D0204548

MONOGRAPHS AND RESEARCH NOTES IN MATHEMATICS

Monomial Algebras

Second Edition

Rafael H. Villarreal

Centro de Investigación
y de Estudios Avanzados del IPN
Mexico City, Mexico

CRC Press is an imprint of the
Taylor & Francis Group, an **informa** business
A CHAPMAN & HALL BOOK

MONOGRAPHS AND RESEARCH NOTES IN MATHEMATICS

Published Titles

Application of Fuzzy Logic to Social Choice Theory, John N. Mordeson, Davender S. Malik and Terry D. Clark

Blow-up Patterns for Higher-Order: Nonlinear Parabolic, Hyperbolic Dispersion and Schrödinger Equations, Victor A. Galaktionov, Enzo L. Mitidieri, and Stanislav Pohozaev

Difference Equations: Theory, Applications and Advanced Topics, Third Edition, Ronald E. Mickens

Dictionary of Inequalities, Second Edition, Peter Bullen

Iterative Optimization in Inverse Problems, Charles L. Byrne

Modeling and Inverse Problems in the Presence of Uncertainty, H. T. Banks, Shuhua Hu, and W. Clayton Thompson

Monomial Algebras, Second Edition, Rafael H. Villarreal

Set Theoretical Aspects of Real Analysis, Alexander B. Kharazishvili

Signal Processing: A Mathematical Approach, *Second Edition*, Charles L. Byrne

Sinusoids: Theory and Technological Applications, Prem K. Kythe

Special Integrals of Gradshetyn and Ryzhik: the Proofs – Volume I, Victor H. Moll

Forthcoming Titles

Actions and Invariants of Algebraic Groups, Second Edition, Walter Ferrer Santos and Alvaro Rittatore

Analytical Methods for Kolmogorov Equations, Second Edition, Luca Lorenzi

Complex Analysis: Conformal Inequalities and the Bierbach Conjecture, Prem K. Kythe

Computational Aspects of Polynomial Identities: Volume I, Kemer's Theorems, 2nd Edition Belov Alexey, Yaakov Karasik, Louis Halle Rowen

Cremona Groups and Icosahedron, Ivan Cheltsov and Constantin Shramov

Geometric Modeling and Mesh Generation from Scanned Images, Yongjie Zhang

Groups, Designs, and Linear Algebra, Donald L. Kreher

Handbook of the Tutte Polynomial, Joanna Anthony Ellis-Monaghan and Iain Moffat

Lineability: The Search for Linearity in Mathematics, Juan B. Seoane Sepulveda, Richard W. Aron, Luis Bernal-Gonzalez, and Daniel M. Pellegrinao

Line Integral Methods and Their Applications, Luigi Brugnano and Felice Iaverno

Microlocal Analysis on R^n and on NonCompact Manifolds, Sandro Coriasco

Forthcoming Titles (continued)

Nonlinear Functional Analysis in Banach Spaces and Banach Algebras: Fixed Point Theory Under Weak Topology for Nonlinear Operators and Block Operators with Applications
Aref Jeribi and Bilel Krichen

Partial Differential Equations with Variable Exponents: Variational Methods and Quantitative Analysis, Vicentiu Radulescu and Dusan Repovs

Practical Guide to Geometric Regulation for Distributed Parameter Systems, Eugenio Aulisa and David S. Gilliam

Reconstructions from the Data of Integrals, Victor Palamodov

Special Integrals of Gradshetyn and Ryzhik: the Proofs – Volume II, Victor H. Moll

Stochastic Cauchy Problems in Infinite Dimensions: Generalized and Regularized Solutions, Irina V. Melnikova and Alexei Filinkov

Symmetry and Quantum Mechanics, Scott Corry

CRC Press
Taylor & Francis Group
6000 Broken Sound Parkway NW, Suite 300
Boca Raton, FL 33487-2742

Printed on acid-free paper
Version Date: 20150202

International Standard Book Number-13: 978-1-4822-3469-5 (Hardback)

Visit the Taylor & Francis Web site at
http://www.taylorandfrancis.com

and the CRC Press Web site at
http://www.crcpress.com

Contents

Preface ix

Preface to First Edition xv

1 Polyhedral Geometry and Linear Optimization 1
- 1.1 Polyhedral sets and cones 1
- 1.2 Relative volumes of lattice polytopes 20
- 1.3 Hilbert bases and TDI systems 28
- 1.4 Rees cones and clutters 39
- 1.5 The integral closure of a semigroup 46
- 1.6 Unimodularity of matrices and normality 48
- 1.7 Normaliz, a computer program 50
- 1.8 Cut-incidence matrices and integrality 51
- 1.9 Elementary vectors and matroids 55

2 Commutative Algebra 61
- 2.1 Module theory 61
- 2.2 Graded modules and Hilbert polynomials 76
- 2.3 Cohen–Macaulay modules 79
- 2.4 Normal rings 86
- 2.5 Valuation rings 92
- 2.6 Krull rings 94
- 2.7 Koszul homology 104
- 2.8 A vanishing theorem of Grothendieck 107

3 Affine and Graded Algebras 111
- 3.1 Cohen–Macaulay graded algebras 111
- 3.2 Hilbert Nullstellensatz 123
- 3.3 Gröbner bases 126
- 3.4 Projective closure 132
- 3.5 Minimal resolutions 134

4 Rees Algebras and Normality **141**
4.1 Symmetric algebras 141
4.2 Rees algebras and syzygetic ideals 142
4.3 Complete and normal ideals 145
4.4 Multiplicities and a criterion of Herzog 159
4.5 Jacobian criterion 165

5 Hilbert Series **171**
5.1 Hilbert–Serre Theorem 171
5.2 a-invariants and h-vectors 177
5.3 Extremal algebras 182
5.4 Initial degrees of Gorenstein ideals 189
5.5 Koszul homology and Hilbert functions 196
5.6 Hilbert functions of some graded ideals 199

6 Stanley–Reisner Rings and Edge Ideals of Clutters **201**
6.1 Primary decomposition 201
6.2 Simplicial complexes and homology 209
6.3 Stanley–Reisner rings 212
6.4 Regularity and projective dimension 226
6.5 Unmixed and shellable clutters 234
6.6 Admissible clutters 246
6.7 Hilbert series of face rings 250
6.8 Simplicial spheres 255
6.9 The upper bound conjectures 258

7 Edge Ideals of Graphs **261**
7.1 Graph theory 261
7.2 Edge ideals and B-graphs 268
7.3 Cohen–Macaulay and chordal graphs 274
7.4 Shellable and sequentially C–M graphs 282
7.5 Regularity, depth, arithmetic degree 293
7.6 Betti numbers of edge ideals 298
7.7 Associated primes of powers of ideals 303

8 Toric Ideals and Affine Varieties **311**
8.1 Binomial ideals and their radicals 311
8.2 Lattice ideals 316
8.3 Monomial subrings and toric ideals 326
8.4 Toric varieties 335
8.5 Affine Hilbert functions 342
8.6 Vanishing ideals over finite fields 345
8.7 Semigroup rings of numerical semigroups 347
8.8 Toric ideals of monomial curves 352

9 Monomial Subrings **365**
9.1 Integral closure of monomial subrings . . . 366
9.2 Homogeneous monomial subrings . . . 372
9.3 Ehrhart rings . . . 380
9.4 The degree of lattice and toric ideals . . . 392
9.5 Laplacian matrices and ideals . . . 396
9.6 Gröbner bases and normal subrings . . . 403
9.7 Toric ideals generated by circuits . . . 410
9.8 Divisor class groups of semigroup rings . . . 416

10 Monomial Subrings of Graphs **423**
10.1 Edge subrings and ring graphs . . . 423
10.2 Incidence matrices and circuits . . . 440
10.3 The integral closure of an edge subring . . . 448
10.4 Ehrhart rings of edge polytopes . . . 454
10.5 Integral closure of Rees algebras . . . 457
10.6 Edge subrings of complete graphs . . . 461
10.7 Edge cones of graphs . . . 467
10.8 Monomial birational extensions . . . 477

11 Edge Subrings and Combinatorial Optimization **483**
11.1 The canonical module of an edge subring . . . 483
11.2 Integrality of the shift polyhedron . . . 484
11.3 Generators for the canonical module . . . 487
11.4 Computing the a-invariant . . . 489
11.5 Algebraic invariants of edge subrings . . . 493

12 Normality of Rees Algebras of Monomial Ideals **499**
12.1 Integral closure of monomial ideals . . . 499
12.2 Normality criteria . . . 505
12.3 Rees cones and polymatroidal ideals . . . 508
12.4 Veronese subrings and the a-invariant . . . 513
12.5 Normalizations of Rees algebras . . . 519
12.6 Rees algebras of Veronese ideals . . . 524
12.7 Divisor class group of a Rees algebra . . . 529
12.8 Stochastic matrices and Cremona maps . . . 531

13 Combinatorics of Symbolic Rees Algebras of Edge Ideals of Clutters **537**
13.1 Vertex covers of clutters . . . 537
13.2 Symbolic Rees algebras of edge ideals . . . 540
13.3 Blowup algebras in perfect graphs . . . 551
13.4 Algebras of vertex covers of graphs . . . 555
13.5 Edge subrings in perfect matchings . . . 558

13.6 Rees cones and perfect graphs 561
13.7 Perfect graphs and algebras of covers 564

14 Combinatorial Optimization and Blowup Algebras 567
14.1 Blowup algebras of edge ideals 568
14.2 Rees algebras and polyhedral geometry 570
14.3 Packing problems and blowup algebras 583
14.4 Uniform ideal clutters 599
14.5 Clique clutters of comparability graphs 610
14.6 Duality and integer rounding problems 615
14.7 Canonical modules and integer rounding 628
14.8 Clique clutters of Meyniel graphs 633

Appendix Graph Diagrams 635
A.1 Cohen–Macaulay graphs 635
A.2 Unmixed graphs 638

Bibliography 639

Notation Index 669

Index 673

Preface

The main purpose of this book is to introduce algebraic, combinatorial, and computational methods to study monomial algebras and their presentation ideals, including Stanley–Reisner rings, monomial subrings, Ehrhart rings, blowup algebras, emphasizing square-free monomials and its corresponding graphs, clutters, or hypergraphs.

Monomial algebras are related to various fields, for instance to numerical semigroups, semigroup rings, algebraic geometry, commutative algebra and combinatorics, integer programming and polyhedral geometry, graph theory and combinatorial optimization. We develop links between the areas shown as the vertices of the following graph.

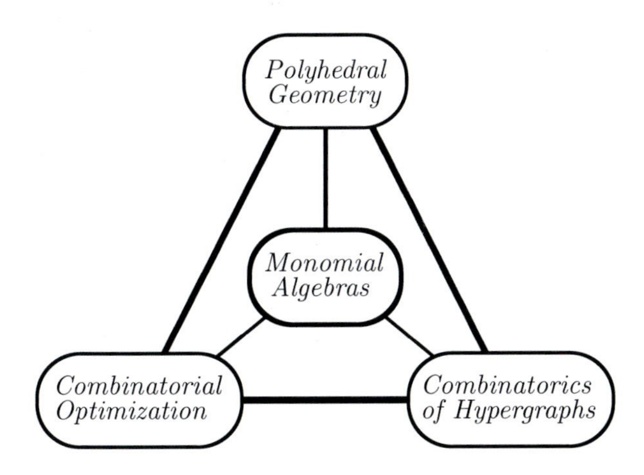

This allows us to solve a variety of problems of monomial algebras using the methods of the other areas and vice versa. An effort has been made to give a unifying presentation of the techniques and notions of these areas. In this book we are interested in the algebraic properties of monomial algebras that can be directly linked to combinatorial structures—such as simplicial complexes, posets, digraphs, graphs, and clutters—and to linear optimization problems. We study various types of affine and graded rings (Cohen–Macaulay, sequentially Cohen–Macaulay, Gorenstein, Artinian, complete intersection, unmixed, normal, reduced) and examine their basic algebraic invariants (type, multiplicity, a-invariant, regularity, Betti numbers, Krull dimension, projective dimension, h-vector, Hilbert polynomial).

In recent years the algebraic properties of blowup algebras have been linked to combinatorial optimization problems of clutters and hypergraphs. In this edition, we include four new chapters (Chapters 1, 11, 13, and 14) to introduce this area of research and to present some of the main advances.

The study of algebraic and combinatorial properties of the edge ideal of clutters and hypergraphs has attracted a great deal of interest in the last two decades [163, 189, 224, 326, 408]. In the present edition we include two chapters with some of the advances in the area (Chapters 6 and 7).

In order to present an up-to-date account of the subject, we have made a full revision of all chapters in the first edition. The chapters have been reorganized and arranged in a different order. In particular Chapters 9 and 12 were originally a single chapter, while Chapters 8 and 10 were originally divided into two chapters for each of them.

This book brings together several areas of pure and applied mathematics. It contains over 550 exercises and over 50 examples, many of them illustrating the use of computer algebra systems. It has extensive indices of terminology and notation.

The contents of this book are as follows. In Chapter 1 we begin by introducing several notions and results coming from polyhedral geometry, combinatorial optimization, and linear programming. Relative volumes of lattice polytope are studied here. We present relations between Hilbert bases, TDI linear systems of inequalities, max-flow min-cut properties of clutters, and normality of affine and Rees semigroups. For instance the max-flow min-cut property is classified in terms of Hilbert bases and the integrality of certain polyhedra. This is used in Chapter 14 to prove that the edge ideal of a clutter is normally torsion-free if and only if the clutter has the max-flow min-cut property. We present a slight generalization of a theorem of Lucchesi and Younger [298] that is useful to detect TDI systems arising from incidence matrices of digraphs. This is used in Chapter 11 to express the a-invariant of the edge subring of a bipartite graph in terms of directed cuts. Elementary integral vectors and matroids are introduced at the end of the chapter. The notion of a matroid will appear in several places in this book in connection with monomial rings and ideals. The computer program we have used to study linear systems of inequalities and to compute Hilbert bases of rational cones is *Normaliz* [68]. The prerequisite for this chapter is a course on linear algebra and familiarity with point set topology.

Chapter 2 discusses certain topics and results on commutative algebra (module theory, normal and graded rings). It includes some detailed proofs, and points the reader to the appropriate references when proofs are omitted. However we make free use of the standard terminology and notation of homological algebra (including Tor and Ext) as described in [363] and [428]. This edition has three new sections on valuation rings, Krull rings, and a vanishing theorem of Grothendieck (Sections 2.5, 2.6, and 2.8).

A number of topics connected to affine and graded algebras are studied in Chapter 3, e.g., Gröbner bases, Hilbert Nullstellensatz and affine varieties, projective closure, minimal resolutions, and Betti numbers. We present two versions of the Noether normalization lemma and some of its applications to affine algebras and to Cohen–Macaulay graded algebras.

In Chapter 4 a thorough presentation of complete and normal ideals is given. Here the systematic use of blowup algebras makes clear their importance for the area. In this chapter we introduce symbolic powers of ideals and normally torsion-free ideals. Then we present a few cases where equality of symbolic and ordinary powers can be described in terms of properties of the associated graded ring. An elegant and useful Cohen–Macaulay criterion due to Herzog is included here.

Chapter 5 deals with the Hilbert series of graded modules and algebras, a topic that is quite useful in Stanley's proof of the upper bound conjecture for simplicial spheres. The h-vector and a-invariant of a graded algebra are defined through the Hilbert–Serre theorem. For Cohen–Macaulay graded algebras we present the main properties of their h-vectors and a-invariants. Some optimal upper bounds for the number of generators in the least degree of Gorenstein and Cohen–Macaulay graded ideals are given, which naturally leads to the notion of an extremal algebra. As an application the Koszul homology of Cohen–Macaulay ideals with pure resolutions is studied using Hilbert function techniques. This edition has a new section on Hilbert functions of a certain type of graded ideals that occur in algebraic coding theory (Section 5.6).

Chapter 6 is an introduction to monomial ideals, Stanley–Reisner rings, and edge ideals of clutters. An understated goal here is to highlight some of the works of T. Hibi, J. Herzog, M. Hochster, G. Reisner, and R. Stanley. We study algebraic and combinatorial properties of edge ideals and simplicial complexes. In particular we examine shellable, unmixed, and sequentially Cohen–Macaulay simplicial complexes and their corresponding edge ideals and invariants (projective dimension, regularity, depth, Hilbert series). As for applications, the proofs of the upper bound conjectures for polytopes are discussed to give a flavor to some of the methods and ideas of the area. Monomial ideals can also be used to solve certain problems of general polynomial ideals using the theory of Gröbner bases.

In Chapter 7 we give an introduction to graph theory and study algebraic properties and invariants of edge ideals using the combinatorial structure of graphs. The study of edge ideals of graphs has been a very active area of research in the last decade because of its connections to graph theory. We present classifications of the following families: unmixed bipartite graphs, Cohen–Macaulay bipartite graphs, Cohen–Macaulay trees, shellable bipartite graphs, sequentially Cohen–Macaulay bipartite graphs. The invariants examined here include regularity, depth, Krull dimension, and multiplicity. Edge ideals of graphs are shown to have the persistence property.

In Chapter 8 we study algebraic and geometric aspects of three special types of polynomial ideals and their quotient rings:

$$\text{TORIC IDEALS} \hookrightarrow \text{LATTICE IDEALS} \hookrightarrow \text{BINOMIAL IDEALS}.$$

These ideals are interesting from a computational point of view and are related to diverse fields, such as combinatorics, algebraic geometry, integer programming, semigroup rings, coding theory, and algebraic statistics.

Chapter 9 deals with point configurations and their lattice polytopes. We consider monomial subrings and binomial ideals associated with them, e.g., Ehrhart rings, Rees algebras, homogeneous subrings, lattice ideals, and toric ideals. Using Hilbert functions, polyhedral geometry, and Gröbner bases, we study normalizations of monomial subrings, initial ideals of toric ideals, normal monomial subrings, primary decompositions and multiplicities of lattice ideals. Algebraic graph theory is used to study matrix ideals of Laplacian matrices. The reciprocity law of Ehrhart for integral polytopes and the Danilov–Stanley formula for canonical modules of monomial subrings will be introduced here. Applications of these results will be given.

In Chapter 10 we study monomial subrings associated with graphs and their toric ideals. We relate the even closed walks and circuits of the vector matroid of a graph with Gröbner bases theory. A description of the integral closure of the edge subring of a multigraph will be presented along with a description of the circuits of its toric ideal. We study the family of graphs whose number of primitive cycles equals its cycle rank. It is shown that this family is precisely the family of ring graphs. These graphs are characterized in algebraic and combinatorial terms. We classify edge subrings of bipartite graphs which are complete intersections. Then we present sharp upper bounds for the multiplicity of edge subrings. Several connections between monomial subrings, graph theory, and polyhedral geometry will occur in this chapter. We study in detail the irreducible representation of an edge cone of a graph and show some applications to graph theory (for instance we show the marriage theorem).

Chapter 11 focuses on edge subrings of connected bipartite graphs and their algebraic invariants. We show how to compute the canonical module and the a-invariant of an edge subring using linear programming techniques and introduce an integral polyhedron whose vertices correspond to minimal generators of the canonical module. The a-invariant of an edge subring will be interpreted in combinatorial optimization terms as the maximum number of edge disjoint directed cuts. We study the Gorenstein property and the type of an edge subring.

Chapter 12 is about Rees algebras of monomial ideals. We study the integral closure of a monomial ideal, and the normality and invariants of its Rees algebra. The normalization of a Rees algebra is examined using the Danilov–Stanley formula, Carathéodory's theorem, and Hilbert bases of Rees cones. Interesting classes of normal ideals, such as ideals of Veronese type and polymatroidal ideals are introduced in the chapter. The divisor class group of a normal Rees algebra is computed using polyhedral geometry.

Chapter 13 shows the interaction between graph theory, combinatorial optimization, commutative algebra, and the theory of blowup algebras. In this chapter, we give a description—using notions from combinatorial optimization—of the minimal generators of the symbolic Rees algebra of the edge ideal of a clutter and show a graph theoretical description of the minimal generators of the symbolic Rees algebra of the ideal of covers of a graph. The minimal sets of generators of symbolic Rees algebras of edge ideals are studied using polyhedral geometry. Indecomposable graphs are related to the strong perfect graph theorem. We give a description—in terms of cliques—of the symbolic Rees algebra and the Simis cone of the edge ideal of a perfect graph.

In Chapter 14 we relate combinatorial optimization with commutative algebra, and present applications to both areas. We establish some links between the algebraic properties of blowup algebras of edge ideals and the combinatorial optimization properties of clutters and polyhedra. A long-standing conjecture of Conforti and Cornuéjols about packing problems is examined from an algebraic point of view. We study max-flow min-cut problems of clutters, packing problems, and integer rounding properties of various systems of linear inequalities to gain insight about the algebraic properties of blowup algebras. Systems with integer rounding properties and clutters with the max-flow min-cut property come from linear optimization problems. The equality between the Rees algebra and the symbolic Rees algebra of an edge ideal is characterized in combinatorial and algebraic terms. A number of properties of clutters and edge ideals are shown to be closed (under taking) minors, Alexander duals, and parallelizations. The max-flow min-cut property of a clutter is characterized in algebraic and combinatorial terms. The structure of ideal uniform clutters is presented here. If a clutter satisfies the max-flow min-cut property, we prove that all invariant factors of its incidence matrix are equal to 1 and that the columns of this matrix form a Hilbert basis. It is shown that the clique clutter of a comparability graph satisfies the max-flow min-cut property. The normality of an ideal is described in terms of the integer rounding property of a linear system and we establish a *duality theorem for monomial subrings*. We show that the Rees algebra of the ideal of covers of a perfect graph is normal and that clique clutters of Meynel graphs are Ehrhart clutters.

This book stresses the use of computational and combinatorial methods in commutative algebra because they have been a major factor in discovering new and interesting results. The main computer algebra programs that we use in this book are *Normaliz* [68] and *Macaulay2* [199]. These programs provide an invaluable tool to study monomial algebras and their algebraic invariants. As a handy reference we include a section summarizing the type of computations that can be done using *Normaliz* (see Section 1.7). We also occasionally use other computer algebra systems, such as *CoCoA* [88], *Maple*™ [80], and *Mathematica*® [431]. *Singular* [200] is another system that can be used for computations in commutative algebra with Gröbner bases. *Porta* [84] and *polymake* [177] are two systems that can be used for computations in convex polyhedra and finite simplicial complexes.

Combinatorial commutative algebra is an extensive area of mathematics. The main references for the area are the books of Bruns and Herzog [65], Hibi [240], Miller and Sturmfels [317], and Stanley [395]. This book emphasizes the use of discrete mathematics and combinatorial optimization methods in combinatorial commutative algebra. There are a number of excellent recent books that offer a complementary view of the subject, namely, Beck and Robins [21], Bruns and Gubelazde [61], De Loera, Hemmecke and Köppe [105], Ene and Herzog [142], Herzog and Hibi [224], Huneke and Swanson [259], and Vasconcelos [414].

Outstanding references for computational and combinatorial aspects that complement and—in some cases—extend some of the material included here are the books of Berge [25, 27], Brøndsted [57], Chvátal [87], Cornuéjols [93], De Loera, Rambau and Santos [106], Diestel [111], Cox, Little and O'Shea [99], Eisenbud [128], Ewald [151], Gitler and Villarreal [189], Godsil and Royle [190], Golumbic [191], Greuel and Pfister [200], Harary [208], Kreuzer and Robbiano [282], Schenck [369], Schrijver [372, 373], Stanley [394, 396], Sturmfels [400], Vasconcelos [413], and Ziegler [438].

We would like to thank Winfried Bruns, Enrique Reyes, and Aron Simis for their helpful comments and corrections. Thanks to executive editor Robert B. Stern for suggesting we prepare an up-to-date second edition of *Monomial Algebras* for the new series in mathematics *Monographs and Research Notes in Mathematics*, Taylor & Francis (Chapman and Hall/CRC Press Group). The support of Marsha Pronin and Samantha White at Taylor & Francis is much appreciated. Finally, we are grateful to Karen Simon for her careful editorial work on the manuscript.

RAFAEL H. VILLARREAL
Cinvestav-IPN
Mexico City, D.F.

Preface to First Edition

Let $R = K[\mathbf{x}] = K[x_1, \ldots, x_n]$ be a polynomial ring in the indeterminates $x_1, \ldots, x_n$, over the field K. Let

$$f_i = x^{v_i} = x_1^{v_{i1}} \cdots x_n^{v_{in}},\ i = 1, \ldots, q,$$

be a finite set of monomials of R. We are interested in studying several algebras and ideals associated with these monomials. Some of these are:

- the monomial subring: $K[f_1, \ldots, f_q] \subset K[\mathbf{x}]$,
- the Rees algebra: $K[\mathbf{x}, f_1 t, \ldots, f_q t] \subset K[\mathbf{x}, t]$, which is also a monomial subring,
- the face ring or Stanley–Reisner ring: $K[\mathbf{x}]/(f_1, \ldots, f_q)$, if the monomials are square-free, and
- the toric ideal: the ideal of relations of a monomial subring.

In the following diagram we stress the most relevant relations between the properties of those algebras that will take place in this text.

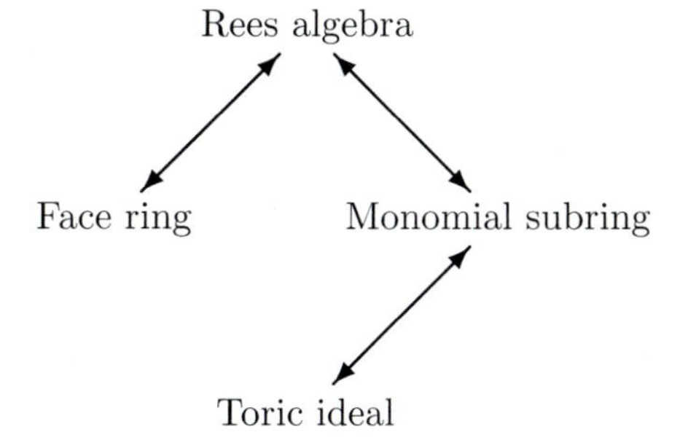

If such monomials are square-free they are indexed by a hypergraph built on the set of indeterminates, which provides a second combinatorial structure in addition to the associated Stanley–Reisner simplicial complex.

This book was written with the aim of providing an introduction to the methods that can be used to study monomial algebras and their presentation ideals, with emphasis on square-free monomials. We have striven to provide methods that are effective for computations.

A substantial part of this volume is dedicated to the case of monomial algebras associated to graphs, that is, those defined by square-free quadratic monomials defining a simple graph. We will systematically use graph theory to study those algebras. Such a systematic treatment is a gap in the literature that we intend to fill. Two outstanding references for graph theory are [48] and [208].

In the text, special attention is paid to providing means to determine whether a given monomial algebra or ideal is Cohen–Macaulay or normal. Those means include diverse characterizations and qualities of those two properties.

Throughout this work base rings are assumed to be Noetherian and modules finitely generated.

An effort has been made to make the book self-contained by including a chapter on commutative algebra (Chapter 2) that includes some detailed proofs and often points the reader to the appropriate references when proofs are omitted. However, we make free use of the standard terminology and notation of homological algebra (including Tor and Ext) as described in [363] and [428].

The first goal is to present basic properties of monomial algebras. For this purpose in Chapter 3 we study affine and graded algebras. The topics include Noether normalizations and their applications, diverse attributes of Cohen–Macaulay graded algebras, Hilbert Nullstellensatz and affine varieties, some Gröbner bases theory, and minimal resolutions.

In Chapter 4 a thorough presentation of complete and normal ideals is given. Here the systematic use of Rees algebras and associated graded rings makes clear their importance for the area.

Chapter 5 deals with the Hilbert series of graded modules and algebras, a topic that is quite useful in Stanley's proof of the upper bound conjecture for simplicial spheres. Here we introduce the h-vector and a-invariant of graded algebras and give several interpretations of the a-invariant when the algebra is Cohen–Macaulay. Some optimal upper bounds for the number of generators in the least degree of Gorenstein and Cohen–Macaulay ideals are presented, which naturally leads to the notion of an extremal algebra. As an application the Koszul homology of Cohen–Macaulay ideals with pure resolutions is studied using Hilbert function techniques.

General monomial ideals and Stanley–Reisner rings are examined in Chapter 6. The first version of this chapter was some notes originally prepared to teach a short course during the *XXVII Congreso Nacional de la Sociedad Matemática Mexicana* in October of 1994. In this course we presented some applications of commutative algebra to combinatorics. We have expanded these notes to include a more complete treatment of shellable and Cohen–Macaulay complexes. The presentation of the last three sections of this chapter, discussing the Hilbert series of face rings and the upper bound conjectures, was inspired by [41, 65] and [393].

Since monomial algebras defined by square-free monomials of degree two have an underlying graph theoretical structure it is natural that some interaction will occur between monomial algebras, graph theory, and polyhedral geometry. We have included three chapters that focus on monomial algebras associated to graphs. One of them is Chapter 7, where we present connections between graphs and ideals and study the Cohen–Macaulay property of the face ring. Another is Chapter 10, where we present a combinatorial description of the integral closure of the corresponding monomial subring and give some applications to graph theory. In Chapter 10 we consider monomial subrings and toric ideals of complete graphs with the aim of computing their Hilbert series, Noether normalizations, and Gröbner bases.

The central topic of Chapters 9 and 12 is the normality of monomial subrings and ideals; some features of toric ideals are presented here.

Chapter 8 is devoted to the study of monomial curves and their toric ideals, where the focus of our attention will be on monomial space curves and monomial curves in four variables. Affine toric varieties and their toric ideals are studied in Chapter 8.

Most of the material in this textbook has been written keeping in mind a typical graduate student with a basic knowledge of abstract algebra and a non-expert who wishes to learn the subject. We hope that this book can be read by people from diverse subjects and fields, such as combinatorics, graph theory, and computer algebra. Various units are accessible to upper undergraduates.

In the last fifteen years a dramatic increase in the number of research articles and books in commutative algebra that stress its connections with computational issues in algebraic geometry and combinatorics has taken place. Excellent references for computational and combinatorial aspects that complement some of the material included here are [99, 128, 413], [65, 395, 400] and [57, 438].

A constant concern during the writing of this text was to give appropriate credits for the proofs and results that were adapted from printed material or communicated to us. We apologize for any involuntary omission and would appreciate any comments and suggestions in this regard.

During the fall of 1999 a course on monomial algebras associated to graphs was given at the University of Messina covering Chapter 7 to Chapter 10 with the support of the *Istituto Nazionale Di Alta Matematica Francesco Severi*. It is a pleasure to thank Vittoria Bonanzinga, Marilena Crupi, Gaetana Restuccia, Rossana Utano, Maurizio Imbesi, Giancarlo Rinaldo,

Fabio Ciolli, and Giovanni Molica for the opportunity to improve those chapters and for their hospitality.

We thank Wolmer V. Vasconcelos for his comments and encouragement to write this book. A number of colleagues and students provided helpful annotations to some early drafts. We are especially grateful to Adrián Alcántar, Joe Brennan, Alberto Corso, José Martínez-Bernal, Susan Morey, Carlos Rentería, Enrique Reyes, and Aron Simis. We are also grateful to Laura Valencia for her competent secretarial assistance.

The *Consejo Nacional de Ciencia y Tecnología* (CONACyT) and the *Sistema Nacional de Investigadores* (SNI) deserve special acknowledgment for their generous support. It should be mentioned that the development of this book was included in the project *Estudios sobre Algebras Monomiales*, which was supported by the CONACyT grant 27931E.

In the homepage "http://www.math.cinvestav.mx/~vila" we will maintain an updated list of corrections.

RAFAEL H. VILLARREAL
Cinvestav-IPN
Mexico City, D.F.

Chapter 1

Polyhedral Geometry and Linear Optimization

In this chapter we introduce several notions and results from polyhedral geometry, combinatorial optimization and linear programming. Excellent general references for these areas are [35, 87, 281, 372, 373, 438]. Then we study relative volumes of lattice polytopes. We present various relations between Hilbert bases, TDI linear systems of inequalities, the max-flow min-cut property of clutters, integral linear systems, and the normality of affine semigroups and Rees semigroups. Elementary integral vectors and matroids are introduced at the end of the chapter.

1.1 Polyhedral sets and cones

An *affine space* or *linear variety* in $\mathbb{R}^n$ is by definition a translation of a linear subspace of $\mathbb{R}^n$. Let A be a subset of $\mathbb{R}^n$. The *affine space generated* by A, denoted by $\text{aff}(A)$, is the set of all *affine combinations* of points in A:

$$\text{aff}(A) = \{a_1v_1 + \cdots + a_rv_r \mid v_i \in A,\, a_i \in \mathbb{R},\, a_1 + \cdots + a_r = 1\}.$$

There is a unique linear subspace V of $\mathbb{R}^n$ such that $\text{aff}(A) = x_0 + V$, for some $x_0 \in \mathbb{R}^n$. The *dimension* of A is defined as $\dim A = \dim_{\mathbb{R}}(V)$.

Definition 1.1.1 An *affine map* is a function between two affine spaces that preserves affine combinations.

A point $x \in \mathbb{R}^n$ is called a *convex combination* of $v_1, \ldots, v_r \in \mathbb{R}^n$ if there are $a_1, \ldots, a_r$ in $\mathbb{R}$ such that $a_i \geq 0$ for all i, $x = \sum_i a_iv_i$ and $\sum_i a_i = 1$. Let A be a subset of $\mathbb{R}^n$. The *convex hull* of A, denoted by $\text{conv}(A)$, is the set of all convex combinations of points in A. If $A = \text{conv}(A)$ we say that A is a *convex set*.

Definition 1.1.2 A *convex polytope* $\mathcal{P} \subset \mathbb{R}^n$ is the convex hull of a finite set of points $v_1, \ldots, v_r$ in $\mathbb{R}^n$, that is, $\mathcal{P} = \text{conv}(v_1, \ldots, v_r)$.

The *inner product* of two vector $x = (x_1, \ldots, x_n)$ and $y = (y_1, \ldots, y_n)$ in $\mathbb{R}^n$ is defined by

$$\langle x, y \rangle = x \cdot y = x_1 y_1 + \cdots + x_n y_n.$$

Definition 1.1.3 Let $x = (x_1, \ldots, x_n)$ be a point in $\mathbb{R}^n$. The *Euclidean norm* of x is defined as

$$\|x\| = \sqrt{\langle x, x \rangle}.$$

We shall always assume that a subset of $\mathbb{R}^n$ has the topology induced by the usual topology of $\mathbb{R}^n$.

Definition 1.1.4 A point a of a set A in $\mathbb{R}^n$ is said to be a *relative interior point* of A if there exists $r > 0$ such that

$$B_r(a) \cap \text{aff}(A) \subset A,$$

where $B_r(a) = \{x | \, \|x - a\| < r\}$.

Let $A \subset \mathbb{R}^n$. The set of relative interior points of A, denoted by $\text{ri}(A)$, is called the *relative interior* of A. We denote the closure of A by $\overline{A}$. The set $\overline{A} \setminus \text{ri}(A)$ is called the *relative boundary* of A and is denoted by $\text{rb}(A)$, points in $\text{rb}(A)$ are called the *relative boundary points* of A. If $\dim(A) = n$, the relative interior of A is sometimes denoted by A^{o}.

An affine space of $\mathbb{R}^n$ of dimension $n-1$ is called an *affine hyperplane*. Given $a \in \mathbb{R}^n \setminus \{0\}$ and $c \in \mathbb{R}$, define the affine hyperplane

$$H(a, c) = \{x \in \mathbb{R}^n | \, \langle x, a \rangle = c\}.$$

Notice that any affine hyperplane of $\mathbb{R}^n$ has this form. The two *closed halfspaces bounded* by $H(a, c)$ are

$$H^+(a, c) = \{x \in \mathbb{R}^n | \, \langle x, a \rangle \geq c\} \text{ and } H^-(a, c) = H^+(-a, -c).$$

If a is a rational vector and c is a rational number, the affine hyperplane $H(a, c)$ (resp. the halfspace $H^+(a, c)$) is called a *rational hyperplane* (resp. *rational halfspace*). If $c = 0$, for simplicity the set H_a will denote the *hyperplane* of $\mathbb{R}^n$ through the origin with normal vector a, that is,

$$H_a := H(a, 0) = \{x \in \mathbb{R}^n | \, \langle x, a \rangle = 0\}.$$

The two *closed halfspaces* bounded by H_a are denoted by

$$H_a^+ = \{x \in \mathbb{R}^n | \, \langle x, a \rangle \geq 0\} \quad \text{and} \quad H_a^- = \{x \in \mathbb{R}^n | \, \langle x, a \rangle \leq 0\}.$$

Definition 1.1.5 A *polyhedral set* or *convex polyhedron* is a subset of $\mathbb{R}^n$ which is the intersection of a finite number of closed halfspaces of $\mathbb{R}^n$. The set $\mathbb{R}^n$ is considered a polyhedron.

The transpose of a matrix A (resp. vector x) will be denoted by A^t or $A^\top$ (resp. x^t or $x^\top$). Often a vector x will denote a column vector or a row vector, from the context the meaning should be clear. Thus, a polyhedral set $\mathcal{Q}$ can be represented as

$$\mathcal{Q} = \{x \in \mathbb{R}^n \mid Ax \leq b\}$$

for some matrix A and for some vector b. As usual, if $a = (a_i)$ and $c = (c_i)$ are vectors in $\mathbb{R}^q$, then $a \leq c$ means $a_i \leq c_i$ for all i. If A and b have rational entries, $\mathcal{Q}$ is called a *rational polyhedron*.

Since the intersection of an arbitrary family of convex sets in $\mathbb{R}^n$ is convex, one derives that any polyhedral set is convex and closed.

Definition 1.1.6 Let $\mathcal{Q}$ be a closed convex set in $\mathbb{R}^n$. A hyperplane H of $\mathbb{R}^n$ is called a *supporting hyperplane* of $\mathcal{Q}$ if $\mathcal{Q}$ is contained in one of the two closed halfspaces bounded by H and $\mathcal{Q} \cap H \neq \emptyset$.

Definition 1.1.7 A *proper face* of a polyhedral set $\mathcal{Q}$ is a set $F \subset \mathcal{Q}$ such that there is a supporting hyperplane $H(a, c)$ satisfying the conditions:

(a) $F = \mathcal{Q} \cap H(a, c) \neq \emptyset$,

(b) $\mathcal{Q} \not\subset H(a, c)$, and $\mathcal{Q} \subset H^+(a, c)$ or $\mathcal{Q} \subset H^-(a, c)$.

The *improper faces* of a polyhedral set $\mathcal{Q}$ are $\mathcal{Q}$ itself and $\emptyset$.

Definition 1.1.8 A proper face F of a polyhedral set $\mathcal{Q} \subset \mathbb{R}^n$ is called a *facet* of $\mathcal{Q}$ if $\dim(F) = \dim(\mathcal{Q}) - 1$.

Proposition 1.1.9 *If $\mathcal{Q}$ is a polyhedral set in $\mathbb{R}^n$ and F_1, F_2 are faces of $\mathcal{Q}$, then their intersection $F = F_1 \cap F_2$ is a face of $\mathcal{Q}$.*

Proof. Let $H_i = H(a_i, c_i)$ be a supporting hyperplane of $\mathcal{Q}$, where $c_i \in \mathbb{R}$ and $0 \neq a_i \in \mathbb{R}^n$, such that $F_i = \mathcal{Q} \cap H_i$ and $\mathcal{Q} \subset H_i^+$ for $i = 1, 2$. Next we prove the equality

$$F = \mathcal{Q} \cap H(a_1 + a_2, c_1 + c_2).$$

The left-hand side is clearly contained in the right-hand side. On the other hand if $x \in \mathcal{Q} \cap H(a_1 + a_2, c_1 + c_2)$, then using $\langle x, a_i \rangle \geq c_i$ one has:

$$c_1 + c_2 = \langle x, a_1 + a_2 \rangle = \langle x, a_1 \rangle + \langle x, a_2 \rangle \geq c_1 + c_2,$$

hence $\langle x, a_i \rangle = c_i$ for $i = 1, 2$ and $x \in F$. As $\mathcal{Q} \subset H^+(a_1 + a_2, c_1 + c_2)$, the set F is a face of $\mathcal{Q}$. $\square$

Definition 1.1.10 A *partially ordered set* or *poset* is a pair $P = (V, \preceq)$, where V is a finite set of vertices and $\preceq$ is a binary relation on V satisfying:

(a) $u \preceq u$, $\forall u \in V$ (reflexivity).

(b) $u \preceq v$ and $v \preceq u$, imply $u = v$ (antisymmetry).

(c) $u \preceq v$ and $v \preceq w$, imply $u \preceq w$ (transitivity).

A poset $P = (V, \preceq)$ with vertex set $V = \{x_1, \ldots, x_n\}$ can be displayed by an *inclusion diagram* , where x_i is joined to x_j by raising a line if $x_i \prec x_j$ and there is no other vertex in between.

Example 1.1.11 The inclusion diagram of the divisors of 12 is:

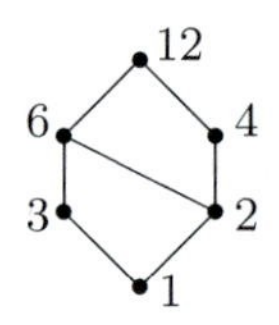

A poset in which any two elements x, y have a greatest lower bound $\inf\{x, y\}$ and a lowest upper bound $\sup\{x, y\}$ is called a *lattice*. A lattice is called *complete* if $\inf S$ and $\sup S$ exist for any subset S. A mapping φ from one lattice $L_1 = (V_1, \preceq)$ onto another lattice $L_2 = (V_2, \preceq)$ is called an *isomorphism* when it is one-to-one and we have $x \preceq y$ if and only if $\varphi(x) \preceq \varphi(y)$ for all $x, y \in V_1$.

Corollary 1.1.12 *Let $\mathcal{Q}$ be a polyhedral set in $\mathbb{R}^n$ and let $\mathcal{F}$ be the set of all faces of $\mathcal{Q}$. The partially ordered set $(\mathcal{F}, \subset)$ is a complete lattice with the lattice operations*

$$\inf \mathcal{G} = \cap\{F \mid F \in \mathcal{G}\} \quad \textit{and} \quad \sup \mathcal{G} = \cap\{G \in \mathcal{F} \mid \forall F \in \mathcal{G}; F \subset G\}.$$

Proof. It follows from Proposition 1.1.9 and from the fact that a convex polyhedron has only finitely many faces [427, Theorem 3.2.1(v)]. $\square$

Definition 1.1.13 The lattice $(\mathcal{F}, \subset)$ is called the *face-lattice* of $\mathcal{Q}$.

Definition 1.1.14 A *polytopal complex* $\mathfrak{C}$ is a finite collection of polytopes in $\mathbb{R}^n$ such that (i) $\emptyset \in \mathfrak{C}$, (ii) if $\mathcal{P} \in \mathfrak{C}$, then all faces of $\mathcal{P}$ are in $\mathfrak{C}$, and (iii) the intersection $\mathcal{P} \cap \mathcal{Q}$ of two polytopes $\mathcal{P}, \mathcal{Q} \in \mathfrak{C}$ is a face of both $\mathcal{P}$ and $\mathcal{Q}$.

Similarly, one can define the notion of a *polyhedral complex* replacing polytope by polyhedron. The *dimension* of a polytopal complex $\mathfrak{C}$, denoted by $\dim(\mathfrak{C})$, is the largest dimension of a polytope in $\mathfrak{C}$.

Definition 1.1.15 Let $\mathcal{P}$ be a convex polytope. All proper faces of $\mathcal{P}$ form the *boundary complex* $\mathfrak{C}(\partial\mathcal{P})$, whose facets are the facets of $\mathcal{P}$.

The boundary complex $\mathfrak{C}(\partial\mathcal{P})$ of a polytope $\mathcal{P}$ of dimension $d+1$ is a pure polytopal complex of dimension d [57, Chapter 2].

Definition 1.1.16 Let $\mathcal{P}$ be a convex polytope of dimension $d+1$. The boundary complex $\mathfrak{C}(\partial\mathcal{P})$ is *shellable* if there is a linear ordering $F_1, \ldots, F_s$ of the facets of $\mathcal{P}$ such that for every $1 \leq i < j \leq s$, there is $1 \leq \ell < j$ satisfying $F_i \cap F_j \subset F_\ell \cap F_j$ and $F_\ell \cap F_j$ is a facet of F_j.

Theorem 1.1.17 [58] *The boundary complex of a polytope is shellable.*

The set of nonnegative real numbers and the set of nonnegative integers are denoted by:

$$\mathbb{R}_+ = \{x \in \mathbb{R} \mid x \geq 0\} \text{ and } \mathbb{N} = \{0, 1, 2, \ldots\}$$

respectively, $\mathbb{R}_+$ is also denoted by $\mathbb{R}_{\geq 0}$. If $\mathcal{A}$ is a set of points in $\mathbb{R}^n$, the cone generated by $\mathcal{A}$, denoted by $\mathbb{R}_+\mathcal{A}$ or cone$(\mathcal{A})$, is defined as

$$\mathbb{R}_+\mathcal{A} = \left\{ \sum_{i=1}^{q} a_i\beta_i \,\middle|\, a_i \in \mathbb{R}_+,\ \beta_i \in \mathcal{A} \text{ for all } i \right\}.$$

The vector space spanned by $\mathcal{A}$ is denoted by $\mathbb{R}\mathcal{A}$. Given a vector $v \in \mathbb{R}^n$, we define:

$$\mathbb{R}_+v := \{\lambda v \mid \lambda \in \mathbb{R}_+\}.$$

Theorem 1.1.18 (Carathéodory's theorem) *Let $v_1, \ldots, v_q$ be a sequence of vectors in $\mathbb{R}^n$ not all of them zero. If $x \in \mathbb{R}_+v_1 + \cdots + \mathbb{R}_+v_q$, then there is a linearly independent set $\mathcal{V} \subset \{v_1, \ldots, v_q\}$ such that $x \in \mathbb{R}_+\mathcal{V}$.*

Proof. By induction on q. The case $q = 1$ is clear. If $q \geq 2$, we can write $x = a_1v_1 + \cdots + a_qv_q$, $a_i \geq 0$ for $i = 1, \ldots, q$. One may assume that $v_1, \ldots, v_q$ are linearly dependent and $a_i > 0$ for all i, otherwise the result follows by induction. There are real numbers $b_1, \ldots, b_q$ such that at least one b_i is positive and $\sum_{i=1}^{q} b_iv_i = 0$. Setting

$$0 < c = \min\{a_i/b_i \mid b_i > 0\} = a_j/b_j,$$

we get $x = x - c(b_1v_1 + \cdots + b_qv_q) = \sum_{i=1}^{q}(a_i - cb_i)v_i$, with $a_i - cb_i \geq 0$ for all i and $a_j - cb_j = 0$. Hence, by induction, the result follows. $\square$

Definition 1.1.19 A set of vectors $\alpha_1, \ldots, \alpha_q \in \mathbb{R}^n$ is *affinely independent* if for every sequence $\lambda_1, \ldots, \lambda_q$ of real numbers satisfying $\sum_{i=1}^{q} \lambda_i\alpha_i = 0$ and $\sum_{i=1}^{q} \lambda_i = 0$, one has $\lambda_i = 0$ for all i.

Corollary 1.1.20 *Let $v_1, \ldots, v_q$ be a set of vectors in $\mathbb{R}^n$ and let x be in its convex hull. Then there exists an affinely independent set $\mathcal{V} \subset \{v_1, \ldots, v_q\}$ such that $x \in \operatorname{conv}(\mathcal{V})$.*

Proof. Consider $\mathcal{B} = \{(v_1, 1), \ldots, (v_q, 1)\}$. Since $(x, 1)$ belongs to $\mathbb{R}_+\mathcal{B}$, the result follows from Carathéodory's theorem and Exercise 1.1.67. $\Box$

Definition 1.1.21 If $C \subset \mathbb{R}^n$ is closed under linear combinations with nonnegative real coefficients, we say that C is a *convex cone*. A *polyhedral cone* is a convex cone which is also a polyhedral set.

Proposition 1.1.22 *If C is a closed convex cone and H is a supporting hyperplane of C, then H is a hyperplane passing through the origin.*

Proof. Let $H = H(a, c)$. Since $0 \in C \subset H^+$, one has $c \leq 0$. If $C \cap H = \{0\}$, then $c = 0$ as required. Assume $C \cap H \neq \{0\}$ and pick $0 \neq z \in C$ such that $\langle z, a \rangle = c$. Using that $tz \in C \subset H^+(a, c)$ for all $t \geq 0$, one derives

$$tc = t\langle z, a \rangle = \langle tz, a \rangle \geq c \;\forall\, t \geq 0 \Rightarrow c = 0. \qquad \Box$$

An affine space of dimension 1 is called a *line*. The following result is quite useful in determining the facets of a polyhedral cone.

Proposition 1.1.23 *Let $\mathcal{A}$ be a finite set of points in $\mathbb{Z}^n$ and let F be a face of $\mathbb{R}_+\mathcal{A}$. The following hold.*

(a) *If $F \neq \{0\}$, then $F = \mathbb{R}_+\mathcal{V}$ for some $\mathcal{V} \subset \mathcal{A}$.*

(b) *If $\dim(F) = 1$ and $\mathbb{R}_+\mathcal{A}$ contains no lines, then $F = \mathbb{R}_+\alpha$ with $\alpha \in \mathcal{A}$.*

(c) *If $\dim(\mathbb{R}_+\mathcal{A}) = n$ and F is a facet defined by the supporting hyperplane H_a, then H_a is generated by a linearly independent subset of $\mathcal{A}$.*

Proof. Let $F = \mathbb{R}_+\mathcal{A} \cap H_a$ with $\mathbb{R}_+\mathcal{A} \subset H_a^-$. Then F is equal to the cone generated by the set $\mathcal{V} = \{\alpha \in \mathcal{A} | \, \langle \alpha, a \rangle = 0\}$. This proves (a). Parts (b) and (c) follow from (a). $\Box$

Most of the notions and results considered thus far make sense if we replace $\mathbb{R}$ by an intermediate field $\mathbb{Q} \subset \mathbb{K} \subset \mathbb{R}$, i.e., we can work in the affine space $\mathbb{K}^n$. However with very few exceptions, like for instance Theorem 1.1.24, we will always work in the Euclidean space $\mathbb{R}^n$ or $\mathbb{Q}^n$.

For convenience we state the *fundamental theorem of linear inequalities*:

Theorem 1.1.24 [372, Theorem 7.1] *Let $\mathbb{Q} \subset \mathbb{K} \subset \mathbb{R}$ be an intermediate field and let $C \subset \mathbb{K}^n$ be a cone generated by $\mathcal{A} = \{\alpha_1, \ldots, \alpha_q\}$. If $\alpha \in \mathbb{K}^n \setminus C$ and $t = \operatorname{rank}\{\alpha_1, \ldots, \alpha_q, \alpha\}$, then there exists a hyperplane H_a containing $t-1$ linearly independent vectors from $\mathcal{A}$ such that $\langle a, \alpha \rangle > 0$ and $\langle a, \alpha_i \rangle \leq 0$ for $i = 1, \ldots, q$.*

Theorem 1.1.25 (Farkas's Lemma) *Let A be an $n \times q$ matrix with entries in a field $\mathbb{K}$ and let $\alpha \in \mathbb{K}^n$. Assume $\mathbb{Q} \subset \mathbb{K} \subset \mathbb{R}$. Then either there exists $x \in \mathbb{K}^q$ with $Ax = \alpha$ and $x \geq 0$, or there exists $v \in \mathbb{K}^n$ with $vA \geq 0$ and $\langle v, \alpha \rangle < 0$, but not both.*

Proof. Let $\mathcal{A} = \{\alpha_1, \ldots, \alpha_q\}$ be the set of column vectors of A. Assume that there is no $x \in \mathbb{K}^q$ with $Ax = \alpha$ and $x \geq 0$, i.e., α is not in $\mathbb{R}_+\mathcal{A}$. By Theorem 1.1.24 there is a hyperplane H_v such that $\langle v, \alpha\rangle < 0$ and $\langle v, \alpha_i\rangle \geq 0$ for all i. Hence $vA \geq 0$, as required. If both conditions hold, then $0 > \langle v, \alpha\rangle = \langle v, Ax\rangle = \langle vA, x\rangle \geq 0$, a contradiction. □

The reader is referred to [438] for other versions of Farkas's lemma. The next result tells us how to separate a point from a cone.

Corollary 1.1.26 *Let $C \subset \mathbb{K}^n$ be a cone generated by $\mathcal{A} = \{a_1, \ldots, a_m\}$. If $\gamma \in \mathbb{K}^n$ and $\gamma \notin C$, then there is a hyperplane H through the origin such that $\gamma \in H^-\backslash H$ and $C \subset H^+$.*

Proof. Let A be the matrix with column vectors $a_1, \ldots, a_m$. By Farkas's lemma (see Theorem 1.1.25) there exists $\mu \in \mathbb{K}^n$ such that $\mu A \geq 0$ and $\langle \gamma, \mu\rangle < 0$. Thus $\langle \mu, a_i\rangle \geq 0$ for all i. If H is the hyperplane through the origin with normal vector μ we get $C \subset H^+$, as required. □

Corollary 1.1.27 *Let $\mathcal{A}$ be a finite set in $\mathbb{Z}^n$, then*

$$\mathbb{Z}\mathcal{A} \cap \mathbb{R}_+\mathcal{A} = \mathbb{Z}\mathcal{A} \cap \mathbb{Q}_+\mathcal{A} \quad \textit{and} \quad \mathbb{Z}^n \cap \mathbb{R}_+\mathcal{A} = \mathbb{Z}^n \cap \mathbb{Q}_+\mathcal{A},$$

where $\mathbb{Z}\mathcal{A}$ is the subgroup of $\mathbb{Z}^n$ spanned by $\mathcal{A}$ and $\mathbb{Q}_+ = \{x \in \mathbb{Q} | \, x \geq 0\}$.

Proof. It follows at once from Theorem 1.1.25. □

Definition 1.1.28 We say that C is a *finitely generated cone* if $C = \mathbb{R}_+\mathcal{A}$, for some finite set $\mathcal{A} = \{v_1, \ldots, v_q\}$.

The proof of the next result yields the *duality theorem for cones*.

Theorem 1.1.29 *If $C \subset \mathbb{R}^n$, then C is a finitely generated cone (resp. finitely generated cone by rational vectors) if and only if C is a polyhedral cone (resp. rational polyhedral cone) in $\mathbb{R}^n$.*

Proof. $\Rightarrow$) Assume that $C \neq (0)$ is a cone generated by $\mathcal{A} = \{\alpha_1, \ldots, \alpha_m\}$. We set $r = \dim(\mathbb{R}_+\mathcal{A})$. Notice that $\mathrm{aff}(C)$ is the real vector space generated by $\mathcal{A}$, because $0 \in C$. If $\mathrm{aff}(C) = C$, then C is a polyhedral cone, because C is the intersection of $n - r$ hyperplanes of $\mathbb{R}^n$ through the origin. Now assume that $C \subsetneq \mathrm{aff}(C)$. Consider the family

$$\mathcal{F} = \{F \,|\, F = H_a \cap C;\ \dim(F) = r - 1;\ C \subset H_a^-\}.$$

By Theorem 1.1.24 the family $\mathcal{F}$ is non-empty. Notice that $\mathcal{F}$ is a finite set because each F in $\mathcal{F}$ is a cone generated by a subset of $\mathcal{A}$; see Proposition 1.1.23. Assume that $\mathcal{F} = \{F_1, \ldots, F_s\}$, where $F_i = H_{a_i} \cap C$. We claim that the following equality holds

$$C = H_{a_1}^- \cap \cdots \cap H_{a_s}^- \cap \mathrm{aff}(C). \tag{1.1}$$

The inclusion "$\subset$" is clear. To show the inclusion "$\supset$", we proceed by contradiction. Assume that there exists $\alpha \notin C$ such that α belongs to the right-hand side of Eq. (1.1). By Theorem 1.1.24 and by reordering the elements of $\mathcal{A}$ if necessary, there is a hyperplane H_a containing linearly independent vectors $\alpha_1, \ldots, \alpha_{r-1}$ such that (i) $\langle a, \alpha \rangle > 0$, and (ii) $\langle a, \alpha_i \rangle \leq 0$ for $i = 1, \ldots, m$. Thus $F = H_a \cap C = H_{a_k} \cap C$ for some $1 \leq k \leq s$. We may assume that $\alpha_1, \ldots, \alpha_r$ form a basis of $\text{aff}(C)$. Since $\alpha \in \text{aff}(C)$ there are scalars $\lambda_1, \ldots, \lambda_r$ with $\alpha = \lambda_1 \alpha_1 + \cdots + \lambda_r \alpha_r$. Using (ii) we get $\langle a, \alpha \rangle = \lambda_r \langle \alpha_r, a \rangle > 0$. Hence $\lambda_r < 0$. On the other hand $\langle a_k, \alpha_r \rangle < 0$ because $C \subset H_{a_k}^-$ and $\dim(F_k) = r - 1$. Therefore

$$\langle a_k, \alpha \rangle = \underbrace{\lambda_r}_{<0} \underbrace{\langle a_k, \alpha_r \rangle}_{<0} > 0,$$

a contradiction to the fact that $\alpha \in H_{a_k}^-$.

$\Leftarrow$) Assume that $C = H_{b_1}^- \cap \cdots \cap H_{b_r}^-$, where $b_1, \ldots, b_r \in \mathbb{R}^n$. Consider the cone C' generated by $b_1, \ldots, b_r$. From the first part of the proof we can write

$$C' = \mathbb{R}_+\{b_1, \ldots, b_r\} = H_{\alpha_1}^- \cap \cdots \cap H_{\alpha_m}^-, \qquad (*)$$

for some set of vectors $\mathcal{A} = \{\alpha_1, \ldots, \alpha_m\}$ in $\mathbb{R}^n$. Next we show the equality $C = \mathbb{R}_+\mathcal{A}$. Notice that $\langle b_i, \alpha_j \rangle \leq 0$ for all i, j, because $b_i \in C'$ for all i. Thus $\mathbb{R}_+\mathcal{A} \subset C$. Assume that there is $\alpha \in C \setminus \mathbb{R}_+\mathcal{A}$. By Corollary 1.1.26, there exists a hyperplane H_a such that $\mathbb{R}_+\mathcal{A} \subset H_a^-$ and $\langle a, \alpha \rangle > 0$. Hence $\langle \alpha_i, a \rangle \leq 0$ for all i, and by Eq. $(*)$ we conclude that $a \in \mathbb{R}_+\{b_1, \ldots, b_r\}$. Therefore, we can write $a = \lambda_1 b_1 + \cdots + \lambda_r b_r$, $\lambda_i \geq 0$ for all i. Since $\alpha \in C$, we have $\langle \alpha, a \rangle = \lambda_1 \langle \alpha, b_1 \rangle + \cdots + \lambda_r \langle \alpha, b_r \rangle \leq 0$, contradicting $\langle a, \alpha \rangle > 0$. Thus $C = \mathbb{R}_+\mathcal{A}$. The respective statement about the rationality character of the representations is left as an exercise. $\square$

Corollary 1.1.30 (Duality theorem for cones) *Let $\mathcal{B} = \{\beta_1, \ldots, \beta_r\}$ be a subset of $\mathbb{R}^n$, and let $\{\alpha_1, \ldots, \alpha_m\}$ and $\{c_1, \ldots, c_s\}$ be subsets of $\mathbb{R}^n \setminus \{0\}$.*

(a) *If $\mathbb{R}_+\mathcal{B} = \cap_{i=1}^m H_{\alpha_i}^-$, then $H_{\beta_1}^- \cap \cdots \cap H_{\beta_r}^- = \mathbb{R}_+\alpha_1 + \cdots + \mathbb{R}_+\alpha_m$.*

(b) *If $\mathbb{R}_+\mathcal{B} = \cap_{i=1}^s H_{c_i}^+$, then $H_{\beta_1}^+ \cap \cdots \cap H_{\beta_r}^+ = \mathbb{R}_+c_1 + \cdots + \mathbb{R}_+c_s$.*

Proof. Part (a) follows from the second part of the proof of Theorem 1.1.29. Part (b) follows using the equality $H_{c_i}^+ = H_{-c_i}^-$ and using part (a). $\square$

By Theorem 1.1.29 a polyhedral cone $C \subsetneq \mathbb{R}^n$ has two representations:

Minkowski representation $C = \mathbb{R}_+\mathcal{B}$ with $\mathcal{B} = \{\beta_1, \ldots, \beta_r\}$ a finite set, and

Implicit representation $C = H_{c_1}^+ \cap \cdots \cap H_{c_s}^+$ for some $c_1, \ldots, c_s \in \mathbb{R}^n \setminus \{0\}$.

From the duality theorem for cones these two representations satisfy:

$$H_{\beta_1}^+ \cap \cdots \cap H_{\beta_r}^+ = \mathbb{R}_+ c_1 + \cdots + \mathbb{R}_+ c_s. \tag{1.2}$$

The *dual cone* of C is defined as

$$C^* := \bigcap_{c \in C} H_c^+ = \bigcap_{a \in \mathcal{B}} H_a^+.$$

By the duality theorem one has $C^{**} = C$. An implicit representation of C is called *irredundant* or *irreducible* if none of the closed half-spaces $H_{c_1}^+, \ldots, H_{c_s}^+$ can be omitted from the intersection.

Remark 1.1.31 The left-hand side of Eq. (1.2) is an irredundant representation of C^* if and only if no proper subset of $\mathcal{B}$ generates C.

Corollary 1.1.32 *Let $C \subset \mathbb{R}^n$ be a finitely generated cone and let*

$$\mathcal{F} = \{F \mid F = H_a \cap C;\ \dim(F) = r - 1;\ C \subset H_a^-\} = \{F_1, \ldots, F_s\} \neq \emptyset,$$

where $F_i = C \cap H_{a_i}$, $\dim(C) = r$. Then $C = \operatorname{aff}(C) \cap H_{a_1}^- \cap \cdots \cap H_{a_s}^-$.

Proof. It follows from the first part of the proof of Theorem 1.1.29. □

One of the fundamental results in polyhedral geometry is the following remarkable decomposition theorem for polyhedra. See [372, Corollary 7.1b] and [427, Theorem 4.1.3] for historical comments and for more information about this result.

Theorem 1.1.33 (Finite basis theorem) *If $\mathcal{Q}$ is a set in $\mathbb{R}^n$, then $\mathcal{Q}$ is a polyhedron (resp. rational polyhedron) if and only if $\mathcal{Q}$ can be expressed as $\mathcal{Q} = \mathcal{P} + C$, where $\mathcal{P}$ is a convex polytope (resp. rational polytope) and C is a finitely generated cone (resp. finitely generated rational cone).*

Proof. $\Rightarrow$) Let $\mathcal{Q} = \{x \mid Ax \leq b\}$ be a polyhedron in $\mathbb{R}^n$, where A is a matrix and b is a vector. Consider the set

$$C' = \left\{ \begin{pmatrix} x \\ \lambda \end{pmatrix} \middle| \, x \in \mathbb{R}^n; \lambda \in \mathbb{R}_+; Ax - \lambda b \leq 0 \right\}.$$

Notice that C' can be written as

$$C' = \left\{ \begin{pmatrix} x \\ \lambda \end{pmatrix} \middle| \, x \in \mathbb{R}^n; \lambda \in \mathbb{R}; \begin{pmatrix} A & -b \\ 0 & -1 \end{pmatrix} \begin{pmatrix} x \\ \lambda \end{pmatrix} \leq 0 \right\},$$

where $-b$ is a column vector. Thus C' is a polyhedral cone in $\mathbb{R}^{n+1}$. Using Corollary 1.1.29 we get that C' can be expressed as

$$C' = \mathbb{R}_+ \left\{ \begin{pmatrix} x_1 \\ \lambda_1 \end{pmatrix}, \ldots, \begin{pmatrix} x_m \\ \lambda_m \end{pmatrix} \right\} \qquad (\lambda_i \geq 0;\ x_i \in \mathbb{R}^n).$$

We may assume that $\lambda_i \in \{0,1\}$ for all i. Set

$$\mathcal{A} = \{x_i |\, \lambda_i = 1\} = \{x_1, \ldots, x_r\}, \quad \mathcal{B} = \{x_i |\, \lambda_i = 0\} = \{x_{r+1}, \ldots, x_m\},$$

$\mathcal{P} = \operatorname{conv}(\mathcal{A})$, and $C = \mathbb{R}_+\mathcal{B}$. Notice that $x \in \mathcal{Q}$ if and only if $(x,1) \in C'$. Thus $x \in \mathcal{Q}$ if and only if $(x,1)$ can be written as

$$\begin{pmatrix} x \\ 1 \end{pmatrix} = \mu_1 \begin{pmatrix} x_1 \\ 1 \end{pmatrix} + \cdots + \mu_r \begin{pmatrix} x_r \\ 1 \end{pmatrix} + \mu_{r+1} \begin{pmatrix} x_{r+1} \\ 0 \end{pmatrix} + \cdots + \mu_m \begin{pmatrix} x_m \\ 0 \end{pmatrix}$$

with $\mu_i \geq 0$ for all i. Consequently $x \in \mathcal{Q}$ if and only if $x \in \mathcal{P}+C$. Therefore we obtain $\mathcal{Q} = \mathcal{P} + C$.

$\Leftarrow$) Assume that $\mathcal{Q}$ is equal to $\mathcal{P} + C$ with $\mathcal{P} = \operatorname{conv}(x_1, \ldots, x_r)$ and $C = \mathbb{R}_+(x_{r+1}, \ldots, x_m)$. Consider the following finitely generated cone

$$C' = \mathbb{R}_+ \left\{ \begin{pmatrix} x_1 \\ 1 \end{pmatrix}, \ldots, \begin{pmatrix} x_r \\ 1 \end{pmatrix}, \begin{pmatrix} x_{r+1} \\ 0 \end{pmatrix}, \ldots, \begin{pmatrix} x_m \\ 0 \end{pmatrix} \right\}.$$

By Corollary 1.1.29 the cone C' is a polyhedron. Thus there exists a matrix A and a vector b such that C' can be written as

$$C' = \left\{ \begin{pmatrix} x \\ \lambda \end{pmatrix} \middle|\, x \in \mathbb{R}^n; \lambda \in \mathbb{R}; Ax + \lambda b \leq 0 \right\}.$$

Since $x \in \mathcal{Q}$ if and only if $(x,1) \in C'$ we conclude that $\mathcal{Q} = \{x |\, Ax \leq -b\}$, that is $\mathcal{Q}$ is a polyhedron. This proof is due to Schrijver [372]. $\Box$

Thus, by the finite basis theorem, a polyhedron has two representations. The computer program PORTA [84] can be used to switch between these two representations. This program is available for Unix and Windows systems. For polyhedral cones one can use *Normaliz* [68].

Corollary 1.1.34 *A set $\mathcal{Q} \subset \mathbb{R}^n$ is a convex polytope if and only if $\mathcal{Q}$ is a bounded polyhedral set.*

Proof. If $\mathcal{Q} = \operatorname{conv}(\alpha_1, \ldots, \alpha_m)$ is a polytope, then by the triangle inequality for all $x \in \mathcal{Q}$ we have $\|x\| \leq \|\alpha_1\| + \cdots + \|\alpha_m\|$. Thus $\mathcal{Q}$ is bounded.

Conversely if $\mathcal{Q}$ is a bounded polyhedron, then by Theorem 1.1.33 we can decompose $\mathcal{Q}$ as $\mathcal{Q} = \mathcal{P} + C$, with $\mathcal{P}$ a polytope and C a finitely generated cone. Notice that $C = \{0\}$, otherwise fixing $p_0 \in \mathcal{P}$ and $c_0 \in C \setminus \{0\}$ we get $p_0 + \lambda c_0 \in \mathcal{Q}$ for all $\lambda > 0$, a contradiction because $\mathcal{Q}$ is bounded. Thus $\mathcal{P} = \mathcal{Q}$, as required. $\Box$

Definition 1.1.35 Let $\mathcal{Q}$ be a polyhedral set and $x_0 \in \mathcal{Q}$. The point x_0 is called a *vertex* or an *extreme point* of $\mathcal{Q}$ if $\{x_0\}$ is a proper face of $\mathcal{Q}$.

Proposition 1.1.36 [57, Theorem 7.2] *If $\mathcal{P} = \operatorname{conv}(\alpha_1, \ldots, \alpha_r) \subset \mathbb{R}^n$ and V is the set of vertices of $\mathcal{P}$, then $V \subset \{\alpha_1, \ldots, \alpha_r\}$ and $\mathcal{P} = \operatorname{conv}(V)$.*

Lemma 1.1.37 [57, Theorem 5.6] *If $\mathcal{Q}$ is a polyhedral set in $\mathbb{R}^n$ and $v \in \mathcal{Q}$ is not a vertex of $\mathcal{Q}$, then there is a face F of $\mathcal{Q}$ such that $v \in \mathrm{ri}(F)$.*

Lemma 1.1.38 *Let $\mathcal{Q}$ be a convex polyhedron in $\mathbb{R}^n$ and $x_0 \in \mathcal{Q}$. Then x_0 is a vertex of $\mathcal{Q}$ if and only if $\mathcal{Q} \setminus \{x_0\}$ is a convex set.*

Proof. $\Rightarrow$) Let $H(a, c)$ be a proper supporting hyperplane of $\mathcal{Q}$ such that $\{x_0\} = \mathcal{Q} \cap H(a, c)$ and $\mathcal{Q} \subset H^+(a, c)$. Take $x, y \in \mathcal{Q} \setminus \{x_0\}$ and consider $z = ty + (1 - t)x$ with $0 < t < 1$. Note that x, y are not in $H(a, c)$, thus

$$\langle z, a\rangle = t\langle y, a\rangle + (1 - t)\langle x, a\rangle > tc + (1 - t)c = c.$$

Hence $\langle z, a\rangle > c$, that is, $z \neq x_0$ and $z \in \mathcal{Q}$, as required.

$\Leftarrow$) Let $\mathcal{Q} = \cap_{i=1}^r H^+(a_i, c_i)$ be a decomposition of $\mathcal{Q}$ as an intersection of closed halfspaces, where $0 \neq a_i \in \mathbb{R}^n$ and $c_i \in \mathbb{R}$ for all i. First observe that x_0 is in $H(a_i, c_i)$ for some i, otherwise if $\langle x_0, a_i\rangle > c_i$ for all i, there is an open ball $B_\delta(x_0)$ in $\mathbb{R}^n$, of radius δ centered at x_0, whose closure lies in $\mathcal{Q}$. Hence taking two antipodal points x_1, x_2 in the boundary of $B_\delta(x_0)$ one obtains $x_0 \in \mathcal{Q} \setminus \{x_0\}$, a contradiction. One may now assume there is $k \geq 1$ such that $x_0 \in H(a_i, c_i)$ for $i \leq k$ and $\langle x_0, a_i\rangle > c_i$ for $i > k$. Set

$$A = H(a_1, c_1) \cap \cdots \cap H(a_k, c_k).$$

We claim that $A = \{x_0\}$. If there is $x_1 \in A \setminus \{x_0\}$, pick $B_\delta(x_0)$ whose closure is contained in $H^+(a_i, c_i)$ for all $i > k$. Since $z = tx_0 + (1 - t)x_1$ is in A for all $t \in \mathbb{R}$, making $t = 1 + \delta/\|x_1 - x_0\|$ one derives $z \in A$ and $\|z - x_0\| = \delta$, thus $z \in \mathcal{Q}$. Note that if $k = r$, then x_1 is already in $\mathcal{Q}$ and in this case we set $z = x_1$. Making $t = -1$ in $z_1 = tz + (1 - t)x_0$ one concludes $z_1 \in A$ and $\|z_1 - x_0\| = \delta$. Altogether z, z_1 are in $\mathcal{Q} \setminus \{x_0\}$ and $x_0 = (z + z_1)/2$, a contradiction because $\mathcal{Q} \setminus \{x_0\}$ is a convex set.

From the equality $\{x_0\} = A = A \cap Q$ one has that $\{x_0\}$ is an intersection of faces. Using Proposition 1.1.9 yields that $\{x_0\}$ is a face, as required. □

Proposition 1.1.39 *If $\mathcal{Q} = C + \mathcal{P} \subset \mathbb{R}^n$ with C a polyhedral cone and $\mathcal{P}$ a polytope, then every vertex of $\mathcal{Q}$ is a vertex of $\mathcal{P}$.*

Proof. Let x be a vertex of $\mathcal{Q}$ and write $x = c + p$ for some $c \in C$ and $p \in \mathcal{P}$. We claim that p is a vertex of $\mathcal{P}$. If p is not a vertex of $\mathcal{P}$, then by Lemma 1.1.37 there is a face F of $\mathcal{P}$ such that $p \in \mathrm{ri}(F)$. Pick $p \neq v \in F$. By [57, Theorem 3.5], there is $y \in F$ with $p \in]y, v[$. Hence $p = \lambda y + (1 - \lambda)v$ with $0 < \lambda < 1$. Note that $x \neq y + c \in \mathcal{Q}$, $x \neq v + c \in \mathcal{Q}$, and

$$x = \lambda(y + c) + (1 - \lambda)(v + c) \qquad (0 < \lambda < 1),$$

thus $\mathcal{Q} \setminus \{x\}$ is not a convex set, a contradiction to Lemma 1.1.38. This proves that p is a vertex of $\mathcal{P}$. Assume $c \neq 0$, then

$$p + \lambda c \neq x = p + c \neq p + 2c \qquad (0 < \lambda < 1).$$

Notice that $x = p+c = \lambda_0(p+\lambda c)+(1-\lambda_0)(p+2c)$ is a convex combination, where $0 < \lambda_0 = 1/(2-\lambda) < 1$. Since $p+\lambda c$ and $p+2c$ are in $\mathcal{Q}\setminus\{x\}$, this shows that $\mathcal{Q}\setminus\{x\}$ is not a convex set, a contradiction. Thus $c = 0$ and x is a vertex of $\mathcal{P}$, as required. □

Theorem 1.1.40 [372, Theorem 8.4] *Let $\mathcal{Q} \subset \mathbb{R}^n$ be a polyhedral set. Then $\mathcal{Q}$ has at least one vertex if and only if $\mathcal{Q}$ does not contain any lines in $\mathbb{R}^n$.*

Proposition 1.1.41 *Let $\mathcal{Q}$ be a polyhedron containing no lines and let $f(x) = \langle x, a\rangle$. If $\max\{f(x)|\, x \in \mathcal{Q}\} < \infty$, then the maximum is attained at a vertex of $\mathcal{Q}$.*

Proof. Let $x_0 \in \mathcal{Q}$ be an optimal solution and let $\lambda = f(x_0)$ be the optimal value. Note that $F = \mathcal{Q}\cap H(a,\lambda)$ is a face of $\mathcal{Q}$ because $\mathcal{Q} \subset H^-(a,\lambda)$ and $x_0 \in F$. Since F contains no lines, by Theorem 1.1.40, the face F contains at least one vertex z_0, which is also a vertex of $\mathcal{Q}$ by transitivity. Thus the optimal value λ is attained at the vertex z_0, as required. □

Theorem 1.1.42 *If $\mathcal{Q} \subset \mathbb{R}^n$ is a polyhedral set containing no lines and $\alpha_1,\dots,\alpha_r$ are the vertices of $\mathcal{Q}$, then there are $\beta_1,\dots,\beta_s \in \mathbb{R}^n$ such that*

$$\mathcal{Q} = \operatorname{conv}(\alpha_1,\dots,\alpha_r) + (\mathbb{R}_+\beta_1 + \cdots + \mathbb{R}_+\beta_s).$$

Proof. It follows from the proof of [427, Theorem 4.1.3]. □

Definition 1.1.43 If a polyhedron $\mathcal{Q}$ in $\mathbb{R}^n$ is represented as

$$\mathcal{Q} = \operatorname{aff}(\mathcal{Q}) \cap H^+(a_1,c_1) \cap \cdots \cap H^+(a_r,c_r) \tag{1.3}$$

with $a_i \in \mathbb{R}^n\setminus\{0\}$, $c_i \in \mathbb{R}$ for all i, and none of the closed halfspaces $H^+(a_1,c_1),\dots,H^+(a_r,c_r)$ can be omitted from the intersection, we say that Eq. (1.3) is an *irreducible representation* of $\mathcal{Q}$.

Theorem 1.1.44 [427, Theorem 3.2.1] *Let $\mathcal{Q}$ be a polyhedral set in $\mathbb{R}^n$ which is not an affine space and let*

$$\mathcal{Q} = \operatorname{aff}(\mathcal{Q}) \cap H^+(a_1,c_1) \cap \cdots \cap H^+(a_r,c_r)$$

be an irreducible representation. If $F_i = \mathcal{Q}\cap H(a_i,c_i)$, $i = 1,\dots,r$, then

(a) $\operatorname{ri}(\mathcal{Q}) = \{x \in \mathcal{Q}|\, \langle x,a_1\rangle > c_1,\dots,\langle x,a_r\rangle > c_r\}$;

(b) $\operatorname{rb}(\mathcal{Q}) = F_1 \cup\cdots\cup F_r$;

(c) *the facets of $\mathcal{Q}$ are precisely the sets $F_1,\dots,F_r$;*

(d) *each face $F \subsetneq \mathcal{Q}$ is the intersection of all F_i such that $F \subset F_i$.*

Corollary 1.1.45 Let $\mathcal{Q}$ be a polyhedron and let β be a vector in $\mathrm{ri}(\mathcal{Q})$. If F is a proper face of $\mathcal{Q}$, then $\beta \notin F$.

Proposition 1.1.46 *Let $\mathcal{Q} = \bigcap_{i=1}^{r} H^+(a_i, c_i)$ be a polyhedral set in $\mathbb{R}^n$ which is not an affine space, where $0 \neq a_i \in \mathbb{R}^n$ and $c_i \in \mathbb{R}$ for all i. If $x_0 \in \mathbb{R}^n$, then x_0 is a vertex of $\mathcal{Q}$ if and only if*

(a) *$\{x_0\} = \cap_{i \in I} H(a_i, c_i)$ for some $I \subset \{1, \ldots, r\}$, and*

(b) *$\langle x_0, a_i \rangle \geq c_i$ for all $i = 1, \ldots, r$.*

Proof. $\Rightarrow$) This direction follows at once from the proof of Lemma 1.1.38.
$\Leftarrow$) Note $x_0 \in \mathcal{Q}$. Since the intersection of faces of $\mathcal{Q}$ is a face (see Proposition 1.1.9), one has that $\{x_0\}$ is a face. □

Corollary 1.1.47 *Let $A = (a_{ij})$ be an $r \times n$ matrix, let $c = (c_i)$ be a column vector and let $a_1, \ldots, a_r$ be the rows of A. If $\mathcal{Q} = \{x \in \mathbb{R}^n \mid Ax \geq c\}$ and $x_0 \in \mathcal{Q}$, then x_0 is a vertex of $\mathcal{Q}$ if and only if there is $J \subset \{1, \ldots, r\}$ with $|J| = n$ such that the set $\{a_i \mid i \in J\}$ is linearly independent and*

$$\{x_0\} = \{(x_i) \in \mathbb{R}^n \mid a_{i1}x_1 + \cdots + a_{in}x_n = c_i \ \ \textit{for all} \ \ i \in J\}.$$

Proof. Assume x_0 is a vertex of $\mathcal{Q}$. Hence, by Proposition 1.1.46, x_0 is the unique solution of a system of linear equations:

$$a_{i1}x_1 + \cdots + a_{in}x_n = c_i \quad (i \in I)$$

for some $I \subset \{1, \ldots, r\}$. Let $[A'|c']$ be the augmented matrix of this system, by Gaussian elimination one obtains that this matrix reduces to

$$\left[\begin{array}{c|c} I_n & c'' \\ 0 & 0 \end{array}\right].$$

Hence the rank of A' is equal to n. Thus there are n linearly independent rows of A' and x_0 is the unique solution of the system

$$a_{i1}x_1 + \cdots + a_{in}x_n = c_i \quad (i \in J \subset I)$$

for some $J \subset I$ with $|J| = n$. The converse is clear. □

Let $\mathcal{Q}$ be a polyhedron in $\mathbb{R}^n$ represented by a system of linear constraints

$$\langle a_i, x \rangle \geq b_i \qquad (i = 1, \ldots, r),$$

where $a_i \in \mathbb{R}^n$ and $b_i \in \mathbb{R}$ for all i. With a slight abuse of language, we will say that the constraints are linearly independent if the corresponding a_i are linearly independent.

Definition 1.1.48 A vector x_0 in $\mathbb{R}^n$ is called a *basic feasible solution* of a system of linear constraints $\langle a_i, x\rangle \geq b_i$, $i = 1, \ldots, r$, if

(a) x_0 satisfies all linear constraints, and

(b) out of the constraints that satisfy $\langle a_i, x_0\rangle = b_i$, there are n of them that are linearly independent.

The next result is a restatement of the corollary above.

Corollary 1.1.49 *Let $\mathcal{Q}$ be a polyhedron in $\mathbb{R}^n$. A vector x_0 in $\mathbb{R}^n$ is a vertex of $\mathcal{Q}$ if and only if x_0 is a basic feasible solution of any system of linear constraints that represents $\mathcal{Q}$.*

This result yields a method to find the vertices of a polyhedron which is by no means the best in practice. See [10, 11] for a thorough discussion on how to find all the vertices of a polyhedron.

Corollary 1.1.50 *If $C \neq \mathbb{R}^n$ is a polyhedral cone of dimension n, then there is a unique irreducible representation*

$$C = H_{a_1}^+ \cap \cdots \cap H_{a_r}^+, \text{ where } a_i \in \mathbb{R}^n \setminus \{0\}.$$

Proof. According to Proposition 1.1.22 a set F is a proper face of C if there is a supporting hyperplane H_a of C such that $F = C \cap H_a \neq \emptyset$ and $C \not\subset H_a$. In particular the facets of C are defined by hyperplanes through the origin. Therefore by Theorem 1.1.44 the irreducible representation of C has the required form and is unique. $\Box$

Proposition 1.1.51 *If $\mathcal{Q} \neq \mathbb{R}^n$ is a rational polyhedral cone of dimension n in $\mathbb{R}^n$, then there are unique (up to sign) $a_1, \ldots, a_r$ in $\mathbb{Z}^n$ with relatively prime entries such that the irreducible representation of $\mathcal{Q}$ is*

$$\mathcal{Q} = H_{a_1}^+ \cap H_{a_2}^+ \cap \cdots \cap H_{a_r}^+$$

Proof. By the finite basis theorem, there are $\alpha_1, \ldots, \alpha_q$ in $\mathbb{Q}^n$ such that $\mathcal{Q} = \mathbb{R}_+\alpha_1 + \cdots + \mathbb{R}_+\alpha_q$. First note that if H_b is a supporting hyperplane of $\mathcal{Q}$ generated by a set of $n-1$ linearly independent vectors in $\{\alpha_1, \ldots, \alpha_q\}$, then by the Gram–Schmidt process H_b has an orthogonal basis of vectors in $\mathbb{Q}^n$, and consequently there is a normal vector a to H_b such that $a \in \mathbb{Q}^n$ and $H_a = H_b$. Hence multiplying a by a suitable integer and then dividing by the greatest common divisor of the entries, one may assume $H_a = H_b$, where a is in $\mathbb{Z}^n$ and has relative prime entries. Observe that a, b are linearly dependent because the orthogonal complement of H_a is one dimensional. It is readily seen that a is uniquely determined up to sign.

To complete the proof use Proposition 1.1.23 (c) to see that any facet of $\mathcal{Q}$ is defined by a supporting hyperplane H_b as above. $\Box$

Definition 1.1.52 A polyhedron containing no lines is called *pointed.*

If F is a face of dimension 1 of a pointed polyhedral cone, then $F = \mathbb{R}_+ v$ for some v (see Proposition 1.1.23). The face F is called an *extremal ray.*

Lemma 1.1.53 [372, Section 8.8] *If C is a pointed polyhedral cone in $\mathbb{R}^n$, then C is generated by non-zero representatives of its extremal rays.*

Theorem 1.1.54 *Let $\mathcal{A}$ be a finite set of non-zero points in $\mathbb{Z}^n$ and let $\mathcal{V}$ be the set of all $\alpha \in \mathcal{A}$ such that $\mathbb{R}_+\alpha$ is a face of $\mathbb{R}_+\mathcal{A}$ of dimension* 1. *If $\mathbb{R}_+\mathcal{A}$ is a pointed cone, then $\mathbb{R}_+\mathcal{A} = \mathbb{R}_+\mathcal{V}$.*

Proof. We set $\mathcal{A} = \{\alpha_1, \ldots, \alpha_q\}$. Let F be a face of $\mathbb{R}_+\mathcal{A}$ of dimension 1. Then, by Proposition 1.1.23, $F = \mathbb{R}_+\alpha_i$ for some i. Thus, the result follows from Lemma 1.1.53. □

Definition 1.1.55 Let A be an $n \times q$ real matrix and let b, c be two real vectors of sizes q and n, respectively. The *primal problem* is defined as

$$\text{Maximize } f(x) = \textstyle\sum_{i=1}^n c_i x_i \qquad (*)$$
$$\text{Subject to } xA \leq b \text{ and } x \geq 0.$$

Its *dual problem* is defined as

$$\text{Minimize } g(y) = \textstyle\sum_{i=1}^q b_i y_i \qquad (**)$$
$$\text{Subject to } Ay \geq c \text{ and } y \geq 0.$$

A solution $x_0 \in \mathbb{R}^n$ of the primal (resp. dual) problem that maximizes the *objective function* f is called and *optimal solution*, the corresponding value $f(x_0)$ of the objective function is called an *optimal value.*

The most important theorem in Linear Programming (LP) theory is:

Theorem 1.1.56 (LP duality theorem, [372, Corollary 7.1.g]) *If the primal problem* $(*)$ *has an optimal solution $(x_1, \ldots, x_n)$, then the dual problem* $(**)$ *has an optimal solution $(y_1, \ldots, y_q)$ such that $\sum_{i=1}^n c_i x_i = \sum_{i=1}^q b_i y_i$.*

Theorem 1.1.57 [372, Corollary 7.1g, p. 92] *Let A be a matrix and b, c vectors. Then*

$$\max\{\langle c, x\rangle \mid xA \leq b\} = \min\{\langle b, y\rangle \mid y \geq 0,\ Ay = c\} \qquad (1.4)$$

provided that at least one of the sets in (1.4) *is non-empty.*

An immediate consequence of the duality theorem is:

Corollary 1.1.58 *Consider the LP primal-dual pair*

$$\max\{\langle x,c\rangle \,|\, xA \leq b\} = \min\{\langle y,b\rangle \,|\, y \geq 0; Ay = c\}.$$

Let x and y be feasible solutions, i.e., $xA \leq b$, $y \geq 0$ and $Ay = c$. Then the following conditions are equivalent:

(a) *x and y are both optimum solutions.*

(b) $\langle x,c\rangle = \langle y,b\rangle$.

(c) $(b - xA)y = 0$.

Condition (c) is called *complementary slackness*. See Exercise 1.1.81 for another form of complementary slackness

Proposition 1.1.59 [281, Proposition 3.3] *Let $\mathcal{Q} = \{x \in \mathbb{R}^n \,|\, xA \leq b\} \neq \emptyset$ be a polyhedral set and let $F \neq \emptyset$ be a proper subset of $\mathcal{Q}$. Then the following conditions are equivalent:*

(a) $F = \{x \in \mathcal{Q} \,|\, A'x = b'\}$ *for some subsystem $xA' \leq b'$ of $xA \leq b$.*

(b) *F is a proper face of $\mathcal{Q}$.*

(c) $F = \{x \in \mathcal{Q} \,|\, \langle x,c\rangle = \delta\}$ *for some vector $0 \neq c \in \mathbb{R}^n$ such that the maximum value $\delta = \max\{\langle x,c\rangle \,|\, x \in \mathcal{Q}\}$ is finite.*

Definition 1.1.60 A *minimal face* of a polyhedron is a face not containing any other face.

Lemma 1.1.61 [372, p. 104] *Let $\mathcal{Q} = \{x \in \mathbb{R}^n \,|\, xA \leq b\} \neq \emptyset$ be a convex polyhedron. A set F is a minimal face of $\mathcal{Q}$ if and only if $\emptyset \neq F \subsetneq \mathcal{Q}$ and F is equal to $F = \{x \in \mathbb{R}^n \,|\, A'x = b'\}$ for some subsystem $xA' \leq b'$ of $xA \leq b$.*

Definition 1.1.62 A rational polyhedron $\mathcal{Q} \subset \mathbb{R}^n$ is called *integral* if $\mathcal{Q}$ is equal to $\mathrm{conv}(\mathbb{Z}^n \cap \mathcal{Q})$.

If $\mathcal{Q}$ is a pointed rational polyhedron, then $\mathcal{Q}$ is integral if and only if $\mathcal{Q}$ has only integral vertices. This is shown below.

Theorem 1.1.63 [281, Theorem 5.12] *If $\mathcal{Q}$ is a rational polyhedron, then $\mathcal{Q}$ is integral if and only if any of the following statements hold:*

(a) *Each face of $\mathcal{Q}$ contains integral vectors.*

(b) *Each minimal face of $\mathcal{Q}$ contains integral vectors.*

(c) $\max\{\langle c,x\rangle \,|\, x \in \mathcal{Q}\}$ *is attained by an integral vector for each c for which the maximum is finite.*

(d) $\max\{\langle c,x\rangle \,|\, x \in \mathcal{Q}\}$ *is an integer for each integral vector c for which the maximum is finite.*

Corollary 1.1.64 *If $\mathcal{Q}$ is a pointed polyhedron, then $\mathcal{Q}$ is integral if and only if all vertices of $\mathcal{Q}$ are integral.*

Proof. Let x_0 be a vertex of $\mathcal{Q}$. There is a supporting hyperplane $H = H(a, c)$ such that $\{x_0\} = H \cap \mathcal{Q}$ and $\mathcal{Q} \subset H^+$. Since x_0 is a convex combination of points in $\mathcal{Q} \cap \mathbb{Z}^n$, it is seen that x_0 is integral. The converse follows from Theorem 1.1.63 and Proposition 1.1.41. □

Proposition 1.1.65 *Let $\mathcal{Q} \subset \mathbb{R}^n$ be a rational polyhedron and let $\mathcal{Q}_I = \operatorname{conv}(\mathcal{Q} \cap \mathbb{Z}^n)$ be its integral hull. Then*

(a) *$\mathcal{Q}_I$ is a polyhedron.*

(b) *If $\mathcal{Q}$ is a pointed polyhedron, then $\mathcal{Q}_I$ is integral.*

Proof. We may assume that $\mathcal{Q}_I$ is non-empty. By the finite basis theorem we can write $\mathcal{Q} = \mathcal{P} + C$, where C is a cone generated by integral vectors $\alpha_1, \ldots, \alpha_s$ and $\mathcal{P}$ is a polytope. Consider the linear map $T\colon \mathbb{R}^s \to \mathbb{R}^n$ induced by $T(e_i) = \alpha_i$. Notice that $B := T([0,1]^s)$ is a polytope whose elements have the form $\lambda_1\alpha_1 + \cdots + \lambda_s\alpha_s$, $0 \leq \lambda_i \leq 1$ for all i. It is not hard to see that $\mathcal{Q}_I = \operatorname{conv}((\mathcal{P}+B)\cap\mathbb{Z}^n)+C$. Since $\mathcal{P}+B$ is bounded, we get that $\mathcal{Q}_I$ is a polyhedron by the finite basis theorem. This proves (a). To prove part (b) observe that the vertices of $\mathcal{Q}_I$ are contained in $(\mathcal{P}+B)\cap\mathbb{Z}^n$, this follows from Proposition 1.1.39. Hence by Corollary 1.1.64, $\mathcal{Q}_I$ is integral. This proof was adapted from [372]. □

Exercises

1.1.66 If $f\colon \mathbb{R}^q \to \mathbb{R}^n$ is an affine map, then $f(x) = Ax + b$, for some $n \times q$ matrix A and some vector b.

1.1.67 Let $\mathcal{A} = \{\alpha_0, \ldots, \alpha_r\}$ be a sequence of distinct vectors in $\mathbb{R}^n$. Prove:

(a) $\mathcal{A}$ is affinely independent $\Leftrightarrow$ $\alpha_1 - \alpha_0, \ldots, \alpha_r - \alpha_0$ are linearly independent $\Leftrightarrow$ $(\alpha_0, 1), \ldots, (\alpha_r, 1)$ are linearly independent.

(b) If $\mathcal{A}$ is contained in an affine hyperplane not containing the origin, then $\mathcal{A}$ is affinely independent if and only if it is linearly independent.

1.1.68 If $V \subset \mathbb{R}^n$, prove that V is an affine space in $\mathbb{R}^n$ if and only if $\lambda_1 V + \lambda_2 V \subset V$ for all λ_1, λ_2 in $\mathbb{R}$ such that $\lambda_1 + \lambda_2 = 1$.

1.1.69 If $C \subset \mathbb{R}^n$, prove that C is a convex cone if and only if $\lambda x + \mu y \in C$ for all $x, y \in C$ and $\lambda, \mu \geq 0$.

1.1.70 Let $\pi_1 \colon \mathbb{R}^2 \to \mathbb{R}$ be the projection $\pi_1(x_1, x_2) = x_1$. Prove that π_1 is not a closed linear map.

Hint Consider the closed set $\{(x, 1/x) \mid x > 0\}$.

1.1.71 Let $\mathcal{Q}$ be a polyhedral set in $\mathbb{R}^n$ and let $f\colon \mathbb{R}^n \to \mathbb{R}^m$ be a linear function. Prove that $f(\mathcal{Q})$ is a polyhedral set.

1.1.72 Let $\mathcal{Q}$ be a polyhedron in $\mathbb{R}^n$ and let $f\colon \mathcal{Q} \to \mathbb{R}$ be a linear function. If f is bounded from above and $\mathcal{Q}$ contains no lines, prove that f attains its maximum at a vertex of $\mathcal{Q}$. Notice that $f(\mathcal{Q})$ is a closed convex set.

1.1.73 Determine the facets of the convex polytope

$$\mathcal{P} = \operatorname{conv}(\pm e_1, \ldots, \pm e_n) \subset \mathbb{R}^n.$$

1.1.74 Let C be a polyhedral cone in $\mathbb{R}^n$ and let C^* be its dual cone. Then C is a pointed cone if and only if $\dim(C^*) = n$.

1.1.75 If $C = \mathbb{R}^n$ and C^* is its dual cone, then $C^* = (0)$.

1.1.76 If $C = H^+_{e_2} \subset \mathbb{R}^2$, then $C = \mathbb{R}_+ e_1 + \mathbb{R}_+(-e_1) + \mathbb{R}_+ e_2$. Use the duality theorem to show that $C^* = \mathbb{R}_+ e_2$.

1.1.77 If $C = \mathbb{R}_+(1,1)$, prove that $C = H^+_{(1,-1)} \cap H^+_{(-1,1)} \cap H^+_{e_1}$ and that the dual cone is given by $C^* = H^+_{(1,1)}$.

1.1.78 If C is a rational cone and $C_I := \operatorname{conv}(C \cap \mathbb{Z}^n)$, then $C_I = C$.

1.1.79 Let $\mathcal{Q} = \{x \mid Ax \leq b\} \neq \emptyset$ be a polyhedron. If $\mathcal{Q} = \mathcal{P} + C$, where $\mathcal{P}$ is a polytope and C is a polyhedral cone, prove that

$$\{y | Ay \leq 0\} = \{y | x + y \in \mathcal{Q},\, \forall\, x \in \mathcal{Q}\}.$$

The cone $C = \{y | Ay \leq 0\}$ is called the *characteristic cone* of $\mathcal{Q}$.

1.1.80 Let $\mathcal{Q} = \{x \mid Ax \leq b\}$ be a polyhedron. The *lineality space* of $\mathcal{Q}$ is the linear space $\text{lin.space}(\mathcal{Q}) = \{x \mid Ax = 0\}$. Prove that $\mathcal{Q}$ is pointed if and only if $\text{lin.space}(\mathcal{Q}) = (0)$.

1.1.81 (Complementary slackness) Consider the LP primal-dual pair

$$\min\{\langle x, c\rangle \mid x \geq 0; xA \geq b\} = \max\{\langle y, b\rangle \mid y \geq 0; Ay \leq c\}.$$

Let x and y be feasible solutions, i.e., $xA \geq b$, $Ay \leq c$ and $x, y \geq 0$. Use Theorem 1.1.56 to show that the following conditions are equivalent:

(a) x and y are both optimum solutions.

(b) $\langle x, c\rangle = \langle y, b\rangle$.

(c) $(b - xA)y = 0$ and $x(c - Ay) = 0$.

1.1.82 (P. Gordan [194]) The system $Ax < 0$ is unsolvable if and only if the system $yA = 0, y \geq 0,\ y \neq 0$ is solvable. The vector inequality $(a_i) < (b_i)$ means that $a_i < b_i$ for all i.

In the following exercises $\mathbb{K}$ will denote an intermediate field $\mathbb{Q} \subset \mathbb{K} \subset \mathbb{R}$.

1.1.83 Let $H \subset \mathbb{K}^n$ be a linear subspace and $\{v_1, v_2, \ldots, v_m\}$ a basis of H. Prove that H has an orthogonal basis using the Gram–Schmidt process:

1. Set $u_1 := v_1$ and $u_2 := v_2 - ((v_2 \cdot u_1)/(u_1 \cdot u_1))u_1$. Verify that $\{u_1, u_2\}$ is an orthogonal set and $\mathbb{K}u_1 + \mathbb{K}u_2 = \mathbb{K}v_1 + \mathbb{K}v_2$.
2. If $u_1, u_2, \ldots, u_k$ $(k < m)$ have been constructed so that they form an orthogonal basis for the linear space generated by $v_1, \ldots, v_k$, setting

$$u_{k+1} = v_{k+1} - \left[\frac{v_{k+1} \cdot u_1}{u_1 \cdot u_1}u_1 + \cdots + \frac{v_{k+1} \cdot u_i}{u_i \cdot u_i}u_i + \cdots + \frac{v_{k+1} \cdot u_k}{u_k \cdot u_k}u_k\right],$$

   verify that $\{u_1, u_2, \ldots, u_k, u_{k+1}\}$ is an orthogonal set and that this set generates $\mathbb{K}v_1 + \mathbb{K}v_2 + \cdots + \mathbb{K}v_k + \mathbb{K}v_{k+1}$.

1.1.84 If $A = (a_{ij})$ is a matrix with entries in $\mathbb{K}$, then $\dim(R_A)$ is equal to $\dim(\mathrm{im}(A))$, where R_A is the row space of A and $\mathrm{im}(A)$ is the image of the linear map defined by A.

1.1.85 If H is a linear subspace of $\mathbb{K}^n$, then $H \oplus H^\perp = \mathbb{K}^n$, where

$$H^\perp = \{x \in \mathbb{K}^n | \, x \cdot h = 0 \text{ for all } h \in H\}.$$

1.1.86 If $A \in M_{n\times q}(\mathbb{K})$, then

$$(\mathrm{im}(A))^\perp = \ker(A^t),\ \mathrm{im}(A) \oplus \ker(A^t) = \mathbb{K}^n, \text{ and } (\mathrm{im}(A^t))^\perp = \ker(A).$$

1.1.87 Let α, β, γ be three vectors in $\mathbb{K}^n$ such that $\langle \alpha, \beta\rangle \leq 0$ and $\langle \beta, \gamma\rangle > 0$. Prove that the vector $\alpha' = \alpha - \frac{\langle \alpha,\beta\rangle}{\langle \gamma,\beta\rangle}\gamma$ is in $\mathbb{K}^n$, $\langle \alpha', \beta\rangle = 0$, and α' belongs to the cone generated by α and γ.

1.1.88 Using the previous exercises show that the proof of Farkas's lemma given in [304, p. 86] is valid for any intermediate field $\mathbb{Q} \subset \mathbb{K} \subset \mathbb{R}$.

1.1.89 Show that the proof of the fundamental theorem of linear inequalities given in [372, Theorem 7.1] works for any intermediate field $\mathbb{Q} \subset \mathbb{K} \subset \mathbb{R}$.

1.1.90 If $A \in M_{n\times q}(\mathbb{K})$ and $\gamma \in \mathbb{K}^n$, then either there exists a vector $x \in \mathbb{K}^q$ with $Ax \leq \gamma$, or there exists a vector $\alpha \in \mathbb{K}^n$ with $\alpha \geq 0$, $\alpha A = 0$ and $\langle \alpha, \gamma\rangle < 0$, but not both.

1.2 Relative volumes of lattice polytopes

In this section we introduce relative volumes and unimodular coverings. Two main references for relative volumes are [21, 146].

Let $\mathcal{A} = \{v_1, \ldots, v_q\}$ be a set of distinct vectors in $\mathbb{Z}^n$ and let

$$\mathcal{P} = \text{conv}(v_1, \ldots, v_q)$$

be the convex hull of $\mathcal{A}$. A point in $\mathbb{Z}^n$ is called a *lattice point* and $\mathcal{P}$ is called a *lattice polytope*. Here a lattice point (resp. lattice polytope) is used as a synonym of integral point (resp. integral polytope).

In the sequel to simplify the exposition and the proofs we set $m = q - 1$ and $\alpha_0 = v_1, \ldots, \alpha_m = v_q$.

If $\mathcal{V} = \{0, \alpha_1 - \alpha_0, \ldots, \alpha_m - \alpha_0\}$ is the image of $\mathcal{A}$ under the translation

$$f\colon \mathbb{R}^n \longrightarrow \mathbb{R}^n, \qquad x \stackrel{f}{\longmapsto} x - \alpha_0,$$

then $\text{aff}(\mathcal{A}) = \alpha_0 + \mathbb{R}\mathcal{V}$. In particular, from this expression we get

$$d = \dim(\mathcal{P}) = \dim_{\mathbb{R}}(\mathbb{R}\mathcal{V}),$$

where $\mathbb{R}\mathcal{V}$ is the linear space spanned by $\mathcal{V}$.

Theorem 1.2.1 [261, Theorem 3.7] *Let $\mathbb{Z}^n$ be the free $\mathbb{Z}$-module of rank n. Then any submodule of $\mathbb{Z}^n$ is free of rank at most n.*

Using this result, and the inclusions $\mathbb{Z}\mathcal{V} \subset \mathbb{Z}^n \cap \mathbb{R}\mathcal{V} \subset \mathbb{R}\mathcal{V}$, it follows that $\mathbb{Z}^n \cap \mathbb{R}\mathcal{V}$ is a free abelian group of rank d. A constructive proof of this fact, which is useful for computing relative volumes, is included below in the proof of Lemma 1.2.3.

Theorem 1.2.2 [332, Theorem II.9, pp. 26-27] *Let A be an integral matrix of rank d. Then there are invertible integral matrices U and Q such that*

$$U^{-1}AQ = \text{diag}\{s_1, \ldots, s_d, 0, \ldots, 0\},$$

$s_i > 0$ for $1 \leq i \leq d$ and s_i divides s_{i+1} for $1 \leq i \leq d-1$.

The matrix $D = \text{diag}\{s_1, \ldots, s_d, 0, \ldots, 0\}$ is called the *Smith normal form* of A and $s_1, \ldots, s_d$ are called the *invariant factors* of A.

Lemma 1.2.3 $\mathbb{R}\mathcal{V} \cap \mathbb{Z}^n = \mathbb{Z}\gamma_1 \oplus \cdots \oplus \mathbb{Z}\gamma_d$ *for some $\gamma_1, \ldots, \gamma_d$ in $\mathbb{Z}^n$.*

Proof. Consider the matrix A of size $n \times m$ whose column vectors correspond to the non-zero vectors in $\mathcal{V}$. By Theorem 1.2.2, there are unimodular integral matrices U and Q, of orders n and m, respectively, such that

$$U^{-1}AQ = D = \text{diag}\{s_1, \ldots, s_d, 0, \ldots, 0\} \qquad (s_i \neq 0; s_i | s_{i+1} \,\forall\, i).$$

Let $u_1, \ldots, u_n$ be the columns of U. We claim that the leftmost d columns of U can serve as $\gamma_1, \ldots, \gamma_d$. We regard some vectors as column vectors.

Take $\alpha \in \mathbb{R}\mathcal{V} \cap \mathbb{Z}^n$, thus $\alpha = A\beta$ for some $\beta \in \mathbb{R}^m$. Set $\beta' = Q^{-1}\beta$ and denote the ith entry of β' by β'_i. Notice the equalities

$$\alpha = U(D\beta') \quad \text{and} \quad U^{-1}\alpha = D\beta' = (s_1\beta'_1, \ldots, s_d\beta'_d, 0, \ldots, 0)^t.$$

Hence $s_i\beta'_i \in \mathbb{Z}$ for all i, and consequently α is in $\mathbb{Z}u_1 + \cdots + \mathbb{Z}u_d$. For the reverse inclusion it suffices to prove that u_i belongs to $\mathbb{R}\mathcal{V}$ for $i = 1, \ldots, d$. Using the equality $(AQ)e_i = (UD)e_i$ (e_i=ith unit vector) we derive

$$Aq_i = U(s_ie_i) = s_i(Ue_i) = s_iu_i \quad (1 \leq i \leq d),$$

where $q_i = Qe_i$ is the ith column of Q. Hence $u_i \in \mathbb{R}\mathcal{V}$ for $i = 1, \ldots, d$. $\square$

Lemma 1.2.4 *There is an affine isomorphism* $\phi\colon \operatorname{aff}(\mathcal{A}) \to \mathbb{R}^d$ *such that* $\phi(\operatorname{aff}(\mathcal{A}) \cap \mathbb{Z}^n) = \mathbb{Z}^d$, *where* $d = \dim(\mathcal{P})$.

Proof. By Lemma 1.2.3 one has $\mathbb{R}\mathcal{V} \cap \mathbb{Z}^n = \mathbb{Z}\gamma_1 \oplus \cdots \oplus \mathbb{Z}\gamma_d$ for some $\gamma_1, \ldots, \gamma_d$ in $\mathbb{Z}^n$. Therefore there is a linear map

$$\psi\colon \mathbb{R}\mathcal{V} \longrightarrow \mathbb{R}^d, \qquad \gamma_i \overset{\psi}{\longmapsto} e_i.$$

Hence $\phi = \psi f\colon \operatorname{aff}(\mathcal{A}) \longrightarrow \mathbb{R}^d$ satisfies the required condition. $\square$

Observe that $\phi(\mathcal{P})$ is an integral polytope of dimension d with a positive Lebesgue measure denoted by $m(\phi(\mathcal{P}))$.

Definition 1.2.5 The *relative volume* of the integral polytope $\mathcal{P}$ is:

$$\operatorname{vol}(\mathcal{P}) := m(\phi(\mathcal{P})).$$

If $\mathcal{P}$ has dimension n, then the relative volume of $\mathcal{P}$ agrees with its usual volume. To see that the relative volume is independent of ϕ we need the following fact about the standard volume.

Theorem 1.2.6 [427, Theorem 6.2.14] *Let* $T\colon \mathbb{R}^n \to \mathbb{R}^n$ *be an affine map given by* $T(x) = Ax + x_0$, *where A is a real matrix of order n and* $x_0 \in \mathbb{R}^n$. *If $\mathcal{P}$ is a bounded convex set in $\mathbb{R}^n$, then $T(\mathcal{P})$ has volume* $|\det(A)|\operatorname{vol}(\mathcal{P})$.

Lemma 1.2.7 *Let* $T\colon \mathbb{Z}^n \to \mathbb{Z}^n$ *be the affine map given by* $T(x) = Ax + \beta$, *where A is an integral matrix of order n and $\beta \in \mathbb{Z}^n$. If T is bijective, then* $\det(A) = \pm 1$.

Proof. Note that the matrix A determines a bijective linear map, thus A has an inverse with integral entries. Indeed, for each i there is a column vector β_i in $\mathbb{Z}^n$ such that $A\beta_i = e_i$. Therefore $A(\beta_1 \cdots \beta_n) = I$, and consequently $\det(A) = \pm 1$. $\square$

Proposition 1.2.8 *The relative volume of $\mathcal{P}$ is independent of ϕ.*

Proof. If ϕ_1 is another affine isomorphism $\phi_1\colon \operatorname{aff}(\mathcal{A}) \to \mathbb{R}^d$ such that $\phi_1(\operatorname{aff}(\mathcal{A}) \cap \mathbb{Z}^n) = \mathbb{Z}^d$, then the affine map $\phi_1\phi^{-1}\colon \mathbb{Z}^d \longrightarrow \mathbb{Z}^d$ is bijective. Using Theorem 1.2.6 and Lemma 1.2.7 one obtains

$$\operatorname{vol}(\phi(\mathcal{P})) = \operatorname{vol}((\phi_1\phi^{-1})(\phi(\mathcal{P}))) = \operatorname{vol}(\phi_1(\mathcal{P})). \qquad \square$$

In practice one can compute volumes of lattice polytopes using *Normaliz* [68] and *Polyprob* [104].

Example 1.2.9 We give an illustration in $\mathbb{R}^3$ of the procedure outlined above to compute relative volumes. We begin by setting

$$\mathcal{A} = \{v_1, v_2, v_3, v_4\} = \{(1,3,1), (1,5,3), (4,9,4), (2,8,5)\}$$

and $\mathcal{P} = \operatorname{conv}(\mathcal{A})$. Then $\mathcal{V} = \{(0,0,0), (0,2,2), (3,6,3), (1,5,4)\}$ and

$$f\colon \mathbb{R}^3 \longrightarrow \mathbb{R}^3, \quad x \stackrel{f}{\longrightarrow} x - \alpha_0,$$

where $\alpha_0 = (1,3,1)$. Consider the matrix

$$A = \begin{bmatrix} 0 & 3 & 1 \\ 2 & 6 & 5 \\ 2 & 3 & 4 \end{bmatrix}.$$

Next using *Maple* [80] to compute the Smith normal form of A we obtain invertible integral matrices U and Q such that $UAQ = D$, where

$$D = \begin{bmatrix} 1 & 0 & 0 \\ 0 & 1 & 0 \\ 0 & 0 & 0 \end{bmatrix}; \quad U = \begin{bmatrix} -2 & 1 & 0 \\ 1 & 0 & 0 \\ 1 & -1 & 1 \end{bmatrix}; \quad U^{-1} = \begin{bmatrix} 0 & 1 & 0 \\ 1 & 2 & 0 \\ 1 & 1 & 1 \end{bmatrix}.$$

By the proof of Lemma 1.2.3, the first two columns of U^{-1} form a $\mathbb{Z}$-basis for $\mathbb{R}\mathcal{V} \cap \mathbb{Z}^3$. Using the affine map $\psi\colon \mathbb{R}\mathcal{V} \to \mathbb{R}^3$ induced by $\psi(0,1,1) = (1,0)$ and $\psi(1,2,1) = (0,1)$, and the map $\phi = \psi f\colon \operatorname{aff}(\mathcal{A}) \to \mathbb{R}$, we get

$$\phi(\mathcal{P}) = \operatorname{conv}((0,0),(2,0),(0,3),(3,1)) \ \text{ and } \ \operatorname{vol}(\mathcal{P}) = m(\phi(\mathcal{P})) = 11/2.$$

Proposition 1.2.10 ([176], [394, Proposition 4.6.30]) *The relative volume of $\mathcal{P}$ is given by:*

$$\operatorname{vol}(\mathcal{P}) = \lim_{i\to\infty} \frac{|\mathbb{Z}^n \cap i\mathcal{P}|}{i^d}.$$

Definition 1.2.11 For an abelian group $(M,+)$ its *torsion subgroup* $T(M)$ is the set of all x in M such that $px = 0$ for some $0 \neq p \in \mathbb{N}$. The group M is *torsion-free* if $T(M) = (0)$.

Lemma 1.2.12 *If* $\mathcal{B} = \{\beta_1, \ldots, \beta_s\} \subset \mathbb{Z}^n$*, then*

(a) $(\mathbb{R}\mathcal{B} \cap \mathbb{Z}^n)/\mathbb{Z}\mathcal{B} = T(\mathbb{Z}^n/\mathbb{Z}\mathcal{B})$*, and* $\quad (*)$

(b) $\mathbb{R}\mathcal{B} \cap \mathbb{Z}^n = \mathbb{Z}\mathcal{B}$ *if and only if* $\mathbb{Z}^n/\mathbb{Z}\mathcal{B}$ *is torsion-free.*

Proof. (a): If $\overline{z} \in T(\mathbb{Z}^n/\mathbb{Z}\mathcal{B}) \subset \mathbb{Z}^n/\mathbb{Z}\mathcal{B}$, then $kz \in \mathbb{Z}\mathcal{B}$ for some $0 \neq k \in \mathbb{N}$, so $z \in \mathbb{Q}\mathcal{B} \subset \mathbb{R}\mathcal{B}$. Since also $z \in \mathbb{Z}^n$, $\overline{z}$ is in the left-hand side of Eq. $(*)$. This shows the inclusion "$\supset$". The other inclusion is left as an exercise.

(b): This follows from (a). $\square$

Lemma 1.2.13 [397, pp. 32-33] *If* $H \subset G$ *are free abelian groups of the same rank* d *with* $\mathbb{Z}$*-bases* $\delta_1, \ldots, \delta_d$ *and* $\gamma_1, \ldots, \gamma_d$ *related by* $\delta_i = \sum_j g_{ij}\gamma_j$*, where* $g_{ij} \in \mathbb{Z}$ *for all* i, j*, then* $|G/H| = |\det(g_{ij})|$.

Theorem 1.2.14 [146] *If* $\mathcal{A} = \{v_1, \ldots, v_q\} \subset \mathbb{Z}^n$ *is a set of vectors lying on an affine hyperplane not containing the origin and* $\mathcal{P} = \mathrm{conv}(\mathcal{A})$ *has dimension* d*, then*

$$\mathrm{vol}(\mathcal{P}) = |T(\mathbb{Z}^n/(v_2 - v_1, \ldots, v_q - v_1))| \lim_{i\to\infty} \frac{|\mathbb{Z}\mathcal{A} \cap i\mathcal{P}|}{i^d}.$$

Proof. As in the beginning of this section for convenience of notation we set $m = q-1$ and $\alpha_0 = v_1, \ldots, \alpha_m = v_q$. Let $x_0 \in \mathbb{Q}^n$ be such that $\langle \alpha_i, x_0\rangle = 1$ for all i. As $\langle \alpha_i - \alpha_0, x_0\rangle = 0$, there is a decomposition $\mathbb{Z}\mathcal{A} = \mathbb{Z}\alpha_0 \oplus \mathbb{Z}\mathcal{V}$ with $\mathcal{V} = \{\alpha_1 - \alpha_0, \ldots, \alpha_m - \alpha_0\}$. Pick $\delta_1, \ldots, \delta_d \in \mathbb{Z}^n$ such that

$$\mathbb{Z}\mathcal{V} = \mathbb{Z}\delta_1 \oplus \cdots \oplus \mathbb{Z}\delta_d. \tag{1.5}$$

Therefore one can write

$$\alpha_i = \alpha_0 + f_{i1}\delta_1 + \cdots + f_{id}\delta_d \qquad (f_{ij} \in \mathbb{Z}). \tag{1.6}$$

Consider the lattice polytope

$$\mathcal{P}_1 = \mathrm{conv}((1, 0, \ldots, 0), (1, f_{11}, \ldots, f_{1d}), \ldots, (1, f_{m1}, \ldots, f_{md})) \subset \mathbb{R}^{d+1}.$$

There is a bijective map $\mathbb{Z}\mathcal{A} \cap i\mathcal{P} \stackrel{\varphi}{\longrightarrow} \mathbb{Z}^{d+1} \cap i\mathcal{P}_1$, namely φ is the restriction of the linear map from $\mathbb{Z}\mathcal{A}$ to $\mathbb{Z}^{d+1}$ that maps each vector into its coordinate vector with respect to the basis $\{\alpha_0, \delta_1, \ldots, \delta_d\}$. Therefore

$$\lim_{i\to\infty} \frac{|\mathbb{Z}\mathcal{A} \cap i\mathcal{P}|}{i^d} = \lim_{i\to\infty} \frac{|\mathbb{Z}^{d+1} \cap i\mathcal{P}_1|}{i^d} = \mathrm{vol}(\mathcal{P}_1).$$

To estimate $\mathrm{vol}(\mathcal{P}_1)$ consider the lattice polytope

$$\mathcal{P}_2 = \mathrm{conv}((0, \ldots, 0), (f_{11}, \ldots, f_{1d}), \ldots, (f_{m1}, \ldots, f_{md})) \subset \mathbb{R}^d$$

and note that applying the translation $\alpha \mapsto \alpha - e_1$ to $\mathcal{P}_1$ gives:

$$\mathrm{vol}(\mathcal{P}_1) \;=\; \mathrm{vol}(\mathrm{conv}((0,\dots,0),(0,f_{11},\dots,f_{1d}),\dots,(0,f_{m1},\dots,f_{md}))).$$

Thus $\mathrm{vol}(\mathcal{P}_1) = \mathrm{vol}(\mathcal{P}_2)$. The next step is to relate $\mathrm{vol}(\mathcal{P}_2)$ with $\mathrm{vol}(\mathcal{P})$ by obtaining a second expression for $\mathrm{vol}(\mathcal{P}_2)$. According to Lemma 1.2.3, there are $\gamma_1,\dots,\gamma_d \in \mathbb{Z}^n$ such that

$$\mathbb{R}\mathcal{V} \cap \mathbb{Z}^n = \mathbb{Z}\gamma_1 \oplus \cdots \oplus \mathbb{Z}\gamma_d. \tag{1.7}$$

Writing

$$\delta_i = g_{i1}\gamma_1 + \cdots + g_{id}\gamma_d \qquad (g_{ij} \in \mathbb{Z};\ i,j = 1,\dots,d), \tag{1.8}$$

from Eqs. (1.6) and (1.8) one derives the matrix equality

$$(f_{ij})(g_{ij}) = (c_{ij}), \tag{1.9}$$

where $\alpha_i - \alpha_0 = c_{i1}\gamma_1 + \cdots + c_{id}\gamma_d$, $1 \le i \le m$. By definition

$$\mathrm{vol}(\mathcal{P}) = \mathrm{vol}(\mathrm{conv}((0,\dots,0),(c_{11},\dots,c_{1d}),\dots,(c_{m1},\dots,c_{md}))).$$

By Eq. (1.9), the linear transformation

$$\sigma\colon \mathbb{Z}^d \longrightarrow \mathbb{Z}^d, \qquad x \overset{\sigma}{\longmapsto} x(g_{ij}),$$

satisfies $\sigma(f_{i1},\dots,f_{id}) = (c_{i1},\dots,c_{id})$. Hence by Theorem 1.2.6 we get

$$\mathrm{vol}(\mathcal{P}) = \mathrm{vol}(\sigma(\mathcal{P}_2)) = |\det(g_{ij})|\mathrm{vol}(\mathcal{P}_2).$$

To finish the proof it suffices to show that $|\det(g_{ij})|$ is the order of the torsion subgroup of $\mathbb{Z}^n/\mathbb{Z}\mathcal{V}$. Now we have

$$T(\mathbb{Z}^n/\mathbb{Z}\mathcal{V}) \overset{\text{(a)}}{=} (\mathbb{R}\mathcal{V} \cap \mathbb{Z}^n)/\mathbb{Z}\mathcal{V} \overset{\text{(b)}}{=} (\mathbb{Z}\gamma_1 \oplus \cdots \oplus \mathbb{Z}\gamma_d)/(\mathbb{Z}\delta_1 \oplus \cdots \oplus \mathbb{Z}\delta_d),$$

where in (a) we use Lemma 1.2.12, and in (b) the identities (1.5) and (1.7). Hence, from the identity (1.8) and Lemma 1.2.13, we get

$$|T(\mathbb{Z}^n/\mathbb{Z}\mathcal{V})| = |\det(g_{ij})|. \qquad \square$$

Proposition 1.2.15 *Let $\alpha_0,\dots,\alpha_n$ be a set of affinely independent points in $\mathbb{R}^n$ and $\Delta = \mathrm{conv}(\alpha_0,\dots,\alpha_n)$. Then the volume of the simplex Δ is*

$$\mathrm{vol}(\Delta) = \frac{\left|\det\begin{pmatrix} \alpha_0 & 1 \\ \vdots & \vdots \\ \alpha_n & 1 \end{pmatrix}\right|}{n!} = \frac{\left|\det\begin{pmatrix} \alpha_1 - \alpha_0 \\ \vdots \\ \alpha_n - \alpha_0 \end{pmatrix}\right|}{n!}.$$

Proof. The result follows using linear algebra or applying the change of variables formula. See Theorem 1.2.6. □

Definition 1.2.16 A set Δ in $\mathbb{R}^n$ is called a *lattice d-simplex* if Δ is the convex hull of a set of $d+1$ affinely independent points in $\mathbb{Z}^n$.

Let Δ be a lattice d-simplex in $\mathbb{R}^n$. The *normalized volume* of Δ is defined as $d!\mathrm{vol}(\Delta)$. If $d = n$, then from Proposition 1.2.15 one has

$$\mathrm{vol}(\Delta) \geq \frac{1}{n!}.$$

Definition 1.2.17 A lattice d-simplex Δ in $\mathbb{R}^n$ is called unimodular if $d!\mathrm{vol}(\Delta) = 1$.

Proposition 1.2.18 *Let* $\Delta = \mathrm{conv}(\alpha_0, \ldots, \alpha_d)$ *be a lattice d-simplex. If*

$$\mathbb{R}(\alpha_1 - \alpha_0, \ldots, \alpha_d - \alpha_0) \cap \mathbb{Z}^n = \mathbb{Z}\gamma_1 \oplus \cdots \oplus \mathbb{Z}\gamma_d$$

for some $\gamma_1, \ldots, \gamma_d \in \mathbb{Z}^n$ *and* $\alpha_i - \alpha_0 = \sum_j c_{ij}\gamma_j$, *then*

$$\mathrm{vol}(\Delta) = \frac{|\det(c_{ij})|}{d!}.$$

Proof. Note $\mathrm{vol}(\Delta) = \mathrm{vol}(\mathrm{conv}(0, c_1, \ldots, c_d))$, where $c_i = (c_{i1}, \ldots, c_{in})$. Hence the formula follows from Proposition 1.2.15. □

Lemma 1.2.19 *If* $\beta_1, \ldots, \beta_m \in \mathbb{Z}^n$, *then*

$$\mathbb{R}(\beta_1, \ldots \beta_m) \cap \mathbb{Z}^n = \mathbb{Z}\beta_1 + \cdots + \mathbb{Z}\beta_m + \mathbb{Z}\delta_1 + \cdots + \mathbb{Z}\delta_s,$$

where $\overline{\delta_1}, \ldots, \overline{\delta_s}$ *generate the torsion subgroup of* $\mathbb{Z}^n/(\beta_1, \ldots, \beta_m)$.

Proof. It is straightforward and will be left as an exercise. □

Lemma 1.2.20 *If* $\alpha_0, \ldots, \alpha_m \in \mathbb{Z}^n$, *then there is an isomorphism of groups*

$$\varphi\colon T\left(\mathbb{Z}^n/(\alpha_1 - \alpha_0, \ldots, \alpha_m - \alpha_0)\right) \longrightarrow T\left(\mathbb{Z}^{n+1}/((\alpha_0, 1), \ldots, (\alpha_m, 1))\right),$$

given by $\varphi(\overline{\alpha}) = \overline{(\alpha, 0)}$. *Here* $T(M)$ *denotes the torsion subgroup of* M.

Proof. The map φ is clearly well-defined and injective. To prove that φ is onto consider an equation $s(\alpha, b) = \lambda_0(\alpha_0, 1) + \cdots + \lambda_m(\alpha_m, 1)$, where $\lambda_i \in \mathbb{Z}$, $0 \neq s \in \mathbb{N}$, $\alpha \in \mathbb{Z}^n$ and $b \in \mathbb{Z}$. Then

$$\begin{aligned} s\alpha &= \lambda_0\alpha_0 + \cdots + \lambda_m\alpha_m, \\ sb &= \lambda_0 + \cdots + \lambda_m. \end{aligned}$$

Hence $s(\alpha - b\alpha_0) = \lambda_1(\alpha_1 - \alpha_0) + \cdots + \lambda_m(\alpha_m - \alpha_0)$. It follows readily that $\varphi(\overline{\alpha - b\alpha_0}) = \overline{(\alpha, b)}$, as required. □

Proposition 1.2.21 *Let* $\underline{\alpha} = \alpha_0, \ldots, \alpha_d$ *be a set of affinely independent vectors in* $\mathbb{Z}^n$ *defining a simplex* Δ*. Then the following are equivalent:*

(a) $d!\mathrm{vol}(\Delta) = 1$.

(b) $\mathbb{R}(\alpha_1 - \alpha_0, \ldots, \alpha_d - \alpha_0) \cap \mathbb{Z}^n = \mathbb{Z}(\alpha_1 - \alpha_0) + \cdots + \mathbb{Z}(\alpha_d - \alpha_0)$.

(c) $\mathbb{Z}^n/\mathbb{Z}\{\alpha_1 - \alpha_0, \ldots, \alpha_d - \alpha_0\}$ *is torsion-free.*

(d) $\mathbb{Z}^{n+1}/\mathbb{Z}\{(\alpha_0, 1), \ldots, (\alpha_d, 1)\}$ *is torsion-free.*

Proof. (a) $\Leftrightarrow$ (b) follows from Proposition 1.2.18 and (b) $\Leftrightarrow$ (c) $\Leftrightarrow$ (d) follows from Lemmas 1.2.19 and 1.2.20. $\square$

Corollary 1.2.22 *If* $\Delta = \mathrm{conv}(\alpha_0, \alpha_1, \ldots, \alpha_d)$ *is a unimodular lattice d-simplex in* $\mathbb{R}^n$*, then* $\Delta \cap \mathbb{Z}^n = \{\alpha_0, \alpha_1, \ldots, \alpha_d\}$.

Proof. Let $\alpha \in \Delta \cap \mathbb{Z}^n$. Then $\alpha = \lambda_0\alpha_0 + \cdots + \lambda_d\alpha_d$, where $\lambda_i \in \mathbb{Q}_+$ for all i and $\lambda_0 + \cdots + \lambda_d = 1$. Using the equality

$$\alpha - \alpha_0 = \lambda_1(\alpha_1 - \alpha_0) + \cdots + \lambda_d(\alpha_d - \alpha_0)$$

and Proposition 1.2.21(c) we get

$$\begin{aligned} \alpha - \alpha_0 &= \eta_1(\alpha_1 - \alpha_0) + \cdots + \eta_d(\alpha_d - \alpha_0) \\ &= \lambda_1(\alpha_1 - \alpha_0) + \cdots + \lambda_d(\alpha_d - \alpha_0), \end{aligned}$$

where $\eta_i \in \mathbb{Z}$ for all i. Since $\alpha_1 - \alpha_0, \ldots, \alpha_d - \alpha_0$ are linearly independent we obtain $\lambda_i = \eta_i$ for all i. Hence exactly one of the λ_i's is equal to 1 and the others are equal to zero, that is, $\alpha \in \{\alpha_0, \alpha_1, \ldots, \alpha_d\}$. $\square$

Definition 1.2.23 A lattice polytope $\mathcal{P} = \mathrm{conv}(\mathcal{A})$ of dimension d is said to have a *unimodular covering with support in* $\mathcal{A}$ if there are simplices $\Delta_1, \ldots, \Delta_m$ of dimension d such that the following two conditions hold

(i) $\mathcal{P} = \cup_{i=1}^m \Delta_i$ and the vertices of Δ_i are contained in $\mathcal{A}$ for all i.

(ii) $d!\mathrm{vol}(\Delta_i) = 1$.

If condition (ii) is replaced by

(ii)' $[\mathbb{Z}\mathcal{A}\colon \mathbb{Z}\mathcal{A}_i] = 1$ for all i,

where $\mathcal{A}_i$ denotes the vertex set of Δ_i, then we say that $\Delta_1, \ldots, \Delta_m$ is a *weakly unimodular covering* of $\mathcal{P}$ with support in $\mathcal{A}$.

For a discussion on the existence of unimodular covers of rational cones see [60] and the references therein.

Proposition 1.2.24 *If $\mathcal{A} = \{v_1, \ldots, v_q\} \subset \mathbb{Z}^n$ and the vectors in $\mathcal{A}$ lie in a hyperplane of $\mathbb{R}^n$ not containing the origin, then any unimodular covering of $\mathcal{P} = \mathrm{conv}(\mathcal{A})$ with support in $\mathcal{A}$ is weakly unimodular.*

Proof. Let $\Delta_i = \mathrm{conv}(\mathcal{A}_i)$ be any unimodular simplex of dimension d with vertex set $\mathcal{A}_i \subset \mathcal{A}$ and $d = \dim(\mathcal{P})$. For convenience of notation assume that $\mathcal{A}_i = \{v_1, \ldots, v_{d+1}\}$. To show the equality $\mathbb{Z}\mathcal{A} = \mathbb{Z}\mathcal{A}_i$ it suffices to prove $\mathbb{Z}\mathcal{A} \subset \mathbb{Z}\mathcal{A}_i$. Take an arbitrary vector α in $\mathbb{Z}\mathcal{A}$ and write:

$$\begin{aligned} \alpha &= \eta_1 v_1 + \cdots + \eta_q v_q & (\eta_i \in \mathbb{Z}), \\ (\alpha, \eta) &= \eta_1(v_1, 1) + \cdots + \eta_q(v_q, 1) & (\eta = \eta_1 + \cdots + \eta_q). \end{aligned}$$

Since $\mathcal{A}$ lies in an affine hyperplane not containing the origin and $\mathcal{A}_i$ is affinely independent, the set $\mathcal{A}_i$ is seen to be linearly independent. Therefore

$$d + 1 = \dim(\mathbb{R}\{(v_1, 1), \ldots, (v_{d+1}, 1)\}) = \dim(\mathbb{R}\{(v_1, 1), \ldots, (v_q, 1)\}).$$

Thus one can write $(\alpha, \eta) = \lambda_1(v_1, 1) + \cdots + \lambda_{d+1}(v_{d+1}, 1)$, $\lambda_i \in \mathbb{Q}$. By Proposition 1.2.21 the group $\mathbb{Z}^{n+1}/\mathbb{Z}\{(v_1, 1), \ldots, (v_{d+1}, 1)\}$ is torsion-free. Hence it follows from Lemma 1.2.12(b) that $\lambda_i \in \mathbb{Z}$ for all i, and consequently α is in $\mathbb{Z}\mathcal{A}_i$. $\square$

Example 1.2.25 Let $\mathcal{A} = \{v_1, \ldots, v_5\}$ be the set of vectors in $\mathbb{Z}^6$ given by

$$\begin{aligned} &v_1 = (1, 0, 1, 0, 0, 1), \quad v_2 = (1, 0, 0, 1, 1, 0), \quad v_3 = (0, 1, 1, 0, 1, 0), \\ &v_4 = (0, 1, 0, 1, 0, 1), \quad v_5 = (1, 0, 0, 1, 0, 1). \end{aligned}$$

Consider $\mathcal{A}_1 = \{v_2, v_3, v_4, v_5\}$, $\mathcal{A}_2 = \{v_1, v_3, v_4, v_5\}$, $\mathcal{A}_3 = \{v_1, v_2, v_3, v_5\}$,

$$\Delta_1 = \mathrm{conv}(\mathcal{A}_1), \qquad \Delta_2 = \mathrm{conv}(\mathcal{A}_2), \qquad \Delta_3 = \mathrm{conv}(\mathcal{A}_3),$$

and $\mathcal{P} = \mathrm{conv}(\mathcal{A})$. The groups $\mathbb{Z}^6/\mathbb{Z}\mathcal{A}$ and $\mathbb{Z}^6/\mathbb{Z}\mathcal{A}_i$ are torsion-free for all i. This readily implies that $\mathbb{Z}\mathcal{A} = \mathbb{Z}\mathcal{A}_i$, $i = 1, 2, 3$. Using the relation $v_1 + v_2 + v_4 = v_3 + 2v_5$ it is not hard to see that $\mathcal{P} = \Delta_1 \cup \Delta_2 \cup \Delta_3$. Thus $\Delta_1, \Delta_2, \Delta_3$ is a weakly unimodular covering. The relative volume of $\mathcal{P}$ can be computed using the procedure described before. It is left to the reader to verify that $\dim(\mathcal{P}) = 3$ and $\mathrm{vol}(\mathcal{P}) = 1/2$.

Exercises

1.2.26 Let Δ be an n-dimensional simplex in $\mathbb{R}^n$ with vertices $\alpha_0, \ldots, \alpha_n$ in $\mathbb{Z}^n$. Prove that Δ is unimodular if and only if any of the following two equivalent conditions hold

(a) $\mathbb{Z}^{n+1} = \mathbb{Z}(\alpha_0, 1) + \cdots + \mathbb{Z}(\alpha_n, 1)$.

(b) $\mathbb{Z}^n = \mathbb{Z}(\alpha_1 - \alpha_0) + \cdots + \mathbb{Z}(\alpha_n - \alpha_0)$.

1.2.27 Construct a lattice simplex of normalized volume greater than 1. Then prove that the converse of Proposition 1.2.24 fails.

1.2.28 If $\mathcal{P}$ is the integral polytope with vertices $(0,0)$ and $(1,3)$:

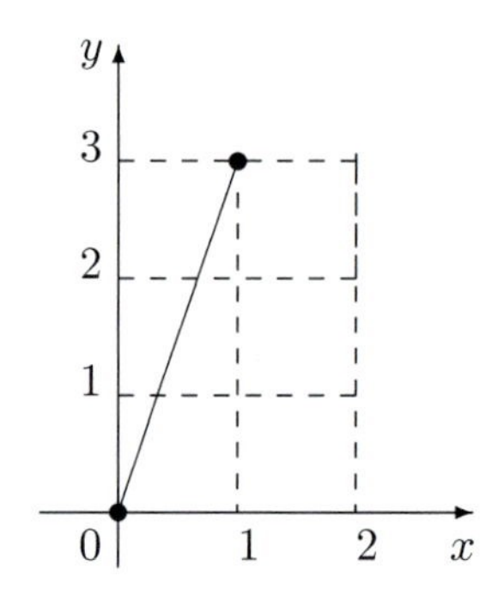

Prove that the relative volume of $\mathcal{P}$ is equal to 1.

1.2.29 If $\mathcal{P} = \text{conv}((3,1),(1,3))$, then $\text{vol}(\mathcal{P}) = 2$.

1.2.30 Consider the set $\mathcal{A} = \{e_1 + e_2,\, e_2 + e_3,\, e_3 + e_4,\, e_1 + e_4\}$ and the polytope $\mathcal{P} = \text{conv}(\mathcal{A})$. Prove that $\dim(\mathcal{P}) = 2$ and $\text{vol}(\mathcal{P}) = 1$.

1.2.31 Let $\mathcal{P}$ be the tetrahedron in $\mathbb{R}^3$ whose vertices are $v_1 = (1,1,0)$, $v_2 = (2,1,-1)$, $v_3 = (2,-5,0)$, $v_4 = (7,1,-8)$. Prove that $\text{vol}(\mathcal{P}) = 2$. Then verify this using *Maple* [80].

1.2.32 If $\mathcal{A} = \{v_1,\ldots,v_q\} \subset \mathbb{Z}^n$, prove that the rank of $\mathbb{R}\mathcal{A} \cap \mathbb{Z}^n$, as a free abelian group, is equal to the dimension of $\mathbb{R}\mathcal{A}$ as a real vector space.

1.3 Hilbert bases and TDI systems

The set of integral points of a rational polyhedral cone form a semigroup that arises in many branches of mathematics, such as combinatorial commutative algebra [65], toric varieties [176], and integer programming [372].

A finite set $\mathcal{H} \subset \mathbb{R}^n$ is called a *Hilbert basis* if

$$\mathbb{Z}^n \cap \mathbb{R}_+\mathcal{H} = \mathbb{N}\mathcal{H},$$

where $\mathbb{N}\mathcal{H}$ is the semigroup spanned by $\mathcal{H}$ consisting of all linear combinations of $\mathcal{H}$ with coefficients in $\mathbb{N}$. Notice that all vectors in a Hilbert basis are integral. A nice introduction to Hilbert bases can be found in [215].

Let $C \subset \mathbb{R}^n$ be a rational polyhedral cone. A finite set $\mathcal{H}$ is called a *Hilbert basis* of C if $C = \mathbb{R}_+\mathcal{H}$ and $\mathcal{H}$ is a Hilbert basis. A Hilbert basis of C is *minimal* if it does not strictly contain any other Hilbert basis of C.

Hilbert bases of rational polyhedral cones always exist. For a pointed cone there is only one minimal Hilbert basis. To prove these two facts, we need Gordan's lemma. Below we give two versions of this lemma.

Definition 1.3.1 Given $a = (a_i) \in \mathbb{R}^n$, we define the *value* of a as:

$$|a| = a_1 + \cdots + a_n.$$

Lemma 1.3.2 (Gordan, version 1) *If $\mathcal{A} = \{v_1, \ldots, v_q\} \subset \mathbb{Z}^n$ and $\mathbb{Z}\mathcal{A}$ is the subgroup generated by $\mathcal{A}$, then there exist $\gamma_1, \ldots, \gamma_m$ in $\mathbb{Z}^n$ such that*

$$\mathbb{Z}\mathcal{A} \cap \mathbb{R}_+\mathcal{A} = \mathbb{N}\gamma_1 + \cdots + \mathbb{N}\gamma_m$$

and $\gamma_i \in [N, M]^n$ for all i, where $N = -q \max_{1 \leq i \leq q} |v_i^-|$ and $M = q \max_{1 \leq i \leq q} |v_i^+|$.

Proof. Recall that $C = \mathbb{Z}\mathcal{A} \cap \mathbb{R}_+\mathcal{A} = \mathbb{Z}\mathcal{A} \cap \mathbb{Q}_+\mathcal{A}$; see Corollary 1.1.27. Let $\beta \in C$. Then one can write

$$\beta = \sum_{i=1}^{q} \left(\frac{x_i}{y_i}\right) v_i,$$

where $x_i \in \mathbb{N}$ and $0 \neq y_i \in \mathbb{N}$. By the division algorithm there are r_i, n_i in $\mathbb{N}$ such that $x_i = n_i y_i + r_i$ and $0 \leq r_i < y_i$. Therefore one can write

$$\beta = \sum_{i=1}^{q} n_i v_i + \sum_{i=1}^{q} a_i v_i,$$

with $a_i \in [0, 1] \cap \mathbb{Q}$. As $\sum_{i=1}^{q} a_i v_i \in C \cap [N, M]^n$, the set $\mathcal{A} \cup (C \cap [N, M]^n)$ is a generating set for C with the required property. Our argument was based on the proof of Gordan's lemma given in [65]. $\Box$

Definition 1.3.3 A semigroup $(\mathcal{S}, +, 0)$ of $\mathbb{Z}^n$ is said to be *finitely generated* if there exists a finite set $\Gamma = \{\gamma_1, \ldots, \gamma_r\} \subset \mathcal{S}$ such that:

$$\mathcal{S} = \mathbb{N}\Gamma := \mathbb{N}\gamma_1 + \cdots + \mathbb{N}\gamma_r.$$

A set of generators Γ of $\mathcal{S}$ is called *minimal* if $\gamma_i \neq 0\ \forall i$ and none of its elements is a linear combination with coefficients in $\mathbb{N}$ of the others.

Remark 1.3.4 There are examples of subsemigroups of $\mathbb{N}^n$, with $n \geq 2$, which are not finitely generated; see Exercise 1.3.34.

Lemma 1.3.5 (Gordan, version 2) *If $\mathbb{R}_+\mathcal{A}$ is a cone in $\mathbb{R}^n$ generated by a finite set $\mathcal{A} \subset \mathbb{Z}^n$, then the semigroup $\mathbb{Z}^n \cap \mathbb{R}_+\mathcal{A}$ is finitely generated.*

Proof. If $\mathcal{A} = \{v_1, \ldots, v_q\}$, consider the finite set of integral points:

$$\{a_1v_1 + \cdots + a_qv_q \,|\, 0 \leq a_i \leq 1\} \cap \mathbb{Z}^n = \{\gamma_1, \ldots, \gamma_r\}.$$

It is left to the reader to prove that $\gamma_1, \ldots, \gamma_r$ is the required set of generators for $\mathbb{Z}^n \cap \mathbb{R}_+\mathcal{A}$. See [372, p. 233]. $\Box$

Let $\mathcal{A}$ be a finite set in $\mathbb{Z}^n$ and let $G = \mathbb{Z}^n$ or $G = \mathbb{Z}\mathcal{A}$. Then, by Gordan's lemma (versions 1 and 2) there exists $\gamma_1, \ldots, \gamma_r \in \mathbb{Z}^n$ such that:

$$G \cap \mathbb{R}_+\mathcal{A} = \mathbb{N}\gamma_1 + \cdots + \mathbb{N}\gamma_r.$$

Computing the γ_i's is in general difficult [69]. Fortunately, the γ_i's can be computed using *Normaliz* [68].

Therefore Hilbert bases of rational polyhedral cones always exist:

Proposition 1.3.6 *Let $\mathcal{A}$ be a finite set in $\mathbb{Z}^n$. Then there exist $\gamma_1, \ldots, \gamma_r$ such that*

$$\mathbb{R}_+\mathcal{A} \cap \mathbb{Z}^n = \mathbb{N}\gamma_1 + \cdots + \mathbb{N}\gamma_r,$$

and $\mathcal{H} = \{\gamma_1, \ldots, \gamma_r\}$ is a Hilbert basis of $\mathbb{R}_+\mathcal{A}$.

Proof. The existence follows from Lemma 1.3.5. That $\mathcal{H}$ is a Hilbert basis of $\mathbb{R}_+\mathcal{A}$ follows from the equality $\mathbb{R}_+\mathcal{A} = \mathbb{R}_+\gamma_1 + \cdots + \mathbb{R}_+\gamma_r$. $\Box$

In Definition 1.1.52 we considered pointed polyhedra. For cones we have the following equivalent definition (see Exercise 1.3.36).

Definition 1.3.7 A polyhedral cone $C = \{x|\, Ax \leq 0\}$ is called *pointed* if the lineality space $\{x|\, Ax = 0\}$ is equal to $\{0\}$.

Lemma 1.3.8 *Let C be a rational polyhedral cone. If C is pointed, then there exists an integral vector b such that $\langle b, x\rangle > 0$ for all $0 \neq x \in C$.*

Proof. There is a rational matrix A such that $C = \{x|\, Ax \leq 0\}$; see Theorem 1.1.29. We may assume that A is integral. If $u_1, \ldots, u_t$ are the rows of A, then $b = -u_1 - \cdots - u_t$ satisfies the required condition. $\Box$

Theorem 1.3.9 [371] *Let $\mathcal{A}$ be a finite set in $\mathbb{Z}^n$ and let $C = \mathbb{R}_+\mathcal{A}$. If C is pointed, then there exists a unique minimal Hilbert basis of C given by*

$$\mathcal{H} = \{x \in C \cap \mathbb{Z}^n \,|\, 0 \neq x \notin \mathbb{N}y_1 + \mathbb{N}y_2;\ \forall y_1, y_2 \in (C \setminus \{0\}) \cap \mathbb{Z}^n\,\}.$$

Proof. Let $\Gamma = \{\gamma_1, \ldots, \gamma_r\}$ be an arbitrary Hilbert basis of C. We claim that $\mathcal{H} \subset \Gamma$. Let $x \in \mathcal{H}$. Since $\mathbb{Z}^n \cap C = \mathbb{N}\Gamma$ we can write $x = \sum_{i=1}^{r} a_i\gamma_i$, $a_i \in \mathbb{N}$. Thus, by construction of $\mathcal{H}$, we get $x = \gamma_i$ for some i, which proves the claim. To finish the proof it suffices to prove $\mathbb{N}\mathcal{H} = \mathbb{N}\Gamma$, because this equality implies $\mathbb{R}_+\mathcal{H} = \mathbb{R}_+\Gamma$ and consequently $\mathbb{Z}^n \cap \mathbb{R}_+\mathcal{H} = \mathbb{N}\mathcal{H}$. We proceed by contradiction by assuming that the set:

$$\mathcal{V} = \{x| \, x \in \mathbb{N}\Gamma = C \cap \mathbb{Z}^n; \, x \notin \mathbb{N}\mathcal{H}\}$$

is not empty. By Lemma 1.3.8 there exists $b \in \mathbb{Z}^n$ such that $\langle b, x\rangle > 0$ for all $0 \neq x \in C$. Let $x_0 \in \mathcal{V}$ such that

$$\langle x_0, b\rangle = \min\{\langle x, b\rangle \,|\, x \in \mathcal{V}\}.$$

Since $x_0 \notin \mathcal{H}$ we can write $x_0 = x_1 + x_2$ with x_1, x_2 in $(C \setminus \{0\}) \cap \mathbb{Z}^n$. Thus, since $\langle x_i, b\rangle < \langle x_0, b\rangle$ for $i = 1, 2$, we have $x_1, x_2 \in \mathbb{N}\mathcal{H}$ and $x_1 + x_2 \in \mathbb{N}\mathcal{H}$, a contradiction. □

Definition 1.3.10 (Edmonds and Giles [125]) A rational system $xA \leq b$ of linear inequalities is called *totally dual integral*, abbreviated TDI, if the minimum in the LP-duality equation

$$\max\{\langle x, c\rangle \,|\, xA \leq b\} = \min\{\langle y, b\rangle \,|\, y \geq 0; Ay = c\} \tag{1.10}$$

has an integral optimum solution y for each integral vector c for which the minimum is finite.

Note that there are rational systems of linear inequalities which define the same polyhedron and such that one system is totally dual integral while the other is not (see Exercise 1.3.39).

TDI systems occur in the theory of Gröbner bases of toric ideals [252], in the theory of perfect graphs [296, 86], and in combinatorial commutative algebra [147, 185, 188]. See Theorems 9.6.21, 13.6.8, and 14.3.6.

Proposition 1.3.11 [372, Corollary 22.1c] *If $xA \leq b$ is a TDI-system and b is integral, then the polyhedron $\{x \,|\, xA \leq b\}$ is integral.*

Proof. By hypothesis $\min\{\langle y, b\rangle \,|\, y \geq 0; Ay = c\}$ has an integral optimum solution y for each integral vector c for which the minimum is finite. Then by the LP duality theorem (see Theorem 1.1.57), we get that

$$\max\{\langle x, c\rangle \,|\, xA \leq b\}$$

is an integer for each integral vector c for which the maximum is finite. Thus the polyhedron $\{x \,|\, xA \leq b\}$ is integral by Theorem 1.1.63. □

Definition 1.3.12 A set $\mathcal{P}$ in $\mathbb{R}^n$ is called a *parallelotope* if it is the image of $[0, 1]^n$ under a non-singular linear transformation, i.e., $\mathcal{P}$ has the form

$$\mathcal{P} = \{\lambda_1 v_1 + \cdots + \lambda_n v_n \,|\, 0 \leq \lambda_i \leq 1\}$$

for some linearly independent vectors $v_1, \ldots, v_n$ in $\mathbb{R}^n$.

Lemma 1.3.13 Let $v_1, \ldots, v_n$ be a basis of $\mathbb{R}^n$ and let

$$\mathcal{P} = \{\lambda_1 v_1 + \cdots + \lambda_n v_n | \, 0 \leq \lambda_i \leq 1\}$$

be a parallelotope. Then $\mathrm{vol}(\mathcal{P}) = |\det[v_1, \ldots, v_n]|$.

Proof. Since $\mathcal{P}$ is the image of $[0,1]^n$ under the linear map T induced by $e_i \mapsto v_i$, the formula follows from Theorem 1.2.6. □

Proposition 1.3.14 *Let A be an integral matrix of order $n \times n$ and let $\mathcal{A} = \{v_1, \ldots, v_n\}$ be the set of columns of A. If $\det(A) \neq 0$, then $\mathcal{A}$ is a Hilbert basis if and only if $|\det(A)| = 1$.*

Proof. $\Rightarrow$) Let $\mathcal{Q} = [0,1]^n$ be the unit cube and let $\mathcal{P}$ be the parallelotope

$$\mathcal{P} = \{\lambda_1 v_1 + \cdots + \lambda_n v_n | \, 0 \leq \lambda_i \leq 1\}.$$

By Lemma 1.3.13 one has $\mathrm{vol}(\mathcal{P}) = |\det(A)|$. As $\mathcal{A}$ is a Hilbert basis and $\mathcal{A}$ is linearly independent, we have

$$(k+1)^n = |k\mathcal{Q} \cap \mathbb{Z}^n| = |k\mathcal{P} \cap \mathbb{Z}^n|, \quad \forall\, k \in \mathbb{N}.$$

Therefore by Proposition 1.2.10 and the fact that $[0,1]^n$ has volume 1, we conclude that $1 = \mathrm{vol}(\mathcal{P}) = |\det(A)|$, as required.

$\Leftarrow$) Let α be a vector in $\mathbb{R}_+\mathcal{A} \cap \mathbb{Z}^n$. As $|\det(A)| = 1$ and since $\mathcal{A}$ is linearly independent, by Cramer's rule, it follows that $\alpha \in \mathbb{N}\mathcal{A}$. Thus $\mathcal{A}$ is a Hilbert basis. □

Notation Let $A \neq (0)$ be an integral matrix. The greatest common divisor of all the non-zero $r \times r$ sub determinants of A will be denoted by $\Delta_r(A)$.

Theorem 1.3.15 [261, Theorem 3.9] *Let A be an integral matrix of rank r and let $d_1, \ldots, d_r$ be the invariant factors of A. Then*

$$d_1 = \Delta_1(A), \; d_2 = \Delta_2(A)\Delta_1(A)^{-1}, \ldots, d_r = \Delta_r(A)\Delta_{r-1}(A)^{-1}.$$

The next result is called the *fundamental structure theorem for finitely generated abelian groups.*

Theorem 1.3.16 [261, pp. 187-188] *Let M be a finitely generated $\mathbb{Z}$-module with a presentation $M \simeq \mathbb{Z}^n/(a_1, \ldots, a_q)$. If A is the matrix of rank r with columns $a_1, \ldots, a_q$ and $d_1, \ldots, d_r$ are the invariant factors of A, then*

$$M \simeq \mathbb{Z}/(d_1) \oplus \mathbb{Z}/(d_2) \oplus \cdots \oplus \mathbb{Z}/(d_r) \oplus \mathbb{Z}^{n-r}.$$

Lemma 1.3.17 *If A is an integral matrix of size $n \times q$ and rank r, then*

$$\Delta_r(A) = |T(\mathbb{Z}^n/(a_1, \ldots, a_q))|,$$

where a_i is the ith column of A. In particular $\Delta_r(A) = 1$ if and only if the quotient group $\mathbb{Z}^n/(a_1, \ldots, a_q)$ is torsion-free.

Proof. Let $d_1, \ldots, d_r$ be the invariant factors of the matrix A. On one hand using Theorem 1.3.15, we get the equality $d_1 d_2 \cdots d_r = \Delta_r(A)$. On the other hand, by Theorem 1.3.16, we obtain an isomorphism

$$T(\mathbb{Z}^n/(a_1, \ldots, a_q)) \simeq \mathbb{Z}/(d_1) \oplus \cdots \oplus \mathbb{Z}/(d_r).$$

Therefore the order of the torsion subgroup is $\Delta_r(A)$, as required. □

Lemma 1.3.18 *Let $\mathcal{A} = \{v_1, \ldots, v_d\} \subset \mathbb{Z}^n$ be a set of linearly independent vectors and let A be the matrix with column vectors $v_1, \ldots, v_d$. Then*

$$d!\mathrm{vol}(\mathrm{conv}(0, v_1, \ldots, v_d)) = \Delta_d(A).$$

Proof. Let $\Gamma = \{\gamma_1, \ldots, \gamma_d\}$ be a set of vectors such that

$$\mathbb{R}\mathcal{A} \cap \mathbb{Z}^n = \mathbb{Z}\gamma_1 \oplus \cdots \oplus \mathbb{Z}\gamma_d, \tag{1.11}$$

see Lemma 1.2.3. Then we can write

$$v_i = c_{i1}\gamma_1 + \cdots + c_{id}\gamma_d \quad (i = 1, \ldots, d) \tag{1.12}$$

where $C = (c_{ij})$ is an integral matrix. By Proposition 1.2.18, we have

$$d!\mathrm{vol}(\mathrm{conv}(0, v_1, \ldots, v_d)) = |\det(C)|.$$

Using Eqs. (1.11) and (1.12), we get that the map

$$T(\mathbb{Z}^n/\mathbb{Z}\mathcal{A}) \longrightarrow T(\mathbb{Z}^d/\mathbb{Z}\{c_1, \ldots, c_d\}), \quad \alpha \mapsto [\alpha]_\Gamma,$$

is an isomorphism of groups, where c_i is the ith row of C and $[\alpha]_\Gamma$ is the coordinate vector of α in the basis Γ. Hence, by Lemma 1.3.17, we get

$$\Delta_d(A) = |T(\mathbb{Z}^n/\mathbb{Z}\mathcal{A})| = |T(\mathbb{Z}^d/\mathbb{Z}\{c_1, \ldots, c_d\})| = |\det(C)|.$$ □

Proposition 1.3.19 *Let $\mathcal{A} = \{v_1, \ldots, v_d\} \subset \mathbb{Z}^n$ be a linearly independent set and let A be the matrix with column vectors $v_1, \ldots, v_d$. Then $\mathcal{A}$ is a Hilbert basis if and only if $\Delta_d(A) = 1$.*

Proof. $\Rightarrow$) Let $C = (c_{ij})$ be as in the proof of Lemma 1.3.18. As $\mathcal{A}$ is a Hilbert basis, it is seen that the rows of C form a Hilbert basis. Then $|\det(C)| = 1$ by Proposition 1.3.14. Therefore $\Delta_d(A) = 1$ by Lemma 1.3.18.

$\Leftarrow$) If $\Delta_d(A) = 1$, then $\mathbb{Z}^n/\mathbb{Z}\mathcal{A}$ is torsion-free. Since the v_i's are linearly independent, it follows readily that $\mathcal{A}$ is a Hilbert basis. □

Theorem 1.3.20 [179] *Let $\mathcal{A} = \{v_1, \ldots, v_q\} \subset \mathbb{Z}^n$ be a Hilbert basis and let $r =$ rank $(\mathbb{Z}\mathcal{A})$. If $\mathbb{R}_+\mathcal{A}$ is pointed, then there exist $\mathcal{H} \subset \mathcal{A}$ such that $\mathcal{H}$ is linearly independent, $\mathcal{H}$ is a Hilbert basis and $|\mathcal{H}| = r$.*

Corollary 1.3.21 *If $\mathcal{A} = \{v_1, \ldots, v_q\} \subset \mathbb{Z}^n$ and $\mathbb{R}_+\mathcal{A}$ is a pointed cone, then $\mathcal{A}$ is a Hilbert basis if and only if $\mathbb{R}_+\mathcal{A} \cap \mathbb{Z}\mathcal{A} = \mathbb{N}\mathcal{A}$ and $\mathbb{Z}^n/\mathbb{Z}\mathcal{A}$ is a torsion-free group.*

Proof. Assume that $\mathcal{A}$ is a Hilbert basis. Let A be the matrix with column vectors $v_1, \ldots, v_q$ and let $r = \text{rank}\ (A)$. Then clearly $\mathbb{R}_+\mathcal{A} \cap \mathbb{Z}\mathcal{A} = \mathbb{N}\mathcal{A}$. By Theorem 1.3.20 and Proposition 1.3.19, we get $\Delta_r(A) = 1$. Thus $\mathbb{Z}^n/\mathbb{Z}\mathcal{A}$ is a torsion-free group. The converse follows readily. Notice that for the converse the hypothesis that the cone $\mathbb{R}_+\mathcal{A}$ is pointed is not needed. $\square$

Let $\mathcal{P} = \{x \,|\, xA \leq b\}$ be a rational polyhedron and let F be a face of $\mathcal{P}$. A column of A is *active* in F if the corresponding inequality in $xA \leq b$ is satisfied with equality for all vectors x in F. A *minimal face* of $\mathcal{P}$ is a face not containing any other face. A face F of $\mathcal{P}$ is minimal if and only if F is an affine subspace [372].

Theorem 1.3.22 [372, Theorem 22.5] *A rational system $xA \leq b$ is TDI if and only if for each minimal face F of the polyhedron $\mathcal{P} = \{x \,|\, xA \leq b\}$, the columns of A which are active in F form a Hilbert basis.*

Proof. $\Rightarrow$) Let $v_1, \ldots, v_q$ be the column vectors of A and let $b = (b_i)$. Assume that $v_1, \ldots, v_r$ are the columns of A which are active in F. Take an integral vector c in $\mathbb{R}_+\{v_1, \ldots, v_r\}$. Then the maximum in the LP-duality equation

$$\max\{\langle x, c\rangle \,|\, xA \leq b\} = \min\{\langle y, b\rangle \,|\, y \geq 0; Ay = c\} \tag{1.13}$$

is attained by each vector of F. Indeed if $x_0 \in F$ and $c = \lambda_1 v_1 + \cdots + \lambda_r v_r$, with $\lambda_i \geq 0$ for all i, then $\langle x_0, c\rangle = \lambda_1 b_1 + \cdots + \lambda_r b_r$ and

$$\min\{\langle y, b\rangle \,|\, y \geq 0; Ay = c\} \leq \lambda_1 b_1 + \cdots + \lambda_r b_r \leq \max\{\langle x, c\rangle \,|\, xA \leq b\}.$$

Thus we have equality everywhere and the maximum is attained at any vector of F. The minimum has an integral optimum solution y. Then by complementary slackness (see Corollary 1.1.58), we get $(b - xA)y = 0$ for any $x \in F$. Therefore $(b_i - \langle x, v_i\rangle)y_i = 0$ for $i = 1, \ldots, q$. If v_i is inactive in F, then $\langle x, v_i\rangle < b_i$ for some $x \in F$. Then $y_i = 0$. This proves that $y_i = 0$ for $i > r$, i.e., $c \in \mathbb{N}\{v_1, \ldots, v_r\}$. Altogether $v_1, \ldots, v_r$ is a Hilbert basis.

$\Leftarrow$) Let c be an integral vector for which the optimum values of Eq. (1.13) are finite. Let F be a face of $\mathcal{P}$ such that each vector in F attains the maximum in Eq. (1.13). We may assume that F is a minimal face of $\mathcal{P}$. By the minimality of F if a vector x_0 in F satisfies $\langle x_0, v_i\rangle = b_i$ for some i, then any other vector x in F satisfies $\langle x_0, v_i\rangle = b_i$, i.e., v_i is active in F. Thus we may assume that there is $1 \leq r \leq q$ such that $\langle x, v_i\rangle = b_i$ for $i \leq r$ and $x \in F$, and $\langle x, v_i\rangle < b_i$ for $i > r$ and $x \in F$. Thus by hypothesis the set $\mathcal{A}_1 = \{v_1, \ldots, v_r\}$ is a Hilbert basis. Let x and $y = (y_i)$ be optimum solutions of Eq. (1.13) with $x \in F$. Then by complementary slackness (see

Corollary 1.1.58), we get $(b - xA)y = 0$, $Ay = c$ and $y \geq 0$. If $y_i > 0$, then $\langle x, v_i \rangle = b_i$, i.e., $i \leq r$. Therefore c is in $\mathbb{R}_+\mathcal{A}_1$. Hence c is in $\mathbb{N}\mathcal{A}_1$ because $\mathcal{A}_1$ is a Hilbert basis. Then we can write $c = \sum_{i=1}^{r} \eta_i v_i$, $\eta_i \in \mathbb{N}$. Consider the vector $y_0 = (\eta_1, \ldots, \eta_r, 0 \ldots, 0)$ in $\mathbb{N}^q$ whose last $q - r$ entries are equal to zero. Thus $c = Ay_0$. Since $b_i - \langle x, v_i \rangle = 0$ for $i \leq r$, we get $(b - xA)y_0 = 0$. Thus the minimum in Eq. (1.13) is attained in y_0, as required. □

Notice that we have in fact shown that Theorem 1.3.22 is valid if we replace "each minimal face F" by "each face F".

Definition 1.3.23 Let A be an integral matrix and let w be an integral vector. The system $xA \leq w$ is said to have the *integer rounding property* if

$$\lceil \min\{\langle y, w \rangle \,|\, y \geq 0;\ Ay = a\} \rceil = \min\{\langle y, w \rangle \,|\, Ay = a;\ y \geq 0\,; y \text{ integral}\}$$

for each integral vector a for which $\min\{\langle y, w \rangle \,|\, y \geq 0;\ Ay = a\}$ is finite.

Theorem 1.3.24 [372, Theorem 22.18] *Let A be an integral matrix of size $n \times q$ with column vectors $v_1, \ldots, v_q$ and $w = (w_i) \in \mathbb{Z}^q$. Then the system $xA \leq w$ has the integer rounding property if and only if the set*

$$\mathcal{H}' = \{(v_1, w_1), \ldots, (v_q, w_q), e_{n+1}\} \subset \mathbb{Z}^{n+1}$$

is a Hilbert basis.

Corollary 1.3.25 *Let A be an integral matrix and let $w = (w_i)$ be an integral vector. The system $xA \leq w$ is TDI if and only if $\{x \,|\, xA \leq w\}$ is an integral polyhedron and the set $\mathcal{H}' \subset \mathbb{Z}^{n+1}$ is a Hilbert basis.*

Proof. It follows from Proposition 1.3.11, Theorem 1.3.24, and using the definition of a TDI system. □

Theorem 1.3.26 *Let A be an integer matrix with column vectors $v_1, \ldots, v_q$ and let $w = (w_i)$ be an integral vector. If the polyhedron $\mathcal{P} = \{x \,|\, xA \leq w\}$ is integral and $\mathcal{H} = \{(v_1, w_1), \ldots, (v_q, w_q)\}$ is a Hilbert basis, then the system $xA \leq w$ is TDI.*

Proof. Let F be a minimal face of $\mathcal{P}$. Recall that a column of A is active in F if the corresponding inequality in $xA \leq w$ is satisfied with equality for all vectors in F. We may assume that $v_1, \ldots, v_r$ are the columns of A which are active in F. Then $\langle x, v_i \rangle = w_i$ for $x \in F$ and $1 \leq i \leq r$.

If $\langle y, v_i \rangle < w_i$ for some $y \in F$, then $\langle x, v_i \rangle < w_i$ for any other $x \in F$. Indeed if $\langle x, v_i \rangle = w_i$ for some $x \in F$, consider the supporting hyperplane of $\mathcal{P}$ given by $H = \{x | \langle x, v_i \rangle = w_i\}$, then $x \in F \cap H \subsetneq F$ because $y \in F$ and $y \notin F \cap H$, a contradiction to the minimality of the face F. Thus we may also assume that $\langle x, v_i \rangle < w_i$ for $x \in F$ and $i > r$.

Since $\mathcal{P}$ is integral, by Theorem 1.1.63, each face of $\mathcal{P}$ contains integral vectors. Pick an integral vector $x_0 \in F$. By Theorem 1.3.22, it suffices to prove that $\mathcal{B} = \{v_1, \dots, v_r\}$ is a Hilbert basis. Take $a \in \mathbb{R}_+\mathcal{B} \cap \mathbb{Z}^n$. Then $a = \lambda_1 v_1 + \cdots + \lambda_r v_r$ with $\lambda_i \geq 0$ for all i. Thus we have

$$\begin{aligned} b = \langle a, x_0 \rangle &= \lambda_1 \langle v_1, x_0 \rangle + \cdots + \lambda_r \langle v_r, x_0 \rangle \\ &= \lambda_1 w_1 + \cdots + \lambda_r w_r. \end{aligned}$$

Hence b is an integer and $(a, b) = \sum_{i=1}^{r} \lambda_i (v_i, w_i)$. As $\mathcal{H}$ is a Hilbert basis, we can write $(a, b) = \sum_{i=1}^{q} \eta_i (v_i, w_i)$, $\eta_i \in \mathbb{N}$ for all i. Therefore on one hand $0 = \langle (a, b), (x_0, -1) \rangle$. On the other hand $\langle (a, b), (x_0, -1) \rangle$ is equal to

$$\sum_{i=1}^{r} \eta_i \underbrace{\langle (v_i, w_i), (x_0, -1) \rangle}_{=0} + \sum_{i=r+1}^{q} \eta_i \underbrace{\langle (v_i, w_i), (x_0, -1) \rangle}_{<0}.$$

Hence $\eta_i = 0$ for $i > r$ and $a = \eta_1 v_1 + \cdots + \eta_r v_r$. Thus $a \in \mathbb{N}\mathcal{B}$. □

The converse is not true in general. However there are some interesting linear systems where the converse holds.

Definition 1.3.27 Let A be an integral matrix and w an integral vector. The system $x \geq 0; xA \leq w$ is TDI if the minimum in the LP-duality equation

$$\max\{\langle a, x \rangle \,|\, x \geq 0; xA \leq w\} = \min\{\langle y, w \rangle \,|\, y \geq 0; Ay \geq a\} \qquad (1.14)$$

has an integral optimum solution y for each integral vector a with finite minimum.

Proposition 1.3.28 *Let A be a nonnegative integral matrix of size $n \times q$ with column vectors $v_1, \dots, v_q$ and let $w = (w_i) \in \mathbb{N}^q$. Then, the system $x \geq 0; xA \leq w$ is TDI if and only if the polyhedron $\mathcal{P} = \{x \,|\, x \geq 0; xA \leq w\}$ is integral and $\mathcal{H} = \{(v_1, w_1), \dots, (v_q, w_q), -e_1, \dots, -e_n\}$ is a Hilbert basis.*

Proof. Assume the system $x \geq 0; xA \leq w$ is TDI. By Proposition 1.3.11, we get that $\mathcal{P}$ is integral. Next we prove that $\mathcal{H}$ is a Hilbert basis. Take $(a, b) \in \mathbb{R}_+\mathcal{H} \cap \mathbb{Z}^{n+1}$, where $a \in \mathbb{Z}^n$ and $b \in \mathbb{Z}$. By hypothesis, the minimum in Eq. (1.14) has an integral optimum solution $y = (y_i)$ such that $\langle y, w \rangle \leq b$. Since $y \geq 0$ and $a \leq Ay$, we can write

$$\begin{aligned} a = y_1 v_1 + \cdots + y_q v_q - \delta_1 e_1 - \cdots - \delta_n e_n \quad (\delta_i \in \mathbb{N}) \implies \\ (a, b) = y_1 (v_1, w_1) + \cdots + y_{q-1} (v_{q-1}, w_{q-1}) \\ +(y_q + b - \langle y, w \rangle)(v_q, w_q) - (b - \langle y, w \rangle) v_q - \delta, \end{aligned}$$

where $\delta = (\delta_i)$. As the entries of A are in $\mathbb{N}$, the vector $-v_q$ can be written as a nonnegative integer combination of $-e_1, \dots, -e_n$. Thus $(a, b) \in \mathbb{N}\mathcal{H}$. This proves that $\mathcal{H}$ is a Hilbert basis. The converse follows readily from Theorem 1.3.26. □

Lemma 1.3.29 *Let A be an integral matrix of size $n \times q$ with column vectors $v_1, \ldots, v_q$ and $w = (w_i) \in \mathbb{Z}^q$. If $\mathcal{P} = \{x \mid xA \leq w\}$ is pointed, then α is a vertex of $\mathcal{P}$ if and only if $H^-_{(\alpha,-1)}$ is a closed halfspace defining a facet of the cone $C = \mathbb{R}_+\{(v_1, w_1), \ldots, (v_q, w_q), e_{n+1}\}$.*

Proof. Since $\mathcal{P}$ is pointed, the rank of A is equal to n and C is a cone of full dimension $n+1$. Using that α is a vertex of $\mathcal{P}$ if and only if α is a basic feasible solution of the system $xA \leq w$ (see Corollary 1.1.49), the proof follows from Theorem 1.1.44. □

Procedure 1.3.30 If $\mathcal{P} = \{x \mid xA \leq w\}$ is pointed, using Corollary 1.3.25 and Lemma 1.3.29 we can check whether a system $xA \leq w$ is TDI. Using *Normaliz* [68] this can be achieved using a single input file of the form:

```
q+1
n+1
v1,w1
  ...
vq,wq
0  1
0
```

The first and third block of the output file contain the Hilbert basis and the support hyperplanes. To verify if $\mathcal{P}$ is integral, by Lemma 1.3.29, it suffices to verify that all rows of the support hyperplanes having its last entry positive are integral and its last entry is equal to 1.

Example 1.3.31 To illustrate Procedure 1.3.30 consider the integer vector $w = (3, 0, 0, 1, 0)$ and the matrix

$$A = \begin{pmatrix} 1 & 1 & 1 & 1 & 1 \\ 0 & 1 & 0 & 1 & 2 \\ 0 & 0 & 1 & 1 & 2 \end{pmatrix}$$

We now verify that the system $xA \leq w$ is not TDI. Using the following input file for *Normaliz*

```
6
4
1 0 0 3
1 1 0 0
1 0 1 0
1 1 1 1
1 2 2 0
0 0 0 1
0
```

We obtain the following output file (we only show the first and third block):

```
6 Hilbert basis elements:          6 support hyperplanes:
 1 0 0 3                             -3  3  3  1
 1 1 0 0                              0  0  0  1
 1 0 1 0                              0  0  1  0
 0 0 0 1                              0  1  0  0
 1 2 2 0                              2 -2  1  0
 1 1 1 0                              2  1 -2  0
```

Thus the polyhedron $\{x|\, xA \leq w\}$ is integral and the rows in the input file do not form a Hilbert basis. For instance the vector $(1,1,1,0)$ is in the cone generated by the rows of the input file but it is not an $\mathbb{N}$-linear combination of the rows.

Corollary 1.3.32 *Let $xA \leq w$ be a system satisfying the integer rounding property and let $\mathcal{A} = \{v_1, \ldots, v_q\}$ be the set of column vectors of A. Then*

(a) *$\mathcal{A}$ is a Hilbert basis.*

(b) *$\mathbb{Z}^n/\mathbb{Z}\mathcal{A}$ is a torsion-free group provided that $\mathbb{R}_+\mathcal{A}$ is a pointed cone.*

Proof. (a): Take $a \in \mathbb{R}_+\mathcal{A} \cap \mathbb{Z}^n$, then we can write $a = \sum_{i=1}^{q} \lambda_i v_i$, for some $\lambda_1, \ldots, \lambda_q$ in $\mathbb{R}_+$. Hence

$$(a, \lceil \textstyle\sum_i \lambda_i w_i \rceil) = \lambda_1(v_1, w_1) + \cdots + \lambda_q(v_q, w_q) + \delta(0,1),$$

where $\delta \geq 0$. Therefore, by Theorem 1.3.24, there are $\lambda_1', \ldots \lambda_q' \in \mathbb{N}$ and $\delta' \in \mathbb{N}$ such that

$$(a, \lceil \textstyle\sum_i w_i \lambda_i \rceil) = \lambda_1'(v_1, w_1) + \cdots + \lambda_q'(v_q, w_q) + \delta'(0,1),$$

Thus $a \in \mathbb{N}\mathcal{A}$. (b): It follows at once from (a) and Corollary 1.3.21. $\square$

Exercises

1.3.33 Prove that any subsemigroup of $\mathbb{N}$ is finitely generated.

1.3.34 Prove that the subsemigroup of $\mathbb{N}^2$ given by $\mathcal{S} = (\mathbb{N}_+ \times \mathbb{N}) \cup \{(0,0)\}$ is not finitely generated, where $\mathbb{N}_+ = \{1, 2, \ldots\}$.

1.3.35 (Gordan's lemma) Let $\mathcal{L}$ be a lattice in $\mathbb{R}^n$, i.e., $\mathcal{L}$ is an additive subgroup of $\mathbb{Z}^n$, and let C be a rational polyhedral cone in $\mathbb{R}^n$. Prove that $\mathcal{L} \cap C$ is a finitely generated semigroup.

Hint Adapt the proof of Lemma 1.3.2.

1.3.36 Let $C = \{x|\, Ax \leq 0\}$ be a rational polyhedral cone. Prove that C contains no lines if and only if $\{x|\, Ax = 0\} = (0)$.

1.3.37 Let A be an integral matrix. If $C = \{x | x \geq 0; Ax = 0\}$, prove that C is a rational polyhedral cone.

1.3.38 If $w = (3, 0, 0, -1, 0)$ and A is the matrix

$$A = \begin{pmatrix} 1 & 1 & 1 & 1 & 1 \\ 0 & 1 & 0 & 1 & 1 \\ 0 & 0 & 1 & 1 & 2 \end{pmatrix},$$

use *Normaliz* to verify that the system $xA \leq w$ is TDI.

1.3.39 Consider the two systems

$$(x_1, x_2, x_3)\begin{pmatrix} 1 & 0 & 1 & 1 \\ 1 & 1 & 0 & 1 \\ 0 & 1 & 1 & 1 \end{pmatrix} \leq \begin{pmatrix} 2 \\ 2 \\ 2 \\ 3 \end{pmatrix}; \ (x_1, x_2, x_3)\begin{pmatrix} 1 & 0 & 1 \\ 1 & 1 & 0 \\ 0 & 1 & 1 \end{pmatrix} \leq \begin{pmatrix} 2 \\ 2 \\ 2 \end{pmatrix}$$

Prove that they define the same polyhedron. Then prove that the first system is TDI but the second is not. Use *Normaliz* to verify these assertions.

1.3.40 Let A be an $n \times d$ integral matrix of rank d whose set of column vectors $\mathcal{A} = \{v_1, \ldots, v_d\}$ form a Hilbert basis. Prove that the system $xA \leq b$ is TDI for each integral vector b.

1.4 Rees cones and clutters

In this section we characterize the max-flow min-cut property of clutters in terms of Hilbert bases of Rees cones and the integrality of polyhedra.

Let A be an $n \times q$ matrix with entries in $\mathbb{N}$, let $\mathcal{A} = \{v_1, \ldots, v_q\}$ be the set of columns of A, and let

$$\mathcal{A}' := \{e_1, \ldots, e_n, (v_1, 1), \ldots, (v_q, 1)\} \subset \mathbb{R}^{n+1},$$

where e_i is the ith unit vector. The cone $\mathbb{R}_+\mathcal{A}'$, generated by $\mathcal{A}'$, is called the *Rees cone* of A or the *Rees cone* of $\mathcal{A}$.

The term Rees cone was coined in [147] because this cone encodes some information about the Rees algebra of the monomial ideal associated to $\mathcal{A}$; see Chapters 12–14. The first aim of this section is to study the irreducible representation, as an intersection of closed halfspaces, of a Rees cone.

We shall always assume that all rows and columns of A are non-zero. The *set covering polyhedron* (in Proposition 13.1.2 we clarify this terminology) is by definition the rational polyhedron:

$$\mathcal{Q}(A) := \{x \in \mathbb{R}^n \mid x \geq \mathbf{0},\ xA \geq \mathbf{1}\},$$

where **0** and **1** are vectors whose entries are equal to 0 and 1, respectively. Often we denote the vectors **0**, **1** simply by 0, 1. As is seen below, one can express the irreducible representation of $\mathbb{R}_+\mathcal{A}'$ in terms of $\mathcal{Q}(A)$.

Some of the facets of a Rees cone are easy to identify. Consider the index set

$$\mathcal{J} = \{i \mid 1 \leq i \leq n \text{ and } \langle e_i, v_j\rangle = 0 \text{ for some } j\} \cup \{n+1\}.$$

Notice that $\mathbb{R}_+\mathcal{A}'$ has dimension $n+1$. It is not hard to see that the set

$$F = \mathbb{R}_+\mathcal{A}' \cap H_{e_i} \quad (1 \leq i \leq n+1)$$

defines a facet of $\mathbb{R}_+\mathcal{A}'$ if and only $i \in \mathcal{J}$. Therefore, by Theorem 1.1.44 and Proposition 1.1.51, the Rees cone has the following unique irreducible representation

$$\mathbb{R}_+\mathcal{A}' = \left(\bigcap_{i\in\mathcal{J}} H_{e_i}^+\right) \bigcap \left(\bigcap_{i=1}^{r} H_{\alpha_i}^+\right) \tag{1.15}$$

such that $0 \neq \alpha_i \in \mathbb{Q}^{n+1}$ and $\langle \alpha_i, e_{n+1}\rangle = -1$ for all i. In many interesting cases, from the viewpoint of commutative algebra, one has the equality $\mathcal{J} = \{1, \ldots, n\}$; see Section 14.2 and Exercise 14.2.34.

Lemma 1.4.1 *Let A be a matrix with entries in $\mathbb{N}$ and let $a = (a_{i1}, \ldots, a_{iq})$ be its ith row. Set $k = \min\{a_{ij} \mid 1 \leq j \leq q\}$. If $a_{ij} > 0$ for all j, then e_i/k is a vertex of $\mathcal{Q}(A)$.*

Proof. Set $x_0 = e_i/k$. Clearly $x_0 \in \mathcal{Q}(A)$, $\langle x_0, v_j\rangle = 1$ for some j, and $\langle x_0, e_\ell\rangle = 0$ for $\ell \neq i$. It is seen that x_0 is a vertex of $\mathcal{Q}(A)$. □

The irreducible representation of a Rees cone can be expressed in terms of the vertices of the set covering polyhedron.

Theorem 1.4.2 *Let V be the vertex set of $\mathcal{Q}(A)$. Then the irreducible representation of the Rees cone is given by*

$$\mathbb{R}_+\mathcal{A}' = \left(\bigcap_{i\in\mathcal{J}} H_{e_i}^+\right) \bigcap \left(\bigcap_{u\in V} H_{(u,-1)}^+\right).$$

Proof. We set $\mathcal{B} = \{e_i \mid i \in \mathcal{J}\} \cup \{(u, -1) \mid u \in V\}$ and $V = \{u_1, \ldots, u_p\}$. First we dualize Eq. (1.15) and use the duality theorem for cones to obtain

$$\begin{aligned}
(\mathbb{R}_+\mathcal{A}')^* &= \{y \in \mathbb{R}^{n+1} \mid \langle y, x\rangle \geq 0, \forall\, x \in \mathbb{R}_+\mathcal{A}'\}\\
&= H_{e_1}^+ \cap \cdots \cap H_{e_n}^+ \cap H_{(v_1,1)}^+ \cap \cdots \cap H_{(v_q,1)}^+\\
&= \sum_{i\in\mathcal{J}} \mathbb{R}_+ e_i + \mathbb{R}_+\alpha_1 + \cdots + \mathbb{R}_+\alpha_r. \qquad (1.16)
\end{aligned}$$

Next we show the equality $(\mathbb{R}_+\mathcal{A}')^* = \mathbb{R}_+\mathcal{B}$. The right-hand side is clearly contained in the left-hand side because a vector α belongs to $\mathcal{Q}(A)$ if and only if $(\alpha, -1)$ is in $(\mathbb{R}_+\mathcal{A}')^*$. To prove the reverse inclusion observe that by Eq. (1.16) it suffices to show that $\alpha_k \in \mathbb{R}_+\mathcal{B}$ for all k. Writing $\alpha_k = (c_k, -1)$ and using $\alpha_k \in (\mathbb{R}_+\mathcal{A}')^*$ gives $c_k \in \mathcal{Q}(A)$. The set covering polyhedron can be written as

$$\mathcal{Q}(A) = \mathbb{R}_+e_1 + \cdots + \mathbb{R}_+e_n + \mathrm{conv}(V),$$

where $\mathrm{conv}(V)$ denotes the convex hull of V, this follows from the structure of polyhedra (see Theorem 1.1.33) by noticing that the characteristic cone of $\mathcal{Q}(A)$ is precisely $\mathbb{R}^n_+$. Thus we can write

$$c_k = \lambda_1 e_1 + \cdots + \lambda_n e_n + \mu_1 u_1 + \cdots + \mu_p u_p,$$

where $\lambda_i \geq 0$, $\mu_j \geq 0$ for all i, j and $\mu_1 + \cdots + \mu_p = 1$. If $1 \leq i \leq n$ and $i \notin \mathcal{J}$, then the ith row of A has all its entries positive. Thus by Lemma 1.4.1 we get that e_i/k_i is a vertex of $\mathcal{Q}(A)$ for some $k_i > 0$. To avoid cumbersome notation we denote e_i and $(e_i, 0)$ simply by e_i, from the context the meaning of e_i should be clear. Therefore from the equalities

$$\sum_{i\notin\mathcal{J}} \lambda_i e_i = \sum_{i\notin\mathcal{J}} \lambda_i k_i \left(\frac{e_i}{k_i}\right) = \sum_{i\notin\mathcal{J}} \lambda_i k_i \left(\frac{e_i}{k_i}, -1\right) + \left(\sum_{i\notin\mathcal{J}} \lambda_i k_i\right) e_{n+1}$$

we conclude that $\sum_{i\notin\mathcal{J}} \lambda_i e_i$ is in $\mathbb{R}_+\mathcal{B}$. From the identities

$$\begin{aligned} \alpha_k &= (c_k, -1) = \lambda_1 e_1 + \cdots + \lambda_n e_n + \mu_1(u_1, -1) + \cdots + \mu_p(u_p, -1) \\ &= \sum_{i\notin\mathcal{J}} \lambda_i e_i + \sum_{i\in\mathcal{J}\setminus\{n+1\}} \lambda_i e_i + \sum_{i=1}^{p} \mu_i(u_i, -1) \end{aligned}$$

we obtain $\alpha_k \in \mathbb{R}_+\mathcal{B}$, as required. Taking duals in $(\mathbb{R}_+\mathcal{A}')^* = \mathbb{R}_+\mathcal{B}$ yields

$$\mathbb{R}_+\mathcal{A}' = \bigcap_{a\in\mathcal{B}} H_a^+. \tag{1.17}$$

Thus, by Remark 1.1.31, the proof reduces to showing that $\beta \notin \mathbb{R}_+(\mathcal{B}\setminus\{\beta\})$ for all $\beta \in \mathcal{B}$. To show this we will assume that $\beta \in \mathbb{R}_+(\mathcal{B}\setminus\{\beta\})$ for some $\beta \in \mathcal{B}$ and derive a contradiction.

Case (I): $\beta = (u_j, -1)$. For simplicity assume $\beta = (u_p, -1)$. Then

$$\begin{aligned} (u_p, -1) &= \sum_{i\in\mathcal{J}} \lambda_i e_i + \sum_{j=1}^{p-1} \mu_j(u_j, -1), \qquad (\lambda_i \geq 0; \mu_j \geq 0) \Rightarrow \\ u_p &= \sum_{i\in\mathcal{J}\setminus\{n+1\}} \lambda_i e_i + \sum_{j=1}^{p-1} \mu_j u_j \qquad (1.18) \\ -1 &= \lambda_{n+1} - (\mu_1 + \cdots + \mu_{p-1}). \qquad (1.19) \end{aligned}$$

To derive a contradiction we claim that $\mathcal{Q}(A) = \mathbb{R}^n_+ + \text{conv}(u_1, \ldots, u_{p-1})$, which is impossible because by Proposition 1.1.39 the vertices of $\mathcal{Q}(A)$ would be contained in $\{u_1, \ldots, u_{p-1}\}$. To prove the claim note that the right-hand side is clearly contained in the left-hand side. For the other inclusion take $\gamma \in \mathcal{Q}(A)$ and write

$$\begin{aligned} \gamma &= \sum_{i=1}^{n} b_i e_i + \sum_{i=1}^{p} c_i u_i \qquad (b_i, c_i \geq 0; \sum_{i=1}^{p} c_i = 1) \\ &\overset{(1.18)}{=} \delta + \sum_{i=1}^{p-1} (c_i + c_p \mu_i) u_i \qquad (\delta \in \mathbb{R}^n_+). \end{aligned}$$

Therefore using the inequality

$$\sum_{i=1}^{p-1} (c_i + c_p \mu_i) = \sum_{i=1}^{p-1} c_i + c_p \left(\sum_{i=1}^{p-1} \mu_i \right) \overset{(1.19)}{=} (1 - c_p) + c_p(1 + \lambda_{n+1}) \geq 1$$

we get $\gamma \in \mathbb{R}^n_+ + \text{conv}(u_1, \ldots, u_{p-1})$. This proves the claim.

Case (II): $\beta = e_k$ for some $k \in \mathcal{J}$. We only consider the subcase $k \leq n$. The subcase $k = n + 1$ can be treated similarly. We can write

$$e_k = \sum_{i \in \mathcal{J} \setminus \{k\}} \lambda_i e_i + \sum_{i=1}^{p} \mu_i (u_i, -1), \qquad (\lambda_i \geq 0; \mu_i \geq 0).$$

From this equality we get $e_k = \sum_{i=1}^{p} \mu_i u_i$. Hence $e_k A \geq (\sum_{i=1}^{p} \mu_i)\mathbf{1} > 0$, a contradiction because $k \in \mathcal{J}$ and $\langle e_k, v_j \rangle = 0$ for some j. □

As a consequence we obtain:

Theorem 1.4.3 [188, Theorem 3.2] *The mapping* $\varphi\colon \mathbb{Q}^n \to \mathbb{Q}^{n+1}$ *given by* $\varphi(\alpha) = (\alpha, -1)$ *induces a bijective mapping*

$$\varphi\colon V \longrightarrow \{\alpha_1, \ldots, \alpha_r\}$$

between the set V *of vertices of* $\mathcal{Q}(A)$ *and the set* $\{\alpha_1, \ldots, \alpha_r\}$ *of vectors that occur in the irreducible representation of* $\mathbb{R}_+\mathcal{A}'$ *given in* Eq. (1.15).

Definition 1.4.4 Let A be a matrix with entries in $\{0, 1\}$. The system $x \geq 0$; $xA \geq 1$ is called *totally dual integral* (TDI) if the maximum in the LP-duality equation

$$\min\{\langle \alpha, x \rangle \,|\, x \geq 0; xA \geq 1\} = \max\{\langle y, 1 \rangle \,|\, y \geq 0; Ay \leq \alpha\}$$

has an integral optimum solution y for each integral vector α with finite maximum.

Proposition 1.4.5 *If the system $x \geq 0$; $xA \geq 1$ is TDI, then $\mathcal{Q}(A)$ has only integral vertices.*

Proof. It follows from Theorem 1.1.63. □

Definition 1.4.6 A *clutter* $\mathcal{C}$ with vertex set $X = \{x_1, \ldots, x_n\}$ is a family of subsets of X, called edges, none of which is included in another. The set of vertices and edges of $\mathcal{C}$ are denoted by $V(\mathcal{C})$ and $E(\mathcal{C})$, respectively.

Let $\mathcal{C}$ be a clutter with vertex set $X = \{x_1, \ldots, x_n\}$ and let $f_1, \ldots, f_q$ be the edges of $\mathcal{C}$. The *incidence matrix* of $\mathcal{C}$ is the $n \times q$ matrix $A = (a_{ij})$ given by $a_{ij} = 1$ if $x_i \in f_j$ and $a_{ij} = 0$ otherwise.

Definition 1.4.7 A clutter $\mathcal{C}$, with incidence matrix A, has the *max-flow min-cut* (MFMC) property if both sides of the LP-duality equation

$$\min\{\langle \alpha, x\rangle \,|\, x \geq 0; xA \geq 1\} = \max\{\langle y, 1\rangle \,|\, y \geq 0; Ay \leq \alpha\} \tag{1.20}$$

have integral optimum solutions x, y for each nonnegative integral vector α.

It turns out that the max-flow min-cut property is equivalent to require that a certain blowup ring has no nilpotent elements; see Theorem 14.3.6.

Proposition 1.4.8 *Let $\mathcal{C}$ be a clutter and let A be its incidence matrix. The system $x \geq 0$; $xA \geq 1$ is* TDI *if and only if $\mathcal{C}$ has the max-flow min-cut property.*

Proof. If follows from Propositions 1.4.5 and 1.1.41. □

Let $\mathcal{C}$ be a clutter with n vertices and let $\mathcal{A} = \{v_1, \ldots, v_q\}$ be the set of columns of its incidence matrix. For use below we denote by $\mathcal{A}'$ the *Rees configuration* of $\mathcal{C}$:

$$\mathcal{A}' = \{e_1, \ldots, e_n, (v_1, 1), \ldots, (v_q, 1)\} \subset \mathbb{N}^{n+1},$$

where e_i is the ith unit vector of $\mathbb{R}^n$

Theorem 1.4.9 [188, Theorem 3.4] *If $\mathcal{C}$ is a clutter and A is its incidence matrix, then $\mathcal{C}$ has the max-flow min-cut property if and only if $\mathcal{Q}(A)$ is an integral polyhedron and $\mathcal{A}'$ is a Hilbert basis of $\mathbb{R}_+\mathcal{A}'$.*

Proof. $\Rightarrow$) By Proposition 1.4.5 the polyhedron $\mathcal{Q}(A)$ is integral. Next we show that $\mathcal{A}'$ is an integral Hilbert basis. Take $(\alpha, \alpha_{n+1}) \in \mathbb{Z}^{n+1} \cap \mathbb{R}_+\mathcal{A}'$. Then $Ay \leq \alpha$ and $\langle y, \mathbf{1}\rangle = \alpha_{n+1}$ for some vector $y \geq 0$. Therefore one concludes that the optimal value of the linear program

$$\max\{\langle y, \mathbf{1}\rangle \,|\, y \geq \mathbf{0};\ Ay \leq \alpha\}$$

is greater than or equal to α_{n+1}. Since $\mathcal{C}$ has the MFMC property, this linear program has an optimal integral solution y_0. By Exercise 1.4.12, there exists an integral vector y_0' such that

$$\mathbf{0} \leq y_0' \leq y_0 \ \text{ and } \ |y_0'| = \alpha_{n+1}.$$

Therefore

$$\begin{pmatrix} \alpha \\ \alpha_{n+1} \end{pmatrix} = \begin{pmatrix} A \\ \mathbf{1} \end{pmatrix} y_0' + \begin{pmatrix} A \\ \mathbf{0} \end{pmatrix} (y_0 - y_0') + \begin{pmatrix} \alpha \\ 0 \end{pmatrix} - \begin{pmatrix} A \\ \mathbf{0} \end{pmatrix} y_0$$

and $(\alpha, \alpha_{n+1}) \in \mathbb{N}\mathcal{A}'$, as required.

$\Leftarrow$) Assume that $\mathcal{C}$ does not satisfy the MFMC property. Then there is an $\alpha_0 \in \mathbb{N}^n$ such that if y_0 is an optimal solution of the linear program:

$$\max\{\langle y, \mathbf{1}\rangle |\ y \geq \mathbf{0};\ Ay \leq \alpha_0\}, \tag{$*$}$$

then y_0 is not integral. We claim that also the optimal value $|y_0| = \langle y_0, \mathbf{1}\rangle$ of this linear program is not integral. If $|y_0|$ is integral, then $(\alpha_0, |y_0|)$ is in $\mathbb{Z}^{n+1} \cap \mathbb{R}_+\mathcal{A}'$. As $\mathcal{A}'$ is a Hilbert basis we get that $(\alpha_0, |y_0|)$ is in $\mathbb{N}\mathcal{A}'$, but this readily yields that the linear program $(*)$ has an integral optimal solution, a contradiction. This completes the proof of the claim.

Now, consider the dual linear program:

$$\min\{\langle x, \alpha_0\rangle |\ x \geq \mathbf{0},\ xA \geq \mathbf{1}\}.$$

By Proposition 1.1.41 the optimal value of this linear program is attained at a vertex x_0 of $\mathcal{Q}(A)$. Then by the duality theorem (Theorem 1.1.56) we get $\langle x_0, \alpha_0\rangle = |y_0| \notin \mathbb{Z}$. Hence x_0 is not integral, a contradiction to the integrality of the set covering polyhedron $\mathcal{Q}(A)$. □

Remark 1.4.10 *Normaliz* [68] computes the irreducible representation and the minimal integral Hilbert basis of a Rees cone. Thus we can effectively use Theorems 1.4.2 and 1.4.9 to determine whether a given clutter has the max-flow min-cut property; see example below.

Example 1.4.11 Let A be the transpose of the matrix:

$$\begin{bmatrix} 1 & 0 & 0 & 0 & 1 \\ 0 & 1 & 0 & 1 & 0 \\ 0 & 0 & 1 & 1 & 1 \\ 1 & 1 & 1 & 0 & 0 \end{bmatrix}$$

and let $\mathcal{C}$ be the clutter with incidence matrix A. Consider the following input file for *Normaliz* that we call "example.in":

```
4
5
 1  0  0  0  1
 0  1  0  1  0
 0  0  1  1  1
 1  1  1  0  0
3
```

Applying *Normaliz*, i.e., typing "normaliz example" in the directory where one has the file normaliz.exe we get the output file "example.out":

```
9 generators of integral closure of Rees algebra:
 1  0  0  0  0  0
 0  1  0  0  0  0
 0  0  1  0  0  0
 0  0  0  1  0  0
 0  0  0  0  1  0
 1  0  0  0  1  1
 0  1  0  1  0  1
 0  0  1  1  1  1
 1  1  1  0  0  1
10 support hyperplanes:
 0   0   1   1   1  -1
 1   0   0   0   0   0
 0   1   0   0   0   0
 0   0   0   0   0   1
 0   0   1   0   0   0
 1   0   0   1   0  -1
 0   0   0   1   0   0
 0   0   0   0   1   0
 0   1   0   0   1  -1
 1   1   1   0   0  -1
semigroup is not homogeneous
```

The first block shows that $\mathcal{A}'$ is a Hilbert basis for the Rees cone. The second block shows the irreducible representation of the Rees cone of A, thus using Theorem 1.4.2 we obtain that $\mathcal{Q}(A)$ is integral. Altogether Theorem 1.4.9 proves that the clutter $\mathcal{C}$ has the max-flow min-cut property.

Exercises

1.4.12 Let $b, \eta_1, \ldots, \eta_q$ be a sequence in $\mathbb{N}$. If $\eta_1 + \cdots + \eta_q \geq b$, then there are $\epsilon_1, \ldots, \epsilon_q \in \mathbb{N}$ such that $0 \leq \epsilon_i \leq \eta_i$ for all i and $\epsilon_1 + \cdots + \epsilon_q = b$.

1.4.13 Let A be the incidence matrix of a clutter $\mathcal{C}$ and let $v_1, \ldots, v_q$ be its column vectors. The system $x \geq 0$; $xA \geq \mathbf{1}$ is TDI if and only if the set

$$\mathcal{A}_u = \{v_i |\, \langle v_i, u \rangle = 1\} \cup \{e_i |\, \langle e_i, u \rangle = 0\}$$

is a Hilbert basis for any vertex u of $\mathcal{Q}(A)$.

1.5 The integral closure of a semigroup

Let $\mathcal{A} = \{v_1, \ldots, v_q\}$ be a set of vectors in $\mathbb{N}^n \setminus \{0\}$. The *integral closure* or *normalization* of the affine semigroup

$$\mathbb{N}\mathcal{A} := \mathbb{N}v_1 + \cdots + \mathbb{N}v_q \subset \mathbb{N}^n,$$

is defined as $\overline{\mathbb{N}\mathcal{A}} := \mathbb{Z}\mathcal{A} \cap \mathbb{R}_+\mathcal{A}$, where $\mathbb{Z}\mathcal{A}$ is the subgroup of $\mathbb{Z}^n$ generated by $\mathcal{A}$. The semigroup $\mathbb{N}\mathcal{A}$ is called *normal* or *integrally closed* if $\overline{\mathbb{N}\mathcal{A}} = \mathbb{N}\mathcal{A}$. The *Rees semigroup* of $\mathcal{A}$ is by definition $\mathbb{N}\mathcal{A}' \subset \mathbb{N}^{n+1}$, where

$$\mathcal{A}' = \{(v_1, 1), \ldots, (v_q, 1), e_1, \ldots, e_n\}.$$

Notice that $e_{n+1} = (v_1, 1) - \langle (v_1, 1), e_1\rangle e_1 - \cdots - \langle (v_1, 1), e_n\rangle e_n$. Hence $\mathbb{Z}\mathcal{A}' = \mathbb{Z}^{n+1}$. As a consequence we get

$$\overline{\mathbb{N}\mathcal{A}'} = \mathbb{R}_+\mathcal{A}' \cap \mathbb{Z}^{n+1}.$$

From this equality we obtain a characterization of the normality of a Rees semigroup in terms of Hilbert bases:

Proposition 1.5.1 $\mathbb{N}\mathcal{A}'$ *is normal if and only if* $\mathcal{A}'$ *is a Hilbert basis.*

The algebraic invariants of semigroup rings of semigroups generated by a Hilbert basis are easier to understand; see [405, Section 6]. This family of semigroups and their cones are studied throughout this book.

Proposition 1.5.2 *If* $\mathcal{A}$ *is a Hilbert basis, then* $\mathbb{N}\mathcal{A}$ *is normal.*

Proof. Since $\mathbb{N}\mathcal{A} \subset \mathbb{R}_+\mathcal{A} \cap \mathbb{Z}\mathcal{A} \subset \mathbb{R}_+\mathcal{A} \cap \mathbb{Z}^n = \mathbb{N}\mathcal{A}$ we obtain the equality $\mathbb{N}\mathcal{A} = \mathbb{R}_+\mathcal{A} \cap \mathbb{Z}\mathcal{A}$. □

Example 1.5.3 If $\mathcal{A} = \{(1, 1, 0), (0, 1, 1), (1, 0, 1)\}$, the affine semigroup $\mathbb{N}\mathcal{A}$ is normal because $\mathcal{A}$ is linearly independent, but $\mathcal{A}$ is not a Hilbert basis because $\mathbb{N}\mathcal{A} \subsetneq \mathbb{Z}^3 \cap \mathbb{R}_+\mathcal{A}$ (see Exercise 1.5.8).

Notation Given $\beta = (\beta_i) \in \mathbb{R}^{n+1}$, we define $\deg_{n+1}(\beta) = \beta_{n+1}$.

Proposition 1.5.4 [147] *Let* $\mathbb{R}_+\mathcal{A}'$ *be the Rees cone of* $\mathcal{A}$ *and let* $\mathcal{B}$ *be its minimal integral Hilbert basis. If* $\beta \in \mathcal{B}$, *then* $\deg_{n+1}(\beta) < n$.

Proof. We set $\beta = (\alpha, b)$, with $\alpha \in \mathbb{N}^n$ and $b = \deg_{n+1}(\beta) \in \mathbb{N}$. Note that $\mathbb{R}_+\mathcal{A}' \cap \mathbb{Z}^{n+1} = \mathbb{Q}_+\mathcal{A}' \cap \mathbb{Z}^{n+1}$ and $\dim(\mathbb{R}_+\mathcal{A}') = n + 1$. Thus, since $(\alpha, b) \in \mathbb{R}_+\mathcal{A}'$, by Carathéodory's theorem (Theorem 1.1.18) we can write

$$(\alpha, b) = \lambda_1(v_{i_1}, 1) + \cdots + \lambda_r(v_{i_r}, 1) + \mu_1 e_{j_1} + \cdots + \mu_s e_{j_s} \quad (\lambda_i, \mu_k \in \mathbb{Q}_+), \ (*)$$

where $\{(v_{i_1},1),\ldots,(v_{i_r},1),e_{j_1},\ldots,e_{j_s}\}$ is a linearly independent subset of $\mathcal{A}'$. By the minimality of $\mathcal{B}$ it is seen that $0 \le \lambda_i < 1$ and $0 \le \mu_k < 1$ for all i,k. From Eq. $(*)$ we get $b = \lambda_1 + \cdots + \lambda_r < r$. Thus $b \le r-1$. If $r \le n$, then $b \le n-1$. Thus from now on we may assume $r = n+1$ and Eq. $(*)$ takes the simpler form

$$(\alpha, b) = \lambda_1(v_{i_1}, 1) + \cdots + \lambda_{n+1}(v_{i_{n+1}}, 1).$$

Consider the cone C generated by $\mathcal{A}'' = \{(v_{i_1},1),\ldots,(v_{i_{n+1}},1)\}$. Since $-e_1$ is not in C, using Exercise 1.5.9, we obtain a point

$$x_0 = (1-\lambda_0)(\alpha, b) + \lambda_0(-e_1) \quad (0 \le \lambda_0 < 1)$$

in the relative boundary of C. By Theorem 1.1.44 the relative boundary of C is the union of its facets. Hence using that any facet of C is an n-dimensional cone generated by a subset of $\mathcal{A}''$ (see Proposition 1.1.23) together with the minimality of $\mathcal{B}$ it follows that we can write (α, b) as:

$$(\alpha, b) = \rho_0 e_1 + \rho_1(v_{j_1}, 1) + \cdots + \rho_n(v_{j_n}, 1) \qquad (0 \le \rho_i < 1\,\forall i),$$

and consequently $b \le n-1$, as required. □

Definition 1.5.5 A matrix A is called *totally unimodular* if each $i \times i$ subdeterminant of A is 0 or ± 1 for all $i \ge 1$.

Examples of totally unimodular matrices include incidence matrices of clutters without odd cycles [27, Chapter 5], incidence matrices of bipartite graphs and digraphs [372, p. 274], and network matrices [372, Chapter 19].

Theorem 1.5.6 (Hoffman, Kruskal) *If A is a totally unimodular matrix, then $\mathcal{Q} = \{x \mid x \ge 0;\ Ax \le b\}$ is integral for each integral vector b.*

Proof. Assume that A has size $n \times q$. Notice that $\mathcal{Q}$ contains no lines, thus it suffices to show that $\mathcal{Q}$ has only integral vertices. Let x_0 be a vertex of $\mathcal{Q}$. By Corollary 1.1.47, $A'x_0 = b'$ for some subsystem $A'x \le b'$ of

$$\begin{pmatrix} A \\ -I_q \end{pmatrix} \le \begin{pmatrix} b \\ 0 \end{pmatrix}$$

such that A' is a square matrix of order q with linearly independent rows. By hypothesis $\det(A') = \pm 1$, hence x_0 is integral because the inverse of A' is an integral matrix by Cramer's rule. □

Corollary 1.5.7 [49] *If $A = (a_{ij})$ is a totally unimodular $\{0,1\}$-matrix of size $n \times q$ with column vectors $v_1,\ldots,v_q$, then $\mathbb{R}_+\mathcal{A}'$ is normal.*

Proof. By the Hoffman–Kruskal theorem, the system $x \ge 0; xA \ge 1$ is TDI. Thus, by Theorem 1.4.9, the set $\mathcal{A}'$ is a Hilbert basis, i.e., the Rees semigroup $\mathbb{R}_+\mathcal{A}'$ is normal. □

Exercises

1.5.8 If $\mathcal{A} = \{(1,1,0),(0,1,1),(1,0,1)\}$, use *Normaliz*, with option 0, to verify that the minimal integral Hilbert basis of $\mathbb{R}_+\mathcal{A}$ is $\mathcal{A} \cup \{(1,1,1)\}$.

1.5.9 Let a be a point of a set $\mathcal{V}$ in $\mathbb{R}^n$ and let x be a point in $\mathrm{aff}(\mathcal{V})$ not in $\mathcal{V}$. If $\lambda_0 = \sup\{\lambda \in [0,1] | \, (1-\lambda)a + \lambda x \in \mathcal{V}\}$, then $x_0 = (1-\lambda_0)a + \lambda_0 x$ is a relative boundary point of $\mathcal{V}$ lying between a and x.

1.5.10 If A is a totally unimodular matrix and A' is obtained from A by adding columns of unit vectors, then the matrix A' is totally unimodular.

1.6 Unimodularity of matrices and normality

In this section, we characterize when a vector belongs to a subgroup of $\mathbb{Z}^n$ and prove Heger's theorem for the existence of solutions of an integer linear system. Then we show the normality of semigroups arising from unimodular matrices and unimodular coverings.

A *minor* of order r (r-minor for short) of a matrix A is defined as the determinant of a square submatrix of A of order r.

Definition 1.6.1 An integral matrix $A \neq (0)$ is t-*unimodular* if all the non-zero r-minors of A have absolute value equal to t, where r is the rank of A. If A is 1-unimodular, we say that A is a *unimodular matrix*.

Theorem 1.6.2 [372, Theorem 19.2] *Let A be an integral matrix of full row rank. Then the polyhedron $\{x | \, x \geq 0; Ax = b\}$ is integral for each integral vector b if and only if A is unimodular.*

Theorem 1.6.3 *Let A be an $n \times q$ integral matrix whose set of column vectors is $\mathcal{A} = \{v_1, \ldots, v_q\}$ and let $b \in \mathbb{Z}^n$ be a column vector such that* $\mathrm{rank}(A) = \mathrm{rank}([A\,b])$. *The following conditions are equivalent:*

(a) $b \in \mathbb{Z}\mathcal{A}$.

(b) *$\mathbb{Z}^n/\mathbb{Z}\mathcal{A}$ and $\mathbb{Z}^n/\mathbb{Z}(\mathcal{A} \cup \{b\})$ have the same invariant factors.*

(c) *The matrices $[A\,0]$ and $[A\,b]$ have the same Smith normal form.*

(d) *$\Delta_r(A) = \Delta_r([A\,b])$, where $r = \mathrm{rank}(A)$ and $\Delta_r(A)$ is the gcd of all the non-zero r-minors of A.*

Proof. There is $k \in \mathbb{N} \setminus \{0\}$ such that $kb = \sum_{i=1}^{q} \lambda_i v_i$, $\lambda_i \in \mathbb{Z}$. Therefore, there is a canonical epimorphism of finite groups

$$\varphi\colon T(\mathbb{Z}^n/\mathbb{Z}\mathcal{A}) \longrightarrow T(\mathbb{Z}^n/\mathbb{Z}\mathcal{B}) \quad (\alpha + \mathbb{Z}\mathcal{A} \stackrel{\varphi}{\longmapsto} \alpha + \mathbb{Z}\mathcal{B}),$$

where $\mathcal{B} = \mathcal{A} \cup \{b\}$. Notice that φ is injective if and only if (a) holds.

The implication (a) $\Rightarrow$ (b) is straightforward. There are invertible integral matrices P_i and Q_i such that

$$D_1 = Q_1[A\ 0]P_1 = \mathrm{diag}(d_1, \ldots, d_r, 0, \ldots, 0),$$
$$D_2 = Q_2[A\ b]P_2 = \mathrm{diag}(e_1, \ldots, e_r, 0, \ldots, 0),$$

are the Smith normal forms of $[A\,0]$ and $[A\,b]$, respectively; that is, d_i, e_i are positive integers satisfying that d_i divides d_{i+1} and e_i divides e_{i+1} for all i. By the fundamental structure theorem for finitely generated abelian groups (see Theorem 1.3.16) there are isomorphisms:

$$T(\mathbb{Z}^n/\mathbb{Z}\mathcal{A}) \simeq (\mathbb{Z}/d_1\mathbb{Z}) \times \cdots \times (\mathbb{Z}/d_r\mathbb{Z}),$$
$$T(\mathbb{Z}^n/\mathbb{Z}\mathcal{B}) \simeq (\mathbb{Z}/e_1\mathbb{Z}) \times \cdots \times (\mathbb{Z}/e_r\mathbb{Z}).$$

Thus (b) $\Leftrightarrow$ (c). Note $\Delta_i(A) = d_1 \cdots d_i$ and $\Delta_i([A\,b]) = e_1 \cdots e_i$ for all i. Hence (c) $\Rightarrow$ (d). To prove (d) $\Rightarrow$ (i) observe $|T(\mathbb{Z}^n/\mathbb{Z}\mathcal{A})| = |T(\mathbb{Z}^n/\mathbb{Z}\mathcal{B})|$ and consequently φ must be injective. □

As an immediate consequence of Theorem 1.6.3 we get:

Theorem 1.6.4 (I. Heger, [372, p. 51]) *Let A be an $n \times q$ integral matrix and let $b \in \mathbb{Z}^n$ be a column vector. If* $\mathrm{rank}(A) = \mathrm{rank}([A\,b])$, *then the system $Ax = b$ has an integral solution if and only if $\Delta_r(A) = \Delta_r([A\,b])$.*

Remark 1.6.5 Let $\mathcal{A} \subset \mathbb{Z}^n$ be a finite set. Note that the finite basis theorem together with Theorem 1.6.3 yield a membership test to decide when a given vector b in $\mathbb{Z}^n$ belongs to $\overline{\mathbb{N}\mathcal{A}} = \mathbb{Z}\mathcal{A} \cap \mathbb{R}_+\mathcal{A}$.

Proposition 1.6.6 *If A is a t-unimodular matrix with columns $v_1, \ldots, v_q$ and $v_{i_1}, \ldots, v_{i_r}$ is a $\mathbb{Q}$-basis for the column space of A, then*

$$\mathbb{Z}\mathcal{A} = \mathbb{Z}v_{i_1} \oplus \cdots \oplus \mathbb{Z}v_{i_r}$$

Proof. For each j one has $\Delta_r(v_{i_1} \cdots v_{i_r}) = \Delta_r(v_{i_1} \cdots v_{i_r}\, v_j) = t$. Then by Theorem 1.6.3, we get $v_j \in \mathbb{Z}v_{i_1} \oplus \cdots \oplus \mathbb{Z}v_{i_r}$, as required. □

Theorem 1.6.7 *If A is a t-unimodular matrix whose set of columns is $\mathcal{A}$, then the affine semigroup $\mathbb{N}\mathcal{A}$ is normal.*

Proof. If $b \in \mathbb{Z}\mathcal{A} \cap \mathbb{R}_+\mathcal{A}$, by Theorem 1.1.18 and Proposition 1.6.6 there are $v_{i_1}, \ldots, v_{i_r}$ linearly independent vectors in $\mathcal{A}$, $r = \mathrm{rank}(A)$, such that

$$b \in \mathbb{R}_+v_{i_1} + \cdots + \mathbb{R}_+v_{i_r} \quad \text{and} \quad b \in \mathbb{Z}v_{i_1} + \cdots + \mathbb{Z}v_{i_r}. \tag{1.21}$$

By comparing the coefficients of b with respect to the two representations obtained from Eq. (1.21), one derives $b \in \mathbb{N}\mathcal{A}$. □

Corollary 1.6.8 *If A is a totally unimodular matrix whose set of columns is $\mathcal{A}$, then $\mathbb{N}\mathcal{A} = \mathbb{Z}^n \cap \mathbb{R}_+\mathcal{A}$; that is, $\mathcal{A}$ is a Hilbert basis.*

Proof. Let $\mathcal{A} = \{v_1, \ldots, v_q\}$ be the set of column vectors of A. Take $b \in \mathbb{Z}^n \cap \mathbb{R}_+\mathcal{A}$, then by Carathéodory's theorem (see Theorem 1.1.18) and after permutation of the v_i's we can write $b = \sum_{i=1}^{r} \eta_i v_i$ with $\eta_i \geq 0$ for all i, where r is the rank of A and $v_1, \ldots, v_r$ are linearly independent.

The submatrix $A' = (v_1 \cdots v_r)$ is totally unimodular. Therefore, by Theorem 1.6.3, the system of equations $A'x = b$ has an integral solution. Thus b is a linear combination of $v_1, \ldots, v_r$ with coefficients in $\mathbb{Z}$. It follows that $\eta_i \in \mathbb{N}$ for all i, that is, $b \in \mathbb{N}\mathcal{A}$. The other inclusion is clear. $\Box$

Proposition 1.6.9 *If $\mathcal{A} = \{v_1, \ldots, v_q\} \subset \mathbb{Z}^n$ lie in a hyperplane of $\mathbb{R}^n$ not containing the origin and $\mathcal{P} = \mathrm{conv}(\mathcal{A})$ has a weakly unimodular covering with support in $\mathcal{A}$, then $\mathbb{N}\mathcal{A}$ is a normal semigroup.*

Proof. Take $0 \neq \beta \in \mathbb{R}_+\mathcal{A} \cap \mathbb{Z}\mathcal{A}$. There exists $\lambda_1 \ldots, \lambda_q \in \mathbb{Q}_+$ such that $\beta = \lambda_1 v_1 + \cdots + \lambda_q v_q$. We set $|\lambda| = \lambda_1 + \cdots + \lambda_q$. Since $\beta/|\lambda|$ is in $\mathcal{P}$ and using that $\mathcal{P}$ has a weakly unimodular covering, there is an affinely independent set $\mathcal{A}_1 \subset \mathcal{A}$ defining a lattice simplex $\Delta = \mathrm{conv}(\mathcal{A}_1)$ such that $\beta/|\lambda| \in \Delta$ and $\mathbb{Z}\mathcal{A}_1 = \mathbb{Z}\mathcal{A}$. Altogether β belongs to $\mathbb{Z}\mathcal{A}_1 \cap \mathbb{R}_+\mathcal{A}_1$. Note that $\mathcal{A}_1$ is a linearly independent set because $\mathcal{A}_1$ lie in a hyperplane not containing the origin. Therefore it follows rapidly that $\beta \in \mathbb{N}\mathcal{A}_1$, as required. $\Box$

Exercises

1.6.10 If $\mathcal{A} = \{v_1, \ldots, v_q\} \subset \mathbb{Z}^n$ and $\mathcal{P} = \mathrm{conv}(\mathcal{A})$ has a unimodular covering with support in $\mathcal{A}$, then $\mathbb{N}(v_1, 1) + \cdots + \mathbb{N}(v_q, 1)$ is normal.

1.7 Normaliz, a computer program

Throughout this book we will frequently use *Normaliz* [68], a program that provides an invaluable effective tool to study monomial subrings and their algebraic invariants. This program computes the following:

- normalization (or integral closure) of an affine semigroup;
- Hilbert basis of a pointed rational cone;
- lattice points of an integral polytope;
- support hyperplanes of a Rees cone;
- generators of the integral closure of the Rees algebra of a monomial ideal $I = (x^{v_1}, \ldots, x^{v_q}) \subset R = K[x_1, \ldots, x_n]$;
- generators of the integral closure of I;

- Hilbert series and Hilbert polynomial of a homogeneous affine semigroup;
- Ehrhart ring (Theorem 9.3.6) and relative volume (Proposition 1.2.10) of a lattice polytope.

It can be used to check the following integer programming properties:

- integer rounding property of any of the linear systems

$$x \geq 0; xA \geq \mathbf{1}, \qquad x \geq 0; xA \leq \mathbf{1}, \qquad xA \leq \mathbf{1},$$

  where A is the matrix with column vectors $v_1, \ldots, v_q$ (Corollary 14.6.9, Theorem 14.6.16, and Proposition 14.6.30);
- integer rounding property and total dual integrality (TDI) of a linear system $xA \leq w$ (Theorem 1.3.24, Procedure 1.3.30);
- perfection of a graph (Remark 13.6.4).

If I is square-free, it can also be used to determine the following:

- minimal primes of I (Remark 13.2.23);
- reducedness of the associated graded ring $\mathrm{gr}_I(R)$ (Example 14.2.20);
- the max-flow min-cut property of a clutter (Example 1.4.11);
- integrality of the set covering polyhedron

$$\mathcal{Q}(A) = \{x \mid xA \geq 1;\ x \geq 0\},$$

  where $A = (v_1, \ldots, v_q)$;
- total dual integrality of the system $x \geq 0$; $xA \geq 1$;
- the generators of the symbolic Rees algebra $R_s(I)$ (Example 13.2.22).

1.8 Cut-incidence matrices and integrality

In this section we introduce incidence matrices of digraphs, and present a generalization of a theorem of Lucchesi and Younger [298] which is very useful to detect TDI systems arising from incidence matrices of digraphs. This theorem will be used in Theorem 11.3.2 to express the a-invariant of the edge subring of a bipartite graph in terms of directed cuts. This illustrates that some deep results in combinatorial optimization can be used to study algebraic invariants of rings.

Next we introduce some more TDI systems which are suited for our purposes (see [281, Chapter 5] and [372, p. 311]).

Definition 1.8.1 Let B be an integral matrix. A system $Bx \geq b$, $x \geq 0$ is *totally dual integral* (TDI) if the maximum in

$$\min\{\langle c, x\rangle |\, x \geq 0;\, Bx \geq b\} = \max\{\langle y, b\rangle |\, y \geq 0;\, yB \leq c\}$$

has an integral optimum solution y for each integral vector c with finite maximum.

Proposition 1.8.2 *If the matrix B is totally unimodular, then the system $Bx \geq b$; $x \geq 0$ is* TDI.

Proof. It follows from Theorem 1.5.6. □

Proposition 1.8.3 *Let B be an integral matrix and let b be an integral vector. If the system $Bx \geq b$; $x \geq 0$ is* TDI, *then $\mathcal{Q} = \{x |\, Bx \geq b; x \geq 0\}$ has only integral vertices.*

Proof. It follows from Theorem 1.1.63. □

Definition 1.8.4 A *digraph* G consists of a finite vertex set $V(G)$ and a family $E(G)$ of ordered pairs of elements of $V(G)$. The pairs $(v_1, v_2) \in E(G)$ are called *directed edges* or *arrows*.

Definition 1.8.5 Let G be a digraph with vertex set $V(G)$ and edge set $E(G)$. Given a family $\mathcal{F}$ of subsets of $V(G)$, the *one-way cut-incidence matrix* of $\mathcal{F}$ is the matrix $B = (b_{X,e})_{X\in\mathcal{F}, e\in E(G)}$, where

$$b_{X,e} = \begin{cases} 1 & \text{if } e \in \delta^+(X) \\ 0 & \text{otherwise.} \end{cases}$$

The *two-way cut-incidence matrix* is $B = (b_{X,e})_{X\in\mathcal{F}, e\in E(G)}$, where

$$b_{X,e} = \begin{cases} 1 & \text{if } e \in \delta^+(X) \\ -1 & \text{if } e \in \delta^-(X) \\ 0 & \text{otherwise.} \end{cases}$$

Here $\delta^+(X) = \{e = (z, w) \in E(G) |\, z \in X,\, w \notin X\}$ is the set of *edges leaving* the vertex set X and $\delta^-(X)$ is the set of *edges entering* the vertex set X. If $\mathcal{F} = \{\{v\} \,|\, v \in V(G)\}$, the matrix B is called the *incidence matrix* of G.

Remark 1.8.6 An interesting case occurs when the one-way and the two-way cut-incidence matrix of the family $\mathcal{F}$ coincide. In this case the rows of the matrix B correspond to directed cuts only. Recall that $\delta^+(X)$ is a *directed cut* of a digraph G if $\emptyset \neq X \subset V(G)$ and $\delta^-(X) = \emptyset$.

It was pointed out to us by Jens Vygen that the next result is a slight generalization of the Lucchesi, Younger theorem [298]. It follows using the technique of proof of [281, Theorem 19.10].

Theorem 1.8.7 [405] *Let G be a digraph and $\mathcal{F}$ a family of subsets of $V(G)$ such that the one-way cut incidence matrix B of $\mathcal{F}$ is equal to the two-way cut incidence matrix of $\mathcal{F}$. If $\mathcal{F}$ satisfies the following three conditions:*

(a) *the rows of B are non-zero,*

(b) *if $X, Y \in \mathcal{F}$ and $X \cup Y \neq V(G)$, then $X \cup Y \in \mathcal{F}$, and*

(c) *if $X, Y \in \mathcal{F}$ and $X \cap Y \neq \emptyset$, then $X \cap Y \in \mathcal{F}$,*

then the system $Bx \geq \mathbf{1}$, $x \geq 0$ is totally dual integral and the rational polyhedron $\{x | \, Bx \geq \mathbf{1}; x \geq 0\}$ is integral.

Proof. Let c be an integral vector such that the maximum in

$$\min\{\langle c, x\rangle | \, x \geq 0; Bx \geq \mathbf{1}\} = \max\{\langle y, \mathbf{1}\rangle | \, y \geq 0; yB \leq c\} \tag{1.22}$$

is attained at $y_0 = (y_{F_1}, \ldots, y_{F_r})$, where $\mathcal{F} = \{F_1, \ldots, F_r\}$. Note $c \geq 0$. Since the set of vectors y for which the maximum in Eq. (1.22) is attained is a face of the polytope $\mathcal{P} = \{y \in \mathbb{R}^r | \, y \geq 0; yB \leq c\}$ and since any face of $\mathcal{P}$ is a compact set, one may assume

$$\sum_{i=1}^{r} y_{F_i}|F_i|^2 = \max\left\{\sum_{i=1}^{r} y_i|F_i|^2 \,\middle|\, \langle y, \mathbf{1}\rangle = \langle y_0, \mathbf{1}\rangle; y \geq 0; yB \leq c\right\}. \tag{1.23}$$

We claim that the family $\mathcal{G} = \{F_i | \, y_{F_i} > 0\}$ is cross-free, that is, for any $F_i, F_j \in \mathcal{G}$, at least one of the four sets $F_i \setminus F_j$, $F_j \setminus F_i$, $F_i \cap F_j$, $V(G) \setminus F_i \cup F_j$ is empty. To prove the claim suppose $F_1 \setminus F_2 \neq \emptyset$, $F_2 \setminus F_1 \neq \emptyset$, $F_1 \cap F_2 \neq \emptyset$, $F_1 \cup F_2 \neq V(G)$ for some $F_1, F_2 \in \mathcal{G}$ (for simplicity we are assuming $i = 1, j = 2$). Set $\epsilon = \min\{y_{F_1}, y_{F_2}\}$ and consider the vector y_0' in $\mathbb{R}^r$ whose entries are given by

$$y'_{F_1} := y_{F_1} - \epsilon, \; y'_{F_2} := y_{F_2} - \epsilon, \; y'_{F_1 \cap F_2} := y_{F_1 \cap F_2} + \epsilon, \; y'_{F_1 \cup F_2} := y_{F_1 \cup F_2} + \epsilon,$$

and $y'_S := y_S$ for all other $S \in \mathcal{F}$. Let $B = (b_{F_i, \alpha_j})$ be the one-way cut-incidence matrix of $\mathcal{F}$, where $\alpha_1, \ldots, \alpha_q$ are the edges of G. Consider any edge $\alpha_j = (z, w)$. For each $1 \leq j \leq q$ one has the following inequality

$$\begin{aligned} b_{F_1,j}(y_{F_1} - \epsilon) + b_{F_2,j}(y_{F_2} - \epsilon) + b_{F_1 \cap F_2,j}(y_{F_1 \cap F_2} + \epsilon) + b_{F_1 \cup F_2,j}(y_{F_1 \cup F_2} + \epsilon) \\ \leq b_{F_1,j}y_{F_1} + b_{F_2,j}y_{F_2} + b_{F_1 \cap F_2,j}y_{F_1 \cap F_2} + b_{F_1 \cup F_2,j}y_{F_1 \cup F_2}. \end{aligned}$$

To show the inequality consider two cases. Case (I): $b_{F_1,j} = b_{F_2,j} = 0$. If $i = 1$ or $i = 2$ and $z \in F_i$, then $w \in F_i$ and $b_{F_1 \cup F_2,j} = 0$. If $i = 1$ or $i = 2$ and $z \notin F_i$, then $w \notin F_i$ and $b_{F_1 \cap F_2,j} = 0$. Hence in this case $y_{F_1 \cap F_2,j} = 0$ and $y_{F_1 \cup F_2,j} = 0$. Case (II): $b_{F_1,j} = 0$ and $b_{F_2,j} = 1$. By the previous considerations one has that either $y_{F_1 \cap F_2,j} = 0$ or $y_{F_1 \cup F_2,j} = 0$.

From the inequality above we obtain $y_0'B \leq y_0B$. Since $\langle y_0', \mathbf{1}\rangle = \langle y_0, \mathbf{1}\rangle$, we obtain a contradiction with the choice of y_0 because

$$\sum_{i=1}^{r} y_{F_i}|F_i|^2 < \sum_{i=1}^{r} y'_{F_i}|F_i|^2.$$

Note that for any numbers $a > b \geq c > d > 0$ with $a + d = b + c$ one has the inequality $a^2 + d^2 > b^2 + c^2$. In our case to show the strict inequality we take $a = |F_1 \cup F_2|$, $b = |F_1| \geq c = |F_2|$ and $d = |F_1 \cap F_2|$. Thus we have shown that the family $\mathcal{G}$ is cross-free.

Let B' be the submatrix of B whose rows correspond to the elements of the family $\mathcal{G}$. Then

$$\max\{\langle y', \mathbf{1}\rangle |\, y' \geq 0; y'B' \leq c\} = \max\{\langle y, \mathbf{1}\rangle |\, y \geq 0; yB \leq c\}. \tag{1.24}$$

The inequality $\leq$ is clear because B' is a submatrix of B. As the maximum in the right-hand side is attained at y_0, where y_0 has zero entries in positions corresponding to rows outside B', the reverse inequality follows. Now the matrix B', being the two-way cut-incidence matrix of a cross-free family $\mathcal{G}$, is a network matrix, hence it is totally unimodular; see [281, Theorem 5.27]. Hence by the Hoffman–Kruskal theorem the polytope

$$\{y' |\, y' \geq 0; y'B' \leq c\}$$

is integral, and so the maximum in the left-hand side of Eq. (1.24) is attained at an integral vector y'. To complete the proof that the system $Bx \geq \mathbf{1}$, $x \geq 0$ is TDI observe that y' can be extended (by adding zero entries) to an integral optimum solution y of the right-hand side of Eq. (1.24). Since the system $Bx \geq \mathbf{1}$, $x \geq 0$ is totally dual integral it follows from Proposition 1.8.3 that the polyhedron $\{x |\, Bx \geq \mathbf{1}; x \geq 0\}$ is integral. □

Corollary 1.8.8 *Let G be a connected digraph and let*

$$\mathcal{F} = \{X \,|\, \emptyset \neq X \subsetneq V(G);\ \delta^-(X) = \emptyset\}.$$

If B is the one-way cut-incidence matrix of $\mathcal{F}$, then $\{x |\, Bx \geq \mathbf{1}; x \geq 0\}$ is a non-empty integral polyhedron.

Proof. It suffices to verify the hypothesis of Theorem 1.8.7. First we prove that $\delta^+(X) \neq \emptyset$ for $X \in \mathcal{F}$. Assume $\delta^+(X) = \emptyset$. Pick $z \in V(G) \setminus X$ and $x \in X$. If $(z, w) \in E(G)$ or $(w, z) \in E(G)$, then $w \notin X$. There is an undirected path $\{z_0 = z, z_1, \ldots, z_r = x\}$, a contradiction since $z_i \notin X$ for all i. Thus (a) is satisfied. Next take $X, Y \in \mathcal{F}$. From the inequality

$$|\delta^-(X)| + |\delta^-(Y)| \geq |\delta^-(X \cap Y)| + |\delta^-(X \cup Y)|,$$

see [281, Lemma 2.1(b)], we get $\delta^-(X\cap Y)=\emptyset$ and $\delta^-(X\cup Y)=\emptyset$. Thus if $X\cap Y\neq\emptyset$ (resp. $X\cup Y\neq V(G)$), then $X\cap Y\in\mathcal{F}$ (resp. $X\cup Y\in\mathcal{F}$). Thus conditions (b) and (c) are satisfied. On the other hand by construction of $\mathcal{F}$ the matrix B is also the two-way cut-incidence matrix of $\mathcal{F}$. □

Lemma 1.8.9 *Let G be a connected bipartite graph with bipartition (V_1,V_2) and let $\mathcal{F}$ be the family*

$$\mathcal{F}=\{A\cup A'\,|\,\emptyset\neq A\subsetneq V_1; N(A)\subset A'\subset V_2\}\cup\{A'\,|\,\emptyset\neq A'\subset V_2\}.$$

If G is regarded as the digraph with all its arrows leaving the vertex set V_2, then the following equality holds $\mathcal{F}=\{X\,|\,\emptyset\neq X\subsetneq V(G);\ \delta^-(X)=\emptyset\}$.

Proof. It follows readily from the definitions. □

Exercises

1.8.10 If A is the incidence matrix of a digraph G, prove that A is totally unimodular.

1.8.11 Let $T\colon\mathbb{R}^q\rightarrow\mathbb{R}^n$ be a linear map and let $\mathcal{Q}'\subset\mathbb{R}^q$ be a rational polyhedron. If $T(\mathbb{Z}^q)\subset\mathbb{Z}^n$, then $T(\mathcal{Q}')$ is a rational polyhedron.

1.9 Elementary vectors and matroids

In this section, we give formulae to compute the circuits of the kernel of an integral matrix and show that the circuits generate the kernel. We introduce the notion of a matroid and explain the relation with vector matroids.

The notion of an *elementary integral vector* or *circuit* occurs in convex analysis [357], in the theory of toric ideals of graphs [34, 400, 417], and in matroid theory [338].

Given $\alpha=(\alpha_1,\ldots,\alpha_q)\in\mathbb{R}^q$, its *support* is $\mathrm{supp}(\alpha)=\{i\,|\,\alpha_i\neq 0\}$. Note that α can be written uniquely as $\alpha=\alpha_+-\alpha_-$, where α_+ and α_- are two nonnegative vectors with disjoint support which are called the *positive* part and the *negative* part of α, respectively. Sometimes we denote the positive and negative part of α by α^+ and α^-, respectively.

Remark 1.9.1 If α is a linear combination of $\beta_1,\ldots,\beta_r\in\mathbb{R}^q$, then one has the inclusion $\mathrm{supp}(\alpha)\subset\mathrm{supp}(\beta_1)\cup\cdots\cup\mathrm{supp}(\beta_r)$.

Definition 1.9.2 Let V be a linear subspace of $\mathbb{Q}^q$. An *elementary vector* of V is a non-zero vector α in V whose support is minimal with respect to inclusion, i.e., $\mathrm{supp}(\alpha)$ does not properly contain the support of any other non-zero vector in V.

The concept of an elementary vector arises in graph theory when V is the kernel of the incidence matrix of a graph G [358, Section 22].

Lemma 1.9.3 *If V is a linear subspace of $\mathbb{Q}^q$ and α, β are two elementary vectors of V with the same support, then $\alpha = \lambda\beta$ for some $\lambda \in \mathbb{Q}$.*

Proof. If $i \in \mathrm{supp}(\alpha)$, then one can write $\alpha_i = \lambda\beta_i$ for some scalar λ. Since $\mathrm{supp}(\alpha - \lambda\beta) \subsetneq \mathrm{supp}(\alpha)$, one concludes $\alpha - \lambda\beta = 0$, as required. $\Box$

Definition 1.9.4 Two vectors $\alpha = (\alpha_i)$ and $\beta = (\beta_i)$ in $\mathbb{Q}^q$ are in *harmony* if $\alpha_i\beta_i \geq 0$ for every i.

Lemma 1.9.5 [357] *Let V be a linear subspace of $\mathbb{Q}^q$. If $0 \neq \alpha \in V$, then there is an elementary vector $\gamma \in V$ in harmony with α such that $\mathrm{supp}(\gamma) \subset \mathrm{supp}(\alpha)$.*

Proof. Let $\beta = (\beta_1, \ldots, \beta_q)$ be an elementary vector of V whose support is contained in $\mathrm{supp}(\alpha)$. By replacing β by $-\beta$ one may assume $\alpha_i\beta_i > 0$ for some i. Consider

$$\lambda_j = \frac{\alpha_j}{\beta_j} = \min_{1\leq i\leq q} \left\{ \frac{\alpha_i}{\beta_i} \,\middle|\, \alpha_i\beta_i > 0 \right\}.$$

Note that $z_i = (\alpha_i - \lambda_j\beta_i)\alpha_i \geq 0$ for all i. Indeed if $\alpha_i\beta_i \leq 0$, then clearly $z_i \geq 0$, and if $\alpha_i\beta_i > 0$ by the minimality of λ_j one has $z_i \geq 0$. Thus the vector $\alpha - \lambda_j\beta$ is in harmony with α and its support is strictly contained in the support of α. If $\alpha - \lambda_j\beta = 0$ we are done, otherwise we apply the same argument with $\alpha - \lambda_j\beta$ playing the role of α. Since being in harmony is an equivalence relation, the result follows by a recursive application of this procedure. $\Box$

Theorem 1.9.6 [357] *If V is a vector subspace of $\mathbb{Q}^q$ and $\alpha \in V\setminus\{0\}$, then α can be written as $\alpha = \sum_{i=1}^{r}\beta_i$ for some elementary vectors $\beta_1, \ldots, \beta_r$ of V with $r \leq \dim V$ such that*

(i) *$\beta_1, \ldots, \beta_r$ are in harmony with α,*

(ii) *$\mathrm{supp}(\beta_i) \subset \mathrm{supp}(\alpha)$ for all i, and*

(iii) *$\mathrm{supp}(\beta_i)$ is not contained in the union of the supports of $\beta_1, \ldots, \beta_{i-1}$ for all $i \geq 2$.*

Proof. By induction on the number of elements in the support of α. If α is not an elementary vector of V, then by Lemma 1.9.5 there is an elementary vector $\alpha_1 \in V$ in harmony with α such that $\mathrm{supp}(\alpha_1) \subset \mathrm{supp}(\alpha)$. Using the proof of Lemma 1.9.5 it follows that there is a positive scalar $\lambda_1 > 0$ such that $\alpha - \lambda_1\alpha_1$ is in harmony with α and the support of $\alpha - \lambda_1\alpha_1$ is strictly contained in $\mathrm{supp}(\alpha)$. Therefore the result follows applying induction. This proof uses the original argument given in [357]. $\Box$

Definition 1.9.7 Let V be a linear subspace of $\mathbb{Q}^q$. An *elementary integral vector* or *circuit* of V is an elementary vector α of V such that $\alpha \in \mathbb{Z}^q$ and the non-zero entries of α are relatively prime.

Corollary 1.9.8 *If V is a linear subspace of $\mathbb{Q}^q$, then there are only a finite number of circuits of N and they generate V as a $\mathbb{Q}$-vector space.*

Proof. It follows from Lemma 1.9.3 and Theorem 1.9.6, respectively. □

Proposition 1.9.9 [134] *Let A be an $n \times q$ integral matrix and let α be a non-zero vector in* $\ker(A)$. *If A has rank n, then α is an elementary vector of* $\ker(A)$ *if and only if there is $0 \neq \lambda \in \mathbb{Q}$ such that*

$$\alpha = \lambda \sum_{k=1}^{n+1} (-1)^k \det[v_{i_1}, \ldots, v_{i_{k-1}}, v_{i_{k+1}}, \ldots, v_{i_{n+1}}] e_{i_k}, \tag{1.25}$$

for some column vectors v_{i_j} of A. Here e_k is the kth unit vector in $\mathbb{R}^q$.

Proof. $\Leftarrow$) Let A' be the submatrix of A consisting of the column vectors $v_{i_1}, \ldots, v_{i_{n+1}}$. Since A' has size $n \times (n+1)$ and rank n, $\ker(A')$ is generated by a non-zero vector. It follows readily that the vector α given by Eq. (1.25) is an elementary vector.

$\Rightarrow$) Let $\text{supp}(\alpha) = \{j_1, \ldots, j_r\}$ and let $v_1, \ldots, v_q$ be the column vectors of A. Since $\alpha \in \ker(A)$ one has

$$\alpha = \lambda_{j_1} e_{j_1} + \cdots + \lambda_{j_r} e_{j_r} \quad \text{and} \quad \lambda_{j_1} v_{j_1} + \cdots + \lambda_{j_r} v_{j_r} = 0, \tag{1.26}$$

where $0 \neq \lambda_{j_k} \in \mathbb{Q}$ and v_{j_k} are columns of A for all k. By the minimality of $\text{supp}(\alpha)$ we may assume that $\{v_{j_1}, \ldots, v_{j_{r-1}}\}$ is linearly independent. Thus, using that $\text{rank}(A) = n$, there are column vectors $v_{j_{r+1}}, \ldots, v_{j_{n+1}}$ such that the set $\mathcal{A}' = \{v_{j_1}, \ldots, v_{j_{r-1}}, v_{j_{r+1}}, \ldots, v_{j_{n+1}}\}$ is linearly independent. After permuting $j_1, \ldots, j_{n+1}$, one can write

$$\mathcal{A}' \cup \{v_{j_r}\} = \{v_{i_1}, \ldots, v_{i_{n+1}}\}$$

with $1 \leq i_1 < \cdots < i_{n+1} \leq q$. Consider the vector

$$\beta = \sum_{k=1}^{n+1} (-1)^k \det[v_{i_1}, \ldots, v_{i_{k-1}}, v_{i_{k+1}}, \ldots, v_{i_{n+1}}] e_{i_k}. \tag{1.27}$$

Note that in Eq. (1.26) and Eq. (1.27) the coefficient of e_{j_r} is non-zero, hence there are scalars $0 \neq \lambda$ and c_{i_k} in $\mathbb{Q}$ such that

$$\alpha - \lambda\beta = \sum_{i_k \neq j_r} c_{i_k} v_{i_k}.$$

Since the support of $\alpha - \lambda\beta$ is strictly contained in the support of β and since β is an elementary vector of $\ker(A)$, one obtains $\alpha - \lambda\beta = 0$, as required. □

Theorem 1.9.10 *Let $A \neq (0)$ be an $n \times q$ integral matrix and let ψ be the $\mathbb{Z}$-linear homomorphism*

$$\psi\colon \mathbb{Z}^q \longrightarrow \mathbb{Z}^n$$

given by $\psi(\alpha) = A(\alpha)$. Then $\ker(\psi)$ is generated as a $\mathbb{Z}$-module by the circuits of $\ker(\psi)$.

Proof. By Theorem 1.2.2, there are invertible integral matrices P and Q such that $PAQ = D$, where D is an $r \times q$ matrix with $r = \operatorname{rank}(A)$. Since A and DQ^{-1} have the same kernel and DQ^{-1} has size $r \times q$ and rank r, we may assume that $n = \operatorname{rank}(A)$.

We set $N = \ker(\psi)$. Let $\alpha_1, \ldots, \alpha_q$ be the column vectors of A and let e_i be the ith unit vector of $\mathbb{Z}^q$. By Proposition 1.9.9, it follows that the set of circuits of N is the set of vectors of the form

$$\frac{\pm 1}{h}\sum_{j=1}^{n+1}(-1)^j \cdot \det(\alpha_{i_1}, \ldots, \alpha_{i_{j-1}}, \alpha_{i_{j+1}}, \ldots, \alpha_{i_{n+1}}) \cdot e_{i_j} = \frac{\pm 1}{h} \cdot v,$$

where $1 \leq i_1 < \cdots < i_{n+1} \leq q$, and h is the greatest common divisor of the entries of the vector v, if v is non-zero. For convenience we introduce the notation $v = v(i_1, \ldots, i_{n+1})$ and

$$[i_1, \ldots, \widehat{i_j}, \ldots, i_{n+1}] = \det(\alpha_{i_1}, \ldots, \alpha_{i_{j-1}}, \alpha_{i_{j+1}}, \ldots, \alpha_{i_{n+1}}).$$

Set d equal to the greatest common divisor of all the non-zero integers having the special form $[i_1, \ldots, \widehat{i_j}, \ldots, i_{n+1}]$ and

$$\mathcal{B} = \left\{ \frac{v(i_1, \ldots, i_{n+1})}{d} \neq 0 \,\middle|\, 1 \leq i_1 < \cdots < i_{n+1} \leq q \right\}.$$

Note the inclusions $\mathbb{Z}\mathcal{B} \subset L \subset N$, where L is the $\mathbb{Z}$-module generated by the circuits of N.

We claim that $\mathbb{Z}\mathcal{B} = N$. First observe that L and N have both rank $q - n$; see Corollary 1.9.8. Since $\mathbb{Z}\mathcal{B}$ and L have equal rank, one concludes that $N/\mathbb{Z}\mathcal{B}$ is a finite group. Thus to show $\mathbb{Z}\mathcal{B} = N$ it suffices to prove that $N/\mathbb{Z}\mathcal{B}$ is torsion-free or equivalently that $\mathbb{Z}^q/\mathbb{Z}\mathcal{B}$ is torsion-free of rank n.

Consider the matrix M whose rows are the vectors of $\mathcal{B}$. By the fundamental theorem for finitely generated abelian groups (see Theorem 1.3.16) the proof reduces to showing that the ideal $I_{q-n}(M)$ of $\mathbb{Z}$ generated by the $(q-n)$-minors of M is equal to $\mathbb{Z}$.

Fix $\Delta = [i_1, \ldots, \widehat{i_j}, \ldots, i_{n+1}]/d$. For simplicity of notation one may consider the case $\Delta = [1, \ldots, n]/d$. Note that Δ occurs in $v(1, \ldots, n, i)/d$ with a $\pm$ sign for $i = n+1, \ldots, q$. Thus the matrix

$$M' = \begin{pmatrix} m_{11} & \cdots & m_{1n} & \Delta & 0 & 0 & \cdots & 0 \\ m_{21} & \cdots & m_{2n} & 0 & \Delta & 0 & \cdots & 0 \\ \vdots & & \vdots & \vdots & \vdots & \vdots & & \vdots \\ m_{(q-n)1} & \cdots & m_{(q-n)n} & 0 & 0 & 0 & \cdots & \Delta \end{pmatrix}$$

whose ith row is the vector $v(1,\ldots,n,i)/d$ is a submatrix of M. Therefore Δ^{q-n} belongs to $I_{q-n}(M)$. If $I_{q-n}(M) \neq \mathbb{Z}$, there is a prime number p dividing Δ^{q-n} for all Δ. Thus p divides Δ for all Δ, and this implies that pd divides d, a contradiction. □

Corollary 1.9.11 *If A is a t-unimodular matrix of size $n \times q$ and α is a circuit of* $\ker(A)$*, then $\alpha = \sum_{i=1}^{q} \epsilon_i e_i$ for some $\epsilon_1, \ldots, \epsilon_q$ in $\{0, \pm 1\}$.*

Proof. It follows readily from the explicit description of the circuits given in Proposition 1.9.9 and from Exercise 1.9.17. □

Matroids Let $M = (E, \mathcal{I})$ be a *matroid* on E, i.e., there is a collection $\mathcal{I}$ of subsets of E satisfying the following three conditions:

(i_1) $\emptyset \in \mathcal{I}$.

(i_2) If $I \in \mathcal{I}$ and $I' \subset I$, then $I' \in \mathcal{I}$.

(i_3) If I_1 and I_2 are in $\mathcal{I}$ and $|I_1| < |I_2|$, then there is an element e of $I_2 \setminus I_1$ such that $I_1 \cup \{e\} \in \mathcal{I}$.

The members of $\mathcal{I}$ are the independent sets of M. It is convenient to write $\mathcal{I}(M)$ for $\mathcal{I}$ and $E(M)$ for E, particularly when several matroids are being considered. A subset of E that is not in $\mathcal{I}$ is called *dependent*. A maximal independent set of M with respect to inclusion is called a *basis*. A minimal dependent set in M is called a *circuit*. The reader is referred to [338] for the general theory of matroids.

Proposition 1.9.12 [338, Proposition 1.1.1, p. 8] *Let E be the set of column labels of an $m \times n$ matrix A over a field K, and let $\mathcal{I}$ be the set of subsets B of E for which the multiset of columns labeled by B is linearly independent in K^m. Then $M[A] = (E, \mathcal{I})$ is a matroid.*

The matroid $M[A]$ obtained as above from the matrix A is called the *vector matroid* of A. Any matroid M arising from a matrix A is called *representable* over the field K.

We are especially interested in vector matroid arising from a nonnegative integral matrix A with column vectors $v_1, \ldots, v_q$. By Proposition 1.9.12 there is a matroid $M[A]$ on $\mathcal{A} = \{v_1, \ldots, v_q\}$ over the field $\mathbb{Q}$ of rational numbers, whose independent sets are the independent subsets of $\mathcal{A}$.

Definition 1.9.13 A *minimal dependent set* or *circuit* of $M[A]$ is a dependent set all of whose proper subsets are independent. A subset B of $\mathcal{A}$ is called a *basis* of $M[A]$ if B is a maximal independent set.

There is a correspondence: Circuits of $\ker(A) \longrightarrow$ Circuits of $M[A]$ given by $\alpha = (\alpha_1, \ldots, \alpha_q) \longrightarrow C(\alpha) = \{v_i | i \in \mathrm{supp}(\alpha)\}$. Thus the set of circuits of the kernel of A is the algebraic realization of the set of circuits of the vector matroid $M[A]$. The circuits of $\ker(A)$ can be used to study the normality of monomial subrings (see Theorem 9.7.1).

Theorem 1.9.14 [338, Corollary 1.2.5, p. 18] *Let $\mathcal{B}$ be a non-empty family of subsets of X. Then $\mathcal{B}$ is the collection of bases of a matroid on X if and only if the following exchange property is satisfied: If B_1 and B_2 are members of $\mathcal{B}$ and $b_1 \in B_1 \setminus B_2$, then there is an element $b_2 \in B_2 \setminus B_1$ such that $(B_1 \setminus \{b_1\}) \cup \{b_2\}$ is in $\mathcal{B}$.*

The family of bases of any matroid satisfies the following *symmetric exchange* property (see [286]):

Theorem 1.9.15 *If B_1 and B_2 are bases of a matroid M and $b_1 \in B_1 \setminus B_2$, then there is an element $b_2 \in B_2 \setminus B_1$ such that both $(B_1 \setminus \{b_1\}) \cup \{b_2\}$ and $(B_2 \setminus \{b_2\}) \cup \{b_1\}$ are bases of M.*

It is well known [338] that all bases of a matroid M have the same number of elements. This common number is called the *rank* of the matroid.

Exercises

1.9.16 Prove that the circuits of the kernel of the matrix M are precisely the row vectors of the matrix A with a $\pm$ sign

$$M = \begin{pmatrix} 4 & 3 & 1 & 0 \\ 0 & 1 & 3 & 4 \end{pmatrix} \qquad A = \begin{pmatrix} 2 & -3 & 1 & 0 \\ 3 & -4 & 0 & 1 \\ 0 & 1 & -3 & 2 \\ 1 & 0 & -4 & 3 \end{pmatrix}.$$

1.9.17 Let A be an $n \times q$ matrix with entries in an integral domain R and let $F_1, \ldots, F_n$ be the rows of A. Assume that $F_i = \sum_{j \neq i} \lambda_j F_j$, for some λ_j's in the field of fractions of R. Consider the R-linear maps

$$\varphi \colon R^q \longrightarrow R^n \qquad \alpha \stackrel{\varphi}{\longmapsto} A\alpha$$
$$\varphi' \colon R^q \longrightarrow R^{n-1} \qquad \alpha \stackrel{\varphi'}{\longmapsto} A'\alpha,$$

where A' is the matrix obtained from A by removing its ith row, prove that $\ker(\varphi) = \ker(\varphi')$.

1.9.18 If $X = \{x_1, \ldots, x_4\}$ and

$$\mathcal{B} = \{\{x_1, x_2\}, \{x_2, x_3\}, \{x_3, x_4\}, \{x_1, x_4\}\},$$

prove that $\mathcal{B}$ satisfies the bases exchange property of matroids.

Chapter 2

Commutative Algebra

In this chapter some basic notions and results from commutative algebra will be introduced. All rings considered in this book are commutative and Noetherian and modules are finitely generated. Our main references are the books of Bruns and Herzog [65], Eisenbud [128], Matsumura [309, 310], and Vasconcelos [413]. Some of the results presented below are just stated without giving proofs, if need be, the reader may locate the missing proofs in those references.

2.1 Module theory

Noetherian modules and localizations Let R be a commutative ring with unit and let M be an R-module. Recall that M is called *Noetherian* if every submodule N of M is finitely generated, that is, $N = Rf_1 + \cdots + Rf_q$, for some $f_1, \ldots, f_q$ in N.

Theorem 2.1.1 *The following conditions are equivalent:*

(a) *M is Noetherian.*

(b) *M satisfies the ascending chain condition for submodules; that is, for every ascending chain of submodules of M*

$$N_0 \subset N_1 \subset \cdots \subset N_n \subset N_{n+1} \subset \cdots \subset M$$

there exists an integer k such that $N_i = N_k$ for every $i \geq k$.

(c) *Any family $\mathcal{F}$ of submodules of M partially ordered by inclusion has a maximal element, i.e., there is $N \in \mathcal{F}$ such that if $N \subset N_i$ and $N_i \in \mathcal{F}$, then $N = N_i$.*

Proof. (a)$\Rightarrow$(b): Consider the submodule $N = \cup_{i\geq 0} N_i$. By hypothesis there are $m_1, \ldots, m_r$ such that $N = Rm_1 + \cdots + Rm_r$. Then, there is k such that $m_i \in N_k$ for all i. It follows that $N_i = N_k$ for all $i \geq k$.

(b)$\Rightarrow$(c): Let $N_1 \in \mathcal{F}$. If N_1 is not maximal, there is $N_2 \in \mathcal{F}$ such that $N_1 \subsetneq N_2$. If N_2 is not maximal, there is $N_3 \in \mathcal{F}$ such that $N_2 \subsetneq N_3$. Applying this argument repeatedly we get that $\mathcal{F}$ has a maximal element.

(c)$\Rightarrow$(a): Let N be a submodule of M and let $\mathcal{F}$ be the family of submodules of N that are finitely generated. By hypothesis $\mathcal{F}$ has a maximal element N'. It follows that $N = N'$. $\Box$

In particular a *Noetherian ring* R is a commutative ring with unit with the property that every ideal of I is finitely generated; that is, given an ideal I of R there exists a finite number of generators $f_1, \ldots, f_q$ such that

$$I = \{a_1 f_1 + \cdots + a_q f_q |\, a_i \in R, \forall i\}.$$

As usual, if I is generated by $f_1, \ldots, f_q$, we write $I = (f_1, \ldots, f_q)$.

Proposition 2.1.2 *If M is a finitely generated R-module over a Noetherian ring R, then M is a Noetherian module.*

Corollary 2.1.3 *If R is a Noetherian ring and I is an ideal of R, then R/I and R^n are Noetherian R-modules. In particular any submodule of R^n is finitely generated.*

Theorem 2.1.4 (Hilbert's basis theorem [9, Theorem 7.5]) *A polynomial ring $R[x]$ over a Noetherian ring R is Noetherian.*

One of the important examples of a Noetherian ring is a polynomial ring over a field k. Often we will denote a polynomial ring in several variables by $k[\mathbf{x}]$ and a polynomial ring in one variable by $k[x]$. The letters k and K will always denote fields.

In this book, unless otherwise stated, by a *ring* (resp. module) we shall always mean a Noetherian ring (resp. finitely generated module).

The *prime spectrum* of a ring R, denoted by $\mathrm{Spec}(R)$, is the set of prime ideals of R. The *minimal primes* of R are the minimal elements of $\mathrm{Spec}(R)$ with respect to inclusion and the maximal ideals of R are the maximal elements of the set of proper ideals of R with respect to inclusion.

Let R be a ring and let $X = \mathrm{Spec}(R)$ be its prime spectrum. Given an ideal I of R, the set of all prime ideals of R containing I will be denoted by $V(I)$. The *minimal primes* of I are the minimal elements of $V(I)$ with respect to inclusion. It is not hard to verify that the pair $(X, \mathcal{Z})$ is a topological space, where $\mathcal{Z}$ is the family of open sets of X, and where U is in $\mathcal{Z}$ iff $U = X \setminus V(I)$, for some ideal I. This topology is called the *Zariski topology* of the prime spectrum of R.

A *local ring* $(R, \mathfrak{m}, k)$ is a Noetherian ring R with exactly one maximal ideal $\mathfrak{m}$, the field $k = R/\mathfrak{m}$ is called the *residue field* of R.

A *homomorphism of rings* is a map $\varphi\colon R \to S$ such that:

(i) $\varphi(a+b) = \varphi(a) + \varphi(b)$, $\forall\, a, b \in A$,

(ii) $\varphi(ab) = \varphi(a)\varphi(b)$, $\forall\, a, b \in A$, and

(iii) $\varphi(1) = 1$.

Let R be a ring and let $\varphi\colon \mathbb{Z} \to R$ be the canonical homomorphism

$$\varphi(a) = a \cdot 1_R,$$

then $\ker(\varphi) = n\mathbb{Z}$, for some $n \geq 0$. The integer n is called the *characteristic* of R and is denoted by $\mathrm{char}(R)$.

Proposition 2.1.5 *Let $(R, \mathfrak{m})$ be a local ring, then either* $\mathrm{char}(R) = 0$ *or* $\mathrm{char}(R) = p^k$, *for some prime number p and some integer $k \geq 1$.*

Proof. Let $n = \mathrm{char}(R) \geq 2$. If $n = pqm$, where p, q are prime numbers with $q \neq p$, then $p \cdot 1_R$ and $q \cdot 1_R$ are both in $\mathfrak{m}$ because they are non invertible, but this is impossible because $1 = ap + bq$ and from the equality

$$1_R = (a \cdot 1_R)(p \cdot 1_R) + (b \cdot 1_R)(q \cdot 1_R),$$

one derives $1_R \in \mathfrak{m}$, a contradiction. □

Modules of fractions and localizations

Let R be a ring, M an R-module, and S a multiplicatively closed subset of R so that $1 \in S$. Then the *module of fractions* of M with respect to S, or the *localization* of M with respect to S, is defined by

$$S^{-1}(M) = \{m/s \mid m \in M,\ s \in S\},$$

where $m/s = m_1/s_1$ if and only if $t(s_1 m - s m_1) = 0$ for some $t \in S$. In particular $S^{-1}R$ has a ring structure given by the usual rules of addition and multiplication, and $S^{-1}M$ is a module over the ring $S^{-1}R$ with the operations:

$$\begin{aligned} m_1/s_1 + m_2/s_2 &= (s_2 m_1 + s_1 m_2)/s_1 s_2 \quad (s_i \in S,\ m_i \in M), \\ (a/s) \cdot (m_1/s_1) &= a m_1/s s_1 \qquad\qquad\quad (a \in R,\ s \in S). \end{aligned}$$

There is a map $\varphi\colon M \to S^{-1}M$, given by $\varphi(m) = m/1$. If $f\colon M \to N$ is a homomorphism of R-modules, then there is an induced homomorphism

$$S^{-1}f\colon S^{-1}M \longrightarrow S^{-1}N$$

of $S^{-1}R$-modules given by $f(m/s) = f(m)/s$.

Example 2.1.6 If f is in a ring R and $S = \{f^i \,|\, i \in \mathbb{N}\}$, then $S^{-1}R$ is usually written R_f. For instance if $R = \mathbb{C}[x]$ is a polynomial ring in one variable over the field $\mathbb{C}$ of complex numbers, then $R_x = \mathbb{C}[x, x^{-1}]$ is the ring of Laurent polynomials.

Definition 2.1.7 Let $\mathfrak{p}$ be a prime ideal of a ring R and $S = R \setminus \mathfrak{p}$. In this case $S^{-1}R$ is written $R_\mathfrak{p}$ and is called the *localization* of R at $\mathfrak{p}$.

Example 2.1.8 Let $\mathfrak{p}$ be a prime ideal of a ring R. The local ring

$$(R_\mathfrak{p}, \mathfrak{p}R_\mathfrak{p}, k(\mathfrak{p}))$$

is the prototype of a local ring, where $k(\mathfrak{p}) = R_\mathfrak{p}/\mathfrak{p}R_\mathfrak{p}$ denotes the residue field of $R_\mathfrak{p}$.

Krull dimension and height

By a *chain* of prime ideals of a ring R we mean a finite strictly increasing sequence of prime ideals

$$\mathfrak{p}_0 \subset \mathfrak{p}_1 \subset \cdots \subset \mathfrak{p}_n,$$

the integer n is called the *length* of the chain. The *Krull dimension* of R, denoted by $\dim(R)$, is the supremum of the lengths of all chains of prime ideals in R. Let $\mathfrak{p}$ be a prime ideal of R, the *height* of $\mathfrak{p}$, denoted by $\mathrm{ht}\,(\mathfrak{p})$ is the supremum of the lengths of all chains of prime ideals

$$\mathfrak{p}_0 \subset \mathfrak{p}_1 \subset \cdots \subset \mathfrak{p}_n = \mathfrak{p}$$

which end at $\mathfrak{p}$. Note $\dim(R_\mathfrak{p}) = \mathrm{ht}\,(\mathfrak{p})$. If I is an ideal of R, then $\mathrm{ht}\,(I)$, the *height* of I, is defined as

$$\mathrm{ht}(I) = \min\{\mathrm{ht}(\mathfrak{p}) |\, I \subset \mathfrak{p} \text{ and } \mathfrak{p} \in \mathrm{Spec}(R)\}.$$

In general $\dim(R/I) + \mathrm{ht}\,(I) \leq \dim(R)$. The difference $\dim(R) - \dim(R/I)$ is called the *codimension* of I and $\dim(R/I)$ is called the *dimension* of I.

Let M be an R-module. The *annihilator* of M is given by

$$\mathrm{ann}_R(M) = \{x \in R |\, xM = 0\},$$

if $m \in M$ the *annihilator* of m is $\mathrm{ann}\,(m) = \mathrm{ann}\,(Rm)$. It is convenient to generalize the notion of annihilator to ideals and submodules. Let N_1 and N_2 be submodules of M, their *ideal quotient* or *colon ideal* is defined as

$$(N_1 \colon {}_R N_2) = \{x \in R |\, xN_2 \subset N_1\}.$$

Let us recall that the *dimension* of an R-module M is

$$\dim(M) = \dim(R/\mathrm{ann}\,(M))$$

and the *codimension* of M is $\mathrm{codim}(M) = \dim(R) - \dim(M)$.

Theorem 2.1.9 *If* $R[x]$ *is a polynomial ring over a Noetherian ring* R, *then* $\dim R[x] = \dim(R) + 1$.

Primary decomposition of modules Let I be an ideal of a ring R. The *radical* of I is

$$\mathrm{rad}\,(I) = \{x \in R \mid x^n \in I \text{ for some } n > 0\},$$

the radical is also denoted by $\sqrt{I}$. In particular $\sqrt{(0)}$, denoted by $\mathfrak{N}_R$ or $\mathrm{nil}(R)$, is the set of *nilpotent elements* of R and is called the *nilradical* of R. A ring is *reduced* if its nilradical is zero. The *Jacobson radical* of R is the intersection of all the maximal ideals of R.

Proposition 2.1.10 *If I is a proper ideal of a ring R, then* $\mathrm{rad}\,(I)$ *is the intersection of all prime ideals containing I.*

Definition 2.1.11 Let M be a module over a ring R. The set of *associated primes* of M, denoted by $\mathrm{Ass}_R(M)$, is the set of all prime ideals $\mathfrak{p}$ of R such that there is a monomorphism ϕ of R-modules:

$$R/\mathfrak{p} \stackrel{\phi}{\hookrightarrow} M.$$

Note that $\mathfrak{p} = \mathrm{ann}\,(\phi(1))$.

Lemma 2.1.12 *If $M \neq 0$ is an R-module, then* $\mathrm{Ass}_R(M) \neq \emptyset$.

Proof. Consider the family of ideals $\mathcal{F} = \{\mathrm{ann}(m) \mid 0 \neq m \in M\}$ ordered by inclusion. By Theorem 2.1.1 this family has a maximal element that we denote by $\mathrm{ann}(m)$. It suffices to show that $\mathrm{ann}(m)$ is prime. Let x, y be two elements of R such that $xy \in \mathrm{ann}(m)$. Assume that $x \notin \mathrm{ann}(m)$. Then $\mathrm{ann}(m) \subset \mathrm{ann}(xm)$ and consequently $\mathrm{ann}(m) = \mathrm{ann}(xm)$. This shows that $y \in \mathrm{ann}(m)$, as required. $\square$

If $M = R/I$, it is usual to say that an associated prime ideal of R/I is an *associated prime* ideal of I and to set $\mathrm{Ass}(I) = \mathrm{Ass}(R/I)$.

Proposition 2.1.13 *Let R be a ring and let S be a multiplicatively closed subset of R. If M is an R-module and $\mathfrak{p}$ is a prime ideal of R with $S \cap \mathfrak{p} = \emptyset$, then $\mathfrak{p}$ is an associated prime of M if and only if $S^{-1}\mathfrak{p}$ is an associated prime of $S^{-1}M$.*

Proof. If $\mathfrak{p}$ is in $\mathrm{Ass}(M)$, then $R/\mathfrak{p} \hookrightarrow M$. Hence $S^{-1}R/S^{-1}\mathfrak{p} \hookrightarrow S^{-1}M$. Thus, $S^{-1}\mathfrak{p}$ is an associated prime of $S^{-1}M$.

For the converse assume $S^{-1}\mathfrak{p} = \mathrm{ann}(m/1)$. Since R is Noetherian, $\mathfrak{p}$ is generated by a finite set $a_1, \ldots, a_n$. Hence for each i there is $s_i \in S$ such that $s_i a_i m = 0$. Set $s = s_1 \cdots s_n$. We claim $\mathfrak{p} = \mathrm{ann}(sm)$. Clearly one has $\mathfrak{p} \subset \mathrm{ann}(sm)$. To show the other containment take $x \in \mathrm{ann}(sm)$, then $xsm = 0$ and $x/1 \in \mathrm{ann}(m/1) = S^{-1}\mathfrak{p}$. Hence $x \in \mathfrak{p}$. $\square$

Definition 2.1.14 Let M be an R-module, the *support* of M, denoted by $\mathrm{Supp}(M)$, is the set of all prime ideals $\mathfrak{p}$ of R such that $M_\mathfrak{p} \neq 0$.

A sequence $0 \to M' \xrightarrow{f} M \xrightarrow{g} M'' \to 0$ of R-modules is called a *short exact sequence* if f is a monomorphism, g is an epimorphism, and $\mathrm{im}(f) = \ker(g)$. A sequence of R modules and homomorphisms:

$$\cdots \longrightarrow M_{i-1} \xrightarrow{f_{i-1}} M_i \xrightarrow{f_i} M_{i+1} \xrightarrow{f_{i+1}} \cdots$$

is said to be *exact* at M_i if $\mathrm{im}(f_{i-1}) = \ker(f_i)$. If the sequence is exact at each M_i it is called an *exact sequence.*

Lemma 2.1.15 *If $0 \to M' \to M \to M'' \to 0$ is a short exact sequence of modules over a ring R, then* $\mathrm{Supp}(M) = \mathrm{Supp}(M') \cup \mathrm{Supp}(M'')$.

Proof. Let $\mathfrak{p}$ be a prime ideal of R. It suffices to observe that from the exact sequence

$$0 \to M'_\mathfrak{p} \to M_\mathfrak{p} \to M''_\mathfrak{p} \to 0,$$

we get $M_\mathfrak{p} \neq 0$ if and only if $M'_\mathfrak{p} \neq 0$ or $M''_\mathfrak{p} \neq 0$. □

Theorem 2.1.16 *If M is an R-module, then there is a filtration of submodules*

$$(0) = M_0 \subset M_1 \subset \cdots \subset M_n = M$$

and prime ideals $\mathfrak{p}_1, \ldots, \mathfrak{p}_n$ of R such that $M_i/M_{i-1} \simeq R/\mathfrak{p}_i$ for all i.

Proof. By Lemma 2.1.12 there is a prime ideal $\mathfrak{p}_1$ and a submodule M_1 of M such that $R/\mathfrak{p}_1 \simeq M_1$. If $M_1 \subsetneq M$, then there is an associated prime ideal $\mathfrak{p}_2$ of M/M_1 such that $R/\mathfrak{p}_2$ is isomorphic to a submodule of M/M_1, i.e., $R/\mathfrak{p}_2 \simeq M_2/M_1$, where M_2 is a submodule of M containing M_1. If $M_2 \subsetneq M$, we pick an associated prime $\mathfrak{p}_3$ of M/M_2 and repeat the argument. Since M is Noetherian a repeated use of this procedure yields the required filtration. □

In general the primes $\mathfrak{p}_1, \ldots, \mathfrak{p}_n$ that occur in a filtration of the type described in the previous result are not associated primes of the module M; see [115] for a careful discussion of filtrations and for some of their applications to combinatorics.

Lemma 2.1.17 *If $0 \to M' \xrightarrow{f} M \xrightarrow{g} M'' \to 0$ is a short exact sequence of modules over a ring R, then* $\mathrm{Ass}(M) \subset \mathrm{Ass}(M') \cup \mathrm{Ass}(M'')$.

Proof. We may assume (without loss of generality) that f is the identity map by replacing M' by $f(M')$. Let $\mathfrak{p}$ be an associated prime of M. There is $0 \neq m \in M$ such that $\mathfrak{p} = \mathrm{ann}(m)$.

Case (I): $Rm \cap M' \neq (0)$. There is $r \in R$ such that $0 \neq rm \in M'$. From the equality $\mathfrak{p} = \mathrm{ann}(m)$ we get $\mathfrak{p} = \mathrm{ann}(rm)$. Thus $\mathfrak{p} \in \mathrm{Ass}(M')$.

Case (II): $Rm \cap M' = (0)$. Notice that $g(m) \neq 0$. From the equality $\mathfrak{p} = \mathrm{ann}(g(m))$ we get $\mathfrak{p} \in \mathrm{Ass}(M'')$. □

Corollary 2.1.18 *If M is an R-module, then $\mathrm{Ass}_R(M)$ is a finite set.*

Proof. Let $\mathfrak{p}_1, \ldots, \mathfrak{p}_n$ be prime ideals as in Theorem 2.1.16. By a repeated use of Lemma 2.1.17 one has $\mathrm{Ass}_R(M) \subset \{\mathfrak{p}_1, \ldots, \mathfrak{p}_n\} \subset \mathrm{Supp}(M)$. □

Let M be an R-module. An element $x \in R$ is a *zero divisor* of M if there is $0 \neq m \in M$ such that $xm = 0$. The set of zero divisors of M is denoted by $\mathcal{Z}(M)$. If x is not a zero divisor on M, we say that x is a *regular element* of M.

Lemma 2.1.19 *If M is an R-module, then*

$$\mathcal{Z}(M) = \bigcup_{\mathfrak{p} \in \mathrm{Ass}_R(M)} \mathfrak{p}.$$

Proof. The right-hand side is clearly contained in the left-hand side by definition of an associated prime. Let r be a zero divisor of M and consider the family $\mathcal{F} = \{\mathrm{ann}(m) |\, 0 \neq m \in M;\, rm = 0\}$. Notice that any maximal element of this family is a prime ideal. □

Proposition 2.1.20 *If M is an R-module, then*

$$\mathrm{Ass}(M) \subset \mathrm{Supp}(M) = V(\mathrm{ann}\,(M)),$$

and any minimal element of $\mathrm{Supp}(M)$ *is in* $\mathrm{Ass}(M)$.

Proof. If $\mathfrak{p}$ is an associated prime of M, then there is a monomorphism $R/\mathfrak{p} \hookrightarrow M$ and thus $0 \neq (R/\mathfrak{p})_\mathfrak{p} \hookrightarrow M_\mathfrak{p}$. Hence $\mathfrak{p}$ is in the support of M, this shows the first containment.

Next, we show $\mathrm{Supp}(M) = V(\mathrm{ann}\,(M))$. Let $\mathfrak{p} \in \mathrm{Supp}(M)$ and let $x \in \mathrm{ann}\,(M)$. If $x \notin \mathfrak{p}$, then $xm = 0$ for all $m \in M$ and $M_\mathfrak{p} = (0)$, which is absurd. Therefore $\mathfrak{p}$ is in $V(\mathrm{ann}\,(M))$. Conversely let $\mathfrak{p}$ be in $V(\mathrm{ann}\,(M))$ and let $m_1, \ldots, m_r$ be a finite set of generators of M. If $M_\mathfrak{p} = (0)$, then for each i there is $s_i \notin \mathfrak{p}$ so that $s_i m_i = 0$, therefore $s_1 \cdots s_r$ is in $\mathrm{ann}\,(M) \subset \mathfrak{p}$, which is impossible. Hence $M_\mathfrak{p} \neq 0$ and $\mathfrak{p}$ is in the support of M.

To prove the last part take a minimal prime $\mathfrak{p}$ in the support of M. As $M_\mathfrak{p} \neq (0)$ there is an associated prime $\mathfrak{p}_1 R_\mathfrak{p}$ of $M_\mathfrak{p}$, where $\mathfrak{p}_1$ is a prime ideal of R contained in $\mathfrak{p}$. Since $M_{\mathfrak{p}_1} \simeq (M_\mathfrak{p})_{\mathfrak{p}_1} \neq (0)$, we get that $\mathfrak{p}_1$ is in the support of M and $\mathfrak{p} = \mathfrak{p}_1$. Therefore using Proposition 2.1.13 one concludes $\mathfrak{p} \in \mathrm{Ass}(M)$. □

Let M be an R-module, the *minimal primes* of M are defined to be the minimal elements of $\text{Supp}(M)$ with respect to inclusion. A minimal prime of M is called an *isolated associated prime* of M. An associated prime of M which is not isolated is called an *embedded prime*.

If R is a ring and I is an ideal, note that the minimal primes of I are precisely the minimal primes of $\text{Ass}_R(R/I)$. In particular the minimal primes of R are precisely the minimal primes of $\text{Ass}_R(R)$.

Definition 2.1.21 Let M be an R-module. A submodule N of M is said to be a $\mathfrak{p}$-*primary submodule* if $\text{Ass}_R(M/N) = \{\mathfrak{p}\}$. An ideal $\mathfrak{q}$ of a ring R called a $\mathfrak{p}$-*primary ideal* if $\text{Ass}_R(R/\mathfrak{q}) = \{\mathfrak{p}\}$.

Proposition 2.1.22 *An ideal $\mathfrak{q} \neq R$ of a ring R is a primary ideal if and only if $xy \in \mathfrak{q}$ and $x \notin \mathfrak{q}$ implies $y^n \in \mathfrak{q}$ for some $n \geq 1$.*

Proof. Assume $\mathfrak{q}$ is a primary ideal. Let $x, y \in R$ such that $xy \in \mathfrak{q}$ and $x \notin \mathfrak{q}$. Hence y is a zero divisor of $R/\mathfrak{q}$ because $y\overline{x} = \overline{0}$ and $\overline{x} \neq 0$. Since $\mathcal{Z}(R/\mathfrak{q}) = \{\mathfrak{p}\}$ and $\text{rad}(\mathfrak{q}) = \mathfrak{p}$, we get $y^n \in \mathfrak{q}$ for some positive integer n. The converse is left as an exercise. $\square$

Definition 2.1.23 Let M be an R-module. A submodule N of M is said to be *irreducible* if N cannot be written as an intersection of two submodules of M that properly contain N.

Proposition 2.1.24 *Let M be an R-module. If $Q \neq M$ is an irreducible submodule of M, then Q is a primary submodule.*

Proof. Assume there are $\mathfrak{p}_1$and $\mathfrak{p}_2$ distinct associated prime ideals of M/Q and pick $r_0 \in \mathfrak{p}_1 \setminus \mathfrak{p}_2$ (or vice versa). There is x_i in $M \setminus Q$ such that $\mathfrak{p}_i = \text{ann}(\overline{x}_i)$, where $\overline{x}_i = x_i + Q$. We claim that

$$(Rx_1 + Q) \cap (Rx_2 + Q) = Q.$$

If z is in the intersection, then $z = \lambda_1 x_1 + q_1 = \lambda_2 x_2 + q_2$, for some $\lambda_i \in R$ and $q_i \in Q$. Note that $r_0 z \in Q$, hence $r_0 \lambda_2 x_2$ is in Q and consequently $r_0 \lambda_2 \in \mathfrak{p}_2$. Thus $\lambda_2 \in \mathfrak{p}_2$ and we get $\lambda_2 x_2 \in Q$. This shows $z \in Q$ and completes the proof of the claim. As Q is irreducible one has $Q = Rx_1 + Q$ or $Q = Rx_2 + Q$, which is a contradiction because $x_i \notin Q$ for $i = 1, 2$. $\square$

Lemma 2.1.25 *If N_1, N_2 are $\mathfrak{p}$-primary submodules of M, then $N_1 \cap N_2$ is a $\mathfrak{p}$-primary submodule.*

Proof. Set $N = N_1 \cap N_2$. There is an inclusion $M/N \hookrightarrow M/N_1 \oplus M/N_2$. Hence using Exercise 2.1.53 we get

$$\text{Ass}(M/N) \subset \text{Ass}(M/N_1 \oplus M/N_2) = \text{Ass}(M/N_1) \cup \text{Ass}(M/N_2) = \{\mathfrak{p}\}.$$

Therefore $\text{Ass}(M/N) = \{\mathfrak{p}\}$. $\square$

Definition 2.1.26 Let M be an R-module and let $N \subsetneq M$ be a proper submodule. An *irredundant primary decomposition* of N is an expression of N as an intersection of submodules, say $N = N_1 \cap \cdots \cap N_r$, such that:

(a) (Submodules are primary) $\mathrm{Ass}_R(M/N_i) = \{\mathfrak{p}_i\}$ for all i.

(b) (Irredundancy) $N \neq N_1 \cap \cdots \cap N_{i-1} \cap N_{i+1} \cap \cdots \cap N_r$ for all i.

(c) (Minimality) $\mathfrak{p}_i \neq \mathfrak{p}_j$ if $N_i \neq N_j$.

Theorem 2.1.27 *Let M be an R-module. If $N \subsetneq M$ is a proper submodule of M, then N has an irredundant primary decomposition*

Proof. First we claim that N is an intersection of a finite number of irreducible submodules of M. Let $\mathcal{F}$ be the family of submodules of M such that the claim is false. Assume $\mathcal{F} \neq \emptyset$. By Theorem 2.1.1, $\mathcal{F}$ has a maximal element that we denote by N. Since N is not irreducible, we can write $N = N_2 \cap N_3$, for some submodules N_2, N_3 strictly containing N. By the maximality of N we get that N_1 and N_2 can be written as an intersection of irreducible submodules, a contradiction. Thus $\mathcal{F} = \emptyset$. Hence the result follows from Proposition 2.1.24 and Lemma 2.1.25. $\square$

Corollary 2.1.28 *If R is a Noetherian ring and I a proper ideal of R, then I has an irredundant primary decomposition $I = \mathfrak{q}_1 \cap \cdots \cap \mathfrak{q}_r$ such that $\mathfrak{q}_i$ is a $\mathfrak{p}_i$-primary ideal and $\mathrm{Ass}_R(R/I) = \{\mathfrak{p}_1, \ldots, \mathfrak{p}_r\}$.*

Proof. Let $(0) = I/I = (\mathfrak{q}_1/I) \cap \cdots \cap (\mathfrak{q}_r/I)$ be an irredundant decomposition of the zero ideal of R/I. Then $I = \mathfrak{q}_1 \cap \cdots \cap \mathfrak{q}_r$ and $\mathfrak{q}_i/I$ is $\mathfrak{p}_i$-primary; that is, $\mathrm{Ass}_R((R/I)/(\mathfrak{q}_i/I)) = \mathrm{Ass}(R/\mathfrak{q}_i) = \{\mathfrak{p}_i\}$. Let us show that $\mathfrak{q}_i$ is a primary ideal. If $xy \in \mathfrak{q}_i$ and $x \notin \mathfrak{q}_i$, then y is a zero-divisor of $R/\mathfrak{q}_i$, but $\mathcal{Z}(R/\mathfrak{q}_i) = \mathfrak{p}_i$, hence $y \in \mathfrak{p}_i = \mathrm{rad}\,(\mathrm{ann}\,(R/\mathfrak{q}_i)) = \mathrm{rad}\,(\mathfrak{q}_i)$ and y^n is in $\mathfrak{q}_i$ for some $n > 0$. $\square$

Corollary 2.1.29 *If M is an R-module, then*

$$\mathrm{rad}(\mathrm{ann}(M)) = \bigcap_{\mathfrak{p} \in \mathrm{Ass}(M)} \mathfrak{p}.$$

Proof. Let $\mathfrak{p}_1, \ldots, \mathfrak{p}_r$ be the minimal elements of $\mathrm{Supp}(M)$. By Proposition 2.1.20 we have $\mathrm{Supp}(M) = V(\mathrm{ann}\,(M))$ and $\mathfrak{p}_i$ is in $\mathrm{Ass}(M)$ for all i. By Corollary 2.1.28 we have that $\mathrm{rad}(\mathrm{ann}(M)) = \mathfrak{p}_1 \cap \cdots \cap \mathfrak{p}_r$. $\square$

Corollary 2.1.30 *If $N \subsetneq M$ and $N = N_1 \cap \cdots \cap N_r$ is an irredundant primary decomposition of N with $\mathrm{Ass}_R(M/N_i) = \{\mathfrak{p}_i\}$, then*

$$\mathrm{Ass}_R(M/N) = \{\mathfrak{p}_1, \ldots, \mathfrak{p}_r\},$$

and $\mathrm{ann}\,(M/N_i)$ is a $\mathfrak{p}_i$-primary ideal for all i.

Proof. There is a natural monomorphism

$$M/(N_1 \cap \cdots \cap N_r) \hookrightarrow (M/N_1) \oplus \cdots \oplus (M/N_r).$$

Hence $\text{Ass}(M/N) \subset \{\mathfrak{p}_1, \ldots, \mathfrak{p}_r\}$. There are natural monomorphisms

$$(N_2 \cap \cdots \cap N_r)/N \hookrightarrow M/N_1; \quad (N_2 \cap \cdots \cap N_r)/N \hookrightarrow M/N.$$

Since $\text{Ass}(M/N_1) = \{\mathfrak{p}_1\}$ we get $\mathfrak{p}_1 \in \text{Ass}(N_2 \cap \cdots \cap N_r/N)$, consequently $\mathfrak{p}_1 \in \text{Ass}(M/N)$. Similarly one can show that any other $\mathfrak{p}_i$ is an associated prime of M/N. This proves the asserted equality.

By Corollary 2.1.29 we have $\text{rad}(\text{ann}(M/N_i)) = \mathfrak{p}_i$. Thus it suffices to show that $I = \text{ann}(M/N_i)$ is a primary ideal. Assume that $xy \in I$ for some $x, y \in R$. If x is not in I, then xM is not contained in N_i. Pick $m \in M$ such that $xm \notin N_i$. Since $y(xm) \subset N_i$, we get that y is a zero divisor of M/N_i, but the zero divisors of this module are precisely the elements of $\mathfrak{p}_i$ according to Lemma 2.1.19. Hence $y^r \in I$ for some r. □

Definition 2.1.31 Let R be a ring and let S be the set of nonzero divisors of R. The ring $S^{-1}R$ is called the *total ring of fractions* of R. If R is a domain, $S^{-1}R$ is the *field of fractions* of R.

Proposition 2.1.32 *Let R be a ring and let K be the total ring of fractions of R. If R is reduced, then K is a direct product of fields.*

Proof. Let $\mathfrak{p}_1, \ldots, \mathfrak{p}_r$ be the minimal primes of R and $S = R \setminus \cup_{i=1}^r \mathfrak{p}_i$. Since R is reduced one has $(0) = \mathfrak{p}_1 \cap \cdots \cap \mathfrak{p}_r$ and $K = S^{-1}R$. Define

$$\phi \colon K \longrightarrow S^{-1}R/S^{-1}\mathfrak{p}_1 \times \cdots \times S^{-1}R/S^{-1}\mathfrak{p}_r$$

by $\phi(x) = (x + S^{-1}\mathfrak{p}_1, \ldots, x + S^{-1}\mathfrak{p}_r)$. As $S^{-1}\mathfrak{p}_1, \ldots, S^{-1}\mathfrak{p}_r$ are maximal ideals and its intersection is zero, it follows from the Chinese remainder theorem that ϕ is an isomorphism (see Exercise 2.1.46). □

Lemma 2.1.33 *Let M be an R-module and L an ideal of R. If $LM = M$, then there is $x \in R$ such that $x \equiv 1 \pmod{L}$ and $xM = (0)$.*

Proof. Let $M = R\alpha_1 + \cdots + R\alpha_n$, $\alpha_i \in M$. As $LM = M$, there are b_{ij} in L such that $\alpha_i = \sum_{j=1}^n b_{ij}\alpha_i$. Set $\alpha = (\alpha_1, \ldots, \alpha_n)$ and $H = (b_{ij}) - I$, where I is the identity matrix. Since $H\alpha^t = 0$ and $H\text{adj}(H) = \det(H)I$, one concludes $\det(H)\alpha_i = 0$ for all i. Hence $xM = (0)$ and $x \equiv 1(\text{mod}\, L)$, where $x = \det(H)$. □

Lemma 2.1.34 (Nakayama) *Let R be a ring and let N be a submodule of an R-module M. If I is an ideal of R contained in the intersection of all the maximal ideals of R such that $M = IM + N$, then $M = N$.*

Proof. Note $I(M/N) = M/N$. By Lemma 2.1.33 there exists an element $x \equiv 1 (\mathrm{mod}\, I)$ such that $x(M/N) = (0)$. To finish the proof note that x is a unit; otherwise x belongs to some maximal ideal $\mathfrak{m}$ and this yields a contradiction because $I \subset \mathfrak{m}$. $\square$

Let M be an R-module. The *minimum number of generators* of M will be denoted by $\mu(M)$. A consequence of Nakayama's lemma is an expression for $\mu(M)$, when the ring R is local (cf. Corollary 3.5.2).

Corollary 2.1.35 *If M is a module over a local ring $(R, \mathfrak{m})$, then*

$$\mu(M) = \dim_k(M/\mathfrak{m}M), \quad \textit{where } k = R/\mathfrak{m}.$$

Proof. Let $\alpha_1, \ldots, \alpha_q$ be a minimal generating set for M and $\overline{\alpha}_i = \alpha_i + \mathfrak{m}M$. After a permutation of the α_i one may assume that $\overline{\alpha}_1, \ldots, \overline{\alpha}_r$ is a basis for $M/\mathfrak{m}M$ as a k-vector space, for some $r \leq q$. Set $N = R\alpha_1 + \cdots + R\alpha_r$. Note the equality $M = N + \mathfrak{m}M$, then by Nakayama's Lemma $N = M$. Therefore $r = q$, as required. $\square$

Modules of finite length An R-module M has *finite length* if there is a *composition series*

$$(0) = M_0 \subset M_1 \subset \cdots \subset M_n = M, \tag{2.1}$$

where M_i/M_{i-1} is a non-zero *simple module* (that is, M_i/M_{i-1} has no proper submodules other than (0)) for all i. Note that M_i/M_{i-1} must be cyclic and thus isomorphic to $R/\mathfrak{m}$, for some maximal ideal $\mathfrak{m}$. The number n is independent of the composition series and is called the *length* of M; it is usually denoted by $\ell_R(M)$ or simply $\ell(M)$.

Proposition 2.1.36 [9, Proposition 6.9] *If $0 \to M' \to M \to M'' \to 0$ is an exact sequence of R-modules of finite length, then*

$$\ell_R(M) = \ell_R(M') + \ell_R(M'').$$

An R-module M is called *Artinian* if M satisfies the descending chain condition for submodules; that is, for every chain of submodules of M

$$\cdots \subset N_{n+1} \subset N_n \subset \cdots \subset N_2 \subset N_1 \subset N_0 = M$$

there exists an integer k such that $N_i = N_k$ for every $i \geq k$. It is easy to verify that M is Artinian if and only if any family $\mathcal{F}$ of submodules of M partially ordered by inclusion has a minimal element, i.e., there is $N \in \mathcal{F}$ such that if $N_i \subset N$ and $N_i \in \mathcal{F}$, then $N = N_i$.

Proposition 2.1.37 *Let M be an R-module. Then $\ell_R(M) < \infty$ if and only if M is Noetherian and Artinian.*

Proof. $\Rightarrow$) Set $n = \ell_R(M)$. If M is not Noetherian or Artinian, pick submodules of M such that

$$(0) = N_0 \subsetneq N_1 \subsetneq \cdots \subsetneq N_n \subsetneq N_{n+1} = M,$$

a contradiction because this chain can be refined to a composition series of length n.

$\Leftarrow$) We construct a finite composition series as follows. Set $M_0 = M$. Consider the family $\mathcal{F}_1$ of proper submodules of M and pick a maximal element M_1, which exists because M is Noetherian. By induction consider the family $\mathcal{F}_i$ of proper submodules of M_i and pick a maximal element M_{i+1}. Notice that this process must stop at the zero module because M is Artinian. $\Box$

Lemma 2.1.38 *Let M be a k-vector space. Then M is Artinian if and only if* $\dim_k(M) = \ell_k(M) < \infty$.

Proof. Assume that M is Artinian and $\dim_k(M) = \infty$. Let $\mathcal{B} \subset M$ be an infinite linearly independent set, say $\mathcal{B} = \{\alpha_1, \ldots, \alpha_i, \ldots, \}$. Then

$$\cdots \subsetneq k(\mathcal{B} \setminus \{\alpha_1, \ldots, \alpha_i\}) \subsetneq \cdots \subsetneq k(\mathcal{B} \setminus \{\alpha_1, \alpha_2\}) \subsetneq k(\mathcal{B} \setminus \{\alpha_1\}) \subsetneq M,$$

a contradiction. The converse is also easy to show. $\Box$

Proposition 2.1.39 *If* $0 \to M' \to M \to M'' \to 0$ *is an exact sequence of modules over a ring R, then* $\dim(M) = \max\{\dim(M'),\ \dim(M'')\}$.

Proof. Set $d = \dim(M)$, $d' = \dim(M')$ and $d'' = \dim(M'')$. First note that $d = \dim(R/\mathfrak{p})$ for some prime $\mathfrak{p}$ containing $\text{ann}\,(M)$, by Proposition 2.1.20 we obtain $M_\mathfrak{p} \neq (0)$. Therefore using Lemma 2.1.15 one has $M'_\mathfrak{p} \neq (0)$ or $M''_\mathfrak{p} \neq (0)$, thus either $\mathfrak{p}$ contains $\text{ann}\,(M')$ or $\mathfrak{p}$ contains $\text{ann}\,(M'')$. This proves $d \leq \max\{d', d''\}$. On the other hand $\text{ann}\,(M)$ is contained in $\text{ann}\,(M') \cap \text{ann}\,(M'')$ and consequently $\max\{d', d''\} \leq d$. $\Box$

Proposition 2.1.40 *If M is an R-module, then M has finite length if and only if every prime ideal in* $\text{Supp}(M)$ *is a maximal ideal.*

Proof. $\Rightarrow$) Let $\{M_i\}_{i=0}^n$ be a composition series as in Eq. (2.1). Notice that $\mathfrak{p}_i$ are maximal ideals for all i. Indeed if N is a simple R-module and $R/\mathfrak{p} \simeq N$ for some prime ideal $\mathfrak{p}$, then $\mathfrak{p}$ is a maximal ideal. Hence from Lemma 2.1.15 we obtain that $\text{Supp}(M)$ is equal to $\{\mathfrak{p}_1, \ldots, \mathfrak{p}_n\}$.

$\Leftarrow$) There is a filtration $(0) = M_0 \subset M_1 \subset \cdots \subset M_n = M$ of submodules and prime ideals $\mathfrak{p}_1, \ldots, \mathfrak{p}_n$ of R such that $M_i/M_{i-1} \simeq R/\mathfrak{p}_i$ for all i (see Theorem 2.1.16). By the proof of Corollary 2.1.18 one has

$$\text{Ass}(M) \subset \{\mathfrak{p}_1, \ldots, \mathfrak{p}_n\} \subset \text{Supp}(M).$$

Hence by Proposition 2.1.20 we get $\text{Ass}(M) = \{\mathfrak{p}_1, \ldots, \mathfrak{p}_n\} = \text{Supp}(M)$. Thus the filtration above is a composition series because $\mathfrak{p}_i$ is a maximal ideal for all i. $\square$

Theorem 2.1.41 *Let R be a ring. Then R is Artinian if and only if*

(a) *R is Noetherian, and*

(b) *every prime ideal of R is maximal.*

Proof. $\Rightarrow$) First we show that $\ell_R(R) < \infty$. Let

$$\mathcal{F} = \{\mathfrak{m}_1\mathfrak{m}_2 \cdots \mathfrak{m}_s |\, \mathfrak{m}_i \in \text{Max}(R),\, \forall\, i\},$$

we do not require $\mathfrak{m}_1, \ldots, \mathfrak{m}_s$ to be distinct. The family $\mathcal{F}$ has a minimal element that we denote by $I = \mathfrak{m}_1 \cdots \mathfrak{m}_r$. We claim that $I = (0)$. Assume that $I \neq (0)$. Clearly $I^2 \in \mathcal{F}$ and $I^2 \subset I$, thus $I = I^2$. Consider the family

$$\mathcal{G} = \{J |\, J \text{ is an ideal; } JI \neq (0)\}.$$

This family is non-empty because $I \in \mathcal{G}$. Since R is Artinian, there exists a minimal element J of $\mathcal{G}$. Then

$$(0) \neq JI = JI^2 = (JI)I \;\Rightarrow\; JI \in \mathcal{G}.$$

Since $IJ \subset J$, we get $J = IJ$. Notice that J is a principal ideal. Indeed since $IJ \neq (0)$, there is $x \in J$ such that $xI \neq (0)$. Hence $(x) \in \mathcal{G}$, and consequently $J = (x)$. If $\mathfrak{m}$ is a maximal ideal of R, then $\mathfrak{m}I \in \mathcal{F}$ and $\mathfrak{m}I \subset I$. Hence $\mathfrak{m}I = I$ and $I \subset \mathfrak{m}$. Thus I is contained in the Jacobson radical of I. From the equality $(x) = J = IJ = Ix$, we obtain that $x = \lambda x$ for some $\lambda \in I$. As $x(1 - \lambda) = 0$, with $1 - \lambda$ a unit of R, we get $x = 0$. A contradiction to $J \neq (0)$. This proves that $I = (0)$. Set $I_k = \mathfrak{m}_1 \cdots \mathfrak{m}_k$ for $1 \leq k \leq r$ and consider the filtration

$$(0) = I = I_r \subset I_{r-1} \subset I_{r-2} \subset \cdots \subset I_1 \subset I_0 = R.$$

Observe that I_i/I_{i+1} is an $R/\mathfrak{m}_{i+1}$ vector space. The vector space I_i/I_{i+1} is Artinian because the subspaces of I_i/I_{i+1} correspond bijectively to the intermediate ideals of $I_{i+1} \subset I_i$. Therefore by Lemma 2.1.38 we get that I_i/I_{i+1} has finite length as an $R/\mathfrak{m}_{i+1}$ vector space and consequently as an R-module. Hence the filtration can be refined to a finite composition series, i.e., $\ell_R(R) < \infty$. Therefore R is Noetherian by Proposition 2.1.37 and (a) holds. To prove (b) take a minimal prime $\mathfrak{p}$ of R. Since $\mathfrak{p}$ contains $I = (0)$ it follows that $\mathfrak{p} = \mathfrak{m}_i$ for some $1 \leq i \leq r$.

$\Leftarrow$) Let $(0) = \mathfrak{q}_1 \cap \cdots \cap \mathfrak{q}_r$ be an irredundant primary decomposition. Then all the associated primes are maximal. By Proposition 2.1.20 we obtain $\text{Ass}_R(R) = \text{Supp}(R)$. Hence R is Artinian by Proposition 2.1.40. $\square$

Definition 2.1.42 A ring is called *semilocal* if it has only finitely many maximal ideals.

Using the Chinese reminder theorem (see Exercise 2.1.46) and the proof of Theorem 2.1.41 we obtain:

Proposition 2.1.43 *If R is an Artinian ring, then R has only a finite number of maximal ideals $\mathfrak{m}_1,\ldots,\mathfrak{m}_s$ and*

$$R \simeq R/\mathfrak{m}_1^{a_1} \times \cdots \times R/\mathfrak{m}_s^{a_s}.$$

Definition 2.1.44 Let R be a ring with total ring of fractions Q. An R-module M has *rank* r if $M \otimes_R Q$ is a free Q-module of rank equal to r.

Lemma 2.1.45 *If M is an R-module of positive rank r, then*

$$\dim(M) = \dim(R).$$

Proof. Let $\mathfrak{p}_1,\ldots,\mathfrak{p}_n$ be the associated primes of R and $S = R \setminus \cup_{i=1}^n \mathfrak{p}_i$. By hypothesis $S^{-1}M \simeq (S^{-1}R)^r$. Since

$$M_{\mathfrak{p}_i} \simeq (S^{-1}M)_{\mathfrak{p}_i} \simeq [(S^{-1}R)^r]_{\mathfrak{p}_i} \simeq [(S^{-1}R)_{\mathfrak{p}_i}]^r \simeq (R_{\mathfrak{p}_i})^r \neq (0),$$

we obtain that all the minimal primes of R are in the support of M and consequently one has $\dim(M) = \dim(R)$. $\Box$

Exercises

2.1.46 (Chinese remainder theorem) Let $I_1,\ldots,I_r$ be ideals of a ring R. If $I_i + I_j = R$ for $i \neq j$, prove:

(a) $I_1 \cap \cdots \cap I_r = I_1 \cdots I_r$,

(b) The rings $R/I_1 \cap \cdots \cap I_r$ and $R/I_1 \times \cdots \times R/I_r$ are isomorphic.

2.1.47 Let I be an ideal of a ring R. Prove that $\mathrm{rad}(I) = I$ if and only if for any $x \in R$ such that $x^2 \in I$ one has $x \in I$.

2.1.48 Let $I_1,\ldots,I_n$ be ideals of a ring R and let $\mathfrak{p}$ be a prime ideal of R. (a) If $I_1 \cdots I_n \subset \mathfrak{p}$, prove that $I_i \subset \mathfrak{p}$ for some i. (b) If $\cap_{i=1}^n I_i \subset \mathfrak{p}$, then $I_i \subset \mathfrak{p}$ for some i. (c) If $\cap_{i=1}^n I_i = \mathfrak{p}$, then $\mathfrak{p} = I_i$ for some i.

2.1.49 Let $I, \mathfrak{p}_1,\ldots,\mathfrak{p}_r$ be ideals of a ring R. If I is contained in $\cup_{i=1}^r \mathfrak{p}_i$ and $\mathfrak{p}_i$ is a prime ideal for $i \geq 3$, prove that $I \subset \mathfrak{p}_i$ for some i (cf. Lemma 3.1.31).

2.1.50 Let R be a ring and let I be an ideal. If $\mathfrak{p}$ is a prime ideal such that $I \subset \mathfrak{p}$, prove that $\mathrm{ht}(I) \leq \mathrm{ht}(I_\mathfrak{p})$. Give an example where the strict inequality holds.

2.1.51 Let R be a ring and let I be an ideal of R. If all the minimal primes of I have the same height and $\mathfrak{p}$ is a prime ideal such that $I \subset \mathfrak{p}$, prove that $\mathrm{ht}(I) = \mathrm{ht}(I_\mathfrak{p})$.

2.1.52 Let I be an ideal of a ring R. Then a prime ideal $\mathfrak{p}$ of R is an associated prime of R/I if and only if $\mathfrak{p} = (I\colon x)$ for some $x \in R$.

2.1.53 If M, N are R-modules, prove $\mathrm{Ass}(M \oplus N) = \mathrm{Ass}(M) \cup \mathrm{Ass}(N)$.

2.1.54 Let M be an R-module and I an ideal of R contained in $\mathrm{ann}_R(M)$. Note that M inherits a structure of R/I-module. Prove that $\mathfrak{p} \in \mathrm{Ass}_R(M)$ if and only if $\mathfrak{p}/I \in \mathrm{Ass}_{R/I}(M)$.

2.1.55 Let M be an R-module and let $\mathfrak{p}$ be a prime ideal of R. Prove that $\mathfrak{p} \in \mathrm{Ass}_R(M)$ if and only if $\mathfrak{p}R_\mathfrak{p} \in \mathrm{Ass}_{R_\mathfrak{p}}(M_\mathfrak{p})$.

2.1.56 Let A be a ring and $\mathfrak{q}$ a proper ideal of A. Prove that $\mathfrak{q}$ is a primary ideal if and only if $\mathcal{Z}(A/\mathfrak{q}) \subset \mathfrak{N}_{A/\mathfrak{q}}$, where $\mathfrak{N}_{A/\mathfrak{q}}$ is the nilradical of $A/\mathfrak{q}$.

2.1.57 Let A be a ring and $\mathfrak{q}$ a proper ideal of A. If every element in $A/\mathfrak{q}$ is either nilpotent or invertible, prove that $\mathfrak{q}$ is a primary ideal.

2.1.58 Let A be a ring and $\mathfrak{q}$ an ideal of A such that $\mathrm{rad}\,(\mathfrak{q}) = \mathfrak{m}$ is a maximal ideal. Show that $\mathfrak{q}$ is a primary ideal.

2.1.59 If $B = A[\mathbf{x}]$ is a polynomial ring over a ring A and $\mathfrak{q}$ a primary ideal of A, then $\mathfrak{q}B$ is a primary ideal of B.

2.1.60 Let M be an R-module and $S \subset T$ two multiplicatively closed subsets of R. Prove $T^{-1}M \simeq T^{-1}(S^{-1}M)$ as $T^{-1}R$ modules.

2.1.61 Let M be an R-module and S a multiplicatively closed subset of R. If $\mathfrak{p}$ is a prime ideal such that $S \cap \mathfrak{p} = \emptyset$, prove $M_\mathfrak{p} \simeq (S^{-1}M)_\mathfrak{p}$.

2.1.62 Let R be a ring and I an ideal. If $x \in R \setminus I$, then there is an exact sequence of R-modules:

$$0 \longrightarrow R/(I\colon x) \stackrel{\psi}{\longrightarrow} R/I \stackrel{\phi}{\longrightarrow} R/(I,x) \longrightarrow 0,$$

where $\psi(\overline{r}) = x\overline{r}$ is multiplication by x and $\phi(\overline{r}) = \overline{r}$.

2.1.63 If $0 \to M' \stackrel{f}{\to} M \stackrel{g}{\to} M'' \to 0$ is an exact sequence of R-modules, prove that the following sequence is also exact

$$0 \longrightarrow S^{-1}M' \stackrel{S^{-1}f}{\longrightarrow} S^{-1}M \stackrel{S^{-1}g}{\longrightarrow} S^{-1}M'' \longrightarrow 0.$$

2.1.64 Let N_1 and N_2 be submodules of an R-module M. If S is a multiplicatively closed subset of R, then

(a) $S^{-1}(N_1 + N_2) = S^{-1}(N_1) + S^{-1}(N_2)$,

(b) $S^{-1}(N_1 \cap N_2) = S^{-1}(N_1) \cap S^{-1}(N_2)$,

(c) $S^{-1}(N_1/N_2) \simeq S^{-1}(N_1)/S^{-1}(N_2)$, as $S^{-1}(R)$-modules,

(d) $S^{-1}(\operatorname{ann}(N_2)) = \operatorname{ann}(S^{-1}(N_2))$, and

(e) $S^{-1}(N_1 \colon N_2) = (S^{-1}(N_1) \colon S^{-1}(N_2))$.

Hint Use that N_2 is a finitely generated R-module.

2.1.65 If I and J are ideals of a ring R and S is a multiplicatively closed subset of R, then

(a) $\sqrt{S^{-1}(I)} = S^{-1}(\sqrt{I})$, and

(b) $S^{-1}(IJ) = S^{-1}(I)S^{-1}(J)$.

2.2 Graded modules and Hilbert polynomials

Let $(H, +)$ be an abelian semigroup. An *H-graded ring* is a ring R together with a decomposition

$$R = \bigoplus_{a \in H} R_a \quad \text{(as a } \mathbb{Z}\text{-module)},$$

such that $R_a R_b \subset R_{a+b}$ for all $a, b \in H$. A *graded ring* is by definition a $\mathbb{Z}$-graded ring.

If R is an H-graded ring and M is an R-module with a decomposition

$$M = \bigoplus_{a \in H} M_a,$$

such that $R_a M_b \subset M_{a+b}$ for all $a, b \in H$, we say that M is an *H-graded module*. An element $0 \neq f \in M$ is said to be *homogeneous* of degree a if $f \in M_a$, in this case we set $\deg(f) = a$. The non-zero elements in R_a are also called *forms* of degree a. Any element $f \in M$ can be written uniquely as $f = \sum_{a \in H} f_a$ with only finitely many $f_a \neq 0$.

A map $\varphi \colon M \to N$ between H-graded modules is *graded* if $\varphi(M_a) \subset N_a$ for all $a \in H$. Let $M = \oplus_{a \in H} M_a$ be an H-graded module and N a *graded submodule*; that is, N is graded with the induced grading $N = \oplus_{a \in H} N \cap M_a$. Then M/N is an H-graded R-module with $(M/N)_a = M_a/N \cap M_a$ for $a \in H$, $R_0 \subset R$ is a subring and M_a is an R_0-module for $a \in H$.

Proposition 2.2.1 [310, p. 92] *Let $M = \oplus_{a\in H} M_a$ be an H-graded module and $N \subset M$ a submodule. Then the following conditions are equivalent:*

(g_1) *N is generated over R by homogeneous elements.*

(g_2) *If $f = \sum_{a\in H} f_a$ is in N, $f_a \in M_a$ for all a, then each f_a is in N.*

(g_3) *N is a graded submodule of M.*

Let $R = K[x_1, \ldots, x_n]$ be a polynomial ring over a field K and let $d_1, \ldots, d_n$ be a sequence in $\mathbb{N}_+$. For $a = (a_i)$ in $\mathbb{N}^n$ we set $x^a = x_1^{a_1} \cdots x_n^{a_n}$ and $|a| = \sum_{i=1}^n a_i d_i$. The *induced* $\mathbb{N}$*-grading* on R is given by:

$$R = \bigoplus_{i=0}^{\infty} R_i, \text{ where } R_i = \bigoplus_{|a|=i} Kx^a.$$

Notice that $\deg(x_i) = d_i$ for all i. The induced grading extends to a $\mathbb{Z}$-grading by setting $R_i = 0$ for $i < 0$. The homogeneous elements of R are called *quasi-homogeneous polynomials*. Let I be a *homogeneous* ideal of R generated by a set $f_1, \ldots, f_r$ of homogeneous polynomials. Setting $\deg(f_i) = \delta_i$, I becomes a *graded ideal* with the grading

$$I_i = I \cap R_i = f_1 R_{i-\delta_1} + \cdots + f_r R_{i-\delta_r}.$$

Hence R/I is an $\mathbb{N}$-graded R-module graded by $(R/I)_i = R_i/I_i$.

Definition 2.2.2 The *standard grading* or *usual grading* of a polynomial ring $K[x_1, \ldots, x_n]$ is the $\mathbb{N}$-grading induced by setting $\deg(x_i) = 1$ for all i.

Hilbert polynomial and multiplicity Let $R = \oplus_{i=0}^{\infty} R_i$ be an $\mathbb{N}$-graded ring. We recall that R is a Noetherian ring if and only if R_0 is a Noetherian ring and $R = R_0[x_1, \ldots, x_n]$ for some $x_1, \ldots, x_n$ in R. If M is a finitely generated $\mathbb{N}$-graded R-module and R_0 is an Artinian local ring, define the *Hilbert function* of M as

$$H(M, i) = \ell_{R_0}(M_i) \qquad (\ell_{R_0} = \text{length w.r.t } R_0).$$

Definition 2.2.3 An $\mathbb{N}$-graded ring $R = \oplus_{i=0}^{\infty} R_i$ is called a *homogeneous ring* if $R = R_0[x_1, \ldots, x_n]$, where $\deg(x_i) = 1$ for all i.

For use below, by convention the zero polynomial has degree -1.

Theorem 2.2.4 (Hilbert [65, Theorem 4.1.3]) *Let $R = \oplus_{i=0}^{\infty} R_i$ be a homogeneous ring and let M be a finitely generated $\mathbb{N}$-graded R-module with $d = \dim(M)$. If R_0 is an Artinian local ring, then there is a unique polynomial $\varphi_M(t) \in \mathbb{Q}[t]$ of degree $d-1$ such that $\varphi_M(i) = H(M, i)$ for $i \gg 0$.*

Definition 2.2.5 The polynomial $\varphi_M(t)$ is called the *Hilbert polynomial* of M. If $\varphi_M(t) = a_{d-1}t^{d-1} + \cdots + a_0$, the *multiplicity* or *degree* of M, denoted by $e(M)$ or $\deg(M)$, is $(d-1)!a_{d-1}$ if $d \geq 1$ and $\ell_{R_0}(M)$ if $d = 0$.

Graded primary decomposition Let $R = K[x_1, \ldots, x_n]$ be a polynomial ring over a field K endowed with a positive grading induced by setting $\deg(x_i) = d_i$ for all i, where d_i is a positive integer for $i = 1, \ldots, n$.

Lemma 2.2.6 [65, Lemma 1.5.6] *If M is an $\mathbb{N}$-graded R-module and $\mathfrak{p}$ is in $\mathrm{Ass}(M)$, then $\mathfrak{p}$ is a graded ideal and there is $m \in M$ homogeneous such that $\mathfrak{p} = \mathrm{ann}(m)$.*

Proposition 2.2.7 [309, p. 63] *Let M be an $\mathbb{N}$-graded R-module and let Q be a $\mathfrak{p}$-primary submodule of M. If $\mathfrak{p}$ is graded and Q^* is the submodule of M generated by the homogeneous elements in Q, then Q^* is again a $\mathfrak{p}$-primary submodule.*

Theorem 2.2.8 *Let M be an $\mathbb{N}$-graded R-module and let N be a proper graded submodule of M. Then N has an irredundant primary decomposition $N = N_1 \cap \cdots \cap N_r$ such that N_i is a graded submodule for all i.*

Proof. Use Lemma 2.2.6, Proposition 2.2.7, and Theorem 2.1.27. $\Box$

Finding primary decompositions of graded ideals in polynomial rings over fields is a difficult task. For a treatment of the principles of primary decomposition consult the book of Vasconcelos [413, Chapter 3].

Notation If $R = \oplus_{i=0}^{\infty} R_i$ is an $\mathbb{N}$-graded ring, we set $R_+ = \oplus_{i \geq 1} R_i$.

Lemma 2.2.9 (Graded Nakayama lemma) *Let R be an $\mathbb{N}$-graded ring and let M be an $\mathbb{N}$-graded R-module. If N is a graded submodule of M and $I \subset R_+$ is a graded ideal of R such that $M = N + IM$, then $N = M$.*

Proof. Since $M/N = I(M/N)$, one may assume $N = (0)$. If $x \in M$ is a homogeneous element of degree r, then a recursive use of the equality $M = IM$ yields that $x \in I^{r+1}M$ and x must be zero. $\Box$

Exercises

2.2.10 Let I be a graded ideal of a polynomial ring R over a field K. Prove that the radical of I is also graded.

2.2.11 Let I be a graded ideal of a polynomial ring $R = K[x_1, \ldots, x_n]$ over a field K. If $\mathfrak{m} = (x_1, \ldots, x_n)$ is the irrelevant maximal ideal of R, prove that $\mathfrak{m} \in \mathrm{Ass}(R/I)$ if and only if $\mathrm{depth}(R/I) = 0$.

Hint Use Lemma 2.1.19.

2.2.12 Let $R = \oplus_{i \in \mathbb{Z}} R_i$ be a graded ring. Prove that R_0 is a subring of R with $1 \in R_0$. A graded ring R with $R_i = 0$ for $i < 0$ is called a *graded R_0-algebra*.

2.2.13 Let $R = K[x_1, x_2]$ be a polynomial ring over a field K with the grading induced by $\deg(x_i) = (-1)^i$. Determine the subring R_0.

2.3 Cohen–Macaulay modules

Here we introduce some special types of rings and modules and present the following fundamental result of dimension theory.

Theorem 2.3.1 (Dimension theorem [310, Theorem 13.4]) *Let $(R, \mathfrak{m})$ be a local ring and let M be an R-module. Set*

$$\delta(M) = \min\{r \mid \text{there are } x_1, \ldots, x_r \in \mathfrak{m} \text{ with } \ell_R(M/(x_1, \ldots x_r)M) < \infty\},$$

then $\dim(M) = \delta(M)$.

Definition 2.3.2 Let $(R, \mathfrak{m})$ be a local ring and let M be an R-module of dimension d. A *system of parameters* (s.o.p for short) of M is a set of elements $\theta_1, \ldots, \theta_d$ in $\mathfrak{m}$ such that $\ell_R(M/(\theta_1, \ldots, \theta_d)M) < \infty$.

Corollary 2.3.3 *Let $(R, \mathfrak{m})$ be a local ring and let M be an R-module of dimension d. If $h_1, \ldots, h_d \in \mathfrak{m}$ is a system of parameters of M, then*

$$\dim M/(h_1, \ldots, h_i)M = d - i \text{ for } 1 \leq i \leq d.$$

Proof. Set $\overline{M} = M/(h_1, \ldots, h_i)M$. First note that $h_{i+1}, \ldots, h_d$ is a s.o.p for $\overline{M}$, hence $\dim(\overline{M}) \leq d - i$ by Theorem 2.3.1. On the other hand if $\overline{\theta}_1, \ldots, \overline{\theta}_r$ is a s.o.p for $\overline{M}$, then $\theta_1, \ldots, \theta_r, h_1, \ldots, h_i$ is a s.o.p for M and again by the dimension theorem one has $d \leq r + i = \dim(\overline{M}) + i$. □

Definition 2.3.4 Let M be an R-module. A sequence $\underline{\theta} = \theta_1, \ldots, \theta_n$ in R is called a *regular sequence* of M or an *M-regular sequence* if $(\underline{\theta})M \neq M$ and $\theta_i \notin \mathcal{Z}(M/(\theta_1, \ldots, \theta_{i-1})M)$ for all i.

Proposition 2.3.5 *Let M be an R-module and let I be an ideal of R such that $IM \neq M$. If $\underline{\theta} = \theta_1, \ldots, \theta_r$ is an M-regular sequence in I, then $\underline{\theta}$ can be extended to a maximal M-regular sequence in I.*

Proof. By induction assume there is an M-regular sequence $\theta_1, \ldots, \theta_i$ in I for some $i \geq r$. Set $\overline{M} = M/(\theta_1, \ldots, \theta_i)M$. If $I \not\subset \mathcal{Z}(\overline{M})$, pick θ_{i+1} in I which is regular on $\overline{M}$. Since

$$(\theta_1) \subset (\theta_1, \theta_2) \subset \cdots \subset (\theta_1, \ldots, \theta_i) \subset (\theta_1, \ldots, \theta_{i+1}) \subset R$$

is an increasing sequence of ideals in a Noetherian ring R, this inductive construction must stop at a maximal M-regular sequence in I. □

Lemma 2.3.6 *Let M be a module over a local ring $(R, \mathfrak{m})$. If $\theta_1, \ldots, \theta_r$ is an M-regular sequence in $\mathfrak{m}$, then $r \leq \dim(M)$.*

Proof. By induction on $\dim(M)$. If $\dim(M) = 0$, then $\mathfrak{m}$ is an associated prime of M and every element of $\mathfrak{m}$ is a zero divisor of M. We claim that $\dim(M/\theta_1 M) < \dim(M)$. If this equality does not hold, there is a saturated chain of prime ideals

$$\operatorname{ann}(M) \subset \operatorname{ann}(M/\theta_1 M) \subset \mathfrak{p}_0 \subset \cdots \subset \mathfrak{p}_d,$$

where d is the dimension of M and $\mathfrak{p}_0$ is minimal over $\operatorname{ann}(M)$. According to Proposition 2.1.20 the ideal $\mathfrak{p}_0$ consists of zero divisors, a contradiction since $\theta_1 \in \operatorname{ann}(M/\theta_1 M) \subset \mathfrak{p}_0$. This proves the claim. Since $\theta_2, \ldots, \theta_r$ is a regular sequence on $M/\theta_1 M$ by induction one derives $r \leq \dim(M)$. $\square$

Proposition 2.3.7 *Let M be an R-module and let I be an ideal of R.*

(a) $\operatorname{Hom}_R(R/I, M) = (0)$ *iff there is $x \in I$ which is regular on M.*

(b) $\operatorname{Ext}_R^r(R/I, M) \simeq \operatorname{Hom}_R(R/I, M/\underline{\theta}M)$, *where $\underline{\theta} = \theta_1, \ldots, \theta_r$ is any M-regular sequence in I.*

Proof. (a) $\Rightarrow$) Assume $I \subset \mathcal{Z}(M)$. Using Lemma 2.1.19 one has $I \subset \mathfrak{p}$ for some $\mathfrak{p} \in \operatorname{Ass}_R(M)$. Hence there is a monomorphism $\psi\colon R/\mathfrak{p} \to M$. To derive a contradiction note that the composition

$$R/I \xrightarrow{\varphi} R/\mathfrak{p} \xrightarrow{\psi} M$$

is a non-zero map, where φ is the canonical map from R/I to $R/\mathfrak{p}$. The converse is left as an exercise.

(b) Consider the exact sequence

$$0 \longrightarrow M \xrightarrow{\theta_1} M \longrightarrow \overline{M} = M/\theta_1 M \longrightarrow 0.$$

According to [363, Theorem 7.3] there is a long exact sequence with natural connecting homomorphisms

$$\cdots \longrightarrow \operatorname{Ext}_R^{r-1}(R/I, M) \xrightarrow{\theta_1} \operatorname{Ext}_R^{r-1}(R/I, M) \longrightarrow \operatorname{Ext}_R^{r-1}(R/I, \overline{M}) \xrightarrow{\partial}$$
$$\operatorname{Ext}_R^{r}(R/I, M) \xrightarrow{\theta_1} \operatorname{Ext}_R^{r}(R/I, M) \longrightarrow \cdots$$

Since θ_1 is in I, using [363, Theorem 7.16] it follows that in the last exact sequence the maps given by multiplication by θ_1 are zero. Hence

$$\operatorname{Ext}_R^{r-1}(R/I, \overline{M}) \simeq \operatorname{Ext}_R^{r}(R/I, M),$$

and the proof follows by induction on r. $\square$

Let $M \neq (0)$ be a module over a local ring $(R, \mathfrak{m})$. The *depth* of M, denoted by $\operatorname{depth}(M)$, is the length of any maximal regular sequence on M which is contained in $\mathfrak{m}$. From Proposition 2.3.7 one derives

$$\operatorname{depth}(M) = \inf\{r \mid \operatorname{Ext}_R^r(R/\mathfrak{m}, M) \neq (0)\}.$$

In general by Lemma 2.3.6 we have $\operatorname{depth}(M) \leq \dim(M)$.

Definition 2.3.8 An R-module M is called *Cohen–Macaulay* (C–M for short) if $\mathrm{depth}(M) = \dim(M)$, or if $M = (0)$.

Lemma 2.3.9 (Depth lemma [413, p. 305]) *If* $0 \to N \to M \to L \to 0$ *is a short exact sequence of modules over a local ring* R*, then*

(a) *If* $\mathrm{depth}(M) < \mathrm{depth}(L)$*, then* $\mathrm{depth}(N) = \mathrm{depth}(M)$.

(b) *If* $\mathrm{depth}(M) = \mathrm{depth}(L)$*, then* $\mathrm{depth}(N) \geq \mathrm{depth}(M)$.

(c) *If* $\mathrm{depth}(M) > \mathrm{depth}(L)$*, then* $\mathrm{depth}(N) = \mathrm{depth}(L) + 1$.

Lemma 2.3.10 *If* M *is a module over a local ring* $(R, \mathfrak{m})$ *and* $z \in \mathfrak{m}$ *is a regular element of* M*, then*

(a) $\mathrm{depth}(M/zM) = \mathrm{depth}(M) - 1$*, and*

(b) $\dim(M/zM) = \dim(M) - 1$.

Proof. As $\mathrm{depth}\, M > \mathrm{depth}\, M/zM$ applying the depth lemma to the exact sequence

$$0 \longrightarrow M \stackrel{z}{\longrightarrow} M \longrightarrow M/zM \longrightarrow 0$$

yields $\mathrm{depth}(M) = \mathrm{depth}(M/zM){+}1$. To prove the second equality first observe the inequality $\dim(M) > \dim(M/zM)$, which follows from the proof Lemma 2.3.6 by making $z = \theta_1$. On the other hand the reverse inequality $\dim(M) \leq \dim(M/zM) + 1$ follows from Theorem 2.3.1. □

Proposition 2.3.11 [65, p. 58] *If* M *is a Cohen–Macaulay* R*-module, then* $S^{-1}M$ *is Cohen–Macaulay for every multiplicatively closed set* S *of* R.

Proposition 2.3.12 [65, p. 58] *Let* M *be an* R*-module and let* $\underline{x}$ *be an* M*-regular sequence. If* M *is Cohen–Macaulay, then* $M/\underline{x}M$ *is Cohen–Macaulay (over* R *or* $R/(\underline{x})$*). The converse holds if* R *is local.*

Proposition 2.3.13 [309, Theorem 30] *If* $M \neq (0)$ *is a Cohen–Macaulay* R*-module over a local ring* R *and* $\mathfrak{p} \in \mathrm{Ass}(M)$*, then* $\dim(R/\mathfrak{p}) = \mathrm{depth}(M)$.

Definition 2.3.14 Let $(R, \mathfrak{m})$ be a local ring of dimension d. A *system of parameters* (s.o.p) of R is a set $\theta_1, \ldots, \theta_d$ generating an $\mathfrak{m}$-primary ideal.

Theorem 2.3.15 *Let* $(R, \mathfrak{m}, k)$ *be a local ring. If*

$$\delta(R) := \min\{\mu(I) = \dim_k(I/\mathfrak{m}I) \,|\, I \textit{ is an ideal of } R \textit{ with } \mathrm{rad}\,(I) = \mathfrak{m}\},$$

then $\dim(R) = \delta(R)$.

Proof. This result follows readily from Theorem 2.3.1. □

Theorem 2.3.16 (Krull principal ideal theorem) *Let I be an ideal of a ring R generated by a sequence $h_1, \ldots, h_r$. Then*

(a) $\mathrm{ht}\,(\mathfrak{p}) \leq r$ *for any minimal prime* $\mathfrak{p}$ *of* I.

(b) *If $h_1, \ldots, h_r$ is a regular sequence, then* $\mathrm{ht}\,(\mathfrak{p}) = r$ *for any minimal prime* $\mathfrak{p}$ *of* I.

Proof. (a): Since $\sqrt{IR_\mathfrak{p}} = \mathfrak{p}R_\mathfrak{p}$ from Theorem 2.3.15 one has the inequality $\mathrm{ht}\,(\mathfrak{p}) = \dim(R_\mathfrak{p}) \leq r$.

(b): Set $J = (h_1, \ldots, h_{i-1})$ and $L = (h_1, \ldots, h_i)$. Assume $\mathrm{ht}\,(P) = i - 1$ for any minimal prime P over J. Since h_i is regular on R/J and L/J is a principal ideal one has $\mathrm{ht}\,(L/J) = 1$, thus there is a prime ideal $\mathfrak{p}_0$ minimal over J such that $J \subset \mathfrak{p}_0 \subset \mathfrak{p}$ and consequently $\mathrm{ht}\,(\mathfrak{p}) \geq i$. Using (a) one gets $\mathrm{ht}\,(\mathfrak{p}) = i$. □

Corollary 2.3.17 *If $(R, \mathfrak{m}, k)$ is a local ring, then* $\dim(R) \leq \dim_k(\mathfrak{m}/\mathfrak{m}^2)$ *and R has finite Krull dimension.*

Proof. Let $x_1, \ldots, x_q$ be a set of elements in $\mathfrak{m}$ whose images in $\mathfrak{m}/\mathfrak{m}^2$ form a basis of this vector space. Then $\mathfrak{m} = (x_1, \ldots, x_q) + \mathfrak{m}^2$. Hence by Nakayama's lemma we get $\mathfrak{m} = (x_1, \ldots, x_q)$. Hence $\dim R \leq q$ by Theorem 2.3.15. □

Remark If R is a Noetherian ring, then $\dim R_\mathfrak{p} < \infty$ for all $\mathfrak{p} \in \mathrm{Spec}(R)$. There are examples of Noetherian rings of infinite Krull dimension [9].

Definition 2.3.18 A local ring $(R, \mathfrak{m}, k)$ is called *regular* if

$$\dim(R) = \dim_k(\mathfrak{m}/\mathfrak{m}^2).$$

A ring R is *regular* if $R_\mathfrak{p}$ is a regular local ring for every $\mathfrak{p} \in \mathrm{Spec}(R)$.

Cohen–Macaulay rings A local ring $(R, \mathfrak{m})$ is called *Cohen–Macaulay* if R is Cohen–Macaulay as an R-module. If R is non local and $R_\mathfrak{p}$ is a C–M local ring for all $\mathfrak{p} \in \mathrm{Spec}(R)$, then we say that R is a *Cohen–Macaulay ring*. An ideal I of R is *Cohen–Macaulay* if R/I is a Cohen–Macaulay R-module.

If R is a Cohen–Macaulay ring and S is a multiplicatively closed subset of R, then $S^{-1}(R)$ is a Cohen–Macaulay ring (see [65, Theorem 2.1.3]).

Proposition 2.3.19 *Let M be a module of dimension d over a local ring $(R, \mathfrak{m})$ and let $\underline{\theta} = \theta_1, \ldots, \theta_d$ be a system of parameters of M. Then M is Cohen–Macaulay if and only if $\underline{\theta}$ is an M-regular sequence.*

Proof. $\Rightarrow$) Let $\mathfrak{p}$ be an associated prime of M. By Proposition 2.3.13 one has $\dim(R/\mathfrak{p}) = d$. We claim that θ_1 is not in $\mathfrak{p}$. If $\theta_1 \in \mathfrak{p}$, then by Nakayama's lemma one has $(M/\theta_1 M)_\mathfrak{p} \neq (0)$. Hence $\mathfrak{p}$ is in the support of $M/\theta_1 M$, a contradiction because by Corollary 2.3.3 one has the equality $\dim(M/\theta_1 M) = d - 1$. Therefore $\theta_1 \notin \mathfrak{p}$ and θ_1 is regular on M. Thus the proof follows by induction, because according to Lemma 2.3.10 one has that $M/\theta_1 M$ is a C–M module of dimension $d - 1$ and the images of $\theta_2, \ldots, \theta_d$ in R/θ_1 form a system of parameters of $M/(\theta_1)M$.

$\Leftarrow$) By Corollary 2.3.3 one has $\dim(M/\underline{\theta}M) = 0$ and $M/(\underline{\theta})M$ is C–M. Therefore M is Cohen–Macaulay by Lemma 2.3.10. $\square$

Lemma 2.3.20 *Let $(R, \mathfrak{m})$ be a local ring and let $(f_1, \ldots, f_r)$ be an ideal of height equal to r. Then there are $f_{r+1}, \ldots, f_d$ in $\mathfrak{m}$ such that $f_1, \ldots, f_d$ is a system of parameters of R.*

Proof. Set $d = \dim(R)$ and $I = (f_1, \ldots, f_r)$. One may assume $r < d$, otherwise there is nothing to prove. Let $\mathfrak{p}_1, \ldots, \mathfrak{p}_s$ be the minimal primes of I. Note that $\mathrm{ht}\,(\mathfrak{p}_i) = r$ for all i by Theorem 2.3.16. Hence if we pick f_{r+1} in $\mathfrak{m} \setminus \cup_{i=1}^{s} \mathfrak{p}_i$, one has the equality $\mathrm{ht}\,(I, f_{r+1}) = r + 1$, and the result follows by induction. $\square$

Definition 2.3.21 Let R be a ring and let I be an ideal of R. If I is generated by a regular sequence we say that I is a *complete intersection*.

Definition 2.3.22 An ideal I of a ring R is called a *set-theoretic complete intersection* if there are $f_1, \ldots, f_r$ in I such that $\mathrm{rad}\,(I) = \mathrm{rad}\,(f_1, \ldots, f_r)$, where $r = \mathrm{ht}\,(I)$.

Definition 2.3.23 An ideal I of a ring R is *height unmixed* or *unmixed* if $\mathrm{ht}\,(I) = \mathrm{ht}\,(\mathfrak{p})$ for all $\mathfrak{p}$ in $\mathrm{Ass}_R(R/I)$.

Proposition 2.3.24 *Let $(R, \mathfrak{m})$ be a Cohen–Macaulay local ring and let I be an ideal of R. If I is a complete intersection, then R/I is Cohen–Macaulay and I is unmixed.*

Proof. Set $d = \dim(R)$ and $r = \mathrm{ht}\,(I)$. By Lemma 2.3.10 R/I is Cohen–Macaulay and $\dim(R/I) = d - r$. Let $\mathfrak{p}$ be an associated prime of R/I, then using Proposition 2.3.13 yields

$$\dim(R) - \mathrm{ht}\,(\mathfrak{p}) \geq \dim(R/\mathfrak{p}) = \mathrm{depth}(R/I) = d - r.$$

As a consequence $\mathrm{ht}\,(\mathfrak{p}) \leq r$ and thus $\mathrm{ht}\,(\mathfrak{p}) = r$. Hence I is unmixed. $\square$

Theorem 2.3.25 *Let R be a Cohen–Macaulay ring and I a proper ideal of R of height r. If I is generated by r elements $f_1, \ldots, f_r$, then I is unmixed.*

Proof. It is enough to prove that I has no embedded primes, because by Krull's theorem all minimal primes of I have height r. Let $\mathfrak{p}$ and $\mathfrak{q}$ be two associated primes of R/I and assume $\mathfrak{p} \subset \mathfrak{q}$, then $IR_{\mathfrak{q}} \subset \mathfrak{p}R_{\mathfrak{q}} \subset \mathfrak{q}R_{\mathfrak{q}}$.

Since $IR_{\mathfrak{q}}$ has height r and is generated by r elements, one obtains (using Lemma 2.3.20) that $f_1/1, \ldots, f_r/1$ is part of a system of parameters of $R_{\mathfrak{q}}$. Therefore by Proposition 2.3.19 the sequence $f_1/1, \ldots, f_r/1$ is a regular sequence and thus Proposition 2.3.24 proves that $IR_{\mathfrak{q}}$ is unmixed. Noticing that $\mathfrak{p}R_{\mathfrak{q}}$ and $\mathfrak{q}R_{\mathfrak{q}}$ are both associated primes of $IR_{\mathfrak{q}}$, one derives $\mathfrak{p}R_{\mathfrak{q}} = \mathfrak{q}R_{\mathfrak{q}}$ and hence $\mathfrak{p} = \mathfrak{q}$. □

Remark 2.3.26 The case $r = 0$ in the statement of Theorem 2.3.25 means that (0) is unmixed.

Theorem 2.3.27 (Unmixedness theorem [310, Theorem 17.6]) *A ring R is Cohen–Macaulay if and only if every proper ideal I of R of height r generated by r elements is unmixed.*

Proposition 2.3.28 *If I is an ideal of height r in a Cohen–Macaulay ring R, then there is a regular sequence $f_1, \ldots, f_r$ in I.*

Proof. By induction assume that $f_1, \ldots, f_s$ is a regular sequence in I such that $s < r$. Note that $(f_1, \ldots, f_s)$ is unmixed by Theorem 2.3.25. If $I \subset \mathcal{Z}(R/(f_1, \ldots, f_s))$, then $\mathrm{ht}\,(I) = s$, which is a contradiction. Therefore there is an element f_{s+1} in I which is regular modulo $(f_1, \ldots, f_s)$. □

Let A be a (Noetherian) ring one says that A is a *catenary* ring if for every pair $\mathfrak{p} \subset \mathfrak{q}$ of prime ideals $\mathrm{ht}\,(\mathfrak{q}/\mathfrak{p})$ is equal to the length of any maximal chain of prime ideals between $\mathfrak{p}$ and $\mathfrak{q}$. If A is a domain, then A is catenary if and only if $\mathrm{ht}\,(\mathfrak{q}/\mathfrak{p}) = \mathrm{ht}\,(\mathfrak{q}) - \mathrm{ht}\,(\mathfrak{p})$ for every pair of prime ideals $\mathfrak{p} \subset \mathfrak{q}$.

Theorem 2.3.29 [65, Theorem 2.1.12] *If R is a Cohen–Macaulay ring, then R is catenary.*

Proposition 2.3.30 *Let A and B be two rings and let $C = A \times B$. If C is a Cohen–Macaulay ring, then A is Cohen–Macaulay.*

Proof. From the ring homomorphism $A \times B \to A$, $(a, b) \mapsto a$, we get $A \times B/\{0\} \times B \simeq A$. It follows that P is a prime ideal of C containing the ideal $\{0\} \times B$ if and only if $P = \mathfrak{p} \times B$ for some prime ideal $\mathfrak{p}$ of A.

Let $I \subsetneq A$ be an ideal and let $I' = I \times B$. We are going to establish a correspondence between the associated primes of I and I'. If $\mathfrak{p}$ is an associated prime ideal of I, there is $z \in A$ such that $\mathfrak{p} = (I \colon z)$. Notice that $\mathfrak{p} \times B = (I' \colon (z, 1))$, thus $\mathfrak{p} \times B$ is an associated prime of I'. Conversely if P is an associated prime of I', then there is a prime ideal $\mathfrak{p}$ containing I and a pair (z_1, z_2) in C such that $P = \mathfrak{p} \times B = (I' \colon (z_1, z_2))$. Thus $\mathfrak{p}$ is an associated prime of I because $\mathfrak{p} = (I \colon z_1)$.

Assume that $I = (x_1, \ldots, x_r)$ is an ideal of A of height r generated by r elements. By Theorem 2.3.27 it suffices to prove that I is unmixed.

Case (I): Assume $r = 0$, that is $I = (0)$. Any associated prime ideal of I' consists of zero divisors only. Since C is C–M, the ideal I' has no embedded primes. Hence I has no embedded primes by the correspondence between associated primes of I and I' established above.

Case (II): Assume $r \geq 1$ and set $s = \mathrm{ht}(I')$. The ideal I' is generated by r elements. Indeed if $(x, y) \in I'$, then

$$(x, y) = (a_1, y)(x_1, 1) + (a_2, 0)(x_2, 0) + \cdots + (a_r, 0)(x_r, 0) \quad (a_i \in A).$$

This means that $I' = ((x_1, 1), (x_2, 0), \ldots, (x_r, 0))$ and by Krull theorem we obtain $s \leq r$. Pick a minimal prime ideal P of I' such that $s = \mathrm{ht}(P)$. There is a prime ideal $\mathfrak{p}$ such that $P = \mathfrak{p} \times B$. It is seen that $\mathfrak{p}$ is a minimal prime of I and by Krull theorem $\mathrm{ht}(\mathfrak{p}) = r$. Thus there is a strictly descending chain of prime ideals

$$\mathfrak{p}_r \subsetneq \cdots \subsetneq \mathfrak{p}_0 = \mathfrak{p} \implies \mathfrak{p}_r \times B \subsetneq \cdots \subsetneq \mathfrak{p}_0 \times B = \mathfrak{p} \times B.$$

Therefore $s = \mathrm{ht}(P) \geq r$, and we obtain that $\mathrm{ht}(I') = r$. Thus since C is Cohen–Macaulay, the ideal I' has no embedded primes and consequently I has no embedded primes either; by the correspondence between associated primes of I and I'. □

Gorenstein rings Let $M \neq (0)$ be a module over a local ring $(R, \mathfrak{m})$ and let $k = R/\mathfrak{m}$ be the residue field of R. The *socle* of M is defined as

$$\mathrm{Soc}(M) = (0\colon {}_M\mathfrak{m}) = \{z \in M \mid \mathfrak{m}z = (0)\},$$

and the *type* of M is defined as $\mathrm{type}(M) = \dim_k \mathrm{Soc}(M/\underline{x}M)$, where $\underline{x}$ is a maximal M-sequence in $\mathfrak{m}$. Observe that the type of M is well-defined because by Proposition 2.3.7 one has:

$$\mathrm{Ext}^r_R(k, M) \simeq \mathrm{Hom}_R(k, M/\underline{x}M) \simeq \mathrm{Soc}(M/\underline{x}M),$$

where $r = \mathrm{depth}(M)$. The ring R is said to be *Gorenstein* if R is a Cohen–Macaulay ring of type 1. An ideal $I \subset R$ is called *Gorenstein* if R/I is a Gorenstein ring. For a thorough study of Gorenstein rings see [18, 65].

Exercises

2.3.31 Let $\phi\colon A \to B$ be an epimorphism of rings. If $\dim(A) = \dim(B) < \infty$ and A is a domain, prove that ϕ is injective.

2.3.32 Let R be a catenary ring. If $S \subset R$ is a multiplicatively closed set and I is an ideal of R, prove that R/I and $S^{-1}R$ are both catenary rings.

2.3.33 Let I be an ideal of a ring R and let $x \in R \setminus I$. If I is unmixed and $I = \mathrm{rad}\,(I)$, then $(I\colon x)$ is unmixed and $\mathrm{ht}(I) = \mathrm{ht}(I\colon x)$.

2.3.34 Let $M = \mathbb{Z}^r \times \mathbb{Z}_{p_1^{\alpha_1}} \times \cdots \times \mathbb{Z}_{p_m^{\alpha_m}}$ be a finitely generated $\mathbb{Z}$-module, where $p_1, \dots, p_m$ are prime numbers. Prove the following:

(a) M has finite length iff $r = 0$.

(b) $\ell(M) = \alpha_1 + \cdots + \alpha_m$, if $\ell(M) < \infty$.

(c) $\text{Ass}(M) \neq \text{Supp}(M)$ if $r \neq 0$.

(d) $\mathcal{Z}(M) = \cup_{i=1}^m (p_i)$.

(e) $\text{Ass}(M) = \{(p_1), \dots, (p_m), (0)\}$ if $r \neq 0$.

(f) $\text{Ass}(M) = \{(p_1), \dots, (p_m)\}$ if $r = 0$.

2.3.35 (Cayley–Hamilton) Let $\varphi\colon M \to M$ be an endomorphism of an R-module M. If I is an ideal of R such that $\varphi(M) \subset IM$, then φ satisfies an equation of the form $\varphi^n + a_{n-1}\varphi^{n-1} + \cdots + a_1\varphi + a_0 = 0 \quad (a_i \in I)$.

2.3.36 If R is a principal ideal domain, prove that R is a regular ring.

2.3.37 Let M be a module over a local ring $(R, \mathfrak{m})$. If $\underline{\theta} = \theta_1, \dots, \theta_r$ is an M-regular sequence in $\mathfrak{m}$, prove that $\underline{\theta}$ can be extended to a system of parameters of M.

2.3.38 Let M be an R-module and let I be an ideal of R. If there is $x \in I$ which is regular on M, show that $\text{Hom}_R(R/I, M) = (0)$.

2.3.39 Let M be an R-module and let I be an ideal of R. If $J \subset I$, then:

$$\text{Hom}_R(R/I, M/JM) \simeq (JM\colon {}_M I)/JM.$$

2.3.40 If A and B are two Cohen–Macaulay rings, prove that $A \times B$ is a Cohen–Macaulay ring.

2.4 Normal rings

Let A and B be two rings. One says that B is an *A-algebra* if there is a homomorphism of rings

$$\varphi\colon A \longrightarrow B.$$

Note that B has an A-module structure (compatible with its ring structure) given by $a \cdot b = \varphi(a)b$ for all $a \in A$ and $b \in B$.

Thus if $A \subset B$ is a *ring extension*, that is, A is a subring of B, then B has in a natural way an A-algebra structure induced by the inclusion map. If $A = K$ is a field and B is a K-algebra, then φ is injective; in this case one may always assume that $K \subset B$ is a ring extension.

Given $F = \{f_1, \dots, f_q\}$ a finite subset of B, we denote the *subring of B generated* by F and $\varphi(A)$ by $A[F] = A[f_1, \dots, f_q]$. Sometimes $A[F]$ is called the *A-subalgebra* or *A-subring* of B *spanned* by F. If $B = A[F]$, one says that B is an A-algebra of *finite type* or a *finitely generated* A-algebra; on the other hand one says that φ is *finite* or that B is a *finite A-algebra* if B is a finitely generated A-module.

A *homomorphism* $\phi\colon B \to C$ of *A-algebras* is a map ϕ which is both a homomorphism of rings and a homomorphism of A-modules. Note that $\phi(a \cdot 1) = a \cdot 1$ for all $a \in A$.

Definition 2.4.1 Let A be a subring of B. An element $b \in B$ is *integral* over A, if there is a monic polynomial $0 \neq f(x) \in A[x]$ such that $f(b) = 0$. The set $\overline{A}$ of all elements $b \in B$ which are integral over A is called the *integral closure* of A in B.

Let A be a subring of B, one says that $A \subset B$ is an *integral extension of rings* or that B is *integral over* A, if b is integral over A for all $b \in B$. A homomorphism of rings $\varphi\colon A \to B$ is called *integral* if B is integral over $\varphi(A)$; in this case one also says that B is *integral over* A.

Proposition 2.4.2 *Let A be a subring of B. If B is a finitely generated A-module, then B is integral over A.*

Proof. There are $\alpha_1, \ldots, \alpha_n \in B$ such that $B = A\alpha_1 + \cdots + A\alpha_n$. Let $\beta \in B$. Then, one can write $\beta\alpha_i = \sum_{j=1}^{n} m_{ij}\alpha_j$, where $m_{ij} \in A$. Set $M = (m_{ij})$, $N = M - \beta I$ and $\alpha = (\alpha_1, \ldots, \alpha_n)$. Here I denotes the identity matrix. Since $N\alpha^\top = 0$, one can use the formula $N\mathrm{adj}(N) = \det(N)I$ to conclude $\alpha_i \det(N) = 0$ for all i. Hence $\det(N) = 0$. To complete the proof note that

$$f(x) = (-1)^n \det(M - xI)$$

is a monic polynomial in $A[x]$ and $f(\beta) = (-1)^n \det(N) = 0$. □

Corollary 2.4.3 *If A is a subring of B, then the integral closure $\overline{A}$ of A in B is a subring of B.*

Proof. Let $\alpha, \beta \in \overline{A}$. If α and β satisfy monic polynomials with coefficients in A of degree m and n, respectively, then $A[\alpha, \beta]$ is a finitely generated A-module with basis $\{\alpha^i\beta^j \,|\, 0 \leq i \leq m-1 \text{ and } 0 \leq j \leq n-1\}$. Hence, by Proposition 2.4.2, $A[\alpha, \beta]$ is integral over A. In particular $\alpha \pm \beta$ and $\alpha\beta$ are integral over A. □

Corollary 2.4.4 *If B is an A-algebra of finite type, then B is integral over A if and only if B is finite over A.*

Proof. $\Rightarrow$) It follows by the arguments given in the proof of Corollary 2.4.3.

$\Leftarrow$) It follows from Proposition 2.4.2. □

Lemma 2.4.5 *Let $A \subset B$ be a ring extension. If B is a domain and $b \in B$ is integral over A, then $A[b]$ is a field if and only if A is a field.*

Proof. $\Rightarrow$) Let $a \in A \setminus \{0\}$ and $c = a^{-1}$ its inverse in $A[b]$. Since $A[b]$ is integral over A, there is an equation

$$c^n + a_{n-1}c^{n-1} + \cdots + a_1c + a_0 = 0 \quad (a_i \in A),$$

multiplying by a^n it follows rapidly that $c \in A$.

$\Leftarrow$) Let $A[x]$ be a polynomial ring and $\psi\colon A[x] \to A[b]$ the epimorphism given by $\psi(f(x)) = f(b)$. As $\ker(\psi)$ is a non-zero prime ideal and $A[x]$ is a principal ideal domain, $\ker(\psi)$ is a maximal ideal and $A[b]$ is a field. $\square$

Proposition 2.4.6 *Let $A \subset B$ be an integral extension of rings. If B is a domain, then A is a field if and only if B is a field.*

Proof. It follows from Lemma 2.4.5. $\square$

Corollary 2.4.7 *Let $A \subset B$ be an integral extension of rings. If P is a prime ideal of B and $\mathfrak{p} = P \cap A$, then P is a maximal ideal of B if and only if $\mathfrak{p}$ is a maximal ideal of A.*

Proof. The result is a direct consequence of Proposition 2.4.6, because $A/\mathfrak{p} \subset B/P$ is an integral extension of rings. $\square$

Corollary 2.4.8 *Let $A \subset B$ be an integral extension of rings. If $P \subset Q$ are two prime ideals of B such that $P \cap A = Q \cap A$, then $P = Q$.*

Proof. Set $\mathfrak{p} = P \cap A$. As $PB_\mathfrak{p} \cap A_\mathfrak{p} = \mathfrak{p}A_\mathfrak{p}$ and $B_\mathfrak{p}$ is integral over $A_\mathfrak{p}$, using Corollary 2.4.7 we get that $PB_\mathfrak{p}$ is maximal and hence $P = Q$. $\square$

Lemma 2.4.9 *If $A \subset B$ is an integral extension of rings and $\mathfrak{p}$ is a maximal ideal of A, then $\mathfrak{p}B \neq B$ and $\mathfrak{p} = P \cap A$ for any maximal ideal P of B containing $\mathfrak{p}B$.*

Proof. If $\mathfrak{p}B = B$, one can write $1 = a_1b_1 + \cdots + a_qb_q$ with $a_i \in \mathfrak{p}$ and $b_i \in B$. Set $S = A[b_1, \ldots, b_q]$, the subring of B generated by $b_1, \ldots, b_q$. Since S is a finitely generated A-module and $\mathfrak{p}S = S$, then by Lemma 2.1.33 there is $x \equiv 1(\operatorname{mod} \mathfrak{p})$ such that $xS = (0)$. As $1 \in B$, one derives $x = 0$ and $1 \in \mathfrak{p}$, which is impossible. Hence $\mathfrak{p}B \neq B$. Note that this part of the proof holds if $\mathfrak{p}$ is a proper ideal of A. If $\mathfrak{p}B \subset P$ is a maximal ideal of B, then $\mathfrak{p} \subset P \cap A$ and thus $\mathfrak{p} = A \cap P$ by the maximality of $\mathfrak{p}$. $\square$

Proposition 2.4.10 (Lying over) *If $A \subset B$ is an integral extension and $\mathfrak{p}$ is a prime ideal of A, then there is a prime ideal P of B such that $\mathfrak{p} = P \cap A$.*

Proof. Since $A_\mathfrak{p} \subset B_\mathfrak{p}$ is an integral extension, then by Lemma 2.4.9 $\mathfrak{p}B_\mathfrak{p} \neq B_\mathfrak{p}$. Note that $P \cap A = \mathfrak{p}$ for any prime ideal P of B such that $PB_\mathfrak{p}$ is prime and contains $\mathfrak{p}B_\mathfrak{p}$ (see Exercise 2.4.22). $\square$

Proposition 2.4.11 *If $A \subset B$ is an integral extension of rings and A is local, then B is semilocal.*

Proof. Let $\mathfrak{m}$ be the maximal ideal of A. By Corollary 2.4.7 every prime ideal of $B/\mathfrak{m}B$ is maximal. Hence, $B/\mathfrak{m}B$ is an Artinian semilocal ring by Proposition 2.1.43. Since the maximal ideals of B are in one to one correspondence with the maximal ideals of $B/\mathfrak{m}B$, the result follows. $\square$

Theorem 2.4.12 (Going-up) *Let $A \subset B$ be an integral ring extension. If $\mathfrak{p} \subset \mathfrak{q}$ are two prime ideals of A and $\mathfrak{p} = P \cap A$ for some P in* Spec(B)*, then there is $Q \in$* Spec(B) *such that $\mathfrak{q} = Q \cap A$ and $P \subset Q$.*

Proof. Since $A/\mathfrak{p} \subset B/P$ is an integral extension, by Proposition 2.4.10 one has $\mathfrak{q}/\mathfrak{p} = (Q/P) \cap (A/\mathfrak{p})$ for some prime ideal Q of B containing P. It follows readily that $\mathfrak{q} = A \cap Q$. $\square$

Proposition 2.4.13 *Let $A \subset B$ be an integral extension of rings. If B is integral over A, then* $\dim(A) = \dim(B)$.

Proof. The formula follows using Theorem 2.4.12, Corollary 2.4.8, and Proposition 2.4.10. $\square$

Let A be an integral domain and let K_A be its field of fractions. The *integral closure* or *normalization* of A, denoted $\overline{A}$, is the set of all $f \in K_A$ satisfying an equation of the form

$$f^n + a_{n-1}f^{n-1} + \cdots + a_1 f + a_0 = 0 \quad (a_i \in A \text{ and } n \geq 1).$$

By Corollary 2.4.3 the integral closure $\overline{A}$ of A is a subring of K_A. If $A = \overline{A}$ we say that A is *integrally closed* or *normal*. If A is not a domain we say that A is *normal* if $A_\mathfrak{p}$ is a normal domain for every prime ideal $\mathfrak{p}$ of A. Any normal ring is a direct product of finitely many normal domains [128].

Proposition 2.4.14 *Let $A \subset B$ be an integral extension of rings. If B is a domain and A is a normal domain, then*

(a) (*going-down*) *if $\mathfrak{p} \subset \mathfrak{q}$ are two prime ideals of A and $\mathfrak{q} = Q \cap A$ for some Q in* Spec(B)*, then there is $P \in$* Spec(B) *such that $\mathfrak{p} = P \cap A$ and $P \subset Q$.*

(b) $\mathrm{ht}\,(I) = \mathrm{ht}\,(I \cap A)$ *for any ideal I of B.*

(c) $\mathrm{ht}\,(I) = \mathrm{ht}\,(IB)$ *for any ideal I of A.*

Proof. For parts (a) and (b) see [309, Theorems 5 and 20]. Notice that, by part (a) and Proposition 2.4.10, $A \subset B$ satisfies the going down and the lying over conditions. Hence, by [259, Proposition B.2.4], $\mathrm{ht}\,(I) = \mathrm{ht}\,(IB)$ for any ideal I of A. This proves (c). $\Box$

Theorem 2.4.15 (Serre's criterion [310, Theorem 23.8]) *A ring A is normal if and only if*

(S_2) $\mathrm{depth}(A_\mathfrak{p}) \geq \inf\{2, \mathrm{ht}\,(\mathfrak{p})\}$ *for all* $\mathfrak{p} \in \mathrm{Spec(A)}$, *and*

(R_1) $A_\mathfrak{p}$ *is regular for all* $\mathfrak{p} \in \mathrm{Spec(A)}$ *with* $\mathrm{ht}\,(\mathfrak{p}) \leq 1$.

Theorem 2.4.16 (Auslander–Buchsbaum [310, Theorem 20.3]) *If $(A, \mathfrak{m})$ is a regular local ring, then A is a unique factorization domain.*

Definition 2.4.17 Let R be a ring and let $G \leq \mathrm{Aut}(R)$ be a group of ring automorphisms of R. The *invariant ring* of G is:

$$R^G = \{r \in R |\, \phi(r) = r, \ \ \forall\, \phi \in G\}.$$

Proposition 2.4.18 *If R is a normal domain and G is a group of ring automorphisms of R, then R^G is a normal domain.*

Proof. Let f be an element in the field of fractions of R^G which is integral over R^G. Since R is normal $f \in R$ and $r_2 f = r_1$ for some $r_1, r_2 \in R^G$. Thus

$$r_2\phi(f) = \phi(r_2)\phi(f) = \phi(r_1) = r_1$$

for all $\phi \in R^G$. Hence $f \in R^G$, as required. $\Box$

Example 2.4.19 If $R = K[x_1, \ldots, x_n]$ is a polynomial ring over a field K and G is a subgroup of $\mathrm{GL}_n(K)$, then for every A in G there is a ring automorphism $\phi_A \colon R \to R$, $f \mapsto f(Ax)$, where $x = (x_1, \ldots, x_n)^t$. Thus G is a group of ring automorphisms of R and $R^G = \{f \in R |\, f(Ax) = f, \ \forall\, A \in G\}$.

Flat and faithfully flat algebras Let $\phi \colon A \to B$ be a homomorphism of rings and consider an exact sequence of A-modules

$$\cdots \longrightarrow N_{i-1} \stackrel{f_{i-1}}{\longrightarrow} N_i \stackrel{f_i}{\longrightarrow} N_{i+1} \stackrel{f_{i+1}}{\longrightarrow} \cdots \qquad (N_\star)$$

If the sequence

$$\cdots \longrightarrow B \otimes_A N_{i-1} \stackrel{1 \otimes f_{i-1}}{\longrightarrow} B \otimes_A N_i \stackrel{1 \otimes f_i}{\longrightarrow} B \otimes_A N_{i+1} \stackrel{1 \otimes f_{i+1}}{\longrightarrow} \cdots \quad (B \otimes_A N_\star)$$

is exact for any $N_\star$, B is called a *flat A-algebra* and ϕ is said to be *flat*. The homomorphism ϕ is said to be *faithfully flat* if any sequence $N_\star$ of A-modules is exact if and only if $B \otimes_A N_\star$ is exact. In a similar way one can define flat and faithfully flat A-modules.

A basic property is that any free A-module (not necessarily finitely generated) is faithfully flat. In particular a polynomial ring $A[\mathbf{x}]$ in several variables with coefficients in A is a faithfully flat A-algebra.

Theorem 2.4.20 [309, Sections 3.H, 5.D and 9.C] *Let $\phi\colon A \to B$ be a flat homomorphism of rings. The following hold:*

(a) $(I_1 \cap I_2)B = I_1B \cap I_2B$, *for any two ideals* I_1, I_2 *of* A.

(b) *The going down theorem holds for* ϕ.

(c) *If* $\mathfrak{q}$ *is a* $\mathfrak{p}$*-primary ideal of* A *such that* $\mathfrak{p}B$ *is prime, then* $\mathfrak{q}B$ *is a* $\mathfrak{p}B$*-primary ideal.*

Theorem 2.4.21 [309, Sections 4.C and 13.B] *Let $\phi\colon A \to B$ be a faithfully flat homomorphism of rings. The following hold:*

(a) *The map* $\phi^*\colon \mathrm{Spec}(B) \to \mathrm{Spec}(A)$, $\phi^*(P) = \phi^{-1}(P)$ *is surjective.*

(b) $IB \cap A = I$ *and* $\mathrm{ht}\,(I) = \mathrm{ht}\,(IB)$ *for any ideal* I *of* A.

Exercises

2.4.22 Let $A \subset B$ be a ring extension and $\mathfrak{p}$ a prime ideal of A. If $S = A \setminus \mathfrak{p}$ and $B_\mathfrak{p} = S^{-1}B$, prove that the map $P \longmapsto PB_\mathfrak{p}$ gives a bijection between the set of prime ideals of P of B such that $P \cap A = \mathfrak{p}$ and the set of prime ideals of $B_\mathfrak{p}$ containing $\mathfrak{p}B_\mathfrak{p}$.

2.4.23 Let K be a field and $\varphi\colon A \to B$ an isomorphism of K-algebras, note $\varphi(r) = r$ for all $r \in K$. If $A = \oplus_{i\geq 0}A_i$ is a graded K-algebra, then B is also a graded K-algebra graded by $B = \oplus_{i\geq 0}\varphi(A_i)$.

2.4.24 Let $\varphi\colon A \to B$ be an isomorphism between two integral domains. If K_A (resp. K_B) is the field of fractions of A (resp. B), then φ extends to an isomorphism $\overline{\varphi}$ from K_A to K_B such that $\overline{\varphi}(\overline{A}) = \overline{B}$.

2.4.25 If A is a unique factorization domain, show that A is normal.

2.4.26 Show an example of an integral extension of rings $A \subset B$, such that A is a field and B is not an integral domain (cf. Lemma 2.4.5).

2.4.27 Let $B = A[\mathbf{x}]$ be a polynomial ring over a ring A and let I be an ideal of A. Prove $(A/I)[\mathbf{x}] \simeq B/IB$, where the left-hand side is a polynomial ring with coefficients in A/I.

2.4.28 If $B = A[\mathbf{x}]$ is a polynomial ring over a ring A and $\mathfrak{q}$ is a $\mathfrak{p}$-primary ideal of A, then $\mathfrak{q}B$ is a $\mathfrak{p}B$-primary ideal of B.

2.4.29 Let $B = A[\mathbf{x}]$ be a polynomial ring over a ring A and let I be an ideal of A. If $I = \cap_{i=1}^r \mathfrak{q}_i$ is a primary decomposition of I, then $IB = \cap_{i=1}^r \mathfrak{q}_iB$ is a primary decomposition of IB.

2.4.30 If $A \subset B$ is an integral extension of rings and A is semilocal, then B is semilocal.

2.4.31 Let A be a domain and let x be an indeterminate over A. Then A is normal if and only if $A[x]$ is normal.

2.5 Valuation rings

In this part we introduce valuation rings. Let A be an integral domain and let K be its field of fractions.

Definition 2.5.1 A is a *valuation ring* of K if for each $0 \neq x \in K$ either $x \in A$ or $x^{-1} \in A$.

Proposition 2.5.2 *Let A be a valuation ring of K. The following hold.*

(a) *If I, J are two ideals of A, then $I \subset J$ or $J \subset I$.*

(b) *A is a local ring.*

(c) *If A' is an intermediate ring $A \subset A' \subset K$, then A' is a valuation ring.*

(d) *A is integrally closed in K.*

Proof. (a) Assume $I \not\subset J$ and pick $x \in I \setminus J$. Let $y \in J$. One has $y^{-1}x \notin A$, because if $y^{-1}x \in A$, then $x = (xy^{-1})y \in J$, a contradiction. Thus $yx^{-1} \in A$ and consequently $y = x(x^{-1}y) \in I$. This proves $J \subset I$, as required. Part (b) follows from (a).

(c) Let $x \in K$ and assume that $x \notin A'$. Then $x \notin A$ and consequently $x^{-1} \in A$. Thus $x^{-1} \in A'$. Hence, A' is a valuation ring of K.

(d) Let $x \in K$ be an integral element over A. There are $a_0, a_1, \ldots, a_{n-1}$ in A such that $x^n + a_{n-1}x^{n-1} + \cdots + a_1x + a_0 = 0$. Assume $x \notin A$. Then $x^{-1} \in A$. Multiplying this equation by x^{1-n} we get

$$x = -(a_{n-1} + a_{n-2}x^{-1} + \cdots + a_1x^{2-n} + a_0x^{1-n}).$$

Thus $x \in A$, a contradiction. Therefore $x \in A$. $\square$

Definition 2.5.3 An abelian group H with a total order $\prec$ is called an *ordered group* if $x \prec y$ and $u \prec v$ implies $x + u \prec y + v$.

Example 2.5.4 Let $G_1, \ldots, G_n$ be abelian ordered groups and let G be its cartesian product. To make G and ordered group for $a = (a_i)$, $b = (b_i)$ in G and $a \neq b$ we define $a \prec b$ if the first non-zero entry of $b - a$ is positive. This defines a total order on G called the lexicographic order.

Let H be an abelian ordered group. We order $H \cup \{\infty\}$ by adding an element ∞ bigger than any element in H and declaring $\infty + x = \infty$ for $x \in H$ and $\infty + \infty = \infty$.

Definition 2.5.5 Let H be an ordered group. A map $\nu\colon K \to H \cup \{\infty\}$ is called a *valuation* of K if ν satisfies:

(1) $\nu(xy) = \nu(x) + \nu(y)$,

(2) $\nu(x+y) \geq \min\{\nu(x), \nu(y)\}$,

(3) $\nu(x) = \infty$ if and only if $x = 0$.

The *valuation ring* of a valuation ν, denoted by A_ν, is the set of all $a \in K$ with $\nu(a) \geq 0$. The *value group* of ν is the subgroup $\nu(K \setminus \{0\})$.

If ν is a valuation of a field K, then A_ν is a valuation ring of K and the ideal $\mathfrak{m}_\nu = \{x \in K | \nu(x) > 0\}$ is its maximal ideal.

Example 2.5.6 Let $p \geq 2$ be a prime number and let $\nu\colon \mathbb{Q} \to \mathbb{Z} \cup \{\infty\}$ be the map defined by

$$\nu\left(\frac{a}{b}\right) = \begin{cases} n & \text{if } \; 0 \neq \dfrac{a}{b} = p^n \dfrac{a_1}{b_1} \text{ and } (a_1 b_1, p) = 1, \\ \infty & \text{if } \; \dfrac{a}{b} = 0. \end{cases}$$

It is easy to see that ν is a valuation of $\mathbb{Q}$ whose valuation ring is $\mathbb{Z}_{(p)}$, the localization of $\mathbb{Z}$ at the prime ideal $\mathfrak{p} = (p)$.

Definition 2.5.7 Let G be an ordered group. A subgroup H of G is called *isolated* if for each $\alpha \in H$ such that $0 \prec \beta \prec \alpha$, one has $\beta \in H$. If H is an isolated subgroup of G and $H \neq G$ we say that H is proper. The *rank* of G is the number of isolated proper subgroups of G.

Definition 2.5.8 Let ν be a valuation of a field K. The *rank of the valuation ring* A_ν is the rank of the value group of ν.

Example 2.5.9 The group $G = \mathbb{Z}^n$ with the lexicographical order has rank n because the proper non-zero isolated subgroups of G have the form:

$$H_i = \{(0, \ldots, 0, x_i, \ldots, x_n) | x_i, \ldots, x_n \in \mathbb{Z}\} \qquad (i = 2, \ldots, n).$$

Definition 2.5.10 A valuation ring whose value group is isomorphic to $\mathbb{Z}$ is called a *discrete valuation ring* (DVR for short).

Theorem 2.5.11 [310, Chapter 4] *A is a discrete valuation ring if and only if any of the following equivalent conditions hold.*

(a) *A is a Noetherian valuation ring.*

(b) *A is a normal Noetherian local ring of dimension* 1.

(c) *A is a Noetherian local ring,* $\dim(A) \geq 1$ *and the maximal ideal of A is principal.*

Theorem 2.5.12 [310, p. 82] *Let A be a Noetherian domain. Then, the following three conditions are equivalent*:

(a) *A is normal.*

(b) *$A_\mathfrak{p}$ is a discrete valuation ring for each prime ideal $\mathfrak{p}$ of A of height* 1.

(c) *All associated primes of non-zero principal ideals of A have height* 1.

Exercises

2.5.13 Let A be a valuation ring of K. Prove that there is an ordered group G and a valuation $\nu\colon K \to G \cup \{\infty\}$ such that the valuation ring of ν is A and G is its value group.

2.5.14 Let $K(x)$ be the field of rational functions in one variable over a field K. For $f(x) \in K[x]$, write $f(x) = x^m g(x)$, where $m \in \mathbb{N}$ and $g(x)$ is a polynomial with $g(0) \neq 0$ and define $\nu(f) = m$. Prove that extend ν to a discrete valuation of $K(x)$ by setting $\nu(f/g) = \nu(f) - \nu(g)$.

2.5.15 Let G be an ordered abelian group and let $K[\{x_i | i \in G\}]$ be a polynomial ring over a field K with quotient field Q. Define $\nu : Q \to G \cup \{\infty\}$ as follows

$$\nu(f) = \begin{cases} k_1 g_1 + \cdots + k_n g_n & \text{if } f = a x_{g_1}^{k_1} \cdots x_{g_n}^{k_n} \text{ is a monomial,} \\ \min\limits_i \{\nu(f_i)\} & \text{if } f = f_1 + \cdots + f_n \text{ is a sum of monomials,} \\ \infty & \text{if } f = 0. \end{cases}$$

Prove that ν is a valuation of Q whose value group is G.

2.5.16 If $(A, \mathfrak{m})$ is a discrete valuation ring, prove that $\mathfrak{m} = (x)$ for some $x \in A$ and that every non-zero ideal of A is a power of $\mathfrak{m}$.

2.6 Krull rings

In this section we study the notion of a Krull ring and introduce its divisor class group. The main references for this section are [161, 309, 310].

Let A be an integral domain with quotient field K and let $Z = X^{(1)}(A)$ be the set of prime ideals of A of height 1.

Definition 2.6.1 An integral domain A is called a *Krull ring* if the following three conditions are satisfied:

(a) $A_\mathfrak{p}$ is a discrete valuation ring for $\mathfrak{p} \in Z$.

(b) $A = \cap_{\mathfrak{p} \in Z} A_\mathfrak{p}$.

(c) Each $0 \neq x \in A$ belongs to at most a finite number of ideals in Z.

Theorem 2.6.2 (Mori–Nagata [161], [309]) *Let A be an integral domain with quotient field K and let L be a finite algebraic field extension of K. If A is Noetherian, then the integral closure A' of A in L is a Krull ring (not necessarily Noetherian). If $\dim(A) = 2$, then A' is Noetherian.*

Corollary 2.6.3 *If A is Noetherian and normal, then A is a Krull ring.*

Theorem 2.6.4 (Krull–Akizuki [309, p. 297]) *Let A be a domain with quotient field K and let L be a finite field extension of K. If A is Noetherian and* $\dim(A) = 1$, *then every subring between A and L is Noetherian.*

Theorem 2.6.5 [309, Theorem 38] *If A is a Noetherian normal domain, then $A = \cap_{\mathfrak{p}\in Z} A_{\mathfrak{p}}$.*

Theorem 2.6.6 [310, Theorem 12.4(iii)] *If A is a Krull ring, then $A[x]$ and $A[\![x]\!]$ are Krull rings.*

Proposition 2.6.7 *Let A be a valuation ring. If* $\dim(A) \geq 2$, *then the formal power series ring $A[\![x]\!]$ is not normal.*

Proof. Let $(0) \subsetneq \mathfrak{p}_1 \subsetneq \mathfrak{p}_2$ be a chain of prime ideals of A. Pick $0 \neq b \in \mathfrak{p}_1$ and $a \in \mathfrak{p}_2 \setminus \mathfrak{p}_1$. Then $ba^{-n} \in A$ for $n > 0$. Indeed if $ba^{-n} \notin A$, then $a^n \in bA \subset \mathfrak{p}_1$ and $a \in \mathfrak{p}_1$, a contradiction. It is not hard to see that there is $f = \sum_{i=1}^{\infty} u_i x^i$ that satisfies the equation $z^2 + az + x = 0$ and such that $u_1 = a^{-1}$ and $u_i \in a^{-2i+1}A$ for $i \geq 2$. Thus $bf \in A[\![x]\!]$ and consequently f is in the field of fractions of $A[\![x]\!]$. Since $a^{-1} \notin A$ we get that $f \notin A[\![x]\!]$. Thus $A[\![x]\!]$ is not normal. □

Corollary 2.6.8 *Let A be a valuation ring. If* $\dim(A) \geq 2$, *then A is not a Krull ring.*

Proof. If A is a Krull ring, then $A[\![x]\!]$ is a Krull ring by Theorem 2.6.6. In particular $A[\![x]\!]$ is normal, a contradiction to Proposition 2.6.7. □

Divisor class groups of Krull rings Let $\mathfrak{p} \in Z$ such that $A_{\mathfrak{p}}$ is a discrete valuation ring. The maximal ideal $\mathfrak{p}A_{\mathfrak{p}}$ of the local ring $A_{\mathfrak{p}}$ is principal and $\cap_{i=1}^{\infty}\mathfrak{p}^i = (0)$ because $A_{\mathfrak{p}}$ is a Noetherian domain. Let $0 \neq x \in A$. Since any non-zero ideal of $A_{\mathfrak{p}}$ is a power of $\mathfrak{p}A_{\mathfrak{p}}$, there is a unique nonnegative integer $n_{\mathfrak{p}}$ satisfying $(x)A_{\mathfrak{p}} = \mathfrak{p}^{n_{\mathfrak{p}}}A_{\mathfrak{p}}$ and $(x)A_{\mathfrak{p}} \neq \mathfrak{p}^{n_{\mathfrak{p}}+1}A_{\mathfrak{p}}$. This defines a valuation $v_{\mathfrak{p}}\colon K^* \longrightarrow \mathbb{Z}$ of K given by

$$v_{\mathfrak{p}}(x) = \begin{cases} n_{\mathfrak{p}} & \text{if } 0 \neq x \in A, \\ v_{\mathfrak{p}}(a) - v_{\mathfrak{p}}(b) & \text{if } x = a/b;\ a, b \in A \setminus \{0\}. \end{cases}$$

The valuation $v_{\mathfrak{p}}$ is called the *associated valuation* of $\mathfrak{p}$.

Lemma 2.6.9 *Let A be a Krull ring and let $0 \neq a \in A$. Then*

$$aA = \bigcap_{\mathfrak{p}\in Z} \mathfrak{p}^{(n_{\mathfrak{p}})},$$

where $n_{\mathfrak{p}} = v_{\mathfrak{p}}(a)$ and $\mathfrak{p}^{(n)} := \mathfrak{p}^n A_{\mathfrak{p}} \cap A$ is the symbolic power of order n.

Proof. From the equality $A = \cap_{\mathfrak{p}\in Z} A_\mathfrak{p}$ we get $aA = \cap_{\mathfrak{p}\in Z}(aA_\mathfrak{p} \cap A)$. As $aA_\mathfrak{p} = \mathfrak{p}^{n_\mathfrak{p}} A_\mathfrak{p}$, we conclude that $\mathfrak{p}^{(n_\mathfrak{p})} = aA_\mathfrak{p} \cap A$ as desired. □

Definition 2.6.10 Let A be a Krull ring and let $D(A)$ be the free abelian group on Z. That is $D(A)$ consists of the formal sums $\sum_{\mathfrak{p}\in Z} n_\mathfrak{p} \cdot \mathfrak{p}$ (with $n_\mathfrak{p} \in \mathbb{Z}$ and $n_\mathfrak{p} = 0$ except for a finite number) with the sum defined by

$$\left(\sum n_\mathfrak{p} \cdot \mathfrak{p}\right) + \left(\sum n'_\mathfrak{p} \cdot \mathfrak{p}\right) = \sum (n_\mathfrak{p} + n'_\mathfrak{p}) \cdot \mathfrak{p}.$$

For each $0 \neq x$ in K, we set

$$\operatorname{div}(x) := \sum_{\mathfrak{p}\in Z} v_\mathfrak{p}(x) \cdot \mathfrak{p}.$$

The function div gives a group homomorphism from K^* to $D(A)$, because $\operatorname{div}(xy) = \operatorname{div}(x) + \operatorname{div}(y)$. The *divisor class group* of A is defined as

$$\operatorname{Cl}(A) := D(A)/F(A),$$

where $F(A)$ is the image of K^* under the homomorphism div.

Lemma 2.6.11 *If A is a unique factorization domain and $0 \neq p \in A$ is a prime element, then* $\operatorname{ht}(p) = 1$.

Proof. Assume there is a prime ideal $\mathfrak{q}$ such that $(0) \subsetneq \mathfrak{q} \subsetneq (p)$. Take $0 \neq x \in \mathfrak{q}$. There is a prime divisor q of x that belongs to $\mathfrak{q}$. Thus $q \in (p)$ and $p = uq$ for some unit $u \in A$, a contradiction. □

The next result is a measure, in terms of $\operatorname{Cl}(A)$, of how far is A from being a unique factorization domain.

Theorem 2.6.12 *An integral domain A is a unique factorization domain if and only if A is a Krull ring and* $\operatorname{Cl}(A) = (0)$.

Proof. $\Rightarrow$) As A is a unique factorization domain the notions of prime element and irreducible element coincide. Hence every prime ideal of A of height 1 is principal. To prove this assertion take $\mathfrak{p} \in Z$. Let $0 \neq x \in \mathfrak{p}$. Notice that some prime factor x_1 of x belongs to $\mathfrak{p}$. Thus $(x_1) = \mathfrak{p}$.

Next we show that A is a Krull ring. Let $\mathfrak{p} \in Z$ and let $x \in A$ such that $\mathfrak{p} = (x)$. Clearly $A_\mathfrak{p}$ is normal and $\dim(A_\mathfrak{p}) = 1$. Let $0 \subsetneq J \subsetneq A_\mathfrak{p}$ be an ideal of $A_\mathfrak{p}$. Set n equal to the smallest integer $n > 0$ such that $x^n \in J$, it follows readily that $J = (x^n)$. This proves that $A_\mathfrak{p}$ is a discrete valuation ring. Now we show the equality

$$A = \bigcap_{\mathfrak{p}\in Z} A_\mathfrak{p}.$$

Let $0 \neq x = a/b$ with $a, b \in A$. We may assume that a and b are relatively prime. Assume that x belongs to the right-hand side of the equality above. To prove that $x \in A$ it suffices to show that b is a unit of A. If b is not a unit pick a prime divisor p of b and set $\mathfrak{p} = (p)$, then $b \in \mathfrak{p}$ and by Lemma 2.6.11 one has $\mathfrak{p} \in Z$. Since $a/b \in A_\mathfrak{p}$ we get $a \in \mathfrak{p}$, a contradiction because a and b are relatively prime. Thus b is a unit and $x \in A$. Since A is a UFD every $0 \neq x \in A$ is contained in at most a finite number of principal prime ideals. Altogether we have shown that A is a Krull ring. It remains to show that $\mathrm{Cl}(A) = (0)$. This follows immediately by observing that if $\mathfrak{p}_0 = (x_0) \in Z$, then as an element of $D(A)$ one has $\mathfrak{p}_0 = \mathrm{div}(x_0)$.

$\Leftarrow$) Let $\mathfrak{p}_0 \in Z$. We claim that $\mathfrak{p}_0$ is a principal ideal. As $\mathrm{Cl}(A) = (0)$, there exists $0 \neq x_0 \in K$ such that

$$\mathrm{div}(x_0) = \sum_{\mathfrak{p} \in Z} v_\mathfrak{p}(x_0) \cdot \mathfrak{p} = \mathfrak{p}_0.$$

Hence $v_{\mathfrak{p}_0}(x_0) = 1$ and $v_\mathfrak{p}(x_0) = 0$ for $\mathfrak{p} \neq \mathfrak{p}_0$. Writing $x_0 = a/b$, with $a, b \in A^*$ we get $v_{\mathfrak{p}_0}(a) = v_{\mathfrak{p}_0}(b) + 1$ and $v_\mathfrak{p}(a) = v_\mathfrak{p}(b)$ for $\mathfrak{p} \neq \mathfrak{p}_0$. By Lemma 2.6.9, we have the equalities

$$aA = \bigcap_{\mathfrak{p} \in Z} \mathfrak{p}^{(v_\mathfrak{p}(a))} \quad \text{and} \quad bA = \bigcap_{\mathfrak{p} \in Z} \mathfrak{p}^{(v_\mathfrak{p}(b))}.$$

Consequently, using $\mathfrak{p}^{(v_{\mathfrak{p}_0}(b)+1)} \subset \mathfrak{p}^{(v_{\mathfrak{p}_0}(b))}$, one has $aA \subset bA$. This proves $x_0 \in A$. Again by Lemma 2.6.9 we get $x_0 A = \mathfrak{p}_0^{(1)} = \mathfrak{p}_0$. This proves that $\mathfrak{p}_0$ is principal.

Let $0 \neq x \in A$ be a non-unit of A. Notice there exists at least one ideal in Z that contains x. Otherwise consider x^{-1}, the inverse of x in K, and observe that $x^{-1} \in \cap_{\mathfrak{p} \in Z} A_\mathfrak{p} = A$, a contradiction because x is not a unit of A. Let $\mathfrak{p}_1 = (x_1), \ldots, \mathfrak{p}_s = (x_s)$ be the ideals in Z that contain x. There are elements $\lambda_1, \ldots, \lambda_s$ in A such that $x = \lambda_1 x_1 = \cdots = \lambda_s x_s$. Notice that $\lambda_1 x_1 \in (x_2)$. Since (x_2) is a prime ideal of height 1 we get $\lambda_1 \in (x_2)$. Using this argument we readily obtain that $x = \lambda x_1^{a_1} \cdots x_s^{a_s}$ with $\lambda \in A$, in addition since $\cap_{i=1}^\infty \mathfrak{p}_k^i = (0)$ for all k we may assume that $x_k^{a_k+1}$ does not divide x for all k. It follows that λ is a unit of A, because if λ is not a unit it must belong to one of the ideals $\mathfrak{p}_1, \ldots, \mathfrak{p}_s$ which contradicts the choice of $a_1, \ldots, a_s$. Clearly the elements $x_1, \ldots, x_s$ are irreducible and we have proved that x is a product of irreducible elements. Finally assume that x has two representations

$$x = y_1 \cdots y_r = z_1 \cdots z_m$$

with y_i and z_j irreducible elements for all i, j. Notice that every irreducible element $z \in A$ is prime. Indeed since z is contained in at least one prime ideal $\mathfrak{p} = (w)$ of height 1, we get $z = \lambda w$. As z is irreducible it follows that

λ is a unit and z is prime. Therefore it follows readily that $r = m$ and there is a permutation σ such that y_i is associated of $z_{\sigma(i)}$ for all i. $\square$

Corollary 2.6.13 *If K is a field and $K[X]$ is a ring of polynomials in any number of variables (finite or infinite), then $K[X]$ is a Krull ring.*

Definition 2.6.14 An A-submodule I of K is called a *fractional ideal* of A if $I \neq (0)$ and $xI \subset A$ for some $0 \neq x \in A$ (or equivalently $0 \neq x \in K$). If I is a fractional ideal of A define $I^{-1} = (A\colon {}_K I) = \{x \in K \mid xI \subset A\}$. We say that a fractional ideal I is *divisorial* if $(I^{-1})^{-1} = I$.

Remark 2.6.15 An A-submodule $I \neq (0)$ of K is a fractional ideal of A if and only if there is an ideal J of A and $z \in K$ such that $I = zJ$. This follows readily using that if $xI \subset A$ for some $x \in K^*$, then $J = xI$ is an ideal of A and $I = x^{-1}J$.

Lemma 2.6.16 *If I is a fractional ideal of A, then I^{-1} is a fractional ideal of A and $II^{-1} \subset A$.*

Definition 2.6.17 A fractional ideal I of A is *invertible* if $II^{-1} = A$.

A fractional ideal of the form xA with $x \in K^*$ is called *principal*. Every principal fractional ideal is invertible and divisorial.

Lemma 2.6.18 *Let I be an invertible fractional ideal of A. If $IJ = A$ for some fractional ideal J of A, then*

$$J = I^{-1} = (A :_K I).$$

In particular if I, J are invertible fractional ideals, then the usual product IJ is invertible and $(IJ)^{-1} = I^{-1}J^{-1}$.

Definition 2.6.19 The *Picard group* of A is defined as

$$\mathrm{Pic}(A) = \left\{ \begin{array}{l} \text{invertible fractional} \\ \text{ideals} \end{array} \right\} \Big/ \left\{ \begin{array}{l} \text{principal fractional} \\ \text{ideals} \end{array} \right\}.$$

Theorem 2.6.20 [310, Theorem 11.3, p. 80] *Let I be a fractional ideal of A. The following conditions are equivalent:*

(a) *I is invertible.*

(b) *I is a projective A-module.*

(c) *I is finitely generated as an A-module and for every maximal ideal $\mathfrak{m}$ of A, the fractional ideal $I_\mathfrak{m} = IA_\mathfrak{m}$ of $A_\mathfrak{m}$ is principal.*

Theorem 2.6.21 *Let $\mathfrak{p} \neq (0)$ be a prime ideal of A. If A is Noetherian and $\mathfrak{p}$ is invertible, then $\mathrm{ht}(\mathfrak{p}) = 1$ and $A_\mathfrak{p}$ is a discrete valuation ring.*

Lemma 2.6.22 *Let I be a fractional ideal of A. Then I^{-1} is a fractional ideal and*

$$(I^{-1})^{-1} = (A\colon {}_K(A\colon {}_K I)) = \bigcap_{I \subset yA} yA.$$

In particular I is divisorial if and only if I is an intersection of principal fractional ideals.

Lemma 2.6.23 *Let $I = x_0 J$ be a fractional ideal of A such that $0 \neq x_0 \in K$ and J is a fractional ideal of A. Then J is divisorial if and only if I is divisorial.*

Proof. Assume J is divisorial. By Lemma 2.6.22 we have

$$J = (J^{-1})^{-1} = (A\colon {}_K(A\colon {}_K J)) = \bigcap_{J \subset xA} xA.$$

Multiplying by x_0 and making the change of variable $z = x_0 x$ we obtain

$$I = x_0 J = \bigcap_{J \subset z x_0^{-1} A} zA = \bigcap_{x_0 J \subset zA} zA = (I^{-1})^{-1}.$$

Therefore I is divisorial. The converse follows similarly. □

Proposition 2.6.24 *Let I, J be two fractional ideals of A. Then every homomorphism of A-modules $f\colon J \to I$ is of the form $f(x) = \alpha x$ for some $\alpha \in (I\colon {}_K J)$. In particular there is an isomorphism of A-modules given by*

$$\mathrm{Hom}_A(J, I) \longrightarrow (I\colon {}_K J); \qquad f \longmapsto \alpha.$$

Proof. Let $0 \neq x, y \in J$. There are $0 \neq a, b \in A$ such that $ax = by$. Therefore the element

$$\frac{f(x)}{x} = \frac{af(x)}{ax} = \frac{bf(y)}{by} = \frac{f(y)}{y} = \alpha$$

is independent of x. Hence $f(x) = \alpha x$ for $x \in J$ and $\alpha \in (I\colon {}_K J)$. □

Corollary 2.6.25 *Let I, J be two fractional ideals of A. If $I \simeq J$ as A-modules, then I is divisorial if and only if J is divisorial.*

Proof. By Proposition 2.6.24 we have $I = \alpha J$ for some $\alpha \in K$. Thus the result follows from Lemma 2.6.23. □

Definition 2.6.26 *Let M be an A-module and let $M^* = \mathrm{Hom}_A(M, A)$ be its dual. We say that M is reflexive if the canonical homomorphism*

$$\varphi\colon M \to M^{**} = \mathrm{Hom}_A(M^*, A); \quad \varphi(m)(f) = f(m); \quad \forall\, m \in M,\ f \in M^*$$

is an isomorphism.

Lemma 2.6.27 *A direct summand of a reflexive module is reflexive.*

Proof. It is left as an exercise. □

Proposition 2.6.28 *If F is a free A-module of finite rank, then F is a reflexive module.*

Corollary 2.6.29 *Every finitely generated projective module is reflexive.*

Proposition 2.6.30 *Let I be a fractional ideal of A. If I is reflexive (as an A-module), then I is divisorial.*

Proof. As I is always contained in $(I^{-1})^{-1}$, it is enough to show that $(I^{-1})^{-1}$ is contained in I. Let $x \in (I^{-1})^{-1}$. The idea is to show that x defines an element ψ of I^{**} and to use that φ is onto to obtain that x belongs to I. Let $f \in I^*$. By Proposition 2.6.24 there exists a unique element α_f in $(A\colon {}_K I)$ such that $f(z) = z\alpha_f$ for all $z \in I$. Consider the map

$$\psi\colon I^* \longrightarrow A; \qquad f \longmapsto x\alpha_f.$$

First let us show that ψ is well-defined, that is we must show $x\alpha_f \in A$. We may assume $\alpha_f \neq 0$. Since $\alpha_f I \subset A$ we obtain $I \subset \alpha_f^{-1}A$. Hence, using

$$(I^{-1})^{-1} = \bigcap_{I \subset yA} yA,$$

we conclude that $x \in \alpha_f^{-1}A$, this proves that $x\alpha_f \in A$. It is not hard to see that ψ is a homomorphism. Hence since φ is onto there exists $x_0 \in I$ such that $\varphi(x_0) = \psi$. Therefore

$$\alpha_f x = \psi(f) = \varphi(x_0)(f) = f(x_0) = x_0\alpha_f \ \forall\, f \in I^* \ \Rightarrow \ x = x_0 \in I. \qquad \square$$

Proposition 2.6.31 [161, Proposition 5.2, Corollary 5.5.] *Let I be a fractional ideal of a Krull ring A. Then I is reflexive if and only if I is divisorial.*

Proposition 2.6.32 *Let A be a Krull ring and let I be an ideal of A. Then I is divisorial if and only if $I = \mathfrak{q}_1 \cap \cdots \cap \mathfrak{q}_s$, where $\mathfrak{q}_i$ is a primary ideal of A of height 1 for all i.*

Example 2.6.33 Let $R = K[x_1, x_2]$ be a polynomial ring over a field K and let $A = K[F]$, the subring of R generated by $F = \{x_1^2, x_1x_2, x_2^2\}$. Let $\mathfrak{p} = x_1R \cap A = (x_1^2, x_1x_2)A$. We claim that $\mathfrak{p}$ is a divisorial prime ideal such that $\mathfrak{p}^2$ is not divisorial. The ring A is a normal Noetherian domain, thus it is a Krull ring. Since $A \subset R$ is an integral extension, by the going down theorem (see Proposition 2.4.14) we have that $\mathfrak{p}$ is a prime ideal of A of height 1. Hence $\mathfrak{p}$ is divisorial.

Now we show that $\mathfrak{p}^2$ is not divisorial. By Proposition 2.6.32 it suffices to observe that from the equality

$$(x_1^2, x_1x_2, x_2^2)A = (\mathfrak{p}^2 \colon {}_A x_1x_2)$$

it follows that the maximal ideal $\mathfrak{m} = (x_1^2, x_1x_2, x_2^2)A$ of A is an embedded associated prime of $\mathfrak{p}^2$.

Definition 2.6.34 The set of divisorial ideals of A is denoted by $\mathrm{Div}(A)$ and the set of principal fractional ideals by Prin(A).

Proposition 2.6.35 *The set* $\mathrm{Div}(A)$ *with the binary operation*

$$\mathrm{Div}(A) \times \mathrm{Div}(A) \longrightarrow \mathrm{Div}(A); \quad (I, J) \longmapsto I * J := ((IJ)^{-1})^{-1}$$

is a monoid whose identity is A.

Definition 2.6.36 Let A be an integral domain and K its field of fractions. An element $x \in K$ is *almost integral* over A if there is $0 \neq a \in A$ such that $ax^n \in A$ for all $n \geq 0$.

Proposition 2.6.37 *Let A be an integral domain and let K be its field of fractions. Then an element $x \in K$ is integral over A if and only if x is almost integral over A.*

Proof. Let $x = c/d$, where c, d are in A and $d \neq 0$. If x is integral over A, then there is an equation

$$x^m + b_1x^{m-1} + \cdots + b_{m-1}x + b_m = 0, \; b_i \in A.$$

Setting $a = d^m$ one has $ax^n \in A$ for all $n > 0$.

Conversely assume there is $0 \neq a \in A$ such that $ax^n \in A$ for all $n > 0$. Since $a^{-1}A$ is a Noetherian A-module and $A[x] \subset a^{-1}A$ one has that $A[x]$ is a finitely generated A-module. As A is a subring of $A[x]$ by Proposition 2.4.2 one derives that x is integral over A. □

Definition 2.6.38 The set of elements $x \in K$ that are almost integral over A is denoted by $\widetilde{A}$ and this set is called the *almost integral closure* of A. If $A = \widetilde{A}$ it is said that A is *completely normal.*

Lemma 2.6.39 [161] *An element $x \in K$ is almost integral over A if and only if there exists a fractional ideal I of A such that $xI \subset I$.*

Corollary 2.6.40 *The set $\widetilde{A}$ is a subring of K and*

$$A \subset \overline{A} \subset \widetilde{A} \subset K.$$

The equality $\overline{A} = \widetilde{A}$ holds if A is Noetherian. In particular if A is completely normal, then A is normal.

Proof. Taking into account Proposition 2.6.37 it suffices to show that $\widetilde{A}$ is a subring of K. Let $x, y \in \widetilde{A}$. By Lemma 2.6.39 there are fractional ideals I, J of A such that $x \in (I :_K I)$, $y \in (J :_K J)$. Observe that

$$xy \in (IJ :_K IJ); \quad x + y \in (I \cap J :_K I \cap J).$$

Since IJ, $I \cap J$ are fractional ideals of A, by Lemma 2.6.39 we get $xy \in \overline{A}$ and $x + y \in \overline{A}$. $\square$

Corollary 2.6.41 *If A is a Krull ring, then A is completely normal.*

Proof. A is an intersection of discrete valuation rings and these rings are normal and Noetherian, hence completely normal. $\square$

Theorem 2.6.42 [289, Theorem 5.19, p. 113] *If A is a valuation ring with $A \neq K$, then A is completely normal if and only if the value group of A has rank* 1.

Remark 2.6.43 By Exercise 2.5.15 there exists a field Q and a valuation ν of Q such that its value group is $\mathbb{Z}\times\mathbb{Z}$. Thus A_ν is not completely normal because the rank of $\mathbb{Z}\times\mathbb{Z}$ is 2. Any valuation ring of Krull dimension greater than or equal to two is normal but not completely normal. See [436].

Theorem 2.6.44 *The set* $\mathrm{Div}(A)$ *with the binary operation*

$$\mathrm{Div}(A) \times \mathrm{Div}(A) \longrightarrow \mathrm{Div}(A); \quad (I, J) \longmapsto I * J := ((IJ)^{-1})^{-1}$$

is a group if and only if A is completely normal.

Proposition 2.6.45 *Let A be a Krull ring and let $I \subsetneq A$ be an ideal of A. Then I is divisorial if and only if I can be written as*

$$I = \mathfrak{p}_1 * \cdots * \mathfrak{p}_r = (A\colon {}_K(A\colon {}_K\mathfrak{p}_1 \cdots \mathfrak{p}_r))$$

for some $\mathfrak{p}_i \in Z$ and this "factorization" is unique.

Corollary 2.6.46 *If A is a Krull ring, then there is an isomorphism from the group $(D(A), +, 0)$ to the group $(\mathrm{Div}(A), *, A)$ given by:*

$$\rho\colon D(A) \longrightarrow \mathrm{Div}(A); \quad a_1\mathfrak{p}_1 + \cdots + a_r\mathfrak{p}_r \longmapsto \mathfrak{p}_1^{a_1} * \cdots * \mathfrak{p}_r^{a_r},$$

where $\mathfrak{p}_1^{a_1}$ means $\mathfrak{p}_1^{a_1} = \mathfrak{p}_1 * \cdots * \mathfrak{p}_1$ (a_1 times) if $a_1 > 0$.

Proof. Clearly ρ is a homomorphism. To prove that ρ is onto take a divisorial fractional ideal I of A. There is $0 \neq a \in A$ such that $aI \subset A$. Notice that aI is a divisorial ideal of A because

$$I = \bigcap_{I\subset xA} xA \implies aI = \bigcap_{aI\subset zA} zA.$$

Thus $(aA)I = aI = ((aI)^{-1})^{-1} = (aA) * I$. By Proposition 2.6.45 we have

$$(aA) * I = \mathfrak{p}_1^{a_1} * \cdots * \mathfrak{p}_r^{a_r} \quad \text{and} \quad aA = \mathfrak{p}_1^{b_1} * \cdots * \mathfrak{p}_r^{b_r} \ \ (a_i, b_i \geq 0).$$

Therefore solving for I we get that I is in the image of ρ. To finish the proof notice that from the uniqueness of the factorization in Proposition 2.6.45 we readily get that ρ is injective. $\square$

Another way of representing the divisor class group of a Krull ring is:

Theorem 2.6.47 *If A is a Krull ring, then* $\mathrm{Cl}(A) \simeq \mathrm{Div}(A)/\mathrm{Prin}(A)$, *where* $\mathrm{Prin}(A)$ *is the group of principal fractional ideals of A.*

Proposition 2.6.48 *If I is an invertible fractional ideal of A, then I is divisorial.*

Proof. Let $x \in (I^{-1})^{-1}$. Then $xI^{-1} \subset A$. Now since I is invertible one has $II^{-1} = A$, then there are $a_i \in I$ and $b_i \in I^{-1}$ such that $1 = \sum_i a_i b_i$. Hence $x = \sum_i a_i(xb_i)$. Notice that $xb_i \in A$ because $b_i \in I^{-1}$, hence $x \in I$. Thus we have shown $(I^{-1})^{-1} \subset I$. To prove the reverse inclusion take $x \in I$. Notice that $xy \in A$ for all $y \in I^{-1}$, thus $x \in (I^{-1})^{-1}$. $\square$

Lemma 2.6.49 *If I, J are two invertible fractional ideals, then*

$$IJ = I * J = ((IJ)^{-1})^{-1}.$$

Proof. By Lemma 2.6.18 one has $(IJ)^{-1} = I^{-1}J^{-1}$. Therefore taking the inverse on both sides gives the required equality. $\square$

As a consequence of the last two results we obtain:

Corollary 2.6.50 *The Picard group of A is a subgroup of the semigroup* $\mathrm{Div}(A)/\mathrm{Prin}(A)$ *of divisor classes.*

Example 2.6.51 Let $R = K[x_1, x_2, \ldots, x_n, \ldots]$ be a polynomial ring in an infinite number of variables over a field K and let $A = K[F]$ be the subring of R generated by $F = \{x_i x_j \mid 1 \leq i \leq j\}$. Consider the ideal

$$\mathfrak{p} = x_1 R \cap A = (x_1^2, x_1x_2, x_1x_3, \ldots)A.$$

We claim that $\mathfrak{p}$ is a divisorial prime ideal of A and that $\mathfrak{p}$ is not invertible. The ring A is a Krull ring [161]. Since $A \subset R$ is an integral extension by the going down theorem one has that $\mathfrak{p}$ is a height 1 prime ideal of A. Therefore $\mathfrak{p}$ is divisorial. Since $\mathfrak{p}$ is not finitely generated it cannot be invertible.

Definition 2.6.52 A is a *Dedekind ring* if every fractional ideal of A is invertible.

Notice that by Remark 2.6.15 A is a Dedekind ring if and only if every ideal $(0) \neq I \subset A$ of A is invertible. Principal ideal domains are Dedekind rings. If S is a multiplicatively closed subset of A and A is a Dedekind ring, then $S^{-1}A$ is a Dedekind ring. If $\dim(A) = 1$, then A is a Dedekind ring if and only if A is a Krull ring.

Dedekind rings occur in algebraic number theory because the ring of algebraic integers in a finite extension of $\mathbb{Q}$ is a Dedekind ring. By the Krull–Akizuki theorem if $A = \mathbb{Z}$, $K = \mathbb{Q}$ and L is a finite extension of $\mathbb{Q}$, then the integral closure of A in L is a Dedekind ring. More generally if A is a Dedekind ring and L is a finite extension of K, then the integral closure $\overline{A}$ of A in L is a Dedekind ring. If $A = \mathbb{Z}$, then $\operatorname{Pic}(\overline{A})$ is finite. See [329].

Theorem 2.6.53 [310, Theorem 11.6] *An integral domain A is a Dedekind ring if and only if any of the following equivalent conditions hold:*

(a) *A is a field or A is a normal Noetherian domain of dimension* 1.

(b) *Each non-zero ideal I of A has a unique factorization $I = \mathfrak{p}_1\mathfrak{p}_2\cdots\mathfrak{p}_s$ into prime ideals.*

Corollary 2.6.54 *If A is a Dedekind ring, then* $\operatorname{Div}(A)$ *is a free abelian group generated by the non-zero prime ideals of A and* $\operatorname{Pic}(A) = \operatorname{Cl}(A)$.

Proof. It follows from Corollary 2.6.46 and Lemma 2.6.49. $\Box$

Exercises

2.6.55 Let A be a domain with quotient field K. Prove that the set given by $G = \{xA \mid x \in K^*\}$ is an ordered abelian group with the usual multiplication of fractional ideals and where G is ordered by $xA \prec yA$ if and only if $xA \supset yA$.

2.6.56 Let $K = \mathbb{Z}_2$ be the field with two elements and let $F = \oplus_{i\in\mathbb{N}}F_i$, where $F_i = K$ for all i. Prove that $F^* = \operatorname{Hom}_K(F, K)$ is uncountable and that F is not a reflexive K-module.

2.7 Koszul homology

Let a be an element of the ring R and let $K_\star(a)$ be the complex defined as

$$K_i = \begin{cases} R & \text{for } i = 0, 1, \\ 0 & \text{otherwise,} \end{cases}$$

with $d_1\colon K_1(a) \to K_0(a)$ being multiplication by a.

Let I be an ideal of R generated by the sequence $\underline{x} = \{x_1, \ldots, x_n\}$. The ordinary *Koszul complex* associated to $\underline{x}$ is defined as

$$K_\star(x; R) = K_\star(x_1) \otimes \cdots \otimes K_\star(x_n).$$

For an R-module M we shall write $K_\star(\underline{x}; M)$ for $K_\star(\underline{x}; R) \otimes M$. The Koszul complex $K_\star(\underline{x}; R)$ is then the exterior algebra complex associated to $E = R^n$ and the map

$$\theta \colon E \longrightarrow R,$$

defined as $\theta(z_1, \ldots, z_n) = z_1 x_1 + \cdots + z_n x_n$. That is, θ defines a differential $\partial = d\theta$ on the exterior algebra $\bigwedge(E)$ of E given in degree r by

$$\partial(e_1 \wedge \cdots \wedge e_r) = \sum_{i=1}^{r} (-1)^{i-1} \theta(e_i) e_1 \wedge \cdots \wedge \widehat{e}_i \wedge \cdots \wedge e_r.$$

From the definition of the differential of $K_\star(\underline{x}; R)$, we get that if w and w' are homogeneous elements of $\bigwedge(E)$, of degrees p and q, respectively, then

$$\partial(w \wedge w') = (-1)^p w \wedge \partial(w') + \partial(w) \wedge w'.$$

This implies that the cycles $\mathcal{Z}(K_\star)$ form a subalgebra of $\bigwedge(E)$, and that the boundaries $\mathcal{B}(K_\star)$ form a two-sided ideal of $\mathcal{Z}(K_\star)$. As a consequence the homology of the Koszul complex, $H_\star(\underline{x})$, inherits a skew commutative R-algebra structure.

One can also see that $H_\star(\underline{x})$ is annihilated by $I = (\underline{x})$. Indeed, if $e \in E$ and $w \in \mathcal{Z}_r(K_\star)$, we have from the last formula $\partial(e \wedge w) = \theta(e) w$.

The ordinary Koszul complex $K_\star(\underline{x}) = K_\star(\underline{x}; R)$ is simply the complex of free modules

$$K_\star(\underline{x}) : \qquad 0 \to \bigwedge^n R^n \to \bigwedge^{n-1} R^n \to \cdots \to \bigwedge^1 R^n \to \bigwedge^0 R^n \to 0,$$

where $\wedge^k R^n$ is the kth exterior power of R^n; thus $\wedge^k R^n$ is a free R-module of rank $\binom{n}{k}$ with basis $\{e_{i_1} \wedge \cdots \wedge e_{i_k} | 1 \leq i_1 < \cdots < i_k \leq n\}$.

Proposition 2.7.1 [410, Theorem 2.3] *If $\underline{x}$ is a regular sequence in R, then the Koszul complex is acyclic; that is, the complex $K_\star(\underline{x})$ is exact.*

Sliding depth Let $(R, \mathfrak{m})$ be a Cohen–Macaulay local ring and let I be an ideal of R generated by $x_1, \ldots, x_n$, denote by $H_\star(\underline{x})$ the homology of the ordinary Koszul complex built on the sequence $\underline{x} = \{x_1, \ldots, x_n\}$.

Definition 2.7.2 (i) (*SD*) I satisfies *sliding depth* if

$$\operatorname{depth} H_i(\underline{x}) \geq \dim(R) - n + i, \ \forall\, i \geq 0.$$

(ii) (*SCM*) I is *strongly Cohen–Macaulay* if $H_i(\underline{x})$ are C–M, $\forall\, i \geq 0$.

(Depths are computed with respect to maximal ideals. It is usual to set depth(0) equal to ∞.)

Remark 2.7.3 (a) The (SD) condition localizes [234], (b) If I satisfies (SD) with respect to some generating set, then it will satisfy (SD) with respect to any other generating set of I. This follows from the isomorphisms:

$$H_i(\{x_1,\ldots,x_n,0\}) \simeq H_i(\underline{x}) \oplus H_{i-1}(\underline{x}), \text{ and}$$

$$H_i(\{x_1,\ldots,x_n,y\}) \simeq H_i(\{x_1,\ldots,x_n,0\}), \text{ where } y \in (\underline{x}).$$

Linkage Let I and J be two ideals in a Cohen–Macaulay local ring R. The ideals I and J are said to be (algebraically) *linked* if there is an R-sequence $\underline{x} = \{x_1,\ldots,x_n\}$ in $I \cap J$ such that

$$I = ((\underline{x})\colon J) \quad \text{and} \quad J = ((\underline{x})\colon I),$$

if in addition I and J are unmixed ideals of the same height n without common components and such that $I \cap J = (\underline{x})$, then I and J are said to be *geometrically linked*.

When I and J are linked we shall write $I \sim J$. We say that J is in the *linkage class* of I if there are ideals $I_1,\ldots,I_m$ such that

$$I \sim I_1 \sim \cdots \sim I_m \sim J.$$

The ideal J is said to be in the *even linkage class* of I if m is odd.

Let R be a Gorenstein ring and let I be a Cohen–Macaulay ideal of R. If J is linked to I, then Peskine and Szpiro [339] showed that J is Cohen–Macaulay. An interesting result of Huneke [253] proves that (SCM) is preserved under even linkage. His method can be adapted to prove the following result.

Proposition 2.7.4 *Let I and J be two ideals in a Gorenstein local ring R of dimension d, and let $\underline{x} = \{x_1,\ldots,x_n\}$ be a generating set for I. Assume that J is evenly linked to I. If I satisfies the condition*

$$(SD_k) \quad \text{depth } H_i(\underline{x};R) \geq d-n+i, \quad 0 \leq i \leq k,$$

then J satisfies the (SD_k) condition as well.

Exercises

2.7.5 Let $R = K[x_1,\ldots,x_6]$ be a polynomial ring over a field K and let $I = I_2(X)$ be the ideal of 2×2-minors of the symmetric matrix:

$$X = \begin{bmatrix} x_1 & x_2 & x_3 \\ x_2 & x_4 & x_5 \\ x_3 & x_5 & x_6 \end{bmatrix}.$$

Prove that I is linked to $I_2(X')$, where X' is the symmetric matrix

$$X' = \begin{bmatrix} x_1 & x_2 & -x_3 \\ x_2 & x_4 & x_5 \\ -x_3 & x_5 & x_6 \end{bmatrix}.$$

In particular if $\text{char}(K) = 2$ the ideal I is self-linked. For other characteristics this is an open question.

2.8 A vanishing theorem of Grothendieck

Let R be a ring and let I be an R-module. We say that I is *injective* if the functor $\text{Hom}_R(\,\cdot\,, I)$ is exact. Note that this functor is always left exact.

Definition 2.8.1 Let R be a ring and let M be an R-module. A complex

$$\mathcal{I}_\star\colon\ 0 \longrightarrow I^0 \xrightarrow{\partial_0} I^1 \xrightarrow{\partial_1} I^2 \xrightarrow{\partial_2} \cdots$$

of injective R-modules is an *injective resolution* of M if $H_i(\mathcal{I}_\star) = 0$ for $i > 0$ and $H_0(\mathcal{I}_\star) = \ker(\partial_0) \cong M$.

The *injective dimension* of M, denoted $\text{inj dim}\, M$, is the smallest integer n for which there exist an injective resolution $\mathcal{I}_\star$ of M with $I^m = 0$ for $m > n$. If there is no such n, the injective dimension of M is infinite.

For the proofs of the next three results and for additional information on Gorenstein rings and injective resolutions see [18, 65, 255, 363].

Theorem 2.8.2 *Let $(R, \mathfrak{m})$ be a local ring and let M be an R-module of finite injective dimension. Then*

$$\dim M \leq \text{inj dim}\, M = \text{depth}\, R.$$

Definition 2.8.3 A local ring R is *Gorenstein* if $\text{inj dim}\, R < \infty$. A ring R is *Gorenstein* if $R_\mathfrak{m}$ is a Gorenstein ring for all $\mathfrak{m}$ maximal ideal of R.

Definition 2.8.4 Let R be a ring and let $N \subset M$ be R-modules. M is an essential extension of N if for any non-zero R-submodule U of M one has $U \cap N \neq 0$. An essential extension M of N is called proper if $N \neq M$.

Proposition 2.8.5 *Let R be a ring. An R-module N is injective if and only if it has no proper essential extensions.*

Definition 2.8.6 Let R be a ring and M an R-module. An injective module E such that $M \subset E$ is an essential extension is called an *injective hull* of M and is denoted by $E = E(M)$ or $E = E_R(M)$.

Let M be an R-module. Then M admits an injective hull and M has a *minimal injective* resolution $E_\star(M)$:

$$0 \to M \stackrel{\partial_{-1}}{\to} E_0(M) \stackrel{\partial_0}{\to} E_1(M) \to \cdots$$
$$\to E_{i-1}(M) \stackrel{\partial_{i-1}}{\to} E_i(M) \stackrel{\partial_i}{\to} E_{i+1}(M) \to \cdots$$

where $E_0(M) = E(M)$ and $E_{i+1} = E(\text{coker}\partial_{i-1})$. Here ∂_{-1} denotes the embedding $M \to E(M)$, and ∂_i is defined in a natural way. Any two minimal injective resolutions of M are isomorphic. If $\mathcal{I}_\star$ is an injective resolution of M, then $E_\star(M)$ is isomorphic to a direct summand of $\mathcal{I}_\star$.

Definition 2.8.7 Let $(R, \mathfrak{m}, K)$ be a local ring R and let $M \neq 0$ be an R-module with depth r. The *type* of M is the number

$$\text{type}(M) = \dim_K \text{Ext}^r_R(K, M).$$

Theorem 2.8.8 *Let R be a local ring. Then R is a Gorenstein ring if and only if R is a Cohen–Macaulay ring of type* 1.

Local cohomology Here we present a vanishing theorem of Grothendieck (Theorem 2.8.12) which is useful to prove Reisner criterion (see Theorem 6.3.12) for Cohen–Macaulay complexes. Our main references for homological algebra and local cohomology are [65, 202, 245, 363].

Let $(R, \mathfrak{m})$ be a local ring and let M be an R-module. Denote by $\Gamma_\mathfrak{m}(M)$ the submodule of M of all the elements with support in $\{\mathfrak{m}\}$, that is,

$$\Gamma_\mathfrak{m}(M) = \{x \in M \,|\, \mathfrak{m}^k x = 0 \text{ for some } k > 0\}.$$

Let $\underline{x} = \{x_1, \ldots, x_n\}$ be a sequence of elements in R generating an $\mathfrak{m}$-primary ideal. Set $\underline{x}^k = \{x_1^k, \ldots, x_n^k\}$. The family $\underline{x}^k$ gives the $\mathfrak{m}$-adic topology on R, hence

$$\Gamma_\mathfrak{m}(M) = \{z \in M \,|\, (\underline{x})^k z = 0 \text{ for some } k \geq 0\}.$$

Since $\text{Hom}_R(R/I, M) = \{x \in M \,|\, Ix = 0\}$ for any ideal I of R, we obtain a natural isomorphism

$$\Gamma_\mathfrak{m}(M) \cong \varinjlim \text{Hom}_R(R/\mathfrak{m}^k, M) \cong \varinjlim \text{Hom}_R(R/(\underline{x}^k), M). \qquad (2.2)$$

Proposition 2.8.9 $\Gamma_\mathfrak{m}(\cdot)$ *is a left exact additive functor.*

Definition 2.8.10 The *local cohomology functors*, denoted by $H^i_\mathfrak{m}(\cdot)$ are the right derived functors of $\Gamma_\mathfrak{m}(\cdot)$.

Remark Let M and I be R-modules. (a) If $\mathcal{I}_\star$ is an injective resolution of M, then $H^i_{\mathfrak{m}}(M) = H^i(\Gamma_{\mathfrak{m}}(\mathcal{I}_\star))$ for $i \geq 0$, (b) $H^0_{\mathfrak{m}}(M) = \Gamma_{\mathfrak{m}}(M)$ and $H^i_{\mathfrak{m}}(M) = 0$ for $i < 0$. If I is injective, then $H^i_{\mathfrak{m}}(I) = 0$ for $i > 0$.

Proposition 2.8.11 *If $(R, \mathfrak{m})$ is a local ring and M is an R-module, then*

$$H^i_{\mathfrak{m}}(M) \cong \varinjlim \operatorname{Ext}^i_R(R/\mathfrak{m}^k, M) \cong \varinjlim \operatorname{Ext}^i_R(R/(\underline{x}^k), M),$$

for $i \geq 0$, where $\underline{x}$ is a sequence in R generating an $\mathfrak{m}$-primary ideal.

Proof. Recall that if $\mathcal{P}_\star$ is a projective resolution of L and $\mathcal{I}_\star$ is an injective resolution of M, then $\operatorname{Ext}^i_R(L, M)$ can be computed as follows:

$$\operatorname{Ext}^i_R(L, M) \cong H^i(\operatorname{Hom}_R(\mathcal{P}_\star, M)) \cong H^i(\operatorname{Hom}_R(L, \mathcal{I}_\star)),$$

see [245, Proposition 8.1]. Assume $\mathcal{I}_\star$ is an injective resolution of M, then

$$H^i_{\mathfrak{m}}(M) \cong H^i(\Gamma_{\mathfrak{m}}(\mathcal{I}_\star)) \text{ and } \Gamma_{\mathfrak{m}}(\mathcal{I}_\star) \cong \varinjlim \operatorname{Hom}_R(R/\mathfrak{m}^k, \mathcal{I}_\star).$$

Therefore

$$\begin{aligned} H^i_{\mathfrak{m}}(M) &\cong H^i(\varinjlim \operatorname{Hom}_R(R/\mathfrak{m}^k, \mathcal{I}_\star)) \cong \varinjlim H^i(\operatorname{Hom}_R(R/\mathfrak{m}^k, \mathcal{I}_\star)) \\ &\cong \varinjlim \operatorname{Ext}^i_R(R/\mathfrak{m}^k, M). \end{aligned}$$

Since

$$\Gamma_{\mathfrak{m}}(M) \cong \varinjlim \operatorname{Hom}_R(R/\mathfrak{m}^k, M) \cong \varinjlim \operatorname{Hom}_R(R/(\underline{x}^k), M),$$

the second isomorphism follows using the same arguments. □

Next we recall the following vanishing theorem.

Theorem 2.8.12 (Grothendieck [202]) *If $(R, \mathfrak{m})$ is a local ring and M is an R-module of depth t and dimension d, then*

(a) $H^i_{\mathfrak{m}}(M) = 0$ *for $i < t$ and $i > d$.*

(b) $H^t_{\mathfrak{m}}(M) \neq 0$ *and* $H^d_{\mathfrak{m}}(M) \neq 0$.

Local cohomology of face rings Let Δ be a simplicial complex and let

$$R = K[\Delta] = K[X_1, \ldots, X_n]/I_\Delta$$

be the Stanley–Reisner ring of Δ, where I_Δ is the ideal of $K[X_1, \ldots, X_n]$ generated by all $X_{i_1} \cdots X_{i_r}$ such that $\{X_{i_1}, \ldots, X_{i_r}\} \notin \Delta$.

Let $\mathfrak{m}$ be the maximal ideal generated by the residue classes x_i of the indeterminates X_i and let $H^i_{\mathfrak{m}}(R)$ be the local cohomology modules of R.

Consider the complex $C^\star$

$$C^\star: \quad 0 \longrightarrow C^0 \longrightarrow C^1 \longrightarrow \cdots \longrightarrow C^n \longrightarrow 0,$$
$$C^t = \bigoplus_{1 \leq i_1 < \cdots < i_t \leq n} R_{x_{i_1} \cdots x_{i_t}},$$

where R_y denotes R localized at $S = \{y^i\}_{i \geq 0}$ and the differentiation map $d_t\colon C^t \to C^{t+1}$ is given on the component

$$R_{x_{i_1} \cdots x_{i_t}} \xrightarrow{d_t} R_{x_{j_1} \cdots x_{j_{t+1}}}$$

to be the natural homomorphism

$$(-1)^{s-1} \cdot \eta : R_{x_{i_1} \cdots x_{i_t}} \longrightarrow R_{(x_{i_1} \cdots x_{i_t}) x_{j_s}}$$

if $\{i_1, \ldots, i_t\} = \{j_1, \ldots, \widehat{j_s}, \ldots, j_{t+1}\}$ and 0 otherwise. If $x = x_{i_1} \cdots x_{i_r}$ is in $K[\Delta]$, then $R_x \neq 0$ iff supp $x \in \Delta$. Hence $H^i(C^\star) = 0$ for $i > \dim K[\Delta]$ (cf. Theorem 2.8.12). Recall that there is an isomorphism

$$H^i_{\mathfrak{m}}(R) \cong H^i(C^\star).$$

The reader should consult [65] for further details and results about the local cohomology of face rings.

Chapter 3

Affine and Graded Algebras

A few topics connected with affine and graded algebras are studied in this chapter, e.g., Gröbner bases, Hilbert Nullstellensatz and affine varieties, projective closure, minimal resolutions, and Betti numbers. We present the affine and graded versions of the Noether normalization lemma and some of its applications to affine algebras and Cohen–Macaulay graded algebras.

As before all base rings considered here are Noetherian and modules are finitely generated.

3.1 Cohen–Macaulay graded algebras

In this section we will emphasize the relationship between graded Cohen–Macaulay rings and their homogeneous Noether normalizations. Then some useful characterizations of those rings will be given. First we introduce affine algebras and affine Noether normalizations.

Definition 3.1.1 Let k be a field and let S be a k-algebra. We say that S is an *affine k-algebra* if $S = k[y_1, \ldots, y_r]$ for some $y_1, \ldots, y_r \in S$.

Definition 3.1.2 Let α and β be in $\mathbb{N}^n$. The *lexicographical order* in $\mathbb{N}^n$ is obtained by declaring $\beta \succ \alpha$ if the last non-zero entry of $\beta - \alpha$ is positive.

Notation The set of positive integers will be denoted by $\mathbb{N}_+$. If $\alpha, \beta \in \mathbb{R}^n$, here $\alpha \cdot \beta$ will denote the usual inner product of α and β.

Lemma 3.1.3 *Let $\alpha_1 \succ \cdots \succ \alpha_m$ be a sequence of m distinct points in $\mathbb{N}^n$ ordered lexicographically. Then, there is $w = (w_1, \ldots, w_n) \in \mathbb{N}_+^n$ such that $w_1 = 1$ and $w \cdot \alpha_1 > w \cdot \alpha_i$ for $i \geq 2$.*

Proof. Let $\alpha_i = (\alpha_{i1}, \ldots, \alpha_{in})$ and $\beta_i = (\alpha_{i1}, \ldots, \alpha_{i(n-1)})$. We proceed by induction on $n \geq 2$. There is k so that $\alpha_{1n} = \cdots = \alpha_{kn}$ and $\alpha_{kn} > \alpha_{in}$ for $i > k$. One may assume $k < m$; for otherwise $\beta_1 \succ \cdots \succ \beta_m$ and one can use induction.

Since $\beta_1 \succ \cdots \succ \beta_k$ by induction there is $w' = (1, w_2, \ldots, w_{n-1})$ such that $w' \cdot \beta_1 > w' \cdot \beta_i$ for $2 \leq i \leq k$. On the other hand for every $i > k$ one can choose $\delta_i \in \mathbb{N}_+$ so that $w' \cdot \beta_1 + \alpha_{1n}\delta_i > w' \cdot \beta_i + \alpha_{in}\delta_i$. Setting $w_n = \max\{\delta_i | \, k < i \leq m\}$, $w = (w', w_n)$, we get $w \cdot \alpha_1 > w \cdot \alpha_i$ for $i \geq 2$. $\Box$

Proposition 3.1.4 *Let $R = k[x_1, \ldots, x_n]$ be a polynomial ring over a field k and let f be a polynomial in $R \setminus k$. Then, there is a change of variables $x_i = x_1^{w_i} + y_i$ for $i \geq 2$ such that*

$$f(x_1, x_1^{w_2} + y_2, \ldots, x_1^{w_n} + y_n) = c_0 x_1^r + c_1 x_1^{r-1} + \cdots + c_{r-1}x_1 + c_r,$$

where $0 \neq c_0 \in k$, $r > 0$, and $c_i \in k[y_2, \ldots, y_n]$ for $i \geq 1$.

Proof. The polynomial f can be written as $f = \sum_{i=1}^m b_i x^{\alpha_i}$, where $0 \neq b_i \in k$ for all i. One may assume that $\alpha_1 \succ \cdots \succ \alpha_m$ are ordered lexicographically. By Lemma 3.1.3 there is $w \in \mathbb{N}_+^n$ such that $x_i = x^{w_i} + y_i$ satisfies the required properties. Note $r = w \cdot \alpha_1$. $\Box$

Lemma 3.1.5 *If $R = k[x_1, \ldots, x_n]$ is an affine k-algebra of dimension n, then R is a polynomial ring.*

Proof. One may assume $R \simeq B/I$, where B is a polynomial ring in n variables with coefficients in the field k and I is an ideal of B. Let

$$I \subset \mathfrak{p}_0 \subset \cdots \subset \mathfrak{p}_n$$

be a chain of prime ideals of B of length n. If $I \neq (0)$, then adding (0) to the chain yields a chain of length $n+1$, which is impossible because $\dim(B) = n$. Hence $I = (0)$. $\Box$

Theorem 3.1.6 *Let $R = k[x_1, \ldots, x_n]$ be a polynomial ring over a field k and let $I \neq R$ be an ideal of R. Then there are $z_1, \ldots, z_n$ in R such that*

(a) *$k[x_1, \ldots, x_n]$ is integral over $k[z_1, \ldots, z_n]$, and*

(b) *$I \cap k[z_1, \ldots, z_n] = z_1 k[z_1, \ldots, z_n] + \cdots + z_g k[z_1, \ldots, z_n]$.*

Proof. The proof is by induction on n. If $n = 1$ and $I = (f(x_1)) \neq 0$, then one sets $z_1 = f(x_1)$. Assume $n \geq 2$ and $I \neq 0$. One may assume that I contains a monic polynomial in x_1. Otherwise take a non-zero polynomial g in I and apply Proposition 3.1.4 to get an isomorphism of k-algebras

$$k[x_1, \ldots, x_n] \xrightarrow{\varphi} k[x_1, y_2 \ldots, y_n]$$

induced by $\varphi(x_1) = x_1$ and $\varphi(x_i) = x_1^{w_i} + y_i$ for $i \geq 2$, such that $\varphi(g)$ is monic in x_1. Let

$$f = x_1^r + c_1 x_1^{r-1} + \cdots + c_{r-1}x_1 + c_r$$

be a polynomial in I with $c_i \in k[x_2, \ldots, x_n]$ and $r > 0$. Set

$$I_1 = I \cap k[x_2, \ldots, x_n].$$

By induction there are $z_2, \ldots, z_n$ such that:

(i) $k[x_2, \ldots, x_n]$ is integral over $k[z_2, \ldots, x_n]$, and

(ii) $I_1 \cap k[z_2, \ldots, z_n] = (z_2, \ldots, z_n)$.

Set $z_1 = f$. It is not hard to see that R is integral over $k[z_1, \ldots, z_n]$ and

$$k[z_1, \ldots, z_n] \cap I = (z_1, \ldots, z_g). \qquad \Box$$

Corollary 3.1.7 *If $R = k[x_1, \ldots, x_n]$ is a polynomial ring over a field k and $I \neq R$ is an ideal of R, then*

$$\dim(R/I) = \dim(R) - \text{ht}\,(I).$$

Proof. One may assume that there are $z_1, \ldots, z_n$ in R and an integer g such that the conditions (a) and (b) of Theorem 3.1.6 are satisfied. We will show that g is equal to the height of I. By Lemma 3.1.5 the z_i's are algebraically independent. Hence Proposition 2.4.14 yields

$$\text{ht}\,(I) = \text{ht}\,(I \cap k[z_1, \ldots, z_n]) = g.$$

Note that there is an integral extension

$$k[z_{g+1}, \ldots, z_n] \simeq k[z_1, \ldots, z_n]/(z_1, \ldots, z_g) \stackrel{\varphi}{\hookrightarrow} k[x_1, \ldots, x_n]/I.$$

Therefore $n - g = \dim(R/I)$. $\Box$

Corollary 3.1.8 (Noether normalization lemma) *If $R = k[x_1, \ldots, x_n]$ is a polynomial ring over a field k and $I \neq R$ is an ideal, then there is an integral extension*

$$k[h_1, \ldots, h_d] \hookrightarrow R/I,$$

where $h_1, \ldots, h_d$ are in R and $d = \dim(R/I)$.

Proof. By Theorem 3.1.6 there is an integral extension

$$k[z_{g+1}, \ldots, z_n] \simeq k[z_1, \ldots, z_n]/(z_1, \ldots, z_g) \stackrel{\varphi}{\hookrightarrow} R/I.$$

To conclude the argument set $h_i = z_{g+i}$ for $i = 1, \ldots, d$. $\Box$

Corollary 3.1.9 *If $R = k[x_1, \ldots, x_n]$ is a polynomial ring over a field k, then R is a catenary ring.*

Proof. Note $\mathrm{ht}\,(\mathfrak{q}/\mathfrak{p}) = \dim(R/\mathfrak{p}) - \dim(R/\mathfrak{q})$ for any two prime ideals $\mathfrak{p} \subset \mathfrak{q}$ in R. Hence by Corollary 3.1.7 we get $\mathrm{ht}\,(\mathfrak{q}/\mathfrak{p}) = \mathrm{ht}\,(\mathfrak{q}) - \mathrm{ht}\,(\mathfrak{p})$. $\square$

Definition 3.1.10 Let $k \subset L$ be a field extension. A subset of L which is algebraically independent and is maximal with respect to inclusions is called a *transcendence basis* of L over k.

Theorem 3.1.11 [262, Theorem 8.35] *If $k \subset L$ is a field extension, then any two transcendence bases have the same cardinality.*

Definition 3.1.12 Let $k \subset L$ be a field extension. The *transcendence degree* of L over k, denoted $\mathrm{trdeg}_k(L)$, is the cardinality of any transcendence basis of L over k.

Corollary 3.1.13 *Let k be a field and let A be a finitely generated k-algebra. If A is a domain with field of fractions L, then*

$$\dim(A) = \mathrm{trdeg}_k(L).$$

Definition 3.1.14 Let k be a field. A *standard algebra* or *homogeneous algebra* is a finitely generated $\mathbb{N}$-graded k-algebra

$$S = \bigoplus_{i=0}^{\infty} S_i = k[y_1, \ldots, y_r]$$

such that $y_i \in S_1$ for all i and $S_0 = k$. If we only require y_i homogeneous and $\deg(y_i) > 0$ for all i, we say that S is a *positively graded* k-algebra.

The *irrelevant maximal ideal* $\mathfrak{m}$ of S is defined by

$$\mathfrak{m} = S_+ = \bigoplus_{i=1}^{\infty} S_i.$$

Definition 3.1.15 Let k be a field and $S = k[y_1, \ldots, y_r]$ a positively graded k-algebra with y_i homogeneous of degree d_i. There is a graded epimorphism

$$\varphi\colon R = k[x_1, \ldots, x_r] \longrightarrow S$$

given by $f(x_1, \ldots, x_r) \stackrel{\varphi}{\longmapsto} f(y_1, \ldots, y_r)$, where R is a polynomial ring graded by $\deg(x_i) = d_i$, the *presentation* of S is the k-algebra $R/\ker(\varphi)$. The graded ideal $\ker(\varphi)$ is called the *ideal of relations* or the *presentation ideal* of S.

Proposition 3.1.16 *If S is a Cohen–Macaulay standard algebra over a field k and L is an ideal of S, then*

$$\dim(S) = \dim(S/L) + \mathrm{ht}\,(L).$$

Proof. One may assume $S = R/I$ and $L = J/I$, where R is a polynomial ring over k and J is an ideal of R containing I. Set $g = \mathrm{ht}\,(J/I)$ and $r = \mathrm{ht}\,(I)$. Let $\mathfrak{p}$ be a prime ideal such that $J \subset \mathfrak{p}$ and $g = \mathrm{ht}\,(\mathfrak{p}/I)$. There is a saturated chain of prime ideals

$$I \subset \mathfrak{p}_0 \subset \mathfrak{p}_1 \subset \cdots \subset \mathfrak{p}_g = \mathfrak{p}.$$

Since $\mathfrak{p}_0$ is a minimal prime ideal of I using Proposition 2.3.13 one obtains $\dim(R/I) = \dim(R/\mathfrak{p}_0)$, that is, $r = \mathrm{ht}\,(\mathfrak{p}_0)$. Hence there is a saturated chain of prime ideals

$$\mathfrak{q}_0 = (0) \subset \mathfrak{q}_1 \subset \cdots \subset \mathfrak{q}_r = \mathfrak{p}_0 \subset \mathfrak{p}_1 \subset \cdots \subset \mathfrak{p}_g = \mathfrak{p}.$$

As R is a catenary domain one obtains $\mathrm{ht}\,(\mathfrak{p}) = r + g$ and consequently $\mathrm{ht}\,(J) \leq r + g$. Therefore, we get

$$\dim(S) \leq \mathrm{ht}\,(L) + \dim(S/L).$$

To conclude the proof note that the reverse inequality holds in general. □

Corollary 3.1.17 *Let R be a positively graded polynomial ring over a field k and I a graded ideal of R. If R/I is Cohen–Macaulay, then I is unmixed.*

Proof. It follows from Propositions 2.3.13 and 3.1.16. □

Definition 3.1.18 Let k be a field and let S be a positively graded k-algebra. A set of homogeneous elements $\underline{\theta} = \{\theta_1, \ldots, \theta_d\}$ is called a *homogeneous system of parameters* (h.s.o.p for short) if $d = \dim(S)$ and $\mathrm{rad}\,(\underline{\theta}) = S_+$.

Corollary 3.1.19 *Let S be a positively graded algebra over a field k and let $h_1, \ldots, h_d$ be a h.s.o.p for S. If $1 \leq i \leq \dim(S)$, then*

$$\dim S/(h_1, \ldots, h_i) = \dim(S) - i.$$

Proof. Set $\overline{S} = S/(h_1, \ldots, h_i)$ and $d = \dim(S)$. Note that the set of images of $h_{i+1}, \ldots, h_d$ in $\overline{S}$ generate an $\overline{\mathfrak{m}}$-primary ideal, where $\mathfrak{m} = S_+$ is the irrelevant maximal ideal of S and $\overline{\mathfrak{m}} = \mathfrak{m}\overline{S}$. Hence, by the graded version of the dimension theorem (Theorem 2.3.15), we get $\dim(\overline{S}) \leq d - i$. On the other hand if $\overline{\theta}_1, \ldots, \overline{\theta}_r$ is a h.s.o.p for $\overline{S}$, then the sequence

$$\theta_1, \ldots, \theta_r, h_1, \ldots, h_i$$

generates an $\mathfrak{m}$-primary ideal of S. Applying the dimension theorem once again one has

$$d \leq r + i = \dim(\overline{S}) + i.$$

Thus $\dim(\overline{S}) = d - i$. $\square$

Proposition 3.1.20 *Let S be a positively graded algebra over a field k and $\underline{\theta} = \theta_1, \ldots, \theta_d$ a h.s.o.p for S. Then S is Cohen–Macaulay if and only if $\underline{\theta}$ is a regular sequence.*

Proof. By Corollary 3.1.19 $\dim S/(\theta_1, \ldots, \theta_i) = d - i$. If S is C–M, then by Proposition 3.1.16 one has $\operatorname{ht}(\theta_1, \ldots, \theta_i) = i$. Assume $\theta_1, \ldots, \theta_{i-1}$ is a regular sequence. Next we show that θ_i is regular on $A = S/(\theta_1, \ldots, \theta_{i-1})$. Otherwise if θ_i is a zero divisor of A, then θ_i belongs to some minimal prime $\mathfrak{p}$ of $(\theta_1, \ldots, \theta_{i-1})$. Since $\operatorname{ht}(\mathfrak{p}) \leq i - 1$ (see Theorem 2.3.16) we obtain $\operatorname{ht}(\theta_1, \ldots, \theta_i) \leq i - 1$, which is impossible. Conversely if $\underline{\theta}$ is a regular sequence, then $\operatorname{depth}(S) = d$ and S is C–M. $\square$

Proposition 3.1.21 *Let S be a positively graded algebra over a field k. If S is Cohen–Macaulay and $\underline{\theta} = \theta_1, \ldots, \theta_q$ is a regular sequence of homogeneous elements in S_+, then $S/(\underline{\theta})$ is Cohen–Macaulay.*

Proof. It suffices to prove the case $q = 1$, which is a direct consequence of Lemma 2.3.10. $\square$

Lemma 3.1.22 *Let $V \neq \{0\}$ be a vector space over an infinite field K. Then V is not a finite union of proper subspaces of V.*

Proof. We proceed by contradiction. Assume that there are proper subspaces $V_1, \ldots, V_m$ of V such that $V = \bigcup_{i=1}^{m} V_i$, where m is the least positive integer with this property. Let

$$v_1 \in V_1 \setminus (V_2 \cup \cdots \cup V_m) \text{ and } v_2 \in V_2 \setminus (V_1 \cup V_3 \cup \cdots \cup V_m).$$

Pick $m + 1$ distinct non-zero scalars $k_0, \ldots, k_m$ in K. Consider the vectors $\beta_i = v_1 - k_i v_2$ for $i = 0, \ldots, m$. By the pigeon-hole principle there are distinct vectors β_r, β_s in V_j for some j. Since $\beta_r - \beta_s \in V_j$ we get $v_2 \in V_j$. Thus $j = 2$ by the choice of v_2. To finish the proof observe that $\beta_r \in V_2$ imply $v_1 \in V_2$, which contradicts the choice of v_1. $\square$

Proposition 3.1.23 *Let $R = k[x_1, \ldots, x_n]$ be a polynomial ring over a field k with a positive grading and let I be a graded ideal of R. Then there are homogeneous polynomials $h_1, \ldots, h_d$ in R_+ such that*

$$\dim R/(I, h_1, \ldots, h_i) = d - i, \quad \textit{for } i = 1, \ldots, d,$$

where $\dim(R/I) = d$. If k is an infinite field and $\deg(x_i) = 1$ for all i, then $h_1, \ldots, h_d$ can be chosen in R_1.

Proof. Assume $d > 0$. Let $\mathfrak{p}_1, \ldots, \mathfrak{p}_r$ be the set of minimal primes of I of height $g = \mathrm{ht}(I)$. We claim that there is a homogeneous polynomial h_1 not in $\cup_{i=1}^r \mathfrak{p}_i$. To show it we use induction on r. Since $\mathfrak{p}_1 \cdots \mathfrak{p}_{r-1} \not\subset \mathfrak{p}_r$ and because $\mathfrak{p}_i$ is graded we can pick $f \in \mathfrak{p}_1 \cdots \mathfrak{p}_{r-1} \cap R_{d_1}$ and $f \notin \mathfrak{p}_r$, $d_1 > 0$. By induction there is $g \in R_{d_2}$ and $g \notin \cup_{i=1}^{r-1} \mathfrak{p}_i$, $d_2 > 0$. Assume $R_i \subset \cup_{i=1}^r \mathfrak{p}_i$ for all $i > 0$, hence $g \in \mathfrak{p}_r$. To complete the proof of the claim consider $h = f^{d_2} - g^{d_1}$ to derive a contradiction.

Note $\dim(R/I) > \dim R/(I, h_1)$, by the choice of h_1. Hence a repeated application of the claim rapidly yields a sequence $h_1, \ldots, h_s$ of homogeneous polynomials in R_+ with $s \leq d$ and such that $\mathrm{ht}(I, h_1, \ldots, h_s) = \dim(R)$. Therefore by Theorem 2.3.15 one concludes $d = s$, as required.

If k is infinite and $\deg(x_i) = 1$ for all i, then there is h_1 in R_1 and not in $\cup_{i=1}^r \mathfrak{p}_i$; for otherwise one has $R_1 = \cup_{i=1}^r (\mathfrak{p}_i)_1$ and since R_1 cannot be a finite union of proper subspaces (see Lemma 3.1.22) we derive $R_+ = \mathfrak{p}_i$ for some i, which is impossible. Thus one may proceed as above to get the required sequence. □

Theorem 3.1.24 (Homogeneous Noether normalization lemma) *Let k be a field and let R be a polynomial ring over k. If R is positively graded and I is a graded ideal of R, then there are homogeneous polynomials $h_1, \ldots, h_d$ in R_+, with $d = \dim(R/I)$, and a natural embedding*

$$A = k[h_1, \ldots, h_d] \stackrel{\varphi}{\hookrightarrow} R/I$$

such that R/I is a finitely generated A-module. Moreover if k is infinite and $\deg(x_i) = 1$ for all i, then $h_1, \ldots, h_d$ can be chosen in R_1.

Proof. We set $R = k[x_1, \ldots, x_n]$ and $\mathcal{B}_i = \{x^a | \deg(x^a) = i\}$. According to Proposition 3.1.23, there are homogeneous polynomials $h_1, \ldots, h_d$ in $\mathfrak{m}$ with $\mathrm{rad}(I, h_1, \ldots, h_d) = \mathfrak{m}$, where $\mathfrak{m} = R_+$. Hence, there is r such that

$$\mathfrak{m}^r \subset (I, h_1, \ldots, h_d).$$

Note that $\mathcal{B}_i \subset (I, h_1, \ldots, h_d)$ for $i \geq s = r\delta$, where δ is the maximum of the degrees of $x_1, \ldots, x_n$. We set $\mathcal{B} = \cup_{i=1}^s \mathcal{B}_i$ and $J = I + \mathcal{B}k[h_1, \ldots, h_d]$. Using induction on i, we now show that $\mathcal{B}_i \subset J$ for $i \geq s$. The inclusion is clear for $i \leq s$. Assume $i > s$ and take x^a in $\mathcal{B}_i$. Let $f_1, \ldots, f_q$ be a homogeneous generating set for I. Since $(I, h_1, \ldots, h_d)$ is graded, we can write $x^a = \sum_{j=1}^q b_j f_j + \sum_{j=1}^d c_j h_j$, where $b_1, \ldots, b_q$ and $c_1, \ldots, c_d$ are homogeneous and $\deg(c_j h_j) = i$. Hence all the monomials in the support of c_j have degree less than i. Thus, by induction, one gets that c_j is in J for all j. Consequently $x^a \in J$.

Observe that the image of $\mathcal{B}$ in R/I generates R/I as an A-module. Next we verify that the canonical map φ from A to R/I is injective. Since R/I is integral over $\varphi(A)$, by Proposition 2.4.13, one obtains that d is equal to $\dim \varphi(A)$. Using $A/\ker(\varphi) \simeq \varphi(A)$ and the fact that A has dimension at most d we get $\ker(\varphi) = 0$. If k is infinite and $\deg(x_i) = 1$, then according to Proposition 3.1.23 one can choose $h_1, \ldots, h_d$ of degree one. $\Box$

Definition 3.1.25 Let R be a positively graded polynomial ring over a field k and $I \neq R$ a graded ideal. A *homogeneous Noether normalization* of $S = R/I$ is an integral extension $k[h_1, \ldots, h_d] \hookrightarrow S$, where $h_1, \ldots, h_d$ are homogeneous polynomials in R_+ and $d = \dim(S)$.

Proposition 3.1.26 (Stanley) *Let $R = k[x_1, \ldots, x_n]$ be a polynomial ring over a field k and let I be a monomial ideal with $d = \dim(R/I)$. Then*

$$A = k[\sigma_1, \ldots, \sigma_d] \hookrightarrow S = R/I$$

is a Noether normalization, where σ_i is the ith *symmetric polynomial.*

Proof. It suffices to prove that S is integral over A. Let $\overline{x}_i = x_i + I$. Since $d = \dim(R/\mathrm{rad}(I))$, one has $\sigma_i \in \mathrm{rad}(I)$ for $i > d$. Hence, from the equality

$$(x - x_1) \cdots (x - x_n) = x^n - \sigma_1 x^{n-1} + \cdots + (-1)^n \sigma_n,$$

we get $\overline{x}_i^n - \overline{\sigma}_1 \overline{x}_i^{n-1} + \cdots + (-1)^d \overline{\sigma}_d \overline{x}_i^{n-d} = \overline{h}$, where $\overline{h}$ is nilpotent in S. Hence if $h^m \in I$, raising the last equality to the mth power yields that $\overline{x}_i$ is integral over A. $\Box$

Proposition 3.1.27 *Let $R = k[x_1, \ldots, x_n]$ be a positively graded polynomial ring over a field k and let $I \neq R$ be a graded ideal of R. If $S = R/I$ is a Cohen–Macaulay ring and $A = k[h_1, \ldots, h_d] \hookrightarrow S$ is a homogeneous Noether normalization of S, then*

$$S = A\overline{x}^{\beta_1} \oplus \cdots \oplus A\overline{x}^{\beta_m}$$

for any set of monomials $\mathcal{B} = \{x^{\beta_1}, \ldots, x^{\beta_m}\}$ whose image in the Artinian ring $\overline{S} = R/(I, h_1, \ldots, h_d)$ is a k-vector space basis of $\overline{S}$.

Proof. Set $\mathbb{M}_i = \{x^\alpha \in R \mid \deg(x^\alpha) = i\}$. First we show that the image of $\mathcal{B}$ generate S as an A-module. It suffices to prove $\mathbb{M}_i \subset J = I + A\mathcal{B}$ for all $i \geq 0$. Let $x^\alpha \in \mathbb{M}_i$. There are homogeneous polynomials $\mu_1, \ldots, \mu_d$ in R and $\lambda_1, \ldots, \lambda_m$ in k such that $x^\alpha = f + \sum_{j=1}^m \lambda_j x^{\beta_j} + \sum_{j=1}^d \mu_j h_j$, where $f \in R_i \cap I$ and $\deg(\mu_j h_j) = i$ if $\mu_j \neq 0$. Since $\deg \mu_j < i$, by induction one obtains that $\mu_j \in J$ for all j. Consequently $x^\alpha \in J$.

We claim that if $c_1 x^{\beta_1} + \cdots + c_m x^{\beta_m}$ belongs to $(I, h_1, \ldots, h_{i-1})$ for some $2 \leq i \leq d$ and $c_1, \ldots, c_m$ are in $k[h_i, \ldots, h_d]$, then $c_j = 0$ for all j. To prove the claim note that $h_1, \ldots, h_d$ is a regular sequence (see Proposition 3.1.20) and use descending induction on i starting with $i = d$.

Next, we show the equality $S = A\overline{x}^{\beta_1} \oplus \cdots \oplus A\overline{x}^{\beta_m}$. Assume that $\sum_{i=1}^{m} a_i x^{\beta_i}$ is in I for some $a_1, \ldots, a_m$ in A. If $a_\ell \neq 0$ for some ℓ, write $a_i = \sum_{j=0}^{s_i} a_{ij} h_1^j$, where a_{ij} are in $k[h_2, \ldots, h_d]$. One may assume $a_{r0} \neq 0$ for some r; otherwise we may factor out h_1 to some power and apply that h_1 is regular on S. Hence $a_{10}x^{\beta_1} + \cdots + a_{m0}x^{\beta_m}$ is in (I, h_1), and applying the claim with $i = 2$ we derive $a_{r0} = 0$, which is a contradiction. Therefore $a_j = 0$ for all j, as required. □

Lemma 3.1.28 *Let k be an infinite field and let S be a standard k-algebra. If S is Cohen–Macaulay, then there exists a h.s.o.p $\underline{\theta} = \{\theta_1, \ldots, \theta_d\}$ such that $\underline{\theta}$ is a regular sequence and θ_i is homogeneous of degree 1 for all i.*

Proof. Let $S = \oplus_{i=0}^{\infty} S_i$ and $\mathrm{Ass}(S) = \{\mathfrak{p}_1, \ldots, \mathfrak{p}_s\}$. Notice that $\mathrm{Ass}(S) = \mathrm{Min}(S)$, because S is Cohen–Macaulay. We may assume $d > 0$, otherwise there is nothing to prove. If $S_1 \subset \mathcal{Z}(S) = \cup_{i=1}^{s} \mathfrak{p}_i$, then $S_1 = \cup_{i=1}^{s} (\mathfrak{p}_i)_1$. Since k is infinite, by Lemma 3.1.22, we obtain that $s = 1$ and $\mathfrak{p}_1 = S_+$, that is, $\dim(S) = 0$, which is a contradiction. Hence there is $\theta_1 \in S_1$ which is regular on S. By Proposition 3.1.16 $\dim S/(h_1) = d - 1$. Applying the depth lemma (see Lemma 2.3.9) to the exact sequence

$$0 \longrightarrow S(-1) \xrightarrow{\theta_1} S \longrightarrow S/(\theta_1) \longrightarrow 0$$

yields depth $S/(\theta_1) = d - 1$. Altogether $S/(\theta_1)$ is C–M and the result follows by induction. □

Lemma 3.1.29 *Let R be an $\mathbb{N}$-graded polynomial ring over a field k and I a graded ideal of R of height g. If I is a complete intersection, then there are homogeneous polynomials $f_1, \ldots, f_g$ such that $I = (f_1, \ldots, f_g)$.*

Proof. Let $h_1, \ldots, h_g$ be a regular sequence generating the ideal I. Using the graded version of Corollary 2.1.35 one has

$$r = \mu(I) = \dim_k(I/\mathfrak{m}I) \leq g,$$

where $\mathfrak{m} = R_+$. On the other hand, since I is graded, there are homogeneous polynomials $f_1, \ldots, f_r$ generating I and such that $\{f_i + \mathfrak{m}I\}_{i=1}^{r}$ is a k-basis for $I/\mathfrak{m}I$. By Theorem 2.3.16 one concludes $r = g$. □

Proposition 3.1.30 *Let R be a positively graded polynomial ring over a field k and let I be a graded ideal of R. If I is a complete intersection, then R/I is Cohen–Macaulay.*

Proof. By Lemma 3.1.29 the ideal I is generated by a sequence $f_1, \ldots, f_g$ consisting of homogeneous polynomials, where g denotes the height of I.

According to Proposition 3.1.23 there is a sequence $h_1, \ldots, h_{n-g}$ consisting of homogeneous polynomials such that

$$\operatorname{rad}(f_1, \ldots, f_g, h_1, \ldots, h_{n-g}) = \mathfrak{m} = R_+.$$

Therefore $\{f_1, \ldots, f_g, h_1, \ldots, h_{n-g}\}$ is a homogeneous system of parameters for R and by Proposition 3.1.20 we derive that $f_1, \ldots, f_g$ is a regular sequence. Finally Lemma 2.3.10 yields that R/I is Cohen–Macaulay. $\Box$

Lemma 3.1.31 *Let $I, \mathfrak{p}_1, \ldots, \mathfrak{p}_m$ be graded ideals of a polynomial ring R over a field k such that $\mathfrak{p}_1, \ldots, \mathfrak{p}_m$ are prime ideals. If $f \in \cup_{i=1}^m \mathfrak{p}_i$ for any $f \in I$ homogeneous, then $I \subset \mathfrak{p}_i$ for some i.*

Proof. By induction on m. If $m = 2$ and $I \not\subset \mathfrak{p}_i$ for $i = 1, 2$, then for each i pick a homogeneous polynomial $f_i \in I \setminus \mathfrak{p}_i$ of degree u_i. Since $f_1^{u_2} + f_2^{u_1}$ is homogeneous we readily derive a contradiction. Hence $I \subset \mathfrak{p}_i$ for some i.

Let G be the set of homogeneous elements in I and suppose $G \subset \cup_{i=1}^m \mathfrak{p}_i$. One may assume $\mathfrak{p}_i \not\subset \mathfrak{p}_j$ for $i \neq j$. If $I \not\subset \mathfrak{p}_i$ for all i, then by induction there is f_1 in $G \setminus \cup_{i=1}^{m-1} \mathfrak{p}_i$. On the other hand since $I\mathfrak{p}_1 \cdots \mathfrak{p}_{m-1} \not\subset \mathfrak{p}_m$ (see Exercise 2.1.48) there is a homogeneous polynomial f_2 in $I\mathfrak{p}_1 \cdots \mathfrak{p}_{m-1}$ and f_2 not in $\mathfrak{p}_m$. Since $f_1^{u_2} + f_2^{u_1}$, $\deg(f_i) = u_i$, is homogeneous we rapidly derive a contradiction. Thus $I \subset \mathfrak{p}_i$ for some i. $\Box$

Proposition 3.1.32 *Let R be a polynomial ring over a field k and let I be a graded ideal of height r, then there is a regular sequence $f_1, \ldots, f_r$ in I of homogeneous polynomials.*

Proof. We will proceed by induction. Assume that $f_1, \ldots, f_s$ is a regular sequence of homogeneous polynomials in I, where $s < r$. Note that the ideal $(f_1, \ldots, f_s)$ is Cohen–Macaulay by Proposition 3.1.30, and hence it is unmixed by Corollary 3.1.17. If all the homogeneous polynomials in I belong to $\mathcal{Z}(R/(f_1, \ldots, f_s))$, then by Lemma 3.1.31 the ideal I must be contained in an associated prime of $(f_1, \ldots, f_s)$ and consequently $\operatorname{ht}(I) \leq s$, which is impossible. Thus there is a homogeneous polynomial f_{s+1} in I such that f_{s+1} is regular modulo $(f_1, \ldots, f_s)$. $\Box$

Tensor product of affine algebras Let A, B be two affine algebras over a field k and consider presentations $A \simeq R_1/I_1$, $B \simeq R_2/I_2$, where $R_1 = k[\mathbf{x}]$, $R_2 = k[\mathbf{y}]$ are polynomial rings in disjoint sets of variables and I_i is an ideal of R_i. Set $R = k[\mathbf{x}, \mathbf{y}]$ and $I = I_1 + I_2$. The map

$$R \longmapsto R_1/I_1 \otimes_k R_2/I_2, \quad x_i \mapsto x_i \otimes 1, \quad y_i \mapsto 1 \otimes y_i,$$

induces a k-algebra homomorphism:

$$\varphi\colon R/I \to R_1/I_1 \otimes_k R_2/I_2, \quad \varphi(\overline{f}(\mathbf{x})\overline{g}(\mathbf{y})) = \overline{f}(\mathbf{x}) \otimes \overline{g}(\mathbf{y}).$$

On the other hand there is a k-bilinear map

$$\psi\colon R_1/I_1 \times R_2/I_2 \to R/I,$$

given by multiplication $\psi(\overline{f}, \overline{g}) = \overline{fg}$. By the universal property of the tensor product there is a map $\overline{\psi}$ that makes the following diagram

$$\begin{array}{ccc} R_1/I_1 \times R_2/I_2 & \xrightarrow{\phi} & R_1/I_1 \otimes_k R_2/I_2 \\ \downarrow \psi & \swarrow \overline{\psi} & \\ R/I & & \end{array}$$

commutative, where ϕ is the canonical map and $\psi = \overline{\psi}\phi$. As a consequence $\overline{\psi}$ is the inverse of φ. Thus we have proved:

Proposition 3.1.33 *If A and B are two affine algebras over a field k with presentations $A \simeq k[\mathbf{x}]/I_1$ and $B \simeq k[\mathbf{y}]/I_2$, then*

$$A \otimes_k B \simeq k[\mathbf{x}, \mathbf{y}]/(I_1 + I_2).$$

Theorem 3.1.34 *If A and B are two standard algebras over a field k, then*

$$\operatorname{depth}(A \otimes_k B) = \operatorname{depth}(A) + \operatorname{depth}(B).$$

Proof. Pick a regular sequence $\underline{g} = g_1, \ldots, g_r$ (resp. $\underline{h} = h_1, \ldots, h_s$) on $A = k[\mathbf{x}]/I_1$ (resp. $B = k[\mathbf{y}]/I_2$) with $\underline{g} \subset (\mathbf{x})$ (resp. $\underline{h} \subset (\mathbf{y})$) and such that $\underline{g}$ (resp. $\underline{h}$) consist of forms, where r is the depth of A and s is the depth of B. As B is a faithfully flat k-algebra, applying the functor $(\cdot) \otimes_k B$ to the injective map

$$0 \longrightarrow k[\mathbf{x}]/(I_1, g_1, \ldots, g_{i-1}) \xrightarrow{g_i} k[\mathbf{x}]/(I_1, g_1, \ldots, g_{i-1})$$

gives a natural commutative diagram

$$\begin{array}{ccc} k[\mathbf{x}]/(I_1, g_1, \ldots, g_{i-1}) \otimes_k B & \xrightarrow{g_i \otimes 1} & k[\mathbf{x}]/(I_1, g_1, \ldots, g_{i-1}) \otimes_k B \\ \downarrow \simeq & & \downarrow \simeq \\ k[\mathbf{x}, \mathbf{y}]/(I_1, I_2, g_1, \ldots, g_{i-1}) & \xrightarrow{g_i} & k[\mathbf{x}, \mathbf{y}]/(I_1, I_2, g_1, \ldots, g_{i-1}) \end{array}$$

such that the map in the first row is injective. Since the vertical arrows are natural isomorphisms by Proposition 3.1.33, one concludes that the map in the second row is also injective. As a consequence $\underline{g}$ is a regular sequence on $k[\mathbf{x}, \mathbf{y}]/(I_1, I_2)$. Similarly if we tensor the injective map

$$0 \longrightarrow k[\mathbf{y}]/(I_2, h_1, \ldots, h_{i-1}) \xrightarrow{h_i} k[\mathbf{y}]/(I_2, h_1, \ldots, h_{i-1})$$

with the faithfully flat k-algebra $k[\mathbf{x}]/(I_1, \underline{g})$ it follows rapidly that $\underline{h}$ is a regular sequence on $k[\mathbf{x}, \mathbf{y}]/(I_1, I_2, \underline{g})$.

Altogether $\underline{g}, \underline{h}$ is a regular sequence on $k[\mathbf{x}, \mathbf{y}]/(I_1, I_2)$. To finish the proof note that $(\mathbf{x}, \mathbf{y})$ is an associated prime ideal of $k[\mathbf{x}, \mathbf{y}]/(I_1, I_2, \underline{g}, \underline{h})$ by Exercise 3.1.41, thus $r + s$ is the length of a maximal regular sequence in $(\mathbf{x}, \mathbf{y})$ and by Proposition 2.3.7 one obtains $\operatorname{depth}(A \otimes_k B) = r + s$. For an alternative proof see [171]. $\square$

Corollary 3.1.35 *If A and B are two standard algebras over a field k, then $A \otimes_k B$ is Cohen–Macaulay if and only if A and B are Cohen–Macaulay.*

Proof. As the dimension is always greater than or equal to the depth by Lemma 2.3.6, the result follows at once using Theorem 3.1.34 together with Exercise 3.1.36. $\square$

A similar statement holds if one replaces Cohen–Macaulay by Gorenstein; indeed the type of the tensor products of two Cohen–Macaulay standard algebras is the product of their types [171].

Exercises

3.1.36 If A and B are standard algebras over a field k, then

$$\dim(A \otimes_k B) = \dim(A) + \dim(B).$$

3.1.37 Let $R = k[x_1, \ldots, x_n]$ be a polynomial ring over a field k and let $A = (a_{ij})$ be an invertible matrix with entries in k. If $y_i = \sum_{j=1}^{n} a_{ij}x_j$ for $i = 1, \ldots, n$, prove that $R = k[y_1, \ldots, y_n]$. In particular $y_1, \ldots, y_n$ are algebraically independent.

3.1.38 Let V be a vector space over an infinite field k and $W, V_1, \ldots, V_m$ vector subspaces of V. If $W \not\subset V_i$ for all i, prove that $W \not\subset \cup_{i=1}^{m} V_i$.

3.1.39 Let $I, I_1, \ldots, I_m$ be ideals of a ring R and let k be an infinite field. If $I \subset \cup_{i=1}^{m} I_i$ and k is a subring of R, then $I \subset I_i$ for some i.

3.1.40 Let R be a polynomial ring over an infinite field k and I a graded ideal of height r. If I is minimally generated by forms of degree $p \geq 1$, prove that there are forms $f_1, \ldots, f_m$ of degree p in I such that $f_1, \ldots, f_r$ is a regular sequence and I is minimally generated by $f_1, \ldots, f_m$.

3.1.41 Let $A = k[\mathbf{x}]/I_1$ and $B = k[\mathbf{y}]/I_2$ be affine algebras over a field k. If $(\mathbf{x})$ (resp. $(\mathbf{y})$) is an associated prime of $k[\mathbf{x}]/I_1$ (resp. $k[\mathbf{y}]/I_2$), then $(\mathbf{x}, \mathbf{y})$ is an associated prime of $k[\mathbf{x}, \mathbf{y}]/(I_1, I_2)$.

3.1.42 Let $k[\mathbf{x}]$ be a polynomial ring and I an ideal (resp. graded). If $k \subset K$ is a field extension, then there is a natural (resp. graded) isomorphism of K-algebras:

$$K[\mathbf{x}]/IK[\mathbf{x}] \simeq k[\mathbf{x}]/I \otimes_k K.$$

3.1.43 Let A be an affine algebra over a field k and let $k \subset K$ be a field extension. Prove:

(a) $\dim(A) = \dim(A \otimes_k K)$.

(b) $\operatorname{depth}(A) = \operatorname{depth}(A \otimes_k K)$ if A is a standard algebra.

3.1.44 Let K be a field and let A be an affine K-algebra. Prove that A is Artinian if and only if $\dim_K(A) < \infty$.

3.2 Hilbert Nullstellensatz

Let K be a field and let $S = K[t_1, \ldots, t_q]$ be a polynomial ring. In what follows $\mathbf{t}$ stands for the set of variables of S, that is, $S = K[\mathbf{t}]$. We define the *affine space* of dimension q over K, denoted by $\mathbb{A}_K^q$, to be the cartesian product $K^q = K \times \cdots \times K$.

Given an ideal $I \subset S$, define the *zero set* or *affine variety* of I as

$$V(I) = \{\alpha \in \mathbb{A}_K^q \mid f(\alpha) = 0, \, \forall f \in I\}.$$

By the Hilbert's basis theorem $V(I)$ is the zero locus of a finite collection of polynomials (see Theorem 2.1.4). Conversely, for any $X \subset \mathbb{A}_K^q$ define $I(X)$, the *vanishing ideal* of X, as the set of polynomials of S that vanish at all points of X. An *affine variety* is the zero set of an ideal. The *dimension* of a variety X is the Krull dimension of its *coordinate ring* $S/I(X)$.

Proposition 3.2.1 (Zariski topology [200, Lemma A.2.4])

(a) $V(1) = \emptyset$ *and* $V(0) = \mathbb{A}_K^q$.

(b) $V(I \cap J) = V(I) \cup V(J) = V(IJ)$, *for all ideals* I *and* J *of* S.

(c) $\cap V(I_\alpha) = V(\cup I_\alpha)$, *where* $\{I_\alpha\}$ *is any family of ideals of* S.

By the previous result the sets in the family

$$\tau = \{\mathbb{A}_K^q \setminus V(I) \mid I \text{ is an ideal of } S\}$$

are the open sets of a topology on $\mathbb{A}_K^q$, called the *Zariski topology*.

Definition 3.2.2 Given $X \subset \mathbb{A}_K^q$, the *Zariski closure* of X, denoted by $\overline{X}$, is the closure of X in the Zariski topology of $\mathbb{A}_K^q$, i.e., $\overline{X}$ is the smallest affine variety of $\mathbb{A}_K^q$ containing X.

Proposition 3.2.3 *If $X \subset \mathbb{A}_K^q$, then $\overline{X} = V(I(X))$.*

Proof. As $V(I(X))$ is closed in the Zariski topology and $X \subset V(I(X))$, taking closures one has $\overline{X} \subset V(I(X))$. For the other inclusion note that $\overline{X}$ is closed; that is, $\overline{X} = V(J)$, where J is an ideal of S. Applying I to $X \subset V(J)$, one gets $I(V(J)) \subset I(X)$. Since one also has the inclusion $J \subset I(V(J))$, by transitivity $J \subset I(X)$. Therefore $V(I(X)) \subset V(J)$ and consequently $V(I(X)) \subset \overline{X}$. $\Box$

Lemma 3.2.4 *Let X and Y be affine varieties in $\mathbb{A}_K^q$. If $I(X) = I(Y)$, then $X = Y$*

Proof. By symmetry it suffices to show the inclusion $X \subset Y$. Take $\alpha \in X$. There are $g_1, \ldots, g_r$ in S such that $Y = V(g_1, \ldots, g_r)$. Clearly $g_i \in I(Y)$ for all i. Then any g_i vanishes at all points of X because $I(Y) = I(X)$. Thus $g_i(\alpha) = 0$ for all i, i.e., $\alpha \in Y$. $\Box$

Definition 3.2.5 An affine variety $X \subset \mathbb{A}_K^q$ is *reducible* if there are affine varieties $X_1 \neq X$ and $X_2 \neq X$ such that $X = X_1 \cup X_2$; otherwise, X is *irreducible*.

Theorem 3.2.6 *Let K be a field and let X be an affine variety of $\mathbb{A}_K^q$, then X is irreducible if and only if $I(X)$ is a prime ideal.*

Proof. $\Rightarrow$) Let I be an ideal of S with $X = V(I)$ and let f, g be polynomials of S such that $fg \in I(X)$. Then $V(I, f) \cup V(I, g) = V(I)$. As X is irreducible, we get that $V(I, f) = V(I)$ or $V(I, g) = V(I)$. Hence $f \in I(X)$ or $g \in I(X)$. Therefore $I(X)$ is a prime ideal.

$\Leftarrow$) Let X_1 and X_2 be affine varieties such that $X = X_1 \cup X_2$. Then $I(X) = I(X_1) \cap I(X_2)$. As $I(X)$ is prime, by Exercise 2.1.48, we get that $I(X) = I(X_1)$ or $I(X) = I(X_2)$. Hence, by Lemma 3.2.4, we have $X = X_1$ or $X = X_2$. Therefore X is irreducible. $\Box$

Proposition 3.2.7 [210, Proposition 1.5] *If $X \subset \mathbb{A}_K^q$ is an affine variety over a field K, then there are unique irreducible affine varieties $X_1, \ldots, X_r$ in $\mathbb{A}_K^q$ such that $X_i \not\subset X_j$ for all $i \neq j$ and $X = \cup_{i=1}^r X_i$.*

Proof. The existence follows using that the Zariski topology is Artinian (Exercise 3.2.16). To show uniqueness assume that $Y_1, \ldots, Y_s$ is another decomposition of X into affine varieties such that $Y_i \not\subset Y_j$ for all $i \neq j$. We have the equality $\cap_{i=1}^r I(X_i) = \cap_{i=1}^s I(Y_i)$. Fix $1 \leq i \leq r$. By Theorem 3.2.6, $I(X_i)$ is a prime ideal of S. Then, by Exercise 2.1.48, we obtain that $I(X_i) \supset I(Y_j)$ for some j. Hence, by Exercise 3.2.15, $X_i \subset Y_j$. Similarly, there is k such that $Y_j \subset X_k$. Altogether, we have $X_i \subset Y_j \subset X_k$. Thus, $i = k$ and $X_i = Y_j$ for some j. This means that $\{X_1, \ldots, X_r\}$ is contained in $\{Y_1, \ldots, Y_s\}$. A symmetric argument shows the reverse inclusion. $\Box$

The irreducible affine varieties $X_1, \dots, X_r$ of Proposition 3.2.7 are called the *irreducible components* of X.

Theorem 3.2.8 *Let $S = K[t_1, \dots, t_q]$ be a polynomial ring over a field K and let $\mathfrak{m}$ be a maximal ideal of S. If K is algebraically closed, then there are $a_1, \dots, a_q \in K$ such that $\mathfrak{m} = (t_1 - a_1, \dots, t_q - a_q)$.*

Proof. There is an integral extension $K[h_1, \dots, h_d] \hookrightarrow S/\mathfrak{m}$, where $d = \dim(S/\mathfrak{m}) = 0$; see Corollary 3.1.8. Hence, the canonical map $\varphi\colon K \to S/\mathfrak{m}$ is an isomorphism because K is algebraically closed. To complete the proof choose $a_i \in K$ so that $\varphi(a_i) = \overline{a}_i = \overline{t}_i = \varphi(t_i)$. $\Box$

Proposition 3.2.9 [200, Lemma 1.8.8] *If I is an ideal of a ring S and $f \in S$, then $f \in \sqrt{I}$ if and only if $(I, 1 - tf) = S[t]$, where t is a new variable.*

Theorem 3.2.10 (Hilbert Nullstellensatz) *Let S be a polynomial ring over an algebraically closed field K and let I be an ideal of S, then*

$$I(V(I)) = \sqrt{I}.$$

Proof. One clearly has $I \subset I(V(I))$, hence $\sqrt{I} \subset I(V(I))$ because $I(V(I))$ is a radical ideal. For the other inclusion take $f \in I(V(I))$ and consider the ideal $J = (I, 1 - ft)$ of the ring $S[t]$, where $S = K[t_1, \dots, t_q]$ and t is a new variable. Next we show the equality $J = S[t]$. If $J \neq S[t]$ by Theorem 3.2.8 one has

$$(I, 1 - ft) \subset (t_1 - a_1, \dots, t_q - a_q, t - a_{q+1}), \quad a_i \in K.$$

Hence $\beta \in V(I)$, where $\beta = (a_1, \dots, a_q)$. Using that $f \in I(V(I))$ gives $f(\beta) = 0$, but this is impossible because $1 - f(\beta)a_{q+1} = 0$. Hence $J = S[t]$. Thus $f \in \sqrt{I}$ by Proposition 3.2.9. $\Box$

Corollary 3.2.11 *Let $X = V(f_1, \dots, f_r)$ be an affine variety, defined by polynomials $f_1, \dots, f_r$ in q variables over an algebraically closed field K. Then $r \geq q - \dim(X)$.*

Proof. By the Nullstellensatz $I(X) = \mathrm{rad}\,(f_1, \dots, f_r)$. Let $\mathfrak{p}$ be a minimal prime of $(f_1, \dots, f_r)$ of height $q - \dim(X)$. Applying Theorem 2.3.16 we get $q - \dim(X) \leq r$. $\Box$

Exercises

3.2.12 Let I and J be two ideals in a polynomial ring S over a field K. If $\mathrm{rad}\,(I) = \mathrm{rad}\,(J)$, prove that $V(I) = V(J)$.

3.2.13 If K is a field and $X \subset \mathbb{A}_K^q$, then $X \subset V(I(X))$ with equality if X is an affine variety.

3.2.14 Let J be an ideal of a polynomial ring S over a field K. Prove that $J \subset I(V(J))$ with equality if J is the ideal of an affine variety.

3.2.15 Let X and Y be affine varieties in $\mathbb{A}_K^q$. Then $X \subset Y$ if and only if $I(X) \supset I(Y)$.

3.2.16 (The Zariski topology is Artinian) For any descending chain

$$X_1 \supset X_2 \supset \cdots \supset X_i \supset \cdots$$

of affine varieties in $\mathbb{A}_K^q$ there is $k \in \mathbb{N} \setminus \{0\}$ such that $X_i = X_k$ for $i \geq k$.

3.2.17 Let I be an ideal of a polynomial ring $S = K[\mathbf{t}]$ over an algebraically closed field K and let $\mathfrak{p}_1, \ldots, \mathfrak{p}_s$ be the minimal primes of I. Prove that Y is an irreducible component of $V(I)$ if and only if $Y = V(\mathfrak{p}_i)$ for some i.

3.2.18 Let K be an algebraically closed field. Then there is a one to one correspondence between affine varieties (resp. irreducible varieties) in $\mathbb{A}_K^q$ and radical ideals (resp. prime ideals) in the polynomial ring $K[t_1, \ldots, t_q]$.

3.2.19 Let I be a monomial ideal in $K[t_1, \ldots, t_q]$ and $X = V(I) \subset \mathbb{A}_K^q$ the associated monomial variety. If K is infinite, prove that X is irreducible if and only if $X = V(t_{i_1}, \ldots, t_{i_r})$.

3.2.20 Let K be an infinite field. Prove: (a) $\mathbb{A}_K^q$ is an irreducible variety, (b) any two non-empty open sets of $\mathbb{A}_K^q$ intersect, (c) any non-empty open set of $\mathbb{A}_K^q$ is dense, (d) $(K^*)^q$ is an open set of $\mathbb{A}_K^q$, (e) $(K^*)^q$ is not an affine variety.

3.2.21 Let $S = K[t_1, \ldots, t_q]$ be a polynomial ring over a field K and let $\mathfrak{m}$ be a maximal ideal of S. If K is uncountable, then $S/\mathfrak{m}$ is a finite extension of K.

3.3 Gröbner bases

In this section we review some basic facts and definitions on Gröbner bases. The reader may consult [1, 99, 142] for a detailed discussion of Gröbner bases and for the missing proofs of this section.

Let K be a field and let $R = K[x_1, \ldots, x_n]$ be a polynomial ring. The *monomials* or *terms* of R are the power products

$$x^a = x_1^{a_1} \cdots x_n^{a_n}, \qquad a = (a_1, \ldots, a_n) \in \mathbb{N}^n.$$

The set of monomials of R is denoted by $\mathbb{M}_n = \{x^a \,|\, a \in \mathbb{N}^n\}$.

Definition 3.3.1 A total order $\succ$ of $\mathbb{M}_n$ is called a *monomial order* or *term order* if

(a) $x^a \succeq 1$ for all $x^a \in \mathbb{M}_n$, and

(b) for all $x^a, x^b, x^c \in \mathbb{M}_n$, $x^a \succ x^b$ implies $x^a x^c \succ x^b x^c$.

Two examples of monomial orders of $\mathbb{M}_n$ are the *lexicographical order* or *lex order* defined as $x^b \succ x^a$ iff the last non-zero entry of $b-a$ is positive, and the *reverse lexicographical order* or *revlex order* given by $x^b \succ x^a$ iff the last non-zero entry of $b - a$ is negative.

In the sequel we assume that a monomial order $\prec$ for $\mathbb{M}_n$ has been fixed. Let f be a non-zero polynomial in R. One can write

$$f = \sum_{i=1}^{r} a_i M_i,$$

with $a_i \in K^* = K \setminus \{0\}$, $M_i \in \mathbb{M}_n$ and $M_1 \succ \cdots \succ M_r$. The *leading term* M_1 of f is denoted by $\mathrm{in}_\prec(f)$ or $\mathrm{lt}_\prec(f)$, or simply by $\mathrm{in}(f)$. The *leading coefficient* a_1 of f and $a_1 M_1$ are denoted by $\mathrm{lc}(f)$ and $\mathrm{lm}(f)$, respectively.

Definition 3.3.2 Let I be an ideal of R. The *initial ideal* of I, denoted by $\mathrm{in}_\prec(I)$ or simply by $\mathrm{in}(I)$, is given by

$$\mathrm{in}_\prec(I) = (\{\mathrm{in}_\prec(f) | \, f \in I\}).$$

Lemma 3.3.3 (Dickson) *If $\{M_i\}_{i=1}^{\infty}$ is a sequence in $\mathbb{M}_n$, then there is an integer k so that M_i is a multiple of some term in the set $\{M_1, \ldots, M_k\}$ for every $i > k$.*

Proof. Let $I \subset K[x_1, \ldots, x_n]$ be the ideal generated by $\{M_i\}_{i=1}^{\infty}$. By the Hilbert's basis theorem I is finitely generated (see Theorem 2.1.4). It is seen that I can be generated by a finite set of terms $M_1, \ldots, M_k$. Hence for each $i > k$, there is $1 \leq j \leq k$ such that M_i is a multiple of M_j. $\square$

Definition 3.3.4 Let $F = \{f_1, \ldots, f_q\} \subset R \setminus \{0\}$ be a set of polynomials in R. One says that f *reduces* to g *modulo* F, denoted $f \to_F g$, if

$$g = f - (au/\,\mathrm{lc}(f_i))f_i$$

for some $f_i \in F$, $u \in \mathbb{M}_n$, $a \in K^*$ such that $a \cdot u \cdot \mathrm{in}_\prec(f_i)$ occurs in f with coefficient a.

Proposition 3.3.5 *The reduction relation "$\longrightarrow_F$" is Noetherian, that is, any sequence of reductions $g_1 \longrightarrow_F \cdots \longrightarrow_F g_i \longrightarrow_F \cdots$ is stationary.*

Proof. Notice that at the ith step of the reduction some term of g_i is replaced by terms of lower degree. Therefore if the sequence above is not stationary, then there is a never ending decreasing sequence of terms in $\mathbb{M}_n$, but this is impossible according to Dickson's lemma. □

Theorem 3.3.6 (Division algorithm [142, Theorem 2.11]) *If $f, f_1, \ldots, f_q$ are polynomials in R, then f can be written as*

$$f = a_1 f_1 + \cdots + a_q f_q + r,$$

where $a_i, r \in R$ and either $r = 0$ or $r \neq 0$ and no term of r is divisible by one of $\mathrm{in}(f_1), \ldots, \mathrm{in}(f_q)$. Furthermore if $a_i f_i \neq 0$, then $\mathrm{in}(f) \geq \mathrm{in}(a_i f_i)$.

Definition 3.3.7 The polynomial r in the division algorithm is called a *remainder* of f with respect to $F = \{f_1, \ldots, f_q\}$.

Definition 3.3.8 Let $I \neq (0)$ be an ideal of R and let $F = \{f_1, \ldots, f_r\}$ be a subset of I. The set F is called a *Gröbner basis* of I if

$$\mathrm{in}_\prec(I) = (\mathrm{in}_\prec(f_1), \ldots, \mathrm{in}_\prec(f_r)).$$

Definition 3.3.9 A Gröbner basis $F = \{f_1, \ldots, f_r\}$ of an ideal I is called a *reduced Gröbner basis* for I if:

(i) $\mathrm{lc}(f_i) = 1 \ \forall i$, and

(ii) none of the terms occurring in f_i belongs to $\mathrm{in}_\prec(F \setminus \{f_i\}) \ \forall i$.

Theorem 3.3.10 [142, Theorem 2.17] *Each ideal I has a unique reduced Gröbner basis.*

Definition 3.3.11 Let $f, g \in R$ and let $[f, g] = \mathrm{lcm}(f, g)$ be its least common multiple. The S-*polynomial* of f and g is given by

$$\mathrm{S}(f, g) = \frac{[\mathrm{in}(f), \mathrm{in}(g)]}{\mathrm{lm}(f)} f - \frac{[\mathrm{in}(f), \mathrm{in}(g)]}{\mathrm{lm}(g)} g,$$

Given a set of generators of a polynomial ideal one can determine a Gröbner basis using the next fundamental procedure:

Theorem 3.3.12 (Buchberger [74]) If $F = \{f_1, \ldots, f_q\}$ is a set of generators of an ideal I of R, then one can construct a Gröbner basis for I using the following algorithm:

Input: F
Output: a Gröbner basis G for I
Initialization: $G := F, \quad B := \{\{f_i, f_j\} \mid f_i \neq f_j \in G\}$
while $B \neq \emptyset$ do

pick any $\{f, g\} \in B$
$B := B \setminus \{\{f, g\}\}$
$r :=$ remainder of $S(f, g)$ with respect to G
if $r \neq 0$ then
  $B := B \cup \{\{r, h\} | \, h \in G\}$
  $G := G \cup \{r\}$

Proposition 3.3.13 *Let I be an ideal of R and let $F = \{f_1, \ldots, f_q\}$ be a Gröbner basis of I. If*

$$\mathcal{B} = \{\overline{u} \,|\, u \in \mathbb{M}_n \text{ and } u \notin (\mathrm{in}(f_1), \ldots, \mathrm{in}(f_q))\},$$

then $\mathcal{B}$ is a basis for the K-vector space R/I.

Proof. First we show that $\mathcal{B}$ is a generating set for R/I. Take $\overline{f} \in R/I$. Since "$\longrightarrow_F$" is Noetherian, we can write $f = \sum_{i=1}^{q} a_i f_i + \sum_{i=1}^{r} \lambda_i u_i$, where $\lambda_i \in K^*$ and such that every u_i is a term which is not a multiple of any of the terms $\mathrm{in}(f_j)$. Accordingly $\overline{u}_i$ is in $\mathcal{B}$ for all i and $\overline{f}$ is a linear combination of the $\overline{u}_i$'s.

To prove that $\mathcal{B}$ is linearly independent assume $h = \sum_{i=1}^{s} \lambda_i u_i \in I$, where $u_i \in \mathcal{B}$ and $\lambda_i \in K$. We must show $h = 0$. If $h \neq 0$, then we can label the u_i's so that $u_1 \succ \cdots \succ u_s$ and $\lambda_1 \neq 0$. Hence $\mathrm{in}(h) = u_1 \in \mathrm{in}(I)$, but this is a clear contradiction because $\mathrm{in}(I) = (\mathrm{in}(f_1), \ldots, \mathrm{in}(f_q))$. Therefore $h = 0$, as required. □

Definition 3.3.14 A monomial in $\mathcal{B}$ is called a *standard monomial* with respect to $f_1, \ldots, f_q$.

Corollary 3.3.15 (Macaulay) *If I is a graded ideal of R, then R/I and $R/\mathrm{in}_{\prec}(I)$ have the same Hilbert function.*

Lemma 3.3.16 [142, Proposition 2.15] *Let f, g be polynomials in R and let $F = \{f, g\}$. If $\mathrm{in}(f)$ and $\mathrm{in}(g)$ are relatively prime, then $\mathrm{S}(f, g) \rightarrow_F 0$.*

Theorem 3.3.17 [74] *Let I be an ideal of R and let $F = \{f_1, \ldots, f_q\}$ be a set of generators of I, then F is a Gröbner basis for I if and only if*

$$\mathrm{S}(f_i, f_j) \longrightarrow_F 0 \quad \textit{for all} \quad i \neq j.$$

Definition 3.3.18 Let I be an ideal of R generated by $F = \{f_1, \ldots, f_r\}$ and consider the homomorphism of R-modules $\varphi\colon R^r \rightarrow I$, $e_i \longmapsto f_i$, where e_i is the ith unit vector. The kernel of φ, denoted by $Z(F)$, is called the *syzygy module* of I with respect to F.

Theorem 3.3.19 [1, Theorem 3.4.1] *Let $G = \{g_1, \ldots, g_r\}$ be a Gröbner basis and write*

$$\mathrm{S}(g_i, g_j) = \frac{[\mathrm{in}(g_i), \mathrm{in}(g_j)]}{\mathrm{lm}(g_i)} \cdot g_i - \frac{[\mathrm{in}(g_i), \mathrm{in}(g_j)]}{\mathrm{lm}(g_j)} \cdot g_j = \sum_{k=1}^{r} a_{ijk} g_k,$$

where $a_{ijk} \in R$ and $\mathrm{in}(\mathrm{S}(g_i, g_j)) \succeq \mathrm{in}(a_{ijk} g_k)$ for all i, j, k. Then the set

$$\left\{ \frac{[\mathrm{in}(g_i), \mathrm{in}(g_j)]}{\mathrm{lm}(g_i)} \cdot e_i - \frac{[\mathrm{in}(g_i), \mathrm{in}(g_j)]}{\mathrm{lm}(g_j)} \cdot e_j - \sum_{k=1}^{r} a_{ijk} e_k \right\}_{1 \leq i, j \leq r}$$

generates the syzygy module of $I = (g_1, \ldots, g_r)$ with respect to G.

Elimination of variables Let $K[x_1, \ldots, x_n, t_1, \ldots, t_q]$ be a polynomial ring over a field K. A useful monomial order is the *elimination order* with respect to the variables $x_1, \ldots, x_n$. This order is given by

$$x^a t^c \succ x^b t^d$$

if and only if $\deg(x^a) > \deg(x^b)$, or both degrees are equal and the last non-zero entry of $(a, c) - (b, d)$ is negative. The elimination order with respect to all variables $x_1, \ldots, x_n, t_1, \ldots, t_q$ is defined accordingly. This order is called the GRevLex order.

Theorem 3.3.20 *Let $B = K[x_1, \ldots, x_n, t_1, \ldots, t_q]$ be a polynomial ring over a field K with a term order $\prec$ such that terms in the x_i's are greater than terms in the t_i's. If I is an ideal of B with a Gröbner basis $\mathcal{G}$, then $\mathcal{G} \cap K[t_1, \ldots, t_q]$ is a Gröbner basis of $I \cap K[t_1, \ldots, t_q]$.*

Proof. Set $S = K[t_1, \ldots, t_q]$ and $I^c = I \cap S$. If $M \in \mathrm{in}(I^c)$, there is $f \in I^c$ with $\mathrm{lt}(f) = M$. Hence $M = \lambda\, \mathrm{lt}(g)$ for some $g \in \mathcal{G}$, because $\mathcal{G}$ is a Gröbner basis. Since $M \in S$ and $x^\alpha \succ t^\beta$ for all α and β we obtain $g \in \mathcal{G} \cap S$, that is, $M \in (\mathrm{in}(\mathcal{G} \cap S))$. Thus $\mathrm{in}(I^c) = (\mathrm{in}(\mathcal{G} \cap S))$, as required. $\square$

Example 3.3.21 Let $\prec$ be the *elimination order* with respect to $x_1, \ldots, x_4$. Using *Macaulay*2 [199], we can compute the reduced Gröbner basis of

$$I = (t_1 - x_1x_2, t_2 - x_1x_3, t_3 - x_1x_4, t_4 - x_2x_3, t_5 - x_2x_4, t_6 - x_3x_4).$$

By Theorem 3.3.20, it follows that $I \cap K[t_1, \ldots, t_6] = (t_3t_4 - t_1t_6, t_2t_5 - t_1t_6)$.

Definition 3.3.22 Let I and J be two ideals of a ring R. The ideal

$$(I \colon J^\infty) = \bigcup_{i \geq 1} (I \colon {}_R J^i)$$

is the *saturation* of I w.r.t J. If $f \in R$, we set $(I \colon (f)^\infty) = (I \colon f^\infty)$.

The saturation can be computed by elimination of variables using the following result.

Proposition 3.3.23 *Let $R[t]$ be a polynomial ring in one variable over a ring R and let I be an ideal of R. If $f \in R$, then*

$$(I\colon f^\infty) = \bigcup_{i\geq 1}(I\colon {}_R f^i) = (I, 1-tf) \cap R.$$

Proof. Let $g \in (I, 1-tf) \cap R$. Then $g = \sum_{i=1}^q a_i f_i + a_{q+1}(1-tf)$, where $f_i \in I$ and $a_i \in R[t]$. Making $t = 1/f$ in the last equation and multiplying by f^m, with m large enough, one derives an equality

$$gf^m = b_1 f_1 + \cdots + b_q f_q,$$

where $b_i \in R$. Hence $gf^m \in I$ and $g \in (I\colon f^\infty)$.

Conversely let $g \in (I\colon f^\infty)$, hence there is $m \geq 1$ such that $gf^m \in I$. Since one can write

$$g = (1 - t^m f^m)g + t^m f^m g \text{ and } 1 - t^m f^m = (1-tf)b,$$

for some $b \in R[t]$, one derives $g \in (I, 1-tf) \cap R$. $\Box$

Exercises

3.3.24 If I and J are two ideals of a polynomial ring R over a field K and let t be a new variable, then

$$I \cap J = (t \cdot I, (1-t) \cdot J) \cap R.$$

3.3.25 Let I be an ideal of a ring R and f a non-zero element of R, then

$$(f)(I\colon f) = I \cap (f).$$

3.3.26 Let $R = k[\mathbf{x}]$ be a polynomial ring, where k is a field. Recall that f is a *binomial* in R if $f = x^\alpha - x^\beta$ for some α, β in $\mathbb{N}^n$. Use Buchberger's algorithm to prove that an ideal of R generated by a finite set of binomials has a Gröbner basis consisting of binomials.

3.3.27 Let I be an ideal of a ring R and $f \in R$. Prove that the following statements are equivalent:

(a) f is regular on R/I.

(b) $(I\colon {}_R f) = I$.

(c) $I = (I, 1-tf) \cap R$, where t is a new variable.

3.3.28 Let $\mathbb{N}^n$ be the set of n-tuples of nonnegative integers endowed with the partial order given by $(a_1, \ldots, a_n) \geq (b_1, \ldots, b_n)$ if and only if $a_i \geq b_i$ for all i. If $A \subset \mathbb{N}^n$, use Dickson's lemma to prove that A has only a finite number of minimal elements.

3.4 Projective closure

Let $S = K[t_1, \ldots, t_q]$ be a polynomial ring over a field K and let u be a new variable. For $f \in S$ of degree e define

$$f^h = u^e f\left(\frac{t_1}{u}, \ldots, \frac{t_q}{u}\right),$$

that is, f^h is the homogenization of the polynomial f with respect to u. The *homogenization* of an ideal $I \subset S$ is the ideal I^h of $S[u]$ given by

$$I^h = (\{f^h | \, f \in I\}).$$

Proposition 3.4.1 *If I is an ideal of S, then* $\sqrt{I^h} = \sqrt{I}^{\,h}$.

Proof. To show the inclusion $\sqrt{I}^{\,h} \subset \sqrt{I^h}$, we need only observe the equality $(f^h)^m = (f^m)^h$ for $f \in S$ and $m \geq 0$. To show the reverse inclusion let $g \in \sqrt{I^h}$. As the ideal $\sqrt{I^h}$ is a graded ideal one may assume that g is homogeneous. Write $g = u^s g_1^h$, where $g_1 \in S$. Since g^m is in I^h, it follows that $g_1 \in \sqrt{I}$, as required. $\square$

Let $\succ$ be the elimination order on the monomials of $S[u]$ with respect to $t_1, \ldots, t_q, u$, this order extends the elimination order with respect to $t_1, \ldots, t_q$ on the monomials of S.

Proposition 3.4.2 *Let I be an ideal of S spanned by a finite set $\mathcal{G}$. Setting $\mathcal{G}^h = \{g^h | \, g \in \mathcal{G}\}$ and $\mathrm{in}\{\mathcal{G}^h\} = \{\mathrm{in}(g^h) | \, g \in \mathcal{G}\}$, the following hold.*

(a) *If $\mathrm{in}(\mathcal{G}^h) = (\mathrm{in}\{\mathcal{G}^h\})$, then $\mathrm{in}(I) = (\{\mathrm{in}(g) | \, g \in \mathcal{G}\})$ and $I^h = (\mathcal{G}^h)$.*

(b) *$\mathcal{G}$ is a Gröbner basis of I if and only if $\mathcal{G}^h$ is a Gröbner basis of I^h.*

Proof. (a): We set $\mathcal{G} = \{g_1, \ldots, g_r\}$ and $\mathrm{in}\{\mathcal{G}\} = \{\mathrm{in}(g) | \, g \in \mathcal{G}\}$. To show the first equality we need only show that $\mathrm{in}(I) \subset (\mathrm{in}\{\mathcal{G}\})$. Let m be a monomial in the ideal $\mathrm{in}(I)$. There is $g \in I$, of degree e, such that $\mathrm{in}(g) = m$. Writing $g = \sum_{i=1}^r f_i g_i$ for some $f_1, \ldots, f_r$ in S, from the equality

$$\frac{g^h}{u^e} = \sum_{i=1}^{r} f_i\left(\frac{t_1}{u}, \ldots, \frac{t_q}{u}\right) g_i\left(\frac{t_1}{u}, \ldots, \frac{t_q}{u}\right) \tag{3.1}$$

we get $u^s g^h \in (\mathcal{G}^h)$ for $s \gg 0$. As $\mathrm{in}(g^h) = \mathrm{in}(g)$, one has $\mathrm{in}(u^s g^h) = u^s m$. Hence $u^s m \in (\mathrm{in}\{\mathcal{G}^h\})$. Using $\mathrm{in}(g_i) = \mathrm{in}(g_i^h)$ yields $m \in (\mathrm{in}\{\mathcal{G}\})$.

To show the second equality it is enough to show that $\{g^h | \, g \in I\} \subset (\mathcal{G}^h)$. By the first equality $\mathcal{G}$ is a Gröbner basis of I. Hence any $g \in I$ can be written as $g = \sum_{i=1}^r f_i g_i$, where $\mathrm{in}(g) \succeq \mathrm{in}(f_i g_i)$ for all i. Notice that $e = \deg(g) \geq \deg(f_i g_i) = \deg(f_i) + \deg(g_i)$. Since Eq. (3.1) holds, it follows that $g^h \in (g_1^h, \ldots, g_r^h)$.

(b) $\Rightarrow$) By part (a) we need only show that $\mathrm{in}(\mathcal{G}^h) \subset (\mathrm{in}\{\mathcal{G}^h\})$. Let $m \in \mathrm{in}(\mathcal{G}^h)$ be a term, then $m = \mathrm{in}(g)$ for some $g \in (\mathcal{G}^h)$. We may assume that g is homogeneous. We can write $g = u^p(m_1 + m_2u^{e_2} + \cdots + m_su^{e_s})$, where $m_1, \ldots, m_s$ are monomials in S such that $m_1 \succ m_2u^{e_2} \succ \cdots \succ m_su^{e_s}$. As all $m_iu^{e_i}$ have the same degree we obtain $0 \leq e_2 \leq \cdots \leq e_s$. It follows that $g' = g(t_1, \ldots, t_q, 1)$ belongs to I and $\mathrm{in}(g') = m_1$. Therefore m_1 belongs to $\mathrm{in}(I) = (\mathrm{in}\{\mathcal{G}\})$. Since $\mathrm{in}(g_i) = \mathrm{in}(g_i^h)$ we obtain that $m \in (\mathrm{in}\{\mathcal{G}^h\})$.

$\Leftarrow$) This implication follows from part (a). $\square$

Projective closure Let K be a field. We define the *projective space* of dimension q over K, denoted by $\mathbb{P}^q_K$, to be the quotient space

$$(K^{q+1} \setminus \{0\})/\sim$$

where two points α, β in $K^{q+1} \setminus \{0\}$ are equivalent under $\sim$ if $\alpha = c\beta$ for some $c \in K$. It is usual to denote the equivalence class of α by $[\alpha]$.

For any set $X \subset \mathbb{P}^q_K$ define $I(X)$, the *vanishing ideal* of X, as the ideal generated by the homogeneous polynomials in $S[u]$ that vanish at all points of X. Conversely, given a homogeneous ideal $I \subset S[u]$ define its *zero set* as

$$V(I) = \{[\alpha] \in \mathbb{P}^q_K \,|\, f(\alpha) = 0, \forall f \in I \text{ homogeneous}\}.$$

A *projective variety* is the zero set of a homogeneous ideal. It is not difficult to see that the members of the family

$$\tau = \{\mathbb{P}^q_K \setminus V(I) \,|\, I \text{ is an ideal of } S[u]\}$$

are the open sets of a topology on $\mathbb{P}^q_K$, called the *Zariski topology*.

Definition 3.4.3 Let $Y \subset \mathbb{A}^q_K$. The *projective closure* of Y is defined as $\overline{Y} := \overline{\varphi(Y)}$, where φ is the map $\varphi\colon \mathbb{A}^q_K \to \mathbb{P}^q_K$, $\alpha \mapsto [(\alpha, 1)]$, and $\overline{\varphi(Y)}$ is the closure of $\varphi(Y)$ in the Zariski topology of $\mathbb{P}^q_K$.

Proposition 3.4.4 *If $Y \subset \mathbb{A}^q_K$, then $\overline{Y} = V(I(\varphi(Y)))$.*

Proof. It is left as an exercise; see Proposition 3.2.3 for a similar formula for the Zariski closure in the affine case. $\square$

Proposition 3.4.5 *Let $Y \subset \mathbb{A}^q_K$ be a set and let $\overline{Y} \subset \mathbb{P}^q_K$ be its projective closure. If $f_1, \ldots, f_r$ is a Gröbner basis of $I(Y)$, then*

$$I(\overline{Y}) = (f_1^h, \ldots, f_r^h).$$

Proof. Note that $I(\overline{Y}) = I(Y)^h$ and use Proposition 3.4.2. $\square$

Corollary 3.4.6 *Let $Y \subset \mathbb{A}^q_K$ be a set and let $\overline{Y} \subset \mathbb{P}^q_K$ be its projective closure. Then the height of $I(Y)$ in S is equal to the height of $I(\overline{Y})$ in $S[u]$.*

Exercises

3.4.7 If I is a prime ideal of a polynomial ring S, prove that I^h is also a prime ideal.

3.4.8 Let $Y = \{(x_1^3, x_1^2, x_1) | \, x_1 \in K\} \subset \mathbb{A}_K^3$ be a monomial curve and let $\overline{Y}$ be its projective closure. If K is an infinite field prove that

$$I(\overline{Y}) = (t_3^2 - t_2u, t_2t_3 - t_1u, t_2^2 - t_1t_3).$$

3.4.9 *Let $S = K[t_1, \ldots, t_q]$ be a polynomial ring over a field K. If I is an ideal of S generated by $f_1, \ldots, f_m$, then $((f_1^h, \ldots, f_m^h)\colon u^\infty) = I^h$.*

3.4.10 Let $Y = \{(x_1^{d_1}, \ldots, x_1^{d_q}) | \, x_1 \in K\}$ be a monomial curve in the affine space $\mathbb{A}_K^q$. If $d_1 > d_2 > \cdots > d_q$ and K is the field of complex numbers, then the projective closure $\overline{Y}$ of Y is equal to

$$\overline{Y} = \left\{[(x_1^{d_1}, x_1^{d_2}u_1^{d_1-d_2}, \ldots, x_1^{d_q}u_1^{d_1-d_q}, u_1^{d_1})] \in \mathbb{P}_K^q \middle| \; u_1, x_1 \in K\right\}.$$

3.5 Minimal resolutions

The aim of this section is to study homogeneous resolutions of positively graded modules over polynomial rings. We shall be interested in the numerical data of these resolutions and in particular in the Betti numbers of these modules.

We begin with a result which is a consequence of the graded Nakayama's lemma (see Lemma 2.2.9).

Proposition 3.5.1 *Let $R = K[\mathbf{x}]$ be a polynomial ring over a field K and let M be an $\mathbb{N}$-graded R-module. If $F = \{f_1, \ldots, f_q\}$ is a set of homogeneous elements of M and $\mathfrak{m} = (\mathbf{x})$, then F is a minimal set of generators for M if and only if the image of F in $M/\mathfrak{m}M$ is a K-basis of $M/\mathfrak{m}M$.*

Corollary 3.5.2 *Let $R = K[\mathbf{x}]$ be a polynomial ring over a field K and let $\mathfrak{m} = (\mathbf{x})$. If M is an $\mathbb{N}$-graded R-module, then*

$$\mu(M) = \dim_{R/\mathfrak{m}}(M/\mathfrak{m}M) = \dim_K(M/\mathfrak{m}M),$$

where $\mu(M)$ is the minimum number of generators of M.

Proof. Let $f_1, \ldots, f_r$ be a minimum set of generators of M. The dimension of $M/\mathfrak{m}M$, as a K-vector space, is less than or equal to $r = \mu(M)$ because the images of $f_1, \ldots, f_r$ generate $M/\mathfrak{m}M$. Hence, by Proposition 3.5.1, we get the required equality. $\square$

Lemma 3.5.3 *Let $R = K[\mathbf{x}]$ be a polynomial ring over a field K and let M be an $\mathbb{N}$-graded module over R with a presentation*

$$0 \longrightarrow \ker(\varphi) \longrightarrow R^q \xrightarrow{\varphi} M \longrightarrow 0, \qquad e_i \overset{\varphi}{\longmapsto} f_i.$$

If $\mathfrak{m} = (\mathbf{x})$ and $F = \{f_1, \ldots, f_q\}$ is a set of homogeneous elements, then F is a minimal set of generators for M if and only if $L = \ker(\varphi) \subset \mathfrak{m}R^q$.

Proof. $\Leftarrow$) Assume $L \subset \mathfrak{m}R^q$. By Proposition 3.5.1 we need only show that the image of F in $M/\mathfrak{m}M$ is linearly independent over K. If

$$\lambda_1 f_1 + \cdots + \lambda_q f_q \in \mathfrak{m}M,$$

for some λ_i's in K, then $(\lambda_i) - (a_i) \in L \subset \mathfrak{m}R^q$, for some a_i's in $\mathfrak{m}$. We are using (λ_i) as a short hand for $(\lambda_1, \ldots, \lambda_q)$. Hence (λ_i) is in $\mathfrak{m}R^q$. Thus, $\lambda_j = 0$ for all j.

$\Rightarrow$) If $L \not\subset \mathfrak{m}R^q$, pick $z = (z_1, \ldots, z_q) \in L$ such that $z \notin \mathfrak{m}R^q$. Then z_i is not in $\mathfrak{m}$ for some i. Since the f_i's are homogeneous this readily implies that f_i is a linear combination of elements in $F \setminus \{f_i\}$, a contradiction. $\Box$

Definition 3.5.4 Let $R = \oplus_{i=0}^{\infty} R_i$ be a positively graded ring and $a \in \mathbb{N}$. The graded R-module obtained by a *shift in the graduation* of R is given by

$$R(-a) = \bigoplus_{i=0}^{\infty} R(-a)_i,$$

where the ith graded component of $R(-a)$ is $R(-a)_i = R_{-a+i}$.

Proposition 3.5.5 *Let $R = \oplus_{i=0}^{\infty} R_i$ be a positively graded polynomial ring over a field K with maximal irrelevant ideal $\mathfrak{m} = R_+$ and M an $\mathbb{N}$-graded R-module. Then there is an exact sequence of graded free modules*

$$\mathbb{F}. \qquad \cdots \longrightarrow \bigoplus_{i=1}^{b_k} R(-d_{ki}) \xrightarrow{\varphi_k} \cdots \longrightarrow \bigoplus_{i=1}^{b_0} R(-d_{0i}) \xrightarrow{\varphi_0} M \longrightarrow 0$$

where φ_k is a degree preserving map and $\operatorname{im}(\varphi_k) \subset \mathfrak{m}R^{b_{k-1}}$ *for* $k \geq 1$.

Proof. Let $f_1, \ldots, f_q$ be a set of homogeneous elements that minimally generate M. Set $d_{0i} = \deg(f_i)$ and $b_0 = q$. There is an exact sequence

$$0 \longrightarrow Z_1 = \ker(\varphi_0) \xrightarrow{i} \bigoplus_{i=1}^{b_0} R(-d_{0i}) \xrightarrow{\varphi_0} M \longrightarrow 0,$$

where φ_0 is a degree preserving homomorphism such that $\varphi_0(e_i) = f_i$ for all i. Note that $Z_1 \subset \mathfrak{m}R^{b_0}$ by Lemma 3.5.3. Since Z_1 is once again a finitely generated graded R-module one may iterate the process to obtain the required exact sequence $\mathbb{F}$. $\Box$

Theorem 3.5.6 (Hilbert syzygy theorem [128]) *Let $R = K[x_1, \ldots, x_n]$ be a polynomial ring over a field K and let M be an $\mathbb{N}$-graded R-module. Then M has a graded free resolution of length at most n.*

Proof. Set $\mathfrak{m} = R_+$. First notice that $\operatorname{Tor}_i^R(M, R/\mathfrak{m}) = 0$ for $i \geq n+1$, because the ordinary Koszul complex $\mathcal{K}_\star(\underline{x})$ is a graded free resolution of $R/\mathfrak{m} = K$ of length n. On the other hand assume that

$$\cdots \longrightarrow F_k \xrightarrow{\varphi_k} \cdots \longrightarrow F_1 \xrightarrow{\varphi_1} F_0 \longrightarrow M \longrightarrow 0$$

is a graded free resolution of M as in Proposition 3.5.5. Applying the functor $(\cdot) \otimes_R R/\mathfrak{m}$ yields the complex

$$\cdots \longrightarrow F_k \otimes R/\mathfrak{m} \xrightarrow{\varphi_k \otimes 1} \cdots \longrightarrow F_1 \otimes R/\mathfrak{m}$$
$$\xrightarrow{\varphi_1 \otimes 1} F_0 \otimes R/\mathfrak{m} \longrightarrow M \otimes R/\mathfrak{m} \longrightarrow 0.$$

Using $\operatorname{im}(\varphi_k) \subset \mathfrak{m}F_{k-1}$ for all $k \geq 1$ one obtains that all the maps $\varphi_k \otimes 1$ are zero. Hence

$$\operatorname{Tor}_i^R(M, R/\mathfrak{m}) \simeq F_i \otimes R/\mathfrak{m}$$

for $i \geq 1$. In particular $F_i \otimes R/\mathfrak{m} \simeq F_i/\mathfrak{m}F_i = 0$ for $i \geq n+1$ and by Nakayama's lemma one obtains $F_i = 0$ for all $i \geq n+1$. $\square$

Corollary 3.5.7 *Let $R = \oplus_{i=0}^{\infty} R_i$ be a polynomial ring of dimension n over a field K and let M be an $\mathbb{N}$-graded R-module. Then there is a unique (up to complex isomorphism) exact sequence of graded modules*

$$0 \longrightarrow \bigoplus_{i=1}^{b_g} R(-d_{gi}) \xrightarrow{\varphi_g} \cdots \longrightarrow \bigoplus_{i=1}^{b_k} R(-d_{ki}) \xrightarrow{\varphi_k} \cdots$$
$$\longrightarrow \bigoplus_{i=1}^{b_1} R(-d_{1i}) \xrightarrow{\varphi_1} \bigoplus_{i=1}^{b_0} R(-d_{0i}) \xrightarrow{\varphi_0} M \longrightarrow 0,$$

such that

(a) $g = \sup\{i \mid \operatorname{Tor}_i^R(M, K) \neq 0\} \leq n$, *and*

(b) $\operatorname{im}(\varphi_k) \subset \mathfrak{m}R^{b_{k-1}}$ *for all* $k \geq 1$, *where* $\mathfrak{m} = R_+$.

Proof. Notice the following isomorphisms:

$$\operatorname{Tor}_j^R(M, K) \simeq \bigoplus_{i=1}^{b_j} K(-d_{ji}) \simeq F_j/\mathfrak{m}F_j,$$

where F_j is the jth free module in the resolution of M. Hence the b_j's and the d_{ji}'s are uniquely determined and so is the length of the resolution. $\square$

Remark 3.5.8 The entries of the matrices φ_k are in $\mathfrak{m}$. This condition is equivalent to require that at each stage we use a minimal generating set.

Remark 3.5.9 If $d_{k1} \leq \cdots \leq d_{kb_k}$ for all k, then

$$d_{11} < d_{21} < \cdots < d_{g1}.$$

Indeed if $\varphi_k(e_i) = \sum_j r_{ij}e_j$, where $r_{ij} \in \mathfrak{m}$ and r_{ij} homogeneous, then $\deg(\varphi_k(e_i)) > d_{(k-1)1}$ because r_{ij} have positive degree. Hence one obtains $d_{k1} > d_{(k-1)1}$.

Definition 3.5.10 The integers $b_0, \ldots, b_g$ are the *Betti numbers* of M. The d_{ji}'s are the *twists*, they indicate a shift in the graduation.

In general the Betti numbers and twists of M may depend on the base field K; see for instance [149, Example 2.10] and [346, Remark 3].

Definition 3.5.11 The R-module $Z_k = \ker(\varphi_{k-1})$ is called the *k-syzygy module* of M.

Definition 3.5.12 The *homogeneous resolution* or *minimal resolution* of M is the unique graded resolution of M by free R-modules described in Corollary 3.5.7.

If M has a minimal free resolution as above, then note that $\mathrm{pd}_R(M)$, the *projective dimension* of M, is equal to g. In our situation the notions of "free resolution" and "projective resolution" coincide because of the theorem of Quillen-Suslin: *if K is a principal ideal domain, then all finitely generated projective $K[x_1, \ldots, x_n]$-modules are free* [363].

Theorem 3.5.13 (Auslander–Buchsbaum [150, Theorem 3.1]) *Let M be an R-module. If R is a regular local ring, then*

$$\mathrm{pd}_R(M) + \mathrm{depth}(M) = \dim(R).$$

Corollary 3.5.14 *If R is a polynomial ring over a field K and I is a graded ideal, then*

$$\mathrm{pd}_R(R/I) \geq \mathrm{ht}\,(I),$$

with equality if and only if R/I is a Cohen–Macaulay R-module.

Proof. The result follows from an appropriate graded version of the Auslander–Buchsbaum formula. □

Proposition 3.5.15 *Let S be a positively graded algebra of dimension d over a field K. If $A = K[h_1, \ldots, h_d] \hookrightarrow S$ is a homogeneous Noether normalization of S, then S is a Cohen–Macaulay ring if and only if S is a free A-module.*

Proof. By Theorem 3.5.13 $\mathrm{pd}_A(S) + \mathrm{depth}(S) = \dim(A)$, where the depth of S is taken with respect to A_+. Assume S is a free A-module, then $\mathrm{depth}(S) = d$ and thus the depth of S with respect to S_+ is also equal d. Therefore S is Cohen–Macaulay.

Conversely assume S Cohen–Macaulay. Note that A is a polynomial ring. Let $\mathfrak{p} \in \mathrm{Spec}\, S$ be a minimal (graded) prime over A_+S. There is an integral extension

$$K = A/A_+ \hookrightarrow S/\mathfrak{p},$$

hence $S/\mathfrak{p}$ is a zero dimensional domain and consequently $\mathfrak{p}$ is a maximal ideal. Therefore $\mathrm{rad}\,(A_+S) = S_+$, that is, $h_1, \ldots, h_d$ is a h.s.o.p for S. Now use Proposition 3.1.20 to conclude that $h_1, \ldots, h_d$ is a regular sequence in S, that is, $\mathrm{depth}\, S = d$. Another application of the Auslander–Buchsbaum formula yields that S is a free A-module. $\square$

Theorem 3.5.16 (Hilbert–Burch [128, Theorem 20.15]) *Let R be a polynomial ring over a field K and let I be a graded Cohen–Macaulay ideal of height two. If*

$$0 \longrightarrow R^{q-1} \stackrel{\varphi}{\longrightarrow} R^q \longrightarrow R \longrightarrow R/I \longrightarrow 0$$

is the minimal resolution of R/I, then I is generated by all the minors of size $q-1$ of the matrix φ.

Example Let $R = \mathbb{Q}[x, y, z, w]$ and let $I = (f_1, f_2, f_3)$, where

$$f_1 = y^2 - xz, \;\; f_2 = x^3 - yzw, \;\; f_3 = x^2y - z^2w.$$

Let us construct the minimal resolution of R/I. Consider the exact sequence

$$0 \longrightarrow \ker(\varphi_1) = Z_1 \stackrel{\mathrm{id}}{\longrightarrow} R(-2) \oplus R^2(-3) \stackrel{\varphi_1}{\longrightarrow} I \longrightarrow 0, \quad e_i \stackrel{\varphi_1}{\longmapsto} f_i.$$

Here Z_1 is the module of syzygies of I. Using *Macaulay* we find that Z_1 is generated as an R-modulo by the column vectors, ψ_1, ψ_2, of the matrix:

$$A = \begin{bmatrix} zw & x^2 \\ y & z \\ -x & -y \end{bmatrix}$$

Notice that $R(-2) \oplus R^2(-3)$ is graded by

$$(R(-2) \oplus R^2(-3))_i = R(-2)_i \oplus R(-3)_i \oplus R(-3)_i.$$

Hence $\psi_1, \psi_2 \in \left[R(-2) \oplus R^2(-3)\right]_4$. Next consider the exact sequence

$$0 \longrightarrow \ker(\varphi_2) = Z_2 \stackrel{\mathrm{id}}{\longrightarrow} R^2(-4) \stackrel{\varphi_2}{\longrightarrow} Z_1 \longrightarrow 0, \quad e_i \stackrel{\varphi_2}{\longmapsto} \psi_i.$$

Since $\psi_1 R + \psi_2 R$ is a free R-module, we have $Z_2 = (0)$. Altogether the minimal homogeneous resolution of R/I is:

$$0 \longrightarrow R^2(-4) \stackrel{\varphi_2}{\longrightarrow} R(-2) \oplus R^2(-3) \stackrel{\varphi_1}{\longrightarrow} R \longrightarrow R/I \longrightarrow 0.$$

In this example R/I is Cohen–Macaulay because $\mathrm{pd}_R(R/I) = 2 = \mathrm{ht}\,(I)$.

Pure and linear resolutions Let $R = \oplus_{i=0}^{\infty} R_i$ be a polynomial ring over a field K with its usual graduation and I a graded ideal of R. Let

$$0 \to \bigoplus_{j=1}^{b_g} R(-d_{gj}) \to \cdots \to \bigoplus_{j=1}^{b_1} R(-d_{1j}) \to R \to S = R/I \to 0 \qquad (3.2)$$

be the minimal graded resolution of S by free R-modules. The ideal I (or the algebra S) has a *pure resolution* if there are constants

$$d_1 < d_2 < \cdots < d_g$$

such that $d_{1j} = d_1, \ldots, d_{gj} = d_g$ for all j. If in addition $d_j = d_1 + j - 1$ for $2 \leq j \leq g$ the resolution is said to be d_1-*linear* .

Theorem 3.5.17 (Herzog–Kühl [230]) *If S is a Cohen–Macaulay algebra with a pure resolution, then*

$$b_i = \prod_{j \neq i} \frac{d_j}{|d_j - d_i|} \qquad (1 \leq i \leq g).$$

Theorem 3.5.18 (Huneke-Miller [256]) *If S is a Cohen–Macaulay algebra with a pure resolution, then the multiplicity $e(S)$ of S is given by*

$$e(S) = d_1 \cdots d_g / g!.$$

Eisenbud and Schreyer proved the following long-standing multiplicity conjecture stated in [237, Conjecture 1, p. 2880].

Theorem 3.5.19 (Huneke–Srinivasan Multiplicity Conjecture [133]) *If S is a Cohen–Macaulay algebra of codimension g, then*

$$\frac{1}{g!} \prod_{i=1}^{g} m_i \leq e(S) \leq \frac{1}{g!} \prod_{i=1}^{g} M_i,$$

where $M_i = \max\{d_{ij} \,|\, j = 1, \ldots, b_i\}$ *and* $m_i = \min\{d_{ij} \,|\, j = 1, \ldots, b_i\}$.

Herzog and Srinivasan [237, Conjecture 2, p. 2881] conjectured that the second inequality holds for non-Cohen–Macaulay algebras. This conjecture was shown by Boij and Söderberg [47]. Since the Betti numbers and the twists are positive integers, these formulas impose restrictions on the numbers that can occur in the resolution of S.

Exercises

3.5.20 Let R be a polynomial ring and let I be a graded Cohen–Macaulay ideal of height 3 with a pure resolution:

$$0 \to R^{b_3}(-(d+a+b)) \to R^{b_2}(-(d+a)) \to R^{b_1}(-d) \to I \to 0,$$

where $a, b \geq 1$. Assume $b_1 = 6$ and $b = 1$. Show that $x = a + 1$ and $y = d + a + 1$ is a solution of the diophantine equation

$$6x(x-1) = y(y-1).$$

Then prove that the integral solutions of this equation are given by

$$x = (X+1)/2 \text{ and } y = (Y+1)/2,$$

where $\sqrt{6}X + Y = \pm(1 \pm \sqrt{6})(5 + 2\sqrt{6})^n$, $n \in \mathbb{Z}$.

3.5.21 Let $R = K[x_1, \ldots, x_n]$ be a polynomial ring over a field K and let I be a graded ideal, then $(x_1, \ldots, x_n)$ is an associated prime of I if and only if $\mathrm{pd}_R(R/I) = n$.

3.5.22 Let $R = K[x, y, z]$ be a polynomial ring and let I be the ideal

$$(x^n + y^n - z^n \mid n \geq 2).$$

If $\mathrm{char}(K) \neq 2$, then $I = (x^2 + y^2 - z^2, x^3 + y^3 - z^3, x^2y^2)$. If $\mathrm{char}(K) \neq 2$, prove that $\mathfrak{q}_1 \cap \mathfrak{q}_2 \cap \mathfrak{q}_3$ is a primary decomposition of I, where

$$\mathfrak{q}_1 = (y^2, x - z),\ \mathfrak{q}_2 = (x^2, y - z),\ \mathfrak{q}_3 = (y^3, z^3, x^2y^2, x^2 + y^2 - z^2).$$

3.5.23 Let $R = K[\mathbf{x}]$ be a polynomial ring over a field K. If I is a homogeneous ideal of R of height r which is generated by r polynomials, then I is generated by r homogeneous polynomials.

Chapter 4

Rees Algebras and Normality

In this part a detailed presentation of complete and normal ideals is given. The systematic use of Rees algebras and associated graded rings will make clear their importance for the area. Some outstanding references for blowup algebras and the normality of Rees algebras are [59, 412, 414]. A general reference for integral closure of ideals, rings, and modules is [259].

4.1 Symmetric algebras

Let M be an R-module. Given $n \geq 0$, we define

$$T^n(M) = \underbrace{M \otimes \cdots \otimes M}_{n-times} \quad \text{and} \quad T^0(M) = R.$$

The *tensor algebra* $T(M)$ of M is the noncommutative graded algebra

$$T(M) = \bigoplus_{n=0}^{\infty} T^n(M),$$

where the product in $T(M)$ is induced by juxtaposition, that is, the product of $x_1 \otimes \cdots \otimes x_m$ and $y_1 \otimes \cdots \otimes y_n$ is $x_1 \otimes \cdots \otimes x_m \otimes y_1 \otimes \cdots \otimes y_n$.

The *symmetric algebra* of M, denoted by $\mathrm{Sym}_R(M)$ or simply $\mathrm{Sym}(M)$, is defined as the quotient algebra

$$\mathrm{Sym}(M) = T(M)/\mathcal{J}$$

where $\mathcal{J}$ is the two-sided ideal generated by all $xy - yx = x \otimes y - y \otimes x$ with x and y running through M. Observe that $\mathrm{Sym}(M)$ is commutative.

Since $\mathcal{J}$ is a graded ideal generated by homogeneous elements of degree two, the symmetric algebra is graded by

$$\mathrm{Sym}_n(M) = T^n(M)/\mathcal{J} \cap T^n(M) \quad \text{and} \quad \mathrm{Sym}_0(M) = R.$$

Note that $\mathrm{Sym}_2(M) = M \otimes M/(x \otimes y - y \otimes x)$, with x and y running through all the elements of M.

Exercises

4.1.1 If M is a free R-module of rank n prove that the symmetric algebra of M is a polynomial ring in n variables with coefficients in R.

4.2 Rees algebras and syzygetic ideals

Let I be an ideal of a ring R generated by $f_1, \ldots, f_q$. The *Rees algebra* of I, denoted by $R[It]$ or $\mathcal{R}(I)$, is the subring of $R[t]$ given by

$$R[It] = R[f_1 t, \ldots, f_q t] \subset R[t],$$

where t is a new variable. Note

$$R[It] = R \oplus It \oplus \cdots \oplus I^n t^n \oplus \cdots \subset R[t].$$

There is an epimorphism of R-algebras

$$\varphi\colon B = R[t_1, \ldots, t_q] \longrightarrow R[It] \longrightarrow 0, \quad t_i \stackrel{\varphi}{\longmapsto} f_i t,$$

where $B = R[\mathbf{t}]$ is a polynomial ring over the ring R. The kernel of φ, denoted by J, is the *presentation ideal* of $R[It]$ with respect to $f_1, \ldots, f_q$. Notice that J is a graded ideal in the t_i-variables: $J = \bigoplus_{i=1}^{\infty} J_i$, where B has the standard grading induced by setting $\deg(t_i) = 1$.

The mapping $\psi\colon R^q \longrightarrow I$ given by $\psi(z_1, \ldots, z_q) = \sum_{i=1}^{q} z_i f_i$ induces an R-algebra epimorphism

$$\beta\colon R[t_1, \ldots, t_q] \longrightarrow \mathrm{Sym}_R(I).$$

Thus, the *symmetric algebra* of I is:

$$\mathrm{Sym}_R(I) \simeq R[t_1, \ldots, t_q]/\ker(\beta),$$

where $\ker(\beta)$ is an ideal of $R[\mathbf{t}]$ generated by linear forms:

$$\ker(\beta) = \left(\left\{ \sum_{i=1}^{q} b_i t_i \,\middle|\, \sum_{i=1}^{q} b_i f_i = 0 \text{ and } b_i \in R \right\} \right).$$

On the other hand the kernel of φ is generated by all forms $F(t_1,\ldots,t_q)$ such that $F(f_1,\ldots,f_q) = 0$. In particular, one may factor φ through $\mathrm{Sym}_R(I)$ and obtain the commutative diagram:

$$\begin{array}{ccc} R[t_1,\ldots,t_q] & \stackrel{\varphi}{\longrightarrow} & R[It] \\ \downarrow^{\beta} & \quad {}^{\alpha}\nearrow & \\ \mathrm{Sym}_R(I) & & \end{array}$$

We say that I is an *ideal of linear type* if α is an isomorphism.

An important module-theoretic obstruction to "$\mathrm{Sym}_R(I) \simeq R[It]$" is given by the following result.

Proposition 4.2.1 (Herzog–Simis–Vasconcelos [233]) *Let I be an ideal of a ring R. If* $\mathrm{Sym}_R(I) \simeq R[It]$, *then for each prime $\mathfrak{p}$ containing I, $I_\mathfrak{p}$ can be generated by* $\mathrm{ht}\,(\mathfrak{p})$ *elements.*

Syzygetic ideals Let I be an ideal of a ring R and let $H_1(I)$ be the first homology module of the Koszul complex $H_\star(\underline{x},R)$ associated to a set $\underline{x} = x_1,\ldots x_n$ of generators of I. In [382] it is pointed out that $H_1(I)$ is related to I/I^2, the *conormal module* of I, by the following exact sequence:

$$H_1(I) \stackrel{f}{\longrightarrow} R^n \otimes (R/I) \stackrel{h}{\longrightarrow} I \otimes (R/I) = I/I^2 \longrightarrow 0. \qquad (*)$$

Here $f([z]) = z \otimes 1$ and $h(e_i \otimes 1) = x_i \otimes 1$. Set $\delta(I) = \ker(f)$.

Definition 4.2.2 The ideal I is called *syzygetic* if $\delta(I) = 0$.

Although $H_1(I)$ may depend on the set of generators for I, $\delta(I)$ depends only on I. Indeed Simis and Vasconcelos [382] proved that

$$\delta(I) \simeq \ker(\mathrm{Sym}_2(I) \to I^2),$$

where $\mathrm{Sym}_2(I) \to I^2$ is the surjection induced by the multiplication map.

Definition 4.2.3 An ideal I of a ring R is said to be *generically a complete intersection* if $IR_\mathfrak{p}$ is a complete intersection for all $\mathfrak{p} \in \mathrm{Ass}_R(R/I)$.

Remark 4.2.4 If I is unmixed and generically a complete intersection, then the R/I-torsion of $H_1(I)$ equals $\delta(I)$. To prove this, notice that $I_\mathfrak{p}$ is syzygetic for all $\mathfrak{p} \in \mathrm{Ass}_R(R/I)$; consequently, $\delta(I)$ is a torsion module.

Computing the presentation ideal of a Rees algebra Let I be a graded ideal of a polynomial ring R over a field k and let $f_1,\ldots,f_q$ be a generating set for I.

Consider the presentation of the Rees algebra:

$$\varphi\colon B=R[t_1,\ldots,t_q]\longrightarrow R[It]\longrightarrow 0,\quad (t_i\longmapsto tf_i).$$

The kernel J of φ can be obtained as follows:

$$J=(t_1-tf_1,\ldots,t_q-tf_q)\cap B.$$

Since J is a graded ideal in the t_i-variables, $J=\oplus_{i\geq 1}J_i$. The relationship between J and the Koszul homology of I is very tight. The exact sequence of Eq. $(*)$ can be made precise:

$$0\longrightarrow J_2/B_1J_1=\delta(I)\longrightarrow H_1(I)\longrightarrow (R/I)^q\longrightarrow I/I^2\longrightarrow 0.$$

In particular one can decide whether I is syzygetic – that is, $J_2=B_1J_1$–or of linear type–that is, $J=J_1B$.

Exercises

4.2.5 Let $A[\mathbf{x}]$ be a polynomial ring over a ring A and let $f_1,\ldots,f_q$ be forms of degree $d\geq 1$ in $A[\mathbf{x}]$. Then there is a graded isomorphism of A-algebras

$$\varphi\colon A[f_1,\ldots,f_q]\longrightarrow A[tf_1,\ldots,tf_q],\ \text{ with } \varphi(f_i)=tf_i,$$

where t is a new variable and both rings have an appropriate grading such that $\deg(f_i)=1$ and $\deg(tf_i)=1$.

4.2.6 Let $I=(f_1,\ldots,f_q)$ be an ideal of a polynomial ring $R=K[\mathbf{x}]$ over a field K. Prove that

$$R[It]\simeq R[t_1,\ldots,t_q]/((f_2t_1-f_1t_2,\ldots,f_qt_1-f_1t_q)\colon f_1^\infty).$$

4.2.7 Let $I=I_2(X)$ be the ideal of 2×2-minors of the symmetric matrix:

$$X=\begin{bmatrix} x_1 & x_2 & x_3\\ x_2 & x_4 & x_5\\ x_3 & x_5 & x_6\end{bmatrix},$$

where the entries of X are indeterminates over a field K. Prove that I is a syzygetic ideal that satisfies sliding depth. See [238].

4.3 Complete and normal ideals

Let R be a ring and let I be an ideal of R, an element $z \in R$ is *integral* over I if z satisfies an equation

$$z^\ell + a_1 z^{\ell-1} + \cdots + a_{\ell-1} z + a_\ell = 0, \ a_i \in I^i,$$

the *integral closure* of I is the set of all elements $z \in R$ which are integral over I. This set will be denoted by $\overline{I}$ or I_a.

Definition 4.3.1 If $I = \overline{I}$, I is said to be *integrally closed* or *complete.* If all the powers I^k are complete, the ideal I is said to be *normal.*

The simplest kinds of integrally closed monomial ideals are the Stanley–Reisner ideals and the powers of face ideals.

Lemma 4.3.2 *Let R be a ring and I an ideal of R. An element $\alpha \in R$ is in the integral closure of I if and only if there is an ideal L of R such that*

(i) *$\alpha L \subset IL$, and*

(ii) *$\alpha^r \operatorname{ann}(L) = (0)$ for some integer $r \geq 0$.*

Proof. $\Rightarrow$) As α is in integral closure of I there is an equation:

$$\alpha^n = a_1 \alpha^{n-1} + \cdots + a_{n-1}\alpha + a_n, \quad \text{where } a_i \in I^i,$$

and $n \geq 1$ is some integer. The required L is obtained by setting

$$L = R\alpha^{n-1} + I\alpha^{n-2} + \cdots + I^{n-2}\alpha + I^{n-1}.$$

Indeed, from the equation above

$$R\alpha^n \subset I\alpha^{n-1} + I^2\alpha^{n-2} + \cdots + I^{n-1}\alpha + I^n = IL,$$

and $\alpha(I^i\alpha^{n-i-1}) = I^i\alpha^{n-i} = I(I^{i-1}\alpha^{n-i}) \subset IL$ for $i \geq 1$. Hence $\alpha L \subset IL$. To finish this part of the proof note $\alpha^{n-1}\operatorname{ann}(L) = (0)$.

$\Leftarrow$) As R is Noetherian, $L = (f_1, \ldots, f_n)$. Then $\alpha f_i = \sum_{j=1}^n b_{ij} f_j$, where $b_{ij} \in I$. Set $B = (b_{ij})$, $C = B - \alpha I_n$ and $f = (f_1, \ldots, f_n)$. Here I_n denotes the identity matrix. As $Cf^t = 0$, one can use the formula

$$C \cdot \operatorname{adj}(C) = \det(C) I_n$$

to conclude $f_i \det(C) = 0$ for all i. Therefore $\det(C)$ is in $\operatorname{ann}(L)$ and by hypothesis $\alpha^r \det(C) = 0$ for some r. Expanding in powers of α gives that α is integral over I. □

Proposition 4.3.3 *If I is an ideal of a ring R, then $\overline{I}$ is an ideal of R.*

Proof. Let $\alpha_i \in \overline{I}$, $i = 1, 2$. By the proof of Lemma 4.3.2, there is an ideal L_i such that $\alpha_i L_i \subset IL_i$, $\alpha_i^{r_i}\mathrm{ann}(L_i) = (0)$ and $I^{r_i} \subset L_i$ for some $r_i \geq 0$. Since α_i is integral over I, one has $\alpha_i^{n_i} \in I$ for some $n_i \geq 1$. Therefore for n large enough β^n is in L_1L_2, where $\beta = \alpha_1 + \alpha_2$. Thus

$$\beta^n \mathrm{ann}(L_1L_2) = (0).$$

On the other hand one clearly has the inclusion $\beta L_1 L_2 \subset IL_1L_2$. Hence, by Lemma 4.3.2, one concludes $\beta \in \overline{I}$.

To finish proving that $\overline{I}$ is an ideal note $x\alpha_1 \in \overline{I}$ for $x \in R$, this follows at once from the definition of $\overline{I}$. □

Proposition 4.3.4 *If I is an ideal of a ring R and S is a multiplicatively closed subset of R, then*

$$\overline{S^{-1}(I)} = S^{-1}(\overline{I}).$$

Proof. To prove $\overline{S^{-1}(I)} \subset S^{-1}(\overline{I})$ take $f \in \overline{S^{-1}(I)}$. One may assume $f = x/1$ with $x \in R$ because the integral closure of an ideal is again an ideal. There is an equation

$$f^n + \frac{a_1}{s_1} f^{n-1} + \cdots + \frac{a_{n-1}}{s_{n-1}} f + \frac{a_n}{s_n} = \frac{0}{1},$$

where $a_i/s_i \in S^{-1}(I)^i = S^{-1}(I^i)$. Thus one may assume $a_i \in I^i$ and $s_i \in S$ for all i. Clearing denominators and multiplying by an appropriate element of S one has an equality (in the ring R) of the form

$$sx^n + t_1 a_1 x^{n-1} + \cdots + t_{n-1}a_{n-1}x + t_n a_n = 0 \quad (s, t_i \in S),$$

multiplying both sides of this equation by s^{n-1} yields sx is in $\overline{I}$. Therefore $f = (xs)/s$ is in $S^{-1}(\overline{I})$. The other inclusion follows readily using that localizations commute with powers of ideals. □

Lemma 4.3.5 *Let A be a domain and $x \in A \setminus \{0\}$ such that A_x is normal. Then A is normal if and only if (x) is a complete ideal.*

Proof. Assume that A is a normal domain. Take $z \in \overline{(x)}$. Since z satisfies an equation of the form

$$z^m + (a_1 x)z^{m-1} + \cdots + (a_{m-1}x^{m-1})z + (a_m x^m) = 0, \quad a_i \in A,$$

dividing by x^m yields that z/x is integral over A and $z \in (x)$.

Conversely assume (x) is complete. Let z be an element of the quotient field of A which is integral over A. As A and A_x have the same field of fractions and A_x is normal one may assume $z = a/x^r$, for some $a \in A$ and $r \geq 1$. To show $z \in A$ it suffices to verify $a \in (x)$. There is an equation

$$z^m + b_1 z^{m-1} + \cdots + b_{m-1}z + b_m = 0, \quad b_i \in A,$$

multiplying by x^{rm} yields the identity

$$a^m + (b_1 x^r)a^{m-1} + \cdots + (b_{m-1}x^{r(m-1)})a + (b_m x^{rm}) = 0,$$

hence $a \in \overline{(x)} = (x)$, as required. $\Box$

Proposition 4.3.6 *If A is a domain and $x \in A \setminus \{0\}$, then*

$$A = A_x \bigcap \left(\bigcap_{\mathfrak{p} \in \operatorname{Ass} A/(x)} A_{\mathfrak{p}} \right).$$

Proof. Note that the left-hand side of the equality is contained in the right-hand side because A is a domain.

Conversely take an element $z \in A_x$ and $z \in A_{\mathfrak{p}}$, for all $\mathfrak{p} \in \operatorname{Ass} A/(x)$. Write $z = a/x^n$, $a \in A$ and $n \geq 1$. It is enough to prove that $a \in (x)$. If a is not in (x) note that

$$((x) \colon a) \subset \mathcal{Z}(A/(x)).$$

Hence $((x) \colon a) \subset \mathfrak{p}$, for some $\mathfrak{p} \in \operatorname{Ass} A/(x)$. Since $z \in A_{\mathfrak{p}}$ one may write $z = a/x^n = b/s$, $b \in A$ and $s \notin \mathfrak{p}$. Therefore $s \in ((x) \colon a) \subset \mathfrak{p}$, which yields a contradiction. $\Box$

Proposition 4.3.7 *If A is an integral domain and S is a multiplicatively closed subset of A, then*

$$\overline{S^{-1}(A)} = S^{-1}(\overline{A}).$$

Proof. Note that A and $S^{-1}(A)$ have the same field of fractions K. First we prove $\overline{S^{-1}(A)} \subset S^{-1}(\overline{A})$. Take any x in K integral over $S^{-1}(A)$. There is an equation

$$x^n + \frac{a_1}{s_1}x^{n-1} + \cdots + \frac{a_{n-1}}{s_{n-1}}x + \frac{a_n}{s_n} = \frac{0}{1},$$

where $a_i \in A$ and $s_i \in S$ for all i. Set $s = s_1 \cdots s_n$. If we multiply by s^n, it follows that $sx \in \overline{A}$ and $x \in S^{-1}(\overline{A})$.

Conversely take $x \in S^{-1}(\overline{A})$. There is $s \in S$ such that sx is integral over A. Hence sx satisfies

$$(sx)^n + a_1(sx)^{n-1} + \cdots + a_{n-1}(sx) + a_n = 0,$$

for some $a_1, \ldots, a_n$ in A, dividing by s^n immediately yields that x is integral over $S^{-1}(A)$, as required. $\Box$

Corollary 4.3.8 *If A is a normal domain and S is a multiplicatively closed subset of A, then $S^{-1}(A)$ is a normal domain.*

Corollary 4.3.9 *Let $R = k[x_1, \ldots, x_n]$ be a polynomial ring over a field k and let K_R be its field of fractions. If $R' = k[x_1^{\pm 1}, \ldots, x_n^{\pm 1}] \subset K_R$ is the ring of Laurent polynomials, then R' is a normal domain.*

Proof. Notice that R' is the localization of R at the multiplicative set of monomials of R. Hence, by Corollary 4.3.8, we get that R' is normal. □

Corollary 4.3.10 *Let A be a domain and $x \in A \setminus \{0\}$. Then A is normal if and only if A_x and $A_\mathfrak{p}$ are normal for every $\mathfrak{p} \in \operatorname{Ass} A/(x)$.*

Proof. If A is normal, by Corollary 4.3.8, A_x and $A_\mathfrak{p}$ are normal for every $x \in A \setminus \{0\}$ and $\mathfrak{p} \in \operatorname{Spec}(A)$. For the converse use Proposition 4.3.6 and observe that A_x and $A_\mathfrak{p}$ have the same quotient field as A. □

To state a useful criterion of normality of Rees algebras, it is convenient to introduce the *extended Rees algebra* of an ideal I in a ring R:

$$A = R[It, u], \ u = t^{-1}.$$

Part of its usefulness is derived from the equality $A_u = R[t, t^{-1}]$.

Proposition 4.3.11 *Let I be an ideal of a ring R and $A = R[It, t^{-1}]$ its extended Rees algebra. If R is a normal domain, then A is normal if and only if $A_\mathfrak{p}$ is normal for each associated prime $\mathfrak{p}$ of $t^{-1}A$.*

Proof. Note $A_{t^{-1}} = R[t, t^{-1}]$. Hence $A_{t^{-1}}$ is a normal domain and one can use Corollary 4.3.10. □

Lemma 4.3.12 *Let A be a ring and x a regular element of A. If $A/(x)$ is reduced and $\mathfrak{p} \in \operatorname{Ass}_A A/(x)$, then $A_\mathfrak{p}$ is a normal domain.*

Proof. Let $\operatorname{Ass}_A A/(x) = \{\mathfrak{p}_1, \ldots, \mathfrak{p}_r\}$. Since $A/(x)$ is reduced

$$(x) = \mathfrak{p}_1 \cap \cdots \cap \mathfrak{p}_r,$$

hence $(x)A_{\mathfrak{p}_i} = \mathfrak{p}_i A_{\mathfrak{p}_i}$ for all i. Note that $\operatorname{ht}(\mathfrak{p}_i) = 1$, by Krull's principal ideal theorem. Therefore the maximal ideal of $A_{\mathfrak{p}_i}$ is generated by a system of parameters, that is, $A_{\mathfrak{p}_i}$ is a regular local ring and consequently $A_{\mathfrak{p}_i}$ is a normal domain by Theorem 2.4.16. □

Proposition 4.3.13 *Let A be a ring and x a regular element of A. If A_x is a normal domain and $A/(x)$ is reduced, then A is a normal domain.*

Proof. As A_x is a domain and x is a regular element on A one obtains that A is a domain. By Lemma 4.3.12 $A_\mathfrak{p}$ is normal for all $\mathfrak{p} \in \operatorname{Ass}_A A/(x)$, hence A must be normal according to Corollary 4.3.10. □

There is another graded algebra associated to an ideal I of a ring R whose properties are related to the normality of the Rees algebra of I, it is called the *associated graded ring* of I and is defined as:

$$\mathrm{gr}_I(R) = R/I \oplus I/I^2 \oplus \cdots \oplus I^i/I^{i+1} \oplus \cdots,$$

with multiplication

$$(a + I^i)(b + I^j) = ab + I^{i+j-1} \quad (a \in I^{i-1},\ b \in I^{j-1}).$$

Given a generating set $f_1, \ldots, f_q$ of I, it is not difficult to verify that $\mathrm{gr}_I(R)$ is a graded algebra over R/I generated by the following elements of degree one:

$$\overline{f}_1 = f_1 + I^2, \ldots, \overline{f}_q = f_q + I^2.$$

Thus one has $\mathrm{gr}_I(R) = (R/I)[\overline{f}_1, \ldots, \overline{f}_q]$.

Lemma 4.3.14 *If I is an ideal of a ring R, then*

$$R[It]/IR[It] \simeq \mathrm{gr}_I(R) \ \text{ and } \ A/t^{-1}A \simeq \mathrm{gr}_I(R),$$

where $A = R[It, t^{-1}]$ is the extended Rees algebra of I.

Proof. It is left as an exercise. □

Theorem 4.3.15 *Let I be an ideal of a ring R. If I is generated by a regular sequence $f_1, \ldots, f_q$, then the epimorphism of graded algebras:*

$$\varphi\colon (R/I)[t_1, \ldots, t_q] \longrightarrow \mathrm{gr}_I(R), \quad \varphi(t_i) = \overline{f}_i = f_i + I^2,$$

is an isomorphism, where $t_1, \ldots, t_q$ are indeterminates over R/I.

Proof. The proof is by induction on q, the case $q = 1$ is easy to show. Let J be the ideal $(f_1, \ldots, f_{q-1})$ and consider the epimorphism:

$$\varphi'\colon (R/J)[t_1, \ldots, t_{q-1}] \longrightarrow \mathrm{gr}_J(R) = (R/J)[\overline{f}_1, \ldots, \overline{f}_{q-1}],$$

$\varphi'(t_i) = \overline{f}_i = f_i + J^2$. Note $(J\colon f_q) = J$ because f_q is regular on R/J; moreover, since φ' is an isomorphism it follows (by induction on m) that one has the following equalities

$$(J^m\colon f_q) = J^m \text{ for all } m \geq 1.$$

Let $F \in R[\mathbf{t}] = R[t_1, \ldots, t_q]$ be a homogeneous polynomial of degree d such that its image in $(R/I)[\mathbf{t}]$ is in the kernel of φ, that is,

$$F(f) = F(f_1, \ldots, f_q) \in I^{d+1}.$$

As φ is graded it suffices to prove $F \in IR[\mathbf{t}]$. To show $F \in IR[\mathbf{t}]$ we proceed by induction on d, the case $d = 0$ is clear. There is $W \in R[\mathbf{t}]$ of degree $d+1$ with $F(f) = W(f)$, write $W = \sum_{i=1}^{q} t_i W_i$, where $W_i = 0$ or W_i is a homogeneous polynomial in $R[\mathbf{t}]$ of degree d. There are polynomials G in $R[t_1, \ldots, t_{q-1}]$ and H in $R[\mathbf{t}]$ of degrees d and $d-1$, respectively, such that

$$F' = F - \sum_{i=1}^{q} f_i W_i = G + t_q H.$$

Observe that $F'(f) = 0$, hence $H(f) \in (J^d \colon f_q) = J^d \subset I^d$ and by induction on d one concludes that H has coefficients in I. It only remains to prove that G has coefficients in I. There is a polynomial $H' \in R[t_1, \ldots, t_{q-1}]$ of degree d with $H(f) = H'(f)$. Set

$$F'' = G + f_q H' \in R[t_1, \ldots, t_{q-1}],$$

noting $F''(f) = F'(f) = 0$ and using that φ' is an isomorphism one derives that F'' has coefficients in J, which implies that G has coefficients in I as required. □

Theorem 4.3.16 *Let I be an ideal of a ring R and $A = R[It, t^{-1}]$ its extended Rees algebra. If R is a normal domain and $\mathrm{gr}_I(R)$ is reduced, then A is normal.*

Proof. As $A/t^{-1}A \simeq \mathrm{gr}_I(R)$ is reduced and $A_{t^{-1}} = R[t, t^{-1}]$ is a normal domain, by Proposition 4.3.13, we get that A is normal. □

Theorem 4.3.17 [236] *Let I be an ideal of a normal domain R. Then the following are equivalent:*

(a) *I is a normal ideal of R.*

(b) *The Rees algebra $R[It]$ is normal.*

(c) *The ideal $IR[It] \subset R[It]$ is complete.*

(d) *The ideal $(t^{-1}) \subset R[It, t^{-1}]$ is complete.*

(e) *The extended Rees algebra $R[It, t^{-1}]$ is normal.*

Proof. (a) ⇒ (b) Set $A = R[It]$. Let $z \in \overline{A} \subset R[t]$ and write $z = \sum_{i=0}^{s} b_i t^i$. It suffices to prove $b_s t^s \in A$. First we prove that $b_s t^s \in \overline{A}$. As z is almost integral over A there is $0 \neq f \in A$ such that $f z^n \in A$ for all $n > 0$. Hence there is $0 \neq f_m \in I^m$ such that $(f_m t^m)(b_s t^s)^n$ is in A for all $n > 0$; that is, $b_s t^s$ is almost integral over A. As R is a Noetherian integral domain using Proposition 2.6.37 one derives that $b_s t^s$ is integral over A. Thus there is an equation of the form

$$(b_s t^s)^m + a_1 (b_s t^s)^{m-1} + \cdots + a_{m-1}(b_s t^s) + a_m = 0,$$

where $a_i = \sum_{j=0}^{r_i} a_{ij}t^j$ and $a_{ij} \in I^j$. Grouping all the terms of t-degree equal to sm one has the equation

$$b_s^m + \sum_{i=1}^{m} a_{i,si} b_s^{m-i} = 0.$$

Thus b_s is integral over I^s and consequently $b_s \in I^s$, as required.

(b) $\Rightarrow$ (c) Let $z = b_0 + b_1 t + \cdots + b_s t^s$ be an element of $R[It]$ which is integral over $IR[It]$. Then z satisfies an equation of the form

$$z^m + a_1 z^{m-1} + \cdots + a_{m-1} z + a_m = 0, \; a_i \in I^i R[It],$$

multiplying by t^m one obtains that tz is integral over $R[It]$. Therefore $tz \in R[It]$, which proves that $z \in IR[It]$.

(c) $\Rightarrow$ (d) Set $B = R[It, t^{-1}]$. Let z be an element of B integral over $t^{-1}B$. As the negative part of the Laurent expansion of z is in $t^{-1}B$, one may assume $z = \sum_{i=0}^{s} b_i t^i$, $s \geq 0$ and $b_i \in I^i$ for all $i \geq 0$. By descending induction on s it suffices to prove that $b_s \in I^{s+1}$. There is an equation

$$z^m + a_1 z^{m-1} + \cdots + a_{m-1} z + a_m = 0, \; a_i \in (t^{-1})^i B,$$

hence zt is almost integral over B. It follows rapidly that $b_s t^{s+1}$ is almost integral over B and thus $b_s t^{s+1}$ is integral over B. A direct calculation yields that $b_s t^{s+1}$ is integral over $IR[It]$, which shows that b_s is in I^{s+1}.

(d) $\Rightarrow$ (e) Set $u = t^{-1}$ and note $R[It, u]_u = R[t, u]$ is a normal domain. Therefore, by Lemma 4.3.5, one concludes that $R[It, u]$ is normal.

(e) $\Rightarrow$ (a) If $z \in \overline{I^r}$, then z satisfies a polynomial equation

$$z^m + a_1 z^{m-1} + \cdots + a_{m-1} z + a_m = 0, \; a_i \in I^{ri},$$

thus multiplying by t^{rm} we get that the element zt^r is integral over the ring $R[It, t^{-1}]$. Hence $z \in I^r$. □

Corollary 4.3.18 *Let I be an ideal of a normal domain R and $R[It]$ its Rees algebra. If $\mathrm{gr}_I(R)$ is reduced, then $R[It]$ is normal.*

Proof. It follows from Theorems 4.3.16 and 4.3.17. □

Theorem 4.3.19 (Huneke [254]) *Let R be a Cohen–Macaulay ring and let I be an ideal of R containing regular elements. If $R[It]$ is Cohen–Macaulay, then $\mathrm{gr}_I(R)$ is Cohen–Macaulay.*

An ideal I of a ring R is a *radical ideal* if $I = \mathrm{rad}\,(I)$. Note: (i) a proper ideal I is radical if and only if I is an intersection of finitely many primes, and (ii) $\overline{I} \subset \mathrm{rad}\,(I)$ and equality occurs if I is a radical ideal.

Corollary 4.3.20 *Let I be a radical ideal of a normal domain R. If I is generated by a regular sequence, then $\mathrm{gr}_I(R)$ is reduced and $R[It]$ is a normal domain.*

Proof. It follows from Theorem 4.3.15 and Corollary 4.3.18. □

Example 4.3.21 Let $R = \mathbb{Q}[x, y]$ be a polynomial ring over the field $\mathbb{Q}$ and let I be the ideal (x^2, y^2). Observe that $\overline{I}$ is equal to (x^2, y^2, xy). Thus $R[It]$ is not normal because I is not even complete.

Definition 4.3.22 Let I be an ideal of a ring R and $\mathfrak{p}_1, \ldots, \mathfrak{p}_r$ the minimal primes of I. Given an integer $n \geq 1$, the nth *symbolic power* of I is defined to be the ideal

$$I^{(n)} = \mathfrak{q}_1 \cap \cdots \cap \mathfrak{q}_r,$$

where $\mathfrak{q}_i$ is the primary component of I^n corresponding to $\mathfrak{p}_i$.

Let $(R, \mathfrak{m})$ be a regular local ring and I an unmixed ideal. An interesting problem is whether $I^{(2)}$ is contained in $\mathfrak{m}I$, although the answer is negative in general, the problem remains open in characteristic zero. For some insight into this problem see [132, 257].

Proposition 4.3.23 *Let I be a proper ideal of a ring R and $S = R \setminus \cup_{i=1}^{r} \mathfrak{p}_i$, where $\mathfrak{p}_1, \ldots, \mathfrak{p}_r$ are the minimal primes of I. Then*

$$I^{(n)} = S^{-1}I^n \cap R \quad \textit{for } n \geq 1.$$

Proof. Let $I^n = \mathfrak{q}_1 \cap \cdots \cap \mathfrak{q}_r \cap \mathfrak{q}_{r+1} \cap \cdots \cap \mathfrak{q}_s$ be a primary decomposition, where $\mathfrak{q}_i$ is the primary component of $\mathfrak{p}_i$ for $i \leq r$ and $\mathfrak{q}_i$ is a primary component of the embedded prime $\mathfrak{p}_i$ for $i > r$. Since $\mathfrak{p}_i \cap S \neq \emptyset$ for $i > r$ one has $S^{-1}\mathfrak{q}_i = S^{-1}R$. Hence

$$S^{-1}I^n = S^{-1}\left(\bigcap_{i=1}^{s} \mathfrak{q}_i\right) = \bigcap_{i=1}^{r} S^{-1}\mathfrak{q}_i.$$

To finish the argument note that $S^{-1}\mathfrak{q}_i \cap R = \mathfrak{q}_i$ and intersect with R the equality above. □

Proposition 4.3.24 *Let I be a radical ideal of a ring R and $\mathfrak{p}_1, \ldots, \mathfrak{p}_r$ the minimal primes of I. Then*

$$I^{(n)} = \mathfrak{p}_1^{(n)} \cap \cdots \cap \mathfrak{p}_r^{(n)} \quad \textit{for } n \geq 1.$$

Proof. Let $I^n = \mathfrak{q}_1 \cap \cdots \cap \mathfrak{q}_r \cap \mathfrak{q}_{r+1} \cap \cdots \cap \mathfrak{q}_s$ be a primary decomposition of I^n, where $\mathfrak{q}_i$ is $\mathfrak{p}_i$-primary for $i \leq r$ and $\mathfrak{q}_i$ is an embedded primary component of I^n for $i > r$. Localizing at $\mathfrak{p}_i$ yields $I^nR_{\mathfrak{p}_i} = \mathfrak{q}_iR_{\mathfrak{p}_i}$ and from $I = \mathfrak{p}_1 \cap \cdots \cap \mathfrak{p}_r$ one obtains:

$$I^nR_{\mathfrak{p}_i} = (IR_{\mathfrak{p}_i})^n = (\mathfrak{p}_iR_{\mathfrak{p}_i})^n = \mathfrak{p}_i^nR_{\mathfrak{p}_i}.$$

Thus $\mathfrak{p}_i^nR_{\mathfrak{p}_i} = \mathfrak{q}_iR_{\mathfrak{p}_i}$ and contracting one has $\mathfrak{p}_i^{(n)} = \mathfrak{q}_i$, as required. □

This result and also Exercise 6.1.25 have interesting generalizations to monomial ideals [92, Theorem 3.7].

Proposition 4.3.25 *Let R be a polynomial ring over a field K and let I be an ideal of R generated by square-free monomials. If $n \geq 1$ and $\mathfrak{p}_1, \ldots, \mathfrak{p}_r$ are the minimal primes of I, then*

$$I^{(n)} = \mathfrak{p}_1^n \cap \cdots \cap \mathfrak{p}_r^n.$$

Proof. From Proposition 6.1.7 one has $\mathfrak{p}_i^n = \mathfrak{p}_i^{(n)}$. Since I is a radical ideal the result follows from Proposition 4.3.24. □

Corollary 4.3.26 *Let R be a polynomial ring over a field K. If I is an ideal of R generated by square-free monomials, then $I^{(n)}$ is integrally closed for $n \geq 1$.*

Proof. Set $J = I^{(n)}$. Let $\mathfrak{p}_1, \ldots, \mathfrak{p}_r$ be the associated primes of I. If $f \in \overline{J}$, then there is $m \geq 1$ so that $f^m \in \mathfrak{p}_i^{nm}$ for all i. By Corollary 4.3.20 $\mathfrak{p}_i^n$ is complete, thus $f \in \mathfrak{p}_i^n$ for all i. Hence Proposition 4.3.25 shows $f \in J$. □

Corollary 4.3.27 *Let R be a polynomial ring over a field K and let I be a monomial ideal. If I is a radical ideal, then $\overline{I^n} \subset I^{(n)}$ for $n \geq 1$.*

Proof. Let $x \in \overline{I^n}$, there is $k \geq 1$ so that $x^k \in (I^n)^k \subset (I^{(n)})^k$, thus x is in $\in \overline{I^{(n)}}$. Note that $I^{(n)}$ is complete by Corollary 4.3.26, hence $x \in I^{(n)}$. □

Definition 4.3.28 An ideal I of a ring R is called *normally torsion-free* if $\mathrm{Ass}(R/I^i)$ is contained in $\mathrm{Ass}(R/I)$ for all $i \geq 1$ and $I \neq R$.

Brodmann [56] showed that when R is a Noetherian ring and I is an ideal of R the sets $\mathrm{Ass}(R/I^n)$ stabilize for large n. If I is a radical ideal which is normally torsion-free, then $\mathrm{Ass}(R/I^n) = \mathrm{Ass}(R/I)$ for all $n \geq 1$; that is, the notion of normally torsion-free is a strong form of stability.

An aspect of symbolic powers that has attracted a lot of attention is to describe when the symbolic and ordinary powers of a given ideal I coincide; see [249, 333] and [185, 188]. Later in this chapter, we present a few cases where equality of symbolic and ordinary powers can be described in terms of properties of the associated graded ring.

In Chapter 14 we characterize normally torsion-free monomial ideals in algebraic and combinatorial optimization terms (see Theorem 14.3.6).

For a certain type of ideals, the next result characterizes normally torsion-free ideals as those ideals whose ordinary and symbolic powers coincide.

Proposition 4.3.29 *Let I be an ideal of a ring R. If I has no embedded primes, then I is normally torsion-free if and only if $I^n = I^{(n)}$ for all $n \geq 1$.*

Proof. $\Rightarrow$) Note $\mathrm{Ass}(R/I^n) = \mathrm{Ass}(R/I)$ for $n \geq 1$, because any associated prime $\mathfrak{p}$ of I is a minimal prime of I^n and thus $\mathfrak{p} \in \mathrm{Ass}(R/I^n)$. As I^n has no embedded primes one concludes that I^n has a unique irredundant minimal primary decomposition and $I^n = I^{(n)}$ for $n \geq 1$.

$\Leftarrow$) Let $\mathfrak{p}_1, \ldots, \mathfrak{p}_r$ be the associated primes of I. Since $\mathfrak{p}_i$ is a minimal prime of I for all i, one derives

$$\mathrm{Ass}(R/I^n) = \mathrm{Ass}(R/I^{(n)}) = \{\mathfrak{p}_1, \ldots, \mathfrak{p}_r\}. \qquad \Box$$

Proposition 4.3.30 *Let R be a polynomial ring with coefficients in a field. If I is a radical monomial ideal of R which is normally torsion-free, then its Rees algebra $R[It]$ is a normal domain.*

Proof. By Proposition 4.3.29 $I^n = I^{(n)}$ for $n \geq 1$. Hence, thanks to Corollary 4.3.27, I^n is integrally closed for $n \geq 1$, i.e., $R[It]$ is normal. $\Box$

Proposition 4.3.31 *Let I be an ideal of a Cohen–Macaulay ring R. If I is generated by a regular sequence, then $I^n = I^{(n)}$ for $n \geq 1$.*

Proof. Let $f_1, \ldots, f_r$ be an R-regular sequence that generates I. By Krull's principal ideal theorem $\mathrm{ht}\,(I) = r$. Hence, by Theorem 2.3.25, I is unmixed. From Theorem 4.3.15 there is an isomorphism of graded rings

$$\varphi\colon B = (R/I)[t_1, \ldots, t_r] \longrightarrow \mathrm{gr}_I(R), \quad t_i \longmapsto f_i + I^2,$$

where B is a polynomial ring over R/I. Hence I^i/I^{i+1} is a free R/I-modulo because it is isomorphic to B_i, the ith graded component of B. Thus

$$\mathrm{Ass}_R(I^i/I^{i+1}) = \mathrm{Ass}_R(R/I).$$

Using the exact sequence $0 \longrightarrow I^i/I^{i+1} \longrightarrow R/I^{i+1} \longrightarrow R/I^i \longrightarrow 0$, it follows by induction that $\mathrm{Ass}_R(R/I^n) \subset \mathrm{Ass}_R(R/I)$ for $n \geq 1$. To finish the proof, we apply Proposition 4.3.29 to conclude the equality between the ordinary and symbolic powers of I. $\Box$

Definition 4.3.32 A proper ideal I of a ring R is said to be *locally a complete intersection* if $IR_\mathfrak{p}$ is a complete intersection for all $\mathfrak{p} \in V(I)$.

Proposition 4.3.33 *Let R be a Cohen–Macaulay ring and I a prime ideal. If I is locally a complete intersection, then $I^n = I^{(n)}$ for all $n \geq 1$.*

Proof. By Proposition 4.3.29 it suffices to prove $\mathrm{Ass}(R/I^n) \subset \mathrm{Ass}(R/I)$ for $n \geq 1$. If $\mathfrak{p}$ is an associated prime of R/I^n, then $\mathfrak{p}R_\mathfrak{p}$ is an associated prime of $R_\mathfrak{p}/I_\mathfrak{p}^n$. Using Proposition 4.3.31 yields that $I_\mathfrak{p}$ is normally torsion-free, that is, the only associated prime of $R_\mathfrak{p}/I_\mathfrak{p}^n$ is $I_\mathfrak{p}$. Therefore $\mathfrak{p}R_\mathfrak{p} = I_\mathfrak{p}$ and $\mathfrak{p} = I$. $\Box$

Proposition 4.3.34 *Let $\mathfrak{p}$ be a prime ideal of a ring R such that $\mathfrak{p}R_\mathfrak{p}$ is a complete intersection, then $\mathfrak{p}^{(n)} = \mathfrak{p}^n$ for all $n \geq 1$ if and only if $\mathrm{gr}_\mathfrak{p}(R)$ is a domain.*

Proof. $\Rightarrow$) Since $\mathrm{gr}_\mathfrak{p}(R) = R[\mathfrak{p}t]/\mathfrak{p}R[\mathfrak{p}t]$, it is enough to show that $\mathfrak{p}R[\mathfrak{p}t]$ is a prime ideal of $R[\mathfrak{p}t]$. Let $x = a_0 + a_1t + \cdots + a_rt^r$ be an element in $R[\mathfrak{p}t]$ and $x' = (a_0/1) + (a_1/1)t + \cdots + (a_r/1)t^r$ the image of x in $R_\mathfrak{p}[\mathfrak{p}R_\mathfrak{p}t]$. Note that a_i is in $\mathfrak{p}^{i+1}$ if and only if $(a_i/1)$ is in $\mathfrak{p}^{i+1}R_\mathfrak{p}$, because the ordinary and symbolic powers of $\mathfrak{p}$ coincide, thus x is in $\mathfrak{p}R[\mathfrak{p}t]$ if and only if x' is in $\mathfrak{p}R_\mathfrak{p}[\mathfrak{p}R_\mathfrak{p}t]$. As a consequence $\mathfrak{p}R[\mathfrak{p}t]$ is prime if and only if $\mathfrak{p}R_\mathfrak{p}[\mathfrak{p}R_\mathfrak{p}t]$ is prime. To finish the argument use Theorem 4.3.15 and the hypothesis that $\mathfrak{p}R_\mathfrak{p}$ is generated by a regular sequence to get that $\mathrm{gr}_{\mathfrak{p}R_\mathfrak{p}}(R_\mathfrak{p})$ is a domain.

$\Leftarrow$) First we prove $\mathrm{Ass}_R(\mathfrak{p}^i/\mathfrak{p}^{i+1}) = \{\mathfrak{p}\}$ for $i \geq 1$. Let $\mathfrak{p}_1$ be an associated prime of $\mathfrak{p}^i/\mathfrak{p}^{i+1}$, that is, $\mathfrak{p}_1 = \mathrm{ann}\,(x + \mathfrak{p}^{i+1})$, for some x in $\mathfrak{p}^i \setminus \mathfrak{p}^{i+1}$. Note $a \in \mathfrak{p}_1$ iff $ax \in \mathfrak{p}^{i+1}$, and since $\mathrm{gr}_\mathfrak{p}(R)$ is a domain, one readily concludes that $a \in \mathfrak{p}_1$ iff $a \in \mathfrak{p}$. Hence $\mathfrak{p}_1 = \mathfrak{p}$. There is an exact sequence:

$$0 \longrightarrow \mathfrak{p}^i/\mathfrak{p}^{i+1} \longrightarrow R/\mathfrak{p}^{i+1} \longrightarrow R/\mathfrak{p}^i \longrightarrow 0.$$

Hence, using induction on $i \geq 1$, Lemma 2.1.17 and the exact sequence above, we get $\mathrm{Ass}(R/\mathfrak{p}^i) \subset \mathrm{Ass}(R/\mathfrak{p}) = \{\mathfrak{p}\}$. Thus $\mathfrak{p}$ is normally torsion-free and by Proposition 4.3.29 one has $\mathfrak{p}^n = \mathfrak{p}^{(n)}$ for all $n \geq 1$. $\Box$

Theorem 4.3.35 [383] *Let R be a normal domain and I a radical ideal which is generically a complete intersection. If I is normally torsion-free, then its Rees algebra $R[It]$ is a normal domain.*

Proof. By Corollary 4.3.18 we need only show that the associated graded ring $\mathrm{gr}_I(R) = R[It]/IR[It]$ is reduced. Let

$$\overline{f} = a_0 + a_1t + \cdots + a_st^s + IR[It]$$

be a nilpotent element of $\mathrm{gr}_I(R)$. Hence $(a_st^s)^m$ is in $IR[It]$ for some $m \geq 1$, by induction it suffices to verify that a_st^s is in $IR[It]$. Let $\mathfrak{p}_1, \ldots, \mathfrak{p}_r$ be the minimal primes of I. Since $IR_{\mathfrak{p}_i} = \mathfrak{p}_iR_{\mathfrak{p}_i}$ is a complete intersection one derives that $\mathrm{gr}_{\mathfrak{p}_iR_{\mathfrak{p}_i}}(R_{\mathfrak{p}_i})$ is reduced (see Corollary 4.3.20). Therefore the image of a_st^s in $\mathrm{gr}_{\mathfrak{p}_iR_{\mathfrak{p}_i}}(R_{\mathfrak{p}_i})$ is zero for all i, and one readily concludes that a_s belongs to the following intersection:

$$(R \cap \mathfrak{p}_1^{s+1}R_{\mathfrak{p}_1}) \cap \cdots \cap (R \cap \mathfrak{p}_r^{s+1}R_{\mathfrak{p}_r}) = \mathfrak{p}_1^{(s+1)} \cap \cdots \cap \mathfrak{p}_r^{(s+1)}.$$

As I is normally torsion-free one can write $I^n = \mathfrak{q}_1 \cap \cdots \cap \mathfrak{q}_r$, where $\mathfrak{q}_i$ is $\mathfrak{p}_i$-primary, localizing yields $I^nR_{\mathfrak{p}_i} = \mathfrak{p}_i^nR_{\mathfrak{p}_i} = \mathfrak{q}_iR_{\mathfrak{p}_i}$ and consequently $\mathfrak{q}_i = \mathfrak{p}_i^{(n)}$. Making n equal to $s+1$ proves that a_s is in I^{s+1}. $\Box$

Gorenstein Rees algebras For use below $R = K[x_1, \ldots, x_n]$ will denote a polynomial ring over a field K and $S = R[It]$ will denote the Rees algebra of an ideal I of R, which is assumed to be Cohen–Macaulay.

Definition 4.3.36 If $(1,t)^m = R \oplus Rt \oplus \cdots \oplus Rt^m \oplus It^{m+1} \cdots$, we say that the canonical module ω_S of S has the *expected form* if $\omega_S = \omega_R(1,t)^m$ for some $m \geq -1$.

Theorem 4.3.37 [235] *If ω_S has the expected form, then $\omega_S \simeq \omega_R(1,t)^{g-2}$ if and only if I is generically a complete intersection.*

Lemma 4.3.38 [412, Page 142] *If S is a regular local ring and J is an ideal of S generated by a regular sequence $h_1, \ldots, h_g$, then the Rees algebra $S[Jt]$ is determinantal:*

$$S[Jt] \simeq S[z_1, \ldots, z_g]/I_2 \begin{pmatrix} z_1 & \cdots & z_g \\ h_1 & \cdots & h_g \end{pmatrix}$$

and its canonical module is $\omega_S(1,t)^{g-2}$.

Proposition 4.3.39 *If I is an ideal of height g generated by square-free monomials and $S = R[It]$ is Gorenstein, then $g = 1$ or $g = 2$.*

Proof. Assume $g > 1$. Then $\omega_S \simeq S = (1,t)^0 = R(1,t)^0 = \omega_R(1,t)^0$ and therefore ω_S has the expected form with $m = 0$. Clearly I is generically a complete intersection because I is the intersection of prime ideals of R; see Proposition 6.1.4. Thus, by Theorem 4.3.37, we get

$$S \simeq \omega_S \simeq \omega_R(1,t)^{g-2} = R \oplus Rt \oplus \cdots \oplus Rt^{g-2} \oplus It^{g-1} \oplus \cdots \tag{4.1}$$

Take a minimal prime $\mathfrak{p}$ of I of height g. Then $S_\mathfrak{p} = R_\mathfrak{p}[I_\mathfrak{p}t]$ is the Rees algebra of the ideal $I_\mathfrak{p}$, which is generated by a regular sequence. Thus localizing the extremes of Eq. (4.1) at $\mathfrak{p}$ and using Lemma 4.3.38 we obtain

$$S_\mathfrak{p} = R_\mathfrak{p}[I_\mathfrak{p}t] \simeq \omega_{R_\mathfrak{p}}(1,t)^{g-2} \simeq \omega_{S_\mathfrak{p}}.$$

Note that it is important to know a priori that the canonical module of $S_\mathfrak{p}$ is $\omega_{R_\mathfrak{p}}(1,t)^{g-2}$. Hence $S_\mathfrak{p}$ is Gorenstein. To finish the proof note that the only Gorenstein determinantal rings that occur in Lemma 4.3.38 are those with $g = 2$; see [65, 71, 73]. □

Descent of normality Let $A \subset B$ be an extension of rings such that B is normal; in general the normality of A is not inherited from B. We discuss some sufficient conditions for A to be normal.

Lemma 4.3.40 *Let $A \subset B$ be an extension of rings. If $B = A \oplus C$ (as A-modules), then $IB \cap A = I$ for every ideal I of A.*

Proof. Let $z \in IB \cap A$ and write $z = \sum_{i=1}^{q} b_i f_i$, where $b_i \in B$ and $f_i \in I$. By hypothesis $b_i = a_i + c_i$, $a_i \in A$ and $c_i \in C$. Since $z \in A$ it follows that $z = \sum_{i=1}^{q} a_i f_i \in I$. This proves the containment $IB \cap A \subset I$, the reverse containment is clear. $\Box$

Proposition 4.3.41 *Let $A \subset B$ be integral domains with field of fractions K_A and K_B, respectively. If $B = A \oplus C$ (as A-modules), then $K_A \cap B \subset A$. In particular if B is normal, then A is normal.*

Proof. By Lemma 4.3.40 one has $IB \cap A = I$ for every ideal I of A. Let $b = a/c \in B$, $a, c \in A$, then $a \in (c)B \cap A = (c)$, hence $a = \lambda c = bc$, with $\lambda \in A$. Therefore $b = \lambda \in A$. $\Box$

Proposition 4.3.42 *Let R be a polynomial ring over a field K and I an ideal of R generated by homogeneous polynomials $f_1, \ldots, f_q$. Assume $\deg(f_i) = d$ for all i. If $R[It]$ is normal, then $K[f_1, \ldots, f_q]$ is normal.*

Proof. Let $\mathfrak{m} = R_+$ be the irrelevant maximal ideal of the polynomial ring R and $A = K[tf_1, \ldots, tf_q]$. Observe that there is a decomposition of A-modules: $R[It] = K[tf_1, \ldots, tf_q] \oplus \mathfrak{m}R[It]$. As $A \simeq K[f_1, \ldots, f_q]$ (as rings), by Proposition 4.3.41, the result follows. $\Box$

Exercises

4.3.43 Let $R = K[x_1, \ldots, x_n]$ be a polynomial ring over a field K and let $\mathfrak{m}$ be the maximal ideal $(x_1, \ldots, x_n)$. If $I = \mathfrak{m}^d$ and $J = (x_1^d, \ldots, x_n^d)$ for some positive integer d, then $\overline{R[Jt]} = R[It]$.

4.3.44 Let I, J be ideals of a normal domain R. Define the *multi Rees algebra* of I, J as $\mathcal{R}(I \oplus J) = R[uI, vJ]$, here u, v are new variables. Then

$$\overline{\mathcal{R}(I \oplus J)} = \oplus \overline{I^n}\ \overline{J^m} u^n v^m.$$

4.3.45 Let R be a ring and $\mathcal{F} = \{I_i\}_{i \in \mathbb{N}}$ a family of ideals in R. It is said that $\mathcal{F}$ is a *filtration* of R, if $I_{i+1} \subset I_i$, $I_0 = R$, and $I_i I_j \subset I_{i+j}$ for all $i, j \in \mathbb{N}$. If I is an ideal of R, prove that the following are filtrations of R:

(a) $I_i = I^i$, the ordinary powers of I,

(b) $I_i = \overline{I^i}$, the integral closure of I^i,

(c) $I_i = I^{(i)}$, the symbolic powers of I, and

(d) $I_n = \oplus_{i \geq n} R_i$, where $R = \oplus_{i \geq 0} R_i$ is a graded ring.

Hint For (b) use the method of proof of Proposition 4.3.3.

4.3.46 Let R be a normal domain and $\mathcal{F} = \{I_i\}_{i\in\mathbb{N}}$ a filtration of R. If I_i is integrally closed for all i, prove that the *Rees algebra*

$$\mathcal{R}(\mathcal{F}) = \bigoplus_{i=0}^{\infty} I_i t^i \subset R[t]$$

of the filtration $\mathcal{F}$ is integrally closed. See [198] for a study of the Cohen–Macaulay and Gorenstein property of $\mathcal{R}(\mathcal{F})$.

4.3.47 Let R be an integral domain. If $\{R_\alpha\}_{\alpha\in\Lambda}$ is a collection of normal subrings of R, then $\cap_{\alpha\in\Lambda} R_\alpha$ is normal.

4.3.48 Let R be a ring and I, J ideals of R. If I and J are integrally closed, prove that $I \cap J$ is integrally closed.

4.3.49 If R is a domain and $I \subset R$ is an ideal, then $\bigoplus_{i=0}^{\infty} \overline{I^i} t^i \subset \overline{R[It]}$ with equality if R is a normal domain.

4.3.50 Let R be a domain and let K be its field of fractions. If $K[x]$ is a polynomial ring in one variable, then $\overline{R}[x] \subset \overline{R[x]} \subset K[x]$ and $\overline{R}[x] = \overline{R[x]}$.

4.3.51 Let F and G be two finite sets of monomials in a polynomial ring R over a field K. If $(F) = (G)$ prove that $R[Ft] = R[Gt]$.

4.3.52 If $F = \{x_1x_2, x_3x_4x_5, x_1x_3, x_2x_4, x_2x_5, x_1x_5\} \subset K[x_1, \ldots, x_5]$ and $I = (F)$, prove that $K[F]$ is not normal and $\mathcal{R}(I)$ is normal.

4.3.53 If $R = \mathbb{Q}[x_1, \ldots, x_7]$ and I is the principal ideal generated by the binomial $f = x_1x_3x_5x_7 - x_2x_4^2x_6$, prove that R/I is a normal domain. Give a description of the irreducible binomials f such that $R/(f)$ is normal.

4.3.54 Let R be a Noetherian ring and a an element in the Jacobson radical of R. If a is regular and $R/(a)$ is an integral domain, prove that R is an integral domain.

4.3.55 Let $R = K[\mathbf{x}]$ be a polynomial ring over a field K. If I is an ideal generated by square-free monomials of degree 2, then I and I^2 are complete.

4.3.56 Let $R = K[x_1, \ldots, x_n]$ be a polynomial ring and let $\mathcal{R}(I)$ be the Rees algebra of an ideal $I = (f_1, \ldots, f_q)$, where K is a field. If f_i is homogeneous of degree $d \geq 1$ for all i. Prove

(a) $\mathcal{R}(I) \simeq \mathcal{R}(I)/\mathfrak{m}\mathcal{R}(I) \oplus \mathfrak{m}\mathcal{R}(I)$ as $\mathcal{R}(I)$ modules, where $\mathfrak{m} = R_+$,

(b) $K[f_1, \ldots, f_q] \simeq \mathcal{R}(I)/\mathfrak{m}\mathcal{R}(I)$ as K-algebras,

(c) if $F = \{f_1, \ldots, f_q\}$ is a set of monomials, then $\mathcal{R}(I) \simeq K[F, t\mathbf{x}]$ as K-algebras, where $\mathbf{x} = \{x_1, \ldots, x_n\}$.

4.4 Multiplicities and a criterion of Herzog

In this section we present an elegant and useful Cohen–Macaulay criterion due to Herzog (Theorem 4.4.13). Once more we recall that all rings considered in this book are Noetherian and modules are finitely generated.

Quasi regular sequences It is useful to generalize Theorem 4.3.15 to modules by introducing a polynomial with coefficients in a module. Let M be an R-module and $R[t_1,\ldots,t_q]$ a polynomial ring over the ring R. Set

$$M[\mathbf{t}] = M[t_1,\ldots,t_q] = M\otimes_R R[t_1,\ldots,t_q],$$

and note that an element of $M[\mathbf{t}]$ can be regarded as a polynomial in the t_i variables with coefficients in M, thus $M[\mathbf{t}]$ is naturally graded. If $f_1,\ldots,f_q$ is a sequence in R and $I=(f_1,\ldots,f_q)$, there is a degree preserving map of additive groups

$$\varphi\colon (M/IM)[t_1,\ldots,t_q] \longrightarrow \mathrm{gr}_I(M) = \bigoplus_{i=0}^{\infty} I^iM/I^{i+1}M,$$

such that $\varphi(\overline{F(\mathbf{t})}) = F(f_1,\ldots,f_q)+I^{i+1}M$, for all $F(\mathbf{t})$ in $M[\mathbf{t}]$ homogeneous of degree i, where the notation $\overline{F(\mathbf{t})}$ means reducing the coefficients of $F(\mathbf{t})$ modulo IM. It is not hard to verify the equivalence between the following two conditions:

(a) φ is injective.

(b) For every homogeneous polynomial F in $M[\mathbf{t}]$ of positive degree n such that $F(\mathbf{f})\in I^{n+1}M$, one has $F\in IM[\mathbf{t}]$.

Definition 4.4.1 If the map φ is an isomorphism and $IM\neq M$, the sequence $f_1,\ldots,f_q$ is called an *M-quasi-regular sequence*.

Theorem 4.4.2 *Let M be an R-module and $\mathbf{f}=f_1,\ldots,f_q$ a sequence in R. If $\mathbf{f}$ is an M-regular sequence, then $\mathbf{f}$ is an M-quasi-regular sequence.*

Proof. It follows adapting the proof of Theorem 4.3.15. $\Box$

Theorem 4.4.3 (Krull's intersection theorem [310, Theorem 8.9]) Let M be an R-module and let I be an ideal of R. Then there is $a\in R$ such that $a\equiv 1 \bmod (I)$ and $a\cdot\left(\bigcap_{i=1}^{\infty} I^iM\right)=0$.

Lemma 4.4.4 *Let $(R,\mathfrak{m})$ be a local ring (resp. $\mathbb{N}$-graded) and M an R-module (resp. $\mathbb{N}$-graded). If N is a submodule of M (resp. graded submodule) and $I\subset\mathfrak{m}$ is an ideal (resp. graded ideal), then $N=\bigcap_{i=1}^{\infty}(N+I^iM)$.*

Proof. Using Krull's intersection theorem one has $\bigcap_{i=1}^{\infty} I^iM' = (0)$ (this equality holds also in the graded case), where $M' = M/N$. Hence the required equality follows at once. $\square$

Proposition 4.4.5 *Let $(R, \mathfrak{m})$ be a local ring (resp. $\mathbb{N}$-graded ring) and M an R-module (resp. $\mathbb{N}$-graded module). If $\mathbf{f} = f_1, \ldots, f_q$ is an M-quasi regular sequence of elements in $\mathfrak{m}$ (resp. homogeneous elements in $\mathfrak{m} = R_+$), then $\mathbf{f}$ is an M-regular sequence.*

Proof. Fix an integer $1 \leq r \leq q$ and set $I = (f_1, \ldots, f_q)$. To begin with we split the ideal I as $I = J + L$, where $J = (f_1, \ldots, f_{r-1})$ and $L = (f_r, \ldots, f_q)$, it is convenient to set $J = (0)$ if $r = 1$. It suffices to show the equality $(JM\colon {}_M f_r) = JM$, because this is equivalent to prove that f_r is not a zero divisor of M/JM. Let $m \in (JM\colon {}_M f_r)$, according to Lemma 4.4.4 one has $\bigcap_{n=1}^{\infty}(JM + L^nM) = JM$, thus the proof reduces to proving by induction on n that $m \in JM + L^nM$ for $n \geq 1$. Since $m \in (JM\colon {}_M f_r)$, there are m_i in M such that

$$mf_r = m_1f_1 + \cdots + m_{r-1}f_{r-1} \quad (m, m_i \in M). \tag{4.2}$$

Consider the polynomial $F = mt_r - m_1t_1 - \cdots - m_{r-1}t_{r-1} \in M[\mathbf{t}]$. Note $\deg_t(F) = 1$ and $F(\mathbf{f}) = 0$. As $\mathbf{f}$ is an M-quasi regular sequence we get $m \in IM = JM + LM$. By induction assume $m \in JM + L^nM$, that is, there is $G \in M[t_r, \ldots, t_q]$ homogeneous of degree n such that

$$m = b_1f_1 + \cdots + b_{r-1}f_{r-1} + G(f_r, \ldots, f_q) \quad (b_i \in M). \tag{4.3}$$

From Eqs. (4.2) and (4.3) we get $f_rG(f_r, \ldots, f_q) = a_1f_1 + \cdots + a_{r-1}f_{r-1}$, $a_i \in M$. Using that $\mathbf{f}$ is an M-quasi regular sequence one may assume (after an induction argument) that $a_i \in I^nM$, and thus $t_rG(t_r, \ldots, t_q)$ is in $IM[\mathbf{t}]$, that is, $G(t_r, \ldots, t_q)$ is in $IM[\mathbf{t}]$. One can write $G = G_1 + G_2$, where G_1, G_2 are homogeneous polynomials in $M[\mathbf{t}]$ of degree n with $G_1 \in JM[\mathbf{t}]$ and $G_2 \in LM[\mathbf{t}]$. Since $G_1(f_r, \ldots, f_q)$ is in JM and $G_2(f_r, \ldots, f_q)$ is in $L^{n+1}M$, from Eq. (4.3) we get that m is in $JM + L^{n+1}M$, as required. $\square$

Lemma 4.4.6 *Let $f(t) \in \mathbb{Q}[t]$ be a polynomial of degree $d-1$ such that $f(n) \in \mathbb{Z}$ for $n \in \mathbb{Z}$, then there are unique integers $a_0, \ldots, a_{d-1}$ such that*

$$f(t) = \sum_{i=0}^{d-1} a_i f_i(t), \quad \textit{where} \quad f_i(t) = \binom{t+i}{i}.$$

Proof. The polynomials $f_i(t)$, $i \in \mathbb{N}$, are a basis for $\mathbb{Q}[t]$ as a $\mathbb{Q}$-vector space. Hence $f(t) = \sum_{i=0}^{d-1} a_i f_i(t)$, for some $a_i \in \mathbb{Q}$. Using the Pascal triangle we get

$$f(t) - f(t-1) = \sum_{i=0}^{d-1} a_i \left[\binom{t+i}{i} - \binom{t+i-1}{i}\right] = \sum_{i=0}^{d-1} a_i \binom{t+i-1}{i-1},$$

thus by induction on the degree it follows that $a_i \in \mathbb{Z}$ for all i. $\square$

Multiplicities Let $(R, \mathfrak{m})$ be a local ring, M a finitely generated R-module and $\mathfrak{q}$ an ideal with $\mathrm{rad}\,(\mathfrak{q}) = \mathfrak{m}$. As $\mathrm{gr}_{\mathfrak{q}}(M)$ is a finitely generated $\mathrm{gr}_{\mathfrak{q}}(R)$-module and $A_0 = R/\mathfrak{q}$ is Artinian, by Theorem 2.2.4 there exists a polynomial $P(t) \in \mathbb{Q}[t]$ of degree $d-1$, where $d = \dim(\mathrm{gr}_{\mathfrak{q}}(M))$, such that

$$P(i) = \ell_{A_0}(\mathfrak{q}^i M/\mathfrak{q}^{i+1}M), \quad \text{for } i \geq n_0.$$

By Lemma 4.4.6 there are integers $a_0, \ldots, a_{d-1}$ such that

$$P(i) = \sum_{j=0}^{d-1} a_j \binom{i+j}{j}, \quad \text{for all } i \geq 0.$$

From the short exact sequences

$$0 \longrightarrow \mathfrak{q}^i M/\mathfrak{q}^{i+1}M \longrightarrow M/\mathfrak{q}^{i+1}M \longrightarrow M/\mathfrak{q}^i M \longrightarrow 0,$$

one derives $\ell(M/\mathfrak{q}^{i+1}M) - \ell(M/\mathfrak{q}^i M) = P(i)$ for $i \geq n_0$. Hence, using the identity of Exercise 4.4.14, we get

$$\begin{aligned} \chi_M^{\mathfrak{q}}(i) := \ell(M/\mathfrak{q}^{i+1}M) &= \ell(M/\mathfrak{q}^{n_0}M) + \sum_{j=n_0}^{i} P(j) \\ &= c_0 + \sum_{j=0}^{d-1} a_j \binom{j+i+1}{j+1}, \end{aligned}$$

for $i \geq n_0$. Thus the function $\chi_M^{\mathfrak{q}}(i) = \ell(M/\mathfrak{q}^{i+1}M)$ is polynomial of degree d, and is called the *Samuel function* of $\mathfrak{q}$ with respect to M. The integer a_{d-1}, is the *multiplicity* of $\mathfrak{q}$ on M and is denoted by $e(\mathfrak{q}, M)$ or simply by $e(\mathfrak{q})$ if $M = R$. Note that $e(\mathfrak{q}, M)/d!$ is the leading coefficient of $\chi_M^{\mathfrak{q}}(i)$.

By [259, Proposition 11.2.1], we have $e(\mathfrak{q}, M) = e(\overline{\mathfrak{q}}, M)$. If R is a polynomial ring with coefficients in a field K, there are efficient methods to compute the multiplicity of a zero dimensional monomial ideal; see [104, 424] and Section 12.5.

We recall the following result in dimension theory that relates the degree of the Samuel function with the dimension of the module.

Theorem 4.4.7 [310, Theorem 13.4] *$\chi_M^{\mathfrak{q}}$ is a polynomial function of degree equal to the dimension of M.*

Definition 4.4.8 Let $(R, \mathfrak{m})$ be a local ring of dimension d, M an R-module and $\mathfrak{q}$ an ideal of R with $\mathrm{rad}\,(\mathfrak{q}) = \mathfrak{m}$. We define

$$e_d(\mathfrak{q}, M) = \begin{cases} e(\mathfrak{q}, M) & \text{if } \dim(M) = d, \\ 0 & \text{if } \dim(M) < d. \end{cases}$$

In order to show how the "multiplicity" behaves under short exact sequences we need to recall a result of E. Artin and D. Rees.

Theorem 4.4.9 (Artin–Rees lemma [310, Theorem 8.5]) *Let M be a module over a ring R. If N is a submodule of M and I an ideal of R, then there is a positive integer c such that*

$$I^n M \cap N = I^{n-c}(I^c M \cap N), \ \forall n > c.$$

Proposition 4.4.10 *Let $(R, \mathfrak{m})$ be a local ring of dimension d and let $\mathfrak{q}$ be an $\mathfrak{m}$-primary ideal of R. If $0 \longrightarrow N \longrightarrow M \longrightarrow N' \longrightarrow 0$ is an exact sequence of R-modules, then $e_d(\mathfrak{q}, M) = e_d(\mathfrak{q}, N) + e_d(\mathfrak{q}, N')$.*

Proof. By Proposition 2.1.39, $\dim(M) = \max\{\dim(N), \dim(N')\}$. Hence, one may assume $d = \dim(M)$. Tensoring with $R/\mathfrak{q}^{n+1}$ the exact sequence above yields an exact sequence

$$0 \to (N \cap \mathfrak{q}^{n+1}M)/\mathfrak{q}^{n+1}N \to N/\mathfrak{q}^{n+1}N \to M/\mathfrak{q}^{n+1}M \to N'/\mathfrak{q}^{n+1}N' \to 0.$$

Taking lengths with respect to $R/\mathfrak{q}$ gives

$$\chi_M^{\mathfrak{q}}(n) = \chi_N^{\mathfrak{q}}(n) + \chi_{N'}^{\mathfrak{q}}(n) - \ell(N \cap \mathfrak{q}^{n+1}M/\mathfrak{q}^{n+1}N). \qquad (*)$$

By Theorem 4.4.9, $N \cap \mathfrak{q}^{n+1}M \subset \mathfrak{q}^{n+1-c}N$, for some integer $c > 0$. Hence

$$\ell(N \cap \mathfrak{q}^{n+1}M/\mathfrak{q}^{n+1}N) \leq \ell(\mathfrak{q}^{n+1-c}N/\mathfrak{q}^{n+1}N) = \chi_N^{\mathfrak{q}}(n) - \chi_N^{\mathfrak{q}}(n-c),$$

as $\chi_N^{\mathfrak{q}}(n) - \chi_N^{\mathfrak{q}}(n-c)$ is a polynomial function of degree at most $d-1$, the result follows by dividing Eq. $(*)$ by n^d and taking limits when n goes to infinity. $\square$

Next we show that the multiplicity is additive (cf. Proposition 8.5.9).

Proposition 4.4.11 *Let $(R, \mathfrak{m})$ be a local ring of dimension d and $\mathfrak{q}$ an $\mathfrak{m}$-primary ideal. If M is a finitely generated R-module and $\mathcal{A}$ is the set of all prime ideals $\mathfrak{p}$ of R with $\dim(R/\mathfrak{p}) = d$, then*

$$e_d(\mathfrak{q}, M) = \sum_{\mathfrak{p} \in \mathcal{A}} \ell(M_{\mathfrak{p}}) e_d(\mathfrak{q}, R/\mathfrak{p}).$$

Proof. Set $\mathcal{B} = \mathcal{A} \cap \operatorname{Supp}(M)$. If $\mathcal{B} = \emptyset$, then $\dim(M) < d$ and $e_d(\mathfrak{q}, M)$ is equal to 0, thus in this case the identity above holds. Hence we may assume $\mathcal{B} \neq \emptyset$, this yields the equality $\dim(M) = d$. By Theorem 2.1.16 there are prime ideals $\mathfrak{p}_1, \ldots, \mathfrak{p}_n$ of R and a filtration of submodules:

$$(0) = M_0 \subset M_1 \subset \cdots \subset M_n = M,$$

such that $M_i/M_{i-1} \simeq R/\mathfrak{p}_i$ for all i. Note $\mathcal{B} = \{\mathfrak{p}_i | \dim(R/\mathfrak{p}_i) = d\}$. To show this equality observe that

$$\mathrm{Ass}(M) \subset \{\mathfrak{p}_1, \ldots, \mathfrak{p}_n\} \subset \mathrm{Supp}(M),$$

see Corollary 2.1.18 and its proof, and recall that the minimal elements of $\mathrm{Supp}(M)$ are in $\mathrm{Ass}(M)$. Using Lemma 4.4.10 and the exact sequences

$$0 \longrightarrow M_{i-1} \longrightarrow M_i \longrightarrow R/\mathfrak{p}_i \longrightarrow 0 \quad (i = 1, \ldots, n),$$

we get $e(\mathfrak{q}, M) = \sum e(\mathfrak{q}, R/\mathfrak{p}_i)$, where the sum is taken over all $\mathfrak{p}_i$, in the multiset $\{\mathfrak{p}_1, \ldots, \mathfrak{p}_n\}$, such that $\dim(R/\mathfrak{p}_i) = d$. Let $\mathfrak{p} \in \mathcal{B}$, then

$$(M_i/M_{i-1})_{\mathfrak{p}} \simeq \begin{cases} (0) & \text{if} \quad \mathfrak{p} \neq \mathfrak{p}_i \\ R_{\mathfrak{p}}/\mathfrak{p}R_{\mathfrak{p}} & \text{if} \quad \mathfrak{p} = \mathfrak{p}_i. \end{cases} \tag{4.4}$$

Since the second module is a field we get that the length of $M_{\mathfrak{p}}$ is equal to the number of times that $\mathfrak{p}$ occurs in the multiset $\{\mathfrak{p}_1, \ldots, \mathfrak{p}_n\}$. Therefore

$$e(\mathfrak{q}, M) = \sum_{\mathfrak{p}_i \in \mathcal{B}} \ell(M_{\mathfrak{p}_i}) e(\mathfrak{q}, R/\mathfrak{p}_i),$$

as required. This proof was adapted from [65]. $\square$

Proposition 4.4.12 *Let $(S, \mathfrak{m})$ be a local ring and M an S-module with a positive rank $r = \mathrm{rank}(M)$. If $\mathfrak{q}$ is an $\mathfrak{m}$-primary ideal, then*

$$e(\mathfrak{q}, M) = e(\mathfrak{q}, S)\mathrm{rank}(M).$$

Proof. First note $d = \dim(M) = \dim(S)$ by Lemma 2.1.45. Let $\mathcal{A}$ be the set of all prime ideals $\mathfrak{p}$ of S with $\dim(S/\mathfrak{p}) = d$. Since M has rank r one has $M_{\mathfrak{p}} \simeq (S_{\mathfrak{p}})^r$ for any associated prime $\mathfrak{p}$ of S, thus $M_{\mathfrak{p}} \simeq (S_{\mathfrak{p}})^r$ for any $\mathfrak{p} \in \mathcal{A}$. Therefore using Proposition 4.4.11 one obtains

$$\begin{aligned} e_d(\mathfrak{q}, M) &= \sum_{\mathfrak{p} \in \mathcal{A}} \ell(M_{\mathfrak{p}}) e_d(\mathfrak{q}, S/\mathfrak{p}) = \sum_{\mathfrak{p} \in \mathcal{A}} \ell(S_{\mathfrak{p}}^r) e_d(\mathfrak{q}, S/\mathfrak{p}) \\ &= \sum_{\mathfrak{p} \in \mathcal{A}} r\ell(S_{\mathfrak{p}}) e_d(\mathfrak{q}, S/\mathfrak{p}) = \mathrm{rank}(M) e_d(\mathfrak{q}, S). \end{aligned}$$

As $e(\mathfrak{q}, M) = e_d(\mathfrak{q}, M)$ and $e(\mathfrak{q}, S) = e_d(\mathfrak{q}, S)$, the proof is complete. $\square$

In Chapter 5 we use the following important Cohen–Macaulay criterion to study the Koszul homology of graded ideals.

Theorem 4.4.13 (Herzog [217]) *Let $(S, \mathfrak{m})$ be a Cohen–Macaulay local ring and let M be a finitely generated S-module with a well-defined and positive rank. If $\mathbf{y} = y_1, \ldots, y_d$ is a system of parameters of S, then*

$$\ell(S/(\mathbf{y})) \cdot \mathrm{rank}(M) \leq \ell(M/(\mathbf{y})M).$$

Furthermore equality holds if and only if M is Cohen–Macaulay.

Proof. By Lemma 2.1.45 one has $d = \dim(M) = \dim(S)$. Let $\mathfrak{q}$ be the $\mathfrak{m}$-primary ideal generated by $\mathbf{y}$. There is a (graded) epimorphism

$$\varphi\colon (M/\mathfrak{q}M)[\mathbf{t}] = (M/\mathfrak{q}M)[t_1,\ldots,t_d] \to \mathrm{gr}_{\mathfrak{q}}(M) = \bigoplus_{i=0}^{\infty} \mathfrak{q}^i M/\mathfrak{q}^{i+1}M \quad (4.5)$$

such that $\varphi(\overline{F}) = F(y_1,\ldots,y_d) + \mathfrak{q}^{i+1}M$, for all F in $M[\mathbf{t}]$ homogeneous of degree i, where $\overline{F}$ means reducing the coefficients of F modulo $\mathfrak{q}M$. By restriction of φ to the ith graded component of $(M/\mathfrak{q}M)[\mathbf{t}]$ gives

$$\ell(M/\mathfrak{q}M)\binom{i+d-1}{d-1} \geq \ell(\mathfrak{q}^i M/\mathfrak{q}^{i+1}M) \quad \text{for } i \geq 0, \quad (4.6)$$

observe that both sides of the inequality are polynomial functions of degree $d-1$ and consequently $\ell(M/\mathfrak{q}M) \geq e(\mathfrak{q}, M)$. On the other hand $\mathbf{y}$ is a regular sequence, because S is Cohen–Macaulay (see Proposition 2.3.19), this forces φ to be an isomorphism when $M = S$ (see Theorem 4.4.2). It follows rapidly that $\ell(S/\mathfrak{q}) = e(\mathfrak{q}, S)$. To finish the first part of the proof note that making use of Proposition 4.4.12 one obtains the required inequality.

At this point one should observe that, by the arguments above, the proof reduces to show that $\ell(M/\mathfrak{q}M) = e(\mathfrak{q}, M)$ if and only if M is Cohen–Macaulay. Next we show both implications.

$\Rightarrow$) Thanks to Propositions 4.4.5 and 2.3.19 it suffices to prove that φ is injective; because this implies that $\mathbf{y}$ is an M-regular sequence and hence M is Cohen–Macaulay. Using Eq. (4.5) we obtain an exact sequence

$$0 \longrightarrow \ker(\varphi)_i \longrightarrow ((M/\mathfrak{q}M)[\mathbf{t}])_i \longrightarrow \mathfrak{q}^i M/\mathfrak{q}^{i+1}M \longrightarrow 0,$$

hence we get

$$\ell(\ker(\varphi)_i) = \ell(M/\mathfrak{q}M)\binom{i+d-1}{d-1} - \ell(\mathfrak{q}^i M/\mathfrak{q}^{i+1}M) \quad \text{for } i \geq 0.$$

As the polynomial functions on the right-hand side have both degree $d-1$ and their leading terms cancel out, one concludes that $\ell(\ker(\varphi)_i)$ grows as a polynomial function of degree at most $d-2$.

Let $F = \sum m_\alpha t^\alpha$ be a homogeneous polynomial in $M[\mathbf{t}]$ of degree p, with m_α in M, and denote $\overline{m}_\alpha = m_\alpha + \mathfrak{q}M$. Assume $\overline{F}$ is in $\ker(\varphi)$ and $\overline{F} \neq 0$. Since $\mathfrak{m}^k \subset \mathfrak{q}$, there is an $s \geq 1$ such that $\mathfrak{m}^{s-1}\overline{F} \neq 0$ and $\mathfrak{m}^s\overline{F} = 0$, thus one may assume $\mathfrak{m}\overline{F} = 0$ and $\overline{F} \neq 0$. Hence there is a graded epimorphism ψ induced by multiplication by $\overline{F}$

$$\psi\colon (S/\mathfrak{m})[\mathbf{t}](-p) \longrightarrow (S/\mathfrak{q})[\mathbf{t}]\overline{F}.$$

We now show that ψ is injective. Let $g = \sum \overline{b}_\beta t^\beta$ be an element in the kernel of ψ, where $b_\beta \in S$ and $\overline{b}_\beta = b_\beta + \mathfrak{m}$. Thus $(\sum \widetilde{b}_\beta t^\beta)(\sum \overline{m}_\alpha t^\alpha) = 0$, where

$\widetilde{b}_\beta = b_\beta + \mathfrak{q}$. One may assume that $\overline{b}_\beta t^\beta$ and $\overline{m}_\alpha t^\alpha$ are the leading terms of g and $\overline{F}$, respectively, w.r.t the lexicographical ordering of the t_i variables. Thus $\widetilde{b}_\beta \overline{m}_\alpha t^{\alpha+\beta}$ is the leading term of $g\overline{F}$ and $b_\beta m_\alpha \in \mathfrak{q}M$. Hence $b_\beta + \mathfrak{q}$ is a zero divisor of $M/\mathfrak{q}M$ and since $\mathrm{Ass}_{R/\mathfrak{q}}(M/\mathfrak{q}M) = \{\mathfrak{m}/\mathfrak{q}\}$, one has that $b_\beta + \mathfrak{q}$ is in $\mathfrak{m}/\mathfrak{q}$, which proves $b_\alpha \in \mathfrak{m}$ and $\overline{b}_\beta = 0$. Altogether one derives $g = 0$ and ψ is injective. Using that ψ is a graded isomorphism yields

$$\binom{i-p+d-1}{d-1} = \ell(((S/\mathfrak{q})[\mathbf{t}]\overline{F})_i) \leq \ell(\ker(\varphi)_i),$$

which is a contradiction because the left-hand side is a polynomial function of degree $d-1$. Therefore $\overline{F} = 0$ and φ is injective, as required.

$\Leftarrow$) As M is Cohen–Macaulay, the sequence $\mathbf{y}$ is an M-regular sequence, an application of Theorem 4.4.2 yields that φ is an isomorphism. Therefore Eq. (4.6) becomes an equality and we get $\ell(M/\mathfrak{q}M) = e(\mathfrak{q}, M)$. □

Exercises

4.4.14 Let $d, m \in \mathbb{N}$. Prove the equality

$$\binom{d+m}{d} = \sum_{j=0}^{m} \binom{j+d-1}{d-1}.$$

4.4.15 Let $f\colon \mathbb{Z} \to \mathbb{Z}$ be a numerical function, f is said to be a *polynomial function* of degree d if there is a polynomial $P(t) \in \mathbb{Q}[t]$ such that $P(i) = f(i)$ for $i \gg 0$. Prove that f is a polynomial function of degree d if and only if the numerical function $g\colon \mathbb{Z} \to \mathbb{Z}$ given by $g(i) = f(i) - f(i-1)$ is a polynomial function of degree $d-1$.

4.5 Jacobian criterion

Here we introduce the Jacobian criterion and present some applications and examples to illustrate its use. Along the way some facts about regular local rings are discussed. This type of rings are used in algebraic geometry to show for instance that the coordinate ring of a smooth irreducible affine variety is normal [100, Proposition 1.0.9].

Let $(R, \mathfrak{m})$ be a local ring. By Nakayama's Lemma it follows readily that R is a regular local ring if and only if $\mathfrak{m}$ is generated by a system of parameters of R. If R is a polynomial ring or a formal power series ring over a field k, then R is a regular ring [310, Theorem 19.5], that is, $R_\mathfrak{p}$ is a regular local ring for all $\mathfrak{p}$ in $\mathrm{Spec}(R)$ or equivalently $R_\mathfrak{m}$ is a regular local ring for any maximal ideal $\mathfrak{m}$ of R (see [310, Theorem 19.3]).

Proposition 4.5.1 *If R is a regular local ring, then R is Cohen–Macaulay.*

Proof. Let $\mathfrak{m}$ be the maximal ideal of R and let $k = R/\mathfrak{m}$ be its residue field. Assume $x_1, \ldots, x_d$ is a set of generators of $\mathfrak{m}$, where $d = \dim(R)$. If $k[t_1, \ldots, t_d]$ is a ring of polynomial over the field k, there is an epimorphism of graded k-algebras

$$\varphi\colon k[\mathbf{t}] = k[t_1, \ldots, t_d] \longrightarrow \mathrm{gr}_{\mathfrak{m}}(R) \tag{4.7}$$

induced by $\varphi(t_i) = x_i + \mathfrak{m}^2$ for all i. It suffices to prove that the map φ is injective. Indeed if φ is injective, then $x_1, \ldots, x_d$ is a regular sequence by Proposition 4.4.5, and hence R is Cohen–Macaulay by Proposition 2.3.19.

If $I = \ker(\varphi) \neq 0$, pick a homogeneous polynomial f in I of degree $s \geq 1$. Using the isomorphism $k[\mathbf{t}](-s) \simeq fk[\mathbf{t}]$, together with Eq. (4.7), one concludes:

$$\dim_k(\mathfrak{m}^i/\mathfrak{m}^{i+1}) = \dim_k k[\mathbf{t}]_i - \dim_k I_i \leq \binom{i+d-1}{d-1} - \binom{i-s+d-1}{d-1}.$$

By Theorem 4.4.7 the left-hand side is a polynomial function of degree $d-1$ while the right-hand side is a polynomial function of degree $d-2$. Therefore I must be zero and φ is injective. □

Corollary 4.5.2 *If $(R, \mathfrak{m})$ is a regular local ring, then R is a domain.*

Proof. By Proposition 4.5.1 $\mathfrak{m}$ is generated by a regular system of parameters. Thus $\mathrm{gr}_{\mathfrak{m}}(R)$ is a domain. Let $x, y \in R$ such that $xy = 0$. If $x \neq 0$ and $y \neq 0$, then by Theorem 4.4.3 there are $r, s \in \mathbb{N}$ with $x \in \mathfrak{m}^r \setminus \mathfrak{m}^{r+1}$ and $y \in \mathfrak{m}^s \setminus \mathfrak{m}^{s+1}$. Multiplying the images $\overline{x}, \overline{y}$ of x, y in $\mathrm{gr}_{\mathfrak{m}}(R)$, one obtains $\overline{xy} = 0$. Hence $\overline{x} = 0$ or $\overline{y} = 0$, which is impossible. This proves R is a domain. □

Proposition 4.5.3 *Let $(R, \mathfrak{m})$ be a regular local ring and $I \neq R$ an ideal of R. If R/I is a regular local ring, then I is a complete intersection.*

Proof. Let $\overline{x}_1, \ldots, \overline{x}_d$ be a generating set of $\overline{\mathfrak{m}} = \mathfrak{m}/I$, where $\overline{x}_i = x_i + I$ and d is the dimension of R/I. Note that the set of images of $x_1, \ldots, x_d$ in $\overline{\mathfrak{m}}/\overline{\mathfrak{m}}^2$ is a basis for $\overline{\mathfrak{m}}/\overline{\mathfrak{m}}^2$ as a vector space over $k = R/\mathfrak{m}$. Hence

$$x_1 + \mathfrak{m}^2, \ldots, x_d + \mathfrak{m}^2$$

are linearly independent in $\mathfrak{m}/\mathfrak{m}^2$. Set $n = \dim(R)$ and $\mathfrak{m}' = (x_1, \ldots, x_d)$. As R is a regular local ring $n = \dim_k(\mathfrak{m}/\mathfrak{m}^2)$, hence using the proof of Corollary 2.1.35 and the equality $\mathfrak{m} = \mathfrak{m}' + I$ one derives $\mathfrak{m} = \mathfrak{m}' + J$, for some ideal $J = (x_{d+1}, \ldots, x_n)$ contained in I.

By Proposition 4.5.1 $x_1, \ldots, x_n$ is a regular sequence, thus it suffices to show $J = I$, to prove this equality observe that R/J is a regular local ring of dimension d, and therefore R/J is a domain by Corollary 4.5.2. Since R/I has also dimension d, we see that the canonical homomorphism $\varphi\colon R/J \to R/I$ is an isomorphism, thus $I = J$, as asserted. □

Definition 4.5.4 Let $R = k[x_1, \ldots x_n]$ be a polynomial ring over a field k and $I = (f_1, \ldots, f_q)$ an ideal. The *Jacobian matrix* of I is the matrix

$$\mathfrak{J} = (\partial f_i / \partial x_j).$$

We denote the Jacobian matrix of I taken modulo an ideal P by

$$(\partial f_i / \partial x_j)(P).$$

Theorem 4.5.5 (Jacobian criterion [309, p. 213]) *Let $B = R/I$ be a quotient ring, where $R = k[x_1, \ldots, x_n]$ is a polynomial ring in n variables over a field k, and let $I = (f_1, \ldots, f_q) \subset R$ be an ideal. If $P \subset R$ is a prime ideal containing I and $\mathfrak{p} = P/I$, one has:*

(a) $\operatorname{rank}(\partial f_i / \partial x_j)(P) \leq \operatorname{ht}(I_P)$.

(b) *If* $\operatorname{rank}(\partial f_i / \partial x_j)(P) = \operatorname{ht}(I_P)$, *then $B_\mathfrak{p}$ is a regular ring.*

(c) *If k is a perfect field and $B_\mathfrak{p}$ is regular, then*

$$\operatorname{rank}(\partial f_i / \partial x_j)(P) = \operatorname{ht}(I_P).$$

Lemma 4.5.6 *Let R be a ring and I an ideal. If I is height unmixed and P is a prime ideal such that $I \subset P$, then I_P is also height unmixed and* $\operatorname{ht}(I) = \operatorname{ht}(I_P)$.

Proof. Let $I = \mathfrak{q}_1 \cap \cdots \cap \mathfrak{q}_r$ be a primary decomposition of I, we may assume $\sqrt{\mathfrak{q}_i} \subset P$ for $i = 1, \ldots s$. Localizing at P gives a primary decomposition $IR_P = \bigcap_{i=1}^{s} \mathfrak{q}_i R_P$. Hence the associated primes of I_P are $\sqrt{\mathfrak{q}_i} R_P$, where $i \leq s$. To finish the proof note $\operatorname{ht}(\sqrt{\mathfrak{q}_i} R_P) = \operatorname{ht}(\sqrt{\mathfrak{q}_i}) = \operatorname{ht}(I)$ for $i \leq s$. $\square$

Definition 4.5.7 Let k be a field and $I = (f_1, \ldots, f_q)$ an ideal of height g of $k[x_1, \ldots, x_n]$. The *Jacobian ideal* J of I is the ideal generated by the $g \times g$ minors of the Jacobian matrix $\mathfrak{J} = (\partial f_i / \partial x_j)$.

Corollary 4.5.8 *Let $R = k[x_1, \ldots, x_n]$ be a polynomial ring over a perfect field k and $I = (f_1, \ldots, f_q) \subset R$ an unmixed ideal. If P is a prime ideal of R containing I and $\mathfrak{p} = P/I$, then $(R/I)_\mathfrak{p}$ is regular if and only if $J/I \not\subset \mathfrak{p}$, where J is the Jacobian ideal of I.*

Proof. $\Rightarrow$) By Theorem 4.5.5 and Lemma 4.5.6 one has

$$\operatorname{rank}(\partial f_i / \partial x_j)(P) = \operatorname{ht}(I),$$

hence there is $f \in J \setminus P$ and $J/I \not\subset P/I$.

$\Leftarrow$) If $(R/I)_\mathfrak{p}$ is not regular, then by Theorem 4.5.5 one has the inequality $\operatorname{rank}(\partial f_i / \partial x_j)(P) < \operatorname{ht}(I)$, hence $J/P = (0)$ and $J/I \subset P/I$, which is impossible. Note that this part of the proof is valid for any field k. $\square$

Proposition 4.5.9 *Let R be a polynomial ring over a field k and I an unmixed ideal of height g. If J is the Jacobian ideal of I and* $\mathrm{ht}\,(J,I)\geq g+2$, *then $(R/I)_\mathfrak{p}$ is regular for any prime $\mathfrak{p}$ of R/I such that* $\mathrm{ht}\,(\mathfrak{p})\leq 1$.

Proof. Let $B=R/I$ and $\mathfrak{p}=P/I$, where P is a prime ideal of R. If $B_\mathfrak{p}$ is not regular, then according to Corollary 4.5.8 one has $J/I\subset\mathfrak{p}$ and hence $(J,I)\subset P$. Since R is a catenary ring and I is unmixed one obtains $\mathrm{ht}\,(P)=g$ if $\mathrm{ht}\,(\mathfrak{p})=0$ and $\mathrm{ht}\,(P)=g+1$ if $\mathrm{ht}\,(\mathfrak{p})=1$. Therefore $\mathrm{ht}\,(J,I)\leq\mathrm{ht}\,(P)\leq g+1$, which is impossible. Thus $B_\mathfrak{p}$ must be regular, as required. $\Box$

Another consequence of the Jacobian criterion is the following radical test. Let k be a field and $I=(f_1,\ldots,f_q)\subset R=k[x_1,\ldots,x_n]$ an ideal of height g. Let J be the Jacobian ideal of I generated by the $g\times g$ minors of the Jacobian matrix $\mathfrak{J}=(\partial f_i/\partial x_j)$.

Theorem 4.5.10 (Vasconcelos [411]) *If k is a perfect field and I is unmixed, then I is a radical ideal if and only if there is an element f in $J\setminus I$ such that $(I\colon f)=I$.*

Proof. $\Rightarrow$) Let $P_1,\ldots,P_r$ be the associated primes of I. If $J\subset\mathcal{Z}(R/I)$, then $J\subset P_i$ for some i because $\mathcal{Z}(R/I)=\cup_{i=1}^r P_i$. Hence by Corollary 4.5.8 one derives that $(R/I)_{P_i}$ is not a regular ring, which is a contradiction because $(R/I)_{P_i}=R_{P_i}/P_iR_{P_i}$ is a field.

$\Leftarrow$) Let $I=\mathfrak{q}_1\cap\cdots\cap\mathfrak{q}_r$ be an irredundant primary decomposition of I and $P_i=\sqrt{\mathfrak{q}_i}$ an associated prime ideal of I. Note $(R/I)_{P_i}\simeq(R/I)_{\mathfrak{p}_i}$ and $I_{P_i}=(\mathfrak{q}_i)_{P_i}$, where $\mathfrak{p}_i=P_i/I$. First we observe that $(R/I)_{\mathfrak{p}_i}$ is a regular ring; otherwise by Corollary 4.5.8 one has $J/I\subset P_i/I$, hence $f\in P_i$, which is impossible because f is regular modulo I. Therefore $(R/I)_{P_i}$ is a regular local ring and thus an integral domain by Theorem 2.4.16. As $I_{P_i}=(\mathfrak{q}_i)_{P_i}$ must be prime, one concludes that $\mathfrak{q}_i$ is also prime and consequently $\mathfrak{q}_i=P_i$. Note that here the hypothesis k perfect is not being used. $\Box$

Example 4.5.11 Let $R=k[x_1,\ldots,x_7]$ and $I=(x_1x_7-x_2x_6,x_3x_5-x_4x_7)$, where k is a field. We now show that $B=R/I$ is a normal ring using Theorem 2.4.15. The Jacobian matrix of I is:

$$\mathfrak{J}=\left(\begin{array}{ccccccc} x_7 & -x_6 & 0 & 0 & 0 & -x_2 & x_1\\ 0 & 0 & x_5 & -x_7 & x_3 & 0 & -x_4\end{array}\right).$$

Note that x_7^2, x_5x_6, x_3x_6, x_4x_6, x_2x_5, x_1x_5, x_2x_4 are in the Jacobian ideal J of I. Hence $\mathrm{ht}\,(J,I)\geq g+2$, where $g=\mathrm{ht}\,(I)=2$. By Proposition 4.5.9, B satisfies (R_1) and, since I is a complete intersection, it follows by Proposition 3.1.30 that B satisfies (S_2). Thus, by Serre's normality criterion, the ring B is normal.

Lemma 4.5.12 *Let R be a ring and I a prime ideal. If $f \in R \setminus I$ and $I_f^{(n)} = I_f^n$ for some $n \geq 1$, then $I^{(n)} = (I^n \colon f^\infty)$.*

Proof. If $z \in (I^n \colon f^\infty)$, then $zf^k \in I^n$ for some $k \geq 1$, hence $z \in I^{(n)}$. Conversely if $z \in I^{(n)}$, then $sz \in I^n$ for some $s \notin I$. Since $s/1 \notin I_f$ and $(s/1)(z/1) \in I_f^n$, then $z/1$ is in the nth symbolic power of I_f. Thus by hypothesis $z/1 \in I_f^n$ and this means that $z \in (I^n \colon f^\infty)$. □

There are a few methods to compute symbolic powers of prime ideals in polynomial rings, see [378] and [413, Chapter 3]. The following is a subtle application of the Jacobian criterion due to Vasconcelos.

Proposition 4.5.13 Let R be a polynomial ring over a field k and I a prime ideal. If f is an element in the Jacobian ideal J of I which is not in the ideal I, then

$$I^{(n)} = (I^n \colon f^\infty) \quad \text{for } n \geq 1,$$

and such element f exist if k is a perfect field.

Proof. Since the localization R_f is Cohen–Macaulay, by Proposition 4.3.33 and Lemma 4.5.12 it is enough to prove that I_f is locally a complete intersection. Let P be a prime ideal such that $I \subset P$ and $f \notin P$. Set $\mathfrak{p} = P/I$. One has

$$(R_f/I_f)_{P_f} \simeq (R_f)_{P_f}/(I_f)_{P_f} \simeq R_P/I_P \simeq (R/I)_P \simeq (R/I)_{\mathfrak{p}}.$$

As $J/I \not\subset \mathfrak{p}$, using Corollary 4.5.8 we conclude that R_P/I_P is a regular ring and consequently $(I_f)_{P_f}$ is a complete intersection (see Proposition 4.5.3). The last assertion follows from Theorem 4.5.10. □

Example 4.5.14 Let $I = (x^3 - yz, y^2 - xz, z^2 - x^2y) \subset \mathbb{Q}[x, y, z]$. The ideal I is prime because it is the kernel of the homomorphism:

$$\varphi\colon \mathbb{Q}[x, y, z] \longrightarrow \mathbb{Q}[t], \quad \text{where } \varphi(h(x, y, z)) = h(t^3, t^4, t^5).$$

Using *CoCoA* [88] and Proposition 4.5.13 we get

$$I^{(2)} = (I^2 \colon f^\infty) = (I^2, \Delta),$$

where $f = 2y^2 + xz$ and $\Delta = -x^5 + 3x^2yz - xy^3 - z^3$. See [280].

Exercises

4.5.15 Let $R = \mathbb{Q}[x, y, z]$ and I the prime ideal generated by

$$f_1 = x^3 - yz, \;\; f_2 = y^2 - xz, \;\; f_3 = z^2 - x^2y.$$

Prove that $I_x = (f_1, f_2)$ and $I^{(n)} = (I^n \colon x^\infty)$ for $n \geq 1$.

4.5.16 Use the following procedure in *CoCoA* [88] to verify the formula for the second symbolic power of Example 4.5.14

```
F1 := x^3 - y * z; -- variables must begin with capital letters
F2 := y^2 - x * z;
F3 := z^2 - x^2 * y;
Jac := Jacobian([F1, F2, F3]);
J := Minors(2, Mat(Jac)); -- 2 × 2 minors of the Jacobian matrix
I := Ideal(F1, F2, F3);
H := Elim(t, I^2 + Ideal(1 - t(2 * y^2 + x * z)));
Print(H);
```

4.5.17 Let $\mathfrak{p}$ be a prime ideal of a ring R such that $\mathfrak{p}R_\mathfrak{p}$ is a complete intersection, then $\mathfrak{p}^{(n)} = \mathfrak{p}^n$ for all $n \geq 1$ if and only if $\mathrm{gr}_\mathfrak{p}(R)$ is a domain.

Hint Use Theorem 4.3.15, Lemma 2.1.17, and Proposition 4.3.29.

4.5.18 Let R be a Cohen–Macaulay ring and I an ideal of R which is height unmixed. If I is a complete intersection, then I is locally a complete intersection.

Hint Use Lemma 2.3.20.

4.5.19 Let R be a Cohen–Macaulay ring and I a prime ideal. If $f \in R \setminus I$ and I_f is a complete intersection, then $I^{(n)} = (I^n \colon f^\infty)$ for $n \geq 1$.

4.5.20 Let $I = I_2(X)$ be the ideal of 2×2-minors of the symmetric matrix:

$$X = \begin{bmatrix} x_1 & x_2 & x_3 \\ x_2 & x_4 & x_5 \\ x_3 & x_5 & x_6 \end{bmatrix},$$

where the entries of X are indeterminates over a field k. Prove that I_{x_1} is a complete intersection. Find a set of generators for $I^{(2)}$ and prove that $\mathrm{depth}(H_1) = 1$, where H_1 is the first Koszul homology module of I.

Chapter 5

Hilbert Series

Hilbert series of graded modules and graded algebras are introduced and studied in this chapter. The h-vector and the a-invariant of a graded algebra are defined using the Hilbert–Serre theorem. For Cohen–Macaulay algebras, we present their main properties. Some features of Cohen–Macaulay and Gorenstein extremal algebras will be presented.

5.1 Hilbert–Serre Theorem

Unless otherwise stated we shall always assume that modules are finitely generated and $\mathbb{N}$-graded. We refer to Section 2.2 for an introduction to graded modules, Hilbert polynomials, and multiplicities.

Let $R = R_0[x_1, \ldots, x_n] = \oplus_{i=0}^{\infty} R_i$ be an R_0-algebra of finite type over an Artinian local ring R_0 with the grading induced by $\deg(x_i) = d_i$, where d_i is a positive integer for $i = 1, \ldots, n$. If

$$M = M_0 \oplus M_1 \oplus \cdots \oplus M_i \oplus \cdots$$

is a finitely generated $\mathbb{N}$-graded module over R, its *Hilbert function* and *Hilbert series* are defined by

$$H(M, i) = \ell(M_i) \ \text{ and } \ F(M, t) = \sum_{i=0}^{\infty} H(M, i) t^i$$

respectively, where $\ell(M_i)$ denotes the length of M_i as an R_0-module, if R_0 is a field $\ell(M_i) = \dim_{R_0}(M_i)$.

Lemma 5.1.1 *If $0 \to M' \to M \to M'' \to 0$ is a degree preserving short exact sequence of $\mathbb{N}$-graded R-modules, then*

(a) $H(M, i) = H(M', i) + H(M'', i)$ *for all i, and*

(b) $F(M, t) = F(M', t) + F(M'', t)$.

Proof. It follows Proposition 2.1.36. □

If $j \in \mathbb{N}$, then $M(-j)$ is the *regrading* of M obtained by a *shift* of the graduation of M; more precisely

$$M(-j) = \bigoplus_{i=0}^{\infty} M(-j)_i,$$

where $M(-j)_i = M_{-j+i}$. Note that we are assuming $M_i = 0$ for $i < 0$. In this way $M(-j)$ becomes an $\mathbb{N}$-graded R-module.

Lemma 5.1.2 $F(M(-j), t) = t^j F(M, t)$.

Proof. Since $M(-j)_i = M_{i-j}$ one has:

$$F(M(-j), t) = t^j \sum_{i=j}^{\infty} \ell(M_{i-j}) t^{i-j} = t^j F(M, t),$$

where the first equality follows using that $M_{i-j} = 0$ for $i = 0, \ldots, j-1$. □

Lemma 5.1.3 *If $z \in R_j$, there is a degree preserving exact sequence*

$$0 \longrightarrow (M/(0\colon z))(-j) \xrightarrow{z} M \xrightarrow{\phi} M/zM \longrightarrow 0 \quad (\phi(m) = m + zM),$$

where $(0\colon z) = \{m \in M \mid zm = 0\}$ and the first map is multiplication by z.

Proof. As the map $\psi\colon M(-j) \to M$, given by $\psi(m) = zm$, is a degree zero homomorphism one has that $(0\colon z)(-j)$ is a graded submodule of $M(-j)$. The exactness of the sequence above follows because ψ induces an exact sequence

$$0 \longrightarrow (0\colon z)(-j) \xrightarrow{\imath} M(-j) \xrightarrow{\psi} M \xrightarrow{\phi} M/zM \longrightarrow 0,$$

where $\imath$ is an inclusion. □

Theorem 5.1.4 (Hilbert–Serre Theorem) *The Hilbert series $F(M, t)$ of M is a rational function that can be written as*

$$F(M, t) = \frac{h(t)}{\prod_{i=1}^{n}(1 - t^{d_i})} \quad \text{for some } h(t) \in \mathbb{Z}[t].$$

If $d_i = 1$ for all i, there is a unique polynomial $h(t) \in \mathbb{Z}[t]$ such that

$$F(M, t) = \frac{h(t)}{(1-t)^d} \quad \text{and} \quad h(1) \neq 0.$$

Proof. If $n = 0$, the Hilbert function of M is zero for $i \gg 0$ and $F(M,t)$ is a polynomial. If $n > 0$, consider the exact sequence of graded modules

$$0 \longrightarrow (0\colon x_n)(-d_n) \longrightarrow M(-d_n) \xrightarrow{x_n} M \longrightarrow M/x_nM \longrightarrow 0.$$

Since the ends of this exact sequence are finitely generated modules over $R_0[x_1, \ldots, x_{n-1}]$ the proof follows readily by induction on n. $\square$

The integer d in the Hilbert–Serre theorem is denoted by $d(M)$. This integer is the Krull dimension of M; see Proposition 5.1.6 and its proof.

Definition 5.1.5 The degree of $F(M,t)$ as a rational function is denoted by $a(M)$; it is called the *a-invariant* of M.

If R has the standard grading, below we see how the a-invariant measures the difference between the Hilbert polynomial and the Hilbert function.

For the rest of this section we shall always assume that the ring R has the standard grading induced by setting $\deg(x_i) = 1$ for $i = 1, \ldots, n$. We denote the Hilbert polynomial of M by $\varphi_M(t)$.

Proposition 5.1.6 *If M is a graded R-module of dimension d. Then, there are integers $a_{-d}, \ldots, a_{-1}$ so that*

$$H(M,i) = \sum_{j=0}^{d-1} a_{-(d-j)} \binom{i+d-j-1}{d-j-1}, \ \forall\, i \geq a(M)+1,$$

where $a(M)$ is the degree of $F(M,t)$ as a rational function.

Proof. By the Hilbert–Serre Theorem there is a polynomial $h(t) \in \mathbb{Z}[t]$ such that

$$F(M,t) = \frac{h(t)}{(1-t)^d}$$

and $h(1) \neq 0$. If $d = 0$, then $a(M)$ is equal to $\deg(h)$ and $H(M,i) = 0$ for $i \geq a(M)+1$. Thus one may assume $d > 0$. Observe that by the division algorithm we can find $e(t) \in \mathbb{Z}[t]$ so that the Laurent expansion of $F(M,t) - e(t)$, in negative powers of $(1-t)$, is equal to

$$F(M,t) - e(t) = \sum_{j=0}^{d-1} \frac{a_{-(d-j)}}{(1-t)^{d-j}}, \ \text{where } a_{-(d-j)} = \frac{(-1)^j h^{(j)}(1)}{j!},$$

where $h^{(j)}$ is the jth derivative of h. Next we expand $(1-t)^{d-j}$ in powers of t to obtain

$$F(M,t) = e(t) + \sum_{i=0}^{\infty} \left[\sum_{j=0}^{d-1} a_{-(d-j)} \binom{i+d-j-1}{d-j-1} \right] t^i = e(t) + \sum_{i=0}^{\infty} \varphi(i) t^i,$$

observe that $\varphi(i) = H(M,i)$ for $i \geq \deg\, e(t)+1$, where the degree of the zero polynomial is set equal to -1. To complete the proof note that $a_{-d}, \ldots, a_{-1}$ are integers by Lemma 4.4.6, and that $d = d(M)$ is the dimension of M by Theorem 2.2.4. □

Remark 5.1.7 The leading coefficient of the Hilbert polynomial $\varphi_M(t)$ of M is equal to $h(1)/(d-1)!$, where $h(t)$ is the polynomial $F(M,t)(1-t)^d$. If $d = 0$, then $h(1) = \ell(M)$ and $\ell(M) = \dim_K(M)$ if R is a polynomial ring over a field K.

Definition 5.1.8 The *index of regularity* of M is the least integer $\ell \geq 0$ such that $H(M,i) = \varphi_M(i)$ for $i \geq \ell$.

Corollary 5.1.9 *Let M be a graded R-module. If*

$$r_0 = \min\{r \in \mathbb{N} \,|\, H(M,i) = \varphi_M(i),\ \forall\ i \geq r\},$$

then $r_0 = 0$ if $a(M) < 0$ and $r_0 = a(M) + 1$ otherwise.

Corollary 5.1.10 *Let S be a standard graded algebra of dimension d over a field K. If S is Cohen–Macaulay and $\underline{h} = \{h_1, \ldots, h_d\}$ is a system of parameters of S consisting of linear forms, then the multiplicity of S is:*

$$e(S) = \ell(S/\underline{h}S).$$

Proof. By Proposition 3.1.20 $\{h_1, \ldots, h_d\}$ is a regular sequence. Thus there are exact sequences of K-vector spaces

$$0 \longrightarrow (S[-1])_i = S_{i-1} \xrightarrow{h_1} S_i \longrightarrow (S/h_1S)_i \longrightarrow 0.$$

Therefore one has the relation $\varphi_{\overline{S}}(i) = \varphi_S(i) - \varphi_S(i-1)$ between Hilbert polynomials for $i \gg 0$, where $\overline{S} = S/h_1S$. Let

$$\varphi_S(t) = \frac{e(S)}{(d-1)!}t^{d-1} + a_{d-2}t^{d-2} + \cdots + a_0,$$

$$\varphi_{\overline{S}}(t) = \frac{e(\overline{S})}{(d-2)!}t^{d-2} + b_{d-3}t^{d-3} + \cdots + b_0.$$

Now, observe that thanks to $\varphi_{\overline{S}}(i) = \varphi_S(i) - \varphi_S(i-1)$ one derives:

$$\varphi_{\overline{S}}(t) = \frac{e(S)(d-1)}{(d-1)!}t^{d-2} + \text{lower order terms},$$

consequently $e(\overline{S}) = e(S)$. Since $\overline{S}$ is Cohen–Macaulay of dimension $d-1$ (see Lemma 2.3.10), the result follows by induction. □

Proposition 5.1.11 *If A, B are standard algebras over a field K, then*

$$\dim(A \otimes_K B) = \dim(A) + \dim(B).$$

Proof. One has the next equality between Hilbert series:

$$\begin{aligned} F(A,t)F(B,t) &= \left[\sum_{i=0}^{\infty}(\dim_K A_i)t^i\right]\left[\sum_{i=0}^{\infty}(\dim_K B_i)t^i\right] \\ &= \sum_{i=0}^{\infty}(\dim_K C_i)t^i = F(A \otimes_K B, t), \end{aligned}$$

where $A \otimes_K B = \bigoplus_{r=0}^{\infty} C_r$ and $C_r = \bigoplus_{i+j=r} A_i \otimes_K B_j$. Thus, Theorem 5.1.4 yields the asserted equality. $\Box$

Using the Noether normalization lemma it is not difficult to prove that the result above holds for arbitrary affine algebras over a field K.

Computation of Hilbert series Let R be a polynomial ring over a field K with a monomial order $\prec$ and let $I \subset R$ be a graded ideal. Since R/I and $R/\operatorname{in}_\prec(I)$ have the same Hilbert function (see Corollary 3.3.15), the actual computation of the Hilbert series of R/I is a two-step process:

- first one finds a Gröbner basis of I using Buchberger's algorithm, and
- second one computes the Hilbert series of $R/\operatorname{in}(I)$ using elimination of variables (see [20, 37]).

Other approaches to compute Hilbert series use minimal resolutions and Stanley decompositions [398]. An ad hoc method to compute Hilbert series of Cohen–Macaulay $\mathbb{N}$-graded algebras is given by Proposition 3.1.27. Some examples are given below.

Example 5.1.12 Let $R = K[x_1, x_2, x_3]$ be a polynomial ring and let I be the ideal $(x_1^2, x_2^2x_3, x_2^3)$. Let us compute the Hilbert series of R/I using elimination. Pick any monomial involving more than one variable, say $x_2^2x_3$. The idea is to eliminate x_2^2 from the monomials containing more than one variable. From the exact sequence of graded modules:

$$0 \longrightarrow R/(x_1^2, x_3, x_2)(-2) \xrightarrow{x_2^2} R/I \longrightarrow R/(x_1^2, x_2^2) \longrightarrow 0,$$

and applying Exercise 5.1.20 to the ends of this sequence, we get

$$F(R/I, t) = t^2\left[\frac{(1-t)^2(1-t^2)}{(1-t)^3}\right] + \frac{(1-t^2)^2}{(1-t)^3} = \frac{-t^4 + 2t^2 + 2t + 1}{(1-t)}.$$

Example 5.1.13 Let $R = \mathbb{Q}[x, y, z, w]$ and $I = (f_1, f_2, f_3)$, where

$$f_1 = y^2 - xz,\ f_2 = x^3 - yzw,\ f_3 = x^2y - z^2w.$$

Using *Macaulay2* [199] one finds that the minimal resolution of R/I is:

$$0 \longrightarrow R^2(-4) \xrightarrow{\varphi_2} R(-2) \oplus R^2(-3) \xrightarrow{\varphi_1} R \longrightarrow R/I \longrightarrow 0.$$

Let us compute the Hilbert series of R/I using its minimal resolution:

$$\begin{aligned} F(R/I, t) &= F(R, t) - F(R(-2) \oplus R^2(-3), t) + F(R^2(-4), t) \\ &= \frac{1}{(1-t)^4} - \frac{t^2}{(1-t)^4} - \frac{2t^3}{(1-t)^4} + \frac{2t^4}{(1-t)^4} \\ &= \frac{1 + 2t + 2t^2}{(1-t)^2}. \end{aligned}$$

Example 5.1.14 Let $R = K[x_1, \ldots, x_5]$ and $I = (x_1x_2, x_3x_4x_5)$. Note that I is a complete intersection. Hence, by Exercise 5.1.20, one has

$$F(R/I, t) = \frac{(1-t^2)(1-t^3)}{(1-t)^5} = \frac{1 + 2t + 2t^2 + t^3}{(1-t)^3} = -1 + \frac{5t^2 - t + 2}{(1-t)^3}.$$

The Laurent expansion of $F(R/I, t) + 1$, in negative powers of $(1-t)$, is

$$\frac{5t^2 - t + 2}{(1-t)^3} = \frac{a_{-3}}{(1-t)^3} + \frac{a_{-2}}{(1-t)^2} + \frac{a_{-1}}{(1-t)},$$

where $a_{-1} = 5$, $a_{-2} = -9$, $a_{-3} = 6$. Therefore the Hilbert polynomial of $S = R/I$ is $\varphi_S(t) = 3t^2 + 2$ because by Proposition 5.1.6 one has:

$$H(S, i) = 6\binom{i+2}{i} - 9\binom{i+1}{i} + 5\binom{i}{i} = 3i^2 + 2 \text{ for } i \geq 1.$$

Exercises

5.1.15 If $n \geq 1$ is an integer, then

$$\frac{1}{(1-t)^n} = \sum_{m=0}^{\infty} \binom{m+n-1}{m} t^m.$$

5.1.16 Let R be a polynomial ring in n variables over a field K. If R has the standard grading, then

$$F(R, t) = \frac{1}{(1-t)^n} \quad \text{and} \quad H(R, m) = \binom{m+n-1}{n-1}.$$

5.1.17 Let K be a field and $R = K[x_1, \ldots, x_n]$ a polynomial ring graded by $\deg(x_i) = d_i \in \mathbb{N}_+$ for $i = 1, \ldots, n$. Then the Hilbert series of R is

$$F(R, t) = 1/\prod_{i=1}^{n}(1 - t^{d_i}).$$

5.1.18 Let $R = K[x, y, z]$ be a polynomial ring over a field K and I is the ideal $(x^a, y^b, z^c, x^{a_1}y^{b_1}z^{c_1})$. If $a \geq a_1$, $b \geq b_1$ and $c \geq c_1$, then

$$e(R/I) = \dim_K(R/I) = abc_1 + (c - c_1)(a_1 b + (a - a_1)b_1).$$

5.1.19 Let $R = K[x, y]$ be a polynomial ring over a field K. If I is the ideal $(x^n, x^a y^b, y^n)$ and $a < n$, $b < n$, then $e(R/I) = n(a + b) - ab$.

5.1.20 Let $R = K[x_1, \ldots, x_n]$ be a polynomial ring over a field K graded by $\deg(x_i) = d_i \in \mathbb{N}_+$ for $i = 1, \ldots, n$. If $f_1, \ldots, f_r$ is a homogeneous regular sequence in R and $a_i = \deg(f_i)$, then

$$F(R/(f_1, \ldots, f_r), t) = \prod_{i=1}^{r}(1 - t^{a_i})/\prod_{i=1}^{n}(1 - t^{d_i}).$$

5.1.21 Let $R = K[x_1, x_2, x_3]$ and let I be the ideal $(x_1^2, x_2^2 x_3, x_2^3)$. Prove that the a-invariant of R/I is 3 and $H(R/I, i) = 4$ for $i \geq 4$.

5.1.22 Let A, B be two standard algebras over a field K and define their *Segre product* as the graded algebra

$$S = A \otimes_S B = (A_0 \otimes_K B_0) \oplus (A_1 \otimes_K B_1) \oplus \cdots \subset A \otimes_K B,$$

where $(A \otimes_K B)_p = \sum_{i+j=p} A_i \otimes_K B_j$. Use Hilbert functions to prove

$$\dim(S) = \dim(A) + \dim(B) - 1.$$

Note $\dim(A_i \otimes_K B_i) = \dim(A_i)\dim(B_i)$.

5.2 a-invariants and h-vectors

Our goal here is to relate the a-invariant of a Cohen–Macaulay $\mathbb{N}$-graded algebra with its minimal resolution and to prove that h-vectors of such algebras are nonnegative. We include Stanley's characterization of Cohen–Macaulay algebras in terms of Hilbert series.

a-invariants of positively graded algebras Let $R = K[x_1, \ldots, x_n]$ be a positively graded polynomial ring over a field K, let I be a graded ideal of R, and let

$$\mathbb{F}_\star : 0 \to \bigoplus_{i=1}^{b_g} R(-d_{gi}) \stackrel{\varphi_g}{\to} \cdots \to \bigoplus_{i=1}^{b_k} R(-d_{ki}) \stackrel{\varphi_k}{\to} \cdots \to \bigoplus_{i=1}^{b_1} R(-d_{1i}) \stackrel{\varphi_1}{\to} R$$

be the minimal resolution of $S = R/I$ (see Section 3.5). From the resolution of S, using Lemmas 5.1.1 and 5.1.2, one can write the Hilbert series as:

$$F(S,t) = \frac{\left(1 + \sum_{k=1}^{g}(-1)^k\left(\sum_{i=1}^{b_k} t^{d_{ki}}\right)\right)}{\prod_{i=1}^{n}(1 - t^{\deg(x_i)})}. \tag{5.1}$$

We begin with a general inequality, that follows from Eq. (5.1), relating $a(S)$, the a-invariant S, with the shifts of $\mathbb{F}_\star$, the minimal resolution of S.

Proposition 5.2.1 $a(S) \leq \max\{d_{ki} | \, 1 \leq k \leq g,\, 1 \leq i \leq b_k\} + a(R)$.

Definition 5.2.2 If $I \subset R$ is a Cohen–Macaulay graded ideal of height g, the *canonical module* of $S = R/I$ is defined as

$$\omega_S = \operatorname{Ext}_R^g(R/I, \omega_R),$$

where $\omega_R = R(-\delta)$ and $\delta = \sum_{i=1}^n \deg(x_i)$.

Proposition 5.2.3 *If $I \subset R$ is a Cohen–Macaulay graded ideal and $a(S)$ is the a-invariant of $S = R/I$, then*

(a) $\max_i\{d_{1i}\} < \cdots < \max_i\{d_{gi}\}$, *and*

(b) $a(S) = \max_{1\leq i\leq b_g}\{d_{gi}\} + a(R) = -\min\{i \,|\, (\omega_S)_i \neq 0\}$.

Proof. (a): We may assume that $d_{k1} \leq \cdots \leq d_{kb_k}$ for $k = 1,\ldots,g$. Since S is Cohen–Macaulay, the height of I is equal g, this follows from Corollary 3.5.14. Recall that by Proposition 3.1.32 there exists a regular sequence $f_1,\ldots,f_g$ inside I. Hence, using Proposition 2.3.7, one has:

$$\operatorname{Ext}_R^i(S, \omega_R) = 0 \ \text{ for } i < g.$$

We set $\delta = \sum_{i=1}^n \deg(x_i)$ and $\omega_R = R(-\delta)$. Therefore dualizing $\mathbb{F}_\star$, the minimal resolution of S, with respect to the canonical module of R, one obtains the (exact) complex:

$$\operatorname{Hom}(\mathbb{F}_\star, \omega_R): \quad 0 \to \operatorname{Hom}(R, \omega_R) \to \cdots \to \operatorname{Hom}\left(\bigoplus_{i=1}^{b_g} R(-d_{gi}), \omega_R\right) \to 0.$$

Observe the following natural isomorphisms:

$$\begin{aligned}\operatorname{Hom}\left(\bigoplus_{i=1}^{b_g} R(-d_{gi}), R(-\delta)\right) &\simeq \operatorname{Hom}\left(\bigoplus_{i=1}^{b_g} R(-d_{gi}), R\right)(-\delta)\\ &\simeq \bigoplus_{i=1}^{b_g} R(d_{g_i})(-\delta)\\ &\simeq \bigoplus_{i=1}^{b_g} R(-(\delta - d_{g_i})).\end{aligned}$$

Since $\operatorname{Hom}(\mathbb{F}_\star, \omega_R)$ is a minimal resolution of $\operatorname{Ext}_R^g(S, \omega_R)$, by Remark 3.5.9, one has $d_{1b_1} < d_{2b_2} < \cdots < d_{gb_g}$.

(b): By part (a), we have $d_{1b_1} < d_{2b_2} < \cdots < d_{gb_g}$. As a consequence, from the expression for $F(S,t)$ given in Eq. (5.1), one concludes the equality $a(S) = d_{gb_g} + a(R)$. In particular one has a surjection

$$\bigoplus_{i=1}^{b_g} R(-(\delta - d_{g_i})) \to \operatorname{Ext}_R^g(S, R(-\delta)) = \omega_S \to 0.$$

To complete the proof note that since $\delta - d_{gb_g} \leq \cdots \leq \delta - d_{g1}$, one has the equalities $\min\{i \mid (\omega_S)_i \neq 0\} = \delta - d_{gb_g} = -a(S)$. □

h-vectors of standard algebras Next we introduce the notion of h-vector of standard algebras through the Hilbert–Serre theorem for Hilbert series (see Theorem 5.1.4).

Definition 5.2.4 Let S be a standard algebra and let $h(t)$ be the (unique) polynomial with integral coefficients such that $h(1) \neq 0$ and

$$F(S,t) = \frac{h(t)}{(1-t)^d},$$

where $d = \dim(S)$. If $h(t) = h_0 + h_1 t + \cdots + h_r t^r$, the *$h$-vector* of S is defined as $h(S) = (h_0, \ldots, h_r)$.

Theorem 5.2.5 [392] *Let S be a standard algebra and let $\theta_1, \ldots, \theta_d$ be a homogeneous system of parameters for S with $a_i = \deg(\theta_i)$. Then S is Cohen–Macaulay if and only if*

$$F(S,t) = \frac{F(S/(\theta_1, \ldots, \theta_d), t)}{\prod_{i=1}^{d}(1 - t^{a_i})}. \tag{5.2}$$

Proof. $\Rightarrow$) Assume S is Cohen–Macaulay. Hence, by Proposition 3.1.20, $\theta_1,\ldots,\theta_d$ is a regular sequence. It is enough to prove the equality

$$F(S/(\theta_1,\ldots,\theta_s),t)=\prod_{i=1}^{s}(1-t^{a_i})F(S,t), \tag{5.3}$$

for $1\leq s\leq d$. We proceed by induction on s. Assume $s=1$. From the exact sequence of graded S-modules

$$0\longrightarrow S[-a_1]\xrightarrow{\theta_1} S\longrightarrow S/(\theta_1)\longrightarrow 0,$$

we obtain $F(S/(\theta_1),t)=F(S,t)-F(S[-a_1],t)$. Since

$$\begin{aligned}F(S[-a_1],t) &= \sum_{i=0}^{\infty}S[-a_1]_it^i=\sum_{i=a_1}^{\infty}S[-a_1]_it^i\\ &= t^{a_1}\sum_{i=a_1}^{\infty}S_{i-a_1}t^{i-a_1}=t^{a_1}F(S,t),\end{aligned}$$

we conclude $F(S/(\theta_1),t)=(1-t^{a_1})F(S,t)$. Now assume $s>1$ and that the equality (5.3) is true for $s-1$. From the exact sequence

$$0\to(S/(\theta_1,\ldots,\theta_{s-1}))[-a_s]\xrightarrow{\theta_s} S/(\theta_1,\ldots,\theta_{s-1})\longrightarrow S/(\theta_1,\ldots,\theta_s)\to 0$$

we obtain $F(S/(\theta_1,\ldots,\theta_s),t)=(1-t^{a_s})F(S/(\theta_1,\ldots,\theta_{s-1}),t)$. Then, the induction hypothesis yields the required equality.

$\Leftarrow$) Let $f(t)=\sum_{i\geq 0}a_it^i$ and $g(t)=\sum_{i\geq 0}b_it^i$ be two formal power series, we say that $f(t)\geq g(t)$ if $a_i\geq b_i$ for all i. From the exact sequence

$$0\longrightarrow(S/\mathrm{ann}\,(\theta_1))[-a_1]\xrightarrow{\theta_1} S\longrightarrow S/(\theta_1)\longrightarrow 0,$$

we obtain $F(S/(\theta_1),t)-(1-t^{a_1})F(S,t)=t^{a_1}F(\mathrm{ann}\,(\theta_1),t)$. Hence, the power series $F(S/(\theta_1),t)$ is greater than or equal to $(1-t^{a_1})F(S,t)$, and by induction it follows that

$$F(S/(\theta_1,\ldots\theta_r),t)\geq\prod_{i=1}^{r}(1-t^{a_i})F(S,t)\ \text{ for } 1\leq r\leq d.$$

Set $S'=S/(\theta_1,\ldots,\theta_{d-1})$. Using the exact sequence

$$0\longrightarrow(S'/\mathrm{ann}\,(\theta_d))[-a_d]\xrightarrow{\theta_d} S'\longrightarrow S'/(\theta_d)=S/(\theta_1,\ldots,\theta_d)\longrightarrow 0,$$

we get $F(S'/(\theta_d),t)=(1-t)^{a_d}F(S',t)+t^{a_d}F(\mathrm{ann}\,(\theta_d),t)$. By hypothesis, we have $F(S'/(\theta_d),t)=\prod_{i=1}^{d}(1-t^{a_i})F(S,t)$. Therefore

$$\underbrace{(1-t)^{a_d}}_{\geq 0}\underbrace{(F(S',t)-(1-t^{a_1})\cdots(1-t^{a_{d-1}})F(S,t))}_{\geq 0}=-t^{a_d}F(\mathrm{ann}\,(\theta_d),t).$$

Hence $F(\mathrm{ann}\,(\theta_d), t) = 0$. This shows that θ_d is a regular element on S'. By induction it follows that $\theta_1, \ldots, \theta_d$ is a regular sequence, Then S is Cohen–Macaulay by Proposition 3.1.20. This proof was adapted from [392]. $\Box$

Theorem 5.2.6 *Let S be a Cohen–Macaulay standard algebra over a field k and $h(S) = (h_i)$ the h-vector of S, then $h_i \geq 0$ for all i.*

Proof. One may assume that k is infinite, otherwise one may change the coefficient field k using the functor $(\cdot) \otimes_k K$, where K is an infinite field extension of k. There is a h.s.o.p $\mathbf{y} = \{y_1, \ldots, y_d\}$ for S, where each y_i is a form in S of degree one. Set $\overline{S} = S/(\mathbf{y})S$. From Theorem 5.2.5 we derive $h_i = H(\overline{S}, i)$ for all i and consequently $h_i \geq 0$. $\Box$

Theorem 5.2.7 (Stanley) *Let K be a field and let S be a positively graded K-algebra of dimension d. If S is a Cohen–Macaulay domain and*

$$F(S, t) = (-1)^d t^a F(S, t^{-1})$$

for some $a \in \mathbb{Z}$, then S is Gorenstein and a is the a-invariant of S.

Proof. By Exercise 5.2.10, $F(S, t) = t^a F(\omega_S, t) = F(\omega_S[-a], t)$. Pick an element $0 \neq x \in \omega_S[-a]$ of degree zero and define

$$\varphi\colon S \longrightarrow \omega_S[-a] \qquad (r \stackrel{\varphi}{\longmapsto} rx),$$

the map φ is injective by Exercise 5.2.12. Therefore the equality above gives that φ is an isomorphism. Thus $\omega_S \simeq S[a]$, as required. $\Box$

Proposition 5.2.8 *Let S be a standard graded algebra and let $M \subset N$ be finitely generated graded S-modules of the same dimension d and multiplicity e. If M and N are Cohen–Macaulay and*

$$H_M(t) = \frac{f(t)}{(1-t)^d}, \qquad H_N(t) = \frac{g(t)}{(1-t)^d}$$

are their Hilbert series, then $\deg f(t) \geq \deg g(t)$.

Proof. There is an exact sequence $0 \to M \to N \to N/M \to 0$ of graded modules. If $M \neq N$, N/M is a module of dimension $< d$ since M and N have the same multiplicity. Since M and N are Cohen–Macaulay, standard depth chasing (see Lemma 2.3.9) implies that N/M is Cohen–Macaulay of dimension $d-1$.

We have the equality of Hilbert series,

$$\frac{g(t)}{(1-t)^d} = \frac{f(t)}{(1-t)^d} + \frac{h(t)}{(1-t)^{d-1}},$$

so $g(t) - f(t) = (1-t)h(t)$. The assertion follows since the h-vectors of these two modules are positive (see Exercise 5.2.9). $\Box$

Exercises

5.2.9 Let R be a polynomial ring over a field K with the usual grading and M a finitely generated $\mathbb{N}$-graded module. If M is Cohen–Macaulay and $h(t) = h_0 + h_1t + \cdots + h_rt^r$ the polynomial in $\mathbb{Z}[t]$ so that $h(1) \neq 0$ and

$$F(M,t) = h(t)/(1-t)^d, \text{ where } d = \dim(M).$$

Prove that $h_i \geq 0$ for all i.

5.2.10 (Stanley) Let K be a field and let S be a positively graded K-algebra. If S is Cohen–Macaulay of dimension d prove the equality:

$$F(\omega_S, t) = (-1)^d F(S, t^{-1}).$$

Hint Note $S \simeq R/I$, with R a polynomial ring. Compute the Hilbert series of S and ω_S using the minimal resolution of R/I as an R-module.

5.2.11 (Stanley) Let K be a field and let S be a positively graded K-algebra. If S is Gorenstein of dimension d and $a(S)$ is its a-invariant, prove:

$$F(S,t) = (-1)^d t^{a(S)} F(S, t^{-1}).$$

Hint $\omega_S \simeq S[a(S)]$.

5.2.12 Let S be a Cohen–Macaulay local domain and let M be a Cohen–Macaulay S-module. If $\dim(S) = \dim(M)$, prove that $\mathcal{Z}(M) = (0)$.

Hint Note $\dim(M) = \dim(S/\mathfrak{p})$ for $\mathfrak{p} \in \mathrm{Ass}(M)$ and $\mathrm{ann}(M) = 0$.

5.2.13 Let K be a field and let S be a positively graded K-algebra. If S is a Cohen–Macaulay domain and the h-vector of S is symmetric, prove that S is a Gorenstein ring.

Hint Use Theorem 5.2.7.

5.2.14 Let R be a polynomial ring over a field K with the standard grading, and let M_1, M_2 be two finitely generated graded R-modules. Then

$$\mathrm{depth}(M_1 \oplus M_2) = \max\{\mathrm{depth}(M_1), \mathrm{depth}(M_2)\}.$$

5.3 Extremal algebras

The concept of an extremal Cohen–Macaulay—or Gorenstein—standard algebra (resp. ideal) has its origins in the works of Sally [366] and Schenzel [370]. Those algebras (resp. ideals) have the smallest possible reduction number (resp. smallest possible number of generators of the least degree). We will study with certain detail those two classes of algebras.

The Cohen–Macaulay case Let R be a polynomial ring over a field k with its usual grading and let $I = \bigoplus_{i=p}^{\infty} I_i$ be a graded ideal with $I_p \neq (0)$. The integer p is the *initial degree* of I. As usual $\mu(I)$ denotes the minimal number of generators of I, and $\mu(I_p)$ stands for the minimal number of generators of I in degree p.

Proposition 5.3.1 *Let R be a polynomial ring over a field k and I a graded ideal in R of height g and initial degree p. If I is Cohen–Macaulay, then*

$$\mu(I_p) = H(I,p) \leq \binom{p+g-1}{p}.$$

Proof. We may assume that k is infinite, since a change of the coefficient field can be easily carried out using the functor $(\cdot) \otimes_k K$, where K is an infinite field extension of k. This change of the coefficient field preserves the hypothesis on I and leaves both the height of I and the dimensions of the vector spaces of forms of a given degree in I unchanged.

We set $S = R/I$. By Lemma 3.1.28, there is a homogeneous system of parameters $\mathbf{y} = \{y_1, \ldots, y_d\}$ for S, where each y_i is a linear form in R. If we tensor the minimal resolution of S with $\overline{R} = R/(\mathbf{y})$ it follows that

$$0 \leq H(\overline{S}, p) = H(\overline{R}, p) - H(I,p),$$

where $\overline{S} = S/(\mathbf{y})S$. To get the desired inequality it suffices to observe that $\overline{R}$ is a polynomial ring in g variables over the field k. $\Box$

Definition 5.3.2 If S is an Artinian positively graded algebra the *socle* of S is given by

$$\mathrm{Soc}(S) = (0 \colon_S S_+).$$

Lemma 5.3.3 [174] *Let $S = R/I$ be a quotient ring of a polynomial ring R over a field k modulo a graded ideal I. Let*

$$\mathbb{F}_\star : \quad 0 \to \bigoplus_{i=1}^{b_g} R(-d_{gi}) \to \cdots \to \bigoplus_{i=1}^{b_1} R(-d_{1i}) \to R \to S = R/I \to 0,$$

be the minimal resolution of S. If S is Artinian, then there is a degree zero isomorphism of graded k-vector spaces $\mathrm{Soc}(S) \simeq \bigoplus_{i=1}^{b_g} k[g - d_{gi}]$.

Proof. Let $R = k[x_1, \ldots, x_g]$ and $\underline{x} = \{x_1, \ldots, x_g\}$. The ordinary Koszul complex $K_\star$ associated to the regular sequence $\underline{x}$ is an R-free resolution of $k = R/(\underline{x})$ and thus

$$\mathrm{Tor}_i^R(S,k) \simeq H_i(K_\star \otimes_R S).$$

On the other hand, we have $\operatorname{Tor}_g^R(S,k) \simeq \bigoplus_{i=1}^{b_g} k[-d_{gi}]$ (see the proof of Corollary 3.5.7). To complete the argument note

$$\begin{aligned} H_g(K_\star \otimes_R S)[g] &\simeq Z_g(K_\star \otimes_R S)[g] \\ &= \left\{ ze_1 \wedge \cdots \wedge e_g \,\middle|\, \sum_{i=1}^{g} (-1)^{i-1} zx_i e_1 \wedge \cdots \wedge \widehat{e_i} \wedge \cdots \wedge e_g = 0 \right\} [g] \\ &\simeq \operatorname{Soc}(S). \end{aligned}$$

Altogether one has $\operatorname{Soc}(S) \simeq \bigoplus_{i=1}^{b_g} k[g - d_{gi}]$, as required. □

Proposition 5.3.4 *If S is a Cohen–Macaulay positively graded k-algebra over a field k with presentation R/I, then the type of S is equal to the last Betti number in the minimal free resolution of R/I as an R-module.*

Proof. By making a change of coefficients using the functor $(\cdot)\otimes_k K$, where K is an infinite field extension of k, one may assume that k is infinite.

By Theorem 3.1.24 there exists a system of parameters $\mathbf{y} = \{y_1, \ldots, y_d\}$ for $S = R/I$ with each y_i a form of degree one of R. Since $\operatorname{Tor}_i(S, R/(\mathbf{y})) = 0$ for $i \geq 1$, the minimal resolution of $S/(\mathbf{y}S)$ as an $R/(\mathbf{y})$-module has the same twists and Betti numbers as the minimal resolution of S over R. Since $S/(\mathbf{y}S)$ is Artinian the result follows from Lemma 5.3.3. □

Corollary 5.3.5 *If R is a polynomial ring over a field k, then a graded ideal I of R is Gorenstein iff R/I is Cohen–Macaulay and the last Betti number in the minimal graded resolution of R/I is equal to 1.*

Corollary 5.3.6 *Let S be a Cohen–Macaulay positively graded k-algebra over a field k and let ω_S be its canonical module. Then*

$$\operatorname{type}(S) = \mu(\omega_S),$$

where $\mu(\omega_S)$ is the minimum number of generators of ω_S.

Proof. It follows from the proofs of Propositions 5.2.3 and 5.3.4. □

Theorem 5.3.7 ([79], [350]) *Let R be a polynomial ring over a field k and let I be a graded ideal in R of height g and initial degree p. If $S = R/I$ is a Cohen–Macaulay ring, then S has a p-linear resolution if and only if the following equality holds*

$$\mu(I_p) = H(I,p) = \binom{p+g-1}{p}. \tag{5.4}$$

Proof. Let $\mathbb{F}_\star$ be the minimal resolution of S as in Lemma 5.3.3. We can order the shifts so that $d_{k1} \leq \cdots \leq d_{kb_k}$ for $k = 1, \ldots, g$; by the minimality of the resolution $d_{11} < d_{21} < \cdots < d_{g1}$ (see Remark 3.5.9). Observe that if the resolution of S is linear, then the Herzog–Kühl formulas of Theorem 3.5.17 give the required equality.

Conversely assume the equality above. Since we may assume that k is infinite, by Lemma 3.1.28 there is a h.s.o.p $\mathbf{y} = \{y_1, \ldots, y_d\}$ for S, where each y_i is a form of degree one in R. Set $\overline{S} = S/(\mathbf{y})S$ and $\overline{R} = R/(\mathbf{y})$. From Theorem 5.2.5 we get that the h-vector of S satisfies $h_p = 0$, and we also get $h_i = H(\overline{S}, i)$ for all i. Therefore $h_i = 0$ for all $i \geq p$. Using Lemma 5.3.3 one obtains a degree zero isomorphism

$$\mathrm{Soc}(\overline{S}) \simeq \bigoplus_{i=1}^{b_g} k[g - d_{gi}]$$

of graded k-vector spaces. In particular, the socle of $\overline{S}$ can only live in degrees $d_{gi} - g$, hence we conclude the inequality $d_{gi} - g \leq p - 1$. On the other hand the minimality of the resolution of $\overline{S}$ gives $p + (g - 1) \leq d_{gi}$ for all i. Altogether we have $d_{gi} = p + g - 1$. Because I is C–M, then by Proposition 5.2.3 we must have $p \leq d_{1b_1} < d_{2b_2} < \cdots < d_{gb_g} = p + g - 1$, which implies that the resolution is p-linear, as required. □

Example 5.3.8 [346] Let $R = k[a, \ldots, f]$ be a polynomial ring over a field k and let I be the ideal

$$I = (abc, abd, ace, adf, aef, bcf, bde, bef, cde, cdf).$$

Then R/I is C–M if $\mathrm{char}(k) \neq 2$ and non C–M otherwise. Thus in the first case R/I has a 3-linear resolution.

Other nice examples of Cohen–Macaulay algebras with linear resolutions include rings of minimal multiplicity [366], the coordinate ring of a variety defined by the submaximal minors of a generic symmetric matrix [264], the coordinate ring of a variety defined by the maximal minors of a generic matrix [123], and some face rings [170].

Cohen–Macaulay rings with linear resolutions have been studied by Sally [366] for the case $p = 2$, and by Schenzel [370] for the general case; more general rings with linear resolutions have been examined in [130, 233].

The Gorenstein case To give a sharp bound for the number of generators in the least degree of a graded Gorenstein ideal is harder than the Cohen–Macaulay case. We deal with this problem in the next section.

The aim here is to introduce and characterize extremal Gorenstein rings in terms of their minimal resolutions and to determine their Betti numbers. Those Betti numbers are our "natural candidates" to bound the initial Betti numbers of graded Gorenstein ideals.

Proposition 5.3.9 *If $S = S_0 + \cdots + S_s$ is a positively graded Artinian Gorenstein algebra over a field k, then the multiplication maps*

$$S_i \times S_{s-i} \to S_s = \mathrm{Soc(S)} \simeq k$$

are a perfect pairing, that is, the homomorphisms:

$$\varphi_i : S_i \to \mathrm{Hom}_k(S_{s-i}, S_s) \simeq S_{s-i}, \quad \varphi_i(r)(x) = rx$$

are isomorphisms of k-vector spaces.

Proof. To show that φ_i is one to one take $r \in \ker(\varphi)$ and assume $r \neq 0$ and $i < s$. Since r is not in the socle of S, one may pick $z \in S$ homogeneous of maximal positive degree such that $zr \neq 0$. Note that $0 < \deg(z) < s - i$, because $rx = 0$ for $x \in S_{s-i}$. Hence zr has degree less than s, using that the socle of S lives only in degree s one obtains a homogeneous element w with $\deg(w) > 0$ such that $wzr = 0$, which contradicts the choice of z. Therefore $r = 0$ and φ is injective. Using the injectivity of φ_i one has $\dim_k S_i \leq \dim_k S_{s-i} \leq \dim_k S_i$, and thus φ_i is an isomorphism. □

Corollary 5.3.10 *Let $(h_0, \ldots, h_s)$ be the h-vector of a standard Gorenstein k-algebra, where k is a field. Then h is symmetric, that is,*

$$h_i = h_{s-i} \ \text{ for all } \ i.$$

Proposition 5.3.11 *Let R be a polynomial ring over a field k with the standard grading and I a graded Gorenstein ideal of initial degree p and height g. If the minimal resolution of $S = R/I$ by free R-modules is:*

$$\mathbb{F}_\star : \quad 0 \to \bigoplus_{i=1}^{b_g} R(-d_{gi}) \to \cdots \to \bigoplus_{i=1}^{b_k} R(-d_{ki}) \to \cdots \to \bigoplus_{i=1}^{b_1} R(-d_{1i}) \to R,$$

then $\bigoplus_{i=1}^{b_{g-k}} R(-(d_g - d_{(g-k)i})) \simeq \bigoplus_{i=1}^{b_k} R(-d_{ki})$ for $1 \leq k \leq g-1$.

Proof. One may always order the shifts so that $d_{k1} \leq \cdots \leq d_{kb_k}$ for $k = 1, \ldots, g$. As $b_g = 1$ and $d_{gi} = d_g$, by Proposition 5.2.3, one has that the a-invariant of S is $a(S) = d_g - n$, where n is the number of variables of R. Since $\mathrm{Hom}(\mathbb{F}_\star, \omega_R)$ is a minimal resolution of

$$\omega_S \simeq \mathrm{Ext}_R^g(S, \omega_R)$$

and using $\omega_S \simeq S(d_g - n)$, one obtains that $\mathbb{F}_\star$ is self dual. □

With the same assumptions and notation of Proposition 5.3.11 one has:

Proposition 5.3.12 *If $(h_0, \ldots, h_s)$ is the h-vector of S, then $s \geq 2(p-1)$ with equality if and only if the minimal resolution of S is of the form*

$$0 \to R(-(2p+g-2)) \to R^{b_{g-1}}(-(p+g-2)) \to \cdots \to R^{b_2}(-(p+1)) \to R^{b_1}(-p) \to R.$$

Proof. As in the proof of Theorem 5.2.6, one may assume that S is Artinian and that h_i is equal to $H(S,i)$. Note that the socle of S lives in degree $d_g - g$ by Lemma 5.3.3. Hence $s \geq d_g - g$. By Proposition 5.3.11 and the minimality of the resolution one concludes

$$s \geq d_g - g = d_{1b_1} + d_{(g-1)1} - g \geq p - g + d_{(g-1)1} \geq 2(p-1),$$

where the last inequality uses $d_{k1} \geq p + k - 1$ for $1 \leq k \leq g-1$. For the second assertion use the chain of inequalities above and compute the h-vector of S using the resolution of S. □

Definition 5.3.13 If $s = 2(p-1)$, S is called an *extremal Gorenstein ring* and its resolution is called *pure and almost linear.*

Proposition 5.3.14 [370] *Let R be a polynomial ring over a field k and I a graded ideal of initial degree p and height g. If $S = R/I$ is an extremal Gorenstein algebra, then for $1 \leq m < g$ the mth Betti number of S is:*

$$b_m = \binom{p+g-1}{p+m-1}\binom{p+m-2}{m-1} + \binom{p+g-1}{m}\binom{p+g-m-2}{p-1} - \binom{g}{m}\binom{p+g-2}{p-1}.$$

Proof. As in the proof of Theorem 5.3.7, one may assume that S is Artinian and that R a polynomial ring in g variables. Set $s = 2(p-1)$. From the minimal resolution of S of Proposition 5.3.12, and using the additivity of Hilbert series (see Lemma 5.1.1), we get

$$\left(\sum_{i=0}^{s} h_i t^i\right)(1-t)^g = 1 + \sum_{i=1}^{g-1}(-1)^i b_i t^{p+i-1} + (-1)^g b_g t^{s+g},$$

where $h_i = H(S,i) = h_{s-i}$. Therefore

$$b_m = \sum_{i=0}^{p+m-1} (-1)^{p-1-i}\binom{g}{p+m-1-i} h_i, \; 1 \leq m \leq g-1$$

Using the identity

$$\sum_{i=0}^{p-1}(-1)^{p-1-i}\binom{g}{p+m-1-i}\binom{g-1+i}{i} = \binom{p+g-1}{p+m-1}\binom{p+m-2}{m-1} \tag{5.5}$$

and the symmetry of the h-vector of S (see Corollary 5.3.10) one has

$$b_m = \binom{p+g-1}{p+m-1}\binom{p+m-2}{m-1} + \sum_{i=p}^{p+m-1} (-1)^{p-1-i}\binom{g}{p+m-1-i} h_{s-i}.$$

Note

$$\sum_{i=p}^{p+m-1} (-1)^{p-1-i}\binom{g}{p+m-1-i} h_{s-i}$$
$$= \sum_{i=0}^{m-1}(-1)^{i+1}\binom{g}{m-1-i}\binom{p+g-3-i}{g-1},$$

since the terms in the right-hand side of the equality are zero for $i > p-1$ the last summation reduces to

$$\sum_{i=0}^{p-2}(-1)^{i+1}\binom{g}{m-1-i}\binom{p+g-3-i}{g-1}$$
$$= \sum_{i=0}^{p-1}(-1)^{p-1-i}\binom{g}{g-m+p-1-i}\binom{g-1+i}{i} - \binom{g}{m}\binom{g+p-2}{p-1}$$
$$= \binom{p+g-1}{m}\binom{p+g-m-2}{p-1} - \binom{g}{m}\binom{g+p-2}{p-1},$$

the last equality follows from Eq. (5.5) making m equal to $g-m$. □

Exercises

5.3.15 Let R be a polynomial ring over a field k and let I be a Cohen–Macaulay graded ideal in R of height g and initial degree p. Let $(h_0, \ldots, h_s)$ be the h-vector of R/I. Prove that $s \geq p-1$, with equality if and only if the minimal resolution of R/I is p-linear. If $s = p-1$, the algebra R/I is called an *extremal Cohen–Macaulay algebra*.

5.3.16 Let R be a polynomial ring over a field k and let I be a C–M graded ideal in R of height g and initial degree p. Prove that the Hilbert series of $S = R/I$ can be written as

$$F(S,t) = \frac{\sum_{i=0}^{p-1}\binom{i+g-1}{g-1}t^i}{(1-t)^d}, \quad \text{where } d = \dim S,$$

if and only if S has a p-linear resolution.

5.3.17 Let R be a polynomial ring over a field k and I a Cohen–Macaulay graded ideal in R of height g and initial degree p. If S has a linear resolution, show that the Betti numbers of $S = R/I$ and its multiplicity are given by

$$b_k = \binom{p+k-2}{k-1} \cdot \binom{p+g-1}{g-k} \quad \text{and} \quad e(S) = \binom{p+g-1}{g}.$$

5.3.18 Let $S = R/I$ be a standard Gorenstein algebra so that I has initial degree p and codimension g. Show that the multiplicity of S satisfies:

$$e(S) \geq e(g,p) = \binom{g+p-1}{g} + \binom{g+p-2}{g},$$

with equality if and only if S is an extremal Gorenstein algebra.

Hint The h-vector $h(S) = (h_0, \ldots, h_s)$ is symmetric and $s \geq 2(p-1)$. Then compute explicitly $h_{p-1} + 2\sum_{i=0}^{p-2} h_i$.

5.3.19 Let R be a polynomial ring over a field k and I a graded Gorenstein ideal of height 4 generated by forms of degree p. Let $h(S) = (h_0, \ldots, h_s)$ be the h-vector of $S = R/I$. Then the minimal resolution of S has the form:

$$0 \to R(-(s+4)) \to R^b(-(s-p+4)) \to \bigoplus_{i=1}^{s-2p+3} R^{a_i}(-(p+i)) \to R^b(-p) \to R,$$

where $r = s - 2p + 3$ and $a_i = a_{r-i+1}$ for $i \geq 1$.

5.3.20 If $x > 1$ is an integer, prove that $x(x+8)$ is never a perfect square.

5.3.21 Let I be a graded Cohen–Macaulay ideal of a polynomial ring R with a minimal resolution:

$$0 \to R^{b_3}(-(p+a+b)) \to R^{b_2}(-(p+a)) \to R^{b_1}(-p) \to R \to R/I.$$

If $b_1 = 9$, use the Herzog–Kühl formulas to prove that I is Gorenstein.

Hint Note $b_3 - b_2 + b_1 - 1 = 0$ and prove that $b_1 b_2 b_3 = 9b_3(b_3+8)$ is a perfect square, then use Exercise 5.3.20.

5.4 Initial degrees of Gorenstein ideals

Assume that R is a polynomial ring over a field and that I is a homogeneous Gorenstein ideal of codimension $g \geq 3$ and initial degree $p \geq 2$. We set

$$e(g,p) = \binom{g+p-1}{g} + \binom{g+p-2}{g} \quad \text{and} \quad \mu_0 = \binom{p+g-1}{g-1} - \binom{p+g-3}{g-1}.$$

By the symmetry of the h-vector of R/I given in Corollary 5.3.10, the multiplicity of R/I satisfies

$$e(R/I) \geq e(g,p),$$

see Exercise 5.3.18. Given codimension g and initial degree p, a graded Gorenstein algebra R/J with multiplicity $e(R/J) = e(g,p)$ is extremal. This has strong structural implications: the minimal free resolution of R/J must be pure and almost linear; hence all the Betti numbers $b_i(R/J)$ are determined, and J is generated by μ_0 forms of degree p (see Section 5.3).

Results in the literature dealing with Cohen–Macaulay ideals (such as [140, 361]) give upper bounds for the minimal number of generators $\mu(I)$ of I in terms of codimension, initial degree, and multiplicity $e(R/I)$. One of the aims below is to elucidate the multiplicity information that is already determined by the codimension and initial degree (see Theorem 5.4.3). In general one expects no such conclusion, but for graded Gorenstein algebras the symmetry of the h-vector $h(R/I)$ can be effectively exploited.

Macaulay's Theorem

First we introduce the binomial expansion, and then state a famous result of Macaulay on the growth of Hilbert functions that will be needed later.

Lemma 5.4.1 *Let h, j be positive integers. Then h can be written uniquely*

$$h = \binom{a_j}{j} + \binom{a_{j-1}}{j-1} + \cdots + \binom{a_i}{i}, \tag{5.6}$$

where $a_j > a_{j-1} > \cdots > a_i \geq i \geq 1$.

Proof. By induction on j. If $j = 1$, then $h = \binom{h}{1}$. Assume $j > 1$, then there is a unique a_j such that $\binom{a_j}{j} \leq h < \binom{a_j+1}{j}$. Let $r = h - \binom{a_j}{j}$. By induction

$$r = h - \binom{a_j}{j} = \binom{a_{j-1}}{j-1} + \cdots + \binom{a_i}{i} \text{ where } a_{j-1} > \cdots > a_i \geq i \geq 1.$$

From the inequality

$$h = \binom{a_j}{j} + \binom{a_{j-1}}{j-1} + \cdots + \binom{a_i}{i} < \binom{a_j+1}{j} = \binom{a_j}{j} + \binom{a_j}{j-1}$$

we obtain $\binom{a_{j-1}}{j-1} < \binom{a_j}{j-1}$ and this implies $a_{j-1} < a_j$. Let us show the uniqueness. Assume h can be written as

$$h = \binom{b_j}{j} + \binom{b_{j-1}}{j-1} + \cdots + \binom{b_k}{k},$$

where $b_j > b_{j-1} > \cdots > b_k \geq k \geq 1$. By induction on j one has

$$\binom{b_j}{j} \leq h < \binom{b_j + 1}{j}.$$

Since a_j is unique we derive $a_j = b_j$. Hence by induction hypothesis $k = i$ and $a_{j-1} = b_{j-1}, \ldots, a_i = b_i$. □

The unique expression for h in Eq. (5.6) is called the *binomial expansion* of h in base j. We set

$$h^{(j)} = \binom{a_j}{j+1} + \binom{a_{j-1}}{j} + \cdots + \binom{a_i}{i+1}, \tag{5.7}$$

$$h^{\langle j\rangle} = \binom{a_j+1}{j+1} + \binom{a_{j-1}+1}{j} + \cdots + \binom{a_i+1}{i+1}, \tag{5.8}$$

sometimes $h^{\langle j\rangle}$ is called the *Macaulay symbol.*

Theorem 5.4.2 (Macaulay [413, Appendix B]) *Let $h\colon \mathbb{N} \to \mathbb{N}$ be a numerical function and k a field. The following conditions are equivalent:*

(*a*) *There exists a homogeneous k-algebra S with $H(S, i) = h(i)$ for $i \geq 0$.*

(*b*) *$h(0) = 1$ and $h(i + 1) \leq h(i)^{\langle i\rangle}$ for all $i \geq 1$.*

Theorem 5.4.3 [318] *If I is a graded Gorenstein ideal of height $g \geq 3$ and initial degree $p \geq 2$, then*

$$\mu(I_p) \leq \mu_0 = \binom{p+g-1}{g-1} - \binom{p+g-3}{g-1},$$

and I is itself extremal if equality holds.

Proof. If either $\mu(I_p) > \mu_0$, or $\mu(I_p) = \mu_0$ and I is not extremal, then by the symmetry of the h-vector $h(R/I)$ there is some $j \geq p$ so that

$$h(R/I) = (h_0, h_1, \ldots, h_{p-1}, h_p, \ldots, h_j, h_{p-1}, \ldots, h_1, h_0),$$

where $h_j = h_p \leq \binom{p+g-3}{g-1} = h_{p-2}$. The idea of the argument is to use the Macaulay bound $h_j^{\langle j\rangle}$ for h_{j+1} to see that such a small value of h_j cannot grow to such a large value of

$$h_{j+1} = h_{p-1} = \binom{p+g-2}{g-1}.$$

Recall that this estimate is calculated from the binomial expansion for h_j:

$$h_j = \binom{a_j}{j} + \binom{a_{j-1}}{j-1} + \cdots + \binom{a_i}{i}, \tag{5.9}$$

where $a_j > a_{j-1} > \cdots > a_i \geq i \geq 1$. Then

$$h_{j+1} \leq h_j^{\langle j \rangle} = \binom{a_j+1}{j+1} + \binom{a_{j-1}+1}{j} + \cdots + \binom{a_i+1}{i+1}. \tag{5.10}$$

We may assume that $h_j > j$, for if not, then $h_j \leq j$ would imply that $a_\ell = \ell$ for all ℓ, and hence $h_j \geq h_{j+1}$, which contradicts our assumption. Notice that by grouping the terms of (5.9) according to the value of $a_\ell - \ell$ the binomial expansion for h_j can be written as

$$\begin{aligned} h_j &= \sum_{n=0}^{r} \left[\binom{j_n + k_n}{j_n} + \binom{j_n - 1 + k_n}{j_n - 1} + \cdots + \binom{j_n - i_n + k_n}{j_n - i_n} \right] \\ &= \sum_{n=0}^{r} \left[\binom{j_n + k_n + 1}{j_n} - \binom{j_n - i_n + k_n}{j_n - i_n - 1} \right], \end{aligned} \tag{5.11}$$

where $a_j - j = k_0 > k_1 > \cdots > k_r \geq 0$, $j = j_0 > j_1 > \cdots > j_r$, $j_r - i_r = i$, and $j_n = j_{n-1} - i_{n-1} - 1$ for $1 \leq n \leq r$. Set $k = k_0$. Since $p \leq j$ and

$$\binom{j+k}{j} \leq h_j \leq \binom{p+g-3}{g-1}$$

it follows that $k \leq g - 2$. From (5.9), (5.10), and (5.11), together with Pascal's identity and

$$\binom{a+b+1}{b+1} = \frac{a+1}{b+1}\binom{a+b+1}{b},$$

we have

$$\begin{aligned} h_{j+1} - h_j &\leq \sum_{n=0}^{r} \left[\binom{j_n + k_n + 1}{j_n + 1} - \binom{j_n - i_n + k_n}{j_n - i_n} \right] \\ &= \sum_{n=0}^{r} \left[\frac{k_n + 1}{j_n + 1} \binom{j_n + k_n + 1}{j_n} - \frac{k_n + 1}{j_n - i_n} \binom{j_n - i_n + k_n}{j_n - i_n - 1} \right]. \end{aligned}$$

On the other hand from the upper bound on h_j and $h_{j+1} = h_{p-1}$ we get

$$\frac{g-1}{p-1} h_j \leq \binom{p+g-3}{g-2} \leq h_{j+1} - h_j.$$

Since $(g-1)/(p-1) > (k+1)/(j+1)$ it follows from (5.11) and the last two inequalities that $F_0 < 0$, where for $0 \leq s \leq r$ we set

$$\begin{aligned} F_s &= \sum_{n=s}^{r} \left[\left(\frac{k_s + 1}{j_s + 1} - \frac{k_n + 1}{j_n + 1} \right) \binom{j_n + k_n + 1}{j_n} \right] \\ &\quad + \sum_{n=s}^{r} \left[\left(\frac{k_n + 1}{j_n - i_n} - \frac{k_s + 1}{j_s + 1} \right) \binom{j_n - i_n + k_n}{j_n - i_n - 1} \right]. \end{aligned}$$

To derive a contradiction we are going to show the following inequalities

$$F_0 > F_1 > \cdots > F_r = \left(\frac{k_r+1}{j_r-i_r} - \frac{k_r+1}{j_r+1}\right)\binom{a_i}{i-1} > 0.$$

Assume $1 \leq s+1 \leq r$. Notice that

$$\begin{aligned} 0 &\leq \sum_{n=s}^{r-1}\left[\binom{j_n-i_n+k_n}{j_n-i_n-1} - \binom{j_{n+1}+k_{n+1}+1}{j_{n+1}}\right] + \binom{j_r-i_r+k_r}{j_r-i_r-1} \\ &= \binom{j_s-i_s+k_s}{j_s-i_s-1} - \sum_{n=s+1}^{r}\left[\binom{j_n+k_n+1}{j_n} - \binom{j_n-i_n+k_n}{j_n-i_n-1}\right]. \end{aligned}$$

Therefore

$$\binom{j_s-i_s+k_s}{j_s-i_s-1} \geq \sum_{n=s+1}^{r}\binom{j_n+k_n+1}{j_n} - \sum_{n=s+1}^{r}\binom{j_n-i_n+k_n}{j_n-i_n-1}.$$

Let A_s denote the first summation and B_s the second in this last inequality; clearly $A_s - B_s > 0$. Also note that

$$\frac{k_s+1}{j_s-i_s} > \frac{k_s+1}{j_s+1} \quad \text{and} \quad \frac{k_s+1}{j_s-i_s} > \frac{k_{s+1}+1}{j_{s+1}+1}.$$

Putting these together, we compute

$$\begin{aligned} F_s - F_{s+1} &= \left(\frac{k_s+1}{j_s-i_s} - \frac{k_s+1}{j_s+1}\right)\binom{j_s-i_s+k_s}{j_s-i_s-1} \\ &\quad + \left(\frac{k_s+1}{j_s+1} - \frac{k_{s+1}+1}{j_{s+1}+1}\right)A_s + \left(\frac{k_{s+1}+1}{j_{s+1}+1} - \frac{k_s+1}{j_s+1}\right)B_s \\ &\geq \left(\frac{k_s+1}{j_s-i_s} - \frac{k_s+1}{j_s+1}\right)(A_s-B_s) + \left(\frac{k_s+1}{j_s+1} - \frac{k_{s+1}+1}{j_{s+1}+1}\right)A_s \\ &\quad + \left(\frac{k_{s+1}+1}{j_{s+1}+1} - \frac{k_s+1}{j_s+1}\right)B_s \\ &= \left(\frac{k_s+1}{j_s-i_s} - \frac{k_{s+1}+1}{j_{s+1}+1}\right)(A_s-B_s) \quad > \quad 0. \end{aligned}$$

Hence $F_0 > F_r > 0$, which contradicts the observation that $F_0 < 0$. □

One might hope for even stronger estimates than suggested by the result above, for instance that $\mu(I) \leq \mu_0$. Next we present a counterexample showing that Theorem 5.4.3 cannot be extended to bound the number of generators in all degrees.

Example 5.4.4 If $I = ((x_1^2, x_2^4, x_3^3, x_4^4)\colon (x_1x_2 - x_3x_4))$, using *Macaulay2* [199] one readily obtains that I is given by:

$(x_1^2,\ x_1x_2x_3 + x_3^2x_4,\ x_3^3,\ x_1x_3^2,\ x_2^4,\ x_4^4,\ x_1x_4^3,\ x_1x_2x_4^2 + x_3x_4^3,\ x_2^3x_3^2,\ x_2^3x_4^3).$

Then R/I is a Gorenstein Artin algebra with h-vector $(1,4,9,13,13,9,4,1)$ and Betti sequence $(1,10,18,10,1)$, whereas the h-vector for an extremal Gorenstein algebra of codimension four and initial degree two is $(1,4,1)$ and the Betti sequence is $(1,9,16,9,1)$; notice that the multiplicity $e(R/I) = 54$ is far greater than the minimal value of six exhibited by an extremal algebra.

In height three one can bound the Betti numbers, thanks to the structure theorem of Buchsbaum and Eisenbud. For initial degree p, the extremal Gorenstein algebra of codimension three has $b_1 = b_2 = 2p+1$.

Theorem 5.4.5 *Let R be a polynomial ring over a field k and let I be a homogeneous Gorenstein ideal of height* 3 *and initial degree p. Then*

$$\mu(I) \leq 2p+1 \quad \textit{and} \quad b_2(R/I) \leq 2p+1.$$

Proof. We may assume, without loss of generality, that R is equal to $k[x_1,x_2,x_3]$ and $A = R/I$ is Artinian and local with socle$(A) = A_\sigma$. Then by [75] the minimal free resolution of A has the form

$$0 \to R(-\sigma-3) \to \bigoplus_{j=1}^{\mu} R(-n_j) \stackrel{Y}{\longrightarrow} \bigoplus_{i=1}^{\mu} R(-m_i) \to R,$$

where Y is an alternating matrix, and the generators $f_1,\ldots,f_\mu$ of I are the maximal pfaffians of Y. If the theorem fails every generator has degree at least $(2p+1)/2 > p$, which is a contradiction since at least one minimal generator has degree p. $\Box$

Conjecture 5.4.6 (Miller–Villarreal) *Let $I \subset R$ be a graded Gorenstein ideal of initial degree p and height g. If $1 \leq i < g$ and γ_i is the* ith *initial Betti number of R/I, then*

$$\gamma_i \leq \binom{p+g-1}{p+i-1}\binom{p+i-2}{i-1} + \binom{p+g-1}{i}\binom{p+g-i-2}{p-1} - \binom{g}{i}\binom{p+g-2}{p-1}.$$

Proposition 5.4.7 ([39], [65, Lemma 4.2.13]) *Let n be a given positive integer and $f,g\colon \mathbb{N} \to \mathbb{N}$ the numerical functions given by*

$$g(x) = n^{\langle x\rangle} \quad \textit{and} \quad f(x) = x^{\langle n\rangle},$$

where $h^{\langle j\rangle}$ is the Macaulay symbol. Then $g(x)$ is non increasing and $f(x)$ is strictly increasing.

There is a short and elegant argument, based on the behavior of the combinatorial functions $g(x) = n^{\langle x \rangle}$ and $f(x) = x^{\langle n \rangle}$, to show the inequality of Theorem 5.4.3. We include the details of this argument in the proof below.

Corollary 5.4.8 *If R is a polynomial ring over a field k and I is a graded Gorenstein ideal of R of height $g \geq 3$ and initial degree $p \geq 2$, then*

$$\mu(I_p) \leq \mu_0 = \binom{p+g-1}{g-1} - \binom{p+g-3}{g-1}.$$

Proof. First we make a change of coefficients using the functor $(\cdot) \otimes_k K$, where K is an infinite field extension of k. Hence we may assume that k is infinite. Let A be an *Artinian reduction* of $S = R/I$, that is,

$$A = S/(h_1, \ldots, h_d),$$

where $\underline{h} = h_1, \ldots, h_d$ is a system of parameters of S with h_i a form of degree 1 for all i. Note that one can write $A = \overline{R}/\overline{I}$, where $\overline{R} = R/(\underline{h})$ is a polynomial ring in g variables and $\overline{I} = I\overline{R}$ is a graded Gorenstein ideal of $\overline{R}$ of codimension g and initial degree p. By Proposition 5.3.9 the Hilbert function of A is symmetric. Since the h-vector $h(S) = (h_i)$ satisfy $h_i = \dim_k(A_i)$ there is some $j \geq p$ so that

$$h(S) = (h_0, h_1, \ldots, h_{p-1}, h_p, \ldots, h_j, h_{p-1}, \ldots, h_1, h_0),$$

thus

$$h_{j+1} = h_{p-1} = \binom{p+g-2}{g-1} \quad \text{and} \quad h_j = h_p \tag{1}$$

The idea of the argument is to use the Macaulay bound

$$h_{j+1} \leq h_j^{\langle j \rangle}, \tag{2}$$

see Theorem 5.4.2. From the inequalities:

$$\begin{aligned} h_{j+1} \quad &\overset{(1)}{=} \quad h_{p-1} = \binom{p+g-2}{p-1} = \binom{p+g-3}{p-2}^{\langle p-2 \rangle} \\ &\overset{(2)}{\leq} \quad h_j^{\langle j \rangle} \overset{(1)}{=} h_p^{\langle j \rangle} = \left[\binom{p+g-1}{p} - \mu(I_p)\right]^{\langle j \rangle} \\ &\overset{\text{use } g(x)}{\leq} \quad \left[\binom{p+g-1}{p} - \mu(I_p)\right]^{\langle p-2 \rangle} \end{aligned}$$

we conclude

$$\binom{p+g-3}{p-2}^{\langle p-2 \rangle} \leq \left[\binom{p+g-1}{p} - \mu(I_p)\right]^{\langle p-2 \rangle}.$$

Using that $f(x)$ is non decreasing yields the required inequality. $\square$

Exercises

5.4.9 Let I be a graded Gorenstein ideal of height 4 and initial degree p, prove $\mu(I_p) \leq (p+1)^2$.

5.4.10 Let I be a graded Gorenstein ideal of height 4 generated by forms of degree p and let R/J be an extremal algebra of the same codimension and initial degree. Prove that their Betti numbers satisfy

$$b_i(R/I) \leq b_i(R/J) \quad \text{for all} \quad i.$$

5.5 Koszul homology and Hilbert functions

The first Koszul homology module of a graded Cohen–Macaulay ideal is studied here using the Cohen–Macaulay criterion of Herzog introduced in Section 4.4. Our approach makes use of Hilbert functions.

Lemma 5.5.1 *Let p and g be positive integers and let*

$$\psi(p) = \binom{p+g-1}{g-1}^2 - 2\binom{2p+g}{g}.$$

If $p \geq 2$ and $g \geq 6$ then $\psi(p) > 0$.

Proof. Notice the equality $6\psi(2) = (g+1)(g^3 - 3g^2 - 13g - 12)$, which is certainly positive for $g \geq 6$. We proceed by induction on p. Assume $\psi(p) > 0$. It is easy to check that $\psi(p+1)$ is greater than

$$2\binom{2p+g}{g}\left[\frac{(p+1)(4(g-2)p^2 + (3g^2 - 3g - 8)p + g^2 - 3g - 2)}{(p+1)^2(2p+1)(2p+2)}\right].$$

Since the right-hand side of this inequality is positive for $p \geq 2$ and $g \geq 3$, the induction step is complete. $\square$

Theorem 5.5.2 [350] *Let $R = \oplus_{i=0}^{\infty} R_i$ be a polynomial ring with its usual graduation over an infinite field k and let I be a Cohen–Macaulay graded ideal of R of height g with a p-linear resolution*

$$0 \to R^{b_g}(-(p+g-1)) \to \cdots \to R^{b_k}(-(p+k-1)) \to \cdots \to R^{b_1}(-p) \to I \to 0.$$

If I is generically a complete intersection with $p \geq 2$ and $g \geq 3$, then $H_1(I)$, the first Koszul homology module of I, is not Cohen–Macaulay.

Proof. The module $H_1 = H_1(I)$ has a well-defined rank equal to $b_1 - g$. Thus, by Theorem 4.4.13, it suffices to prove that for some homogeneous system of parameters $\mathbf{y}$ of $S = R/I$, one has

$$\ell(H_1(I)/\mathbf{y}H_1(I)) > \text{rank}(H_1(I)) \cdot \ell(S/\mathbf{y}S).$$

We start by making a specialization to the case of a polynomial ring of dimension g. Since k is infinite according to Theorem 3.1.24, there exists a system of parameters $\mathbf{y} = \{y_1, \ldots, y_d\}$ for S with each y_i a form of degree one of R. We make two observations: (i) Because $\operatorname{Tor}_i(S, R/(\mathbf{y})) = 0$ for $i \geq 1$, it is clear that the minimal resolution of $\overline{S} = S/(\mathbf{y}S)$ as an $\overline{R}$ $(= R/(\mathbf{y}))$-module has the same twists and Betti numbers as the minimal resolution of S over R. (ii) With H_1 Cohen–Macaulay it is easy to see that the first Koszul homology module of $I \otimes \overline{R}$ over $\overline{R}$ is precisely $H_1(I) \otimes \overline{R}$. We may then in the sequel assume that S is zero dimensional.

We will compare the integer r=rank(original $H_1(I)$)$\cdot\ell(S)$, with partial Hilbert sums contributing to $\ell(H_1)$. First notice (see Exercise 5.3.17) that the length of S and its Betti numbers can be calculated using:

$$b_k = \binom{p+k-2}{k-1} \cdot \binom{p+g-1}{g-k} \quad \text{and} \quad \ell(S) = \binom{p+g-1}{g}. \qquad (5.12)$$

From the Koszul complex we obtain the exact sequences

$$0 \to B_1 \to Z_1 \to H_1 \to 0,$$

$$0 \to Z_2 \to R^{\binom{b_1}{2}}(-2p) \to B_1 \to 0,$$

where Z_i and B_i are the modules of cycles and boundaries defining $H_i(I)$. To simplify notation we set $\ell(M)_i = H(M, i)$, the dimension of the ith component of the graded module M. One may write

$$\ell(H_1)_i = \ell(Z_1)_i - \ell(B_1)_i = \ell(Z_1)_i - \binom{b_1}{2}\ell(R(-2p))_i + \ell(Z_2)_i,$$

and therefore

$$\ell(H_1) \geq \sum_{i=0}^{2p} \ell(H_1)_i \geq \sum_{i=0}^{2p} \ell(Z_1)_i - \binom{b_1}{2} \sum_{i=0}^{2p} \ell(R(-2p))_i. \qquad (5.13)$$

From the minimal resolution of S we obtain

$$\ell(S)_i = \ell(R)_i - b_1\ell(R(-p))_i + \ell(Z_1)_i \geq 0.$$

Hence $\ell(Z_1)_i \geq b_1\ell(R(-p))_i - \ell(R)_i$ and $\ell(Z_1)_i = 0$ for $0 \leq i \leq p$, which leads to

$$\sum_{i=0}^{2p} \ell(Z_1)_i \geq b_1 \sum_{i=p+1}^{2p} \ell(R(-p))_i - \sum_{i=p+1}^{2p} \ell(R)_i. \qquad (5.14)$$

Finally by (5.13) and (5.14) we have

$$\ell(H_1) \geq b_1 \sum_{i=p+1}^{2p} \ell(R(-p))_i - \sum_{i=p+1}^{2p} \ell(R)_i - \binom{b_1}{2}. \qquad (5.15)$$

The proof now reduces to showing that the right-hand side of (5.15) is greater than $\operatorname{rank}(H_1(I)) \cdot \ell(S) = (b_1 - g) \cdot \ell(S)$; that is, we must show that the following inequality holds for $p \geq 2$ and $g \geq 3$

$$f(p,g) = b_1 \sum_{i=p+1}^{2p} \ell(R(-p))_i - \sum_{i=p+1}^{2p} \ell(R)_i - \binom{b_1}{2} - (b_1 - g) \cdot \ell(S) > 0.$$

It is not hard to see that $f(p,g)$ simplifies to

$$f(p,g) = \binom{b_1+1}{2} + (1+g)\binom{p+g-1}{p-1} - \binom{2p+g}{g}.$$

Observe that from this equality we obtain the inequality:

$$2f(p,g) > \binom{p+g-1}{p}^2 - 2\binom{2p+g}{g}.$$

It is easy to check that $f(p,g) > 0$ for $g \in \{3,4,5\}$ and $p \geq 2$. The required inequality is now a direct consequence of Lemma 5.5.1. $\Box$

The result above complements one of Ulrich [404] in which the twisted conormal module $\omega_S \otimes_R I = \omega_S \otimes_R I/I^2$ was shown not to be Cohen–Macaulay; either result implies that I is not in the linkage class of a complete intersection.

Corollary 5.5.3 *Let I be a graded Cohen–Macaulay ideal of height $g \geq 3$ in a polynomial ring R. If I is generically a complete intersection and has a p-linear resolution with $p \geq 2$, then I is not in the linkage class of a complete intersection.*

Proof. If I is in the linkage class of a complete intersection, then I is a strongly Cohen–Macaulay ideal, a contradiction because $H_1(I)$ is not Cohen–Macaulay. See [253] and [412, Corollary 4.2.4]. $\Box$

Exercises

5.5.4 Let I be a graded ideal of a polynomial ring R over a field K with a pure resolution:

$$0 \to R^3(-15) \to R^{27}(-10) \to R^{25}(-9) \to I \to 0.$$

Prove that $H_1(I)$ is not Cohen–Macaulay.

5.6 Hilbert functions of some graded ideals

This section is about the behavior of the Hilbert function in some special cases of interest in algebraic coding theory [349, 348] (see Section 8.6).

Throughout this section $R = K[x_1, \ldots, x_n] = \oplus_{i=0}^{\infty} R_i$ is a polynomial ring over a field K with the standard grading induced by setting $\deg(x_i) = 1$ for $i = 1, \ldots, n$ and $\mathfrak{m} = (x_1, \ldots, x_n)$ is the irrelevant maximal ideal of R.

Proposition 5.6.1 *Let I be a graded ideal of R. If K is infinite and $\mathfrak{m}$ is not in $\mathrm{Ass}(R/I)$, then there is $h_1 \in R_1$ such that $h_1 \notin \mathcal{Z}(R/I)$.*

Proof. Let $\mathfrak{p}_1, \ldots, \mathfrak{p}_m$ be the associated primes of R/I. As R/I is graded, $\mathfrak{p}_1, \ldots, \mathfrak{p}_m$ are graded ideals by Lemma 2.2.6. We proceed by contradiction. Assume that R_1, the degree 1 part of R, is contained in $\mathcal{Z}(R/I)$. Thanks to Lemma 2.1.19, one has that $\mathcal{Z}(R/I) = \cup_{i=1}^{m} \mathfrak{p}_i$. Hence

$$R_1 \subset (\mathfrak{p}_1)_1 \cup (\mathfrak{p}_2)_1 \cup \cdots \cup (\mathfrak{p}_m)_1 \subset R_1,$$

where $(\mathfrak{p}_i)_1$ is the homogeneous part of degree 1 of the graded ideal $\mathfrak{p}_i$. Since K is infinite, from Lemma 3.1.22, we get $R_1 = (\mathfrak{p}_i)_1$ for some i. Hence, $\mathfrak{p}_i = \mathfrak{m}$, a contradiction. $\Box$

The hypothesis that $\mathfrak{m} \notin \mathrm{Ass}\,(R/I)$ is equivalent to require that R/I has positive depth. This follows from Lemma 2.1.19 (see Exercise 2.2.11).

Theorem 5.6.2 *Let I be a graded ideal of R. If $\mathrm{depth}(R/I) > 0$, and H_I is the Hilbert function of R/I, then $H_I(i) \leq H_I(i+1)$ for $i \geq 0$.*

Proof. Case (I): If K is infinite, by Proposition 5.6.1, there exists $h \in R_1$ a non-zero divisor of R/I. The homomorphism of K-vector spaces

$$(R/I)_i \longrightarrow (R/I)_{i+1}, \quad \overline{z} \mapsto \overline{hz}$$

is injective, therefore $H_I(i) = \dim_K(R/I)_i \leq \dim_K(R/I)_{i+1} = H_I(i+1)$.

Case (II): If K is finite, consider the algebraic closure $\overline{K}$ of K. We set

$$\overline{R} = R \otimes_K \overline{K}, \qquad \overline{I} = I\overline{R}.$$

Hence, from [392, Lemma 1.1], one has that $H_I(i) = H_{\overline{I}}(i)$. This means that the Hilbert function does not change when the base field is extended from K to $\overline{K}$. Applying Case (I) to $H_{\overline{I}}(i)$ we obtain the result. $\Box$

Lemma 5.6.3 *Let I be a graded ideal of R. The following hold.*

(a) *If $R_i = I_i$ for some i, then $R_\ell = I_\ell$ for all $\ell \geq i$.*

(b) *If $\dim R/I \geq 2$, then $\dim_K(R/I)_i > 0$ for $i \geq 0$.*

Proof. (a): It suffices to prove the case $\ell = i+1$. As $I_{i+1} \subset R_{i+1}$, we need only show $R_{i+1} \subset I_{i+1}$. Take a non-zero monomial x^a in R_{i+1}. Then, $x^a = x_1^{a_1} \cdots x_n^{a_n}$ with $a_j > 0$ for some j. Thus, $x^a \in R_1 R_i$. As $R_1 I_i \subset I_{i+1}$, we get $x^a \in I_{i+1}$.

(b): If $\dim_K(R/I)_i = 0$ for some i, then $R_i = I_i$. Thus, by (a), $H_I(j)$ vanishes for $j \geq i$, a contradiction because the Hilbert polynomial of R/I has degree $\dim(R/I) - 1 \geq 1$; see Theorem 2.2.4. □

Theorem 5.6.4 [178] *Let I be a graded ideal with* $\operatorname{depth}(R/I) > 0$.

(i) *If* $\dim(R/I) \geq 2$, *then* $H_I(i) < H_I(i+1)$ *for* $i \geq 0$.

(ii) *If* $\dim(R/I) = 1$, *then there is an integer r and a constant c such that*

$$1 = H_I(0) < H_I(1) < \cdots < H_I(r-1) < H_I(i) = c \quad \text{for } i \geq r.$$

Proof. Consider the algebraic closure $\overline{K}$ of K. Notice that $|\overline{K}| = \infty$. As in the proof of Theorem 5.6.2, we make a change of coefficients using the functor $(\cdot) \otimes_K \overline{K}$. Hence we may assume that K is infinite. By Proposition 5.6.1, there is $h \in R_1$ a non-zero divisor of R/I. From the exact sequence

$$0 \longrightarrow (R/I)[-1] \xrightarrow{h} R/I \longrightarrow R/(h,I) \longrightarrow 0,$$

we get $H_I(i+1) - H_I(i) = H_S(i+1)$, where $S = R/(h,I)$.

(i): If $H_S(i+1) = 0$, then, by Lemma 5.6.3, $\dim_K(S) < \infty$. Hence S is Artinian (see Exercise 3.1.44). Thus $\dim(S) = 0$, a contradiction.

(ii): Let $r \geq 0$ be the first integer such that $H_I(r) = H_I(r+1)$, thus $S_{r+1} = (0)$ and $R_{r+1} = (h,I)_{r+1}$. Then, by Lemma 5.6.3, $S_k = (0)$ for $k \geq r+1$. Hence, the Hilbert function of R/I is constant for $k \geq r$ and strictly increasing on $[0, r-1]$. □

Exercises

5.6.5 Let $\mathbb{P}^{n-1}$ be a projective space over a field K and let $[P_1], \ldots, [P_m]$ be a set of distinct points in $\mathbb{P}^{n-1}$, then there exists $F \in K[x_1, \ldots, x_n]$ a form of degree $m-1$ such that $F(P_1) \neq 0$ and $F(P_i) = 0$ for $2 \leq i \leq m$.

5.6.6 Let X be a finite set in a projective space $\mathbb{P}^{n-1}$ over a field K. If $I = I(X) \subset R$ is the vanishing ideal of X and H_I is the Hilbert function of $R/I(X)$, then $H_I(i) = |X|$ for all $i \geq |X| - 1$.

Chapter 6

Stanley–Reisner Rings and Edge Ideals of Clutters

This chapter is intended as an introduction to the study of *combinatorial commutative algebra*, some of the topics were chosen in order to motivate the subject. Part of our exposition was inspired by [41, 393]. An understated goal here is to highlight some of the works of T. Hibi, J. Herzog, M. Hochster, G. Reisner, and R. Stanley.

This chapter deals with monomial ideals, Stanley–Reisner rings and edge ideals of clutters. We study algebraic and combinatorial properties of edge ideals of clutters and simplicial complexes. In particular we examine shellable, unmixed, and sequentially Cohen–Macaulay simplicial complexes and their corresponding edge ideals and invariants (projective dimension, regularity, depth, Hilbert series). As applications, the proofs of the upper bound conjectures for convex polytopes and simplicial spheres are discussed in order to give a flavor of some of the methods and ideas of the area.

6.1 Primary decomposition

General monomial ideals will be examined first, subsequently we specialize to the case of square-free monomial ideals and edge ideals of clutters, and present some of their relevant properties.

Let $R = K[\mathbf{x}] = K[x_1, \ldots, x_n]$ be a polynomial ring over a field K. To make notation simpler, we will use the following multi-index notation:

$$x^a := x_1^{a_1} \cdots x_n^{a_n} \quad \text{for } a = (a_1, \ldots, a_n) \in \mathbb{N}^n.$$

Definition 6.1.1 An ideal I of R is called a *monomial ideal* if there is $\mathcal{A} \subset \mathbb{N}_+^n$ such that I is generated by $\{x^a | a \in \mathcal{A}\}$. If I is a monomial ideal the quotient ring R/I is called a *monomial ring*.

Note that, by Dickson's lemma (see Lemma 3.3.3), a monomial ideal I is always minimally generated by a unique finite set of monomials. This unique set of generators of I is denoted by $G(I)$.

Definition 6.1.2 A *face ideal* is an ideal $\mathfrak{p}$ of R generated by a subset of the set of variables, that is, $\mathfrak{p} = (x_{i_1}, \ldots, x_{i_k})$ for some variables x_{i_j}.

Definition 6.1.3 A monomial f in R is called *square-free* if $f = x_{i_1} \ldots x_{i_r}$ for some $1 \leq i_1 < \cdots < i_r \leq n$. A *square-free monomial ideal* is an ideal generated by square-free monomials.

Any square-free monomial ideal is a finite intersection of face ideals:

Theorem 6.1.4 *Let $I \subset R$ be a monomial ideal. The following hold.*

(i) *Every associated prime of I is a face ideal.*

(ii) *If I is square-free, then $I = \cap_{i=1}^s \mathfrak{p}_i$, where $\mathfrak{p}_1, \ldots, \mathfrak{p}_s$ are the associated primes of I. In particular $\mathfrak{p}_i$ is a minimal prime of I for all i.*

Proof. (i): By induction on the number of variables that occur in $G(I)$. Set $\mathfrak{m} = (x_1, \ldots, x_n)$. Let $\mathfrak{p}$ be an associated prime of I. If $\text{rad}(I) = \mathfrak{m}$, then $\mathfrak{p} = \mathfrak{m}$. Hence we may assume $\text{rad}(I) \neq \mathfrak{m}$. Pick a variable x_1 not in $\text{rad}(I)$ and consider the ascending chain of ideals

$$I_0 = I \text{ and } I_{i+1} = (I_i \colon x_1) \qquad (i \geq 0).$$

Since R is Noetherian, one has $I_k = (I_k \colon x_1)$ for some k. There are two cases to consider. If $\mathfrak{p}$ is an associated prime of (I_i, x_1) for some i, then by induction $\mathfrak{p}$ is a face ideal because one can write $(I_i, x_1) = (I_i', x_1)$, where I_i' is an ideal minimally generated by a finite set of monomials in the variables $x_2, \ldots, x_n$ (cf. Exercise 6.1.26). Assume we are in the opposite case. By Lemma 5.1.3 for each i there is an exact sequence

$$0 \longrightarrow R/(I_i \colon x_1) \stackrel{x_1}{\longrightarrow} R/I_i \longrightarrow R/(I_i, x_1) \longrightarrow 0,$$

hence making a recursive application of Lemma 2.1.17 one obtains that $\mathfrak{p}$ is an associated prime of I_i for all i. Since x_1 is regular on R/I_k one concludes that I_k is an ideal minimally generated by monomials in the variables $x_2, \ldots, x_n$, thus by induction $\mathfrak{p}$ is a face ideal.

(ii): We only have to check $\cap_{i=1}^s \mathfrak{p}_i \subset I$ because I is contained in any of its associated primes. Take a monomial f in $\cap_{i=1}^s \mathfrak{p}_i$ and write $f = x_{i_1}^{a_1} \cdots x_{i_r}^{a_r}$, where $i_1 < \cdots < i_r$ and $a_i > 0$ for all i. By Corollary 2.1.29, $f^k \in I$ for some $k \geq 1$. Then, using that I is generated by square-free monomials, we obtain $x_{i_1} \cdots x_{i_r} \in I$. Hence $f \in I$. To finish the proof observe that, by part (i), $\cap_{i=1}^{s} \mathfrak{p}_i$ is a monomial ideal because the intersection of monomial ideals is again a monomial ideal (see Exercise 6.1.22). $\square$

Definition 6.1.5 Let $x^a = x_1^{a_1} \cdots x_n^{a_n}$ be a monomial in R. The *support* of x^a is given by $\operatorname{supp}(x^a) := \{x_i \mid a_i > 0\}$.

Proposition 6.1.6 [169] *Let $I \subset R$ be a monomial ideal. If $S = R/I$, then there is a polynomial ring R' and a square-free monomial ideal I' of R' such that $S = S'/(\underline{h})$, where $S' = R'/I'$ and $\underline{h}$ is a regular sequence on S' of forms of degree one.*

Proof. Let $G(I) = \{f_1, \ldots, f_r\}$ be the set of monomials that minimally generate I. Assume that one of the variables, say x_1, occurs in at least one of the monomials in $G(I)$ with multiplicity greater than 1. We may assume that the elements of $G(I)$ can be written as

$$f_1 = x_1^{a_1} g_1, \ldots, f_s = x_1^{a_s} g_s,$$

where $a_i \geq 1$ for all i, $a_1 \geq 2$, $x_1 \notin \operatorname{supp}(g_i)$ for all $i \leq s$ and $x_1 \notin \operatorname{supp}(f_i)$ for $i > s$. We set

$$I' = (x_0 x_1^{a_1 - 1} g_1, \ldots, x_0 x_1^{a_s - 1} g_s, f_{s+1}, \ldots, f_r) \subset R' = R[x_0],$$

where x_0 is a new variable. We claim that $x_0 - x_1$ is a non-zero divisor of $S' = R'/I'$. On the contrary assume that $x_0 - x_1$ belongs to an associated prime $\mathfrak{p}$ of I' and write $\mathfrak{p} = (I' \colon h)$, for some monomial h. Since $\mathfrak{p}$ is a face ideal, one has $x_i h \in I'$ for $i = 0, 1$. Hence using $h \notin I'$ gives $x_1 h = x_0 x_1^{a_i - 1} g_i h_1$ for some i and consequently $h = x_0 x_1^{a_i - 2} g_i h_1$. Hence

$$x_0 h = x_0^2 x_1^{a_i - 2} g_i h_1 = \begin{cases} x_0 x_1^{a_j - 1} g_j M & \text{or,} \\ f_j M & \text{for some } j > s. \end{cases}$$

In both cases one obtains $x_0 \in \operatorname{supp}(M)$ and $h \in I'$, a contradiction. Since

$$S'/(x_0 - x_1) = S$$

one can repeat the construction to obtain the asserted monomial ideal I'. This proof is due to R. Fröberg. $\square$

The ideal I' constructed above is called the *polarization* of I. Thus any monomial ring is a deformation by linear forms of a monomial ring with square-free relations. Note that I is Cohen–Macaulay (resp. Gorenstein) iff I' is Cohen–Macaulay (resp. Gorenstein).

Proposition 6.1.7 *A monomial ideal $\mathfrak{q} \subset R$ is primary if and only if, after permutation of the variables, $\mathfrak{q}$ has the form:*

$$\mathfrak{q} = (x_1^{a_1}, \ldots, x_r^{a_r}, x^{b_1}, \ldots, x^{b_s}),$$

where $a_i \geq 1$ and $\cup_{i=1}^{s} \mathrm{supp}(x^{b_i}) \subset \{x_1, \ldots, x_r\}$.

Proof. If $\mathrm{Ass}(R/\mathfrak{q}) = \{\mathfrak{p}\}$, then by permuting the variables $x_1, \ldots, x_n$ and using Theorem 6.1.4 one may assume that $\mathfrak{p}$ is equal to $(x_1, \ldots, x_r)$. Since $\mathrm{rad}\,(\mathfrak{q}) = \mathfrak{p}$, the ideal $\mathfrak{q}$ is minimally generated by a set of the form:

$$\{x_1^{a_1}, \ldots, x_r^{a_r}, x^{b_1}, \ldots, x^{b_s}\}.$$

Let $x_j \in \mathrm{supp}(x^{b_i})$, then $x^{b_i} = x_j x^c$, where x^c is a monomial not in $\mathfrak{q}$. Since $\mathfrak{q}$ is primary, a power of x_j is in $\mathfrak{q}$. Thus $x_j \in (x_1, \ldots, x_r)$ and consequently $1 \leq j \leq r$, as required.

For the converse note that any associated prime $\mathfrak{p}$ of $R/\mathfrak{q}$ can be written as $\mathfrak{p} = (\mathfrak{q} \colon f)$, for some monomial f. It follows readily that $(x_1, \ldots, x_r)$ is the only associated prime of $\mathfrak{q}$. □

Corollary 6.1.8 *If $\mathfrak{p}$ is a face ideal, then $\mathfrak{p}^n$ is a primary ideal for all n.*

Proposition 6.1.9 *If $I \subset R$ is a monomial ideal, then I has an irredundant primary decomposition $I = \mathfrak{q}_1 \cap \cdots \cap \mathfrak{q}_r$, where $\mathfrak{q}_i$ is a primary monomial ideal for all i and* $\mathrm{rad}\,(\mathfrak{q}_i) \neq \mathrm{rad}\,(\mathfrak{q}_j)$ *if $i \neq j$.*

Proof. Let $G(I) = \{f_1, \ldots, f_q\}$ be the set of monomials that minimally generate I. We proceed by induction on the number of variables that occur in the union of the supports of $f_1, \ldots, f_q$.

One may assume that one of the variables in $\cup_{i=1}^{q} \mathrm{supp}(f_i)$, say x_n, satisfy $x_n^i \notin I$ for all i, otherwise I is a primary ideal and there is nothing to prove. Next we permute the f_i in order to find integers $0 \leq a_1 \leq \cdots \leq a_q$, with $a_q \geq 1$, such that f_i is divisible by $x_n^{a_i}$ but by not higher power of x_n. If we apply this procedure to $(I, x_n^{a_q})$, instead of I, note that one must choose a variable different from x_n.

Since $(I \colon x_n^{a_q})$ is generated by monomials in less than n variables and because of the equality

$$I = (I, x_n^{a_q}) \cap (I \colon x_n^{a_q})$$

one may apply the argument above recursively to the two monomial ideals occurring in the intersection— and use induction—to obtain a decomposition of I into primary monomial ideals $\mathfrak{q}_1', \ldots, \mathfrak{q}_s'$ of R. Finally we remove redundant primary ideals from $\mathfrak{q}_1', \ldots, \mathfrak{q}_s'$ and group those primary ideals with the same radical. □

Even for monomial ideals a minimal irredundant primary decomposition is not unique. What is unique is the number of terms in such a decomposition and the primary components that correspond to minimal primes [9].

Example 6.1.10 If $I = (x^2, xy) \subset K[x, y]$, then

$$I = (x) \cap (x^2, xy, y^2) = (x) \cap (x^2, y),$$

are two minimal irredundant primary decompositions of I.

The computation of a primary decomposition of a monomial ideal can be carried out by successive elimination of powers of variables, as described in the proof of Proposition 6.1.9. Now we illustrate this procedure with a specific ideal.

Example 6.1.11 If $R = K[x, y, z]$ and $I = (yz^2, x^2z, x^3y^2)$.

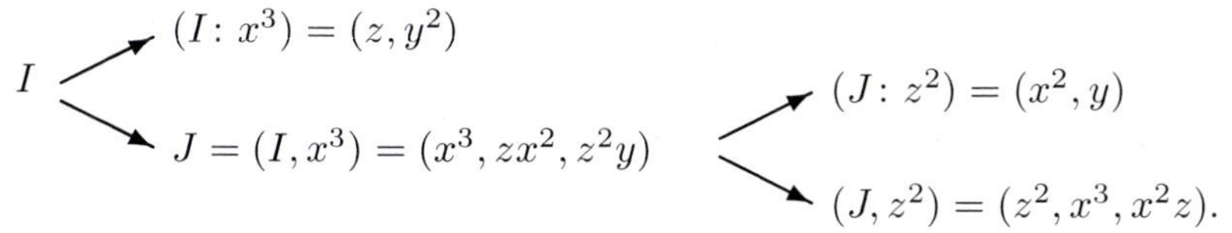

$$I \nearrow (I\colon x^3) = (z, y^2)$$

$$I \searrow J = (I, x^3) = (x^3, zx^2, z^2y) \nearrow (J\colon z^2) = (x^2, y)$$

$$J \searrow (J, z^2) = (z^2, x^3, x^2z).$$

Thus $I = (z, y^2) \cap (x^2, y) \cap (z^2, x^3, x^2z)$.

Corollary 6.1.12 *If $I \subset R$ is a monomial ideal, then there is a primary decomposition $I = \mathfrak{q}_1 \cap \cdots \cap \mathfrak{q}_m$, such that $\mathfrak{q}_i$ is generated by powers of variables for all i.*

Proof. Let $\mathfrak{q}$ be a primary ideal. Then, by Proposition 6.1.7, $\mathfrak{q}$ is minimally generated by a set of monomials $x_1^{a_1}, \ldots, x_r^{a_r}, f_1, \ldots, f_s$ such that

$$\bigcup_{i=1}^{s} \text{supp}(f_i) \subset \{x_1, \ldots, x_r\},$$

where $a_i > 0$ for all i. Note that if $f_1 = x_1^{b_1} \cdots x_r^{b_r}$ and $b_1 > 0$, then $a_1 > b_1$ and one has a decomposition:

$$\mathfrak{q} = (x_1^{b_1}, x_2^{a_2}, \ldots, x_r^{a_r}, f_2, \ldots, f_s) \cap (x_1^{a_1}, x_2^{a_2}, \ldots, x_r^{a_r}, x_2^{b_2} \cdots x_r^{b_r}, f_2, \ldots, f_s),$$

where in the first ideal of the intersection we have lower the degree of $x_1^{a_1}$ and have eliminated f_1, while in the second ideal we have eliminated the variable x_1 from f_1. Applying the same argument repeatedly it follows that one can write $\mathfrak{q}$ as an intersection of primary monomial ideals such that for each of those ideals the only minimal generators that contain x_1 are pure powers of x_1. Therefore, by induction, $\mathfrak{q}$ is the intersection of ideals generated by powers of variables. Hence the result follows from Proposition 6.1.9. $\Box$

Proposition 6.1.13 *Let I be an ideal of $R = K[x_1, \ldots, x_n]$ generated by monomials in the variables $x_1, \ldots, x_r$ with $r < n$. If*

$$I = I_1 \cap \cdots \cap I_s$$

is an irredundant decomposition of I into monomial ideals, then none of the ideals I_i can contain a monomial in $K[x_{r+1}, \ldots, x_n]$.

Proof. Set $X = \{x_1, \ldots, x_n\}$ and $X' = X \setminus \{x_1, \ldots, x_r\}$. Assume some of the I_i's contain monomials in $K[X']$. One may split the I_i's into two sets so that $I_1, \ldots, I_m$ do not contain monomials in $K[X']$ (note m could be zero), while $I_{m+1}, \ldots, I_s$ contain monomials in the set of variables X'. For $i \geq m+1$ pick a monomial g_i in I_i whose support is contained in X'. Since the decomposition of I is irredundant there is a monomial $f \in \cap_{i=1}^m I_i$ and $f \notin \cap_{i=m+1}^s I_i$, where we set $f = 1$ if $m = 0$. To derive a contradiction consider $f_1 = f g_{m+1} \cdots g_s$. As $f_1 \in I$, we get $f \in I$ which is absurd. $\Box$

Corollary 6.1.14 *Let $I \subset R = K[x_1, \ldots, x_n]$ be an ideal generated by monomials in the variables $x_1, \ldots, x_r$ with $r < n$. If*

$$I = \mathfrak{q}_1 \cap \cdots \cap \mathfrak{q}_s$$

is an irredundant primary decomposition into monomial ideals, then $\mathfrak{q}_i$ is generated by monomials in $K[x_1, \ldots, x_r]$.

Proof. It follows from Propositions 6.1.7 and 6.1.13. $\Box$

Definition 6.1.15 An ideal I of a ring R is called *irreducible* if I cannot be written as an intersection of two ideals of R that properly contain I.

Proposition 6.1.16 *If $I \subset R = K[x_1, \ldots, x_n]$ is a monomial ideal, then I is irreducible if and only if up to permutation of the variables*

$$I = (x_1^{a_1}, \ldots, x_r^{a_r}),$$

where $a_i > 0$ for all i.

Proof. $\Rightarrow$) Since I must be primary (see Proposition 2.1.24) from the proof of Corollary 6.1.12 one derives that I is generated by powers of variables.

$\Leftarrow$) If I is reducible, then $I = I_1 \cap I_2$ for some ideals I_1 and I_2 that properly contain I. Pick $f \in I_1 \setminus I$ (resp. $g \in I_2 \setminus I$) with the smallest possible number of terms. We can write

$$f = \lambda_1 x^{\gamma_1} + \cdots + \lambda_s x^{\gamma_s}; \qquad 0 \neq \lambda_i \in K \text{ for all } i.$$

Let $1 \leq k \leq r$ be an integer and let $x_k^{b_i}$ be the maximum power of x_k that divides x^{γ_i}, i.e., we can write $x^{\gamma_i} = x_k^{b_i} x^{\delta_i}$, where x_k does not divide x^{δ_i}. After permuting terms we may assume that $b_1 \geq \cdots \geq b_s$. Note that $x_k^{a_k}$ does not divide x^{γ_i} for all i, k, otherwise we can find a polynomial in $I_1 \setminus I$, namely $f - \lambda_i x^{\gamma_i}$, with less than s terms. Thus $b_i < a_k$ for all i. We claim that $b_1 = \cdots = b_s$. We proceed by contradiction assuming that $b_p > b_{p+1}$ for some p. From the equality

$$x_k^{a_k - b_p} f = x_k^{a_k - b_p}(\lambda_1 x^{\gamma_1} + \cdots + \lambda_p x^{\gamma_p}) + x_k^{a_k - b_p}(\lambda_{p+1} x^{\gamma_{p+1}} + \cdots + \lambda_s x^{\gamma_s})$$

we obtain that the polynomial $x_k^{a_k-b_p}(\lambda_{p+1}x^{\gamma_{p+1}}+\cdots+\lambda_s x^{\gamma_s})$ is in $I_1\setminus I$ and has fewer terms than f, a contradiction. This completes the proof of the claim. Therefore we can write $f=x_1^{c_1}\cdots x_r^{c_r}f_1$ and $g=x_1^{d_1}\cdots x_r^{d_r}g_1$, where f_1 and f_2 are in $R'=K[x_{r+1},\ldots,x_n]$. Setting $e_i=\max\{c_i,d_i\}$ we get that $h=x_1^{e_1}\cdots x_r^{e_r}f_1g_1$ is in $I_1\cap I_2$, i.e., $h\in I$, a contradiction because $e_i<a_i$ for all i and $f_1g_1\in R'$. Thus I is irreducible. □

Theorem 6.1.17 *If $I\subset R$ is a monomial ideal, then there is a unique irredundant decomposition $I=\mathfrak{q}_1\cap\cdots\cap\mathfrak{q}_r$ such that $\mathfrak{q}_i$ is an irreducible monomial ideal.*

Proof. The existence follows from Corollary 6.1.12. For the uniqueness assume one has two irredundant decompositions:

$$\mathfrak{q}_1\cap\cdots\cap\mathfrak{q}_r=\mathfrak{q}_1'\cap\cdots\cap\mathfrak{q}_s',$$

where $\mathfrak{q}_i$ and $\mathfrak{q}_j'$ are irreducible for all i,j. Using the arguments given in the proof of Proposition 6.1.16 one concludes that for each i, there is σ_i such that $\mathfrak{q}_{\sigma_i}\subset\mathfrak{q}_i'$ and vice versa for each j there is π_j such that $\mathfrak{q}_{\pi_j}'\subset\mathfrak{q}_j$. Therefore $r=s$ and $\mathfrak{q}_i=\mathfrak{q}_{\rho_i}'$ for some permutation ρ. □

For further information on the primary decomposition of more general monomial ideals, e.g., monomial ideals obtained from a regular sequence or with coefficients in a ring other than a field, consult [122, 213, 214].

Example 6.1.18 Let $I=(xy^2,x^2y)\subset K[x,y]$ and $V(I)$ the affine variety defined by I. From the primary decomposition

$$I=(x)\cap(y)\cap(x^2,y^2)$$

one has $V(I)=\{(0,0)\}\cup V(x)\cup V(y)=V(x)\cup V(y)\subset\mathbb{A}_K^2$, where (x^2,y^2) corresponds to $\{(0,0)\}$ which is embedded in $V(x)\cup V(y)$. For this reason (x^2,y^2) is said to be an embedded primary component. If K is infinite, then the coordinate axes $V(x)$ and $V(y)$ are the irreducible components of $V(I)$.

Monomial ideals form a lattice Let L be a *lattice*, that is, L is a poset in which any two elements x, y have a greatest lower bound or a *meet* $x\wedge y$ and a lowest upper bound or a *join* $x\vee y$. The lattice L is *distributive* if

(a) $x\wedge(y\vee z)=(x\wedge y)\vee(x\wedge z)$, and

(b) $x\vee(y\wedge z)=(x\vee y)\wedge(x\vee z)$, $\forall\, x,y,z$ in L.

Proposition 6.1.19 *If L is the family of monomial ideals of R, order by inclusion, then L is a distributive lattice under the operations $I\wedge J=I\cap J$ and $I\vee J=I+J$.*

Proof. It follows from Exercises 6.1.22 and 6.1.23. □

Exercises

6.1.20 If I is a monomial ideal of R, then any associated prime $\mathfrak{p}$ of R/I can be written as $\mathfrak{p} = (I\colon f)$, for some monomial f.

6.1.21 If L is a lattice, then the following conditions are equivalent:

(a) $x \wedge (y \vee z) = (x \wedge y) \vee (x \wedge z)$, $\forall\, x, y, z$ in L.

(b) $x \vee (y \wedge z) = (x \vee y) \wedge (x \vee z)$, $\forall\, x, y, z$ in L.

6.1.22 Let I and J be two ideals generated by finite sets of monomials F and G, respectively, prove that the intersection $I \cap J$ is generated by the set

$$\{\mathrm{lcm}(f, g) |\, f \in F \text{ and } g \in G\}.$$

6.1.23 If I, J, L are monomial ideals, prove the equalities

$$\begin{array}{rcl} I \cap (J + L) & = & (I \cap J) + (I \cap L), \\ I + (J \cap L) & = & (I + J) \cap (I + L). \end{array}$$

6.1.24 If I and J are two monomial ideals, then $(I\colon J)$ is a monomial ideal.

6.1.25 If $\mathfrak{q}_1, \ldots, \mathfrak{q}_r$ are primary monomial ideals of R with non-comparable radicals and I is an ideal such that $I = \mathfrak{q}_1 \cap \cdots \cap \mathfrak{q}_r$, then

$$I^{(n)} = \mathfrak{q}_1^n \cap \cdots \cap \mathfrak{q}_r^n.$$

Hint Proceed as in the proof of Proposition 4.3.24 (cf. [92, Theorem 3.7]).

6.1.26 Let $R' = k[x_2, \ldots, x_n]$ and $R = R'[x_1]$ be polynomial rings over a field k. If I' is an ideal of R' and $\mathfrak{p} \in \mathrm{Ass}_R(R/(I', x_1))$, then

(a) $\mathfrak{p} = x_1 R + \mathfrak{p}' R$, where $\mathfrak{p}'$ is a prime ideal of R', and

(b) $\mathfrak{p}'$ is an associated prime of R'/I'.

6.1.27 If $I = (x_1^3, x_2^2 x_3^2, x_1 x_2^2 x_3) \subset K[x_1, x_2, x_3]$, show that $I = \mathfrak{q}_1 \cap \mathfrak{q}_2$ is a primary decomposition of I, where $\mathfrak{q}_1 = (x_1^3, x_2^2)$ and $\mathfrak{q}_2 = (x_1^3, x_1 x_3, x_3^2)$. Prove that $I^2 = I^{(2)}$ and $I^3 \neq I^{(3)}$.

Hint If $z = x_1^3 x_2^4 x_3^3$ and $a = x_1 + x_2^2 + x_3$, then $az \in I^3$ and $z \notin I^3$.

6.1.28 If $I = (x_1 x_2^2, x_1^2 x_2, x_1 x_3^2, x_1^2 x_3, x_2 x_3^2, x_2^2 x_3, x_1 x_2 x_3)$, find a primary decomposition of I. Notice that (x_1, x_2, x_3) is an associated prime of I.

6.1.29 Show that $I = (x_1^2, x_1 x_2)$ is a non-primary ideal with a prime radical.

6.1.30 If $n \geq 5$ prove that the number of monomials of degree n in n variables whose support is at least three is given by

$$\mu(n) = \binom{2n-1}{n-1} - n - (n-1)\binom{n}{2}.$$

6.2 Simplicial complexes and homology

A *finite simplicial complex* consists of a finite set V of vertices and a collection Δ of subsets of V called *faces* or *simplices* such that

(i) If $v \in V$, then $\{v\} \in \Delta$.

(ii) If $F \in \Delta$ and $G \subset F$, then $G \in \Delta$.

Let Δ be a simplicial complex and F a face of Δ. Define the *dimensions* of F and Δ by

$$\dim(F) = |F| - 1 \quad \text{and} \quad \dim(\Delta) = \sup\{\dim(F) |\, F \in \Delta\}$$

respectively. A face of dimension q is sometimes referred to as a q-face or as a q-simplex.

Example 6.2.1 A triangulation of a disk with a whisker attached.

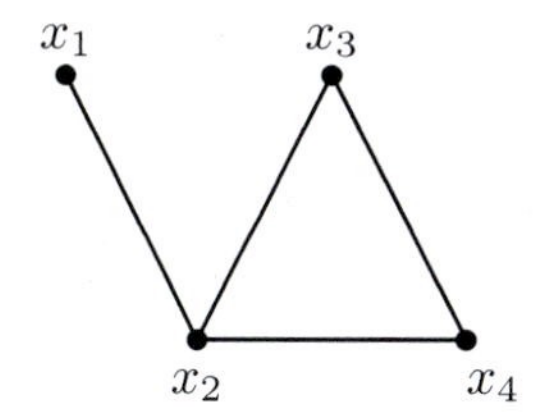

0-simplices: x_1, x_2, x_3, x_4
1-simplices: $x_1x_2, x_2x_3, x_3x_4, x_2x_4$
2-simplices: $x_2x_3x_4$

Let F be a q-simplex in Δ with vertices $v_0, v_1, \ldots, v_q$. We say that two total orderings of the vertices

$$v_{i_0} < \cdots < v_{i_q} \quad \text{and} \quad v_{j_0} < \cdots < v_{j_q}$$

are *equivalent* if $(i_0, \ldots, i_q)$ is an even permutation of $(j_0, \ldots, j_q)$. This is an equivalence relation because the set of even permutations forms a group, and for $q > 1$, it partitions the total orderings of $v_0, \ldots, v_q$ into two equivalence classes. An *oriented q-simplex* of Δ is a q-simplex F with a choice of one of these equivalence classes. If $v_0, \ldots, v_q$ are the vertices of F, the oriented simplex determined by the ordering $v_0 < \cdots < v_q$ will be denoted by $[v_0, \ldots, v_q]$.

Suppose A is a commutative ring with unit. Let $C_q(\Delta)$ be the free A-module with basis consisting of the oriented q-simplices in Δ, modulo the relations

$$[v_0, v_1, v_2, \ldots, v_q] + [v_1, v_0, v_2, \ldots, v_q].$$

In particular $C_q(\Delta)$ is defined for any field K and $\dim_K C_q(\Delta)$ equals the number of q-simplices of Δ. For $q \geq 1$ we define the homomorphism

$$\partial_q \colon C_q(\Delta) \longrightarrow C_{q-1}(\Delta)$$

induced by

$$\partial_q([v_0, v_1, \ldots, v_q]) = \sum_{i=0}^{q} (-1)^i [v_0, v_1, \ldots, v_{i-1}, \widehat{v}_i, v_{i+1}, \ldots, v_q],$$

where $\widehat{v}_i$ means that the symbol v_i is to be deleted. Since $\partial_q \partial_{q+1} = 0$ we obtain the chain complex $C(\Delta) = \{C_q(\Delta), \partial_q\}$, which is called *the oriented chain complex* of Δ. The *augmented oriented chain complex* of Δ is the complex

$$0 \longrightarrow C_d(\Delta) \xrightarrow{\partial_d} C_{d-1}(\Delta) \longrightarrow \cdots \longrightarrow C_0(\Delta) \xrightarrow{\epsilon} A \longrightarrow 0,$$

where $d = \dim(\Delta)$ and $\epsilon(v) = 1$ for every vertex v of Δ. This chain complex is denoted by $C_\star(\Delta)$ and we set $\partial_0 = \epsilon$ and $C_{-1}(\Delta) = A$.

Setting $Z_q(\Delta, A) = \ker(\partial_q)$, $B_q = \operatorname{im}(\partial_{q+1})$ and

$$\widetilde{H}_q(\Delta; A) = Z_q(\Delta, A)/B_q(\Delta, A), \text{ for } q \geq 0,$$

the elements of $Z_q(\Delta, A)$ and $B_q(\Delta, A)$ are called *cycles* and *boundaries*, respectively, and $\widetilde{H}_q(\Delta; A)$ is the qth *reduced simplicial homology group* of Δ with coefficients in A.

Remark 6.2.2 Note that if $\Delta \neq \emptyset$, then $\widetilde{H}_i(\Delta; A) = 0$ for $i < 0$.

Let $C_\star(\Delta)$ be the augmented chain complex of Δ over the ring A. The q-reduced singular cohomology group with coefficients in A is defined as

$$\widetilde{H}^q(\Delta; A) = \widetilde{H}^q(\operatorname{Hom}_A(C_\star(\Delta), A)).$$

If K is a field, there are canonical isomorphisms

$$\widetilde{H}^q(\Delta; K) \cong \operatorname{Hom}_K(\widetilde{H}_q(\Delta, K), K) \text{ and}$$
$$\widetilde{H}_q(\Delta; K) \cong \operatorname{Hom}_K(\widetilde{H}^q(\Delta, K), K).$$

Thus, in particular, we have $\widetilde{H}_q(\Delta; K) \cong \widetilde{H}^q(\Delta; K)$.

Proposition 6.2.3 *If Δ is a non-empty simplicial complex, then $\widetilde{H}_0(\Delta; A)$ is a free A-module of rank $r - 1$, where r is the number of connected components of Δ.*

Proof. Let $V = \{x_1, \ldots, x_n\}$ be the vertex set of Δ and let $\Delta_1, \ldots, \Delta_r$ be its connected components. Denote the vertex set of Δ_i by V_i and pick a vertex in each one of the V_i; one may assume $x_i \in V_i$ for $i = 1, \ldots, r$.

Set $\beta_i = x_i - x_r$ and $\overline{\beta}_i = \beta_i + \operatorname{im}(\partial_1)$ for $i = 1, \ldots, r-1$. We claim that the set $\overline{\mathcal{B}} = \{\overline{\beta}_1, \ldots, \overline{\beta}_{r-1}\}$ is an A-basis for $\widetilde{H}_0(\Delta; A) = \ker(\partial_0)/\operatorname{im}(\partial_1)$.

First we prove that $\overline{\mathcal{B}}$ is linearly independent. Assume $\sum_{i=1}^{r-1} a_i\beta_i$ is in $\text{im}(\partial_1)$ for some a_i in A, by making an appropriate grouping of terms one can write

$$\sum_{i=1}^{r-1} a_i\beta_i = \sum_{x_i,x_j\in V_1} b_{ij}(x_i - x_j) + \cdots + \sum_{x_i,x_j\in V_r} b_{ij}(x_i - x_j),$$

for some b_{ij} in A. Since $C_0(\Delta)$ is a free A-module with basis V and the V_i are mutually disjoint one has

$$a_kx_k = \sum_{x_i,x_j\in V_k} b_{ij}(x_i - x_j) \ \text{ for } \ 1 \leq k \leq r-1,$$

applying ∂_0 to both sides of this equality we obtain $a_k = 0$ for all k.

Next we show that $\overline{\mathcal{B}}$ is a set of generators. Let x_{i_1} and x_{j_1} be two vertices of Δ, it suffices to prove that $x_{i_1} - x_{j_1} + \text{im}(\partial_1)$ is in the submodule generated by $\overline{\mathcal{B}}$ because $\ker(\partial_0)$ is generated by the elements of the form $x_i - x_j$. Assume $x_{i_1} \in V_m$ and $x_{j_1} \in V_s$. There is a path $x_{i_1}, \ldots, x_{i_t}, x_m$ in Δ_m and a path $x_{j_1}, \ldots, x_{j_p}, x_s$ in Δ_s such that every two consecutive vertices of those paths form a 1-face of Δ. From the equalities

$$\begin{array}{rcl}(x_{i_1} - x_{i_2}) + (x_{i_2} - x_{i_3}) + \cdots + (x_{i_t} - x_m) & = & x_{i_1} - x_m \\ (x_{j_1} - x_{j_2}) + (x_{j_2} - x_{j_3}) + \cdots + (x_{j_p} - x_s) & = & x_{j_1} - x_s,\end{array}$$

one concludes $x_{i_1} - x_{j_1} + \text{im}(\partial_1) = (x_m - x_r) - (x_s - x_r) + \text{im}(\partial_1)$. □

Definition 6.2.4 The *reduced Euler characteristic* $\widetilde{\chi}(\Delta)$ of a simplicial complex Δ of dimension d is given by

$$\widetilde{\chi}(\Delta) = -1 + \sum_{i=0}^{d}(-1)^i f_i = -1 + \chi(\Delta),$$

where f_i is the number of i-faces in Δ and $\chi(\Delta)$ is its *Euler characteristic.*

Exercises

6.2.5 Let Δ be a simplicial complex of dimension d and f_i the number of i-faces in Δ. If K is a field, then

$$\sum_{i=-1}^{d} (-1)^i \text{rank}\, \widetilde{H}_i(\Delta; K) = -1 + \sum_{i=0}^{d}(-1)^i f_i.$$

6.2.6 Let Δ be a simplicial complex, prove that there is an exact sequence

$$0 \longrightarrow \widetilde{H}_0(\Delta; A) \longrightarrow H_0(\Delta; A) \longrightarrow A \longrightarrow 0.$$

Then, show $H_0(\Delta; A) \simeq \widetilde{H}_0(\Delta, A) \oplus A$.

6.2.7 Let Δ be a simplicial complex with vertex set $V = \{x_1, \ldots, x_n\}$ and

$$\cdots \xrightarrow{\partial_1} C_0(\Delta) \xrightarrow{\partial_0} C_{-1}(\Delta) = A \longrightarrow 0,$$

the right end of the augmented chain complex over a ring A. Prove that $\ker(\partial_0)$ is generated by the cycles of the form $x_i - x_j$.

6.2.8 Let Δ be a simplicial complex and f_i the number of i-faces of Δ. If

$$\cdots \longrightarrow C_1(\Delta) \xrightarrow{\partial_1} C_0(\Delta) \xrightarrow{\partial_0} C_{-1}(\Delta) = K \longrightarrow 0,$$

is the augmented chain complex of Δ over a field K, then $\dim_K(Z_1)$ is equal to $f_1 - f_0 + r$; where r is the number of connected components of Δ and $Z_1 = \ker(\partial_1)$.

6.3 Stanley–Reisner rings

In this section we study algebraic and combinatorial properties of edge ideals of clutters and simplicial complexes. In particular we examine shellable, unmixed, and sequentially Cohen–Macaulay simplicial complexes and their corresponding edge ideals.

Let $R = K[x_1, \ldots, x_n]$ be a polynomial ring over a field K. If I is an ideal of R generated by square-free monomials, the *Stanley–Reisner simplicial complex* Δ_I associated to I has vertex set $V = \{x_i | \, x_i \notin I\}$ and its faces are defined by

$$\Delta_I = \{\{x_{i_1}, \ldots, x_{i_k}\} | \; i_1 < \cdots < i_k \,, \; x_{i_1} \cdots x_{i_k} \notin I\}.$$

Conversely if Δ is a simplicial complex with vertex set V contained in $\{x_1, \ldots, x_n\}$, the *Stanley–Reisner ideal* I_Δ is defined as

$$I_\Delta = (\{x_{i_1} \cdots x_{i_r} | \; i_1 < \cdots < i_r, \; \{x_{i_1}, \ldots, x_{i_r}\} \notin \Delta\}),$$

and its *Stanley–Reisner ring* $K[\Delta]$ is defined as the quotient ring R/I_Δ.

Definition 6.3.1 Let $I \subset R$ be a square-free monomial ideal. The quotient ring R/I is called a *face ring*.

Example 6.3.2 A simplicial complex Δ of dimension 1 and its Stanley–Reisner ideal:

x_2

x_1 x_3

x_4

$$I_\Delta = (x_1x_3, x_1x_2x_4, x_2x_3x_4)$$

Definition 6.3.3 A face F of a simplicial complex Δ is said to be a *facet* if F is not properly contained in any other face of Δ.

Proposition 6.3.4 *If Δ is a simplicial complex with vertices $x_1, \ldots, x_n$, then the primary decomposition of the Stanley–Reisner ideal of Δ is:*

$$I_\Delta = \bigcap_F \mathfrak{p}_F,$$

where the intersection is taken over all facets F of Δ, and $\mathfrak{p}_F$ denotes the face ideal generated by all x_i such that $x_i \notin F$.

Proof. Let F be a face of Δ and $\mathfrak{p}_F$ the face ideal generated by the x_i such that $x_i \notin F$. Note that $\mathfrak{p}_F$ is a minimal prime of I_Δ if and only if F is a facet. Therefore the result follows from Theorem 6.1.4. □

Corollary 6.3.5 *Let Δ be a simplicial complex of dimension d on the vertex set $V = \{x_1, \ldots, x_n\}$ and let K be a field. Then*

$$\dim K[\Delta] = d + 1 = \max\{s \mid x_{i_1} \cdots x_{i_s} \notin I_\Delta \text{ and } i_1 < \cdots < i_s\}.$$

Proof. Let F be a facet of Δ with $d+1$ vertices. By Proposition 6.3.4 the face ideal $\mathfrak{p}_F$ generated by the variables not in F has height equal to the height of I_Δ. Hence $\mathrm{ht}(I_\Delta) = n - d - 1$, and thanks to Proposition 3.1.16 one has $\dim K[\Delta] = d + 1$. □

Definition 6.3.6 Let K be a field. A simplicial complex Δ is said to be *Cohen–Macaulay* over K if the Stanley–Reisner ring $K[\Delta]$ is a Cohen–Macaulay ring.

Corollary 6.3.7 *A Cohen–Macaulay simplicial complex Δ is pure; that is, all its maximal faces have the same dimension.*

Proof. It follows from Proposition 6.3.4 and Corollary 3.1.17. □

Definition 6.3.8 A pure d-dimensional complex Δ is called *strongly connected* or *connected in codimension* 1 if each pair of facets F, G can be connected by a sequence of facets $F = F_0, F_1, \ldots, F_s = G$, such that $\dim(F_{i-1} \cap F_i) = d - 1$ for $1 \leq i \leq s$.

Theorem 6.3.9 [43, Proposition 11.7] *Every Cohen–Macaulay complex is strongly connected.*

Definition 6.3.10 Let Δ be a simplicial complex. For $F \in \Delta$, define $\mathrm{lk}(F)$, the *link* of F, as

$$\mathrm{lk}(F) := \{H \in \Delta \mid H \cap F = \emptyset \text{ and } H \cup F \in \Delta\}.$$

Theorem 6.3.11 (Hochster [65, Theorem 5.3.8]) *Let Δ be a simplicial complex. If K is a field, then the Hilbert series of the local cohomology modules of $K[\Delta]$ with respect to the fine grading is given by*

$$F(H_{\mathfrak{m}}^{p}(K[\Delta]),\mathbf{t}) = \sum_{F\in\Delta} \dim_K \widetilde{H}_{p-|F|-1}(\mathrm{lk}\ F;K) \prod_{x_j\in F} \frac{t_j^{-1}}{1-t_j^{-1}}.$$

Theorem 6.3.12 (Reisner [346]) *Let Δ be a simplicial complex. If K is a field, then the following conditions are equivalent:*

(a) *Δ is Cohen–Macaulay over K.*

(b) *$\widetilde{H}_i(\mathrm{lk}\ F;K)=0$ for $F\in\Delta$ and $i<\dim \mathrm{lk}\ F$.*

Proof. Let $H_{\mathfrak{m}}^{i}(K[\Delta])$ be the local cohomology modules of $K[\Delta]$. Recall that $H_{\mathfrak{m}}^{i}(K[\Delta]) \cong H^i(C^\star)$; see Section 2.8 for the definition of $C^\star$. By Theorem 2.8.12, Δ is C–M if and only if $H^i(C^\star)=0$ for $i<d+1$, where $d=\dim(\Delta)$.

$\Rightarrow$) If Δ is C–M, then by Hochster's theorem $\widetilde{H}_{i-|F|-1}(\mathrm{lk}(F);K)=0$ for $F\in\Delta$ and $i<d+1$. By Corollary 6.3.7 we get that Δ is pure. If $F\in\Delta$, there is a face F_1 of dimension d containing F, since $F_1\setminus F\in \mathrm{lk}(F)$ it follows that $\dim \mathrm{lk}(F)=|F_1\setminus F|-1=d-|F|$. Hence $\widetilde{H}_i(\mathrm{lk}(F);K)=0$ for all $F\in\Delta$ and all $i<\dim \mathrm{lk}(F)$.

$\Leftarrow$) Using $\dim \mathrm{lk}(F)\leq d-|F|$, the hypothesis $\widetilde{H}_i(\mathrm{lk}(F);K)=0$ for $F\in\Delta$ and $i<\dim \mathrm{lk}(F)$ implies $\widetilde{H}_{i-|F|-1}(\mathrm{lk}(F);K)=0$ for $F\in\Delta$ and $i<d+1$. Hence by Hochster's theorem $H^i(C^\star)=0$ for $i<d+1$. Thus Δ is Cohen–Macaulay. This proof was adapted from [65]. The original proof can be found in [346]. $\Box$

In particular a Cohen–Macaulay complex Δ, being the link of its empty face, must satisfy $\widetilde{H}_i(\Delta;K)=0$ for $i<\dim\Delta$.

In general the Cohen–Macaulay property of a face ring may depend on the characteristic of the base field; classical examples of this dependence are triangulations of the projective plane [346]. See the exercises at the end of this section.

Example 6.3.13 If Δ is a discrete set with n vertices

x_5, x_6, x_4, x_7, x_3, x_n, x_2, x_1

then Δ is Cohen–Macaulay and $I_\Delta=(x_ix_j|1\leq i<j\leq n)$.

Corollary 6.3.14 *If Δ is a 1-dimensional simplicial complex, then Δ is connected if and only if Δ is Cohen–Macaulay.*

Proof. By Reisner's theorem Δ is Cohen–Macaulay iff $\widetilde{H}_0(\Delta; K) = 0$, because the link of a vertex of Δ consists of a discrete set of vertices. To complete the argument apply Proposition 6.2.3. $\square$

Proposition 6.3.15 [250] *If K is a field, then a simplicial complex Δ is Cohen–Macaulay over K if and only if* $\mathrm{lk}(F)$ *is Cohen–Macaulay over K for all $F \in \Delta$.*

Proof. $\Rightarrow$) Let F be a fixed face in Δ and $\Delta' = \mathrm{lk}(F)$. If $G \in \Delta'$, using the equality

$$\mathrm{lk}_{\Delta'}(G) = \mathrm{lk}_{\Delta}(G \cup F)$$

and Reisner's theorem we get $\widetilde{H}_i(\mathrm{lk}_{\Delta'}(G), K) = 0$ for $i < \dim \mathrm{lk}_{\Delta'}(G)$, thus Δ' is Cohen–Macaulay.

$\Leftarrow$) Since all the links are Cohen–Macaulay, it suffices to take $F = \emptyset$ and observe $\Delta = \mathrm{lk}(\emptyset)$ to conclude that Δ is Cohen–Macaulay. $\square$

Definition 6.3.16 The *q-skeleton* of a simplicial complex Δ is the simplicial complex Δ^q consisting of all p-simplices of Δ with $p \leq q$.

Proposition 6.3.17 Let Δ be a simplicial complex of dimension d and Δ^q its q-skeleton. If Δ is Cohen–Macaulay over a field K, then Δ^q is Cohen–Macaulay for $q \leq d$.

Proof. Set $\Delta' = \Delta^q$ and take $F \in \Delta'$ with $\dim(F) < q \leq d$. Note

$$\mathrm{lk}_{\Delta'}(F) = (\mathrm{lk}_{\Delta}(F))^{q-|F|}.$$

Since Δ is pure there is a facet F' of Δ of dimension d containing F. Hence $F' \setminus F$ is a face of $\mathrm{lk}_{\Delta}(F)$ of dimension $d - |F|$. It follows that the dimension of $\mathrm{lk}_{\Delta'}(F)$ is $q - |F|$. Using Proposition 6.3.15 we get

$$\widetilde{H}_i(\mathrm{lk}_{\Delta}(F); K) = 0 \ \text{ for } i < \dim(\mathrm{lk}_{\Delta}(F)) = d - |F|,$$

and consequently

$$\widetilde{H}_i((\mathrm{lk}_{\Delta}(F))^{q-|F|}; K) = \widetilde{H}_i(\mathrm{lk}_{\Delta}(F); K) = 0 \ \text{ for } i < q - |F| \leq d - |F|.$$

Observe that the equality between the homology modules holds because the augmented oriented chain complexes of $(\mathrm{lk}_{\Delta}(F))^{q-|F|}$ and $\mathrm{lk}_{\Delta}(F)$ are equal up to degree $q - |F|$ and thus their homologies have to agree up to degree one less. Hence by Reisner's theorem Δ^q is Cohen–Macaulay. $\square$

Example 6.3.18 [170] Let $R = K[x_1, \ldots, x_n]$ be a polynomial ring over a field K and Δ the $(p-2)$-skeleton of a simplex of dimension $n-1$, where $p \geq 2$. Then I_Δ is a Cohen–Macaulay ideal and the face ring

$$K[\Delta] = R/(\{x_{i_1} \cdots x_{i_p} \mid 1 \leq i_1 < \cdots < i_p \leq n\})$$

has a p-linear resolution by Theorem 5.3.7.

Face rings with pure and linear resolutions have been studied in [66] and [170, 172]; special results have been obtained when the corresponding Stanley–Reisner ideal is generated in degree two.

Definition 6.3.19 Let Δ_1, Δ_2 be two simplicial complexes with vertex sets V_1, V_2. The join $\Delta_1 * \Delta_2$ is the complex on the vertex set $V_1 \cup V_2$ with faces $F_1 \cup F_2$, where $F_i \in \Delta_i$ for $i = 1, 2$.

Proposition 6.3.20 *If Δ_1 and Δ_2 are simplicial complexes, then their join $\Delta_1 * \Delta_2$ is Cohen–Macaulay if and only if Δ_i is Cohen–Macaulay for $i = 1, 2$.*

Proof. Since one has the equality $I_{\Delta_1 * \Delta_2} = I_{\Delta_1} + I_{\Delta_2}$, there is a graded isomorphism of K-algebras

$$K[\Delta_1 * \Delta_2] \simeq K[\Delta_1] \otimes_K K[\Delta_2],$$

see Proposition 3.1.33. To complete the proof use Corollary 3.1.35. □

Given a simplicial complex Δ and $\Delta_1, \ldots, \Delta_s$ sub complexes of Δ, we define their *union*

$$\bigcup_{i=1}^{s} \Delta_i$$

as the simplicial complex with vertex set $\cup_{i=1}^{s} V_i$ and such that F is a face of $\cup_{i=1}^{s} \Delta_i$ if and only if F is a face of Δ_i for some i. The intersection can be defined similarly.

Definition 6.3.21 A simplicial complex Δ of dimension d is called *pure shellable* if Δ is pure and the facets (maximal faces) of Δ can be order $F_1, \ldots, F_s$ such that

$$\overline{F}_i \bigcap \left(\bigcup_{j=1}^{i-1} \overline{F}_j \right)$$

is pure of dimension $d-1$ for all $i \geq 2$. Here $\overline{F}_i = \{\sigma \in \Delta \mid \sigma \subset F_i\}$. If Δ is pure shellable, $F_1, \ldots, F_s$ is called a *shelling*.

The next definition of *shellable* is due to Björner and Wachs [44] and is usually referred to as *non-pure shellable*, although in this book we will drop the adjective "non-pure."

Definition 6.3.22 A simplicial complex Δ is *shellable* if the facets (maximal faces) of Δ can be ordered $F_1, \ldots, F_s$ such that for all $1 \leq i < j \leq s$, there exists some $v \in F_j \setminus F_i$ and some $\ell \in \{1, \ldots, j-1\}$ with $F_j \setminus F_\ell = \{v\}$. We call $F_1, \ldots, F_s$ a *shelling* of Δ. For a fixed shelling of Δ, if $F, F' \in \Delta$ then we write $F < F'$ to mean that F appears before F' in the ordering.

If Δ is a pure simplicial complex, then Δ is shellable if and only if Δ is pure shellable (Exercise 6.3.60).

Theorem 6.3.23 *Let Δ be a simplicial complex. If Δ is pure shellable, then Δ is Cohen–Macaulay over any field K.*

Proof. Let $V = \{x_1, \ldots, x_n\}$ be the vertex set of Δ and let $R = K[V]$ be a polynomial ring over a field K. Assume that $F_1, \ldots, F_s$ is a shelling of Δ such that $\overline{F}_i \cap (\cup_{j=1}^{i-1} \overline{F}_j)$ is pure of dimension $d-1$ for all $i \geq 2$, where d is the dimension of Δ. Set

$$\sigma = \left\{ x_i \in F_s \;\middle|\; F_s \setminus \{x_i\} \in \overline{F}_s \bigcap \left(\bigcup_{j=1}^{s-1} \overline{F}_j \right) \right\},$$

one may assume $\sigma = \{x_1, \ldots, x_r\}$. There is a short exact sequence

$$0 \longrightarrow R/(I\colon f) \xrightarrow{f} R/I \longrightarrow R/(I, f) \longrightarrow 0, \qquad (*)$$

where $f = x_1 \cdots x_r$ and $I = I_\Delta$. First we show the equality

$$R/(I\colon f) = K[\overline{F}_s],$$

note that $K[\overline{F}_s]$ is a polynomial ring in $d+1$ variables. By Proposition 6.3.4 one can write $I = \cap_{i=1}^s \mathfrak{p}_i$, where $\mathfrak{p}_i$ is generated by $V \setminus F_i$ for all i. Hence

$$(I\colon f) = \bigcap_{i=1}^{s} (\mathfrak{p}_i\colon f) = \bigcap_{\sigma \subset F_i} (\mathfrak{p}_i\colon f).$$

Observe that if $\sigma \subset F_i$ for some i, then $i = s$; otherwise if $i < s$, then σ belongs to $\overline{F}_s \cap (\cup_{j=1}^{s-1} \overline{F}_j)$, but since this simplicial complex is pure of dimension $d-1$, there is $v \in \sigma$ such that $\sigma \subset F_s \setminus \{v\}$, a contradiction. Hence $i = s$ and $(I\colon f) = (\mathfrak{p}_s\colon f) = \mathfrak{p}_s$, as asserted. Next, we show

$$R/(I, f) = K[\overline{F}_1 \cup \cdots \cup \overline{F}_{s-1}].$$

Note that $f \in \mathfrak{p}_i$ for $i = 1, \ldots, s-1$; otherwise if $f \notin \mathfrak{p}_i$ for some $i < s$, then $\sigma \subset F_i$ and the previous argument yields a contradiction. If $\mathfrak{p}$ is a minimal prime of (I, f) different from $\mathfrak{p}_1, \ldots, \mathfrak{p}_{s-1}$, then $\mathfrak{p}_s \subset \mathfrak{p}$ and $x_i \in \mathfrak{p}$ for some $x_i \in \sigma$; thus by construction of σ one has $F_s \setminus \{x_i\} \subset F_k$ for some $k < s$.

Therefore $V \setminus F_k \subset \{x_i\} \cup (V \setminus F_s) \subset \mathfrak{p}$, and by the minimality of $\mathfrak{p}$ one derives $\mathfrak{p}_k = \mathfrak{p}_s$, which is impossible. Altogether we conclude that $\mathfrak{p}_1, \ldots, \mathfrak{p}_{s-1}$ are the minimal primes of (I, f) and $(I, f) = \bigcap_{i=1}^{s-1} \mathfrak{p}_i$, as required.

Note that the two ends of Eq. $(*)$ are Cohen–Macaulay R-modules of dimension $d+1$. Hence, using induction on s and Eq. $(*)$, by the depth lemma it follows that R/I is Cohen–Macaulay. Our proof follows ideas of Hibi [239]. $\square$

Definition 6.3.24 A simplicial complex Δ is *constructible* if Δ can be obtained by the following recursive procedure:

(i) any simplex is constructible,

(ii) if Δ_1, Δ_2 are constructible complexes of dimension d, and if $\Delta_1 \cap \Delta_2$ is constructible of dimension $d-1$, then $\Delta_1 \cup \Delta_2$ is constructible.

Proposition 6.3.25 *If Δ is a pure shellable complex of dimension d, then Δ is constructible.*

Proof. We proceed by induction on d. If $d = 0$, then Δ is a discrete set of vertices, which is constructible. Assume $d > 0$ and that Δ has at least two facets. Let $F_1, \ldots, F_r$ be a shelling of Δ. Thus, the complex

$$\Delta_m = (\overline{F}_1 \cup \cdots \cup \overline{F}_m) \cap \overline{F}_{m+1}$$

is pure of dimension $d-1$ for $1 \leq m \leq r-1$. Next we prove that Δ_m is pure shellable. Consider the set

$$\begin{aligned} \mathcal{F} &= \{F_i \cap F_{m+1} \mid \dim(F_i \cap F_{m+1}) = d-1;\ i \leq m\} \\ &= \{F_{\ell_1} \cap F_{m+1}, \ldots, F_{\ell_s} \cap F_{m+1}\} \end{aligned}$$

where $\ell_1 < \cdots < \ell_s < m+1$ and $|\mathcal{F}| = s$. We claim that a face of the form

$$F = F_{\ell_i} \cap F_{\ell_j} \cap F_{m+1}$$

has dimension $d-2$ for $\ell_j < \ell_i < m+1$. Clearly $\dim(F) \leq d-2$, otherwise one has $F_{\ell_i} \cap F_{m+1} = F_{\ell_j} \cap F_{m+1}$, a contradiction because $|\mathcal{F}| = s$. By Exercise 6.3.59, there are $v_i \in F_{m+1} \setminus F_{\ell_i}$ and $v_j \in F_{m+1} \setminus F_{\ell_j}$ such that

$$\begin{aligned} F_{m+1} \cap F_{\ell_i} &\subset F_{m+1} \cap F_{k_1} = F_{m+1} \setminus \{v_i\}; \quad (k_1 < m+1), \\ F_{m+1} \cap F_{\ell_j} &\subset F_{m+1} \cap F_{k_2} = F_{m+1} \setminus \{v_j\}; \quad (k_2 < m+1). \end{aligned}$$

By dimension considerations we get

$$\begin{aligned} F_{m+1} \cap F_{\ell_i} &= F_{m+1} \setminus \{v_i\}, \\ F_{m+1} \cap F_{\ell_j} &= F_{m+1} \setminus \{v_j\}. \end{aligned}$$

Hence $F = F_{m+1} \setminus \{v_i, v_j\}$, consequently $\dim(F)$ is equal to $d-2$, as claimed. It follows readily that $\mathcal{F}$ gives a shelling of Δ_m because the complex

$$((\overline{F}_{\ell_1} \cap \overline{F}_{m+1}) \cup \cdots \cup (\overline{F}_{\ell_k} \cap \overline{F}_{m+1})) \cap (\overline{F}_{\ell_{k+1}} \cap \overline{F}_{m+1}).$$

is pure of dimension $d-2$ for $1 \leq k \leq s-1$. Therefore by induction hypothesis Δ_m is constructible for $1 \leq m \leq r-1$. Hence, using induction on m, it follows that the complex

$$\overline{F}_1 \cup \cdots \cup \overline{F}_m \cup \overline{F}_{m+1}$$

is constructible for $1 \leq m \leq r-1$, as required. $\Box$

Definition 6.3.26 Let $R = K[x_1, \ldots, x_n]$. A graded R-module M is called *sequentially Cohen–Macaulay* (over K) if there exists a finite filtration of graded R-modules

$$0 = M_0 \subset M_1 \subset \cdots \subset M_r = M$$

such that each M_i/M_{i-1} is Cohen–Macaulay, and the Krull dimensions of the quotients are increasing:

$$\dim(M_1/M_0) < \dim(M_2/M_1) < \cdots < \dim(M_r/M_{r-1}).$$

A filtration with these properties is called a C–M filtration of M.

As first observed by Stanley [395, p. 87], shellable implies sequentially Cohen–Macaulay. In [207] Haghighi, Terai, Yassemi, and Zaare-Nahandi introduce and study the notion of a sequentially S_r module which is very reminiscent of the definition of a sequentially Cohen–Macaulay module.

Theorem 6.3.27 *If Δ is a shellable simplicial complex, then its associated Stanley–Reisner ring R/I_Δ is sequentially Cohen–Macaulay.*

If Δ is a simplicial complex and v is a vertex of Δ, the *deletion* of v, denoted by $\mathrm{del}_\Delta(v)$, is the subcomplex consisting of the faces of Δ that do not contain v. The *deletion of a face* is defined similarly.

Definition 6.3.28 [43] A simplicial complex Δ is *vertex decomposable* if either Δ is a simplex, or $\Delta = \emptyset$, or Δ contains a vertex v, called a *shedding vertex*, such that both the link $\mathrm{lk}_\Delta(v)$ and the deletion $\mathrm{del}_\Delta(v)$ are vertex-decomposable, and such that every facet of $\mathrm{del}_\Delta(v)$ is a facet of Δ.

Definition 6.3.29 Let Δ be a simplicial complex. The *pure i-skeleton* of Δ is defined as:

$$\Delta^{[i]} \quad = \quad \langle \{F \in \Delta | \dim(F) = i\} \rangle; \quad -1 \leq i \leq \dim(\Delta),$$

where $\langle \mathcal{F} \rangle$ denotes the subcomplex generated by $\mathcal{F}$. Notice that $\Delta^{[i]}$ is always pure of dimension i.

We say that a simplicial complex Δ is *sequentially Cohen–Macaulay* if its Stanley–Reisner ring has this property.

Theorem 6.3.30 [121, Theorem 3.3] *A simplicial complex Δ is sequentially Cohen–Macaulay if and only if the pure i-skeleton $\Delta^{[i]}$ is Cohen–Macaulay for $-1 \leq i \leq \dim(\Delta)$.*

Corollary 6.3.31 *A simplicial complex Δ is Cohen–Macaulay if and only if Δ is sequentially Cohen–Macaulay and Δ is pure.*

Proposition 6.3.32 *The following hold for simplicial complexes:*

(I) *pure shellable $\Rightarrow$ constructible $\Rightarrow$ Cohen–Macaulay $\Rightarrow$ pure.*

(II) *vertex decomposable $\Rightarrow$ shellable $\Rightarrow$ sequentially Cohen–Macaulay.*

Proof. (I): The first implication is Proposition 6.3.25. The second implication follows using Exercise 6.3.62 together with the depth lemma. The third implication follows at once from Corollary 3.1.17.

(II): The first implication is shown in [45, Theorem 11.3]. The second implication is Theorem 6.3.27. $\square$

Definition 6.3.33 A *clutter* $\mathcal{C}$ is a finite ground set X together with a family E of subsets of X such that if $f_1, f_2 \in E$, then $f_1 \not\subset f_2$. The ground set X is called the *vertex set* of $\mathcal{C}$ and E is called the *edge set* of $\mathcal{C}$; they are denoted by $V(\mathcal{C})$ and $E(\mathcal{C})$, respectively.

Clutters are simple hypergraphs (see below) and they are called *Sperner families* in the literature. One example of a clutter is a graph with the vertices and edges defined in the usual way.

Definition 6.3.34 A *hypergraph* $\mathcal{H}$ is a pair (V, E) such that V is a finite set and E is a subset of the set of all subsets of V. The elements of E are called *edges* and the elements of V are called vertices. A hypergraph is *simple* if $f_1 \not\subset f_2$ for any two edges f_1, f_2.

An excellent reference for hypergraph theory is the 3-volume book of Schrijver on combinatorial optimization [373].

Definition 6.3.35 Let $\mathcal{C}$ be a clutter with vertex set $X = \{x_1, \ldots, x_n\}$. The *edge ideal* of $\mathcal{C}$, denoted by $I(\mathcal{C})$, is the ideal of R generated by all monomials $x_e = \prod_{x_i \in e} x_i$ such that $e \in E(\mathcal{C})$.

Edge ideals of graphs, clutters, and hypergraphs were introduced in [416], [188], and [206], respectively. The assignment

$$\mathcal{C} \longmapsto I(\mathcal{C})$$

establishes a one-to-one correspondence between the family of clutters and the family of square-free monomial ideals. Edge ideals of clutters are in one-to-one correspondence with simplicial complexes via the Stanley–Reisner correspondence

$$I(\mathcal{C}) \longmapsto \Delta_{I(\mathcal{C})}.$$

We shall be interested in studying the relationships between the algebraic and combinatorial properties of $\mathcal{C}$, $\Delta_{I(\mathcal{C})}$ and $R/I(\mathcal{C})$.

Let $\mathcal{C}$ be a clutter with vertex set $X = \{x_1, \ldots, x_n\}$. A subset F of X is called *independent* or *stable* if $e \not\subset F$ for any $e \in E(\mathcal{C})$. The dual concept of a stable vertex set is a *vertex cover*, i.e., a subset C of X is a vertex cover if and only if $X \setminus C$ is a stable vertex set. A *minimal vertex cover* is a vertex cover which is minimal with respect to inclusion. The number of vertices in any smallest vertex cover, denoted by $\alpha_0(\mathcal{C})$, is called the *vertex covering number*. The *vertex independence number*, denoted by $\beta_0(\mathcal{C})$, is the number of vertices in any largest independent set of vertices.

Notice that the Stanley–Reisner complex of $I(\mathcal{C})$ is given by

$$\Delta_{I(\mathcal{C})} = \Delta_{\mathcal{C}},$$

where $\Delta_{\mathcal{C}}$ is the simplicial complex whose faces are the independent vertex sets of $\mathcal{C}$. Thus

$$K[\Delta_{\mathcal{C}}] = R/I(\mathcal{C}),$$

where $K[\Delta_{\mathcal{C}}]$ is the Stanley–Reisner ring of $\Delta_{\mathcal{C}}$.

Definition 6.3.36 The simplicial complex $\Delta_{\mathcal{C}}$ whose faces are the independent vertex sets of $\mathcal{C}$ is called the *independence complex* of $\mathcal{C}$.

Lemma 6.3.37 *Let C be a set of vertices of a clutter $\mathcal{C}$ and let $\mathfrak{p}$ be the face ideal of R generated by C. The following are equivalent:*

(a) *C is a minimal vertex cover of $\mathcal{C}$.*

(b) *$\mathfrak{p}$ is a minimal prime of $I(\mathcal{C})$.*

(c) *$V(\mathcal{C}) \setminus C$ is a maximal face of $\Delta_{\mathcal{C}}$.*

Proof. Note that $I(\mathcal{C}) \subset \mathfrak{p}$ if and only if C is a vertex cover of $\mathcal{C}$. Taking this into account, the proof follows readily from the fact that any minimal prime of $I(\mathcal{C})$ is a face ideal; see Theorem 6.1.4. □

Definition 6.3.38 The clutter of minimal vertex covers of $\mathcal{C}$, denoted by $\mathcal{C}^\vee$ or $b(\mathcal{C})$ is called the *Alexander dual clutter* or *blocker* of $\mathcal{C}$. The edge ideal of $\mathcal{C}^\vee$, denoted by $I_c(\mathcal{C})$, is called the *ideal of covers* of $\mathcal{C}$. The ideal $I_c(\mathcal{C})$ is also called the *Alexander dual* of $I(\mathcal{C})$ and is also denoted by $I(\mathcal{C})^\vee$.

Theorem 6.3.39 *Let $e_1,\ldots,e_q$ and $c_1,\ldots,c_s$ be the edges and minimal vertex covers of a clutter $\mathcal{C}$. Then, there is a duality given by:*

$$\begin{array}{rclcl} I(\mathcal{C}) & = & (x_{e_1},x_{e_2},\ldots,x_{e_q}) & = & (c_1)\cap(c_2)\cap\cdots\cap(c_s) \\ & & \updownarrow & & \updownarrow \\ I_c(\mathcal{C}) & = & (e_1)\cap(e_2)\cap\cdots\cap(e_q) & = & (x_{c_1},x_{c_2},\ldots,x_{c_s}), \end{array}$$

where $x_{e_k}=\prod_{x_i\in e_k}x_i$ and $x_{c_k}=\prod_{x_i\in c_k}x_i$ for $1\leq k\leq s$. In particular the heights of $I(\mathcal{C})$ and $I_c(\mathcal{C})$ are $\min_i\{|c_i|\}$ and $\min_i\{|e_i|\}$, respectively.

Proof. If $\mathfrak{p}_i$ is the prime ideal generated by c_i, then $\mathfrak{p}_1,\ldots,\mathfrak{p}_s$ are the minimal primes of $I(\mathcal{C})$ (see Lemma 6.3.37). Thus, by Theorem 6.1.4, we get $I(\mathcal{C})=\cap_{i=1}^s\mathfrak{p}_i$. As $\mathcal{C}$ is a clutter, we have $(\mathcal{C}^\vee)^\vee=\mathcal{C}$ and $I_c(\mathcal{C}^\vee)=I(\mathcal{C})$. Hence, the minimal vertex covers of $\mathcal{C}^\vee$ are $e_1,\ldots,e_q$. Thus, by the previous argument, we have $I(\mathcal{C}^\vee)=\cap_{i=1}^q(e_i)$. □

The survey article [218] explains the role of Alexander duality to prove combinatorial and algebraic theorems.

Definition 6.3.40 Let Δ be a simplicial complex on the vertex set V, the *Alexander dual* $\Delta^\vee$ of Δ is the simplicial complex given by

$$\Delta^\vee=\{G\subset V|\,V\setminus G\notin\Delta\}.$$

Theorem 6.3.41 (Eagon–Reiner [124]) *If Δ is a simplicial complex, then the Alexander dual $\Delta^\vee$ is Cohen–Macaulay if and only if the Stanley–Reisner ideal I_Δ has a linear resolution.*

This result can be rephrased as:

Theorem 6.3.42 *Let $\mathcal{C}$ be a clutter. Then, $R/I(\mathcal{C})$ is Cohen–Macaulay if and only if $I_c(\mathcal{C})$ has a linear resolution.*

This result has been generalized [220] replacing linear resolution by the notion of *componentwise linear ideal*, and Cohen–Macaulay by *sequentially Cohen–Macaulay*. Next, we introduce the result of Herzog and Hibi that link these two notions.

Definition 6.3.43 Let (I_d) denote the ideal generated by all degree d elements of a homogeneous ideal I. Then I is called *componentwise linear* if (I_d) has a linear resolution for all d.

If I is a square-free monomial ideal we write $I_{[d]}$ for the ideal generated by all the square-free monomial ideals of degree d in I.

Theorem 6.3.44 [220] *Let I be a square-free monomial ideal of R. Then* (a) *R/I is sequentially Cohen–Macaulay if and only if $I^\vee$ is componentwise linear.* (b) *I is componentwise linear if and only if $I_{[d]}$ has a linear resolution for all $d\geq 0$.*

Definition 6.3.45 *A monomial ideal I has linear quotients if the monomials that minimally generate I can be ordered $g_1, \ldots, g_q$ such that for all $1 \leq i \leq q-1$, $((g_1, \ldots, g_i) : g_{i+1})$ is generated by linear forms $x_{i_1}, \ldots, x_{i_t}$.*

Recall that a clutter is called *uniform* if all its edges have the same cardinality.

Proposition 6.3.46 [155, Lemma 5.2] *If I is an edge ideal of a uniform clutter and I has linear quotients, then I has a linear resolution.*

Theorem 6.3.47 [228, Theorem 1.4(c)] *If $\mathcal{C}$ is a uniform clutter, then $\Delta_\mathcal{C}$ is shellable if and only if $I_c(\mathcal{C})$ has linear quotients.*

Exercises

6.3.48 Let R be a polynomial ring over a field K. Prove that the family of ideals of R generated by square-free monomials is a sublattice of the lattice of monomial ideals (cf. Proposition 6.1.19).

6.3.49 A monomial ideal I is square-free monomials iff any of the following conditions hold:

(a) I is an intersection of prime ideals.

(b) $\mathrm{rad}\,(I) = I$.

(c) A monomial f is in I iff $x_1 \cdots x_r \in I$, where $\mathrm{supp}(f) = \{x_i\}_{i=1}^r$.

6.3.50 If Δ is the simplicial complex:

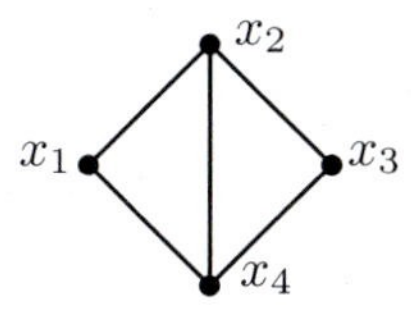

then $I_\Delta = (x_1, x_2) \cap (x_1, x_3) \cap (x_1, x_4) \cap (x_2, x_3) \cap (x_3, x_4)$.

6.3.51 Let Δ be a Cohen–Macaulay simplicial complex and $F \in \Delta$. If $\mathrm{lk}(F)$ is not a discrete set of vertices, prove that $\mathrm{lk}(F)$ is connected.

6.3.52 Let Δ be a simplicial complex on the vertex set V and I its Stanley–Reisner ideal. Take $x \in V$ and set $J = (I\colon x)$.
(a) Prove that $J = \cap_{x \notin \mathfrak{p}_i} \mathfrak{p}_i$, where $\mathrm{Ass}(I) = \{\mathfrak{p}_1, \ldots, \mathfrak{p}_r\}$.
(b) Find a relation between Δ and Δ_J.

6.3.53 Let R be a polynomial ring over a field K and I an ideal minimally generated by square-free monomials $f_1, \ldots, f_q$. If R/I is Cohen–Macaulay and x is a variable not in I such that $x \in \cup_{i=1}^q \mathrm{supp}(f_i)$, then $R/(I\colon x)$ is Cohen–Macaulay.

6.3.54 Let Δ be a pure simplicial complex on the vertex set V and $I = I_\Delta$. Assume that x is a variable in $\cup_{i=1}^q \text{supp}(f_i)$, where the f_i's are monomials that minimally generate I. If $(I\colon x)$ and (I, x) are Cohen–Macaulay, then I is Cohen–Macaulay.

Hint Show $\text{ht}(I) = \text{ht}(I\colon x) = \text{ht}(I, x)$.

6.3.55 Let Δ be a simplicial complex and let $\sigma \in \Delta$, define the *star* of the face σ as

$$\text{star}(\sigma) = \{G \in \Delta \,|\, \sigma \cup G \in \Delta\}.$$

Prove that the facets of $\text{star}(\sigma)$ are the facets of Δ that contain σ.

6.3.56 Let $R = K[\mathbf{x}]$ be a polynomial ring over a field K and Δ a simplicial complex with vertex set $\mathbf{x}$. If $\sigma = \{x_1, \ldots, x_r\} \in \Delta$, prove the equality

$$K[\text{star}(\sigma)] = R/(I_\Delta\colon f),$$

where $f = \prod_{i=1}^r x_i$ and $\text{star}(\sigma)$ is the star of σ.

6.3.57 If Δ is a Cohen–Macaulay complex and F a face of Δ, then $\text{star}(F)$ is a Cohen–Macaulay complex.
Hint $\text{star}(F) = \overline{F} * \text{lk}(F)$.

6.3.58 A simplicial complex Δ of dimension d is pure shellable if and only if Δ is pure and the facets of Δ can be listed $F_1, \ldots, F_s$ such that for every $1 \leq i < j \leq s$, there is $1 \leq \ell < j$ with $F_i \cap F_j \subset F_\ell \cap F_j$ and $\dim(F_\ell \cap F_j) = d - 1$.

6.3.59 A simplicial complex Δ is pure shellable if and only if Δ is pure and the facets of Δ can be ordered $F_1, \ldots, F_s$ such that for all $1 \leq j < i \leq s$, there is $v \in F_i \setminus F_j$ and $k < i$ with $F_i \cap F_j \subset F_i \cap F_k = F_i \setminus \{v\}$.

6.3.60 A simplicial complex Δ is pure shellable if and only if Δ is pure and the facets of Δ can be listed $F_1, \ldots, F_s$ such that for all $1 \leq j < i \leq s$, there exist some $v \in F_i \setminus F_j$ and some $k \in \{1, \ldots, i-1\}$ with $F_i \setminus F_k = \{v\}$.

6.3.61 Let R be a ring and I_1, I_2 ideals of R. Prove that there is an exact sequence of R-modules:

$$0 \longrightarrow R/(I_1 \cap I_2) \stackrel{\varphi}{\longrightarrow} R/I_1 \oplus R/I_2 \stackrel{\phi}{\longrightarrow} R/(I_1 + I_2) \longrightarrow 0,$$

where $\varphi(\overline{r}) = (\overline{r}, -\overline{r})$ and $\phi(\overline{r}_1, \overline{r}_2) = \overline{r_1 + r_2}$.

6.3.62 Let $R = K[\mathbf{x}]$ be a polynomial ring over a field K and Δ_1, Δ_2 simplicial complexes whose vertex sets are contained in $\mathbf{x}$. Prove that there is an exact sequence of R-modules:

$$0 \longrightarrow K[\Delta_1 \cup \Delta_2] \longrightarrow K[\Delta_1] \oplus K[\Delta_2] \longrightarrow K[\Delta_1 \cap \Delta_2] \longrightarrow 0.$$

6.3.63 Prove that any constructible simplicial complex is Cohen–Macaulay.

6.3.64 (Reisner) Let $R = K[a, \ldots, f]$ be a polynomial ring over a field K and let Δ be the following triangulation of the real projective plane $\mathbb{P}^2$:

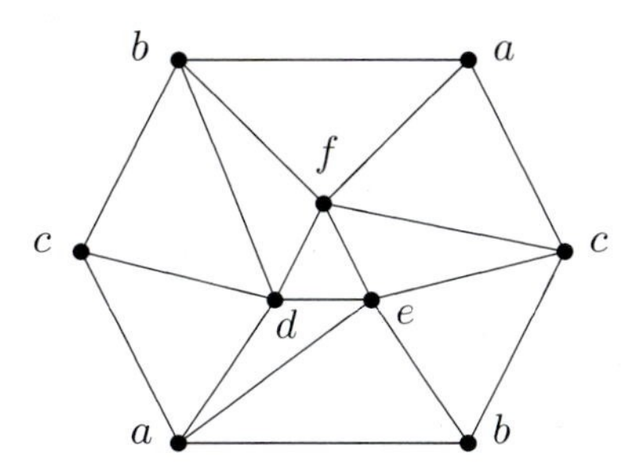

Show that $I_\Delta = (abc, abd, ace, adf, aef, bcf, bde, bef, cde, cdf)$. The ring $K[\Delta]$ is Cohen–Macaulay and has a 3-linear resolution if K has characteristic other than 2, and is not Cohen–Macaulay and has a non-linear resolution otherwise. Prove that Δ is not pure shellable and $\dim \Delta = 2$.

The figure below gives a subdivision of the minimal triangulation of the projective plane given above (cf. Reisner [346, Remark 3]).

6.3.65 (N. Terai) If K is a field and Δ is the following triangulation of the real projective plane $\mathbb{P}^2$

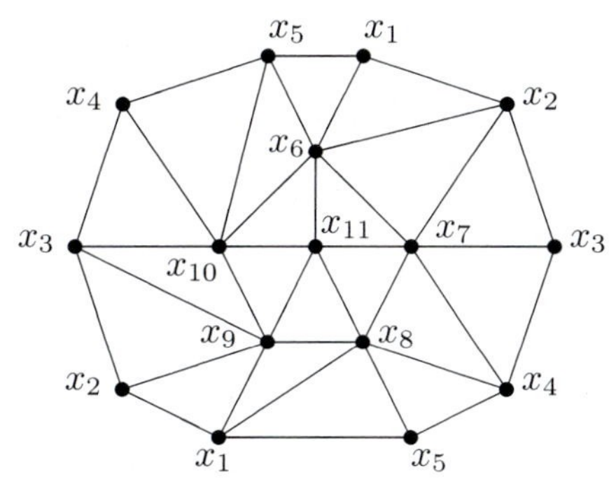

then the link of any vertex is a cycle and $K[\Delta]$ is Cohen–Macaulay if and only if $\mathrm{char}(K) \neq 2$.

Hint $\widetilde{H}_1(\Delta; K) \simeq \widetilde{H}_1(\mathbb{P}^2; K) = K/2K$ and use Theorem 6.3.12.

6.3.66 [228] Let Δ be a simplicial complex with vertex set X and facets $F_1, \ldots, F_s$. Then, Δ is shellable if and only if the ideal $(x_{F_1^c}, \ldots, x_{F_s^c})$ has linear quotients, where $F_i^c = X \setminus F_i$ and $x_{F^c} = \prod_{x_i \in F^c} x_i$.

6.3.67 Let Δ be a simplicial complex and let v be a vertex. The following are equivalent: (i) v is a shedding vertex, i.e., every facet of $\mathrm{del}_\Delta(v)$ is a facet of Δ, (ii) No face of $\mathrm{lk}_\Delta(v)$ is a facet of $\mathrm{del}_\Delta(v)$.

6.3.68 Let Δ be a simplicial complex on the vertex set $V = \{x_1, \ldots, x_n\}$ and $\Delta^\vee$ its Alexander dual. Prove that the Stanley–Reisner ideal of $\Delta^\vee$ is equal to the ideal of minimal covers of $I = I_\Delta$, that is:

$$I_{\Delta^\vee} = (\{x_{i_1} \cdots x_{i_r} \mid (x_{i_1}, \ldots, x_{i_r}) \text{ is a minimal prime of } I\}) = I^\vee.$$

6.4 Regularity and projective dimension

Let $R = K[x_1, \ldots, x_n]$ be a polynomial ring over a field K with the standard grading and let $\mathcal{C}$ be a clutter with vertex set $X = \{x_1, \ldots, x_n\}$. In this section we study the regularity, depth, and projective dimension of $R/I(\mathcal{C})$, where $I = I(\mathcal{C})$ is the edge ideal of $\mathcal{C}$. There are several well-known results relating these invariants. We collect some of them here for ease of reference.

The first result is a basic relation between the dimension and the depth (see Lemma 2.3.6):

$$\mathrm{depth}\, R/I(\mathcal{C}) \leq \dim R/I(\mathcal{C}). \tag{6.1}$$

The deviation from equality in the above relationship can be quantified using the projective dimension, as is seen in a formula discovered by Auslander and Buchsbaum (see Theorem 3.5.13):

$$\mathrm{pd}_R(R/I(\mathcal{C})) + \mathrm{depth}\, R/I(\mathcal{C}) = \dim(R). \tag{6.2}$$

Another invariant of interest also follows from a closer inspection of a minimal projective resolution of R/I. Consider the minimal graded free resolution of $M = R/I$ as an R-module:

$$\mathbb{F}_\star : \quad 0 \to \bigoplus_j R(-j)^{b_{gj}} \to \cdots \to \bigoplus_j R(-j)^{b_{1j}} \to R \to R/I \to 0.$$

The *Castelnuovo–Mumford regularity* of M (*regularity* of M for short) is defined as

$$\mathrm{reg}(M) = \max\{j - i \mid b_{ij} \neq 0\}.$$

The a-invariant, the regularity, and the depth of M are closely related.

Theorem 6.4.1 [413, Corollary B.4.1] *$a(M) \leq \mathrm{reg}(M) - \mathrm{depth}(M)$, with equality if M is Cohen–Macaulay.*

An excellent reference for the regularity of graded ideals is the book of Eisenbud [129]. There are methods to compute the regularity of R/I avoiding the construction of a minimal graded free resolution; see [31, 32] and [200, p. 614]. These methods work for any homogeneous ideal over an arbitrary field.

Theorem 6.4.2 [401] *Let $\mathcal{C}$ be a clutter. If* $\mathrm{ht}(I(\mathcal{C})) \geq 2$, *then*

$$\operatorname{reg} I(\mathcal{C}) = 1 + \operatorname{reg} R/I(\mathcal{C}) = \operatorname{pd} R/I_c(\mathcal{C}).$$

This formula also holds for edge ideals of height one:

Corollary 6.4.3 *If* $\mathrm{ht}(I(\mathcal{C})) = 1$, *then* $\operatorname{reg} R/I(\mathcal{C}) = \operatorname{pd} R/I_c(\mathcal{C}) - 1$.

Proof. We set $I = I(\mathcal{C})$. The formula clearly holds if I is a principal ideal. Assume that I is not principal. By permuting variables, we may assume that the primary decomposition of I has the form

$$I = (x_1) \cap \cdots \cap (x_r) \cap \mathfrak{p}_1 \cap \cdots \cap \mathfrak{p}_m,$$

where $L = \mathfrak{p}_1 \cap \cdots \cap \mathfrak{p}_m$ is an edge ideal of height at least 2. Notice that $I = fL$, where $f = x_1 \cdots x_r$. Then, the Alexander dual of I is

$$I^\vee = (x_1, \ldots, x_r) + L^\vee.$$

The multiplication map $L[-r] \overset{f}{\to} fL$ induces an isomorphism of graded R-modules. Thus $\operatorname{reg}(L[-r]) = r + \operatorname{reg}(L) = \operatorname{reg}(I)$. By the Auslander–Buchsbaum formula, one has the equality $\operatorname{pd}(R/I^\vee) = r + \operatorname{pd}(R/L^\vee)$. Therefore, using Theorem 6.4.2, we get

$$\operatorname{reg}(R/I) = \operatorname{reg}(R/L) + r = (\operatorname{pd}(R/L^\vee) - 1) + r = \operatorname{pd}(R/I^\vee) - 1. \quad \Box$$

Proposition 6.4.4 [434, Lemma 7] *Let $R_1 = K[\mathbf{x}]$ and $R_2 = K[\mathbf{y}]$ be two polynomial rings over a field K and let $R = K[\mathbf{x}, \mathbf{y}]$. If I_1 and I_2 are edge ideals of R_1 and R_2, respectively, then*

$$\operatorname{reg} R/(I_1R + I_2R) = \operatorname{reg}(R_1/I_1) + \operatorname{reg}(R_2/I_2).$$

Proof. By abuse of notation, we will write I_i in place of I_iR for $i = 1, 2$ when it is clear from context that we are using the generators of I_i but extending to an ideal of the larger ring. Let $\mathbf{x} = \{x_1, \ldots, x_n\}$ and $\mathbf{y} = \{y_1, \ldots, y_m\}$ be two disjoint sets of variables. Notice that

$$(I_1 + I_2)^\vee = I_1^\vee I_2^\vee = I_1^\vee \cap I_2^\vee$$

where $I_i^\vee$ is the Alexander dual of I_i. Hence, by Theorem 6.4.2 and using the Auslander–Buchsbaum formula, we need only show the equality

$$\operatorname{depth}(R/(I_1^\vee \cap I_2^\vee)) = \operatorname{depth}(R_1/I_1^\vee) + \operatorname{depth}(R_2/I_2^\vee) + 1.$$

The equality $\text{depth}(R/(I_1^\vee + I_2^\vee)) = \text{depth}(R_1/I_1^\vee) + \text{depth}(R_2/I_2^\vee)$ holds thanks to Theorem 3.1.34. Thus the proof reduces to showing the equality

$$\text{depth}(R/(I_1^\vee \cap I_2^\vee)) = \text{depth}(R/(I_1^\vee + I_2^\vee)) + 1. \tag{6.3}$$

We may assume that $\text{depth}(R/I_1^\vee) \geq \text{depth}(R/I_2^\vee)$. There is an exact sequence of graded R-modules:

$$0 \longrightarrow R/(I_1^\vee \cap I_2^\vee) \stackrel{\varphi}{\longrightarrow} R/I_1^\vee \oplus R/I_2^\vee \stackrel{\phi}{\longrightarrow} R/(I_1^\vee + I_2^\vee) \longrightarrow 0, \tag{6.4}$$

where $\varphi(\overline{r}) = (\overline{r}, -\overline{r})$ and $\phi(\overline{r}_1, \overline{r}_2) = \overline{r_1 + r_2}$. From the inequality

$$\begin{aligned} \text{depth}(R/I_1^\vee \oplus R/I_2^\vee) &= \max\{\text{depth}(R/I_i^\vee)\}_{i=1}^2 = \text{depth}(R/I_1^\vee) \\ &= \text{depth}(R_1/I_1^\vee) + m \\ &> \text{depth}(R_1/I_1^\vee) + \text{depth}(R_2/I_2^\vee) \\ &= \text{depth}(R/(I_1^\vee + I_2^\vee)) \end{aligned}$$

and applying the depth lemma to Eq. (6.4), we obtain Eq. (6.3). $\square$

Definition 6.4.5 Let S be a set of vertices of a clutter $\mathcal{C}$. The *induced subclutter* on S, denoted by $\mathcal{C}[S]$, is the maximal subclutter of $\mathcal{C}$ with vertex set S. A clutter of the form $\mathcal{C}[S]$ for some $S \subset V(\mathcal{C})$ is called an *induced subclutter* of $\mathcal{C}$.

Thus, the vertex set of $\mathcal{C}[S]$ is S and the edges of $\mathcal{C}[S]$ are exactly the edges of $\mathcal{C}$ contained in S. Notice that $\mathcal{C}[S]$ may have isolated vertices, i.e., vertices that do not belong to any edge of $\mathcal{C}[S]$. If $\mathcal{C}$ is a discrete clutter, i.e., all the vertices of $\mathcal{C}$ are isolated, we set $I(\mathcal{C}) = 0$ and $\alpha_0(\mathcal{C}) = 0$.

Proposition 6.4.6 *If $\mathcal{D}$ is an induced subclutter of $\mathcal{C}$, then*

$$\text{reg}(R/I(\mathcal{D})) \leq \text{reg}(R/I(\mathcal{C})).$$

Proof. There is $S \subset V(\mathcal{C})$ such that $\mathcal{D} = \mathcal{C}[S]$. Let $\mathfrak{p}$ be the prime ideal of R generated by S. By duality (see Theorem 6.3.39), we have

$$I_c(\mathcal{C}) = \bigcap_{e \in E(\mathcal{C})} (e) \Longrightarrow I_c(\mathcal{C})_\mathfrak{p} = \bigcap_{e \in E(\mathcal{C})} (e)_\mathfrak{p} = \bigcap_{e \in E(\mathcal{D})} (e)_\mathfrak{p} = I_c(\mathcal{D})_\mathfrak{p}.$$

Therefore, using Theorem 6.4.2 and Exercise 6.4.30, we get

$$\begin{aligned} \text{reg}(R/I(\mathcal{C})) &= \text{pd}(R/I_c(\mathcal{C})) - 1 \\ &\geq \text{pd}(R_\mathfrak{p}/I_c(\mathcal{C})_\mathfrak{p}) - 1 = \text{pd}(R_\mathfrak{p}/I_c(\mathcal{D})_\mathfrak{p}) - 1 \\ &= \text{pd}(R'/I_c(\mathcal{D})) - 1 = \text{pd}(R/I_c(\mathcal{D})) - 1 = \text{reg}(R/I(\mathcal{D})), \end{aligned}$$

where $R' = K[S]$. Thus, $\text{reg}(R/I(\mathcal{C})) \geq \text{reg}(R/I(\mathcal{D}))$. $\square$

Definition 6.4.7 An *induced matching* in a clutter $\mathcal{C}$ is a set of pairwise disjoint edges $f_1, \ldots, f_r$ such that the only edges of $\mathcal{C}$ contained in $\cup_{i=1}^r f_i$ are $f_1, \ldots, f_r$. The *induced matching number*, denoted by $\mathrm{im}(\mathcal{C})$, is the number of edges in the largest induced matching.

Corollary 6.4.8 *Let $\mathcal{C}$ be a clutter and let $f_1, \ldots, f_r$ be an induced matching of $\mathcal{C}$ with $d_i = |f_i|$ for $i = 1, \ldots, r$. Then*

$$\left(\sum_{i=1}^{r} d_i\right) - r \leq \mathrm{reg}(R/I(\mathcal{C})).$$

Proof. Let $\mathcal{D} = \mathcal{C}[\cup_{i=1}^r f_i]$. Notice that $I(\mathcal{D}) = (x_{f_1}, \ldots, x_{f_r})$. Thus $I(\mathcal{D})$ is a complete intersection and the regularity of $R/I(\mathcal{D})$ is the degree of its h-polynomial. The Hilbert series of $R/I(\mathcal{D})$ is given by

$$F(t) = \frac{\prod_{i=1}^r (1 + t + \cdots + t^{d_i - 1})}{(1-t)^{n-r}}.$$

Thus, the degree of the h-polynomial equals $(\sum_{i=1}^r d_i) - r$. Therefore, the inequality follows from Proposition 6.4.6. $\Box$

Corollary 6.4.9 *If $\mathcal{C}$ is a clutter and $R/I_c(\mathcal{C})$ is Cohen–Macaulay, then* $\mathrm{im}(\mathcal{C}) = 1$.

Proof. Let r be the induced matching number of $\mathcal{C}$ and let d be the cardinality of any edge of $\mathcal{C}$. Using Theorem 6.4.2 and Corollary 6.4.8, we obtain $d - 1 \geq r(d-1)$. Thus $r = 1$, as required. $\Box$

Lemma 6.4.10 [128, Corollary 20.19] *If $0 \to N \to M \to L \to 0$ is a short exact sequence of graded finitely generated R-modules, then*

(a) $\mathrm{reg}(N) \leq \max(\mathrm{reg}(M), \mathrm{reg}(L) + 1)$.

(b) $\mathrm{reg}(M) \leq \max(\mathrm{reg}(N), \mathrm{reg}(L))$.

(c) $\mathrm{reg}(L) \leq \max(\mathrm{reg}(N) - 1, \mathrm{reg}(M))$.

The following result was shown by Kalai and Meshulam for square-free monomial ideals and by Herzog for arbitrary monomial ideals.

Proposition 6.4.11 [269, 219] *If I_1, I_2 are monomial ideals of R. Then*

(a) $\mathrm{reg}\, R/(I_1 + I_2) \leq \mathrm{reg}(R/I_1) + \mathrm{reg}(R/I_2)$,

(b) $\mathrm{reg}\, R/(I_1 \cap I_2) \leq \mathrm{reg}(R/I_1) + \mathrm{reg}(R/I_2) + 1$.

Corollary 6.4.12 *If* $\mathcal{C}_1,\ldots,\mathcal{C}_s$ *are clutters on the vertex set* X*, then*

$$\operatorname{reg}(R/I(\cup_{i=1}^{s}\mathcal{C}_i)) \leq \operatorname{reg}(R/I(\mathcal{C}_1)) + \cdots + \operatorname{reg}(R/I(\mathcal{C}_s)).$$

Proof. The set of edges of $\mathcal{C} = \cup_{i=1}^{s}\mathcal{C}_i$ is $\cup_{i=1}^{s}E(\mathcal{C}_i)$. By Proposition 6.4.11, it suffices to notice the equality $I(\cup_{i=1}^{s}\mathcal{C}_i) = \sum_{i=1}^{s} I(\mathcal{C}_i)$. □

A clutter $\mathcal{C}$ is called *co-CM* if $I_c(\mathcal{C})$ is Cohen–Macaulay. A co-CM clutter is uniform because Cohen–Macaulay clutters are unmixed.

Corollary 6.4.13 *If* $\mathcal{C}_1,\ldots,\mathcal{C}_s$ *are co-CM clutters with vertex set* X*, then*

$$\operatorname{reg}(R/I(\cup_{i=1}^{s}\mathcal{C}_i)) \leq (d_1-1) + \cdots + (d_s-1),$$

where d_i *is the number of elements in any edge of* $\mathcal{C}_i$*.*

Proof. By Theorem 6.4.2, we get that $\operatorname{reg} R/I(\mathcal{C}_i) = d_i - 1$ for all i. Thus the result follows from Corollary 6.4.12. □

Definition 6.4.14 Let $\mathcal{C}$ be a clutter. We will denote the cardinality of a smallest maximal independent set of $\mathcal{C}$ by $\beta_0'(\mathcal{C})$. This number is called the *small independence number* of $\mathcal{C}$.

Lemma 6.4.15 *Let* $\mathcal{C}$ *be a clutter and let* $\Delta = \Delta_{\mathcal{C}}$ *be its independence complex. Then* $\Delta^{[i]} = \Delta^i$ *for* $i \leq \beta_0'(\mathcal{C}) - 1$*.*

Proof. First we prove the inclusion $\Delta^{[i]} \subset \Delta^i$. Let F be a face of $\Delta^{[i]}$. Then F is contained in a face of Δ of dimension i, and so F is in Δ^i. Conversely, let F be a face of Δ^i. Then

$$\dim(F) \leq i \leq \beta_0'(\mathcal{C}) - 1 \implies |F| \leq i+1 \leq \beta_0'(\mathcal{C}).$$

Since $\beta_0'(\mathcal{C})$ is the cardinality of any smallest maximal independent set of $\mathcal{C}$, we can extend F to an independent set of $\mathcal{C}$ with $i+1$ vertices. Thus F is in $\Delta^{[i]}$. □

Let us recall the following expression for the depth of $K[\Delta_{\mathcal{C}}]$; a simple proof can be found in [172].

Theorem 6.4.16 [388] *Let* $\mathcal{C}$ *be a clutter and let* $\Delta = \Delta_{\mathcal{C}}$ *be its independence complex. Then*

$$\operatorname{depth} R/I(\mathcal{C}) = 1 + \max\{i \mid K[\Delta^i] \text{ is Cohen–Macaulay}\},$$

where $\Delta^i = \{F \in \Delta \mid \dim(F) \leq i\}$ *is the* i*-skeleton and* $-1 \leq i \leq \dim(\Delta)$*.*

Definition 6.4.17 Let $\mathcal{C}$ be a clutter. The cardinality of a largest minimal vertex cover of $\mathcal{C}$, denoted by $\operatorname{bight}(I(\mathcal{C}))$, is called the *big height* of $I(\mathcal{C})$.

Theorem 6.4.18 [326] *Let $\mathcal{C}$ be a clutter with n vertices. Then*

(a) $\operatorname{reg} R/I_c(\mathcal{C}) \geq \operatorname{bight}(I(\mathcal{C})) - 1$,

(b) $\operatorname{pd}_R(R/I(\mathcal{C})) \geq \operatorname{bight}(I(\mathcal{C}))$,

(c) $\operatorname{depth} R/I(\mathcal{C}) \leq n - \operatorname{bight}(I(\mathcal{C}))$,

with equality everywhere if $R/I(\mathcal{C})$ is sequentially Cohen–Macaulay.

Proof. (a): We set $\alpha_0'(\mathcal{C}) = \operatorname{bight}(I(\mathcal{C}))$. Note that $\beta_0'(\mathcal{C}) = n - \alpha_0'(\mathcal{C})$. By Theorem 6.4.2 and the Auslander–Buchsbaum formula, the proof reduces to showing: $\operatorname{depth} R/I(\mathcal{C}) \leq \beta_0'(\mathcal{C})$, with equality if $R/I(\mathcal{C})$ is sequentially Cohen–Macaulay.

First we show $\operatorname{depth} R/I(\mathcal{C}) \leq \beta_0'(\mathcal{C})$. Assume Δ^i is Cohen–Macaulay for some $-1 \leq i \leq \dim(\Delta)$, where Δ is the independence complex of $\mathcal{C}$. By Theorem 6.4.16, it suffices to prove that $1 + i \leq \beta_0'(\mathcal{C})$. We can pick a maximal independent set F of $\mathcal{C}$ with $\beta_0'(\mathcal{C})$ vertices. Since Δ^i is Cohen–Macaulay all maximal faces of Δ have dimension i. If $1 + i > \beta_0'(\mathcal{C})$, then F is a maximal face of Δ^i of dimension $\beta_0'(\mathcal{C}) - 1$, a contradiction.

Assume that $R/I(\mathcal{C})$ is sequentially Cohen–Macaulay. By Lemma 6.4.15 $\Delta^{[i]} = \Delta^i$ for $i \leq \beta_0'(\mathcal{C}) - 1$. Then by Theorem 6.3.30, the ring $K[\Delta^i]$ is Cohen–Macaulay for $i \leq \beta_0'(\mathcal{C}) - 1$. Therefore, applying Theorem 6.4.16, we get that the depth of $R/I(\mathcal{C})$ is at least $\beta_0'(\mathcal{C})$. Consequently, in this case one has the equality $\operatorname{depth} R/I(\mathcal{C}) = \beta_0'(\mathcal{C})$.

(b): It follows from the proof of (a).

(c): It follows from (b) and the Auslander–Buchsbaum formula. $\square$

The inequality in part (a) of Theorem 6.4.18 also follows directly from the definition of regularity because $\operatorname{reg}(I_c(\mathcal{C}))$ is an upper bound for the largest degree of a minimal generator of $I_c(\mathcal{C})$.

Definition 6.4.19 Let I be a monomial ideal of R. The *big height* of I, denoted by $\operatorname{bight}(I)$, is $\max\{\operatorname{ht}(\mathfrak{p}) \mid \mathfrak{p} \in \operatorname{Ass}(R/I)\}$.

The following result underlines the advantage of viewing a square-free monomial ideal as the edge ideal of a clutter. Herzog pointed out that this result holds for any sequentially Cohen–Macaulay module, as is seen below.

Corollary 6.4.20 [326] *If I is a monomial ideal and R/I is sequentially Cohen–Macaulay, then* $\operatorname{pd}_R(R/I) = \operatorname{bight}(I)$.

Proof. Let $I' \subset R'$ be the polarization of $I \subset R$. Since R/I and R'/I' have the same projective dimension and $\operatorname{bight}(I) = \operatorname{bight}(I')$, we may assume that I is a square-free monomial ideal. As any square-free monomial ideal is the edge ideal of a clutter. The formula follows from Theorem 6.4.18. $\square$

This formula can be applied to a wide variety of square-free monomial ideals as is seen below.

Definition 6.4.21 A clutter $\mathcal{C}$ is called *sequentially Cohen–Macaulay* if $R/I(\mathcal{C})$ is sequentially Cohen–Macaulay.

Theorem 6.4.22 *The following clutters are sequentially Cohen–Macaulay:*

(a) [432] *graphs with no chordless cycles of length other than* 3 *or* 5*,*

(b) [228] *clutters whose ideal of covers has linear quotients,*

(c) [211] *clutters of paths of length* t *of directed rooted trees,*

(d) [155] *totally balanced clutters,*

(e) [205] *uniform admissible clutters whose covering number is* 3.

A clutter $\mathcal{C}$ is called *shellable* if $\Delta_{\mathcal{C}}$ is shellable. The clutters of parts (a)–(e) are in fact shellable, and the clutters of part (a) are in fact vertex decomposable; see [113, 228, 407, 409, 433, 432]. The families of clutters in (a)–(e) will be studied later in this book (see Chapter 7 and the index).

Theorem 6.4.23 *If* M *is a sequentially Cohen–Macaulay* R*-module, then*

$$\mathrm{pd}_R(M) = \max\{\mathrm{ht}(\mathfrak{p})|\, \mathfrak{p} \in \mathrm{Ass}_R(M)\}.$$

Proof. By Theorem 3.5.13 and [413, Proposition A.7.3] one has

$$\mathrm{pd}_R(M) = \dim(R) - \mathrm{depth}(M) = \sup\{r \,|\, \mathrm{Ext}^r_R(M,R) \neq 0\}, \tag{6.5}$$

for the second equality see also [65, Exercise 3.1.24, p. 96]. Since M is sequentially Cohen–Macaulay, there is a filtration of graded R-modules

$$0 = M_0 \subset M_1 \subset \cdots \subset M_r = M$$

such that each M_i/M_{i-1} is C–M and $d_1 < \cdots < d_r$, where d_i is equal to $\dim(M_i/M_{i-1})$ for $i = 1, \ldots, r$. By [231, Proposition 2.5] one has

$$\mathrm{Ass}_R(M_i/M_{i-1}) = \{\mathfrak{p} \in \mathrm{Ass}_R(M) \,|\, \dim(R/\mathfrak{p}) = d_i\}.$$

In particular $\mathrm{Ass}_R(M) = \cup_i \mathrm{Ass}_R(M_i/M_{i-1})$. Hence

$$n - d_1 = \max\{\mathrm{ht}(\mathfrak{p})|\, \mathfrak{p} \in \mathrm{Ass}_R(M)\},$$

where $n = \dim(R)$. By [231, Proposition 2.2], $\mathrm{Ext}^{n-d_i}_R(M,R)$ is C–M of dimension d_i for al i, $\mathrm{Ext}^j_R(M,R) = (0)$ for $j \notin \{n-d_1, \ldots, n-d_r\}$ and

$$\mathrm{Ext}^{n-d_i}_R(\mathrm{Ext}^{n-d_i}_R(M,R),R) \simeq M_i/M_{i-1}$$

for all i. Hence, by Eq. (6.5), we get $\mathrm{pd}_R(M) = n - d_1$. $\Box$

Definition 6.4.24 The number of maximal independent sets of $\mathcal{C}$, denoted by arith-deg$(I(\mathcal{C}))$, is called the *arithmetic degree* of $I(\mathcal{C})$.

The *arithmetic rank* of an ideal I, denoted by $\text{ara}(I)$, is the least number of elements of R which generate the ideal I up to radical. By Krull's principal ideal theorem $\text{ara}(I) \geq \text{ht}(I)$. If equality occurs, I is called a *set-theoretic complete intersection.*

Theorem 6.4.25 [401] *If* $\text{ht}(I(\mathcal{C})) \geq 2$, *then* $\text{reg}(I(\mathcal{C})) \leq \text{arith-deg}(I(\mathcal{C}))$.

Theorem 6.4.26 [301, Proposition 3] $\text{pd}(R/I(\mathcal{C})) \leq \text{ara}(I(\mathcal{C}))$.

There are some instances where the equality $\text{pd}(R/I) = \text{ara}(I)$ holds; see [15, 16, 143, 279] and the references therein. Barile [15] conjectured that the equality holds for edge ideals of forests. This conjecture was recently shown by Kimura and Terai [278]. Since edge ideals of forests are sequentially Cohen–Macaulay (see Theorem 6.5.25), by Theorem 6.4.18, it follows that $\text{bight}(I) = \text{ara}(I)$ if I is the edge ideal of a forest [278].

Conjecture 6.4.27 (Eisenbud–Goto [130]) If $\mathfrak{p} \subset (x_1, \ldots, x_n)^2$ is a prime graded ideal, then $\text{reg}(R/\mathfrak{p}) \leq \deg(R/\mathfrak{p}) - \text{codim}(R/\mathfrak{p})$.

The following gives a partial answer to the monomial version of the Eisenbud–Goto regularity conjecture.

Theorem 6.4.28 [401] *Let I be the edge ideal of a clutter $\mathcal{C}$. If $\Delta_{\mathcal{C}}$ is connected in codimension* 1, *then* $\text{reg}(R/I) \leq \deg(R/I) - \text{codim}(R/I)$.

The *multiplicity* or *degree* of the edge-ring $R/I(\mathcal{C})$ is equal to the number of independent sets of $\mathcal{C}$ with $\beta_0(\mathcal{C})$ vertices (see Exercise 6.7.10).

Theorem 6.4.29 *If $\mathcal{C}$ is a k-uniform clutter, then* $\deg(R/I(\mathcal{C})) \leq k^{\alpha_0(\mathcal{C})}$.

Proof. Let I be the edge ideal of $\mathcal{C}$. We set $r = \alpha_0(\mathcal{C})$. One may assume that K is an infinite field by considering a field extension $K \subset L$, with L an infinite field, and using the functor $(\cdot) \otimes_K L$. It is not hard to see that I is minimally generated by forms $f_1, \ldots, f_q$ of degree k such that $f_1, \ldots, f_r$ is a regular sequence (see Exercise 3.1.40). Consider the subideal $I' = (f_1, \ldots, f_r)$. There is a graded epimorphism $R/I' \to R/I \to 0$, where $\dim(R/I') = \dim(R/I)$. Therefore $\deg(R/I) \leq \deg(R/I')$. On the other hand by Exercise 5.1.20 the Hilbert series of R/I' is given by

$$F(R/I', t) = \frac{(1 + t + \cdots + t^{k-1})^r}{(1 - t)^{n-r}}.$$

Making $t = 1$ in the numerator and using Remark 5.1.7 yields the required bound. $\Box$

Exercises

6.4.30 Let $R = K[x_1, \ldots, x_n]$ and I an ideal of R. If $I \subset (x_1, \ldots, x_{n-1})$, and $R' = R/(x_n) \cong K[x_1, \ldots, x_{n-1}]$, then $\mathrm{reg}(R/I) = \mathrm{reg}(R'/I)$ and $\mathrm{pd}_R(R/I) = \mathrm{pd}_{R'}(R'/I)$. Similarly, if $x_n \in I$ and $I' = I/(x_n)$, then $\mathrm{reg}(R/I) = \mathrm{reg}(R'/I')$ and $\mathrm{pd}_R(R/I) = \mathrm{pd}_{R'}(R'/I') + 1$.

6.4.31 Let $\mathcal{C}$ be a clutter. If $\alpha_0'(\mathcal{C}) := \mathrm{bight}(I(\mathcal{C}))$, then

$$\alpha_0'(\mathcal{C}) = \max\{|e| : e \in E(\mathcal{C}^\vee)\} \text{ and } \alpha_0'(\mathcal{C}^\vee) = \max\{|e| : e \in E(\mathcal{C})\}.$$

6.4.32 Let $\mathcal{C}$ be a clutter. If $I(\mathcal{C})$ has linear quotients, then

$$\mathrm{reg}\, R/I(\mathcal{C}) = \max\{|e| : e \in E(\mathcal{C})\} - 1.$$

6.4.33 Let $I \subset R$ be a monomial ideal and let $I' \subset R'$ be its polarization. Prove that the following equalities hold:

$$\mathrm{pd}(R/I) = \mathrm{pd}(R'/I'),\ \mathrm{reg}(R/I) = \mathrm{reg}(R'/I') \text{ and } \mathrm{bight}(I) = \mathrm{bight}(I').$$

6.5 Unmixed and shellable clutters

In this section we study unmixed, Cohen–Macaulay and shellable clutters using some notions that come from combinatorial optimization. We are interested in determining what families of clutters have these properties.

Let $\mathcal{C}$ be a clutter with vertex set $X = \{x_1, \ldots, x_n\}$, let $I = I(\mathcal{C})$ be its edge ideal, and let $\Delta_{\mathcal{C}}$ be its independence complex. We shall always assume that $\mathcal{C}$ has no *isolated vertices*, i.e., each vertex occurs in at least one edge.

Definition 6.5.1 If $\Delta_{\mathcal{C}}$ is pure (resp. Cohen–Macaulay, shellable), we say that $\mathcal{C}$ is *unmixed* (resp. *Cohen–Macaulay*, *shellable*).

The following notions of *contraction*, *deletion*, and *minor* come from combinatorial optimization [373].

Definition 6.5.2 For $x_i \in X$, the *contraction* $\mathcal{C}/x_i$ and *deletion* $\mathcal{C} \setminus x_i$ are the clutters constructed as follows: both have $X \setminus \{x_i\}$ as vertex set, $E(\mathcal{C}/x_i)$ is the set of minimal elements of $\{e \setminus \{x_i\} | \, e \in E(\mathcal{C}\}$, minimal w.r.t to inclusion, and $E(\mathcal{C} \setminus x_i)$ is the set $\{e | \, x_i \notin e \in E(\mathcal{C})\}$.

The edge ideals of a deletion and a contraction have a nice algebraic interpretation. For $x_i \in X$, define the *contraction* and *deletion* of I as the ideals:

$$(I \colon x_i) \text{ and } I^c = I \cap K[x_1, \ldots, \widehat{x}_i, \ldots, x_n],$$

respectively. Notice that the clutter associated to the square-free monomial ideal $(I \colon x_i)$ (resp. I^c) is the contraction $\mathcal{C}/x_i$ (resp. deletion $\mathcal{C} \setminus x_i$), i.e., the edge ideals of $\mathcal{C}/x_i$ and $\mathcal{C} \setminus x_i$ are $(I \colon x_i)$ and I^c, respectively.

Definition 6.5.3 A *minor* of a clutter $\mathcal{C}$ is a clutter obtained from $\mathcal{C}$ by a sequence of deletions and contractions in any order.

A *c-minor* of I is any ideal generated by the set of monomials obtained from $G(I)$, the minimal set of generators of I, by making any sequence of variables equal to 1 in the monomials of $G(I)$. If a c-minor I' contains a variable x_i and we remove this variable from I', we still consider the new ideal a c-minor of I. A *c-minor* of $\mathcal{C}$ is any clutter that corresponds to a c-minor of I.

Proposition 6.5.4 *Let $C_1, \ldots, C_s$ be the minimal vertex covers of a clutter $\mathcal{C}$. Then the primary decomposition of $I(\mathcal{C})$ is $(C_1) \cap (C_2) \cap \cdots \cap (C_s)$, and the facets of $\Delta_{\mathcal{C}}$ are $X \setminus C_1, \ldots, X \setminus C_s$.*

Proof. This follows from Theorem 6.3.39. □

Corollary 6.5.5 *$\Delta_{\mathcal{C}}$ is pure if and only if all minimal vertex covers of $\mathcal{C}$ have the same cardinality.*

Definition 6.5.6 A *perfect matching of König type* of $\mathcal{C}$ is a collection $e_1, \ldots, e_g$ of pairwise disjoint edges whose union is X and such that g is the height of $I(\mathcal{C})$.

A set of pairwise disjoint edges is called *independent* or a *matching* and a set of independent edges of $\mathcal{C}$ whose union is X is called a *perfect matching*. A clutter $\mathcal{C}$ satisfies the *König property* if the maximum number of independent edges of $\mathcal{C}$ equals the height of $I(\mathcal{C})$. For uniform clutters, it is easy to check that if $\mathcal{C}$ has the König property and a perfect matching, then the perfect matching is of König type.

A vertex x of $\mathcal{C}$ is called *isolated* if x does not occur in any edge of $\mathcal{C}$. A clutter with a perfect matching of König type has the König property. Next we show the converse to be true for unmixed clutters.

Lemma 6.5.7 *If $\mathcal{C}$ is an unmixed clutter with the König property and without isolated vertices, then $\mathcal{C}$ has a perfect matching of König type.*

Proof. Let X be the vertex set of $\mathcal{C}$. By hypothesis there are $e_1, \ldots, e_g$ independent edges of $\mathcal{C}$, where g is the height of $I(\mathcal{C})$. If $e_1 \cup \cdots \cup e_g \subsetneq X$, pick $x_r \in X \setminus (e_1 \cup \cdots \cup e_g)$. Since the vertex x_r occurs in some edge of $\mathcal{C}$, there is a minimal vertex cover C containing x_r. Thus using that $e_1, \ldots, e_g$ are mutually disjoint we conclude that C contains at least $g + 1$ vertices, a contradiction. □

Proposition 6.5.8 *Let $\mathcal{C}$ be an unmixed clutter with a perfect matching $e_1, \ldots, e_g$ of König type and let $C_1, \ldots, C_r$ be any collection of minimal vertex covers of $\mathcal{C}$. If $\mathcal{C}'$ is the clutter associated to $I' = \cap_{i=1}^{r}(C_i)$, then $\mathcal{C}'$ has a perfect matching $e_1', \ldots, e_g'$ of König type such that:* (a) *$e_i' \subset e_i$ for all i, and* (b) *every vertex of $e_i \setminus e_i'$ is isolated in $\mathcal{C}'$.*

Proof. We denote the minimal set of generators of the ideal $I = I(\mathcal{C})$ by $G(I)$. There are monomials $x^{v_1}, \ldots, x^{v_g}$ in $G(I)$ so that $\text{supp}(x^{v_i}) = e_i$ for $i = 1, \ldots, g$. Since x^{v_i} is in I and $I \subset I'$, there is $e_i' \subset e_i$ such that e_i' is an edge of $\mathcal{C}'$. Let x be any vertex in $e_i \setminus e_i'$. If x is not isolated in $\mathcal{C}'$, there would a minimal vertex cover C_k of $\mathcal{C}'$ containing x. As C_k contains a vertex of e_j' for each $1 \leq j \leq g$ and since $e_1', \ldots, e_g'$ are pairwise disjoint, we get that C_k contains at least $g+1$ vertices, a contradiction. Thus (a) and (b) are satisfied. Clearly g is the height of I' by construction of I'. Let X' be the vertex set of $\mathcal{C}'$. To finish the proof we need only show that $X' = e_1' \cup \cdots \cup e_g'$. Let $x \in X'$, then $x \in e_i$ for some i and x belongs to at least one edge of $\mathcal{C}'$. By part (b) we get that $x \in e_i'$, as required. $\Box$

Corollary 6.5.9 *Let $\mathcal{C}$ be an unmixed clutter. If $\mathcal{C}$ has a perfect matching $\{e_i\}_{i=1}^g$ of König type, then $\mathcal{C}/x_j$ has a perfect matching $\{e_i'\}_{i=1}^g$ of König type such that:* (a) *$e_i' \subset e_i$ for all i, and* (b) *every vertex of $e_i \setminus e_i'$ is isolated in $\mathcal{C}/x_j$.*

Proof. Let $C_1, \ldots, C_s$ be the minimal vertex covers of $\mathcal{C}$. Since $I(\mathcal{C})$ is equal to $\cap_{i=1}^s (C_i)$, one has $(I(\mathcal{C})\colon x_j) = \cap_{x_j \notin C_i}(C_i)$ for any vertex $x_j \notin I(\mathcal{C})$. The clutter associated to $(I(\mathcal{C})\colon x_j)$ is the contraction $\mathcal{C}/x_j$. Hence from Proposition 6.5.8, we get that $\mathcal{C}/x_j$ has a perfect matching $e_1', \ldots, e_g'$ satisfying (a) and (b). $\Box$

Lemma 6.5.10 *Let $\mathcal{C}$ be an unmixed clutter with a perfect matching $\{e_i\}_{i=1}^g$ of König type and let $I = I(\mathcal{C})$. If $e_1 = \{x_1, \ldots, x_r\}$ and $C_1, \ldots, C_s$ are the minimal vertex covers of $\mathcal{C}$, then*

$$\bigcap_{x_1 \in C_i} (C_i) = (((\cdots(((I\colon x_2)\colon x_3)\colon x_4)\cdots)\colon x_{r-1})\colon x_r),$$

Proof. Let I' be the ideal on the right-hand side of the equality. If $x^{v_1}, \ldots, x^{v_q}$ generate I and we make $x_i = 1$ for $i = 2, \ldots, r$ in $x^{v_1}, \ldots, x^{v_q}$, we obtain a generating set for I'. Notice that $I' = (I\colon x_2 \cdots x_r)$ by the definition of the colon operation.

"$\supset$": Take a $x^a = x_1^{a_1} x_{r+1}^{a_{r+1}} \cdots x_n^{a_n}$ in I'. We may assume $a_1 = 0$, otherwise x^a is already in the left-hand side. Then $x_2 \cdots x_r x_{r+1}^{a_{r+1}} \cdots x_n^{a_n}$ is in I. Let C_i be any minimal vertex cover of $\mathcal{C}$ containing x_1. Observe that C_i cannot contain x_j for $2 \leq j \leq r$. Indeed if $x_j \in C_i$ for some $2 \leq j \leq r$, then C_i would contain $\{x_1, x_j\}$ plus at least one vertex of each edge in the collection $e_2, \ldots, e_g$, a contradiction because C_i has exactly g vertices. Hence, using that $x_2 \cdots x_r x_{r+1}^{a_{r+1}} \cdots x_n^{a_n}$ is in I, we get that $x_{r+1}^{a_{r+1}} \cdots x_n^{a_n}$ is in (C_i). Consequently x^a is in the left-hand side of the equality.

"$\subset$": Let x^a be a minimal generator in the left-hand side of the equality. Then $x^a \in (C_i)$ if $x_1 \in C_i$. If $x_1 \notin C_i$, then $x_2 \cdots x_r \in (C_i)$ since C_i covers e_1. Thus $x^a x_2 \cdots x_r \in (C_i)$ for all i, and so $x^a x_2 \cdots x_r \in \cap_{i=1}^s (C_i) = I$. Thus x^a is in the right-hand side of the equality. $\Box$

Definition 6.5.11 We say x_i is a *free variable* (resp. *free vertex*) of I (resp. $\mathcal{C}$) if x_i only appears in one of the monomials of $G(I)$ (resp. in one of the edges of $\mathcal{C}$), where $G(I)$ denotes the minimal set of generators of $I = I(\mathcal{C})$ consisting of monomials.

Theorem 6.5.12 [325] *Let $\mathcal{C}$ be a clutter with a perfect matching $e_1, \ldots, e_g$ of König type. If all c-minors of $\mathcal{C}$ have a free vertex and $\mathcal{C}$ is unmixed, then $\Delta_{\mathcal{C}}$ is pure shellable.*

Proof. The proof is by induction on the number of vertices. We may assume that $\mathcal{C}$ is a non-discrete clutter, i.e., it contains an edge with at least two vertices. Let z be a free vertex of $\mathcal{C}$ and let $C_1, \ldots, C_s$ be the minimal vertex covers of $\mathcal{C}$. We may also assume that $z \in e_m$ for some $e_m = \{z_1, \ldots, z_r\}$, with $r \geq 2$. For simplicity of notation assume that $z = z_1$ and $m = g$. Consider the clutters $\mathcal{C}_1$ and $\mathcal{C}_2$ associated with

$$I_1 = \bigcap_{z_1 \notin C_i} (C_i) \quad \text{and} \quad I_2 = \bigcap_{z_1 \in C_i} (C_i) \tag{6.6}$$

respectively. By Proposition 6.5.8, the clutter $\mathcal{C}_2$ has a perfect matching $e_1', \ldots, e_g'$ of König type such that: (a) $e_i' \subset e_i$ for all i, and (b) every vertex x of $e_i \setminus e_i'$ is isolated in $\mathcal{C}_2$, i.e., x does not occur in any edge of $\mathcal{C}_2$. In particular all vertices of $e_g \setminus \{z_1\}$ are isolated vertices of $\mathcal{C}_2$. Similar statements hold for $\mathcal{C}_1$ because of Proposition 6.5.8. By Lemma 6.5.10 and Corollary 6.5.9 we get

$$I_1 = (I \colon z_1) \quad \text{and} \quad I_2 = (((\cdots(((I \colon z_2) \colon z_3) \colon z_4) \cdots) \colon z_{r-1}) \colon z_r),$$

that is, $\mathcal{C}_1 = \mathcal{C}/z_1$ and $\mathcal{C}_2 = \mathcal{C}/\{z_2, \ldots, z_r\}$. Hence the ideals I_1 and I_2 are c-minors of I. The number of vertices of $\mathcal{C}_i$ is less than that of $\mathcal{C}$ for $i = 1, 2$. Thus $\Delta_{\mathcal{C}_1}$ and $\Delta_{\mathcal{C}_2}$ are shellable by the induction hypothesis. Consider the clutter $\mathcal{C}_i'$ whose edges are the edges of $\mathcal{C}_i$ and whose vertex set is X. The minimal vertex covers of $\mathcal{C}_i'$ are exactly the minimal vertex covers of $\mathcal{C}_i$. Thus it follows that $\Delta_{\mathcal{C}_i'}$ is shellable for $i = 1, 2$. Let $F_1, \ldots, F_p$ be the facets of $\Delta_{\mathcal{C}}$ that contain z_1 and let $G_1, \ldots, G_t$ be the facets of $\Delta_{\mathcal{C}}$ that do not contain z_1. Notice that the edge ideals of $\mathcal{C}_i$ and $\mathcal{C}_i'$ coincide, the vertex set of $\mathcal{C}_i'$ is equal to the vertex set of $\mathcal{C}$, and $I = I_1 \cap I_2$. Hence from Eq. (6.6) we get that $F_1, \ldots, F_p$ are the facets of $\Delta_{\mathcal{C}_1'}$ and $G_1, \ldots, G_t$ are the facets of $\Delta_{\mathcal{C}_2'}$. By the induction hypothesis we may assume $F_1, \ldots, F_p$ is a shelling of $\Delta_{\mathcal{C}_1'}$ and $G_1, \ldots, G_t$ is a shelling of $\Delta_{\mathcal{C}_2'}$. We now prove that

$$F_1, \ldots, F_p, G_1, \ldots, G_t$$

is a shelling of $\Delta_{\mathcal{C}}$. We need only show that given G_j and F_i there is $v \in G_j \setminus F_i$ and F_ℓ such that $G_j \setminus F_\ell = \{v\}$. We can write

$$G_j = X \setminus C_j \quad \text{and} \quad F_i = X \setminus C_i,$$

where C_j (resp. C_i) is a minimal vertex cover of $\mathcal{C}$ containing z_1 (resp. not containing z_1). Notice that $z_2,\ldots,z_r$ are not in C_j because $e_1,\ldots,e_g$ is a perfect matching and $|C_j| = g$. Thus $z_2,\ldots,z_r$ are in G_j. Since $z_1 \in F_i$ and F_i cannot contain the edge e_g, there is a z_k so that $z_k \notin F_i$ and $k \neq 1$. Set $v = z_k$ and $F_\ell = (G_j \setminus \{z_k\}) \cup \{z_1\}$. Clearly F_ℓ is an independent vertex set because z_1 is a free vertex in e_g and G_j is an independent vertex set. Thus F_ℓ is a facet because $\mathcal{C}$ is unmixed. To complete the proof observe that $G_j \setminus F_\ell = \{z_k\}$. $\square$

In what follows, we set $x_e = \prod_{x_i\in e} x_i$ for any $e \subset X$. Next we give a characterization of the unmixed property of $\mathcal{C}$. This characterization can be formulated combinatorially or algebraically.

Theorem 6.5.13 [325] *Let $\mathcal{C}$ be a clutter with a perfect matching $e_1,\ldots,e_g$ of König type. Then the following conditions are equivalent:*

(a) *$\mathcal{C}$ is unmixed.*

(b) *For any two edges $e \neq e'$ and for any two distinct vertices $x \in e$, $y \in e'$ contained in some e_i, $(e \setminus \{x\}) \cup (e' \setminus \{y\})$ contains an edge.*

(c) *For any two edges $e \neq e'$ and for any $T \subset e_i$ such that x_T divides $x_e x_{e'}$, $\operatorname{supp}(x_e x_{e'}/x_T)$ contains an edge.*

(d) *For any two edges $e \neq e'$ and for any e_i, $(x_e x_{e'} \colon x_{e_i}) \subset I(\mathcal{C})$.*

(e) *$I(\mathcal{C}) = (I(\mathcal{C})^2 \colon x_{e_1}) + \cdots + (I(\mathcal{C})^2 \colon x_{e_g})$.*

Proof. (a) $\Rightarrow$ (c): We may assume $i = 1$. Let T be a subset of e_1 such that x_T divides $x_e x_{e'}$. If $T \subset e$, then e' is an edge contained in $S = \operatorname{supp}(x_e x_{e'}/x_T)$ and there is nothing to show. The proof is similar if $T \subset e'$. So we can define $T_1 = e \cap T$ and $T_2 = T \setminus T_1$ and we may assume neither T_1 nor T_2 is empty. Note that $T_1 \subset e$ and $T_2 \subset e'$. In fact, $T_2 \subset T \cap e'$, but equality does not necessarily hold. Notice that $S = (e \setminus T_1) \cup (e' \setminus T_2)$. If S does not contain an edge, its complement contains a minimal vertex cover C. We use c to denote complement. Then

$$C \subset X \setminus S = S^c = (e \setminus T_1)^c \cap (e' \setminus T_2)^c = (e^c \cup T_1) \cap (e'^c \cup T_2).$$

Now $C \cap e \neq \emptyset$, so there is an $x \in C \cap e$. Then $x \in e^c \cup T_1$. This forces $x \in T_1$. Similarly there is a $y \in C \cap e'$, and so $y \in e'^c \cap T_2$. Thus $y \in T_2$. By the definition of T_2, $x \neq y$. To derive a contradiction pick $z_k \in e_k \cap C$ for $k \geq 2$ and notice that $x, y, z_2,\ldots,z_g$ is a set of $g+1$ distinct vertices in C, which is impossible because $\mathcal{C}$ is unmixed.

(c) $\Rightarrow$ (b): Let $x \in e$ and $y \in e'$ be two distinct vertices contained in some e_i. Let $T = \{x, y\}$. Then x_T divides $x_e x_{e'}$ and

$$S = \operatorname{supp}(x_e x_{e'}/x_T) \subset (e \setminus \{x\}) \cup (e' \setminus \{y\}).$$

By (c), S contains an edge. Thus $(e \setminus \{x\}) \cup (e' \setminus \{y\})$ contains an edge.

(b) $\Rightarrow$ (a): Let C be a minimal vertex cover of $\mathcal{C}$. Since the matching is perfect, there is a partition $C = (C \cap e_1) \cup \cdots \cup (C \cap e_g)$. Hence it suffices to prove that $|C \cap e_i| = 1$ for all i. We proceed by contradiction. For simplicity of notation assume $i = 1$ and $|C \cap e_1| \geq 2$. Pick $x \neq y$ in $C \cap e_1$. Since C is minimal, there are edges e, e' such that

$$e \cap (C \setminus \{x\}) = \emptyset \text{ and } e' \cap (C \setminus \{y\}) = \emptyset. \tag{6.7}$$

Clearly $x \in e$, $y \in e'$, and $e \neq e'$ because $y \notin e$. Then by hypothesis the set $S = (e \setminus \{x\}) \cup (e' \setminus \{y\})$ contains an edge e''. Take $z \in e'' \cap C$, then $z \in e \setminus \{x\}$ or $z \in e' \setminus \{y\}$, which is impossible by Eq. (6.7).

(c) $\Rightarrow$ (d): Let $x^a \in (x_e x_{e'} \colon x_{e_i})$ be a monomial generator of the colon ideal. Then $x^a x_{e_i} = m x_e x_{e'}$ for some monomial m. Let $T \subset e_i$ be maximal such that x_T divides $x_e x_{e'}$. Then $x_{e_i \setminus T}$ divides m, and $x^a = (m/x_{e_i \setminus T})(x_e x_{e'}/x_T)$. Since $\text{supp}(x_e x_{e'}/x_T)$ contains an edge, we have $x_e x_{e'}/x_T \in I$. Thus $x^a \in I(\mathcal{C})$ as desired.

(d) $\Rightarrow$ (c): Suppose $T \subset e_i$ is such that x_T divides $x_e x_{e'}$. Then

$$(x_e x_{e'}/x_T) x_{e_i} = x_e x_{e'} x_{e_i \setminus T},$$

and so $(x_e x_{e'}/x_T) \in (x_e x_{e'} \colon x_{e_i}) \subset I(\mathcal{C})$. Thus $(x_e x_{e'}/x_T)$ is a multiple of a monomial generator of $I(\mathcal{C})$. Hence $\text{supp}(x_e x_{e'}/x_T)$ contains an edge.

(e) $\Rightarrow$ (d): If equality in (e) holds, then $(I(\mathcal{C})^2 \colon x_{e_i}) = I(\mathcal{C})$ for all i. Hence from $(I(\mathcal{C})^2 \colon x_{e_i}) \subset I(\mathcal{C})$ we get condition (d).

(d) $\Rightarrow$ (e): We set $I = I(\mathcal{C})$. It suffices to show $(I^2 \colon x_{e_i}) = I$ for all i. Since I is contained in $(I^2 \colon x_{e_i})$, we need only show the inclusion $(I^2 \colon x_{e_i}) \subset I$. Take $x^a \in (I^2 \colon x_{e_i})$, then $x^a x_{e_i} = m x_e x_{e'}$ for some edges e, e' of $\mathcal{C}$ and some monomial m. If $e \neq e'$, then by hypothesis $x^a \in (x_e x_{e'} \colon x_{e_i}) \subset I$, i.e., $x^a \in I$. If $e = e'$, then $x^a x_{e_i} = m x_e^2$. Thus x_e divides x^a because x_{e_i} is a square-free monomial, but this means that $x^a \in I$, as required. $\square$

Definition 6.5.14 A matrix A with entries in $\{0, 1\}$ is *balanced* if A has no square submatrix of odd order with exactly two 1's in each row and column.

Definition 6.5.15 Let A be the incidence matrix of a clutter $\mathcal{C}$. A clutter $\mathcal{C}$ has a *special cycle* of length r if there is a square submatrix of A of order $r \geq 3$ with exactly two 1's in each row and column. A clutter with no special odd cycles is called *balanced* and a clutter with no special cycles is called *totally balanced*.

This definition of special cycle is equivalent to the usual definition of *special cycle* in hypergraph theory [8, 226] (see Exercise 6.5.34).

The next result classifies all unmixed balanced clutters because balanced clutters have the König property (Corollary 14.3.11). In particular it gives a classification of all unmixed bipartite graphs (cf. Theorem 7.4.19).

Corollary 6.5.16 [325] *A clutter $\mathcal{C}$ with the König property is unmixed if and only if there is a perfect matching $e_1, \ldots, e_g$ of König type such that for any two edges $e \neq e'$ and for any two distinct vertices $x \in e$, $y \in e'$ contained in some e_i, one has that $(e \setminus \{x\}) \cup (e' \setminus \{y\})$ contains an edge.*

Proof. $\Rightarrow$) Assume that $\mathcal{C}$ is unmixed. By Theorem 6.5.13 it suffices to observe that any unmixed clutter with the König property and without isolated vertices has a perfect matching of König type; see Lemma 6.5.7.

$\Leftarrow$) This implication follows at once from Theorem 6.5.13. □

Faridi [155] introduced the notion of a leaf for a simplicial complex Δ. Precisely, a facet F of Δ is a *leaf* if F is the only facet of Δ, or there exists a facet $G \neq F$ such that $F \cap F' \subset F \cap G$ for all facets $F' \neq F$. Δ is called a *simplicial forest* if every non-empty subcollection, i.e., a subcomplex whose facets are also facets of Δ, of Δ contains a leaf.

A graph G is called *chordal* if every cycle C_r of G of length $r \geq 4$ has a chord in G. A *chord* of C_r is an edge joining two non-adjacent vertices of C_r. A chordal graph is called *strongly chordal* if every cycle C_r of even length at least six has a chord that divides C_r into two odd length paths.

A *clique* of a graph G is a set of vertices inducing a complete subgraph. The *clique clutter* of G, denoted by $\mathrm{cl}(G)$, is the clutter on $V(G)$ whose edges are the maximal cliques of G (maximal with respect to inclusion).

If all the minors of a clutter $\mathcal{C}$ have free vertices, we say that $\mathcal{C}$ has the *free vertex property.* Note that if $\mathcal{C}$ has the free vertex property, then so do all of its minors.

Theorem 6.5.17 *A clutter $\mathcal{C}$ is totally balanced if and only if any of the following equivalent conditions hold:*

(a) [226, Theorem 3.2] *$\mathcal{C}$ is the clutter of the facets of a simplicial forest.*

(b) [389, Corollary 3.1] *$\mathcal{C}$ has the free vertex property.*

(c) [152] *$\mathcal{C}$ is the clique clutter of a strongly chordal graph.*

Theorem 6.5.18 [325] *Let $\mathcal{C}$ be a clutter with a perfect matching $e_1, \ldots, e_g$ of König type. If $\mathcal{C}$ has no special cycles of length 3 or 4 and $\mathcal{C}$ is unmixed, then for any two edges f_1, f_2 of $\mathcal{C}$ and for any e_i, one has that $f_1 \cap e_i \subset f_2 \cap e_i$ or $f_2 \cap e_i \subset f_1 \cap e_i$.*

Proof. For simplicity assume $i = 1$. We proceed by contradiction. Assume there are $x_1 \in f_1 \cap e_1 \setminus f_2 \cap e_1$ and $x_2 \in f_2 \cap e_1 \setminus f_1 \cap e_1$. As $\mathcal{C}$ is unmixed, by Theorem 6.5.13(b) there is an edge e of $\mathcal{C}$ such that

$$e \subset (f_1 \setminus \{x_1\}) \cup (f_2 \setminus \{x_2\}) = (f_1 \cup f_2) \setminus \{x_1, x_2\}.$$

Since $e \not\subset e_1$, there is $x_3 \in e \setminus e_1$. Then either $x_3 \in f_1$ or $x_3 \in f_2$. Without loss of generality we may assume $x_3 \in f_1 \setminus e_1$. For use below we denote the incidence matrix of $\mathcal{C}$ by A.

Case(I): $x_3 \in f_2$. Then the matrix

$$\begin{array}{cccc} & f_1 & f_2 & e_1 \\ x_1 & 1 & 0 & 1 \\ x_2 & 0 & 1 & 1 \\ x_3 & 1 & 1 & 0 \end{array}$$

is a submatrix of A, a contradiction.

Case(II): $x_3 \notin f_2$. Notice that $e \not\subset f_1$, otherwise $e = f_1$ which is impossible because $x_1 \in f_1 \setminus e$. Thus there is $x_4 \in e \setminus f_1$ and $x_4 \in (e \cap f_2) \setminus f_1$.

Subcase(II.a): $x_4 \in e_1$. Then the matrix

$$\begin{array}{cccc} & f_1 & e & e_1 \\ x_1 & 1 & 0 & 1 \\ x_3 & 1 & 1 & 0 \\ x_4 & 0 & 1 & 1 \end{array}$$

is a submatrix of A, a contradiction.

Subcase(II.b): $x_4 \notin e_1$. Then the matrix

$$\begin{array}{ccccc} & f_1 & e & f_2 & e_1 \\ x_1 & 1 & 0 & 0 & 1 \\ x_2 & 0 & 0 & 1 & 1 \\ x_3 & 1 & 1 & 0 & 0 \\ x_4 & 0 & 1 & 1 & 0 \end{array}$$

is a submatrix of A, a contradiction. $\square$

Proposition 6.5.19 *Let $\mathcal{C}$ be an unmixed clutter with no special cycles of length 3 or 4. If $e_1, \ldots, e_g$ is a perfect matching of $\mathcal{C}$ of König type, then e_i has a free vertex for all i.*

Proof. Fix an integer i in $[1, g]$. We may assume that e_i has at least one non-free vertex. Consider the set: $\mathcal{F} = \{f \in E(\mathcal{C}) |\, e_i \cap f \neq \emptyset;\, f \neq e_i\}$. By Theorem 6.5.18, the edges of $\mathcal{F}$ can be listed as $f_1, \ldots, f_r$ so that they satisfy the inclusions $f_1 \cap e_i \subset f_2 \cap e_i \subset \cdots \subset f_r \cap e_i \subsetneq e_i$. Thus any vertex of $e_i \setminus (f_r \cap e_i)$ is a free vertex of e_i. $\square$

Theorem 6.5.20 *Let $\mathcal{C}$ be an unmixed clutter with a perfect matching $e_1, \ldots, e_g$ of König type. If $\mathcal{C}$ has no special cycles of length 3 or 4, then $\Delta_{\mathcal{C}}$ is pure shellable.*

Proof. All hypotheses are preserved under contractions, i.e., under c-minors. This follows from Corollary 6.5.9 and the fact that the incidence matrix of a contraction of $\mathcal{C}$ is a submatrix of the incidence matrix of $\mathcal{C}$. Thus by Proposition 6.5.19 any c-minor has a free vertex and the result follows from Theorem 6.5.12. $\square$

Theorem 6.5.21 [325] *Let $\mathcal{C}$ be a clutter with a perfect matching $e_1, \ldots, e_g$ of König type. If for any two edges f_1, f_2 of $\mathcal{C}$ and for any edge e_i of the perfect matching, one has that $f_1 \cap e_i \subset f_2 \cap e_i$ or $f_2 \cap e_i \subset f_1 \cap e_i$, then $\Delta_{\mathcal{C}}$ is pure shellable.*

Proof. First we show that $\Delta_{\mathcal{C}}$ is pure or equivalently that $\mathcal{C}$ is unmixed. It suffices to verify condition (b) of Theorem 6.5.13. Let $f_1 \neq f_2$ be two edges and let $x \in f_1$, $y \in f_2$ be two distinct vertices contained in some e_i. For simplicity we assume $i = 1$. Set $B = (f_1 \setminus \{x\}) \cup (f_2 \setminus \{y\})$. Then $f_2 \cap e_1 \subset f_1 \cap e_1$ or $f_1 \cap e_1 \subset f_2 \cap e_1$. In the first case we have that $f_2 \subset B$. Indeed let $z \in f_2$. If $z \neq y$, then $z \in f_2 \setminus \{y\} \subset B$, and if $z = y$, then $z \in f_2 \cap e_1 \subset f_1 \cap e_1$ and $z \neq x$, i.e., $z \in f_1 \setminus \{x\} \subset B$. In the second case $f_1 \subset B$. This proves that $\mathcal{C}$ is unmixed.

Next we show that $\Delta_{\mathcal{C}}$ is shellable. Notice that (i) e_i has a free vertex for all i, which follows from the proof of Proposition 6.5.19. Thus, as $\mathcal{C}$ is unmixed, by Theorem 6.5.12 we need only show that any c-minor has a free vertex. By (i) it suffices to show that our hypotheses are closed under contractions. Let x be a vertex of $\mathcal{C}$ and let $\mathcal{C}' = \mathcal{C}/x$. By Corollary 6.5.9, we get that $\mathcal{C}/x$ has a perfect matching $e_1', \ldots, e_g'$ satisfying: (a) $e_i' \subset e_i$ for all i, and (b) every vertex of $e_i \setminus e_i'$ is isolated in $\mathcal{C}'$. Let e, e' be two edges of $\mathcal{C}'$ and let e_i' be an edge of the perfect matching of $\mathcal{C}'$. There are edges f, f' of $\mathcal{C}$ such that one of the following is satisfied: $e = f$ and $e' = f' \setminus \{x\}$, $e = f \setminus \{x\}$ and $e' = f'$, $e = f \setminus \{x\}$ and $e' = f' \setminus \{x\}$, $e = f$ and $e' = f'$. We may assume $f \cap e_i \subset f' \cap e_i$. To finish the proof we now show that $e \cap e_i' \subset e' \cap e_i'$. Take $z \in e \cap e_i'$. Then $z \in f \cap e_i$ and consequently $z \in f' \cap e_i'$. Since $x \notin e_i'$, one has $z \neq x$. It follows that $z \in e' \cap e_i'$. $\Box$

Let G be a graph and let V be its vertex set. For use below consider the graph $G \cup W(V)$ obtained from G by adding new vertices $\{y_i \,|\, x_i \in V\}$ and new edges $\{\{x_i, y_i\} \,|\, x_i \in V\}$.

Corollary 6.5.22 *If G is a graph and $H = G \cup W(V)$, then Δ_H is pure shellable.*

Proof. It follows at once from Theorem 6.5.21. Indeed if $V = \{x_1, \ldots, x_n\}$, then $\{x_1, y_1\}, \ldots, \{x_n, y_n\}$ is a perfect matching of H satisfying the ordering condition in Theorem 6.5.21. $\Box$

Lemma 6.5.23 *Let $\mathcal{C}$ be a clutter with minimal vertex covers $C_1, \ldots, C_s$. If $\Delta_{\mathcal{C}}$ is shellable, $A \subset V_{\mathcal{C}}$ is a set of vertices, and*

$$I' = \bigcap_{C_i \cap A = \emptyset} (C_i),$$

then $\Delta_{I'}$ is shellable with respect to the linear ordering of the facets of $\Delta_{I'}$ induced by the shelling of the simplicial complex $\Delta_{\mathcal{C}}$.

Proof. Let $H_1, \ldots, H_s$ be a shelling of $\Delta_{\mathcal{C}}$. We may assume that H_i is equal to $V_{\mathcal{C}} \setminus C_i$ for all i. Let H_i and H_j be two facets of $\Delta_{I'}$ with $i < j$, i.e., $A \cap C_i = \emptyset$ and $A \cap C_j = \emptyset$. By the shellability of $\Delta_{\mathcal{C}}$, there is an $x \in H_j \setminus H_i$ and an $\ell < j$ such that $H_j \setminus H_\ell = \{x\}$. It suffices to prove that $C_\ell \cap A = \emptyset$. If $C_\ell \cap A \neq \emptyset$, pick $z \in C_\ell \cap A$. Then $z \notin C_i \cup C_j$ and $z \in H_i \cap H_j$. Since $z \notin H_\ell$ (otherwise $z \notin C_\ell$, a contradiction), we get $z \in H_j \setminus H_\ell$, i.e., $z = x$, a contradiction because $x \notin H_i$. □

Lemma 6.5.24 *Let x_n be a free variable of $I(\mathcal{C}) = (x^{v_1}, \ldots, x^{v_{q-1}}, x^{v_q})$, and let $x^{v_q} = x_n x^u$. Then the following hold.*

(a) *If $\mathcal{C}_1$ is the clutter associated to $(x^{v_1}, \ldots, x^{v_{q-1}})$, then C is a minimal vertex cover of $\mathcal{C}$ containing x_n if and only if $C \cap \operatorname{supp}(x^u) = \emptyset$ and $C = \{x_n\} \cup C'$ for some minimal vertex cover C' of $\mathcal{C}_1$.*

(b) *If $\mathcal{C}_2$ is the clutter associated to $(x^{v_1}, \ldots, x^{v_{q-1}}, x^u)$, then C is a minimal vertex cover of $\mathcal{C}$ not containing x_n if and only if C is a minimal vertex cover of $\mathcal{C}_2$.*

Proof. (a) Assume that C is a minimal vertex cover of $\mathcal{C}$ containing x_n. If $C \cap \operatorname{supp}(x^u) \neq \emptyset$, then $C \setminus \{x_n\}$ is a vertex cover of $\mathcal{C}$, a contradiction. Thus $C \cap \operatorname{supp}(x^u) = \emptyset$. Hence it suffices to notice that $C' = C \setminus \{x_n\}$ is a minimal vertex cover of $\mathcal{C}_1$. The converse also follows readily.

(b) Assume that C is a minimal vertex cover of $\mathcal{C}$ not containing x_n. Let x^a be a minimal generator of $I(\mathcal{C}_2)$, then either x^u divides x^a or $x^a = x^{v_i}$ for some $i < q$. Then clearly $C \cap \operatorname{supp}(x^a) \neq \emptyset$ because $C \cap A \neq \emptyset$, where $A = \operatorname{supp}(x^u)$. Thus C is a vertex cover of $\mathcal{C}_2$. To prove that C is minimal take $C' \subsetneq C$. We must show that there is an edge of $\mathcal{C}_2$ not covered by C'. As C is a minimal vertex cover of $\mathcal{C}$, there is x^{v_i} such that $\operatorname{supp}(x^{v_i}) \cap C' = \emptyset$. If x^{v_i} is a minimal generator of $\mathcal{C}_2$ there is nothing to prove, otherwise x^u divides x^{v_i} and the edge A of $\mathcal{C}_2$ is not covered by C'. The converse also follows readily. □

If all c-minors have a free vertex and $\mathcal{C}$ is unmixed, then $\Delta_{\mathcal{C}}$ is pure shellable (see Theorem 6.5.12). The next result complements this fact.

Theorem 6.5.25 [409] *If the clutter $\mathcal{C}$ has the free vertex property, then the independence complex $\Delta_{\mathcal{C}}$ is shellable.*

Proof. We proceed by induction on the number of vertices of $\mathcal{C}$. Let x_n be a free variable of $I = I(\mathcal{C}) = (x^{v_1}, \ldots, x^{v_{q-1}}, x^{v_q})$. We may assume that x_n occurs in x^{v_q}. Hence we can write $x^{v_q} = x_n x^u$ for some x^u such that $x_n \notin \operatorname{supp}(x^u)$. For use below we set $A = \operatorname{supp}(x^u)$. Consider the ideals $J = (x^{v_1}, \ldots, x^{v_{q-1}})$ and $L = (J, x^u)$. Then $J = I(\mathcal{C}_1)$ and $L = I(\mathcal{C}_2)$, where $\mathcal{C}_1$ and $\mathcal{C}_2$ are the clutters defined by the ideals J and L, respectively. Notice that J and L are minors of the ideal I obtained by setting $x_n = 0$

and $x_n = 1$, respectively. The vertex set of $\mathcal{C}_i$ is $V_{\mathcal{C}_i} = X \setminus \{x_n\}$ for $i = 1, 2$. Thus $\Delta_{\mathcal{C}_1}$ and $\Delta_{\mathcal{C}_2}$ are shellable by the induction hypothesis. Let $F_1, \ldots, F_r$ be the facets of $\Delta_{\mathcal{C}}$ that contain x_n and let $G_1, \ldots, G_s$ be the facets of $\Delta_{\mathcal{C}}$ that do not contain x_n. Set $C_i = X \setminus G_i$ and $C_i' = C_i \setminus \{x_n\}$ for $i = 1, \ldots, s$. Then $C_1, \ldots, C_s$ is the set of minimal vertex covers of $\mathcal{C}$ that contain x_n, and by Lemma 6.5.24(a) $C_1', \ldots, C_s'$ is the set of minimal vertex covers of $\mathcal{C}_1$ that do not intersect A. One has the equality $G_i = V_{\mathcal{C}_1} \setminus C_i'$ for all i. Hence, by the shellability of $\Delta_{\mathcal{C}_1}$ and using Lemma 6.5.23, we may assume that $G_1, \ldots, G_s$ is a shelling for the simplicial complex generated by $G_1, \ldots, G_s$. By Lemma 6.5.24(b) one has that C is a minimal vertex cover of $\mathcal{C}$ not containing x_n if and only if C is a minimal vertex cover of $\mathcal{C}_2$. Thus, F is a facet of $\Delta_{\mathcal{C}}$ that contains x_n, i.e., $F = F' \cup \{x_n\}$ if and only if F' is a facet of $\Delta_{\mathcal{C}_2}$. By induction we may also assume that $F_1' = F_1 \setminus \{x_n\}, \ldots, F_r' = F_r \setminus \{x_n\}$ is a shelling of $\Delta_{\mathcal{C}_2}$. We now prove that

$$F_1, \ldots, F_r, G_1, \ldots, G_s \quad \text{with } F_i = F_i' \cup \{x_n\}$$

is a shelling of $\Delta_{\mathcal{C}}$. We need only show that given G_j and F_i there is $a \in G_j \setminus F_i$ and F_ℓ such that $G_j \setminus F_\ell = \{a\}$. We can write

$$G_j = X \setminus C_j \quad \text{and} \quad F_i = X \setminus C_i,$$

where C_j (resp. C_i) is a minimal vertex cover of $\mathcal{C}$ containing x_n (resp. not containing x_n). Recall that $A = \text{supp}(x^u)$ is an edge of $\mathcal{C}_2$. Notice the following: (i) $C_j = C_j' \cup \{x_n\}$ for some minimal vertex cover C_j' of $\mathcal{C}_1$ such that $A \cap C_j' = \emptyset$, and (ii) C_i is a minimal vertex cover of $\mathcal{C}_2$. From (i) we get that $A \subset G_j$. Observe that $A \not\subset F_i$, otherwise $A \cap C_i = \emptyset$, a contradiction because C_i must cover the edge $A = \text{supp}(u)$. Hence there is $a \in A \setminus F_i$ and $a \in G_j \setminus F_i$. Since $C_j' \cup \{a\}$ is a vertex cover of $\mathcal{C}$, there is a minimal vertex cover C_ℓ of $\mathcal{C}$ contained in $C_j' \cup \{a\}$. Clearly $a \in C_\ell$ because C_ℓ has to cover x^u and $C_j' \cap A = \emptyset$. Thus $F_\ell = X \setminus C_\ell$ is a facet of $\Delta_{\mathcal{C}}$ containing x_n. To finish the proof we now prove that $G_j \setminus F_\ell = \{a\}$. We know that $a \in G_j$. If $a \in F_\ell$, then $a \notin C_\ell$, a contradiction. Thus $a \in G_j \setminus F_\ell$. Conversely take $z \in G_j \setminus F_\ell$. Then $z \notin C_j' \cup \{x_n\}$ and $z \in C_\ell \subset C_j' \cup \{a\}$. Hence $z = a$, as required. □

According to Theorem 6.5.17, a totally balanced clutter satisfies the free vertex property. Thus, we obtain:

Corollary 6.5.26 *If $\mathcal{C}$ is a totally balanced clutter, then $\Delta_{\mathcal{C}}$ is shellable.*

By Theorem 6.5.17, a clutter $\mathcal{C}$ is a simplicial forest if and only if $\mathcal{C}$ is a totally balanced clutter. Thus, we also obtain:

Corollary 6.5.27 *If $\mathcal{C}$ is the clutter of facets of a simplicial forest, then $\Delta_{\mathcal{C}}$ is shellable.*

Definition 6.5.28 [155] Let Δ be a simplicial complex. The *facet ideal* of Δ, denoted by $I(\Delta)$, is the edge ideal of the clutter of facets of Δ.

Facet ideals were introduced and studied by Faridi in a series of papers [153, 154, 155, 156, 157].

Corollary 6.5.29 [155] *Let $I = I(\Delta)$ be the facet ideal of a simplicial forest. Then $R/I(\Delta)$ is sequentially Cohen–Macaulay.*

Proof. If $\Delta = \langle F_1, \ldots, F_s \rangle$, then $I(\Delta)$ is also the edge ideal of the clutter $\mathcal{C}$ whose edge set is $E_{\mathcal{C}} = \{F_1, \ldots, F_s\}$. Now apply Corollary 6.5.27 and Theorem 6.3.27. $\square$

Exercises

6.5.30 Prove that a clutter with a perfect matching of König type has the König property.

6.5.31 Let $\mathcal{C}$ be a uniform clutter. If $\mathcal{C}$ has the König property and a perfect matching, then the perfect matching is of König type.

6.5.32 Consider the clutter $\mathcal{C}$ whose edges are

$$e_1 = \{x_1, x_2\},\ e_2 = \{x_3, x_4, x_5, x_6\},\ e_3 = \{x_7, x_8, x_9\},\ f_4 = \{x_1, x_3\},$$
$$f_5 = \{x_2, x_4\},\ f_6 = \{x_5, x_7\},\ f_7 = \{x_6, x_8\}.$$

Prove that $\mathcal{C}$ has the König property and that $\mathcal{C}$ has no perfect matching of König type.

6.5.33 [185, Proposition 5.8] If $\mathcal{C}$ is a uniform totally balanced clutter, there is a partition $X^1, \ldots, X^d$ of X such that any edge of $\mathcal{C}$ intersects any X^i in exactly one vertex.

6.5.34 Let $\mathcal{H}$ be a hypergraph and let A be its incidence matrix. Recall that a *cycle* (resp. *special cycle*) of $\mathcal{H}$ is a sequence $x_1e_1x_2e_2 \cdots x_re_rx_1$ of r distinct vertices x_i and r distinct edges e_j ($r \geq 3$) such that $\{x_i, x_{i+1}\} \subset e_i$ (resp. $\{x_1, \ldots, x_r\} \cap e_i = \{x_i, x_{i+1}\}$) for $i = 1, \ldots, r$ ($x_{r+1} = x_1$). The value r is the length of the cycle. Prove that $\mathcal{H}$ has a special cycle of length r if and only if there is a square submatrix of A of order $r \geq 3$ with exactly two 1's in each row and column.

6.5.35 If $e_1 = \{x_1, x_2, x_4\}, e_2 = \{x_2, x_3, x_5\}, e_3 = \{x_1, x_3, x_6\}$, the clutter with edges e_1, e_2, e_3 has a special cycle $x_1e_1x_2e_2x_3e_3x_1$ of order 3.

6.5.36 [8] If $\mathcal{C}$ is a totally balanced clutter with n vertices, then $\mathcal{C}$ has at most $\binom{n}{2} + n$ edges.

6.6 Admissible clutters

Let $X^1,\ldots,X^d$ and $e_1,\ldots,e_g$ be two partitions of a finite set X such that $|e_i \cap X^j| \leq 1$ for all i,j. The variables of $K[X]$ are linearly ordered by: $x \prec y$ iff ($x \in X^i$, $y \in X^j$, $i<j$) or ($x,y \in X^i$, $x \in e_k$, $y \in e_\ell$, $k<\ell$).

Let $e \subset X$ such that $|e|=k$ and $|e \cap X^i| \leq 1$ for all i. There are unique integers $1 \leq i_1 < \cdots < i_k \leq d$ and integers $j_1,\ldots,j_k \in [1,g]$ such that

$$\emptyset \neq e \cap X^{i_1} = \{x_1\},\ \emptyset \neq e \cap X^{i_2} = \{x_2\}, \ldots,\ \emptyset \neq e \cap X^{i_k} = \{x_k\}$$

and $x_1 \in e_{j_1},\ldots,x_k \in e_{j_k}$. We say that e is an *admissible* set if one has $i_1=1, i_2=2,\ldots,i_k=k$ and $j_1 \leq \cdots \leq j_k$.

We can represent an admissible set $e=\{x_1,\ldots,x_k\}$ as $e=x_{j_1}^1\cdots x_{j_k}^k$, i.e., $x_i=x_{j_i}^i$ and $x_{j_i}^i \in X^i \cap e_{j_i}$ for all i. A monomial x^a is admissible if $\mathrm{supp}(x^a)$ is admissible. A clutter $\mathcal{C}$ is called *admissible* if $e_1,\ldots,e_g$ are edges of $\mathcal{C}$, e_i is admissible for all i, and all other edges are admissible sets not contained in any of the e_i's. We can think of $X^1,\ldots,X^d$ as color classes that color the edges.

Lemma 6.6.1 *If $\mathcal{C}$ is an admissible clutter, then $e_1,\ldots,e_g$ is a perfect matching of König type.*

Proof. It suffices to prove that $g = \mathrm{ht}\, I(\mathcal{C})$. Clearly $\mathrm{ht}\, I(\mathcal{C}) \geq g$ because any minimal vertex cover of $\mathcal{C}$ must contain at least one vertex of each e_i and the e_i's form a partition of X. For each $1 \leq i \leq g$ there is $y_i = x_i^1$ so that $e_i \cap X^1 = \{y_i\}$. Since the e_i's form a partition we have the equality

$$(e_1 \cap X^1) \cup \cdots \cup (e_g \cap X^1) = X^1.$$

Thus $|X^1|=g$. To complete the proof notice that X^1 is a vertex cover of $\mathcal{C}$ because all edges of $\mathcal{C}$ are admissible. This shows $\mathrm{ht}\, I(\mathcal{C}) \leq g$. $\square$

Admissible clutters with two color classes X^1, X^2 are special types of bipartite graphs (see Section 7.4).

Example 6.6.2 Let $\mathcal{C}$ be the Cohen–Macaulay admissible balanced clutter with color classes X^1,X^2,X^3 and edges e_1,e_2,e_3,f_1,f_2,f_3.

		X^1	X^2	X^3
e_1	$=$	x_1	y_1	
e_2	$=$	x_2	y_2	z_2
e_3	$=$	x_3	y_3	

		X^1	X^2	X^3
f_1	$=$	x_1	y_2	z_2
f_2	$=$	x_1	y_3	
f_3	$=$	x_2	y_3	

The edges e_1,e_2,e_3 form a perfect matching of König type of $\mathcal{C}$.

Example 6.6.3 Consider the clutter $\mathcal{C}$ whose edge ideal is generated by:

$$a_1b_1c_1d_1g_1h_1k_1,\quad a_2b_2c_2d_2g_2h_2k_2,\quad a_3b_3c_3d_3g_3h_3k_3,$$
$$a_4b_4c_4d_4g_4h_4k_4,\quad a_1b_1c_1d_1g_2h_3k_4,\quad a_1b_2c_3d_4g_2h_3k_4,$$

where $a_i, b_i, \ldots$ are variables. This clutter is balanced because its incidence matrix is totally unimodular and $I(\mathcal{C})$ is Cohen–Macaulay. However, we cannot order its vertices so that it becomes an admissible uniform clutter.

Definition 6.6.4 If $e_1, \ldots, e_g$ are admissible subsets of X, the clutter $\mathcal{C}$ on X whose set of edges is:

$$E(\mathcal{C}) = \left\{ e \subset X \;\middle|\; \begin{array}{l} e_i \not\subset e \text{ for } i = 1, \ldots, g,\ e \text{ is admissible,} \\ e \not\subset e' \text{ for any admissible set } e' \neq e \end{array} \right\} \cup \{e_1, \ldots, e_g\}$$

is called a *complete admissible clutter* (cf. Definition 14.5.11).

The edges of this clutter are the maximal admissible sets with respect to inclusion. By Lemma 6.6.1, $e_1, \ldots, e_g$ is a perfect matching of König type. In Section 14.5 we show an optimization property of complete admissible uniform clutters (see Theorems 14.3.6 and 14.5.12).

Proposition 6.6.5 *If $\mathcal{C}$ is a complete admissible clutter, then $\mathcal{C}$ is unmixed.*

Proof. To show that $\mathcal{C}$ is unmixed it suffices to verify condition (b) of Theorem 6.5.13. Let $e \neq e'$ be two edges of $\mathcal{C}$ and let $x \neq y$ be two vertices such that $\{x, y\} \subset e_i$ for some e_i, $x \in e$, and $y \in e'$. Since e, e', e_i are admissible we can write

$$e = \{x_1, \ldots, x_k\},\ e' = \{y_1, \ldots, y_{k'}\},\ e_i = \{z_1, \ldots, z_{k''}\},$$

where $x_i \in X^i$, $y_i \in X^i$, $z_i \in X^i$. There are i_1, i_2 such that $x = x_{i_1}$, $y = y_{i_2}$, $x = z_{i_1}$, and $y = z_{i_2}$. We may assume $i_1 < i_2$. One has $i_1 < k$, because if $k = i_1$, then $e \subsetneq e \cup \{z_{i_1+1}, \ldots, z_{i_2}\}$ and the right-hand side is admissible, a contradiction. Set $f = \{y_1, \ldots, y_{i_1}, x_{i_1+1}, \ldots, x_k\}$. Then

$$f \subset e \setminus \{x\} \cup e' \setminus \{y\}.$$

Thus to finish the proof we need only show that f is an edge of $\mathcal{C}$. Since $y_{i_2} \in e_i$ and $x_{i_1} \in e_i$, then $y_{i_1} \in e_\ell$ for some $\ell \leq i$ and $x_{i_1+1} \in e_t$ for some $i \leq t$. Hence f is admissible. Next we show that f is maximal. Assume that f is not maximal. Then there exists an admissible subset f' that properly contains f. Then there is $z \in f' \cap X^{k+1}$ and since $f \cup \{z\} \subset f'$, we get that $e \cup \{z\} = \{x_1, \ldots, x_k, z\}$ is admissible, but $e \subsetneq e \cup \{z\}$, a contradiction. Hence f is maximal. □

Theorem 6.6.6 [325] *If $\mathcal{C}$ is a complete admissible uniform clutter, then the simplicial complex generated by the edges of $\mathcal{C}$ is pure shellable.*

Proof. Order the variables of $K[X]$ as in the beginning of Section 6.6. We can represent the edges of $\mathcal{C}$ as $F_i = x_{i_1}^1 x_{i_2}^2 \cdots x_{i_d}^d$, where $x_{i_j}^i \in X^i \cap e_{i_j}$. Then, we order the edges of $\mathcal{C}$ lexicographically, that is

$$F_i = x_{i_1}^1 x_{i_2}^2 \cdots x_{i_d}^d < F_j = x_{j_1}^1 x_{j_2}^2 \cdots x_{j_d}^d$$

if the first non-zero entry of $(j_1, j_2, \ldots, j_d) - (i_1, i_2, \ldots, i_d) = \mathbf{j} - \mathbf{i}$ is positive. Under this order, we show that $\mathcal{C}$ is shellable.

Suppose F_i and F_j are edges of $\mathcal{C}$ with $F_i < F_j$. Suppose the first non-zero entry of $\mathbf{j}-\mathbf{i}$ is $j_t - i_t$. Then $1 \leq i_t < j_t$. Let $F_k = F_j \setminus \{x_{j_t}^t\} \cup \{x_{i_t}^t\}$ and let $v = x_{j_t}^t$. Since $j_1 = i_1 \leq \cdots \leq j_{t-1} = i_{t-1} \leq i_t < j_t \leq j_{t+1} \leq \cdots \leq j_d$, F_k is maximal admissible, $v \in F_j \setminus F_i$, $F_k < F_j$ and $F_j \setminus F_k = \{v\}$. $\Box$

The next example illustrates the construction of the lexicographical shelling used in the proof of Theorem 6.6.6.

Example 6.6.7 Let $\mathcal{C}$ be the complete admissible uniform clutter with color classes $X^1 = \{x_1, x_2, x_3\}$, $X^2 = \{y_1, y_2, y_3\}$, $X^3 = \{z_1, z_2, z_3\}$. Then the shelling of the simplicial complex generated by the edges of $\mathcal{C}$ is:

$$\begin{array}{llllll}
F_1 = \{x_1, y_1, z_1\} & < & F_2 = \{x_1, y_1, z_2\} & < & F_3 = \{x_1, y_1, z_3\} & < \\
F_4 = \{x_1, y_2, z_2\} & < & F_5 = \{x_1, y_2, z_3\} & < & F_6 = \{x_1, y_3, z_3\} & < \\
F_7 = \{x_2, y_2, z_2\} & < & F_8 = \{x_2, y_2, z_3\} & < & F_9 = \{x_2, y_3, z_3\} & < \\
F_{10} = \{x_3, y_3, z_3\}. & & & & &
\end{array}$$

Lemma 6.6.8 *If $\mathcal{C}$ is a complete admissible uniform clutter, then the simplicial complex $\Delta_{\mathcal{C}^\vee}$ generated by $\{X \setminus F | \, F \in E(\mathcal{C})\}$ is pure shellable.*

Proof. Let $F_1, \ldots F_r$ be the shelling of the edges of $\mathcal{C}$ defined in Theorem 6.6.6. Let $G_1 = X \setminus F_1, \ldots, G_r = X \setminus F_r$ be the facets of $\Delta_{\mathcal{C}^\vee}$. We claim that $G_1, \ldots, G_r$ is the desired shelling. Suppose $G_i < G_j$. Then $F_i < F_j$. Using the notation defined in Theorem 6.6.6, let $v = x_{j_t}^t$ and define $u = x_{i_t}^t$. Then $u \in G_j \setminus G_i$ and $G_j \setminus G_k = \{u\}$ as required. $\Box$

Theorem 6.6.9 [325] *If $\mathcal{C}$ is a complete admissible d-uniform clutter, then the face ring $R/I(\mathcal{C})$ is a Cohen–Macaulay ring with a d-linear resolution and $|E(\mathcal{C})| = \binom{d+g-1}{g-1}$.*

Proof. Consider the clutter $\mathcal{C}^\vee$ of minimal vertex covers of $\mathcal{C}$. By Lemma 6.6.8 and Exercise 6.6.14 we have that $\Delta_{\mathcal{C}^\vee}$ is pure shellable. Now recall that the Stanley–Reisner ideal of $\Delta_{\mathcal{C}^\vee}$ is $I(\mathcal{C}^\vee)$ and that $I(\mathcal{C}^\vee)$ is the Alexander dual of $I(\mathcal{C})$. Thus $R/I(\mathcal{C}^\vee)$ is Cohen–Macaulay, and by Theorem 6.3.41 the ideal $I(\mathcal{C})$ has a linear resolution. Since the Alexander dual of a complete admissible uniform clutter is also a complete admissible uniform clutter and since $(\mathcal{C}^\vee)^\vee = \mathcal{C}$ it follows that $R/I(\mathcal{C})$ is Cohen–Macaulay. The formula for the number of edges of $\mathcal{C}$ follows from the explicit formula given in

Theorem 3.5.17 for the Betti numbers of a Cohen–Macaulay ideal with a pure resolution; see also Exercise 5.3.17. $\Box$

Let $\mathcal{C}$ be a complete admissible uniform clutter. For each edge of $\mathcal{C}$, $e = x_{j_1}^1 x_{j_2}^2 \cdots x_{j_d}^d$, consider all pairs $(x_{j_i}^i, x_{j_k}^k)$ with $i < k$ and consider the union of all these pairs with $e = x_{j_1}^1 x_{j_2}^2 \cdots x_{j_d}^d$ running through all edges of $\mathcal{C}$. This defines a poset $P = (X, \prec)$ on X whose comparability graph G is defined by all the unordered pairs $\{x_{j_i}^i, x_{j_k}^k\}$.

Corollary 6.6.10 *If G' is the complement of the comparability graph G defined above, then $R/I(G')$ is Cohen–Macaulay.*

Proof. Notice that $\Delta_{G'} = \{\mathcal{K}_r \mid \mathcal{K}_r \text{ is a clique of } G\} = \mathcal{O}(P)$, where $\mathcal{O}(P)$ is the order complex of P. Since the maximal faces of $\mathcal{O}(P)$ are precisely the edges of $\mathcal{C}$, by Theorem 6.6.6, we obtain that $\mathcal{O}(P)$ is a pure shellable complex whose Stanley–Reisner ring is equal to $R/I(G')$. Hence $R/I(G')$ is Cohen–Macaulay by Theorem 6.3.23. $\Box$

Let $\mathcal{C}$ be a clutter and let $x^{v_1}, \ldots, x^{v_q}$ be the minimal set of generators of $I(\mathcal{C})$. Consider the ideal $I^* = (x^{w_1}, \ldots, x^{w_q})$, where $v_i + w_i = (1, \ldots, 1)$. Following the terminology of matroid theory we call I^* the *dual* of I. If I^* has linear quotients and all x^{w_i} have the same degree, then I^* has a linear resolution (see Proposition 6.3.46).

Corollary 6.6.11 *If $\mathcal{C}$ is a complete admissible uniform clutter, then $I(\mathcal{C})^*$ has linear quotients.*

Proof. Let $x^{v_1}, \ldots, x^{v_q}$ be the minimal set of generators of $I = I(\mathcal{C})$. We set $F_i = \operatorname{supp}(x^{v_i})$ for $i = 1, \ldots, q$. By Theorem 6.6.6, we may assume that $F_1, \ldots, F_q$ is a shelling for the simplicial complex $\langle F_1, \ldots, F_q \rangle$ generated by the F_i's. Thus, according to Exercise 6.3.66, the ideal $I^* = (x_{F_1^c}, \ldots, x_{F_q^c})$ has linear quotients, where $F_k^c = X \setminus F_k$ and $x_{F_k^c} = \prod_{x_i \in F_k^c} x_i$. $\Box$

Exercises

6.6.12 Consider the following clutter with edges e_1, e_2, f_1, f_2 and color classes X^1, X^2, X^3

$$
\begin{array}{ccccc}
 & & X^1 & X^2 & X^3 \\
e_1 & = & x_1 & y_1 & \\
e_2 & = & & y_2 & z_2
\end{array}
\qquad\qquad
\begin{array}{ccccc}
 & & X^1 & X^2 & X^3 \\
f_1 & = & & y_1 & z_2 \\
f_2 & = & x_1 & y_2 &
\end{array}
$$

Prove that this clutter is unmixed, is not Cohen–Macaulay, has a perfect matching e_1, e_2 of König type, and the height of $I(\mathcal{C})$ is two.

6.6.13 Prove that the uniform admissible clutters with three color classes

$$X^1 = \{x_1, \dots, x_g\},\ X^2 = \{y_1, \dots, y_g\},\ X^3 = \{z_1, \dots, z_g\}$$

are, up to permutation of variables, exactly the clutters with a perfect matching $e_i = \{x_i, y_i, z_i\}$ for $i = 1, \dots, g$ such that all edges of $\mathcal{C}$ have the form $\{x_i, y_j, z_k\}$, with $1 \leq i \leq j \leq k \leq g$.

6.6.14 [325] If $\mathcal{C}$ is a complete admissible uniform clutter, then the blocker $\mathcal{C}^\vee$ of $\mathcal{C}$ is also a complete admissible uniform clutter.

6.7 Hilbert series of face rings

For a Stanley–Reisner ring $K[\Delta]$ there are formulas for its Hilbert function and Hilbert series in terms of the combinatorial data of the complex Δ. If $I(\mathcal{C})$ is the edge ideal of a clutter $\mathcal{C}$, its Hilbert series can be obtained from the so-called edge induced polynomial of $\mathcal{C}$ [193, 347].

Hilbert series with the fine and standard grading Let K be a field. Note that the polynomial ring $R = K[x_1, \dots, x_n]$ can be endowed with a *fine $\mathbb{Z}^n$-grading* as follows. For $a = (a_1, \dots, a_n) \in \mathbb{Z}^n$, set

$$R_a = \begin{cases} Kx^a, & \text{if } a_i \geq 0 \text{ for } i = 1, \dots, n, \\ 0, & \text{if } a_i < 0 \text{ for some } i. \end{cases}$$

Let $I \subset R$ be an ideal generated by monomials. Since I is $\mathbb{Z}^n$-graded, the quotient ring R/I inherits the $\mathbb{Z}^n$-grading given by $(R/I)_a = R_a/I_a$ for all $a \in \mathbb{Z}^n$. In particular Stanley–Reisner rings have a *fine grading*.

Let M be a $\mathbb{Z}^n$-graded R-module. Each homogeneous component M_a of M is an R_0-module. Define the *Hilbert function* $H(M, a) = \ell(M_a)$, provided that the length $\ell(M_a)$ of M_a is finite for all a, and call

$$F(M, \mathbf{t}) = \sum_{a \in \mathbb{Z}^n} H(M, a)\mathbf{t}^a$$

the *Hilbert–Poincaré series* of M. Here $\mathbf{t} = (t_1, \dots, t_n)$, where the t_i are indeterminates and $\mathbf{t}^a = t_1^{a_1} \cdots t_n^{a_n}$ for $a = (a_1, \dots, a_n) \in \mathbb{Z}^n$.

By induction on n it follows that the polynomial ring $R = K[x_1, \dots, x_n]$ with the fine grading has Hilbert–Poincaré series:

$$F(R, \mathbf{t}) = \sum_{a \in \mathbb{N}^n} \mathbf{t}^a = \prod_{i=1}^{n} \frac{1}{1 - t_i}.$$

Let Δ be a simplicial complex with vertices $X_1, \dots, X_n$, permitting an abuse of notation we also denote by X_i the residue class of x_i in $K[\Delta]$.

Thus $K[\Delta] = K[X_1, \ldots, X_n]$. The *support* of $a \in \mathbb{Z}^n$, denoted by $\mathrm{supp}(a)$, is defined as the set $\{X_i | \, a_i > 0\}$.

Let $X^a \in K[\Delta]$ and let $\mathrm{supp}(a) = \{X_{i_1}, \ldots, X_{i_m}\}$. Since I_Δ is generated by square free monomials we have

$$X^a \neq 0 \Leftrightarrow X_{i_1} \cdots X_{i_m} \neq 0 \Leftrightarrow x_{i_1} \cdots x_{i_m} \notin I_\Delta \Leftrightarrow \mathrm{supp}(a) \in \Delta.$$

Hence the non-zero monomials X^a form a K-basis of $K[\Delta]$. Therefore

$$F(K[\Delta], \mathbf{t}) = \sum_{\substack{a \in \mathbb{N}^n \\ \mathrm{supp}(a) \in \Delta}} \mathbf{t}^a = \sum_{F \in \Delta} \sum_{\substack{a \in \mathbb{N}^n \\ \mathrm{supp}(a) = F}} \mathbf{t}^a$$

Let $F \in \Delta$. If $F = \emptyset$, then $\sum_{\mathrm{supp}\; a = F} \mathbf{t}^a = 1$, and if $F \neq \emptyset$, then

$$\sum_{\substack{a \,\in\, \mathbb{N}^n \\ \mathrm{supp}\; a \subset F}} \mathbf{t}^a = \prod_{x_i \in F} \frac{1}{1 - t_i} \Longrightarrow \sum_{\substack{a \,\in\, \mathbb{N}^n \\ \mathrm{supp}\; a = F}} \mathbf{t}^a = \prod_{x_i \in F} \frac{t_i}{1 - t_i}.$$

Altogether we obtain that the expression for $F(K[\Delta], \mathbf{t})$ simplifies to

$$F(K[\Delta], \mathbf{t}) = \sum_{F \in \Delta} \prod_{x_i \in F} \frac{t_i}{1 - t_i}, \tag{6.8}$$

where the product over an empty index set is equal to 1.

Definition 6.7.1 Given a simplicial complex Δ of dimension d its *f-vector* is the $(d+1)$-tuple:

$$f(\Delta) = (f_0, \ldots, f_d),$$

where f_i is the number of i-faces of Δ. Note $f_{-1} = 1$.

To compute the Hilbert series of $K[\Delta]$, as a standard $\mathbb{N}$-graded algebra, note that for $i \in \mathbb{Z}$ we have

$$K[\Delta]_i = \bigoplus_{a \in \mathbb{Z}^n \; |a| = i} K[\Delta]_a$$

where $|a| = a_1 + \cdots + a_n$ for $a = (a_1, \ldots, a_n)$. Observe that the Hilbert series of $K[\Delta]$ with the fine grading specializes to the Hilbert series of $K[\Delta]$ with the $\mathbb{Z}$-grading, that is, if $t_i = t$ for all i, then

$$F(K[\Delta], \mathbf{t}) = F(K[\Delta], t).$$

Thus, by Eq. (6.8), we obtain a formula for the Hilbert series of $K[\Delta]$:

Theorem 6.7.2 [395] *If Δ is a simplicial complex and $f(\Delta) = (f_0, \ldots, f_d)$ is its f-vector, then the Hilbert series of $K[\Delta]$ is given by*

$$F(K[\Delta], z) = \sum_{i=-1}^{d} \frac{f_i z^{i+1}}{(1-z)^{i+1}}, \quad d = \dim(\Delta).$$

The Hilbert function of $K[\Delta]$ can be read off from its Hilbert series:

Proposition 6.7.3 *If Δ is a simplicial complex of dimension d and (f_i) is its f-vector, then the Hilbert function of $K[\Delta]$ is:*

$$H(K[\Delta], j) = \sum_{i=0}^{d} \binom{j-1}{i} f_i, \quad \textit{for } j \geq 1 \textit{ and } H(K[\Delta], 0) = 1.$$

A result of Kruskal–Katona

Kruskal and Katona [283] showed that $(f_0, f_1, \ldots, f_d) \in \mathbb{Z}^{d+1}$ is the f-vector of some d-dimensional simplicial complex if and only if

$$0 < f_{i+1} \leq f_i^{(i+1)}, \ 0 \leq i \leq d-1,$$

where $f_i^{(i+1)}$ is defined according to Eq. (5.7), cf. [395, Theorem 2.1, p. 55].

Simplicial complexes and their h-vectors Next we derive formulas for the h-vector of $K[\Delta]$. If Δ is Cohen–Macaulay, we present some numerical constraints for the h-vector which are central for applications of commutative algebra to combinatorics [391] (see Theorem 6.7.7).

Definition 6.7.4 The *h-vector of a simplicial complex* Δ, denoted by $h(\Delta)$,, is defined as the h-vector of the standard graded algebra $K[\Delta]$.

Theorem 6.7.5 [395, p. 58] *Let Δ be a simplicial complex of dimension d with h-vector (h_i) and f-vector (f_i). Then $h_k = 0$ for $k > d+1$ and*

$$h_k = \sum_{i=0}^{k} (-1)^{k-i} \binom{d+1-i}{k-i} f_{i-1} \quad \textit{for} \quad 0 \leq k \leq d+1. \tag{6.9}$$

Proof. The idea is to write the Hilbert series of $K[\Delta]$ in two ways. By Theorems 6.7.2 and 5.1.4 there is a polynomial $h(t) = h_0 + h_1 t + \cdots + h_r t^r$ with integral coefficients so that $h(1) \neq 0$ and satisfying

$$F(K[\Delta], t) = \sum_{i=-1}^{d} \frac{f_i t^{i+1}}{(1-t)^{i+1}} = \frac{h(t)}{(1-t)^{d+1}}. \tag{6.10}$$

Comparing the series yields the asserted equalities. $\square$

Theorem 6.7.6 *Let Δ be a simplicial complex of dimension d with h-vector (h_i) and f-vector (f_i). Then*

$$f_{k-1} = \sum_{i=0}^{k} \binom{d+1-i}{k-i} h_i, \quad \text{for } 1 \le k \le d+1, \tag{6.11}$$

$f_d = \sum_{i=0}^{d+1} h_i$, $h_1 = f_0 - (d+1)$ and $h_{d+1} = (-1)^d \sum_{i=-1}^{d} (-1)^i f_i$.

Proof. By the substitution $t = z/(1+z)$ in the equality

$$\frac{h_0 + h_1 t + \cdots + h_{d+1} t^{d+1}}{(1-t)^{d+1}} = \sum_{i=0}^{d+1} \frac{f_{i-1} t^i}{(1-t)^i},$$

we obtain $\sum_{i=0}^{d+1} h_i z^i (1+z)^{d+1-i} = \sum_{i=0}^{d+1} f_{i-1} z^i$. Hence

$$f_{k-1} = \sum_{i=0}^{d+1} \binom{d+1-i}{k-i} h_i = \sum_{i=0}^{k} \binom{d+1-i}{k-i} h_i, \quad \text{for } 1 \le k \le d+1.$$

From Eqs. (6.9) and (6.11) we derive the formulas for f_d, h_1 and h_{d+1}. □

Computation of the h-vector of face rings

To compute the h-vector (of face rings) the following procedure of Stanley [393] can be used. Consider the 3-dimensional convex polytope $\mathcal{P}$ consisting of two solid tetrahedrons joined by a 2-face:

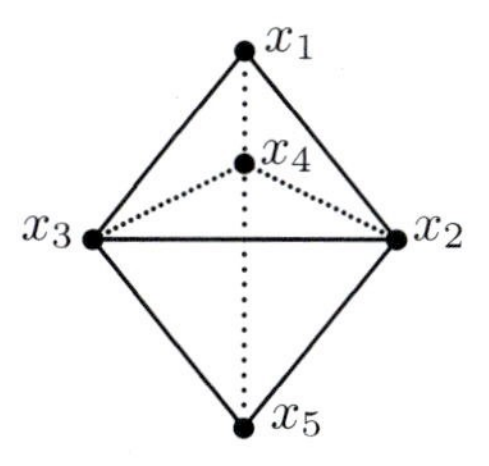

one has $f(\mathcal{P}) = (5, 9, 6)$. Write down the f-vector on a diagonal, and put a 1 to the left of f_0.

$$\begin{array}{cccc} 1 & 5 & & \\ & & 9 & \\ & & & 6 \end{array}$$

Complete this array constructing a *table*, by placing below a pair of consecutive entries their difference, and placing a 1 on the left edge:

$$\begin{array}{ccccccccc} & & & 1 & & 5 & & & \\ & & 1 & & 4 & & 9 & & \\ & 1 & & 3 & & 5 & & 6 & \\ \hline h(\mathcal{P}) = & (1 & & 2 & & 2 & & 1) & \end{array}$$

After completing the table the next row of differences will be the h-vector. Hence the boundary simplicial complex $\Delta = \Delta(\mathcal{P})$ is 2-dimensional, $K[\Delta]$ has multiplicity 6, and $I_\Delta = (x_1x_5, x_2x_3x_4)$.

Now we present a result that plays a role in the proof of the upper bound conjecture for spheres [391].

Theorem 6.7.7 *Let Δ be a simplicial complex of dimension d with n vertices. If $K[\Delta]$ is Cohen–Macaulay and K is an infinite field, then the h-vector of Δ satisfies*

$$0 \leq h_i(\Delta) \leq \binom{i+n-d-2}{i} \quad \textit{for} \quad 0 \leq i \leq d+1. \tag{6.12}$$

Proof. Set $S = K[\Delta] = R/I$, where $R = K[x_1, \ldots, x_n]$ and $I = I_\Delta$. By Lemma 3.1.28 there exists a regular homogeneous system of parameters $\underline{\theta} = \theta_1, \ldots, \theta_{d+1}$ for S such that each θ_i has degree one. Since $A = S/(\underline{\theta})S$ is Artinian, in this case Theorem 5.2.5 says that $h_i(\Delta) = H(A, i)$.

Note that $S/(\underline{\theta})S \simeq \overline{R}/\overline{I}$, where $\overline{R} = R/(\underline{\theta})$ is a polynomial ring in $n-d-1$ variables and $\overline{I}$ is the image of I in $\overline{R}$. Therefore we have

$$h_i(\Delta) = H(A, i) = H(\overline{R}/\overline{I}, i) \leq H(\overline{R}, i) = \binom{i+n-d-2}{i}. \qquad \Box$$

Example 6.7.8 Let Δ be a 2-dimensional complex with $f(\Delta) = (5, 6, 2)$. Thus $K[\Delta]$ is not Cohen–Macaulay because its h vector has a negative entry:

$$\begin{array}{ccccccccc} & & & 1 & & 5 & & & \\ & & 1 & & 4 & & 6 & & \\ & 1 & & 3 & & 2 & & 2 & \\ \hline h(\Delta) = & (1 & & 2 & & -1 & & 0). & \end{array}$$

Exercises

6.7.9 Let $S = K[\Delta]$ be the Stanley–Reisner ring of a simplicial complex Δ over a field K and $\varphi(t)$ the Hilbert polynomial of S. If $\widetilde{\chi}(\Delta)$ is the reduced Euler characteristic, prove:

(a) $\varphi(i) = \dim_K(S_i)$ for all $i \geq 1$, and

(b) $\varphi_S(0) = 1$ if and only if $\widetilde{\chi}(\Delta) = 0$.

6.7.10 Let Δ be a simplicial complex of dimension d and $(f_0, \ldots, f_d)$ its f-vector. Prove that f_d is equal to $\deg(K[\Delta])$, the degree of $K[\Delta]$.

6.7.11 Let $S = K[\Delta]$ be the Stanley–Reisner ring of a simplicial complex Δ with vertices $x_1, \ldots, x_n$ and $A = S/(x_1^2, \ldots, x_n^2)$. Prove that Δ is pure if and only if all the non-zero elements of the socle of A have the same degree.

6.7.12 If Δ is a connected simplicial complex having all its vertices of degree two, then Δ is a cycle.

6.7.13 [23] Let $R = K[x_1, \ldots, x_n]$ be a polynomial ring over a field K. If I is a monomial Gorenstein ideal and $\operatorname{rad}(I) = (x_1, \ldots, x_n)$, prove that I is generated by $x_1^{a_1}, \ldots, x_n^{a_n}$ for some $a_1, \ldots, a_n$ in $\mathbb{N}_+$.

6.7.14 Give an example of a monomial Gorenstein ideal which is not a complete intersection.

6.7.15 Let $R = K[x_1, \ldots, x_n]$ be a polynomial ring over a field K and let Δ be a simplicial complex with vertices $x_1, \ldots, x_n$. Associated to the Stanley–Reisner ideal I_Δ define another ideal

$$I'(\Delta) = (I_\Delta, x_1y_1, \ldots, x_ny_n) \subset K[x_1, \ldots, x_n, y_1, \ldots, y_n],$$

and denote by $S(\Delta)$ the Stanley–Reisner complex corresponding to this ideal $I'(\Delta)$. Prove the following:

(a) The simplicial complex $S(\Delta)$ is pure shellable.

(b) The f-vector of Δ is related to the h-vector of $S(\Delta)$ by

$$f_i(\Delta) = h_{i+1}(S(\Delta)) \quad \text{for all} \quad i.$$

6.7.16 Let k and n be two positive integers. A *k-partition* of n is an element $\alpha = (\alpha_1, \ldots, \alpha_k)$ in $\mathbb{N}_+^k$ so that $n = \alpha_1 + \cdots + \alpha_k$. Prove that the number of k-partitions of n is equal to $\binom{n-1}{k-1}$.

6.8 Simplicial spheres

Let Δ be a finite simplicial complex with vertices $x_1, \ldots, x_n$ and let e_i be the ith unit vector in $\mathbb{R}^n$. Given a face $F \in \Delta$, we set

$$|F| = \operatorname{conv}\{e_i \mid x_i \in F\},$$

where "conv" is the convex hull. Define the *geometric realization* of Δ as

$$|\Delta| = \bigcup_{F \in \Delta} |F|.$$

Then $|\Delta|$ is a topological space with the induced usual topology of $\mathbb{R}^n$. Note that there is a canonical isomorphism $\widetilde{H}_i(\Delta, A) \cong \widetilde{H}_i(|\Delta|; A)$ for all i, where $\widetilde{H}_i$ are the corresponding reduced homology modules.

Definition 6.8.1 A *q-simplex* is a polytope generated by a set of $q+1$ affinely independent points. A polytope is *simplicial* if everyone of its proper faces is a simplex.

Let $\mathcal{P}$ be a polytope of dimension $d+1$ and denote the number of faces of dimension i of $\mathcal{P}$ by f_i, thus f_0 is the number of vertices of $\mathcal{P}$. The *f-vector* of $\mathcal{P}$ is the vector $f(\mathcal{P}) = (f_0, \ldots, f_d)$ and we set $f_{-1} = 1$. The *boundary complex* $\Delta(\mathcal{P})$ of $\mathcal{P}$ is by definition the "abstract simplicial complex" whose vertices are the vertices of $\mathcal{P}$ and whose faces are those sets of vertices that span a proper face of $\mathcal{P}$, thus $\Delta(\mathcal{P})$ has dimension d. Notice that $\mathcal{P}$ and $\Delta(\mathcal{P})$ have the same f-vector and accordingly one defines the *h-vector* of $\mathcal{P}$ as the h-vector of $\Delta(\mathcal{P})$.

The geometric realization $|\Delta(\mathcal{P})|$ is therefore homeomorphic to a sphere S^d of dimension d. This suggests the following more general concept.

Definition 6.8.2 A simplicial complex Δ of dimension d is a *simplicial sphere* if $|\Delta| \simeq S^d$, in this case Δ is said to be a *triangulation* of S^d.

Remark 6.8.3 There are simplicial d-spheres with $d = 3$ and $n = 8$ which are not the boundary complex of a simplicial $(d+1)$-polytope [393]. Moreover it follows from a result of Steinitz (a graph is the 1-skeleton of a 3-polytope in $\mathbb{R}^3$ if and only if it is a 3-connected planar graph) that such an example does not exist for $d = 2$; see [438, Theorem 4.1].

Theorem 6.8.4 [65, Chapter 5] *Let Δ be a simplicial complex of dimension d. If $|\Delta| \cong S^d$, then*

$$\widetilde{H}_i(\mathrm{lk}\, F; K) \cong \begin{cases} K & \text{for } i = \dim(\mathrm{lk}(F)), \\ 0 & \text{otherwise.} \end{cases}$$

Proposition 6.8.5 *If Δ is a simplicial sphere, then Δ is Cohen–Macaulay and the following equality holds*

$$\widetilde{\chi}(\mathrm{lk}\ F) = (-1)^{\dim\, \mathrm{lk}\ F} \quad \text{for} \quad F \in \Delta.$$

Proof. If follows from Reisner's criterion and Theorem 6.8.4. □

Remark 6.8.6 If $\mathcal{P}$ is a simplicial d-polytope, then $\Delta(\mathcal{P})$ is a simplicial sphere of dimension $d-1$. Hence, from Proposition 6.8.5, we obtain the Euler's formula

$$\widetilde{\chi}(\mathcal{P}) = \widetilde{\chi}(\Delta(\mathcal{P})) = \sum_{i=-1}^{d-1} (-1)^i f_i = (-1)^{d-1} = -1 + \chi(\mathcal{P}),$$

where $f = (f_0, \ldots, f_{d-1})$ is the f-vector of $\mathcal{P}$. The Euler's formula is valid for any d-polytope, not just the simplicial ones (see [57, Theorem 16.1]).

Corollary 6.8.7 *If Δ is a simplicial sphere, then $K[\Delta]$ is Gorenstein.*

Proof. It follows from [395, Theorem 5.1] and Theorem 6.8.4. □

Theorem 6.8.8 *If Δ is a simplicial sphere of dimension d, then its h-vector satisfies $h_i = h_{d+1-i}$ for $i = 0, \ldots, d+1$.*

Proof. Since $K[\Delta]$ is a Gorenstein ring, its h-vector is symmetric according to Proposition 5.3.9 and Corollary 5.3.10. □

Remark 6.8.9 If Δ is a simplicial sphere, then $h_i = h_{d+1-i}$ and one can recover Euler's formula (make $i = 0$ and use Theorem 6.7.6). Those relations are the *Dehn–Sommerville equations.*

Corollary 6.8.10 [314] *If $h = (h_0, \ldots, h_{d+1})$ is the h-vector of a simplicial polytope of dimension $d+1$, then $h_i = h_{d+1-i}$ for $0 \leq i \leq d+1$.*

Proof. It follows from Theorem 6.8.8. □

The next two conjectures were raised around 1993 in connection with the problem of bounding the Betti numbers of graded Gorenstein ideals [318].

Conjecture 6.8.11 *Let Δ be a simplicial sphere and I the Stanley–Reisner ideal of Δ. If I has initial degree p and height g, then for $1 \leq i < g$ the ith Betti number β_i of $K[\Delta]$ satisfies:*

$$\beta_i \leq \binom{p+g-1}{p+i-1}\binom{p+i-2}{i-1} + \binom{p+g-1}{i}\binom{p+g-i-2}{p-1} - \binom{g}{i}\binom{p+g-2}{p-1}.$$

Conjecture 6.8.12 *Let Δ be a simplicial sphere and I its Stanley–Reisner ideal. If I has initial degree p and height g, then*

$$\mu(I) \leq \binom{p+g-1}{g-1} - \binom{p+g-3}{g-1}.$$

Exercises

6.8.13 If Δ is a simplicial sphere on n vertices whose geometric realization $|\Delta|$ is isomorphic to the unit circle S^1, then

(a) Δ is a cycle of length.

(b) $\mu(I_\Delta) = \binom{n}{2} - n$ for $n \geq 4$.

6.9 The upper bound conjectures

In this section we present the main steps in the proof of the upper bound conjecture for simplicial spheres. An excellent reference for a detailed proof of this conjecture is [65].

Let $\mathcal{P}$ be a polytope of dimension $d+1$ and let $f_i = f_i(\mathcal{P})$ be the number of faces of dimension i of $\mathcal{P}$. In 1893 Poincaré proved the *Euler characteristic formula*

$$\sum_{i=0}^{d}(-1)^i f_i = 1 + (-1)^d, \tag{6.13}$$

see [57, Theorem 16.1]. In [393] there are some historical comments about this formula. It follows from Proposition 6.8.5 that this formula also holds for simplicial spheres. Are there any optimal bounds for $f_i(\mathcal{P})$? As it will be seen there is a positive answer to this question which is valid in a more general setting.

In order to formulate the upper bound theorem for simplicial spheres and the upper bound theorem for convex polytopes we need to introduce some results on *cyclic polytopes*, the reader is referred to [57] for a detailed discussion on this topic.

Consider the *monomial curve* $\Gamma \subset \mathbb{R}^{d+1}$ given parametrically by

$$\Gamma = \{(\tau, \tau^2, \ldots, \tau^{d+1}) | \, \tau \in \mathbb{R}\}.$$

A *cyclic polytope*, denoted by $C(n, d+1)$, is the convex hull of any n distinct points in Γ such that $n > d+1$. The f-vector of $C(n, d+1)$ depends only on n and d and not on the points chosen, and $\dim C(n, d+1) = d+1$.

The cyclic polytope $C(n, d+1)$ is simplicial and has the remarkable property that its f-vector satisfies:

$$f_i(C(n, d+1)) = \binom{n}{i+1} \quad \text{for} \quad 0 \leq i < \left\lfloor \frac{d+1}{2} \right\rfloor, \tag{6.14}$$

this means that $C(n, d+1)$ has the highest possible number of i-faces when i is within the specified rank. Hence for any polytope $\mathcal{P}$ of dimension $d+1$ with n vertices we have

$$f_i(\mathcal{P}) \leq \binom{n}{i+1} = f_i(C(n, d+1)),$$

for all $0 \leq i < \lfloor (d+1)/2 \rfloor$. The *upper bound conjecture for polytope* (UBCP for short) states that

$$f_i(\mathcal{P}) \leq f_i(C(n, d+1)) \quad \text{for} \quad 0 \leq i \leq d.$$

This conjecture was posed by Motzkin [327] in 1957 and proved by McMullen [314] in 1970. It was motivated by the performance of the simplex algorithm in linear programming.

Proposition 6.9.1 *If $h = (h_0, \ldots, h_{d+1})$ is the h-vector of a cyclic polytope $C(n, d+1)$, then*

$$h_k = \binom{n-d+k-2}{k}, \qquad 0 \leq k \leq \lfloor (d+1)/2 \rfloor.$$

Proof. Let $0 \leq k \leq \lfloor (d+1)/2 \rfloor$. One has

$$\begin{aligned} h_k = \sum_{i=0}^{k}(-1)^{k-i}\binom{d+1-i}{k-i} f_{i-1} &= \sum_{i=0}^{k}(-1)^{2(k-i)}\binom{k-d-2}{k-i}\binom{n}{i} \\ &= \binom{n-d+k-2}{k}. \end{aligned}$$

The second equality follows from Exercise (6.9.5). □

Since a cyclic polytope $C(n, d+1)$ is simplicial of dimension $d+1$ its boundary complex $\Delta(C(n, d+1))$ is a d-sphere. In 1964 V. Klee pointed out that the upper bound conjecture for polytopes may as well be made for simplicial spheres. This conjecture was proved by Stanley [391] in 1975 using techniques from commutative and homological algebra. Next we prove the *upper bound theorem for simplicial spheres*.

Theorem 6.9.2 [391] *Let Δ be simplicial complex of dimension d with n vertices. If $|\Delta| \cong S^d$, then $f_i(\Delta) \leq f_i(\Delta(C(n, d+1)))$ for $i = 0, \ldots, d$.*

Proof. Consider the following four conditions:

(a) $h_i(\Delta) \leq h_i(\Delta(C(n, d+1)))$ for $0 \leq i \leq \lfloor (d+1)/2 \rfloor$.

(b) $h_i(\Delta(C(n, d+1))) = h_{d+1-i}(\Delta(C(n, d+1)))$ for $0 \leq i \leq d+1$.

(c) $h_i(\Delta) \leq \binom{i+n-d-2}{i}$ all i.

(d) $h_i(\Delta) = h_{d+1-i}(\Delta)$ for $0 \leq i \leq d+1$.

We claim that (a), (b) and (d) imply the theorem. To show it notice that if $\lfloor (d+1)/2 \rfloor < i \leq d+1$, then $0 \leq d+1-i \leq \lfloor (d+1)/2 \rfloor$ and

$$h_i(\Delta) = h_{d+1-i}(\Delta) \leq h_{d+1-i}(\Delta(C(n, d+1))) = h_i(\Delta(C(n, d+1))),$$

hence $h_i(\Delta) \leq h_i(\Delta(C(n, d+1)))$ for $0 \leq i \leq d+1$. Therefore

$$\begin{aligned} f_{k-1}(\Delta) &= \sum_{i=0}^{d+1}\binom{d+1-i}{k-i} h_i(\Delta) \\ &\leq \sum_{i=0}^{d+1}\binom{d+1-i}{k-i} h_i(\Delta(C(n, d+1))) \\ &= f_{k-1}(\Delta(C(n, d+1))) \quad \text{for } 1 \leq k \leq d+1, \end{aligned}$$

and the proof of the claim is completed.

By Proposition 6.9.1 (c) $\Rightarrow$ (a), and by Lemma 6.8.10 (b) is satisfied due to the fact that $C(n, d+1)$ is a simplicial polytope. Hence the theorem is reduced to prove (c) and (d). To complete the proof notice that (c) follows from Lemma 6.7.7, and (d) follows from Theorem 6.8.8. $\square$

As a particular case of the upper bound theorem for spheres we obtain the *upper bound theorem for convex polytopes*:

Theorem 6.9.3 [314] *Let $\mathcal{P}$ be a convex polytope of dimension d with n vertices, then $f_i(\mathcal{P}) \leq f_i(C(n,d))$ for $0 \leq i \leq d-1$.*

Proof. By pulling the vertices $\mathcal{P}$ can be transformed into a simplicial polytope with the same number of vertices as $\mathcal{P}$ and at least as many faces of higher dimension; see [315] and [438, Lemma 8.24]. Hence, we may assume that $\mathcal{P}$ is simplicial. Since $\mathcal{P}$ is simplicial, the boundary complex $\Delta(\mathcal{P})$ is a simplicial sphere, that is, $|\Delta(\mathcal{P})| \cong S^{d-1}$. Using Theorem 6.9.2 we obtain

$$f_i(\mathcal{P}) = f_i(\Delta(\mathcal{P})) \leq f_i(\Delta(C(n,d))) = f_i(C(n,d)), \quad \text{for } i = 0, \ldots, d. \quad \square$$

Theorem 6.9.4 [57, Corollary 18.3] *If $\mathcal{P}$ is a simplicial d-polytope with n vertices, then*

$$f_j(\mathcal{P}) \leq \varphi_{d-j-1}(d,n) \qquad (j = 1, \ldots, d-1),$$

where

$$\varphi_{d-j-1}(d,n) = \sum_{i=0}^{\lfloor d/2 \rfloor} \binom{i}{d-j-1}\binom{n-d+i-1}{i} + \sum_{i=0}^{\lfloor (d-1)/2 \rfloor} \binom{d-i}{d-j-1}\binom{n-d+i-1}{i}.$$

Exercises

6.9.5 If $r \in \mathbb{R}$ and $k \in \mathbb{N}$, then

$$\binom{r}{k} = (-1)^k \binom{-r+k-1}{k}.$$

6.9.6 [57, p. 79] Prove that a d-polytope $\mathcal{P}$ is simplicial if and only if each facet of $\mathcal{P}$ is a simplex.

Chapter 7

Edge Ideals of Graphs

In this chapter we give an introduction to graph theory and study algebraic properties and invariants of edge ideals using the combinatorial structure of graphs. We present classifications of the following families: unmixed bipartite graphs, Cohen–Macaulay bipartite graphs, Cohen–Macaulay trees, shellable bipartite graphs, sequentially Cohen–Macaulay bipartite graphs. The invariants examined here include regularity, depth, Krull dimension, and multiplicity. Edge ideals are shown to have the persistence property.

For those readers interested in a comprehensive view of graph theory itself, we recommend the books of Bollobás [48], Diestel [111], and Harary [208], which we take as our main references for the subject.

7.1 Graph theory

A *graph* G is an ordered pair of disjoint finite sets (V, E) such that E is a subset of the set of unordered pairs of V. The set V is the set of *vertices* and the set E is called the set of *edges*. The number of vertices of a graph is its *order*. An edge $e = \{u, v\}$, with $u, v \in V$, is said to join the vertices u and v. When working with several graphs it is convenient to write $V(G)$ and $E(G)$ for the vertex set and edge set of G, respectively. Sometimes we will also write V_G and E_G for the vertex set and edge set of G, respectively.

Let G be a graph on the vertex set V. If $e = \{u, v\}$ is an edge of G one says that the vertices u and v are *adjacent* or neighboring vertices of G. In this case it is also usual to say that the edge e is incident with u and v. The *degree* of a vertex v in V denoted by $\deg(v)$, is the number of edges incident with v. A vertex with degree zero is called an *isolated vertex*.

Note that by definition a graph does not contain *loops*, a pair $\{v, v\}$ ("an edge joining a vertex to itself"); neither does it contain a pair $\{u, v\}$ that occurs several times ("that is, several edges joining the same two vertices").

If we allow these type of relations as edges we will call G a *multigraph*. Most results on graphs carry over to multigraphs in a natural way. There are areas in graph theory (such as plane duality [111]) where multigraphs arise more naturally than graphs. Terminology introduced earlier for graphs can be used correspondingly for multigraphs.

Given two graphs G and H, a mapping φ, from $V(G)$ to $V(H)$ is called a *homomorphism* if $\{\varphi(u), \varphi(v)\} \in E(H)$ whenever $\{u, v\} \in E(G)$. Two graphs G and H are *isomorphic* if there is a bijective map ψ from $V(G)$ to $V(H)$ such that $\{u, v\} \in E(G)$ if and only if $\{\psi(u), \psi(v)\} \in E(H)$. A map taking graphs as arguments is called a *graph invariant* if it assigns equal values to isomorphic graphs. The number of vertices and the number of edges are two simple examples of graph invariants.

Let H and G be two graphs, then H is called a *subgraph* of G if $V(H) \subset V(G)$ and $E(H) \subset E(G)$. A subgraph H is called an *induced subgraph* if H contains all the edges $\{u, v\} \in E(G)$ with $u, v \in V(H)$. In this case H is said to be the subgraph induced by $V(H)$. An induced subgraph is denoted by $H = G[V(H)]$ or $H = \langle V(H) \rangle$. It is also denoted by $H = G_{V(H)}$. A *spanning subgraph* is a subgraph H of G containing all the vertices of G.

A *walk* of *length* n in G is an alternating sequence of vertices and edges, written as $w = \{v_0, z_1, v_1, \ldots, v_{n-1}, z_n, v_n\}$, where $z_i = \{v_{i-1}, v_i\}$ is the edge joining the vertices v_{i-1} and v_i. A walk may also be written $\{v_0, \ldots, v_n\}$ with the edges understood, or $\{z_1, z_2, \ldots, z_n\}$ with the vertices understood. If $v_0 = v_n$, the walk w is called a *closed walk*. A *path* is a walk with all its vertices distinct.

We say that G is *connected* if for every pair of vertices u and v there is a path from u to v. Notice that G has a vertex disjoint decomposition $G = \cup_{i=1}^{r} G_i$ where $G_1, \ldots, G_r$ are the maximal (with respect to inclusion) connected subgraphs of G, the G_i's are called the *connected components* of G. A component is called *even* (resp. *odd*) if its order is even (resp. odd).

A *cycle* of *length* n is a closed path $\{v_0, \ldots, v_n\}$ in which $n \geq 3$. A cycle is *even* (resp. odd) if its length is even (resp. odd). We denote by C_n the graph consisting of a cycle with n vertices, C_3 will be called a *triangle*, C_4 a *square* and so on. If all the vertices of G are isolated, G is called a *discrete graph*. A *forest* is an acyclic graph and a *tree* is a connected forest. The *complete graph* $\mathcal{K}_n$ has every pair of its n vertices adjacent. For instance:

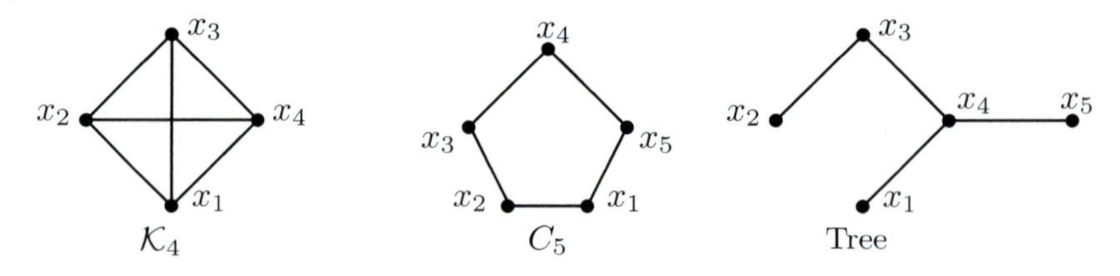

A *coloring* of the vertices of a graph G is an assignment of colors to the vertices of G in such a way that adjacent vertices have distinct colors. The *chromatic number* of a graph G, denoted by $\chi(G)$, is the minimal number of colors in a coloring of G. There are algebraic methods to compute the chromatic number [165] (see Theorem 7.7.19).

A *clique* of a graph G is a set of vertices that induces a complete subgraph. We will also call a complete subgraph of G a clique. A *maximal clique* of G is a clique which is maximal with respect to inclusion.

The *clique number* of G, denoted by $\omega(G)$, is the size of the largest complete subgraph of G. The clique number and the chromatic number are related by the inequality

$$\omega(G) \leq \chi(G).$$

A graph is called *perfect* if for every induced subgraph H, the chromatic number of H is equal to the size of the largest complete subgraph of H, i.e., $\omega(H) = \chi(H)$ for every induced subgraph of G. This notion was introduced by Berge [25, Chapter 16]. An excellent reference for the theory of perfect graphs is the book of Golumbic [191]; see also [93, 373].

A graph G is *bipartite* if its vertex set $V(G)$ can be partitioned into two disjoint subsets V_1 and V_2 such that every edge of G has one vertex in V_1 and one vertex in V_2. The pair (V_1, V_2) is called a *bipartition* of G. If G is connected such a bipartition is uniquely determined.

A bipartite graph G with bipartition (V_1, V_2) is called a *complete bipartite graph* if $\{u, v\} \in E(G)$ for all $u \in V_1$ and $v \in V_2$. If V_1 and V_2 have m and n vertices, respectively, we denote a complete bipartite graph by $\mathcal{K}_{m,n}$. A *star* is a complete bipartite graph of the form $\mathcal{K}_{1,n}$. For instance:

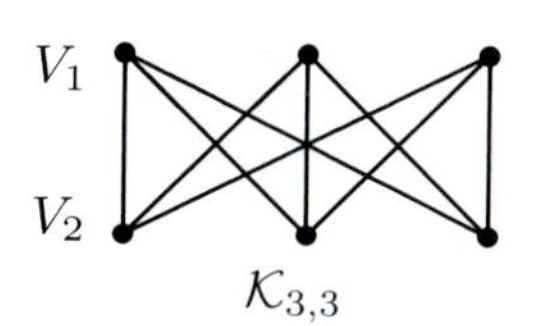

$\mathcal{K}_{3,3}$

Definition 7.1.1 The *distance* $d(u, v)$ between two vertices u and v of a graph G is defined to be the minimum of the lengths of all possible paths from u to v. If there is no path joining u and v, then $d(u, v) = \infty$.

Proposition 7.1.2 *A graph G is bipartite if and only if all the cycles of G are even. In particular every forest is a bipartite graph.*

Proof. $\Rightarrow$) Let (V_1, V_2) be a bipartition of G. If $\{v_0, \ldots, v_n\}$ is a cycle of G, one may assume $v_0 \in V_1$. Since $v_1 \in V_2$, it follows at once that v_i is in V_1 if and only if i is even, thus n must be even.

$\Leftarrow$) It suffices to prove that each connected component of G is bipartite. Thus one may assume that G is connected. Pick a vertex $v_0 \in V$. Set

$$V_1 = \{v \in V(G) | d(v, v_0) \text{ is even}\} \ \text{ and } \ V_2 = V(G) \setminus V_1.$$

It follows that no two vertices of V_i are adjacent for $i = 1, 2$, otherwise G would contain an odd cycle. Therefore (V_1, V_2) is a bipartition of G and the graph G is bipartite. $\square$

If e is an edge, we denote by $G\setminus\{e\}$ the spanning subgraph of G obtained by deleting e and keeping all the vertices of G. The removal of a vertex v from a graph G results in a subgraph $G \setminus \{v\}$ of G consisting of all the vertices in G except v and all the edges not incident with v. For instance:

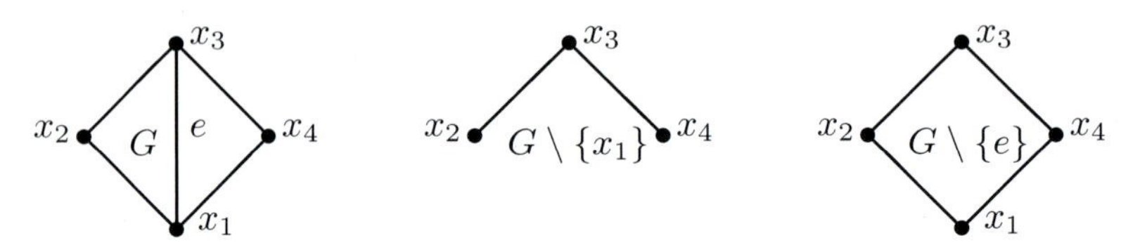

Given a set of vertices A, by $G \setminus A$, we mean the graph formed from G by deleting all the vertices in A, and all edges incident to a vertex in A.

Let G be a graph. A vertex v (resp. an edge e) of G is called a *cutvertex* or *cutpoint* (resp. a *bridge*) if the number of connected components of $G\setminus\{v\}$ (resp. $G \setminus \{e\}$) is larger than that of G. A maximal connected subgraph of G without cut vertices is called a *block*. A graph G is 2-*connected* if $|V(G)| > 2$ and G has no cut vertices. Thus a block of G is either a maximal 2-connected subgraph, a bridge, or an isolated vertex. By their maximality, different blocks of G intersect in at most one vertex, which is then a cutvertex of G. Therefore every edge of G lies in a unique block, and G is the union of its blocks.

Definition 7.1.3 A set of edges in a graph G is called *independent* or a *matching* if no two of them have a vertex in common.

Definition 7.1.4 Let A be a set of vertices of a graph G. The *neighbor set* of A, denoted by $N_G(A)$ or simply by $N(A)$ if G is understood, is the set of vertices of G that are adjacent with at least one vertex of A.

In particular, if G is a graph and $v \in V(G)$, the *neighbor set* of v, denoted by $N_G(v)$ or $N(v)$, is the set of vertices of G adjacent to v. Notice that if C_r is a cycle of a graph G, then $C_r \subset N_G(C_r)$.

Proposition 7.1.5 *Let G be a bipartite graph with bipartition (V_1, V_2) and let m, n be the number of vertices in V_1, V_2, respectively. If $|A| \leq |N_G(A)|$ for all $A \subset V_1$, then there are m independent edges in G.*

Proof. By induction on m. If $m = 1$, then there is at least one vertex in V_2 connected to V_1, thus there is one independent edge. Assume $m \geq 2$.

Case (I): Assume that $|A| < |N_G(A)|$ for all $A \subset V_1$ with $|A| < m$. Let $x_1 \in V_1$ and $y_1 \in V_2$ be two adjacent vertices. Consider the subgraph

$H = G \setminus \{x_1, y_1\}$ obtained from G by removing x_1 and y_1. Note that for every $A \subset V_1 \setminus \{x_1\}$ one has $N_H(A) = N_G(A)$ if $y_1 \notin N_G(A)$ and $N_H(A) = N_G(A) \setminus \{y_1\}$ otherwise. Hence by induction there are independent edges $z_2, \ldots, z_m$ in H, which together with $z_1 = \{x_1, y_1\}$ yield the required number of independent edges.

Case (II): Assume that $|A| = |N_G(A)|$ for some $A \subset V_1$ with $|A| < m$. Set $r = |A|$. Consider the subgraph $H = G \setminus (V_1 \setminus A)$. As $N_H(S) = N_G(S)$ for all $S \subset A$, by induction there are independent edges $z_1, \ldots, z_r$ in H. On the other hand consider $F = G \setminus (A \cup N_G(A))$. If $S \subset V_1 \setminus A$, then

$$|N_G(A \cup S)| \geq |A \cup S| = |S| + |A|,$$

since $N_G(S \cup A) = N_G(A) \cup N_F(S)$ and $r = |N_G(A)|$ we get $|N_F(S)| \geq |S|$. Applying induction there are independent edges $w_1, \ldots, w_{m-r}$ in F. Putting altogether it is seen that there are m independent edges in G, as required. This proof is due to Halmos and Vaughn. $\Box$

Corollary 7.1.6 *Let G be a bipartite graph with bipartition (V_1, V_2). If there is an integer $p \geq 0$ such that $|A| - p \leq |N_G(A)|$ for all $A \subset V_1$, then there are $m - p$ independent edges in G, where $m = |V_1|$.*

Proof. By Proposition 7.1.5 the assertion holds for $p = 0$. Consider the graph F obtained from G by adding a new vertex y to V_2 and joining every vertex of V_1 to y. Let $A \subset V_1$, since $N_F(A) = N_G(A) \cup \{y\}$ one concludes $|N_F(A)| \geq |A| - (p-1)$. Hence by induction there are $m-p+1$ independent edges in F, it follows that there are $m - p$ independent edges in G. $\Box$

Let G be a graph with vertex set V. A subset $C \subset V$ is a *minimal vertex cover* of G if: (i) every edge of G is incident with at least one vertex in C, and (ii) there is no proper subset of C with the first property. If C satisfies condition (i) only, then C is called a *vertex cover* of G.

The following example illustrates the notion of a minimal vertex cover:

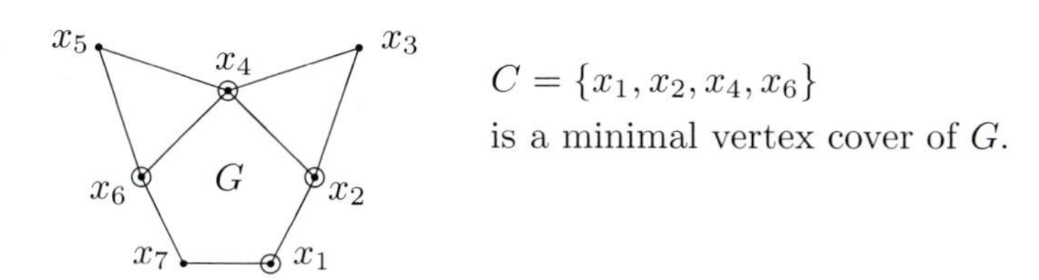

It is convenient to regard the empty set as a minimal vertex cover for a graph with all its vertices isolated. A set of vertices of G is called *independent* or *stable* if no two of them are adjacent. Notice the following duality: a set of vertices in G is a maximal independent set (with respect to inclusion) if and only if its complement is a minimal vertex cover for G.

Definition 7.1.7 Let G be a graph. The *vertex covering number*, denoted by $\alpha_0(G)$, is the number of vertices in any smallest vertex cover. The *matching number*, denoted by $\beta_1(G)$, is the number of edges in any largest independent set of edges. The *vertex independence number*, denoted by $\beta_0(G)$, is the number of vertices in any largest independent set of vertices.

In [26] the integer $\alpha_0(G)$ (resp. $\beta_0(G)$) is denoted by $\tau(G)$ (resp. $\alpha(G)$) and is called the *transversal number* (resp. *stability number*) of G.

Theorem 7.1.8 (König) *If G is a bipartite graph, then $\beta_1(G) = \alpha_0(G)$.*

Proof. Let (V_1, V_2) be a bipartition of G. Set $r = \beta_1(G)$ and $m = |V_1|$. Since there are r independent edges in G one obtains $\alpha_0(G) \geq r$.

There exists a subset $A \subset V_1$ such that $|N_G(A)| \leq |A| - (m-r)$, otherwise if $|N_G(A)| > |A| - (m-r)$ for all $A \subset V_1$, then by Corollary 7.1.6 one would have $r+1$ independent edges which is impossible. Hence $N_G(A) \cup (V_1 \setminus A)$ is a vertex cover of G with at most r elements, thus we get $\alpha_0(G) \leq r$. □

A pairing by an independent set of edges of all the vertices of a graph G is called a *perfect matching* or a 1-*factor*. Thus G has a perfect matching if and only if G has an even number of vertices and there is a set of independent edges containing all the vertices.

Next we present a criterion for the existence of a perfect matching.

Theorem 7.1.9 (Marriage theorem) *Let G be a bipartite graph. Then the following are equivalent*

(a) *G has a perfect matching.*

(b) *$|A| \leq |N_G(A)|$ for all $A \subset V_G$ independent set of vertices.*

Proof. Let (V_1, V_2) be a bipartition of G. Set $r = \beta_1(G)$. It is easy to see that (a) implies (b). To prove that (b) implies (a), we begin by noticing that $|V_1| = |V_2|$ because V_i is an independent set for $i = 1, 2$. By König theorem there are $z_1, \ldots, z_r$ independent edges and a minimal vertex cover A of G with r elements. Hence $z_i \cap A$ has exactly one vertex for any i and A is an independent set. Using that $V_G \setminus A$ is an independent set of vertices and from the equality $N_G(V_G \setminus A) = A$ we conclude that $|V_G \setminus A| \leq |N_G(V_G \setminus A)| = |A|$. It follows that $|V_1| = r$, and thus $z_1, \ldots, z_r$ yields a perfect matching. □

What happens if the graph G is not bipartite? The answer is given by the following theorem of Tutte. A very readable and comprehensive reference about matchings in finite graphs is the book of Lovász and Plummer [297].

Theorem 7.1.10 (Tutte; see [111, Theorem 2.2.1]) *A graph G has a perfect matching if and only if $c_0(G \setminus S) \leq |S|$ for all $S \subset V_G$, where $c_0(G)$ denotes the number of odd components (connected components with an odd number of vertices) of G.*

A *directed graph* or *digraph* $\mathcal{D}$ consists of a finite set $V(\mathcal{D})$ of vertices together with a prescribed collection $E(\mathcal{D})$ of ordered pairs of distinct points called edges or arrows. An *oriented graph* is a digraph having no-cycles of length two. In other words an oriented graph is a graph together with an orientation of its edges. A *tournament* is a complete oriented graph. Any tournament has a spanning directed path [14, Theorem 1.4.5].

Exercises

7.1.11 Let G be a graph with vertex set $V_G = \{x_1, \ldots, x_n\}$ and edge set E_G, prove Euler's identity: $2|E_G| = \sum_{i=1}^{n} \deg(x_i)$.

7.1.12 If G is a connected bipartite graph, prove that G has a unique bipartition.

7.1.13 If u, v are two vertices at maximum distance in a connected graph G, then u, v are not cutpoints.

7.1.14 If G is a regular bipartite graph (all vertices have the same degree), then G has a perfect matching.

7.1.15 An *edge cover* in a graph G is a set of edges collectively incident with each vertex of G. The *edge covering number* of G denoted by $\alpha_1(G)$, is the number of edges in any smallest edge cover in G. Prove:

(a) If G has no isolated vertices, then $\beta_1(G) + \alpha_1(G) = |V_G|$.

(b) For any graph G, $\beta_1(G) \leq \alpha_0(G)$. If G is bipartite, $\alpha_1(G) = \beta_0(G)$.

7.1.16 Let G be a bipartite graph with minimum degree r. Then G is the union of r edge disjoint edge covers.

7.1.17 Define a 2-*coloration* of a set A as a surjective mapping from A onto a set of two elements. A 2-coloration of a graph is a 2-coloration of its vertex set. Prove that a graph is connected if for any 2-coloration of the graph there exists a heterochromatic edge.

7.1.18 If G is a bipartite graph with bipartition (V_1, V_2), prove that the mapping from $V(G)$ to $V(\mathcal{K}_2)$ that sends all the vertices from V_i to vertex i is a homomorphism from G to $\mathcal{K}_2$.

7.1.19 Prove that the clique number and the chromatic number of a graph G satisfy $\omega(G) \leq \chi(G)$.

7.1.20 Let G be a connected perfect graph with vertex set $V(G)$. Then there are cliques $K_1, \ldots, K_s$ of G such that $V(G)$ is the disjoint union of $K_1, \ldots, K_s$ and $\alpha_0(G) = \alpha_0(G[K_1]) + \cdots + \alpha_0(G[K_s])$.

7.1.21 Prove that any tournament has a spanning directed path.

7.2 Edge ideals and B-graphs

In this section we introduce edge ideals and edge rings of graphs. We will compare several families of graphs and examine some of their numerical invariants. Some bounds for the Krull dimension are given.

Let G be a graph with vertices $x_1, \ldots, x_n$ and let $R = K[x_1, \ldots, x_n]$ be a polynomial ring over a field K. Here we are permitting an abuse of notation by using x_i to denote a vertex of G and also a variable of R.

Definition 7.2.1 The *edge ideal* $I(G)$ associated to the graph G is the ideal of R generated by the set of all square-free monomials x_ix_j such that x_i is adjacent to x_j. If all the vertices of G are isolated we set $I(G) = (0)$. The ring $R/I(G)$ is called the *edge ring* of G.

Edge ideals are algebraic representations of graphs. These ideals occur in computational chemistry [291], physics [51], and combinatorics [395]. The non-zero edge ideals are precisely those ideals generated by square-free monomials of degree two.

A prime example of an edge ideal comes from the theory of posets:

Definition 7.2.2 Given a poset $\mathcal{P}$ with vertex set $X = \{x_1 \ldots, x_n\}$, its *order complex*, denoted by $\Delta(\mathcal{P})$, is the simplicial complex on X whose faces are the *chains* (linearly ordered sets) in $\mathcal{P}$, thus

$$K[\Delta(\mathcal{P})] = K[X]/(x_ix_j \,|\, x_i \not\sim x_j)$$

is the Stanley–Reisner ring of $\Delta(\mathcal{P})$. Here $x_i \not\sim x_j$ means that x_i is not comparable to x_j.

The next result establishes a one-to-one correspondence between the minimal vertex covers of G and the minimal primes of $I(G)$.

Proposition 7.2.3 *A prime ideal $\mathfrak{p} \subset R$ is a minimal prime of $I(G)$ if and only if $\mathfrak{p}$ is generated by a minimal vertex cover of G.*

Proof. It follows at once from Lemma 6.3.37 and Theorem 6.1.4. □

Corollary 7.2.4 *If G is a graph, then the vertex covering number $\alpha_0(G)$ is equal to the height of the edge ideal $I(G)$.*

Proof. It follows at once from Proposition 7.2.3. □

Corollary 7.2.5 *Let G be a graph with n vertices and let $I(G)$ be its edge ideal. Then $n = \alpha_0(G) + \beta_0(G) = \mathrm{ht}(I(G)) + \dim R/I(G)$.*

Proof. It follows from Corollary 7.2.4. □

Definition 7.2.6 Let $I \subset R$ be a graded ideal. The quotient ring R/I is called *Cohen–Macaulay* (C–M for short) if $\operatorname{depth}(R/I) = \dim(R/I)$. The ideal I is called *Cohen–Macaulay* if R/I is Cohen–Macaulay.

Definition 7.2.7 A graph G is called *Cohen–Macaulay* over the field K if $R/I(G)$ is a Cohen–Macaulay ring.

Definition 7.2.8 A graph G is called *unmixed graph* or *well-covered* if any two minimal vertex covers of G have the same cardinality.

Proposition 7.2.9 *If G is a Cohen–Macaulay graph, then G is unmixed.*

Proof. By Theorem 6.1.4 $I(G)$ is the intersection of its minimal primes. Therefore $I(G)$ is an unmixed ideal by Corollary 3.1.17, thus G is unmixed by the correspondence between minimal covers and minimal primes. □

Definition 7.2.10 A vertex v is critical if $\alpha_0(G \setminus \{v\}) < \alpha_0(G)$. A graph G is called *vertex-critical* if all its vertices are critical.

Proposition 7.2.11 *Let G be a graph. If $\alpha_0(G \setminus \{v\}) < \alpha_0(G)$ for some vertex v, then $\alpha_0(G \setminus \{v\}) = \alpha_0(G) - 1$.*

Proof. Set $r = \alpha_0(G \setminus \{v\})$ and note that $r + 1 \leq \alpha_0(G)$. Pick a minimal vertex cover C of $G \setminus \{v\}$ with r vertices. Since $\alpha_0(G \setminus \{v\}) < \alpha_0(G)$, v is not an isolated vertex and $N_G(v) \neq C$. Thus $C \cup \{v\}$ is a minimal vertex cover of G. Hence $\alpha_0(G) \leq r + 1$ and one has the asserted equality. □

Proposition 7.2.12 *Let G be a graph without isolated vertices. Then G is vertex critical if and only if any of the following equivalent conditions hold.*

(a) *Any vertex is in some minimal vertex cover with $\alpha_0(G)$ vertices.*

(b) *$\alpha_0(G) = \alpha_0(G \setminus \{v\}) + 1$ for any vertex v of G.*

(c) *$\beta_0(G) = \beta_0(G \setminus \{v\})$ for any vertex v of G.*

Proof. It follows readily from Proposition 7.2.11. □

Proposition 7.2.13 *Let I be an ideal of a catenary local domain $(R, \mathfrak{m})$ and let $x \in \mathfrak{m}$. If $\operatorname{ht}(I) < \operatorname{ht}(I, x)$, then $\operatorname{ht}(I, x) = \operatorname{ht}(I) + 1$.*

Proof. Let $\mathfrak{p}$ be a minimal prime of I such that $\operatorname{ht}(I) = \operatorname{ht}(\mathfrak{p})$. Note that the image of $(x) + \mathfrak{p}$ in $R/\mathfrak{p}$ is a principal ideal, hence by Krull's principal ideal theorem (see Theorem 2.3.16) there is a minimal prime ideal $\mathfrak{q}$ of $(x)+\mathfrak{p}$ such that $\operatorname{ht}(\mathfrak{q}/\mathfrak{p}) \leq 1$. As R is a catenary domain one has

$$\operatorname{ht}(I, x) - \operatorname{ht}(I) \leq \operatorname{ht}(\mathfrak{q}) - \operatorname{ht}(\mathfrak{p}) = \operatorname{ht}(\mathfrak{q}/\mathfrak{p}) \leq 1.$$

Therefore $\operatorname{ht}(I, x) \leq \operatorname{ht}(I) + 1$, and the required equality follows. □

Definition 7.2.14 A family $\mathcal{F}$ consisting of independent sets of a graph G is said to be a *maximal independent cover* of G if $V(G) = \bigcup_{C\in\mathcal{F}} C$ and $|C| = \beta_0(G)$ for all $C \in \mathcal{F}$. A graph G without isolated vertices having a maximal independent cover is called a *B-graph* [26].

Proposition 7.2.15 [26] *If G is a B-graph, then G is vertex critical.*

Proof. Let v be a vertex of G and let $e = \{v, w\}$ be an edge of G. Since w is contained in an independent set A of G with $\beta_0(G)$ vertices, then v is in $V(G) \setminus A$. Hence $\alpha_0(G \setminus \{v\}) < \alpha_0(G)$. $\square$

Proposition 7.2.16 [26] *If G is an unmixed graph, then G has a maximal independent cover.*

Proof. The result follows readily because any two maximal independent sets have the same cardinality. $\square$

Let us summarize some of the results obtained thus far. The following implications hold for any graph without isolated vertices:

$$\text{C–M} \implies \text{unmixed} \implies B\text{-graph} \implies \text{vertex-critical}$$

Proposition 7.2.17 *If G is a vertex critical graph and $G\setminus\{v\}$ is not vertex critical for all $v \in V(G)$, then G has a maximal independent cover.*

Proof. We use induction on the number of vertices of G. If G has two vertices or $\alpha_0(G) = 1$, then for both cases the assumptions on G imply that G has a single edge and the result is clear. Thus one may assume G has at least three vertices and $\alpha_0(G) \geq 2$.

Assume G has no maximal independent cover. Let $C_1, \ldots, C_r$ be the set of all minimal vertex covers of G with $\alpha_0(G)$ vertices. Pick a vertex $v \in V(G)$ such that v is in C_i for all i. By hypothesis $V(G) = \cup_{i=1}^{r} C_i$ because G is vertex-critical (Proposition 7.2.12). Hence $\deg_G(x) \geq 2$ for all $x \in V(G)$ adjacent to v, indeed if $\deg_G(x) = 1$, then $x \in C_j$ for some j and $C_j \setminus \{x\}$ is a vertex cover of G which is impossible. Thus $G \setminus \{v\}$ has no isolated vertices. Since $\alpha_0(G \setminus \{v\})$ is equal to $\alpha_0(G) - 1$, the set $C_i \setminus \{v\}$ is a minimal vertex cover of $G \setminus \{v\}$ with $\alpha_0(G \setminus \{v\})$ vertices for all i. Thus, from the equality $V(G) \setminus \{v\} = (C_1 \setminus \{v\}) \cup \cdots \cup (C_r \setminus \{v\})$ we conclude that $G \setminus \{v\}$ is vertex critical, a contradiction. $\square$

Remark 7.2.18 Let G be a graph and let A be the intersection of all minimal vertex covers of G with $\alpha_0(G)$ vertices. Note that G has a maximal independent cover if and only if $A = \emptyset$.

Proposition 7.2.19 *Let G be a graph and let A be the intersection of all the minimal vertex covers of G with $\alpha_0(G)$ vertices. If G is vertex critical, then the graph $G \setminus A$ is vertex-critical, has a maximal independent cover, and $\beta_0(G) = \beta_0(G \setminus A)$.*

Proof. Let $A = \{v_1, \ldots, v_s\}$. If $A = \emptyset$, then G has a maximal independent cover and there is nothing to prove. Assume $A \neq \emptyset$. By the arguments in the proof of Proposition 7.2.17 the graph $G \setminus \{v_1\}$ is vertex-critical and $\beta_0(G) = \beta_0(G \setminus \{v_1\})$. Since the intersection of all the minimal vertex covers of $G \setminus \{v_1\}$ with $\alpha_0(G \setminus \{v_1\})$ vertices is equal to $A \setminus \{v_1\}$, the result follows by induction. □

Theorem 7.2.20 [186] *If G is a B-graph, then $\beta_0(G) \leq \alpha_0(G)$.*

Let G be a graph. In what follows Δ_G is the independence complex of G and $K[\Delta_G]$ is the Stanley–Reisner ring of Δ_G over a field K.

Theorem 7.2.21 (Erdös–Gallai [144]) *If G is a vertex-critical graph with n vertices, then $\beta_0(G) = \dim R/I(G) = \dim K[\Delta_G] \leq \lfloor \frac{n}{2} \rfloor$.*

Proof. By induction on n. The case $n = 2$ is clear. Assume $n \geq 3$. One may assume that G has no maximal independent cover, otherwise the result follows directly from Theorem 7.2.20 and $\alpha_0(G) + \beta_0(G) = n$. By Proposition 7.2.17 there is $v \in V(G)$ such that $G \setminus \{v\}$ is vertex-critical. Thus by the induction hypothesis $\beta_0(G \setminus \{v\}) \leq (n-1)/2$. Since $\beta_0(G)$ is $\beta_0(G \setminus \{v\})$ (see Proposition 7.2.12), the assertion follows. □

Theorem 7.2.22 *Let G be a graph and let A be the intersection of all the minimal vertex covers of G with $\alpha_0(G)$ vertices. If G is vertex critical, then* $\dim K[\Delta_G] \leq n - |A|/2$.

Proof. It follows from Proposition 7.2.19 and Theorem 7.2.21. □

Example 7.2.23 Given n, k positive integers such that $n \geq 2(k-1)$, let G be the following graph G on n vertices.

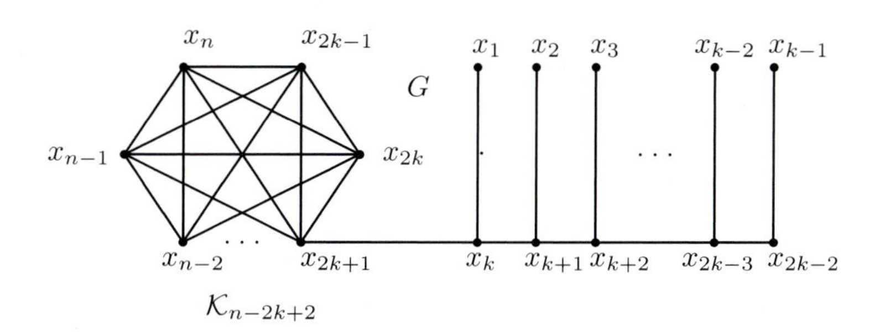

Using Proposition 7.3.2 it follows that G is Cohen–Macaulay. Notice that $\beta_0(G) = k$. In particular if $n = 2k$ or $n = 2k+1$ one has $\beta_0(G) = \lfloor \frac{n}{2} \rfloor$.

Lemma 7.2.24 *If G is a bipartite graph with n vertices, then $\alpha_0(G) \leq \lfloor \frac{n}{2} \rfloor$.*

Proof. Let (V_1, V_2) be a bipartition of G. Since $n = |V_1| + |V_2|$, then $|V_i| \leq \lfloor \frac{n}{2} \rfloor$ for some i. Thus $\alpha_0(G)) \leq \frac{n}{2}$. □

Example 7.2.25 Let G be the graph below and let I be its edge ideal.

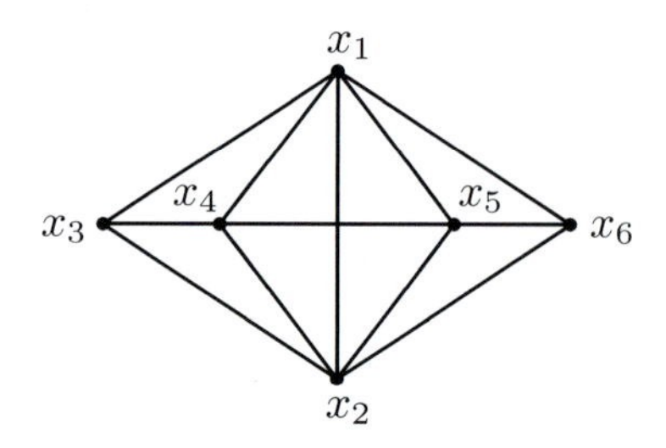

Using the following simple procedure for *Macaulay*2

```
R=ZZ/101[x1,x2,x3,x4,x5,x6]
I=ideal(x1*x2, x1*x3, x1*x4, x1*x5, x1*x6, x2*x3,
            x2*x4, x2*x5, x2*x6, x3*x4, x4*x5, x5*x6)
primaryDecomposition I
```

we obtain that the minimal vertex covers of G are:

```
{x6, x5, x4, x3, x1}, {x6, x5, x4, x3, x2},
{x5, x3, x2, x1}, {x5, x4, x2, x1}, {x6, x4, x2, x1}.
```

The graph G is vertex critical, $\beta_0(G) = 2$, and $\alpha_0(G) = 4$. Thus the set $A = \{x_1, x_2\}$ is the intersection of the minimal vertex cover of G with four vertices. Hence the bound in Theorem 7.2.22 gives $\beta_0(G) \leq 2$ and the Erdös–Gallai bound gives $\beta_0(G) \leq 3$.

Exercises

7.2.26 If I is the ideal $(x_1x_2x_3 - 1) \cap (x_1, x_2, x_3)$ of the polynomial ring $K[x_1, x_2, x_3]$, then $I = (x_1x_2x_3^2 - x_3, x_1x_2^2x_3 - x_2, x_1^2x_2x_3 - x_1)$, $\text{ht}(I, x_i)$ is strictly greater than $\text{ht}(I) + 1$ and $V(I, x_i) = \{0\}$ for all i.

7.2.27 Let $\mathcal{C}$ be a clutter with vertex set $X = \{x_1, \ldots, x_n\}$ and let I be the ideal $I = I(\mathcal{C}) + (x_1^2, \ldots, x_n^2) \subset R$, where $R = K[X]$ is a polynomial ring over a field K. Prove the following: (a) the standard monomials of R/I are in one-to-one correspondence with the independent sets of $\mathcal{C}$, (b) the Hilbert series of R/I is a polynomial $a_0 + a_1t + \cdots + a_dt^d$ of degree $d = \beta_0(\mathcal{C})$, where a_i is the number of independent sets of $\mathcal{C}$ of size i. This polynomial is called the *independence polynomial* of $\mathcal{C}$.

7.2.28 Let G be a graph. Construct a graph H by appending an isolated vertex x to G. Prove that $R/I(G)[x] = R[x]/I(H)$, $\alpha_0(G) = \alpha_0(H)$ and $\beta_0(H) = \beta_0(G) + 1$.

7.2.29 Let G be a graph. If e is an edge of G such that $\alpha_0(G\setminus\{e\}) < \alpha_0(G)$, then $\alpha_0(G) = \alpha_0(G \setminus \{e\}) + 1$.

7.2.30 A graph G is *edge-critical* if $\alpha_0(G \setminus \{e\}) < \alpha_0(G)$ for all $e \in E(G)$. If G is an edge-critical graph, then G has a maximal independent cover.

7.2.31 [145] If G is an edge-critical graph, then $q \leq (\alpha_0(G) + \alpha_0(G)^2)/2$ where q is the number of edges of G.

7.2.32 If G is a graph, then there is a spanning subgraph H of G such that H is edge-critical and $\beta_0(H) = \beta_0(G)$.

7.2.33 Prove that every edge-critical bipartite graph is unmixed.

7.2.34 Prove that a cycle with nine vertices is an edge-critical graph which is not unmixed.

7.2.35 Let G be a graph with q edges. If G has a maximal independent cover, then $q \leq \alpha_0(G)^2$.

7.2.36 Prove that the complete bipartite graph $G = \mathcal{K}_{g,g}$ is an unmixed graph with $\alpha_0(G)^2$ edges and has a maximal independent cover.

7.2.37 Let G be a graph with n vertices. If G is an unmixed graph without isolated vertices such that $2\alpha_0(G) = n$, then G has a perfect matching.

7.2.38 Prove that the following graph is neither edge-critical nor vertex-critical, is not unmixed, and $\{x_1, x_4, x_6, x_7\}, \{x_2, x_3, x_5, x_6\}$ is a maximal independent cover.

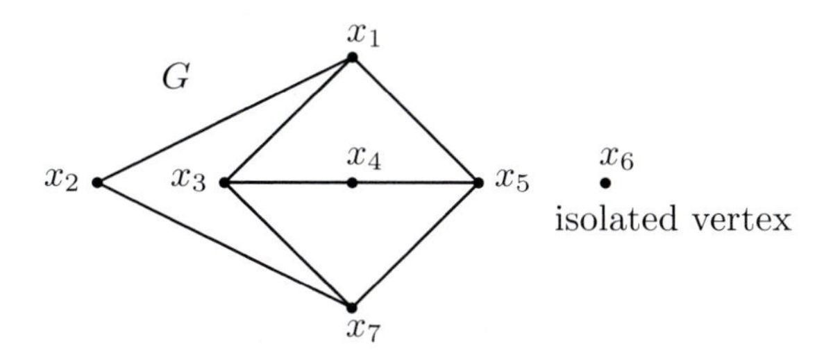

7.2.39 (N. Terai) Let $R = K[a, \ldots, j]$ be a polynomial ring on 10 variables over a field K and let

$$I = (abc, abd, acd, bcd, abe, acf, adg, bch, bdi, cdj)$$

be the edge ideal of the clutter which consists of: the boundary of a tetrahedron (abc, abd, acd, bcd) and a new triangle on each edge of the tetrahedron using a new vertex (e.g., $ab \to abe$). Prove that this ideal is unmixed of height 3 (in fact Cohen–Macaulay).

7.2.40 Let G be a bipartite graph with n vertices. If G has no isolated vertices and G has a maximal independent cover, then n is an even integer.

7.2.41 Prove that the graph G shown below is vertex-critical, $\alpha_0(G) = 3$, $\beta_0(G) = 2$, and G does not have a maximal independent cover.

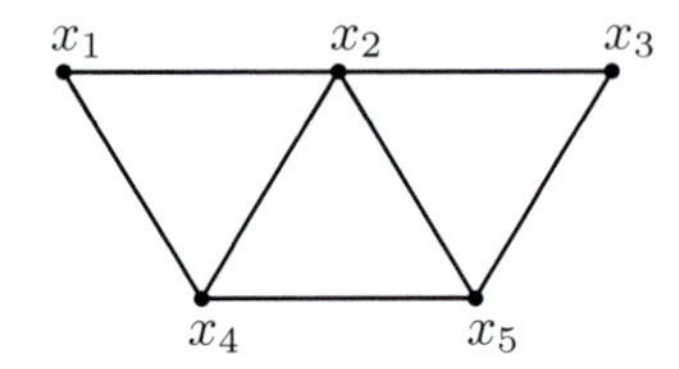

7.2.42 (Dilworth's theorem) Using König's theorem, prove that in any finite poset the cardinality of any largest antichain equals the cardinality of any smallest chain partition.

7.2.43 Prove that in a finite poset the cardinality of any longest chain equals the cardinality of any smallest antichain partition.

7.3 Cohen–Macaulay and chordal graphs

In this section we study Cohen–Macaulay graphs, chordal graphs, and the Cohen–Macaulay property of the ideal of covers of a graph. A classification of Cohen–Macaulay trees is presented.

Constructions of Cohen–Macaulay graphs We begin by showing how large classes of Cohen–Macaulay graphs can be produced and give some obstructions for a graph to be Cohen–Macaulay.

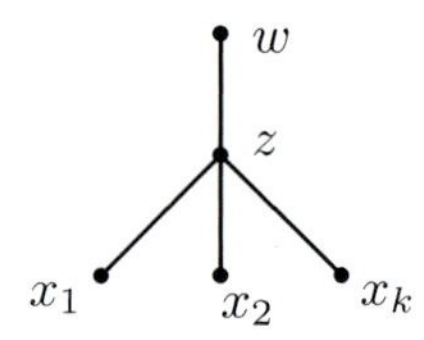

First construction Let H be a graph with vertex set $V(H) = \{x_1, \ldots, x_n, z, w\}$ and J its edge ideal. Assume that z is adjacent to w with $\deg(z) \geq 2$ and $\deg(w) = 1$. We label the vertices of H such that $x_1, \ldots, x_k, w$ are the vertices of H adjacent to z, as shown in the figure. The next two results describe how the Cohen–Macaulay property of H relates to that of the two subgraphs $G = H \setminus \{z, w\}$ and $F = G \setminus \{x_1, \ldots, x_k\}$. One has the equalities

$$J = (I, x_1 z, \ldots, x_k z, zw) \text{ and } (I, x_1, \ldots, x_k) = (L, x_1 \ldots, x_k),$$

where I and L are the edge ideals of G and F, respectively.

Assume H is unmixed with height of J equal to $g + 1$. Since z is not isolated, there is a minimal prime $\mathfrak{p}$ over I containing $\{x_1, \ldots, x_k\}$ and such that $\mathrm{ht}(I) = \mathrm{ht}(\mathfrak{p}) = g$. It is not difficult to prove that $k < n$ and $\deg(x_i) \geq 2$ for $i = 1, \ldots, k$.

Proposition 7.3.1 [416] *If H is a Cohen–Macaulay graph, then F and G are Cohen–Macaulay graphs.*

Proof. Set $A = K[x_1, \ldots, x_n]$ and $R = A[z, w]$. By Proposition 3.1.23 there exists a homogeneous system of parameters $\{f_1, \ldots, f_d\}$ for A/I, where $f_i \in A_+$ for all i. Because of the hypothesis and the equalities

$$z(z - w) + zw = z^2 \text{ and } w(w - z) + zw = w^2,$$

the set $\{f_1, \ldots, f_d, z - w\}$ is a regular system of parameters for R/J. Hence $f_1, \ldots f_d$ is a regular sequence on R/I, that is, G is Cohen–Macaulay. To show the corresponding property for F we consider the exact sequence

$$0 \longrightarrow R/(I, x_1, \ldots, x_k, w)[-1] \stackrel{z}{\longrightarrow} R/J \stackrel{\psi}{\longrightarrow} R/(I, z) \longrightarrow 0$$

where the first map is multiplication by z and ψ is induced by a projection. The exactness of this sequence follows from Theorem 6.1.4. Taking depths with respect to the maximal ideal of R, by the depth lemma one has

$$n - g + 1 \leq \text{depth } R/(I, x_1, \ldots, x_k, w), \text{ where } g = \text{ht}(I).$$

Since $(I, x_1, \ldots, x_k, w) = (L, x_1, \ldots, x_k, w)$, F is Cohen–Macaulay. $\Box$

Proposition 7.3.2 *If F and G are Cohen–Macaulay graphs and $x_1, \ldots, x_k$ are in some minimal vertex cover of G, then H is a Cohen–Macaulay graph.*

Proof. Consider the exact sequence of Proposition 7.3.1. Since the ends of this sequence have R-depth equal to $\dim(R/J)$, by the depth lemma H is a Cohen–Macaulay graph. $\Box$

Corollary 7.3.3 *If G is Cohen–Macaulay and $\{x_1, \ldots, x_k\}$ is a minimal vertex cover of G, then H is Cohen–Macaulay.*

Proof. Note that in this case F is C–M because $I(F) = (0)$. $\Box$

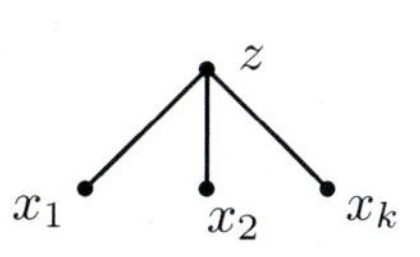

Second construction For the discussion of the second construction we change our notation. Let H be a graph on the vertex set $V(H) = \{x_1, \ldots, x_n, z\}$. Let $\{x_1, \ldots, x_k\}$ be the vertices of H adjacent to z, as shown in the figure. Taking into account the first construction one may assume $\deg(x_i) \geq 2$ for $i = 1, \ldots, k$ and $\deg(z) \geq 2$. Setting $G = H \setminus \{z\}$ and $F = G \setminus \{x_1, \ldots, x_k\}$, notice that the ideals J, I, and L associated to H, G, and F, respectively are related by the equalities $J = (I, x_1 z, \ldots x_k z)$ and $(I, x_1, \ldots, x_k) = (L, x_1, \ldots, x_k)$.

Proposition 7.3.4 [416] *If H is a Cohen–Macaulay graph, then F is a Cohen–Macaulay graph.*

Proof. We set $R = K[x_1, \ldots, x_n, z]$, $A = K[x_1, \ldots, x_n]$ and $\mathrm{ht}(J) = g + 1$. The polynomial $f = z - x_1 - \cdots - x_k$ is regular on R/J because it is clearly not contained in any associated prime of J. Therefore there is a sequence $\{f, f_1, \ldots, f_m\}$ regular on R/J so that $\{f_1, \ldots, f_m\} \subset A_+$, where $m = n - g - 1$. Observe that $\{f_1, \ldots, f_m\}$ is in fact a regular sequence on A/I, which gives $\mathrm{depth}(A/I) \geq n - g - 1$. Next, we use the exact sequence

$$0 \longrightarrow R/(I, x_1, \ldots, x_k)[-1] \stackrel{z}{\longrightarrow} R/J \longrightarrow R/(I, z) \longrightarrow 0$$

and $\mathrm{ht}(I, x_1, \ldots, x_k) = g + 1$ to conclude that F is Cohen–Macaulay. □

Proposition 7.3.5 *Assume that $x_1, \ldots, x_k$ are not contained in any minimal vertex cover for G and* $\mathrm{ht}(I, x_1, \ldots, x_k) = \mathrm{ht}(I) + 1$. *If F and G are Cohen–Macaulay, then H is Cohen–Macaulay.*

Proof. The assumption on $\{x_1, \ldots, x_k\}$ forces $\mathrm{ht}(J) = \mathrm{ht}(I) + 1$. From the exact sequence

$$0 \longrightarrow R/(I, x_1, \ldots, x_k)[-1] \stackrel{z}{\longrightarrow} R/J \longrightarrow R/(I, z) \longrightarrow 0$$

we obtain that H is Cohen–Macaulay. □

Corollary 7.3.6 *If G is Cohen–Macaulay and $\{x_1, \ldots, x_{k-1}\}$ is a minimal vertex cover for G, then H is Cohen–Macaulay.*

Connected components of Cohen–Macaulay graphs An important property of the family of Cohen–Macaulay graphs is their additivity with respect to connected components.

Lemma 7.3.7 *Let $K[\mathbf{x}]$ and $K[\mathbf{y}]$ be two polynomial rings over a field K. If I_1 and I_2 are graded ideals of $K[\mathbf{x}]$ and $K[\mathbf{y}]$, respectively, then*

$$\mathrm{depth}(K[\mathbf{x}]/I_1) + \mathrm{depth}(K[\mathbf{y}]/I_2) = \mathrm{depth}(K[\mathbf{x}, \mathbf{y}]/(I_1 + I_2)).$$

Proof. The formula is a direct consequence of Proposition 3.1.33 and Theorem 3.1.34. □

Proposition 7.3.8 *A graph G is Cohen–Macaulay if and only if all its connected components are Cohen–Macaulay.*

Proof. The result follows from Lemma 7.3.7. □

Whisker graphs in Cohen–Macaulay trees In this part we introduce whisker graphs, present a characterization of all Cohen–Macaulay trees, and construct a family of graphs containing all Cohen–Macaulay trees.

Proposition 7.3.9 *Let $R = K[x_1, \ldots, x_n, y_1, \ldots, y_n]$ be a polynomial ring over a field K and $M \subset \{(i,j)|\ 1 \leq i < j \leq n\}$. Then, the ideal*

$$I = (\{x_iy_i, y_jy_\ell|\ i = 1, \ldots, n \text{ and } (j,\ell) \in M\}),$$

is Cohen–Macaulay.

Proof. The polarization of $(x_i^2, x_jx_\ell|\, 1 \leq i \leq n,\, (j,\ell) \in M)$ is equal to I. Hence, the result follows applying Proposition 6.1.6. □

Definition 7.3.10 Let G_0 be a graph on the vertex set $Y = \{y_1, \ldots, y_n\}$ and take a new set of variables $X = \{x_1, \ldots, x_n\}$. The *whisker graph* or *suspension* of G_0, denoted by $G_0 \cup W(Y)$, is the graph obtained from G_0 by attaching to each vertex y_i a new vertex x_i and the edge $\{x_i, y_i\}$. The edge $\{x_i, y_i\}$ is called a *whisker.*

Proposition 7.3.11 *Let G_0 be a graph and let $G = G_0 \cup W(Y)$ be its whisker graph. Then the multiplicity of $R/I(G)$ is $\sum_{i=-1}^{d} f_i$, where f_i is the number of i-faces of the Stanley–Reisner complex of $I(G_0)$.*

Proof. Let $Y = \{y_1, \ldots, y_n\}$ be the vertex set of G_0. By Proposition 7.3.9, the set of linear forms $\{y_i - x_i\}_{i=1}^n$ is a regular system of parameters for $R/I(G)$. The reduction of $R/I(G)$ modulo these forms is

$$S = K[Y]/(y_1^2, \ldots, y_n^2, I(G_0)),$$

whose K-vector space basis is formed by the standard monomials, that is, by the faces of $\Delta_{I(G_0)}$. To complete the proof notice that the degree of $R/I(G)$ is the degree of S by Corollary 5.1.10. □

Proposition 7.3.12 *Let G_0 and G_1 be two copies of the same graph on disjoint sets of vertices $\{y_1, \ldots, y_n\}$ and $\{x_1, \ldots, x_n\}$, respectively. If $G_1^\vee$ is the clutter of minimal vertex covers of G_1 and*

$$L = (\{x_{i_1} \cdots x_{i_s} |\, \{x_{i_1}, \ldots, x_{i_s}\} \in E(G_1^\vee)\}),$$

then $((\mathbf{z})\colon I(G)) = (\mathbf{z}, L)$, where $\mathbf{z} = \{x_iy_i\}_{i=1}^n$ and $G = G_0 \cup W(Y)$.

Proof. Take $\{x_{i_1}, \ldots, x_{i_r}\} \in E(G_1^\vee)$ and $y_ky_\ell \in I(G_0)$. Since $i_s \in \{k, \ell\}$ for some s we obtain $x_{i_1} \cdots x_{i_r}y_ky_\ell \in (\mathbf{z})$. This shows $(\mathbf{z}, L) \subset ((\mathbf{z})\colon I(G))$. Conversely, assume M is a monomial in $((\mathbf{z})\colon I(G)) \setminus (\mathbf{z})$. Let $\{y_k, y_\ell\}$ be any edge in G_0. Since $My_ky_\ell = x_ty_tM_1$, either x_k divides M or x_ℓ divides M; in either case, $M \in L$. □

Corollary 7.3.13 *The type of* $K[X,Y]/I(G_0 \cup W(Y))$ *is* $|E(G_0^\vee)|$.

Example 7.3.14 Let $I(G_0) = (y_1y_3, y_1y_4, y_2y_4, y_2y_5, y_3y_5, y_3y_6, y_4y_6)$ and let G be the whisker graph of G_0. The ideal $((\mathbf{z})\colon I(G))$ is generated by

$$\{x_iy_i\}_{i=1}^6 \cup \{x_2x_3x_4,\ x_3x_4x_5,\ x_1x_2x_5x_6,\ x_1x_2x_3x_6,\ x_1x_4x_5x_6\}.$$

The last five monomials correspond to the minimal primes of $I(G_0)$, the type of $R/I(G)$ is 5 and its multiplicity is 17.

Lemma 7.3.15 *Let G be a tree. If v and w are two adjacent vertices of degree at least two, then there is a minimal vertex cover of G containing both v and w.*

Proof. Let x be the edge joining v and w. Since $G \setminus \{x\}$ has exactly two components, say G_1 and G_2, there are minimal vertex covers A and B for G_1 and G_2, respectively, so that $v \in A$ and $w \in B$. Therefore $A \cup B$ is the required cover for G. □

Theorem 7.3.16 [416] *Let G be a tree. Then, G is Cohen–Macaulay if and only if $|V(G)| \leq 2$ or $2 < |V(G)| = 2r$ and G has a perfect matching $\{x_1, y_1\}, \ldots, \{x_r, y_r\}$ so that* $\deg(x_i) = 1$, $\deg(y_i) \geq 2$ *for $i = 1, \ldots, r$.*

Proof. $\Rightarrow$) Let (V_1, V_2) be a bipartition of G. Since V_1, V_2 are minimal vertex covers of G, and G is unmixed, we have $r = |V_1| = |V_2| = \alpha_0(G)$. By Theorem 7.1.8 there are r independent edges. Therefore, we may assume $V_1 = \{z_i\}_{i=1}^r$, $V_2 = \{w_i\}_{i=1}^r$, and $\{z_i, w_i\} \in E(G)$ for $i = 1, \ldots, r$. To complete the proof it suffices to show that for each i either z_i or w_i has degree one. Assume $\deg(z_i) \geq 2$ and $\deg(w_i) \geq 2$ for some i. By Lemma 7.3.15 there is a minimal vertex cover for T, say A, containing z_i and w_i. Since $\{z_j, w_j\} \bigcap A \neq \emptyset$ for $j \neq i$ we conclude $|A| \geq r+1$, which is a contradiction.

$\Leftarrow$) The sufficiency follows from Proposition 7.3.9. □

The significance of the notion of a whisker graph lies partly in the next restatement of Theorem 7.3.16.

Theorem 7.3.17 *If G is a graph, then G is a Cohen–Macaulay tree if and only if $G = G_0 \cup W(Y)$ for some tree G_0 with vertex set Y.*

Corollary 7.3.18 *If G is a tree, then G is Cohen–Macaulay if and only if G is unmixed.*

Proof. If G is Cohen–Macaulay, then G is unmixed by Proposition 7.2.9. If G is unmixed, using the proof above it follows that $I(G)$ is a Cohen–Macaulay ideal of the form described in Proposition 7.3.9. □

Corollary 7.3.19 *The only Cohen–Macaulay cycles are the triangle and the pentagon.*

Proof. Let $C_n = \{x_0, x_1, \ldots, x_n = x_0\}$ be a Cohen–Macaulay cycle of length n. The cases $n = 4$, 6, and 7 can be treated separately, so we assume $n \geq 8$. By Proposition 7.3.4 $C_n \setminus \{x_0, x_1, x_{n-1}\}$ is a Cohen–Macaulay path of length $n - 3$ which is impossible by Theorem 7.3.16 □

The next result generalizes the Cohen–Macaulay criterion for trees given in Theorem 7.3.16 and is a generalization of Faridi's characterization of unmixed simplicial trees [156].

Theorem 7.3.20 [325] *Let $\mathcal{C}$ be a clutter with the König property and with no special cycles of length* 3 *or* 4*. The following conditions are equivalent:*

(a) *$\mathcal{C}$ is unmixed.*

(b) *There is a perfect matching $e_1, \ldots, e_g$, $g = \operatorname{ht} I(\mathcal{C})$, such that e_i has a free vertex for all i, and for any two edges f_1, f_2 of $\mathcal{C}$ and for any edge e_i, one has that $f_1 \cap e_i \subset f_2 \cap e_i$ or $f_2 \cap e_i \subset f_1 \cap e_i$.*

(c) *$R/I(\mathcal{C})$ is Cohen–Macaulay.*

(d) *$\Delta_{\mathcal{C}}$ is a pure shellable simplicial complex.*

Proof. Using Lemma 6.5.7, Theorems 6.5.18, and 6.5.21, and Proposition 6.5.19 it follows readily that conditions (a) and (b) are equivalent. Since (a) is equivalent to (b), from Theorem 6.5.20 we get that (b) implies (d). That (d) implies (c) and (c) implies (a), were shown in Theorem 6.3.23 and Corollary 6.3.7. □

Simplicial complexes of edge ideals Given a graph G on the vertex set V one defines a simplicial complex $\operatorname{Simp}(G)$ on the same set of vertices:

$$\operatorname{Simp}(G) = \{F \mid \exists\, H \text{ a subgraph of } G \text{ such that } F = V(H) \text{ and } H \simeq \mathcal{K}_r\},$$

Conversely a simplicial complex Δ defines a graph $\operatorname{Skel}(\Delta)$, the 1-skeleton of Δ, on the same set of vertices:

$$\operatorname{Skel}(\Delta) = \{F \in \Delta \mid \dim(F) \leq 1\}.$$

In general one has $\Delta \subset \operatorname{Simp}(\operatorname{Skel}(\Delta))$.

Proposition 7.3.21 *Let Δ be a simplicial complex and I_Δ its Stanley–Reisner ideal. Then $\Delta = \operatorname{Simp}(\operatorname{Skel}(\Delta))$ if and only if I_Δ is generated by square-free monomials of degree two.*

Proof. Let $V = \{x_1, \ldots, x_n\}$ be the vertex set of Δ. Assume I_Δ is generated by square-free monomials of degree two. Take $F \in \mathrm{Simp}(\mathrm{Skel}(\Delta))$, for simplicity set $F = \{x_1, \ldots, x_r\}$, that is, $\{x_i, x_j\} \in \Delta$ for all $1 \leq i < j \leq r$. If $F \notin \Delta$, then $x_1 \cdots x_r$ is in I_Δ, and by assumption $x_i x_j$ is in I_Δ for some $1 \leq i < j \leq r$. Hence $\{x_i, x_j\} \notin \Delta$, which is a contradiction. This shows $\Delta = \mathrm{Simp}(\mathrm{Skel}(\Delta))$. The converse also follows readily. □

Simplicial complexes associated to edge ideals have been studied in the literature. They are called *flag complexes.* (See [395, Chapter III] and the references there for a series of amusing related problems).

Definition 7.3.22 The *complement* $\overline{G}$ of a graph G has vertex set $V(G)$ and two vertices are adjacent in $\overline{G}$ if and only if they are not adjacent in G. The complement of G is also denoted by G'.

Proposition 7.3.23 *Let G be a graph and let Δ_G be its independence complex. Then $\overline{G} \subset \Delta_G$, with equality if and only if $\overline{G}$ has no triangles.*

Proof. If $x = \{v_1, v_2\}$ is an edge of $\overline{G}$, then $\{v_1, v_2\}$ is a stable set of G and $x \in \Delta_G$. Hence $\overline{G} \subset \Delta_G$. Assume $\overline{G}$ has no triangles. If $A = \{v_1, \ldots, v_n\}$ is a stable set of G, then $n = 1$ or $n = 2$; otherwise if $n > 2$, then $\overline{G}$ would contain the triangle $\{v_1, v_2, v_3\}$. Therefore A is either a point or an edge of $\overline{G}$, thus $\overline{G} = \Delta_G$. The converse also follows readily. □

Chordal graphs The use of graph theoretical methods in commutative algebra is implicit in Lyubeznik thesis [299] and explicit in the work of Fröberg [172]. In both cases the central notion is that of a chordal graph.

Let G be a graph and let $\overline{G}$ be its complement. Recall that $\overline{G}$ is said to be *chordal* if every cycle C_n in $\overline{G}$ of length $n \geq 4$ has a chord in $\overline{G}$. A *chord* of C_n is an edge joining two non-adjacent vertices of C_n. Chordal graphs have been extensively studied [112, 228, 403].

Lemma 7.3.24 *Let G be a connected graph and let d be the minimum degree among the vertices of G. If G is not a complete graph, then there is $S \subset V(G)$ such that $G \setminus S$ is disconnected and $d = |S|$.*

Proof. Let v be a vertex of G of degree d and $S = N_G(v)$ the neighbor set of v. Since G is not a complete graph and v has minimum degree there is a vertex $w \notin S \cup \{v\}$. Thus $G \setminus S$ has at least two components. □

Proposition 7.3.25 (Dirac [112]) *If H is a chordal non-complete graph, then it can be constructed out of two smaller disjoint chordal graphs H_1 and H_2 by identifying two (possibly empty) complete subgraphs of the same size in H_1 and H_2.*

Proof. One may assume that H is connected and has n vertices. Since H has at least one vertex of degree less than $n-1$, by Lemma 7.3.24 there is a minimal set of vertices S such that $H \setminus S$ is disconnected and $|S| \leq n-2$. Hence $H \setminus S = F_1 \cup F_2$, where F_1 and F_2 are disjoint subgraphs of H. Consider the induced subgraphs

$$H_1 = H[V(F_1) \cup S] \quad \text{and} \quad H_2 = H[V(F_2) \cup S].$$

Notice that $H_1 \cup H_2 = H$ and $H_1 \cap H_2 = H[S]$.

Next we show that $H[S]$ is a complete graph. Assume there are vertices x_1 and x_2 in S not connected by an edge of H. Since $H \setminus (S \setminus \{x_i\})$ is connected for $i = 1, 2$, there are paths p_1 and p_2 in H_1 and H_2, respectively, joining x_1 with x_2, whose only vertices in S are x_1 and x_2. If p_1 and p_2 are the shortest paths with this property, then $p_1 \cup p_2$ is a cycle of length greater than or equal to four without a chord, a contradiction. $\Box$

Theorem 7.3.26 [300] *Let G be a graph and let $\overline{G}$ be its complement. Then $I_c(G)$ is Cohen–Macaulay if and only if $\overline{G}$ is a chordal graph.*

Proposition 7.3.27 *If $I_c(G)$ is Cohen–Macaulay, then $\mu(I_c(G)) \leq g + 1$. where $\mu(I_c(G))$ is the minimum number of generators of $I_c(G)$ and g is the height of $I(G)$.*

Proof. By the Hilbert–Burch Theorem (see Theorem 3.5.16) $I_c(G)$ is generated by the $m-1$ minors of an $m \times (m-1)$ matrix A with homogeneous entries, where $m = \mu(I_c(G))$. Take f in $I_c(G)$ of degree g, since any $m-1$ minor of A has degree at least $m-1$, one obtains $\deg(f) = g \geq m-1$. $\Box$

Exercises

7.3.28 Show that a cycle C_n of length $n \geq 3$ is unmixed if and only if $n = 3, 4, 5, 7$.

7.3.29 Let G_1, G_2 be two graphs with vertex set V. If their complements are triangle free and they have the same number of edges, then the Hilbert series of $K[V]/I(G_1)$ and $K[V]/I(G_2)$ are identical.

7.3.30 If G is a graph and $\mathrm{ht}(I(G)) = |V(G)| - 2$, then

(a) G is Cohen–Macaulay if and only if Δ_G is connected, and

(b) if $I(G)$ is not Cohen–Macaulay, then $I(G)$ is unmixed if and only if Δ_G has no isolated vertices.

7.3.31 If G is a Cohen–Macaulay graph with q edges, then $q \leq (g^2 + g)/2$, where g is the height of $I(G)$.

7.3.32 If G is a Cohen–Macaulay graph and u, v are two vertices of degree 1, prove that they cannot have an adjacent vertex in common.

7.3.33 Let G be a connected graph and let $\{v_1, \ldots, v_n\}$ be a path with $n \geq 4$. If $\deg(v_1) = 1$, $\deg(v_i) = 2$ for $i = 2, \ldots n-1$ and $\deg(v_n) \geq 3$, prove that G cannot be unmixed.

7.3.34 Prove that $H = \text{Skel}(\text{Simp}(H))$, for any graph H.

7.3.35 Let T be a tree with $2r \geq 4$ vertices. Then T is Cohen–Macaulay if and only if $\text{ht}\, I(T) = r$ and T has exactly r vertices of degree 1.

7.3.36 Prove that the ideal $I = I_\Delta$ of Example 6.3.65 is generated by

$$\begin{array}{llllllll} x_1x_3, & x_1x_4, & x_1x_7, & x_1x_{10}, & x_1x_{11}, & x_2x_4, & x_2x_5, & x_2x_8, \\ x_2x_{10}, & x_2x_{11}, & x_3x_5, & x_3x_6, & x_3x_8, & x_3x_{11}, & x_4x_6, & x_4x_9, \\ x_4x_{11}, & x_5x_7, & x_5x_9, & x_5x_{11}, & x_6x_8, & x_6x_9, & x_7x_9, & x_7x_{10}, \\ x_8x_{10}. & & & & & & & \end{array}$$

Let $R_i = K[x_1, \ldots, \widehat{x}_i, \ldots, x_{11}]$ and $G_i = G \setminus \{x_i\}$. Then $I = I(G)$ for some graph G. Prove that the f-vector of Δ_{G_i} is $(10, 25, 15)$ or $(10, 24, 14)$, the h-vector of $R_i/I(G_i)$ is $(1, 7, 8, -1)$ or $(1, 7, 7, -1)$, and G_i is not a C–M graph for all i.

7.3.37 Let $I = (x_iy_j \,|\, 1 \leq i \leq j \leq n) \subset R = K[X, Y]$. Prove that R/I has type n and its a-invariant is $-(n-1)$.

7.4 Shellable and sequentially C–M graphs

In this section we classify all sequentially Cohen–Macaulay bipartite graphs (they are precisely the shellable bipartite graphs) and give a recursive procedure to verify if a bipartite graph is shellable. Then we present some structure theorems for unmixed and Cohen–Macaulay bipartite graphs.

Definition 7.4.1 Let G be a graph whose independence complex is Δ_G. We say G is a *shellable graph* if Δ_G is a shellable simplicial complex.

To prove that a graph G is shellable, it suffices to prove each connected component of G is shellable.

Lemma 7.4.2 [409] *Let G_1 and G_2 be two graphs with disjoint sets of vertices and let $G = G_1 \cup G_2$. Then G_1 and G_2 are shellable if and only if G is shellable.*

Proof. $\Rightarrow$) Let $F_1, \ldots, F_r$ and $H_1, \ldots, H_s$ be the shellings of Δ_{G_1} and Δ_{G_2}, respectively. Then if we order the facets of Δ_G as

$$F_1 \cup H_1, \ldots, F_1 \cup H_s;\ F_2 \cup H_1, \ldots, F_2 \cup H_s;\ \ldots;\ F_r \cup H_1, \ldots, F_r \cup H_s$$

we get a shelling of Δ_G. Indeed if $F' < F$ are two facets of Δ_G we have two cases to consider. Case (i): $F' = F_i \cup H_k$ and $F = F_j \cup H_t$, where $i < j$. Because Δ_{G_1} is shellable there is $v \in F_j \setminus F_i$ and $\ell < j$ with $F_j \setminus F_\ell = \{v\}$. Hence $v \in F \setminus F'$, $F_\ell \cup H_t < F$, and $F \setminus (F_\ell \cup H_t) = \{v\}$. Case (ii): $F' = F_k \cup H_i$ and $F = F_k \cup H_j$, where $i < j$. This case follows from the shellability of Δ_{G_2}.

$\Leftarrow$) Note that if F is a facet of Δ_G, then $F' = F \cap V_{G_1}$, respectively, $F'' = F \cap V_{G_2}$, is a facet of Δ_{G_1}, respectively, Δ_{G_2}. We now show that G_1 is shellable and omit the similar proof for the shellability of G_2. Let $F_1, \ldots, F_t$ be a shelling of Δ_G, and consider the subsequence

$$F_{i_1}, \ldots, F_{i_s} \quad \text{with } 1 = i_1 < i_2 < \cdots < i_s$$

where $F_1 \cap V_{G_2} = F_{i_j} \cap V_{G_2}$ for $i_j \in \{i_1, \ldots, i_s\}$, but $F_1 \cap V_{G_2} \neq F_k \cap V_{G_2}$ for any $k \in \{1, \ldots, t\} \setminus \{i_1, \ldots, i_s\}$. We then claim that

$$F_1' = F_{i_1} \setminus V_{G_2},\ F_2' = F_{i_2} \setminus V_{G_2}, \ldots, F_s' = F_{i_s} \setminus V_{G_2}$$

is a shelling of Δ_{G_1}. We first show that this is a complete list of facets; indeed, each $F_j' = F_{i_j} \cap V_{G_1}$ is a facet of Δ_{G_1}, and furthermore, for any facet $F \in \Delta_{G_1}$, $F \cup (F_1 \cap V_{G_2})$ is a facet of Δ_G, and hence $F \cup (F_1 \cap V_{G_2}) = F_{i_j}$ for some $i_j \in \{i_1, \ldots, i_s\}$.

Because the F_i's form a shelling, if $1 \leq k < j \leq s$, there exists v in $F_{i_j} \setminus F_{i_k} = (F_{i_j} \setminus V_{G_2}) \setminus (F_{i_k} \setminus V_{G_2}) = F_j' \setminus F_k'$ such that $\{v\} = F_{i_j} \setminus F_\ell$ for some $1 \leq \ell < i_j$. It suffices to show that F_ℓ is among $F_{i_1}, \ldots, F_{i_s}$. Now because $F_{i_j} \cap V_{G_2} \subset F_{i_j}$ and $v \notin F_{i_j} \cap V_{G_2}$, we must have $F_{i_j} \cap V_{G_2} \subset F_\ell$. So, $F_\ell \cap V_{G_2} \supset F_{i_j} \cap V_{G_2}$. But $F_\ell \cap V_{G_2}$ is a facet of Δ_{G_2}, so we must have $F_\ell \cap V_{G_2} = F_{i_j} \cap V_{G_2}$. So $F_\ell = F_{i_r}$ for some $r < j$, and hence, $\{v\} = F_j' \setminus F_r'$, as desired. $\square$

If F is a face of a simplicial complex Δ, the *link* of F is defined to be $\mathrm{lk}_\Delta(F) = \{G \mid G \cup F \in \Delta,\ \ G \cap F = \emptyset\}$. When $F = \{x\}$, then we shall abuse notation and write $\mathrm{lk}_\Delta(x)$ instead of $\mathrm{lk}_\Delta(\{x\})$.

Lemma 7.4.3 *Let x be a vertex of G and let $G' = G \setminus (\{x\} \cup N_G(x))$. Then*

$$\Delta_{G'} = \mathrm{lk}_{\Delta_G}(x),$$

and F is a facet of $\Delta_{G'}$ if and only if $x \notin F$ and $F \cup \{x\}$ is a facet of Δ_G.

Proof. If $F \in \mathrm{lk}_{\Delta_G}(x)$, then $x \notin F$, and $F \cup \{x\} \in \Delta_G$ implies that $F \cup \{x\}$ is an independent set of G. So $(F \cup \{x\}) \cap N_G(x) = \emptyset$. But this means that

$F \subset V_{G'}$ because $V_{G'} = V_G \setminus (\{x\} \cup N_G(x))$. Thus $F \in \Delta_{G'}$ since F is also an independent set of the smaller graph G'.

Conversely, if $F \in \Delta_{G'}$, then F is an independent set of G' that does not contain any of the vertices of $\{x\} \cup N_G(x)$. But then $F \cup \{x\}$ is an independent set of G, i.e., $F \cup \{x\} \in \Delta_G$. So $F \in \text{lk}_{\Delta_G}(x)$.

The last statement follows readily from the fact that F is a facet of $\text{lk}_{\Delta_G}(x)$ if and only if $x \notin F$ and $F \cup \{x\}$ is a facet of Δ_G. $\square$

Theorem 7.4.4 [409] *Let x be a vertex of a graph G and let G' be the graph $G' = G \setminus (\{x\} \cup N_G(x))$. If G is shellable, then G' is shellable.*

Proof. Let $F_1, \ldots, F_s$ be a shelling of Δ_G. Suppose the subsequence

$$F_{i_1}, F_{i_2}, \ldots, F_{i_t} \quad \text{with } i_1 < i_2 < \cdots < i_t$$

is the list of all the facets with $x \in F_{i_j}$. Setting $H_j = F_{i_j} \setminus \{x\}$ for each $j = 1, \ldots, t$, Lemma 7.4.3 implies that the H_j's are the facets of $\Delta_{G'}$.

We claim that $H_1, \ldots, H_t$ is a shelling of $\Delta_{G'}$. As the F_i's form a shelling, if $1 \leq k < j \leq t$, there is $v \in F_{i_j} \setminus F_{i_k} = (F_{i_j} \setminus \{x\}) \setminus (F_{i_k} \setminus \{x\}) = (H_j \setminus H_k)$ such that $\{v\} = F_{i_j} \setminus F_\ell$ for some $1 \leq \ell < i_j$. It suffices to show that F_ℓ is among the list $F_{i_1}, \ldots, F_{i_t}$. But because $x \in F_{i_j}$ and $x \neq v$, we must have $x \in F_\ell$. Thus $F_\ell = F_{i_k}$ for some $k \leq j$. But then $\{v\} = F_{i_j} \setminus F_\ell = H_j \setminus H_k$. So, the H_i's form a shelling of $\Delta_{G'}$. $\square$

Let G be a graph and let $S \subset V_G$. The graph $G \cup W_G(S)$ obtained from G by adding new vertices $\{y_i \mid x_i \in S\}$ and new edges $\{\{x_i, y_i\} \mid x_i \in S\}$ is called the *whisker graph* of G over S. The edges $\{x_i, y_i\}$ are called *whiskers*.

Corollary 7.4.5 [409] *Let G be a graph and let $S \subset V_G$. If $G \cup W_G(S)$ is shellable, then $G \setminus S$ is shellable.*

Proof. We may assume that $S = \{x_1, \ldots, x_s\}$. Set $G_0 = G \cup W_G(S)$ and $G_i = G_{i-1} \setminus (\{y_i\} \cup N_G(y_i))$ for $i = 1, \ldots, s$. Notice that $G_s = G \setminus S$. Hence, by repeatedly applying Theorem 7.4.4, the graph $G \setminus S$ is shellable. $\square$

Lemma 7.4.6 *Let G be a bipartite graph with bipartition $\{x_1, \ldots, x_m\}$, $\{y_1, \ldots, y_n\}$. If G is shellable and G has no isolated vertices, then there is $v \in V_G$ with* $\deg(v) = 1$.

Proof. Let $F_1, \ldots, F_s$ be a shelling of Δ_G. We may assume that $F_i = \{y_1, \ldots, y_n\}$, $F_j = \{x_1, \ldots, x_m\}$ and $i < j$. Then there is $x_k \in F_j \setminus F_i$ and F_ℓ with $\ell \leq j - 1$ such that $F_j \setminus F_\ell = \{x_k\}$. For simplicity assume that $x_k = x_1$. Then $\{x_2, \ldots, x_m\} \subset F_\ell$ and there is y_t in F_ℓ for some $1 \leq t \leq n$. Since $\{y_t, x_2, \ldots, x_m\}$ is an independent set of G, we get that y_t can only be adjacent to x_1. Thus $\deg(y_t) = 1$ because G has no isolated vertices. $\square$

Theorem 7.4.7 [409] *Let G be a graph and let $\{x_1, x_2\}$ be an edge of G with* $\deg(x_1) = 1$. *If $G_i = G \setminus (\{x_i\} \cup N_G(x_i))$ for $i = 1, 2$, then G is shellable if and only if G_1 and G_2 are shellable.*

Proof. If G is shellable, then G_1 and G_2 are shellable by Theorem 7.4.4. So it suffices to prove the reverse direction. Let $F_1', \ldots, F_r'$ be a shelling of Δ_{G_1} and let $H_1', \ldots, H_s'$ be a shelling of Δ_{G_2}. It suffices to prove that

$$F_1' \cup \{x_1\}, \ldots, F_r' \cup \{x_1\}, H_1' \cup \{x_2\}, \ldots, H_s' \cup \{x_2\}$$

is a shelling of Δ_G. One first shows that this is the complete list of facets of Δ_G using Lemma 7.4.3. Indeed, take any facet F of Δ_G. If $x_2 \in F$, then $x_1 \notin F$ because $\{x_1, x_2\}$ is an edge of G, and by Lemma 7.4.3, $F \setminus \{x_2\} = H_i'$ for some i. On the other hand, if $x_2 \notin F$, we must have $x_1 \in F$, because if not, then $\{x_1\} \cup F$ is larger independent set of G because x_1 is only adjacent to x_2. Again, by Lemma 7.4.3, we have $F \setminus \{x_1\} = F_i'$ for some i. Let $F' < F$ be two facets of Δ_G. Consider the case in which $F' = F_i' \cup \{x_1\}$ and $F = H_j' \cup \{x_2\}$. Since $H_j' \cup \{x_1\}$ is an independent set of G, it is contained in a facet of Δ_G, i.e., $H_j' \cup \{x_1\} \subset F_\ell' \cup \{x_1\}$ for some ℓ. Hence $(H_j' \cup \{x_2\}) \setminus (F_\ell' \cup \{x_1\}) = \{x_2\}$, $x_2 \in F \setminus F'$, and $F_\ell' \cup \{x_1\} < F$. The remaining two cases follow readily from the shellability of Δ_{G_1} and Δ_{G_2}. $\Box$

The last two results yield a recursive procedure to verify if a bipartite graph is shellable. The next result is the first combinatorial characterization of all shellable bipartite graphs.

Corollary 7.4.8 [409] *Let G be a bipartite graph. Then G is shellable if and only if there are adjacent vertices x and y with* $\deg(x) = 1$ *such that the bipartite graphs $G \setminus (\{x\} \cup N_G(x))$ and $G \setminus (\{y\} \cup N_G(y))$ are shellable.*

Proof. By Lemma 7.4.2 it suffices to verify the statement when G is connected. By Lemma 7.4.6 there exists a vertex of x_1 with $\deg(x_1) = 1$. Now apply the previous theorem. $\Box$

A clutter is called *vertex decomposable* if its independence complex is vertex decomposable. The following major result of Woodroofe gives a sufficient condition for vertex decomposability of graphs and is an extension of the fact that chordal graphs are shellable [409].

Theorem 7.4.9 [432, Theorem 1.1] *If G is a graph with no chordless cycles of length other than 3 or 5, then G is vertex decomposable (hence shellable and sequentially Cohen–Macaulay).*

Definition 7.4.10 A clutter $\mathcal{C}$ is called *sequentially Cohen–Macaulay* if $R/I(\mathcal{C})$ is sequentially Cohen–Macaulay.

We now give a sequentially Cohen–Macaulay analog of Theorem 7.4.4.

Theorem 7.4.11 [409] *Let x be a vertex of a graph G and let G' be the graph $G' = G \setminus (\{x\} \cup N_G(x))$. If G is sequentially Cohen–Macaulay, then G' is sequentially Cohen–Macaulay.*

Proof. Let $F_1, \ldots, F_s$ be the facets of $\Delta = \Delta_G$, and let $F_1, \ldots, F_r$ be the facets of Δ that contain x. Set $\Gamma = \Delta_{G'}$; by Lemma 7.4.3, the facets of Γ are $F_1' = F_1 \setminus \{x\}, \ldots, F_r' = F_r \setminus \{x\}$.

Consider the pure simplicial complexes

$$\begin{aligned} \Delta^{[k]} &= \langle \{F \in \Delta \mid \dim(F) = k\} \rangle; \quad -1 \leq k \leq \dim(\Delta), \\ \Gamma^{[k]} &= \langle \{F \in \Gamma \mid \dim(F) = k\} \rangle; \quad -1 \leq k \leq \dim(\Gamma), \end{aligned}$$

where $\langle \mathcal{F} \rangle$ denotes the subcomplex generated by the set of faces $\mathcal{F}$. Recall that H is a face of $\langle \mathcal{F} \rangle$ if and only if H is contained in some F in $\mathcal{F}$. Take a facet F_i' of Γ of dimension $d = \dim(\Gamma)$. Then $F_i' \cup \{x\} \in \Delta^{[d+1]}$ and consequently $\{x\} \in \Delta^{[k+1]}$ for $k \leq d$. Because the facets of Γ are $F_1' = F_1 \setminus \{x\}, \ldots, F_r' = F_r \setminus \{x\}$, we have the equality

$$\Gamma^{[k]} = \mathrm{lk}_{\Delta^{[k+1]}}(x)$$

for $k \leq d$. By Theorem 6.3.30, the complex Δ is sequentially Cohen–Macaulay if and only if $\Delta^{[k]}$ is Cohen–Macaulay for $-1 \leq k \leq \dim(\Delta)$. Because $\Delta^{[k]}$ is Cohen–Macaulay, by Proposition 6.3.15 $\mathrm{lk}_{\Delta^{[k]}}(F)$ is Cohen–Macaulay for any $F \in \Delta^{[k]}$. Thus, $\Gamma^{[k]} = \mathrm{lk}_{\Delta^{[k+1]}}(x)$ is Cohen–Macaulay for any $-1 \leq k \leq \dim(\Gamma) \leq \dim(\Delta) - 1$. Therefore Γ is sequentially Cohen–Macaulay by Theorem 6.3.30, as required. □

The following result is due to Francisco and Hà.

Corollary 7.4.12 [159, Theorem 4.1] *Let G be a graph and let $S \subset V_G$. If the whisker graph $G \cup W_G(S)$ of G over S is sequentially Cohen–Macaulay, then $G \setminus S$ is sequentially Cohen–Macaulay.*

Proof. We may assume that $S = \{x_1, \ldots, x_s\}$. Set $G_0 = G \cup W_G(S)$ and $G_i = G_{i-1} \setminus (\{y_i\} \cup N_G(y_i))$ for $i = 1, \ldots, s$ where y_i is the degree 1 vertex adjacent to x_i. Notice that $G_s = G \setminus S$. Hence by Theorem 7.4.11 the graph $G \setminus S$ is sequentially Cohen–Macaulay. □

Theorem 7.4.13 [229] *If I is the edge ideal of a graph, then I has linear quotients if and only if I has a linear resolution if and only if each power of I has a linear resolution*

Lemma 7.4.14 *Let G be a bipartite graph with bipartition $\{x_1, \ldots, x_m\}$, $\{y_1, \ldots, y_n\}$. If G is sequentially Cohen–Macaulay and G is not a discrete graph, then there is $v \in V_G$ with* $\deg(v) = 1$.

Proof. We may assume that $m \leq n$ and that G has no isolated vertices. Let J be the Alexander dual of $I = I(G)$ and let $L = J_{[n]}$ be the monomial ideal generated by the square-free monomials of J of degree n. We may assume that L is generated by $g_1, \ldots, g_q$, where $g_1 = y_1y_2 \cdots y_n$ and $g_2 = x_1 \cdots x_m y_1 \cdots y_{n-m}$. Consider the linear map

$$R^q \xrightarrow{\varphi} R \quad (e_i \mapsto g_i).$$

The kernel of this map is generated by syzygies of the form

$$(g_j/\gcd(g_i, g_j))e_i - (g_i/\gcd(g_i, g_j))e_j.$$

Since the vector $\alpha = x_1 \cdots x_m e_1 - y_{n-m+1} \cdots y_n e_2$ is in $\ker(\varphi)$ and since $\ker(\varphi)$ is generated by linear syzygies (see Theorem 6.3.44), there is a linear syzygy of L of the form $x_j e_1 - z e_k$, where z is a variable, $k \neq 1$. Hence $x_j(y_1 \cdots y_n) = z(g_k)$ and $g_k = x_j y_1 \cdots y_{i-1} y_{i+1} \cdots y_n$ for some i. Because the support of g_k is a vertex cover of G, we get that the complement of the support of g_k, i.e., $\{y_i, x_1, \ldots, x_{j-1}, x_{j+1}, \ldots, x_m\}$, is an independent set of G. Thus y_i can only be adjacent to x_j, i.e., $\deg(y_i) = 1$. □

Proposition 7.4.15 *If G is a C–M bipartite graph, then $G \setminus \{u\}$ is C–M for some vertex u of G.*

Proof. By Lemma 7.4.14, there is an edge $\{u, v\}$ of the graph G such that $\deg(v) = 1$. Then by Proposition 7.3.1 we obtain that $G \setminus \{u\}$ is Cohen–Macaulay. □

Theorem 7.4.16 [409] *Let G be a bipartite graph. Then G is shellable if and only if G is sequentially Cohen–Macaulay.*

Proof. Assume that G is sequentially Cohen–Macaulay. The proof is by induction on the number of vertices of G. By Lemma 7.4.14 there is a vertex x_1 of G of degree 1. Let x_2 be the vertex of G adjacent to x_1. For $i = 1, 2$ consider the subgraph $G_i = G \setminus (\{x_i\} \cup N_G(x_i))$. By Theorem 7.4.11 G_1 and G_2 are sequentially Cohen–Macaulay. Hence Δ_{G_1} and Δ_{G_2} are shellable by the induction hypothesis. Therefore Δ_G is shellable by Theorem 7.4.7. The converse follows at once from Theorem 6.3.27. □

Van Tuyl [407] has shown that the independence complex Δ_G must be vertex decomposable for any bipartite graph G whose edge ring $R/I(G)$ is sequentially Cohen–Macaulay. Thus Theorem 7.4.16 remains valid if we replace shellable by vertex decomposable.

As we saw in Corollary 7.4.8, one can verify recursively that a bipartite graph is shellable. The above theorem, therefore, implies the same for sequentially Cohen–Macaulay bipartite graphs.

Lemma 7.4.17 The complete bipartite graph $\mathcal{K}_{m,n}$ is Cohen–Macaulay if and only if $m = n = 1$.

Proof. By Exercise 6.3.51 it is enough to observe that the independence complex of $\mathcal{K}_{m,n}$ is disconnected for $m + n \geq 3$. □

Lemma 7.4.18 *Let G be an unmixed bipartite graph and let $I(G)$ be its edge ideal. If $I(G)$ has height g, then there are disjoint sets of vertices $V_1 = \{x_1, \ldots, x_g\}$ and $V_2 = \{y_1, \ldots, y_g\}$ such that:* (i) *$\{x_i, y_i\}$ is an edge of G for all i, and* (ii) *every edge of G joins V_1 with V_2.*

Proof. The statement follows by using that g is equal to the maximum number of independent edges of G; see Theorem 7.1.8. □

The following is a combinatorial characterization of all unmixed bipartite graphs. Unmixed graphs are also called *well-covered* [340].

Theorem 7.4.19 [421] *Let G be a bipartite graph. Then G is unmixed if and only if there is a perfect matching $e_1, \ldots, e_g$ such that for any two edges $e \neq e'$ and for any two distinct vertices $x \in e$, $y \in e'$ contained in some e_i, one has that $(e \setminus \{x\}) \cup (e' \setminus \{y\})$ is an edge.*

Proof. It follows at once from Corollary 6.5.16 because, by Theorem 7.1.8, bipartite graphs satisfy the König property. □

This result can be reformulated as:

Theorem 7.4.20 [421] *Let G be a bipartite graph without isolated vertices. Then G is unmixed if and only if G has a bipartition $V_1 = \{x_1, \ldots, x_g\}$, $V_2 = \{y_1, \ldots, y_g\}$ such that:* (a) *$\{x_i, y_i\} \in E(G)$ for all i, and* (b) *if $\{x_i, y_j\}$ and $\{x_j, y_k\}$ are in $E(G)$ and i, j, k are distinct, then $\{x_i, y_k\} \in E(G)$.*

The following nice result of Herzog and Hibi classifies the family of Cohen–Macaulay bipartite graphs. This family is contained in the class of uniform admissible clutters studied in [160, 205, 325].

Theorem 7.4.21 [222] *Let G be a bipartite graph without isolated vertices. Then G is a Cohen–Macaulay graph if and only if there is a bipartition $V_1 = \{x_1, \ldots, x_g\}$, $V_2 = \{y_1, \ldots, y_g\}$ of G such that:* (i) *$\{x_i, y_i\} \in E(G)$ for all i,* (ii) *if $\{x_i, y_j\} \in E(G)$, then $i \leq j$, and* (iii) *if $\{x_i, y_j\}$ and $\{x_j, y_k\}$ are in $E(G)$ and $i < j < k$, then $\{x_i, y_k\} \in E(G)$.*

Proof. By Proposition 7.3.8 we may assume that G is connected.

$\Leftarrow$) The proof is by induction on g. The case $g = 1$ is clear. Assume $g \geq 2$. Let $S = \{x_{r_1}, \ldots, x_{r_s}\}$ be the set of all vertices of G adjacent to y_g. Consider the subgraph $G' = G \setminus S$ obtained from G be removing the vertices in S. We claim that $y_{r_1}, \ldots, y_{r_s}$ are isolated vertices of G'. Indeed if y_{r_j} is not isolated, there is an edge $\{x_i, y_{r_j}\}$ in G' with $i < r_j$. Hence by condition (iii) we get that $\{x_i, y_g\}$ is an edge of G and x_i must be a vertex in S, a contradiction. Thus by induction the graphs $G' \setminus \{y_{r_1}, \ldots, y_{r_s}\}$ and $G \setminus \{x_g, y_g\}$ are Cohen–Macaulay. Applying Proposition 7.3.2 we get that G is Cohen–Macaulay.

$\Rightarrow$) By Lemma 7.4.18 there is a bipartition (V_1, V_2) satisfying (i). By Lemma 7.4.14 we may assume that $\deg(x_g) = 1$ and $\deg(y_g) \geq 2$. Since G is Cohen–Macaulay we get that $G \setminus \{y_g\}$ is Cohen–Macaulay; see Proposition 7.3.1. Using induction on G it is seen that there is a bipartition of G satisfying (i) and (ii). To complete the proof notice that G must also satisfy (iii) according to Theorem 7.4.20. □

The following result characterizes all bipartite graphs with a perfect matching that satisfies condition (ii) of Theorem 7.4.21.

Theorem 7.4.22 [325, Theorem 4.3] *Let G be a bipartite graph with a perfect matching $\{e_i\}_{i=1}^{g}$ such that $e_i = \{x_i, y_i\}$ for all i and let $\Gamma = \Delta_G^{[g]}$ be the pure g-skeleton of Δ_G. Then, Γ is pure shellable if and only if we can order $e_1, \ldots, e_g$ such that $\{x_i, y_j\} \in E(G)$ implies $i \leq j$.*

Directed graphs in sequentially C–M graphs Let G be an unmixed bipartite graph without isolated vertices. Then, by Theorem 7.4.20, there is a bipartition V_1, V_2 of G such that

(1) if $V_1 = \{x_1, \ldots, x_g\}$ and $V_2 = \{y_1, \ldots, y_g\}$, where $g = \alpha_0(G)$,

(2) for $i = 1, \ldots, g$, (after relabeling) $\{x_i, y_i\}$ is an edge of G.

Following Carrá Ferro and Ferrarello [78], we define a directed graph $\mathcal{D}$ with vertex set V_1 as follows: (x_i, x_j) is a directed edge of $\mathcal{D}$ if $i \neq j$ and $\{x_i, y_j\}$ is an edge of G. We say that a cycle $\mathcal{C}$ of $\mathcal{D}$ is *oriented* if all the arrows of $\mathcal{C}$ are oriented in the same direction. Recall that $\mathcal{D}$ is called *transitive* if for any two (x_i, x_j), (x_j, x_k) in $E_{\mathcal{D}}$ with i, j, k distinct, we have that $(x_i, x_k) \in E_{\mathcal{D}}$.

Example 7.4.23 If $G = \mathcal{C}_4$ is a four cycle with edge set

$$\{\{x_1, y_1\}, \{x_2, y_2\}, \{x_1, y_2\}, \{x_2, y_1\}\},$$

then $\mathcal{D}$ has two vertices x_1, x_2 and two arrows (x_1, x_2), (x_2, x_1) forming an oriented cycle of length two.

Lemma 7.4.24 [208, Theorem 16.3(4), p. 200] *Let $\mathcal{D}$ be the directed graph described above. $\mathcal{D}$ is acyclic, i.e., $\mathcal{D}$ has no oriented cycles, if and only if there is a linear ordering of the vertex set V_1 such that all the edges of $\mathcal{D}$ are of the form (x_i, x_j) with $i < j$.*

Theorem 7.4.25 *Let G be a bipartite graph satisfying* (1) *and* (2). *If G is sequentially Cohen–Macaulay, then the directed graph $\mathcal{D}$ is acyclic.*

Proof. We proceed by induction on the number of vertices of G. Assume that $\mathcal{D}$ has an oriented cycle $\mathcal{C}_r$ with vertices $\{x_{i_1},\ldots,x_{i_r}\}$. This means that the graph G has a cycle

$$\mathcal{C}_{2r}=\{y_{i_1},x_{i_1},y_{i_2},x_{i_2},y_{i_3},\ldots,y_{i_{r-1}},x_{i_{r-1}},y_{i_r},x_{i_r}\}$$

of length $2r$. By Lemma 7.4.14, the graph G has a vertex v of degree 1. Notice that $v\notin\{x_{i_1},\ldots,x_{i_r},y_{i_1},\ldots,y_{i_r}\}$. Furthermore, if w is the vertex adjacent to v, we also have $w\notin\{x_{i_1},\ldots,x_{i_r},y_{i_1},\ldots,y_{i_r}\}$. Hence by Theorem 7.4.11 the graph $G'=G\setminus(\{v\}\cup N_G(v))$ is sequentially Cohen–Macaulay and $\mathcal{D}_{G'}$ has an oriented cycle, a contradiction to the induction hypotheses. Thus $\mathcal{D}$ has no oriented cycles, as required. $\Box$

Cohen–Macaulay trees can also be described in terms of $\mathcal{D}$:

Theorem 7.4.26 *Let G be a tree satisfying* (1) *and* (2). *Then G is a Cohen–Macaulay tree if and only if $\mathcal{D}$ is a tree such that every vertex x_i of $\mathcal{D}$ is either a source (i.e., has only arrows leaving x_i) or a sink (i.e., has only arrows entering x_i).*

Proof. $\Rightarrow$) Since a tree is bipartite, $\mathcal{D}$ is both acyclic and transitive. Suppose there is a vertex x_i that is not a sink or source. i.e., there is an arrow entering x_i and one leaving x_i. Suppose the arrow entering x_i originates at x_j, and the arrow leaving x_i goes to x_k. Note that $x_j\neq x_k$ because otherwise we would have a cycle in the acyclic graph $\mathcal{D}$. Because $\mathcal{D}$ is transitive, the directed edge (x_j,x_k) also belongs to $\mathcal{D}$. But then the induced graph on the vertices $\{x_j,y_i,x_i,y_k\}$ in G forms the cycle $\mathcal{C}_4$, contradicting the fact that G is a tree.

$\Leftarrow$) The hypotheses on $\mathcal{D}$ imply $\mathcal{D}$ is acyclic and transitive, so apply Exercise 7.4.36(d). $\Box$

Computing the type Assume that G is a Cohen–Macaulay bipartite graph. Note that $\{x_i-y_i\}_{i=1}^g$ is a regular system of parameters for $R/I(G)$, where $R=K[\mathbf{x},\mathbf{y}]$ and $R/I(G)$ modulo $(\{x_i-y_i\}_{i=1}^g)$ reduces to:

$$A=K[\mathbf{x}]/(x_1^2,\ldots,x_g^2,I(H)),$$

where H is a graph on the vertex set V_1. As a K-vector space, A is generated by the image of 1 and the images of all the monomials $x_{i_1}\cdots x_{i_r}$ such that $\{x_{i_1},\ldots,x_{i_r}\}$ is an independent set of H. If $H^\vee$ denotes the clutter of minimal vertex covers of H, then the type of $R/I(G)$ is equal to $|E(H^\vee)|$, because $\mathrm{Soc}(A)$ is generated by the images of all the monomials $x_{i_1}\cdots x_{i_r}$ such that $\{x_{i_1},\ldots,x_{i_r}\}$ is a maximal independent set of H.

Example 7.4.27 Let G be the Cohen–Macaulay bipartite graph whose edge ideal is

$$I(G) = (x_1y_1, x_1y_2, x_1y_3, x_1y_4, x_2y_2, x_2y_3, x_2y_4, x_3y_3, x_4y_4).$$

The graph H has edges $x_1x_2, x_1x_3, x_1x_4, x_2x_3, x_2x_4$. Since this graph has only three minimal vertex covers, the type of $R/I(G)$ is equal to three.

Exercises

7.4.28 Let $\mathcal{K}_{m,n}$ be the complete bipartite graph. Prove that if $m, n \geq 2$, then $\mathcal{K}_{m,n}$ is not shellable. If $m = 1$ and $n \geq 1$, prove that $\mathcal{K}_{m,n}$ is shellable.

7.4.29 If G is a chordal graph, then the ideal $I(G)^\vee$ has linear quotients.

7.4.30 [149] Prove that a bipartite graph G is Cohen–Macaulay if and only if Δ_G is pure shellable.

7.4.31 Prove that no even cycle can be sequentially Cohen–Macaulay.

7.4.32 [167, Proposition 4.1] Prove that C_3 and C_5 are the only sequentially Cohen–Macaulay cycles.

7.4.33 Let G be a Cohen–Macaulay bipartite graph. Show that the h-vector of $R/I(G)$ has length at most $g = \operatorname{ht} I(G)$. Can we replace Cohen–Macaulay by unmixed?

7.4.34 (Ravindra [344]) A connected bipartite graph G is unmixed if and only if there is a perfect matching such that for every edge $\{x, y\}$ in the perfect matching, the induced subgraph $G[N_G(x) \cup N_G(y)]$ is a complete bipartite graph.

7.4.35 A bipartite graph G without isolated vertices is unmixed if and only if G has a perfect matching $\{x_1, y_1\}, \ldots, \{x_g, y_g\}$ such that $(I(G)^2 \colon x_iy_i)$ is equal to $I(G)$ for $i = 1, \ldots, g$.

7.4.36 Let G be a connected bipartite graph with bipartition $V_1 = \{x_i\}_{i=1}^g$ and $V_2 = \{y_i\}_{i=1}^g$ such that $\{x_i, y_i\} \in E(G)$ for all i and $g \geq 2$. Define a directed graph $\mathcal{D}$ with vertex set V_1 as follows: (x_i, x_j) is an edge of $\mathcal{D}$ if $i \neq j$ and $\{x_i, y_j\}$ is an edge of G. Prove that the following hold.

(a) If G is a square, then $\mathcal{D}$ has two vertices x_1, x_2 and two arrows (x_1, x_2), (x_2, x_1) forming a cycle of length two.

(b) $\mathcal{D}$ is *acyclic*, i.e., $\mathcal{D}$ has no directed cycles, if and only if there is a linear ordering of the vertex set V_1 such that all the edges of $\mathcal{D}$ are of the form (x_i, x_j) with $i < j$.

(c) $\mathcal{D}$ is called *transitive* if for any two (x_i, x_j), (x_j, x_k) in $E(\mathcal{D})$ with i, j, k distinct, we have that $(x_i, x_k) \in E(\mathcal{D})$. The digraph $\mathcal{D}$ is transitive if and only if G is unmixed.

(d) (Carrá Ferro–Ferrarello [78]) G is Cohen–Macaulay if and only if $\mathcal{D}$ is acyclic and transitive.

(e) If G is a Cohen–Macaulay graph, then $\mathcal{D}$ has at least one source x_i and at least one sink x_j.

(f) If every vertex of $\mathcal{D}$ is either a source or a sink, then G is Cohen–Macaulay.

(g) If $\mathcal{D}$ is transitive and has a cycle, then $\mathcal{D}$ has a cycle of length 2.

(h) $\mathcal{D}$ is transitive and acyclic if and only if $\mathcal{D}$ is transitive and has no cycles of length 2.

(i) $\mathcal{D}$ has no cycles of length 2 if and only if for any edge $\{x_i, y_j\} \in E(G)$ with $i \neq j$, one has that $\{x_j, y_i\}$ is not an edge of G.

(j) (Zaare-Nahandi [435]) G is Cohen–Macaulay if and only if the following conditions hold: (1) The induced subgraph $G[N(x_i) \cup N(y_i)]$ is a complete bipartite graph for all i, and (2) if x_i is adjacent to y_j for $i \neq j$, then x_j is not adjacent to y_i.

(k) [435] G is Cohen–Macaulay if and only if G is unmixed and G has a unique perfect matching.

7.4.37 Let G be a connected bipartite graph satisfying the conditions of Theorem 7.4.21. Prove that $\deg(y_1) = 1$ and $\deg(x_g) = 1$.

7.4.38 Let G be a Cohen–Macaulay bipartite graph. If G is not a discrete graph, prove that G has at least two vertices of degree one.

7.4.39 Prove that the following is the full list of Cohen–Macaulay connected bipartite graphs with eight vertices.

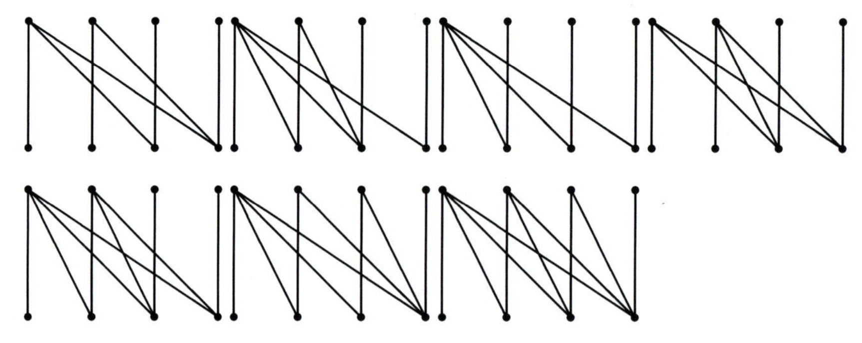

7.4.40 Let $I = (x_i y_j \,|\, 1 \leq i \leq j \leq n)$, draw a picture of the bipartite graph G such that $I = I(G)$ and prove that I is a Cohen–Macaulay ideal.

7.5 Regularity, depth, arithmetic degree

In this section we study the regularity and depth of edge ideals of graphs. We introduce a device to estimate the regularity of edge ideals and show some upper bounds for the arithmetic degree.

Let G be a simple graph with vertex set $X = \{x_1, \ldots, x_n\}$ and let $R = K[X]$ be a polynomial ring over a field K. As usual we denote the vertex set and edge set of G by $V(G)$ and $E(G)$, respectively.

Lemma 7.5.1 [274, Lemma 2.2] *Let G be a graph and let $\mathrm{im}(G)$ be its induced matching number. Then $\mathrm{im}(G) \leq \mathrm{reg}(R/I(G))$.*

Proof. It follows at once from Corollary 6.4.8. □

Theorem 7.5.2 [326] *Let $\mathcal{F}$ be a family of graphs containing any discrete graph and let $\beta\colon \mathcal{F} \to \mathbb{N}$ be a function satisfying that $\beta(G) = 0$ for any discrete graph G, and such that given $G \in \mathcal{F}$, with $E(G) \neq \emptyset$, there is $x \in V(G)$ such that the following two conditions hold:*

(i) *$G \setminus \{x\}$ and $G \setminus (\{x\} \cup N_G(x))$ are in $\mathcal{F}$.*

(ii) *$\beta(G \setminus (\{x\} \cup N_G(x))) < \beta(G)$ and $\beta(G \setminus \{x\}) \leq \beta(G)$.*

Then $\mathrm{reg}(R/I(G)) \leq \beta(G)$ for any $G \in \mathcal{F}$.

Proof. The proof is by induction on the number of vertices. Let G be a graph in $\mathcal{F}$. If G is a discrete graph, then $I(G) = (0)$, $\mathrm{reg}(R) = 0$ and $\beta(G) = 0$. Assume that G has at least one edge. There is a vertex $x \in V(G)$ such that the induced subgraphs $G_1 = G \setminus \{x\}$ and $G_2 = G \setminus (\{x\} \cup N_G(x))$ satisfy (i) and (ii). There is an exact sequence of graded R-modules

$$0 \longrightarrow R/(I(G)\colon x)[-1] \stackrel{x}{\longrightarrow} R/I(G) \longrightarrow R/(x, I(G)) \longrightarrow 0.$$

Notice that $(I(G)\colon x) = (N_G(x), I(G_2))$ and $(x, I(G)) = (x, I(G_1))$. The graphs G_1 and G_2 have fewer vertices than G. It follows directly from the definition of regularity that $\mathrm{reg}(M[-1]) = 1 + \mathrm{reg}(M)$ for any graded R-module M. Therefore applying the induction hypothesis to G_1 and G_2, and using conditions (i) and (ii) and Exercise 6.4.30, we get

$$\begin{aligned} \mathrm{reg}(R/(I(G)\colon x)[-1]) &= \mathrm{reg}(R'/I(G_2)) + 1 \leq \beta(G_2) + 1 \leq \beta(G), \\ \mathrm{reg}(R/(x, I(G))) &\leq \beta(G_1) \leq \beta(G) \end{aligned}$$

where $R' = K[V(G_2)]$. Hence, by Lemma 6.4.10, we get that the regularity of $R/I(G)$ is bounded by the maximum of the regularities of $R/(x, I(G))$ and $R/(I(G)\colon x)[-1]$. Thus $\mathrm{reg}(R/I(G)) \leq \beta(G)$, as required. □

Definition 7.5.3 A vertex x of a graph G is called *simplicial* if the subgraph $G[N_G(x)]$ induced by the neighbor set $N_G(x)$ is a complete subgraph.

Lemma 7.5.4 [403] *Let H be a chordal graph and $\mathcal{K}$ a complete subgraph of H. If $\mathcal{K} \neq H$, then there is $x \notin V(\mathcal{K})$ such that the subgraph $H[N_H(x)]$ induced by the neighbor set $N_H(x)$ of x is a complete subgraph.*

Proof. By induction on $V(H)$. If H is complete, any $x \notin V(\mathcal{K})$ would satisfy the requirement. Otherwise, using Proposition 7.3.25, $H = H_1 \cup H_2$, where H_1, H_2 are chordal graphs smaller than H such that $H_1 \cap H_2$ is a complete subgraph. It follows that $\mathcal{K}$ is a subgraph of H_i for some i, say $i = 1$. As $H_1 \cap H_2$ is a proper complete subgraph of H_2, by induction x can be chosen in $H_2 \setminus (H_1 \cap H_2)$. □

Let G be a graph. We let $\beta'(G)$ be the cardinality of any smallest maximal matching of G. Hà and Van Tuyl proved that the regularity of $R/I(G)$ is bounded from above by the matching number of G and Woodroofe improved this result showing that $\beta'(G)$ is an upper bound for the regularity.

For edge ideals whose resolutions are k-steps linear and for ideals of covers of graphs some new upper bounds for the regularity are given by Dao, Huneke, and Schweig [103]. There are also upper bounds given by Banerjee [13] for the regularity of powers of edge ideals of graphs whose complement does not have any induced four cycle.

Theorem 7.5.5 *Let G be a graph and let $R = K[V(G)]$. Then*

(a) [206, Corollary 6.9] $\mathrm{reg}(R/I(G)) = \mathrm{im}(G)$ *for any chordal graph G.*

(b) ([206, Theorem 6.7], [434]) $\mathrm{reg}(R/I(G)) \leq \beta'(G)$.

(c) [407, Theorem 3.3] $\mathrm{reg}(R/I(G)) = \mathrm{im}(G)$ *if G is bipartite and $R/I(G)$ is sequentially Cohen–Macaulay.*

Proof. (a) Let $\mathcal{F}$ be the family of chordal graphs and let G be a chordal graph with $E(G) \neq \emptyset$. By Lemma 7.5.1 and Theorem 7.5.2 it suffices to prove that there is $x \in V(G)$ such that $\mathrm{im}(G_1) \leq \mathrm{im}(G)$ and $\mathrm{im}(G_2) < \mathrm{im}(G)$, where G_1 and G_2 are the subgraphs $G \setminus \{x\}$ and $G \setminus (\{x\} \cup N_G(x))$, respectively. The inequality $\mathrm{im}(G_1) \leq \mathrm{im}(G)$ is clear because any induced matching of G_1 is an induced matching of G. We now show the other inequality. By Lemma 7.5.4, there is $y \in V(G)$ such that $G[N_G(y) \cup \{y\}]$ is a complete subgraph. Pick $x \in N_G(y)$ and set $f_0 = \{x, y\}$. Consider an induced matching $f_1, \ldots, f_r$ of G_2 with $r = \mathrm{im}(G_2)$. We claim that $f_0, f_1, \ldots, f_r$ is an induced matching of G. Let e be an edge of G contained in $\cup_{i=0}^r f_i$. We may assume that $e \cap f_0 \neq \emptyset$ and $e \cap f_i \neq \emptyset$ for some $i \geq 1$, otherwise $e = f_0$ or $e = f_i$ for some $i \geq 1$. Then $e = \{y, z\}$ or $e = \{x, z\}$ for some $z \in f_i$. If $e = \{y, z\}$, then $z \in N_G(y)$ and $x \in N_G(y)$. Hence $\{z, x\} \in E(G)$ and $z \in N_G(x)$, a contradiction because the vertex set of G_2 is disjoint from $N_G(x) \cup \{x\}$. If $e = \{x, z\}$, then $z \in N_G(x)$, a contradiction. This completes the proof of the claim. Hence $\mathrm{im}(G_2) < \mathrm{im}(G)$.

(b) Let $\mathcal{F}$ be the family of all graphs and let G be a graph with $E(G) \neq \emptyset$. By Theorem 7.5.2 it suffices to prove that there is $x \in V(G)$ such that $\beta'(G_1) \leq \beta'(G)$ and $\beta'(G_2) < \beta'(G)$, where G_1 and G_2 are the subgraphs $G \setminus \{x\}$ and $G \setminus (\{x\} \cup N_G(x))$, respectively. We leave the proof of this as an exercise.

(c) Let $\mathcal{F}$ be the family of all bipartite graphs G such that $R/I(G)$ is sequentially Cohen–Macaulay, and let $\beta\colon \mathcal{F} \to \mathbb{N}$ be the function $\beta(G) = \mathrm{im}(G)$. Let G be a graph in $\mathcal{F}$ with $E(G) \neq \emptyset$. By Lemma 7.5.1 and Theorem 7.5.2 it suffices to observe that, according to Corollary 7.4.8 and Theorem 7.4.16, there are adjacent vertices x_1 and x_2 with $\deg(x_2) = 1$ such that the bipartite graph $G \setminus (\{x_i\} \cup N_G(x_i))$ is sequentially Cohen–Macaulay for $i = 1, 2$. Thus conditions (i) and (ii) of Theorem 7.5.2 are satisfied. □

Corollary 7.5.6 [437] *If G is a forest, then* $\mathrm{reg}(R/I(G)) = \mathrm{im}(G)$.

Proof. Any forest is a bipartite graph. Thus, by Theorem 6.5.25, Δ_G is shellable and $R/I(G)$ is sequentially Cohen–Macaulay. Hence the result follows from Theorem 7.5.5. □

A graph G is *weakly chordal* if every induced cycle in both G and its complement $\overline{G}$ has length at most 4.

Theorem 7.5.7 [434] *If G is weakly chordal, then* $\mathrm{reg}(R/I(G)) = \mathrm{im}(G)$.

Proof. It is shown in [77] that a weakly chordal graph G can be covered by $\mathrm{im}(G)$ co-CM graphs. Thus, the result follows from Corollary 6.4.13. □

If G is an unmixed graph, Kummini [285] showed that $\mathrm{reg}(R/I(G))$ is equal to the induced matching number of G. This result was later extended to all very well-covered graphs [303] (a graph G is *very well-covered* if it is unmixed without isolated vertices and $2\,\mathrm{ht}(I(G)) = |V(G)|$).

If G is claw-free and $\overline{G}$, the complement of G, has no induced 4-cycles, then $\mathrm{reg}(R/I(G)) \leq 2$ with equality if $\overline{G}$ is not chordal [331], in this case $\mathrm{reg}(R/I(G)) = \mathrm{im}(G) + 1$. Formulas for the regularity of ideals of mixed products are given in [260]. The regularity and depth of lex segment edge ideals are computed in [143]. The regularity and other algebraic properties of edge ideal of Ferrers graphs are studied in detail in [96].

Corollary 7.5.8 *Let G be a bipartite graph without isolated vertices. If G has n vertices, then* $\mathrm{depth}\, K[\Delta_G] \leq \lfloor \frac{n}{2} \rfloor$.

Proof. Let (V_1, V_2) be a bipartition of G with $|V_1| \leq |V_2|$. Note $2|V_1| \leq n$ because $|V_1| + |V_2| = n$. Since V_1 is a maximal independent set of vertices one has $\beta_0'(G) \leq |V_1|$. Therefore using Theorem 6.4.18 we conclude

$$\mathrm{depth}\, K[\Delta_G] \leq \beta_0'(G) \leq |V_1| \leq n/2. \qquad \square$$

The bound above does not extend to non-bipartite graphs. The depth of powers of edge ideals of trees is studied in [324].

Proposition 7.5.9 *Let $\mathcal{K}_r$ be the complete graph with r vertices and let $G_{r,i}$ be the graph obtained by attaching i lines at each vertex of $\mathcal{K}_r$. Then*

$$\operatorname{depth}(R/I(G_{r,i})) = (r-1)i + 1.$$

Proof. Let $R = K[x_1, \ldots, x_n]$ be a polynomial ring and let $x_1, \ldots, x_n$ be the vertices of $G_{r,i}$. Note that $G_{r,i}$ has $n = r(i+1)$ vertices. We proceed by induction on r. If $r = 1$, then $G_{1,i}$ is a star and clearly $K[\Delta_{G_{1,i}}]$ has depth 1. Assume $r \geq 2$. Set $G = G_{r,i}$ and $I = I(G)$. Let x_1 be any vertex of $\mathcal{K}_r$ and consider the exact sequence

$$0 \longrightarrow R/(I\colon x_1) \xrightarrow{x_1} R/I \longrightarrow R/(x_1, I) \longrightarrow 0. \qquad (*)$$

It is not hard to see that $(I\colon x_1)$ is a face ideal generated by $i + (r-1)$ variables. Hence $\operatorname{depth}(R/(I\colon x_1)) = (r-1)i + 1$. On the other hand note the equality $(I, x_1) = (x_1) + I(G_{r-1,i})$. Since x_1 is adjacent in G to exactly i-vertices of degree 1 and those vertices do not occur in (I, x_1), by induction hypothesis we derive $\operatorname{depth}(R/(x_1, I)) = i + [(r-2)i + 1] = (r-1)i + 1$. Altogether the ends of the exact sequence $(*)$ have depth equal to $(r-1)i+1$. Thus, by the depth Lemma 2.3.9, we get $\operatorname{depth}(R/I) = (r-1)i + 1$. □

Example 7.5.10 Let $K = \mathbb{Q}$ and let $G = G_{4,3}$ be the following graph

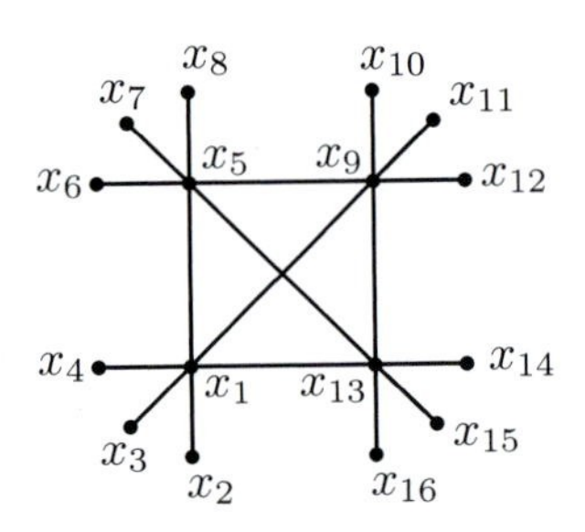

Then, by Proposition 7.5.9, $\operatorname{depth} R/I(G) = 10$ and $\operatorname{pd}(R/I(G)) = 6$.

Theorem 7.5.11 [189, Theorem 2.7.14] *Let G be a graph without isolated vertices and let $\beta_0'(\mathcal{C})$ be the cardinality of a smallest maximal independent set of G. If G has n vertices, then*

$$\operatorname{depth} K[\Delta_G] \leq \beta_0'(G) \leq \frac{n + \beta_0(G)n - 2\beta_0(G) - \beta_0(G)^2}{(n - \beta_0(G))},$$
$$\beta_0(G) \leq \alpha_0(G)[1 + (\beta_0(G) - \beta_0'(G))].$$

Lemma 7.5.12 *Let G be a graph and let $G^\vee$ be its blocker. Then the arithmetic degree of G is given by* $\operatorname{arith-deg}(I(G)) = |E(G^\vee)|$.

Proof. It suffices to recall that a set of vertices of G is a maximal stable set if and only if its complement is a minimal vertex cover. $\Box$

The basic problem of enumeration and counting of the stable sets in a graph—using an algebraic perspective—is addressed in [110]. The following estimate is due to E. Sperner (cf. [168, Theorem 3.1]).

Theorem 7.5.13 [390] *If G is a graph with n vertices, then*

$$\text{arith-deg}(I(G)) = |E(G^\vee)| \leq \binom{n}{\lfloor n/2 \rfloor}.$$

Theorem 7.5.14 [212] *If G is a graph, then $|E(G^\vee)| \leq 2^{\alpha_0(G)}$.*

Proof. We may assume G has no isolated vertices. We set $\mathcal{C} = G^\vee$. Let C be a fixed vertex cover with $\alpha_0(G)$ vertices. For $0 \leq i \leq \alpha_0(G)$ consider:

$$\mathcal{C}_i = \{C' \,|\, C' \in \mathcal{C} \text{ and } |C \cap C'| = i\}.$$

We claim that the following inequality holds: $|\mathcal{C}_i| \leq \binom{\alpha_0(G)}{i}$ for all $i \geq 0$. We are going to show that if C' and C'' are two vertex covers in $\mathcal{C}_i$ such that $C \cap C' = C \cap C''$, then $C' = C''$. Using $C \cap C' = C \cap C''$ we obtain that the neighbors sets satisfy: $(*)$ $N_G(C \setminus C') = N_G(C \setminus C'')$. Since C' and C'' are vertex covers of G one has:

$$N_G(C \setminus C') \cup (C \cap C') \subset C' \quad \text{and} \quad N_G(C \setminus C'') \cup (C \cap C'') \subset C''.$$

Observe that $N_G(C \setminus C') \cup (C \cap C')$ and $N_G(C \setminus C'') \cup (C \cap C'')$ are vertex covers of G, because C is a vertex cover of G. From the minimality of C' and C'' one concludes the equalities:

$$N_G(C \setminus C') \cup (C \cap C') = C' \quad \text{and} \quad N_G(C \setminus C'') \cup (C \cap C'') = C'',$$

which together with $(*)$ allow us to derive the equality $C' = C''$, as required. Therefore since there are only $\binom{\alpha_0(G)}{i}$ subsets in C with i vertices, one obtains the inequality $|\mathcal{C}_i| \leq \binom{\alpha_0(G)}{i}$, as claimed. Hence

$$|E(G^\vee)| = |\mathcal{C}| = \sum_{i=0}^{\alpha_0(G)} |\mathcal{C}_i| \leq \sum_{i=0}^{\alpha_0(G)} \binom{\alpha_0(G)}{i} = 2^{\alpha_0(G)}. \qquad \Box$$

Exercises

7.5.15 A graph whose complement is chordal is called *co-chordal.* A complex Δ is called a quasi-forest if Δ is the clique complex of a chordal graph. Prove that an edge ideal $I(\mathcal{C})$ of a clutter $\mathcal{C}$ has regularity 2 if and only if $\Delta_{\mathcal{C}}$ is the independence complex of a co-chordal graph.

7.5.16 Let G be the complement of a cycle of length six. Use *Macaulay*2, to show that $\mathrm{reg}(R/I(G)) = 2$ and $\mathrm{im}(G) = 1$.

7.5.17 Let $\mathcal{K}_r$ be a complete graph and let G be the graph obtained by attaching i_j lines at each vertex v_j of $\mathcal{K}_r$. If $i_j \leq i_{j+1}$ for all j, then

$$\mathrm{depth}(R/I(G)) = i_1 + \cdots + i_{r-1} + 1.$$

7.5.18 Let $\gamma_n = \sup\{|E(G^\vee)| \mid G \text{ is a graph with } n \text{ vertices}\}$. If $\{F_n\}$ is the Fibonacci sequence, $F_0 = F_1 = 1$ and $F_n = F_{n-1} + F_{n-2}$, then $\gamma_n \leq F_n$ for $n \geq 1$. If P_n is the path $\{x_1, \ldots, x_n\}$, $n \geq 3$, then $F_n = |E(P_n^\vee)|$.

7.5.19 Find $e(G)$, the multiplicity of $R/I(G)$, and arith-deg$(I(G))$, the arithmetic degree of $I(G)$ if G is a disjoint union of stars.

7.5.20 Let G be the graph which is the disjoint union of k complete graphs with two vertices. Then, $|E(G^\vee)| = e(G) = 2^{\alpha_0(G)}$ and $k = \alpha_0(G)$.

7.6 Betti numbers of edge ideals

Let G be a graph and $I(G) \subset R$ its edge ideal. An interesting problem is to express some of the first initial Betti numbers and invariants of $I(G)$ in terms of the graph theoretical data of G (see [138, 141, 276, 321, 362, 437] and the references therein). Kimura [277] has studied the projective dimension of edge ideals of unmixed bipartite graphs.

Let R be a polynomial ring over a field K with the usual grading and I a graded ideal of R. The minimal graded resolution of R/I by free R-modules can be expressed as:

$$0 \longrightarrow \bigoplus_{i=1}^{c_g} R^{b_{gi}}(-d_{gi}) \xrightarrow{\varphi_g} \cdots \longrightarrow \bigoplus_{i=1}^{c_1} R^{b_{1i}}(-d_{1i}) \xrightarrow{\varphi_1} R \longrightarrow R/I \longrightarrow 0,$$

where all the maps are degree preserving and the d_{ij} are positive integers. One may assume $d_{j1} < \cdots < d_{jc_j}$ for all $j = 1, \ldots, g$. The integer g is equal to $\mathrm{pd}_R(R/I)$, the *projective dimension* of R/I. To simplify notation set

$$F_j = \bigoplus_{i=1}^{c_j} R^{b_{ji}}(-d_{ji}).$$

The rank of F_j is the jth *Betti number* of I. From the minimality of the resolution $d_{11} < \cdots < d_{g_1}$; see Remark 3.5.9. We define the jth *initial Betti number* of I as $\gamma_j = b_{j1}$ and the jth *initial virtual Betti number* of I as

$$v_j = \dim_K (F_j)_{d_{11}+j-1}.$$

Note that some of the initial virtual Betti numbers may be zero.

For a general monomial ideal I there is an explicit and elegant description of a minimal set of generators for the first module of syzygies of I due to Eliahou [137]; see also [64].

Definition 7.6.1 The *line graph* of G, denoted by $L(G)$, has vertex set $E = E(G)$ with two vertices of $L(G)$ adjacent whenever the corresponding edges of G have exactly one common vertex.

Proposition 7.6.2 *If G is a graph with vertices $x_1, \ldots, x_n$ and edge set $E(G)$, then the number of edges of the line graph $L(G)$ is given by*

$$|E(L(G))| = \sum_{i=1}^{n} \binom{\deg(x_i)}{2} = -|E(G)| + \sum_{i=1}^{n} \frac{\deg^2 x_i}{2}. \tag{7.1}$$

Proof. To prove the first equality note that each vertex x_i of G of degree d_i contributes with $\binom{d_i}{2}$ edges to the number of edges of $L(G)$. The second equality follows from the Euler's identity $2|E(G)| = \sum_{i=1}^{n} \deg(x_i)$. □

Let us now present a formula, in graph theoretical terms, for the second initial Betti number of $I(G)$.

Proposition 7.6.3 [138] *Let $I \subset R$ be the edge ideal of a graph G, let V be the vertex set of G, and let $L(G)$ be the line graph of G. If*

$$\cdots \longrightarrow R^c(-4) \oplus R^b(-3) \longrightarrow R^q(-2) \stackrel{\psi}{\longrightarrow} R \longrightarrow R/I \longrightarrow 0$$

is the minimal graded resolution of R/I. Then $b = |E(L(G))| - N_t$, where N_t is the number of triangles of G and c is the number of unordered pairs of edges $\{f, g\}$ such that $f \cap g = \emptyset$ and f and g cannot be joined by an edge.

Proof. Let $f_1, \ldots, f_q$ be the set of all $x_i x_j$ such that $\{x_i, x_j\}$ is an edge of G. We may assume $\psi(e_i) = f_i$. Let Z_1' be the set of elements in $\ker(\psi)$ of degree 3. We regard the f_i's as the vertices of $L(G)$. Every edge $e = \{f_i, f_j\}$ in $L(G)$ determines a syzygy $\mathrm{syz}(e) = x_j e_i - x_i e_j$, where $f_i = x_i z$ and $f_j = x_j z$ for some z, x_i, x_j in V. By Theorem 3.3.19 the set of those syzygies generate Z_1'. Given a triangle $C_3 = \{z_1, z_2, z_3\}$ in G we set

$$\phi(C_3) = \{z_1 e_j - z_3 e_i, z_1 e_j - z_2 e_k, z_2 e_k - z_3 e_i\},$$

where $f_i = z_1 z_2$, $f_j = z_2 z_3$, and $f_k = z_1 z_3$. Notice that $\phi(C_3) \cap \phi(C_3') = \emptyset$ if $C_3 \neq C_3'$. For every triangle C_3 in G choose an element $\rho(C_3) \in \phi(C_3)$. It is not hard to show that the set

$$B = \{\mathrm{syz}(e) | \, e \in L(G)\} \setminus \{\rho(C_3) | \, C_3 \text{ is a triangle in } G\}$$

is a minimal generating set for Z_1'. Thus $b = |E(L(G))| - N_t$. The proof of the expression for c is left as an exercise. □

The 2nd Betti number of a Stanley–Reisner ideal (resp. monomial ideal) is independent of the ground field [244] (resp. [64, 137]); moreover for edge ideals the 3rd and 4th Betti numbers and the 5th and 6th Betti numbers are also independent of the field [243] and [274], respectively.

An application to Hilbert series Let Δ be a simplicial complex of dimension d and $f = (f_0, \ldots, f_d)$ its f-vector, where f_i is the number of faces of dimension i in Δ and $f_{-1} = 1$. By Theorem 6.7.2 the Hilbert series of the Stanley–Reisner ring $S = R/I_\Delta$ can be expressed as

$$F(t) = f_{-1} + \frac{f_0 t}{(1-t)} + \cdots + \frac{f_d t^{d+1}}{(1-t)^{d+1}}. \tag{7.2}$$

We now use this formula to compute a particular instance.

Corollary 7.6.4 *Let G be a graph with q edges, $V = \{x_1, \ldots, x_n\}$ its vertex set and $F(t)$ the Hilbert series of $S = R/I(G)$. If S has dimension 3, then $F(t)(1-t)^3$ is equal to*

$$1 + gt + \left(\binom{g+1}{2} - q\right)t^2 + \left(\frac{2g + 3g^2 + g^3 - 6q - 6gq - 6N + 3v}{6}\right)t^3,$$

where $g = n - 3$, N is the number of triangles of G, and $v = \sum_{i=1}^{n} \deg^2 x_i$.

Proof. Let $\Delta = \Delta_G$ be the Stanley–Reisner complex of $I = I(G)$. The first entries of the f-vector of Δ are $f_{-1} = 1$, $f_0 = n$, $f_1 = \frac{n(n-1)}{2} - q$. From the resolution above and Proposition 6.7.3, $\dim(R/I)_3 = \binom{n+2}{n-1} - qn + b$ and $f_2 = \binom{n+2}{n-1} - qn + b - 2f_1 - n$. The desired formula follows by substitution of the f_i's in Eq. (7.2) and using Eq. (7.1). □

The number of triangles of a graph Let G be a graph with q edges and vertex set $V = \{x_1, \ldots, x_n\}$. As the number of edges of the line graph $L(G)$ of G is given by the formula

$$|E(L(G))| = \sum_{i=1}^{n} \binom{\deg(x_i)}{2} = -q + \sum_{i=1}^{n} \frac{\deg^2 x_i}{2},$$

observe that Proposition 7.6.3 provides a method to compute the number of triangles of a graph by computing syzygies with *Macaulay*2. An alternative method using algebraic graph theory is recalled next.

The *adjacency matrix* of G is the $n \times n$ matrix $A = (a_{ij})$ with entries

$$a_{ij} = \begin{cases} 1 & \text{if } x_i \text{ is adjacent } x_j, \\ 0 & \text{otherwise.} \end{cases}$$

It follows directly that A is symmetric and that its trace is equal to zero. If

$$f(x) = x^n + c_1x^{n-1} + c_2x^{n-2} + c_3x^{n-3} + \cdots + c_n,$$

is the characteristic polynomial of A, then $c_1 = 0$, $-c_2$ is the number of edges of G, and $-c_3$ is twice the number of triangles of G. For an interpretation of all the c_i's see [40].

Example 7.6.5 Let G be the graph whose complement is the union of two disjoint copies of a path of length two. The edge ideal of G is equal to:

$$I(G) = (x_1x_4, x_1x_5, x_1x_6, x_2x_3, x_2x_4, x_2x_5, x_2x_6, x_3x_4, x_3x_5, x_3x_6, x_5x_6),$$

and part of its minimal resolution is:

$$\cdots \rightarrow R^{24}(-3) \rightarrow R^{11}(-2) \rightarrow R \rightarrow R/I(G) \rightarrow 0.$$

Hence the number of triangles of G is equal to 6. The same number is obtained noticing that $f(x) = x^6 - 11x^4 - 12x^3 + 3x^2 + 4x - 1$ is the characteristic polynomial of the adjacency matrix of G.

Linear resolutions of edge ideals First we introduce d-trees and state an interesting result of Fröberg.

Definition 7.6.6 A *d-tree* is defined inductively: (i) a complete graph $\mathcal{K}_{d+1}$ is a d-tree, (ii) let G be a d-tree and H a subgraph of G with $H \simeq \mathcal{K}_d$, if v is a new vertex connected to all vertices in H, then $G \cup \{v\}$ is a d-tree.

Theorem 7.6.7 [172] *Let G be a simple graph, let Δ_G be its independence complex, and let $\overline{G}$ be the complement of G. The following hold.*

(a) *$K[\Delta_G]$ has a 2-linear resolution if and only if $\overline{G}$ is a chordal graph.*

(b) *If G is Cohen–Macaulay over a field K, then $K[\Delta_G]$ has a 2-linear resolution if and only if $\overline{G}$ is a d-tree, where $d = \dim \Delta_G$.*

The problem of determining the simplicial complexes Δ whose Stanley–Reisner ideal I_Δ has a pure resolution was solved by Bruns and Hibi [66, 67].

Corollary 7.6.8 *Let G be a connected Cohen–Macaulay bipartite graph. If the number of edges of G is $g(g+1)/2$, where $g = \operatorname{ht} I(G)$, then*

$$I(G) = (x_iy_j | \, 1 \leq i \leq j \leq g).$$

Proof. The ring $R/I(G)$ has a linear resolution by Theorem 5.3.7. Hence, according to Theorem 7.6.7, the 1-skeleton of Δ_G is a $(g-1)$-tree, which implies that there is a vertex w of G of degree g. We now use Exercise 7.4.37 to conclude that $\deg(x_i) = 1$ and $\deg(y_j) = 1$, for some i, j. To finish the proof consider the graph $G \setminus \{w\}$ and use induction. □

Exercises

7.6.9 Let G be a graph and let $k = \max\{\deg(v) | v \in V(G)\}$. If G is connected, then G is k-regular, i.e., its vertices have degree k, if and only if k is a characteristic value of the adjacency matrix of G.

7.6.10 Let G be a graph with vertex set $\mathbf{x}$ and $F_G(t)$ the Hilbert series of $K[\mathbf{x}]/I(G)$, where K is a field. For any vertex $x \in V(G)$ one has:

$$F_G(t) = F_{G\setminus\{x\}}(t) + \frac{t}{1-t} F_{G\setminus(N(x)\cup\{x\})}(t), \quad \text{with } F_{\emptyset}(t) = 1.$$

7.6.11 If $\mathcal{K}_{r,s}$ is a complete bipartite graph, then

$$F_{\mathcal{K}_{r,s}}(t) = \frac{1}{(1-t)^r} + \frac{1}{(1-t)^s} - 1.$$

7.6.12 [426] Let P_n be a path graph with vertices $\mathbf{x} = x_1, \ldots, x_n$. Prove that the Hilbert series and the Hilbert function of $K[\mathbf{x}]/I(P_n)$ are given by

$$\begin{aligned} F_{P_n}(t) &= \sum_{j=0}^{\lfloor \frac{n+1}{2} \rfloor} \binom{n-j+1}{j} \sum_{i=0}^{\infty} \binom{j+i-1}{i} t^{j+i} \\ H_{L_n}(m) &= \sum_{j=0}^{\min(m, \lfloor \frac{n+1}{2} \rfloor)} \binom{n-j+1}{j} \binom{m-1}{m-j}. \end{aligned}$$

7.6.13 Let $G = C_7$ be a heptagon. Prove the formula:

$$F_G(t) = \frac{-t^3 + 3t^2 + 4t + 1}{(1-t)^3}.$$

What is the number of minimal primes of $I(G)$?

7.6.14 Let $G = C_7$ be a heptagon and let Δ_G be its independence complex. Prove that Δ_G is a triangulation of a Möbius band whose reduced Euler characteristic is $\widetilde{\chi}(\Delta_G) = -1$.

7.6.15 If G is a graph and $I(G) \subset R$ is its edge ideal, then $a(R/I(G)) \leq 0$.

7.6.16 If G is a graph and H is its whisker graph, then $\beta_0(G)$ equals the induced matching number of H.

7.6.17 Let Δ be a simplicial complex of dimension d with n vertices and $f = (f_0, \ldots, f_d)$ its f-vector. If K is a field and $K[\Delta]$ is a Cohen–Macaulay ring with a 2-linear resolution, then

$$f_k = \binom{d+1}{k+1} + (n-d-1)\binom{d}{k}, \quad 0 \leq k \leq d.$$

7.6.18 Let I be the ideal $(abc, abd, ace, adf, aef, bcf, bde, bef, cde, cdf)$ of the polynomial ring $K[a, \ldots, f]$. Prove that the third Betti number of I depends on the base field, and that I is equal to its Alexander dual.

7.7 Associated primes of powers of ideals

In this section we show that the sets of associated primes of the powers of the edge ideal of a graph form an ascending chain. Two excellent references for the general theory of asymptotic prime divisors in commutative Noetherian rings are [259] and [313].

Let G be a simple *graph* with finite vertex set $X = \{x_1, \ldots, x_n\}$ and without isolated vertices, let $R = K[x_1, \ldots, x_n]$ be a polynomial ring over a field K and let $I = I(G)$ be the edge ideal of G.

In [56], Brodmann showed that there exists a positive integer N_1 such that $\mathrm{Ass}(R/I^k) = \mathrm{Ass}(R/I^{N_1})$ for all $k \geq N_1$. A minimal such N_1 is called the *index of stability* of I.

Proposition 7.7.1 [81, Corollary 4.3] *If G is a connected non-bipartite graph with n vertices, s leaves, and the smallest odd cycle of G has length $2k+1$, then $N_1 \leq n - k - s$.*

Definition 7.7.2 [165] *An ideal $I \subset R$ has the persistence property if*

$$\mathrm{Ass}(R/I^k) \subset \mathrm{Ass}(R/I^{k+1})$$

for all $k \geq 1$, i.e., the sets $\mathrm{Ass}(R/I^k)$ form an ascending chain.

There are some interesting cases where associated primes are known to form ascending chains [36, 163, 165, 232, 326]. The first listed is quite general, but has applications to square-free monomial ideals.

Theorem 7.7.3 ([313, Proposition 3.9], [342, 343]) *If R is a Noetherian ring and I is an ideal, then the sets $\mathrm{Ass}(R/\overline{I^k})$ form an ascending chain. In particular if I is normal, then I has the persistence property.*

Definition 7.7.4 Following Schrijver [373], the *duplication* of a vertex x_i of a graph G means extending its vertex set X by a new vertex x_i' and replacing $E(G)$ by

$$E(G) \cup \{(e \setminus \{x_i\}) \cup \{x_i'\} | \, x_i \in e \in E(G)\}.$$

The duplication of the vertex x_i is denoted by G^{x_i}. The *deletion* of x_i is the subgraph $G \setminus \{x_i\}$. A graph obtained from G by a sequence of deletions and duplications of vertices is called a *parallelization* of G.

These two operations commute. If $a = (a_i)$ is a vector in $\mathbb{N}^n$, we denote by G^a the graph obtained from G by successively deleting any vertex x_i with $a_i = 0$ and duplicating $a_i - 1$ times any vertex x_i if $a_i \geq 1$.

Lemma 7.7.5 *Any parallelization of G has the form G^a for some a.*

Proof. It suffices to notice that, by Exercise 7.7.23, any parallelization of G can be obtained by duplications and deletions of vertices of G. $\Box$

The subring $K[G] = K[x_ix_j \mid \{x_i, x_j\}$ is an edge of $G] \subset R$ is called the *edge subring* of G. This subring will be studied in detail in Chapter 10.

Lemma 7.7.6 *Let G be a graph and let $K[G]$ be its edge subring. Then, G^a has a perfect matching if and only if $x^a \in K[G]$.*

Proof. We may assume that $a = (a_1, \ldots, a_n)$ and $a_i \geq 1$ for all i, because if a has zero entries we can use the induced subgraph on the vertex set $\{x_i | \, a_i > 0\}$. The vertex set of G^a is

$$X^a = \{x_1^1, \ldots, x_1^{a_1}, \ldots, x_i^1, \ldots, x_i^{a_i}, \ldots, x_n^1, \ldots, x_n^{a_n}\}$$

and the edges of G^a are exactly those pairs of the form $\{x_i^{k_i}, x_j^{k_j}\}$ with $i \neq j$, $k_i \leq a_i$, $k_j \leq a_j$, for some edge $\{x_i, x_j\}$ of G. We can regard x^a as an ordered multiset

$$x^a = x_1^{a_1} \cdots x_n^{a_n} = \underbrace{(x_1 \cdots x_1)}_{a_1} \cdots \underbrace{(x_n \cdots x_n)}_{a_n}$$

on the set X; that is, we can identify the monomial x^a with the multiset

$$X_a = \{\underbrace{x_1, \ldots, x_1}_{a_1}, \ldots, \underbrace{x_n, \ldots, x_n}_{a_n}\}$$

on X in which each variable is uniquely identified with an integer between 1 and $|a|$. This integer is the position, from left to right, of x_i in X_a. There is a bijective map

$$\begin{array}{ccccccccc}
1 & 2 & \cdots & a_1 & \cdots & a_1 + \cdots + a_{n-1} + 1 & \cdots & a_1 + \cdots + a_n \\
\downarrow & \downarrow & \cdots & \downarrow & \cdots & \downarrow & \cdots & \downarrow \\
x_1 & x_1 & \cdots & x_1 & \cdots & x_n & \cdots & x_n \\
\downarrow & \downarrow & \cdots & \downarrow & \cdots & \downarrow & \cdots & \downarrow \\
x_1^1 & x_1^2 & \cdots & x_1^{a_1} & \cdots & x_n^1 & \cdots & x_n^{a_n}.
\end{array}$$

Hence if G^a has a perfect matching, then the perfect matching induces a factorization of x^a in which each factor corresponds to an edge of G, i.e., $x^a \in K[G]$. Conversely, if $x^a \in K[G]$ we can factor x^a as a product of monomials corresponding to edges of G and this factorization induces a perfect matching of G^a. $\Box$

Given an edge $f = \{x_i, x_j\}$ of G, we denote by G^f or $G^{\{x_i, x_j\}}$ the graph obtained from G by successively duplicating the vertices x_i and x_j, i.e., $G^f = G^{\mathbf{1}+e_i+e_j}$, where e_i is the ith unit vector and $\mathbf{1} = (1, \ldots, 1)$. Recall that def$(G)$, the *deficiency* of G, is given by def$(G) = |V(G)| - 2\beta_1(G)$, where $\beta_1(G)$ is the matching number of G. Hence $\operatorname{def}(G)$ is the number of vertices left uncovered by any maximum matching.

Notation In what follows $F = \{f_1, \ldots, f_q\}$ denotes the set of all monomials x_ix_j such that $\{x_i, x_j\} \in E(G)$. We use f^c as an abbreviation for $f_1^{c_1} \cdots f_q^{c_q}$, where $c = (c_i) \in \mathbb{N}^q$.

Lemma 7.7.7 *Let G be a graph and let $a \in \mathbb{N}^n$ and $c \in \mathbb{N}^q$. Then*

(a) $x^a = x^\delta f^c$, *where* $|\delta| = \operatorname{def}(G^a)$ *and* $|c| = \beta_1(G^a)$.

(b) x^a *belongs to* $I(G)^k \setminus I(G)^{k+1}$ *if and only if* $k = \beta_1(G^a)$.

(c) $(G^a)^f = (G^a)^{\{x_i,x_j\}}$ *for any edge* $f = \{x_i^{k_i}, x_j^{k_j}\}$ *of* G^a.

Proof. Parts (a) and (b) follow using the bijective map used in the proof of Lemma 7.7.6. To show (c) we use the notation used in the proof of Lemma 7.7.6. We now prove the inclusion $E((G^a)^f) \subset E((G^a)^{\{x_i,x_j\}})$. Let y_i and y_j be the duplications of $x_i^{k_i}$ and $x_j^{k_j}$, respectively. We also denote the duplications of x_i and x_j by y_i and y_j, respectively. The common vertex set of $(G^a)^f$ and $(G^a)^{\{x_i,x_j\}}$ is $V(G^a) \cup \{y_i, y_j\}$. Let e be an edge of $(G^a)^f$. If $e = \{y_i, y_j\}$ or $e \cap \{y_i, y_j\} = \emptyset$, then clearly e is an edge of $(G^a)^{\{x_i,x_j\}}$. Thus, we may assume that $e = \{y_i, x_\ell^{k_\ell}\}$. Then $\{x_i^{k_i}, x_\ell^{k_\ell}\} \in E(G^a)$, so $\{x_i, x_\ell\} \in E(G)$. Hence $\{x_i, x_\ell^{k_\ell}\}$ is in $E(G^a)$, so $e = \{y_i, x_\ell^{k_\ell}\}$ is an edge of $(G^a)^{\{x_i,x_j\}}$. This proves the inclusion "$\subset$". The other inclusion follows using similar arguments (arguing backwards). $\square$

Theorem 7.7.8 (Berge [297, Theorem 3.1.14]) *Let G be a graph. Then*

$$\operatorname{def}(G) = \max\{c_0(G \setminus S) - |S| \mid S \subset V(G)\},$$

where $c_0(G)$ denotes the number of odd components (components with an odd number of vertices) of a graph G.

The theorem of Berge is equivalent to a classical result of Tutte that describes perfect matchings [297] (see Theorem 7.1.10).

Theorem 7.7.9 *Let G be a graph. Then* $\operatorname{def}(G^f) = \delta$ *for all $f \in E(G)$ if and only if* $\operatorname{def}(G) = \delta$ *and* $\beta_1(G^f) = \beta_1(G) + 1$ *for all $f \in E(G)$.*

Proof. Assume that $\operatorname{def}(G^f) = \delta$ for all $f \in E(G)$. In general, $\operatorname{def}(G) \geq \operatorname{def}(G^f)$ for any $f \in E(G)$. We proceed by contradiction. Assume that $\operatorname{def}(G) > \delta$. Then, by Berge's theorem, there is an $S \subset V(G)$ such that $c_0(G \setminus S) - |S| > \delta$. We set $r = c_0(G \setminus S)$ and $s = |S|$. Let $H_1, \ldots, H_r$ be the odd components of $G \setminus S$.

Case (I): $|V(H_k)| \geq 2$ for some $1 \leq k \leq r$. Pick an edge $f = \{x_i, x_j\}$ of H_k. Consider the parallelization H_k' obtained from H_k by duplicating the vertices x_i and x_j, i.e., $H'_k = H^f_k$. The odd connected components of $G^f \setminus S$ are $H_1, H_2, \ldots, H_{k-1}, H'_k, H_{k+1} \ldots, H_r$. Thus

$$c_0(G^f \setminus S) - |S| > \delta = \mathrm{def}(G^f).$$

This contradicts Berge's theorem when applied to G^f.

Case (II): $|V(H_k)| = 1$ for $1 \leq k \leq r$. Notice that $S \neq \emptyset$ because G has no isolated vertices. Pick $f = \{x_i, x_j\}$ an edge of G with $\{x_i\} = V(H_1)$ and $x_j \in S$. Let y_i and y_j be the duplications of x_i and x_j, respectively. The odd components of $G^f \setminus (S \cup \{y_j\})$ are $H_1, \ldots, H_r, \{y_i\}$. Thus

$$c_0(G^f \setminus (S \cup \{y_j\})) - |S \cup \{y_j\}| = c_0(G \setminus S) - |S| > \delta = \mathrm{def}(G^f).$$

This contradicts Berge's theorem when applied to G^f. Therefore $\mathrm{def}(G) = \mathrm{def}(G^f)$ for all $f \in E(G)$. Hence $\beta_1(G^f) = \beta_1(G) + 1$ for all $f \in E(G)$. The converse follows using the definition of $\mathrm{def}(G)$ and $\mathrm{def}(G^f)$. □

Corollary 7.7.10 *Let G be a graph. Then G has a perfect matching if and only if G^f has a perfect matching for every edge f of G.*

Proof. Assume that G has a perfect matching. Let $f_1, \ldots, f_{n/2}$ be a set of edges of G that form a perfect matching of $V(G)$, where n is the number of vertices of G. If $f = \{x_i, x_j\}$ is any edge of G and y_i, y_j are the duplications of the vertices x_i and x_j, respectively, then clearly $f_1, \ldots, f_{n/2}, \{y_i, y_j\}$ form a perfect matching of $V(G^f)$. Conversely, if G^f has a perfect matching for all $f \in E(G)$, then $\mathrm{def}(G^f) = 0$ for all $f \in E(G)$. Hence, by Theorem 7.7.9, we get that $\mathrm{def}(G) = 0$, so G has a perfect matching. □

If $I \neq (0)$ is an ideal of a commutative Noetherian domain, Ratliff showed that $(I^{k+1} \colon I) = I^k$ for all large k [342, Corollary 4.2] and that equality holds for all k when I is normal [342, Proposition 4.7]. The next lemma shows that equality holds for all k when I is an edge ideal.

Lemma 7.7.11 *Let I be the edge ideal of a graph G. Then $(I^{k+1} \colon I) = I^k$ for $k \geq 1$.*

Proof. Let $f_1, \ldots, f_q$ be the set of all $x_i x_j$ such that $\{x_i, x_j\} \in E(G)$. Given $c = (c_i) \in \mathbb{N}^q$, we set $f^c = f_1^{c_1} \cdots f_q^{c_q}$. The colon ideal $(I^{k+1} \colon I)$ is a monomial ideal. Clearly $I^k \subset (I^{k+1} \colon I)$. To show the other inclusion take a monomial x^a in $(I^{k+1} \colon I)$. Then, $f_i x^a \in I^{k+1}$ for all i. We may assume that $f_i x^a \notin I^{k+2}$, otherwise $x^a \in I^k$ as required. Thus $x^{a+e_i+e_j} \in I^{k+1} \setminus I^{k+2}$ for any $e_i + e_j$ such that $\{x_i, x_j\} \in E(G)$. Hence, by Lemma 7.7.7(b), $\beta_1(G^{a+e_i+e_j}) = k+1$ for any $\{x_i, x_j\} \in E(G)$, that is, $(G^a)^{\{x_i, x_j\}}$ has a maximum matching of size $k+1$ for any edge $\{x_i, x_j\}$ of G. With the notation of the proof of Lemma 7.7.6, for any edge $\{x_i^{k_i}, x_j^{k_j}\}$ of G^a we have

$$(G^a)^{\{x_i^{k_i}, x_j^{k_j}\}} = (G^a)^{\{x_i, x_j\}},$$

see Lemma 7.7.7(c). Then, $(G^a)^f$ has a maximum matching of size $k+1$ for any edge f of G^a. As a consequence

$$\operatorname{def}((G^a)^f) = (|a|+2) - 2(k+1) = |a| - 2k$$

for any edge f of G^a. Hence, by Theorem 7.7.9, $\operatorname{def}(G^a) = |a| - 2k$. Using Lemma 7.7.7(a), we can write $x^a = x^\delta f^c$, where $|\delta| = \operatorname{def}(G^a)$ and $|c| = \beta_1(G^a)$. Taking degrees in $x^a = x^\delta f^c$ gives $|a| = |\delta| + 2|c| = (|a|-2k) + 2|c|$, that is, $|c| = k$. Then $x^a \in I^k$ and the proof is complete. $\Box$

Proposition 7.7.12 *Let $I = I(G)$ be the edge ideal of a graph G and let $\mathfrak{m} = (x_1, \ldots, x_n)$. If $\mathfrak{m} \in \operatorname{Ass}(R/I^k)$, then $\mathfrak{m} \in \operatorname{Ass}(R/I^{k+1})$.*

Proof. As $\mathfrak{m}$ is an associated prime of R/I^k, there is $x^a \notin I^k$ such that $\mathfrak{m}x^a \subset I^k$. By Lemma 7.7.11 there is an edge $\{x_i, x_j\}$ of G such that $x_i x_j x^a \notin I^{k+1}$. Then, $x_\ell(x_i x_j x^a) \in I^{k+1}$ for $\ell = 1, \ldots, n$; that is, $\mathfrak{m}$ is an associated prime of R/I^{k+1}. $\Box$

Lemma 7.7.13 [204, Lemma 3.4] *Let I be a square-free monomial ideal in $S = K[x_1, \ldots, x_m, x_{m+1}, \ldots, x_r]$ such that $I = I_1 S + I_2 S$, where $I_1 \subset S_1 = K[x_1, \ldots, x_m]$ and $I_2 \subset S_2 = K[x_{m+1}, \ldots, x_r]$. Then $\mathfrak{p} \in \operatorname{Ass}(S/I^k)$ if and only if $\mathfrak{p} = \mathfrak{p}_1 S + \mathfrak{p}_2 S$, where $\mathfrak{p}_1 \in \operatorname{Ass}(S_1/I_1^{k_1})$ and $\mathfrak{p}_2 \in \operatorname{Ass}(S_2/I_2^{k_2})$ with $(k_1 - 1) + (k_2 - 1) = k - 1$.*

If $\mathfrak{p}_i$ is an ideal of R, the generators of $\mathfrak{p}_i$ will generate a prime ideal in any ring that contains those variables. We will abuse notation by denoting the ideal generated by the generators of $\mathfrak{p}_i$ in any other ring by $\mathfrak{p}_i$ as well.

Theorem 7.7.14 [305] *Let G be a graph and let $I = I(G)$ be its edge ideal. Then*

$$\operatorname{Ass}(R/I^k) \subset \operatorname{Ass}(R/I^{k+1})$$

for all k. That is, I has the persistence property.

Proof. Take $\mathfrak{p}$ in $\operatorname{Ass}(R/I^k)$. We set $\mathfrak{m} = (x_1, \ldots, x_n)$. We may assume that $\mathfrak{p} = (x_1, \ldots, x_r)$. By Proposition 7.7.12, we may assume that $\mathfrak{p} \subsetneq \mathfrak{m}$. Write $I_\mathfrak{p} = (I_2, I_1)_\mathfrak{p}$, where I_2 is the ideal of R generated by all $x_i x_j$, $i \neq j$, whose image, under the canonical map $R \to R_\mathfrak{p}$, is a minimal generator of $I_\mathfrak{p}$, and I_1 is the ideal of R generated by all x_i whose image is a minimal generator of $I_\mathfrak{p}$, which correspond to the isolated vertices of the graph associated to $I_\mathfrak{p}$. The minimal generators of I_2 and I_1 lie in $S = K[x_1, \ldots, x_r]$, and the two sets of variables occurring in the minimal generating sets of I_1 and I_2 (respectively) are disjoint and their union is $\{x_1, \ldots, x_r\}$.

If $I_2 = (0)$, then $\mathfrak{p}$ is a minimal prime of I so it is an associated prime of R/I^{k+1}. Thus, we may assume $I_2 \neq (0)$. An important fact is that localization preserves associated primes; that is $\mathfrak{p} \in \operatorname{Ass}(R/I^k)$ if and only if $\mathfrak{p}R_\mathfrak{p} \in \text{Ass}(R_\mathfrak{p}/(I_\mathfrak{p}R_\mathfrak{p})^k)$. Hence, $\mathfrak{p}$ is in $\text{Ass}(R/I^k)$ if and only if $\mathfrak{p}$ is in $\text{Ass}(R/(I_1, I_2)^k)$ if and only if $\mathfrak{p}$ is in $\text{Ass}(S/(I_1, I_2)^k)$. By Proposition 7.7.12 and Lemma 7.7.13, $\mathfrak{p}$ is an associated prime of $S/(I_1, I_2)^{k+1}$. Hence, we can argue backwards to conclude that $\mathfrak{p}$ is an associated prime of R/I^{k+1}. □

If G is a graph with loops, then its edge ideal $I(G)$ has the persistence property [353].

Corollary 7.7.15 *If I is a square-free monomial ideal and $(I^{k+1} \colon I) = I^k$ for $k \geq 1$. Then, I has the persistence property.*

Proof. As in Proposition 7.7.12, we first show that $\mathfrak{m} \in \text{Ass}(R/I^k)$ implies $\mathfrak{m} \in \text{Ass}(R/I^{k+1})$. Assume $\mathfrak{m} \in \text{Ass}(R/I^k)$. Then there is a monomial $x^a \notin I^k$ with $x_i x^a \in I^{k+1}$ for all i. By the hypothesis, $x^a \notin (I^{k+1} \colon I)$, so there is a square-free monomial generator e of I (which can be viewed as the edge of a clutter associated to I) with $ex^a \notin I^{k+1}$. But $x_i e x^a = e(x_i x^a) \in I^{k+1}$ for all i, so $\mathfrak{m} \in \text{Ass}(R/I^{k+1})$.

Recall that since I is finitely generated, $(I^{k+1} \colon I)_\mathfrak{p} = (I_\mathfrak{p}^{k+1} \colon I_\mathfrak{p})$. Thus $(I_\mathfrak{p}^{k+1} \colon I_\mathfrak{p}) = I_\mathfrak{p}^k$. The remainder of the argument now follows from localization, as in the proof of Theorem 7.7.14. □

Lemma 7.7.16 *If I is a monomial ideal, then* $\text{Ass}(I^{k-1}/I^k) = \text{Ass}(R/I^k)$.

Proof. Suppose that $\mathfrak{p} \in \text{Ass}(R/I^k)$. Then $\mathfrak{p} = (I^k : c)$ for some monomial $c \in R$. But since $\mathfrak{p}$ is necessarily a monomial prime, generated by a subset of the variables, then if $xc \in I^k$ for a variable $x \in \mathfrak{p}$, then $c \in I^{k-1}$ and so $\mathfrak{p} \in \text{Ass}(I^{k-1}/I^k)$. The other inclusion is automatic. □

Conjecture 7.7.17 (Persistence problem) If I is a square-free monomial ideal, then I has the persistence property.

The next counterexample for this conjecture is due to Kaiser, Stehlík, and Škrekovski [268] (cf. Exercise 7.7.25).

Example 7.7.18 [268] Let $I^\vee$ be the Alexander dual of the edge ideal

$$\begin{aligned} I = \; & (x_1x_2, x_2x_3, x_3x_4, x_4x_5, x_5x_6, x_6x_7, x_7x_8, x_8x_9, x_9x_{10}, x_1x_{10}, \\ & x_2x_{11}, x_8x_{11}, x_3x_{12}, x_7x_{12}, x_1x_9, x_2x_8, x_3x_7, x_4x_6, x_1x_6, x_4x_9, \\ & x_5x_{10}, x_{10}x_{11}, x_{11}x_{12}, x_5x_{12}). \end{aligned}$$

Using *Macaulay*2, one can readily verify that $\text{Ass}(R/(I^\vee)^3)$ is not a subset of $\text{Ass}(R/(I^\vee)^4)$ (see Exercise 7.7.27).

Theorem 7.7.19 [165] *If G is a simple graph, then the chromatic number of G is the minimal k such that $(x_1 \cdots x_n)^{k-1} \in I_c(G)^k$.*

Note that one can use *Normaliz* [68] inside *Macaulay*2 to compute the integral closure of a monomial ideal and also the normalization of the Rees algebra of a monomial ideal.

Procedure 7.7.20 The following simple procedure for *Macaulay*2 decides whether $\mathrm{Ass}(R/I^3)$ is contained in $\mathrm{Ass}(R/I^4)$ and whether we have the equality $\mathrm{Ass}(R/I^3) = \mathrm{Ass}(R/I^4)$. It also computes $\overline{I^4}$ and decides whether $\mathrm{Ass}(R/I^4)$ is equal to $\mathrm{Ass}(R/\overline{I^4})$.

```
R=QQ[x1,x2,x3,x4,x5,x6,x7,x8,x9];
load "normaliz.m2";
I=monomialIdeal(x1*x2,x2*x3,x1*x3,x3*x4,x4*x5,x5*x6,
x6*x7,x7*x8,x8*x9,x5*x9);
isSubset(ass(I^3),ass(I^4))
ass(I^3)==ass(I^4)
(intCl4,normRees4)=intclMonIdeal I^4;
intCl4'=substitute(intCl4,R);
ass(monomialIdeal(intCl4'))==ass(I^4)
```

Exercises

7.7.21 Let I be the square-free monomial ideal generated by

$$x_1x_2x_5,\ x_1x_3x_4,\ x_1x_2x_6,\ x_1x_3x_6,\ x_1x_4x_5,\ x_2x_3x_4,\ x_2x_3x_5,$$
$$x_2x_4x_6,\ x_3x_5x_6,\ x_4x_5x_6.$$

Use *Normaliz* [68] and *Macaulay*2 [199], to show that I is a non-normal ideal such that $(I^2 : I) = I$ and $(I^3 : I) \neq I^2$. Then, prove that I has the persistence property and that the index of stability of I is equal to 3.

7.7.22 Let $I = I(G)$ be the edge ideal of a graph G. Prove that the sets $\mathrm{Ass}(I^{k-1}/I^k)$ form an ascending chain for $k \geq 1$.

7.7.23 Let G be a graph and let x_i be a vertex. If $y \in V(G^{x_i})$, then $(G^{x_i})^y = (G^{x_i})^{x_j}$ for some $x_j \in V(G)$.

7.7.24 If $I = (I_1, \ldots, I_s)$, where the I_i are square-free monomial ideals in disjoint sets of variables, then $\mathfrak{p} \in \mathrm{Ass}(R/I^k)$ if and only if $\mathfrak{p} = (\mathfrak{p}_1, \ldots, \mathfrak{p}_s)$, where $\mathfrak{p}_i \in \mathrm{Ass}(R/I_i^{k_i})$ with $(k_1 - 1) + \cdots + (k_s - 1) = k - 1$.

7.7.25 [326] Let $I = (x_1x_2^2x_3, x_2x_3^2x_4, x_3x_4^2x_5, x_4x_5^2x_1, x_5x_1^2x_2)$ be the ideal obtained by taking the product of consecutive edges of a pentagon. Then $\mathfrak{m} \in \mathrm{Ass}(R/I^k)$ for $k = 1, 4$, $\mathfrak{m} \notin \mathrm{Ass}(R/I^k)$ for $k = 2, 3$, $\mathfrak{m} = (x_1, \ldots, x_5)$.

7.7.26 [323, Lemma 2.3] If I is a monomial ideal of R and $a \in I/I^2$ is a regular element of the associated graded ring $\mathrm{gr}_I(R)$. Then the sets $\mathrm{Ass}(I^{k-1}/I^k)$ form an ascending chain.

7.7.27 Use the following procedure for *Macaulay*2 to verify that the ideal $I^\vee$ of Example 7.7.18 does not have the persistence property.

```
R=QQ[x1,x2,x3,x4,x5,x6,x7,x8,x9,x10,x11,x12]
I=monomialIdeal(x1*x2,x2*x3,x3*x4,x4*x5,x5*x6,x6*x7,x7*x8,
x8*x9,x9*x10,x1*x10,x2*x11,x8*x11,x3*x12,x7*x12,x1*x9,x2*x8,
x3*x7,x4*x6,x1*x6,x4*x9,x5*x10,x10*x11,x11*x12,x5*x12)
J=dual I
isSubset(ass(J^3),ass(J^4))
```

7.7.28 Consider the graph G of the figure below, where vertices are labeled with i instead of x_i. The duplication of the vertices x_3 and x_4 of G is shown below. If $f = \{x_3, x_4\}$, prove that deficiencies of G and G^f are not equal.

7.7.29 Load the package EdgeIdeals [166] in *Macaulay*2 [199] to verify that the chromatic number $\chi(G)$ of the graph G of Example 7.7.18 is equal to 4.

7.7.30 Pick any planar graph and use EdgeIdeals [166] to verify the four color theorem (*any planar graph has chromatic number at most* 4).

Chapter 8

Toric Ideals and Affine Varieties

In this chapter we study algebraic and geometric aspects of three important types of polynomial ideals and their quotient rings:

$$\text{TORIC IDEALS} \hookrightarrow \text{LATTICE IDEALS} \hookrightarrow \text{BINOMIAL IDEALS}.$$

These ideals are interesting from a computational point of view and are related to diverse fields, such as, numerical semigroups [173, 341], semigroup rings [180], commutative algebra and combinatorics [61, 317, 395], algebraic geometry [176, 317], linear algebra and polyhedral geometry [354, 420], integer programming [400], graph theory and combinatorial optimization [405], algebraic coding theory [348, 367], and algebraic statistics [142].

8.1 Binomial ideals and their radicals

Ideals generated by binomials are an interesting family of polynomial ideals. In this section we study the radical of those ideals.

Definition 8.1.1 Let $S = K[t_1, \ldots, t_q]$ be a polynomial ring over a field K. A *binomial* of S is an element of the form $f = t^\alpha - t^\beta$, for some α, β in $\mathbb{N}^q$. An ideal generated by binomials is called a *binomial ideal.*

A polynomial with at most two terms, say $\lambda t^a - \mu t^b$, where $\lambda, \mu \in K$ and $a, b \in \mathbb{N}^q$, is called a *non-pure binomial.* Accordingly an ideal generated by non-pure binomials is called a *non-pure binomial ideal.*

The essence of the Eisenbud–Sturmfels paper [134] is that the associated primes, the primary components and the radical of a non-pure binomial ideal are non-pure binomial ideals if K is algebraically closed. Binomial primary decompositions can be recovered from the theory of mesoprimary decomposition of congruences [267].

Computing a primary decomposition of a binomial ideal over a field of characteristic zero can be done using "Binomials"[266], a package for the computer algebra system *Macaulay*2 [199].

To study binomial ideals over arbitrary fields we introduce the notion of congruence that comes from semigroup theory, this allows us to link general semigroup rings with binomial ideals (see Proposition 8.1.6).

An *equivalence relation* on a set $\mathcal{S}$ is a subset R of $\mathcal{S}\times\mathcal{S}$ which is reflexive: $(a,a)\in R$, symmetric: $(a,b)\in R\Rightarrow(b,a)\in R$, and transitive: $(a,b)\in R$ and $(b,c)\in R\Rightarrow(a,c)\in R$. As usual we write $a\sim b$ if $(a,b)\in R$.

Definition 8.1.2 A *congruence* in a commutative semigroup with identity $(\mathcal{S},+)$ is an equivalence relation $\sim$ on $\mathcal{S}$ compatible with $+$.

Example 8.1.3 Let $\mathcal{L}\subset\mathbb{Z}^q$ be a subgroup. The relation, $a\sim_{\mathcal{L}} b$ if $a-b$ is in $\mathcal{L}$, defines a congruence in the additive semigroup $\mathbb{N}^q$.

Given a congruence $\sim$ on a semigroup $\mathcal{S}$, it is usual to denote the quotient semigroup of $\mathcal{S}$ by $\mathcal{S}/\sim:=\{\overline{a}\mid a\in\mathcal{S}\}$, where $\overline{a}$ is the equivalence class of a.

Lemma 8.1.4 *If $\sim$ is a congruence on a semigroup $(\mathcal{S},+)$ and $a\sim b$, then $ra\sim rb$ for $r\in\mathbb{N}$.*

Proof. By induction on r. If $(r-1)a\sim(r-1)b$, then $(r-1)a+b\sim rb$. Since $a\sim b$, we get $ra\sim b+(r-1)a$. Thus $ra\sim rb$. □

Lemma 8.1.5 *Let $\mathcal{S}$ be a semigroup with a congruence $\sim$ and let $a,b\in\mathcal{S}$. If m and n are relatively prime positive integers such that $ma\sim mb$ and $na\sim nb$, then $ra\sim rb$ for all $r>mn-m-n$.*

Proof. By Lemma 8.7.8 we can write $r=\eta_1m+\eta_2n$ for some η_1,η_2 in $\mathbb{N}$. As $n\overline{a}=n\overline{b}$ and $m\overline{a}=m\overline{b}$, it follows that $r\overline{a}=r\overline{b}$ as required. □

The radical of a binomial ideal Let $S=K[t_1,\ldots,t_q]$ be a polynomial ring over a field K and let $I=(t^{a_1}-t^{b_1},\ldots,t^{a_s}-t^{b_s})$ be a binomial ideal of S. We set $\varrho_0:=\{(a_1,b_1),\ldots,(a_s,b_s)\}$. The intersection of all congruences in $(\mathbb{N}^q,+)$ that contain ϱ_0 is called the *congruence generated* by ϱ_0 and it will be denoted by ϱ in what follows.

According to [180] we can construct ϱ as follows:

$$\begin{aligned}
\varrho_1 &= \varrho_0\cup\{(a,a)\mid a\in\mathbb{N}^q\}\cup\{(b,a)\mid (a,b)\in\varrho_0\},\\
\varrho_2 &= \{(a+c,b+c)\mid (a,b)\in\varrho_1,\ c\in\mathbb{N}^q\},\\
\varrho &= \{(a,b)\mid \exists a_0,\ldots,a_r\in\mathbb{N}^q \text{ with } a_0=a, a_r=b, (a_i,a_{i+1})\in\varrho_2\,\forall i\}.
\end{aligned}$$

Since ϱ is clearly a congruence contained in any congruence containing ϱ_0, we have that ϱ is the congruence generated by ϱ_0.

Proposition 8.1.6 [180, Corollary 7.3] $I = (\{t^a - t^b | \, (a, b) \in \varrho\})$.

Proof. It follows rapidly from the construction of ϱ. $\square$

Let $\sim$ be a congruence in $\mathbb{N}^q$. We associate to $\sim$ an equivalence relation on the monomials in S, by $t^a \sim t^b$ if $a \sim b$. Note that this relation is compatible with the product, i.e., $t^a \sim t^b$ implies $t^c t^a \sim t^c t^b$ for all $c \in \mathbb{N}^q$.

Definition 8.1.7 A non-zero polynomial $f = \sum_\alpha \lambda_\alpha t^\alpha$ in S, with $\lambda_\alpha \in K$, is called *simple* (with respect to $\sim$) if all its monomials, *i.e.*, those t^α with non-zero coefficient λ_α, are equivalent under $\sim$.

Given any polynomial $f \in S \setminus \{0\}$, we can group together its monomials by equivalence classes under $\sim$, thereby obtaining a decomposition

$$f = h_1 + \cdots + h_m$$

with the property that each summand h_i is simple, and that no monomial in h_i is equivalent with a monomial in h_j if $j \neq i$. Such a decomposition of f as a sum of maximal simple subpolynomials is unique up to order. We will refer to the h_i's as the *simple components* of f (with respect to $\sim$). These concepts were introduced by Eliahou [135].

Lemma 8.1.8 *Let $\sim$ be a congruence in $\mathbb{N}^q$. If J is an ideal of S generated by a set $\mathcal{G}$ of simple polynomials and $f \in J$, then every simple component of f also belongs to J.*

Proof. Each generator $g \in \mathcal{G}$ is simple. Moreover, $t^c g$ remains simple for any $c \in \mathbb{N}^q$, since the relation $\sim$ on monomials is compatible with the product. Let f be any element in the ideal J. Then, f is a finite linear combination of polynomials of the form $t^{c_i} g_i$, with $g_i \in \mathcal{G}$ simple and $c_i \in \mathbb{N}^q$, which are simple. Hence, every simple component of f is also a linear combination of some $t^{c_i} g_i$ and therefore belongs to J. $\square$

Definition 8.1.9 Let $a, b \in \mathbb{N}^q$ and let p be a positive prime number. If $(p^r a, p^r b) \in \varrho$ for some integer $r \geq 0$, we say that a and b are *p-equivalent* modulo ϱ and write $a \sim_p b$.

The radical of a binomial ideal is generated by binomials as is seen below. If K is algebraically closed, then the radical of a non-pure binomial ideal is a non-pure binomial ideal [134, Theorem 3.1] (see Exercise 8.1.15).

Theorem 8.1.10 [180] *If* $p = \operatorname{char}(K) \neq 0$, *then*

$$\operatorname{rad}(I) = (\{t^a - t^b | \, a \sim_p b\}).$$

Proof. "$\supset$": If $a \sim_p b$, then $t^{p^r a} - t^{p^r b} \in I$ by Proposition 8.1.6. Thus $(t^a - t^b)^{p^r}$ belongs to I and $t^a - t^b \in \mathrm{rad}(I)$.

"$\subset$": Take $f \in \mathrm{rad}(I)$, we can write $f = \sum_{i=1}^m \lambda_i t^{\alpha_i}$, with $\lambda_i \in K$ for all i. For $r \gg 0$ we have $f^{p^r} \in I$. From the equality $f^{p^r} = \sum_{i=1}^m \mu_i t^{p^r \alpha_i}$, $\mu_i = \lambda_i^{p^r}$, and using Lemma 8.1.8 we may assume that f^{p^r} is a simple component with respect to the congruence ϱ. Note that $\sum_{i=1}^m \mu_i = 0$, hence

$$f^{p^r} = \mu_2(t^{p^r\alpha_2} - t^{p^r\alpha_1}) + \cdots + \mu_m(t^{p^r\alpha_m} - t^{p^r\alpha_1}),$$

with $(p^r\alpha_i, p^r\alpha_1) \in \varrho$ for all $i \geq 2$, as required. □

Lemma 8.1.11 *Let $a, b \in \mathbb{N}^q$. If there is an integer $s > 0$ such that (ra, rb) is in ϱ for all $r \geq s$, then $f = t^a - t^b \in \mathrm{rad}(I)$.*

Proof. Set $r = 2s + 1$. We claim $f^r \in I$. From the binomial expansion:

$$f^r = \sum_{i=0}^{s} \left[(-1)^i \binom{r}{i} (t^a)^{r-i}(t^b)^i + (-1)^{r-i}\binom{r}{r-i}(t^a)^i(t^b)^{r-i}\right]. \quad (8.1)$$

Note $s \leq r - i$ for $0 \leq i \leq s$. Hence, from $((r-i)a, (r-i)b) \in \varrho$, we get

$$(ra, (r-i)b + ia) \in \varrho \ \text{ and } \ ((r-i)a + ib, rb) \in \varrho.$$

Thus, as $(ra, rb) \in \varrho$, we get $((r-i)b + ia, (r-i)a + ib)) \in \varrho$. Therefore any summand in Eq. (8.1) is in I, as required. □

Following [180], two elements a, b in $\mathbb{N}^q$ that satisfy the conditions of Lemma 8.1.11 are called *asymptotically equivalent* modulo ϱ.

Theorem 8.1.12 [180] *If* $\mathrm{char}(K) = 0$, *then*

$$\mathrm{rad}(I) = (\{t^a - t^b \mid \exists\, 0 \neq s \in \mathbb{N} \ \text{ such that } \ (ra, rb) \in \varrho; \forall\, r \geq s\}).$$

Proof. By Lemma 8.1.11 we need only show the inclusion "$\subset$". Take $0 \neq f \in \mathrm{rad}\,(I)$. We can write $f = \lambda_1 t^{\alpha_1} + \cdots + \lambda_m t^{\alpha_m}$ with $0 \neq \lambda_i \in K$ for all i and $\lambda_1 + \cdots + \lambda_m = 0$. If α_1 is asymptotically equivalent modulo ϱ to α_2, from the equality

$$f_1 = f - \lambda_2(t^{\alpha_2} - t^{\alpha_1}) = \lambda_3(t^{\alpha_3} - t^{\alpha_1}) + \cdots + \lambda_m(t^{\alpha_m} - t^{\alpha_1})$$

and Lemma 8.1.11 we get $f_1 \in \mathrm{rad}(I)$. Thus we may either assume that α_i is not asymptotically equivalent modulo ϱ to α_j for $i \neq j$ or that $f = 0$. To complete the proof we assume $f \neq 0$ and derive a contradiction.

Let E be the set of prime numbers $p > 0$ such that α_i is p-equivalent modulo ϱ to α_j. By Lemma 8.1.5 the set E is finite. Consider the ring $A_0 = \mathbb{Z}[\lambda_1, \ldots, \lambda_m] \subset K$ and the set $\{\mathfrak{m}_i\}_{i \in \mathcal{I}}$ of maximal ideals of A_0. Since

A_0 is a Hilbert ring, using [65, Corollary A.18], it follows that $(0) = \cap_{i\in\mathcal{I}}\mathfrak{m}_i$ and $A_0/\mathfrak{m}_i$ is a finite field for $i \in \mathcal{I}$. If $\mathcal{I}_0$ is the subset of $\mathcal{I}$ consisting of all i such that $\mathrm{char}(A_0/\mathfrak{m}_i)$ is not in E, then $(0) = \cap_{i\in\mathcal{I}_0}\mathfrak{m}_i$. Indeed take $a \in \cap_{i\in\mathcal{I}_0}\mathfrak{m}_i$, then $(\prod_{p\in E} p)a = 0$ because $(\prod_{p\in E} p)a \in \mathfrak{m}_i$ for all $i \in \mathcal{I}$, thus $a = 0$ because $\mathrm{char}(A_0) = 0$.

Let $i \in \mathcal{I}_0$, we set $\mathfrak{m} = \mathfrak{m}_i$ and $p = \mathrm{char}(A_0/\mathfrak{m})$. Note $p \notin E$. Pick r sufficiently large such that f^{p^r} is in I. We can write $f^{p^r} = \sum_{i=1}^n \mu_i t^{\beta_i}$, with $\mu_i \in A_0 \setminus \{0\}$ and $t^{\beta_1}, \ldots, t^{\beta_n}$ distinct. Since the simple components of f^{p^r}, with respect to ϱ, are again in I using that $\sum_{i=1}^n \mu_i = 0$ (on every component of f^{p^r}) we readily see that

$$f^{p^r} = \eta_1(t^{\gamma_1} - t^{\delta_1}) + \cdots + \eta_s(t^{\gamma_s} - t^{\delta_s}) \qquad ((\gamma_i, \delta_i) \in \varrho)$$

with $\eta_i \in A_0$ for all i. If $h \in A_0[\mathbf{t}]$, we denote by $\overline{h}$ the image of h under the canonical map $A_0[\mathbf{t}] \to (A_0/\mathfrak{m})[\mathbf{t}]$. If $\overline{f} \neq \overline{0}$, then from the equation

$$\overline{f^{p^r}} = \overline{\eta_1}(t^{\gamma_1} - t^{\delta_1}) + \cdots + \overline{\eta_s}(t^{\gamma_s} - t^{\delta_s}) \qquad ((\gamma_i, \delta_i) \in \varrho)$$

we get $\alpha_i \sim_p \alpha_j$ modulo ϱ for some $i \neq j$, a contradiction because p is not in E. Thus $f \in \mathfrak{m}[\mathbf{t}]$. Therefore $\lambda_1, \ldots, \lambda_m$ are in the maximal ideal $\mathfrak{m}_i$ for every $i \in \mathcal{I}_0$ and consequently $\lambda_i = 0$ for all i, a contradiction. This proof was adapted from [180]. $\square$

Exercises

8.1.13 Let $I = (t^{a_1} - t^{b_1}, \ldots, t^{a_s} - t^{b_s}) \subset S$ and let $\mathcal{L}$ be the subgroup of $\mathbb{Z}^q$ generated by $a_1 - b_1, \ldots, a_s - b_s$. Prove the following:

(a) The congruence ϱ generated by $\varrho_0 = \{(a_1, b_1), \ldots, (a_s, b_s)\}$ is contained in the congruence $\varrho' = \{(a, b) |\, a, b \in \mathbb{N}^q;\ a - b \in \mathcal{L}\}$.

(b) $\varrho = \varrho'$ if t_i is not a zero divisor of S/I for all i.

(c) If $\mathrm{char}(K) = p > 0$, then I is a radical ideal if and only if $\mathbb{N}^q/\varrho$ is a semigroup that is p-torsion-free.

8.1.14 Let $I = (y^2z - x^2t, z^4 - xt^3) \subset \mathbb{Q}[x, y, z, t]$. Prove that I is a complete intersection whose primary decomposition is given by $I = \mathfrak{q}_1 \cap \mathfrak{q}_2 \cap \mathfrak{q}_3$, where

$$\mathfrak{q}_1 = (y^2z - x^2t, z^4 - xt^3, xz^3 - y^2t^2, x^3z^2 - y^4t, y^6 - x^5z),$$

$\mathfrak{q}_2 = (t^3, zt^2, z^2t, z^3, y^2z - x^2t)$ and $\mathfrak{q}_3 = (x, z)$. Prove that $\mathfrak{q}_1$ is prime.

8.1.15 (I. Swanson) If $K = \mathbb{Z}_2(t)$ is a field of rational functions in one variable and $R = K[x, y]$ is a polynomial ring, prove that the radical of the ideal $(x^2 + t, y^2t + 1)$ is equal to $(x^2 + t, x + y + 1)$.

8.2 Lattice ideals

In this section we study lattice ideals and their radicals. We present some classifications of lattice and toric ideals. In Chapter 9 we will study primary decompositions and the degree of lattice ideals.

The class of lattice ideals has been studied in several places; see for instance [22, 134, 251, 337] and the references therein.

Let $S = K[t_1, \ldots, t_q]$ be a polynomial ring over a field K and let $\mathcal{L}$ be a *lattice* in $\mathbb{Z}^q$; that is, $\mathcal{L}$ is an additive subgroup of $\mathbb{Z}^q$. The *rank* of $\mathcal{L}$, denoted by $\operatorname{rank}(\mathcal{L})$, is the nonnegative integer r such that $\mathcal{L} \simeq \mathbb{Z}^r$. A *partial character* is a group homomorphism from $\mathcal{L}$ to the multiplicative group $K^* = K \setminus \{0\}$.

Definition 8.2.1 Let $\mathcal{L}$ be a lattice in $\mathbb{Z}^q$ and let $\rho\colon \mathcal{L} \to K^*$ be a partial character. The *lattice ideal* of $\mathcal{L}$ relative to ρ is the ideal:

$$I_\rho(\mathcal{L}) = (\{t^{a_+} - \rho(a)t^{a_-} \,|\, a \in \mathcal{L}\}) \subset S.$$

If ρ is the *trivial character* $\rho(a) = 1$ for a in $\mathcal{L}$, we denote $I_\rho(\mathcal{L})$ simply by $I(\mathcal{L})$. If no partial character is specified we shall always assume that ρ is the trivial character.

Theorem 8.2.2 [134] *Let $\mathcal{L}$ be a lattice in $\mathbb{Z}^q$ of rank r and let $\rho\colon \mathcal{L} \to K^*$ be a partial character. The following hold.*

(i) *$I_\rho(\mathcal{L})$ contains no monomials.*

(ii) $\dim(S/I_\rho(\mathcal{L})) = q - r$, *that is,* $\operatorname{ht}(I(\mathcal{L})) = \operatorname{rank}(\mathcal{L})$.

(iii) *t_i is a non-zero divisor of $S/I_\rho(\mathcal{L})$ for all i.*

(iv) *If $\mathbb{Z}^q/\mathcal{L}$ is torsion-free, then $I_\rho(\mathcal{L})$ is prime.*

Proof. Let $\alpha_1, \ldots, \alpha_r$ be a $\mathbb{Z}$-basis of $\mathcal{L}$ and let A be the $r \times q$ matrix with rows $\alpha_1, \ldots, \alpha_r$. Consider unimodular integral matrices $P = (p_{ij})$ and $Q = (q_{ij})$ of orders q and r, respectively, and make a change of basis

$$\alpha_i' = \sum_{j=1}^{r} q_{ij}\alpha_j \ \text{ and } \ e_i' = \sum_{j=1}^{q} p_{ij}e_j.$$

By Theorem 1.2.2, $QAP^{-1} = \operatorname{diag}\{d_1, \ldots, d_r, 0, \ldots, 0\}$ for some P and Q. By the change of basis theorem the new basis is related by $\alpha_i' = d_ie_i'$ for $i = 1\ldots, r$, where $e_i = e_i'P^{-1}$. Thus $\mathcal{L} = \langle d_1e_1', \ldots, d_re_r'\rangle$. Note that the isomorphism $\phi\colon \mathbb{Z}^q \to \mathbb{Z}^q$, $\alpha \to \alpha P^{-1}$, induces a K-algebra isomorphism

$$\varphi\colon S' = K[t_1^{\pm 1}, \ldots, t_q^{\pm 1}] \longrightarrow S' \quad (t^\alpha \xrightarrow{\varphi} t^{\phi(\alpha)}).$$

Consider the ideal $I' = (t^{\alpha} - \rho(\alpha) | \alpha \in \mathcal{L}) \subset S'$. If $\alpha \in \mathcal{L}$, then we can write $\alpha = \sum_{i=1}^{r} a_i \alpha_i'$ for some $a_1, \ldots, a_r$ in $\mathbb{Z}$ and $t^{\alpha} - \rho(\alpha)$ maps under φ to $t_1^{a_1 d_1} \cdots t_r^{a_r d_r} - \rho(\alpha_1')^{a_1} \cdots \rho(\alpha_r')^{a_r}$. Therefore, setting $h_i = t_i^{d_i} - \rho(\alpha_i')$ for $i = 1, \ldots, r$, it follows that $\varphi(I')$ is equal to $(h_1, \ldots, h_r)$. This expression for $\varphi(I')$ is useful to study the structure of $I_\rho(\mathcal{L})$ as is seen below.

(i): Let $\overline{K}$ be the algebraic closure of K and write $\rho(\alpha_i') = \gamma_i^{d_i}$ for some γ_i in $\overline{K}$. If $I_\rho(\mathcal{L})$ contains a monomial, then $\varphi(I')$ contains a Laurent monomial since $I_\rho(\mathcal{L}) \subset I'$, a contradiction because $t_i^{d_i} - \gamma_i^{d_i} = 0$ if $t_i = \gamma_i$.

(ii): As $h_1, \ldots, h_r$ form a regular sequence, we get $\mathrm{ht}(\varphi(I')) = r$. Since S'/I' is isomorphic to $S'/\varphi(I')$, by Exercise 8.2.31, it follows that the Krull dimension of $S/I_\rho(\mathcal{L})$ is equal to $q - r$.

(iii): Let $\mathcal{G} = \{g_1, \ldots, g_s\}$ be a Gröbner basis for $I_\rho(\mathcal{L})$ with respect to a term order $\prec$. For simplicity we set $i = 1$. Assume that there is $0 \neq h \in S$ such that $t_1 h \in I_\rho(\mathcal{L})$. By the division algorithm (see Theorem 3.3.6) we can write $h = h_1 g_1 + \cdots + h_s g_s + f$, where no term in f is divisible by $\mathrm{in}_\prec(g_i)$ for all i; that is, the distinct terms $t^{a_1}, \ldots, t^{a_r}$ occurring in f are standard monomials. To show that t_1 is not a zero divisor of $S/I_\rho(\mathcal{L})$, it suffices to prove that $f = 0$. Assume that $f \neq 0$. Let $\sim$ be the congruence given by $a \sim b$ if $a - b \in \mathcal{L}$. Since $I_\rho(\mathcal{L})$ cannot contain monomials (see part (i)), we can decompose $t_1 f$ into simple components with respect to $\sim$ and apply Lemma 8.1.8 to derive that $(e_1 + a_i) - (e_1 + a_j) \in \mathcal{L}$ for some $i \neq j$. Hence $t^{a_i} - \rho(a_i - a_j) t^{a_j}$ belongs to $I_\rho(\mathcal{L})$ and either t^{a_i} or t^{a_j} is not standard, a contradiction.

(iv): If $\mathbb{Z}^q/\mathcal{L}$ is torsion-free, then $d_i = 1$ for $i = 1, \ldots, r$. Thus $\varphi(I')$ is prime because $h_i = t_i - \rho(\alpha_i')$ for all i. Hence I' is also prime. By part (iii) it follows that $I' \cap S = I_\rho(\mathcal{L})$. Thus $I_\rho(\mathcal{L})$ is prime as well. $\Box$

Definition 8.2.3 The *saturation* of a lattice $\mathcal{L} \subset \mathbb{Z}^q$, denoted by $\mathcal{L}_s$, is the lattice consisting of all $\alpha \in \mathbb{Z}^q$ such that $\eta\alpha \in \mathcal{L}$ for some $\eta \in \mathbb{Z} \setminus \{0\}$.

From the proof of Theorem 8.2.2 we obtain the following expression for the saturation (cf. Lemma 8.2.5).

Corollary 8.2.4 *If $\mathcal{L}$ is a lattice in $\mathbb{Z}^q$, then $\mathcal{L}_s = \mathbb{Z}\{e_1', \ldots, e_r'\}$*

Proof. Since $\alpha_i' = d_i e_i'$ for $i = 1, \ldots, r$, we need only show that $\mathcal{L}_s$ is contained in $\mathbb{Z}\{e_1', \ldots, e_r'\}$. Take $\alpha \in \mathcal{L}_s$, then $\alpha = \sum_{i=1}^{r} (a_i d_i / n) e_i'$ with $n, a_i \in \mathbb{Z}; n \neq 0$. Using that the set $\{e_1', \ldots, e_q'\}$ is a $\mathbb{Z}$-basis for $\mathbb{Z}^q$ we obtain that all the coefficients $a_i d_i / n$ are in $\mathbb{Z}$, thus α is in $\mathbb{Z}\{e_1', \ldots, e_r'\}$. $\Box$

The next result gives another method to compute the saturation.

Lemma 8.2.5 *If $\mathcal{L}$ is a lattice of rank r in $\mathbb{Z}^q$, then there is an integral matrix A of size $(q - r) \times q$ and* $\mathrm{rank}(A) = q - r$ *such that $\mathcal{L} \subset \ker_{\mathbb{Z}}(A)$, with equality if and only if $\mathbb{Z}^q/\mathcal{L}$ is torsion-free. Furthermore* $\ker_{\mathbb{Z}}(A)$ *is the saturation of $\mathcal{L}$.*

Proof. We may assume that $\mathcal{L} = \mathbb{Z}\alpha_1 \oplus \cdots \oplus \mathbb{Z}\alpha_r$, where $\alpha_1, \ldots, \alpha_q$ is a basis of the vector space $\mathbb{Q}^q$. Consider the hyperplane H_i of $\mathbb{Q}^q$ generated by $\alpha_1, \ldots, \widehat{\alpha}_i, \ldots, \alpha_q$. Note that the subspace of $\mathbb{Q}^q$ generated by $\alpha_1, \ldots, \alpha_r$ is equal to $H_{r+1} \cap \cdots \cap H_q$. There is a normal vector $w_i \in \mathbb{Z}^q$ such that

$$H_i = \{\alpha \in \mathbb{Q}^q | \langle \alpha, w_i \rangle = 0\}.$$

It is not hard to see that the matrix A with rows $w_{r+1}, \ldots, w_q$ is the matrix with the required conditions because by construction $\alpha_i \in H_j$ for $i \neq j$ and consequently $w_{r+1}, \ldots, w_q$ are linearly independent. In particular we have the equality $\text{rank}(\mathcal{L}) = \text{rank}(\ker_{\mathbb{Z}}(A))$. Since $\ker_{\mathbb{Z}}(A)/\mathcal{L}$ is a finite group it follows that $\mathcal{L}_s$, the saturation of $\mathcal{L}$, is equal to $\ker_{\mathbb{Z}}(A)$. □

Lemma 8.2.6 *Let $\mathcal{L}$ be a lattice in $\mathbb{Z}^q$. If $f = t^a - \lambda t^b \in S$ with $0 \neq \lambda \in K$, then $f \in I_\rho(\mathcal{L})$ if and only if $a - b \in \mathcal{L}$ and $\lambda = \rho(a-b)$.*

Proof. Assume that $f \in I_\rho(\mathcal{L})$. Consider the congruence in $\mathbb{N}^q$ given by $a \sim_{\mathcal{L}} b$ if $a - b \in \mathcal{L}$. By Theorem 8.2.2, $I_\rho(\mathcal{L})$ contains no monomials. Hence, by Lemma 8.1.8, f is simple, i.e., $a - b \in \mathcal{L}$. As $t^a - \rho(a-b)t^b$ is in $I_\rho(\mathcal{L})$, we get $\lambda = \rho(a-b)$. The converse is clear. □

Proposition 8.2.7 *Let $I_\rho(\mathcal{L}) \subset S$ be a lattice ideal and let $\prec$ be a term order. Then the elements in the reduced Gröbner basis of $I_\rho(\mathcal{L})$ are of the form $t^{a_+} - \rho(a)t^{a_-}$ with $a \in \mathcal{L}$.*

Proof. Let $\mathcal{G}$ be a finite set of generators of $I_\rho(\mathcal{L})$ consisting of polynomials of the form $\lambda_1 t^a - \lambda_2 t^b$ with $\lambda_i \in K^*$ for $i = 1, 2$ and let $f, g \in \mathcal{G}$. As $I_\rho(\mathcal{L})$ contains no monomials (see Theorem 8.2.2), the S-polynomial $\mathrm{S}(f,g)$ and the remainder of $\mathrm{S}(f,g)$ with respect to $\mathcal{G}$ are both also of this form. Then it follows that the output of the Buchberger's algorithm (see Theorem 3.3.12) is a Gröbner basis of $I_\rho(\mathcal{L})$ consisting of polynomials of the same form. Hence, by Lemma 8.2.6 and Theorem 8.2.2(iii), the elements in the reduced Gröbner basis of $I_\rho(\mathcal{L})$ are of the form $t^{a_+} - \rho(a)t^{a_-}$ with $a \in \mathcal{L}$. □

Theorem 8.2.8 *If I is a non-pure binomial ideal of S, then I is a lattice ideal of the form $I_\rho(\mathcal{L})$ for some lattice $\mathcal{L}$ and some partial character ρ if and only if $I \subsetneq S$ and t_i is a non-zero divisor of S/I for all i.*

Proof. $\Rightarrow$) This part follows from items (i) and (iii) of Theorem 8.2.2.

$\Leftarrow$) Let $\mathcal{L}$ be the set of $a \in \mathbb{Z}^q$ such that $t^{a_+} - \lambda t^{a_-} \in I$ for some $\lambda \in K^*$ and let a, b be two points in $\mathcal{L}$. Then $t^{b_+} - \lambda_1 t^{b_-} \in I$ for some λ_1 in K^*. We claim that $\mathcal{L}$ is a lattice. Since $t^{a_-} - \lambda^{-1}t^{a_+} \in I$, we get $-a \in \mathcal{L}$. Notice that $t^{a_+ + b_+} - \lambda\lambda_1 t^{a_- + b_-}$ is in I. Factoring the gcd of $t^{a_+ + b_+}$ and $t^{a_- + b_-}$, we get $a + b \in \mathcal{L}$. Given $a \in \mathcal{L}$ there is a unique $\lambda \in K^*$ such that $t^{a_+} - \lambda t^{a_-} \in I$. Setting $\rho(a) = \lambda$, we get a partial character. Thus I is the lattice ideal $I_\rho(\mathcal{L})$. This part was adapted from [134]. □

Let $\mathcal{L} \subset \mathbb{Z}^q$ be a lattice. We associate to $\mathcal{L}$ an equivalence relation on the monomials in $S = K[t_1, \ldots, t_q]$, by $t^\alpha \sim_{\mathcal{L}} t^\beta$ if $\alpha - \beta \in \mathcal{L}$.

Lemma 8.2.9 *Let $I_\rho(\mathcal{L})$ be a lattice ideal and let I be an ideal generated by a subset $\{t^{a_i} - \rho(a_i - b_i)t^{b_i}\}_{i=1}^r$ of $I_\rho(\mathcal{L})$. If $\mathcal{G} = \mathbb{Z}\{a_i - b_i\}_{i=1}^r$ and $f \in I$, then every simple component of f with respect to $\sim_{\mathcal{G}}$ also belongs to I.*

Proof. It follows from Lemma 8.1.8 because the relation, $a \sim_{\mathcal{G}} b$ if $a - b \in \mathcal{G}$ defines a congruence in $\mathbb{N}^q$ and $t^{a_i} - \rho(a_i - b_i)t^{b_i}$ is simple for all i. □

Lemma 8.2.10 *Let a, α_i, b, β_i be in $\mathbb{N}^q$ for $i = 1, \ldots, r$, let $a - b$ be in the lattice $\mathcal{L} = \mathbb{Z}\{\alpha_1 - \beta_1, \ldots, \alpha_r - \beta_r\}$ and let $\rho\colon \mathcal{L} \to K^*$ be a partial character. Then there is $t^\delta \in S$ such that*

$$t^\delta(t^a - \rho(a - b)t^b) \in (t^{\alpha_1} - \rho(\alpha_1 - \beta_1)t^{\beta_1}, \ldots, t^{\alpha_r} - \rho(\alpha_r - \beta_r)t^{\beta_r}).$$

Proof. We set $f = t^a - \rho(a - b)t^b$ and $g_i = t^{\alpha_i} - \rho(\alpha_i - \beta_i)t^{\beta_i}$ for $i = 1, \ldots, r$. There are integers $n_1, \ldots, n_r$ such that

$$(t^a/t^b) - \rho(a - b) = \left(t^{\alpha_1}/t^{\beta_1}\right)^{n_1} \cdots \left(t^{\alpha_r}/t^{\beta_r}\right)^{n_r} - \rho(a - b). \qquad (*)$$

We may assume that $n_i \geq 0$ for all i by replacing, if necessary, t^{α_i}/t^{β_i} by its inverse. Writing $t^{\alpha_i}/t^{\beta_i} = ((t^{\alpha_i}/t^{\beta_i}) - \rho(\alpha_i - \beta_i)) + \rho(\alpha_i - \beta_i)$ and using the binomial theorem, from Eq. $(*)$, it follows that $t^\delta f \in (g_1, \ldots, g_r)$ for some monomial t^δ. □

Lemma 8.2.11 *A lattice $\mathcal{L} \subset \mathbb{Z}^q$ is generated by $a_1, \ldots, a_m$ if and only if*

$$I_\rho(\mathcal{L}) = ((t^{a_1^+} - \rho(a_1)t^{a_1^-}, \ldots, t^{a_m^+} - \rho(a_m)t^{a_m^-})\colon (t_1 \cdots t_q)^\infty).$$

Proof. $\Rightarrow$) We set $I = (t^{a_1^+} - \rho(a_1)t^{a_1^-}, \ldots, t^{a_m^+} - \rho(a_m)t^{a_m^-})$. "$\subset$": Take $\alpha \in \mathcal{L}$. By Lemma 8.2.10 there is $t^\delta \in S$ such that $t^\delta(t^{\alpha^+} - \rho(\alpha)t^{\alpha^-}) \in I$. Thus $t^{\alpha^+} - \rho(\alpha)t^{\alpha^-} \in (I\colon (t_1 \cdots t_q)^\infty)$. "$\supset$": This inclusion follows readily from Theorem 8.2.8.

$\Leftarrow$) This part follows from Lemma 8.2.9 □

The next result can be used to compute the *presentation ideal* of a subring generated by rational functions.

Proposition 8.2.12 *Let $F = \{f_1/g_1, \ldots, f_q/g_q\} \subset K(\mathbf{x})$ be a finite set of rational functions, with $f_i, g_i \in K[\mathbf{x}] = K[x_1, \ldots, x_n]$ and $g_i \neq 0$ for all i, and let φ be the homomorphism of K-algebras*

$$\varphi\colon S = K[t_1, \ldots, t_q] \longrightarrow K[F], \text{ induced by } \varphi(t_i) = f_i/g_i.$$

Then $\ker(\varphi) = (g_1t_1 - f_1, \ldots, g_qt_q - f_q, yg_1 \cdots g_q - 1) \cap S$, *where y is a new variable. If $g_i = 1$ for all i, then* $\ker(\varphi) = (t_1 - f_1, \ldots, t_q - f_q) \cap S$.

Proof. We set $J = (g_1t_1 - f_1, \ldots, g_qt_q - f_q, yg_1 \cdots g_q - 1)$ and $h_i = f_i/g_i$ for $i = 1, \ldots, q$. Let us show the first equality. The second equality will follow by similar arguments. To show that $\ker(\varphi) \subset J \cap S$ take $f \in \ker(\varphi)$. Making $t_i = (t_i - h_i) + h_i$ and using the binomial theorem we can write:

$$f = f(t_1, \ldots, t_q) = \textstyle\sum_\gamma a_\gamma (t_1 - h_1)^{\gamma_1} \cdots (t_q - h_q)^{\gamma_q} + \sum_\beta a_\beta h_1^{\beta_1} \cdots h_q^{\beta_q},$$

with $a_\gamma \in K[F]$ for all $\gamma = (\gamma_1, \ldots, \gamma_q)$ and all γ's in $\mathbb{N}^q \setminus \{0\}$. As $f(h_1, \ldots, h_q) = 0$, we get $\sum_\beta a_\beta h_1^{\beta_1} \cdots h_q^{\beta_q} = 0$. Therefore multiplying the equality above by an appropriate positive power of $g_1 \cdots g_q$ we get

$$(g_1 \cdots g_q)^s f = \textstyle\sum_\gamma b_\gamma (g_1t_1 - f_1)^{\gamma_1} \cdots (g_qt_q - f_q)^{\gamma_q}, \tag{8.2}$$

where b_γ is polynomial in $K[\mathbf{x}]$ for all γ. Making $z = yg_1 \cdots g_q - 1$ we get $g_1 \cdots g_q = (z + 1)/y$. Thus from Eq. (8.2), we obtain that $(z + 1)^s f \in J$. Thus $f \in J \cap S$. Conversely let $f \in J \cap S$. Then

$$f = f(t_1, \ldots, t_q) = a_1(g_1t_1 - f_1) + \cdots + a_q(g_qt_q - f_q) + b(g_1 \cdots g_q y - 1).$$

Hence $f(h_1, \ldots, h_q) = b'(g_1 \cdots g_q y - 1)$. The polynomial f is independent of y. Making $y = 1/g_1 \cdots g_q$, we get $f(h_1, \ldots, h_q) = 0$, i.e., $f \in \ker(\varphi)$. □

Let us illustrate how to compute presentation ideals using Gröbner bases and elimination of variables.

Example 8.2.13 Let $F = \{x_ix_j \mid 1 \leq i < j \leq 4\}$ and let $\varphi\colon K[\mathbf{t}] \to K[F]$ be the map of K-algebras induced by $\varphi(t_{ij}) = x_ix_j$, where $\mathbf{t}$ is the set of variables $\{t_{ij} \mid 1 \leq i < j \leq 4\}$. By Proposition 8.2.12 one has

$$P_F = \ker(\varphi) = L \cap K[t_{ij}],\ L = (t_{ij} - x_ix_j \mid 1 \leq i < j \leq 4).$$

Using *Macaulay*2 [199] we get that the reduced Gröbner basis of L with respect to the elimination ordering in the first variables $x_1, \ldots, x_4$ is:

$t_{14}t_{23} - t_{12}t_{34}$, $t_{13}t_{24} - t_{12}t_{34}$, $x_4^2t_{23} - t_{24}t_{34}$, $x_4^2t_{13} - t_{14}t_{34}$,
$x_4^2t_{12} - t_{14}t_{24}$, $x_3t_{24} - x_4t_{23}$, $x_3t_{14} - x_4t_{13}$, $x_3x_4 - t_{34}$,
$x_3^2t_{12} - t_{13}t_{23}$, $x_2t_{34} - x_4t_{23}$, $x_2t_{14} - x_4t_{12}$, $x_2t_{13} - x_3t_{12}$,
$x_2x_4 - t_{24}$, $x_2x_3 - t_{23}$, $x_1t_{34} - x_4t_{13}$, $x_1t_{24} - x_4t_{12}$,
$x_1t_{23} - x_3t_{12}$, $x_1x_4 - t_{14}$, $x_1x_3 - t_{13}$, $x_1x_2 - t_{12}$,
$x_3t_{12}t_{34} - x_4t_{13}t_{23}$.

Therefore $P_F = (t_{14}t_{23} - t_{12}t_{34}, t_{13}t_{24} - t_{12}t_{34})$ by Theorem 3.3.20.

The ring of *Laurent polynomials* in the variables $x_1, \ldots, x_n$, denoted by $K[\mathbf{x}^{\pm 1}]$ or $K[x_1^{\pm 1}, \ldots, x_n^{\pm 1}]$, is generated as a K-vector space by the set of *Laurent monomials*, i.e., by the set of all x^a with $a \in \mathbb{Z}^n$.

Definition 8.2.14 Let $\mathcal{A} = \{v_1, \ldots, v_q\}$ be a set of points in $\mathbb{Z}^n$. The *monomial subring* generated or spanned by $F = \{x^{v_1}, \ldots, x^{v_q}\}$ is:

$$K[F] := \bigcap_{D \in \mathcal{F}} D,$$

where $\mathcal{F}$ is the family of all subrings D of R such that $K \cup F \subset D$.

The elements of $K[F]$ have the form $\sum c_a (x^{v_1})^{a_1} \cdots (x^{v_q})^{a_q}$ with $c_a \in K$ and all but a finite number of c_a's are zero. As a K-vector space $K[F]$ is generated by the set of monomials of the form x^a, with a in the semigroup $\mathbb{N}\mathcal{A}$ generated by $\mathcal{A}$. That is $K[F] = K[x^a | \, a \in \mathbb{N}\mathcal{A}]$. This means that $K[F]$ coincides with $K[\mathbb{N}\mathcal{A}]$, the *semigroup ring* of the semigroup $\mathbb{N}\mathcal{A}$ (see [180]).

Definition 8.2.15 An ideal $P \subset S$ is called a *toric ideal* if there is a finite set $F = \{x^{v_1}, \ldots, x^{v_q}\}$ in $K[\mathbf{x}^{\pm 1}]$ such that P is the kernel of the epimorphism of K-algebras:

$$\varphi\colon\ S = K[t_1, \ldots, t_q] \longrightarrow K[F] \longrightarrow 0, \text{ induced by } \varphi(t_i) = x^{v_i}.$$

The ideal P is called the *toric ideal* of $K[F]$ and is denoted by P_F.

Proposition 8.2.16 [400, p. 32] *If $F = \{x^{v_1}, \ldots, x^{v_q}\}$ is a set of Laurent monomials in $K[\mathbf{x}^{\pm 1}]$, then the corresponding toric ideal is given by*

$$P_F = (x^{v_1^-} t_1 - x^{v_1^+}, \ldots, x^{v_q^-} t_q - x^{v_q^+}, y x_1 \cdots x_n - 1) \cap S.$$

Proof. It follows adapting the proof of Proposition 8.2.12. □

This result gives a method to compute toric ideals. Another method can be found in [109]. In [38] some algorithms are devised, that in many respects improve existing algorithms for the computation of toric ideals.

Lemma 8.2.17 *Let $B = K[y_1, \ldots, y_n, t_1, \ldots, t_q]$ be a polynomial ring over a field K. If I is a binomial ideal of B, then the reduced Gröbner basis of I with respect to any term order $\prec$ consists of binomials. Furthermore $I \cap K[t_1, \ldots, t_q]$ is a binomial ideal.*

Proof. Let $\mathcal{B}$ be a finite set of generators of I consisting of binomials and let $f, g \in \mathcal{B}$. Since the S-polynomial $\mathrm{S}(f, g)$ is again a binomial and the remainder of $\mathrm{S}(f, g)$ with respect to $\mathcal{B}$ is also a binomial, the output of the Buchberger's algorithm (Theorem 3.3.12) is a Gröbner basis of I consisting of binomials. Hence the reduced Gröbner basis of I consists of binomials.

If $\prec$ is the lex order $y_1 \succ \cdots \succ y_n \succ t_1 \succ \cdots \succ t_q$ and $K[\mathbf{t}]$ is the ring $K[t_1, \ldots, t_q]$, then by elimination theory (see Theorem 3.3.20) $\mathcal{G} \cap K[\mathbf{t}]$ is a Gröbner basis of $I \cap K[\mathbf{t}]$. Hence $I \cap K[\mathbf{t}]$ is a binomial ideal. □

Corollary 8.2.18 *The reduced Gröbner basis of any toric ideal $P \subset S$ consists of binomials of the form $t^{a_+} - t^{a_-}$ with respect to any term order.*

Proof. By Proposition 8.2.16 and Lemma 8.2.17, P is a binomial ideal. Therefore, by Theorem 8.2.8, P is a lattice ideal because P is prime. Hence the result follows at once from Proposition 8.2.7. $\Box$

Definition 8.2.19 Let $F = \{x^{v_1}, \ldots, x^{v_q}\}$ be a set of Laurent monomials. The *associated matrix* of the monomial subring $K[F]$, denoted by A, is the matrix whose columns are the vectors $v_1, \ldots, v_q$.

Corollary 8.2.20 *If A is the associated matrix of a monomial subring $K[F]$, then the toric ideal P_F is the lattice ideal of* $\ker_{\mathbb{Z}}(A)$.

Proof. By Corollary 8.2.18, P_F is generated by binomials of the form $t^{a_+} - t^{a_-}$. Since a binomial $t^{a_+} - t^{a_-}$ is in P_F if and only if $A(a_+) = A(a_-)$, we get that P_F is generated by the set of all $t^{a_+} - t^{a_-}$ such that $Aa = 0$, i.e., P_F is the lattice ideal of $\ker_{\mathbb{Z}}(A)$. $\Box$

Corollary 8.2.21 [216] *Let $K[F]$ be a Laurent monomial subring and let A be its associated matrix. Then* $\dim K[F] = \operatorname{rank}(A)$.

Proof. The toric ideal P of $K[F]$ is the lattice ideal of $\mathcal{L} = \ker_{\mathbb{Z}}(A)$ by Corollary 8.2.20. Since the rank of $\mathcal{L}$ is $q - \operatorname{rank}(A)$, using Theorem 8.2.2, we get that $\dim K[F] = \operatorname{rank}(A)$. $\Box$

Toric ideals form a special class of lattice ideals.

Theorem 8.2.22 *Let $\mathcal{L} \subset \mathbb{Z}^q$ be a lattice. The following are equivalent:*

(a) *$I(\mathcal{L})$ is a toric ideal.*

(b) *$I(\mathcal{L})$ is a prime ideal.*

(c) *$\mathbb{Z}^q/\mathcal{L}$ is torsion-free.*

(d) $\mathcal{L} = \ker_{\mathbb{Z}}(A)$ *for some integral matrix A.*

Proof. (a) $\Rightarrow$ (b): If $I(\mathcal{L})$ is a toric ideal, then $S/I(\mathcal{L}) \simeq K[F]$ for some finite set of Laurent monomials of $K[\mathbf{x}^{\pm 1}]$. Since $K[F]$ is an integral domain, the ideal $I(\mathcal{L})$ is prime.

(b) $\Rightarrow$ (c): Assume $n\alpha \in \mathcal{L}$ for some integer $n > 0$ and $\alpha \in \mathbb{Z}^q$. First we assume that either $p = \operatorname{char}(K) = 0$ or that $p \neq 0$ is relatively prime to n. Since $t^{n\alpha^+} - t^{n\alpha^-}$ is in $I(\mathcal{L})$ and $n \cdot 1_K \neq 0$, from the equation

$$x^n - y^n = (x - y)(x^{n-1} + x^{n-2}y + \cdots + xy^{n-2} + y^{n-1})$$

with $x = t^{\alpha^+}$ and $y = t^{\alpha^-}$ we get $t^{\alpha^+} - t^{\alpha^-} \in I(\mathcal{L})$. Thus $\alpha \in \mathcal{L}$. If $p > 0$ and $\gcd(p, n) = p$, we write $n = p^r m$, where p and m are relatively prime. Since $t^{m\alpha^+} - t^{m\alpha^-} \in I(\mathcal{L})$, by the previous argument we obtain $\alpha \in \mathcal{L}$.

(c) $\Rightarrow$ (d): Let A be as in Lemma 8.2.5. Thus $\ker_{\mathbb{Z}}(A)/\mathcal{L}$ is a finite group, which must be zero because it is torsion-free.

(d) $\Rightarrow$ (a): Let $v_1, \ldots, v_q$ be the columns of A. If $F = \{x^{v_1}, \ldots, x^{v_q}\}$, by Corollary 8.2.20, P_F is the lattice ideal of $\mathcal{L} = \ker_{\mathbb{Z}}(A)$. $\Box$

Given a non-pure binomial $g = t^\alpha - \lambda t^\beta$, with $\lambda \in K^*$, we set $\widehat{g} = \alpha - \beta$. If $\mathcal{B} \subset \mathbb{Z}^q$, we denote by $\langle \mathcal{B} \rangle$ or $\mathbb{Z}\mathcal{B}$ the subgroup of $\mathbb{Z}^q$ generated by $\mathcal{B}$.

Lemma 8.2.23 *Let $\mathcal{L} \subset \mathbb{Z}^q$ be a lattice. If $g_1, \ldots, g_r$ is a set of non-pure binomials that generate $I_\rho(\mathcal{L})$, then $\mathcal{L} = \mathbb{Z}\{\widehat{g}_1, \ldots, \widehat{g}_r\}$. In particular if I is a lattice ideal, there are unique $\mathcal{L}$ and ρ such that $I = I_\rho(\mathcal{L})$.*

Proof. Consider the lattice $\mathcal{G} = \mathbb{Z}\{\widehat{g}_1, \ldots, \widehat{g}_r\}$. First we show the inclusion $\mathcal{L} \subset \mathcal{G}$. Take $0 \neq a \in \mathcal{L}$. Then $f = t^{a^+} - \rho(a) t^{a^-}$ is in $I_\rho(\mathcal{L}) = (g_1, \ldots, g_r)$. By Lemma 8.2.9 any simple component of f, with respect to $\sim_{\mathcal{G}}$, is also in $(g_1, \ldots, g_r)$. Since t^{a^+} and t^{a^-} are not in $I_\rho(\mathcal{L})$, then f simple with respect to $\sim_{\mathcal{G}}$, i.e., $a = a^+ - a^- \in \mathcal{G}$. Thus $\mathcal{L} \subset \mathcal{G}$. To show the other inclusion notice that, by Lemma 8.2.6, $\widehat{g}_i \in \mathcal{L}$ for all i, i.e., $\mathcal{G} \subset \mathcal{L}$. $\Box$

Lemma 8.2.24 *Let I be an ideal generated by a set of binomials $g_1, \ldots, g_r$ of S and let $h_1, \ldots, h_s$ be a set of binomials of S that generate* $\mathrm{rad}(I)$. *If* $\mathrm{char}(K) = 0$, *then* $\mathbb{Z}\{\widehat{g}_1, \ldots, \widehat{g}_r\} = \mathbb{Z}\{\widehat{h}_1, \ldots, \widehat{h}_s\}$.

Proof. We begin by writing the binomials g_i and h_i as $g_i = t^{\alpha_i} - t^{\beta_i}$ and $h_i = t^{\gamma_i} - t^{\delta_i}$. We set $\mathcal{L}_1 = \mathbb{Z}\{\widehat{g}_1, \ldots, \widehat{g}_r\}$ and $\mathcal{L}_2 = \mathbb{Z}\{\widehat{h}_1, \ldots, \widehat{h}_s\}$.

"$\subset$" Since $g_i \in \mathrm{rad}(I)$, then by Lemma 8.2.9, any simple component of g_i with respect to $\mathcal{L}_2$ belongs to $\mathrm{rad}(I)$. Therefore $\alpha_i \sim_{\mathcal{L}_2} \beta_i$ otherwise $\mathrm{rad}(I)$ would contain t^{α_i}, which is impossible. This proves that $\mathcal{L}_1 \subset \mathcal{L}_2$.

"$\supset$" Since $h_i = t^{\gamma_i} - t^{\delta_i}$ is in $\mathrm{rad}(I)$, then $h_i^{p^m}$ is in I for $m \gg 0$ and for an arbitrary odd prime number p. We claim that $t^{p^m \gamma_i} \sim_{\mathcal{L}_1} t^{p^m \delta_i}$. Consider the following equality which is obtained from the binomial theorem

$$h_i^{p^m} = \sum_{k=0}^{p^m} (-1)^k \binom{p^m}{k} (t^{\gamma_i})^{p^m - k} (t^{\delta_i})^k.$$

If $t^{p^m \gamma_i}$ and $t^{p^m \delta_i}$ are not in the same simple component of $h_i^{p^m}$ with respect to $\mathcal{L}_1$, then there is $C \subset \{1, \ldots, p^m - 1\}$, $C \neq \emptyset$, such that the polynomial

$$f = t^{p^m \gamma_i} + \sum_{k \in C} (-1)^k \binom{p^m}{k} (t^{\gamma_i})^{p^m - k} (t^{\delta_i})^k$$

is a simple component of $h_i^{p^m}$ with respect to $\mathcal{L}_1$. By Lemma 8.2.9, $f \in I$. Hence, since $\mathrm{char}(K) = 0$, we get

$$f(1, \ldots, 1) = 0 = 1 + \sum_{k \in C} (-1)^k \binom{p^m}{k},$$

a contradiction to $\binom{p^m}{k} \equiv 0 \bmod(p)$ for $1 \leq k \leq p^m - 1$; see Exercise 8.3.43. Therefore $t^{p^m\gamma_i} \sim_{\mathcal{L}_1} t^{p^m\delta_i}$. Thus $p^m(\gamma_i - \delta_i) \in \mathcal{L}_1$. If we pick another odd prime number $q \neq p$ and $t \gg 0$, repeating the previous argument, we get $q^t(\gamma_i - \delta_i) \in \mathcal{L}_1$, and hence $\gamma_i - \delta_i \in \mathcal{L}_1$, as required. $\Box$

Theorem 8.2.25 *Let I be a binomial ideal and let $I = \mathfrak{q}_1 \cap \cdots \cap \mathfrak{q}_m$ be an irredundant primary decomposition of I. If $\mathfrak{q}_1, \ldots, \mathfrak{q}_m$ are binomial ideals and* $\mathrm{char}(K) = 0$, *then* $\mathrm{rad}(I) = I$.

Proof. There are binomials $g_1, \ldots, g_r$ and $h_1, \ldots, h_s$ that generate I and $\mathrm{rad}(I)$, respectively. Say $g_i = t^{\alpha_i} - t^{\beta_i}$ and $h_i = t^{\gamma_i} - t^{\delta_i}$. Since $\mathrm{rad}(I)$ is equal to $\mathrm{rad}(\mathfrak{q}_1) \cap \cdots \cap \mathrm{rad}(\mathfrak{q}_m)$, it suffices to prove the case $m = 1$. Thus, let us assume that I is primary and show that $h_i = t^{\gamma_i} - t^{\delta_i}$ belongs to I for all $i = 1, \ldots, s$. By Lemma 8.2.24, we can write $\gamma_i - \delta_i = \sum_{j=1}^{r} \eta_j(\alpha_j - \beta_j)$ with $\eta_j \in \mathbb{Z}$ for all j, and by Lemma 8.2.10 there is a monomial t^γ such that $t^\gamma h_i \in I$. If $h_i \notin I$, then $(t^\gamma)^\ell \in I$ for some $\ell \geq 1$ because I is primary, but this is impossible. Thus $h_i \in I$, as required. $\Box$

Corollary 8.2.26 *Let I be a binomial ideal without embedded primes. If* $\mathrm{char}(K) = 0$ *and* $\mathrm{rad}(I)$ *is prime, then* $\mathrm{rad}(I) = I$ *and I is a prime ideal.*

Proof. Set $\mathfrak{p} = \mathrm{rad}(I)$. Since I has no embedded primes, by Exercise 2.1.48, one has $\mathrm{Ass}(S/I) = \{\mathfrak{p}\}$. Thus I is a $\mathfrak{p}$-primary ideal and I is a radical ideal by Theorem 8.2.25. Thus $I = \mathfrak{p}$ and I is a prime ideal. $\Box$

Theorem 8.2.27 *Let $I(\mathcal{L})$ be the lattice ideal of $\mathcal{L}$ and let $p =$* $\mathrm{char}(K)$.

(a) *If $p = 0$, then* $\mathrm{rad}(I(\mathcal{L})) = I(\mathcal{L})$.

(b) *If $p \neq 0$, then* $\mathrm{rad}(I(\mathcal{L})) = (t^a - t^b \,|\, p^r(a - b) \in \mathcal{L}$ *for some* $r \in \mathbb{N})$.

Proof. (a) Assume $t^{ra} - t^{rb} \in I(\mathcal{L})$ for $r \geq s_0 > 0$. By Theorem 8.1.12 and Exercise 8.1.13 it suffices to prove that $a - b \in \mathcal{L}$. Note that the relation "$\alpha \sim \beta$ if and only if $\alpha - \beta \in \mathcal{L}$" is a congruence in $\mathbb{N}^q$, thus $r(a - b) \in \mathcal{L}$ for $r \geq s_0$ and consequently $a - b \in \mathcal{L}$.

(b) It follows from Theorem 8.1.10 and Lemma 8.1.8. $\Box$

Definition 8.2.28 An ideal $I \subset S$ is called a *binomial set-theoretic complete intersection* if there are binomials $f_1, \ldots, f_r$ in I such that $\mathrm{rad}\,(I)$ is equal to $\mathrm{rad}\,(f_1, \ldots, f_r)$, where $r = \mathrm{ht}\,(I)$.

Theorem 8.2.29 [293] *Let $I(\mathcal{L}) \subset S$ be a lattice ideal of height r and let I be an ideal of S generated by binomials $g_1, \ldots, g_r$. If* $\mathrm{rad}(I(\mathcal{L})) = \mathrm{rad}(I)$ *and* $\mathrm{char}(K) = 0$, *then* $I(\mathcal{L}) = I$.

Proof. By Theorem 8.2.27 the ideal $I(\mathcal{L})$ is radical. Let $I = \mathfrak{q}_1 \cap \cdots \cap \mathfrak{q}_s$ be a primary decomposition of I. Since I is an ideal of height r generated by r elements, by the unmixedness theorem I has no embedded primes; see Theorem 2.3.27. Hence $\operatorname{rad}(\mathfrak{q}_i) = \mathfrak{p}_i$ is a minimal prime of both I and $I(\mathcal{L})$. Take a binomial h in $I(\mathcal{L})$. By Lemma 8.2.24 and Lemma 8.2.10, there is a monomial t^α so that $t^\alpha h \in I$. Thus $t^\alpha h \in \mathfrak{q}_i$ for all i. It suffices to prove that h belongs to $\mathfrak{q}_i$ for all i. If $h \notin \mathfrak{q}_i$ for some i, then $(t^\alpha)^\ell \in \mathfrak{q}_i$ and consequently $\mathfrak{p}_i$ must contain a variable, a contradiction to Theorem 8.2.8 because none of the variables of S is a zero divisor of $S/I(\mathcal{L})$. $\Box$

Exercises

8.2.30 If $\mathcal{L}$ is a lattice of $\mathbb{Z}^q$ and $I_\rho(\mathcal{L}) \subset S$ is a lattice ideal. Prove that

$$I_\rho(\mathcal{L}) = (\{t^a - \rho(a) |\, a \in \mathcal{L}\}) \cap S$$

and $(\{t^a - \rho(a) |\, a \in \mathcal{L}\}) = I_\rho(\mathcal{L})S'$, where $S' = K[t_1^\pm, \ldots, t_q^\pm]$.

8.2.31 Let T be the set of monomials of S. If $L = I_\rho(\mathcal{L})$ is a lattice ideal of S and $\mathfrak{p} \in \operatorname{Ass}_S(S/L)$, then $T \cap \mathfrak{p} = \emptyset$. Prove that $\operatorname{ht}(L)$ is equal to $\operatorname{ht}(T^{-1}(L))$ and that $T^{-1}(L) = (\{t^a - \rho(a) |\, a \in \mathcal{L}\}) \subset K[t_1^\pm, \ldots, t_q^\pm]$.

8.2.32 Let I be a binomial ideal of $S = K[t_1, \ldots, t_q]$. Then

(a) I is a lattice ideal of S if and only if $I = (I\colon (t_1 \cdots t_q)^\infty)$.

(b) If $I = (t^{a_1} - t^{b_1}, \ldots, t^{a_s} - t^{b_s})$, then $(I\colon (t_1 \cdots t_q)^\infty) = I(\mathcal{L})$, where $\mathcal{L}$ is the lattice $\mathcal{L} = \mathbb{Z}\{a_i - b_i\}_{i=1}^s$ in $\mathbb{Z}^q$ generated by $\{a_i - b_i\}_{i=1}^s$.

8.2.33 Let $S = K[t_1]$ be a polynomial ring in one variable. Prove that the lattice ideals of S are the ideals of the form $(t_1^m - 1)$, where $m \geq 0$.

8.2.34 Let $I \subset K[t_1, \ldots, t_q]$ be a binomial ideal generated by $t^{c_i^+} - t^{c_i^-}$, $i = 1, \ldots, m$, and let $\mathcal{L} \subset \mathbb{Z}^q$ be the lattice generated by $c_1, \ldots, c_s$. Prove that $I(\mathcal{L}) = IK[\mathbf{t}^{\pm 1}] \cap S$.

8.2.35 Let $K \subset F$ be a field extension, let $S = K[\mathbf{t}]$ and $B = F[\mathbf{t}]$ be polynomial rings and let I be an ideal of S. Then the following hold. (a) $IB \cap S = I$. (b) If $\cap_{i=1}^r \mathfrak{q}_i$ is a primary decomposition of IB, then $\cap_{i=1}^r (\mathfrak{q}_i \cap S)$ is a primary decomposition of I such that $\operatorname{rad}(\mathfrak{q}_i \cap S) = \operatorname{rad}(\mathfrak{q}_i) \cap S$. (c) If I is the lattice ideal of $\mathcal{L}$ in S, then IB is the lattice ideal of $\mathcal{L}$ in B.

8.2.36 Let $\mathcal{L}$ be a lattice in $\mathbb{Z}^q$ and let $\rho\colon \mathcal{L} \to K^*$ be a partial character. If $I_\rho(\mathcal{L})$ is prime and K is algebraically closed, then $\mathbb{Z}^q/\mathcal{L}$ is torsion-free.

8.2.37 Let $\mathcal{L}$ be the lattice $2\mathbb{Z}$ and let $\rho\colon \mathcal{L} \to \mathbb{Q}^*$ be the partial character given by $\rho(2p) = (-1)^p$. If $K = \mathbb{Q}$, prove that $I_\rho(\mathcal{L}) = (t_1^2 + 1)$.

8.3 Monomial subrings and toric ideals

Let $R = K[\mathbf{x}] = K[x_1, \ldots, x_n] = \oplus_{i\geq 0} R_i$ be a polynomial ring over a field K with the standard grading. As usual, $\mathbf{x}$ denotes the set of indeterminates $x_1, \ldots, x_n$ of the ring R. In this section we focus on monomial subrings generated by monomials in the polynomial ring R.

Let $F = \{f_1, \ldots, f_q\}$ be a finite set of monomials in R with $f_i = x^{v_i}$ and $f_i \neq 1$ for all i. The set F has a corresponding set of vectors in $\mathbb{N}^n$:

$$F = \{x^{v_1}, \ldots, x^{v_q}\} \longleftrightarrow \mathcal{A} = \{v_1, \ldots, v_q\}.$$

The monomial subring $K[F]$ is a graded subring of R with the grading given by $K[F]_i = K[F] \cap R_i$. There is a graded epimorphism of K-algebras

$$\varphi\colon\ S = K[t_1, \ldots, t_q] \longrightarrow K[F] \longrightarrow 0, \text{ induced by } \varphi(t_i) = f_i,$$

where $S = \oplus_{i\geq 0} S_i$ is a positively graded polynomial ring with the grading induced by setting $\deg(t_i) = \deg(x^{v_i}) = |v_i|$. We also denote S by $K[\mathbf{t}]$.

The kernel of φ, denoted by P_F, is the so called *toric ideal* of $K[F]$ with respect to $f_1, \ldots, f_q$. We also denote the toric ideal of $K[F]$ by $I_{\mathcal{A}}$. In this case we say that $I_{\mathcal{A}}$ is the *toric ideal* of $\mathcal{A}$.

Let A be the $n \times q$ matrix with column vectors $v_1, \ldots, v_q$. This matrix is called the *associated matrix* of $K[F]$. Closely related to the map φ is the homomorphism $\psi\colon \mathbb{Z}^q \longrightarrow \mathbb{Z}^n$ determined by the matrix A in the standard bases of $\mathbb{Z}^q$ and $\mathbb{Z}^n$. Indeed, we have $\varphi(t^\alpha) = x^{\psi(\alpha)}$ for all $\alpha \in \mathbb{N}^q$.

Recall that given a binomial $g = t^\alpha - t^\beta$, we set $\widehat{g} = \alpha - \beta$.

Proposition 8.3.1 *If $g_1, \ldots, g_r$ is a set of binomials that generate the toric ideal P_F, then $\widehat{g}_1, \ldots, \widehat{g}_r$ generate* $\ker_{\mathbb{Z}}(A)$.

Proof. By Corollary 8.2.20, P_F is the lattice ideal of $\ker_{\mathbb{Z}}(A)$. Thus the result follows from Lemma 8.2.23. $\square$

Definition 8.3.2 A binomial $t^\alpha - t^\beta \in P_F$ is called *primitive* if there is no other binomial $t^\gamma - t^\delta \in P_F$ such that t^γ divides t^α and t^δ divides t^β.

Lemma 8.3.3 *If f is a binomial in the reduced Gröbner basis of P_F with respect to some term order $\prec$, then f is a primitive binomial.*

Proof. Let $\mathcal{G}$ be the reduced Gröbner basis of P_F. One may assume $f = t^{\alpha_+} - t^{\alpha_-}$ and $\alpha_+ \succ \alpha_-$. Assume $g = t^{\beta_+} - t^{\beta_-} \neq f$ is a binomial in P_F such that t^{β_+} divides t^{α_+} and t^{β_-} divides t^{α_-}. Note $\beta_+ \neq \alpha_+$ and $\beta_- \neq \alpha_-$, otherwise $f = g$. If $\beta_+ \succ \beta_-$, then $t^{\beta_+} \in \mathrm{in}(\mathcal{G} \setminus \{f\})$ and consequently $t^{\alpha_+} \in \mathrm{in}(\mathcal{G} \setminus \{f\})$, a contradiction. If $\beta_- \succ \beta_+$, then $t^{\beta_-} \in \mathrm{in}(\mathcal{G} \setminus \{f\})$ and consequently $t^{\alpha_-} \in \mathrm{in}(\mathcal{G} \setminus \{f\})$, a contradiction. $\square$

According to a result of [400], the set $\mathcal{G}_{P_F}$ of all primitive binomials in a toric ideal P_F is finite and is called a *Graver basis* of P_F.

Definition 8.3.4 The *universal Gröbner basis* of a toric ideal P_F is the union of all reduced Gröbner basis of P_F. It is denoted by $\mathcal{U}_{P_F}$.

Definition 8.3.5 If α is a circuit of $\ker(A)$ we call the binomial $t^{\alpha_+} - t^{\alpha_-}$ a *circuit* of P_F. The set of circuits of P_F is denoted by $\mathcal{C}_{P_F}$.

Proposition 8.3.6 [400] $\mathcal{C}_{P_F} \subset \mathcal{U}_{P_F} \subset \mathcal{G}_{P_F}$.

Proof. The second inclusion follows from Lemma 8.3.3. To show the first inclusion take a circuit $f = t^{\alpha_+} - t^{\alpha_-}$ of the toric ideal P_F. We may assume that $f = t_1^{\alpha_1} \cdots t_r^{\alpha_r} - t_{r+1}^{-\alpha_{r+1}} \cdots t_s^{-\alpha_s}$, $\alpha = (\alpha_i) = \alpha^+ - \alpha^-$. We reorder the set of variables $\mathbf{t}$ as $\mathbf{t} \setminus \{t_1, \ldots, t_s\}, t_1, \ldots, t_s$. Consider the lex order: $t^a \succ t^b$ if the first non-zero entry of $a - b$ is positive. Let $\mathcal{G}$ be the reduced Gröbner basis of P_F with respect to $\succ$ and let $g = t^{\beta_+} - t^{\beta_-}$ be a binomial in $\mathcal{G}$ so that $\mathrm{in}(g) = t^{\beta_+}$ divides $\mathrm{in}(f)$. Clearly $\mathrm{supp}(t^{\beta_+}) \subset \mathrm{supp}(t^{\alpha_+})$. By the choice of term order it follows that $\mathrm{supp}(t^{\beta_-}) \subset \mathrm{supp}(t^{\alpha})$. Thus $\mathrm{supp}(\beta) \subset \mathrm{supp}(\alpha)$. Since α is a circuit we get equality. Thus using Lemma 1.9.3 we obtain that $\beta = \lambda\alpha$ for some $\lambda \in \mathbb{Q}$, consequently $\alpha = \beta$ because the non-zero entries of α (resp. β) are relatively prime. $\Box$

Corollary 8.3.7 *If in every circuit $t^{a_+} - t^{a_-}$ of P_F both monomials t^{a_+} and t^{a_-} are square-free, then $\mathcal{C}_{P_F} = \mathcal{U}_{P_F} = \mathcal{G}_{P_F}$.*

Proof. By Proposition 8.3.6 it suffices to show the inclusion $\mathcal{G}_{P_F} \subset \mathcal{C}_{P_F}$. Let $f = t^{\beta_+} - t^{\beta_-} \in \mathcal{G}_{P_F}$, i.e., f is primitive. By Lemma 1.9.5 there is a circuit α, in harmony with β, whose support is contained in the support of β. Since t^{α_+} and t^{α_-} are square-free, we conclude that t^{α_+} (resp. t^{α_-}) divides t^{β_+} (resp. t^{β_-}). Hence $f = t^{\alpha_+} - t^{\alpha_-}$ because f is primitive. $\Box$

Corollary 8.3.8 *If A is t-unimodular, then $\mathcal{C}_{P_F} = \mathcal{U}_{P_F} = \mathcal{G}_{P_F}$.*

Proof. By Corollary 1.9.11 in every circuit $t^{a_+} - t^{a_-}$ of P_F both terms t^{a^+} and t^{a^-} are square-free. Thus the equality follows from Corollary 8.3.7. $\Box$

Definition 8.3.9 A *universal Gröbner basis* of a polynomial ideal I is a finite set $\mathcal{U} \subset I$ which is a Gröbner basis of I with respect to all term orders.

Proposition 8.3.10 *$\mathcal{G}_{P_F}$ is a universal Gröbner basis of P_F.*

Proof. It follows from Proposition 8.3.6. $\Box$

Theorem 8.3.11 *The following are equivalent:*

(a) *In every circuit $t^{a_+} - t^{a_-}$ of P_F both monomials are square-free.*

(b) *Every initial monomial ideal of P_F is square-free.*

Proof. That (a) implies (b) follows from Corollary 8.3.7. Conversely let f be a circuit of P_F. We may assume that $f = t_r^{a_r} \cdots t_s^{a_s} - t_{s+1}^{a_{s+1}} \cdots t_q^{a_q}$. Consider the lex order $\succ$ with $t_1 \succ \cdots \succ t_q$ and consider the weight vector $\omega = e_1 + \cdots + e_{r-1} \in \mathbb{N}^q$. We define the term order: $a \succ_\omega b$ if and only if $\omega \cdot a > \omega \cdot b$ or $\omega \cdot a = \omega \cdot b$ and $a \succ b$. Then there is $g = t^{b^+} - t^{b^-}$, in the reduced Gröbner basis of P_F, such that t^{b^+} is square-free, $\mathrm{in}(g) = t^{b^+}$, and t^{b^+} divides t^{a^+}. Hence $\mathrm{supp}(t^{b^+})$ is contained in $\mathrm{supp}(t^{a^+})$. Consequently, by definition of $\succ_\omega$, we have that $\mathrm{supp}(t^{b^-})$ is contained in $\{t_r, \ldots, t_q\}$. Since a is a circuit we get that a and b have the same support. Thus, by Lemma 1.9.3, $a = \lambda b$ for some rational number λ. Using that a (resp. b) has relatively prime non-zero entries we obtain that $a = \pm b$ and $t^{b^+} = t^{a^+}$, i.e., t^{a^+} is square-free. A similar argument shows that t^{a^-} is also square-free. □

Corollary 8.3.12 [17, Theorem 4] *If* $\mathrm{char}(K) = 0$ *and* P_F *is a binomial set theoretic complete intersection, then* P_F *is a complete intersection.*

Proof. Set $r = \dim S/P_F$. By hypothesis, there are $g_1, \ldots, g_{q-r}$ binomials of S such that $\mathrm{rad}(g_1, \ldots, g_{q-r}) = P_F$. Since the ideal $(g_1, \ldots, g_{q-r})$ is C–M (see Proposition 2.3.24), the result follows from Corollary 8.2.26. □

Lemma 8.3.13 [134] *Let* R *be a ring and let* $x_1, \ldots, x_n \in R$. *If* I *is an ideal of* R, *then the radical of* I *satisfies*

$$\mathrm{rad}(I) = \mathrm{rad}(I\colon (x_1 \cdots x_n)^\infty) \cap \mathrm{rad}(I, x_1) \cap \cdots \cap \mathrm{rad}(I, x_n).$$

Proof. The inclusion "$\subset$" is clear. To show the inclusion "$\supset$" take a prime $\mathfrak{p} \supset I$, it suffices to show that $\mathfrak{p}$ contains one of the ideals on the right-hand side. If $\mathrm{rad}(I\colon (x_1 \cdots x_n)^\infty) \not\subset \mathfrak{p}$, there is $f \in R \setminus \mathfrak{p}$ such that $f(x_1 \cdots x_n)^s$ is in I for some $s \geq 1$. Hence $x_i \in \mathfrak{p}$ for some i and $\mathfrak{p} \supset \mathrm{rad}(I, x_i)$. □

Proposition 8.3.14 *Let* $\mathfrak{p}$ *be a prime ideal of a ring* R *and let* $x_1, \ldots, x_n$ *be a sequence in* $R \setminus \mathfrak{p}$. *If* $I \subset \mathfrak{p}$ *is an ideal, then* $\mathrm{rad}(I) = \mathfrak{p}$ *if and only if* $\mathfrak{p} = \mathrm{rad}(I\colon (x_1 \cdots x_n)^\infty)$ *and* $\mathrm{rad}(I, x_i) = \mathrm{rad}(\mathfrak{p}, x_i)$ *for all* i.

Proof. If J is any ideal of R and $y_1, \ldots, y_n \in R$, then by Lemma 8.3.13 $\mathrm{rad}(J)$ is equal to $\mathrm{rad}(J\colon (y_1 \cdots y_n)^\infty) \cap \mathrm{rad}(J, y_1) \cap \cdots \cap \mathrm{rad}(J, y_n)$. Hence the result follows by applying this equality to I and $\mathfrak{p}$, together with the fact that $x_1 \cdots x_n$ is regular on $R/\mathfrak{p}$. □

Proposition 8.3.15 [139] *Let* I *be an ideal generated by a set of binomials* $g_1, \ldots, g_r$ *in the toric ideal* P_F. *If* $\mathrm{char}(K) = p \neq 0$ *(resp.* $\mathrm{char}(K) = 0$) *and* $\mathcal{L}$ *is the lattice* $\mathbb{Z}\{\widehat{g}_i\}_{i=1}^r$, *then the following conditions are equivalent:*

(a$_1$) $P_F = \mathrm{rad}(I\colon (t_1 \cdots t_q)^\infty)$.

(a$_2$) $p^u \ker(\psi) \subset \mathcal{L}$ *for some* $u \in \mathbb{N}$ *(resp.* $\ker(\psi) = \mathcal{L}$).

Proof. (a_1) $\Rightarrow$ (a_2): First we consider the case $\text{char}(K) = p \neq 0$. Setting $z = t_1 \cdots t_q$, by hypothesis, there exists $u \geq 0$ such that $f^{p^u} \in (I\colon z^\infty)$ for all $f \in P_F$. Let $\gamma \in \ker(\psi)$. Write $\gamma = \alpha - \beta$ with $\alpha, \beta \in \mathbb{N}^q$ and $g = t^\alpha - t^\beta$. Thus $g \in P_F$ and $\widehat{g} = \alpha - \beta = \gamma$. The lattice $\mathcal{L}$ defines a congruence in $\mathbb{N}^q$ given by $a \sim_{\mathcal{L}} b$ if $a - b \in \mathcal{L}$. We know that $t^\delta g^{p^u} \in I$ for some $\delta \in \mathbb{N}^q$ and using that $p = \text{char}(K)$, one has $t^\delta g^{p^u} = t^{\alpha p^u + \delta} - t^{\beta p^u + \delta}$. Since the simple components of $t^\delta g^{p^u}$ with respect to $\mathcal{L}$ belong to I and hence to P_F (see Lemma 8.2.9), and since $t^\delta t^{\alpha p^u}$ does not belong to I because it does not belong to P_F, it follows that $t^\delta g^{p^u}$ is simple, *i.e.*, $\delta + p^u\gamma \sim_{\mathcal{L}} p^u\beta + \delta$. Hence $p^u\alpha - p^u\beta \in \mathcal{L}$. This shows the inclusion $p^u\ker(\psi) \subset \mathcal{L}$.

Next consider the case $\text{char}(K) = 0$. By Lemma 8.2.11 one has the equality $(I\colon z) = I(\mathcal{L})$. Thus $P_F = \text{rad}(I(\mathcal{L}))$. As P_F is the lattice ideal of $\ker(\psi)$ (Corollary 8.2.20), by Lemmas 8.2.23 and 8.2.24 we get $\mathcal{L} = \ker(\psi)$.

(a_1) $\Leftarrow$ (a_2): It suffices to prove the inclusion $P_F \subset \text{rad}(I\colon z^\infty)$, the other inclusion is clear. Assume $\text{char}(K) = p > 0$. Let $f = t^c - t^e$ be a binomial in P_F. By (a_2) and Lemma 8.2.10, we get that $t^\gamma(t^{p^u c} - t^{p^u e}) = t^\gamma f^{p^u}$ is in I for some monomial t^γ. Thus f is in $\text{rad}(I\colon z^\infty)$. If $\text{char}(K) = 0$, the same argument works if we replace p^u by 1. □

Theorem 8.3.16 [139] *Let $I \subset P_F$ be an ideal generated by a finite set of binomials $g_1, \ldots, g_r$. If* $\text{char}(K) = p \neq 0$ *(resp.* $\text{char}(K) = 0$*), then* $\text{rad}(I) = P_F$ *if and only if*

(a) $p^u\ker(\psi) \subset \mathbb{Z}\{\widehat{g}_i\}_{i=1}^r$ *for some $u \in \mathbb{N}$ (resp.* $\ker(\psi) = \mathbb{Z}\{\widehat{g}_i\}_{i=1}^r$*), and*

(b) $\text{rad}(I, t_i) = \text{rad}(P_F, t_i)$ *for all i.*

Proof. It is a consequence of Propositions 8.3.14 and 8.3.15. □

Next we give a result that holds over a field of arbitrary characteristic. It is a consequence of the proofs of Theorem 8.2.25 and Corollary 8.2.26.

Proposition 8.3.17 *Let $I \subset S$ be an ideal, and let $\{g_i\}_{i=1}^r$, $\{h_i\}_{i=1}^s$ be sets of binomials that generate I,* $\text{rad}(I)$*, respectively. If $\mathbb{Z}\{\widehat{g}_i\}_{i=1}^r$ is equal to $\mathbb{Z}\{\widehat{h}_i\}_{i=1}^s$, I has no embedded primes, and* $\text{rad}(I)$ *is prime, then* $I = \text{rad}(I)$.

Corollary 8.3.18 *A toric ideal P_F of height r is a complete intersection if and only if there are binomials $g_1, \ldots, g_r$ in P_F such that*

(a) $\ker(\psi) = \mathbb{Z}\{\widehat{g}_1, \ldots, \widehat{g}_r\}$*, and*

(b) $(g_1, \ldots, g_r, t_i) = (P_F, t_i)$ *for all i.*

Proof. $\Rightarrow$) It follows readily from Proposition 8.3.1.

$\Leftarrow$) By Theorem 8.3.16 we get $\text{rad}(g_1, \ldots, g_r) = P_F$. Notice that the ideal $I = (g_1, \ldots, g_r)$ is a complete intersection because $r = \text{ht}(P_F)$ and I is quasi-homogeneous. Hence applying Proposition 8.3.17 yields $I = P_F$. □

Given a subset $I \subset S$ we denote its zero set in $\mathbb{A}_K^q$ by $V(I)$, and given a subset $X \subset \mathbb{A}_K^q$ we denote its vanishing ideal in S by $I(X)$.

Corollary 8.3.19 *Let I be an ideal generated by binomials $g_1,\ldots,g_r$ in P_F. If* $\mathrm{char}(K)=p\neq 0$ (resp. $\mathrm{char}(K)=0$) *and* $p^m\ker(\psi)\subset\mathbb{Z}\{\widehat{g}_i\}_{i=1}^r$ *for some* $m\in\mathbb{N}$ (resp. $\ker(\psi)=\mathbb{Z}\{\widehat{g}_i\}_{i=1}^r$), *then* $V(I)\subset V(P_F)\cup V(t_1\cdots t_q)$.

Proof. By Proposition 8.3.15, there is a monomial t^δ and an integer N such that $t^\delta P_F^N\subset I$. It follows that

$$V(I)\subset V(P_F^N)\cup V(t^\delta)\subset V(P_F)\cup V(t_1\cdots t_q). \qquad \Box$$

Lemma 8.3.20 *Let I be a binomial ideal of S such that $V(I,t_i)=\{0\}$ for all i. If $\mathfrak{p}$ is a prime ideal containing (I,t_m) for some $1\leq m\leq q$, then $\mathfrak{p}=(t_1,\ldots,t_q)$.*

Proof. Let $h_1,\ldots,h_r$ be a set of binomials that generate I. For simplicity of notation assume that $m=1$. We may assume that $t_1,\ldots,t_k$ are in $\mathfrak{p}$ and $t_i\notin\mathfrak{p}$ for $i>k$. If $t_i\in\mathrm{supp}(h_j)$ for some $1\leq i\leq k$, say $h_j=t^{a_j}-t^{b_j}$ and $t_i\in\mathrm{supp}(t^{a_j})$, then $t^{b_j}\in\mathfrak{p}$ and there is $1\leq\ell\leq k$ such that t_ℓ is in the support of t^{b_j}. Thus, $h_j\subset(t_1,\ldots,t_k)$. Hence, for each $1\leq j\leq r$, either

(i) $\mathrm{supp}(h_j)\cap\{t_1,\ldots,t_k\}=\emptyset$ or (ii) $h_j\in(t_1,\ldots,t_k)$.

Consider the point $c=(c_i)\in\mathbb{A}_K^q$, with $c_i=0$ for $i\leq k$ and $c_i=1$ for $i>k$. If (i) occurs, then $h_j(c)=(t^{a_j}-t^{b_j})(c)=1-1=0$. If (ii) occurs, then $h_j(c)=(t^{a_j}-t^{b_j})(c)=0-0=0$. Clearly the polynomial t_1 vanishes at c. Hence, $c\in V(I,t_1)=\{0\}$. Therefore, $k=q$. Thus, $\mathfrak{p}$ contains all the variables of S, i.e., $\mathfrak{p}=(t_1,\ldots,t_q)$. $\Box$

Corollary 8.3.21 [139] *Let I be an ideal generated by binomials $g_1,\ldots,g_r$ in P_F. If $F=\{x_1^{d_1},\ldots,x_1^{d_q}\}$ and* $\mathrm{char}(K)=p\neq 0$ *(resp.* $\mathrm{char}(K)=0$*), then* $\mathrm{rad}(I)=P_F$ *if and only if*

(a) $p^m\ker(\psi)\subset\mathbb{Z}\{\widehat{g}_i\}_{i=1}^r$ *for some $m\in\mathbb{N}$ (resp.* $\ker(\psi)=\mathbb{Z}\{\widehat{g}_i\}_{i=1}^r$)

(b) $V(g_1,\ldots,g_r,t_i)=\{0\}$, *for all i.*

Proof. $\Rightarrow$) By Theorem 8.3.16 condition (a) holds and $\mathrm{rad}(I,t_i)$ is equal to $\mathrm{rad}(P_F,t_i)$ for all i. On the other hand $\mathrm{rad}(P_F,t_i)=(t_1,\ldots,t_q)$, because the height of P_F is $q-1$ and t_i is regular on S/P_F. Hence $V(I,t_i)=\{0\}$.

$\Leftarrow$) By Proposition 8.3.15 one has $P_F=\mathrm{rad}(I\colon z^\infty)$, where $z=t_1\cdots t_q$. Hence there is monomial t^δ and an integer N such that $t^\delta P_F^N\subset I$.

To prove $\mathrm{rad}(I)=P_F$, it suffices to show that every prime ideal $\mathfrak{p}$ containing I also contains P_F. If $\mathfrak{p}$ contains no variables, then the inclusions $t^\delta P_F^N\subset I\subset\mathfrak{p}$ imply that $P_F\subset\mathfrak{p}$, as desired. If $\mathfrak{p}$ contains at least one variable, then $\mathfrak{p}$ contains all the variables by Lemma 8.3.20 and consequently contains P_F, as desired. $\Box$

Proposition 8.3.22 *Let $I \subset S$ be a graded binomial ideal. The following hold.* (a) *If $V(I, t_i) = \{0\}$ for all i, then* $\mathrm{ht}(I) = q - 1$. (b) *If I is a lattice ideal and* $\mathrm{ht}(I) = q - 1$, *then $V(I, t_i) = \{0\}$ for all i.*

Proof. (a) As I is graded, all associated prime ideal of S/I are graded. Thus, all associated prime ideals of S/I are contained in $\mathfrak{m} = (t_1, \ldots, t_q)$. If $\mathrm{ht}(I) = q$, then $\mathfrak{m}$ would be the only associated prime of S/I, that is, $\mathfrak{m}$ is the radical of I, a contradiction because I cannot contain a power of t_i for any i. Thus, $\mathrm{ht}(I) \leq q - 1$. On the other hand, by Lemma 8.3.20, the ideal (I, t_q) has height q. Hence, $q = \mathrm{ht}(I, t_q) \leq \mathrm{ht}(I) + 1$ (here we use the fact that I is graded). Altogether, we get $\mathrm{ht}(I) = q - 1$.

(b) Let $\mathcal{L}$ be the lattice that defines I and $g_1, \ldots, g_r$ a set of homogeneous binomials that generate I. By Lemma 8.2.23, we get $\mathcal{L} = \langle \widehat{g}_1, \ldots, \widehat{g}_r \rangle$. Notice that $q - 1 = \mathrm{ht}(I) = \mathrm{rank}(\mathcal{L})$. Given two integers $1 \leq i, k \leq q$, the vector space $\mathbb{Q}^q$ is generated by $e_k, \widehat{g}_1, \ldots, \widehat{g}_r$. Hence, as $\mathcal{L}$ is homogeneous with respect to $\omega = (\omega_1, \ldots, \omega_q)$, there are positive integers r_i and r_k such that $r_i e_i - r_k e_k \in \mathcal{L}$ and $r_i \omega_i - r_k \omega_k = 0$. By Lemma 8.2.10, there is t^δ such that $t^\delta(t_i^{r_i} - t_k^{r_k})$ is in I. Hence, by Theorem 8.2.8, $t_i^{r_i} - t_k^{r_k}$ is in I. Therefore, $V(I, t_i) = \{0\}$ for all i. $\Box$

Lemma 8.3.23 *Let $I \subset S$ be a graded binomial ideal. If $V(I, t_i) = \{0\}$ for all i and I is a complete intersection, then I is a lattice ideal.*

Proof. By Proposition 8.3.22(a), the height of I is $q - 1$. It suffices to prove that t_i is a non-zero divisor of S/I for all i (see Theorem 8.2.8). If t_i is a zero divisor of S/I for some i, there is an associated prime ideal $\mathfrak{p}$ of S/I containing (I, t_i). Hence, using Lemma 8.3.20, we get that $\mathfrak{p} = \mathfrak{m}$, a contradiction because I is a complete intersection of height $q - 1$ and all associated prime ideals of I have height equal to $q - 1$ (see Proposition 2.3.24). $\Box$

Theorem 8.3.24 [293] *If $L \subset S$ is a graded lattice ideal and $V(L, t_i) = \{0\}$ for all i, then L is a complete intersection if and only if there are homogeneous binomials $h_1, \ldots, h_{q-1}$ in L satisfying the following conditions:*

(i) $\mathcal{L} = \langle \widehat{h}_1, \ldots, \widehat{h}_{q-1} \rangle$, *where $\mathcal{L}$ is the lattice that defines L.*

(ii) $V(h_1, \ldots, h_{q-1}, t_i) = \{0\}$ *for all i.*

(iii) $h_i = t^{a_i^+} - t^{a_i^-}$ *for $i = 1, \ldots, q - 1$.*

Proof. As L is graded, By Proposition 8.3.22, the height of L is $q - 1$.

$\Rightarrow$) By Lemma 3.1.29 and Exercise 3.3.26, L is generated by homogeneous binomials $h_1, \ldots, h_{q-1}$. Then, by Lemma 8.2.23 and Theorem 8.2.8, (i) and (iii) hold. From the equality $(L, t_i) = (h_1, \ldots, h_{q-1}, t_i)$, it follows readily that (ii) holds.

$\Leftarrow$) We set $I = (h_1, \ldots, h_{q-1})$. By hypothesis $I \subset L$. Thus, we need only show the inclusion $L \subset I$. Let $g_1, \ldots, g_m$ be a generating set of L consisting of binomials, then $\widehat{g}_i \in \mathcal{L}$ for all i. Using condition (i) and Lemma 8.2.10, for each i there is a monomial t^{γ_i} such that $t^{\gamma_i} g_i \in I$. Hence, $t^\gamma L \subset I$, where t^γ is equal to $t^{\gamma_1} \cdots t^{\gamma_m}$. By (ii) and Proposition 8.3.22, the height of I is $q-1$. This means that I is a complete intersection. As $t^\gamma L \subset I$, to show the inclusion $L \subset I$, it suffices to notice that by (ii), Lemma 8.3.23 and Theorem 8.2.8 t_i is a non-zero divisor of S/I for all i. □

Remark 8.3.25 The result remains valid if we remove condition (iii), i.e., condition (iii) is redundant. In both implications of the theorem the set $h_1, \ldots, h_{q-1}$ is shown to generate L.

Definition 8.3.26 The *circuit ideal* of the matrix A is given by

$$I = (\{t^{\alpha_+} - t^{\alpha_-} \,|\, \alpha \text{ is a circuit of } \ker(A)\}).$$

The next lemma is useful in induction arguments.

Lemma 8.3.27 *Let A' be the matrix obtained from A by removing its ith column. Then the inclusion map $j\colon \mathbb{Z}^{q-1} \to \mathbb{Z}^q$ given by*

$$(a_1, \ldots, a_{i-1}, a_{i+1}, \ldots, a_q) \xrightarrow{j} (a_1, \ldots, a_{i-1}, 0, a_{i+1}, \ldots, a_q)$$

induces a bijection between the circuits of $\ker(A')$ *and the circuits of* $\ker(A)$ *whose ith entry is equal to zero.*

Proof. Since the inclusion map j takes $\ker(A')$ into $\ker(A)$, the result follows readily. □

Proposition 8.3.28 [134] *If I is the circuit ideal of A, then* $\mathrm{rad}\,(I) = P_F$.

Proof. We proceed by induction on q, the number of columns of A. If $q = 1$ or $q = 2$, then P_F is a principal ideal and the result is clear. According to Theorems 8.3.16 and 1.9.10 it suffices to prove that $\mathrm{rad}\,(I, t_i)$ is equal to $\mathrm{rad}\,(P_F, t_i)$ for all i. For simplicity of notation we assume $i = 1$. Let $\gamma_1, \ldots, \gamma_r$ be the circuits of $\ker(A)$. We set $f_i = t^{\gamma_i^+} - t^{\gamma_i^-}$ for $i = 1, \ldots, r$ and $f_0 = f_{r+1} = 0$. If $\mathfrak{p}$ is a prime ideal containing (I, t_1), then after relabeling $t_1, \ldots, t_q$ and $f_1, \ldots, f_r$ there are $t_1, \ldots, t_s \in \mathfrak{p}$ and $0 \leq k \leq r$ such that (i) $(f_{k+1}, \ldots, f_r) \subset (t_1, \ldots, t_s)$ and (ii) $t_i \notin \cup_{j=1}^k \mathrm{supp}(f_j)$ for $i = 1, \ldots, s$.

We claim that $\mathfrak{p}$ contains (P_F, t_1). Take $f = t^{\alpha^+} - t^{\alpha^-}$ in P_F. Setting $\alpha = \alpha^+ - \alpha^-$, by Theorem 1.9.6, we can write $p\alpha = \sum_{i=1}^r n_i \gamma_i$ for some $p \in \mathbb{N}_+$ and $n_1, \ldots, n_r \in \mathbb{N}$, such that for each i either $n_i = 0$ or $n_i > 0$ and γ_i is in harmony with α and $\mathrm{supp}(\gamma_i) \subset \mathrm{supp}(\alpha)$.

Case (I): If $t_i \in \mathrm{supp}(f)$ for some $1 \leq i \leq s$, then $t_i \in \mathrm{supp}(f_j)$ for some j such that $n_j > 0$. Thus by (ii) one has $j \geq k+1$ and $f_j \in (t_1, \ldots, t_s)$. From the equality $p\alpha = \sum_{i=1}^r n_i \gamma_i$ it follows that $f \in (t_1, \ldots, t_s) \subset \mathfrak{p}$.

Case (II): If $t_i \notin \operatorname{supp}(f)$ for $1 \leq i \leq s$. Let φ' (resp. ψ') be the restriction of φ (resp. ψ) to $K[t_{s+1}, \ldots, t_q]$ (resp. $\mathbb{Z}e_{s+1} + \cdots + \mathbb{Z}e_q$). We set $F' = \{x^{v_{s+1}}, \ldots, x^{v_q}\}$. Note that $\gamma_1, \ldots, \gamma_k$ are the circuits of $\ker(\psi')$ and $P'_{F'} = K[t_{s+1}, \ldots, t_q] \cap P_F$ is the toric ideal of $K[F']$; see Exercise 8.3.33 and Lemma 8.3.27. Hence by induction one has

$$\operatorname{rad}(I') = P'_{F'} = K[t_{s+1}, \ldots, t_q] \cap P_F,$$

where I' is a circuit ideal generated by $f_1, \ldots, f_k$. Since $f \in P'_{F'}$, then there is $m \geq 1$ such that $f^m \in I' \subset \mathfrak{p}$. Thus $f \in \mathfrak{p}$.

Altogether $\mathfrak{p}$ contains (P_F, t_1). As $\operatorname{rad}(P_F, t_1)$ is the intersection of the minimal primes of (P_F, t_1) one concludes $\operatorname{rad}(P_F, t_1) \subset \operatorname{rad}(I, t_1)$. Since the other inclusion is clear one has equality. □

Exercises

8.3.29 If the matrix A is t-unimodular, then the reduced Gröbner basis of P_F with respect to any term order consists of square-free binomials, i.e., of binomials $t^{a^+} - t^{a^-}$ with t^{a^+} and t^{a^-} square-free.

8.3.30 Let $\mathcal{B}$ be a set of binomials of $S = K[t_1, \ldots, t_q]$ and let f be a binomial. Prove that dividing f by $\mathcal{B}$ produces a residue $h = t^\gamma - t^\delta$ which is independent of the field K as long as we follow identical steps in the division process when changing fields.

8.3.31 Let F be a finite set of monomials in a polynomial ring $\mathbb{Q}[x_1, \ldots, x_n]$ and let P be the toric ideal of $\mathbb{Q}[F]$. Consider the toric ideal P_F of $K[F]$ over a field K. If $\mathcal{G}$ is a Gröbner basis of P consisting of binomials with respect to a fixed term order, prove that $\mathcal{G}$ is also a Gröbner basis for P_F.

8.3.32 Use the previous exercise to draw some conclusion if P is a toric ideal minimally generated by a Gröbner basis over a field K.

8.3.33 Let $F = \{f_1, \ldots, f_q\}$ be a set of monomials of a polynomial ring R over a field K. Consider the homomorphism

$$\varphi\colon S = K[t_1, \ldots, t_q] \longrightarrow K[F] \longrightarrow 0, \text{ induced by } \varphi(t_i) = f_i.$$

Prove that $\ker(\varphi) \cap K[t_1, \ldots, t_r]$ is a binomial ideal which is equal to $\ker(\varphi')$, where φ' is the restriction of φ to $K[t_1, \ldots, t_r]$.

8.3.34 If I is an ideal of $S = K[t_1, \ldots, t_q]$ and $\operatorname{rad}(I) = P_F$, prove that $\operatorname{rad}(IS') = P_F S'$, where $S' = K[t_1^{\pm 1}, \ldots, t_q^{\pm 1}]$.

8.3.35 Let A be an $n \times q$ integral matrix. If $\mathcal{L}$ is a subgroup of $\ker_{\mathbb{Z}}(A)$, prove that $\mathrm{T}(\mathbb{Z}^q/\mathcal{L}) = \mathrm{T}(\ker_{\mathbb{Z}}(A)/\mathcal{L})$, where T denotes torsion.

8.3.36 Let A be an $n \times q$ integral matrix with non-zero columns. If $\mathcal{L}$ is a subgroup of $\ker_{\mathbb{Z}}(A)$ and $p \geq 2$ a prime number, then $p^u \ker_{\mathbb{Z}}(A) \subset \mathcal{L}$ for some $u \in \mathbb{N}$ if and only if $\ker_{\mathbb{Z}}(A)/\mathcal{L}$ is a finite p-group or $\ker_{\mathbb{Z}}(A)/\mathcal{L} = (0)$.

8.3.37 Let $d_1, \ldots, d_q$ be a sequence of relatively prime integers and ψ the linear map from $\mathbb{Z}^q$ to $\mathbb{Z}$ induced by $\psi(e_i) = d_i$. If $\mathcal{L}$ is a subgroup of $\ker(\psi)$, then $\mathrm{T}(\mathbb{Z}^q/\mathcal{L}) \subset \ker(\psi)/\mathcal{L}$, with equality if $\ker(\psi)/\mathcal{L}$ is a finite group.

8.3.38 Let A be an $n \times q$ integer matrix. If $\mathcal{L}$ is a subgroup of $\ker_{\mathbb{Z}}(A)$, then $\mathcal{L} = \ker_{\mathbb{Z}}(A)$ if and only if $\mathbb{Z}^q/\mathcal{L}$ is torsion-free of rank equal to $\mathrm{rank}(A)$.

8.3.39 Let P be the toric ideal of the subring $K[x_1^6, x_1^8, x_1^9]$ and let I be the ideal generated by g_1, g_2, where $g_1 = t_1^3 - t_3^2$, $g_2 = t_2^9 - t_3^8$. Then

(a) $P = (g_1, t_2^3 - t_1 t_3^2)$ for any field K, (b) $\mathbb{Z}^3/\langle \widehat{g}_1, \widehat{g}_2 \rangle \simeq \mathbb{Z} \times \mathbb{Z}_3$,

(c) $\mathrm{rad}(I) = P$ if $\mathrm{char}(K) = 3$, and $\mathrm{rad}(I) \neq P$ if $\mathrm{char}(K) = 2$,

(d) $\Gamma = \{(x^6, x^8, x^9) | \, x \in K\} = \{(0,0,0), (1,1,1)\} = V(I)$, if $K = \mathbb{Z}_2$.

8.3.40 Let $d_1, \ldots, d_n$ be a sequence of relatively prime integers and let ψ be the linear map from $\mathbb{Z}^n$ to $\mathbb{Z}$ induced by $\psi(e_i) = d_i$, then the set

$$L = \left\{ \frac{d_j}{\gcd(d_i, d_j)} e_i - \frac{d_i}{\gcd(d_i, d_j)} e_j \,\middle|\, 1 \leq i < j \leq n \right\}$$

is a generating set for $\ker(\psi)$, where e_i is the ith unit vector.

8.3.41 Let $d_1, \ldots, d_n$ be a sequence of non-zero integers and let ψ be the linear map from $\mathbb{Z}^n$ to $\mathbb{Z}$ induced by $\psi(e_i) = d_i$. Prove that the set

$$L = \left\{ \left(\frac{d_j}{d}\right) e_i - \left(\frac{d_i}{d}\right) e_j \,\middle|\, 1 \leq i < j \leq n \right\}$$

is a generating set for $\ker(\psi)$, where $d = \gcd(d_1, \ldots, d_n)$.

8.3.42 Let $k \geq 1$ be an integer and let p be a prime number. Consider the expansion of k in base p: $k = \sum_{i=0}^{m_1} a_i p^i$, where $a_{m_1} \neq 0$, the a_i's are integers and $0 \leq a_i \leq p-1$ for all i. If p^{u_1} is the largest power of p dividing $k!$ (the factorial of k), prove the equalities

$$u_1 = \sum_{i=1}^{m_1} a_i \left(\frac{p^i - 1}{p - 1} \right) \quad \text{and} \quad u_1 = \frac{p^m - 1}{p - 1} \quad \text{if} \quad k = p^m.$$

8.3.43 If $p > 1$ is a prime number and $m \geq 1$ is an integer. Prove that p divides the binomial coefficient $\binom{p^m}{k}$ for $1 \leq k \leq p^m - 1$.

8.4 Toric varieties

In this section we use linear algebra methods to characterize when a toric set is an affine toric variety in terms of the existence of certain roots in the base field and a vanishing condition.

First we fix some notation. Let K be a field and let A be an $n \times q$ matrix with entries in $\mathbb{N}$ and with non-zero columns $v_1, \ldots, v_q$. The *toric set parameterized* by $F = \{x^{v_1}, \ldots, x^{v_q}\} \subset K[\mathbf{x}]$ is the set given by

$$\Gamma = \left\{(a_1^{v_{11}} \cdots a_n^{v_{n1}}, \ldots, a_1^{v_{1q}} \cdots a_n^{v_{nq}}) \in \mathbb{A}_K^q \;\middle|\; a_1, \ldots, a_n \in K\right\},$$

where $v_i = (v_{1i}, \ldots, v_{ni})$ for all i. Similarly one can define the toric set Γ parameterized by $x^{v_1}, \ldots, x^{v_q}$, with $v_1, \ldots, v_q$ in $\mathbb{Z}^n$, as the set of all points $(a_1^{v_{11}} \cdots a_n^{v_{n1}}, \ldots, a_1^{v_{1q}} \cdots a_n^{v_{nq}})$ in $\mathbb{A}_K^q$ that are well-defined for some a_i's in K. As usual the toric ideal of $K[F]$ is denoted by P_F. Recall that P_F is a prime ideal of the polynomial ring $S = K[t_1, \ldots, t_q]$ over the field K.

There are unimodular integral square matrices $U = (u_{ij})$ and $Q = (q_{ij})$ of orders n and q, respectively, such that

$$D = UAQ = \operatorname{diag}(\lambda_1, \ldots, \lambda_s, 0, \ldots, 0),$$

where s is the rank of A and $\lambda_1, \ldots, \lambda_s$ are the invariant factors of A; that is, λ_i divides λ_{i+1} and $\lambda_i > 0$ for all i (see Theorem 1.2.2). For use below we set $U^{-1} = (f_{ij})$ and $Q^{-1} = (b_{ij})$. In what follows e_i will denote the ith unit vector in $\mathbb{Z}^q$.

Lemma 8.4.1 $\ker_{\mathbb{Z}}(A) = \mathbb{Z}q_{s+1} \oplus \cdots \oplus \mathbb{Z}q_q$, *where q_i is the ith column of the matrix Q.*

Proof. Let $x \in \mathbb{Z}^q$. Make the change of variables $y = Q^{-1}x$. As $D = UAQ$, it follows that $Ax = 0$ if and only if $Dy = 0$. Set $y = (y_1, \ldots, y_q)$.

First note that $q_i \in \ker_{\mathbb{Z}}(A)$ for $i \geq s+1$, because $DQ^{-1}q_i = De_i$. If x is in $\ker_{\mathbb{Z}}(A)$, then $\lambda_i y_i = 0$ for $i = 1, \ldots, s$. Thus $x = Qy = \sum_{i=s+1}^{q} y_i q_i$. To complete the proof observe that the columns of Q are a basis for $\mathbb{Z}^q$. $\Box$

Proposition 8.4.2 $\ker_{\mathbb{Z}}(A) = \mathbb{Z}\{w_j - e_j\}_{j=1}^{q}$, *where $w_j = \sum_{i=1}^{s} b_{ij} q_i$ and q_i is the ith column of Q.*

Proof. Note that $e_j = \sum_{i=1}^{q} b_{ij} q_i$ for all j, because $QQ^{-1} = I$. Hence

$$w_j = \sum_{i=1}^{s} b_{ij} q_i = e_j - \sum_{i=s+1}^{q} b_{ij} q_i \qquad (j = 1, 2, \ldots, q).$$

Then, using Lemma 8.4.1, we obtain $w_j - e_j \in \ker_{\mathbb{Z}}(A)$ for $j = 1, \ldots, q$. By Lemma 8.4.1, it suffices to observe that from the equality above one has

$$\begin{aligned}\sum_{j=1}^{q} q_{jk}(e_j - w_j) &= \sum_{j=1}^{q} q_{jk}\left(\sum_{i=s+1}^{q} b_{ij}q_i\right) = \sum_{i=s+1}^{q} q_i\left(\sum_{j=1}^{q} q_{jk}b_{ij}\right)\\ &= \sum_{i=s+1}^{q} q_i\delta_{ik} = q_k,\end{aligned}$$

for all $k \geq s+1$, where $\delta_{ik} = 1$ if $i = k$ and $\delta_{ik} = 0$ otherwise. □

Definition 8.4.3 An *affine toric variety* V is defined as the zero set of a toric ideal, that is, $V = V(P_F)$ for some toric ideal P_F.

Theorem 8.4.4 [354] *Let $\Gamma \subset K^q$ be the toric set parameterized by F. Then $\Gamma = V(P_F)$ if and only if the following two conditions hold:*

(a) *If $(a_i) \in V(P_F)$ and $a_i \neq 0\ \forall i$, then $a_1^{q_{1i}} \cdots a_q^{q_{qi}}$ has a λ_i-root in K for $i = 1, \ldots, s$, where $\lambda_1, \ldots, \lambda_s$ are the invariant factors of A,*

(b) *$V(P_F, t_i) \subset \Gamma$ for $i = 1, \ldots, q$.*

Proof. $\Leftarrow$) One invariably has $\Gamma \subset V(P_F)$. To prove the other inclusion take a point $a = (a_1, \ldots, a_q)$ in $V(P_F)$, by condition (b) one may assume $a_i \neq 0$ for all i. Thus using (a) there are $x_1', \ldots, x_s'$ in K such that

$$(x_i')^{\lambda_i} = a_1^{q_{1i}} \cdots a_q^{q_{qi}} = a^{q_i} \qquad (i = 1, \ldots, s). \tag{8.3}$$

For convenience of notation we extend the definition of x_i' by putting $x_i' = 1$ for $i = s+1, \ldots, n$ and $x' = (x_1', \ldots, x_n')$. Set

$$x_j = (x_1')^{u_{1j}} \cdots (x_n')^{u_{nj}} \qquad (j = 1, \ldots, n). \tag{8.4}$$

We claim that $x^{v_k} = x_1^{v_{1k}} \cdots x_n^{v_{nk}} = a_k$ for $k = 1, \ldots, q$. Setting $U^{-1} = (f_{ij})$ and comparing columns in the equality $U^{-1}D = AQ$ one has:

$$\lambda_i f_i = \sum_{j=1}^{q} q_{ji} v_j \qquad (i = 1, 2, \ldots, s), \tag{8.5}$$

where $f_i = (f_{1i}, \ldots, f_{ni})$ and $v_j = (v_{1j}, \ldots, v_{nj})$ denote the ith and jth columns of U^{-1} and A. Comparing columns in $A = (U^{-1}D)Q^{-1}$ we get:

$$v_k = \sum_{j=1}^{s} \lambda_j b_{jk} f_j \qquad (k = 1, 2, \ldots, q), \tag{8.6}$$

where $Q^{-1} = (b_{ij})$. Using $UU^{-1} = I$ and Eq.(8.4) we rapidly conclude:

$$x^{f_k} = x'_k \qquad (k = 1, \dots, n). \tag{8.7}$$

From Proposition 8.4.2 we derive $Aw_j = Ae_j = v_j$ for $j = 1, \dots, q$, where

$$w_j = \sum_{i=1}^{s} b_{ij} q_i = \left(\sum_{\ell=1}^{s} q_{1\ell} b_{\ell j}, \dots, \sum_{\ell=1}^{s} q_{n\ell} b_{\ell j} \right) \qquad (j = 1, \dots, q). \tag{8.8}$$

Hence $Aw_j^+ = A(e_j + w_j^-)$, that is, $t^{w_j^+} - t^{e_j + w_j^-}$ belongs to the toric ideal P_F. Using that $a \in V(P_F)$ yields $a^{w_j^+} = a^{e_j + w_j^-}$, thus

$$a^{w_j} = a^{e_j} = a_j \qquad (j = 1, \dots, q). \tag{8.9}$$

Therefore putting everything together:

$$\begin{aligned} x^{v_k} &\overset{8.6}{=} x^{\sum_{j=1}^{s} \lambda_j b_{jk} f_j} = (x^{f_1})^{\lambda_1 b_{1k}} \cdots (x^{f_s})^{\lambda_s b_{sk}} \overset{8.7}{=} (x'_1)^{\lambda_1 b_{1k}} \cdots (x'_s)^{\lambda_s b_{sk}} \\ &\overset{8.3}{=} (a^{q_1})^{b_{1k}} \cdots (a^{q_s})^{b_{sk}} = a^{q_1 b_{1k} + \cdots + q_s b_{sk}} \overset{8.8}{=} a^{w_k} \overset{8.9}{=} a_k \end{aligned}$$

for $k = 1, \dots, q$. Thus $a \in \Gamma$, as required.

$\Rightarrow$) It is clear that (b) holds because $V(P_F, t_i) \subset V(P_F)$. To prove (a) take $(a_i) \in V(P_F)$ with $a_i \neq 0$ for all i, then by definition of Γ there are $x_1, \dots, x_n$ in K such that $a_j = x^{v_j}$ for all j. Therefore by Eq. (8.5) one has:

$$(x^{f_i})^{\lambda_i} = x^{\lambda_i f_i} = x^{q_{1i} v_1} \cdots x^{q_{qi} v_q} = a_1^{q_{1i}} \cdots a_q^{q_{qi}}. \qquad \square$$

The proof of Theorem 8.4.4 works for arbitrary Laurent monomials. Another classification, that uses polyhedral geometry, of the affine toric varieties that are parameterized by Laurent monomials is given in [272].

Corollary 8.4.5 *$V(P_F) \subset \Gamma \cup V(t_1 \cdots t_q)$ if K is algebraically closed.*

Proof. Let $a = (a_i) \in V(P_F)$ such that $a_i \neq 0$ for all i. Since K is algebraically closed condition (a) above holds. Therefore one may proceed as in the first part of the proof of Theorem 8.4.4 to get $a \in \Gamma$. $\square$

Next we present another consequence that can be used to prove that monomial curves over arbitrary fields are affine toric varieties.

Corollary 8.4.6 *If the columns of A generate $\mathbb{Z}^n$ as $\mathbb{Z}$-module, then the equality $\Gamma = V(P_F)$ holds if and only if $V(P_F, t_i) \subset \Gamma$ for all i.*

Proof. Since $\mathbb{Z}v_1 + \cdots + \mathbb{Z}v_q = \mathbb{Z}^n$, one has $\lambda_i = 1$ for all i, thus condition (a) holds. Therefore Γ is an affine toric variety if and only if (b) holds. $\square$

Corollary 8.4.7 *If K is algebraically closed, then $\Gamma = V(P_F)$ if and only if $V(P_F, t_i) \subset \Gamma$ for all i.*

Proof. If K is algebraically closed, then (a) is satisfied, thus Γ is a toric variety if and only if $V(P_F, t_i) \subset \Gamma$ for all i. □

Proposition 8.4.8 [139] *If A has only one row and $v_1, \ldots, v_q$ are relatively prime positive integers, then $\Gamma = V(P_F)$ over any field K.*

Proof. As $\mathbb{Z} = \mathbb{Z}v_1 + \cdots + \mathbb{Z}v_q$ by Corollary 8.4.6 it suffices to show $V(P_F, t_i) \subset \Gamma$. Let $a \in V(P_F, t_i)$ since all the binomials $t_i^{v_j} - t_j^{v_i}$ vanish on a one obtains $a = 0$ and $a \in \Gamma$. □

A natural question is whether a toric set Γ can be a variety but not a toric variety (see Exercise 8.4.18):

Example 8.4.9 Let K be the field $\mathbb{Z}_3$ and let A be the matrix $(2, 4)$. Then

$$\Gamma = \{(0,0)\} \cup \{(1,1)\} = V(t_1 - t_2, t_2^2 - t_2).$$

On the other hand $P_F = (t_1 - t_2^2)$ and $(1,2) \in V(P_F)$. Thus $\Gamma \neq V(P_F)$.

Affine toric varieties are fully parameterized by monomials when K is algebraically closed or when the variety is normal.

Theorem 8.4.10 [271, 272] *Let $F = \{x^{v_1}, \ldots, x^{v_q}\}$ be a set of Laurent monomials such that K is algebraically closed or $K[F]$ is normal. Then there exists a toric set Γ_G parameterized by $G = \{x^{\epsilon_1}, \ldots, x^{\epsilon_q}\}$ with $\epsilon_i \in \mathbb{Z}^n$ such that $V(P_G) = \Gamma_G$ and $P_F = P_G$, where P_G is the toric ideal of $K[G]$.*

Theorem 8.4.11 (Combinatorial Nullstellensatz [7]) *Let $S = K[t_1, \ldots, t_q]$ be a polynomial ring over a field K, let $f \in S$, and let $c = (c_i) \in \mathbb{N}^q$. Suppose that the coefficient of t^c in f is non-zero and $\deg(f) = c_1 + \cdots + c_q$. If $A_1, \ldots, A_q$ are subsets of K, with $|A_i| > c_i$ for all i, then there are $a_1 \in A_1, \ldots, a_q \in A_q$ such that $f(a_1, \ldots, a_q) \neq 0$.*

Lemma 8.4.12 *Let f be a polynomial in $K[x_1, \ldots, x_n]$. If K is an infinite field and $f(a) = 0$ for all $a \in K^n$, then f is the zero polynomial.*

Proof. It follows readily from Theorem 8.4.11. □

Corollary 8.4.13 *Let $R = K[\mathbf{x}]$ be a polynomial ring over a field K and let $X \subset K^q$ be a set parameterized by $t_i = f_i(\mathbf{x})$, where $F = \{f_1, \ldots, f_q\} \subset R$. If K is infinite and $I(X)$ is the vanishing ideal of X, then*

(a) *$I(X)$ is equal to P_F, the presentation ideal of $K[F]$, and*

(b) *the Zariski closure of X is equal to $V(P_F)$.*

Proof. (a): As $P_F = (t_1 - f_1, \ldots, t_q - f_q) \cap K[t_1, \ldots, t_q]$, one has $P_F \subset I(X)$. Conversely take $f \in I(X)$ and consider the revlex ordering such that $t_i > x_j$ for all i, j. By the division algorithm $f(t_1, \ldots, t_q) = \sum_i g_i(t_i - f_i) + r$, where r is a polynomial in R. Hence for any $a \in K^n$ one has

$$f(f_1(a), \ldots, f_q(a)) = \textstyle\sum_i g_i(a, x)(f_i(a) - f_i) + r = 0.$$

Thus $r(a) = 0$ for any $a \in K^n$. Therefore r is the zero polynomial because K is an infinite field (see Lemma 8.4.12).

(b): By part (a) and Proposition 3.2.3 we get $\overline{X} = V(P_F)$. □

Affine normal toric varieties Let $\mathcal{A} = \{v_1, \ldots, v_q\} \subset \mathbb{Z}^n$ be a point configuration such that $\mathbb{R}_+\mathcal{A} \cap \mathbb{Z}\mathcal{A} = \mathbb{N}\mathcal{A}$. Hence the monomial subring

$$K[x^{v_1}, \ldots, x^{v_q}] \subset K[x_1^{\pm 1}, \ldots, x_n^{\pm 1}]$$

is normal (Corollary 9.1.3). Pick a $\mathbb{Z}$-basis $\mathcal{B} = \{w_1, \ldots, w_r\}$ of $\mathbb{Z}\mathcal{A}$, where $r = \operatorname{rank}(\mathbb{Z}\mathcal{A})$, and consider the linear map $T\colon \mathbb{Z}\mathcal{A} \longrightarrow \mathbb{Z}^r$, $w_i \mapsto e_i$. For $1 \leq i \leq q$ we can write $v_i = u_{1i}w_1 + \cdots + u_{ri}w_r$. Notice that $T(v_i) = u_i$, where $u_i = (u_{1i}, \ldots, u_{ri})$. Setting $\mathcal{H} = \{u_1, \ldots, u_q\}$, we have $\mathbb{Z}\mathcal{H} = \mathbb{Z}^r$.

Lemma 8.4.14 $\mathbb{Z}^r \cap \mathbb{R}_+\mathcal{H} = \mathbb{N}\mathcal{H}$.

Proof. The equality follows readily by observing that T can be extended to an isomorphism from $\mathbb{Q}\mathcal{A}$ onto $\mathbb{Q}^r$. □

Let P and P_1 be the toric ideals of $K[x^{v_1}, \ldots, x^{v_q}]$ and $K[x^{u_1}, \ldots, x^{u_q}]$, respectively.

Lemma 8.4.15 $P = P_1$ *and* $K[x^{v_1}, \ldots, x^{v_q}] \simeq K[x^{u_1}, \ldots, x^{u_q}]$.

Proof. Let $K[t_1, \ldots, t_q]$ be a polynomial ring over the field K. Consider the following ring maps

$$\begin{array}{ccc} K[t_1, \ldots, t_q] & \xrightarrow{\ \varphi_1\ } & K[x^{u_1}, \ldots, x^{u_q}] \\ \Big\downarrow \varphi & \nearrow \overline{\varphi}_1 & \\ K[x^{v_1}, \ldots, x^{v_q}] & & \end{array}$$

induced by $\varphi_1(t_i) = x^{u_i}$ and $\varphi(t_i) = x^{v_i}$. Since the rings $K[x^{v_1}, \ldots, x^{v_q}]$ and $K[x^{u_1}, \ldots, x^{u_q}]$ have dimension r it suffices to prove that $\ker(\varphi) \subset \ker(\varphi_1)$ and to observe that $P = \ker(\varphi)$, $P_1 = \ker(\varphi_1)$. Let $f = t_1^{a_1} \cdots t_q^{a_q} - t_1^{c_1} \cdots t_q^{c_q}$ be a binomial in $\ker(\varphi)$. Then

$$(x^{v_1})^{a_1} \cdots (x^{v_q})^{a_q} = (x^{v_1})^{c_1} \cdots (x^{v_q})^{c_q}$$

and $\sum_{i=1}^{q}(a_i - c_i)v_i = 0$. Applying T we get $\sum_{i=1}^{q}(a_i - c_i)u_i = 0$. Thus

$$(x^{u_1})^{a_1}\cdots(x^{u_q})^{a_q} = (x^{u_1})^{c_1}\cdots(x^{u_q})^{c_q}.$$

This proves that $f \in \ker(\varphi_1)$, as required. □

Lemma 8.4.16 *The map* $\phi\colon (K^*)^r \to (K^*)^q \cap V(P)$, $x \mapsto (x^{u_1}, \ldots, x^{u_q})$, *is bijective, where* $K^* = K \setminus \{0\}$.

Proof. As $e_i \in \mathbb{Z}\mathcal{H} = \mathbb{Z}^r$ for $1 \leq i \leq r$, there is $\alpha_i \in \mathbb{N}\mathcal{H}$ such that $e_i + \alpha_i = \beta_i$ and $\beta_i \in \mathbb{N}\mathcal{H}$. Then the monomial $x^\alpha = x^{\alpha_1}\cdots x^{\alpha_r}$ satisfies that $x^\alpha \in \mathbb{N}\mathcal{H}$ and $x_i x^\alpha = x^{\gamma_i} \in \mathbb{N}\mathcal{H}$. Thus we can write

$$x^\alpha = (x^{u_1})^{r_1}\cdots(x^{u_q})^{r_q};\ x^{\gamma_i} = (x^{u_1})^{s_{i1}}\cdots(x^{u_q})^{s_{iq}};\ (r_i \in \mathbb{N};\ s_{ij} \in \mathbb{N}).$$

Setting $h_i(t_1, \ldots, t_q) = t_1^{r_1}\cdots t_q^{r_q}$ and $h_0(t_1, \ldots, t_q) = t_1^{s_{i1}}\cdots t_q^{s_{iq}}$, we have

$$x_i h_i(x^{u_1}, \ldots, x^{u_q}) = h_0(x^{u_1}, \ldots, x^{u_q}) \tag{8.10}$$

for $1 \leq i \leq r$. If $\phi(x_1, \ldots, x_r) = \phi(y_1, \ldots, y_r)$, then $x^{u_i} = y^{u_i}$ for all i. Thus, from Eq. (8.10), $x_i = y_i$ for all i. This proves that ϕ is injective.

To prove that ϕ is onto take $a = (a_i) \in V(P) \cap (K^*)^q$. Let A_u be the matrix with column vectors $u_1, \ldots, u_q$. There are q_{ij} in $\mathbb{Z}$ such that $e_j = \sum_{i=1}^{q} q_{ij}u_i$ for $j = 1, \ldots, r$. Thus $A_u(q_j) = e_j$ for $j = 1, \ldots, r$, where $q_j = (q_{1j}, \ldots, q_{qj})$. Hence, setting $w_i = \sum_{j=1}^{r} u_{ji}q_j$ for $i = 1, \ldots, q$, we get

$$A_u(w_i) = u_i = A_u(e_i) \quad (i = 1, \ldots, q)$$

Hence $Aw_i^+ = A(e_j + w_i^-)$, that is, $t^{w_i^+} - t^{e_i + w_i^-}$ is in the toric ideal P. Using that $a \in V(P)$ yields $a^{w_i^+} = a^{e_i + w_i^-}$, thus $a^{w_i} = a^{e_i} = a_i$ for $i = 1, \ldots, q$. Setting $x_j = a^{q_j} = a_1^{q_{1j}}\cdots a_q^{q_{qj}}$ for $j = 1, \ldots, r$, we get

$$x^{u_i} = x_1^{u_{1i}}\cdots x_r^{u_{ri}} = (a^{q_1})^{u_{1i}}\cdots(a^{q_r})^{u_{ri}} = a^{w_i} = a_i$$

for $i = 1, \ldots, q$. Thus a is in the image of ϕ, as required. □

Theorem 8.4.17 *If K is algebraically closed, then*

(a) *$V(P)$ contains the torus $(K^*)^r$ as a dense open subset, and*

(b) *there is an action of the torus $(K^*)^r$ on the toric variety $V(P)$:*

$$\begin{array}{ccc} (K^*)^r \times V(P) & \longrightarrow & V(P) \\ (x, (y_1, \ldots, y_q)) & \longmapsto & (x^{u_1}y_1, \ldots, x^{u_q}y_q) \end{array}$$

Proof. The set $\Gamma^* := \operatorname{im}(\phi) = (K^*)^r \cap V(P)$ is an open subset of $V(P)$ because the set $V(t_1 \cdots t_q)$ is closed in the Zariski topology of K^q, i.e., the set $(K^*)^q = K^q \setminus V(t_1 \cdots t_q)$ is open in K^q. By Hilbert Nullstellensatz (see Theorem 3.2.10) $V(P)$ is irreducible. Hence Γ^* is dense in $V(P)$. Thus (a) follows from Lemma 8.4.16. Part (b) is left as an exercise. □

Exercises

8.4.18 Let Γ be a toric set parameterized by F. If K is an infinite field and $\Gamma = V(J)$ for some ideal J, then $J \subset P_F$ and $\Gamma = V(P_F)$.

8.4.19 (Rational quartic curve in $\mathbb{P}^3$) Let K be an algebraically closed field and let Γ be the curve parameterized by $F = \{x_1^4, x_1^3x_2, x_1x_2^3, x_2^4\}$. Then $\mathrm{rad}\,(t_2t_3 - t_1t_4,\, t_3^3 - t_2t_4^2,\, t_2^3 - t_1^2t_3) = P_F$ and Γ is an affine toric variety.

8.4.20 Prove that Corollaries 8.4.5 and 8.4.7 are valid assuming condition (a) of Theorem 8.4.4, instead of assuming K algebraically closed.

8.4.21 Let $S = K[t_1, t_2, t_3, t_4]$ be a polynomial ring over a field K. Then

$$V(t_1t_3 - t_2t_4) = \{(x_1x_2, x_2x_3, x_3x_4, x_1x_4)|\; x_1, \ldots, x_4 \in K\}.$$

8.4.22 (Segre variety) Let m, n be two positive integers. Prove that

$$\Gamma = \{(a_ib_j) \in K^{mn}|\, a_i \in K,\, b_j \in K\,\forall i, j\} \subset \mathbb{A}_K^{mn}$$

is an affine toric variety over any field K.

8.4.23 (Veronese variety) Let d be a positive integer and let $\mathcal{A}$ be the set of k-partitions of d consisting of all $a \in \mathbb{N}^k$ such that $|a| = d$. If K is an algebraically closed field and $F = \{x^a|\, a \in \mathcal{A}\}$, then the toric set Γ parameterized by F is an affine toric variety.

8.4.24 Let $\mathcal{A} = \{v_1, \ldots, v_q\} \subset \mathbb{N}^n$ and let K be a field. Prove that $K[x^{v_1}, \ldots, x^{v_q}]$ is the coordinate ring of an affine toric variety, in the sense of [176], if and only if $\mathbb{R}_+\mathcal{A} \cap \mathbb{Z}^n = \mathbb{N}\mathcal{A}$ and $\mathrm{rank}(\mathcal{A}) = n$.

8.4.25 Let $\mathcal{A} = \{v_1, \ldots, v_q\} \subset \mathbb{Z}^n$ and let K be a field. Prove that if $K[F] = K[x^{v_1}, \ldots, x^{v_q}]$ is normal and $\mathrm{rank}(\mathcal{A}) = n$, then $K[F]$ is isomorphic to the coordinate ring of an affine toric variety, in the sense of [176].

8.4.26 An affine variety $X \subset K^q$ is called a *set-theoretic complete intersection* if $I(X)$ is a set-theoretic complete intersection. If K is algebraically closed, then X is a set-theoretic complete intersection if and only if X can be defined by $q - \dim(X)$ polynomials in q variables with coefficients in K.

8.4.27 Let $K = \mathbb{Z}_2$ and let $X \subset K^3$ be a space curve given parametrically by $t_1 = x_1^7$, $t_2 = x_1^8$, $t_3 = x_1^9$. If $F = \{x_1^7, x_1^8, x_1^9\}$, prove that $P_F \neq I(X)$ and $P_F = (t_1^5 - t_2t_3^3, t_2^2 - t_1t_3, t_1^4t_2 - t_3^4)$.

8.4.28 [272] Let $F = \{x_1^2x_3,\, x_2^4x_3^2, x_1x_2x_3\}$ and let $K = \mathbb{R}$ be the field of real numbers. Prove that $V(P_F)$ is never fully parameterized by Laurent monomials in the sense of Theorem 8.4.10.

8.5 Affine Hilbert functions

In this section, we introduce the notion of degree via affine Hilbert functions and show additivity of the degree. For lattice ideals the degree turns out to be independent of the base field and the partial character.

Let $S = K[t_1, \ldots, t_q]$ be a polynomial ring over a field K and let I be an ideal of S. The vector space of polynomials in S (resp. I) of degree at most i is denoted by $S_{\leq i}$ (resp. $I_{\leq i}$). The functions

$$H_I^a(i) = \dim_K(S_{\leq i}/I_{\leq i}) \quad \text{and} \quad H_I(i) = H_I^a(i) - H_I^a(i-1)$$

are called the *affine Hilbert function* and the *Hilbert function* of S/I. Let $u = t_{s+1}$ be a new variable and let $I^h \subset S[u]$ be the *homogenization* of I, where $S[u]$ is given the standard grading.

Lemma 8.5.1 *If $\mathfrak{m} = (t_1, \ldots, t_q)$ is the irrelevant maximal ideal of S and $I \subset S$ is a graded ideal of dimension zero, then*

$$\ell_{S_\mathfrak{m}}(S_\mathfrak{m}/I_\mathfrak{m}) = \ell_S(S/I) = \ell_{S/I}(S/I) = \dim_K(S/I).$$

Proof. Let $\overline{\mathfrak{m}}$ be the image of $\mathfrak{m}$ in the quotient ring $\overline{S} = S/I$ and let r be the least positive integer such that $\overline{\mathfrak{m}}^r = (\overline{0})$. From the ascending chain

$$(\overline{0}) = \overline{\mathfrak{m}}^r \subset \overline{\mathfrak{m}}^{r-1} \subset \cdots \subset \overline{\mathfrak{m}}^2 \subset \overline{\mathfrak{m}} \subset \overline{S}$$

and using that $\overline{S}/\overline{\mathfrak{m}} = K$, we get

$$\ell(\overline{S}) = \sum_{i=0}^{r-1} \ell_{\overline{S}}(\overline{\mathfrak{m}}^i/\overline{\mathfrak{m}}^{i+1}) = \sum_{i=0}^{r-1} \ell_{\overline{S}/\overline{\mathfrak{m}}}(\overline{\mathfrak{m}}^i/\overline{\mathfrak{m}}^{i+1}) = \sum_{i=0}^{r-1} \dim_K(\overline{\mathfrak{m}}^i/\overline{\mathfrak{m}}^{i+1}).$$

As the last summation is equal to $\dim_K(\overline{S})$, we get that $\ell(\overline{S})$ is equal to $\dim_K(\overline{S})$. The equality $\ell_{S_\mathfrak{m}}(S_\mathfrak{m}/I_\mathfrak{m}) = \ell_S(S/I)$ follows readily by localizing at the maximal ideal $\mathfrak{m}$. $\square$

Remark 8.5.2 If $S = \oplus_{i=0}^{\infty} S_i$ has the standard grading and $I \subset S$ is a graded ideal, then

$$H_I^a(i) = \textstyle\sum_{k=0}^{i} \dim_K(S_k/I_k)$$

where $I_k = I \cap S_k$. Thus, one has $H_I(i) = \dim_K(S_i/I_i)$ for all i. There are isomorphisms of K-vector spaces

$$\overline{S}/\mathfrak{m}^{i+1}\overline{S} \simeq S/(\mathfrak{m}^{i+1} + I) \simeq S_{\leq i}/I_{\leq i},$$

where $\overline{S} = S/I$ and $\mathfrak{m} = (t_1, \ldots, t_q)$. By Lemma 8.5.1, one has

$$\ell_S(S/(\mathfrak{m}^{i+1} + I)) = \dim_K(S/(\mathfrak{m}^{i+1} + I)).$$

Thus, H_I^a is the Samuel or Hilbert–Samuel function of $\overline{S}$ with respect to $\mathfrak{m}$ (see Section 4.4 and [413, Definition B.3.1]).

Definition 8.5.3 The *graded reverse lexicographical order* (GRevLex for short) on the monomials of S is defined as $t^b \succ t^a$ if $\deg(t^b) > \deg(t^a)$ or $\deg(t^b) = \deg(t^a)$ and the last non-zero entry of $b - a$ is negative.

Lemma 8.5.4 *Let I be an ideal of S. Then the following hold:*

(a) $\dim(S[u]/I^h) = \dim(S/I) + 1$.

(b) $H_I^a(i) = H_{I^h}(i)$ *for* $i \geq 0$.

Proof. (a): Let $\prec$ be the GRevLex order and let $\mathcal{G}$ be a Gröbner basis of I. Then $\{g^h | \, g \in \mathcal{G}\}$ is a Gröbner basis of I^h and $\mathrm{in}_\prec(g) = \mathrm{in}_\prec(g^h)$ for $g \in \mathcal{G}$ (see Proposition 3.4.2). Since $\dim(S/I) = \dim(S/\mathrm{in}_\prec(I))$ and $\dim(S[u]/I) = \dim(S[u]/\mathrm{in}_\prec(I^h))$, the equality follows.

(b): Fix $i \geq 0$. The mapping $S[u]_i \to S_{\leq i}$ induced by mapping $u \mapsto 1$ is a K-linear surjection. Consider the induced composite K-linear surjection $S[u]_i \to S_{\leq i} \to S_{\leq i}/I_{\leq i}$. An easy check shows that this has kernel I_i^h. Hence, we have an isomorphism of K-vector spaces

$$S[u]_i/I_i^h \simeq S_{\leq i}/I_{\leq i} \;\Rightarrow\; H_I^a(i) = H_{I^h}(i). \qquad \square$$

The degree of affine algebras and lattice ideals Let d be the Krull dimension of S/I. By Lemma 8.5.4 and Proposition 5.1.6, there are unique polynomials $h_I^a(t) = \sum_{i=0}^{d} a_i t^i \in \mathbb{Q}[t]$ and $h_I(t) = \sum_{i=0}^{d-1} c_i t^i \in \mathbb{Q}[t]$ of degrees d and $d-1$, respectively, such that $h_I^a(i) = H_I^a(i)$ and $h_I(i) = H_I(i)$ for $i \gg 0$. By convention, the zero polynomial has degree -1. Notice that $d!\, a_d = (d-1)!\, c_{d-1}$ for $d \geq 1$. If $d = 0$, then $H_I^a(i) = \dim_K(S/I)$ for $i \gg 0$.

Definition 8.5.5 The integer $d!\, a_d$, denoted by $\deg(S/I)$ or $e(S/I)$, is called the *degree* or *multiplicity* of S/I.

Proposition 8.5.6 $\deg(S/I) = \deg(S[u]/I^h)$.

Proof. From Lemma 8.5.4(a), $\dim(S[u]/I^h)$ is equal to $\dim(S/I) + 1$. Hence, the equality follows from Lemma 8.5.4(b). $\square$

Lemma 8.5.7 *Let F be a field extension of K, let $B = F[t_1, \ldots, t_q]$ be a polynomial ring, and let I be an ideal of S, then* $\deg(S/I) = \deg(B/IB)$.

Proof. Let $\mathcal{G}$ be a Gröbner basis of I w.r.t the GRevLex order. Thanks to Buchberger's criterion, $\mathcal{G}$ is a Gröbner basis for IB. By Proposition 3.4.2, I^h and IB^h are both generated by $\mathcal{G}^h = \{g^h | \, g \in \mathcal{G}\}$, and $\mathcal{G}^h$ is a Gröbner basis for IB^h. Hence to show the equality $\deg(S/I) = \deg(B/IB)$ notice that, by Lemma 8.5.4, one has

$$H_I^a(i) = H_{I^h}(i) = \dim_K(S[u]/I^h)_i = \dim_F(B[u]/IB^h)_i = H_{IB^h}(i) = H_{IB}^a(i) \text{ for } i \geq 0. \qquad \square$$

For a lattice ideal $I_\rho(\mathcal{L})$ the degree depends only on the lattice $\mathcal{L}$ and is independent of the base field K and the character ρ. In Section 9.4 we show effective methods to compute the degree of any lattice ideal.

Proposition 8.5.8 *Let $\rho\colon \mathcal{L} \to K^*$ be a partial character of a lattice $\mathcal{L}$ in $\mathbb{Z}^q$ and let $I_{\mathbb{Q}}(\mathcal{L})$ be the lattice ideal over the field $\mathbb{Q}$. Then*
$$\deg(S/I_\rho(\mathcal{L})) = \deg(S/I(\mathcal{L})) = \deg(\mathbb{Q}[t_1,\ldots,t_q]/I_{\mathbb{Q}}(\mathcal{L})).$$

Proof. First we show the equality on the left. The ideal $I_\rho(\mathcal{L})$ contains no monomials (Theorem 8.2.2). Let $\mathcal{G} = \{g_1,\ldots,g_m\}$ be the reduced Gröbner basis of $I(\mathcal{L})$ w.r.t the GRevLex order $\succ$. We can write $g_i = t^{a_i^+} - t^{a_i^-}$ with $\mathrm{in}_\prec(g_i) = t^{a_i^+}$. We set $h_i = t^{a_i^+} - \rho(a_i)t^{a_i^-}$ and $\mathcal{H} = \{h_1,\ldots,h_m\}$. Next, we show that $\mathcal{H}$ is a Gröbner basis of $I_\rho(\mathcal{L})$. Let $f \neq 0$ be any non-pure binomial of $I_\rho(\mathcal{L})$. We claim that f reduces to zero w.r.t $\mathcal{H}$. By the Division Algorithm, $f = \sum_{i=1}^m f_ih_i + g$, where $\mathrm{in}_\prec(f) \succ \mathrm{in}_\prec(f_ih_i)$ for all i, and g is a non-pure binomial in $I_\rho(\mathcal{L})$ such that none of the terms of g is divisible by any of the terms $t^{a_1^+},\ldots,t^{a_m^+}$. Then we can write
$$g = \mu(t^a - \lambda t^b) = \mu t^\delta(t^{u_+} - \lambda t^{u_-}),$$
with $\mu,\lambda \in K^*$ and $a - b = u_+ - u_-$. As $t^{u_+} - \lambda t^{u_-}$ is in $I_\rho(\mathcal{L})$, by Lemma 8.2.6, $u = u_+ - u_-$ is in $\mathcal{L}$ and $\lambda = \rho(u)$. If $g \neq 0$, we obtain that $t^{u_+} - t^{u_-}$, being in $I(\mathcal{L})$, has one of its terms in $(t^{a_1^+},\ldots,t^{a_m^+}) = \mathrm{in}_\prec(I(\mathcal{L}))$, a contradiction. Thus, g must be zero, i.e., f reduces to zero w.r.t $\mathcal{H}$. This proves the claim. In particular, $I_\rho(\mathcal{L})$ is generated by $\mathcal{H}$. Note that the S-polynomial of h_i and h_j is a binomial; thus, by the claim, it reduces to zero with respect to $\mathcal{H}$. Therefore by Buchberger's criterion, $\mathcal{H}$ is a Gröbner basis of $I_\rho(\mathcal{L})$. Hence, $S/I(\mathcal{L})$ and $S/I_\rho(\mathcal{L})$ have the same degree.

Now we show the equality on the right. Let $\succ$ be the GRevLex order on S and $S_{\mathbb{Q}} = \mathbb{Q}[t_1,\ldots,t_q]$, and on the extensions $S[u]$ and $S_{\mathbb{Q}}[u]$. Let $\mathcal{G}_{\mathbb{Q}}$ be the reduced Gröbner basis of $I_{\mathbb{Q}}(\mathcal{L})$. We set $I = I(\mathcal{L})$ and $I_{\mathbb{Q}} = I_{\mathbb{Q}}(\mathcal{L})$. Notice that $\mathbb{Z} \subset K$ if $\mathrm{char}(K) = 0$ and $\mathbb{Z}_p \subset K$ if $p = \mathrm{char}(K) > 0$. In the second case, one has a map $\mathbb{Z} \mapsto K$, $1 \mapsto 1_K$. Hence $S_{\mathbb{Z}} = \mathbb{Z}[t_1,\ldots,t_q]$ embeds into S if $\mathrm{char}(K) = 0$ and $S_{\mathbb{Z}}$ maps into S if $p = \mathrm{char}(K) > 0$. If $\mathcal{G}$ denotes the image of $\mathcal{G}_{\mathbb{Q}}$ under either of these two maps, by Buchberger's criterion, it is seen that $\mathcal{G}$ is a Gröbner basis of I. Hence, by Proposition 3.4.2, $\mathcal{G}_{\mathbb{Q}}^h$ and $\mathcal{G}^h$ are Gröbner basis of $I_{\mathbb{Q}}^h$ and I^h, where $\mathcal{G}^h$ is the set of f^h with $f \in \mathcal{G}$. Thus, the standard monomials of $S_{\mathbb{Q}}[u]/I_{\mathbb{Q}}^h$ and $S[u]/I^h$ are the "same." Therefore, these two rings have the same Hilbert function and the same degree. Thus, by Proposition 8.5.6, the result follows. $\square$

Proposition 8.5.9 (Additivity of the degree) *If I is an ideal of S and $I = \mathfrak{q}_1 \cap \mathfrak{q}_2 \cap \cdots \cap \mathfrak{q}_m$ is an irredundant primary decomposition, then*
$$\deg(S/I) = \sum_{\mathrm{ht}(\mathfrak{q}_i)=\mathrm{ht}(I)} \deg(S/\mathfrak{q}_i).$$

Proof. We may assume that $\mathfrak{p}_1, \ldots, \mathfrak{p}_r$ are the associated primes of I of height $\mathrm{ht}(I)$ and that $\mathrm{rad}(\mathfrak{q}_i) = \mathfrak{p}_i$ for $i = 1, \ldots, r$. The proof is by induction on m. We set $J = \cap_{i=2}^m \mathfrak{q}_i$. There is an exact sequence of K-vector spaces

$$0 \to S_{\leq i}/(\mathfrak{q}_1 \cap J)_{\leq i} \stackrel{\varphi}{\to} S_{\leq i}/(\mathfrak{q}_1)_{\leq i} \oplus S/(J)_{\leq i} \stackrel{\phi}{\to} S_{\leq i}/(\mathfrak{q}_1 + J)_{\leq i} \to 0,$$

where $\varphi(\overline{f}) = (\overline{f}, -\overline{f})$ and $\phi(\overline{f}_1, \overline{f}_2) = \overline{f_1 + f_2}$. Hence

$$H^a_{\mathfrak{q}_1}(i) + H^a_J(i) = H^a_I + H^a_{\mathfrak{q}_1+J}(i). \tag{8.11}$$

As the decomposition of I is irredundant, by Exercise 8.5.10, one has $\dim(S/(\mathfrak{q}_1 + J)) < \dim(S/J)$. If $r = 1$, then $\dim(S/J) < \dim(S/I)$. Hence, from Eq. (8.11), we get that $\deg(S/I) = \deg(S/\mathfrak{q}_1)$. If $r > 1$, then $\dim(S/J) = \dim(S/I) = \dim(S/\mathfrak{q}_1)$. Hence, from Eq. (8.11), we get that $\deg(S/I)$ is $\deg(S/\mathfrak{q}_1) + \deg(S/J)$. Therefore, by induction, we get the required formula. $\square$

Exercises

8.5.10 Let $I \subset S$ be an ideal and let $I = \mathfrak{q}_1 \cap \mathfrak{q}_2 \cap \cdots \cap \mathfrak{q}_m$ be an irredundant primary decomposition. If $J = \cap_{i=2}^m \mathfrak{q}_i$, then $\dim(S/(\mathfrak{q}_1 + J)) < \dim(S/J)$.

8.5.11 Let $I \subset S$ be an ideal. If $\mathfrak{q}$ is a primary component of I of height $\mathrm{ht}(I)$, then $\mathfrak{q}^h$ is a primary component of I^h of height $\mathrm{ht}(I^h)$.

8.6 Vanishing ideals over finite fields

Let $K = \mathbb{F}_q$ be a finite field with q elements and X a subset of a projective space $\mathbb{P}^{s-1}$ over K. Let $S = K[t_1, \ldots, t_s]$ be a polynomial ring over the finite field K with the standard grading $S = \oplus_{d=0}^\infty S_d$ and let $I(X)$ be the *vanishing ideal* of X which is the ideal of S generated by the homogeneous polynomials of S that vanish at all points of X.

The following family arises in algebraic coding theory [348].

Proposition 8.6.1 [330] *$I(X)$ is a lattice ideal if and only if*

$$X = \{[(x^{v_1}, \ldots, x^{v_s})] \,|\, x_i \in K^* = K \setminus \{0\}\ \forall i\} \subset \mathbb{P}^{s-1}, \tag{8.12}$$

for some monomials $x^{v_i} := x_1^{v_{i1}} \cdots x_n^{v_{in}}$, $i = 1, \ldots, s$.

In what follows we assume that X is a subset of $\mathbb{P}^{s-1}$ *parameterized by monomials* as in Eq. (8.12).

The following formula allows us to compute the algebraic invariants of $I(X)$ (degree, regularity, Hilbert function) using *Macaulay*2 [199].

Theorem 8.6.2 [348] $I(X)=(\{t_i-x^{v_i}z\}_{i=1}^{s}\cup\{x_i^{q-1}-1\}_{i=1}^{n})\cap S$.

Lemma 8.6.3 *The degree of $S/I(X)$ is $|X|$.*

Proof. For $[P]\in X$, let $I_{[P]}$ be its vanishing ideal. Since $I(X)=\cap_{[P]\in X}I_{[P]}$ and since $\deg(S/I_{[P]})=1$ (see Exercise 8.6.10), the lemma follows from the additivity of the degree (see Proposition 8.5.9). $\Box$

In this context it is usual to denote the Hilbert function of $S/I(X)$ simply by H_X.

Proposition 8.6.4 *There is an integer $r\geq 0$ such that*

$$1=H_X(0)<H_X(1)<\cdots<H_X(r-1)<H_X(d)=|X| \quad \textit{for } d\geq r.$$

Proof. As $S/I(X)$ is Cohen–Macaulay of dimension 1, by Theorem 2.2.4, its Hilbert polynomial has degree 0. Hence, by Lemma 8.6.3, $H_X(d)=|X|$ for $d\gg 0$. Consequently the result follows readily from Theorem 5.6.4. $\Box$

Recall from Chapter 5 that the integer r is the index of regularity of $S/I(X)$ and that by Theorem 6.4.1 this number is the so-called Castelnuovo–Mumford regularity of $S/I(X)$. For these reasons r is simply called the *regularity* and is denoted by $\operatorname{reg}(S/I(X))$.

Definition 8.6.5 Let $X=\{[P_1],\ldots,[P_m]\}$ and let $f_0(t_1\ldots,t_s)=t_1^d$, where $d\geq 1$. The linear map of K-vector spaces:

$$\operatorname{ev}_d\colon S_d\rightarrow K^{|X|},\qquad f\mapsto\left(\frac{f(P_1)}{f_0(P_1)},\ldots,\frac{f(P_m)}{f_0(P_m)}\right)$$

is called an *evaluation map*. Notice that $\ker(\operatorname{ev}_d)=I(X)_d$. The image of ev_d, denoted by $C_X(d)$, is a *linear code* and $C_X(d)$ is called a *parameterized code* of order d.

The algebraic invariants of $S/I(X)$ are closely related to the so-called basic parameters of parameterized codes [120, 348].

Definition 8.6.6 The *basic parameters* of the linear code $C_X(d)$ are:

(b$_1$) $\dim_K C_X(d)=H_X(d)$, the *dimension*,

(b$_2$) $|X|=\deg(S/I(X))$, the *length*,

(b$_3$) $\delta_X(d)=\min\{\|v\|\colon 0\neq v\in C_X(d)\}$, the *minimum distance*, where $\|v\|$ is the number of non-zero entries of v.

Lemma 8.6.7 $\delta_X(d)=1$ *for* $d\geq\operatorname{reg}(S/I(X))$.

Proof. As $C_X(d)$ is a linear subspace of $K^{|X|}$ of dimension equal to $|X|$ for $d \geq \text{reg}(S/I(X))$, we get the equality $C_X(d) = K^{|X|}$. Thus $\delta_X(d)$ is equal to 1 for $d \geq \text{reg}(S/I(X))$. $\Box$

A good linear code should have large $|X|$ together with $\dim_K C_X(d)/|X|$ and $\delta_X(d)/|X|$ as large as possible. From Lemma 8.6.7 we conclude that the potentially good codes $C_X(d)$ can occur only if $1 \leq d < \text{reg}(S/I(X))$.

Problem 8.6.8 Find explicit formulas, in terms of the numerical data n, s, q, d, and the combinatorics of $x^{v_1}, \ldots, x^{v_s}$, for the *basic parameters*:

(a) $H_X(d)$, (b) $\deg(S/I(X))$, (c) $\delta_X(d)$, (d) $\text{reg}(S/I(X))$.

Formulas for (a)-(d) are known when X is a projective torus [367] and when X is parameterized by $x_1^{d_1}, \ldots, x_s^{d_s}, 1$, where $d_i \in \mathbb{N}_+$ for al i [192, 292].

Example 8.6.9 If $X = \{[(x_1^{90}, x_2^{36}, x_3^{20}, 1)] | \, x_i \in \mathbb{F}_{181}^* \text{ for } i = 1, 2, 3\}$, then $\text{reg}(S/I(X)) = 13$. The basic parameters of the family $\{C_X(d)\}_{d \geq 1}$ are:

d	1	2	3	4	5	6	7	8	9	10	11	12	13
$\|X\|$	90	90	90	90	90	90	90	90	90	90	90	90	90
$H_X(d)$	4	9	16	25	35	45	55	65	74	81	86	89	90
$\delta_X(d)$	45	36	27	18	9	8	7	6	5	4	3	2	1

Exercises

8.6.10 Let X be a finite subset of a projective space $\mathbb{P}^{s-1}$ over a field K. Let $[P]$ be a point in X, with $P = (\alpha_1, \ldots, \alpha_s)$ and $\alpha_k \neq 0$ for some k, and let $I_{[P]}$ be the ideal generated by the homogeneous polynomials of S that vanish at $[P]$. Prove that $I_{[P]}$ is a prime ideal of height $s - 1$,

$$I_{[P]} = (\{\alpha_k t_i - \alpha_i t_k | \, k \neq i \in \{1, \ldots, s\}),$$

$\deg(S/I_{[P]}) = 1$ and that $I(X) = \bigcap_{[Q] \in X} I_{[Q]}$.

8.7 Semigroup rings of numerical semigroups

Let $\mathcal{S} \neq (0)$ be a *semigroup* of $(\mathbb{N}, +)$, that is, $\mathcal{S}$ is a subset of $\mathbb{N}$ which is closed under addition and $0 \in \mathcal{S}$. Note that there exists a sequence $d_1 < \cdots < d_q$ of positive integers such that

$$\mathcal{S} = d_1\mathbb{N} + \cdots + d_q\mathbb{N}.$$

If $\gcd(d_1, \ldots, d_q) = 1$, $\mathcal{S}$ is called a *numerical semigroup* [360]. Let $K[x]$ be a polynomial ring in one variable over a field K, the semigroup ring of $\mathcal{S}$, denoted by $K[\mathcal{S}]$, is equal to the monomial subring $K[x^{d_1}, \ldots, x^{d_q}]$.

Lemma 8.7.1 *If $\mathcal{S}$ is a numerical semigroup, then $c+\mathbb{N}\subset\mathcal{S}$ for some c.*

Proof. There are integers $r_1,\ldots,r_q$ such that $1=r_1d_1+\cdots+r_qd_q$. The integer $c=(|r_1|d_1+\cdots+|r_q|d_q)d_q$ satisfies the required property. $\Box$

Definition 8.7.2 Let $\mathcal{S}$ be a numerical semigroup. The *Frobenius number* of $\mathcal{S}$, denoted by $g(\mathcal{S})$, is the largest integer not in $\mathcal{S}$. The semigroup $\mathcal{S}$ is called *symmetric* if $g(\mathcal{S})-z\in\mathcal{S}$ for all $z\notin\mathcal{S}$.

Proposition 8.7.3 *A numerical semigroup $\mathcal{S}$ is symmetric if and only if*

$$(g(\mathcal{S})+1)/2=|\mathbb{N}\setminus\mathcal{S}|.$$

Proof. We set $c=g(\mathcal{S})$. Consider the map $\varphi\colon\mathbb{N}\setminus\mathcal{S}\rightarrow[0,c]\cap\mathbb{N}$, $z\mapsto c-z$. Note that $\mathcal{S}$ is symmetric if and only if $\varphi(\mathbb{N}\setminus\mathcal{S})=\mathcal{S}\cap[0,c)$. Because of the decomposition $[0,c]=(\mathbb{N}\setminus\mathcal{S})\cup(\mathcal{S}\cap[0,c))$, one obtains that $\mathcal{S}$ is symmetric if and only if $c+1=2|\mathbb{N}\setminus\mathcal{S}|$. $\Box$

Proposition 8.7.4 [173] *Let $\mathcal{S}$ be a numerical semigroup. If*

$$T(\mathcal{S})=\{y\in\mathbb{N}\setminus\mathcal{S}\,|\,y+s\in\mathcal{S},\ \forall s\in\mathcal{S},\ s>0\},$$

then $\mathcal{S}$ is symmetric if and only if $T(\mathcal{S})=\{g(\mathcal{S})\}$.

Proof. $\Rightarrow$) We set $c=g(\mathcal{S})$. Let $y\in T(\mathcal{S})$ and assume $y<c$, then $0<c-y\in\mathcal{S}$ and $y+(c-y)=c$ is in $\mathcal{S}$, which is impossible, hence $y=c$.

$\Leftarrow$) Let $z\notin\mathcal{S}$, $z>0$, one must show $c-z\in\mathcal{S}$. If $c-z$ is not in $\mathcal{S}$, choose the least positive integer z such that $c-z$ is not in $\mathcal{S}$. Since $c-z\neq c$, by definition of $T(\mathcal{S})$ there is $s\in\mathcal{S}$, $s>0$ with $(c-z)+s$ not in $\mathcal{S}$, hence $z-s>0$, which contradicts the choice of z. $\Box$

The following nice result is due to Fröberg. It gives an expression for the Cohen–Macaulay type for the semigroup ring of a numerical semigroup.

Theorem 8.7.5 (Fröberg [173]) *Let $\mathcal{S}$ be a numerical semigroup. If*

$$T(\mathcal{S})=\{y\in\mathbb{N}\setminus\mathcal{S}\,|\,y+s\in\mathcal{S},\ \forall s\in\mathcal{S},\ s>0\},$$

then $|T(\mathcal{S})|$ is equal to the Cohen–Macaulay type of $K[\mathcal{S}]$.

Proof. We set $D=K[\mathcal{S}]$. Let $d_1<\cdots<d_q$ be a minimal generating set of $\mathcal{S}$ with $d_1,\ldots,d_q$ relatively prime and $s\in\mathcal{S}$, $s>0$. Consider the mapping

$$\varphi\colon T(\mathcal{S})\rightarrow\operatorname{Soc}(D/x^sD),\quad y\mapsto x^{y+s}+x^sD.$$

The type of $K[\mathcal{S}]$ is, by definition, equal to $\dim_K(\operatorname{Soc}(D/x^sD))$ (see the last paragraph of Section 2.3). Thus it is enough to show that the image of $T(\mathcal{S})$ is a K-vector space basis of $\mathrm{Soc}(D/x^sD)$, because φ is injective. Note that $\mathrm{Soc}(D/x^sD)$ has a K-basis consisting of monomials of the form $x^w + x^sD$, with $w - s \notin \mathcal{S}$ and $x^{d_i}(x^w + x^sD) = x^sD$ for all i, write $d_i + w = s + s_i$, where s_i is in $\mathcal{S}$.

We claim $w - s \in T(\mathcal{S})$. It follows that $(w-s)+s'$ is in $\mathcal{S}$ for all $s' \in \mathcal{S}$, $s' > 0$. Observe that $w - s > 0$; otherwise from $w - s = s_1 - d_1 = s_2 - d_2$ we derive $s_1 = 0$ and $d_2 = s_2 + d_1$, hence $d_2 = \mu d_1$, which is impossible because d_1 and d_2 are part of a minimal generating set of $\mathcal{S}$. Hence $w - s \in T(\mathcal{S})$ and $\varphi(w - s) = w$. To finish note that $\varphi(T(\mathcal{S}))$ is linearly independent. $\square$

Theorem 8.7.6 [287] *Let $\mathcal{S}$ be a numerical semigroup. Then the ring $K[\mathcal{S}]$ is Gorenstein if and only if $\mathcal{S}$ is symmetric.*

Proof. It follows from Proposition 8.7.4 and Theorem 8.7.5. $\square$

Proposition 8.7.7 *Let $\mathcal{S}$ be a numerical semigroup. If $K[\mathcal{S}] \subset K[x]$ has the induced grading, then the Frobenius number of $\mathcal{S}$ is equal to the degree, as a rational function, of the Hilbert series of $K[\mathcal{S}]$.*

Proof. Let $c = g(\mathcal{S})$ be the Frobenius number of $\mathcal{S}$. Since $K[\mathcal{S}]_i = Kx^i$ if $i \in \mathcal{S}$ and $K[\mathcal{S}]_i = (0)$ otherwise, one has

$$F(K[\mathcal{S}], z) = \sum_{i\in\mathcal{S}} z^i = f(z) + \sum_{i=c+1}^{\infty} z^i = f(z) + \frac{1}{(1-z)} - \sum_{i=0}^{c} z^i$$

where $f(z)$ is a polynomial with coefficients in $\{0,1\}$ of degree at most $c-1$. Hence $\deg(F(K[\mathcal{S}], z)) = c$. $\square$

Arithmetical symmetric semigroups Let $\mathcal{S}$ be a numerical semigroup of $\mathbb{N}$. The *problem of Frobenius* consists in determining $g(\mathcal{S})$, the Frobenius number of $\mathcal{S}$. This problem can be explained as follows. Given a sufficient supply of coins of various denominations find the largest amount that cannot be formed with these coins. The problem of computing $g(\mathcal{S})$ has been examined by several authors [173, 359, 365, 374]. In [364, 365] the famous formula of Scarf and Shallcross [368], for the Frobenius number, is expressed in algebraic terms using homological algebra and lattice theory.

Lemma 8.7.8 *Let $u, v \in \mathbb{N}_+$. If $\gcd(u, v) = 1$ and $\mathcal{S} = u\mathbb{N} + v\mathbb{N}$, then $g(\mathcal{S}) = uv - u - v$.*

Proof. We set $c = uv - u - v$. It is easy to see that c is not in $\mathcal{S}$. Hence, to show that $c = g(\mathcal{S})$, it suffices to show that $c + i \in \mathcal{S}$ for $i \in \mathbb{N}_+$. One can write $1 = \mu_1 u + \mu_2 v$ for some μ_1, μ_2 in $\mathbb{Z}$, hence

$$c + i = (v - 1 + i\mu_1)u + (-1 + i\mu_2)v.$$

By the division algorithm we can write $c+i=b_1u+b_2v$, where $0\leq b_2<u$. From the identity

$$c-b_1u-b_2v=(-b_1-1)u+(u-1-b_2)v=-i\notin\mathcal{S},$$

one concludes $b_1\geq 0$ and consequently $c+i\in\mathcal{S}$. □

Lemma 8.7.9 *Let $q\geq 3$ and $v\geq 1$ be integers and let $\mathcal{S}$ be the semigroup*

$$\mathcal{S}=d_1\mathbb{N}+\cdots+d_q\mathbb{N},$$

where $d_i=d_1+(i-1)v$ and $d_1=r+k(q-1)$ for some $k,r\in\mathbb{N}$. If $\gcd(d_1,v)=1$ and $2\leq r\leq q$, then $g(\mathcal{S})$ is equal to $(k+v)d_1-v$.

Proof. Setting $c=(k+v)d_1-v$, we will show that $c+i\in\mathcal{S}$ for $i\geq 1$. One may assume $1\leq i\leq d_1$. Notice

$$c+i=(k+1)d_1+\lambda+i,$$

where $\lambda=vd_1-d_1-v$. Consider $U=d_1\mathbb{N}+v\mathbb{N}$. By Lemma 8.7.8, λ is equal to $g(U)$. Hence $\lambda+i=ad_1+bv$, for some a,b in $\mathbb{N}$. One has

$$vd_1+i=(a+1)d_1+(b+1)v.$$

Hence $b\leq(q-1)(k+a+1)$ and we can write $b=j(k+a+1)+\ell$, for some $1\leq\ell\leq k+a+1$ and $0\leq j\leq q-2$. Therefore the equality

$$c+i=(k+a+1)d_1+bv=(k+a+1-\ell)d_{j+1}+\ell d_{j+2},$$

gives $c+i$ in $\mathcal{S}$. It remains to show $c\notin\mathcal{S}$. If $c\in\mathcal{S}$, then $c=\sum_{i=1}^q a_id_i$, for some a_i's in $\mathbb{N}$. Setting $s=\sum_{i=1}^q a_i$, note that one can rewrite c as:

$$c=(k+1)d_1+\lambda=a_1d_1+\cdots+a_qd_q=sd_1+pv,$$

for some $p\in\mathbb{N}$. Since $\lambda\notin U$, one derives $s\leq k$ and $c\leq sd_q\leq kd_q$, which is a contradiction since $c=kd_q+(r-1)v$. □

Theorem 8.7.10 [148, 265] *Let $r,k,q,v\in\mathbb{N}$ and let*

$$d_1=r+k(q-1)\quad\textit{and}\quad d_i=d_1+(i-1)v$$

be an arithmetical sequence. If $2\leq r\leq q\neq 2$ and $\gcd(d_1,v)=1$, then $r=2$ if and only if the semigroup $\mathcal{S}=d_1\mathbb{N}+\cdots+d_q\mathbb{N}$ is symmetric.

Proof. $\Rightarrow$) Let $d_1=2+(q-1)k$. Set $c=(k+v)d_1-v=kd_q+v$ and $U=d_1\mathbb{N}+v\mathbb{N}$. Assume $z\notin\mathcal{S}$. First consider the case $z\in U$, write $z=rd_1+sv$, $r,s\in\mathbb{N}$. By the Euclidean algorithm and Lemma 8.7.9 one can write $z=ad_q+bv$, where a,b are integers so that $0\leq a\leq k$

and $1 \leq b < d_1$. Writing $a = k - i$, where $0 \leq i \leq k$, we claim that $1 \leq b \leq (q-1)i + 1$. If $b > (q-1)i + 1$, then $z = x_1 d_1 + x_2 v$, with $x_1 = k + v - i$ and $x_2 = b - (q-1)i - 2$. Since $0 \leq x_2 \leq (q-1)x_1$, it is not hard to show that $z \in \mathcal{S}$, which is impossible. Hence $1 \leq b \leq (q-1)i + 1$. Set $y_1 = i$ and $y_2 = (q-1)i - b + 1$. Since $0 \leq y_2 \leq (q-1)y_1$ from the equality $c - z = y_1 d_1 + y_2 b$, one gets $c - z \in \mathcal{S}$. Now consider the case $z \notin U$. One can write $z = ad_1 + bv$, with $1 \leq b < d_1$ and $a < 0$. Set $w_1 = k+1$ and $w_2 = d_1 - b - 1$. Using $c - z = w_1 d_1 - (1+a)d_1 + w_2 v$ and $0 \leq w_2 \leq (q-1)w_1$, gives $c - z \in \mathcal{S}$. Therefore $\mathcal{S}$ is symmetric.

$\Leftarrow$) Assume that $\mathcal{S}$ is symmetric and keep the same notations as above. As $v \notin \mathcal{S}$ one has $c - v \in \mathcal{S}$ and $c - v = \sum_{i=1}^{q} b_i d_i$, where $b_i \in \mathbb{N}$ for all i. Set $a = \sum_{i=1}^{q} b_i$ and notice that

$$c - v = (k+1)d_1 + \lambda - v = ad_1 + q_1 v,$$

for some $q_1 \in \mathbb{N}$. Therefore $a \leq k$, otherwise one obtains $\lambda \in U$. From the inequality $c - v \leq ad_q \leq kd_q$, it follows that $r = 2$. $\square$

Exercises

8.7.11 Let $\mathcal{S}$ be a numerical semigroup of $\mathbb{N}$. Prove that there is a unique minimal set of generators $\{d_1, \ldots, d_q\}$ of $\mathcal{S}$ with $\gcd(d_1, \ldots, d_q) = 1$.

8.7.12 Let $\mathcal{S} \neq (0)$ be a semigroup of $\mathbb{N}$. Prove that the following conditions are equivalent:

(a) $\mathcal{S}$ is a numerical semigroup.

(b) $c + \mathbb{N} \subset \mathcal{S}$ for some $c \in \mathcal{S}$.

(c) $|\mathbb{N} \setminus \mathcal{S}| < \infty$.

8.7.13 Prove that the semigroup $\mathcal{S} = 4\mathbb{N} + 5\mathbb{N} + 6\mathbb{N}$ is symmetric.

8.7.14 If $\mathcal{S} = 5\mathbb{N} + 6\mathbb{N} + 7\mathbb{N} + 8\mathbb{N}$ and K is a field, prove that $K[\mathcal{S}]$ is a Gorenstein ring which is not a complete intersection.

8.7.15 Let $\mathcal{S} \subset \mathbb{N}$ be a numerical semigroup with Frobenius number c and let K be a field. If $D = K[\mathcal{S}] \subset K[x]$ has the induced grading, prove that the degree of the Hilbert series of $D/x^{c+1}D$ is $2c + 1$.

8.7.16 Let K be a field and let $D' = K[t_1, t_2]/(t_1^{d_2} - t_2^{d_1}, t_1^{a_1} t_2^{a_2})$, where t_i has degree $d_i \in \mathbb{N}_+$. If $\gcd(d_1, d_2) = 1$ and $c + 1 = a_1 d_1 + a_2 d_2$, prove that the Hilbert series of D' is

$$\frac{(1 - z^{c+1})(1 - z^{d_1 d_2})}{(1 - z^{d_1})(1 - z^{d_2})},$$

where c is the Frobenius number of the proper semigroup $\mathcal{S} = d_1\mathbb{N} + d_2\mathbb{N}$. Note $D' \simeq D/x^{c+1}D$, where $D = K[x^{d_1}, x^{d_2}]$.

8.8 Toric ideals of monomial curves

In this section we study toric ideals of monomial curves. Using binary trees, we characterize when the toric ideal of an affine monomial curve over an arbitrary field is a complete intersection. If the base field has positive characteristic, we show that 1-dimensional graded lattice ideals are set theoretic complete intersections.

Let $\mathcal{S} = \mathbb{N}d_1 + \cdots + \mathbb{N}d_q$ be a numerical semigroup of $\mathbb{N}$ generated by a sequence $\underline{d} = \{d_1, \ldots, d_q\}$ of relatively prime positive integers.

As usual $S = K[t_1, \ldots, t_q]$ denotes a polynomial ring over a field K. The kernel of the homomorphism of K-algebras

$$\varphi\colon S = K[t_1, \ldots, t_q] \longrightarrow K[\mathcal{S}] = K[x^{d_1}, \ldots, x^{d_q}],$$

induced by $\varphi(t_i) = x^{d_i}$, is called the *toric ideal* of $K[\mathcal{S}]$ and will be denoted by P. The map φ is graded if we endow S and $K[x]$ with the gradings induced by setting $\deg(t_i) = d_i$ and $\deg(x) = 1$.

The corresponding monomial curve in $\mathbb{A}_K^q$ parameterized by $t_i = x^{d_i}$, $i = 1, \ldots, q$, will be denoted by Γ. If $q = 3$, Γ is called a *monomial space curve.* In this context the ideal P is also called the toric ideal of Γ.

Lemma 8.8.1 *P is a binomial graded prime ideal of S of dimension* 1 *and $\Gamma = V(P)$. If K is an infinite field, then $I(\Gamma) = P$.*

Proof. As $K[x]$ is an integral domain, we get that P is prime. Since $K[x]$ is integral over $K[\mathcal{S}]$, by Proposition 2.4.13, we have $\mathrm{ht}(P) = q - 1$. By Corollary 8.2.18, the toric ideal P is generated by binomials. According to Proposition 8.4.8, if $\gcd(\underline{d}) = 1$, Γ is an affine toric variety, that is $\Gamma = V(P)$. If K is infinite, by Corollary 8.4.13, we get $I(\Gamma) = P$. □

Definition 8.8.2 A binomial $f_i = t_i^{m_i} - \prod_{j\neq i} t_j^{a_{ij}} \in P$ is called *critical* with respect to t_i or *t_i-critical* if m_i is the least positive integer such that

$$m_i d_i \in \sum_{j\neq i} d_j \mathbb{N}.$$

The notion of a critical binomial was first introduced by Eliahou [135], and later studied in [4]. This notion—and some of the results of this section—can be extended to binomial and lattice ideals [335].

Definition 8.8.3 A set $\{f_1, \ldots, f_q\}$ is called a *full set* of critical binomials if f_i is a t_i-critical binomial for all i.

The interest in studying ideals generated by critical binomials comes from a famous result of Herzog [216] showing that the toric ideal of any monomial space curve is generated by a full set of critical binomials (see Theorem 8.8.8). Herzog's result is no longer true for toric ideals of monomial curves in higher dimensions.

Example 8.8.4 [135] Let m be a positive integer. If $d_1 = m^2$, $d_2 = 2m^2-1$, $d_3 = 3m^2 + m$, $d_4 = 4m^2 + m - 1$, then $\mu(P) \geq m$, where $\mu(P)$ is the minimum number of generators of P.

If $\mathcal{S}$ is symmetric and generated non-redundantly by four integers, then P is minimally generated by 3 or 5 elements [53]; thus even in this case P is not generated by critical binomials. A main open problem is to determine when a given full set of critical binomials generates P. If the toric ideal of a monomial curve in $\mathbb{A}_K^4$ is generated by a full set of critical binomials, produced by any critical-binomial-algorithm, then one can predict how these generators should look [4].

In what follows, for simplicity of notation, f_i will stand for a t_i-critical binomial for $i = 1, \ldots, q$. Note that we do not assume $\pm f_1, \ldots, \pm f_q$ distinct.

Theorem 8.8.5 [216] *Let $\mathcal{B} = \{f_1, f_2, f_3\}$ be a full set of critical binomials and let $I = (\mathcal{B})$. If $f_1 = t_1^{a_1} - t_2^{a_2}t_3^{a_3}$, $f_2 = t_2^{b_2} - t_1^{b_1}t_3^{b_3}$, $f_3 = -t_1^{c_1}t_2^{c_2} + t_3^{c_3}$ and $b_i > 0, c_i > 0$ for all i, then*

(a) *$a_i > 0$ for all i and $a_1 = b_1 + c_1, b_2 = a_2 + c_2, c_3 = a_3 + b_3$.*

(b) *$\mathcal{B}$ is a Gröbner basis for I w.r.t the revlex order $t_1 \succ t_2 \succ t_3$.*

(c) *The ideal $I = (f_1, f_2, f_3)$ is equal to the toric ideal P of $K[\mathcal{S}]$.*

Proof. (a) If $a_2 = 0$, then $a_3 \geq c_3$. As $f_1 + t_3^{a_3-c_3}f_3 = t_1^{a_1} - t_1^{c_1}t_2^{c_2}t_3^{a_3-c_3}$, one obtains a contradiction, hence $a_2 > 0$, by a similar argument $a_3 > 0$. We set $\alpha = (-a_1, a_2, a_3)$, $\beta = (b_1, -b_2, b_3)$, $\gamma = (c_1, c_2, -c_3)$. Consider the vector $(v_1, v_2, v_3) = \alpha + \beta + \gamma$. We now show that $v_i \geq 0$ for all i. Note $a_1 > b_1$, otherwise using $t_1^{b_1-a_1}t_3^{b_3}f_1 + f_2$ one derives a contradiction. Hence, using $t_3^{b_3}f_1 + t_1^{a_1-b_1}f_2$, one has $a_3 + b_3 \geq c_3$ and $v_3 \geq 0$. Similarly, one shows that $v_1, v_2 \geq 0$. As $v_1d_1 + v_2d_2 + v_3d_3 = 0$, we get $v_i = 0$ for all i.

(b) It follows from part (a) and Buchberger's criterion (Theorem 3.3.17).

(c) Let f be a binomial in P and assume we order the monomials with the revlex ordering $t_1 \succ t_2 \succ t_3$. As f_1, f_2, f_3 are binomials, by the division algorithm (Theorem 3.3.6), we can write

$$f = h_1f_1 + h_2f_2 + h_3f_3 + r, \quad \text{where } r = t_1^{r_1}t_2^{r_2}t_3^{r_3} - t_1^{s_1}t_2^{s_2}t_3^{s_3} \in P,$$

$r_1, s_1 < a_1$, $r_2, s_2 < b_2$ and such that none of the monomials occurring in r is divisible by the leading term of f_3. It suffices to show that $r = 0$.

Assume $r \neq 0$. As $t_3^{r_3}$ or $t_3^{s_3}$ is a factor of r, one may assume that $s_3 = 0$. If $s_1 \leq r_1$ or $s_2 \leq r_2$ we get a contradiction with the minimality of b_2 or a_1, respectively; hence $s_1 > r_1$ and $s_2 > r_2$. From the equality $r = t_1^{r_1}t_2^{r_2}h$, where $h = t_3^{r_3} - t_1^{s_1-r_1}t_2^{s_2-r_2}$, we get $h \in P$ and $r_3 \geq c_3$. Using the identity

$$h - t_3^{r_3-c_3}f_3 = t_3^{r_3-c_3}t_1^{c_1}t_2^{c_2} - t_1^{s_1-r_1}t_2^{s_2-r_2}$$

and the minimality of b_2 and a_1 allows us to conclude $s_1 - r_1 > c_1$ and $s_2 - r_2 > c_2$, a contradiction because $t_1^{s_1}t_2^{s_2}$ is not divisible by $t_1^{c_1}t_2^{c_2}$. $\square$

Proposition 8.8.6 *If $f_1 = t_1^{a_1} - t_2^{a_2} = -f_2$, $f_3 = t_3^{c_3} - t_1^{c_1}t_2^{c_2}$ is a full set of critical binomials, then (f_1, f_3) is equal to the toric ideal P of $K[\mathcal{S}]$.*

Proof. Let f be a binomial in P and assume we order the terms with the revlex order $t_3 \succ t_1 \succ t_2$. As f_1, f_3 are binomials, by the division algorithm, we can write $f = h_1 f_1 + h_3 f_3 + r$, where $r = t_1^{r_1}t_2^{r_2}t_3^{r_3} - t_1^{s_1}t_2^{s_2}t_3^{s_3} \in P$, such that $r_1, s_1 < a_1$ and $r_3, s_3 < c_3$.

Assume $r \neq 0$. As $t_2^{r_2}$ or $t_2^{s_2}$ is a factor of r, one may assume that $s_2 = 0$. Note $s_1 > r_1$ and $s_3 > r_3$. From the equality $r = t_1^{r_1}t_3^{r_3}h$, where $h = t_2^{r_2} - t_1^{s_1-r_1}t_3^{s_3-r_3}$, we get $h \in P$ and $r_2 \geq a_2$. Therefore the identity

$$h + t_2^{r_2-a_2}f_1 = t_1^{a_1}t_2^{r_2-a_2} - t_1^{s_1-r_1}t_3^{s_3-r_3}$$

contradicts the choice of c_3 because one has the inequality $a_1 > s_1 - r_1$. $\Box$

Definition 8.8.7 The *support of a binomial* $f = t^\alpha - t^\beta$, denoted by $\mathrm{supp}(f)$, is defined as $\mathrm{supp}(t^\alpha) \cup \mathrm{supp}(t^\beta)$, where $\mathrm{supp}(t^\alpha) = \{t_i \mid \alpha_i > 0\}$.

The next result also holds for graded lattice ideals [335].

Theorem 8.8.8 [216] *If $\{f_1, f_2, f_3\}$ is a full set of critical binomials, then the ideal (f_1, f_2, f_3) is equal to the toric ideal P of $K[\mathcal{S}]$.*

Proof. Let $f_1 = t_1^{a_1} - t_2^{a_2}t_3^{a_3}$, $f_2 = t_2^{b_2} - t_1^{b_1}t_3^{b_3}$, $f_3 = -t_1^{c_1}t_2^{c_2} + t_3^{c_3}$. By Theorem 8.8.5(c) one may assume $\mathrm{supp}(f_i) \neq \{t_1, t_2, t_3\}$ for at least two f_i. Hence one may assume $a_3 = 0$ and the cases to consider are the following: (i) $b_1 = 0$, (ii) $b_3 = 0$, (iii) $c_1 = 0$, and (iv) $c_2 = 0$.

(i) Note $b_3 \geq c_3$, if $b_3 = c_3$, then f_2 is a critical binomial w.r.t t_3 and by Proposition 8.8.6 one concludes that P is generated by $\{f_1, f_2\}$. If $b_3 > c_3$, from the identity $f_2 + t_3^{b_3-c_3}f_3 = t_2^{b_2} - t_3^{b_3-c_3}t_1^{c_1}t_2^{c_2}$, we get $c_2 = 0$. Thus

$$f_2' = f_2 + t_3^{b_3-c_3}f_3 = t_2^{b_2} - t_3^{b_3-c_3}t_1^{c_1} \text{ and}$$
$$f_3' = f_3 + t_1^{c_1-a_1}f_1 = t_3^{c_3} - t_1^{c_1-a_1}t_2^{a_2}.$$

Applying Theorem 8.8.5(a) to f_1, f_2', f_3' yield $c_1 = a_1$. Hence, f_3 is t_1-critical. Using Proposition 8.8.6, we get that P is generated by $\{f_2, f_3\}$.

(ii) As $a_2 \geq b_2$ and $b_1 \geq a_1$, from $f_1 - t_2^{a_2-b_2}f_2 = t_1^{a_1} - t_2^{a_2-b_2}t_1^{b_1}$ we get $f_1 = -f_2$ and $P = (f_1, f_3)$ by Proposition 8.8.6.

One may now assume $b_1 > 0$ and $b_3 > 0$. (iii) Using $t_2^{c_2-b_2}f_2 + f_3$ we readily see that this case cannot occur. (iv) Using $t_2^{a_2-b_2}f_2 + f_1$ we derive a contradiction. $\Box$

Binary trees and complete intersections

Definition 8.8.9 A *binary tree* is a connected directed rooted tree such that: (i) two edges leave the root and every other vertex has either degree

1 or 3, (ii) if a vertex has degree 3, then one edge enters the vertex and the other two edges leave the vertex, and (iii) if a vertex has degree 1, then one edge enters the vertex. The vertices of degree 1 are called *terminal*.

Proposition 8.8.10 *If G is a binary tree with q terminal vertices, then the number of non-terminal vertices of G is $q-1$.*

Proof. It follows by induction on q. $\square$

Definition 8.8.11 A binary tree G is said to be *labeled by* $[q] := \{1,\ldots,q\}$ if its terminal vertices are labeled by $\{1\}$, ..., $\{q\}$. Extending this definition, we will also consider binary trees with q terminal vertices labeled by arbitrary finite subsets of $\mathbb{N}$ with q elements.

If G is a binary tree labeled by $[q]$ and $\mathbf{v}$ is a non-terminal vertex of G, consider $\mathbf{v}_1$ and $\mathbf{v}_2$, the two vertices of G such that

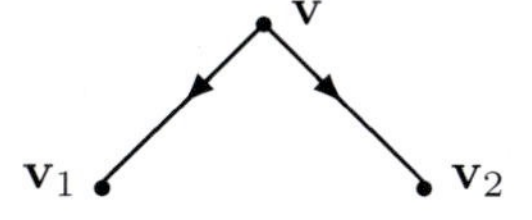

is a subgraph of G, and denote by G_1, resp. G_2, the subtree of G whose root is $\mathbf{v}_1$, resp. $\mathbf{v}_2$. We denote by $\ell_1[\mathbf{v}]$ and $\ell_2[\mathbf{v}]$ the two disjoint subsets of $[n]$ formed by the union of the labels of the terminal vertices of G_1 and G_2, respectively.

Definition 8.8.12 Let $\mathcal{B} = \{g_1,\ldots,g_{q-1}\}$ be a set of binomials of S with $g_i = t^{\alpha_i} - t^{\beta_i}$, $\text{supp}(t^{\alpha_i}) \cap \text{supp}(t^{\beta_i}) = \emptyset$, and $\alpha_i \neq 0$, $\beta_i \neq 0$ for all $i = 1,\ldots,q-1$, and let G be a binary tree labeled by $[q]$. We say that G is *compatible* with $\mathcal{B}$ if, denoting by $\mathcal{F}$ the set of non-terminal vertices of G, there is a bijection

$$\mathcal{B} \xrightarrow{f} \mathcal{F}$$

such that $\text{supp}(t^{\alpha_i}) \subset \ell_1[f(g_i)]$ and $\text{supp}(t^{\beta_i}) \subset \ell_2[f(g_i)]$ for all $i \in [q-1]$.

Example 8.8.13 The following binary tree G labeled by $[5]$ is compatible with $\mathcal{B} = \{g_1 = t_1^2t_2^4 - t_4t_5\,,\ g_2 = t_1 - t_2t_3\,,\ g_3 = t_4^4 - t_5^2\,,\ g_4 = t_2 - t_3^7\}$

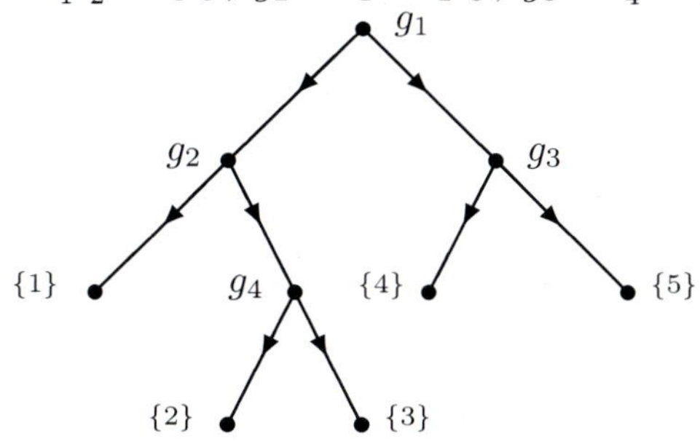

and if $\mathbf{v}$ is the root of G, then $\ell_1[\mathbf{v}] = \{1,2,3\}$ and $\ell_2[\mathbf{v}] = \{4,5\}$.

Theorem 8.8.14 *Let $\mathcal{B} = \{g_1, \ldots, g_{q-1}\}$ be a set of binomials of S such that $g_i = t^{\alpha_i} - t^{\beta_i}$, $\mathrm{supp}(t^{\alpha_i}) \cap \mathrm{supp}(t^{\beta_i}) = \emptyset$, and $\alpha_i \neq 0$, $\beta_i \neq 0$ for all $i = 1, \ldots, q-1$. Then the following two conditions are equivalent:*

(1) *$V(\mathcal{B}, t_i) = \{0\}$ for all $i = 1, \ldots, q$.*

(2) *There is a binary tree G labeled by $[q]$ which is compatible with $\mathcal{B}$.*

Proof. (1) $\Rightarrow$ (2): Set $V_1 := \{1\}$, ..., $V_q := \{q\}$ and consider the partition $\mathcal{F}_1 := \{V_1, \ldots, V_q\}$ of $[q]$. Let us show that there exist $V_{q+1}, \ldots, V_{2q-1}$, subsets of $[q]$, and $\mathcal{F}_2, \ldots, \mathcal{F}_q$, partitions of $[q]$, such that, reindexing $g_1, \ldots, g_{q-1}$ if necessary, the following assertions hold for all $i \in [q-1]$:

(a) $V_{q+i} = V_j \cup V_k$ for some $V_j, V_k \in \mathcal{F}_i$, $j \neq k$.

(b) $\mathrm{supp}(t^{\alpha_i}) \subset V_j$ and $\mathrm{supp}(t^{\beta_i}) \subset V_k$.

(c) $\mathcal{F}_{i+1} = (\mathcal{F}_i \setminus \{V_j, V_k\}) \cup \{V_{q+i}\}$.

Consider the digraph G with $2q-1$ vertices, denoted by $\mathbf{v}_1, \ldots, \mathbf{v}_{2q-1}$, where we connect $\mathbf{v}_{q+i}$ with $\mathbf{v}_j$ and $\mathbf{v}_k$ as follows:

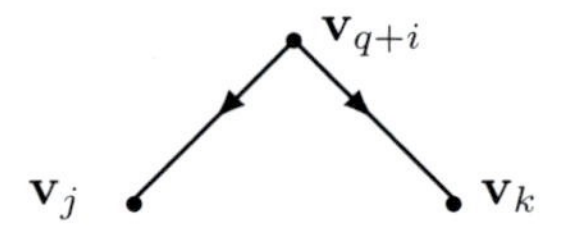

whenever $V_{q+i} = V_j \cup V_k$ in (a), it is not hard to see that G is a binary tree labeled by $[q]$. The root of G is $\mathbf{v}_{2q-1}$, and the set of its non-terminal vertices is $\mathcal{F} := \{\mathbf{v}_{q+1}, \ldots, \mathbf{v}_{2q-1}\}$. Moreover, by construction, for $i \in \{1, \ldots, q-1\}$, one has that $\ell_1[\mathbf{v}_{q+i}] = V_j$ and $\ell_2[\mathbf{v}_{q+i}] = V_k$ for V_j and V_k in (a). Hence, by (b), G is compatible with $\mathcal{B}$ via the map $f : \mathcal{B} \to \mathcal{F}$, $g_i \mapsto \mathbf{v}_{q+i}$, and (2) will follow.

Let us first construct V_{q+1} and $\mathcal{F}_2$ satisfying (a), (b), and (c). We first claim that for all $i \in [q]$, there exists an element $g_j \in \mathcal{B}$ such that either $\mathrm{supp}(t^{\alpha_j}) \subset V_i$ or $\mathrm{supp}(t^{\beta_j}) \subset V_i$ because otherwise, we have that the ith unit vector e_i of $\mathbb{A}_K^q$ belongs to $V(\mathcal{B}, t_1, \ldots, t_{i-1}, t_{i+1}, \ldots, t_q)$ which is $\{0\}$ by (1). Since $|\mathcal{F}_1| = q$ and $|\mathcal{B}| = q-1$, by the pigeonhole principle there exists an element in $\mathcal{B}$, say g_1, and $V_j, V_k \in \mathcal{F}_1$ with $j \neq k$, such that $\mathrm{supp}(t^{\alpha_1}) \subset V_j$ and $\mathrm{supp}(t^{\beta_1}) \subset V_k$. Setting $V_{q+1} := V_j \cup V_k$ and $\mathcal{F}_2 := (\mathcal{F}_1 \setminus \{V_j, V_k\}) \cup \{V_{q+1}\}$, (a), (b), and (c) hold for $i = 1$.

Assume now that for $i \in [\![2, q-1]\!]$, we have constructed $V_{q+1}, \ldots, V_{q+i-1}$ and $\mathcal{F}_2, \ldots \mathcal{F}_i$ such that (a), (b), and (c) hold, and let us construct V_{q+i} and $\mathcal{F}_{i+1}$ satisfying (a), (b), and (c).

Observe first that for all $j \leq i-1$, $\mathrm{supp}(g_j)$ is contained in some element of $\mathcal{F}_i$. Set $\mathcal{B}_i := \mathcal{B} \setminus \{g_1, \ldots, g_{i-1}\}$. We claim that for each $V_k \in \mathcal{F}_i$, there exists $g_j \in \mathcal{B}_i$ such that either $\mathrm{supp}(t^{\alpha_j}) \subset V_k$ or $\mathrm{supp}(t^{\beta_j}) \subset V_k$. In order to prove this, we show that if there exists an element in $\mathcal{F}_i$, say $V_s =$

$\{i_1, \ldots, i_m\}$, that does not satisfy the claim, then $\alpha := e_{i_1} + \cdots + e_{i_m}$ belongs to $V(\mathcal{B})$, which is a contradiction by (1). Take $g_j \in \mathcal{B}$. If $g_j \in \mathcal{B}_i$, then $\text{supp}(t^{\alpha_j}) \not\subset V_s$ and $\text{supp}(t^{\beta_j}) \not\subset V_s$ by definition of V_s, and hence $g_j(\alpha) = 0$. If $g_j \notin \mathcal{B}_i$, i.e., if $j \leq i-1$, then $\text{supp}(g_j)$ is contained in some element of $\mathcal{F}_i$, say V_t. If $t = s$, i.e., if $\text{supp}(g_j) \subset V_s$, then $g_j(\alpha) = 1 - 1 = 0$. Otherwise, since $\mathcal{F}_i$ is a partition of $[q]$ and $V_s, V_t \in \mathcal{F}_i$, one has that $V_s \cap V_t = \emptyset$, and hence $\text{supp}(g_j) \cap V_s = \emptyset$. Thus, $g_j(\alpha) = 0$, and the claim is proved.

We have proved that for each $V_k \in \mathcal{F}_i$, there exists $g_j \in \mathcal{B}_i$ such that either $\text{supp}(t^{\alpha_j}) \subset V_k$ or $\text{supp}(t^{\beta_j}) \subset V_k$. Since $|\mathcal{F}_i| = q - i + 1$ and $|\mathcal{B}_i| = q - i$, and using that $\mathcal{F}_i$ is a partition of $[q]$, we get by the pigeonhole principle that there exists an element in $\mathcal{B}_i$, say g_i, and $V_j, V_k \in \mathcal{F}_i$ such that $\text{supp}(t^{\alpha_i}) \subset V_j$ and $\text{supp}(t^{\beta_i}) \subset V_k$. Setting $V_{q+i} := V_j \cup V_k$ and $\mathcal{F}_{i+1} = (\mathcal{F}_i \setminus \{V_j, V_k\}) \cup \{V_{q+i}\}$, the statements (a), (b), and (c) hold, and we are done.

(2) $\Rightarrow$ (1): The proof is by induction on q. The result is clear if $q = 2$. Denoting by $\mathbf{v}$ the root of G, we may assume without loss of generality, that $\ell_1[\mathbf{v}] = [r]$ and $\ell_2[\mathbf{v}] = [\![r+1, q]\!]$ for some $r \in \{1, \ldots, q-1\}$. Then, if G_1 and G_2 are the two connected components of the digraph $G \setminus \{\mathbf{v}\}$ obtained from G by removing the vertex $\mathbf{v}$, one has that G_1 and G_2 are binary trees (for convenience we regard an isolated vertex as a binary tree), labeled by $[r]$ and $[\![r+1, q]\!]$, respectively. Reindexing the g_i's if necessary, we may also assume that G_1 is compatible with $\mathcal{B}_1 := \{g_2, \ldots, g_r\}$, G_2 is compatible with $\mathcal{B}_2 := \{g_{r+1}, \ldots, g_{q-1}\}$, and $g_1 = t^{\alpha_1} - t^{\beta_1}$ with $\text{supp}(t^{\alpha_1}) \subset [r]$ and $\text{supp}(t^{\beta_1}) \subset [\![r+1, q]\!]$. Then, $\text{supp}(g_i) \subset [r]$ if $i = 2, \ldots, r$, and $\text{supp}(g_i) \subset [\![r+1, q]\!]$ if $i = r+1, \ldots, q-1$. Moreover, applying the induction hypothesis, one has that $\{0\} = V(\mathcal{B}_1, t_i) \subset K^r$ for $i = 1, \ldots, r$, and $\{0\} = V(\mathcal{B}_2, t_i) \subset K^{q-r}$ for $i = r+1, \ldots, q$. Fix $i \in [q]$ and take $a \in V(\mathcal{B}, t_i)$. The result will be proved if we show that $a = 0$. By symmetry, we may assume that $1 \leq i \leq r$. The vector $a = (a_1, \ldots, a_q)$ can be decomposed as $a = b + c$, where $b = (a_1, \ldots, a_r, 0, \ldots, 0)$. Then $b \in V(\mathcal{B}_1, t_i)$, and hence $b = 0$. On the other hand, $g_1(a) = 0$ implies that $a_j = 0$ for some $j \in \{r+1, \ldots, q\}$. Thus $c \in V(\mathcal{B}_2, t_j)$ which is $\{0\}$, and hence $a = 0$, as required. $\square$

Complete intersections Let $\underline{d} = \{d_1, \ldots, d_q\}$ be a set of distinct positive integers and let $P \subset S$ be the corresponding toric ideal defined at the beginning of the section. The exact sequence

$$0 \longrightarrow \ker(\psi) \longrightarrow \mathbb{Z}^q \xrightarrow{\psi} \mathbb{Z} \longrightarrow 0; \quad e_i \overset{\psi}{\longmapsto} d_i$$

is related to P as follows. If $g = t^a - t^b$ is a binomial, then $g \in P$ if and only if $\widehat{g} = a - b \in \ker(\psi)$.

Definition 8.8.15 Let G be a binary tree labeled by $[q]$, and consider a set of vectors in $\mathbb{Z}^q$, $W = \{w_1, \ldots, w_{q-1}\}$. We say that G is *compatible* with W if G is compatible with the set of binomials $\{t^{w_i^+} - t^{w_i^-};\ i = 1, \ldots, q-1\}$.

Theorem 8.8.16 *Let G be a binary tree labeled by $[q]$ and denote by $\mathcal{F}$ the set of its non-terminal vertices. The following two conditions are equivalent:*

(1) *There exist vectors $w_1, \ldots, w_{q-1} \in \mathbb{Z}^q$ such that G is compatible with $W = \{w_1, \ldots, w_{q-1}\}$, and $\ker(\psi) = \mathbb{Z}W$.*

(2) *For all $\mathbf{v} \in \mathcal{F}$,*

$$\frac{\gcd(d_j\,;j \in \ell_1[\mathbf{v}])\gcd(d_j\,;j \in \ell_2[\mathbf{v}])}{\gcd(d_j\,;j \in \ell_1[\mathbf{v}] \cup \ell_2[\mathbf{v}])} \in \bigcap_{i=1}^{2} \mathbb{N}\{d_j\,;j \in \ell_i[\mathbf{v}]\}.$$

Proof. Let $\mathbf{v}$ be the root of G and let G_1 and G_2 be the two connected components of the digraph $G \setminus \{\mathbf{v}\}$ obtained from G by removing $\mathbf{v}$. We may assume that $\ell_1[\mathbf{v}] = [r]$ and $\ell_2[\mathbf{v}] = [\![r+1, q]\!]$ for some $r \in \{1, \ldots, q-1\}$. Then G_1 and G_2 are binary trees (for convenience we regard an isolated vertex as a binary tree) labeled by $[r]$ and $[\![r+1, q]\!]$. The result is clear if $q = 2$, and we will prove both implications by induction on q.

(1) $\Rightarrow$ (2): Reindexing the w_i's if necessary, we may assume that w_{q-1} is the element of W associated to $\mathbf{v}$ through the map that makes G compatible with W, and that $W_1 = \{w_1, \ldots, w_{r-1}\}$ and $W_2 = \{w_r, \ldots, w_{q-2}\}$ are the set of vectors in W such that G_i is compatible with W_i for $i = 1, 2$.

There is a decomposition $\mathbb{Z}^q = \mathbb{Z}^r \oplus \mathbb{Z}^{q-r}$, where $\mathbb{Z}^r := \mathbb{Z}^r \times \{0\}^{q-r}$ and $\mathbb{Z}^{q-r} := \{0\}^r \times \mathbb{Z}^{q-r}$. Consider the linear maps

$$\overline{\psi}_1 \colon \mathbb{Z}^q \longrightarrow \mathbb{Z}, \qquad \overline{\psi}_1(e_i) = \begin{cases} d_i & \text{if } 1 \le i \le r, \\ 0 & \text{if } r < i \le q, \end{cases}$$

and $\overline{\psi}_2 = \psi - \overline{\psi}_1$. Let ψ_1 (resp. ψ_2) be the restriction of $\overline{\psi}_1$ (resp. $\overline{\psi}_2$) to $\mathbb{Z}^r$ (resp. $\mathbb{Z}^{q-r}$). It is not hard to see that $\ker(\psi_1) = \mathbb{Z}W_1$ and $\ker(\psi_2) = \mathbb{Z}W_2$. Hence, setting

$$\begin{array}{lll} d = \gcd(d_1, \ldots, d_q), & d' = \gcd(d_1, \ldots, d_r), & d'' = \gcd(d_{r+1}, \ldots, d_q), \\ \underline{d} = \{d_1, \ldots, d_q\}, & \underline{d}' = \{d_1, \ldots, d_r\}, & \underline{d}'' = \{d_{r+1}, \ldots, d_q\}, \end{array}$$

and using induction we need only show $(d'd'')/d \in \mathbb{N}\underline{d}' \cap \mathbb{N}\underline{d}''$. For $1 \le j \le r$ and $r+1 \le k \le q$ we can write

$$\frac{d_k}{d}e_j - \frac{d_j}{d}e_k = \lambda_{jk}^1 w_1 + \cdots + \lambda_{jk}^{q-1} w_{q-1},$$

for some $\lambda_{kj}^1, \ldots, \lambda_{kj}^{q-1}$ in $\mathbb{Z}$. We can write the last vector in the set W as

$$w_{q-1} = (a_1, \ldots, a_r, -a_{r+1}, \ldots, -a_q) = w_{q-1}^+ - w_{q-1}^-.$$

Hence, we get $-(d_j/d)e_k = \lambda_{jk}^r w_r + \cdots + \lambda_{jk}^{q-1} w_{q-1}$. This implies

$$(d_j/d)d_k \quad = \quad \lambda_{jk}^{q-1}(a_{r+1}d_{r+1} + \cdots + a_q d_q).$$

Set $h = a_{r+1}d_{r+1} + \cdots + a_q d_q$. If we fix k and vary j, we get

$$\gcd((d_1/d)d_k, \ldots, (d_r/d)d_k) = \mu_k h \quad (\mu_k \in \mathbb{Z}) \;\Rightarrow\; d_k d' = \mu_k h d.$$

Therefore varying k yields

$$\gcd(d_{r+1}d', \ldots, d_q d') = \gcd(\mu_{r+1}hd, \ldots, \mu_q hd) = hd\mu \quad (\mu \in \mathbb{Z}).$$

As a consequence $(d'd'')/d = (hd\mu)/d \in \mathbb{N}\underline{d}''$. A symmetric argument gives $(d'd'')/d \in \mathbb{N}\underline{d}'$, as required.

(2) $\Rightarrow$ (1): By induction hypothesis there are

$$W_1 = \{w_1, \ldots, w_{r-1}\} \text{ and } W_2 = \{w_r, \ldots, w_{q-2}\}$$

such that G_i is compatible with W_i and $\ker(\psi_i) = \mathbb{Z}W_i$ for $i = 1, 2$. The result will be proved if we give $w_{q-1} \in \mathbb{Z}^q$ such that $\mathrm{supp}(w_{q-1}^+) \subset [r]$, $\mathrm{supp}(w_{q-1}^-) \subset [\![r+1, q]\!]$, and $\ker(\psi) = \mathbb{Z}W$ for $W = W_1 \cup W_2 \cup \{w_{q-1}\}$.

By hypothesis there are where $a_1, \ldots, a_q$ in $\mathbb{N}$ such that

$$\frac{d'd''}{d} = \frac{\gcd(d_1, \ldots, d_r)\gcd(d_{r+1}, \ldots, d_q)}{\gcd(d_1, \ldots, d_q)} = \sum_{i=1}^{r} a_i d_i = \sum_{i=r+1}^{q} a_i d_i.$$

Setting $w_{q-1} := (a_1, \ldots, a_r, -a_{r+1}, \ldots, -a_q)$ and $W := W_1 \cup W_2 \cup \{w_{q-1}\}$, we get $\mathrm{supp}(w_{q-1}^+) \subset [r]$ and $\mathrm{supp}(w_{q-1}^-) \subset [\![r+1, q]\!]$. Thus G is compatible with W. To complete the proof it remains to prove that $\mathbb{Z}W = \ker(\psi)$. Clearly $\mathbb{Z}W \subset \ker(\psi)$. To prove the reverse inclusion define

$$\sigma_{jk} = (d_j/d)e_k - (d_k/d)e_j; \qquad j, k \in [q].$$

By Exercise 8.3.41 the set $\{\sigma_{jk} | \, j, k \in [q]\}$ generates $\ker(\psi)$. Thus we need only show that $\sigma_{jk} \in \mathbb{Z}W$ for all $j, k \in [q]$. If $j, k \in [r]$ or $j, k \in [\![r+1, q]\!]$, then $\sigma_{jk} \in \ker(\psi_1) \subset \mathbb{Z}W$ or $\sigma_{jk} \in \ker(\psi_2) \subset \mathbb{Z}W$. Assume $j \in [r]$ and $k \in [\![r+1, q]\!]$. From the equalities

$$\begin{aligned} S_1 &= \sum_{i=1}^{r} a_i \left(\frac{d_i}{d'} e_j - \frac{d_j}{d'} e_i \right) = \frac{d''}{d} e_j - \frac{d_j}{d'} \sum_{i=1}^{r} a_i e_i, \\ S_2 &= \sum_{i=r+1}^{q} a_i \left(\frac{d_i}{d''} e_k - \frac{d_k}{d''} e_i \right) = \frac{d'}{d} e_k - \frac{d_k}{d''} \sum_{i=r+1}^{q} a_i e_i \end{aligned}$$

we conclude

$$\frac{d_k}{d''} S_1 - \frac{d_j}{d'} S_2 = \left(\frac{d_k}{d} e_j - \frac{d_j}{d} e_k \right) - \frac{d_j d_k}{d'd''} w_{q-1}.$$

Since $S_i \in \ker(\psi_i) \subset \mathbb{Z}W$ we obtain $\sigma_{jk} \in \mathbb{Z}W$, as required. $\square$

Notation Given a binomial $g = t^a - t^b$, we set $\widehat{g} = a - b$.

Theorem 8.8.17 *Let K be an arbitrary field and let $\mathcal{B} = \{g_1, \ldots, g_{q-1}\}$ be a set of binomials in the toric ideal P. Then $P = (\mathcal{B})$ if and only if*

(a) $\ker(\psi) = \mathbb{Z}\{\widehat{g}_1, \ldots, \widehat{g}_{q-1}\}$ *and*

(b) $V(g_1, \ldots, g_{q-1}, t_i) = \{0\}$ *for* $i = 1, \ldots, q$.

Proof. If $P = (\mathcal{B})$, then (a) follows at once from Proposition 8.3.1, and (b) follows from Corollary 8.3.21(b). Conversely, if (a) and (b) hold then by Corollary 8.3.21, one has $\mathrm{rad}(\mathcal{B}) = P$. Let $\{h_1, \ldots, h_s\}$ be a set of generators of P consisting of binomials. Notice that $\ker(\psi) = \mathbb{Z}\{\widehat{g}_1, \ldots, \widehat{g}_r\}$ by (a), and $\ker(\psi) = \mathbb{Z}\{\widehat{h}_1, \ldots, \widehat{h}_s\}$ by Proposition 8.3.1. Thus using that $(\mathcal{B})$ is a quasi homogeneous complete intersection and applying Proposition 8.3.17, $(\mathcal{B})$ is a radical ideal, and hence $P = (\mathcal{B})$. □

Theorem 8.8.18 [33] *The toric ideal P is a complete intersection if and only if there is a binary tree G labeled by $[q]$ such that, for all non-terminal vertex $\mathbf{v}$ of G, one has that*

$$\frac{\gcd(d_j,\, j \in \ell_1[\mathbf{v}])\gcd(d_j,\, j \in \ell_2[\mathbf{v}])}{\gcd(d_j,\, j \in \ell_1[\mathbf{v}] \cup \ell_2[\mathbf{v}])} \in \mathbb{N}\{d_j,\, j \in \ell_1[\mathbf{v}]\} \cap \mathbb{N}\{d_j,\, j \in \ell_2[\mathbf{v}]\}.$$

Proof. $\Rightarrow$) There are binomials $g_1, \ldots, g_{q-1}$ such that $P = (g_1, \ldots, g_{q-1})$. We may assume that $g_i = t^{\alpha_i} - t^{\beta_i}$ and $\mathrm{supp}(t^{\alpha_i}) \cap \mathrm{supp}(t^{\beta_i}) = \emptyset$ for all i. By Theorem 8.8.17(b) and Theorem 8.8.14 there exists a binary tree G labeled by $[q]$ which is compatible with $\{g_1, \ldots, g_{q-1}\}$. Then G is compatible with $W = \{\widehat{g}_1, \ldots, \widehat{g}_r\}$ and $\ker(\psi) = \mathbb{Z}\{\widehat{g}_1, \ldots, \widehat{g}_r\}$ (see Theorem 8.8.17(a)). Thus applying Theorem 8.8.16 we obtain the required conditions.

$\Leftarrow$) By Theorem 8.8.16, there is $W = \{w_1, \ldots, w_{q-1}\} \subset \mathbb{Z}^q$ such that W is compatible with G and $\ker(\psi) = \mathbb{Z}W$. Setting $g_i := t^{w_i^+} - t^{w_i^-}$, one has that G is compatible with $\{g_1, \ldots, g_{q-1}\}$. Hence, by Theorem 8.8.14, we get $V(g_1, \ldots, g_{q-1}, t_i) = \{0\}$ for $i = 1, \ldots, q$. Therefore, by Theorem 8.8.17, we deduce the equality $P = (g_1, \ldots, g_{q-1})$. □

There is a characterization of complete intersection semigroups of $\mathbb{N}$ given in [107] (see also [158] for a generalization of this description to semigroups of arbitrary dimension). The complete intersection property of toric ideals of monomial curves has been studied in [30] from a computational point of view (showing an efficient algorithm that checks this property). In the area of complete intersection toric ideals there are some other papers (see the introductions of [182, 184, 322] and the references therein).

Remark 8.8.19 If P is a complete intersection and G is a binary tree labeled by $[q]$ such that the conditions of Theorem 8.8.18 hold, then

(i) The generators $\{g_1, \ldots, g_{q-1}\}$ of P and their degrees $D_1, \ldots, D_{q-1}$ can be obtained as shown in the proofs of Theorems 8.8.16 and 8.8.18.

(ii) The Frobenius number $g(\mathcal{S})$ of a numerical semigroup $\mathcal{S} = \mathbb{N}\underline{d}$ can be expressed entirely in terms of $\{d_1, \ldots, d_q\}$.

This last assertion is a consequence of the following. Recall that the quasi-homogeneous Hilbert series of S/P is

$$F(S/P, z) = \frac{f(z)}{(1 - z^{d_1}) \cdots (1 - z^{d_q})}$$

for some polynomial $f \in \mathbb{Z}[z]$; see Theorem 5.1.4. When $d_1, \ldots, d_q$ are relatively prime, by Proposition 8.7.7 and its proof, one can write

$$F(S/P, z) = \frac{h(z)}{1 - z}$$

for some polynomial h of degree $g(\mathcal{S}) + 1$. If P is a complete intersection, one can write $f(z) = (1 - z^{D_1}) \cdots (1 - z^{D_{q-1}})$ where $D_1, \ldots, D_{q-1}$ are the degrees of the minimal generators of P (see Exercise 5.1.20), and hence $g(\mathcal{S}) = D_1 + \cdots + D_{q-1} - (d_1 + \cdots + d_q)$. Denoting by $\{\mathbf{v}_1, \ldots, \mathbf{v}_{q-1}\}$ the set of non-terminal vertices of G and using (i), we get :

$$g(\mathcal{S}) = \left(\sum_{i=1}^{q-1} \frac{\gcd(d_j,\, j \in \ell_1[\mathbf{v}_i]) \gcd(d_j,\, j \in \ell_2[\mathbf{v}_i])}{\gcd(d_j,\, j \in \ell_1[\mathbf{v}_i] \cup \ell_2[\mathbf{v}_i])} \right) - \left(\sum_{i=1}^{q} d_i \right).$$

Example 8.8.20 Let K be a field, and consider $d_1 = 16$, $d_2 = 27$, $d_3 = 45$, and $d_4 = 56$. The corresponding toric ideal $P \subset K[t_1, t_2, t_3, t_4]$ is a complete intersection because using the following binary tree labeled by [4]

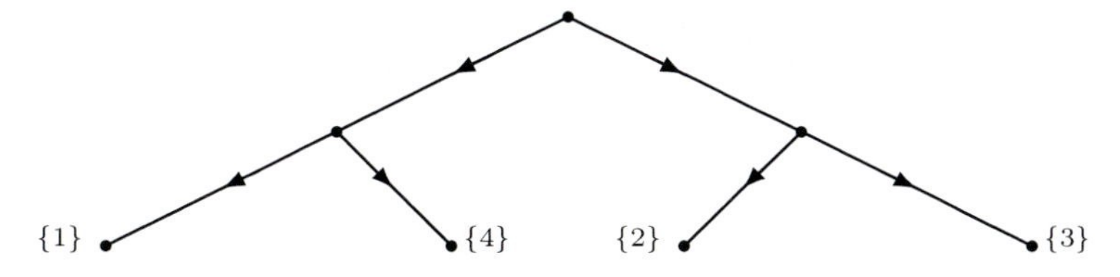

the arithmetical conditions in Theorem 8.8.18 are satisfied:

$$112 = \frac{(16)(56)}{\gcd(16, 56)} \in 16\mathbb{N} \cap 56\mathbb{N}; \qquad 2(56) \overset{(1)}{=} 7(16)$$

$$135 = \frac{(27)(45)}{\gcd(27, 45)} \in 27\mathbb{N} \cap 45\mathbb{N}; \qquad 3(45) \overset{(2)}{=} 5(27)$$

$$72 = \frac{\gcd(16, 56) \gcd(27, 45)}{\gcd(16, 27, 45, 56)} \in (16, 56)\mathbb{N} \cap (27, 45)\mathbb{N};$$

$$1(16) + 1(56) \overset{(3)}{=} 1(27) + 1(45).$$

Moreover, the equalities (1), (2) and (3) provide, by Remark 8.8.19(i), a set of minimal generators of P:

$$g_1 = t_4^2 - t_1^7, \quad g_2 = t_3^3 - t_2^5, \quad g_3 = t_1 t_4 - t_2 t_3.$$

Finally, by Remark 8.8.19(ii), the Frobenius number of the numerical semigroup $\mathcal{S} = \mathbb{N}\{16, 27, 45, 56\}$ is

$$g(\mathcal{S}) = 112 + 135 + 72 - (16 + 27 + 45 + 56) = 175\,.$$

Monomial curves in positive characteristic Let K be a field and let $P \subset S$ be the toric ideal of a monomial curve. In characteristic zero it is an open problem whether P is a set theoretic complete intersection. Several authors have studied this problem; see [135, 302, 319, 402]. The solution of the case $n = 3$ is treated in [54].

The next result gives a nice family of binomial set theoretic complete intersections. We show this result using a theorem of Katsabekis, Morales, and Thoma [270, Theorem 4.4(2)].

Proposition 8.8.21 [293] *If K is a field of positive characteristic and L is a graded lattice ideal of S of dimension* 1, *then L is a binomial set theoretic complete intersection.*

Proof. Let $\mathcal{L}$ be the homogeneous lattice of $\mathbb{Z}^q$ such that $L = I(\mathcal{L})$. Notice that $\mathcal{L}$ is a lattice of rank $q - 1$ because $\mathrm{ht}(L) = \mathrm{rank}(\mathcal{L})$. Thus, there is an isomorphism of groups $\psi\colon \mathbb{Z}^q/\mathcal{L}_s \to \mathbb{Z}$, where $\mathcal{L}_s$ is the saturation of $\mathcal{L}$ consisting of all $a \in \mathbb{Z}^q$ such that $da \in \mathcal{L}$ for some $0 \neq d \in \mathbb{Z}$. For each $1 \leq i \leq q$, we set $a_i = \psi(e_i + \mathcal{L}_s)$, where e_i is the ith unit vector in $\mathbb{Z}^q$. Following [270], the multiset $A = \{a_1, \ldots, a_q\}$ is called the configuration of vectors associated to $\mathcal{L}$. Recall that $q - 1 = \mathrm{rank}(\mathcal{L})$. Hence, as $\mathcal{L}$ is homogeneous with respect to $d = (d_1, \ldots, d_q)$, there are positive integers r_i and r_k such that $r_ie_i - r_ke_k \in \mathcal{L}$ and $r_id_i - r_kd_k = 0$. Thus, $r_ia_i = r_ka_k$ and a_i has the same sign as a_k. This means that $a_1, \ldots, a_q$ are all positive or all negative. It follows that A is a full configuration in the sense of [270, Definition 4.3]. Thus, $I(\mathcal{L})$ is a binomial set theoretic complete intersection by [270, Theorem 4.4(2)] and its proof. $\square$

Corollary 8.8.22 [319] *Let $P \subset S$ be the toric ideal of a monomial curve. If* $\mathrm{char}(K) > 0$, *then P is a binomial set theoretic complete intersection.*

Proof. By Lemma 8.8.1 P is a 1-dimensional graded lattice ideal. Thus, the result follows at once from Proposition 8.8.21. $\square$

Remark 8.8.23 Working over an algebraically closed field K of characteristic zero Eliahou [136] proved that the toric ideal P of a monomial curve in $\mathbb{A}_K^n$ is the radical of an ideal generated by n binomials.

Exercises

8.8.24 Let P be the toric ideal of $K[x^{d_1}, x^{d_2}, x^{d_3}]$, where K is a field and d_1, d_2, d_3 are relatively prime positive integers. Prove that P is a complete intersection if and only if P is Gorenstein.

8.8.25 Let $f_1 = t_1^{a_1} - t_2^{a_2}t_3^{a_3}, f_2 = t_2^{b_2} - t_1^{b_1}t_3^{b_3}, f_3 = -t_1^{c_1}t_2^{c_2} + t_3^{c_3}$ be a full set of critical binomials. If $b_i > 0, c_i > 0$ for all i, prove the equalities $d_1 = a_2b_3 + a_3b_2$, $d_2 = a_1b_3 + a_3b_1$, $d_3 = a_1b_2 - a_2b_1$.

8.8.26 (a) If $d_1 = 7$, $d_2 = 8$, $d_3 = 9$, $d_4 = 17$. Prove that

$$f_1 = t_1^5 - t_3^2t_4,\ f_2 = t_2^2 - t_1t_3,\ f_3 = t_3^4 - t_1^4t_2,\ f_4 = t_4 - t_2t_3$$

is a set of critical binomials that generate a height 3 prime ideal of type 2.

(b) If $d_1 = 204$, $d_2 = 855$, $d_3 = 1216$, $d_4 = 1260$. Prove that

$$f_1 = t_1^{47} - t_2^4t_3^3t_4^2,\ f_2 = t_2^8 - t_1^{15}t_4^3,\ f_3 = t_3^6 - t_1^{19}t_2^4,\ f_4 = t_4^5 - t_1^{13}t_3^3$$

is a set of critical binomials that generate a height 3 prime ideal of type 3.

(c) If $d_1 = 9$, $d_2 = 12$, $d_3 = 18$, $d_4 = 19$. Prove that

$$f_1 = -f_3 = t_1^2 - t_3,\ f_2 = t_2^3 - t_3^2,\ f_4 = t_4^3 - t_1t_2t_3^2$$

is a complete intersection prime ideal.

8.8.27 Let $f_1 = t_1^{a_1} - t_2^{a_2}t_3^{a_3}$, $f_2 = t_2^{b_2} - t_1^{b_1}t_3^{b_3}$, $f_3 = -t_1^{c_1}t_2^{c_2} + t_3^{c_3}$ be a full set of critical binomials and $I = (f_1, f_2, f_3)$. If $b_i > 0, c_i > 0$ for all i and $S = K[t_1, t_2, t_3]$ has the grading induced by $\deg(t_i) = d_i$, then the minimal resolution of S/I is given by

$$\begin{array}{l} 0 \to S(-d_2b_2 - d_3a_3) \oplus S(-d_3c_3 - d_2a_2) \to \\ \qquad S(-d_1a_1) \oplus S(-d_2b_2) \oplus S(-d_3c_3) \to S \to S/I \to 0 \end{array}$$

and $\max\{d_2a_2 + d_3c_3 - (d_1 + d_2 + d_3), d_2b_2 + d_3a_3 - (d_1 + d_2 + d_3)\}$ is the Frobenius number of $\mathcal{S} = \mathbb{N}\{d_1, d_2, d_3\}$.

Hint Use Theorem 8.8.5(b) and Proposition 8.7.7.

8.8.28 Let K be a field, let P be the toric ideal of $K[t^{d_1}, \ldots, t^{d_q}]$ and let

$$I = (\{t_i^{d_j} - t_j^{d_i} \mid 1 \leq i < j \leq q\}).$$

If $\gcd(d_1, \ldots, d_q) = 1$ and Γ is the monomial curve in $\mathbb{A}_K^q$ parameterized by $t_i = x^{d_i}$, then $\mathrm{rad}(I) = P$ and $\Gamma = V(I)$.

8.8.29 Let K be an algebraically closed field and let X be the curve in $\mathbb{A}_K^q$ given parametrically by $t_i = f_i(x)$, where $f_i(x) \in K[x] \setminus K$ for $i = 1, \ldots, q$. Use the extension theorem [99, Chapter 3] to prove that $V(P) = X$, where P is the presentation ideal of $K[f_1(x), \ldots, f_q(x)]$.

8.8.30 Prove that part (c) of Theorem 8.8.5 is valid for arbitrary distinct positive integers d_1, d_2, d_3.

8.8.31 [216] If P is the toric ideal of $K[x^{d_1}, x^{d_2}, x^{d_3}]$, prove that P is a set theoretic complete intersection.

8.8.32 A rational polyhedral cone $C \subset \mathbb{R}^n$ is *unimodular* if there exists $\Gamma = \{\gamma_1, \ldots, \gamma_n\}$ a $\mathbb{Z}$-basis of $\mathbb{Z}^n$ such that $C = \mathbb{R}_+\mathcal{B}$ for some $\mathcal{B} \subset \Gamma$. A lattice polytope $\mathcal{P} \subset \mathbb{R}^n$ with vertex set $\mathcal{A}$ is *smooth* [61, p. 371] if the cone $\mathbb{R}_+(\mathcal{A} - v)$ is unimodular for any $v \in \mathcal{A}$, where $\mathcal{A} - v = \{a - v | \, a \in \mathcal{A}\}$. Let $\mathcal{P} \subset \mathbb{R}^n$ be a lattice polytope of dimension n and let

$$K[\mathcal{P}] := K[\{x^a t | \, a \in \mathbb{Z}^n \cap \mathcal{P}\}] \subset K[x_1, \ldots, x_n, t]$$

be its *polytopal subring*, where $K[x_1, \ldots, x_n, t]$ is a polynomial ring over a field K. An open problem posed by Bøgvad is whether for a smooth normal polytope $\mathcal{P}$ the toric ideal of $K[\mathcal{P}]$ is generated by quadrics (see [61, 400]). If $n = 1$, prove that $\mathcal{P}$ is smooth, $K[\mathcal{P}]$ is normal, and $\deg(K[\mathcal{P}]) = \mathrm{vol}(\mathcal{P})$.

Hint Use Theorem 9.3.5, Lemma 9.3.7, and Theorem 9.3.25.

8.8.33 If $\mathcal{P} = \mathrm{conv}(v_1, v_1 + (s - 1)) \subset \mathbb{R}$, with $v_1 \in \mathbb{N}_+$ and $s \geq 3$ an integer, prove that the toric ideal P of $K[\mathcal{P}]$ is the ideal $I_2(M)$ generated by the 2×2 minors of the following $2 \times (s - 1)$ generic Hankel matrix

$$M = \begin{pmatrix} t_1 & t_2 & \cdots & t_{s-1} \\ t_2 & t_3 & \cdots & t_s \end{pmatrix}.$$

Hint The ideal $I_2(M)$ is a prime ideal of height $s - 2$ [128, p. 612].

8.8.34 If $\mathcal{P} = \mathrm{conv}((-1, -1), (1, 0), (0, 1)) \subset \mathbb{R}^2$, prove that the polytope $\mathcal{P}$ is not smooth, $K[\mathcal{P}]$ is normal, and the toric ideal P of $K[\mathcal{P}]$ is generated by $t_1t_2t_3 - t_4^3$. Then prove that the point $[(0, 0, 1, 0)]$ is a singular point of the projective toric variety $V(P) \subset \mathbb{P}^3$ defined by P.

Hint Use the Jacobian criterion of [210, 5.8, p. 37].

Chapter 9

Monomial Subrings

This chapter deals with point configurations and their lattice polytopes. We consider various subrings and ideals associated with them, e.g., Ehrhart rings, Rees algebras, homogeneous subrings, lattice and toric ideals, matrix and Laplacian ideals of graphs.

Throughout we shall use the following symbology and terminology which is consistent with the notation introduced thus far:

K; R	field; Laurent polynomial ring $K[x_1^{\pm 1},\ldots,x_n^{\pm 1}]$
$\mathcal{A}; F$	$\mathcal{A}=\{v_1,\ldots,v_q\}\subset\mathbb{Z}^n$; $F=\{x^v\in R\,\vert\, v\in\mathcal{A}\}$
If $\mathcal{A}\subset\mathbb{N}^n$	set $R=K[\mathbf{x}]=K[x_1,\ldots,x_n]$, $I=(F)\subset R$
A	$n\times q$ matrix with column vectors $v_1,\ldots,v_q$
$\mathcal{P}:=\mathrm{conv}(\mathcal{A})\subset\mathbb{R}^n$	lattice polytope
$R[t]$	the polynomial ring $K[x_1,\ldots,x_n,t]$
$A(\mathcal{P})$	Ehrhart ring $K[\{x^at^i\,\vert\, a\in\mathbb{Z}^n\cap i\mathcal{P}, i\in\mathbb{N}\}]$
$K[\mathcal{P}]$	polytopal subring $K[\{x^at\,\vert\, a\in\mathcal{P}\cap\mathbb{Z}^n\}]$
$R[Ft]$	Rees algebra $R[x^{v_1}t,\ldots,x^{v_q}t]$
$K[F]$	monomial subring $K[x^{v_1},\ldots,x^{v_q}]$
S	polynomial ring $K[\mathbf{t}]=K[t_1,\ldots,t_q]$
$P_F\subset S$; $I(\mathcal{L})\subset S$	toric ideal of $K[F]$; lattice ideal of $\mathcal{L}\subset\mathbb{Z}^q$
$\mathrm{in}_\prec(P_F)$	initial ideal of P_F with respect to $\prec$
$K[Ft]$	monomial subring $K[x^{v_1}t,\ldots,x^{v_q}t]$
$\overline{K[Ft]}$	integral closure of $K[Ft]$.

Following [146, 189, 383], we shall be interested in comparing these monomial subrings and their algebraic properties and invariants. One has the following inclusions:

$$K[Ft]\subset K[\mathcal{P}]\subset A(\mathcal{P})\subset\overline{R[Ft]}\quad\text{and}\quad\overline{K[Ft]}\subset A(\mathcal{P}).$$

Using Hilbert functions, polyhedral geometry and Gröbner bases, we will study normalizations of monomial subrings as well as initial ideals, special generating sets, primary decompositions and multiplicities of lattice and toric ideals. The reciprocity law of Ehrhart for integral polytopes and the Danilov–Stanley formula for canonical modules of monomial subrings will be introduced here. Applications of these results will be given.

9.1 Integral closure of monomial subrings

Here we present a description of the integral closure of a monomial subring using polyhedral geometry and integer programming techniques.

Normality of monomial subrings If A is an integral domain with field of fractions K_A, recall that the *normalization* or *integral closure* of A is the subring $\overline{A}$ consisting of all the elements of K_A which are integral over A.

Normal monomial subrings arise in the theory of toric varieties [100]. The next result is essentially shown in the book of Fulton [176, pp. 29–30].

Theorem 9.1.1 *If $\mathcal{A} \subset \mathbb{Z}^n$ is a finite set of points and $\mathcal{S} = \mathbb{Z}\mathcal{A} \cap \mathbb{R}_+\mathcal{A}$, then the following hold:*

(a) $K[\mathcal{S}] := K[\{x^a \mid a \in \mathcal{S}\}]$ *is normal,*

(b) $\overline{K[F]} = K[\mathcal{S}]$, *where* $F = \{x^a \mid a \in \mathcal{A}\}$.

Proof. (a): By Theorem 1.1.29, there are non-zero vectors $a_1, \ldots, a_p$ in $\mathbb{Z}^n$ such that $\mathbb{R}_+\mathcal{A} = H_{a_1}^+ \cap \cdots \cap H_{a_p}^+$. We set $\mathcal{S}_i = \mathbb{Z}\mathcal{A} \cap H_{a_i}^+$. Since $K[\mathcal{S}]$ is equal to $K[\mathcal{S}_1] \cap \cdots \cap K[\mathcal{S}_p]$ and since the intersection of normal domains is a normal domain, it suffices to show that $K[\mathcal{S}_i]$ is normal for all i. If $\mathbb{Z}\mathcal{A} \subset H_{a_i}$, then $\mathcal{S}_i = \mathbb{Z}\mathcal{A} \simeq \mathbb{Z}^r$. Hence, $K[\mathcal{S}_i] \simeq K[\mathbb{Z}^r] = K[x_1^{\pm 1}, \ldots, x_r^{\pm 1}]$ which is normal by Corollary 4.3.9. Thus, we may assume that $\mathbb{Z}\mathcal{A} \not\subset H_{a_i}$. Setting $r = \operatorname{rank}(\mathbb{Z}\mathcal{A})$ and $\mathcal{L} = \mathbb{Z}\mathcal{A} \cap H_a$, one has $\operatorname{rank}(\mathcal{L}) = r - 1$. This follows by noticing that $\mathbb{Q}\mathcal{L} = \mathbb{Q}\mathcal{A} \cap H_a$, and using the equality

$$n = \dim_{\mathbb{Q}}(\mathbb{Q}\mathcal{A} + H_a) = \dim_{\mathbb{Q}}(\mathbb{Q}\mathcal{A}) + \dim_{\mathbb{Q}}(H_a) - \dim_{\mathbb{Q}}(\mathbb{Q}\mathcal{A} \cap H_a)$$

together with the fact that the ranks of $\mathbb{Z}\mathcal{A}$ and $\mathcal{L}$ are equal to $\dim_{\mathbb{Q}}(\mathbb{Q}\mathcal{A})$ and $\dim_{\mathbb{Q}}(\mathbb{Q}\mathcal{L})$, respectively. The quotient group $H = \mathbb{Z}\mathcal{A}/\mathcal{L}$ is torsion-free of rank 1. Thus, H is a free abelian group of rank 1 and we can write $H = \mathbb{Z}\overline{\alpha}$ for some is $0 \neq \alpha \in \mathbb{Z}\mathcal{A}$ such that $\langle \alpha, a_i \rangle > 0$. As a consequence, we get that $\mathcal{S}_i = \mathcal{L} \oplus \mathbb{N}\alpha \simeq \mathbb{Z}^{r-1} \oplus \mathbb{N}$ and $K[\mathcal{S}_i] \simeq K[\mathbb{Z}^{r-1} \oplus \mathbb{N}]$. Therefore $\mathcal{S}_i$ is normal because $K[\mathbb{Z}^{r-1} \oplus \mathbb{N}]$ is equal to $K[x_1^{\pm 1}, \ldots, x_{r-1}^{\pm 1}, x_r]$ and this ring is normal again by Corollary 4.3.9.

(b): Since $K[F] \subset K[\mathcal{S}]$, taking integral closures and using part (a) gives the inclusion $\overline{K[F]} \subset K[\mathcal{S}]$. To show the reverse inclusion, note the equality $\mathbb{Z}\mathcal{A} \cap \mathbb{R}_+\mathcal{A} = \mathbb{Z}\mathcal{A} \cap \mathbb{Q}_+\mathcal{A}$ (see Corollary 1.1.27). A straightforward calculation shows that x^α is in the field of fractions of $K[F]$ if $\alpha \in \mathbb{Z}\mathcal{A}$, and x^α is an integral element over $K[F]$ if $\alpha \in \mathbb{Q}_+\mathcal{A}$. Hence $K[\mathcal{S}] \subset \overline{K[F]}$. $\square$

Theorem 9.1.2 *Let $\mathcal{A}$ be a finite set of points in $\mathbb{Z}^n$. If $\mathbb{R}_+\mathcal{A}$ is a pointed cone, then the integral closure of $K[F]$ in $R = K[\mathbf{x}^{\pm 1}]$ is equal to*

$$K[\{x^a |\, a \in \mathbb{Z}^n \cap \mathbb{R}_+\mathcal{A}\}].$$

Proof. We set $B = K[\{x^a|\, a \in \mathbb{Z}^n \cap \mathbb{R}_+\mathcal{A}\}]$. Let $\overline{K[F]}$ be the integral closure of $K[F]$ in R. To show $\overline{K[F]} \subset B$, take $z \in \overline{K[F]}$. One can write $z = d_1 x^{\gamma_1} + \cdots + d_r x^{\gamma_r}$, where $d_i \in K \setminus \{0\}$ for all i and $\gamma_1, \ldots, \gamma_r$ distinct non-zero points in $\mathbb{Z}^n$. Write $\gamma_i = n_i \beta_i$, with n_i equal to the gcd of the entries of γ_i (observe that $\beta_1, \ldots, \beta_r$ are not necessarily distinct). Consider the cone C spanned by $\mathcal{A}$ and $\{\beta_1, \ldots, \beta_r\}$. It suffices to show that $\beta_i \in \mathbb{R}_+\mathcal{A}$ for all i. Assume on the contrary that C is not equal to $\mathbb{R}_+\mathcal{A}$. By Theorem 1.1.54 one may assume that $\mathbb{R}_+\beta_1$ is a face of C not contained in $\mathbb{R}_+\mathcal{A}$. Let H_a be a hyperplane so that $H_a \cap C = \mathbb{R}_+\beta_1$ and $C \subset H_a^-$. There is $1 \leq \ell \leq r$ such that

$$n_\ell = \sup_{1 \leq j \leq r} \{n_j |\, \gamma_j = n_j \beta_1\}.$$

Since z is integral over $K[F]$ it satisfies a monic polynomial f of degree m with coefficients in $K[F]$. The monomials that occur in the expansion of $f(z)$ as a sum of monomials are of the form $x^\alpha (x^{\gamma_{i_1}})^{m_{i_1}} \cdots (x^{\gamma_{i_t}})^{m_{i_t}}$, where $m_{i_j} > 0$ for all j, $m \geq \sum_{j=1}^t m_{i_j}$ and $\alpha \in \mathbb{R}_+\mathcal{A}$. To derive a contradiction we claim that the term $x^{m\gamma_\ell}$ occurs only once in the expansion of $f(z)$ as a sum of monomials. Assume the equality

$$(x^{\gamma_\ell})^m = x^\alpha (x^{\gamma_{i_1}})^{m_{i_1}} \cdots (x^{\gamma_{i_t}})^{m_{i_t}},$$

where $m_{i_j} > 0$ for all j, $m \geq \sum_{j=1}^t m_{i_j}$, and $\alpha \in \mathbb{R}_+\mathcal{A}$. As $\gamma_\ell = n_\ell \beta_1$, one has $\langle m\gamma_\ell, a\rangle = 0$ and from this equality one rapidly derives

$$\langle \alpha, a\rangle = \langle \beta_{i_1}, a\rangle = \cdots = \langle \beta_{i_t}, a\rangle = 0.$$

Hence $\alpha, \beta_{i_i}, \ldots, \beta_{i_t} \in H_a \cap C = \mathbb{R}_+\beta_1$. Note $\beta_1 \notin \mathbb{R}_+\mathcal{A}$. Thus $\alpha = 0$ and $\beta_{i_j} = \beta_1$ for all j. Therefore $t = 1$ and $\gamma_\ell = \gamma_{i_1}$, as claimed. Altogether we get $f(z) \neq 0$, which is impossible.

Now we show the inclusion $B \subset \overline{K[F]}$. By Corollary 1.1.27 one has the equality $\mathbb{Z}^n \cap \mathbb{R}_+\mathcal{A} = \mathbb{Z}^n \cap \mathbb{Q}_+\mathcal{A}$. A straightforward calculation shows that x^a is an integral element over $K[F]$ if $a \in \mathbb{Q}_+\mathcal{A} \cap \mathbb{Z}^n$. Hence $B \subset \overline{K[F]}$. $\square$

Corollary 9.1.3 (a) *$K[F]$ is normal if and only if $\mathbb{N}\mathcal{A} = \mathbb{Z}\mathcal{A} \cap \mathbb{R}_+\mathcal{A}$.*

(b) *$\overline{K[F]} = K[x^{\gamma_1}, \ldots, x^{\gamma_r}]$ for some $\gamma_1, \ldots, \gamma_r$ in $\mathbb{Z}\mathcal{A} \cap \mathbb{Q}_+\mathcal{A}$.*

(c) *$\overline{K[F]} = x^{\beta_1} K[F] + \cdots + x^{\beta_m} K[F]$ for some $x^{\beta_1}, \ldots, x^{\beta_m}$ in $\overline{K[F]}$.*

(d) *If A is t-unimodular, then $K[F]$ is normal.*

Proof. (a): It suffices to observe the equality $\overline{K[F]} = K[\{x^\alpha | \alpha \in \mathbb{N}\mathcal{A}\}]$ and to use the description of $\overline{K[F]}$ given in Theorem 9.1.1(b).

(b): It follows from Corollary 1.1.27, Lemma 1.3.2, and Theorem 9.1.1.

(c): By part (b) $\overline{K[F]}$ is generated, as a $K[F]$-algebra, by a finite set of Laurent monomials. As $\overline{K[F]}$ is integral over $K[F]$, by Corollary 2.4.4, we get that $\overline{K[F]}$ is finite over $K[F]$ and the result follows readily.

(d): It follows from Theorem 1.6.7 and part (a). □

Definition 9.1.4 A decomposition $K[F] = \bigoplus_{i=0}^{\infty} K[F]_i$ of the K-vector space $K[F]$ is an *admissible grading* if $K[F]$ is a positively graded K-algebra with respect to this decomposition and each component $K[F]_i$ has a finite K-basis consisting of monomials.

Theorem 9.1.5 (Danilov, Stanley [65, Theorem 6.3.5]) *Let $\mathcal{A} \subset \mathbb{Z}^n$ be a finite set of points and let $\mathrm{ri}(\mathbb{R}_+\mathcal{A})$ be the relative interior of $\mathbb{R}_+\mathcal{A}$. If $K[F]$ is normal and $\mathbb{R}_+\mathcal{A}$ is a pointed cone, then the canonical module $\omega_{K[F]}$ of $K[F]$, with respect to an admissible grading, can be expressed as*

$$\omega_{K[F]} = (\{x^a \,|\, a \in \mathbb{N}\mathcal{A} \cap \mathrm{ri}(\mathbb{R}_+\mathcal{A})\}). \tag{9.1}$$

The formula above represents the canonical module of $K[F]$ as an ideal of $K[F]$ generated by Laurent monomials. For a comprehensive treatment of the Danilov–Stanley theorem, see [61, 65, 102].

Theorem 9.1.6 (Hochster [248], [65, Theorem 6.3.5]) *If $\mathcal{A} \subset \mathbb{Z}^n$ is a finite set of points and $K[F]$ is normal, then $K[F]$ is Cohen–Macaulay.*

Lemma 9.1.7 *Let $\mathcal{A} \subset \mathbb{Z}^n$ be a finite set. If $K[F]$ is a normal positively graded K-algebra and $\omega_{K[F]}$ is the canonical module of $K[F]$, then*

$$a(K[F]) = -\min\{\, i \,|\, (\omega_{K[F]})_i \neq 0\}. \tag{9.2}$$

Proof. Let P_F be the toric ideal of $K[F]$. Recall that $K[F]$ is isomorphic, as a K-algebra, to $K[t_1, \ldots, t_q]/P_F$, where $K[t_1, \ldots, t_q]$ has the grading induced by $\deg(t_i) = \deg(x^{v_i})$. By Theorem 9.1.6, the algebra $K[F]$ is Cohen–Macaulay. Therefore by Proposition 5.2.3 the formula follows. □

Theorem 9.1.8 *If $\mathcal{A} \subset \mathbb{N}^n$ and $K[\underline{h}] := K[h_1, \ldots, h_d] \hookrightarrow K[F]$ is a homogeneous Noether normalization of $K[F]$, then $\overline{K[F]}$ is a free $K[\underline{h}]$-module whose generators have degree at most $\sum_{i=1}^{d} \deg(h_i)$.*

Proof. Let $R = \oplus_{i \geq 0} R_i$ be the standard grading of R and endow $\overline{K[F]}$ with the grading $K[F]_i = K[F] \cap R_i$. The composite $K[\underline{h}] \hookrightarrow K[F] \hookrightarrow \overline{K[F]}$ is a homogeneous Noether normalization. Since $\overline{K[F]}$ is Cohen–Macaulay (see Theorem 9.1.6), by Proposition 3.1.27, we get a direct sum decomposition

$$\overline{K[F]} = K[\underline{h}]\overline{x}^{\beta_1} \oplus \cdots \oplus K[\underline{h}]\overline{x}^{\beta_m}.$$

Therefore the Hilbert series of $\overline{K[F]}$ can be expressed as

$$F(\overline{K[F]}, t) = \sum_{i=1}^{m} F(K[\underline{h}]x^{\beta_i}, t) = \left(\sum_{i=1}^{m} t^{\deg \beta_i}\right) \Big/ \prod_{i=1}^{d}(1 - t^{\deg h_i}).$$

Applying Theorem 9.1.5 one derives that $a(\overline{K[F]})$ is negative. Using that $a(\overline{K[F]})$ is equal to the degree of $F(\overline{K[F]}, t)$ as a rational function (see Proposition 5.2.3) one concludes $\deg x^{\beta_i} \leq \sum_{i=1}^{m} \deg h_i$. □

Applications of the Danilov–Stanley theorem Next we introduce some techniques to compute the canonical module and the a-invariant of a wide class of monomial subrings.

Let $\mathcal{A}$ be a finite set of points in $\mathbb{Z}^n$. The *dual cone* of $\mathbb{R}_+\mathcal{A}$ is the polyhedral cone given by

$$(\mathbb{R}_+\mathcal{A})^* = \{x \mid \langle x, y\rangle \geq 0;\ \forall\, y \in \mathbb{R}_+\mathcal{A}\}.$$

A set $\mathcal{H} \subset \mathbb{Z}^n \setminus \{0\}$ is called an *integral basis* of $(\mathbb{R}_+\mathcal{A})^*$ if $(\mathbb{R}_+\mathcal{A})^* = \mathbb{R}_+\mathcal{H}$.

Theorem 9.1.9 *Let $c_1, \ldots, c_r$ be an integral basis of $(\mathbb{R}_+\mathcal{A})^*$ and let $b = (b_i)$ be the vector given by $b_i = 0$ if $\mathbb{R}_+\mathcal{A} \subset H_{c_i}$ and $b_i = -1$ if $\mathbb{R}_+\mathcal{A} \not\subset H_{c_i}$. Assume there is $x_0 \in \mathbb{Q}^n$ such that $\langle x_0, v_i\rangle = 1$ for all i. If $\mathbb{N}\mathcal{A} = \mathbb{Z}^n \cap \mathbb{R}_+\mathcal{A}$ and B is the matrix with column vectors $-c_1, \ldots, -c_r$, then*

(a) $\omega_{K[F]} = (\{x^a \mid a \in \mathbb{Z}^n \cap \{x \mid xB \leq b\}).$

(b) $a(K[F]) = -\min\,\{\langle x_0, x\rangle \mid\, x \in \mathbb{Z}^n \cap \{x \mid xB \leq b\}\}.$

Proof. $K[F]$ is a standard graded K-algebra with the grading induced by declaring that a monomial $x^a \in K[F]$ has degree i if and only if $\langle a, x_0\rangle = i$. We set $\mathcal{H} = \{c_1, \ldots, c_r\}$. As $\mathbb{R}_+\mathcal{H} = (\mathbb{R}_+\mathcal{A})^* = H_{v_1}^+ \cap \cdots \cap H_{v_q}^+$, by duality (see Corollary 1.1.30), we have the equality

$$\mathbb{R}_+\mathcal{A} = H_{c_1}^+ \cap \cdots \cap H_{c_r}^+. \tag{9.3}$$

Observe that $\mathbb{R}_+\mathcal{A} \cap H_{c_i}$ is a proper face if $b_i = -1$ and it is an improper face otherwise. From Eq. (9.3) we get that each facet of $\mathbb{R}_+\mathcal{A}$ has the form $\mathbb{R}_+\mathcal{A} \cap H_{c_i}$ for some i. The relative boundary of the cone $\mathbb{R}_+\mathcal{A}$ is the union of its facets (see Theorem 1.1.44). Hence, using that $\mathcal{H}$ is an integral basis, we obtain the equality

$$\mathbb{Z}^n \cap (\mathbb{R}_+\mathcal{A})^{\mathrm{o}} = \mathbb{Z}^n \cap \{x \mid xB \leq b\}. \tag{9.4}$$

Now, part (a) follows readily from Eq. (9.1) of Theorem 9.1.5 and Eq. (9.4). Part (b) follows from Eq. (9.2) of Lemma 9.1.7 and part (a). □

Definition 9.1.10 If x^a is a monomial of $K[\mathbf{x}]$, we set $\log(x^a) = a$. The vector a is called the *exponent vector* of x^a. Given a set F of monomials, the *log set* of F, denoted $\log(F)$, consists of all $\log(x^a)$ with $x^a \in F$.

Example 9.1.11 If $F = \{x_1, \ldots, x_4, x_1x_2x_5, x_2x_3x_5, x_3x_4x_5, x_1x_4x_5\}$ and $\mathcal{A} = \log(F)$, then $\mathcal{A}$ is a Hilbert basis and $\langle x_0, v\rangle = 1$ for $v \in \mathcal{A}$, where $x_0 = (1,1,1,1,-1)$. An integral basis for $(\mathbb{R}_+\mathcal{A})^*$ is given by

$$\{e_1, e_2, e_3, e_4, e_5, (0,1,0,1,-1), (1,0,1,0,-1)\}.$$

Then, using Theorem 9.1.9, it is easy to verify that $\omega_{K[F]}$ is generated by the set of all x^a such that $a = (a_i)$ satisfies the system of linear inequalities:

$$a_i \geq 1\,\forall\, i; \quad a_1 + a_3 - a_5 \geq 1; \quad a_2 + a_4 - a_5 \geq 1. \qquad (*)$$

The only vertex of the polyhedron defined by Eq. $(*)$ is $v_0 = (1,1,1,1,1)$. Thus, by Theorem 9.1.9, the a-invariant of $K[F]$ is $-\langle x_0, v_0\rangle = -3$.

Let $\mathbb{R}_+\mathcal{A} = \mathrm{aff}(\mathbb{R}_+\mathcal{A}) \cap H^-_{a_1} \cap \cdots \cap H^-_{a_r}$ be an irreducible representation of $\mathbb{R}_+\mathcal{A}$ such that $a_i \in \mathbb{Z}^n$ and the non-zero entries of a_i are relatively prime for all i. The existence of this representation follows from Theorem 1.1.29.

Definition 9.1.12 If C is the integral matrix with rows $a_1, \ldots, a_r$, the polyhedron

$$\mathcal{Q} = \mathrm{aff}(\mathbb{R}_+\mathcal{A}) \cap \{x \in \mathbb{R}^n \,|\, Cx \leq -\mathbf{1}\}$$

is called the *shift polyhedron* of $\mathbb{R}_+\mathcal{A}$ relative to C.

If $\mathbb{Z}^n \cap \mathrm{ri}(\mathbb{R}_+\mathcal{A}) \neq \emptyset$, then $\mathcal{Q} \neq \emptyset$. The shift polyhedron will be used to give a technique to compute the canonical module of $K[F]$.

Example 9.1.13 A shift polyhedron of a cone is a shift of the cone as drawn below.

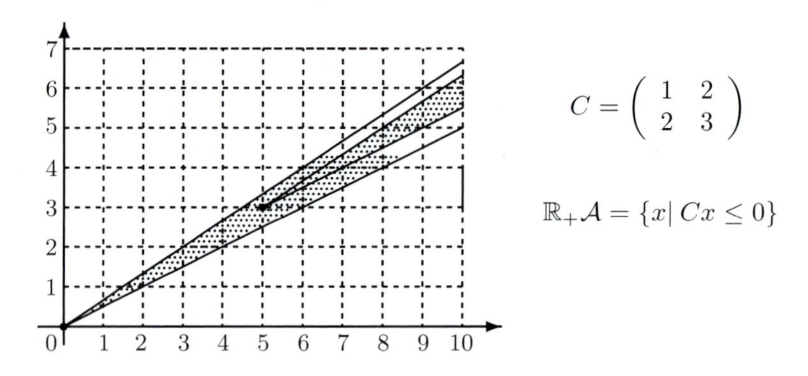

Here the shift polyhedron of the cone $\mathbb{R}_+\mathcal{A}$ is $\mathcal{Q} = \{x \,|\, Cx \leq -\mathbf{1}\}$.

Proposition 9.1.14 *If $\mathcal{Q}$ is the shift polyhedron of $\mathbb{R}_+\mathcal{A}$ with respect to C and $\mathbb{R}_+\mathcal{A}$ is a pointed cone, then $\mathcal{Q}$ is a pointed polyhedron,*

$$\mathbb{Z}^n \cap \mathcal{Q} = \mathbb{Z}^n \cap \operatorname{ri}(\mathbb{R}_+\mathcal{A}), \tag{9.5}$$

and $\operatorname{conv}(\mathbb{Z}^n \cap \operatorname{ri}(\mathbb{R}_+\mathcal{A}))$ *is an integral polyhedron if* $\mathbb{Z}^n \cap \operatorname{ri}(\mathbb{R}_+\mathcal{A}) \neq \emptyset$. *If $\mathcal{Q}$ is integral, then* $\mathcal{Q} = \operatorname{conv}(\mathbb{Z}^n \cap \operatorname{ri}(\mathbb{R}_+\mathcal{A}))$.

Proof. It is not hard to see that the lineality spaces of $\mathbb{R}_+\mathcal{A}$ and $\mathcal{Q}$ are equal. As $\mathbb{R}_+\mathcal{A}$ is pointed, so is $\mathcal{Q}$ by Exercise 1.1.80. To prove Eq. (9.5) it suffices to note that a point α is in $\operatorname{ri}(\mathbb{R}_+\mathcal{A})$ if and only if $\alpha \in \mathbb{R}_+\mathcal{A}$ and $\langle \alpha, a_i\rangle < 0$ for $i = 1, \ldots, r$, see Theorem 1.1.44. The third assertion follows by taking convex hulls in Eq. (9.5) and observing that $\operatorname{conv}(\mathbb{Z}^n \cap \mathcal{Q})$ is the integer hull of $\mathcal{Q}$, which is a polyhedron by Proposition 1.1.65(a) and is integral by Proposition 1.1.65(b). The last assertion follows by taking convex hulls in Eq. (9.5) and using that $\mathcal{Q}$ is integral. □

Theorem 9.1.15 *Let $K[F]$ be a monomial subring such that* (a) *there is* $0 \neq x_0 \in \mathbb{Q}^n$ *such that* $\langle x_0, v_i\rangle = 1$ *for all i,* (b) $\mathbb{N}\mathcal{A} = \mathbb{Z}^n \cap \mathbb{R}_+\mathcal{A}$, *and* (c) *the shift polyhedron $\mathcal{Q}$, relative to C, of the cone $\mathbb{R}_+\mathcal{A}$ is integral. Then*

(i) *$K[F]$ is a normal standard graded K-algebra whose canonical module and a-invariant are given by*

$$\omega_{K[F]} = (\{x^a \mid a \in \mathbb{Z}^n \cap \mathcal{Q}\}) \text{ and } a(K[F]) = -\min\{\langle x_0, x\rangle \mid x \in \mathcal{Q}\}.$$

(ii) *x^β is a minimal generator of $\omega_{K[F]}$ for any vertex β of $\mathcal{Q}$.*

Proof. (i): $K[F]$ is graded as follows: a monomial x^a with $a \in \mathbb{N}\mathcal{A}$ has degree i if and only if $\langle a, x_0\rangle = i$. Using (a), it is seen that $K[F]$ is a standard graded K-algebra. That $K[F]$ is normal follows from (b). The canonical module is the ideal of $K[F]$ given by

$$\begin{aligned} \omega_{K[F]} &\stackrel{(9.1)}{=} (\{x^a \mid a \in \mathbb{N}\mathcal{A} \cap \operatorname{ri}(\mathbb{R}_+\mathcal{A})\}) \\ &\stackrel{(b)}{=} (\{x^a \mid a \in \mathbb{Z}^n \cap \operatorname{ri}(\mathbb{R}_+\mathcal{A})\}) \qquad (9.6) \\ &\stackrel{(9.5)}{=} (\{x^a \mid a \in \mathbb{Z}^n \cap \mathcal{Q}\}). \qquad (9.7) \end{aligned}$$

To show the formula for the a-invariant observe that a monomial x^a of $K[F]$ has degree $\langle x_0, a\rangle$. Thus from Eq. (9.7) and Lemma 9.1.7 we obtain

$$\min\{\langle x_0, x\rangle \mid x \in \mathcal{Q}\} \leq -a(K[F]).$$

As the minimum is attained at a vertex β of $\mathcal{Q}$ (see Proposition 1.1.41), to show equality, it suffices to observe that β is integral (see Corollary 1.1.64).

(ii): By condition (c), β is integral. If $\mathcal{F}$ is a proper face of $\mathbb{R}_+\mathcal{A}$, then $\beta \notin \mathcal{F}$ because β is in $\operatorname{ri}(\mathbb{R}_+\mathcal{A})$ (see Theorem 1.1.44). Then, by Eq. (9.6), we get $x^\beta \in \omega_{K[F]}$. There are $c \in \mathbb{Q}^n$ and $b \in \mathbb{Q}$ such that

(1) $\{\beta\} = \{x|\langle x, c\rangle = b\} \cap \mathcal{Q}$ and $\mathcal{Q} \subset \{x|\langle x, c\rangle \leq b\}$.

Then, by definition of $\mathcal{Q}$, one has $v_i + \beta \in \mathcal{Q}$ for all i. Thus

$$\langle v_i + \beta, c\rangle = \langle v_i, c\rangle + b \leq b \Rightarrow \langle v_i, c\rangle \leq 0 \quad (i = 1, \ldots, q).$$

Assume there is $\alpha \in \mathcal{Q}$ and $\eta_1, \ldots, \eta_q$ in $\mathbb{N}$ such that

$$\beta = \eta_1 v_1 + \cdots + \eta_q v_q + \alpha,$$

then $b = \langle \beta, c\rangle = \eta_1\langle v_1, c\rangle + \cdots + \eta_q\langle v_q, c\rangle + \langle \alpha, c\rangle \leq \langle \alpha, c\rangle \leq b$. Hence $\langle \alpha, c\rangle = b$ and by (1) we get $\alpha = \beta$. Thus x^β is a minimal generator of the canonical module $\omega_{K[F]}$. □

Exercises

9.1.16 If $K[F] = K[x^3, y^3, x^2y]$, prove that $\overline{K[F]} = K[F][xy^2]$.

9.1.17 Prove that the ring $K[F]$ is normal if and only if any element $\alpha \in \mathbb{Z}\mathcal{A}$ satisfying $r\alpha \in \mathbb{N}\mathcal{A}$ for some integer $r \geq 1$ belongs to $\mathbb{N}\mathcal{A}$.

9.1.18 If $\mathcal{S} = \mathbb{R}_+\mathcal{A} \cap \mathbb{Z}^n$, then $K[\mathcal{S}] = K[\{x^\alpha|\alpha \in \mathcal{S}\}]$ is normal.

9.1.19 The shift polyhedron $\mathcal{Q}$ of $\mathbb{R}_+\mathcal{A}$, relative to C, can be written as $\mathcal{Q} = \mathbb{R}_+\mathcal{A} + \mathcal{P}$, i.e., $\mathbb{R}_+\mathcal{A}$ is the characteristic cone of $\mathcal{Q}$.

9.1.20 Let $\mathcal{A} \subset \mathbb{N}^n$ be a finite set. If $K[F]$ is normal and F' is the set of all $x^{v_i} \in F$ such that x_1 is not in the support of x^{v_i}, then $K[F']$ is normal.

9.1.21 If there is $x_0 \in \mathbb{Q}^n$ such that $\langle x_0, v\rangle = 1$ for all $v \in \mathcal{A}$, then $\mathbb{R}_+\mathcal{A}$ is a pointed cone.

9.2 Homogeneous monomial subrings

The class of homogeneous monomial subrings will be studied here. Examples in this class include Rees algebras of ideals generated by monomials of the same degree and polytopal subrings.

Definition 9.2.1 The subring $K[F]$ (resp. $\mathcal{A}$) is said to be *homogeneous* if there is $x_0 \in \mathbb{Q}^n$ such that $\langle v_i, x_0\rangle = 1$ for $i = 1, \ldots, q$.

Proposition 9.2.2 *$K[F]$ is homogeneous if and only if $K[F]$ is a standard graded algebra with the grading*

$$K[F]_i = \sum_{|a|=i} K(x^{v_1})^{a_1} \cdots (x^{v_q})^{a_q}, \text{ where } a = (a_1, \ldots, a_q) \in \mathbb{N}^q.$$

Proof. $\Rightarrow$) There is a vector $x_0 \in \mathbb{Q}^n$ such that $\langle v_i, x_0\rangle = 1$ for all i. One clearly has $K[F]_i K[F]_j \subset K[F]_{i+j}$ for all i, j. Thus, it suffices to prove $K[F] = \bigoplus_{i=0}^{\infty} K[F]_i$. First let us prove that $K[F]_i \cap K[F]_j = \{0\}$ for $i \neq j$. If this intersection is not zero, one has an equality

$$(x^{v_1})^{a_1}\cdots(x^{v_q})^{a_q} = (x^{v_1})^{b_1}\cdots(x^{v_q})^{b_q},$$

with $a = (a_1, \ldots, a_q) \in \mathbb{N}^q$, $b = (b_1, \ldots, b_q) \in \mathbb{N}^q$, $|a| = i$, $|b| = j$. Hence

$$i = |a| = \langle a_1v_1 + \cdots + a_qv_q, x_0\rangle = \langle b_1v_1 + \cdots + b_qv_q, x_0\rangle = |b| = j,$$

a contradiction. To finish the proof assume $f_{i_1} + \cdots + f_{i_r} = 0$, with f_{i_j} in $K[F]_{i_j}$ for all j and $i_1 < \cdots < i_r$. If $f_{i_r} \neq 0$, using that the monomials of R form a K-basis, one has $K[F]_{i_r} \cap K[F]_{i_j} \neq \{0\}$ for some $j < r$, a contradiction. Thus $f_{i_r} = 0$. By induction $f_{i_j} = 0$ for all j, as required.

$\Leftarrow$) Set $V = \mathrm{aff}_{\mathbb{Q}}(v_1, \ldots, v_q)$ and $r = \dim(V)$. To begin with we claim that 0 is not in V. Otherwise if $0 \in V$, write $0 = \lambda_1v_1 + \cdots + \lambda_qv_q$ with $\lambda_i \in \mathbb{Q}$ for all i and $\sum_{i=1}^{q}\lambda_i = 1$. There is $0 \neq p \in \mathbb{Z}$ such that $p\lambda_i \in \mathbb{Z}$ for all i. One may assume, by reordering the v_i, that $p\lambda_i \geq 0$ for $i \leq s$ and $p\lambda_i \leq 0$ for $i > s$. As $K[F]$ is graded, from the equality

$$(x^{v_1})^{p\lambda_1}\cdots(x^{v_s})^{p\lambda_s} = (x^{v_{s+1}})^{-p\lambda_{s+1}}\cdots(x^{v_q})^{-p\lambda_q},$$

one derives $p\sum_{i=1}^{q}\lambda_i = 0$, a contradiction. Thus $0 \notin V$, and consequently $r < n$. It is well known that a linear variety V in $\mathbb{Q}^n$ of dimension r can be written as an intersection of hyperplanes $V = \bigcap_{i=1}^{n-r} H(y_i, c_i)$, where $0 \neq y_i \in \mathbb{Q}^n$ and $c_i \in \mathbb{Q}$ for all i; see [427, Corollary 1.4.2]. Since $0 \notin V$, one has $c_j \neq 0$ for some j. To finish the proof we set $x_0 = y_j/c_j$. $\square$

Corollary 9.2.3 *P_F is a graded ideal with respect to the standard grading of S if and only if $K[F]$ is homogeneous.*

Proof. As P_F is generated by a finite set of binomials (Corollary 8.2.18), the result follows readily from Proposition 9.2.2. $\square$

Proposition 9.2.4 *If ψ, φ are the maps of K-algebras defined by*

$$\begin{array}{ccc} K[t_1,\ldots,t_q] & \xrightarrow{\varphi} & K[F] \\ \downarrow \psi & \overset{\overline{\varphi}}{\nearrow} & \\ K[Ft] & & \end{array} \qquad \begin{array}{ccc} t_i & \xrightarrow{\varphi} & x^{v_i} \\ \downarrow \psi & \overset{\overline{\varphi}}{\nearrow} & \\ x^{v_i}t & & \end{array}$$

then there is a unique epimorphism $\overline{\varphi}$ such $\varphi = \overline{\varphi}\psi$. In addition $\overline{\varphi}$ is an isomorphism if and only if $K[F]$ and $K[Ft]$ have the same Krull dimension.

Proof. To show the existence of $\overline{\varphi}$ it suffices to prove $\ker(\psi) \subset \ker(\varphi)$. This inclusion follows at once because any binomial in $\ker(\psi)$ clearly belongs to $\ker(\varphi)$, and $\ker(\psi)$ is a binomial ideal.

Note that $\ker(\psi)$ and $\ker(\varphi)$ are both prime ideals. The map $\overline{\varphi}$ is an isomorphism if and only if $\ker(\psi) = \ker(\varphi)$; thus to finish the proof we need only observe that the last equality holds if and only if $K[F]$ and $K[Ft]$ have the same dimension. $\Box$

Corollary 9.2.5 *The map $\overline{\varphi}\colon K[Ft] \to K[F]$ is an isomorphism if and only if $K[F]$ is homogeneous.*

Proof. $\Rightarrow$) As $\ker(\varphi) = \ker(\psi)$, by Corollary 9.2.3, $K[F]$ is homogeneous because $\ker(\psi)$ is homogeneous in the standard grading of S.

$\Leftarrow$) By Corollary 9.2.3, $\ker(\varphi)$ is homogeneous in the standard grading of S. Note that any homogeneous binomial in $\ker(\varphi)$ is also in $\ker(\psi)$, thus one has $\ker(\varphi) = \ker(\psi)$ and $\overline{\varphi}$ is an isomorphism. $\Box$

Remark 9.2.6 The subring $K[Ft]$ is always homogeneous because $(v_i, 1)$ lies in the hyperplane $x_{n+1} = 1$ for all i. If $K[F]$ is homogeneous, there is $x_0 \in \mathbb{Q}^n$ such that $\langle x_0, v_i\rangle = 1$ for all i. Then $K[F] = \oplus_{i\in\mathbb{N}} K[F]_i$ is a standard graded K-algebra with ith graded component defined by

$$K[F]_i = \sum_{\langle a, x_0\rangle = i} K\{x^a | x^a \in K[F]\}.$$

The isomorphism of K-algebras $\overline{\varphi}$ is degree preserving, that is, $\overline{\varphi}$ maps $K[Ft]_i$ into $K[F]_i$ for all $i \in \mathbb{N}$. In particular $e(K[F]) = e(K[Ft])$.

Lemma 9.2.7 *There is $n_0 \in \mathbb{N}_+$ such that $(x^\alpha)^{n_0} \in K[F]$ for $x^\alpha \in \overline{K[F]}$.*

Proof. By Lemma 1.3.2 and Theorem 9.1.1, there are $\gamma_1, \ldots, \gamma_r$ in the semigroup $\mathbb{Z}\mathcal{A} \cap \mathbb{Q}_+\mathcal{A}$ such that $\overline{K[F]} = K[x^{\gamma_1}, \ldots, x^{\gamma_r}]$. For each i there is a positive integer n_i such that $x^{n_i\gamma_i} \in K[F]$. The integer $n_0 = \mathrm{lcm}(n_1, \ldots, n_r)$ satisfies the required condition, because any $x^\alpha \in \overline{K[F]}$ can be written as $x^\alpha = x^{a_1\gamma_1} \cdots x^{a_r\gamma_r}$, for some $a_1, \ldots, a_r \in \mathbb{N}$. $\Box$

Lemma 9.2.8 *There is x^γ such that $x^\gamma\overline{K[F]} \subset K[F]$.*

Proof. By Lemma 1.3.2 and Theorem 9.1.1, there are $\gamma_1, \ldots, \gamma_r$ in the semigroup $\mathbb{Z}\mathcal{A} \cap \mathbb{Q}_+\mathcal{A}$ such that $\overline{K[F]} = K[x^{\gamma_1}, \ldots, x^{\gamma_r}]$. For each i, we can write $x^{\gamma_i} = x^{\beta_i}/x^{\delta_i}$, where x^{β_i} and x^{δ_i} are in $K[F]$. By Lemma 9.2.7 there is $0 \neq n_0 \in \mathbb{N}$ such that $x^{n_0\alpha} \in K[F]$ for all $x^\alpha \in \overline{K[F]}$. We set $x^\gamma = x^{n_0\delta_1} \cdots x^{n_0\delta_r}$. Take x^α in $\overline{K[F]}$ and write $x^\alpha = (x^{\gamma_1})^{a_1} \cdots (x^{\gamma_r})^{a_r}$, where $a_i \in \mathbb{N}$ for all i. By the division algorithm, for each i there are q_i, c_i in $\mathbb{N}$ such that $a_i = q_i n_0 + c_i$ and $0 \leq c_i < n_0$. From the equality

$$x^\gamma x^\alpha = \underbrace{\prod_{i=1}^{r}(x^{n_0\gamma_i})^{q_i}}_{\text{is in } K[F]} \underbrace{\prod_{i=1}^{r}(x^{\delta_i+\gamma_i})^{c_i}}_{\text{is in } K[F]} \underbrace{\prod_{i=1}^{r}(x^{\delta_i})^{n_0-c_i}}_{\text{is in } K[F]}$$

and using that $\delta_i + \gamma_i = \beta_i$ for all i, we get $x^\gamma x^\alpha \in K[F]$. As x^α was an arbitrary monomial in $\overline{K[F]}$, we get $x^\gamma \overline{K[F]} \subset K[F]$. □

The *normalized Ehrhart ring* of $\mathcal{P} = \mathrm{conv}(\mathcal{A})$ is the graded algebra

$$A_{\mathcal{P}} = \bigoplus_{i=0}^{\infty}(A_{\mathcal{P}})_i \subset R[t],$$

where the ith component is given by

$$(A_{\mathcal{P}})_i = \sum_{\alpha\in\mathbb{Z}\mathcal{A}\cap i\mathcal{P}} Kx^\alpha t^i.$$

To prove that $A_{\mathcal{P}}$ is a graded algebra note that because of the convexity of the polytope $\mathcal{P}$ one has

- $(A_{\mathcal{P}})_i(A_{\mathcal{P}})_j \subset (A_{\mathcal{P}})_{i+j}$ and
- $(A_{\mathcal{P}})_i \cap (A_{\mathcal{P}})_j = \{0\}$ for $i \neq j$.

Proposition 9.2.9 *$\overline{K[Ft]} \subset A_{\mathcal{P}}$, with equality if $K[F]$ is homogeneous.*

Proof. We set $\mathcal{B}' = \{(v_1, 1), \ldots, (v_q, 1)\} \subset \mathbb{Z}^{n+1}$. First we show:

$$\mathbb{Z}\mathcal{B}' \cap \mathbb{R}_+\mathcal{B}' \subset \{(\alpha, i)|\, \alpha \in \mathbb{Z}\mathcal{A} \cap i\mathcal{P} \text{ and } i \in \mathbb{N}\}. \qquad (*)$$

Take $z = (\alpha, i)$ in $\mathbb{Z}\mathcal{B}' \cap \mathbb{R}_+\mathcal{B}'$. There are n_i's in $\mathbb{Z}$ and λ_i's in $\mathbb{R}_+$ such that

$$z = n_1(v_1, 1) + \cdots + n_q(v_q, 1) = \lambda_1(v_1, 1) + \cdots + \lambda_q(v_q, 1).$$

Note $\alpha = 0$ if $i = 0$, and $\alpha/i \in \mathcal{P}$ if $i \geq 1$. Hence z is in the right-hand side of Eq. $(*)$. Thus, by Theorem 9.1.1, we get $\overline{K[Ft]} \subset A_{\mathcal{P}}$.

Assume that $K[F]$ is homogeneous. Let x_0 be a vector in $\mathbb{Q}^n$ such that $\langle v_j, x_0\rangle = 1$ for all j. By Theorem 9.1.1 it suffices to prove that equality holds in Eq. $(*)$. Take $z = (\alpha, i)$ in the right-hand side of Eq. $(*)$ and write

$$\alpha = n_1v_1 + \cdots + n_qv_q = i(\lambda_1v_1 + \cdots + \lambda_qv_q),$$

where $n_j \in \mathbb{Z}$, $\lambda_j \geq 0$, and $\sum_j \lambda_j = 1$. Hence $\langle \alpha, x_0\rangle = \sum_j n_j = i$, and

$$\begin{aligned} z = (\alpha, i) &= n_1(v_1, 1) + \cdots + n_q(v_q, 1)\\ &= i\lambda_1(v_1, 1) + \cdots + i\lambda_q(v_q, 1). \end{aligned}$$

Thus $z \in \mathbb{Z}\mathcal{B}' \cap \mathbb{R}_+\mathcal{B}'$, as required. □

Corollary 9.2.10 *If $K[F]$ is homogeneous, then $K[F]$ is normal if and only if $K[Ft] = A_{\mathcal{P}}$.*

Proof. It follows from Corollary 9.2.5 and Proposition 9.2.9. □

The *normalized Ehrhart function* of $\mathcal{P}$ is defined as

$$E'(i) = \dim_K(A_{\mathcal{P}})_i = |\mathbb{Z}\mathcal{A} \cap i\mathcal{P}|, \quad i \in \mathbb{N}.$$

Proposition 9.2.11 *If $K[F]$ is a homogeneous subring of dimension d, then E' is a polynomial function of degree $d-1$.*

Proof. As the normalized Ehrhart ring $A_{\mathcal{P}}$ is equal to $\overline{K[Ft]}$, thanks to Corollary 9.1.3 one has that $A_{\mathcal{P}}$ is a finitely generated K-algebra. Hence $A_{\mathcal{P}}$ is a finitely generated graded $K[Ft]$-module.

Using Theorem 2.2.4 and Proposition 2.4.13 one concludes that the Hilbert function of $A_{\mathcal{P}}$ is a polynomial function of degree $d-1$. □

Example 9.2.12 Let $F = \{x_1^4, x_1^3x_2, x_1x_2^3, x_2^4\}$ and let $\mathcal{P}$ be the convex hull of $\mathcal{A} = \log(F)$. As $K[F]$ is homogeneous, using Proposition 9.2.9 and *Normaliz* [68], we get that the normalized Ehrhart ring is

$$A_{\mathcal{P}} = K[x_1^4t, x_1^3x_2t, x_1x_2^3t, x_2^4t, x_1^2x_2^2t].$$

The subrings $K[F]$ and $A_{\mathcal{P}}$ have different Hilbert functions, but they have the same Hilbert polynomial, which is equal to $4t+1$.

Proposition 9.2.13 *If $K[F]$ is homogeneous, then $\overline{K[F]}$ is a positively graded K-algebra and its multiplicity is equal to the multiplicity of $K[F]$.*

Proof. We set $\mathcal{P} = \text{conv}(\mathcal{A})$. As $K[F]$ is a homogeneous subring, using Proposition 9.2.9, we get $\overline{K[Ft]} = A_{\mathcal{P}}$. Hence $\overline{K[F]} = \bigoplus_{i=0}^{\infty} \overline{K[F]}_i$ is a graded K-algebra whose ith component is given by

$$\overline{K[F]}_i = \sum_{\alpha \in \mathbb{Z}\mathcal{A} \cap i\mathcal{P}} Kx^{\alpha}.$$

Recall that $K[F]$ is graded as in Proposition 9.2.2. Let h and $\overline{h}$ be the Hilbert functions of $K[F]$ and $\overline{K[F]}$, respectively. Since $K[F]$ and $\overline{K[F]}$ have the same dimension d, one can write

$$h(i) = a_0 i^{d-1} + \text{terms of lower degree,}$$
$$\overline{h}(i) = c_0 i^{d-1} + \text{terms of lower degree,}$$

for $i \gg 0$. One clearly has $a_0 \leq c_0$ because $K[F]_i \subset \overline{K[F]}_i$ for all i. On the other hand by Lemma 9.2.8 there is $x^\gamma \in K[F]$ of degree m such that $x^\gamma \overline{K[F]} \subset K[F]$. Thus for each i there is a one-to-one map

$$\overline{K[F]}_i \xrightarrow{x^\gamma} K[F]_{i+m}.$$

Hence $\overline{h}(i) \leq h(i+m)$ for all i, and consequently $c_0 \leq a_0$. Altogether $a_0 = c_0$. Therefore

$$e(K[F]) = a_0(d-1)! = c_0(d-1)! = e(\overline{K[F]}). \qquad \square$$

Corollary 9.2.14 [72] *If $K[F]$ is homogeneous and Cohen–Macaulay, then*

$$a(\overline{K[F]}) \leq a(K[F]).$$

Proof. As $\overline{K[F]}$ is normal, by Theorem 9.1.6, $\overline{K[F]}$ is Cohen–Macaulay. Hence, the inequality follows from Propositions 5.2.8 and 9.2.13. $\square$

Definition 9.2.15 A binomial $t^\alpha - t^\beta$ in $K[t_1,\ldots,t_q]$ is said to *have a square-free term* if at least one of its two terms t^α, t^β is square-free.

The following result gives a necessary condition for the normality of a homogeneous subring $K[F]$ in terms of its toric ideal.

Theorem 9.2.16 [385] *Let $\mathcal{B}$ be a finite set of binomials in the toric ideal P_F of $K[F]$. If $K[F]$ is a normal homogeneous subring and P_F is minimally generated by $\mathcal{B}$, then every element of $\mathcal{B}$ has a square-free term.*

Proof. Recall that the toric ideal of $K[F]$ is the kernel of the epimorphism

$$\varphi\colon S = K[t_1,\ldots,t_q] \longrightarrow K[F] \qquad (t_i \stackrel{\varphi}{\longmapsto} x^{v_i})$$

of K algebras. Since $K[F]$ is homogeneous, P_F is a graded ideal in the standard grading of S (see Corollary 9.2.3). We set $\mathcal{B} = \{g_1,\ldots,g_\ell\}$ and $f_i = x^{v_i}$. Let g be a binomial in $\mathcal{B}$. We proceed by contradiction assuming that g has no square free-term. After permuting variables one can write

$$g = t_1^{a_1}\cdots t_r^{a_r} - t_{r+1}^{a_{r+1}}\cdots t_s^{a_s},$$

with $a_i \geq 1$ for all i and $2 \leq a_1 = \max\{a_i\}_{i=1}^r \leq a_s = \max\{a_i\}_{i=r+1}^s$. From the equality $f_1^{a_1}\cdots f_r^{a_r} = f_{r+1}^{a_{r+1}}\cdots f_s^{a_s}$, we obtain $f_1\cdots f_r/f_s \in K[F]$ because $K[F]$ is normal. Hence there exists a binomial h_1 in P_F of the form

$$h_1 = t_1\cdots t_r - t_s t^\alpha,$$

with $r = \deg(h_1) = \deg(t^\alpha)+1 < \deg(g)$. Note the equality

$$g - t_1^{a_1-1}\cdots t_r^{a_r-1}h_1 = t_s h_2, \tag{9.8}$$

where h_2 is a binomial in P_F with $\deg(h_2) < \deg(g)$ or $h_2 = 0$. Writing

$$h_j = \sum_{g_i \neq g} c_{ij} g_i \qquad (j = 1,2),$$

and using Eq. (9.8) we conclude $g = \sum_{g_i \neq g} c_i g_i$, $c_i \in S$, a contradiction because P_F is minimally generated by $g_1, \ldots, g_\ell$. $\Box$

A full converse to Proposition 9.2.16 is not true even if the monomials have the same positive total degree. Notice however that there is a partial converse due to Sturmfels (see Theorem 9.6.16).

Example 9.2.17 Let $F = \{x_1x_2, x_2x_3, x_3x_4, x_1x_4, x_1^2, x_2^2, x_3^2, x_4^2\} \subset K[\mathbf{x}]$. Using *Macaulay*2 [199], one obtains that P_F is minimally generated by

$$\mathcal{B} = \{t_4^2 - t_5t_8, t_3^2 - t_7t_8, t_1t_3 - t_2t_4, t_2^2 - t_6t_7,$$
$$t_1^2 - t_5t_6, t_3t_4t_6 - t_1t_2t_8, t_2t_3t_5 - t_1t_4t_7\}.$$

Using *Normaliz* [68], we obtain $\overline{K[F]} = K[F][x_1x_3, x_2x_4]$.

The next result complements [400, Theorem 13.14].

Theorem 9.2.18 *If* $K[F]$ *is normal and homogeneous, then the toric ideal* P_F *is generated by homogeneous binomials of degree at most* $\operatorname{rank}(A)$.

Proof. Let $\mathcal{B} = \{g_1, \ldots, g_\ell\}$ be a minimal generating set of P_F consisting of binomials. Set $r = \max_i\{\deg(g_i)\}$.

Claim (I): If $\deg(g_i) = r$ for some i and $r > \operatorname{rank}(A)$, then g_i is a linear combination of two binomials in P_F of degree strictly less than r. By Theorem 9.2.16 we may assume that g_i can be written as:

$$g_i = t_1 \cdots t_r - t_{r+1}^{a_{r+1}} \cdots t_m^{a_m} \quad \text{with} \quad a_j \geq 1 \text{ for all } j.$$

One has $v_1 + \cdots + v_r = a_{r+1}v_{r+1} + \cdots + a_m v_m$. Applying Carathéodory's theorem (see Theorem 1.1.18) to both sides of this equality and permuting variables if necessary, we obtain an equation:

$$b_1v_1 + \cdots + b_{r_1}v_{r_1} = c_{r+1}v_{r+1} + \cdots + c_{m_1}v_{m_1}$$

with $b_j, c_k \in \mathbb{N} \setminus \{0\}$ for all j, k, $r_1 \leq \operatorname{rank}(A)$, $m_1 - r \leq \operatorname{rank}(A)$, and $m_1 \leq m$. Thus the binomial $f = t_1^{b_1} \cdots t_{r_1}^{b_{r_1}} - t_{r+1}^{c_{r+1}} \cdots t_{m_1}^{c_{m_1}}$ belongs to P_F.

Case (A): $b_i = 1$ for all i. We can write $g_i - (t_{r_1+1} \cdots t_r)f = t_{r+1}h$ for some h. Hence g_i is a linear combination of f and h, which in this case have degree less than r.

Case (B): $c_i = 1$ for all i. Notice that $\deg(f) = m_1 - r \leq \operatorname{rank}(A) < r$. Pick t^γ so that $t_{r+1}^{a_{r+1}} \cdots t_m^{a_m} = t^\gamma(t_{r+1} \cdots t_{m_1})$. We can write $g_i - t^\gamma f = t_1 h$ for some h. Thus g_i is a linear combination of f and h, which in this case also have degree less than r.

Case (C): $\max_i\{c_i\} \geq \max_i\{b_i\} \geq 2$. After permutation of variables we may assume $c_{r+1} = \max_i\{c_i\} \geq \max_i\{b_i\} = b_1 \geq 2$. From the equality

$$f_1^{b_1} \cdots f_{r_1}^{b_{r_1}} = f_{r+1}^{c_{r+1}} \cdots f_{m_1}^{c_{m_1}} \qquad (f_i = x^{v_i}),$$

we get $f_1 \cdots f_{r_1}/f_{r+1} \in K[F]$ because $K[F]$ is normal. Hence there is h_1 in P_F of the form $h_1 = t_1 \cdots t_{r_1} - t_{r+1}t^{\alpha}$, with $\deg(h_1) = \deg(t^{\alpha})+1 < \deg(g_i)$. Note that $r_1 = \deg(h_1)$. We can write

$$g_i - t_{r_1+1} \cdots t_r h_1 = t_{r+1} h_2,$$

where h_2 is a binomial in P_F with $\deg(h_2) < \deg(g)$ or $h_2 = 0$. Thus once again g_i is a linear combination of binomials of degree less than r. The remaining cases follow by symmetry. This completes the proof of the claim.

Applying Claim (I) to each binomial g_i of degree r, it follows that P_F is minimally generated by binomials of degree less than r. Thus, by induction, it follows that P_F is generated by binomials of degree at most rank (A). □

Exercises

9.2.19 If $K[F]$ is a homogeneous monomial subring over a field K, use the Danilov–Stanley formula to show $a(\overline{K[F]}) < 0$.

9.2.20 Prove that the following conditions are equivalent:

(a) $K[F] = K[x^{v_1}, \ldots, x^{v_q}]$ is homogeneous.

(b) $0 \notin \text{aff}_{\mathbb{Q}}(v_1, \ldots, v_q) \subset \mathbb{Q}^n$.

(c) $\dim_{\mathbb{Q}} \mathbb{Q}\{(v_1, 1), \ldots, (v_q, 1)\} < \dim_{\mathbb{Q}} \mathbb{Q}\{(v_1, 1), \ldots, (v_q, 1), (\mathbf{0}, 1)\}$.

(d) $\dim_{\mathbb{Q}} \mathbb{Q}\{(v_1, 1), \ldots, (v_q, 1)\} = \dim_{\mathbb{Q}} \mathbb{Q}\{v_1, \ldots, v_q\}$.

9.2.21 Is the integral closure of a homogeneous subring homogeneous?

9.2.22 Is the monomial subring $K[x^2, xy, y^3] \subset K[x, y]$ homogeneous?

9.2.23 Let $K[F]$ be a normal homogeneous subring. If $\text{rank}(A) = 2$, then P_F is generated by homogeneous binomials of degree at most 2.

9.2.24 Consider the monomial subring $\mathbb{Q}[F] = \mathbb{Q}[x_1^3 x_2, x_2 x_3^3, x_1^2 x_2^2, x_1^3 x_3]$. Prove that $\overline{\mathbb{Q}[F]}$ is equal to $\mathbb{Q}[F][x_1^2 x_2 x_3, x_1 x_2 x_3^2]$. Prove that the toric ideal of $\mathbb{Q}[F]$ is equal to $(t_1^5 t_2 - t_3^3 t_4^3)$.

9.2.25 If $K[F]$ is a homogeneous and h (resp. E') is the Hilbert function of $K[F]$ (resp. normalized Ehrhart function of $\mathcal{P}$), then

(a) $h(i) \leq E'(i)$ for all $i \geq 0$, and

(b) $h(i) = E'(i)$ for all $i \geq 0$ if and only if $K[F]$ is normal.

9.2.26 If $K[F]$ is a two-dimensional homogeneous monomial subring, prove that the Hilbert polynomials of $K[Ft]$ and $\overline{K[Ft]}$ are equal.

9.2.27 Let $K[F]$ be a Cohen–Macaulay homogeneous monomial subring. If $\dim(K[F]) = 2$ and $a(K[F]) < 0$, prove that $K[F]$ is normal.

9.3 Ehrhart rings

In this section we introduce the Ehrhart ring and the polytopal subring of a lattice polytope. We study some of its properties.

Let $\mathcal{A} = \{v_1, \ldots, v_q\}$ be a set of points in $\mathbb{Z}^n$ and let $\mathcal{P}$ be the *lattice polytope* $\mathcal{P} = \mathrm{conv}(\mathcal{A}) \subset \mathbb{R}^n$. The set $\mathcal{A}$ is called a *point configuration.* A point configuration $\mathcal{A}$ is called *normal* if $\mathbb{N}\mathcal{A} = \mathbb{Z}\mathcal{A} \cap \mathbb{R}_+\mathcal{A}$, and it is called *homogeneous* if $\mathcal{A}$ lies on an affine hyperplane of $\mathbb{R}^n$ not containing the origin. In [106] by a "point configuration" they mean a finite multiset of vectors in $\mathbb{R}^n$. In [106] they also introduce "vector configurations" and "homogeneous point configurations."

To link the lattice polytope $\mathcal{P}$ and the point configuration $\mathcal{A}$ to ring theory consider a Laurent polynomial ring $R = K[x_1^{\pm 1}, \ldots, x_n^{\pm 1}]$ over a field K. If $\mathcal{A} \subset \mathbb{N}^n$, we set $R = K[x_1, \ldots, x_n]$. There is an isomorphism between the group $\mathbb{Z}^n$ and the multiplicative group of monomials of R given by

$$a = (a_1, \ldots, a_n) \longleftrightarrow x^a = x_1^{a_1} \cdots x_n^{a_n}.$$

In this correspondence:

$$\mathcal{A} = \{v_1, \ldots, v_q\} \longleftrightarrow F = \{x^{v_1}, \ldots, x^{v_q}\}.$$

The monomial subring $K[\mathcal{P}] = K[\{x^a t |\, a \in \mathcal{P} \cap \mathbb{Z}^n\}] \subset R[t]$ is called the *polytopal subring* of $\mathcal{P}$. Notice the equality

$$K[\mathcal{P}] = K[\{x^a t |\, (a, 1) \in \mathbb{Q}_+\{(v_1, 1), \ldots, (v_q, 1)\} \cap \mathbb{Z}^{n+1}\}],$$

which readily implies that $K[Ft] \subset K[\mathcal{P}]$ is an integral extension. The polytope $\mathcal{P}$ is said to be *normal* if $K[\mathcal{P}]$ is normal [62] (cf. [100, 203]).

Proposition 9.3.1 $\dim K[\mathcal{P}] = \dim(\mathcal{P}) + 1$.

Proof. We set $V = \mathbb{R}(v_2 - v_1) + \cdots + \mathbb{R}(v_q - v_1)$. From the equality

$$\mathrm{aff}(\mathcal{P}) = \mathrm{aff}(v_1, \ldots, v_q) = v_1 + V,$$

we get $\dim(\mathcal{P}) = \dim_{\mathbb{R}}(V)$. Let B be the matrix with rows $(v_1, 1), \ldots, (v_q, 1)$ and let B' be the matrix with rows $(v_2 - v_1, 0), \ldots, (v_q - v_1, 0)$. Using Corollary 8.2.21, we obtain

$$\dim K[Ft] = \mathrm{rank}(B) = \mathrm{rank}(B') + 1 = \dim(\mathcal{P}) + 1.$$

Since $K[Ft] \subset K[\mathcal{P}]$ is an integral extension, by Proposition 2.4.13, we get that $\dim K[Ft]$ is equal to $\dim K[\mathcal{P}]$.

Lemma 9.3.2 *Let $\mathcal{A} = \{v_1, \ldots, v_{d+1}\}$ be an affinely independent set in $\mathbb{Z}^n$. If $\mathcal{P} = \mathrm{conv}(\mathcal{A})$ is a unimodular d-simplex, then $\mathcal{P} \cap \mathbb{Z}^n = \mathcal{A}$ and $K[\mathcal{P}]$ is a polynomial ring.*

Proof. By Corollary 1.2.22 we have $\mathcal{P}\cap\mathbb{Z}^n=\mathcal{A}$. Hence $K[\mathcal{P}]$ is equal to $K[x^{v_1}t,\ldots,x^{v_{d+1}}t]$. Thus $K[\mathcal{P}]$ has dimension $d+1$ and by Lemma 3.1.5 $K[\mathcal{P}]$ is a polynomial ring. $\Box$

Proposition 9.3.3 [62] *Let $\{\mathcal{S},\mathcal{S}_i\}$ be a family of finitely generated semigroups of $\mathbb{Z}^n$ such that $\mathcal{S}=\cup_i\mathcal{S}_i$. If $\mathbb{Z}\mathcal{S}=\mathbb{Z}\mathcal{S}_i$ and $K[\mathcal{S}_i]$ is normal for all i, then $K[\mathcal{S}]$ is normal.*

Proof. By Corollary 9.1.3 it suffices to show that $\mathbb{Z}\mathcal{S}\cap\mathbb{R}_+\mathcal{S}$ is contained in $\mathcal{S}$. Take $a\in\mathbb{Z}\mathcal{S}\cap\mathbb{R}_+\mathcal{S}$. There is an integer $s\geq 1$ such that $sa\in\mathcal{S}$, thus $sa\in\mathcal{S}_i$ for some i. As $\mathbb{Z}\mathcal{S}=\mathbb{Z}\mathcal{S}_i$ and $K[\mathcal{S}_i]$ is normal we get $a\in\mathcal{S}_i\subset\mathcal{S}$, as required. $\Box$

Definition 9.3.4 A collection $\{\Delta_\ell\}$ of unimodular lattice simplices in $\mathbb{R}^n$ is called a *unimodular covering* of $\mathcal{P}$ if $\mathcal{P}=\cup_\ell\Delta_\ell$ and $\dim(\Delta_\ell)=d$ for all ℓ, where d is the dimension of $\mathcal{P}$.

Theorem 9.3.5 [62] *If $\mathcal{P}$ has a unimodular covering, then $K[\mathcal{P}]$ is normal.*

Proof. Let $\{\Delta_\ell\}$ be a unimodular covering of $\mathcal{P}$. Consider the semigroups $\mathcal{S}$ and $\mathcal{S}_\ell$ of $\mathbb{Z}^{n+1}$ defined as:

$$\mathcal{S}=\mathbb{N}\{(\alpha,1)\,|\,\alpha\in\mathbb{Z}^n\cap\mathcal{P}\}\quad\text{and}\quad\mathcal{S}_\ell=\mathbb{N}\{(\alpha,1)\,|\,\alpha\in\mathbb{Z}^n\cap\Delta_\ell\}.$$

By Lemma 9.3.2 $K[\mathcal{S}_\ell]$ is a polynomial ring. In particular $K[\mathcal{S}_\ell]$ is normal for all ℓ. Hence, by Proposition 9.3.3, we need only show that $\mathcal{S}=\cup_\ell\mathcal{S}_\ell$ and $\mathbb{Z}\mathcal{S}=\mathbb{Z}\mathcal{S}_\ell$. One clearly has the inclusion $\cup_\ell\mathcal{S}_\ell\subset\mathcal{S}$, because $\mathcal{P}=\cup_\ell\Delta_\ell$. To show the reverse inclusion take $0\neq z\in\mathcal{S}$. Then

$$z=(\beta,s)=n_1(\beta_1,1)+\cdots+n_r(\beta_r,1),$$

where $n_i\in\mathbb{N}$ and $\beta_i\in\mathbb{Z}^n\cap\mathcal{P}$ for all i. Note $\beta/s\in\mathcal{P}$, where $\beta=\sum_i n_i\beta_i$ and $s=\sum_i n_i$. Hence $\beta/s\in\Delta_\ell$ for some ℓ. There is an affinely independent set $\Gamma=\{\gamma_0,\ldots,\gamma_d\}$ in $\mathbb{Z}^n$ such that $\Delta_\ell=\operatorname{conv}(\Gamma)$. Thus $\beta/s=\sum_{i=0}^d\mu_i\gamma_i$, for some μ_i's satisfying $\mu_i\geq 0$ and $\sum_i\mu_i=1$. Hence

$$z=(\beta,s)=s\mu_0(\gamma_0,1)+\cdots+s\mu_d(\gamma_d,1)\in\mathbb{R}_+\{(\gamma_0,1),\ldots,(\gamma_d,1)\}.$$

As Δ_ℓ is unimodular, by Corollary 1.1.27 and Proposition 1.2.21, one has

$$z\in\mathbb{Z}\{(\gamma_0,1),\ldots,(\gamma_d,1)\}$$

Hence using that $K[\mathcal{S}_\ell]=K[x^{\gamma_0}t,\ldots,x^{\gamma_d}t]$ is normal yields $z\in\mathcal{S}_\ell$. To show the equality $\mathbb{Z}\mathcal{S}=\mathbb{Z}\mathcal{S}_\ell$ note that $\mathbb{Z}\mathcal{S}$ and $\mathbb{Z}\mathcal{S}_\ell$ are free groups of rank $d+1$. This follows from the assumption that $\mathcal{P}$ and Δ_ℓ have the same dimension. Since Δ_ℓ is unimodular, by Proposition 1.2.21, we get that $\mathbb{Z}^{n+1}/\mathbb{Z}\mathcal{S}_\ell$ is torsion-free. Thus, $\mathbb{Z}\mathcal{S}=\mathbb{Z}\mathcal{S}_\ell$. $\Box$

Recall that the *Ehrhart ring* of $\mathcal{P}$ is:

$$A(\mathcal{P})=K[x^\alpha t^i|\alpha\in\mathbb{Z}^n\cap i\mathcal{P}]\subset R[t].$$

Theorem 9.3.6 *The following assertions hold:*

(a) $A(\mathcal{P}) = K[\{x^\alpha t^i \mid (\alpha, i) \in \mathbb{Z}^{n+1} \cap \mathbb{R}_+\mathcal{B}'\}]$, $\mathcal{B}' = \{(\alpha, 1) \mid \alpha \in \mathbb{Z}^n \cap \mathcal{P}\}$,

(b) $A(\mathcal{P}) = K[\{x^\alpha t^i \mid (\alpha, i) \in \mathbb{Z}^{n+1} \cap \mathbb{R}_+\mathcal{B}\}]$, $\mathcal{B} = \{(v_i, 1)\}_{i=1}^{n}$,

(c) $A(\mathcal{P})$ *is a finitely generated* K*-algebra and a normal domain,*

(d) $K[Ft] \subset A(\mathcal{P})$ *is a finitely generated integral extension,*

(e) $\overline{K[Ft]} = A(\mathcal{P}) \Leftrightarrow A(\mathcal{P})$ *is contained in the field of fractions of* $K[Ft]$,

(f) $\dim(K[Ft]) = \dim(\overline{K[Ft]}) = \dim(A(\mathcal{P})) = \dim(\mathcal{P}) + 1$,

(g) $K[Ft] = A(\mathcal{P})$ *if and only if* $\mathcal{B} = \{(v_i, 1)\}_{i=1}^{q}$ *is a Hilbert basis.*

Proof. (a): "$\subset$" Take a monomial $x^\alpha t^i$ in $A(\mathcal{P})$, then $\alpha \in \mathbb{Z}^n \cap i\mathcal{P}$ and $i \in \mathbb{N}$. Thus we can write $(\alpha, i) = \sum_{j=1}^{q} i\lambda_j(v_j, 1)$, with $\sum_{j=1}^{q} \lambda_j = 1$ and $\lambda_j \geq 0$ for all j. This proves that (α, i) is in the semigroup $\mathbb{Z}^{n+1} \cap \mathbb{R}_+\mathcal{B}'$.

"$\supset$" Take a monomial $x^\alpha t^i$ with (α, i) in $\mathbb{Z}^{n+1} \cap \mathbb{R}_+\mathcal{B}'$. Any vector of the form $(\beta, 1)$ with $\beta \in \mathbb{Z}^n \cap \mathcal{P}$ can be written as $(\beta, 1) = \sum_j \mu_j(v_j, 1)$ for some μ_j's in $\mathbb{R}_+$. Hence we can write $(\alpha, i) = \sum_{j=1}^{q} \rho_j(v_j, 1)$, with $\rho_j \in \mathbb{R}_+$ for all j. Consequently $\alpha \in \mathbb{Z}^n \cap i\mathcal{P}$. Thus $x^\alpha t^i \in A(\mathcal{P})$.

(b): This follows using part (a).

(c): By Proposition 1.3.6 there is a finite Hilbert basis $\mathcal{H} \subset \mathbb{Z}^n$ such that $\mathbb{R}_+\mathcal{B} \cap \mathbb{Z}^n = \mathbb{N}\mathcal{H}$ and $\mathbb{R}_+\mathcal{B} = \mathbb{R}_+\mathcal{H}$. Thus, by Proposition 1.5.2, $\mathbb{N}\mathcal{H}$ is normal. Using part (b) and Corollary 9.1.3, we get that $A(\mathcal{P})$ is normal.

(d): If $x^a t^i$ is in $A(\mathcal{P})$, then $(x^a t^i)^s \in K[Ft]$ for some $0 \neq s \in \mathbb{N}$. Hence $A(\mathcal{P})$ is integral over $K[Ft]$. As a result, using (c) together with Corollary 2.4.4, we obtain that $A(\mathcal{P})$ is a finitely generated $K[Ft]$-module.

(e): This follows directly from (c) and (d).

(f): The first two equalities follow from Proposition 2.4.13 because the extensions $K[Ft] \subset \overline{K[Ft]} \subset A(\mathcal{P})$ are integral. The third equality follows from Proposition 9.3.1.

(g): This follows from part (a) because $K[Ft] = K[\mathbb{N}\mathcal{B}]$. $\square$

Multiplicity of monomial subrings The subrings $K[Ft]$, $\overline{K[Ft]}$ and $A(P)$ are graded as follows. We denote the multiplicity by $e(\cdot)$ or $\deg(\cdot)$.

Notice that $K[Ft]$ is homogeneous because $(v_i, 1)$ lies in the hyperplane $x_{n+1} = 1$ for all i. The rings $K[Ft]$ and $\overline{K[Ft]}$ are graded K-algebras with ith graded components given by

$$K[Ft]_i = \sum_{(a,i)\in\mathbb{N}\mathcal{B}} Kx^a t^i \quad \text{and} \quad \overline{K[Ft]}_i = \sum_{(a,b)\in\mathbb{Z}\mathcal{B}\cap i\mathcal{Q}} Kx^a t^b, \tag{9.9}$$

respectively, where $\mathcal{B} = \{(v_1, 1), \ldots, (v_q, 1)\}$ and $\mathcal{Q} = \mathrm{conv}(\mathcal{B})$.

The Ehrhart ring $A(\mathcal{P})$ is a normal finitely generated graded K-algebra with ith component given by

$$A(\mathcal{P})_i = \sum_{\alpha\in\mathbb{Z}^n\cap i\mathcal{P}} Kx^\alpha t^i.$$

Since $A(\mathcal{P})$ is a finitely generated module over the homogeneous ring $K[Ft]$, by Proposition 5.1.6, we obtain that its Hilbert function:

$$\chi_{\mathcal{P}}(i) := |\mathbb{Z}^n \cap i\mathcal{P}| = c_d i^d + \cdots + c_1 i + c_0 \quad (i \gg 0)$$

is a polynomial function of degree $d = \dim(\mathcal{P})$ such that $c_i \in \mathbb{Q}$ for all i and $d!c_d$ is an integer, which is the *multiplicity* of $A(\mathcal{P})$.

The function $\chi_{\mathcal{P}}$ is the *Ehrhart function* of $\mathcal{P}$. The Hilbert polynomial of $A(\mathcal{P})$, denoted by $E_{\mathcal{P}}(x)$, is called the *Ehrhart polynomial* of $\mathcal{P}$. Recall that the relative volume of $\mathcal{P}$ is given by

$$\mathrm{vol}(\mathcal{P}) = \lim_{i\to\infty}\frac{|\mathbb{Z}^n \cap i\mathcal{P}|}{i^d},$$

where $d = \dim(\mathcal{P})$; see Proposition 1.2.10. Hence $\mathrm{vol}(\mathcal{P})$ is equal to c_d, the leading coefficient of the Ehrhart function of $\mathcal{P}$. For this reason $d!c_d$ is called the *normalized volume* of $\mathcal{P}$.

Altogether one has:

Lemma 9.3.7 $e(A(\mathcal{P})) = d!c_d = d!\mathrm{vol}(\mathcal{P})$.

By the Hilbert–Serre theorem (Theorem 5.1.4) the Hilbert series of $A(\mathcal{P})$ is a rational function that can be written uniquely as:

$$F(A(\mathcal{P}), x) := \sum_{i=0}^{\infty} |\mathbb{Z}^n \cap i\mathcal{P}| x^i = \frac{h_0 + h_1 x + \cdots + h_s x^s}{(1-x)^{d+1}},$$

with $h_0 + h_1 + \cdots + h_s \neq 0$, $h_i \in \mathbb{Z}$ for all i and $d = \dim(\mathcal{P})$. This series is called the *Ehrhart series* of $\mathcal{P}$.

Lemma 9.3.8 (a) $a(A(\mathcal{P})) = s - (d+1) < 0$,

(b) $E_{\mathcal{P}}(i) = \chi_{\mathcal{P}}(i)$ *for all integers* $i \geq 0$,

(c) $h_i \geq 0$ *for all* i,

(d) $h_0 + h_1 + \cdots + h_s = d!\mathrm{vol}(\mathcal{P})$.

Proof. (a): This follows from Theorem 9.1.5. (b): This follows from Corollary 5.1.9. (c): By Theorem 9.3.6, $A(\mathcal{P})$ is a normal finitely generated K-algebra. Hence, by Theorem 9.1.6, $A(\mathcal{P})$ is Cohen–Macaulay. Therefore $h_i \geq 0$ for all i; see Exercise 5.2.9. (d): This follows from Remark 5.1.7. $\square$

Proposition 9.3.9 *The Ehrhart function of $\mathcal{P}$ can be expressed as:*

$$E_{\mathcal{P}}(i) = e_0\binom{i+d}{d} - e_1\binom{i+d-1}{d-1} + \cdots + (-1)^{d-1}e_{d-1}\binom{i+1}{1} + (-1)^d e_d,$$

where $e_i = h^{(i)}(1)/i!$ for all i and $h(x) = h_0 + h_1x + \cdots + h_sx^s$.

Proof. This follows from Proposition 5.1.6 and its proof. □

Definition 9.3.10 $e_0, \ldots, e_d$ are called the *Hilbert coefficients*.

Theorem 9.3.11 (Reciprocity law of Ehrhart [127]) *If $E^+_{\mathcal{P}}$ is the function $E^+_{\mathcal{P}}(i) = |\mathbb{Z}^n \cap \mathrm{ri}(i\mathcal{P})|$, where $\mathrm{ri}(i\mathcal{P})$ is the relative interior of $i\mathcal{P}$, then*

$$E^+_{\mathcal{P}}(i) = (-1)^d E_{\mathcal{P}}(-i) \quad \forall\, i \geq 1.$$

Proof. By Exercise 5.2.10 and [394, Proposition 4.2.3], we have

$$F(\omega_{A(P)}, t) = (-1)^{d-1}F(A(P), t^{-1}) = (-1)^d \sum_{i\geq 1} E_{\mathcal{P}}(-i)t^i.$$

Therefore the required identity follows by comparing coefficients. □

Example 9.3.12 To illustrate the reciprocity law consider the polytope $\mathcal{P}$:

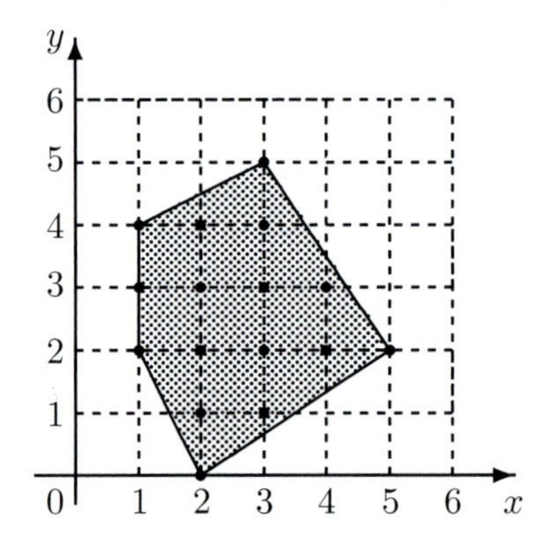

$|\mathbb{Z}^2 \cap \mathcal{P}| = 16$

$|\mathbb{Z}^2 \cap \partial\mathcal{P}| = 6$

$E_{\mathcal{P}}(x) = 12x^2 + 3x + 1$

$E^+_{\mathcal{P}}(1) = (-1)^2 E_{\mathcal{P}}(-1) = 10$

Proposition 9.3.13 *If $E_{\mathcal{P}}(x) = c_dx^d + c_{d-1}x^{d-1} + \cdots + c_1x + c_0$ is the Ehrhart polynomial and $F_1, \ldots, F_s$ are the facets of $\mathcal{P}$, then $c_d = \mathrm{vol}(\mathcal{P})$, $c_{d-1} = \frac{1}{2}\sum_{i=1}^{s}\mathrm{vol}(F_i)$ and $c_0 = 1$.*

Proof. By Lemma 9.3.7, $c_d = \mathrm{vol}(\mathcal{P})$. Since $E_{\mathcal{P}}(0) = \chi_{\mathcal{P}}(0) = 1$, we get $c_0 = 1$. Now we show the formula for c_{d-1}. Notice the equalities

$$i\mathcal{P} = \mathrm{ri}(i\mathcal{P}) \cup \partial(i\mathcal{P}); \quad \partial(\mathcal{P}) = F_1 \cup \cdots \cup F_s,$$

where $\partial(\mathcal{P})$ is the relative boundary. Using the reciprocity law and the inclusion-exclusion principle [2, p. 38, Formula 2.12] we get

$$\begin{aligned} |\partial(i\mathcal{P}) \cap \mathbb{Z}^n| &= E_{\mathcal{P}}(i) - E^+(i) = E_{\mathcal{P}}(i) - (-1)^d E_{\mathcal{P}}(-i) \\ &= 2c_{d-1}i^{d-1} + \text{terms of lower degree} \\ &= E_{F_1}(i) + \cdots + E_{F_s}(i) + h(i), \end{aligned}$$

where h is a polynomial of degree at most $d-2$. Comparing leading terms, the required formula follows. □

In general some of the coefficients $c_0, \ldots, c_d$ of $E_{\mathcal{P}}(x)$ may be negative; see [21, Example 3.22] and [241]. It is unknown whether the coefficients are nonnegative if the vertices of $\mathcal{P}$ have $\{0,1\}$-entries.

Theorem 9.3.14 *If $\mathcal{P} \subset \mathbb{R}^2$ is a lattice polytope of dimension 2, then the Ehrhart polynomial is given by:*

$$E_{\mathcal{P}}(x) = \text{area}(\mathcal{P})x^2 + \frac{|\mathbb{Z}^2 \cap \partial\mathcal{P}|}{2}x + 1.$$

Proof. Let $E_{\mathcal{P}}(x) = c_2x^2 + c_1x + c_0$ be the Ehrhart polynomial of $\mathcal{P}$. By Proposition 9.3.13, $c_2 = \text{area}(\mathcal{P})$ and $c_0 = 1$. Writing $\mathcal{P} = \text{ri}(\mathcal{P}) \cup \partial(\mathcal{P})$, where $\partial(\mathcal{P})$ is the boundary of $\mathcal{P}$, we obtain by the reciprocity law:

$$\begin{aligned} |\partial(\mathcal{P}) \cap \mathbb{Z}^2| &= |\mathcal{P} \cap \mathbb{Z}^2| - |\text{ri}(\mathcal{P}) \cap \mathbb{Z}^2| \\ &= E_{\mathcal{P}}(1) - E_{\mathcal{P}}^+(1) = E_{\mathcal{P}}(1) - E_{\mathcal{P}}(-1) \\ &= (c_2 + c_1 + c_0) - (c_2 - c_1 + c_0) = 2c_1. \end{aligned}$$

Thus one has $c_1 = \frac{|\mathbb{Z}^2 \cap \partial(\mathcal{P})|}{2}$, as required. □

Corollary 9.3.15 (Pick's Formula) *If $\mathcal{P} \subset \mathbb{R}^2$ and $\dim(\mathcal{P}) = 2$, then*

$$|\mathbb{Z}^2 \cap \mathcal{P}| = \text{area}(\mathcal{P}) + \frac{|\mathbb{Z}^2 \cap \partial\mathcal{P}|}{2} + 1.$$

Proof. It follows making $x = 1$ in the formula of Theorem 9.3.14. □

Lemma 9.3.16 *If $v_1, \ldots, v_q$ have $\{0,1\}$-entries, then $\mathcal{P} \cap \mathbb{Z}^n = \mathcal{A}$.*

Proof. Take $\alpha \in \mathcal{P} \cap \mathbb{Z}^n$. Then we can write $\alpha = \sum_j \lambda_j v_j$ with $\sum_j \lambda_j = 1$ and $\lambda_j \geq 0$ for all j. We may assume $\lambda_j > 0$ for all j and $v_1 \neq 0$. We set $v_j = (v_{j1}, \ldots, v_{jn})$. As the ith entry of v_1 is 1 for some i, one has

$$1 \leq \lambda_1 v_{1i} + \lambda_2 v_{2i} + \cdots + \lambda_q v_{qi} \leq \lambda_1 + \cdots + \lambda_q = 1.$$

Thus $q = 1$ and $\lambda_1 = 1$, i.e., $\alpha = v_1$ and $\alpha \in \mathcal{A}$. □

Corollary 9.3.17 *If $v_1, \ldots, v_q$ have $\{0,1\}$-entries, then $E_{\mathcal{P}}(-1) = 0$.*

Proof. By Lemma 9.3.16 and Theorem 9.3.11, $(-1)^d E_{\mathcal{P}}(-1) = E_{\mathcal{P}}^{+}(1) = 0$. Thus $E_{\mathcal{P}}(-1) = 0$, as required. $\Box$

Corollary 9.3.18 If $x^{v_1}, \ldots, x^{v_q}$ are square-free, then $K[Ft] = K[\mathcal{P}]$.

Proof. It follows from Lemma 9.3.16. $\Box$

Proposition 9.3.19 *If $\mathcal{A} = \{v_1, \ldots, v_q\} \subset \mathbb{Z}^n$ and $\mathcal{P} = \mathrm{conv}(\mathcal{A})$, then the multiplicities of $A(\mathcal{P})$ and $K[Ft]$ are related by*

$$e(A(\mathcal{P})) = |T(\mathbb{Z}^n/(v_2 - v_1, \ldots, v_q - v_1))|\, e(K[Ft]).$$

Proof. We set $\mathcal{B} = \{(v_1, 1)\}_{i=1}^{q}$ and $\mathcal{Q} = \mathrm{conv}(\mathcal{B})$. Note that there is a bijective map $\mathbb{Z}^n \cap i\mathcal{P} \to \mathbb{Z}^{n+1} \cap i\mathcal{Q}$, $\alpha \mapsto (\alpha, i)$, and $d = \dim(\mathcal{P}) = \dim(\mathcal{Q})$. Thus $\mathrm{vol}(\mathcal{P}) = \mathrm{vol}(\mathcal{Q})$. Since $\mathcal{B}$ lies in the affine hyperplane $x_{n+1} = 1$, applying Theorem 1.2.14 gives

$$\mathrm{vol}(\mathcal{P}) = |T(\mathbb{Z}^{n+1}/((v_2 - v_1, 0), \ldots, (v_q - v_1, 0)))| \lim_{i \to \infty} \frac{|\mathbb{Z}\mathcal{B} \cap i\mathcal{Q}|}{i^d}. \tag{9.10}$$

On the other hand, one has:

$$e(K[Ft]) = e(\overline{K[Ft]}) = d! \lim_{i \to \infty} \frac{|\mathbb{Z}\mathcal{B} \cap i\mathcal{Q}|}{i^d}. \tag{9.11}$$

The first equality follows from Proposition 9.2.13, while the second equality follows by definition of multiplicity noticing that $\overline{K[Ft]}$ is graded as in Eq. (9.9) and using Theorem 9.3.6(f). Hence the asserted equality follows from Eq. (9.10), Eq. (9.11), and observing that the map

$$T(\mathbb{Z}^n/(v_2 - v_1, \ldots, v_q - v_1)) \longrightarrow T(\mathbb{Z}^{n+1}/((v_2 - v_1, 0), \ldots, (v_q - v_1, 0)))$$

given by $\overline{v} \longmapsto \overline{(v, 0)}$ is bijective. $\Box$

Lemma 9.3.20 *Let $P^h \subset S[u]$ be the homogenization of $P = P_F$. Then*

(a) *P^h is the toric ideal of $K[F'] := K[z, x^{v_2 - v_1} z, \ldots, x^{v_q - v_1} z, x^{-v_1} z]$, where z is a new variable.*

(b) *The toric ideal of $K[x^{v_1} z, \ldots, x^{v_q} z, z]$ is the toric ideal P^h of $K[F']$.*

Proof. (a): The toric ideal of $K[F']$, denoted by P', is the kernel of the map $\varphi\colon S[u] \to K[F']$ induced by $t_i \mapsto x^{v_i - v_1} z$ for $i = 1, \ldots, q+1$, where $v_{q+1} = 0$ and $t_{q+1} = u$. Let $\mathcal{G}$ be the reduced Gröbner basis of P with respect to the GRevLex order $\prec$. First, we show the inclusion $P^h \subset P'$. Take an element f of $\mathcal{G}$. By Proposition 3.4.2, it suffices to show that f^h is in P'. We can write $f = t^{a^+} - t^{a^-}$ with $\mathrm{in}_\prec(f) = t^{a^+}$. Thus, $|a^+| \geq |a^-|$ and $f^h = t^{a^+} - t^{a^-}u^{|a|}$, where $a = a^+ - a^-$. We set $a = (a_1, \ldots, a_q)$. Using

$$0 = a_1 v_1 + \cdots + a_q v_q = a_2(v_2 - v_1) + \cdots + a_q(v_q - v_1) + |a| v_1,$$

we get that $f^h \in P'$, as required. Let $\mathcal{G}'$ be the reduced Gröbner basis of P' with respect to the GRevLex order. Next we show the inclusion $P' \subset P^h$. Take an element f' of $\mathcal{G}'$. It suffices to show that f' is in P^h. As f' is homogeneous in the standard grading of $S[u]$, we can write $f' = t^{c^+} - t^{c^-}u^{|c|}$ with $\mathrm{in}_\prec(f') = t^{c^+}$ and $c = (c_1, \ldots, c_q)$. Since $f' \in P'$, we get

$$\begin{aligned}(z^{c_1^+})(x^{v_2 - v_1}z)^{c_2^+} \cdots (x^{v_q - v_1}z)^{c_q^+} \\ = (z^{c_1^-})(x^{v_2 - v_1}z)^{c_2^-} \cdots (x^{v_q - v_1}z)^{c_q^-}(x^{-v_1}z)^{|c|}.\end{aligned}$$

Hence, $c_2(v_2 - v_1) + \cdots + c_q(v_q - v_1) = -|c| v_1$. Thus, $c_1 v_1 + \cdots + c_q v_q = 0$, that is, the binomial $f = t^{c^+} - t^{c^-}$ is in P. As $f' = f^h$, we get $f' \in P^h$.

(b): The mapping that sends z to $x^{v_1}z$ induces an isomorphism of K-algebras $K[F'] \to K[F'']$. So this part is a consequence of (a). □

Recall that the multiplicity of an affine algebra S/I is also denoted by $\deg(S/I)$ (see Definition 8.5.5). Thus, by Lemma 9.3.7, one can conveniently restate Proposition 9.3.19 as follows:

Proposition 9.3.21 *Let $\mathcal{B} = \{\beta_1, \ldots, \beta_m\}$ be a set of points of $\mathbb{Z}^n$ and let $\mathcal{Q} = \mathrm{conv}(\mathcal{B})$ be its convex hull. If $r = \dim(\mathcal{Q})$, then*

$$|T(\mathbb{Z}^n/\langle \beta_1 - \beta_m, \ldots, \beta_{m-1} - \beta_m \rangle)| \deg(K[x^{\beta_1}z, \ldots, x^{\beta_m}z]) = r!\mathrm{vol}(\mathcal{Q}).$$

Theorem 9.3.22 *Let P be the toric ideal of $K[F] = K[x^{v_1}, \ldots, x^{v_q}]$, let A be the matrix whose columns are $v_1, \ldots, v_q$ and let r be the rank of A. Then*

$$|T(\mathbb{Z}^n/\langle v_1, \ldots, v_q \rangle)| \deg(S/P) = r!\mathrm{vol}(\mathrm{conv}(v_1, \ldots, v_q, 0)).$$

Proof. By Proposition 8.5.6, $\deg(S/P) = \deg(S[u]/P^h)$. On the other hand, by Lemma 9.3.20(b), P^h is the toric ideal of the monomial subring:

$$K[F''] = K[x^{v_1}z, \ldots, x^{v_q}z, z],$$

where z is a new variable. Hence $S[u]/P^h \simeq K[F'']$. Therefore, setting $m = q + 1$, $\beta_i = v_i$ for $i = 1, \ldots, m - 1$, $\beta_m = 0$ and $\mathcal{B} = \{\beta_1, \ldots, \beta_m\}$, the result follows readily from Proposition 9.3.19. □

Corollary 9.3.23 *If $v_1, \ldots, v_q$ are positive integers and P is the toric ideal of $K[x^{v_1}, \ldots, x^{v_q}]$, then $\gcd(v_1, \ldots, v_q)\deg(S/P) = \max\{v_1, \ldots, v_q\}$.*

Proof. We may assume that $v_1 \leq \cdots \leq v_q$. The order of the group $T(\mathbb{Z}/\mathbb{Z}\{v_1, \ldots, v_q\})$ is equal to $\gcd(v_1, \ldots, v_q)$. Then, by Theorem 9.3.22, we get that $\gcd(v_1, \ldots, v_q)\deg(S/P)$ is $\mathrm{vol}([0, v_q]) = v_q$. $\Box$

Example 9.3.24 Let P be the toric ideal of the monomial subring

$$K[H] = K[x_2x_1^{-1}, x_3x_2^{-1}, x_4x_3^{-1}, x_1x_4^{-1}, x_5x_2^{-1}, x_3x_5^{-1}, x_4x_5^{-1}].$$

We employ the notation of Theorem 9.3.22. The rank of A is 4 and the height of P is 3. Using *Normaliz*, we get

$$4!\mathrm{vol}(\mathrm{conv}(v_1, \ldots, v_7, 0)) = 11.$$

As the group $\mathbb{Z}^5/\mathbb{Z}\{v_1, \ldots, v_7\}$ is torsion- free, by Theorem 9.3.22, we have that $\deg(K[t_1, \ldots, t_7]/P) = 11$.

Notation If B is an integral matrix, $\Delta_i(B)$ will denote the greatest common divisor of all the non-zero $i \times i$ minors of B.

Theorem 9.3.25 [146] *If $\mathcal{A} = \{v_1, \ldots, v_q\} \subset \mathbb{Z}^n$ and B is the matrix whose columns are the vectors in $\mathcal{B} = \{(v_i, 1)\}_{i=1}^q$, then the following hold:*

(a) $e(A(\mathcal{P})) = \Delta_r(B)e(K[Ft])$, *where* $r = \mathrm{rank}(B)$.

(b) $\Delta_r(B) = 1$ *if and only if* $\overline{K[Ft]} = A(\mathcal{P})$.

(c) $K[Ft] = A(\mathcal{P})$ *if and only if* $\mathbb{R}_+\mathcal{B} \cap \mathbb{Z}\mathcal{B} = \mathbb{N}\mathcal{B}$ *and* $T(\mathbb{Z}^n/\mathbb{Z}\mathcal{B}) = (0)$.

Proof. (a): It follows at once by successively applying Proposition 9.3.19, Lemma 1.2.20 and Lemma 1.3.17.

(b): $\Rightarrow$) By Theorem 9.3.6, $K[Ft] \subset A(\mathcal{P})$ is an integral extension of rings and $A(\mathcal{P})$ is normal. Thus $\overline{K[Ft]} \subset A(\mathcal{P})$. For the other inclusion, by Theorem 9.3.6(e), it suffices to prove that $A(\mathcal{P})$ is contained in the field of fractions of $K[Ft]$. Let $x^\alpha t^i \in A(\mathcal{P})_i$, that is, $\alpha \in \mathbb{Z}^n \cap i\mathcal{P}$ and $i \in \mathbb{N}$. Hence the system of equations $By = \alpha'$ has a rational solution, where α' is the column vector $(\alpha, i)^\top$. Thus, the augmented matrix $[B\,\alpha']$ has rank r. In general $\Delta_r([B\,\alpha'])$ divides $\Delta_r(B)$, so in this case they are equal because $\Delta_r(B) = 1$. By Theorem 1.6.3, the linear system $By = \alpha'$ has an integral solution. Hence $x^\alpha t^i$ is in the field of fractions of $K[Ft]$, as required.

(b): $\Leftarrow$) By hypothesis $\overline{K[Ft]} = A(\mathcal{P})$. Hence, taking multiplicities and using Proposition 9.2.13 and part (a), we get $\Delta_r(B) = 1$.

(c): By Theorem 9.3.6(g), $K[Ft] = A(\mathcal{P})$ if and only if $\mathcal{B} = \{(v_i, 1)\}_{i=1}^q$ is a Hilbert basis. Hence this part follows readily from Corollary 1.3.21 because $\mathbb{R}_+\mathcal{B}$ is a pointed cone (see Exercise 9.1.21). $\Box$

Corollary 9.3.26 *If $\Delta_r(B) = 1$, then $K[Ft]$ is normal if and only if $a(K[Ft]) < 0$ and the Hilbert polynomial of $K[Ft]$ is equal to $E_\mathcal{P}(x)$.*

Proof. $\Rightarrow$) It follows from Theorems 9.1.5 and 9.3.25.

$\Leftarrow$) Since the a-invariants of $K[Ft]$ and $A(\mathcal{P})$ are both negative, then $\dim_K K[Ft]_i$ is equal to $\dim_K A(\mathcal{P})_i$ for $i \in \mathbb{N}$. Hence, noticing that $K[Ft]_i \subset A(\mathcal{P})_i$, we get $K[Ft]_i = A(\mathcal{P})_i$ for $i \in \mathbb{N}$. Therefore $K[Ft] = A(\mathcal{P})$ and consequently $K[Ft]$ is normal. □

Definition 9.3.27 If $K[Ft] = A(\mathcal{P})$ and A is the incidence matrix of a clutter $\mathcal{C}$, we say that $\mathcal{C}$ is an *Ehrhart clutter.* If $K[Ft] = A(\mathcal{P})$, we say that $\mathcal{A}$ is an *Ehrhart point configuration.*

There are many interesting families of Ehrhart point configurations that will be presented later; see Theorems 9.3.29, 9.3.31, and Section 14.8.

Corollary 9.3.28 *If the subgroup* $L = (\{v_i - v_j \,|\, i, j = 1, \ldots, q\})$ *of* $\mathbb{Z}^n$ *has rank* n*, then* $\overline{K[Ft]} = A(\mathcal{P})$ *if and only if* $L = \mathbb{Z}^n$.

Proof. It follows from Theorem 9.3.25 and Lemma 1.2.20. □

Theorem 9.3.29 *If* $\mathcal{P}$ *has a unimodular covering with support in* $\mathcal{A}$*, then* $K[Ft] = A(\mathcal{P})$.

Proof. Let $\Delta = \mathrm{conv}(v_{i_1}, \ldots, v_{i_{d+1}})$ be any simplex in the unimodular covering of $\mathcal{P}$, where $d = \dim(\mathcal{P})$. Note that $\mathbb{Z}^{n+1}/\mathbb{Z}\{(v_{i_j}, 1)\}_{j=1}^{d+1}$ is torsion-free by Proposition 1.2.21. Hence, by Lemma 1.3.17 and Theorem 9.3.25, we obtain the equality $\overline{K[Ft]} = A(\mathcal{P})$. To finish the proof note that $K[Ft]$ is normal by Exercise 1.6.10. □

Proposition 9.3.30 [146] *If* A *is a unimodular matrix and the vectors* $v_1, \ldots, v_q$ *lie on a hyperplane not containing the origin, then* $K[Ft] = A(\mathcal{P})$.

Proof. Let B be the matrix with column vectors $(v_1, 1), \ldots, (v_q, 1)$. By Corollary 9.2.5, we have $K[F] \simeq K[Ft]$. Hence $r = \mathrm{rank}(A) = \mathrm{rank}(B)$. Using Corollary 9.1.3(d), one obtains that $K[Ft]$ is normal. As $\Delta_r(A) = 1$, applying Theorem 9.3.25 yields the asserted equality. □

The following result is a useful generalization of [383, Theorem 7.1].

Theorem 9.3.31 [146] *If the Rees algebra* $R[Ft]$ *is a normal domain and* $v_1, \ldots, v_q$ *lie on a hyperplane*

$$b_1x_1 + \cdots + b_nx_n = 1 \qquad (b_i > 0 \; \forall i),$$

then $K[Ft] = A(\mathcal{P})$.

Proof. The inclusion $K[Ft] \subset A(\mathcal{P})$ is clear. To show the reverse inclusion take $x^\alpha t^i \in A(\mathcal{P})$. By Theorem 9.3.6(b), one can write

$$(\alpha, i) = \lambda_1(v_1, 1) + \cdots + \lambda_q(v_q, 1) \qquad (\lambda_j \geq 0 \ \forall j). \tag{9.12}$$

Hence $(\alpha, i) \in \mathbb{Z}^{n+1} \cap \mathbb{R}_+\mathcal{A}' = \mathbb{Z}\mathcal{A}' \cap \mathbb{R}_+\mathcal{A}$, where

$$\mathcal{A}' = \{(e_1, 0), \ldots, (e_n, 0), (v_1, 1), \ldots, (v_q, 1)\}$$

and e_i is the ith unit vector in $\mathbb{R}^n$. By the normality of $R[Ft]$ one has $(\alpha, i) \in \mathbb{N}\mathcal{A}'$. Thus one can write

$$(\alpha, i) = m_1(e_1, 0) + \cdots + m_n(e_n, 0) + n_1(v_1, 1) + \cdots + n_q(v_q, 1), \tag{9.13}$$

where m_i, n_j are in $\mathbb{N}$ for all i, j. From Eqs. (9.12) and (9.13):

$$\begin{aligned} \alpha &= m_1e_1 + \cdots + m_ne_n + n_1v_1 + \cdots + n_qv_q \\ &= \lambda_1v_1 + \cdots + \lambda_qv_q, \\ i &= n_1 + \cdots + n_q. \end{aligned}$$

Taking the inner product of α with $b = (b_1, \ldots, b_n)$ yields

$$\langle \alpha, b \rangle = m_1b_1 + \cdots + m_nb_n + i$$

and $\langle \alpha, b \rangle = \lambda_1\langle v_1, b \rangle + \cdots + \lambda_q\langle v_q, b \rangle = i$. Hence $m_i = 0$ for all i. Since $x^\alpha t^i = (x^{v_1}t)^{n_1} \cdots (x^{v_q}t)^{n_q}$, we get $x^\alpha t^i \in K[Ft]$, as required. $\square$

Proposition 9.3.32 *If $x^{v_1}, \ldots, x^{v_q}$ are monomials of degree $k \geq 2$ in the polynomial ring $R = K[x_1, \ldots, x_n]$ and $R[Ft]$ is normal, then*

$$e(R[Ft]) = n!\mathrm{vol}(\mathcal{P}'),$$

where $\mathcal{P}' = \mathrm{conv}(\mathcal{A}')$ and $\mathcal{A}' = \{(v_1, 1), \ldots, (v_q, 1), e_1, \ldots, e_n\}$.

Proof. The affine hyperplane $x_1 + \cdots + x_n - (k-1)x_{n+1} = 1$ contains $\mathcal{A}'$. Then $R[Ft]$ is a standard graded K-algebra and $\dim(\mathcal{P}') = n$. Thus, using Theorem 9.3.25, we get an isomorphism $R[Ft] \simeq A(\mathcal{P}')$, as graded K-algebras. By Lemma 9.3.7 the multiplicity of $A(\mathcal{P}')$ is $n!\mathrm{vol}(\mathcal{P}')$. $\square$

Proposition 9.3.33 [385] *If $R[Ft]$ is normal and there is $x_0 \in \mathbb{N}_+^n$ such that $\langle x_0, v_i \rangle = k$, $i = 1, \ldots, q$, for some $k \geq 2$, then the following hold:*

(a) *The torsion subgroup of $\mathbb{Z}^n/\mathbb{Z}\mathcal{A}$ is cyclic and its order divides k.*

(b) *If $\mathbb{Z}^n/\mathbb{Z}\mathcal{A}$ is a finite group and $x_0 = (1, \ldots, 1)$, then $\mathbb{Z}^n/\mathbb{Z}\mathcal{A} \simeq \mathbb{Z}_k$.*

Proof. We set $\mathcal{B} = \{(v_1, 1)\}_{i=1}^q$. There is an exact sequence of finite groups

$$0 \longrightarrow T(\mathbb{Z}^{n+1}/\mathbb{Z}\mathcal{B}) \stackrel{\varphi}{\longrightarrow} T(\mathbb{Z}^n/\mathbb{Z}\mathcal{A}) \stackrel{\psi}{\longrightarrow} \mathbb{Z}_k, \qquad (*)$$

where φ and ψ are given by $\varphi(\overline{(\alpha, b)}) = \overline{\alpha}$ and $\psi(\overline{\alpha}) = \overline{\langle \alpha, x_0 \rangle}$, $\alpha \in \mathbb{Z}^n$, $b \in \mathbb{Z}$. Consider the matrix B whose columns are the vectors in the set $\mathcal{B}$. By Theorems 9.3.31 and 9.3.25 we get $\Delta_r(B) = 1$, where r is the rank of $\mathbb{Z}\mathcal{B}$. Since $\Delta_r(B)$ is the order of $T(\mathbb{Z}^{n+1}/\mathbb{Z}\mathcal{B})$, part (a) follows using Eq. $(*)$. To prove (b), note that $\mathbb{Z}^n/\mathbb{Z}\mathcal{A}$ is a torsion group and the ith unit vector $\overline{e}_i$ maps into the element $\overline{1}$ under the map ψ. Hence ψ is onto, and ψ gives the required isomorphism. □

Exercises

9.3.34 Let $\mathcal{P}$ be a lattice polytope in $\mathbb{Z}^n$. Then $K[\mathcal{P}] = A(\mathcal{P})$ if and only if $k\mathcal{P} \cap \mathbb{Z}^n = \mathcal{P} \cap \mathbb{Z}^n + \cdots + \mathcal{P} \cap \mathbb{Z}^n$ (k-times) for any $k \in \mathbb{N}_+$. If this holds $\mathcal{P}$ is said to satisfy the *integer decomposition property* (cf. Definition 14.6.3).

9.3.35 Let $F = \{x^{v_1}, \ldots, x^{v_q}\} \subset K[\mathbf{x}]$ be a set of square-free monomials of degree k. If $\mathcal{P} = \mathrm{conv}(v_1, \ldots, v_q)$, then $K[\mathcal{P}] \simeq K[F]$ as graded K-algebras.

9.3.36 Let $\mathcal{P} = \mathrm{conv}((0,0), (0,1), (1,0), (1,1))$ be the unit square. In the figure below we show the integral points of $4\mathcal{P}$. Verify the following:

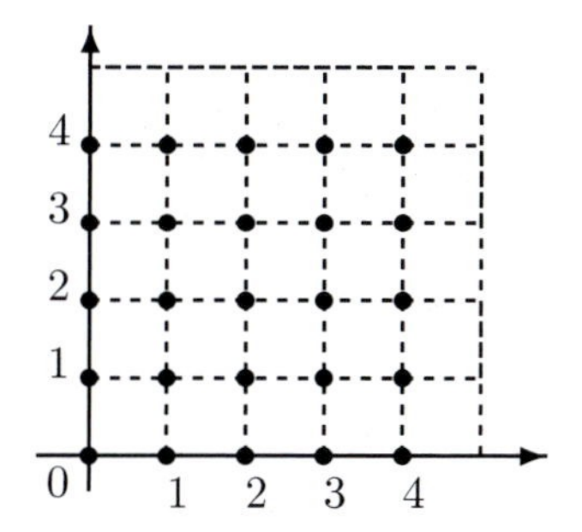

$$|\mathbb{Z}^2 \cap 4\mathcal{P}| = 25$$
$$E_{\mathcal{P}}(x) = (x+1)^2$$
$$\mathrm{vol}(\mathcal{P}) = 1$$
$$A(\mathcal{P}) = K[Ft]$$
$$F(A(\mathcal{P}), x) = (1+x)/(1-x)^3$$

9.3.37 Let A be an integral matrix of size $n \times n$ with $\det(A) \neq 0$ and let $\mathcal{A} = \{v_1, \ldots, v_n\}$ be the set of columns of A. Consider the nth unit cube $\mathcal{Q} = [0,1]^n$ and the parallelotope

$$\mathcal{P} = \{\lambda_1 v_1 + \cdots + \lambda_n v_n | 0 \leq \lambda_i \leq 1\}.$$

If $\mathcal{A}$ is a Hilbert basis, prove that the Ehrhart polynomials of $\mathcal{Q}$ and $\mathcal{P}$ are equal to $(x+1)^n$. Prove that the normalized volume of $\mathcal{P}$ is equal to $n!$.

9.3.38 If $\mathcal{P} = \text{conv}((0,0,0),(12,1,0),(0,1,1),(1,0,1))$, use *Normaliz* [68] to show that the Ehrhart polynomial of $\mathcal{P}$ is $E_{\mathcal{P}}(x) = 13/6x^3+x^2-1/6x+1$.

9.3.39 Let $R = K[x_1,\ldots,x_n]$ be a polynomial ring. If $\mathfrak{m} = (x_1,\ldots,x_n)$, prove that $e(K[\mathfrak{m}^d])$, the multiplicity of $K[\mathfrak{m}^d]$, is equal to d^{n-1}.

9.3.40 Let $\mathcal{P}$ be the convex hull of $\mathcal{A} = \{(3,1),(1,3)\}$ in $\mathbb{R}^2$, let K be the real numbers field and let $F = \{x_1^3x_2, x_1x_2^3\} \subset K[x_1,x_2]$. Prove that $\mathcal{A}$ is a weakly unimodular covering of $\mathcal{P}$ and $\overline{K[Ft]} \neq A(\mathcal{P})$.

9.3.41 If $K[F]$ is homogeneous and the lattice polytope $\mathcal{P}$ has a weakly unimodular covering with support in $\mathcal{A}$, then $K[F]$ is normal.

9.3.42 If $F = \{x^{v_1},\ldots,x^{v_q}\}$ is a set of monomials in $R = K[x_1,\ldots,x_n]$ and P' is the toric ideal of $R[Ft]$, then

$$\deg(K[x_1,\ldots,x_n,t_1,\ldots,t_q]/P') = (n+1)!\text{vol}(\mathcal{P}'),$$

where $\mathcal{P}' = \text{conv}(\mathcal{A}' \cup \{0\})$ and $\mathcal{A}' = \{(v_1,1),\ldots,(v_q,1),e_1,\ldots,e_n\}$.

9.3.43 If $\{(v_1,1),\ldots,(v_q,1)\} \subset \mathbb{N}^{n+1}$ is a Hilbert basis with $|v_i| = d$ for all i and $\text{rank}(v_1,\ldots,v_q) = n$, prove that $\mathbb{Z}^n/\mathbb{Z}\{v_1,\ldots,v_q\} \simeq \mathbb{Z}_d$.

9.4 The degree of lattice and toric ideals

In this section we give formulas to compute the degree of a lattice ideal in terms of the torsion of certain factor groups of $\mathbb{Z}^q$ and in terms of relative volumes of polytopes. Then we study primary decompositions of lattice ideals over an arbitrary field. In this section we will use the Eisenbud–Sturmfels theory of binomial ideals over algebraically closed fields [134].

Let $\mathcal{L} \subset \mathbb{Z}^q$ be a lattice and let e_i be the ith unit vector in $\mathbb{Z}^{q+1}$. For $a = (a_i) \in \mathbb{Z}^q$ define the *value* of a as $|a| = \sum_{i=1}^q a_i$, and the *homogenization* of a with respect to e_{q+1} as $a^h = (a,0) - |a|e_{q+1}$ if $|a| \geq 0$ and $a^h = (-a)^h$ if $|a| < 0$. The *homogenization* of $\mathcal{L}$, denoted by $\mathcal{L}^h$, is the lattice of $\mathbb{Z}^{q+1}$ generated by all a^h such that $a \in \mathcal{L}$.

Lemma 9.4.1 *Let $I(\mathcal{L}) \subset S$ be a lattice ideal. Then the following hold.*

(a) *$I(\mathcal{L})^h \subset S[u]$ is a lattice ideal.*

(b) *If $\mathcal{L} = \mathbb{Z}\{b_1,\ldots,b_r\}$, then $\mathcal{L}^h = \mathbb{Z}\{b_1^h,\ldots,b_r^h\}$.*

(c) *$I(\mathcal{L}^h) = I(\mathcal{L})^h$.*

Proof. It is left as an exercise. □

Theorem 9.4.2 [294, 335] *If $\mathcal{L} \subset \mathbb{Z}^q$ is a lattice of rank r and $I_\rho(\mathcal{L})$ is its lattice ideal with respect to a partial character ρ, then*

(a) *If $r < q$, there is an integer matrix A of size $(q-r) \times q$ and rank $q-r$ such that we have the containment of rank r lattices $\mathcal{L} \subset \ker_{\mathbb{Z}}(A)$, with equality if and only if $\mathbb{Z}^q/\mathcal{L}$ is torsion-free.*

(b) *If $r < q$ and $v_1, \ldots, v_q$ are the columns of A, then*

$$\deg(S/I_\rho(\mathcal{L})) \;=\; \frac{|T(\mathbb{Z}^q/\mathcal{L})|(q-r)!\mathrm{vol}(\mathrm{conv}(0, v_1, \ldots, v_q))}{|T(\mathbb{Z}^{q-r}/\langle v_1, \ldots, v_q\rangle)|}.$$

(c) *If $r = q$, then* $\deg(S/I_\rho(\mathcal{L})) = |\mathbb{Z}^q/\mathcal{L}|$.

Proof. (a): This follows at once from Lemma 8.2.5.

(b): By Proposition 8.5.8, we may assume that K is algebraically closed of characteristic zero and that $\rho(a) = 1$ for $a \in \mathcal{L}$, i.e., $I_\rho(\mathcal{L}) = I(\mathcal{L})$. Let P be the toric ideal of $K[x^{v_1}, \ldots, x^{v_q}]$. By Theorem 9.3.22, we need only show the equality

$$\deg(S/I(\mathcal{L})) = |T(\mathbb{Z}^q/\mathcal{L})| \deg(S/P).$$

Let $\mathcal{L}_s$ be the saturation of $\mathcal{L}$. By part (a), $\mathcal{L}_s$ is equal to $\ker_{\mathbb{Z}}(A)$. We set $c = |T(\mathbb{Z}^q/\mathcal{L})|$. Notice that $T(\mathbb{Z}^q/\mathcal{L}) = \mathcal{L}_s/\mathcal{L}$. According to [134, Corollaries 2.2 and 2.5], there are distinct partial characters $\rho_1, \ldots, \rho_c$ of $\mathcal{L}_s$, extending the trivial character $\rho(a) = 1$ for a in $\mathcal{L}$, such that the minimal primary decomposition of $I(\mathcal{L})$ is given by

$$I(\mathcal{L}) = I_{\rho_1}(\mathcal{L}_s) \cap \cdots \cap I_{\rho_c}(\mathcal{L}_s),$$

By part (a), P is a minimal prime of $I(\mathcal{L})$. Thus we may assume $\rho_1(a) = 1$ for a in $\mathcal{L}_s$, i.e., P is equal to the lattice ideal $I_{\rho_1}(\mathcal{L}_s)$. By additivity of the degree (see Proposition 8.5.9), we get

$$\deg(S/I(\mathcal{L})) = \deg(S/I_{\rho_1}(\mathcal{L}_s)) + \cdots + \deg(S/I_{\rho_c}(\mathcal{L}_s)).$$

Therefore, it suffices to recall that the degree is independent of the character, i.e., $\deg(S/P) = \deg(S/I_{\rho_k}(\mathcal{L}_s))$ for $k = 1, \ldots, c$ (see Proposition 8.5.8).

(c): Let $t^{c_1^+} - t^{c_1^-}, \ldots, t^{c_m^+} - t^{c_m^-}$ be a set of generators of $I(\mathcal{L})$. By Lemma 8.2.23, $\mathcal{L}$ is generated by $c_1, \ldots, c_m$. We may assume that $|c_i| \geq 0$ for all i. Then, by Lemma 9.4.1(b), $\mathcal{L}^h$ is generated by $c_1^h, \ldots, c_m^h$. Let C and C^h be the matrices with rows $c_1, \ldots, c_m$ and $c_1^h, \ldots, c_m^h$, respectively. Notice that C^h is obtained from C by adding the column vector given by $b = (-|c_1|, \ldots, -|c_m|)^\top$. Since b is a linear combination of the columns of C, using the fundamental theorem of finitely generated abelian group (Theorem 1.3.16), we get that the groups $\mathbb{Z}^q/\mathcal{L}$ and $\mathbb{Z}^{q+1}/\mathcal{L}^h$ have the same torsion. Thus, $|\mathbb{Z}^q/\mathcal{L}|$ is equal to $|T(\mathbb{Z}^{q+1}/\mathcal{L}^h)|$. By Proposition 8.5.6, $\deg(S/I(\mathcal{L})) = \deg(S[u]/I(\mathcal{L})^h)$, and by Lemma 9.4.1(c), $I(\mathcal{L})^h = I(\mathcal{L}^h)$. Altogether it suffices to show that the degree of $S[u]/I(\mathcal{L}^h)$ is equal to $|T(\mathbb{Z}^{q+1}/\mathcal{L}^h)|$. Since $S[u]/I(\mathcal{L}^h)$ has dimension 1, we get $q = \operatorname{rank}(\mathcal{L}^h)$. By part (a), there is a row vector $A' = (v_1', \ldots, v_{q+1}')$ such that $\mathcal{L}^h \subset \ker_{\mathbb{Z}}(A')$. Since the $\mathbb{Q}$-vector spaces generated by $\mathcal{L}^h$ and $\ker_{\mathbb{Z}}(A')$ are equal and have dimension q and since $\langle \alpha, \mathbf{1}\rangle = 0$ for all α in $\mathcal{L}^h$, it follows readily, using orthogonal complements, that $v_1' = v_i'$ for all i. Thus, by part (b), the required equality follows. □

Corollary 9.4.3 *If $I(\mathcal{L})$ is a lattice ideal of dimension 1 which is homogeneous with respect to a positive vector $\mathbf{d} = (d_1, \ldots, d_q)$, then*

$$\gcd(d_1, \ldots, d_q)\deg(S/I(\mathcal{L})) = \max\{d_1, \ldots, d_q\}|T(\mathbb{Z}^q/\mathcal{L})|.$$

Proof. Let A be the $1 \times q$ matrix $(d_1, \ldots, d_q)$. By hypothesis $\mathcal{L} \subset \ker_{\mathbb{Z}}(A)$. The order of $T(\mathbb{Z}/\mathbb{Z}\{d_1, \ldots, d_q\})$ is equal to $\gcd(d_1, \ldots, d_q)$. Therefore, by Theorem 9.4.2, we get

$$\deg(S/I(\mathcal{L})) = \frac{T(\mathbb{Z}^q/\mathcal{L})\mathrm{vol}(\mathrm{conv}(0, d_1, \ldots, d_q))}{T(\mathbb{Z}/\mathbb{Z}\{d_1, \ldots, d_q\})} = \frac{|T(\mathbb{Z}^q/\mathcal{L})|\max\{d_1, \ldots, d_q\}}{\gcd(d_1, \ldots, d_q)},$$

as required. □

Corollary 9.4.4 *Let $\mathcal{L} \subset \mathbb{Z}^q$ be a lattice of rank $q-1$ and let $\alpha_1, \ldots, \alpha_{q-1}$ be a $\mathbb{Z}$-basis of $\mathcal{L}$. If $\alpha_i = (\alpha_{i,1}, \ldots, \alpha_{i,q})$, for $i = 1, \ldots, q-1$, and*

$$n_i = (-1)^i \det\begin{pmatrix} \alpha_{1,1} & \cdots & \alpha_{1,i-1} & \alpha_{1,i+1} & \cdots & \alpha_{1,q} \\ \vdots & & \vdots & \vdots & \vdots & \\ \alpha_{q-1,1} & \cdots & \alpha_{q-1,i-1} & \alpha_{q-1,i+1} & \cdots & \alpha_{q-1,q} \end{pmatrix} \quad \text{for } 1 \leq i \leq q,$$

then $\deg(S/I(\mathcal{L})) = \max\{n_1, \ldots, n_q, 0\} - \min\{n_1, \ldots, n_q, 0\}$.

Proof. Let B be the $(q-1) \times q$ matrix with rows $\alpha_1, \ldots, \alpha_{q-1}$ and let A be the $1 \times q$ matrix $(n_1, \ldots, n_q)$. We set $r = q-1$. The order of $T(\mathbb{Z}^q/\mathcal{L})$ is equal to $\gcd(n_1, \ldots, n_q)$, the gcd of the $r \times r$ minors of B. The order of $T(\mathbb{Z}/\mathbb{Z}\{n_1, \ldots, n_q\})$ is also equal to $\gcd(n_1, \ldots, n_q)$. Since $\alpha_i \in \ker(A)$ for all i, we obtain that $\mathcal{L} \subset \ker(A)$. Hence, by Theorem 9.4.2, we get that $\deg(S/I(\mathcal{L}))$ is equal to $\mathrm{vol}(\mathrm{conv}(0, n_1, \ldots, n_q))$ which is equal to $\max\{n_1, \ldots, n_q, 0\} - \min\{n_1, \ldots, n_q, 0\}$. □

Normaliz [68] computes *normalized volumes* of lattice polytopes using polyhedral geometry. Thus we can compute the degree of any lattice ideal using Theorem 9.4.2.

Example 9.4.5 Let K be the field of rational numbers and let $\mathcal{L}$ be the lattice generated by

$$a_1 = (2,1,1,1,-1,-1,-1,-2), \qquad a_2 = (1,1,-1,-1,1,1,-1,-1),$$
$$a_3 = (2,-1,1,-2,1,-1,1,-1), \qquad a_4 = (5,-5,0,0,0,0,0,0).$$

We use the notation of Theorem 9.4.2. The lattice $\mathcal{L}$ has rank 4 and $a_1,\ldots,a_4,e_1,e_3,e_4,e_5$ form a $\mathbb{Q}$-basis of $\mathbb{Q}^8$. In this case we obtain the matrix

$$A = \begin{pmatrix} 4 & 4 & 0 & 0 & 0 & -1 & 1 & 6 \\ 0 & 0 & 1 & 0 & 0 & 1 & 0 & 0 \\ 0 & 0 & 0 & 4 & 0 & 7 & 9 & -6 \\ 0 & 0 & 0 & 0 & 2 & -3 & -3 & 2 \end{pmatrix}$$

whose columns are denoted by $v_1,\ldots,v_8$. Let P be the toric ideal of $K[x^{v_1},\ldots,x^{v_8}]$. Therefore, by Theorem 9.4.2, we get

$$\deg(S/I(\mathcal{L})) = \frac{5(4)!\mathrm{vol}(\mathrm{conv}(0,v_1,\ldots,v_8))}{|T(\mathbb{Z}^4/\langle v_1,\ldots,v_8\rangle)|} = \frac{(5)(200)}{8} = 125.$$

The normalized volume of the polytope $\mathrm{conv}(0,v_1,\ldots,v_8)$ was computed using *Normaliz* [68].

Theorem 9.4.6 [335] *Let $I(\mathcal{L})$ be a lattice ideal of S over an arbitrary field K of characteristic p, let c be the number of associated primes of $I(\mathcal{L})$, and for $p>0$, let G be the unique largest subgroup of $T(\mathbb{Z}^q/\mathcal{L})$ whose order is relatively prime to p. Then*

(a) *All associated primes of $I(\mathcal{L})$ have height equal to* $\mathrm{rank}(\mathcal{L})$.

(b) $|T(\mathbb{Z}^q/\mathcal{L})| \geq c$ *if* $p=0$ *and* $|G|\geq c$ *if* $p>0$, *with equality if K is algebraically closed.*

(c) $\deg(S/I(\mathcal{L})) \geq |T(\mathbb{Z}^q/\mathcal{L})|$ *if* $p=0$ *and* $\deg(S/I(\mathcal{L}))\geq |G|$ *if* $p>0$.

Proof. Let $\overline{K}$ be the algebraic closure of K and let $\overline{S} = \overline{K}[t_1,\ldots,t_q]$ be the corresponding polynomial ring with coefficients in $\overline{K}$. Thus, we have an integral extension $S\subset \overline{S}$ of normal domains. We set $I=I(\mathcal{L})$ and $\overline{I} = I\overline{S}$, where the latter is the extension of I to $\overline{S}$. The ideal $\overline{I}$ is the lattice ideal of $\mathcal{L}$ in $\overline{S}$ (see Exercise 8.2.35(c)). Hence, as $\overline{K}$ is algebraically closed, by [134, Corollaries 2.2 and 2.5] $\overline{I}$ has a unique irredundant primary decomposition

$$\overline{I} = \overline{\mathfrak{q}}_1 \cap \cdots \cap \overline{\mathfrak{q}}_{c_1}, \tag{$\dagger$}$$

where $c_1 = |T(\mathbb{Z}^q/\mathcal{L})|$ if $p=0$ and $c_1=|G|$ if $p>0$. Notice that $c_1 = |T(\mathbb{Z}^q/\mathcal{L})|$ if p is relatively prime to $|T(\mathbb{Z}^q/\mathcal{L})|$. Furthermore, also by [134, Corollaries 2.2 and 2.5], one has that if $\overline{\mathfrak{p}}_i = \mathrm{rad}(\overline{\mathfrak{q}}_i)$ for $i=1,\ldots,c_1$, then $\overline{\mathfrak{p}}_1, \ldots, \overline{\mathfrak{p}}_{c_1}$ are the associated primes of $\overline{I}$ and $\text{ht}(\overline{\mathfrak{p}}_i) = \text{rank}(\mathcal{L})$ for $i = 1, \ldots, c_1$. Hence, by Exercise 8.2.35(b), one has a primary decomposition

$$I = (\overline{\mathfrak{q}}_1 \cap S) \cap \cdots \cap (\overline{\mathfrak{q}}_{c_1} \cap S) \qquad (\ddagger)$$

such that $\text{rad}(\overline{\mathfrak{q}}_i \cap S) = \text{rad}(\overline{\mathfrak{q}}_i) \cap S = \overline{\mathfrak{p}}_i \cap S$. We set $\mathfrak{p}_i = \overline{\mathfrak{p}}_i \cap S$ and $\mathfrak{q}_i = \overline{\mathfrak{q}}_i \cap S$ for $i = 1, \ldots, c_1$.

(a): Since S is a normal domain and $S \subset \overline{S}$ is an integral extension, we get $\text{ht}(\mathfrak{p}_i) = \text{ht}(\overline{\mathfrak{p}}_i) = \text{rank}(\mathcal{L})$ for all i (see Proposition 2.4.14).

(b): By Eq. ($\ddagger$), the associated primes of I are contained in $\{\mathfrak{p}_1, \ldots, \mathfrak{p}_{c_1}\}$. Thus, $c_1 \geq c$, which proves the first part. Now, assume that $K = \overline{K}$. By (a), we may assume that $\mathfrak{p}_1, \ldots, \mathfrak{p}_c$ are the minimal primes of I. Consequently I has a unique minimal primary decomposition $I = \mathfrak{Q}_1 \cap \cdots \cap \mathfrak{Q}_c$ such that $\mathfrak{Q}_i$ is $\mathfrak{p}_i$-primary and $\text{ht}(\mathfrak{p}_i) = \text{rank}(\mathcal{L})$ for $i = 1, \ldots, c$. As $I = \overline{I}$, from Eq. ($\dagger$), we get that $c_1 = c$.

(c): Using Lemma 8.5.7, Eq. ($\dagger$) that was stated at the beginning of the proof, and the additivity of the degree (Proposition 8.5.9), we get

$$\deg(S/I) = \deg(\overline{S}/\overline{I}) = \textstyle\sum_{i=1}^{c_1} \deg(\overline{S}/\overline{\mathfrak{q}}_i) \geq c_1. \qquad \square$$

Exercises

9.4.7 Let $\mathcal{L} = \langle (2,-1,-1), (-3,1,-1) \rangle$. Prove that $I(\mathcal{L})$ is a non-graded lattice ideal of height 2 given by

$$I(\mathcal{L}) = ((t_1^2 - t_2t_3, t_2 - t_1^3t_3)\colon (t_1t_2t_3)^\infty) = (t_1^2 - t_2t_3, t_1t_3^2 - 1).$$

Then, apply Corollary 9.4.4 with $v_1 = -2$, $v_2 = -5$ and $v_3 = 1$, to show that $K[t_1, t_2, t_3]/I(\mathcal{L})$ has degree 6.

9.5 Laplacian matrices and ideals

In this section we use algebraic graph theory to study pure binomial matrices and their matrix ideals. We are interested in relating the combinatorics of a graph with the algebraic invariants and properties of the Laplacian ideal associated to the Laplacian matrix of the graph.

If $\mathbf{d} = (d_1, \ldots, d_q) \in \mathbb{N}_+^q$, then $S = \oplus_{k=0}^{\infty} S_k$ has a grading induced by setting $\deg(t_i) = d_i$ for $i = 1, \ldots, q$. A graded ideal of $S = K[t_1, \ldots, t_q]$ is an ideal which is graded with respect to some vector $\mathbf{d}$ in $\mathbb{N}_+^q$.

Definition 9.5.1 A *matrix ideal* is an ideal of the form

$$I(L) = (t^{a_i^+} - t^{a_i^-} \mid i = 1, \ldots, m) \subset S,$$

where the a_i's are the columns of an integer matrix L of size $q \times m$.

Note that $I(L)$ is graded if and only if $\mathbf{d}L = 0$ for some $\mathbf{d} \in \mathbb{N}_+^q$.

Example 9.5.2 If L is the matrix

$$L = \begin{pmatrix} 2 & -1 & -1 \\ 0 & 1 & -1 \\ 0 & -1 & 1 \end{pmatrix},$$

then $I(L) = (t_1^2 - 1,\, t_2 - t_1t_3,\, t_3 - t_1t_2)$.

Proposition 9.5.3 [335] *Let $I = I(L)$ be a graded matrix ideal and let $\mathcal{L}$ be the lattice spanned by the columns of L. Suppose that $V(I, t_i) = \{0\}$ for all i. Then we have the following.*

(a) *$I = I(\mathcal{L})$ if I is unmixed, or else $I = I(\mathcal{L}) \cap \mathfrak{q}$, if I is not unmixed, where $\mathfrak{q}$ is an $\mathfrak{m}$-primary component of I and $\mathfrak{m} = (t_1, \ldots, t_q)$.*

(b) $(\gcd\{d_i\}_{i=1}^q) \deg(S/I) = \max_i\{d_i\}|T(\mathbb{Z}^q/\mathcal{L})|$.

Proof. By Lemma 8.2.11, one has $(I\colon (t_1 \cdots t_q)^\infty) = I(\mathcal{L})$. By considering a primary decomposition of I, part (a) follows from Lemma 8.3.20 and Proposition 8.3.22. Part (b) follows from Corollary 9.4.3 and part (a). $\square$

Definition 9.5.4 Let $a_{i,j} \in \mathbb{N}$, $i, j = 1, \ldots, q$, and let L be a $q \times q$ matrix of the following special form:

$$L = \begin{pmatrix} a_{1,1} & -a_{1,2} & \cdots & -a_{1,q} \\ -a_{2,1} & a_{2,2} & \cdots & -a_{2,q} \\ \vdots & \vdots & \cdots & \vdots \\ -a_{q,1} & -a_{q,2} & \cdots & a_{q,q} \end{pmatrix}.$$

L is called a *pure binomial* matrix (*PB* matrix) if $a_{j,j} > 0$ for all j, and for each row and each column of L at least one off-diagonal entry is non-zero.

Definition 9.5.5 Let $B = (b_{i,j})$ be a $q \times q$ real matrix. The *underlying digraph* G_B of B has vertex set $\{t_1, \ldots, t_q\}$, with an arc from t_i to t_j iff $b_{i,j} \neq 0$. If $b_{i,i} \neq 0$, we put a loop at t_i. A digraph is *strongly connected* if for any two vertices t_i and t_j there is a directed path form t_i to t_j and a direct path from t_j to t_i.

Theorem 9.5.6 (Duality Theorem [335]) *Let L be a PB matrix of size $q \times q$ such that $L\mathbf{c}^\top = 0$ for some $\mathbf{c} \in \mathbb{N}_+^q$. If G_L is strongly connected, then* $\operatorname{rank}(L) = q - 1$ *and there is $\mathbf{d}$ in $\mathbb{N}_+^q$ such that $\mathbf{d}L = 0$.*

Proof. Let L be a PB matrix as in Definition 9.5.4. First we treat the case $\mathbf{c} = \mathbf{1} = (1, \ldots, 1)$. Let L be a PB matrix as in Definition 9.5.4. We can write $L = D - A$, where $D = \operatorname{diag}(a_{1,1}, \ldots, a_{q,q})$ and A is the matrix whose

i, j entry is $a_{i,j}$ if $i \neq j$ and whose diagonal entries are equal to zero. We set $\delta_i = a_{i,i}$ for $i = 1, \ldots, q$. By hypothesis $L\mathbf{1}^\top = 0$, hence $\operatorname{rank}(L) \leq q - 1$. There exists a non-zero vector $\mathbf{d} \in \mathbb{Z}^q$ such that $\mathbf{d}L = 0$. Therefore

$$\mathbf{d}D = \mathbf{d}A = \mathbf{d}D(D^{-1}A).$$

Since $L\mathbf{1}^\top = 0$, we get that $D\mathbf{1}^\top = A\mathbf{1}^\top$ or equivalently $(D^{-1}A)\mathbf{1}^\top = \mathbf{1}^\top$. Thus, as the entries of $D^{-1}A$ are nonnegative, the matrix $B := D^{-1}A$ is stochastic. It is well-known that the spectral radius $\rho(B)$ of a stochastic matrix B is equal to 1 [28, Theorem 5.3], where $\rho(B)$ is the maximum of the moduli of the eigenvalues of B. As the diagonal entries of B are zero and $\delta_i > 0$ for all i, the underlying digraph G_B of B is equal to the digraph obtained from G_L by removing all loops of G. Since G_L is strongly connected so is G_B, and by the Perron–Frobenius Theorem for nonnegative matrices [190, Theorem 8.8.1, p. 178], $\rho(B) = 1$ and 1 is a simple eigenvalue of B (i.e., the eigenspace of B relative to $\rho(B) = 1$ is 1-dimensional), and if z is an eigenvector for $\rho(B) = 1$, then no entries of z are zero and all have the same sign. Applying this to $z = \mathbf{d}D$, we get that $d_i \neq 0$ for all i and all entries of $\mathbf{d}$ have the same sign. Hence, $\ker(L^\top) = (\mathbf{d}^\top)$ for any non-zero vector $\mathbf{d}$ such that $\mathbf{d}L = 0$ and L has rank $q - 1$.

The general case follows by considering the matrix $\widetilde{L} = L\operatorname{diag}(c_1, \ldots, c_q)$, where $\mathbf{c} = (c_1, \ldots, c_q)$. Notice that $\widetilde{L}\mathbf{1}^\top = 0$ because $L\mathbf{c}^\top = 0$. $\square$

Lemma 9.5.7 *Let L be an integer matrix of size $q \times q$ with rows $\ell_1, \ldots, \ell_q$ and let $\operatorname{adj}(L) = (L_{i,j})$ be the adjoint matrix of L. Suppose $L\mathbf{c}^\top = 0$ for some $\mathbf{c}$ in $\mathbb{N}_+^q$. The following hold.*

(a) *If $\operatorname{rank}(\ell_1, \ldots, \widehat{\ell_i}, \ldots, \ell_q) = q - 1$ and $L_{i,i} \geq 0$ for all i, then $\mathbf{d}L = 0$ for some $\mathbf{d} \in \mathbb{N}_+^q$.*

(b) *If $L_{i,i} > 0$ for all i, then $\mathbf{d}L = 0$ for some $\mathbf{d} \in \mathbb{N}_+^q$.*

(c) *If L is a PB matrix and $\operatorname{rank}(\ell_1, \ldots, \widehat{\ell_i}, \ldots, \ell_q) = q - 1$ for all i, then $\mathbf{d}L = 0$ for some $\mathbf{d} \in \mathbb{N}_+^q$.*

Proof. (a): Let L_i be the ith column of $\operatorname{adj}(L)$. Since $\ell_1, \ldots, \widehat{\ell_i}, \ldots, \ell_q$ are linearly independent, we get $L_i \neq 0$. The vector $\mathbf{c}$ generates $\ker_{\mathbb{Q}}(L)$ because L has rank $q - 1$ and $L\mathbf{c}^\top = 0$. Then, because of the equality $L\operatorname{adj}(L) = 0$, we can write $L_i = \mu_i\mathbf{c}$ for some $\mu_i \in \mathbb{Q}$. Notice that $\mu_i > 0$ because $L_{i,i} \geq 0$ and $\mathbf{c} \in \mathbb{N}_+^q$. Hence, all entries of $\operatorname{adj}(L)$ are positive integers. If $\mathbf{d}$ is any row of $\operatorname{adj}(L)$, we get $\mathbf{d}L = 0$ because $\operatorname{adj}(L)L = 0$.

(b): For any i, the vectors are $\ell_1, \ldots, \widehat{\ell_i}, \ldots, \ell_q$ are linearly independent because $L_{i,i} > 0$. Thus this part follows from (a).

(c): Let L be as in Definition 9.5.4. (c_1): First we treat the case $\mathbf{c} = \mathbf{1}$. By part (a) it suffices to show that $L_{i,i} \geq 0$ for all i. Let $H_{i,i}$ be the submatrix of L obtained by eliminating the ith row and ith column. By the Gershgorin Circle Theorem, every (possibly complex) eigenvalue λ of $H_{i,i}$ lies within at least one of the discs $\{z \in \mathbb{C} \mid \|z - a_{j,j}\| \leq r_j\}$, $j \neq i$, where $r_j = \sum_{u\neq i,j} |-a_{j,u}| \leq a_{j,j}$ since $L\mathbf{1}^\top = 0$ and $a_{i,j} \geq 0$ for all i, j. If $\lambda \in \mathbb{R}$, we get $|\lambda - a_{j,j}| \leq a_{j,j}$, and consequently $\lambda \geq 0$. If $\lambda \notin \mathbb{R}$, then since $H_{i,i}$ is a real matrix, its conjugate $\overline{\lambda}$ must also be an eigenvalue of $H_{i,i}$. Since $\det(H_{i,i})$ is the product of the $q-1$ (possibly repeated) eigenvalues of $H_{i,i}$, we get $L_{i,i} \geq 0$. (c$_2$): Now, we treat the general case. Let B be the $q \times q$ diagonal matrix $\mathrm{diag}(c_1, \ldots, c_q)$, where $\mathbf{c} = (c_1, \ldots, c_q)$, and let $\widetilde{L} = LB$. Notice that $\widetilde{L}\mathbf{1}^\top = 0$ because $L\mathbf{c}^\top = 0$, and $\widetilde{L}$ is a PB matrix because L is a PB matrix. Let $(\widetilde{L}_{i,j})$ be the adjoint matrix of $\widetilde{L}$. Since $c_i > 0$ for all i, by the multilinearity of the determinant, it follows that $L_{i,j} \neq 0$ if and only if $\widetilde{L}_{i,j} \neq 0$. Hence, any set of $q-1$ rows of $\widetilde{L}$ is linearly independent. Therefore, applying case (c$_1$) to $\widetilde{L}$, we obtain that there is $\mathbf{d} \in \mathbb{N}_+^q$ such that $\mathbf{d}\widetilde{L} = 0$. Then, $\mathbf{d}L = 0$. □

Theorem 9.5.8 [335] *Let L be a PB matrix of size $q \times q$ such that $L\mathbf{c}^\top = 0$ for some $\mathbf{c} \in \mathbb{N}_+^q$. The following conditions are equivalent:*

(a) *G_L is strongly connected.*

(b) *$V(I(L), t_i) = \{0\}$ $\forall\, i$, where $V(I(L), t_i)$ is the zero set of $(I(L), t_i)$.*

(c) *$L_{i,j} > 0$ $\forall\, i, j$; $\mathrm{adj}(L) = (L_{i,j})$ is the adjoint of L.*

Proof. (a) $\Rightarrow$ (b): Let L be as in Definition 9.5.4. For $1 \leq k \leq q$, let f_k be the binomial defined by the kth column of L. Set $I = I(L)$ and fix i such that $1 \leq i \leq q$. Clearly $\{0\}$ is contained in $V(I, t_i)$ because I is graded by Theorem 9.5.6. To show the reverse containment, let $\alpha = (\alpha_1, \ldots, \alpha_q)$ be a point in $V(I, t_i)$. Consider the set $E(t_i)$ of all arrows (t_i, t_k) leaving t_i. If $(t_i, t_k) \in E(t_i)$, i.e., $a_{i,k} \neq 0$, we claim that $\alpha_k = 0$. We can write

$$f_k = t_k^{a_{k,k}} - \prod_{j\neq k} t_j^{a_{j,k}}.$$

Using that $f_k(\alpha) = 0$, we get $\alpha_k^{a_{k,k}} = \prod_{j\neq k} \alpha_j^{a_{j,k}}$. Since $a_{i,k} \neq 0$ and using that $\alpha_i = 0$, we obtain that $\alpha_k = 0$, as claimed. Let ℓ be an integer in $[q] := \{1, \ldots, q\}$. Since the digraph G_L is strongly connected, there is a directed path $\{v_1, \ldots, v_r\}$ joining t_i and t_ℓ, i.e., $v_1 = t_i$, $v_r = t_\ell$ and $(v_j, v_{j+1}) \in E(G_L)$ for all j. There is a permutation π of $[q]$ such that $\pi(1) = i$, $\pi(r) = \ell$ and $v_j = t_{\pi(j)}$ for $j = 1, \ldots, r$. Applying the claim successively for $j = 2, \ldots, r$, we obtain $\alpha_{\pi(2)} = 0, \alpha_{\pi(3)} = 0, \ldots, \alpha_{\pi(r)} = 0$. Thus, $\alpha_\ell = 0$. This proves that $\alpha = 0$.

(b) $\Rightarrow$ (a): We proceed by contradiction. Assume that G_L is not strongly connected. Without loss of generality we may assume that there is no directed path from t_1 to t_q. Let W be the set of all vertices t_i such that there is a directed path from t_1 to t_i, the vertex t_1 being included in W. The set W is non-empty because G_L has no sources or sinks by definition of a PB matrix, and the vertex t_q is not in W. Consider the vector $\alpha \in K^q$ defined as $\alpha_i = 1$ if $t_i \notin W$ and $\alpha_i = 0$ if $t_i \in W$. To derive a contradiction it suffices to show that all binomials of $I(L)$ vanish at α. Let L be as in Definition 9.5.4 and let $f_k = t_k^{a_{k,k}} - \prod_{j \neq k} t_j^{a_{j,k}}$ be the binomial defined by the k-columns of L. If $t_k \in W$, there is a directed path $\mathcal{P}$ from t_1 to t_k. Then t_j is part of the path $\mathcal{P}$ for some j such that $a_{j,k} > 0$ because the last arrow of the path $\mathcal{P}$ has the form (t_j, t_k). Thus, since $t_j \in W$, $f_k(\alpha) = 0$. If $t_k \notin W$, then t_j is not in W for any j such that $a_{j,k} > 0$, because if $a_{j,k} > 0$, the pair (t_j, t_k) is an arc of G_L. Thus, $f_k(\alpha) = 0$.

(a) $\Rightarrow$ (c): By the proof of Lemma 9.5.7(c), one has that $L_{i,i} \geq 0$ for all i. By Theorem 9.5.6 and using the proof of Lemma 9.5.7(a), we get that $L_{i,j} > 0$ for all i, j.

(c) $\Rightarrow$ (a): We proceed by contradiction. Assume that G_L is not strongly connected. We may assume that there is no directed path from t_1 to t_q. Let W be as above and let $W^c = \{t_i | \, t_i \notin W\}$ be its complement. We can write $W^c = \{t_{\ell_1}, \ldots, t_{\ell_r}\}$. Consider the $r \times r$ submatrix B obtained from L by fixing rows $\ell_1, \ldots, \ell_r$ and columns $\ell_1, \ldots, \ell_r$. Notice that by the arguments above $W^c = \cup_{t_k \in W^c}(\text{supp}(f_k))$. Hence any column of B extends to a column of L by adding 0's only, and consequently since $\mathbf{d}L = 0$ for some $\mathbf{d} \in \mathbb{N}^s$, the rows of B are linearly dependent. Hence $\det(B) = 0$. By permuting rows and columns, L can be brought to the form

$$L' = \left(\begin{array}{cc} C & 0 \\ C' & C'' \end{array} \right)$$

where C and C'' are square matrices of orders r and $q - r$, respectively, and $\det(C) = 0$. Hence the adjoint of L' has a zero entry, and so does the adjoint of L, a contradiction. $\square$

Laplacian matrices Let G be a connected graph, where $V = \{t_1, \ldots, t_q\}$ is the set of vertices, E is the set of edges, and let w be a weight function

$$w \colon E \to \mathbb{N}_+, \qquad e \mapsto w_e.$$

Let $E(t_i)$ be the set of edges incident to t_i. The *Laplacian matrix* $L(G)$ of G is the $q \times q$ matrix whose (i, j)-entry $L(G)_{i,j}$ is given by

$$L(G)_{i,j} := \begin{cases} \sum_{e \in E(t_i)} w_e & \text{if } i = j, \\ -w_e & \text{if } i \neq j \text{ and } e = \{t_i, t_j\} \in E, \\ 0 & \text{otherwise.} \end{cases}$$

Notice that $L(G)$ is symmetric and $\mathbf{1}L(G) = 0$.

The notion of a Laplacian matrix can be extended to weighted digraphs; see [95] and the references there. Let $G = (V, E, w)$ be a weighted digraph without loops and with vertices $t_1, \ldots, t_q$, let $w(t_i, t_j)$ be the weight of the directed arc from t_i to t_j and let $A(G)$ be the adjacency matrix of G given by $A(G)_{i,j} = w(t_i, t_j)$. The Laplacian matrix of G is given by $L(G) = D^+(G) - A(G)$, where $D^+(G)$ is the diagonal matrix with the out-degrees of the vertices of G in the diagonal entries. Note that $L(G)\mathbf{1}^\top = 0$ and that the Laplacian matrix of a digraph may not be symmetric. If t_i is a sink, i.e., there is no arc of the form (t_i, t_j), then the ith row of $L(G)$ is zero. Thus, the rank of $L(G)$ may be much less than $q - 1$.

Definition 9.5.9 The matrix ideal I_G of $L(G)$ is called the *Laplacian ideal* of G. If $\mathcal{L}$ is the lattice generated by the columns of $L(G)$, the group

$$K(G) := T(\mathbb{Z}^q/\mathcal{L})$$

is called the *critical group* or the *sandpile group* of G.

The structure, as a finite abelian group, of $K(G)$ is only known for a few families of graphs [5, 295].

Theorem 9.5.10 (the Matrix-Tree Theorem [396, Theorem 9.8]) *If G is regarded as a multigraph (where each edge e occurs w_e times), then the number of spanning trees of G is the (i, j)-entry of the adjoint matrix of $L(G)$ for any (i, j).*

Remark 9.5.11 The order of $T(\mathbb{Z}^q/\mathcal{L})$ is the gcd of all $(q-1)$-minors of $L(G)$. Thus the order of $K(G)$ is the number of spanning trees of G.

Proposition 9.5.12 [335] *Let G be a connected graph, $I_G \subset S$ its Laplacian ideal, and $\mathcal{L}$ the lattice spanned by the columns of $L(G)$. Then:*

(a) $V(I_G, t_i) = \{0\}$ *for all i and* $\mathrm{rank}(L(G)) = q - 1$.

(b) $\deg(S/I_G) = \deg(S/I(\mathcal{L})) = |K(G)|$.

(c) *If $G = \mathcal{K}_q$ is a complete graph, then* $\deg(S/I_G) = q^{q-2}$.

(d) $\mathrm{Hull}(I_G) = I(\mathcal{L})$, *where* $\mathrm{Hull}(I_G)$ *is the intersection of the isolated primary components of I_G.*

Proof. The underlying digraph $G_{L(G)}$ of the Laplacian matrix $L(G)$ is strongly connected because $L(G)$ is symmetric and G is connected. Hence, by Theorems 9.5.6 and 9.5.8, (a) holds. Parts (b) and (d) follow from (a) and Proposition 9.5.3. Part (c) follows from (b) and the fact that a complete graph with q vertices has q^{q-2} spanning trees [396, p. 143]. $\square$

Exercises

9.5.13 Let L be the following PB matrix and $\mathrm{adj}(L)$ its adjoint:

$$L = \begin{pmatrix} 4 & -1 & -1 & -1 & -1 \\ -1 & 4 & -1 & -1 & -1 \\ 0 & -1 & 2 & -1 & 0 \\ -1 & -1 & -1 & 4 & -1 \\ -1 & 0 & 0 & 0 & 1 \end{pmatrix}.$$

Prove that $V(I(L), t_i) = \{0\}$ for all i and that G_L is strongly connected.

9.5.14 Let G be a connected graph and let $I_G \subset S$ be its Laplacian ideal. Prove that the following hold.

(a) If $\deg_G(t_i) \geq 3$ for all i, then I_G is not a lattice ideal.

(b) If $\deg_G(t_i) \geq 2$ for all i, then I_G is not a complete intersection.

9.5.15 Let G be the following weighted digraph and L its Laplacian matrix. Prove that the underlying digraph G_L is not strongly connected, the matrix ideal $I(L^\top)$ is graded but $I(L)$ is not.

$$L = L(G) = \begin{pmatrix} 5 & -4 & 0 & -1 \\ 0 & 1 & -1 & 0 \\ 0 & -1 & 1 & 0 \\ -3 & 0 & -1 & 4 \end{pmatrix}$$

9.5.16 Let G be the complete graph on 3 vertices. Prove that

$$L(G) = \begin{pmatrix} 2 & -1 & -1 \\ -1 & 2 & -1 \\ -1 & -1 & 2 \end{pmatrix},$$

$I_G = (t_1^2 - t_2t_3,\, t_2^2 - t_1t_3,\, t_3^2 - t_1t_2)$, $V(I_G, t_i) = \{0\}$ for all i, I_G is not a complete intersection, is a lattice ideal, is not a prime ideal, has degree 3 and dimension 1.

9.5.17 Use the package "Binomials" [266] and the following procedure for *Macaulay*2 [199] to verify that over $\mathbb{C}$ the ideal $(t_1^2 - t_2t_3, t_2^2 - t_3t_1, t_3^2 - t_4^2)$ is a complete intersection vanishing ideal and is not a lattice ideal.

```
load "Binomial.m2"
R=QQ[t1,t2,t3,t4]
I=ideal(t1^2-t2*t3,t2^2-t3*t1,t3^2-t4^2)
binomialAssociatedPrimes I
saturate(I,t1*t2*t3*t4)==I
```

9.6 Gröbner bases and normal subrings

Let $R = K[x_1^{\pm 1}, \ldots, x_n^{\pm 1}]$ be a Laurent polynomial ring over a field K and consider a finite set $F = \{x^{v_i}\}_{i=1}^q$ of distinct monomials with $v_i \neq 0$ for all i. There is a homomorphism of K-algebras:

$$S = K[t_1, \ldots, t_q] \xrightarrow{\varphi} K[F] \quad (t_i \xrightarrow{\varphi} x^{v_i}),$$

where S is a polynomial ring. The kernel of φ, denoted by $P := P_F$, is the *toric ideal* of $K[F]$. In what follows A will denote the $n \times q$ matrix with column vectors $v_1, \ldots, v_q$ and $\mathcal{A}$ will denote the set $\{v_1, \ldots, v_q\}$.

The main reference for this section is the book of Sturmfels on Gröbner bases and convex polytopes [400].

For the rest of this section we assume that $\prec$ is a fixed term order for the set of monomials of $K[t_1, \ldots, t_q]$. We denote the initial ideal of P by $\mathrm{in}_\prec(P)$ or simply by $\mathrm{in}(P)$. The Stanley–Reisner complex of the square-free monomial ideal $\mathrm{rad}\,(\mathrm{in}_\prec(P))$ will be denoted by Δ.

The following result was pointed out by Sturmfels. It can be shown using essentially "line shellings" of polytopes (as explained in [438, pp. 240–242]).

Theorem 9.6.1 [106, Theorem 9.5.10] Δ *is pure shellable.*

Corollary 9.6.2 (Sturmfels) $\mathrm{rad}(\mathrm{in}_\prec(P))$ *is a Cohen–Macaulay ideal.*

Proof. The simplicial complex Δ is pure shellable by Theorem 9.6.1. Hence Δ is Cohen–Macaulay by Theorem 6.3.23. □

Lemma 9.6.3 *If* $\sigma = \{t_1, \ldots, t_r\}$ *is a maximal face of* Δ, *then* $v_1, \ldots, v_r$ *are linearly independent.*

Proof. If $v_1, \ldots, v_r$ are linearly dependent, then after permutation of the variables we can write $\lambda_1 v_1 + \cdots + \lambda_j v_j = \lambda_{j+1} v_{j+1} + \cdots + \lambda_r v_r$, $\lambda_i \in \mathbb{N}$ and $\lambda_r \neq 0$. Thus if $f = t_1^{\lambda_1} \cdots t_j^{\lambda_j} - t_{j+1}^{\lambda_{j+1}} \cdots t_r^{\lambda_r}$, then $\mathrm{in}(f) \in \mathrm{in}(P)$, a contradiction because $\mathrm{rad}(\mathrm{in}(P)) \subset (t_{r+1}, \ldots, t_q)$. □

Definition 9.6.4 [400, p. 4] Let I be an ideal of S and let $\omega = (\omega_1, \ldots, \omega_q)$ be a vector in $\mathbb{R}^q$. If $f = \lambda_1 t^{a_1} + \cdots + \lambda_s t^{a_s}$, define $\mathrm{in}_\omega(f)$, the *initial form* of f relative to ω, as the sum of all terms $\lambda_i t^{a_i}$ such that $\langle \omega, a_i \rangle$ is maximal. The ideal generated by $\{\mathrm{in}_\omega(f) | \, f \in I\}$ is denoted by $\mathrm{in}_\omega(I)$.

If $\omega \in \mathbb{N}^q$, then $\mathrm{in}_\omega(f)$ is the leading coefficient in the x-variable of the univariate polynomial $h(x) = f(xt_1^{\omega_1}, \ldots, xt_q^{\omega_q}) \in S[x]$.

If $\omega \geq 0$, the following rule defines a term order $\prec_\omega$ for S: $t^a \prec_\omega t^b$ if and only if $\langle \omega, a \rangle < \langle \omega, b \rangle$, or $\langle \omega, a \rangle = \langle \omega, b \rangle$ and $t^a \prec t^b$.

Proposition 9.6.5 [400, Proposition 1.8] *Let I be an ideal of S. If $\omega \geq 0$, then* $\mathrm{in}(\mathrm{in}_\omega(I)) = \mathrm{in}_{\prec_\omega}(I)$. *In addition if* $\mathrm{in}_\omega(I)$ *is a monomial ideal, then* $\mathrm{in}_\omega(I) = \mathrm{in}_{\prec_\omega}(I)$.

Proposition 9.6.6 *Let I be an ideal of S and let $\mathcal{G}$ be the reduced Gröbner basis of I. If $\omega \in \mathbb{R}_+^q$, then* $\mathrm{in}(I) = \mathrm{in}_\omega(I)$ *if and only if* $\mathrm{in}(g) = \mathrm{in}_\omega(g)$ *for all $g \in \mathcal{G}$.*

Proof. $\Rightarrow$) Let $g \in \mathcal{G}$. There are $t^{a_s}, \ldots, t^{a_s}$ distinct monomials in S such that $g = \lambda_1 t^{a_1} + \cdots + \lambda_s t^{a_s}$ and $\mathrm{in}_\omega(g) = \lambda_1 t^{a_1} + \cdots + \lambda_r t^{a_r}$, where $r \leq s$ and $0 \neq \lambda_i \in k$ for all i. Since $\mathcal{G}$ is reduced there is j satisfying $\mathrm{in}(g) = t^{a_j} \in \mathrm{in}(I)$ and $t^{a_i} \notin \mathrm{in}(I)$ for $i \neq j$. Using that $\mathrm{in}(I)$ is a monomial ideal we get $t^{a_i} \in \mathrm{in}(I)$ for $i = 1, \ldots, r$. Thus $r = 1$ and $j = 1$, that is, $\mathrm{in}(g) = \mathrm{in}_\omega(g)$.

$\Leftarrow$) First let us prove $\mathrm{in}(I) = \mathrm{in}_{\prec_\omega}(I)$. Clearly $\mathrm{in}(I) \subset \mathrm{in}_\omega(I)$. Hence taking initial ideals with respect to $\prec$ in both sides and using the previous proposition we get $\mathrm{in}(I) \subset \mathrm{in}_{\prec_\omega}(I)$. To show the reverse inclusion take $t^a \in \mathrm{in}_{\prec_\omega}(I)$. Assume $t^a \notin I$, otherwise $t^a \in \mathrm{in}(I)$. Since the standard monomials with respect to $\prec_\omega$ are a K-basis for S/I (see Proposition 3.3.13), we have $t^a + I = (\lambda_1 t^{a_1} + I) + \cdots + (\lambda_s t^{a_s} + I)$, $\lambda_i \in K^*$, where $t^{a_i} \notin \mathrm{in}_{\prec_\omega}(I)$. Therefore $t^a \in \mathrm{in}(I)$, as required.

Let $\mathrm{in}_\omega(f)$ with $f \in I$. There are $t^{a_s}, \ldots, t^{a_s}$ distinct monomials in S and non-zero scalars $\lambda_1, \ldots, \lambda_s$ such that $f = \lambda_1 t^{a_1} + \cdots + \lambda_r t^{a_r} + \cdots + \lambda_s t^{a_s}$, where $\mathrm{in}_\omega(f) = \lambda_1 t^{a_1} + \cdots + \lambda_r t^{a_r}$ and $t^{a_1} \succ \cdots \succ t^{a_r}$. Note $\mathrm{in}_{\prec_\omega}(f) = t^{a_1}$, hence $t^{a_1} \in \mathrm{in}(I)$. There is $g \in \mathcal{G}$ such that

$$\lambda_1 t^{a_1} = \lambda_1 t^\delta \mathrm{in}(g) = \lambda_1 t^\delta \mathrm{in}_\omega(g) = \lambda_1 \mathrm{in}_\omega(t^\delta g).$$

Set $h = f - \lambda_1 t^\delta g$. As $\mathrm{in}_\omega(h) = \lambda_2 t^{a_2} + \cdots + \lambda_r t^{a_r}$, we can repeat the same argument to get $\lambda_i t^{a_i} \in \mathrm{in}(I)$ for $2 \leq i \leq r$, that is, $\mathrm{in}_\omega(f) \in \mathrm{in}(I)$. □

The next result shows that $\mathrm{in}(P)$ can be represented by a weight vector.

Proposition 9.6.7 [400, Proposition 1.11] $\mathrm{in}(P) = \mathrm{in}_\omega(P)$ *for some non-negative integer weight vector $\omega \in \mathbb{N}^q$.*

A fundamental result linking Gröbner basis theory with "regular triangulations" of point configurations is the following result of Sturmfels.

Theorem 9.6.8 [399, 400] *If* $\mathrm{in}(P) = \mathrm{in}_\omega(P)$, *then*

$$\Delta = \{\sigma \mid \exists\, c \in \mathbb{R}^n \text{ such that } \langle v_i, c\rangle = \omega_i \text{ if } t_i \in \sigma \;\&\; \langle v_i, c\rangle < \omega_i \text{ if } t_i \notin \sigma\}.$$

In what follows ω denotes a nonnegative integer vector that represents the initial ideal of the toric ideal P, that is, $\mathrm{in}(P) = \mathrm{in}_\omega(P)$.

Lemma 9.6.9 *If $\sigma = \{t_1, \ldots, t_r\}$ is a face of Δ, then*

$$\mathrm{in}(P) = \mathrm{in}_{\omega'}(P) = \mathrm{in}_{\prec_{\omega'}}(P)$$

for some $\omega' = (\omega'_i) \in \mathbb{R}^q$ such that $\omega'_i = 0$ if $t_i \in \sigma$ and $\omega'_i > 0$ if $t_i \notin \sigma$.

Proof. According to Theorem 9.6.8 there is $c \in \mathbb{R}^q$ such that $\langle v_i, c\rangle = \omega_i$ if $t_i \in \sigma$ and $\langle v_i, c\rangle < \omega_i$ if $t_i \notin \sigma$. Let $g = t^a - t^b$ be an element in the reduced Gröbner basis of P with $t^a \succ t^b$. Note $\mathrm{in}_\omega(g) = t^a$, otherwise if $\mathrm{in}_\omega(g) = t^b$ or $\mathrm{in}_\omega(g) = g$, the using $\mathrm{in}(P) = \mathrm{in}_\omega(P)$ we get that t^b is not a standard monomial, a contradiction. Thus $\langle \omega, a\rangle > \langle \omega, b\rangle$.

Consider the vector $\omega' = (\omega_i) - (\langle v_i, c\rangle)$. By Proposition 9.6.5 and Proposition 9.6.6 we need only show the inequality $\langle \omega', a\rangle > \langle \omega', b\rangle$. If $a = (a_i)$ and $b = (b_i)$, from $a_1v_1 + \cdots + a_qv_q = b_1v_1 + \cdots + b_qv_q$, we get $a_1\langle c, v_1\rangle + \cdots + a_q\langle c, v_q\rangle = b_1\langle c, v_1\rangle + \cdots + b_q\langle c, v_q\rangle$. Hence

$$\langle \omega', a\rangle - \langle \omega', b\rangle = \langle \omega, a\rangle - \langle \omega, b\rangle > 0. \qquad \Box$$

Lemma 9.6.10 *If $\mathrm{in}(P)$ is square-free and $0 \neq v \in \mathbb{N}\mathcal{A}$, then there is σ' a face of Δ such that $v \in \mathbb{N}\mathcal{A}'$, where $\mathcal{A}' = \{v_i | \, t_i \in \sigma'\}$.*

Proof. Let $\{t^{\beta_1} - t^{\gamma_1}, \ldots, t^{\beta_p} - t^{\gamma_p}\}$ be the reduced Gröbner basis of the toric ideal P, where $t^{\beta_i} \succ t^{\gamma_i}$ for all i. There is $\lambda = (\lambda_i)$ in $\mathbb{N}^q$ such that v is equal to $\lambda_1v_1 + \cdots + \lambda_qv_q$. Consider the monomial $h_0 = t^\lambda$ and its support $\sigma_0 = \mathrm{supp}(h)$. If $\sigma_0 \in \Delta$, set $\sigma' = \sigma_0$. Assume $\sigma_0 \notin \Delta$, then $h_0 \in \mathrm{in}(P)$ and we may assume $h_0 = t^{\delta_1}t^{\beta_1}$. Set $h_1 = t^{\delta_1}t^{\gamma_1}$ and $\sigma_1 = \mathrm{supp}(h_1)$. Note

$$f^\lambda := f_1^{\lambda_1} \cdots f_q^{\lambda_q} = f^{\delta_1}f^{\beta_1} = f^{\delta_1}f^{\gamma_1}.$$

Thus one may assume $\sigma_1 \notin \Delta$, otherwise we set $\sigma' = \sigma_1$. Since $h_0 \succ h_1$ and using that $\prec$ is Noetherian, the existence of σ' follows by induction. $\Box$

Lemma 9.6.11 *Let $\mathcal{A}' = \{v_1, \ldots, v_r\}$ and let π be the projection map*

$$\pi \colon \mathbb{Z}^q \longrightarrow \mathbb{Z}^{q-r} \qquad (\pi(a_i) = (a_{r+1}, \ldots, a_q)).$$

If $\mathcal{L} = \pi(\ker(A) \cap \mathbb{Z}^q)$, then $\mathbb{Z}^{q-r}/\mathcal{L} \simeq \mathbb{Z}\mathcal{A}/\mathbb{Z}\mathcal{A}'$.

Proof. It suffices to note that the linear map $\mathbb{Z}^{q-r} \stackrel{\psi}{\longrightarrow} \mathbb{Z}\mathcal{A}/\mathbb{Z}\mathcal{A}'$ given by $\psi(a_{r+1}, \ldots, a_q) = a_{r+1}v_{r+1} + \cdots + a_qv_q + \mathbb{Z}\mathcal{A}'$ is an epimorphism whose kernel is equal to $\mathcal{L}$. $\Box$

For use below consider the epimorphism of K-algebras:

$$S = K[t_1, \ldots, t_q] \stackrel{\phi}{\longrightarrow} S' = K[t_{r+1}, \ldots, t_q]$$

induced by $\phi(t_i) = 1$ for $i \leq r$ and $\phi(t_i) = t_i$ otherwise. Note that ϕ is related to the map π by $\phi(t_i) = t^{\pi(e_i)}$ for all i.

Lemma 9.6.12 *Let I be a monomial ideal of S and let $\mathfrak{p} = (t_{r+1}, \ldots, t_q)$ be a minimal prime of I, then*

(a) $\mathrm{Ass}(S'/I') = \{\mathfrak{p}'\}$, *where $I' = \phi(I)$ and $\mathfrak{p}' = \mathfrak{p} \cap S'$,*

(b) $\dim(S'/I') = 0$, *and*

(c) $I' = \mathfrak{p}'$, *if I is generated by square-free monomials.*

Proof. (a): Let $\mathfrak{q}'$ be an associated prime of S'/I' and set $\mathfrak{q} = \mathfrak{q}'S$. Note that $I \subset I'S$ and $\mathfrak{q}'$ is generated by variables in $V = \{t_{r+1}, \ldots, t_q\}$. Since $I \subset \mathfrak{q}$, there is a minimal prime $\mathfrak{q}_i$ of I contained in $\mathfrak{q}$, thus $\mathfrak{q}_i$ is generated by variables in V. Hence by the minimality of $\mathfrak{p}$ we get $\mathfrak{q} = \mathfrak{p}$ and $\mathfrak{q}' = \mathfrak{p}'$.

(b): From (a) the radical of I' is equal to the irrelevant maximal ideal of S', thus S'/I' is Artinian. Part (c) follows by noticing that I' is also a square-free monomial ideal and applying (a). $\square$

Definition 9.6.13 Let I be a monomial ideal of S and let $\mathfrak{p}$ be a minimal prime of I. The *multiplicity* of $\mathfrak{p}$ in I is: $m_I(\mathfrak{p}) = \ell_{S_\mathfrak{p}}(S_\mathfrak{p}/I_\mathfrak{p})$.

Proposition 9.6.14 *Let I be a monomial ideal of S and $\mathfrak{p} = (t_{r+1}, \ldots, t_q)$ a minimal prime of I. If $I' = \phi(I)$, then $m_I(\mathfrak{p}) = \dim_K(S'/I')$.*

Proof. Since $S_\mathfrak{p}/I_\mathfrak{p} \simeq (S')_\mathfrak{p}/(I')_\mathfrak{p}$, the result follows from Lemma 8.5.1 $\square$

Proposition 9.6.15 *If $\mathcal{L}$ is a lattice in $\mathbb{Z}^{q-r}$ of rank $q - r$, then*

$$\dim_K(S'/I(\mathcal{L})) = |\mathbb{Z}^{q-r}/\mathcal{L}|.$$

Proof. It follows at once from Theorem 9.4.2 (c) because $\dim_K(S'/I(\mathcal{L}))$ is equal to $\deg S'/I(\mathcal{L})$. $\square$

For use below recall that the map $\sigma \mapsto (\sigma^c)$ induces a bijection between the maximal faces of Δ and the minimal primes of $\mathrm{rad}(\mathrm{in}(P))$, where σ^c is equal to $\{t_i | \, t_i \notin \sigma\}$ (see Proposition 6.3.4).

Theorem 9.6.16 [400] *If the initial ideal* $\mathrm{in}(P)$ *is generated by square-free monomials, then $K[F]$ is normal.*

Proof. Let $\alpha \in \mathbb{R}_+\mathcal{A} \cap \mathbb{Z}\mathcal{A}$. By Corollary 1.1.27, $p\alpha \in \mathbb{N}\mathcal{A}$ for some $0 \neq p \in \mathbb{N}$. By Lemma 9.6.10 there is σ', a maximal face of Δ, such that $p\alpha$ is in $\mathbb{N}\mathcal{A}'$, where $\mathcal{A}' = \{v_i | \, t_i \in \sigma'\}$. We may assume that $\sigma' = \{t_1, \ldots, t_r\}$. Consider the prime ideal $\mathfrak{p}$ of S generated by $\{t_{r+1}, \ldots, t_q\}$. As $\mathfrak{p}$ is a minimal prime of the initial ideal $\mathrm{in}(P)$, using Lemma 9.6.12(c), we get

$$\phi(\mathrm{in}(P)) = \mathfrak{p}' = (t_{r+1}, \ldots, t_q) \subset S'.$$

By Lemma 9.6.9 we may assume that $\text{in}(P) = \text{in}_\omega(P) = \text{in}_{\prec_\omega}(P)$ for some $\omega = (\omega_i) \in \mathbb{R}^q$ such that $\omega_i = 0$ if $t_i \in \sigma'$ and $\omega_i > 0$ if $t_i \notin \sigma'$. Therefore

$$\mathfrak{p}' = \phi(\text{in}(P)) = \text{in}_\omega(\phi(P)) = \text{in}_{\prec_\omega}(\phi(P)). \tag{9.14}$$

Indeed the first equality follows from Proposition 9.6.6 and noticing that $\text{in}_\omega(\phi(P)) \subset \mathfrak{p}'$. The second equality follows at once from Proposition 9.6.5 because $\text{in}_\omega(\phi(P))$ is a monomial ideal by the first equality. If $t^a - t^b$ is a binomial in the reduced Gröbner basis of $\phi(P)$ with respect to $\prec_\omega$, then t^a or t^b is a standard monomial. Thus by Eq. (9.14) $t^a = 1$ or $t^b = 1$. Hence

$$\phi(P) = (t_{r+1} - 1, \ldots, t_q - 1) \subset S'.$$

Let $\mathcal{L}$ be the lattice $\pi(\ker(A) \cap \mathbb{Z}^q)$. As a result $S'/\phi(P)$ is Artinian, $\mathcal{L}$ has rank $q-r$ and $\dim_K(S'/\phi(P)) = 1$. Since $\phi(P) = I(\mathcal{L})$, using Lemma 9.6.11 and Proposition 9.6.15 we conclude $\mathbb{Z}\mathcal{A} = \mathbb{Z}\mathcal{A}'$. By Lemma 9.6.3 the set $\mathcal{A}'$ is linearly independent, thus $\alpha \in \mathbb{N}\mathcal{A}$, as required. This proof was adapted from [400]. □

Proposition 9.6.17 *Let I be an ideal of S. The following hold.*

(a) [336] *If* $\text{in}_\prec(I)$ *is square-free, then* $\text{rad}(I) = I$.

(b) [142, Corollary 6.9] *If I is graded and* $\text{in}_\prec(I)$ *is Cohen–Macaulay (resp. Gorenstein), then I is Cohen–Macaulay (resp. Gorenstein).*

Proposition 9.6.18 *Let $I = I(\mathcal{L}) \subset S$ be a standard graded lattice ideal. If the initial ideal* $\text{in}_\prec(I)$ *is square-free, then I is a prime ideal and S/I is normal and Cohen–Macaulay.*

Proof. By Theorem 9.4.6 and Proposition 9.6.17 all associated prime ideals of I have height $r = \text{rank}(\mathcal{L})$ and I is a radical ideal. Then I has an irredundant primary decomposition $I = \mathfrak{p}_1 \cap \cdots \cap \mathfrak{p}_m$, where $\mathfrak{p}_i$ is a prime ideal of height r for all i. Let $\mathcal{L}_s$ be the saturation of $\mathcal{L}$ consisting of all $a \in \mathbb{Z}^q$ such that $pa \in \mathcal{L}$ for some $0 \neq p \in \mathbb{N}$ and let $I(\mathcal{L}_s)$ be its lattice ideal. Since $\text{rank}(\mathcal{L})$ is equal to $\text{rank}(\mathcal{L}_s)$, by Theorem 8.2.2, we get that r is also the height of $I(\mathcal{L}_s)$. As $\mathbb{Z}^q/\mathcal{L}_s$ is torsion-free, by Theorem 8.2.22, $I(\mathcal{L}_s)$ is a prime toric ideal. Then we may assume that $\mathfrak{p}_1 = I(\mathcal{L}_s)$. We claim that $\text{in}_\prec(I) = \text{in}_\prec(I(\mathcal{L}_s))$. Clearly $\text{in}_\prec(I) \subset \text{in}_\prec(I(\mathcal{L}_s))$ because $I \subset I(\mathcal{L}_s)$. To show the reverse inclusion take any element f in the reduced Gröbner basis of $I(\mathcal{L}_s)$. It suffices to show that $\text{in}_\prec(f) \in \text{in}_\prec(I)$. By Lemma 8.3.3, we can write $f = t^{a^+} - t^{a^-}$ for some $a = a^+ - a^-$ in $\mathcal{L}_s$. We may assume that $\text{in}_\prec(f) = t^{a^+}$. There is $p \in \mathbb{N}_+$ such that $pa \in \mathcal{L}$. The binomial $g = t^{pa^+} - t^{pa^-}$ is in $I = I(\mathcal{L})$ and $\text{in}_\prec(g) = t^{pa^+}$. Thus $t^{pa^+} \in \text{in}_\prec(I)$ and since this ideal is square-free we get that $t^{a^+} \in \text{in}_\prec(I)$. This proves the claim. Hence $\deg(S/I)$ is $\deg(S/I(\mathcal{L}_s))$ because S/I and $S/I(\mathcal{L}_s)$ have the same Hilbert function. Therefore, by additivity of the degree, we get that $m = 1$. Consequently, by Theorems 9.6.16 and 9.1.6, S/I is normal and Cohen–Macaulay. $\Box$

Interesting families of graded ideals with square-free initial ideals are given in [196]. For these families the corresponding simplicial complexes associated to the initial ideals are vertex decomposable and hence shellable.

Regular triangulations of cones Consider the primary decomposition of the radical of in(P) as a finite intersection of face ideals:

$$\mathrm{rad}\,(\mathrm{in}(P)) = \mathfrak{p}_1 \cap \mathfrak{p}_2 \cap \cdots \cap \mathfrak{p}_r,$$

see Theorem 6.1.4. Recall that, by Proposition 6.3.4, the facets of Δ are given by $\sigma_i = \{t_j | \, t_j \notin \mathfrak{p}_i\}$, or equivalently by $\mathcal{A}_i = \{v_j | \, t_j \notin \mathfrak{p}_i\}$ if one identifies t_i with v_i.

According to [400, Theorem 8.3], the family of cones (resp. family of simplices if $\mathcal{A}$ is homogeneous)

$$\{\mathbb{R}_+\mathcal{A}_1, \ldots, \mathbb{R}_+\mathcal{A}_r\} \quad (\text{resp. } \{\mathrm{conv}(\mathcal{A}_1), \ldots, \mathrm{conv}(\mathcal{A}_r)\})$$

is a *regular triangulation* of the cone $\mathbb{R}_+\mathcal{A}$ (resp. the polytope conv$(\mathcal{A})$) in the sense of [400, pp. 63-64] (resp. in the sense of [438, pp. 129-130]). This means that $\mathbb{R}_+\mathcal{A}_1, \ldots, \mathbb{R}_+\mathcal{A}_r$ (resp. conv$(\mathcal{A}_1), \ldots,$ conv$(\mathcal{A}_r)$) are obtained by projection onto the first n coordinates of the lower facets of

$$Q' = \mathbb{R}_+\{(v_1, \omega_1), \ldots, (v_q, \omega_q)\} \ (\text{resp. } \mathrm{conv}((v_1, \omega_1), \ldots, (v_q, \omega_q))),$$

a facet of the cone Q' is *lower* if it has a normal vector with negative last entry, i.e., a lower facet of Q' has the form

$$\mathcal{F} = \{x \in Q' | \, \langle \alpha, x \rangle = 0\}, \quad \langle \alpha, x \rangle \geq 0 \text{ valid for } Q', \quad \alpha_{n+1} < 0,$$

see Example 9.6.19. Furthermore all regular triangulations of $\mathbb{R}_+\mathcal{A}$ (resp. conv$(\mathcal{A})$ if $\mathcal{A}$ is homogeneous) arise in this way, i.e., the regular triangulations of $\mathbb{R}_+\mathcal{A}$ (resp. conv$(\mathcal{A})$ if $\mathcal{A}$ is homogeneous) are in one to one correspondence with the radicals of the monomial initial ideals of the toric ideal P; see [400, Theorem 8.3] and [106, Theorem 9.4.5]. The simplicial complex Δ and also its set of facets $\{\mathcal{A}_1, \ldots, \mathcal{A}_r\}$ is called a *regular triangulation* of the cone $\mathbb{R}_+\mathcal{A}$ (resp. polytope conv$(\mathcal{A})$ if $\mathcal{A}$ is homogeneous). For a thorough study of triangulations we refer to [106] and [438].

The regular triangulation $\{\mathbb{R}_+\mathcal{A}_1, \ldots, \mathbb{R}_+\mathcal{A}_r\}$ is called *weakly unimodular* (resp. *unimodular*) if $\mathbb{Z}\mathcal{A}_i = \mathbb{Z}\mathcal{A}$ for all i (resp. the simplex conv$(\mathcal{A}_i)$ is unimodular for all i). If $\mathcal{A}$ is homogeneous, this regular triangulation is weakly unimodular if and only if the monomial ideal in(P) is square-free; see [400, Corollary 8.9].

Example 9.6.19 The facets of the cone generated by

$$(1,1,0,0,1),(0,1,1,0,1),(0,0,1,1,1),(1,0,0,1,1),$$
$$(1,0,0,0,0),(0,1,0,0,0),(0,0,1,0,0),(0,0,0,1,0).$$

are given by the system of linear inequalities

$$x_i \geq 0,\ i=1,\ldots,5 \quad x_1+x_3-x_5 \geq 0,\ x_2+x_4-x_5 \geq 0,$$

Thus the lower facets are defined by the last two inequalities.

Example 9.6.20 Let $R = \mathbb{Q}[x_1, x_2, x_3, y_1, y_2, y_3]$ be a polynomial ring with the default monomial order $\prec$ in *Macaulay*2 [199], let $F = \{f_1, \ldots, f_{10}\}$ be the set of monomials

$$f_1 = y_1y_2y_3,\ f_2 = x_1x_2y_1,\ f_3 = x_2x_3y_1,\ f_4 = x_1x_3y_1,\ f_5 = x_1x_2y_2$$
$$f_6 = x_2x_3y_2,\ f_7 = x_1x_3y_2,\ f_8 = x_1x_2y_3,\ f_9 = x_2x_3y_3,\ f_{10} = x_1x_3y_3,$$

and let $\mathcal{A} = \{v_1, \ldots, v_{10}\}$ be the set of exponent vectors of $f_1, \ldots, f_{10}$. The initial ideal of P is $\mathrm{in}(P) = (t_4t_5,\ t_3t_5,\ t_4t_6,\ t_4t_8,\ t_3t_8,\ t_4t_9,\ t_7t_8,\ t_6t_8,\ t_7t_9)$. The primary decomposition of the initial ideal $\mathrm{in}(P)$ gives the following regular triangulation of $\mathrm{conv}(\mathcal{A})$:

$$\mathcal{A}_1 = \{v_1, v_2, v_5, v_8, v_9, v_{10}\},\ \mathcal{A}_2 = \{v_1, v_2, v_5, v_6, v_9, v_{10}\},$$
$$\mathcal{A}_3 = \{v_1, v_2, v_3, v_6, v_9, v_{10}\},\ \mathcal{A}_4 = \{v_1, v_2, v_5, v_6, v_7, v_{10}\},$$
$$\mathcal{A}_5 = \{v_1, v_2, v_3, v_6, v_7, v_{10}\},\ \mathcal{A}_6 = \{v_1, v_2, v_3, v_4, v_7, v_{10}\}.$$

It is easy to verify that $\mathbb{Z}\mathcal{A} = \mathbb{Z}\mathcal{A}_i$ for all i, i.e., the regular triangulation is weakly unimodular. However notice that this triangulation is not unimodular in the sense that the simplex $\Delta_i = \mathrm{conv}(\mathcal{A}_i)$ has normalized volume equal to 2 for all i, i.e., $\mathrm{vol}(\Delta_i) = 2/5!$ and $\dim(\Delta_i) = 5$ for all i. The Ehrhart function of $\mathcal{P} = \mathrm{conv}(\mathcal{A})$ is given by

$$|\mathbb{Z}^6 \cap i\mathcal{P}| = \frac{1}{10}i^5 + \frac{1}{2}i^4 + \frac{11}{6}i^3 + \frac{7}{2}i^2 + \frac{46}{15}i + 1 = \frac{12}{5!}i^5 + \cdots$$

Thus the normalized volume of $\mathcal{P}$ is equal to 12.

The notion of weakly unimodular regular triangulation is related to TDI systems and to Hilbert bases [252].

Proposition 9.6.21 [252] *Let $\mathcal{A} = \{v_1, \ldots, v_q\}$ and let A be the matrix with column vectors $v_1, \ldots, v_q$. If $\mathbb{Z}\mathcal{A} = \mathbb{Z}^n$, then the system $xA \leq \omega$ is TDI if and only if Δ is a weakly unimodular regular triangulation of $\mathbb{R}_+\mathcal{A}$.*

Proof. It follows from Theorems 1.3.22 and 9.6.8, and Proposition 1.3.14. We leave the details as an exercise. $\Box$

Conjecture 9.6.22 [119] If A is the incidence matrix of a uniform clutter $\mathcal{C}$ that satisfies the max-flow min-cut property, then the rational polyhedral cone $\mathbb{R}_+\{v_1, \ldots, v_q\}$ has a weakly unimodular regular triangulation.

This conjecture holds for some interesting classes of uniform clutters coming from combinatorial optimization (see Theorem 14.4.17).

Proposition 9.6.23 [400] *If A is a t-unimodular matrix, then the regular triangulation $\{\mathbb{R}_+\mathcal{A}_1, \ldots, \mathbb{R}_+\mathcal{A}_r\}$ of $\mathbb{R}_+\mathcal{A}$ is weakly unimodular.*

Proof. From Proposition 1.6.6, we get $\mathbb{Z}\mathcal{A}_i = \mathbb{Z}\mathcal{A}$ for all i, that is, the triangulation is weakly unimodular. □

Exercises

9.6.24 Let $\mathcal{A} = \{v_1, \ldots, v_q\} \subset \mathbb{N}^n$ be a homogeneous configuration. Then $\mathcal{F} = \mathrm{conv}(v_1, \ldots, v_s)$ is a face of $\mathrm{conv}(\mathcal{A})$ if and only if $\mathcal{F}' = \mathbb{R}_+\{v_1, \ldots, v_s\}$ is a face of $\mathbb{R}_+\mathcal{A}$.

9.6.25 Let $xA \leq c$ be a rational system. Assume that $\mathcal{Q} = \{x \mid xA \leq c\}$ is pointed and $\mathrm{rank}(A) < \mathrm{rank}\,((v_1, c_1), \ldots (v_q, c_q))$, where the v_i's are the columns of A. Prove that the map $\alpha \mapsto (\alpha, -1)$ gives a bijection between the vertices of $\mathcal{Q}$ and the lower facets of the cone spanned by the (v_i, c_i)'s.

9.7 Toric ideals generated by circuits

In this section we give sufficient conditions for the normality of $K[F]$ and classify when the toric ideal of a homogeneous normal monomial subring is generated by circuits.

Theorem 9.7.1 [49] *If each circuit of* $\ker(A)$ *has a positive or negative part with entries consisting of* 0's *and* 1's, *then* $K[F]$ *is normal.*

Proof. The proof is by induction on q, the number of generators of $K[F]$. The case $q = 1$ is clear. Assume $q \geq 2$ and that the result holds for monomial subrings with less than q generators. To simplify the notation we set $f_i = x^{v_i}$ for $i = 1, \ldots, q$. Using Theorem 9.1.1 one has:

$$\overline{K[F]} = K[\{x^a \mid a \in \mathbb{Z}\mathcal{A} \cap \mathbb{R}_+\mathcal{A}\}].$$

Let $z = x^a \in \overline{K[F]}$ be a minimal generator of the integral closure of $K[F]$. One can write $z = x^a = f_1^{\lambda_1} \cdots f_q^{\lambda_q}$ for some $\lambda_1, \ldots, \lambda_q \in \mathbb{Z}$. There is a positive integer m such that

$$z^m = f_1^{m\lambda_1} \cdots f_q^{m\lambda_q} = f_1^{b_1} \cdots f_q^{b_q}, \tag{9.15}$$

for some $b_1,\ldots,b_q \in \mathbb{N}$. First we assume that $b_i = \lambda_i = 0$ for some i. Consider the matrix A' obtained from A by removing its ith column, then Lemma 8.3.27 and the induction hypothesis yield $z \in K[F \setminus \{f_i\}]$. Hence, we may assume that for each i either $\lambda_i \neq 0$ or $b_i > 0$.

If $\ker(A) = (0)$, the columns of A are linearly independent and $K[F]$ is normal because it is a polynomial ring. Assume $\ker(A) \neq (0)$ and take a circuit $0 \neq \alpha \in \ker(A)$, one can write

$$\alpha = \underbrace{(e_{i_1} + \cdots + e_{i_k})}_{\alpha_+} - \underbrace{(\epsilon_1 e_{j_1} + \cdots + \epsilon_t e_{j_t})}_{\alpha_-},$$

where e_i is the ith unit vector, $\epsilon_i \in \mathbb{N}$ for all i, and $i_1,\ldots,i_k,j_1,\ldots,j_t$ distinct. Thus one has the equality

$$\sum_{\ell=1}^{k} v_{i_\ell} = \sum_{\ell=1}^{t} \epsilon_\ell v_{j_\ell}. \tag{9.16}$$

If $b_{i_r} = 0$ for some $1 \leq r \leq k$, then using Eq. (9.16) one can rewrite Eq. (9.15) as $z^m = f_1^{m\mu_1} \cdots f_q^{m\mu_q} = f_1^{b_1} \cdots f_q^{b_q}$, where $\mu_i \in \mathbb{Z}$ for all i and $\mu_{i_r} = b_{i_r} = 0$, which by induction yields that $z \in K[F \setminus \{f_{i_r}\}]$. It remains to consider the case $b_{i_r} > 0$ for all $1 \leq r \leq k$. One may assume $b_{i_1} \leq \cdots \leq b_{i_k}$. Using Eqs. (9.15) and (9.16) we obtain

$$ma \overset{(9.15)}{=} \sum_{i=1}^{q} b_i v_i = b_{i_1} \sum_{\ell=1}^{k} v_{i_\ell} + \sum_{i \neq i_1} c_i v_i \overset{(9.16)}{=} b_{i_1} \sum_{\ell=1}^{t} \epsilon_\ell v_{j_\ell} + \sum_{i \neq i_1} c_i v_i, \tag{9.17}$$

where the c_i's are nonnegative integers. Therefore, using Eq. (9.16) and (9.17), we can rewrite Eq. (9.15) as $z^m = f_1^{m\delta_1} \cdots f_q^{m\delta_q} = f_1^{d_1} \cdots f_q^{d_q}$, where $\delta_i \in \mathbb{Z}$, $d_i \in \mathbb{N}$ for all i, and $d_{i_1} = \delta_{i_1} = 0$. Hence by the induction hypothesis one obtains $z \in K[F \setminus \{f_{i_1}\}]$. □

A non-zero binomial $t^a - t^b$ is said to have a *square-free term* if t^a is square-free or t^b is square-free. If t^a and t^b are both not square-free monomials we say that the binomial $t^a - t^b$ has *nonsquare-free terms.*

Definition 9.7.2 A binomial $g = t_1^{a_1} \cdots t_q^{a_q} - t_1^{b_1} \cdots t_q^{b_q}$ is called *balanced* if the following holds: $\max\{a_1,\ldots,a_q\} = \max\{b_1,\ldots,b_q\}$. If g is not balanced it is called *unbalanced.* Let g be an unbalanced binomial of the form:

$$g = t_1^{b_1} \cdots t_r^{b_r} - t_{r+1}^{b_{r+1}} \cdots t_s^{b_s}, \quad b_i \geq 1\,\forall i,$$

where $1 \leq m_1 = \max\{b_1,\ldots,b_r\} < \max\{b_{r+1},\ldots,b_s\} = m_2$. A *connector* of g is a binomial: $t_{i_1} \cdots t_{i_j} - t_{i_{j+1}}^{c_{j+1}} \cdots t_{i_m}^{c_m}$, $c_i \geq 1\,\forall i$, with a square-free term $t_{i_1} \cdots t_{i_j}$ such that $\{i_1,\ldots,i_j\} \subset \{1,\ldots,r\}$ and the intersection of $\{i_{j+1},\ldots,i_m\}$ with $\{r+1,\ldots,s\}$ is non-empty.

Theorem 9.7.3 [308] *If $K[F]$ is homogeneous and normal and P_F is its toric ideal, then the following are equivalent:*

(a) *P_F is generated by a finite set of circuits.*

(b) *P_F is generated by a finite set of circuits with a square-free term.*

(c) *Every unbalanced circuit of P_F has a connector which is a linear combination (with coefficients in $K[\mathbf{t}]$) of circuits of P_F with a square-free term.*

Proof. As $K[F]$ is homogeneous, we get that any binomial $t^a - t^b$ in P_F is homogeneous with respect to the standard grading of $K[\mathbf{t}] = K[t_1, \ldots, t_q]$ induced by setting $\deg(t_i) = 1$ for all i, a fact that will be used repeatedly below without any further notice.

(a) $\Rightarrow$ (b): The toric ideal P_F is minimally generated by a finite set $\mathcal{B}$ of circuits. Thus, by Theorem 9.2.16, each binomial of $\mathcal{B}$ is a circuit with a square-free term.

(*b*) $\Rightarrow$ (*c*): Let $g = t_1^{b_1} \cdots t_r^{b_r} - t_{r+1}^{b_{r+1}} \cdots t_s^{b_s}$, $b_i \geq 1\,\forall i$, be an unbalanced circuit of A, where $1 \leq m_1 = \max\{b_1, \ldots, b_r\} < \max\{b_{r+1}, \ldots, b_s\} = m_2$. We may assume $m_2 = b_{r+1}$. Then

$$(x^{v_1} \cdots x^{v_r}/x^{v_{r+1}})^{m_1} \in K[F]. \tag{9.18}$$

The element $x^{v_1} \cdots x^{v_r}/x^{v_{r+1}}$ is in the field of fractions of $K[F]$ and by Eq. (9.18) it is integral over $K[F]$. Hence, by the normality of $K[F]$, the element $x^{v_1} \cdots x^{v_r}/x^{v_{r+1}}$ is in $K[F]$. Since $K[F]$ is generated as a K-vector space by Laurent monomials of the form x^a, with $a \in \mathbb{N}\mathcal{A}$, it is not hard to see that there is a monomial t^γ such that $t_1 \cdots t_r - t_{r+1}t^\gamma \in P_F$. This is a connector of g and by hypothesis it is a linear combination of circuits of P_F with a square-free term.

(c) $\Rightarrow$ (a): By Theorem 9.2.16, the toric ideal P_F is minimally generated by a finite set $\mathcal{B} = \{f_1, \ldots, f_m\}$ consisting of binomials with a square-free term. We will show, by induction on the degree, that each one of the f_i's is a linear combination of circuits. The degree is taken with respect to the standard grading of $K[\mathbf{t}]$.

Let f be a binomial in $\mathcal{B}$. We may assume that $f = t_1 \cdots t_p - t_{p+1}^{a_{p+1}} \cdots t_\ell^{a_\ell}$, $a_i \geq 1\,\forall i$, $\ell \leq q$. Assume that f is not a circuit. Then by Lemma 1.9.5 there is a circuit in P_F (permuting variables if necessary) of the form

$$g = t_1^{b_1} \cdots t_r^{b_r} - t_{p+1}^{b_{p+1}} \cdots t_s^{b_s}, \quad b_i \geq 1\,\forall i,$$

with $r < p$ or $s < \ell$. Set $m_1 = \max\{b_1, \ldots, b_r\}$ and $m_2 = \max\{b_{p+1}, \ldots, b_s\}$.

We claim that there exist binomials h and h_1 (we allow $h = h_1$ or $h = 0$) in P_F of degree less than $\deg(f) = p$ and a binomial h_2 which is a linear combination of circuits of P_F such that f is in the ideal of $K[\mathbf{t}]$ generated by g, h, h_1, h_2. To prove this we consider the following two cases.

Case (A): $r = p$ and $s < \ell$. Then

$$g = t_1^{b_1} \cdots t_p^{b_p} - t_{p+1}^{b_{p+1}} \cdots t_s^{b_s}. \tag{9.19}$$

Subcase (A_1): $b_i = 1$ for $i = 1, \ldots, p$. Then we can write

$$f - g = t_{p+1}^{b_{p+1}} \cdots t_s^{b_s} - t_{p+1}^{a_{p+1}} \cdots t_\ell^{a_\ell} = t_{p+1}h,$$

for some binomial $0 \neq h \in P_F$ (P_F is prime) with $\deg(h) < \deg(f) = p$.

Subcase (A_2): $b_i = 1$ for $i = p+1, \ldots, s$. Then $g = t_1^{b_1} \cdots t_p^{b_p} - t_{p+1} \cdots t_s$. By subcase ($A_1$), we may assume that $b_i \geq 2$ for some $1 \leq i \leq p$. Then, on the one hand, by the homogeneity of g, $p + 1 \leq \sum_{i=1}^{p} b_i = s - p$, so $2p+1 \leq s$. On the other hand, by the homogeneity of f, $p \geq \ell - p \geq s - p + 1$, so $2p - 1 \geq s$. This is a contradiction. So this case cannot occur.

Subcase (A_3): $b_i \geq 2$ for some $1 \leq i \leq p$, $b_{p+j} \geq 2$ for some $1 \leq j \leq s-p$, and $m_1 \geq m_2$. Then $f = t_1 \cdots t_p - t_{p+1}^{a_{p+1}} \cdots t_\ell^{a_\ell}$, $g = t_1^{b_1} \cdots t_p^{b_p} - t_{p+1}^{b_{p+1}} \cdots t_s^{b_s}$. For simplicity of notation we may assume that $m_1 = b_1$. Using that $g \in P_F$ and $m_1 \geq m_2 \geq 2$, we get $(x^{v_{p+1}} \cdots x^{v_s}/x^{v_1})^{m_2} \in K[F]$. Hence, by the normality of $K[F]$, there is t^γ such that $h_1 = t_{p+1} \cdots t_s - t_1 t^\gamma$ is in P_F. The binomial h_1 is non-zero and has degree less than $\deg(f)$ because $s - p < \ell - p \leq p$; the second inequality follows from the homogeneity of f. Let $t^\delta = t_{p+1}^{a_{p+1}} \cdots t_\ell^{a_\ell} / t_{p+1} \cdots t_s$. We have

$$f + h_1 t^\delta = f + (t_{p+1} \cdots t_s - t_1 t^\gamma) t^\delta = t_1 \cdots t_p - t_1 t^\gamma t^\delta = t_1 h, \tag{9.20}$$

where $0 \neq h \in P_F$ and $\deg(h) < p$.

Subcase (A_4): $b_i \geq 2$ for some $1 \leq i \leq p$, $b_{p+j} \geq 2$ for some $1 \leq j \leq s-p$, and $m_1 < m_2$. Since g is an unbalanced circuit, by hypothesis g has a connector

$$h_2 = t_{i_1} \cdots t_{i_k} - t_{i_{k+1}} t^\gamma, \quad i_1 < \cdots < i_k,$$

with $i_{k+1} \in \{p+1, \ldots, s\}$, $\{i_1, \ldots, i_k\} \subset \{1, \ldots, p\}$ and such that h_2 is a linear combination of circuits of P_F. Set $t^\delta = t_1 \cdots t_p / t_{i_1} \cdots t_{i_k}$. If $f = h_2 t^\delta$, then $t^\delta = 1$ and $f = h_2$. If $f \neq h_2 t^\delta$, then we can write

$$f - h_2 t^\delta = t_{i_{k+1}} t^\gamma t^\delta - t_{p+1}^{a_{p+1}} \cdots t_\ell^{a_\ell} = t_{i_{k+1}} h, \tag{9.21}$$

with $0 \neq h \in P_F$ and $\deg(h) < p$.

Case (B): $r < p$ and $s \leq \ell$. In this case

$$f = t_1 \cdots t_p - t_{p+1}^{a_{p+1}} \cdots t_\ell^{a_\ell}, \quad g = t_1^{b_1} \cdots t_r^{b_r} - t_{p+1}^{b_{p+1}} \cdots t_s^{b_s}.$$

Subcase (B_1): $b_i = 1$ for $i = 1, \ldots, r$. Then

$$f - g t_{r+1} \cdots t_p = t_{p+1}^{b_{p+1}} \cdots t_s^{b_s} t_{r+1} \cdots t_p - t_{p+1}^{a_{p+1}} \cdots t_\ell^{a_\ell} = t_{p+1} h, \tag{9.22}$$

where $0 \neq h \in P_F$ and $\deg(h) < p$.

Subcase (B$_2$): $b_i = 1$ for $i = p+1, \ldots, s$. Let $t^\gamma = t_{p+1}^{a_{p+1}} \cdots t_\ell^{a_\ell}/t_{p+1} \cdots t_s$. Then we have

$$f - gt^\gamma = t_1 \cdots t_p - t_1^{b_1} \cdots t_r^{b_r} t^\gamma = t_1 h \tag{9.23}$$

where $0 \neq h \in P_F$ and $\deg(h) < p$.

Subcase (B$_3$): $b_i \geq 2$ for some $1 \leq i \leq r$, $b_j \geq 2$ for some $p+1 \leq j \leq s$, and $m_1 \leq m_2$. We may assume $m_2 = b_{p+1}$. Using that $g \in P_F$ and that $m_2 \geq m_1 \geq 2$, we get $(x^{v_1} \cdots x^{v_r}/x^{v_{p+1}})^{m_1} \in K[F]$. Hence, by the normality of $K[F]$, there is t^γ such that $h_1 = t_1 \cdots t_r - t_{p+1}t^\gamma$ is in P_F. The binomial h_1 is non-zero and has degree less than $\deg(f)$ because $r < p$. Then we have

$$\begin{aligned} f - h_1 t_{r+1} \cdots t_p &= f - (t_1 \cdots t_r - t_{p+1}t^\gamma)t_{r+1} \cdots t_p \\ &= t_{p+1}t^\gamma t_{r+1} \cdots t_p - t_{p+1}^{a_{p+1}} \cdots t_\ell^{a_\ell} = t_{p+1}h, \end{aligned} \tag{9.24}$$

where $0 \neq h \in P_F$ and $\deg(h) < p$.

Subcase (B$_4$): $b_i \geq 2$ for some $1 \leq i \leq r$, $b_j \geq 2$ for some $p+1 \leq j \leq s$, and $m_1 > m_2$. Since g is an unbalanced circuit, by hypothesis g has a connector

$$h_2 = t_{i_{k+1}} \cdots t_{i_{k+t}} - t_{i_d}t^\gamma, \quad i_{k+1} < \cdots < i_{k+t},$$

with $\{i_{k+1}, \cdots, i_{k+t}\} \subset \{p+1, \ldots, s\}$, $i_d \in \{1, \ldots, r\}$, and such that h_2 is a linear combination of circuits of P_F. Set $t^\delta = t_{p+1}^{a_{p+1}} \cdots t_\ell^{a_\ell}/t_{i_{k+1}} \cdots t_{i_{k+t}}$. If $f = -h_2 t^\delta$, then $t^\delta = 1$ and $f = -h_2$. If $f \neq -h_2 t^\delta$, then we can write

$$f + h_2 t^\delta = t_1 \cdots t_p - t_{i_d} t^\gamma t^\delta = t_{i_d} h, \tag{9.25}$$

with $0 \neq h \in P_F$ and $\deg(h) < p$. This completes the proof of the claim.

We are now ready to show that each f_i in $\mathcal{B}$ is a linear combination (with coefficients in $K[\mathbf{t}]$) of circuits. We proceed by induction on $\deg(f_i)$. Let $p = \min\{\deg(f_i) |\, 1 \leq i \leq m\}$ be the initial degree of P_F. If f_i is a binomial in $\mathcal{B}$ of degree p, then either f_i is a circuit or f_i is not a circuit and by the claim f_i is a linear combination of circuits (notice that in this case $h = h_1 = 0$ because there are no non-zero binomials in P_F of degree less than p). Let d be an integer greater than p and let f_k be a binomial of $\mathcal{B}$ of degree d (if any). Assume that each f_i of degree less than d is a linear combination of circuits. If f_k is a circuit there is nothing to prove. If f_k is not a circuit, then by the claim (or more precisely by Eqs. (9.19)–(9.25)) we can write

$$f_k = \lambda g + \mu h + \mu_1 h_1 + \mu_2 h_2, \tag{9.26}$$

where $\lambda, \mu, \mu_1, \mu_2$ are monomials, h, h_1 are binomials in P_F of degree less than $d = \deg(f_k)$, h_2 is a linear combination of circuits, and g is a circuit. Since P_F is a graded ideal with respect to the standard grading of $K[t_1, \ldots, t_q]$, we get that h and h_1 are linear combinations of binomials in

$\mathcal{B}$ of degree less than d. Therefore by Eq. (9.26) and the induction hypothesis, we conclude that f_k is a linear combination of circuits. Therefore the ideal P_F is generated by a finite set of circuits. $\Box$

The following result will be used below to show a class of toric ideals generated by circuits. See Theorem 10.3.14 for an application of this result.

Corollary 9.7.4 *Let $K[F]$ be a homogeneous and normal subring. If each circuit of P_F with nonsquare-free terms is balanced, then P_F is generated by a finite set of circuits with a square-free term.*

Proof. The circuits of P_F satisfy condition (c) of Theorem 9.7.3. Indeed, let f be an unbalanced circuit of P_F. Then f has a square-free term by hypothesis. Thus f is a circuit with a square-free term and it is a connector of f. Hence the result follows from Theorem 9.7.3. $\Box$

The best known examples of toric ideals generated by circuits come from configurations whose matrix A is t-unimodular (see Corollary 8.3.8). Other interesting examples of toric ideals generated by circuits are the phylogenetic ideals studied in [82]. As noted in [328], these phylogenetic ideals actually represent the family of cut ideals of cycles.

Corollary 9.7.5 *If $K[F]$ is a homogeneous subring such that each circuit of P_F has a square-free term, then $K[F]$ is normal and P_F is generated by a finite set of circuits with a square-free term.*

Proof. The normality of $K[F]$ follows from Theorem 9.7.1. Since the circuits of P_F satisfy condition (c) of Theorem 9.7.3, we get that P_F is generated by a finite set of circuits with a square-free term. $\Box$

Exercises

9.7.6 Let $F = \{x_1x_2, x_2x_3, x_3x_4, x_1x_4, x_1x_3, x_1^2, x_2^2, x_3^2, x_4^2\}$. Use *Macaulay2* and the procedure below to show that P_F is minimally generated by

$$\begin{array}{llllll} t_3t_4 - t_5t_9, & t_4t_5 - t_3t_6, & t_3t_5 - t_4t_8, & t_2t_5 - t_1t_8, & t_4^2 - t_6t_9, & t_5^2 - t_6t_8, \\ t_1t_5 - t_2t_6, & t_1t_2 - t_5t_7, & t_1t_3 - t_2t_4, & t_2^2 - t_7t_8, & t_3^2 - t_8t_9, & t_1^2 - t_6t_7. \end{array}$$

Prove that $\mathbb{Q}[F]$ is not normal. Notice that P_F is minimally generated by "circuits" with square-free part.

```
KK=ZZ/31991
R=KK[x_1..x_4,t_1..t_9,MonomialOrder=>Eliminate 4]
I=ideal(t_1-x_1*x_2,t_2-x_2*x_3,t_3-x_3*x_4,t_4-x_1*x_4,
t_5-x_1*x_3,t_6-x_1^2,t_7-x_2^2,t_8-x_3^2,t_9-x_4^2)
J= ideal selectInSubring(1,gens gb I)
mingens J
```

9.7.7 We set $F = \{x^{v_1}, \ldots, x^{v_7}\}$, where the v_i's are given by:

$$\begin{array}{lll} v_1 = (0,1,1,0,1,0), & v_2 = (1,0,0,1,1,0), & v_3 = (1,0,1,0,0,1), \\ v_4 = (1,0,0,1,0,1), & v_5 = (0,1,1,0,0,1), & v_6 = (0,1,0,1,1,0), \\ v_7 = (0,1,0,1,0,1). & & \end{array}$$

Let A be the matrix with column vectors $v_1, \ldots, v_7$. Prove that the circuits of $\ker(A)$ are given by the rows of the matrix:

$$\left[\begin{array}{rrrrrrr} -1 & 1 & 0 & -1 & 1 & 0 & 0 \\ -1 & 0 & 1 & -1 & 0 & 1 & 0 \\ -1 & 1 & 1 & -2 & 0 & 0 & 1 \\ 0 & -1 & 1 & 0 & -1 & 1 & 0 \\ 1 & -1 & 1 & 0 & -2 & 0 & 1 \\ 1 & 1 & -1 & 0 & 0 & -2 & 1 \\ 0 & 1 & 0 & -1 & 0 & -1 & 1 \\ 1 & 0 & 0 & 0 & -1 & -1 & 1 \\ 0 & 0 & -1 & 1 & 1 & 0 & -1 \end{array}\right]$$

Then show that $K[F]$ is a normal monomial subring using the criterion of Theorem 9.7.1.

9.7.8 In the proof of Theorem 9.7.3 (from (c) to (a)), show that the subcases (A_3), (B_1), and (B_3) cannot occur.

9.8 Divisor class groups of semigroup rings

In this section we present an algorithm to compute the divisor class group of a Krull semigroup ring of the form $K[\mathbb{N}\mathcal{A}]$, where K is a field and $\mathbb{N}\mathcal{A}$ is a subsemigroup of $\mathbb{N}^n$ generated by a finite set of vectors.

Krull semigroup rings In this part the semigroups are commutative, satisfy the cancellation law, have a unit, and the total ring of quotients is torsion-free. The semigroups will be written additively.

Let $\mathcal{S}$ be a semigroup and let $\sim$ be the equivalence relation on $\mathcal{S} \times \mathcal{S}$ given by:

$$(a, b) \sim (c, d) \quad \text{if and only if } a + d = c + b.$$

We denote each equivalence class $[(a, b)]$ by $a - b$ or simply by a if $b = 0$. Let $\langle \mathcal{S} \rangle$ be the set of equivalence classes with the operation

$$(a - b) + (c - d) = (a + c) - (b + d).$$

It follows readily that $\langle \mathcal{S} \rangle$ is a group which is called the *total quotient group* of $\mathcal{S}$ or the *group of differences* of $\mathcal{S}$.

Definition 9.8.1 $\mathcal{S}$ is called *normal* if $s \in \langle \mathcal{S} \rangle$ and $ns \in \mathcal{S}$ for some $n \geq 1$ implies $s \in \mathcal{S}$. $\mathcal{S}$ is called *completely normal* if $s \in \langle \mathcal{S} \rangle$, $a \in \mathcal{S}$ and $a+ns \in \mathcal{S}$ for all $n \geq 1$ implies $s \in \mathcal{S}$.

If $\mathcal{S} = \mathbb{N}\mathcal{A}$ is a semigroup of $\mathbb{N}^n$ generated by a finite set $\mathcal{A}$, then the group of differences of $\mathcal{S}$ is $\langle \mathcal{S} \rangle = \mathbb{Z}\mathcal{A}$ and $\mathcal{S}$ is normal if and only if $\mathbb{N}\mathcal{A} = \mathbb{Z}\mathcal{A} \cap \mathbb{R}_+\mathcal{A}$.

Remark 9.8.2 If $\mathcal{S}$ is completely normal, then $\mathcal{S}$ is normal. The converse holds if $\mathcal{S}$ satisfies the ascending chain condition on ideals, thus in particular when $\mathcal{S}$ is finitely generated (see [83, 161]).

Definition 9.8.3 A *discrete valuation* of an abelian group G is a homomorphism of groups $\nu\colon G \to \mathbb{Z}$. The set $\{x \in G \mid \nu(x) \geq 0\}$ is the *valuation semigroup* of ν. The *residue group* of ν is $\ker(\nu)$. We say that ν is *normalized* if $\nu(G) = \mathbb{Z}$.

Proposition 9.8.4 If $\nu \neq 0$ is a valuation of a group G, then the valuation semigroup of ν is isomorphic to $\ker(\nu) \times \mathbb{N}$.

Proof. There is $0 \neq n_0 \in \mathbb{N}$ such that $\nu(G) = n_0\mathbb{Z}$ and we can write $n_0 = \nu(x_0)$ for some $x_0 \in G$. The following mapping gives the required isomorphism

$$\varphi\colon \ker(\nu) \times \mathbb{N} \to \{x \in G \mid \nu(x) \geq 0\}, \quad (x, n) \mapsto x + nx_0. \qquad \square$$

Definition 9.8.5 $\mathcal{S}$ is a *Krull semigroup* if there is a family $(\nu_i)_{i\in\mathcal{I}}$ of discrete valuations of $\langle \mathcal{S} \rangle$ such that $\mathcal{S}$ is the intersection of the valuation semigroups of the ν_i's and for each $s \in \mathcal{S}$ the set $\{i \in \mathcal{I} \mid \nu_i(s) > 0\}$ is finite.

Notation Let $\mathfrak{F} = \bigoplus_{i\in I} \mathbb{Z}e_i$ be the free abelian group with basis $\{e_i \mid i \in I\}$. For each $i \in I$ let π_i the projection $\pi_i\colon \mathfrak{F} \to \mathbb{Z}$ given by $\pi_i(a_ke_k)_{k\in I} = a_i$. The *positive* part of $\mathfrak{F}$ is: $\mathfrak{F}_+ = \{x \in \mathfrak{F} \mid \pi_i(x) \geq 0 \ \forall\ i \in I\}$. Notice that $\mathfrak{F}_+ = \bigoplus_{i\in I} \mathbb{N}e_i$ and if $\mathcal{S}$ is a subsemigroup of $\mathfrak{F}$, then $\mathbb{Z}\mathcal{S}$ (the subgroup generated by $\mathcal{S}$) is equal to $\langle \mathcal{S} \rangle$ (the group of differences of $\mathcal{S}$).

Proposition 9.8.6 [83] *$\mathcal{S}$ is a Krull semigroup if and only if $\mathcal{S} \cong G \times \mathcal{S}_1$, where G is a group and $\mathcal{S}_1$ is a subsemigroup of a free group $\mathfrak{F} = \bigoplus_{i\in I} \mathbb{Z}e_i$ such that $\mathcal{S}_1 = \langle \mathcal{S}_1 \rangle \cap \mathfrak{F}_+$.*

Corollary 9.8.7 *If $\mathcal{S}$ is a semigroup of $\mathfrak{F}$ such that $\mathcal{S} = \langle \mathcal{S} \rangle \cap \mathfrak{F}_+$, then $\mathcal{S}$ is a Krull semigroup.*

Proof. Notice that $\mathcal{S} \simeq \{0\} \times \mathcal{S}$ and apply Proposition 9.8.6. $\square$

Let $\mathcal{S}$ be a semigroup and let A be a ring. The *semigroup ring* of $\mathcal{S}$ with coefficients in A is denoted by $A[\mathcal{S}]$. The elements of $A[\mathcal{S}]$ can be written in the form $\sum_{s\in\mathcal{S}} r_s X^s$, $r_s \in A$, where $r_s = 0$ except for a finite number, with addition and multiplication defined as for polynomials. See [180, p. 64] for a formal construction of a semigroup ring.

Theorem 9.8.8 [83] *$A[\mathcal{S}]$ is a Krull ring if and only if A is a Krull ring and $\mathcal{S}$ is a Krull semigroup such that $\langle\mathcal{S}\rangle$ satisfies the ascending chain condition on cyclic subgroups.*

Corollary 9.8.9 *Let $\mathcal{A} \subset \mathbb{N}^n$ be a finite set. Then $K[\mathbb{N}\mathcal{A}]$ is a Krull ring if and only if $\mathbb{N}\mathcal{A}$ is a Krull semigroup.*

The divisor class group of a Krull semigroup

Definition 9.8.10 A subsemigroup $\mathcal{S}$ of $\mathfrak{F}$ is called *reduced* in $\mathfrak{F}$ if the projection $\pi_i|_{\langle\mathcal{S}\rangle}\colon \langle\mathcal{S}\rangle \to \mathbb{Z}$ is onto for all $i \in I$.

Theorem 9.8.11 [83] *Let $\mathfrak{F} = \bigoplus_{i\in I} \mathbb{Z}e_i$ be a free abelian group and let $\mathcal{S}$ be a subsemigroup of $\mathfrak{F}$ such that $\mathcal{S} = \langle\mathcal{S}\rangle \cap \mathfrak{F}_+$. If any of the following two equivalent conditions hold, then $\mathrm{Cl}(K[\mathcal{S}]) \cong \mathfrak{F}/\langle\mathcal{S}\rangle$.*

(a) *$\mathcal{S}$ is a reduced subsemigroup of $\mathfrak{F}$, and for all $i, j \in I$ with $i \neq j$, there exists $s \in \mathcal{S}$ such that $\pi_j(s) > 0 = \pi_i(s)$.*

(b) *$T_i = \{x_j + \langle\mathcal{S}\rangle \mid i \neq j \in I\}$ generates $\mathfrak{F}/\langle\mathcal{S}\rangle$ as a subsemigroup $\forall\, i \in I$.*

Next we state a particular case of Theorem 9.8.11 that can be used to compute the divisor class group of an affine semigroup.

Theorem 9.8.12 [83] *Let $\mathfrak{F} = \mathbb{Z}e_1 \oplus \cdots \oplus \mathbb{Z}e_r$ be a free abelian group of rank r and let $\mathcal{S} = \mathbb{N}w_1 + \cdots + \mathbb{N}w_q$ be a subsemigroup of $\mathfrak{F}$ such that the following two conditions hold:*

(a) *$\mathcal{S} = \mathbb{Z}\mathcal{S} \cap \mathfrak{F}_+$, where $\mathbb{Z}\mathcal{S}$ is the subgroup of $\mathfrak{F}$ generated by $\mathcal{S}$.*

(b) *$\pi_i(\mathbb{Z}\mathcal{S}) = \mathbb{Z}$ for $i = 1, \ldots, r$ and for each pair $i \neq j$ in $\{1, \ldots, r\}$ there exists $w \in \mathcal{S}$ such that $\pi_i(w) > 0 = \pi_j(w)$.*

Then $\mathrm{Cl}(K[\mathcal{S}]) \cong \mathfrak{F}/\mathbb{Z}\mathcal{S}$ for any field K.

Divisor class groups of affine semigroups Let $\mathcal{A} = \{v_1, \ldots, v_q\}$ be a finite set of distinct vectors in $\mathbb{N}^n \setminus \{0\}$ and let $\mathbb{N}\mathcal{A}$ be the affine semigroup of $\mathbb{N}^n$ generated by $\mathcal{A}$. If $R = K[x_1, \ldots, x_n]$ is a polynomial ring over a field K and $F = \{x^{v_1}, \ldots, x^{v_q}\}$, then the elements in the monomial subring $K[F]$ are expressions of the form:

$$\sum c_a\, (x^{v_1})^{a_1} \cdots (x^{v_q})^{a_q}\,,$$

where $a = (a_1, \ldots, a_q) \in \mathbb{N}^q$, $c_a \in K$ and $c_a = 0$ except for a finite number. Therefore $K[F]$ is generated, as a K-vector space, by the set of all x^a such that $a \in \mathbb{N}\mathcal{A}$, i.e., $K[F]$ is the semigroup ring of $\mathbb{N}\mathcal{A}$:

$$K[F] = K[\mathbb{N}\mathcal{A}] = K[\{x^a | \, a \in \mathbb{N}\mathcal{A}\}].$$

Theorem 9.8.13 *The following conditions are equivalent:*

(a) $K[F]$ *is normal.* (b) $K[\mathbb{N}\mathcal{A}]$ *is a Krull ring.* (c) $\mathbb{N}\mathcal{A} = \mathbb{R}_+\mathcal{A} \cap \mathbb{Z}\mathcal{A}$.

Proof. Notice that $K[F]$ is a Noetherian domain because $K[F]$ is a finitely generated K-algebra. Thus, by the Mori–Nagata theorem (Theorem 2.6.2), part (a) is equivalent to part (b). From Corollary 9.1.3, we obtain that part (a) is equivalent to part (c). □

Corollary 9.8.14 $\mathbb{N}\mathcal{A}$ *is a Krull semigroup if and only if* $\mathbb{N}\mathcal{A}$ *is equal to* $\mathbb{R}_+\mathcal{A} \cap \mathbb{Z}\mathcal{A}$.

Proof. It follows from Theorem 9.8.13 and Corollary 9.8.9. □

Theorem 9.8.15 $\mathbb{Z}\mathcal{A} \cap \mathbb{R}_+\mathcal{A}$ *is a finitely generated Krull semigroup.*

Proof. By the finite basis theorem there are $a_1, \ldots, a_k$ in $\mathbb{Q}^n$ such that

$$\mathbb{R}_+\mathcal{A} = H^+_{a_1} \cap \cdots \cap H^+_{a_k}. \tag{$\dagger$}$$

For $1 \leq i \leq k$ consider the valuation $\nu_i \colon \mathbb{Z}\mathcal{A} \to \mathbb{Z}$ given by $\nu_i(x) = \langle a_i, x \rangle$. We may assume that $a_i \in \mathbb{Z}^n$ for all i because $H^+_a = H^+_{ra}$ for $r > 0$. Hence

$$\mathbb{Z}\mathcal{A} \cap \mathbb{R}_+\mathcal{A} = \bigcap_{i=1}^{k} \{x \in \mathbb{Z}\mathcal{A} \mid \nu_i(x) \geq 0\}.$$

Hence $\mathbb{Z}\mathcal{A} \cap \mathbb{R}_+\mathcal{A}$ is a Krull semigroup because it is the intersection of the valuation semigroups of $\nu_1, \ldots, \nu_k$. By Gordan's lemma the semigroup is finitely generated. □

Proposition 9.8.16 *There exists an irreducible representation:*

$$\mathbb{R}_+\mathcal{A} = \mathbb{R}\mathcal{A} \cap H^+_{\ell_1} \cap \cdots \cap H^+_{\ell_r}, \tag{9.27}$$

with $\ell_i \in \mathbb{Q}^n$ *for all* i *and such that the following conditions are satisfied:*

(a) $\langle \ell_i, v_j \rangle \in \mathbb{N}$ *for all* i, j.

(b) $\mathbb{Z} = \langle \ell_i, v_1 \rangle \mathbb{Z} + \cdots + \langle \ell_i, v_q \rangle \mathbb{Z}$ *for all* i.

Proof. The decomposition of $\mathbb{R}_+\mathcal{A}$ given in Eq. (†) can be brought to the form of Eq. (9.27), where $\ell_i \in \mathbb{Q}^n$ for all i and $\mathbb{R}\mathcal{A}$ is the vector space generated by $\mathcal{A}$. By multiplying each ℓ_i by an appropriate positive integer, we can choose the ℓ_i's such that condition (a) holds. Let $1 \leq i \leq r$ be a fixed integer. We claim that there exists $1 \leq j \leq q$ such that $\langle \ell_i, v_j\rangle > 0$. Otherwise $\langle \ell_i, v_j\rangle = 0$ for all j and one has $\mathbb{R}_+\mathcal{A} \subset H_{\ell_i}$. By the irreducibility of the representation (see Theorem 1.1.44), the set

$$\mathbb{R}_+\mathcal{A} \cap H_{\ell_i} = \mathbb{R}_+\mathcal{A}$$

is a facet (proper face of maximum dimension), a contradiction. Let m be the greatest common divisor of the non-zero elements of the set

$$\{\langle \ell_i, v_1\rangle, \ldots, \langle \ell_i, v_q\rangle\}.$$

We can write $\langle \ell_i, v_k\rangle = \lambda_k m$. Setting $\ell_i' = \ell_i/m$, conditions (a) and (b) are satisfied if we substitute ℓ_i by ℓ_i', this follows from the fact that the non-zero elements of the set $\lambda_1, \ldots \lambda_q$ are relatively prime. □

For the rest of this section we assume that $\mathbb{R}_+\mathcal{A}$ has an irreducible representation satisfying the conditions of Proposition 9.8.16. Consider the homomorphism of $\mathbb{Z}$-modules $\varphi\colon \mathbb{Z}^n \to \mathbb{Z}^r$ given by:

$$\varphi(v) = \langle \ell_1, v\rangle e_1 + \cdots + \langle \ell_r, v\rangle e_r.$$

In what follows w_i will denote the vector of $\mathbb{N}^r$ defined as

$$w_i := \varphi(v_i) = \langle \ell_1, v_i\rangle e_1 + \cdots + \langle \ell_r, v_i\rangle e_r \quad (i = 1, \ldots, q),$$

$\mathcal{S}$ will denote the semigroup $\mathcal{S} := \mathbb{N}w_1 + \cdots + \mathbb{N}w_q$, and $\mathfrak{F}$ will denote the free abelian group $\mathfrak{F} := \mathbb{Z}e_1 \oplus \cdots \oplus \mathbb{Z}e_r$.

Proposition 9.8.17 $\mathbb{N}\mathcal{A} \simeq \mathcal{S}$.

Proof. Clearly φ induces a homomorphism from $\mathbb{N}\mathcal{A}$ onto $\mathcal{S}$ that we also denote by φ. Thus it suffices to prove that φ is an injective map. Let $\alpha = a_1v_1 + \cdots + a_qv_q$ be a vector in $\ker(\varphi)$. It follows readily that $\langle \alpha, \ell_i\rangle = 0$ for $i = 1, \ldots, r$. Therefore α belongs to the set

$$\mathbb{R}_+\mathcal{A} \cap H_{\ell_1} \cap \cdots \cap H_{\ell_r} = \{0\}$$

and consequently $\alpha = 0$. The previous equality follows using that the vector 0 is a vertex of the pointed cone $\mathbb{R}_+\mathcal{A}$, hence $\{0\}$ is an intersection of facets. From the irreducibility of the representation of the cone one obtains that its facets have the form $\mathbb{R}_+\mathcal{A} \cap H_{\ell_i}$ for $i = 1, \ldots, r$ (see Theorem 1.1.44). □

Corollary 9.8.18 $K[\mathbb{N}\mathcal{A}] \simeq K[\mathcal{S}]$.

Proof. By Proposition 9.8.17 we have $\mathbb{N}\mathcal{A} \simeq \mathcal{S}$. Hence applying [180, Theorem 7.2, p. 69] we obtain the required isomorphism. □

Theorem 9.8.19 *If* $K[\mathbb{N}\mathcal{A}]$ *is a Krull ring , then*

$$\mathrm{Cl}(K[\mathbb{N}\mathcal{A}]) \simeq \mathfrak{F}/\langle\mathcal{S}\rangle = (\mathbb{Z}e_1 \oplus \cdots \oplus \mathbb{Z}e_r)/(\mathbb{Z}w_1 + \cdots + \mathbb{Z}w_q)\,.$$

In particular $\mathrm{rank}(\mathfrak{F}/\langle\mathcal{S}\rangle) = r - \mathrm{rank}(\langle\mathcal{S}\rangle)$.

Proof. Taking into account Theorem 9.8.12 and Corollary 9.8.18 it suffices to prove that the following two conditions are satisfied:

(a) $\mathcal{S} = \mathbb{Z}\mathcal{S} \cap \mathfrak{F}_+$.

(b) $\pi_i(\mathbb{Z}\mathcal{S}) = \mathbb{Z}$ for $i = 1, \ldots, r$ and for every $i \neq j$ in $\{1, \ldots, r\}$ there exists $w \in \mathcal{S}$ such that $\pi_i(w) > 0 = \pi_j(w)$.

First we prove (a). Since $\langle \ell_j, v_i\rangle \in \mathbb{N}$ for all i, j, we get that $w_i \in \mathfrak{F}_+$ and clearly $w_i \in \mathbb{Z}\mathcal{S}$ for all i. Therefore $\mathcal{S} \subset \mathbb{Z}\mathcal{S} \cap \mathfrak{F}_+$. Conversely let $\alpha \in \mathbb{Z}\mathcal{S} \cap \mathfrak{F}_+$. We can write

$$\alpha = a_1w_1 + \cdots + a_qw_q = \eta_1e_1 + \cdots + \eta_re_r \quad (a_i \in \mathbb{Z};\ \eta_j \in \mathbb{N}\,\forall\, i, j).$$

Substituting the w_i's and matching the coefficients of the e_i's yields:

$$\eta_i \;\; = \;\; a_1\langle \ell_i, v_1\rangle + \cdots + a_q\langle \ell_i, v_q\rangle, \quad i = 1, \ldots, r.$$

Consider the vector $\beta = a_1v_1 + \cdots + a_qv_q \in \mathbb{Z}\mathcal{A}$. Clearly $\varphi(\beta) = \alpha$ because $\varphi(v_i) = w_i$ for $= 1, \ldots, q$. Using these equations we obtain

$$\langle \beta, \ell_i\rangle \;\; = \;\; a_1\langle \ell_i, v_1\rangle + \cdots + a_q\langle \ell_i, v_q\rangle = \eta_i \geq 0, \quad i = 1, \ldots, r.$$

Therefore $\beta \in \mathbb{R}\mathcal{A} \cap H^+_{\ell_1} \cap \cdots \cap H^+_{\ell_r} = \mathbb{R}_+\mathcal{A}$ and as a consequence we get $\beta \in \mathbb{Z}\mathcal{A} \cap \mathbb{R}_+\mathcal{A}$. By the normality of $K[\mathbb{N}\mathcal{A}]$ we conclude that $\beta \in \mathbb{N}\mathcal{A}$. This means that we can rewrite β as $\beta = \sum_{i=1}^q c_iv_i$ with $c_i \in \mathbb{N}$ for all i. Hence $\varphi(\beta) = \sum_{i=1}^q c_iw_i = \alpha$. Thus $\alpha \in \mathcal{S}$, as desired.

Next we prove (b). Since $\pi_i(w_j) = \langle \ell_i, v_j\rangle$ for all i, j, one has

$$\begin{aligned}\pi_i(\mathbb{Z}\mathcal{S}) &= \pi_i(\mathbb{Z}w_1 + \cdots + \mathbb{Z}w_q) = \mathbb{Z}\pi_i(w_1) + \cdots + \mathbb{Z}\pi_i(w_q)\\ &= \mathbb{Z}\langle \ell_i, v_1\rangle + \cdots + \mathbb{Z}\langle \ell_i, v_q\rangle = \mathbb{Z},\end{aligned}$$

where the last equality is satisfied thanks to the choice of $\ell_1, \ldots, \ell_r$ (see Proposition 9.8.16). To prove the second part of condition (b) it suffices to prove that for $i \neq j$ there exists $w_k = \varphi(v_k) \in \mathcal{S}$ such that

$$\langle \ell_i, v_k\rangle = \pi_i(w_k) > 0 = \pi_j(w_k) = \langle \ell_j, v_k\rangle.$$

By the irreducibility of the representation of $\mathbb{R}_+\mathcal{A}$, the sets $F_i = H_{\ell_i} \cap \mathbb{R}_+\mathcal{A}$ and $F_j = H_{\ell_j} \cap \mathbb{R}_+\mathcal{A}$ are distinct facets. By Proposition 1.1.23, F_i is a cone generated by a subset of $\mathcal{A}$ and since $F_j \not\subset F_i$ there exists $v_k \in F_j \setminus F_i$. Therefore $\langle \ell_j, v_k\rangle = 0$ and $\langle \ell_i, v_k\rangle > 0$, as desired. □

Algorithm To compute the divisor class group of the semigroup ring $K[\mathbb{N}\mathcal{A}]$ proceed as follows:

(1) First, using *Normaliz* [68], compute the integral closure of $K[\mathbb{N}\mathcal{A}]$ in its field of fractions to determine whether $K[\mathbb{N}\mathcal{A}]$ is a Krull ring.

(2) Second, using *PORTA* [84], compute an irreducible representation of $\mathbb{R}_+\mathcal{A}$ and adjust this representation (if necessary) to satisfy conditions (a) and (b) of Proposition 9.8.16.

(3) Third, using (2) determine the vectors $w_1, \ldots, w_q$. Then, using *Maple* [80], diagonalize (over the integers) the matrix B whose rows are $w_1, \ldots, w_q$, to obtain a "diagonal matrix" $D = \operatorname{diag}\{d_1, \ldots, d_s\}$, where $d_1, \ldots, d_s$ are the invariant factors of $\mathbb{Z}^r/\mathbb{Z}\{w_1, \ldots, w_q\}$.

(4) Fourth, use Theorem 9.8.19 together with the fundamental theorem for finitely generated abelian groups (see Theorem 1.3.16), to obtain:

$$\operatorname{Cl}(K[\mathbb{N}\mathcal{A}]) \simeq \mathbb{Z}^r/\mathbb{Z}\{w_1, \ldots, w_q\} \simeq \mathbb{Z}^t \oplus \mathbb{Z}/\mathbb{Z}d_1 \oplus \cdots \oplus \mathbb{Z}/\mathbb{Z}d_s,$$

where $t = r - \operatorname{rank}(B) = r - s$.

Example 9.8.20 Let $\mathcal{A} = \{(a_1, \ldots, a_d) \in \mathbb{N}^n |\, a_1 + \cdots + a_n = d\}$ be the set of ordered partitions of an integer $d \geq 2$. Assume $\mathcal{A} = \{v_1, \ldots, v_q\}$. Notice that $\mathbb{R}\mathcal{A} = \mathbb{R}^n$ and the irreducible representation of $\mathbb{R}_+\mathcal{A}$ is:

$$\mathbb{R}_+\mathcal{A} = \mathbb{R}_+^n = H_{e_1}^+ \cap \cdots \cap H_{e_n}^+.$$

In this case $r = n$ and $\ell_i = e_i$ for $i = 1, \ldots, n$. It is easy to see that $\varphi(v_i) = w_i = v_i$ for all i. Thus $\mathcal{S} = \mathbb{Z}\mathcal{A}$. Let B be the matrix whose rows are $w_1, \ldots, w_q$. Using elementary operations we obtain that the Smith normal form of B is $\operatorname{diag}\{1, \ldots, 1, d\}$, where the number of 1's is $n-1$. By Theorem 9.8.19 we get $\operatorname{Cl}(K[\mathbb{N}\mathcal{A}]) \simeq \mathbb{Z}^n/\mathbb{Z}\mathcal{A} \simeq \mathbb{Z}_d$.

Exercises

9.8.21 A semigroup $(0) \neq \mathcal{S}$ of $\mathbb{N}^n$ is called *full* if $\mathbb{N}^n \cap \mathbb{Z}\mathcal{S} = \mathcal{S}$. If $\mathcal{S} \subset \mathbb{N}^n$ is a full semigroup, prove that $\mathcal{S}$ is a finitely generated semigroup.

9.8.22 [248] Let $\mathcal{A} \subset \mathbb{N}^n$ be a finite set. If $\mathbb{N}\mathcal{A}$ is a normal semigroup of $\mathbb{N}^n$, then there is a full semigroup $\mathcal{S} \subset \mathbb{N}^r$ such that $\mathbb{N}\mathcal{A} \simeq \mathcal{S}$ (as semigroups). In particular $K[\mathbb{N}\mathcal{A}] \simeq K[\mathcal{S}]$ (as rings).

9.8.23 If $\mathcal{S} \subset \mathbb{N}^n$ is a full semigroup, prove that $\mathbb{R}_+\mathcal{S} \cap \mathbb{Z}\mathcal{S} = \mathbb{N}\mathcal{S}$, that is, $\mathcal{S}$ is a normal semigroup.

9.8.24 Let F be the set of square-free monomials of degree d of a polynomial ring over a field K and let $K[F]$ be the monomial subring spanned by F. Find a formula for $\operatorname{Cl}(K[F])$.

Chapter 10

Monomial Subrings of Graphs

In this chapter we study monomial subrings associated to graphs and their toric ideals. We relate the even closed walks and circuits of the vector matroid of a graph with Gröbner bases. A description of the integral closure of the edge subring of a multigraph will be presented along with a description of the circuits of its toric ideal. The Smith normal form and the invariant factors of the incidence matrix of a graph are fully determined. As an application we compute the multiplicity of edge subrings in terms of relative volumes. The family of ring graphs is studied here. These graphs are characterized in algebraic and combinatorial terms.

Several interesting connections between monomial subrings, polyhedral geometry and graph theory will occur in this chapter. We study in detail the irreducible representation of an edge cone of a graph and show some applications to graph theory, e.g., we show the marriage theorem.

10.1 Edge subrings and ring graphs

Let $R = K[x_1, \ldots, x_n]$ be a polynomial ring over a field K with the standard grading $R = \bigoplus_{i=0}^{\infty} R_i$ induced by $\deg(x_i) = 1$ for all i and let G be a graph on the vertex set $X = \{x_1, \ldots, x_n\}$.

Definition 10.1.1 The *edge subring* of the graph G, denoted by $K[G]$, is the K-subalgebra of R given by:

$$K[G] = K[\{x_i x_j \mid x_i \text{ is adjacent to } x_j\}] \subset R.$$

To obtain a presentation of the edge subring of G note that $K[G]$ is a standard K-algebra with the normalized grading

$$K[G]_i = K[G] \cap R_{2i}$$

and consider the set $F = \{f_1, \ldots, f_q\}$ of all monomials x_ix_j such that x_i is adjacent to x_j. Since the elements in $K[G]$ are polynomial expressions in F with coefficients in K, there is a graded epimorphism of K-algebras

$$\varphi\colon S = K[t_1, \ldots, t_q] \longrightarrow K[G], \quad t_i \longmapsto f_i,$$

where S is a polynomial ring graded by $\deg(t_i) = 1$ for all i. The kernel of φ, denoted by $P(G)$, is a graded ideal of S called the *toric ideal* of $K[G]$ with respect to $f_1, \ldots, f_q$.

Definition 10.1.2 Let $w = \{x_0, x_1, \ldots, x_r = x_0\}$ be an even closed walk of G such that $f_i = x_{i-1}x_i$. The binomial $t_w = t_1 \cdots t_{r-1} - t_2 \cdots t_r$ is called the *binomial associated* to w.

Remark 10.1.3 $t_w \in P(G)$ because $f_1f_3 \cdots f_{r-1} = f_2f_4 \cdots f_r$.

Definition 10.1.4 A closed walk of even length will be called a *monomial walk*. A monomial x_ix_j of R is said to be an *edge generator* if $\{x_i, x_j\}$ is an edge of G.

Notation In what follows $\mathcal{I}_s$ denotes the set of all non-decreasing sequences $\alpha = (i_1, \cdots, i_s)$ of length s. If $a_1, \ldots, a_q$ is a sequence and $\alpha = (i_1, \cdots, i_s)$ is in $\mathcal{I}_s$, then we set $a_\alpha = a_{i_1} \cdots a_{i_s}$.

Proposition 10.1.5 [417] *If G is a graph and $P = P(G)$ is the toric ideal of the edge subring $K[G]$, then*

(a) $P = (\{t_w | \, w \text{ is an even closed walk}\})$, *and*

(b) $P = (\{t_w \,|\, w \text{ is an even cycle}\})$ *if G is bipartite.*

Proof. (a): We set $\mathcal{B} = \{t_w | \, w \text{ is an even closed walk}\}$. As P is a graded ideal in the standard grading of $S = K[t_1, \ldots, t_q]$, we can write

$$P = S \cdot \left(\bigcup_{s=2}^{\infty} P_s \right).$$

One clearly has $(\mathcal{B}) \subset P$. To prove the other inclusion we use induction on s. First recall that P is a binomial ideal by Corollary 8.2.18. If $s = 2$, it is enough to note that the binomials in P_2 come from squares. Assume

$P_{s-1} \subset (\mathcal{B})$. To show $P_s \subset (\mathcal{B})$, take $t_\alpha - t_\beta$ in P_s, where $\alpha = (i_1, \ldots, i_s)$ and $\beta = (j_1, \ldots, j_s)$. Define G_1 as the subgraph of G having vertex set

$$V_1 = \{x_i \in V \,|\, x_i \text{ divides } f_\alpha\}$$

and edge set

$$E_1 = \{\{x, y\} \in E(G) \,|\, f_\ell = xy \text{ for some } \ell \in \{i_1, \ldots, i_s, j_1, \ldots, j_s\}\}.$$

Note that if $f_{i_1} \cdots f_{i_m} = f_{j_1} \cdots f_{j_m}$ for some $m < s$ and for some ordering of the generators then $t_\alpha - t_\beta \in (\mathcal{B})$; to prove it, notice that $t_\alpha - t_\beta$ can be written as

$$\begin{aligned} t_\alpha - t_\beta = t_{i_{m+1}} \cdots t_{i_s} (t_{i_1} \cdots t_{i_m} - t_{j_1} \cdots t_{j_m}) \\ + t_{j_1} \cdots t_{j_m} (t_{i_{m+1}} \cdots t_{i_s} - t_{j_{m+1}} \cdots t_{j_s}) \end{aligned}$$

and use induction hypothesis. Now we can assume $f_{i_1} \cdots f_{i_m} \neq f_{j_1} \cdots f_{j_m}$ for $m < s$ and for any re-ordering of the f_i's. Observe that this forces G_1 to be connected. Take $x_0 \in V_1$. Since $f_\alpha = f_\beta$, it is easy to check that after re-ordering we can write $f_{i_k} = x_{2k-2}x_{2k-1}$ and $f_{j_k} = x_{2k-1}x_{2k}$, where $1 \leq k \leq s$ and $x_0 = x_{2s}$. Therefore the monomial walk $w = \{x_0, \ldots, x_{2n}\}$ satisfies $t_w = t_\alpha - t_\beta$ and the induction is complete.

(b): Assume that G is bipartite. This part can be proven similarly to part (a) if we notice that in this case the condition $f_{i_1} \cdots f_{i_m} \neq f_{j_1} \cdots f_{j_m}$ for $m < s$ and for any re-ordering of the f_i's implies that the graph G_1 is an even cycle. □

Corollary 10.1.6 *If G is a graph and $f = t^\alpha - t^\beta$ is a primitive binomial in $P(G)$, then $f = t_w$ for some even closed walk w of G.*

Proof. It follows from the proof of Proposition 10.1.5. □

Corollary 10.1.7 *If G is a bipartite graph and $f = t^\alpha - t^\beta$ is a primitive binomial in $P(G)$, then $f = t_w$ for some even cycle w of G.*

Proof. It follows from the proof of Proposition 10.1.5. □

Proposition 10.1.8 *If G is a graph and $f = t^\alpha - t^\beta$ is a primitive binomial in $P(G)$, then the entries of α satisfy $\alpha_i \leq 2$ for all i.*

Proof. By Corollary 10.1.6 one can write

$$f = t_w = t_1 t_3 \cdots t_{\ell-1} - t_2 t_4 \ldots t_\ell,$$

where ℓ is even, and w is an even closed walk in G of length ℓ:

$$w = \{x_0, x_1, \ldots, x_{\ell-1}, x_\ell = x_0\}$$

such that $f_1 f_3 \cdots f_{\ell-1} = f_2 f_4 \cdots f_\ell$ and $f_i = x_{i-1}x_i$ for all i. Note that there may be some repetitions of the variables t_i and accordingly some repetitions in the monomials f_i. As usual we are identifying the vertices of G with variables. It suffices to verify that x_0 occurs at most three times (including the end vertices) in the closed walk w. If x_0 occurs more than three times in w, one can write

$$w = \{x_0, x_1, \ldots, x_{t_1} = x_0, x_{t_1+1}, \ldots, x_{t_1+t_2} = x_0, x_{t_1+t_2+1}, \ldots, x_\ell = x_0\},$$

note that t_1 must be odd; otherwise using $w_1 = \{x_0, x_1, \ldots, x_{t_1} = x_0\}$ one has that $t_{w_1} = t^\gamma - t^\delta$, where t^γ divides $t_1 t_3 \cdots t_{\ell-1}$ and t^δ divides $t_2 t_4 \ldots t_\ell$, a contradiction because f is primitive. By a similar argument t_2 must be odd. Thus $t_1 + t_2$ is even, to get a contradiction note that the closed walk

$$w' = \{x_{t_1+t_2} = x_0, x_{t_1+t_2+1}, \ldots, x_\ell = x_0\}$$

is of even length, and apply the previous argument. □

In [352] there are characterizations of when a binomial of a toric ideal of $K[G]$ is primitive, minimal, indispensable, or fundamental in terms of the even closed walks.

Lemma 10.1.9 *Let G be a graph and let $P = P(G)$ be the toric ideal of $K[G]$. If f is a polynomial in any reduced Gröbner bases of P, then*

(a) *f is a primitive binomial and $f = t_w$ for some even closed walk w of the graph G.*

(b) *If G is bipartite, then f is primitive and $f = t_w$ for some even cycle w of the graph G.*

Proof. Let $\mathcal{B}$ be a reduced Gröbner basis of P and take $f \in \mathcal{B}$. Using Corollary 8.2.18 and the fact that normalized reduced Gröbner bases are uniquely determined, we obtain $f = t^\alpha - t^\beta$. By Lemma 8.3.3 f is primitive, thus by Corollaries 10.1.6 and 10.1.7 f has the required form. □

Proposition 10.1.10 *If G is a graph, then the set*

$$\{t_w | \, t_w \textit{ is primitive and } w \textit{ is an even closed walk}\}$$

is a universal Gröbner basis of $P(G)$.

Proof. It follows from Lemma 10.1.9. □

Proposition 10.1.11 *If G is a bipartite graph, then the set*

$$\{t_w | \, w \textit{ is an even cycle}\}$$

is a universal Gröbner basis of $P(G)$.

Proof. It follows from Lemma 10.1.9. □

Definition 10.1.12 A *chord* of a cycle c of a graph G is any edge of G joining two non-adjacent vertices of c. A cycle without chords is called *primitive*.

Proposition 10.1.13 *If a is a cycle and $t_a \in (t_c|\, c$ is a cycle, $c \neq a)$, then a has a chord.*

Proof. If $t_a = t^\alpha - t^\beta \in (t_c|\, c$ is a cycle, $c \neq a)$, then each term of t_a must be divisible by a term of t_c for some $c \neq a$. This means that $t^\alpha = t^\delta t^{\alpha_1}$ and $t^\beta = t^\gamma t^{\alpha_2}$, where t^{α_i} are terms of some binomial of the form t_{c_i} with $c_i \neq a$ and c_i a cycle. Hence (after permuting variables) $x_1 \cdots x_r = x^\theta(x_1 \cdots x_s)$, where $x_1, \ldots, x_r$ and $x_1, \ldots, x_s$ are the vertices of the cycles a and c_1. It is seen that a must have a chord. □

Proposition 10.1.14 *If G is a bipartite graph, then $P(G)$ is a prime ideal minimally generated by the set $\{t_c|\, c$ is a primitive cycle of $G\}$.*

Proof. It follows from Proposition 10.1.13. □

Rees algebras of edge ideals We will look closely at the toric ideal of the Rees algebra of an edge ideal, and then show a few applications.

Let $I = I(G)$ be the edge ideal of a graph G, let $F = \{f_1, \ldots, f_q\}$ be the set of edge generators of I, and let $R[It] = R[f_1t, \ldots, f_qt] \subset R[t]$ be the Rees algebra of I. There is an epimorphism

$$\phi : R[t_1, \ldots, t_q] \to R[It], \text{ induced by } t_i \to f_it.$$

We set $J = \ker(\phi)$ and $B = R[t_1, \cdots, t_q]$, notice that $J = \oplus_{i=1}^{\infty} J_i$ is a graded ideal (in the t_i-variables) of $B = \oplus_{i=0}^{\infty} B_i$. The ideal J is the *presentation ideal* or *toric ideal* of $R[It]$ with respect to F. Recall that the ideal I is said to be of *linear type* if $J = J_1B$.

Theorem 10.1.15 [417] *Let $I = I(G)$ be the edge ideal of a graph G and let J be the toric ideal of $R[It]$. Then $J_s = B_1J_{s-1} + RP_s$ for $s \geq 2$ and*

$$J = BJ_1 + B \cdot \left(\bigcup_{s=2}^{\infty} P_s \right),$$

where $P_s = \{t_\alpha - t_\beta|\ f_\alpha = f_\beta, \text{ for some } \alpha,\, \beta \in \mathcal{I}_s\}$.

Corollary 10.1.16 [417] *Let G be a connected graph and let I be its edge ideal. Then I is an ideal of linear type if and only if G is a tree or G has a unique cycle of odd length.*

Proof. $\Rightarrow$) If I is an ideal of linear type, then by Proposition 4.2.1, G is a tree or G has a unique cycle of odd length.

$\Leftarrow$) Assume that G is either a tree or G has a unique cycle of odd length. We claim that $RP_s = (0)$ for $s \geq 2$, we proceed by induction on n. If $n \leq 3$ this is easily verified, assume the claim true for graphs with less than n vertices. Using induction on s we now show $RP_s = (0)$ for $s \geq 2$, notice that $RP_2 = (0)$ because G has no squares. Assume $RP_{s-1} = (0)$ and $RP_s \neq (0)$, take a non-trivial relation $f_{i_1} \cdots f_{i_s} = f_{j_1} \cdots f_{j_s}$. Since a tree has a vertex of degree one, by induction hypothesis it follows that G must be a cycle. Therefore, using induction again, we may write $\{f_{i_1}, \ldots, f_{i_s}\} = \{g_1, \ldots, g_s\}$, where $\gcd(g_i, g_j) = 1$ for $i \neq j$ and $g_1 \cdots g_k = x_1 \cdots x_n$. Notice that this equality cannot occur if n is odd, hence $RP_s = (0)$ and the proof of the claim is complete. Hence, by Proposition 10.1.15, I is of linear type. $\Box$

Arbitrary square-free monomial ideals of linear type have been studied in [6, 162, 376] using simplicial complexes. These papers show a deep interplay between graph theory and the defining equations of Rees algebras. An interesting family of square-free monomial ideals of linear type is given in [376]. This family includes edge ideals of totally balanced clutters and facet ideals of disjoint simplicial cycles of odd lengths. This family is closed under the operation of adding M-elements.

Let $I \subset R$ be a square-free monomial ideal minimally generated by monomials $f_1, \ldots, f_q$. The *line graph* of I, denoted by $L(I)$, is the graph whose vertex set is $f_1, \ldots, f_q$ and where $\{f_i, f_j\}$ is an edge of $L(I)$ if and only if f_i and f_j have at least one variable in common.

Recently Fouli and Lin [162] have shown that if the line graph $L(I)$ is a disjoint union of trees and graphs with a unique cycle of odd length, then I is of linear type. A generalization of this fact is given in [6].

A dimension formula Let us describe the *cycle space* of a graph G over the field $\mathbb{F} = \mathbb{Z}_2$. We denote the edge set and vertex set of G by

$$E(G) = \{f_1, \ldots, f_q\} \quad \text{and} \quad X = \{x_1, \ldots, x_n\},$$

respectively. Let C_0 and C_1 denote the vector spaces over $\mathbb{F}$ of 0-chains and 1-chains, respectively. Recall that a 0-chain of G is a formal linear combination

$$\sum a_i x_i$$

of points and a 1-chain is a formal linear combination

$$\sum b_i f_i$$

of edges, where $a_i \in \mathbb{F}$ and $b_i \in \mathbb{F}$. The boundary operator ∂ is the linear transformation defined by

$$\partial(f_k) = \partial(\{x_i, x_j\}) = x_i + x_j.$$

A *cycle vector* is a 1-chain of the form $t_1 + \cdots + t_r$ where $t_1, \ldots, t_r$ are the edges of a cycle of G. The *cycle space* $\mathcal{Z}(G)$ of G over $\mathbb{F}$ is equal to $\ker(\partial)$.

The vectors in $\mathcal{Z}(G)$ can be regarded as a set of edge-disjoint cycles. A *cycle basis* for G is a basis for $\mathcal{Z}(G)$ which consist entirely of cycle vectors, such a basis can be constructed as follows:

Remark 10.1.17 If G is connected, then G has a spanning tree T. The subgraph of G consisting of T and any edge of G not in T has exactly one cycle, the collection of all the cycle vectors obtained in this way form a cycle basis for G. See [208] for details (cf. Exercise 6.2.8). In particular

$$\dim_{\mathbb{F}} \mathcal{Z}(G) = q - n + 1.$$

Definition 10.1.18 If G is a graph, the number $\dim_{\mathbb{F}} \mathcal{Z}(G)$ is called the *cycle rank* of G and is denoted by $\operatorname{rank}(G)$.

Lemma 10.1.19 *Let G be a connected graph and $\mathcal{Z}_e(G)$ the subspace of $\mathcal{Z}(G)$ of all cycle vectors of G with an even number of terms. Then*

$$\dim_{\mathbb{F}} \mathcal{Z}_e(G) = \begin{cases} q - n + 1 & \text{if } G \text{ is bipartite, and} \\ q - n & \text{otherwise.} \end{cases}$$

Proof. Let $c_1, \ldots, c_l, c_{l+1}, \ldots, c_m$ be a cycle basis for G, where $c_1, \ldots, c_l$ are even cycles and $c_{l+1}, \ldots, c_m$ odd cycles. It is clear that a basis for $\mathcal{Z}_e(G)$ is given by $\{c_1, \ldots, c_l, c_{l+1} + c_m, \ldots, c_{m-1} + c_m\}$. □

Proposition 10.1.20 [417] *If G is a connected graph, then*

$$\operatorname{ht}(P(G)) = \dim_{\mathbb{F}} \mathcal{Z}_e(G).$$

Proof. Let G be a connected graph with q edges and n vertices. Assume that $I(G)$ is minimally generated by the monomials $f_1, \ldots, f_q$. There is a spanning tree T of G so that (after re-ordering) $I(T) = (f_1, \ldots, f_{n-1})$. Since $I(T)$ is an ideal of linear type

$$\dim K[G] = \operatorname{tr.deg}_K K[G] \geq n - 1.$$

If G is bipartite notice that $f_k \in k(f_1, \ldots, f_{n-1})$ for $k \geq n$; to prove it we write $f_k = xz$ and observe that the graph $T \cup \{x, z\}$ has a unique cycle of even length. This shows the equality $\dim K[G] = n - 1$.

If G is not bipartite, then for some $f_k = xz$, $k \geq n$, the subgraph $T \cup \{x, z\}$ has a unique cycle of odd length. By Corollary 10.1.16 the ideal $(f_1, \ldots, f_{n-1}, f_k)$ is of linear type, hence $\dim K[G] \geq n$; to show equality recall that $\operatorname{tr.deg}_K K[G] \leq n$. To finish the proof use Lemma 10.1.19. □

Corollary 10.1.21 *If G is a connected graph with n vertices and $K[G]$ is its edge subring, then*

$$\dim(K[G]) = \begin{cases} n & \text{if } G \text{ is not bipartite, and} \\ n-1 & \text{otherwise.} \end{cases}$$

Proof. It follows from the proof of Proposition 10.1.20. □

Definition 10.1.22 The number of primitive cycles of a graph G, denoted by $\operatorname{frank}(G)$, is called the *free rank* of G.

Proposition 10.1.23 *If G is a graph, then $\mathcal{Z}(G)$ is generated by cycle vectors of primitive cycles. In particular* $\operatorname{rank}(G) \leq \operatorname{frank}(G)$.

Proof. Let $\mathbf{c}_1, \ldots, \mathbf{c}_r$ be a cycle basis for the cycle space of G and let $c_1, \ldots, c_r$ be the corresponding cycles of G. It suffices to notice that if some c_j has a chord, we can write $\mathbf{c}_j = \mathbf{c}'_j + \mathbf{c}''_j$, where $\mathbf{c}'_j$ and $\mathbf{c}''_j$ are cycle vectors of cycles of length smaller than that of c_j. □

Corollary 10.1.24 *Let G be a graph. The following are equivalent:*

(a) $\operatorname{rank}(G) = \operatorname{frank}(G)$.

(b) *The set of cycle vectors of primitive cycles is a basis for $\mathcal{Z}(G)$.*

(c) *The set of cycle vectors of primitive cycles is linearly independent.*

Proof. (a) $\Rightarrow$ (b): By Proposition 10.1.23 there is a basis $\mathcal{B}$ of $\mathcal{Z}(G)$ consisting of cycle vectors of primitive cycles. By hypothesis $\operatorname{rank}(G)$ is $\operatorname{frank}(G)$. Thus $\mathcal{B}$ is the set of all cycle vectors of primitive cycles and $\mathcal{B}$ is a basis. That (b) implies (c) and (c) implies (a) are also easy to prove. □

The family of graphs satisfying the equality $\operatorname{rank}(G) = \operatorname{frank}(G)$ can be constructed as is seen later in this section.

Ring graphs In this part we characterize ring graphs in algebraic and combinatorial terms. The reader is referred to Section 7.1 for the notions of cutvertex, bridge, and block.

Lemma 10.1.25 *Let G be a graph and let $G_1, \ldots, G_r$ be its blocks. Then* $\operatorname{rank}(G) = \operatorname{frank}(G)$ *if and only if* $\operatorname{rank}(G_i) = \operatorname{frank}(G_i)$ *for all i.*

Proof. $\Rightarrow$) Let G_i be any block of G. We may assume $|V(G_i)| > 2$, otherwise $\operatorname{rank}(G_i) = \operatorname{frank}(G_i) = 0$. If c is a primitive cycle of G_i, then by the maximality condition of a block one has that c is also a primitive cycle of G. Thus by Corollary 10.1.24 the set of cycle vectors of primitive cycles of G_i is linearly independent and $\operatorname{rank}(G_i) = \operatorname{frank}(G_i)$.

$\Leftarrow$) Let $\mathcal{B}_i$ and $\mathcal{B}$ be the set of cycle vectors of primitive cycles of G_i and G, respectively. As $\cup_{i=1}^{r}\mathcal{B}_i$ is linearly independent, by Corollary 10.1.24 it suffices to prove that $\cup_{i=1}^{r}\mathcal{B}_i = \mathcal{B}$. In the first part of the proof we have already observed that $\cup_{i=1}^{r}\mathcal{B}_i \subset \mathcal{B}$. To prove the equality take any cycle vector $\mathbf{c}$ of a primitive cycle c of G. Since c is a 2-connected subgraph, it must be contained in some block of G, i.e., in some G_i. Thus c is a primitive cycle of G_i, so $\mathbf{c}$ is in $\mathcal{B}_i$. □

Definition 10.1.26 Given a graph H, we call a path $\mathcal{P}$ an *H-path* if $\mathcal{P}$ is non-trivial and meets H exactly in its ends.

Theorem 10.1.27 ([111, Proposition 3.1.2], [208, Theorem 5.10]) *Let G be a graph. Then the following three conditions are equivalent*:

(a) *G is 2-connected.*

(b) *G can be constructed from a cycle by successively adding H-paths to graphs H already constructed.*

(c) *Every pair of vertices of G is joined by at least 2 vertex-disjoint paths.*

This result suggests the following more restrictive notion:

Definition 10.1.28 A graph G is called a *ring graph* if each block of G which is not a bridge can be constructed from a cycle by successively adding H-paths of length at least 2 that meet graphs H already constructed in two adjacent vertices.

Families of ring graphs include forests and cycles.

Remark 10.1.29 Let G be a 2-connected ring graph and let c be a fixed primitive cycle of G, then G can be constructed from c by successively adding H-paths of length at least 2 that meet graphs H already constructed in two adjacent vertices.

Definition 10.1.30 A graph H is called a *subdivision* of a graph G if H arises from G by replacing edges by paths. That is H is obtained by iteratively choosing an edge (u, v), introducing a new vertex w, deleting edge (u, v), and adding edges (u, w) and (w, v).

Example 10.1.31 A complete bipartite graph $\mathcal{K}_{2,3}$ and a subdivision

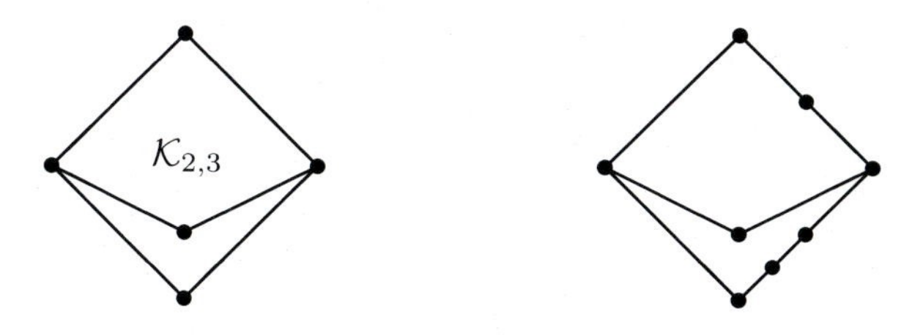

Lemma 10.1.32 [2, Lemma 7.78, p. 387] *Let G be a graph with vertex set V. If G is 2-connected and $\deg(v) \geq 3$ for all $v \in V$, then G contains a subdivision of $\mathcal{K}_4$ as a subgraph.*

Lemma 10.1.33 *Let G be a graph. If* $\text{rank}(G) = \text{frank}(G)$ *and x, y are two non-adjacent vertices of G, then there are at most two vertex disjoint paths joining x and y.*

Proof. Assume that there are three vertex disjoint paths joining x and y:

$$\mathcal{P}_1 = \{x, x_1, \ldots, x_r, y\}, \qquad \mathcal{P}_2 = \{x, z_1, \ldots, z_t, y\},$$
$$\mathcal{P}_3 = \{x, y_1, \ldots, y_s, y\},$$

where r, s, t are greater than or equal to 1. We may assume that the sum of the lengths of the $\mathcal{P}_i$'s is minimal. Consider the cycles

$$\begin{aligned} c_1 &= \{x, x_1, \ldots, x_r, y, z_t, \ldots, z_1, x\}, \\ c_2 &= \{x, z_1, \ldots, z_t, y, y_s, \ldots, y_1, x\}, \\ c_3 &= \{x, x_1, \ldots, x_r, y, y_s, \ldots, y_1, x\}. \end{aligned}$$

Thus we are in the following situation:

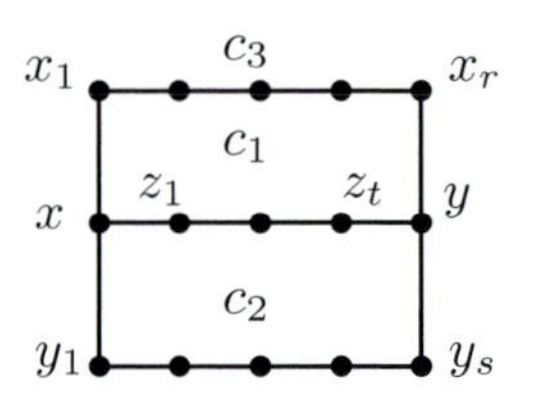

Observe that, by the choice of the $\mathcal{P}_i$'s, a chord of the cycle c_1 (resp. c_2, c_3) must join x_i and z_j (resp. z_i and y_j, x_i and y_j) for some i, j. Therefore we can write

$$\mathbf{c_1} = \sum_{i=1}^{n_1} \mathbf{a}_i, \quad \mathbf{c_2} = \sum_{i=1}^{n_2} \mathbf{b}_i, \quad \mathbf{c_3} = \sum_{i=1}^{n_3} \mathbf{d}_i$$

where $\mathbf{a}_1, \ldots, \mathbf{a}_{n_1}, \mathbf{b}_1, \ldots, \mathbf{b}_{n_2}, \mathbf{d}_1, \ldots, \mathbf{d}_{n_3}$ are distinct cycle vectors of primitive cycles of G and $\mathbf{c}_i$ is the cycle vector corresponding to c_i. Thus from the equality $\mathbf{c}_3 = \mathbf{c}_1 + \mathbf{c}_2$. we get that the set of cycle vectors of primitive cycles is linearly dependent, a contradiction to Corollary 10.1.24. $\Box$

Definition 10.1.34 A graph G has the *primitive cycle property* (PCP) if any two primitive cycles intersect in at most one edge.

Lemma 10.1.35 *Let G be a graph. If* $\text{rank}(G) = \text{frank}(G)$, *then G has the primitive cycle property.*

Proof. Let c_1, c_2 be two distinct primitive cycles. Assume that c_1 and c_2 intersect in at least two edges. Thus c_1 and c_2 must intersect in two non-adjacent vertices. Hence there are two non-adjacent vertices x, y in c_1 and a path $\mathcal{P} = \{x, x_1, \ldots, x_r, y\}$ of length at least two that intersect c_1 in exactly the vertices x, y:

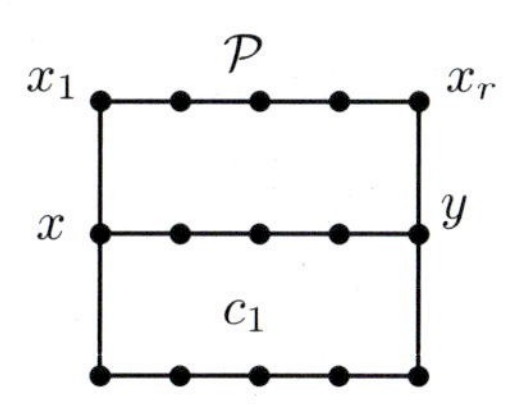

This contradicts Lemma 10.1.33. Hence c_1 and c_2 have at most one edge in common. □

Lemma 10.1.36 *Let G be a graph. If G satisfies* PCP *and G does not contain a subdivision of $\mathcal{K}_4$ as a subgraph, then for any two non-adjacent vertices x, y there are at most two vertex disjoint paths joining x and y.*

Proof. Assume that there are three vertex disjoint paths joining x and y:

$$\mathcal{P}_1 = \{x, x_1, \ldots, x_r, y\},\ \mathcal{P}_2 = \{x, z_1, \ldots, z_t, y\},\ \mathcal{P}_3 = \{x, y_1, \ldots, y_s, y\},$$

where r, s, t are greater than or equal to 1. We may assume that the sum of the lengths of the $\mathcal{P}_i$'s is minimal. Consider the cycles

$$c_1 = \{x, x_1, \ldots, x_r, y, z_t, \ldots, z_1, x\}, \quad c_2 = \{x, z_1, \ldots, z_t, y, y_s, \ldots, y_1, x\},$$
$$c_3 = \{x, x_1, \ldots, x_r, y, y_s, \ldots, y_1, x\}.$$

Thus we are in the following situation:

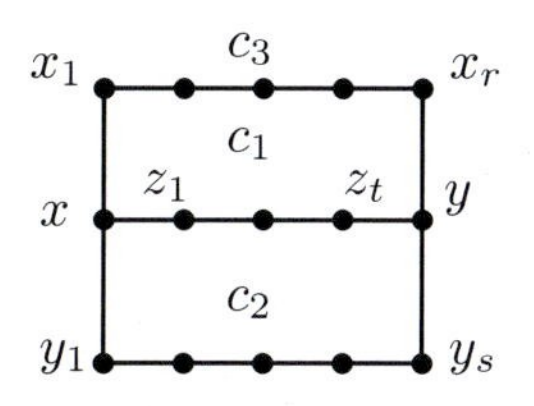

Observe that, by the choice of the $\mathcal{P}_i$'s, a chord of the cycle c_1 (resp. c_2, c_3) must join x_i and z_j (resp. z_i and y_j, x_i and y_j) for some i, j. Using that G does not contain a subdivision of $\mathcal{K}_4$ as a subgraph, it is seen that the cycles c_1 and c_3 are primitive. Thus, since c_1 and c_3 have at least two edges in common, we obtain that G does not satisfy PCP, a contradiction. □

Lemma 10.1.37 *Let G be a graph. If* $\operatorname{rank}(G) = \operatorname{frank}(G)$*, then G does not contain a subdivision of $\mathcal{K}_4$ as a subgraph.*

Proof. Assume there is a subgraph $H \subset G$ which is a subdivision of $\mathcal{K}_4$. If $\mathcal{K}_4$ is a subgraph of G, then G has four distinct triangles whose cycle vectors are linearly dependent, a contradiction to Corollary 10.1.24. If $\mathcal{K}_4$ is not a subgraph of G, then H is a strict subdivision of $\mathcal{K}_4$, i.e., H has more than four vertices. It follows that there are two vertices x, y in $V(H)$ which are non-adjacent in G. Notice that x, y can be chosen in $\mathcal{K}_4$ before subdivision. Therefore there are at least three non-adjacent paths joining x and y, a contradiction to Lemma 10.1.33. □

Theorem 10.1.38 [184] *Let G be a graph. The following are equivalent:*

(a) *G is a ring graph.*

(b) $\operatorname{rank}(G) = \operatorname{frank}(G)$.

(c) *G satisfies* PCP *and does not contain a subdivision of $\mathcal{K}_4$ as a subgraph.*

Proof. (a) $\Rightarrow$ (b): By induction on the number of vertices it is not hard to see that any ring graph G satisfies the equality $\operatorname{rank}(G) = \operatorname{frank}(G)$.

(b) $\Rightarrow$ (c): It follows at once from Lemmas 10.1.35 and 10.1.37.

(c) $\Rightarrow$ (a): Let $G_1, \ldots, G_r$ be the blocks of G. The proof is by induction on the number of vertices of G. If each G_i is either a bridge or an isolated vertex, then G is a forest and consequently a ring graph. Hence by Lemma 10.1.25 we may assume that G is 2-connected and that G is not a cycle. We claim that G has at least one vertex of degree 2. If $\deg(v) \geq 3$ for all $v \in V(G)$, then by Lemma 10.1.32 there is a subgraph $H \subset G$ which is a subdivision of $\mathcal{K}_4$, which is impossible. Let $v_0 \in V(G)$ be a vertex of degree 2 as claimed. By the primitive cycle property there is a unique primitive cycle $c = \{v_0, v_1, \ldots, v_s = v_0\}$ of G containing v_0. The graph $H = G \setminus \{v_0\}$ satisfies PCP and does not have a subdivision of $\mathcal{K}_4$ as a subgraph. Consequently H is a ring graph. Thus we may assume that c is not a triangle, otherwise G is a ring graph because it can be obtained by adding the H-path $\{v_2, v_0, v_1\}$ to H.

Next we claim that if $1 \leq i < j < k \leq s-1$, then v_i and v_k cannot be in the same connected component of $H \setminus \{v_j\}$. Otherwise there is a path of $H \setminus \{v_j\}$ that joins v_i with v_k. It follows that there is a path $\mathcal{P}$ of $H \setminus \{v_j\}$ with at least three vertices that joins a vertex of $\{v_{j+1}, \ldots, v_{s-1}\}$ with a vertex of $\{v_1, \ldots, v_{j-1}\}$ and such that $\mathcal{P}$ intersects c exactly in its ends, but this contradicts Lemma 10.1.36. This proves the claim. In particular v_i is a cutvertex of H for $i = 2, \ldots, s-2$ and v_{i-1}, v_{i+1} are in different connected components of $H \setminus \{v_i\}$. For each $1 \leq i \leq s-2$ there is a block K_i of H such that $\{v_i, v_{i+1}\}$ is an edge of K_i. Notice that if $1 \leq i < j < k \leq s-1$, then v_i, v_j, v_k cannot lie in some K_ℓ. Indeed if the three vertices lie in some

K_ℓ, then there is a path $\mathcal{P}'$ in $K_\ell \setminus \{v_j\}$ that joins v_i and v_k. Since $\mathcal{P}'$ is also a path in $H \setminus \{v_j\}$, we get that v_i and v_k are in the same connected component of $H \setminus \{v_j\}$, but this contradicts the last claim. In particular $V(K_\ell)$ intersects the cycle c in exactly the vertices $v_\ell, v_{\ell+1}$ for $1 \leq \ell \leq s-2$.

Observe that at least one of the edges of c not containing v_0 is not a bridge of H. To show this pick $x \notin c$ such that $\{x, v_k\}$ is an edge of H. We may assume that $v_{k+1} \neq v_0$ (or $v_{k-1} \neq v_0$). Since $G' = G \setminus \{v_k\}$ is connected, there is a path $\mathcal{P}$ of G' joining x and v_{k+1} (or v_{k-1}). This readily yields a cycle of H containing an edge of c which is not a bridge of H. Hence at least one of the blocks $K_1, \ldots, K_{s-2}$, say K_i, contains vertices outside c.

Next we show that two distinct blocks B_1, B_2 of H cannot intersect outside c. We proceed by contradiction assuming that $V(B_1) \cap V(B_2) = \{z\}$ for some z not in c. Let $H_1, \ldots, H_t$ be the connected components of $H \setminus \{z\}$. Notice that $t \geq 2$ because $\{z\}$ is the intersection of two different blocks of H. We may assume that $\{v_1, \ldots, v_{s-1}\}$ are contained in H_1. Consider the subgraph H_1' of $G \setminus \{z\}$ obtained from H_1 by adding the vertex v_0 and the edges $\{v_0, v_1\}$, $\{v_0, v_{s-1}\}$. It follows that the connected components of $G \setminus \{z\}$ are $H_1', H_2, \ldots, H_t$, which is impossible because G is 2-connected.

Let K_i be a block of H that contains vertices outside of c for some $1 \leq i \leq s-2$. By induction hypothesis K_i is a ring graph. Thus by Remark 10.1.29 we can construct K_i starting with a primitive cycle c_1 that contains the edge $\{v_i, v_{i+1}\}$, and then adding appropriate paths. Suppose that $\mathcal{P}_1, \ldots, \mathcal{P}_m$ is the sequence of paths added to c_1 to obtain K_i. If we remove the path $\mathcal{P}_m$ from G and use the fact that distinct blocks of H cannot intersect outside c, then again by induction hypothesis we obtain a ring graph. It follows that G is a ring graph as well. $\square$

A graph G that can be embedded in the (Euclidean) plane $\mathbb{R}^2$ is called a *planar graph*. The point set of a plane graph is compact. Therefore exactly one face of G (a region of $\mathbb{R}^2 \setminus G$) is unbounded. It is called the unbounded face of G (outer face, exterior face). A polygonal arc is a subset of $\mathbb{R}^2$ which is the union of a finite number of straight line segments

$$\{p + \lambda(q - p) \mid 0 \leq \lambda \leq 1\} \quad (p, q \in \mathbb{R}^2)$$

and is homeomorphic to the closed unit interval $[0, 1]$.

A ring graph is planar by construction. Thus an immediate consequence of Theorem 10.1.38 is:

Corollary 10.1.39 *Let G be a graph. If* $\mathrm{rank}(G) = \mathrm{frank}(G)$, *then G is planar.*

Lemma 10.1.40 [320] *If G is a planar graph, then G has a representation in the plane such that all edges are simple polygonal arcs.*

Proposition 10.1.41 [111, Proposition 4.2.5] *If G is a 2-connected planar graph, then every face of G is bounded by a cycle.*

Theorem 10.1.42 (Euler's formula; see [111, Theorem 4.2.7]) *Let G be a connected planar graph with n vertices, q edges, and r faces. Then*

$$n - q + r = 2$$

Proposition 10.1.43 (Kuratowski; see [48, p. 14, Theorem 17]) *A graph is planar if and only if it does not contain a subdivision of $\mathcal{K}_5$ or a subdivision of $\mathcal{K}_{3,3}$ as a subgraph.*

Definition 10.1.44 A planar graph is *outerplanar* if it can be embedded in the plane so that all its vertices lie on some face; it is usual to choose this face to be the exterior.

Definition 10.1.45 Two graphs H_1 and H_2 are called *homeomorphic* if there exists a graph G such that both H_1 and H_2 are subdivisions of G.

Theorem 10.1.46 [208, Theorem 11.10] *A graph is outerplanar if and only if it has no subgraph homeomorphic to $\mathcal{K}_4$ or $\mathcal{K}_{2,3}$ except $\mathcal{K}_4 \setminus \{e\}$, where e is an edge.*

Proposition 10.1.47 *If G is outerplanar, then* $\mathrm{rank}(G) = \mathrm{frank}(G)$.

Proof. By Theorem 10.1.38(c) it suffices to prove that G satisfies PCP and G does not contain a subdivision of $\mathcal{K}_4$ as a subgraph. If G contains a subdivision H of $\mathcal{K}_4$ as a subgraph, then G contains a subgraph, namely H, homeomorphic to $\mathcal{K}_4$, but this is impossible by Theorem 10.1.46. To finish the proof we now show that G has the PCP property. Let $c_1 = \{x_1, x_2, \ldots, x_m = x_1\}$ and $c_2 = \{y_1, y_2, \ldots, y_n = y_1\}$ be two distinct primitive cycles having at least one common edge. We may assume that $x_i = y_i$ for $i = 1, 2$ and $x_3 \neq y_3$. Notice that $y_3 \notin c_1$ because otherwise $\{y_2, y_3\} = \{x_2, y_3\}$ is a chord of c_1. We need only show that $\{x_1, x_2\} = c_1 \cap c_2$, because this implies that c_1 and c_2 cannot have more than one edge in common. Assume that $\{x_1, x_2\} \subsetneq c_1 \cap c_2$. Let r be the minimum integer such that y_r belong to $(c_1 \cap c_2) \setminus \{x_1, x_2\}$. Notice that $y_r \neq x_3$ because otherwise $\{x_2, x_3\}$ is a chord of c_2. Hence c_1 together with the path $\{x_2 = y_2, y_3, \ldots, y_r\}$ give a subgraph H of G which is a subdivision of $\mathcal{K}_{2,3}$, a contradiction to Theorem 10.1.46. $\square$

Proposition 10.1.48 *Let G be a bipartite graph and let $G_1, \ldots, G_r$ be their blocks, then*

$$K[G] \simeq K[G_1] \otimes_K \cdots \otimes_K K[G_r] \text{ and } P(G) = (P(G_1), \ldots, P(G_r)).$$

Proof. Since G is bipartite it follows from Exercise 10.1.71. □

Definition 10.1.49 A graph G is called a *complete intersection* (over K) if the toric ideal $P(G)$ of $K[G]$ is a complete intersection.

The complete intersection property of the edge subring of a bipartite graph was first studied in [114], and later in [379].

Proposition 10.1.50 [379] *If G is a bipartite graph, then G is a complete intersection if and only if* $\mathrm{rank}(G) = \mathrm{frank}(G)$.

Proof. Let $G_1, \ldots, G_r$ be the blocks of G. By Proposition 10.1.48 G is a c.i. if and only if G_i is a c.i. for all i. On the other hand by Exercise 10.1.72 G_i is a c.i. if and only if $\mathrm{rank}(G_i) = \mathrm{frank}(G_i)$. To finish the proof apply Lemma 10.1.25. □

Corollary 10.1.51 *If G is a bipartite graph and G is a complete intersection, then G is a planar graph.*

Proof. It follows from Proposition 10.1.50 and Corollary 10.1.39. □

Corollary 10.1.52 *If G is a bipartite graph, then G is a complete intersection if and only if G is a ring graph.*

Proof. By Proposition 10.1.50 G is a complete intersection if and only if $\mathrm{rank}(G) = \mathrm{frank}(G)$ and the result follows from Theorem 10.1.38. □

Theorem 10.1.53 [273] *The edge subring $K[G]$ of a bipartite graph G is a complete intersection if and only if G is planar and any two primitive cycles of G have at most one common edge.*

Proof. $\Rightarrow$) It follows from Lemma 10.1.35 and Corollary 10.1.51.
$\Leftarrow$) See [273, Theorem 3.5]. □

Let us summarize some of the results obtained thus far. The following implications hold for any graph G:

$$\begin{array}{ccccccc} \text{outerplanar} & \Rightarrow & \text{ring graph} & \Leftrightarrow & \text{PCP} + \text{contains no} & & \\ & & \Updownarrow & & \text{subdivision of } \mathcal{K}_4 & \Rightarrow & \text{planar} \\ & & \text{rank} = \text{frank} & & \text{as a subgraph} & & \end{array}$$

If G is bipartite, then

$$\text{ring graph} \iff \text{complete intersection} \iff \text{PCP} + \text{planar}.$$

If G is not bipartite, the complete intersection property of $P(G)$ is hard to characterize in combinatorial terms. In [29] a polynomial time algorithm is presented that checks whether a given graph is a complete intersection.

Definition 10.1.54 Let $g_1 = t^{\alpha_1} - t^{\beta_1}, \ldots, g_r = t^{\alpha_r} - t^{\beta_r}$ be homogeneous binomials of degree at least 2 in the polynomial ring $S = K[t_1, \ldots, t_q]$. We say that $\mathcal{B} = \{g_1, \ldots, g_r\}$ is a *foliation* if the following conditions hold:

(a) t^{α_i} and t^{β_i} are square-free for all i,

(b) $\mathrm{supp}(t^{\alpha_i}) \cap \mathrm{supp}(t^{\beta_i}) = \emptyset$ for all i, and

(c) $|(\cup_{i=1}^{j} C_i) \cap C_{j+1}| = 1$ for $2 \leq j < r$, where $C_i = \mathrm{supp}(g_i)$ for all i.

Proposition 10.1.55 *If $\mathcal{B} = \{g_1, \ldots, g_r\}$ is a foliation, then the binomial ideal $I = (\mathcal{B})$ is a complete intersection.*

Proof. By the constructive nature of $\mathcal{B}$ we can order the variables $t_1, \ldots, t_q$ so that the leading terms of $g_1, \ldots, g_r$, with respect to the lex order $\prec$, are relatively prime. Since

$$\dim(S/I) = \dim(S/(\mathrm{in}_\prec(g_1), \ldots, \mathrm{in}_\prec(g_r))),$$

we obtain that the height of I is equal to r, as required.

Corollary 10.1.56 *If G is a 2-connected bipartite graph with at least four vertices, then the toric ideal $P(G)$ is a complete intersection if and only if it is generated by a foliation.*

Exercises

10.1.57 [376] A clutter $\mathcal{C}$ is called of *linear type* if its edge ideal $I(\mathcal{C})$ is of linear type. If all connected components of a clutter are of linear type, then the clutter is of linear type.

10.1.58 [376] If $\mathcal{C}$ is a clutter of linear type, then so are all minors of $\mathcal{C}$.

10.1.59 Let $R = \mathbb{Q}[x_1, \ldots, x_7]$ and let I be the ideal of R generated by $f_1 = x_1x_2x_3, f_2 = x_2x_4x_5, f_3 = x_5x_6x_7, f_4 = x_3x_6x_7$. Use *Macaulay*2 to verify that the toric ideal of $R[It]$ is minimally generated by the binomials

$$x_3t_3 - x_5t_4, x_6x_7t_1 - x_1x_2t_4, x_6x_7t_2 - x_2x_4t_3, x_4x_5t_1 - x_1x_3t_2, x_4t_1t_3 - x_1t_2t_4.$$

Prove that Theorem 10.1.15 does not extend to uniform square-free monomial ideals.

10.1.60 Let G be a graph. Then G is planar (resp. outerplanar) if and only if each block of G is planar (resp. outerplanar).

10.1.61 Let G be a ring graph. Prove that G is outerplanar if and only if G does not contain a subdivision of $\mathcal{K}_{2,3}$ as a subgraph.

10.1.62 Let G be a ring graph with $n \geq 3$ vertices and q edges. If G is 2-connected and has no triangles, then $q \leq 2(n-2)$.

10.1.63 Let G be a graph with q edges and $P(G)$ the toric ideal of the edge subring $K[G]$. If f is a primitive binomial in $P(G)$ prove:

$$\deg(f) \leq \begin{cases} q & \text{if } q \text{ is even, and} \\ q-1 & \text{otherwise.} \end{cases}$$

10.1.64 Let G be a bipartite graph with bipartition (X, Y). Prove that the largest degree of a binomial which is part of a minimal set of homogeneous generators of $P(G)$ is less than or equal to $\min(|X|, |Y|)$.

10.1.65 If G is a connected graph, then G has a *spanning tree* T, that is, there exists a tree T which is a spanning subgraph of G.

10.1.66 [208] Prove that a graph G is bipartite if and only if every cycle in some cycle basis of G is even.

10.1.67 If G is a connected non-bipartite graph with n vertices, prove that there is a connected subgraph H of G with n vertices and n edges and with a unique cycle of odd length.

10.1.68 If G is a connected graph, prove that e is a bridge if and only if e is not on any cycle of G.

10.1.69 Let G be a connected graph and let $f_1, \ldots, f_q$ be the edge generators of $K[G]$. If f_r correspond to a bridge of G and

$$f^a = f_1^{a_1} \cdots f_q^{a_q} = f_1^{b_1} \cdots f_q^{b_q} = f^b$$

for some $a, b \in \mathbb{N}^q$ such that $\text{supp}(a) \cap \text{supp}(b) = \emptyset$ and $a_r > 0$, prove that a_r is an even integer.

10.1.70 The graph below is planar but not outerplanar and it satisfies $\text{rank}(G) = \text{frank}(G) = 3$. The regions of this embedding are not bounded by primitive cycles. Thus, the converse of Proposition 10.1.47 fails.

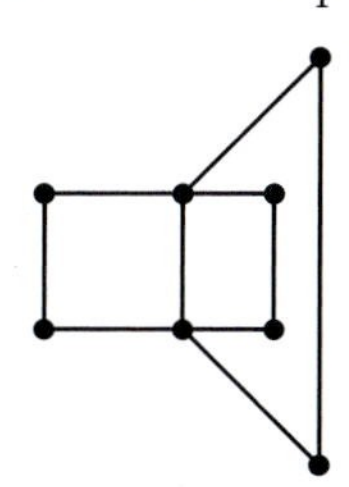

10.1.71 Let G be a graph that can be written as $G = G_1 \cup G_2$, where G_1 and G_2 are subgraphs with at most one vertex in common. If G_1 is bipartite, then $P(G) = (P(G_1), P(G_2))$ and $K[G] \cong K[G_1] \otimes_K K[G_2]$.

10.1.72 [379] Let G be a connected bipartite graph with n vertices and q edges. Prove that $\text{frank}(G) = q - n + 1$ if and only if the toric ideal $P(G)$ of $K[G]$ is a complete intersection.

10.1.73 Let G be a bipartite graph. If G is a subdivision of $\mathcal{K}_5$ or a subdivision $\mathcal{K}_{3,3}$ prove that the binomials of $P(G)$ that correspond to primitive cycles do not form a regular sequence.

10.1.74 Let G be a bipartite graph. Use Kuratowski's theorem (see Theorem 10.1.43) to prove that if the toric ideal of $K[G]$ is a complete intersection, then G is a planar graph.

10.1.75 If G is a ring graph and H is an induced subgraph of G, then H is a ring graph.

10.1.76 Prove that the complete graph $\mathcal{K}_4$ has the primitive cycle property and $\text{rank}(G) \neq \text{frank}(G)$. Thus, the converse of Lemma 10.1.35 fails.

10.2 Incidence matrices and circuits

Let G be a simple graph with vertex set $V = \{x_1, \ldots, x_n\}$ and with edge set $E = \{f_1, \ldots, f_q\}$, where every edge f_i is an unordered pair of distinct vertices $f_i = \{x_{i_j}, x_{i_k}\}$. The *incidence matrix* $A_G = (a_{ij})$ associated to G is the $n \times q$ matrix defined by

$$a_{ij} = \begin{cases} 1 & \text{if } x_i \in f_j, \text{ and} \\ 0 & \text{if } x_i \notin f_j. \end{cases}$$

Notice that each column of A_G has exactly two 1's and the rest of its entries equal to zero. If $f_i = \{x_{i_j}, x_{i_k}\}$ define $v_i = e_{i_j} + e_{i_k}$, where e_i is the ith canonical vector in $\mathbb{R}^n$. Thus the columns of A_G are precisely the vectors $v_1, \ldots, v_q$ regarded as column vectors. In what follows we denote the set $\{v_1, \ldots, v_q\}$ by $\mathcal{A}$.

Example 10.2.1 Consider a triangle G with vertices x_1, x_2, x_3. The incidence matrix of G is:

$$A_G = \begin{pmatrix} 1 & 0 & 1 \\ 1 & 1 & 0 \\ 0 & 1 & 1 \end{pmatrix},$$

with the vectors $v_1 = e_1 + e_2$, $v_2 = e_2 + e_3$, and $v_3 = e_1 + e_3$ corresponding to the edges $f_1 = \{x_1, x_2\}$, $f_2 = \{x_2, x_3\}$, and $f_3 = \{x_1, x_3\}$.

Proposition 10.2.2 *If A is the incidence matrix of a bipartite graph G, then A is totally unimodular.*

Proof. By induction on the number of columns of A. Let (V_1, V_2) be the bipartition of G and let B be a square submatrix of A. If a column of B has at most one entry equal to 1, then by induction $\det(B) = 0, \pm 1$. Thus we may assume that all the columns of B have two entries equal to 1. Hence

$$\sum_{x_i \in V_1} F_i = \sum_{x_i \in V_2} F_i = (1, \ldots, 1),$$

where $x_1, \ldots, x_n$ are the vertices of G and $F_1, \ldots, F_r$ are the rows of B. Therefore the rows of B are linearly dependent and $\det(B) = 0$. $\square$

Theorem 10.2.3 [187] *Let G be a simple bipartite graph with n vertices and q edges and let $A = (a_{ij})$ be its incidence matrix. If $e_1, \ldots, e_n$ are the first n unit vectors in $\mathbb{R}^{n+1}$ and C is the matrix*

$$C = \begin{pmatrix} a_{11} & \ldots & a_{1q} & e_1 & \cdots & e_n \\ \vdots & \vdots & \vdots & & & \\ a_{n1} & \ldots & a_{nq} & & & \\ 1 & \ldots & 1 & & & \end{pmatrix}$$

obtained from A by adjoining a row of 1's *and the column vectors $e_1, \ldots, e_n$, then C is totally unimodular.*

Proof. Suppose that $\{1, \ldots, m\}$ and $\{m+1, \ldots, n\}$ is the bipartition of the graph G. Let C' be the matrix obtained by deleting the last $n - m$ columns from C. It suffices to show that C' is totally unimodular. First one successively subtracts the rows $1, 2, \ldots, m$ from the row $n+1$. Then one reverses the sign in the rows $m+1, \ldots, n$. These elementary row operations produce a new matrix C''. The matrix C'' is the incidence matrix of a directed graph, namely, consider G as a directed graph, and add one more vertex $n+1$, and add the edges $(i, n+1)$ for $i = 1, \ldots, m$. The matrix C'', being the incidence matrix of a directed graph, is totally unimodular; see Exercise 1.8.10. As the last m column vectors of C'' are

$$e_1 - e_{n+1}, \ldots, e_m - e_{n+1},$$

one can successively pivot on the first non-zero entry of $e_i - e_{n+1}$ for $i = 1, \ldots, m$ and reverse the sign in the rows $m+1, \ldots, n$ to obtain back the matrix C'. Here a pivot on the entry c'_{st} means transforming column t of C'' into the sth unit vector by elementary row operations. Since pivoting preserves total unimodularity [338, Lemma 2.2.20] one derives that C' is totally unimodular, and hence so is C. In the next specific example we display C' and C''. $\square$

Example 10.2.4 To illustrate the constructions in the proof above consider the complete bipartite graph $G = \mathcal{K}_{2,3}$. In this case the two $\mathbb{Z}$-row equivalent matrices C' and C'' are

$$\begin{bmatrix} 1 & 1 & 1 & 0 & 0 & 0 & 1 & 0 \\ 0 & 0 & 0 & 1 & 1 & 1 & 0 & 1 \\ 1 & 0 & 0 & 1 & 0 & 0 & 0 & 0 \\ 0 & 1 & 0 & 0 & 1 & 0 & 0 & 0 \\ 0 & 0 & 1 & 0 & 0 & 1 & 0 & 0 \\ 1 & 1 & 1 & 1 & 1 & 1 & 0 & 0 \end{bmatrix}, \quad \begin{bmatrix} 1 & 1 & 1 & 0 & 0 & 0 & 1 & 0 \\ 0 & 0 & 0 & 1 & 1 & 1 & 0 & 1 \\ -1 & 0 & 0 & -1 & 0 & 0 & 0 & 0 \\ 0 & -1 & 0 & 0 & -1 & 0 & 0 & 0 \\ 0 & 0 & -1 & 0 & 0 & -1 & 0 & 0 \\ 0 & 0 & 0 & 0 & 0 & 0 & -1 & -1 \end{bmatrix},$$

respectively. Note that pivoting in the last two columns of C'' amounts to adding row 1 to row 6, and then adding row 2 to row 6.

Remark 10.2.5 Let G be a graph, let A_G be the incidence matrix of G, and let B be a square submatrix of A_G. In [201] it is shown that either $\det(B) = 0$ or $\det(B) = \pm 2^k$, for some integer k such that $0 \leq k \leq \tau_0$, where τ_0 is the maximum number of vertex disjoint odd cycles in G. Moreover for any such value of k there exists a minor equal to $\pm 2^k$.

The number of bipartite (resp. non-bipartite) connected components of a graph G will be denoted by c_0 (resp. c_1). Thus $c = c_0 + c_1$ is the total number of components of G.

Lemma 10.2.6 [201] *If G is a graph with n vertices and A_G is its incidence matrix, then* $\mathrm{rank}(A_G) = n - c_0$.

Proof. Let $G_1, \ldots, G_c$ be the connected components of G, and n_i the number of vertices of G_i. After permuting the vertices we may assume that A_G is a "diagonal" matrix

$$A_G = \mathrm{diag}(A_{G_1}, \ldots, A_{G_c}),$$

where A_{G_i} is the incidence matrix of G_i. By Proposition 10.1.20 the rank of A_{G_i} is equal to $n_i - 1$ if G_i is bipartite, and is equal to n_i otherwise. Hence the rank of A_G is equal to $n - c_0$. $\Box$

The incidence matrix of G is denoted by A_G or simply by A. Regarding the v_i's as column vectors, we define the matrix B as:

$$B = \begin{pmatrix} v_1 & \cdots & v_q \\ 1 & \cdots & 1 \end{pmatrix}.$$

Observe that $\mathrm{rank}(B) = \mathrm{rank}(A_G)$ because the last row of B is a linear combination of the first n rows. Recall that $\Delta_r(B)$ denotes the gcd of all non-zero r-minors of B.

For simplicity we keep the notation introduced above throughout the rest of this section. We shall assume, if need be, that the graph G has no isolated vertices.

Lemma 10.2.7 *If G is a unicyclic graph with a unique odd cycle and r is the rank of B, then $\Delta_r(B) = 1$.*

Proof. If G is an odd cycle of length n, then the matrix B' obtained from B by deleting any of its first n rows has determinant ± 1. The matrix B' is in fact totally unimodular by Theorem 10.2.3. We proceed by induction. If G is not an odd cycle, then G has a vertex x_j of degree 1. Thus the jth row of A has exactly one entry equal to 1, say $a_{jk} = 1$. Consider the matrix C obtained from B by deleting the jth row and the kth column. Thus using induction $\Delta_{r-1}(C) = 1$, which gives $\Delta_r(B) = 1$. □

Lemma 10.2.8 *Let G be a graph with connected components $G_0, \ldots, G_m$. If G_0 is non-bipartite, then there exists a subgraph H of G with connected components $H_0, \ldots, H_m$ such that H_0 is a spanning unicyclic subgraph of G_0 with a unique odd cycle and H_i is a spanning tree of G_i for all $i \geq 2$.*

Proof. The existence of $H_2, \ldots, H_m$ is clear because any connected graph has a spanning tree. Take a spanning tree T_0 of G_0. For each line e of G_0 not in T_0, the graph $T_0 \cup \{e\}$ has exactly one cycle. According to Remark 10.1.17 the set $Z(T)$ of such cycles form a basis for the cycle space of G_0. Since a graph is bipartite if and only if every cycle in some cycle basis is even (see Exercise 10.1.66), there is an edge e of G_0 such that $T_0 \cup \{e\}$ is a unicyclic connected graph with a unique odd cycle. □

Proposition 10.2.9 *If G is a graph with exactly one non-bipartite connected component and $r = \mathrm{rank}(B)$, then $\Delta_r(B) = 1$.*

Proof. Let $G_0, \ldots, G_m$ be the components of G, with G_0 non-bipartite, and let H be as in Lemma 10.2.8. If A' is the incidence matrix of H, then using that G and H have the same number of non-bipartite components one has $\mathrm{rank}(A) = \mathrm{rank}(A')$. Therefore $\Delta_r(B) = 1$ by Lemma 10.2.7. □

Proposition 10.2.10 *Let G be a non-bipartite graph and let A be its incidence matrix. If A has rank r, then $\Delta_r(A) = 2\Delta_r(B)$.*

Proof. We set $\mathcal{A} = \{v_1, \ldots, v_q\}$ and $\mathcal{A}' = \{(v_1, 1), \ldots, (v_q, 1)\}$. It suffices to prove that there is an exact sequence of groups

$$0 \longrightarrow T(\mathbb{Z}^{n+1}/\mathbb{Z}\mathcal{A}') \xrightarrow{\varphi} T(\mathbb{Z}^n/\mathbb{Z}\mathcal{A}) \xrightarrow{\psi} \mathbb{Z}_2 \longrightarrow 0.$$

For $\beta = (a_1, \ldots, a_n) \in \mathbb{Z}^n$ and $b \in \mathbb{Z}$, define

$$\varphi(\overline{\beta, b}) = \overline{\beta} \quad \text{and} \quad \psi(\overline{\beta}) = \overline{a_1 + \cdots + a_n}.$$

It is not hard to verify that φ is injective and that $\mathrm{im}(\varphi) = \ker(\psi)$.

To show that ψ is onto consider a non-bipartite component G_1 of G. If $x_1, \ldots, x_s$ are the vertices of G_1 and $v_1', v_2', \ldots, v_m'$ are the columns of the incidence matrix of G_1, then $\mathbb{Z}^s/(v_1', v_2', \ldots, v_m')$ is a finite group. It follows that for any $1 \leq j \leq s$, the element $\overline{e_j}$ is in the torsion subgroup of $\mathbb{Z}^n/\mathbb{Z}\mathcal{A}$, hence $\psi(\overline{e_j}) = \overline{1}$, as required. □

Corollary 10.2.11 [201] *If G is a graph and c_0 (resp. c_1) is the number of bipartite (resp. non-bipartite) components, then*

$$\mathbb{Z}^n/\mathbb{Z}\mathcal{A} \simeq \mathbb{Z}^{n-r} \times \mathbb{Z}_2^{c_1} = \mathbb{Z}^{c_0} \times \mathbb{Z}_2^{c_1},$$

where $r = n - c_0$ is the rank of A_G.

Proof. Let $G_1, \ldots, G_c$ be the components of G and let $\mathcal{A}_i$ be the set of vectors in $\mathcal{A}$ corresponding to the edges of G_i. Since the vectors in $\mathcal{A}_i$ can be regarded as vector in $\mathbb{Z}^{\eta_i}$, where η_i is the number of vertices in G_i, there is a canonical isomorphism

$$\mathbb{Z}^n/\mathbb{Z}\mathcal{A} \simeq \mathbb{Z}^{\eta_1}/\mathbb{Z}\mathcal{A}_1 \oplus \cdots \oplus \mathbb{Z}^{\eta_c}/\mathbb{Z}\mathcal{A}_c.$$

To finish the proof apply Propositions 10.2.9 and 10.2.10 to derive

$$\mathbb{Z}^{\eta_i}/\mathbb{Z}\mathcal{A}_i \simeq \begin{cases} \mathbb{Z}_2 & \text{if} \quad G_i \text{ is non-bipartite,} \\ \mathbb{Z} & \text{if} \quad G_i \text{ is bipartite.} \end{cases}$$

In the second isomorphism we are using the fact that incidence matrices of bipartite graphs are totally unimodular; see Proposition 10.2.2. □

Corollary 10.2.12 *If c_1 is the number of non-bipartite components of G and r is the rank of B, then*

$$\Delta_r(B) = \begin{cases} 2^{c_1 - 1} & \text{if } c_1 \geq 1, \\ 1 & \text{if } c_1 = 0. \end{cases}$$

Proof. It follows from Proposition 10.2.10 and Corollary 10.2.11. □

Definition 10.2.13 A square matrix U, with entries in a commutative ring R, is called *unimodular* if $\det(U)$ is a unit of R.

Theorem 10.2.14 [201] *If G is a graph with n vertices and A_G is the incidence matrix of G, then there are unimodular integral matrices U, U' such that*

$$S = UA_GU' = \begin{pmatrix} D & 0 \\ 0 & 0 \end{pmatrix},$$

where $D = \operatorname{diag}(1, 1, \ldots, 1, 2, 2, \ldots, 2)$, $n - c$ is the number of 1's and c_1 is the number of 2's.

Proof. It follows using Corollary 10.2.11 and the fundamental structure theorem for finitely generated abelian groups. See Theorem 1.3.16. □

The matrix S is called the *Smith normal form* of A_G and the numbers in the diagonal matrix D are the *invariant factors* of A_G.

Circuits of the kernel of an incidence matrix We give a geometric description of the elementary integral vectors or circuits of the kernel of the incidence matrix of a graph (see Section 10.3 for the multigraph case).

Notation For $\alpha = (\alpha_1, \ldots, \alpha_q) \in \mathbb{N}^q$ and $f_1, \ldots, f_q$ in a commutative ring with identity we set $f^\alpha = f_1^{\alpha_1} \cdots f_q^{\alpha_q}$. The *support* of f^α is defined as the set $\text{supp}(f^\alpha) = \{f_i \mid \alpha_i \neq 0\}$.

Let G be a graph on the vertex set V with edge set E and incidence matrix A. We set $N = \ker(A)$, the kernel of A in $\mathbb{Q}^q$. Note that a vector $\alpha \in N \cap \mathbb{Z}^q$ determine the subgraph G_α of G having vertex set

$$V_\alpha = \{x \in V \mid x \text{ divide } f^{\alpha_+}\}$$

and edge set $E_\alpha = \{\{x, y\} \in E \mid xy \in \text{supp}(f^{\alpha_+}) \cup \text{supp}(f^{\alpha_-})\}$, where $f_1, \ldots, f_q$ are the square-free monomials corresponding to the edges of G, that is, the edge generators of G.

The next result can be generalized to multigraphs; see Section 10.3.

Proposition 10.2.15 [417] *Let G be a graph with incidence matrix A and let N be the kernel of A in $\mathbb{Q}^q$. Then a vector $0 \neq \alpha \in \mathbb{Z}^q \cap N$ is an elementary vector of N if and only if*

(i) *G_α is an even cycle, or*

(ii) *G_α is a connected graph consisting of two odd cycles with at most one common vertex joined by a path.*

Proof. Assume $\alpha \in \mathbb{Z}^q \cap N$ is an elementary vector of N so that G_α is not an even cycle. From the equation $f^{\alpha_+} = f^{\alpha_-}$ we obtain a walk in G_α

$$w = \{x_0, x_1, \ldots, x_n = x_0, x_{n+1}, \ldots, x_\ell\},$$

so that $x_1, \ldots, x_{\ell-1}$ are distinct vertices, n is odd and x_ℓ is in $\{x_1, \ldots, x_{\ell-2}\}$. The walk w can be chosen so that $x_i x_{i+1} \in \text{supp}(f^{\alpha_+})$ if i is even and $x_i x_{i+1} \in \text{supp}(f^{\alpha_-})$ otherwise. We claim that x_ℓ is not in $\{x_1, \ldots, x_{n-1}\}$. Assume $x_\ell = x_m$ for some $1 \leq m \leq n-1$. If m and $\ell - n$ are odd then the monomial walk

$$w_1 = \{x_0, x_1, \ldots, x_m, x_{\ell-1}, x_{\ell-2}, \ldots, x_n = x_0\}$$

yields a relation $f^\delta = f^\gamma$ with $\text{supp}(\delta - \gamma)$ properly contained in $\text{supp}(\alpha)$, which is a contradiction. If m is odd and $\ell - m$ is even then we can consider the walk

$$w_1 = \{x_n, x_{n+1}, \ldots, x_\ell = x_m, x_{m+1}, \ldots, x_n\}$$

and derive a contradiction; the remaining cases are treated similarly. We may now assume $x_\ell = x_m$ for $n \leq m \leq \ell - 2$. The cycle

$$C_1 = \{x_m, \ldots, x_\ell = x_m\}$$

must be odd, otherwise C_1 would give a relation $f^\delta = f^\gamma$ with $\mathrm{supp}(\delta - \gamma)$ a proper subset of $\mathrm{supp}(\alpha)$. The walk w gives a relation $f^\delta = f^\gamma$ with

$$\mathrm{supp}(f^\delta) \subset \mathrm{supp}(f^{\alpha_+}) \ \text{ and } \ \mathrm{supp}(f^\gamma) \subset \mathrm{supp}(f^{\alpha_-}).$$

Hence $\mathrm{supp}(\delta - \gamma)$ is a subset of $\mathrm{supp}(\alpha)$, and since the support of α is minimal one has the equality $\mathrm{supp}(\delta - \gamma) = \mathrm{supp}(\alpha)$. Therefore

$$V_\alpha = \{x_1, \ldots, x_\ell\}$$

and G_α is as required. The converse follows from Corollary 10.1.16. □

Definition 10.2.16 A *circuit* of a graph G is either an even cycle or a subgraph consisting of two odd cycles with at most one common vertex joined by a path.

A family of connected graphs is called *matroidal* if given any graph G, the subgraphs of G isomorphic to a member of the family are the circuits of a matroid on the edge set of G. Thus Proposition 10.2.15 implies that the family of circuits is matroidal as was shown in [387].

Proposition 10.2.17 *If A is the incidence matrix of a connected simple graph G, then the circuits of the vector matroid $M[A]$ consists of even cycles, two edge disjoint odd cycles meeting at exactly one vertex, and two vertex disjoint odd cycles joined by an arbitrary path.*

Proof. It follows from Proposition 10.2.15 (see Section 1.9). □

Circuits of a graph and Gröbner bases Let G be a graph. We study the toric ideal $P(G)$ of the edge subring $K[G]$ from the view point of Gröbner bases. For general results on Gröbner bases of toric ideals a standard reference is [400, Chapter 4].

Proposition 10.2.18 *If G is a graph with incidence matrix A and P is the toric ideal of $K[G]$, then*

$$P = (\{t^{\alpha_+} - t^{\alpha_-} \mid \alpha \in \mathbb{Z}^q \text{ and } A\alpha = 0\}).$$

Proof. The result is a consequence of Corollary 8.2.20. □

Corollary 10.2.19 *Let G be a graph with incidence matrix A. If $\lambda = (\lambda_i)$ is a circuit of $N = \ker(A)$, then $|\lambda_i| \leq 2$ for all i.*

Proof. Let λ be a circuit of N. Since G_λ is an even cycle or two edge disjoint cycles joined by a path we obtain a circuit α with $\mathrm{supp}(\alpha) = \mathrm{supp}(\lambda)$ and so that $|\alpha_i| \leq 2$ for all i. Because any two elementary vectors with the same support are dependent we obtain $\lambda = \alpha$. □

Theorem 10.2.20 *Let G be a connected graph with q edges and let P be the toric ideal of $K[G]$. Then the total degree of a polynomial in any reduced Gröbner basis of P is less than or equal to $2q \dim Z_e(G)$.*

Proof. We set $N = \{\alpha \in \mathbb{Q}^q \mid A\alpha = 0\}$, where A is the incidence matrix of G. Assume F is the reduced Gröbner basis of P and take $f = f(t) \in F$. Using Corollary 8.2.18 and Corollary 10.2.18, together with the fact that normalized reduced Gröbner bases are uniquely determined, we obtain

$$f(t) = t^{\alpha} - t^{\beta}.$$

Notice that $\mathrm{supp}(\alpha) \cap \mathrm{supp}(\beta) = \emptyset$, for otherwise take i in the intersection and write $f(t) = t_i(t^a - t^b)$. Since $t^a - t^b$ reduces with respect to F it follows that $f(t)$ reduces with respect to $F \setminus \{f\}$, which is impossible. We can now write $f(t) = t^{\alpha_+} - t^{\alpha_-}$ for some $\alpha \in N \cap \mathbb{Z}^q$. By Theorem 1.9.6 we can write

$$\alpha = a_1\lambda_1 + \cdots + a_r\lambda_r$$

for some $a_i \in \mathbb{Q}^+$, and for some circuit λ_i each in harmony with α such that $\mathrm{supp}(\lambda_i) \subset \mathrm{supp}(\alpha)$ for all i and $r \leq \dim N$. Hence

$$\alpha_+ = a_1(\lambda_1)_+ + \cdots + a_r(\lambda_r)_+ \quad \text{and} \quad \alpha_- = a_1(\lambda_1)_- + \cdots + a_r(\lambda_r)_-.$$

If $a_i > 1$ for some i then α_{i+} and α_{i-} are componentwise larger than λ_{i+} and λ_{i-}, respectively; note that this is not possible by the previous argument. Hence $0 < a_i \leq 1$. Using Corollary 10.2.19 it now follows that the entries of α satisfy $|\alpha_i| \leq 2 \dim Z_e(G)$ for all i, which gives the asserted inequality. □

Exercises

10.2.21 Let G be a cycle of length n and A_G its incidence matrix. Prove

$$\det(A_G) = \begin{cases} \pm 2 & \text{if } n \text{ is odd} \\ 0 & \text{if } n \text{ is even.} \end{cases}$$

10.2.22 Let U be a square matrix with entries in a commutative ring R. Prove that U is unimodular if and only if U is invertible.

10.2.23 Let G be a graph with incidence matrix A and $N = \ker(A) \subset \mathbb{Q}^q$. Give an example to show that the set of circuits of N does not determine the presentation ideal of $K[G]$.

10.2.24 Let G be a graph with incidence matrix A and $N = \ker(A)$ the kernel of A in $\mathbb{Q}^q$. Let $\lambda_1, \ldots, \lambda_s$ be the circuits of N. Define the *elementary zonotope* of N as $\mathcal{E}_N = [0, \lambda_1] + \cdots + [0, \lambda_s]$. Prove that $\mathcal{E}_N$ is a polytope.

10.2.25 If N is the kernel of the incidence matrix of a graph G and P is the toric ideal of $K[G]$, prove that $\{t^{\alpha_+} - t^{\alpha_-} \,|\, \alpha \in \mathcal{E}_N \cap \mathbb{Z}^q\}$ is a universal Gröbner basis for P.

10.2.26 (I. Gitler) Prove that the complete graph $\mathcal{K}_5$ is not a circuit and consists of two edge disjoint odd cycles of length 5.

10.2.27 (K. Truemper) Consider the matrices

$$V = \begin{pmatrix} 1 & 0 & 0 & 1 \\ 0 & 1 & 0 & 1 \\ 0 & 0 & 1 & 1 \\ 1 & 1 & 1 & 1 \end{pmatrix}, \quad A = \begin{pmatrix} 0 & 0 & 1 & 0 \\ 0 & 0 & 1 & 0 \\ 0 & 1 & 0 & 0 \\ 0 & 1 & 0 & 0 \\ 1 & 0 & 0 & 0 \\ 1 & 0 & 0 & 0 \\ 1 & 0 & 0 & 1 \\ 0 & 1 & 0 & 1 \\ 0 & 0 & 1 & 1 \end{pmatrix}, \quad B = \begin{pmatrix} A \\ 1\,1\,1\,1 \end{pmatrix}.$$

Note $\det(V) = -2$. Prove that A is totally unimodular but B is not. Thus Theorem 10.2.3 does not extend to k-hypergraphs with $k > 2$.

10.3 The integral closure of an edge subring

Let G be a multigraph. In this section we unfold a construction for the integral closure of $K[G]$ and study the circuits of the toric ideal of $K[G]$.

Let G a *multigraph* with vertex set $X = \{x_1, \ldots, x_n\}$, i.e., G is obtained from a simple graph by allowing multiple edges and loops. Thus the edges of G have the form $\{x_i, x_j\}$. If $e = \{x_i, x_j\}$ is an edge of G, its characteristic vector is given by $v_e = e_i + e_j$, where e_i is the ith unit vector in $\mathbb{R}^n$. Notice that if e is a loop, i.e., if $i = j$, then $v_e = 2e_i$. The *incidence matrix* of G, denoted by A, is the matrix whose column vectors are the characteristic vectors of the edges and loops of G.

Let $v_1, \ldots, v_q$ be the characteristic vectors of the edges and loops of G and let $R = K[x_1, \ldots, x_n]$ be a polynomial ring over a field K. In what follows $\mathcal{A}$ denotes the set $\{v_1, \ldots, v_q\}$ and F denotes the set $\{f_1, \ldots, f_q\}$ of all monomials x_ix_j in R such that $\{x_i, x_j\}$ is an edge of G. The *edge subring* of G is given by

$$K[G] = K[f_1, \ldots, f_q] \subset R.$$

We may assume that $f_i = x^{v_i}$ for $i = 1, \ldots, q$. One has the following description of the integral closure of $K[G]$:

$$\overline{K[G]} = K[\{x^a \,|\, a \in \mathbb{R}_+\mathcal{A} \cap \mathbb{Z}\mathcal{A}\}],$$

where $\mathbb{R}_+\mathcal{A}$ is the cone in $\mathbb{R}^n$ generated by $\mathcal{A}$ and $\mathbb{Z}\mathcal{A}$ is the subgroup of $\mathbb{Z}^n$ generated by $\mathcal{A}$. See Theorem 9.1.1.

Proposition 10.3.1 *If G is a bipartite graph, then $K[G]$ is normal and Cohen–Macaulay.*

Proof. By Proposition 10.2.2 the incidence matrix of a bipartite graph is totally unimodular. Then, by Corollary 1.6.8, we have $\mathbb{N}\mathcal{A} = \mathbb{Z}^n \cap \mathbb{R}_+\mathcal{A}$. Thus $K[G] = \overline{K[G]}$, that is $K[G]$ is normal. Hence, by Theorem 9.1.6, $K[G]$ is Cohen–Macaulay. $\Box$

Definition 10.3.2 A *bow-tie* of G is an induced submultigraph w of G consisting of two odd cycles with at most one common vertex

$$Z_1 = \{z_0, z_1, \ldots, z_r = z_0\} \text{ and } Z_2 = \{z_s, z_{s+1}, \ldots, z_t = z_s\}$$

joined by a path $\{z_r, \ldots, z_s\}$. In this case we set $M_w = z_1 \cdots z_r z_{s+1} \cdots z_t$. We regard a loop of G as an odd cycle.

Remark 10.3.3 A bow-tie w, as above, always defines an even closed walk:

$$w = \{z_0, z_1, \ldots, z_r, z_{r+1}, \ldots, z_s, z_{s+1}, \ldots, z_t = z_s, z_{s-1}, \ldots, z_r = z_0\}.$$

Lemma 10.3.4 *If w is a bow-tie of a multigraph G, then M_w is in the integral closure of $K[G]$.*

Proof. Let w be a bow-tie. With the notation above. If $f_i = z_{i-1}z_i$, then

$$z_1^2 \cdots z_r^2 = f_1 \cdots f_r, \qquad z_s^2 \cdots z_{t-1}^2 = f_{s+1} \cdots f_t,$$

which together with the identities

$$M_w = \prod_{i \text{ odd}} f_i \prod_{\substack{i \text{ even} \\ r < i \le s}} f_i^{-1} \qquad \text{and} \qquad M_w^2 = f_1 \cdots f_r f_{s+1} \cdots f_t$$

gives $M_w \in \overline{K[G]}$. $\Box$

Remark 10.3.5 Putting the two equations above together one obtains

$$\prod_{\substack{i \text{ odd} \\ r < i \le s}} f_i^2 \prod_{\substack{i \text{ odd} \\ i \notin (r,s]}} f_i = \prod_{\substack{i \text{ even} \\ r < i \le s}} f_i^2 \prod_{\substack{i \text{ even} \\ i \notin (r,s]}} f_i. \tag{10.1}$$

If $r = s$, that is Z_1 and Z_2 intersect at a vertex, then

$$\prod_{i \text{ odd}} f_i = \prod_{i \text{ even}} f_i \tag{10.2}$$

If $s = r + 1$, that is Z_1 and Z_2 are joined by an edge $\{z_r, z_s\}$, then

$$\prod_{i \text{ odd}} f_i = f_s^2 \prod_{\substack{i \text{ even} \\ i \neq s}} f_i. \tag{10.3}$$

Example 10.3.6 A bow-tie formed with a triangle and a pentagon joined by a path of length 2:

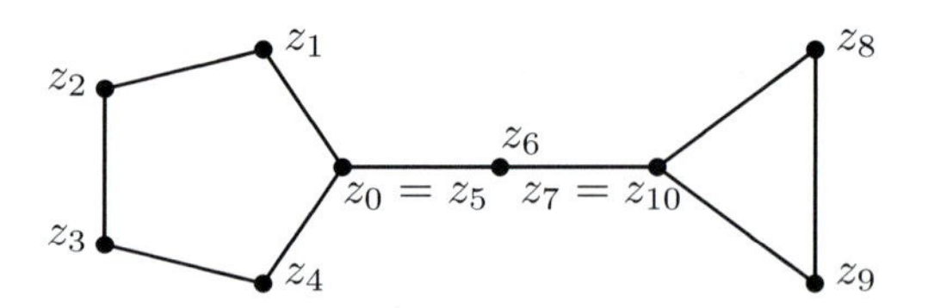

Example 10.3.7 A bow-tie w formed with two loops (odd cycles of length 1) joined by a path of length 2:

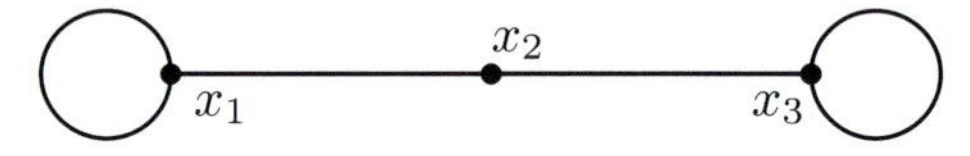

We can write w as an even closed walk $w = \{x_1, x_1, x_2, x_3, x_3, x_2, x_1\}$. In this example $M_w = x_1x_3$.

The *cone generated* by $\mathcal{A}$ is given by $\mathbb{R}_+\mathcal{A} = \mathbb{R}_+v_1 + \cdots + \mathbb{R}_+v_q$. This cone will be studied in detail in Section 10.7 when G is a graph.

Definition 10.3.8 A multigraph having just one cycle is called *unicyclic.*

Theorem 10.3.9 [384] *If G is a connected multigraph, then*

$$\overline{K[G]} = K[\{f_1, \ldots, f_q\} \cup \{M_w |\, w \textit{ is a bow-tie}\}].$$

Proof. We set $\mathcal{B} = \{f_1, \ldots, f_q\} \cup \{M_w |\, w \text{ is a bow-tie}\}$. According to Theorem 9.1.1 and Corollary 1.1.27 one has:

$$\overline{K[G]} = K[\{x^a |\, a \in \mathbb{Z}\mathcal{A} \cap \mathbb{R}_+\mathcal{A}\}] = K[\{x^a |\, a \in \mathbb{Z}\mathcal{A} \cap \mathbb{Q}_+\mathcal{A}\}].$$

Hence, by Lemma 10.3.4, it suffices to show that $\overline{K[G]} \subset K[\mathcal{B}]$. Let x^a be a minimal (irreducible) generator of $\overline{K[G]}$. Then $a \in \mathbb{Q}_+\mathcal{A}$ and $a \in \mathbb{Z}\mathcal{A}$. By Carathéodory's Theorem (see Theorem 1.1.18), and the irreducibility of x^a, we can write

$$a = \lambda_1 v_1 + \cdots + \lambda_r v_r, \tag{10.4}$$

where $v_1, \ldots, v_r$ are linearly independent vectors in $\mathcal{A}$ and $\lambda_i \in \mathbb{Q}_+ \cap (0, 1)$ for $i = 1, \ldots, r$. Permitting an abuse of notation, we will also denote the edges of G by $v_1, \ldots, v_q$, and refer to v_i as an edge of G.

Let G_a be the multigraph whose edges are $v_1, \ldots, v_r$ and let

$$(G_a)_1, \ldots, (G_a)_s$$

be the components of G_a. For each i consider the graph H_i obtained from $(G_a)_i$ by removing loops and by replacing multiple edges by single edges.

Each H_i is either a tree or a unicyclic graph with a unique odd cycle because $v_1, \ldots, v_r$ are linearly independent. If H_i is a unicyclic graph with a unique odd cycle, then $(G_a)_i$ has no loops or multiple edges because $v_1, \ldots, v_r$ are linearly independent and the rank of the incidence matrix of $(G_a)_i$ is equal to the number of vertices of $(G_a)_i$. Thus in this case $(G_a)_i = H_i$. If H_i is a tree, then $(G_a)_i$ is a tree with at most one loop. This follows using that the rank of the incidence matrix of $(G_a)_i$ is equal to $|V((G_a)_i)| - 1$ if $(G_a)_i$ is a tree. Since $0 < \lambda_i < 1$, by Eq. (10.4), the multigraph G_a has no vertices of degree 1. Therefore the connected components of G_a are odd cycles. Hence, as $0 < \lambda_i < 1$ for all i and the components of G_a are odd cycles, we get that the entries of a are in $\{0, 1\}$. The multigraph G_a has at least two components, i.e., two odd cycles Z_1 and Z_2, because $a \in \mathbb{Z}\mathcal{A}$. Thus, by the irreducibility of x^a, the components of G_a are Z_1 and Z_2. As the graph is connected, there is a bow-tie w whose odd cycles are Z_1 and Z_2 and such that $x^a = M_w$. So $x^a \in \mathcal{B}$ as required. This proof was adapted from [263]. $\square$

There are two special types of bow-ties that can be omitted in the description of the integral closure of $K[G]$.

Proposition 10.3.10 *Let w be a bow-tie of G with two odd cycles Z_1, Z_2 such that either Z_1 and Z_2 meet at exactly one vertex or Z_1 and Z_2 are disjoint and are connected by an edge, then M_w is in $K[G]$.*

Proof. It is left as an exercise. $\square$

Definition 10.3.11 A multigraph G has the *odd cycle condition* if every two vertex disjoint odd cycles can be joined by at least one edge.

The *odd cycle condition* comes from graph theory [175]. This notion was used to study the normality of edge subrings [242, 383, 384] and the circuits of a graph [358, 417].

Corollary 10.3.12 *Let G be a connected multigraph. Then $K[G]$ is normal if and only if G satisfies the odd cycle condition.*

Proof. It follows from Theorem 10.3.9 and Proposition 10.3.10. $\square$

Normality and the circuits of a multigraph Let $S = K[t_1, \ldots, t_q]$ be a polynomial ring over a field K. Recall that the toric ideal of $K[G]$, denoted by $P(G)$, is the kernel of the epimorphism of K-algebras:

$$S = K[t_1, \ldots, t_q] \longrightarrow K[G] \text{ induced by } t_i \mapsto x^{v_i}.$$

Lemma 10.3.13 *If $f = t^a - t^b$ is a circuit of $P(G)$, then f has a square-free term or f has nonsquare-free terms and $\max_i\{a_i\} = \max_i\{b_i\} = 2$.*

Proof. If G is a simple graph, the result was shown in Corollary 10.2.19. In the case of a multigraph it follows using an identical argument as the one given in the proof of this corollary. □

Theorem 10.3.14 *Let G be a multigraph. If $K[G]$ is normal, then $P(G)$ is generated by circuits with a square-free term.*

Proof. It follows at once applying Lemma 10.3.13 and Corollary 9.7.4. □

The converse of this result is not true (see Exercise 10.3.25).

Definition 10.3.15 A sub-multigraph H of G is called a *circuit* of G if H has one of the following forms:

(a) H is an even cycle.

(b) H consists of two odd cycles intersecting in exactly one vertex; a loop is regarded as an odd cycle of length 1.

(c) H consists of two vertex disjoint odd cycles joined by a path.

The circuits of G are in one-to-one correspondence with the circuits of $P(G)$ as we now explain. (See Section 10.2 for a detailed discussion when G is a graph.) Any circuit H of G can be regarded as an even closed walk

$$w = \{w_0, w_1, \ldots, w_r, w_0\},$$

where r is even, $w_0, w_1, \ldots, w_r$ are the vertices of H (we allow repetitions) and $\{w_i, w_{i+1}\}$ is an edge (we allow loops) of G for all i. Then the binomial $t_w = t_1t_3 \cdots t_{r-1} - t_2t_4 \cdots t_r$ is in $P(G)$, where $f_i = w_{i-1}w_i$ and t_i maps to f_i for all i.

Remark 10.3.16 The circuits of $P(G)$ with a square-free term correspond to the following types of circuits of G: (a) Even cycles, (b) two odd cycles intersecting in exactly one vertex, (c) two vertex disjoint odd cycles joined by an edge. Thus if $K[G]$ is normal, by Theorem 10.3.14 we obtain a precise graph theoretical description of a generating set of circuits for $P(G)$.

Toric ideals of edge subrings of oriented graphs were studied in [184]. In this case the toric ideal is also generated by circuits and the circuits correspond to the cycles of the graph. In [182] they completely characterize the graphs that are complete intersections for all orientation.

Exercises

10.3.17 Prove that $K[G]$ is normal for every connected graph G with six vertices, where K is a field.

10.3.18 Let G be a graph and let $G_1, \ldots, G_r$ be its connected components. Then $K[G]$ is normal if and only if $K[G_i]$ is normal for each i.

10.3.19 Let G be a graph and let $G_1, \ldots, G_r$ be the connected components of G. If K is a field, prove that $K[G]$ is normal if and only if G_i satisfies the odd cycle condition for all i.

10.3.20 Let G_1 and G_2 be two finite sets of monomials in disjoint sets of indeterminates. Then $K[G_1 \cup G_2] \simeq K[G_1] \otimes_K K[G_2]$ and

$$\overline{K[G_1 \cup G_2]} \simeq \overline{K[G_1]} \otimes_K \overline{K[G_2]} \simeq \overline{K[G_1]}\ \overline{K[G_2]},$$

is the subring generated by the generators of $\overline{K[G_1]}$ and $\overline{K[G_2]}$.

10.3.21 Prove that the smallest connected graph G such that $K[G]$ is non-normal is the graph consisting of two disjoint triangles joined by a 2-path (see Example 10.5.1).

10.3.22 Le G be a graph and let w be one of the following bow-ties of G with cycles Z_1 and Z_2

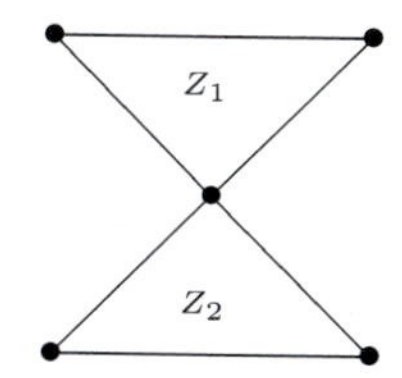

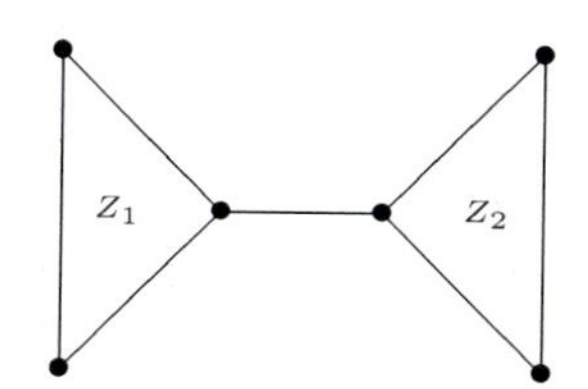

prove that M_w is in $K[G]$.

10.3.23 If G is a graph and $K[G]$ is normal, then $P(G)$ is generated by the binomials that correspond to the circuits of the incidence matrix of G.

10.3.24 Let $F = \{x_1x_2, x_2x_3, x_3x_4, x_1x_4, x_1^2, x_2^2, x_3^2, x_4^2\} \subset K[\mathbf{x}]$. Find all the bow-ties of the multigraph defined by F. Use Theorem 10.3.9 to show that $\overline{K[F]} = K[F][x_1x_3, x_2x_4]$.

10.3.25 (A. Thoma) Let G be the graph whose edge subring is

$$K[G] = K[x_1x_2, x_2x_3, x_1x_3, x_1x_4, x_5x_4, x_5x_6, x_6x_4, x_5x_7, x_6x_7].$$

Prove that the toric ideal is a complete intersection generated by circuits but that $K[G]$ is not normal.

10.4 Ehrhart rings of edge polytopes

Let G be a graph on the vertex set $V = \{x_1, \ldots, x_n\}$. The *edge polytope* of G is the lattice polytope

$$\mathcal{P} = \operatorname{conv}(\mathcal{A}) \subset \mathbb{R}^n,$$

where $\mathcal{A} = \{v_1, \ldots, v_q\}$ is the set of vectors in $\mathbb{N}^n$ of the form $e_i + e_j$ such that x_i is adjacent to x_j. Let $R = K[x_1, \ldots, x_n]$ be a polynomial ring over a field K. Recall that the Ehrhart ring of $\mathcal{P}$ is

$$A(\mathcal{P}) = K[\{x^\alpha t^i | \alpha \in \mathbb{Z}^n \cap i\mathcal{P}\}] \subset R[t],$$

the *edge subring* of G is the K-subalgebra

$$K[G] = K[x^{v_1}, \ldots, x^{v_q}] \subset R,$$

the polytopal subring is given by

$$K[\mathcal{P}] = K[x^\alpha t | \, \alpha \in \mathbb{Z}^n \cap \mathcal{P}] \subset R[t],$$

and the incidence matrix of G, denoted by A, is the matrix with column vector $v_1, \ldots, v_q$. In what follows B denotes the matrix

$$B = \begin{pmatrix} v_1 & \cdots & v_q \\ 1 & \cdots & 1 \end{pmatrix}.$$

As was earlier observed $\operatorname{rank}(B) = \operatorname{rank}(A)$. We shall keep these notations throughout the rest of this section. If need be, we assume that the graph G has no isolated vertices.

Proposition 10.4.1 $\dim(\mathcal{P}) = n - c_0 - 1$, *where c_0 denotes the number of bipartite components of G.*

Proof. By Lemma 10.2.6 and Proposition 9.3.1 one has the equalities

$$\dim K[G] = \operatorname{rank}(A) = n - c_0 \quad \text{and} \quad \dim K[\mathcal{P}] = \dim(\mathcal{P}) + 1.$$

Thanks to Lemma 9.3.16 one has that $K[\mathcal{P}]$ is equal to $K[x^{v_1}t, \ldots, x^{v_q}t]$. Thus $K[G] \simeq K[\mathcal{P}] \subset A(\mathcal{P})$. Consequently $\mathcal{P}$ has dimension $n - c_0 - 1$. $\Box$

Theorem 10.4.2 $\overline{K[\mathcal{P}]} = A(\mathcal{P})$ *if and only if the graph G has at most one non-bipartite connected component.*

Proof. $\Rightarrow$) If G has two non-bipartite components G_1, G_2, take

$$Z_1 = \{x_0, x_1, \ldots, x_r = x_0\}, \qquad Z_2 = \{x_{r+1}, x_{r+2} \ldots, x_{r+s} = x_{r+1}\}$$

two odd cycles of lengths r, $s-1$ in G_1, G_2, respectively. Setting

$$\beta = e_1 + \cdots + e_r + e_{r+1} + \cdots + e_{r+s-1},$$

note the equality

$$\beta = \frac{m}{2}\left[\left(\frac{e_r + e_1}{m} + \sum_{i=1}^{r-1} \frac{e_i + e_{i+1}}{m}\right) + \left(\frac{e_{r+s-1} + e_{r+1}}{m} + \sum_{i=r+1}^{r+s-2} \frac{e_i + e_{i+1}}{m}\right)\right]$$

where $m = r + s - 1$. Hence $\beta \in \mathbb{Z}^n \cap p\mathcal{P}$, where $p = m/2$, that is, $x^\beta t^p$ belongs to $A(\mathcal{P}) = \overline{K[\mathcal{P}]}$. Hence (β, p) is in the field of fractions of the polytopal subring $K[\mathcal{P}]$. There are integers $\lambda_1, \ldots, \lambda_q$ such that

$$e_1 + e_2 + \cdots + e_{r+s-1} + pe_{n+1} = \lambda_1(v_1, 1) + \cdots + \lambda_q(v_q, 1).$$

Since Z_1 and Z_2 are in different connected components we obtain an equality $e_1 + \cdots + e_r = \sum_j \lambda_j v_j$, where the sum is taken over all j such that the vector v_j corresponds to an edge of G_1. Setting $\mathbf{1} = (1, \ldots, 1)$ and taking inner products in the last equality one has $r = \langle e_1 + \cdots + e_r, \mathbf{1}\rangle = 2\sum_j \lambda_j$, a contradiction because r is odd.

$\Leftarrow$) By Lemma 9.3.16 $K[\mathcal{P}] = K[x^{v_1}t, \ldots, x^{v_q}t]$. If G is bipartite then, by Proposition 9.3.30, $\overline{K[\mathcal{P}]} = A(\mathcal{P})$. If G is not bipartite, the equality $\overline{K[\mathcal{P}]} = A(\mathcal{P})$ follows from Theorem 9.3.25 and Proposition 10.2.9. □

Corollary 10.4.3 [242, p. 423] *If G is connected, then $\overline{K[\mathcal{P}]} = A(\mathcal{P})$.*

Theorem 10.4.4 *The multiplicities of $A(\mathcal{P})$ and $K[G]$ are related by*

$$e(A(\mathcal{P})) = \mathrm{vol}(\mathcal{P})d! = \begin{cases} 2^{c_1 - 1}e(K[G]) & \text{if } G \text{ is non-bipartite,} \\ e(K[G]) & \text{if } G \text{ is bipartite,} \end{cases}$$

where $d = \dim(\mathcal{P})$ and c_1 is the number of non-bipartite components of G.

Proof. By Theorem 9.3.25 one has $e(A(\mathcal{P})) = \Delta_r(B)e(K[\mathcal{P}])$, where r is the rank of A. From Corollary 10.2.11 one derives

$$\Delta_r(A) = |T(\mathbb{Z}^n/\mathbb{Z}\mathcal{A})| = 2^{c_1}.$$

Thus the result follows readily from Proposition 10.2.10 and using the fact that A is totally unimodular if $c_1 = 0$. □

Lemma 10.4.5 *Let G be a connected non-bipartite graph with n vertices and let $\mathcal{B} = \{v_1, \ldots, v_n\}$ be a set of linearly independent columns of A. If $K[G]$ is normal and $0 \neq \beta \in \mathbb{R}_+\mathcal{B} \subset \mathbb{R}^n$, then β can be written as*

$$\beta = \mu_1 v_1' + \cdots + \mu_n v_n' \quad (\mu_i \geq 0),$$

such that $v_1', \ldots, v_n'$ are linearly independent columns of A defining a spanning subgraph of G which is unicyclic connected and non-bipartite.

Proof. For simplicity we identify the edges of G with the columns of A. There are nonnegative real numbers $\lambda_1, \ldots, \lambda_n$ such that

$$\beta = \lambda_1 v_1 + \cdots + \lambda_n v_n \quad (\lambda_i \geq 0). \tag{10.5}$$

Consider the subgraph H of G defined by the edges $v_1, \ldots, v_n$ and the subgraph H' of G defined by the set of edges $\{v_i | \lambda_i > 0\}$. Using that $v_1, \ldots, v_n$ are linearly independent one rapidly derives that H is a spanning subgraph of G whose connected components are unicyclic and non-bipartite (cf. [400, Lemma 9.5]). Let $H_1, \ldots, H_s$ be the components of H and Z_i the unique odd cycle of H_i, for $i = 1, \ldots, s$. First assume that H' has at most one odd cycle, say Z_1. For each cycle Z_i with $i \geq 2$ there is v_{j_i} in Z_i with $\lambda_{j_i} = 0$. Construct a graph K by removing $v_{j_1}, \ldots, v_{j_s}$ from H and then adding an edge connecting Z_1 with Z_i for each $i \geq 2$, such construction is possible thanks to Corollary 10.3.12. This new graph K is a spanning connected subgraph of G, is non-bipartite and has a unique cycle Z_1. Thus in this case Eq. (10.5) is itself the required expression for β. Next we assume that H' has at least two odd cycles Z_1, Z_2. Let v be an edge joining Z_1 with Z_2. In this case, using Remark 10.3.5, it is seen that we can rewrite

$$\beta = \mu_1 v_1' + \cdots + \mu_n v_n' \quad (\mu_i \geq 0),$$

where $v_1', \ldots, v_n'$ are linearly independent edges such that the graph H'' determined by the set $\{v_i' | \, \mu_i > 0\}$ has one cycle less than H'. Thus, by induction, the result follows. □

For connected graphs the next result was shown in [242].

Theorem 10.4.6 *If $K[G]$ is normal, then $\mathcal{P}$ has a unimodular covering with support in the vertices of $\mathcal{P}$ if and only if G has at most one non-bipartite component.*

Proof. $\Rightarrow$) By Theorem 9.3.29 $K[Ft] = A(\mathcal{P})$. Hence using Theorem 10.4.4 we obtain that the number of non-bipartite components of G is at most 1.

$\Leftarrow$) Let $G_1, \ldots, G_c$ be the connected components of the graph G and let η_i be the number of vertices of G_i.

Assume that G_1 is non-bipartite and $G_2, \ldots, G_c$ are bipartite. We denote the set of columns of the incidence matrix of G by $\mathcal{A} = \{v_1, \ldots, v_q\}$. To construct a unimodular covering of $\mathcal{P}$ consider any subgraph H of G with connected components $H_1, \ldots, H_c$ such that H_i is a spanning tree of G_i for $i > 1$ and H_1 is a unicyclic connected spanning subgraph of G_1 with a unique odd cycle. The edge polytope Δ_H of H is a simplex of dimension $d = n - c$. From Proposition 10.2.9 and 1.2.21 Δ_H is a simplex of normalized volume equal to 1.

We claim that the family of all simplices Δ_H is a unimodular covering of $\mathcal{P}$. Let $\beta \in \mathcal{P}$. By Carathéodory's Theorem (see Theorem 1.1.18) one can write (after permutation of edges) $\beta = \sum_{i=1}^{m} \lambda_i v_i$ with $\lambda_i \geq 0$, where $m = n-(c-1)$ and $v_1, \ldots, v_m$ are linearly independent. Note that $v_1, \ldots, v_m$ define a subgraph H of G. For each i there is a subgraph H_i of G_i such that $H = H_1 \cup \cdots \cup H_c$. Since $v_1, \ldots, v_m$ are linearly independent, H_i is bipartite for $i > 1$ and hence the number of edges of H_i is at most $\eta_i - 1$ for all $i > 1$. On the other hand H_1 has at most η_1 edges. Altogether the number of edges of H is at most $\eta_1 + (\eta_2 - 1) + \cdots + (\eta_c - 1)$, but since the number of edges of H is exactly this number one concludes that the number of edges of H_i (resp. H_1) is $\eta_i - 1$ (resp. η_1) for $i > 1$ (resp. $i = 1$). Hence $H_2, \ldots, H_c$ are trees. Since G_1 is connected and non-bipartite applying Lemma 10.4.5 we can rewrite β such that H_1 is a spanning subgraph of G_1 which is connected unicyclic and non-bipartite. Therefore $\beta \in \Delta_H$, as required.

If G_1 is bipartite the proof follows using similar arguments as above. Indeed is suffices to make $G_1 = \emptyset$ and proceed as above. □

Exercise

10.4.7 Let G be an arbitrary graph and let $\mathcal{P}$ be its edge polytope. Prove that $K[\mathcal{P}] = A(\mathcal{P})$ if and only if G has at most one non-bipartite connected component and this component satisfies the odd cycle condition.

10.5 Integral closure of Rees algebras

Let G be a graph with vertex set $X = \{x_1, \ldots, x_n\}$, let $R = K[x_1, \ldots, x_n]$ be a polynomial ring over a field K and let $I(G)$ be the edge ideal of G. As usual we denote the Rees algebra of $I(G)$ by $R[I(G)t]$.

Example 10.5.1 (Hochster's configuration) Let G be the graph

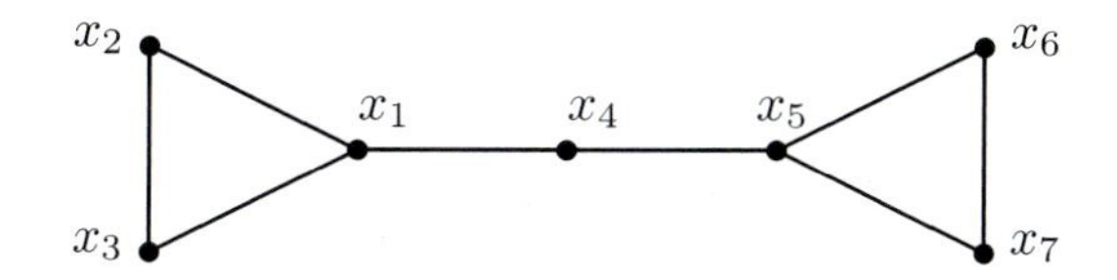

Note that $f = (x_1x_2x_3)(x_5x_6x_7)$ satisfies $f \in \overline{I(G)^3} \setminus I(G)^3$.

Definition 10.5.2 A *Hochster configuration of order k* of G consists of two odd cycles C_{2r+1} and C_{2s+1} satisfying the following conditions:

(i) $C_{2r+1} \cap N_G(C_{2s+1}) = \emptyset$ and $k = r + s + 1$.

(ii) No chord of either C_{2r+1} or C_{2s+1} is an edge of G.

Proposition 10.5.3 *If G has a Hochster configuration of order k, then* $\overline{I(G)^k} \neq I(G)^k$.

Proof. We set $I = I(G)$. The monomial obtained by multiplying all the variables in the two cycles of the configuration is an element of $\overline{I^k} \setminus I^k$. Indeed, let

$$\delta = z_1 \cdots z_{2r+1} \text{ and } \gamma = w_1 \cdots w_{2s+1}$$

be the monomials corresponding to the two cycles in the configuration. By definition, we have $r + s + 1 = k$. Therefore $\delta\gamma \in I^{k-1} \setminus I^k$. On the other hand $\delta^2 \in I^{2r+1}$ and $\gamma^2 \in I^{2s+1}$, thus $(\delta\gamma)^2 \in I^{2(r+s+1)} = I^{2k}$. This shows that $\delta\gamma \in \overline{I^k}$. $\square$

Definition 10.5.4 The *cone* $C(G)$, over the graph G, is obtained by adding a new vertex t to G and joining every vertex of G to t.

Proposition 10.5.5 *$R[I(G)t]$ is isomorphic to $K[C(G)]$.*

Proof. Let $R[I(G)t] = K[\{x_1, \ldots, x_n, tf_i | 1 \leq i \leq q\}]$ be the Rees algebra of $I(G) = (f_1, \ldots, f_q)$, where $f_1, \ldots, f_q$ are the monomials corresponding to the edges of G. Let

$$K[C(G)] = K[\{tx_i, f_j | 1 \leq i \leq n, 1 \leq j \leq q\}]$$

be the monomial subring of the cone. $R[I(G)t]$ and $K[C(G)]$ are integral domains of the same dimension by Corollary 8.2.21 and Proposition 10.1.20. It follows that there is an isomorphism

$$\varphi \colon R[I(G)t] \longrightarrow K[C(G)], \text{ induced by } \varphi(x_i) = tx_i \text{ and } \varphi(tf_i) = f_i.$$

That is $R[I(G)t] \overset{\varphi}{\simeq} K[C(G)]$. $\square$

Corollary 10.5.6 *Let G be a connected graph and $I = I(G)$ its edge ideal. Then $K[G]$ is normal if and only if the Rees algebra $R[It]$ is normal.*

Proof. $\Rightarrow$) Let $C(G)$ be the cone over G. By Proposition 10.5.5 $R[I(G)t]$ is isomorphic to $K[C(G)]$. Let M_w be a bow-tie of $C(G)$ with edge disjoint cycles Z_1 and Z_2 joined by a path P. It is enough to verify that M_w is in $K[C(G)]$. If $t \notin Z_1 \cup Z_2 \cup P$, then w is a bow-tie of G and $M_w \in K[G]$. Assume $t \in Z_1 \cup Z_2$, say $t \in Z_1$. If $Z_1 \cap Z_2 \neq \emptyset$, then $M_w \in K[C(G)]$. On the other hand if $Z_1 \cap Z_2 = \emptyset$, then $M_w \in K[C(G)]$ because in this case Z_1 and Z_2 are joined by the edge $\{t, z\}$, where z is any vertex in Z_2. It remains to consider the case $t \notin Z_1 \cup Z_2$ and $t \in P$, since G is connected there is a path in G joining Z_1 with Z_2. Therefore $M_w = M_{w_1}$ for some bow-tie w_1 of G and $M_w \in K[G]$.

$\Leftarrow$) Conversely if $R[I(G)t]$ is normal, then by Proposition 4.3.42 we obtain that $K[G]$ is normal. $\square$

Example 10.5.7 Let $F = \{x_1x_2, x_3x_4x_5, x_1x_3, x_2x_4, x_2x_5, x_1x_5\}$ and let $I = (F)$. Using *Normaliz* [68] we get that $R[It]$ is normal but $K[F]$ is not.

Proposition 10.5.8 *Let G be a graph. Then $R[I(G)t]$ is normal if and only if either* (i) *G is bipartite, or* (ii) *exactly one of the components, say G_1, is non-bipartite and $K[G_1]$ is a normal domain.*

Proof. It follows adapting the proof of Corollary 10.5.6. □

Corollary 10.5.9 [383] *The edge ideal of a graph G is normal if and only if G admits no Hochster configurations.*

Remark 10.5.10 (a) Let G be a graph consisting of two disjoint triangles. Note that G itself is a Hochster configuration but G is not contained in a bow-tie, since the two triangles cannot be joined by a path.

(b) If two disjoint odd cycles Z_1, Z_2 of a graph G are in the same connected component, then the two cycles form a Hochster configuration if and only if Z_1, Z_2 cannot be connected by an edge of G and Z_1, Z_2 have no chords in G

Integral closure of Rees algebras Consider the endomorphism φ of the field $K(x_1, \ldots, x_n, t)$ defined by $x_i \mapsto x_it$, $t \mapsto t^{-2}$. If $I = I(G)$ is the edge ideal of G, then by Proposition 10.5.5 it induces an isomorphism

$$R[It] \stackrel{\varphi}{\longmapsto} K[C(G)], \quad x_i \mapsto tx_i, \quad x_ix_jt \mapsto x_ix_j.$$

We thank Wolmer Vasconcelos for showing us how to get the description of the integral closure of $R[It]$ given below. First we describe the integral closure of $K[C(G)]$.

If $Z_1 = \{x_1, x_2, \ldots, x_r = x_1\}$ and $Z_2 = \{z_1, z_2, \ldots, z_s = z_1\}$ are two edge disjoint odd cycles in $C(G)$. Then one has

$$M_w = x_1 \cdots x_r z_1 \cdots z_s$$

for some bow-tie w of $C(G)$, this follows readily because $C(G)$ is a connected graph. In particular M_w is in the integral closure of $K[C(G)]$. Observe that if t occurs in Z_1 or Z_2, then $M_w \in K[C(G)]$ because in this case either Z_1 and Z_2 meet at a point, or Z_1 and Z_2 are joined by an edge in $C(G)$; see Proposition 10.3.10. Thus in order to compute the integral closure of $K[C(G)]$ the only bow-ties that matter are those defining the set:

$$\mathcal{B} = \{M_w |\, w \text{ is a bow-tie in } C(G) \text{ such that } t \notin \operatorname{supp}(M_w)\}.$$

Therefore by Theorem 10.3.9 we have proved:

Proposition 10.5.11 $\overline{K[C(G)]} = K[C(G)][\mathcal{B}]$.

To obtain a description of the integral closure of the Rees algebra of I note that using φ together with the equality

$$(tx_1)M_w = (tz_1)\prod_{i=1}^{\frac{r+1}{2}}(x_{2i-1}x_{2i})\prod_{i=1}^{\frac{s-1}{2}}(z_{2i}z_{2i+1}),$$

where $x_{r+1} = x_1$, we see that M_w is mapped back in the Rees algebra in the element

$$M_w t^{\frac{r+s}{2}} = x_1x_2\cdots x_r z_1 z_2\cdots z_s t^{\frac{r+s}{2}}.$$

If $\mathcal{B}'$ is the set of all monomials of this form, then one obtains:

Proposition 10.5.12 $\overline{R[It]} = R[It][\mathcal{B}']$.

Exercises

10.5.13 Let G be a graph. Prove that $I(G)^2$ is integrally closed.

10.5.14 If G is a graph consisting of two disjoint triangles, then $K[G]$ is normal and the Rees algebra of $I(G)$ is not normal.

10.5.15 Let G be a graph. If H is the whisker graph of G or the cone of G and $I(G)$ is normal, then $I(H)$ is normal.

10.5.16 Let G be a graph and let $m = (r+s)/2$ be the least positive integer such that G has two vertex disjoint odd cycles of lengths r and s. If $I = I(G)$ is the edge ideal of G, prove that I^i is integrally closed for $i < m$.

10.5.17 If G is any non-discrete graph with n vertices and $C(G)$ its cone, prove that the edge subring $K[C(G)]$ has dimension $n+1$.

10.5.18 Let G be a simple graph with vertex set $X = \{x_1,\ldots,x_n\}$ and let $I \subset R = K[X]$ be its edge ideal. If φ, ψ are the maps of K-algebras defined by the diagram

$$\begin{array}{ccc} R[t_1,\ldots,t_q] & \xrightarrow{\ \varphi\ } & R[It] \\ \downarrow{\scriptstyle\psi} & & \\ K[C(G)] & & \end{array} \qquad \begin{array}{ccc} t_i & \overset{\varphi}{\dashrightarrow} & Tf_i \\ \downarrow{\scriptstyle\psi} & & \\ f_i & & \end{array} \qquad \begin{array}{ccc} x_i & \overset{\varphi}{\dashrightarrow} & x_i \\ \downarrow{\scriptstyle\psi} & & \\ xt_i & & \end{array}$$

where $C(G)$ is the cone of G and $f_1,\ldots,f_q$ are the edge generators, then $\ker(\varphi) = \ker(\psi)$.

10.5.19 Let $R = K[x_1,x_2,x_3,y_1,y_2,y_3,y_4]$ and $I = I_1J_3 + I_3J_1$, where I_r (resp. J_r) denote the ideal of R generated by the square-free monomials of degree r in the x_i variables (resp. y_i variables). If $z = x_1x_2x_3y_1^2y_2y_3y_4$, then $z^2 \in I^4$ and $z \in \overline{I^2} \setminus I^2$.

10.5.20 Let G be a graph. Explain the differences between a circuit of G and a Hochster configuration of G.

10.5.21 Let G be a bipartite graph and $I(G)$ its edge ideal. Prove that the toric ideal of the Rees algebra of $I(G)$ has a universal Gröbner basis consisting of square-free binomials.

10.6 Edge subrings of complete graphs

In this section we treat two special types of edge subrings, attached to complete graphs, and compute their Hilbert series and a-invariant.

Edge subrings of bipartite graphs One result that simplifies the study of an edge subring associated to a complete bipartite graph is the fact that its toric ideal is a determinantal prime ideal. Let us begin with a classical formula for determinantal ideals.

Theorem 10.6.1 [73, Theorem 2.5] *Let $T = (t_{ij})$ be an $m \times n$ matrix of indeterminates over a field K and $I_r(T)$ the ideal generated by the r-minors of T. If $m \leq n$ and $1 \leq r \leq m+1$, then*

$$\text{height}(I_r(T)) = (m-r+1)(n-r+1).$$

Proposition 10.6.2 *Let K be a field and let $\mathcal{K}_{m,n}$ be the complete bipartite graph. Then, the toric ideal P of $K[\mathcal{K}_{m,n}]$ is generated by the 2×2 minors of a generic $m \times n$ matrix.*

Proof. Let $V_1 = \{x_1, \ldots, x_m\}$ and $V_2 = \{y_1, \ldots, y_n\}$ be the bipartition of $\mathcal{K}_{m,n}$ and let

$$S = K[\{t_{ij} | \, 1 \leq i \leq m, \ 1 \leq j \leq n\}]$$

be a polynomial ring over a field K in the indeterminates t_{ij} and consider the graded homomorphism of K-algebras

$$\varphi \colon S \longrightarrow K[\mathcal{K}_{m,n}] = K[x_i y_j\text{'s}], \quad \varphi(t_{ij}) = x_i y_j.$$

We claim that the kernel of φ is generated by the 2×2 minors of a generic $m \times n$ matrix. We set $T = (t_{ij})$. It is clear that the ideal $I_2(T)$, generated by the 2×2 minors of T, is contained in $P = \ker(\varphi)$. On the other hand, since the graph is bipartite, one has that the dimension of $K[\mathcal{K}_{m,n}]$ is equal to $m+n-1$; see Corollary 10.1.21. Therefore

$$\text{height}(P) = (mn) - (m+n-1) = (m-1)(n-1) = \text{height}(I_2(T)),$$

the last equality uses Theorem 10.6.1. Since they are both prime ideals, we have $I_2(T) = P$. □

By Proposition 10.3.1 the ring $K[\mathcal{K}_{m,n}]$ is normal and Cohen–Macaulay. The Gorensteiness of $K[\mathcal{K}_{m,n}]$ has been dealt with in great detail in [71].

Proposition 10.6.3 [418] *Let S/P be the presentation of the edge subring $K[\mathcal{K}_{m,n}]$. If $n \geq m$, then the Hilbert series of S/P is given by*

$$F(S/P,z) = \frac{\sum_{i=0}^{m-1} \binom{m-1}{i}\binom{n-1}{i} z^i}{(1-z)^{n+m-1}}.$$

Corollary 10.6.4 *If $n \geq m$, then the a-invariant and the multiplicity of $K[\mathcal{K}_{m,n}]$ are given by:*

$$a(K[\mathcal{K}_{m,n}]) = -n \quad \text{and} \quad e(K[\mathcal{K}_{m,n}]) = \binom{m+n-2}{m-1}.$$

Proof. It follows using Remark 5.1.7. □

Proposition 10.6.5 *Let G be a spanning connected subgraph of $\mathcal{K}_{m,n}$. If $n \geq m$, then the a-invariant of $K[G]$ satisfies*

$$a(K[G]) \leq a(K[\mathcal{K}_{m,n}]) = -n.$$

Proof. From Corollary 10.6.4 one has $a(K[\mathcal{K}_{m,n}]) = -n$. Hence one may proceed as in the proof of Proposition 12.4.13 to rapidly derive the asserted inequality. □

An application of Dedekind–Mertens formula If R is a commutative ring and $f = f(t) \in R[t]$ is a polynomial, say

$$f = a_0 + \cdots + a_m t^m,$$

the *content* of f is the R-ideal $(a_0, \ldots, a_m)$. It is denoted by $c(f)$. Given another polynomial g, the *Gaussian ideal* of f and g is the R-ideal

$$c(fg).$$

This ideal bears a close relationship to the ideal $c(f)c(g)$, one aspect of which is expressed in the classical lemma of Gauss: If R is a principal ideal domain, then

$$c(fg) = c(f)c(g).$$

In general, these two ideals are different but one aspect of their relationship is given by a formula due to Dedekind–Mertens (see [126] and [334]):

$$c(fg)c(g)^m = c(f)c(g)^{m+1}. \tag{10.6}$$

We consider the ideal $c(fg)$ in the case when f, g are generic polynomials. It turns out that some aspects of the theory of Cohen–Macaulay rings show up very naturally when we closely examine $c(fg)$.

One path to our analysis, and its applications to Noether normalizations, starts by multiplying both sides of (10.6) by $c(f)^m$; we get

$$c(fg)[c(f)c(g)]^m = c(f)c(g)[c(f)c(g)]^m. \tag{10.7}$$

It will be shown that this last formula is sharp in terms of the exponent $m = \deg(f)$.

To make this connection, we recall the notion of a *reduction* of an ideal. Let R be a ring and I an ideal. A *reduction* of I is an ideal $J \subset I$ such that, for some nonnegative integer r, the equality

$$I^{r+1} = JI^r$$

holds. The smallest such integer is the *reduction number* $r_J(I)$ of I *relative* to J, and the *reduction number* $r(I)$ of I is the smallest reduction number among all reductions J of I.

Thus (10.7) says that $J = c(fg)$ is a reduction for $I = c(f)c(g)$, and that the reduction number is at most $\min\{\deg(f), \deg(g)\}$.

The following result solves the question of Noether normalizations for monomial subrings of complete bipartite graphs.

Theorem 10.6.6 [126] *Let $R = K[x_0, \dots, x_m, y_0, \dots, y_n]$ be a polynomial ring over a field K and let $h_q = \sum_{i+j=q} x_i y_j$. Then*

$$A = K[h_0, h_1, \dots, h_{m+n}] \hookrightarrow S = K[\{x_i y_j | \, 0 \leq i \leq m, \ 0 \leq j \leq n\}]$$

is a Noether normalization of S.

Proof. Let $R[t]$ be a polynomial ring in a new variable t and $f, g \in R[t]$ the generic polynomials

$$f = f(t) = \sum_{i=0}^{m} x_i t^i \quad \text{and} \quad g = g(t) = \sum_{j=0}^{n} y_j t^j.$$

Set $I = c(f)c(g)$ and $J = c(fg)$. Note that J is a reduction of I by Eq. (10.7) and more precisely $JI^m = I^{m+1}$. Therefore

$$R[Jt] \hookrightarrow R[It]$$

is an integral extension; indeed $R[It]$ is generated as an $R[Jt]$-module by a finite set of elements of t-degree at most m. On the other hand, as I and J are generated by homogeneous polynomials of the same degree, one has an integral embedding:

$$R[Jt] \otimes_R R/\mathfrak{m} = K[h_0, \dots, h_{m+n}] \hookrightarrow R[It] \otimes R/\mathfrak{m} = K[x_i y_j\text{'s}],$$

where $\mathfrak{m} = (x_0 \dots, x_m, y_0, \dots, y_n)$. To complete the proof note that the dimension of $K[x_i y_j\text{'s}]$ is equal to $n + m + 1$, by Corollary 10.1.21. $\square$

Theorem 10.6.7 [97]*Let $R = K[x_0, \ldots, x_m, y_0, \ldots, y_n]$ be a polynomial ring over a field K and let $f, g \in R[t]$ be the generic polynomials*

$$f = \sum_{i=0}^{m} x_i t^i \quad \text{and} \quad g = \sum_{j=0}^{n} y_j t^j$$

with $n \geq m$. If $I = c(f)c(g)$ and $J = c(fg)$, then $r_J(I) = m$.

Proof. By Theorem 10.6.6 there is a Noether normalization:

$$A = K[h_0, h_1, \ldots, h_{m+n}] \hookrightarrow S = K[x_i y_j\text{'s}],$$

where $h_q = \sum_{i+j=q} x_i y_j$. Set $I = (f_1, \ldots, f_s)$. We note that the ideal $I = (x_i y_j\text{'s})$ is the edge ideal associated to a complete bipartite graph G which is the join of two discrete graphs, one with $m+1$ vertices and another with $n+1$ vertices. Since $K[G]$ is Cohen–Macaulay (see Corollary 10.3.12), using Proposition 3.1.27 we get

$$K[G] = Af^{\beta_1} \oplus \cdots \oplus Af^{\beta_k}, \;\; \beta_i \in \mathbb{N}^s.$$

By Corollary 10.6.4 the a-invariant of $K[G]$ is $-n-1$, we can compute the Hilbert series of $K[G]$ using the decomposition above to get that

$$\max_{1 \leq i \leq k} \{\deg(f^{\beta_i})\} = m,$$

where the degree is taken with respect to the normalized grading.

On the other hand assume $r = r_J(I)$ and $JI^r = I^{r+1}$, since we already know the inequality $r \leq m$. It suffices to prove $r \geq m$. Note

$$K[G] = \sum_{\alpha \in \mathcal{I}} Af^{\alpha},$$

where $\mathcal{I} = \{\alpha \mid \alpha = (v_1, \ldots, v_s) \in \mathbb{N}^s \text{ and } |\alpha| = \sum_{i=1}^{s} v_i \leq r\}$. If $r < m$, this rapidly leads to a contradiction because, by Exercise 10.6.14, there is $\mathcal{J} \subset \mathcal{I}$ such that $K[G]$ can be written as

$$K[G] = \bigoplus_{\alpha \in \mathcal{J}} Af^{\alpha}. \qquad \square$$

Edge subrings of complete graphs Let $R = K[x_1, \ldots, x_n]$ be a ring of polynomials over a field K and let $\mathcal{K}_n$ be the complete graph on the vertex set $X = \{x_1, \ldots, x_n\}$. For $i < j$, we set $f_{ij} = x_i x_j$. Let

$$K[\mathcal{K}_n] = K[\{f_{ij} | \, 1 \leq i < j \leq n\}]$$

be the K-subring of R spanned by the f_{ij}. Consider the homomorphism

$$\psi\colon A_n = K[\{t_{ij} | \, 1 \leq i < j \leq n\}] \to K[\mathcal{K}_n], \;\; \text{induced by } t_{ij} \to f_{ij}.$$

The ideal $P_n = \ker(\psi)$ is the *presentation ideal* or *toric ideal* of $K[\mathcal{K}_n]$.

Proposition 10.6.8 [418] *If the terms in A_n are ordered lexicographically by $t_{12} \prec \cdots \prec t_{1n} < t_{23} \prec \cdots \prec t_{2n} \prec \cdots \prec t_{(n-1)n}$, then the set*

$$\mathcal{B} = \{t_{ij}t_{k\ell} - t_{i\ell}t_{jk},\, t_{ik}t_{j\ell} - t_{i\ell}t_{jk} \,|\, 1 \leq i < j < k < \ell \leq n\}.$$

is a minimal generating set for P_n and $\mathcal{B}$ is a reduced Gröbner basis for P_n.

Example 10.6.9 If $K[\{t_{ij} \,|\, 1 \leq i < j \leq 5\}]$ has the lex ordering

$$t_{45} \prec t_{34} \prec t_{35} \prec t_{23} \prec t_{24} \prec t_{25} \prec t_{12} \prec t_{13} \prec t_{14} \prec t_{15}.$$

Then the reduced Gröbner basis for the toric ideal P of $K[\mathcal{K}_5]$ is equal to:

$$\begin{array}{lll} t_{35}t_{24} - t_{45}t_{23}, & t_{34}t_{25} - t_{45}t_{23}, & t_{45}t_{23}t_{13} - t_{34}t_{35}t_{12}, \\ t_{25}t_{13} - t_{35}t_{12}, & t_{35}t_{14} - t_{45}t_{13}, & t_{23}t_{14} - t_{34}t_{12}, \\ t_{34}t_{15} - t_{45}t_{13}, & t_{23}t_{15} - t_{35}t_{12}, & t_{24}t_{15} - t_{45}t_{12}. \\ t_{24}t_{13} - t_{34}t_{12} & t_{25}t_{14} - t_{45}t_{12}. & \end{array}$$

Theorem 10.6.10 [418] *Let $\mathcal{K}_n$ be the complete graph on n vertices and A_n/P_n be the presentation of $K[\mathcal{K}_n]$. If $n \geq 3$, then the Hilbert series $F_n(z)$ of A_n/P_n satisfies:*

$$(1-z)^n F_n(z) = 1 + \frac{n(n-3)}{2} z + \sum_{m=2}^{\lfloor \frac{n}{2} \rfloor} \binom{n}{2m} z^m$$

Corollary 10.6.11 *The multiplicity $e(K[\mathcal{K}_n])$, of the ring $K[\mathcal{K}_n]$, is equal to $2^{n-1} - n$.*

Theorem 10.6.12 *Let G be a connected graph with n vertices and q edges. Then the following are sharp upper bounds for the multiplicity of $K[G]$:*

$$e(K[G]) \leq \begin{cases} 2^{n-1} - n, & \text{if } G \text{ is non-bipartite,} \\ \binom{m+r-2}{m-1}, & \text{if } G \text{ is a spanning subgraph of } \mathcal{K}_{m,r}. \end{cases}$$

Proof. It follows from Corollary 10.6.4 and Corollary 10.6.11. □

It is easy to produce examples where those bounds are very lose. The extreme cases being the unicyclic connected non-bipartite graphs and the trees, because in both cases the multiplicity of $K[G]$ is equal to 1. In our case adding edges to G increases the volume of the edge polytope (see Lemma 9.3.16); hence the multiplicity is an increasing function of the edges. For additional information on the multiplicity of edge subrings of bipartite planar graphs, see [189] and the references therein.

Exercises

10.6.13 Let $R[t]$ be a polynomial ring over a ring R. If $f = f(t)$ and $g = g(t)$ are two polynomials in $R[t]$, prove the Dedekind–Mertens formula

$$c(fg)c(g)^m = c(f)c(g)^{m+1} \quad (m = \deg(f)).$$

10.6.14 Let M be a graded free module over a polynomial ring A such that $M = \oplus_{i=1}^r Af_i$, where f_i is homogeneous for all i. If M is generated by homogeneous elements $g_1, \ldots, g_s$, then $M = \oplus_{j=1}^r Ag_{i_j}$, for some $i_1, \ldots, i_r$.

10.6.15 Let $A = K[x_1, \ldots, x_m]$ and $R = K[y_1, \ldots, y_n]$ be two polynomial rings over a field K, prove that the Segre product

$$(A_0 \otimes_K R_0) \oplus (A_1 \otimes_K R_1) \oplus \cdots \subset A \otimes_K R$$

is isomorphic to $S = K[x_iy_j\text{'s}] \subset K[x_i\text{'s}, y_j\text{'s}]$.

10.6.16 Let $R = K[x_1, \ldots, x_n]$ be a polynomial ring over a field K and $\mathfrak{m} = (x_1, \ldots, x_n)$. Prove that the toric ideal of the Rees algebra $R[\mathfrak{m}t]$ is the ideal generated by the 2-minors of the following matrix of indeterminates:

$$X = \begin{pmatrix} x_1 & x_2 & \cdots & x_n \\ t_1 & t_2 & \cdots & t_n \end{pmatrix}.$$

10.6.17 Let $R = K[x_1, \ldots, x_n]$ be a polynomial ring over a field K and $I = (x_1, \ldots, x_n)^k$. Set $s = -\lceil n/k \rceil + n$, and write $n = qk + r$, for some $q, r \in \mathbb{N}$ such that $0 \leq k < r$. By a "counting degrees" argument prove:

$$\begin{aligned} x_1^{k-1} \cdots x_n^{k-1} &\in I^s \qquad \text{if } r = 0, \\ x_1^{k-1} \cdots x_q^{k-1} x_{q+1}^{r-1} x_{q+2}^{k-1} \cdots x_n^{k-1} &\in I^s \qquad \text{if } r > 0. \end{aligned}$$

10.6.18 Let $R = K[x_1, \ldots, x_n]$ be a polynomial ring over a field K and $\mathfrak{m} = (x_1, \ldots, x_n)$. If $I = \mathfrak{m}^k$ and $J = (x_1^k, \ldots, x_n^k)$, then J is a reduction of I and the reduction number of I relative to J is given by

$$r_J(I) = -\left\lceil \frac{n}{k} \right\rceil + n, \quad \text{where } \lceil x \rceil \text{ is the ceiling of } x.$$

10.6.19 Let S be a standard algebra over an infinite field K and

$$A = K[h_1, \ldots, h_d] \hookrightarrow S$$

a Noether normalization of S with $h_i \in S_1$ for all i. If $b_1, \ldots, b_t$ is a minimal set of homogeneous generators of S as an A-module, prove

$$(h_1, \ldots, h_d)S_r = S_{r+1},$$

where $r = \max_i\{\deg(b_i)\}$, and that $r \geq 0$ is the minimum integer where equality occurs. The *reduction number* of S, denote by $r(S)$, is the minimum r taken over all Noether normalizations. See [413, Chapter 9] for a study of several degrees of complexity associated to a graded module.

10.6.20 Let S be a standard algebra over an infinite field K and $a(S)$ its a-invariant. If S is Cohen–Macaulay, then $r(S) = a(S) + \dim(S)$.

10.6.21 Let $\phi\colon B \to S$ be a graded epimorphism of standard K-algebras, where K is an infinite field. If $\dim(B) = \dim(S)$ and

$$A = K[h_1, \ldots, h_d] \stackrel{\imath}{\hookrightarrow} B$$

is a Noether normalization of B with $h_i \in S_1$ for all i, prove that S is integral over $A' = K[\phi(h_1), \ldots, \phi(h_d)]$ and that the composite

$$A = K[h_1, \ldots, h_d] \stackrel{\imath}{\hookrightarrow} B \stackrel{\phi}{\longrightarrow} S$$

is a Noether normalization of S.

10.6.22 Let $\phi\colon B \to S$ be a graded epimorphism of standard K-algebras, where K is an infinite field. If $\dim(B) = \dim(S)$, prove $r(B) \geq r(S)$.

10.7 Edge cones of graphs

In this section we give explicit combinatorial descriptions of the edge cone of a graph. In Chapter 11 these descriptions will be used to compute the a-invariant and the canonical module of an edge subring. The main tools used to show these descriptions are linear algebra (Farkas's Lemma, incidence matrices of graphs, Carathéodory's Theorem), graph theory and polyhedral geometry (finite basis theorem, facet structure of polyhedra).

First we fix the notation that will be used throughout this section and adapt the terminology and notation of Chapter 1 for the case of edge cones.

Let G be a simple graph with vertex set $V = \{x_1, \ldots, x_n\}$ and edge set $E = \{f_1, \ldots, f_q\}$. Every edge f_i is an unordered pair of distinct vertices $f_i = \{x_{i_j}, x_{i_k}\}$. If $f_i = \{x_{i_j}, x_{i_k}\}$ define $v_i = e_{i_j} + e_{i_k}$, where e_i is the ith unit vector in $\mathbb{R}^n$. The incidence matrix of G, denoted by A, is the matrix of size $n \times q$ whose columns are precisely the vectors $v_1, \ldots, v_q$.

We set $\mathcal{A}_G$ (or simply $\mathcal{A}$ if G is understood) equal to the set $\{v_1, \ldots, v_q\}$ of row vectors of the transpose of the incidence matrix of G. Since v_i represents an edge of G, sometimes v_i is called an edge or an *edge vector*. The *edge cone* of G, denoted by $\mathbb{R}_+\mathcal{A}$, is the cone generated by $\mathcal{A}$. Notice that $\mathbb{R}_+\mathcal{A} \neq (0)$ if G is not a discrete graph. By Corollary 10.2.11 one has

$$n - c_0(G) = \text{rank }(A) = \dim \mathbb{R}_+\mathcal{A},$$

where $c_0(G)$ is the number of bipartite components of G.

Lemma 10.7.1 *If x_i is not an isolated vertex of the graph G, then the set $F = H_{e_i} \cap \mathbb{R}_+\mathcal{A}$ is a proper face of the edge cone.*

Proof. Note $F \neq \emptyset$ because $0 \in F$, and $\mathbb{R}_+\mathcal{A} \subset H_{e_i}^+$. Since x_i is not an isolated vertex $\mathbb{R}_+\mathcal{A} \not\subset H_{e_i}$. □

Let $\mathfrak{A}$ be an independent set of vertices of G. The supporting hyperplane of the edge cone defined by

$$\sum_{x_i \in \mathfrak{A}} x_i = \sum_{x_i \in N(\mathfrak{A})} x_i$$

will be denoted by $H_{\mathfrak{A}}$, where $N(\mathfrak{A})$ is the neighbor set of $\mathfrak{A}$. The sum over an empty set is defined to be zero.

Lemma 10.7.2 *If $\mathfrak{A}$ is an independent set of vertices of the graph G and $F = \mathbb{R}_+\mathcal{A} \cap H_{\mathfrak{A}}$, then either F is a proper face of the edge cone or $F = \mathbb{R}_+\mathcal{A}$.*

Proof. It suffices to prove $\mathbb{R}_+\mathcal{A} \subset H_{\mathfrak{A}}^-$. Take an edge $\{x_j, x_\ell\}$ of G. If $\{x_j, x_\ell\} \cap \mathfrak{A} \neq \emptyset$, then $e_j + e_\ell$ is in $H_{\mathfrak{A}}$, else $e_j + e_\ell$ is in $H_{\mathfrak{A}}^-$. □

Definition 10.7.3 The *support* of a vector $\beta = (\beta_i) \in \mathbb{R}^n$ is defined as

$$\text{supp}(\beta) = \{\beta_i \mid \beta_i \neq 0\}.$$

Lemma 10.7.4 [419] *Let $G_1, \ldots, G_r$ be the components of G. If G_1 is a tree with at least two vertices and $G_2, \ldots, G_r$ are unicyclic non-bipartite graphs, then* $\ker(A_G^\top) = (\beta)$ *for some $\beta \in \mathbb{R}^n$ with* $\text{supp}(\beta) = \{1, -1\}$ *such that $V(G_1) = \{v_i \in V \mid \beta_i = \pm 1\}$.*

Proof. If G is a tree, using induction on the number of vertices it is not hard to show that $\ker(A^\top) = (\beta)$, where $\text{supp}(\beta) = \{1, -1\}$. On the other hand if $G = G_i$ for some $i \geq 2$, then $A^\top$ is non-singular and $\ker(A^\top) = (0)$. The general case follows readily by writing $A^\top$ as a "diagonal" matrix

$$A^\top = \text{diag}(A_{G_1}^\top, \ldots, A_{G_m}^\top),$$

where A_{G_i} is the incidence matrix of the graph G_i. Observe the equality $V(G_1) = \{x_i \in V \mid \beta_i = \pm 1\}$. □

To determine whether $H_{\mathfrak{A}}$ defines a facet of $\mathbb{R}_+\mathcal{A}$ consider the subgraph $L = L_1 \cup L_2$, where L_1 is the subgraph of G with vertex set and edge set

$$V(L_1) = \mathfrak{A} \cup N(\mathfrak{A}) \text{ and } E(L_1) = \{z \in E \mid z \cap \mathfrak{A} \neq \emptyset\}$$

respectively, and $L_2 = G[S]$ is the subgraph of G induced by $S = V \setminus V(L_1)$. The vectors in $\mathcal{A} \cap H_{\mathfrak{A}}$ correspond precisely to the edges of L.

Remark 10.7.5 Let F be a facet of $\mathbb{R}_+\mathcal{A}$ defined by the hyperplane H_a. If the rank of A is n, then it is not hard to see that H_a is generated by a linearly independent subset of $\mathcal{A}$. See Proposition 1.1.23.

Theorem 10.7.6 [419] *If* $\mathrm{rank}(A) = n$*, then* F *is a facet of* $\mathbb{R}_+\mathcal{A}$ *if and only if either*

(*a*) $F = H_{e_i} \cap \mathbb{R}_+\mathcal{A}$ *for some* i*, where all the connected components of* $G \setminus \{x_i\}$ *are non-bipartite graphs, or*

(*b*) $F = H_{\mathfrak{A}} \cap \mathbb{R}_+\mathcal{A}$ *for some independent set* $\mathfrak{A} \subset V$ *such that* L_1 *is a connected bipartite graph, and the connected components of* L_2 *are non-bipartite graphs.*

Proof. $\Rightarrow$) Assume that F is a facet of $\mathbb{R}_+\mathcal{A}$. By Remark 10.7.5 we may assume that there is $a \in \mathbb{R}^n$ such that

(i) $F = \mathbb{R}_+\mathcal{A} \cap H_a$, $\langle a, v_i\rangle \leq 0\ \forall\, i \in \{1, \ldots, q\}$, and

(ii) $v_1, \ldots, v_{n-1}$ are linearly independent vectors in H_a.

Consider the subgraph D of G whose edges correspond to $v_1, \ldots, v_{n-1}$ and its vertex set is the union of the vertices in the edges of D. Let Γ be the transpose of the incidence matrix of D, and $D_1, \ldots, D_r$ the connected components of D. Without loss of generality one may assume

$$\Gamma = \begin{bmatrix} \Gamma_1 & 0 & \ldots & 0 \\ 0 & \Gamma_2 & \ldots & 0 \\ \vdots & \vdots & \ddots & \vdots \\ 0 & 0 & \ldots & \Gamma_r \end{bmatrix}, \quad \mathrm{rank}(\Gamma) = n - 1,$$

where Γ_i is the transpose of the incidence matrix of D_i. We set n_i and q_i equal to the number of vertices and edges of D_i, respectively. Let Θ be the matrix with rows $v_1, \ldots, v_{n-1}$. We consider two cases.

Case (I): Assume Θ has a zero column. Hence it has exactly one zero column and D has vertex set $V(D) = V \setminus \{x_i\}$, for some i. This implies that $\sum_{i=1}^r n_i = \sum_{i=1}^r q_i = n - 1$. Since $n_i \leq q_i$ for all i, one obtains $n_i = q_i$ for all i. Note that $c_0(D) = 0$, by Lemma 10.2.6. Hence $D_1, \ldots, D_r$ are unicyclic non-bipartite graphs. As any component of $G \setminus \{x_i\}$ contains a component of D, we are in case (a).

Case (II): Assume that all the columns of Θ are non-zero, that is, D has vertex set equal to V. Note that in this case

$$n = \sum_{i=1}^r n_i = 1 + \sum_{i=1}^r q_i. \tag{10.8}$$

Since rank $(\Gamma) = n - c_0(D) = n - 1$, we get that D has exactly one bipartite component, say D_1. Hence $n_1 - 1 \leq q_1$ and $n_i \leq q_i$ for all $i \geq 2$, which together with Eq. (10.8) yields $q_1 = n_1 - 1$ and $n_i = q_i$ for all $i \geq 2$. Altogether D_1 is a tree and $D_2, \ldots, D_r$ are unicyclic non-bipartite graphs.

By Lemma 10.7.4, $\ker(\Gamma) = (\beta)$ and $\operatorname{supp}(\beta) \subset \{1, -1\}$ for some $\beta \in \mathbb{R}^n$. We may harmlessly assume $a = \beta$, because $H_a = H_\beta$ and $a \in (\beta)$. Set

$$\mathfrak{A} = \{x_i \in V \mid a_i = 1\} \text{ and } B = \{x_i \in V \mid a_i = -1\}.$$

Note that $\mathfrak{A} \neq \emptyset$, because D has no isolated vertices and its vertex set is V. We will show that $\mathfrak{A}$ is an independent set in G and $B = N(\mathfrak{A})$, where the neighbor set of $\mathfrak{A}$ is taken w.r.t G.

On the contrary if $\{x_i, x_j\}$ is an edge of G for some x_i, x_j in $\mathfrak{A}$, then $v_k = e_i + e_j$ and by (i) we get $\langle a, v_k\rangle \leq 0$, which is impossible because $\langle a, v_k\rangle = 2$. This proves that $\mathfrak{A}$ is an independent set.

Next we show $N(\mathfrak{A}) = B$. If $x_i \in N(\mathfrak{A})$, then $v_k = e_i + e_j$ for some x_j in $\mathfrak{A}$, using (i) we obtain $\langle a, v_k\rangle = a_i + 1 \leq 0$ and $a_i = -1$, hence $x_i \in B$. Conversely if $x_i \in B$, since D has no isolated vertices, there is $1 \leq k \leq n-1$ so that $v_k = e_i + e_j$, for some j, by (ii) we obtain $\langle a, v_k\rangle = -1 + a_j = 0$, which shows that $x_j \in \mathfrak{A}$ and $x_i \in N(\mathfrak{A})$.

From the proof of Lemma 10.7.4 we obtain $V(D_1) = \mathfrak{A} \cup N(\mathfrak{A})$. Therefore $v_1, \ldots, v_{n-1}$ are in $H_{\mathfrak{A}}$ and $H_a = H_{\mathfrak{A}}$. Observe that L_1 (see notation before Remark 10.7.5) cannot contain odd cycles because $\mathfrak{A}$ is independent and every edge of L_1 contains a vertex in $\mathfrak{A}$. Hence L_1 is bipartite, that L_1 is connected follows from the fact that D_1 is a spanning tree of L_1. Note that every connected component of L_2 is non-bipartite because it contains some D_i, with $i \geq 2$.

$\Leftarrow$) Assume that F is as in (a). Since the vectors in $\mathcal{A} \cap H_{e_i}$ correspond precisely to the edges of $G \setminus \{x_i\}$ one obtains that F is a facet of the edge cone. If F is as in (b), a similar argument shows that F is also a facet. $\square$

Theorem 10.7.7 *If G is a connected graph and F is a facet of the edge cone of G, then either*

(a) $F = \mathbb{R}_+\mathcal{A} \cap \{x \in \mathbb{R}^n \mid x_i = 0\}$ *for some* $1 \leq i \leq n$, *or*

(b) $F = \mathbb{R}_+\mathcal{A} \cap H_{\mathfrak{A}}$ *for some independent set* $\mathfrak{A}$ *of* G.

Proof. By Theorem 10.7.6 one may assume that G is bipartite and $n \geq 3$. Notice that $F = \mathbb{R}_+\mathcal{B}$, where $\mathcal{B} = \{v_i \in \mathcal{A} \mid \langle v_i, a\rangle = 0\}$. Since $\dim(F)$ is equal to $n-2$, one may assume that $v_1, \ldots, v_{n-2}$ are linearly independent vectors in F. There is $0 \neq a \in \mathbb{R}^n$ such that

(i) $F = \mathbb{R}_+\mathcal{A} \cap H_a$, $\langle a, v_i\rangle = 0 \ \forall i \in \{1, 2, \ldots, n-2\}$, and

(ii) $\langle a, v_i\rangle \leq 0$ for $i \in \{1, 2, \ldots, q\}$.

Let D be the subgraph of G whose edges correspond to $v_1, \ldots, v_{n-2}$ and its vertex set is the union of the vertices in the edges of D. Setting $k = |V(D)|$, by Corollary 10.2.11 one has $n - 2 = \operatorname{rank}(A_D) = k - c_0(D)$, where A_D is the incidence matrix of D and $c_0(D)$ is the number of bipartite components of D. Thus $0 \leq n - k = 2 - c_0(D)$. This shows that either $c_0(D) = 1$ and $k = n - 1$ or $c_0(D) = 2$ and $k = n$.

Case (I): Assume $c_0(D) = 1$ and $k = n - 1$. Set $V(D) = \{x_1, \ldots, x_{n-1}\}$. As D is a tree with $n - 2$ edges and $\langle v_i, a\rangle = 0$ for $i = 1, \ldots, n-2$, applying Lemma 10.7.4, one may assume $a = (a_1, \ldots, a_{n-1}, a_n)$, where $a_i = \pm 1$ for $1 \leq i \leq n - 1$.

Subcase(I.a): Assume that $a_n = 0$. Set $\mathfrak{A} = \{x_i \in V \,|\, a_i = 1\}$ and $B = \{x_i \in V \,|\, a_i = -1\}$. Note that $\emptyset \neq \mathfrak{A} \subset V(D)$, because D is a tree with at least two vertices. We will show that $\mathfrak{A}$ is an independent set of G and $B = N_G(\mathfrak{A})$. If $\mathfrak{A}$ is not an independent set of G, there is an edge $\{x_i, x_j\}$ of G for some x_i, x_j in $\mathfrak{A}$. Thus $v_k = e_i + e_j$ and by (ii) we get $\langle a, v_k\rangle \leq 0$, which is impossible because $\langle a, v_k\rangle = 2$. Next we will show $N_G(\mathfrak{A}) = B$. If $x_i \in N(\mathfrak{A})$, then $v_k = e_i + e_j$ for some x_j in $\mathfrak{A}$, using (ii) we obtain $\langle a, v_k\rangle = a_i + 1 \leq 0$ and $a_i = -1$, hence $x_i \in B$. Conversely if $x_i \in B$, since D has no isolated vertices, there is $1 \leq k \leq n - 2$ so that $v_k = e_i + e_j$, for some j, by (i) we obtain $\langle a, v_k\rangle = -1 + a_j = 0$, which shows that $x_j \in \mathfrak{A}$ and $x_i \in N(\mathfrak{A})$. Therefore $H_a = H_{\mathfrak{A}}$ and F is as in (b).

Subcase(I.b): Assume $a_n > 0$. In this case consider $a' = -e_n$. Then:

(c) $\langle v_i, a'\rangle = 0$ for $i = 1, \ldots, n - 2$.

(d) If $v_j \in \mathbb{R}(v_1, \ldots, v_{n-2}) \Rightarrow \langle v_j, a'\rangle = \langle v_j, a\rangle = 0$.

(e) If $v_j \notin \mathbb{R}(v_1, \ldots, v_{n-2})$, then $\langle v_j, a'\rangle = -1$ and $\langle v_j, a\rangle < 0$.

To prove (e) assume $v_j \notin \mathbb{R}(v_1, \ldots, v_{n-2})$, then $v_j = e_k + e_n$, because otherwise the "edge" v_j added to the tree D form a graph with a unique even cycle, a contradiction. Thus $\langle v_j, a'\rangle = -1$. On the other hand $\langle v_j, a\rangle < 0$, because if $\langle v_j, a\rangle = 0$, then H_a contains the linearly independent vectors $v_1, \ldots, v_{n-2}, v_j$, a contradiction to $\dim(F) = n - 2$.

Thus given $1 \leq j \leq q$, then $\langle v_j, a\rangle = 0$ if and only if $\langle v_j, a'\rangle = 0$. Note

$$\mathbb{R}_+\mathcal{A} \cap H_{a'} = \mathbb{R}_+\mathcal{B}', \tag{10.9}$$

where $\mathcal{B}' = \{v_i \in \mathcal{A} \,|\, \langle v_i, a'\rangle = 0\}$. Hence using $F = \mathbb{R}_+\mathcal{B}$ and Eq. (10.9) we get $F = \mathbb{R}_+\mathcal{A} \cap H_{a'}$, as required. The case $a_n < 0$ can be treated similarly.

Case (II): Assume $c_0(D) = 2$ and $k = n$. Let D_1 and D_2 be the two components of D and set $V_1 = V(D_1)$ and $V_2 = V(D_2)$.

Using Lemma 10.7.4 we can relabel the vertices of D to write $a = rb + sc$, where $0 \neq r \geq s \geq 0$ are rational numbers and $b = (b_1, \ldots, b_m, 0, \ldots, 0)$, $c = (0, \ldots, 0, c_{m+1}, \ldots, c_n)$, where $V_1 = \{x_1, \ldots, x_m\}$, $b_i = \pm 1$ for $i \leq m$, and $c_i = \pm 1$ for $i > m$. Set $a' = b$. Note the following:

(f) $\langle v_i, a'\rangle = 0$ for $i = 1, \ldots, n - 2$.

(g) If $v_j \in \mathbb{R}(v_1, \ldots, v_{n-2}) \Rightarrow \langle v_j, a'\rangle = \langle v_j, a\rangle = 0$.

(h) If $v_j \notin \mathbb{R}(v_1, \ldots, v_{n-2})$, then $\langle v_j, a\rangle < 0$ and $\langle v_j, a'\rangle = -1$.

Let us prove assertion (h). Assume $v_j \notin \mathbb{R}(v_1,\ldots,v_{n-2})$. Since $\dim(F)$ is equal to $= n-2$, by (ii) we have $\langle v_j, a\rangle < 0$. Observe that if an "edge" v_k has vertices in V_1 (resp. V_2), then $\langle v_k, a\rangle = 0$. Indeed if we add the edge v_k to the tree D_1 (resp. D_2) we get a graph with a unique even cycle and this implies that $v_1,\ldots,v_{n-2},v_k$ are linearly dependent, that is, $\langle v_k, a\rangle = 0$. Thus $v_j = e_i + e_\ell$ for some $x_i \in V_1$ and $x_\ell \in V_2$. From the inequality

$$\langle v_j, a\rangle = r\langle v_j, b\rangle + s\langle v_j, c\rangle = rb_i + sc_\ell < 0$$

we obtain $b_i = -1 = \langle v_j, a'\rangle$, as required. In particular note that $H_{a'}$ is a supporting hyperplane of $\mathbb{R}_+\mathcal{A}$ because $\mathbb{R}_+\mathcal{A} \subset H_{a'}^-$. We may now proceed as in case (I) to obtain the equality $F = \mathbb{R}_+\mathcal{A} \cap H_a = \mathbb{R}_+\mathcal{A} \cap H_{a'}$. Using that D_1 has no isolated vertices, one may proceed as in case (I) to prove that $H_{a'} = H_{\mathfrak{A}}$ for the independent set of vertices given by $\mathfrak{A} = \{x_i \mid b_i = 1\}$. □

Theorem 10.7.8 *If G is a connected graph, then*

$$\mathbb{R}_+\mathcal{A} = \left(\bigcap_{\mathfrak{A}\in\mathcal{F}} H_{\mathfrak{A}}^-\right) \bigcap \left(\bigcap_{i=1}^{n} H_{e_i}^+\right),$$

where $\mathcal{F}$ is the family of all the independent sets of vertices of G and $H_{e_i}^+$ is the closed halfspace $\{x \in \mathbb{R}^n \mid x_i \geq 0\}$.

Proof. Set $C = \mathbb{R}_+\mathcal{A}$. From Theorem 10.7.7 and Corollary 1.1.32 we get

$$C = \operatorname{aff}(C) \cap \left(\bigcap_{\mathfrak{A}\in\mathcal{F}} H_{\mathfrak{A}}^-\right) \bigcap \left(\bigcap_{i=1}^{n} H_{e_i}^+\right).$$

We may assume that G is bipartite, otherwise $\operatorname{aff}(C) = \mathbb{R}^n$ and there is nothing to prove. If (V_1, V_2) is the bipartition of G, then $\operatorname{aff}(\mathbb{R}_+\mathcal{A})$ is the hyperplane $\sum_{x_i\in V_1} x_i = \sum_{x_i\in V_2} x_i$, because $\dim(\mathbb{R}_+\mathcal{A}) = n-1$. Since $H_{V_1}^- \cap H_{V_2}^- = H_{V_1} = \operatorname{aff}(C)$, we obtain the required equality. □

Studying the bipartite case For connected bipartite graphs we will present more precise results about the irreducible representations of edge cones. Let G be a connected bipartite graph with bipartition (V_1, V_2). To avoid repetitions we continue using the notation introduced at the beginning of the section.

Proposition 10.7.9 *If $\mathfrak{A}$ is an independent set of G such that $\mathfrak{A} \neq V_i$ for $i = 1, 2$, then $F = \mathbb{R}_+\mathcal{A} \cap H_{\mathfrak{A}}$ is a proper face of the edge cone.*

Proof. Assume $N(\mathfrak{A}) = V_2$. Take any $x_i \in V_1 \setminus \mathfrak{A}$ and any $x_j \in V_2$ adjacent to x_i, then $e_i + e_j \notin H_{\mathfrak{A}}$. This means that $\mathbb{R}_+\mathcal{A} \not\subset H_{\mathfrak{A}}$. Thus we may assume $N(\mathfrak{A}) \neq V_i$ for $i = 1, 2$.

Case (I): $N(\mathfrak{A}) \cap V_i \neq \emptyset$ for $i = 1, 2$. If the vertices in $N(\mathfrak{A}) \cap V_i$ for $i = 1, 2$ are only adjacent to vertices in $\mathfrak{A}$, then pick vertices $x_i \in N(\mathfrak{A}) \cap V_i$ and note that there is no path between x_1 and x_2, a contradiction. Thus there must be a vector in the edge cone which is not in $H_{\mathfrak{A}}$.

Case (II): $A \subsetneq V_1$. If the vertices in $N(\mathfrak{A})$ are only adjacent to vertices in $\mathfrak{A}$. Then a vertex in $\mathfrak{A}$ cannot be joined by a path to a vertex in $V_2 \setminus N(\mathfrak{A})$, a contradiction. As before we obtain $\mathbb{R}_+\mathcal{A} \not\subset H_{\mathfrak{A}}$. □

Proposition 10.7.10 *Let $\mathcal{F}$ be the family of all independent sets $\mathfrak{A}$ of G such that $H_{\mathfrak{A}} \cap \mathbb{R}_+\mathcal{A}_G$ is a facet. If $\mathfrak{A}$ is in $\mathcal{F}$ and $V_i \cap \mathfrak{A} \neq \emptyset$ for $i = 1, 2$, then $H_{\mathfrak{A}}^-$ is redundant in the following expression of the edge cone*

$$\mathbb{R}_+\mathcal{A} = \operatorname{aff}(\mathbb{R}_+\mathcal{A}) \cap \left(\bigcap_{\mathfrak{A}\in\mathcal{F}} H_{\mathfrak{A}}^-\right) \bigcap \left(\bigcap_{i=1}^{n} H_{e_i}^+\right).$$

Proof. Set $\mathcal{A} = \{v_1, \ldots, v_q\}$. One can write $\mathfrak{A} = \mathfrak{A}_1 \cup \mathfrak{A}_2$ with $\mathfrak{A}_i \subset V_i$ for $i = 1, 2$. There are $v_1, \ldots, v_{n-2}$ linearly independent vectors in $H_{\mathfrak{A}} \cap \mathbb{R}_+\mathcal{A}$, where n is the number of vertices of G. Consider the subgraph D of G whose edges correspond to $v_1, \ldots, v_{n-2}$ and its vertex set is the union of the vertices in those edges. Note that D cannot be connected. Indeed there is no edge of D connecting a vertex in $N_G(\mathfrak{A}_1)$ with a vertex in $N_G(\mathfrak{A}_2)$ because all the vectors $v_1, \ldots, v_{n-2}$ satisfy the equation

$$\sum_{x_i \in \mathfrak{A}} x_i = \sum_{x_i \in N_G(\mathfrak{A})} x_i.$$

Hence D is a spanning subgraph of G with two components D_1, D_2, which are trees, such that $V(D_i) = \mathfrak{A}_i \cup N_G(\mathfrak{A}_i)$, $i = 1, 2$. Therefore $H_{\mathfrak{A}_i}$ is a proper support hyperplane defining a facet $F_i = H_{\mathfrak{A}_i} \cap \mathbb{R}_+\mathcal{A}$, that is $\mathfrak{A}_1, \mathfrak{A}_2$ are in $\mathcal{F}$. Since $H_{\mathfrak{A}_1}^- \cap H_{\mathfrak{A}_2}^-$ is contained in $H_{\mathfrak{A}}^-$ the proof is complete. □

Proposition 10.7.11 *If $\mathfrak{A}_2 \subsetneq V_2$, $F = H_{\mathfrak{A}_2} \cap \mathbb{R}_+\mathcal{A}$ is a facet of the edge cone of G and $\mathcal{A}' = \mathcal{A} \cup \{0\}$, then*

$$H_{\mathfrak{A}_2}^- \cap \operatorname{aff}(\mathcal{A}') = \begin{cases} H_{\mathfrak{A}_1}^- \cap \operatorname{aff}(\mathcal{A}'), & \text{where } \mathfrak{A}_1 = V_1 \setminus N(\mathfrak{A}_2) \neq \emptyset, \text{ or} \\ H_{e_i}^+ \cap \operatorname{aff}(\mathcal{A}'), & \text{for some } x_i \text{ with } G \setminus \{x_i\} \text{ connected.} \end{cases}$$

Proof. Let us assume that G has n vertices $x_1, \ldots, x_n$ and that V_1 is the set of the first m vertices of G. There are $v_1, \ldots, v_{n-2}$ linearly independent vectors in the hyperplane $H_{\mathfrak{A}_2}$. Consider the subgraph D of G whose edges correspond to $v_1, \ldots, v_{n-2}$ and its vertex set is the union of the vertices in those edges. As G is connected, either D is a tree with $n - 1$ vertices or D is a spanning subgraph of G with two connected components.

Assume that D is a tree (the other case is left as an exercise), write $V(D) = V \setminus \{x_i\}$ for some i. Note $\langle v_j, v_{\mathfrak{A}_2}\rangle = -\langle v_j, e_i\rangle$, $j = 1, \ldots, q$, where

$$v_{\mathfrak{A}_2} = \sum_{x_i \in \mathfrak{A}_2} e_i - \sum_{x_i \in N(\mathfrak{A}_2)} e_i.$$

Indeed if the "edge" v_j has vertices in $V(D)$, then both sides of the equality are zero, otherwise write $v_j = e_i + e_\ell$. Observe $x_i \notin \mathfrak{A}_2$ and $x_\ell \in N(\mathfrak{A}_2)$ because as $H_{\mathfrak{A}_2}$ is a facet it cannot contain v_j, thus both sides of the equality are equal to -1. As a consequence since $\mathrm{aff}(\mathcal{A}') = \mathbb{R}(v_1, \ldots, v_{n-2}, v_j)$ for some $v_j = e_i + e_\ell$ we rapidly obtain $\langle \alpha, v_{\mathfrak{A}_2}\rangle = -\langle \alpha, e_i\rangle$, $\forall\, \alpha \in \mathrm{aff}(\mathcal{A}')$. Therefore $H^-_{\mathfrak{A}_2} \cap \mathrm{aff}(\mathcal{A}') = H^+_{e_i} \cap \mathrm{aff}(\mathcal{A}')$, as required. □

From the proof of Proposition 10.7.11 we get:

Lemma 10.7.12 *Let $F = H_{\mathfrak{A}} \cap \mathbb{R}_+\mathcal{A}$ be a facet of $\mathbb{R}_+\mathcal{A}$ with $\mathfrak{A} \subsetneq V_1$.*

(a) *If $N(\mathfrak{A}) = V_2$, then $\mathfrak{A} = V_1 \setminus \{x_i\}$ for some $x_i \in V_1$ and $F = H_{e_i} \cap \mathbb{R}_+\mathcal{A}$.*

(b) *If $N(\mathfrak{A}) \subsetneq V_2$, then $F = H_{V_2 \setminus N(\mathfrak{A})} \cap \mathbb{R}_+\mathcal{A}$ and $N(V_2 \setminus N(\mathfrak{A})) = V_1 \setminus \mathfrak{A}$.*

Proposition 10.7.13 *Let $\mathfrak{A} \subsetneq V_1$. Then $F = H_{\mathfrak{A}} \cap \mathbb{R}_+\mathcal{A}$ is a facet of $\mathbb{R}_+\mathcal{A}$ if and only if*

(a) *$G[\mathfrak{A} \cup N(\mathfrak{A})]$ is connected with vertex set $V \setminus \{v\}$ for some $v \in V_1$, or*

(b) *$G[\mathfrak{A} \cup N(\mathfrak{A})]$ and $G[(V_2 \setminus N(\mathfrak{A})) \cup (V_1 \setminus \mathfrak{A})]$ are connected and their union is a spanning subgraph of G.*

Moreover any facet has the form $F = H_{\mathfrak{A}} \cap \mathbb{R}_+\mathcal{A}$ for some $\mathfrak{A} \subsetneq V_i$, $i \in \{1, 2\}$.

Proof. The first statement follows readily from Lemma 10.7.12 and using part of the proof of Theorem 10.7.8. The last statement follows combining Theorem 10.7.8 with Proposition 10.7.10. □

Remark 10.7.14 In Proposition 10.7.13 the case (a) is included in case (b). To see this fact, take $N(\mathfrak{A}) = V_2$ and note that $G[(V_2 \setminus N(\mathfrak{A})) \cup (V_1 \setminus \mathfrak{A})]$ must consist of a point. The condition in case (a) is equivalent to require $G \setminus \{v\}$ connected and in this case $F = H_{e_i} \cap \mathbb{R}_+\mathcal{A}$, where $v = x_i$ correspond to the unit vector e_i.

Lemma 10.7.15 *Let F be a facet of $\mathbb{R}_+\mathcal{A}$. If $F = H_{\mathfrak{A}} \cap \mathbb{R}_+\mathcal{A} = H_B \cap \mathbb{R}_+\mathcal{A}$ with $\mathfrak{A} \subsetneq V_1$ and $B \subsetneq V_1$, then $\mathfrak{A} = B$.*

Proof. It follows from Lemma 10.7.12 and Proposition 10.7.13. □

Putting together the previous results we get the following canonical way of representing the edge cone. The uniqueness follows from Lemma 10.7.12 and Lemma 10.7.15.

Theorem 10.7.16 *There is a unique irreducible representation*

$$\mathbb{R}_+\mathcal{A} = \mathrm{aff}(\mathbb{R}_+\mathcal{A}) \cap (\cap_{i=1}^{r} H^-_{\mathfrak{A}_i}) \cap (\cap_{i\in\mathcal{I}} H^+_{e_i})$$

such that $\mathfrak{A}_i \subsetneq V_1$ *for all* i *and* $\mathcal{I} = \{i | x_i \in V_2$ *is not cut vertex* $\}$.

Corollary 10.7.17 *There is an irreducible representation*

$$\mathbb{R}_+\mathcal{A} = \mathrm{aff}(\mathbb{R}_+\mathcal{A}) \cap (\cap_{i=1}^{r} H^-_{\mathfrak{A}_i}) \cap (\cap_{i=1}^{s} H^-_{B_i})$$

such that $\mathfrak{A}_i \subsetneq V_1$ *for all* i *and* $B_i \subsetneq V_2$ *for all* i.

Lemma 10.7.18 $\mathbb{Z}^n \cap \mathbb{R}_+\mathcal{A} = \mathbb{N}\mathcal{A}$. *In particular if* $(\beta_1, \ldots, \beta_n)$ *is an integral vector in the edge cone, then* $\sum_{i=1}^{n} \beta_i$ *is an even integer.*

Proof. Since G is bipartite, by Proposition 10.2.2 its incidence matrix is totally unimodular, the lemma follows from Corollary 1.6.8. □

As an application we recover the marriage theorem (Theorem 7.1.9). For a generalized version of the marriage theorem; see Proposition 13.5.3. Recall that a pairing off of all the vertices of a graph G is called a *perfect matching*.

Theorem 10.7.19 (Marriage Theorem) *If* G *is a connected bipartite graph, then* G *has a perfect matching if and only if* $|\mathfrak{A}| \leq |N(\mathfrak{A})|$ *for every independent set of vertices* $\mathfrak{A}$ *of* G.

Proof. G has a perfect matching if and only if the vector $\beta = (1, \ldots, 1)$ is in $\mathbb{N}\mathcal{A}$. By Lemma 10.7.18 β is in $\mathbb{N}\mathcal{A}$ if and only if $\beta \in \mathbb{R}_+\mathcal{A}$. Thus the result follows from Theorem 10.7.8. □

Exercises

10.7.20 Let G be a graph with n vertices and let A be its incidence matrix. Prove that if $\mathrm{rank}(A) = n$ and $\mathfrak{A}$ is an independent set of G, then the sets $F = \mathbb{R}_+\mathcal{A} \cap H_{\mathfrak{A}}$ and $F = H_{e_i} \cap \mathbb{R}_+\mathcal{A}$ are proper faces of the edge cone.

10.7.21 Let G be a connected bipartite graph and let $\mathcal{Q}$ be the polyhedron defined as the set of $x = (x_i) \in \mathbb{R}^{n+1}$ satisfying

$$\begin{array}{l} x_i \geq 1 \text{ for } i = 1, \ldots, n+1, \text{ and} \\ -x_{n+1} + \sum_{v_i \in C} x_i \geq 1 \text{ for any minimal vertex cover } C \text{ of } G. \end{array}$$

If G is unmixed and $v_0 \geq 3$, then $a = (1, \ldots, 1)$ and $b = (1, \ldots, 1, v_0 - 1)$ are the vertices of $\mathcal{Q}$.

10.7.22 [72] Let $G = \mathcal{K}_n$ be the complete graph on n vertices and let $\mathcal{A}$ be the set of column vectors of its incidence matrix. Then a vector $x \in \mathbb{R}^n$ is in $\mathbb{R}_+\mathcal{A}$ if and only if $x = (x_1, \ldots, x_n)$ satisfies

$$\begin{array}{rcl} -x_i & \leq & 0, \quad i = 1, \ldots, n \\ x_i - \sum_{j \neq i} x_j & \leq & 0, \quad i = 1, \ldots, n. \end{array}$$

In addition if $n \geq 4$, these equations define all the facets of $\mathbb{R}_+\mathcal{A}$.

10.7.23 If G is a triangle with vertices $\{x_1, x_2, x_3\}$, then the edge cone of G has three facets defined by

$$x_1 \leq x_2 + x_3, \quad x_2 \leq x_1 + x_3, \quad x_3 \leq x_1 + x_2.$$

Note that $x_1 \geq 0$, $x_2 \geq 0$ and $x_3 \geq 0$ define proper faces of dimension 1.

10.7.24 If G is an arbitrary graph, then

$$\mathbb{R}_+\mathcal{A} = \left(\bigcap_{\mathfrak{A} \in \mathcal{F}} H_{\mathfrak{A}}^-\right) \bigcap \left(\bigcap_{i=1}^{n} H_{e_i}^+\right),$$

where $\mathcal{F}$ is the set of all the independent sets of vertices of G.

10.7.25 If $G_1, \ldots, G_r$ are the components of a graph G, then

$$\mathbb{R}_+\mathcal{A} = \left(\bigcap_{\mathfrak{A}} H_{\mathfrak{A}}^-\right) \bigcap \left(\bigcap_{i=1}^{n} H_{e_i}^+\right),$$

where the first intersection is taken over all the independent sets of vertices of the components G_i of G.

10.7.26 Let $G = \mathcal{K}_{m,n}$ be the complete bipartite graph with $m \leq n$. If $V_1 = \{x_1, \ldots, x_m\}$ and $V_2 = V \setminus V_1$ is the bipartition of G, then a vector $z \in \mathbb{R}^{m+n}$ is in $\mathbb{R}_+\mathcal{A}$ if and only if $z = (x_1, \ldots, x_m, y_1, \ldots, y_n)$ satisfies

$$\begin{array}{rcll} x_1 + \cdots + x_m & = & y_1 + \cdots + y_n, & \\ -x_i & \leq & 0, & i = 1, \ldots, m, \\ -y_i & \leq & 0, & i = 1, \ldots, n. \end{array}$$

In addition if $m \geq 2$, the inequalities define all the facets of $\mathbb{R}_+\mathcal{A}$.

10.7.27 If $G = \mathcal{K}_{1,3}$ is the star with vertices $\{v, x_1, x_2, x_3\}$ and center x, then the edge cone of G has three facets defined by

$$x_i \geq 0, \quad (i = 1, 2, 3)$$

and $x = 0$ defines a proper face of dimension 1.

10.8 Monomial birational extensions

Let $R = K[x_1, \ldots, x_n]$ be a polynomial ring over a field K. In this section we consider a finite set of distinct monomials $F = \{x^{v_1}, \ldots, x^{v_q}\} \subset R$ of degree $d \geq 2$ and consider the integer matrices:

$$A = (v_1, \ldots, v_q) \text{ and } B = \begin{pmatrix} v_1 & \cdots & v_q \\ 1 & \cdots & 1 \end{pmatrix},$$

where the v_i's are regarded as column vectors, A is sometimes called the *log-matrix* of F. In the sequel let $\mathbf{x}_d$ denote the set of *all* monomials of degree d in R. Then $K[\mathbf{x}_d]$ is the dth Veronese subring $R^{(d)}$ of R.

If L is an integer matrix with r rows, we denote by $\mathbb{Z}L$ (resp. $\mathbb{Q}L$) the subgroup of $\mathbb{Z}^r$ (resp. subspace of $\mathbb{Q}^r$) generated by the columns of L.

Definition 10.8.1 An extension $D_1 \subset D_2$ of integral domains is said to be *birational* if D_1 and D_2 have the same field of fractions.

Proposition 10.8.2 *Let $F_1 = \{x^{\beta_1}, \ldots, x^{\beta_r}\}$ be a set monomials of R such that $F \subset F_1$. Then the ring extension $K[F] \subset K[F_1]$ is birational if and only if $\mathbb{Z}A = \mathbb{Z}L$, where L is the matrix with column vectors $\beta_1, \ldots, \beta_r$.*

Proof. It is left as an exercise. □

Following [385] an aim here is to study the birationality of the extension $K[F] \subset R^{(d)}$ in terms of the matrices:

(a) the *linear syzygy matrix* $\mathrm{Syz}_\ell(F)$ whose columns are the set of vectors of the form $x_ie_j - x_ke_\ell$ such that $x_ix^{v_j} = x_kx^{v_\ell}$,

(b) the *numerical linear syzygy matrix* L obtained from $\mathrm{Syz}_\ell(F)$ by making the substitution $x_i = 1$ for all i,

(c) the matrix M whose columns are the set of vectors $e_i - e_k \in \mathbb{R}^n$ such that $e_i - e_k = v_j - v_\ell$, i.e., $M = AL$,

(d) the *Jacobian matrix* $\Theta(F) = (\partial f_i/\partial x_j)$.

Notice that the matrices in the first row of the diagram:

$$\begin{array}{ccc} \Theta(F) & \mathrm{Syz}_\ell(F) & \mathcal{M} := \Theta(F)^\top \mathrm{Syz}_\ell(F) \\ \downarrow & \downarrow & \downarrow \\ A & L & M := AL \end{array}$$

specialize to the matrices in the second row by making $x_i = 1$ for all i. The matrices $\mathrm{Syz}_\ell(F)$ and L have size $q \times r$, while the matrices $\mathcal{M}$ and M have size $n \times r$.

Proposition 10.8.3 [385] *If* $\mathrm{rank}(M) = n - 1$, *then* $K[F] \subset K[\mathbf{x}_d]$ *is a birational extension.*

Proof. Let $a = (a_i) \in \mathbb{N}^n$ such that $|a| = \sum_i a_i = d$. It suffices to prove that $a \in \mathbb{Z}A$. Let $w_1, \ldots, w_r$ be the column vectors of the matrix M. Each w_m has the form $w_m = e_i - e_k = v_j - v_\ell$ for some $i \neq k$ and $j \neq \ell$. Hence $\operatorname{rank}(A) = n$ because $v_1 \notin \mathbb{Q}M$. Therefore we can write

$$\lambda a = \lambda_1 w_1 + \cdots + \lambda_r w_r + \mu v_1 \quad (\lambda, \mu, \lambda_i \in \mathbb{Z}).$$

Taking inner product with $\mathbf{1} = (1, \ldots, 1)$ yields

$$\begin{aligned} \lambda d = \lambda|a| = \lambda_1|w_1| + \cdots + \lambda_r|w_r| + \mu|v_1| = \mu d \;\Rightarrow\; \lambda = \mu \\ \Rightarrow\; \lambda(a - v_1) = \lambda_1 w_1 + \cdots + \lambda_r w_r. \end{aligned} \tag{10.10}$$

Consider the digraph $\mathcal{D}$ with vertex set $X = \{x_1, \ldots, x_n\}$ such that (x_i, x_k) is a directed edge iff $w_m = e_i - e_k$ for some m. The incidence matrix of $\mathcal{D}$ is M, thus M is totally unimodular by Exercise 1.8.10 and consequently $\mathbb{Z}^n/\mathbb{Z}M$ is torsion-free. Hence from Eq. (10.10) we get $a - v_1 \in \mathbb{Z}M$ and $a \in \mathbb{Z}M + v_1 \subset \mathbb{Z}A$, as required. □

Remark 10.8.4 Let $\mathcal{D}$ be the digraph in the proof of Proposition 10.8.3. Then according to [190, Theorem 8.3.1] we have

$$\operatorname{rank}(M) = n - c,$$

where c is the number of connected components of $\mathcal{D}$. In particular M has rank $n - 1$ if and only if $\mathcal{D}$ is connected.

A generalization of the next result is given in Theorem 10.8.12.

Corollary 10.8.5 *Let G be a connected non-bipartite graph with n vertices and let $K[G]$ be its edge subring. Then $K[G] \subset K[\mathbf{x}_2]$ is birational.*

Proof. There is a connected subgraph H of G with n vertices and n edges and with a unique cycle of odd length. This follows using that any connected graph has a spanning tree together with the fact that a graph is bipartite if and only if every cycle in some cycle basis is even; see Exercise 10.1.67. By induction on n it is not hard to see that the numerical matrix M associated to H has rank $n-1$. Thus $K[H] \subset K[\mathbf{x}_2]$ is birational by Proposition 10.8.3, which implies the birationality of $K[G] \subset K[\mathbf{x}_2]$. □

Proposition 10.8.6 [385] *$K[F] \subset K[\mathbf{x}_d]$ is a birational extension if and only if $n = \operatorname{rank}(A)$ and $\Delta_n(A) = d$.*

Proof. Assume the given extension is birational. Then $\mathbb{Z}\log(\mathbf{x}_d) = \mathbb{Z}A$ by Proposition 10.8.2. By Proposition 9.3.33(b) one has $\mathbb{Z}^n/\mathbb{Z}\log(\mathbf{x}_d) \simeq \mathbb{Z}_d$. Therefore, $\Delta_n(A) = d$. Conversely, if $\Delta_n(A) = d$, the surjection

$$\mathbb{Z}^n/\mathbb{Z}A \longrightarrow \mathbb{Z}_d, \qquad \overline{a} \longmapsto \overline{|a|},$$

is an isomorphism. Since $\log(\mathbf{x}_d)$ maps to zero under this map, one obtains $\mathbb{Z}A = \mathbb{Z}\log(\mathbf{x}_d)$, as required. □

Proposition 10.8.7 [385] *The following conditions are equivalent*

(a) $K[F] \subset K[\mathbf{x}_d]$ *is birational.*

(b) $\mathbb{Z}^n/\mathbb{Z}(\{v_i - v_j | 1 \leq i < j \leq q\})$ *is torsion-free of rank* 1.

(c) $n = \mathrm{rank}(A)$ *and*

$$\mathbb{Z}(\{v_i - v_j | 1 \leq i < j \leq q\}) = \mathbb{Z}(\{e_i - e_j | 1 \leq i < j \leq n\}).$$

Proof. If $n = \mathrm{rank}(A)$, then there is an exact sequence of finite groups

$$0 \longrightarrow T(\mathbb{Z}^{n+1}/\mathbb{Z}B) \stackrel{\varphi}{\longrightarrow} T(\mathbb{Z}^n/\mathbb{Z}A) \stackrel{\psi}{\longrightarrow} \mathbb{Z}_d \longrightarrow 0, \qquad (*)$$

where the maps φ and ψ are given by

$$\begin{array}{rcll} \varphi(\overline{(\alpha, b)}) & = & \overline{\alpha} & (\alpha \in \mathbb{Z}^n, \quad b \in \mathbb{Z}), \\ \psi(\overline{\alpha}) & = & \overline{\langle \alpha, x_0 \rangle}. & \end{array}$$

Pick a basis $\{v_1, \ldots, v_n\}$ of the column space of A and notice that

$$\{v_1 - v_n, v_2 - v_n, \ldots, v_{n-1} - v_n, v_n\}$$

is also a basis because $|v_i| = d$ for all i. Therefore one has the equality

$$\mathbb{Q}(\{v_i - v_j | 1 \leq i < j \leq q\}) = \mathbb{Q}(\{e_i - e_j | 1 \leq i < j \leq n\}).$$

The equivalences now follow using Lemma 1.2.20, Proposition 9.3.33, the exact sequence $(*)$, together with the fact that the group

$$\mathbb{Z}^n/\mathbb{Z}(\{e_i - e_j | 1 \leq i < j \leq n\})$$

is torsion-free of rank 1, see the proof of Proposition 10.8.3. $\Box$

Proposition 10.8.8 [385] *If* $\mathrm{rank}(L) = q - 1$ *and* $\mathrm{rank}(A) = n$, *then* $\mathrm{rank}(M) = n - 1$. *In particular* $K[F] \subset K[\mathbf{x}_d]$ *is birational.*

Proof. There are linear maps

$$\mathbb{Q}^r \stackrel{L}{\longrightarrow} \mathbb{Q}^q \stackrel{A}{\longrightarrow} \mathbb{Q}^n.$$

Hence we have a linear map

$$\mathrm{im}(L) \stackrel{A_1}{\longrightarrow} \mathrm{im}(AL) = \mathrm{im}(M) \longrightarrow 0,$$

where A_1 is the restriction of A to $\mathrm{im}(L)$. By hypothesis $\dim(\mathrm{im}(A)) = n$ and $\dim(\mathrm{im}(L)) = q - 1$. Hence

$$\begin{array}{rcl} q - 1 & = & \dim(\mathrm{im}(L)) = \dim(\ker(A_1)) + \dim(\mathrm{im}(M)), \\ q - n & = & \dim(\ker(A)) \geq \dim(\ker(A_1)). \end{array}$$

Therefore $\dim(\mathrm{im}(M)) \geq n - 1$. Since $\mathrm{im}(M)$ is generated by vectors of the form $e_i - e_j$ it follows that $\mathrm{im}(M)$ has rank n simply because e_1 is not in the linear space generated by $\{e_i - e_j | 1 \leq i < j \leq n\}$. $\Box$

Lemma 10.8.9 *If m is a non-zero minor of $\mathrm{Syz}_\ell(F)$ of order s for some $s \geq 1$, then*

$$m = \pm x_1^{a_1} \cdots x_n^{a_n} \qquad (a_i \in \mathbb{N}).$$

Proof. By induction on s. The case $s = 1$ is clear because the entries of $\mathrm{Syz}_\ell(F)$ are monomials. Assume $s \geq 2$. Let L_1 be a submatrix of $\mathrm{Syz}_\ell(F)$ of size $s \times s$. We may assume that L_1 is obtained using the first s rows and columns of $\mathrm{Syz}_\ell(F)$. Let $g_1, \ldots, g_q$ be the rows obtained from $\mathrm{Syz}_\ell(F)$ by fixing the first s columns. Thus L_1 is the matrix with rows $g_1, \ldots, g_s$.

Case (I): If the vector g_i has a non-zero entry for some $s + 1 \leq i \leq q$, say the jth entry of g_i is non-zero, then expanding the determinant of L_1 along the jth column and using the induction hypothesis we get that

$$\det(L_1) = \pm x_1^{a_1} \cdots x_n^{a_n} \qquad (a_i \in \mathbb{N}).$$

Case (II): Now we assume that g_i is the zero vector for $s + 1 \leq i \leq q$, but this implies that the linear syzygy matrix of $F' = \{x^{v_1}, \ldots, x^{v_s}\}$ has rank at least s, a contradiction because from the minimal free resolution of $I' = (x^{v_1}, \ldots, x^{v_s})$ one has $\mathrm{rank}(\mathrm{Syz}_\ell(F')) \leq s - 1$; see Exercise 10.8.17. □

Theorem 10.8.10 [385] *If* $\mathrm{rank}(A) = n$ *and* $\mathrm{rank}(\mathrm{Syz}_\ell(F)) = q - 1$, *then* $K[F] \subset K[\mathbf{x}_d]$ *is birational.*

Proof. By Lemma 10.8.9 the matrix L obtained from $\mathrm{Syz}_\ell(F)$ by making $x_i = 1$ for all i has also rank $q - 1$. Therefore by Propositions 10.8.8 and 10.8.3 we get that $K[F] \subset K[\mathbf{x}_d]$ is birational. □

Corollary 10.8.11 [385] *If* $I = (F)$, $\mathrm{rank}(A) = n$, *and* I *has a linear presentation, then* $K[F] \subset K[\mathbf{x}_d]$ *is birational.*

Proof. It follows at once from Theorem 10.8.10 because in this case the rank of $\mathrm{Syz}_\ell(F)$ is $q - 1$. □

Monomials of degree two The birational theory of monomials of degree two can be fully established. Theorem 10.8.12 gives a complete answer for the birationality of subrings of multigraphs.

In what follows we assume that $\deg(x^{v_i}) = 2$ for all i. Consider the multigraph G on the vertex set $X = \{x_1, \ldots, x_n\}$ whose edges (resp. loops) are the pairs $\{x_i, x_j\}$ (resp. $\{x_\ell, x_\ell\}$) such that $x_i x_j \in F$ (resp. $x_\ell^2 \in F$) for some $i \neq j$. Notice that, in our situation, the log-matrix A of F is the incidence matrix of G and the monomial subring $K[F]$ is the edge subring $K[G]$ of the multigraph G.

Theorem 10.8.12 *$K[G] \subset K[\mathbf{x}_2]$ is a birational extension if and only if the following two conditions hold:*

(a) *G is connected.*

(b) *G is bipartite and has at least one loop or G is non-bipartite.*

Proof. $\Leftarrow$) Since G is connected, there is a spanning tree T of G containing all the vertices of G; see Exercise 10.1.65.

Case (I): G is a bipartite and x_n is a loop of G. We may then regard T as a tree with a loop at x_n. Notice that T has exactly $n-1$ simple edges plus a loop. The incidence matrix A_T of T has order n, is non-singular, and we may assume that the last column of A_T has the form $(0,0,\ldots,0,2)^\top$. Consider the matrix A_T' obtained from A_T by removing the last column. The matrix A_T' is totally unimodular because it is the incidence matrix of a bipartite graph; see Proposition 10.2.2. Therefore $\det(A_T) = \pm 2$ and $\text{rank}(A) = n$. From Lemma 10.8.6 we obtain that $K[T] \subset K[\mathbf{x}_2]$ is birational. Hence $K[G] \subset K[\mathbf{x}_2]$ is birational as well.

Case (II): G is non-bipartite. This case follows from Corollary 10.8.5. For use below notice that in Cases (I) and (II) we have that $\text{rank}(A) = n$ and $\mathbb{Z}^n/\mathbb{Z}A \simeq \mathbb{Z}_2$.

$\Rightarrow$) Let $G_1,\ldots,G_r$ be the connected components of G and let A_i be the incidence matrix of G_i. Thus we have $A = \text{diag}(A_1,A_2,\ldots,A_r)$. Notice that for each i the multigraph G_i is either non-bipartite or bipartite with at least one loop, for otherwise $\text{rank}(A_i) < n_i$ for some i, where n_i is the number of vertices of G_i, a contradiction because $n = n_1 + \cdots + n_r$ and $n = \text{rank}(A) = \sum_j \text{rank}(A_j)$. Therefore by the observation at the end of the proof of Case (II) we have $\text{rank}(A_i) = n_i$ and $\mathbb{Z}^{n_i}/\mathbb{Z}A_i \simeq \mathbb{Z}_2$. This means that the Smith normal form of A_i is $\text{diag}(1,\ldots,1,2,\mathbf{0})$ and consequently the Smith normal form of A has exactly r entries equal to 2 and the remaining entries equal to 1 and 0. Hence $\mathbb{Z}^n/\mathbb{Z}A \simeq \mathbb{Z}_2^r$ and $|\mathbb{Z}^n/\mathbb{Z}A| = |\Delta_n(A)| = 2^r$. By the birationality of the extension, using Lemma 10.8.6, we get $r = 1$, i.e., G is connected, as required. □

Corollary 10.8.13 *If G is a connected graph, then $K[G] \subset K[\mathbf{x}_2]$ is a birational extension if and only if G is non-bipartite.*

Proof. If G is bipartite, $\dim K[G] = \text{rank}(A) = n-1$, hence $K[G] \subset K[\mathbf{x}_2]$ cannot be birational. The converse follows from Theorem 10.8.12. □

Exercises

10.8.14 Let A be an integral matrix of size $n \times q$ with equal column sum d. Prove the equalities

$$I_n(A) = I_n\begin{pmatrix} A \\ A_1 + \cdots + A_n \end{pmatrix} = I_n\begin{pmatrix} A \\ d\mathbf{1} \end{pmatrix} = dI_n\begin{pmatrix} A \\ \mathbf{1} \end{pmatrix},$$

where $A_1, \ldots, A_n$ are the rows of A and $I_n(A)$ is the ideal of $\mathbb{Z}$ generated by the n-minors of A.

Hint $d\mathbf{1} = (d, \ldots, d) = A_1 + \cdots + A_n$.

10.8.15 Let $F = \{x^{v_1}, \ldots, x^{v_q}\}$ and $G = \{x^{\beta_1}, \ldots, x^{\beta_r}\}$ be two sets of monomials of R such that $F \subset G$. Prove that $K[F] \subset K[G]$ is a birational extension if and only if $\mathbb{Z}\log(\mathrm{F}) = \mathbb{Z}\log(G)$.

10.8.16 Let F be a set of monomials of $R = K[x_1, \ldots, x_n]$ of the same degree d such that $\mathbb{Z}^n/\mathbb{Z}\log(F)$ is a finite group. Prove that $\overline{K[Ft]} = A(P)$ if and only if $\mathbb{Z}^n/\mathbb{Z}\log(F) \simeq \mathbb{Z}_d$.

10.8.17 Let $I = (x^{v_1}, \ldots, x^{v_q})$ be a monomial ideal of a polynomial ring R over a field K. Prove that if

$$\cdots \longrightarrow R^{b_2} \xrightarrow{\varphi} R^q \longrightarrow I \longrightarrow 0$$

is the minimal free resolution of I, then $\mathrm{rank}(\varphi) = q - 1$ and the columns of the matrix φ have the form $x^{a_i}e_j - x^{a_k}e_\ell$ (cf. Theorem 3.3.19 and [137]).

Chapter 11

Edge Subrings and Combinatorial Optimization

Let G be a connected bipartite graph. We present an approach to compute the canonical module of the edge subring $K[G]$ using linear programming techniques and study the Gorenstein property and the type of $K[G]$. The a-invariant of $K[G]$ will be interpreted in combinatorial optimization terms as the maximum number of edge disjoint directed cuts.

11.1 The canonical module of an edge subring

Let G be a connected bipartite graph on the vertex set $V = \{x_1, \ldots, x_p\}$ and let $R = K[x_1, \ldots, x_p] = \oplus_{i=0}^{\infty} R_i$ be a polynomial ring over a field K with the standard grading induced by $\deg(x_i) = 1$. The *edge subring* of G is the K-subalgebra

$$K[G] = K[\{x_i x_j \mid x_i \text{ is adjacent to } x_j\}] \subset R,$$

we grade $K[G]$ with the *normalized grading* $K[G]_i = K[G] \cap R_{2i}$. Recall that, by Proposition 10.3.1, $K[G]$ is normal and Cohen–Macaulay.

The set of vectors $v_k = e_i + e_j \in \mathbb{R}^p$ such that x_i is adjacent to x_j will be denoted by $\mathcal{A} = \{v_1, \ldots, v_q\}$. Note that $\mathcal{A}$ is the set of column vectors of the incidence matrix of G. As G is bipartite, by Lemma 10.7.18, we have

$$\mathbb{N}\mathcal{A} = \mathbb{Z}^p \cap \mathbb{R}_+\mathcal{A}. \tag{11.1}$$

Thus, by a formula of Danilov–Stanley (see Theorem 9.1.5), the canonical module $\omega_{K[G]}$ is the ideal given by

$$\begin{aligned} \omega_{K[G]} &= (\{x^a \mid a \in \mathbb{N}\mathcal{A} \cap \mathrm{ri}(\mathbb{R}_+\mathcal{A})\}) \\ &\overset{(11.1)}{=} (\{x^a \mid a \in \mathbb{Z}^p \cap \mathrm{ri}(\mathbb{R}_+\mathcal{A})\}) \subset K[G], \qquad (11.2) \end{aligned}$$

where $\mathrm{ri}(\mathbb{R}_+\mathcal{A})$ is the interior of $\mathbb{R}_+\mathcal{A}$ relative to the affine hull of $\mathbb{R}_+\mathcal{A}$. Recall, from Section 10.7, that $\mathbb{R}_+\mathcal{A}$ is called the *edge cone* of G.

11.2 Integrality of the shift polyhedron

In what follows G will denote a connected bipartite graph with $p = m + n$ vertices and bipartition (V_1, V_2). We will assume that the vertices in V_1 are $x_1, \ldots, x_m$ and the vertices in V_2 are $x_{m+1}, \ldots, x_{n+m}$, where $2 \le m \le n$.

Consider the family

$$\mathcal{F} = \{\mathfrak{A} \cup \mathfrak{A}' \mid \emptyset \ne \mathfrak{A} \subsetneq V_1; N(\mathfrak{A}) \subset \mathfrak{A}' \subset V_2\} \cup \{\mathfrak{A}' \mid \emptyset \ne \mathfrak{A}' \subset V_2\},$$

where $N(\mathfrak{A})$ is the neighbor set of $\mathfrak{A}$. For each $Y = \mathfrak{A} \cup \mathfrak{A}' \in \mathcal{F}$ we associate the following vector

$$\beta_Y = \sum_{x_i \in \mathfrak{A}} e_i - \sum_{x_i \in \mathfrak{A}'} e_i \in \mathbb{R}^{m+n},$$

note that if $\mathfrak{A} = \emptyset$ the vector β_Y is a $\{0, -1\}$-vector. Let C' be the matrix whose rows are the vectors in $\{\beta_Y\}_{Y \in \mathcal{F}}$. According to Corollary 10.7.17 the edge cone of G can be written as:

$$\mathbb{R}_+\mathcal{A} = \mathrm{aff}(\mathbb{R}_+\mathcal{A}) \cap \{x \mid C'x \le 0\}.$$

The *shift polyhedron* of the edge cone of G, with respect to C', is the rational polyhedron:

$$\mathcal{Q} = \mathrm{aff}(\mathbb{R}_+\mathcal{A}) \cap \{x \mid C'x \le -\mathbf{1}\},$$

see Definition 9.1.12. From the finite basis theorem (see Theorem 1.1.33) the shift polyhedron can be written as the sum of a unique cone and a polytope. In our case:

$$\mathcal{Q} = \mathbb{R}_+\mathcal{A} + \mathrm{conv}(\beta_1, \ldots, \beta_r),$$

where $\beta_1, \ldots, \beta_r$ are the vertices of $\mathcal{Q}$. Recall that $\mathcal{Q}$ is an *integral polyhedron* if $\mathcal{Q} = \mathrm{conv}(\mathbb{Z}^p \cap \mathcal{Q})$. As $\mathcal{Q}$ is a pointed polyhedron, it is integral if and only if $\beta_1, \ldots, \beta_r$ are integral vectors; see Corollary 1.1.64.

A reason for introducing the shift polyhedron is that its integral points define the canonical module of $K[G]$.

Proposition 11.2.1 $\mathbb{Z}^p \cap \mathcal{Q} = \mathbb{Z}^p \cap \mathrm{ri}(\mathbb{R}_+\mathcal{A})$.

Proof. It follows from Eq. (11.2) and Corollary 1.1.45. $\Box$

Thus, the shift polyhedron is a bridge which allows us to use combinatorial optimization techniques to study the edge subring $K[G]$.

Theorem 11.2.2 *The shift polyhedron* $\mathcal{Q} = \mathrm{aff}(\mathbb{R}_+\mathcal{A}) \cap \{x \mid C'x \leq -\mathbf{1}\}$ *is integral.*

Proof. If G is regarded as the digraph with all its arrows leaving the vertex set V_2, then it is seen that one has the equality $C'A = -B'$, where A is the incidence matrix of G and B' is the one-way cut-incidence matrix of the family $\mathcal{F}$; see Definition 1.8.5. Let b be any vector in $\mathbb{R}^p$, $p = m + n$, such that the following maximum is finite

$$\max\{\langle x, b\rangle \mid x \in \mathcal{Q}\}. \tag{11.3}$$

According to Theorem 1.1.63 it suffices to prove that the maximum in Eq. (11.3) is attained by an integral vector. As $\mathcal{Q} \subset \mathbb{R}_+\mathcal{A}$, any vector $x \in \mathcal{Q}$ can be written as $x = A\widetilde{x}$ for some $\widetilde{x} \geq 0$, $\widetilde{x} \in \mathbb{R}^q$, where q is the number of edges of G. Hence

$$\{\langle b, x\rangle \mid x \in \mathcal{Q}\} = \{\langle bA, \widetilde{x}\rangle \mid \widetilde{x} \geq 0; B'\widetilde{x} \geq \mathbf{1}\}.$$

Therefore

$$\max\{\langle x, b\rangle \mid x \in \mathcal{Q}\} = \max\{\langle bA, \widetilde{x}\rangle \mid \widetilde{x} \geq 0; B'\widetilde{x} \geq \mathbf{1}\}. \tag{11.4}$$

By Corollary 1.8.8 and Lemma 1.8.9 the polyhedron

$$\mathcal{Q}' = \{\widetilde{x} \in \mathbb{R}^q \mid \widetilde{x} \geq 0; B'\widetilde{x} \geq \mathbf{1}\}$$

is integral. Hence, by Theorem 1.1.63, the maximum in the right-hand side of Eq. (11.4) is attained by an integral vector $\widetilde{x}_0 \in \mathcal{Q}'$. Consequently the maximum in Eq. (11.3) is attained by the integral vector $A\widetilde{x}_0 \in \mathcal{Q}$. $\Box$

Example 11.2.3 Consider the following bipartite graph G and make G a digraph with "edges" $v_1, \ldots, v_6$ as shown below.

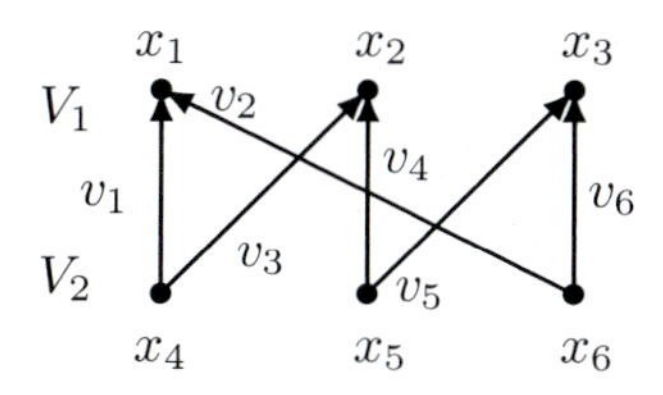

The family $\mathcal{F}$ consists of the subsets that occur in the directed cuts:

$$\begin{array}{rr}
\delta^+(\{x_1,x_4,x_6\})=\{v_3,v_6\}, & \delta^+(\{x_5\})=\{v_4,v_5\},\\
\delta^+(\{x_2,x_4,x_5\})=\{v_1,v_5\}, & \delta^+(\{x_6\})=\{v_2,v_6\},\\
\delta^+(\{x_3,x_5,x_6\})=\{v_2,v_4\}, & \delta^+(\{x_4,x_5\})=\{v_1,v_3,v_4,v_5\},\\
\delta^+(\{x_1,x_2,x_4,x_5,x_6\})=\{v_5,v_6\}, & \delta^+(\{x_4,x_6\})=\{v_1,v_2,v_3,v_6\},\\
\delta^+(\{x_1,x_3,x_4,x_5,x_6\})=\{v_3,v_4\}, & \delta^+(\{x_5,x_6\})=\{v_2,v_4,v_5,v_6\},\\
\delta^+(\{x_2,x_3,x_4,x_5,x_6\})=\{v_1,v_2\}, & \delta^+(\{x_4,x_5,x_6\})=\{v_1,\dots,v_6\}.\\
\delta^+(\{x_4\})=\{v_1,v_3\}, &
\end{array}$$

As pointed out in the proof of Theorem 11.2.2 the one-way cut-incidence matrix B' of $\mathcal{F}$ is related to the incidence matrix A of G and to the matrix C' by the equality $C'A=-B'$. In this example one has:

$$C'=\begin{bmatrix}
1&0&0&-1&0&-1\\
0&1&0&-1&-1&0\\
0&0&1&0&-1&-1\\
1&1&0&-1&-1&-1\\
1&0&1&-1&-1&-1\\
0&1&1&-1&-1&-1\\
0&0&0&-1&0&0\\
0&0&0&0&-1&0\\
0&0&0&0&0&-1\\
0&0&0&-1&-1&0\\
0&0&0&-1&0&-1\\
0&0&0&0&-1&-1\\
0&0&0&-1&-1&-1
\end{bmatrix}\quad\text{and}\quad B'=\begin{bmatrix}
0&0&1&0&0&1\\
1&0&0&0&1&0\\
0&1&0&1&0&0\\
0&0&0&0&1&1\\
0&0&1&1&0&0\\
1&1&0&0&0&0\\
1&0&1&0&0&0\\
0&0&0&1&1&0\\
0&1&0&0&0&1\\
1&0&1&1&1&0\\
1&1&1&0&0&1\\
0&1&0&1&1&1\\
1&1&1&1&1&1
\end{bmatrix}$$

The polyhedron $\mathcal{Q}'=\{\widetilde{x}|B'\widetilde{x}\geq\mathbf{1};\widetilde{x}\geq 0\}$ is integral and has two vertices $(0,1,1,0,1,0)$ and $(1,0,0,1,0,1)$ that map under A onto the vector $\mathbf{1}$ of $\mathcal{Q}$.

Corollary 11.2.4 *If G is a connected bipartite graph and $\mathcal{Q}$ is the shift polyhedron of the edge cone of G, with respect to C', then*

$$\mathcal{Q}=\text{conv}(\mathbb{Z}^p\cap\text{ri}(\mathbb{R}_+\mathcal{A})).$$

Proof. The polyhedron $\mathcal{Q}$ is integral by Theorem 11.2.2. Hence the result follows from Proposition 9.1.14. □

Exercises

11.2.5 Let $T\colon\mathbb{R}^q\to\mathbb{R}^n$ be a linear map and let $\mathcal{Q}'\subset\mathbb{R}^q$ be an integral polyhedron. If $T(\mathbb{Z}^q)\subset\mathbb{Z}^n$, then $T(\mathcal{Q}')$ is an integral polyhedron.

11.2.6 Prove that, in the proof Theorem 11.2.2, the linear transformation $\widetilde{x}\mapsto A\widetilde{x}$ maps $\mathcal{Q}'$ onto $\mathcal{Q}$. Prove that $\dim(\mathcal{Q}')=q$ and $\dim(\mathcal{Q})=m+n-1$. Give a shorter proof of Theorem 11.2.2 using Exercise 11.2.5.

11.3 Generators for the canonical module

Let S be a standard graded K-algebra over a field K. The *a-invariant* of S, denoted by $a(S)$, is the degree, as a rational function, of the Hilbert series of S. If S is Cohen–Macaulay and ω_S is the canonical module of S, then

$$a(S) = -\min\{\, i \,|\, (\omega_S)_i \neq 0\},$$

see Proposition 5.2.3. In our situation, since G is bipartite, $S = K[G]$ is a normal Cohen–Macaulay standard K-algebra. Thus this formula applies.

Theorem 11.3.1 [405] *If $\mathcal{Q}$ is the shift polyhedron of the edge cone of a connected bipartite graph G, then the a-invariant of $K[G]$ is given by*

$$a(K[G]) = -\min\left\{ \frac{|x|}{2} \,\middle|\, x \in \mathcal{Q} \right\}.$$

Proof. Let C' be the matrix defining the shift polyhedron. Note that any row of C' defines a proper face of the edge cone, except the row with the first m entries equal to 0 and the last n entries equal to -1. Thus from Corollary 1.1.45 and Theorem 1.1.44 we have $\mathbb{Z}^p \cap \mathcal{Q} = \mathbb{Z}^p \cap \mathrm{ri}(\mathbb{R}_+\mathcal{A})$. Hence the inequality "$\leq$" follows from the Danilov–Stanley formula of Eq. (11.2). By Proposition 1.1.41 the minimum above is attained at a vertex β of $\mathcal{Q}$. Hence it suffices to observe that β has integral entries by Theorem 11.2.2 and Corollary 1.1.64. $\square$

Theorem 11.3.2 [405, Proposition 4.2] *Let G be a connected bipartite graph. If G is regarded as the digraph with all its arrows leaving the vertex set V_2, then the following three numbers are equal:*

(a) $-a(K[G])$, *minus the a-invariant of $K[G]$.*

(b) *The minimum cardinality of an edge set that contains at least one edge of each directed cut.*

(c) *The maximum number of edge disjoint directed cuts.*

Proof. Let A be the incidence matrix of G and let C' be the matrix defining the shift polyhedron. As $C'A = -B'$, then by Theorem 1.8.7 and LP duality (Theorem 1.1.56) one has that the optimum values in the equality

$$\min\{\langle \mathbf{1}, \widetilde{x}\rangle \,|\, \widetilde{x} \geq 0; B'\widetilde{x} \geq \mathbf{1}\} = \max\{\langle y, \mathbf{1}\rangle \,|\, y \geq 0; yB' \leq \mathbf{1}\}$$

are attained by integral vectors. By looking at B' as a one-way cut-incidence matrix it follows that the two numbers in (b) and (c) are equal. In general, for digraphs, it is known that the numbers in (b) and (c) are equal. See for instance [281, Theorem 19.10]. Noticing that

$$\langle \mathbf{1}, A\widetilde{x}\rangle/2 = \langle \mathbf{1}, \widetilde{x}_1 v_1 + \cdots + \widetilde{x}_q v_q\rangle/2 = (2\widetilde{x}_1 + \cdots + 2\widetilde{x}_q)/2 = \langle \mathbf{1}, \widetilde{x}\rangle,$$

where $v_1, \ldots, v_q$ are the column vectors of A, from the equalities

$$\begin{aligned}&\min\{\langle \mathbf{1}, x\rangle/2 \,|\, x \in \operatorname{aff}(\mathbb{R}_+\mathcal{A}); C'x \leq -\mathbf{1}\} = \\ &\qquad \min\{\langle \mathbf{1}, A\widetilde{x}\rangle/2 \,|\, \widetilde{x} \geq 0; B'\widetilde{x} \geq \mathbf{1}\} = \min\{\langle \mathbf{1}, \widetilde{x}\rangle \,|\, \widetilde{x} \geq 0; B'\widetilde{x} \geq \mathbf{1}\},\end{aligned}$$

and Theorem 11.3.1 we get that the numbers in (a) and (b) are equal. □

Proposition 11.3.3 *If G is a connected bipartite graph and β is a vertex of the shift polyhedron $\mathcal{Q}$, then x^β is a minimal generator of $\omega_{K[G]}$.*

Proof. By Eq. (11.2), Corollary 1.1.45, and Theorem 11.2.2 we get that x^β is in $\omega_{K[G]}$. There are $c \in \mathbb{Q}^p$ and $b \in \mathbb{Q}$ such that

$$\text{(i)} \ \ \{\beta\} = \{x | \langle x, c\rangle = b\} \cap \mathcal{Q} \ \ \text{and} \ \ \text{(ii)} \ \mathcal{Q} \subset \{x | \langle x, c\rangle \leq b\}.$$

If $v_1, \ldots, v_q$ are the columns of the incidence matrix of G, then by definition of $\mathcal{Q}$ one has $v_i + \beta \in \mathcal{Q}$ for all i. Thus

$$\langle v_i + \beta, c\rangle = \langle v_i, c\rangle + b \leq b \Rightarrow \langle v_i, c\rangle \leq 0 \quad (i = 1, \ldots, q).$$

Assume there are $\alpha \in \mathcal{Q}$, $\eta_1, \ldots, \eta_q \in \mathbb{N}$ such that $\beta = (\sum_{i=1}^q \eta_i v_i) + \alpha$, then

$$b = \langle \beta, c\rangle = \eta_1\langle v_1, c\rangle + \cdots + \eta_q\langle v_q, c\rangle + \langle \alpha, c\rangle \leq \langle \alpha, c\rangle \leq b.$$

Hence $\langle \alpha, c\rangle = b$ and by (i) we get $\alpha = \beta$. Thus x^β is a minimal generator of the canonical module $\omega_{K[G]}$. □

Definition 11.3.4 Let G be a bipartite graph with incidence matrix A and let C' be the matrix defining the shift polyhedron. If $C'A = -B'$, the polyhedron

$$\mathcal{Q}' = \{\widetilde{x} \in \mathbb{R}^q \,|\, \widetilde{x} \geq 0; B'\widetilde{x} \geq \mathbf{1}\}$$

is called the *blocking polyhedron*.

Theorem 11.3.5 *If G is a connected bipartite graph, then $\mathcal{Q}'$ is integral.*

Proof. It follows from Corollary 1.8.8 and Lemma 1.8.9. □

Proposition 11.3.6 *Let G be a connected bipartite graph and let $\mathcal{Q}$ be the shift polyhedron with respect to C'. If x^β is a minimal generator of $\omega_{K[G]}$ and A is the incidence matrix of G, then there is a vertex $\widetilde{\alpha}$ of the blocking polyhedron $\mathcal{Q}'$ such that $A\widetilde{\alpha} = \beta$.*

Proof. Let $v_1, \ldots, v_q$ be the column vectors of the matrix A and let C' be the matrix defining $\mathcal{Q}$. As A is totally unimodular (see Proposition 10.2.2), using Carathéodory's theorem (see Theorem 1.1.18) and Heger's theorem (see Theorem 1.6.4), after permuting the v_i's, we can write $\beta = \sum_{i=1}^r \eta_i v_i$

with $\eta_i \in \mathbb{N} \setminus \{0\}$ and $r \leq p-1 \leq q$, where $p = m+n$. We claim that $\eta_i = 1$ for all i. Assume $\eta_i > 1$. Take any row v of C'. Observe that $\langle v, v_k \rangle$ is equal to 0 or -1 for any v_k. The vector

$$\beta' = \eta_1 v_1 + \cdots + \eta_{i-1} v_{i-1} + (\eta_i - 1) v_i + \eta_{i+1} v_{i+1} + \cdots + \eta_r v_r$$

satisfies $\langle \beta', v \rangle \leq -1$ because $\langle \beta, v \rangle \leq -1$. Thus $\beta' \in \mathcal{Q}$, a contradiction because $\beta' = \beta - v_i$ and x^β is minimal. Thus $\eta_i = 1$. Notice that the vector $\widetilde{\alpha} = e_1 + \cdots + e_r$ satisfies $A\widetilde{\alpha} = \beta$, and from $C'A = -B'$ we get $\widetilde{\alpha} \in \mathcal{Q}'$. Consider the linear program

$$\begin{aligned} &\min \widetilde{x}_{r+1} + \cdots + \widetilde{x}_q \\ &\text{subject to } B'\widetilde{x} \geq \mathbf{1} \text{ and } \widetilde{x} \geq 0 \end{aligned}$$

and notice that 0 is the optimum value of this linear program because $\widetilde{\alpha}$ is in $\mathcal{Q}'$. By Theorem 11.3.5 there is an integral vertex $\widetilde{\gamma}$ of $\mathcal{Q}'$ where the minimum is attained. Hence $\widetilde{\gamma}_i = 0$ for $i > r$. By Exercise 11.3.8 the vector γ has $\{0,1\}$-entries. If $\widetilde{\gamma}_k = 0$ for some $1 \leq k \leq r$, then $A\widetilde{\gamma} = \sum_{i \neq k} \epsilon_i v_i \in \mathcal{Q}$, where $\epsilon_i \in \{0,1\}$ for all $i \neq k$, a contradiction to the minimality of x^β. Thus $\widetilde{\alpha} = \widetilde{\gamma}$, as required. Observe that the last part of the argument works even if $q = r$, which is the case of a tree. □

Theorem 11.3.7 *If G is a connected bipartite graph with incidence matrix A and Q' is the blocking polyhedron, then the canonical module $\omega_{K[G]}$ of $K[G]$ is generated by the set*

$$\{x^{A\widetilde{\alpha}} \mid \widetilde{\alpha} \text{ is a vertex of } \mathcal{Q}'\}.$$

Proof. It follows by recalling that the blocking polyhedron $\mathcal{Q}'$ is integral and using Proposition 11.3.6. □

Exercise

11.3.8 *If B is a $\{0,1\}$-matrix, then any integral vertex of the rational polyhedron $\mathcal{Q} = \{x \mid x \geq 0; Bx \geq \mathbf{1}\}$ is a $\{0,1\}$-vector.*

11.4 Computing the a-invariant

Let G be a bipartite graph. In order to use the results of Section 11.3 in an efficient way, we introduce other representations of the shift polyhedron. For each independent set of vertices $\mathfrak{A}$ of G consider the vector

$$\alpha_{\mathfrak{A}} = \sum_{x_i \in \mathfrak{A}} e_i - \sum_{x_i \in N(\mathfrak{A})} e_i.$$

By Corollary 10.7.17, there exists an irreducible representation of the edge cone of G, as an intersection of closed half-spaces, of the form:

$$\mathbb{R}_+\mathcal{A} = \operatorname{aff}(\mathbb{R}_+\mathcal{A}) \cap H^-_{\alpha_{\mathfrak{A}_1}} \cap \cdots \cap H^-_{\alpha_{\mathfrak{A}_r}} \cap H^-_{-e_{i_1}} \cap \cdots \cap H^-_{-e_{i_s}}, \tag{11.5}$$

where for each i either $\mathfrak{A}_i \subsetneq V_1$ or $\mathfrak{A}_i \subsetneq V_2$ and none of the half-spaces can be omitted from the intersection. Let us denote by C the matrix whose rows are the vectors $\alpha_{\mathfrak{A}_1}, \ldots, \alpha_{\mathfrak{A}_r}, -e_{i_1}, \ldots, -e_{i_s}$ and by C' the matrix defining the shift polyhedron, as defined in Section 11.2.

The next result says that the shift polyhedrons $\mathcal{Q}$ and $\mathcal{Q}_1$ with respect to C' and C, respectively, are equal.

Theorem 11.4.1 *If G is a connected bipartite simple graph, then the shift polyhedron $\mathcal{Q}$ of the edge cone of G, with respect to C', is given by*

$$\mathcal{Q} = \operatorname{aff}(\mathbb{R}_+\mathcal{A}) \cap \{x \mid Cx \leq -\mathbf{1}\}.$$

Proof. It follows from Exercises 11.4.7, 11.4.8, and Lemma 10.7.15. □

Remark 11.4.2 If $CA = -B$ and $C'A = -B'$, where A is the incidence matrix of G, then B and B' define the same blocking polyhedron.

There are linear programming techniques to convert the description of a rational polyhedron given by a "finite basis" into an irreducible representation as intersection of closed half-spaces and vice versa. These techniques have been converted into very efficient routines in several programming environments; see for instance PORTA [84]. Thus, one can effectively compute a generating set for the ideal $\omega_{K[G]}$ (see Example 11.4.4 for an illustration).

Remark 11.4.3 To compute the vertices of a shift polyhedron of an edge cone using PORTA we need a "valid" point. Note that if $\mathcal{A} = \{v_1, \ldots, v_q\}$ is the set of column vectors of the incidence matrix of G, then the point $\alpha = \sum_{i=1}^{q} v_i = (\deg(x_1), \ldots, \deg(x_{n+m}))$ is *valid*, that is, $\alpha \in \mathcal{Q}$:

Example 11.4.4 Consider the following bipartite simple graph G:

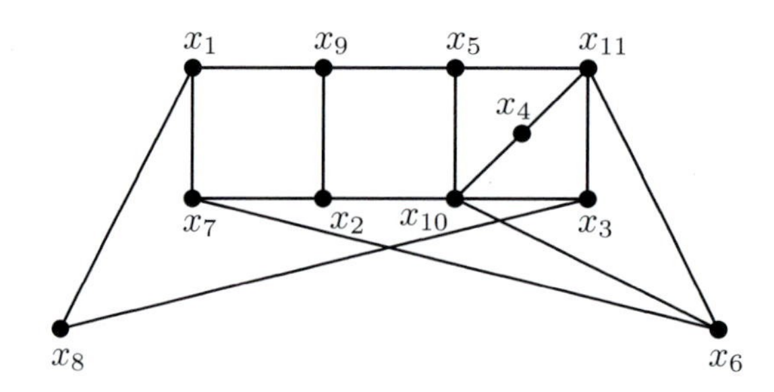

with bipartition $V_1 = \{x_1, \ldots, x_6\}$ and $V_2 = \{x_7, \ldots, x_{11}\}$. The incidence matrix of G, denoted by A, is the transpose of the following matrix. We will display the data as input files for PORTA.

```
CONE_SECTION (columns of A)                corresponding monomial
(  1) 1 0 0 0 0 0 1 0 0 0 0                      x1x7
(  2) 1 0 0 0 0 0 0 1 0 0 0                      x1x8
(  3) 0 1 0 0 0 0 1 0 0 0 0                      x2x7
(  4) 0 1 0 0 0 0 0 0 1 0 0                      x2x9
(  5) 0 0 1 0 0 0 0 1 0 0 0                      x3x8
(  6) 0 0 1 0 0 0 0 0 0 1 0                      x3x10
(  7) 0 0 1 0 0 0 0 0 0 0 1                      x3x11
(  8) 0 0 0 1 0 0 0 0 0 1 0                      x4x10
(  9) 0 0 0 0 1 0 0 0 0 1 0                      x5x10
( 10) 0 0 0 0 0 1 0 0 0 1 0                      x6x10
( 11) 0 0 0 0 0 1 0 0 0 0 1                      x6x11
( 12) 0 0 0 0 1 0 0 0 1 0 0                      x5x9
( 13) 0 0 0 1 0 0 0 0 0 0 1                      x4x11
( 14) 1 0 0 0 0 0 0 0 1 0 0                      x1x9
( 15) 0 1 0 0 0 0 0 0 0 1 0                      x2x10
( 16) 0 0 0 0 0 1 1 0 0 0 0                      x6x7
( 17) 0 0 0 0 1 0 0 0 0 0 1                      x5x11
```

Applying PORTA to this input file we obtain an irreducible representation of $\mathbb{R}_+\mathcal{A}$, which immediately yields the following representation of $\mathcal{Q}$.

```
VALID POINT: 3 3 3 2 3 3 3 2 3 5 4 (see previous Remark)
INEQUALITIES_SECTION
      -xi<= -1 (for i=7,...,11)
( 1)  x1+x2+x3+x4+x5+x6-x7-x8-x9-x10-x11   == 0
( 1)  x1+x3+x4+x5+    x6-x7-x8-x9-x10-x11 <= -1
( 2)  x1+x2+x4+x5+    x6-x7-x8-x9-x10-x11 <= -1
( 3)  x1+x2+x3+x5+    x6-x7-x8-x9-x10-x11 <= -1
( 4)  x1+x2+x3+x4+    x6-x7-x8-x9-x10-x11 <= -1
( 5)  x1+x2+x3+x4+x5    -x7-x8-x9-x10-x11 <= -1
( 6)            x4                -x10-x11 <= -1
( 7)     x2                -x7  -x9 -x10     <= -1
( 8)            x4+   x6-x7       -x10-x11 <= -1
( 9)            x4+x5          -x9-x10-x11 <= -1
(10)         x3+x4          -x8   -x10-x11 <= -1
(11)  x1+x2            -x7-x8-x9-x10       <= -1
(12)         x3+x4   +x6-x7-x8    -x10-x11 <= -1
(13)         x3+x4+x5       -x8-x9-x10-x11 <= -1
(14)     x2   +x4+x5+x6-x7    -x9-x10-x11 <= -1
(15)  x1                -x7-x8-x9          <= -1
(16)     x2+x3+x4+x5+x6-x7-x8-x9-x10-x11 <= -1
```

Using PORTA we get that the vertices of the shift polyhedron are:

```
(1) 1 2 1 1 1 1 1 1 1 1 3        (5) 1 1 1 1 1 1 1 1 1 2 1
(2) 2 1 1 1 1 1 1 1 1 1 3        (6) 1 1 1 1 1 1 1 1 2 1 1
(3) 2 1 1 1 1 1 1 1 1 3 1        (7) 1 1 1 1 1 1 2 1 1 1 1
(4) 1 1 1 1 1 1 1 1 1 1 2
```

From the equality $CA = -B$ (Remark 11.4.2), we get that the corresponding blocking polyhedron is defined by

```
INEQUALITIES_SECTION
xi>=0 (for i=1,...,17)
( 1) x15+x3+x4>=1
( 2) x5+x6+x7>=1
( 3) x13+x8>=1
( 4) x12+x17+x9>=1
( 5) x10+x11+x16>=1
( 6) x1+x16+x3>=1
( 7) x2+x5>=1
( 8) x12+x14+x4>=1
( 9) x10+x15+x6+x8+x9>=1
(10) x11+x13+x17+x7>=1
(11) x10+x11+x15+x17+x6+x7+x9>=1
(12) x1+x10+x12+x14+x16+x6+x8+x9>=1
(13) x1+x15+x17+x3+x6+x7+x9>=1
(14) x10+x11+x14+x15+x4+x6+x7>=1
(15) x10+x11+x15+x17+x2+x9>=1
(16) x10+x12+x16+x5+x6+x8+x9>=1
(17) x1+x15+x17+x2+x3+x9>=1
(18) x10+x11+x14+x15+x2+x4>=1
(19) x1+x14+x6+x7>=1
(20) x12+x16+x3+x4+x5>=1
(21) x1+x14+x2>=1
```

Using PORTA we get that the blocking polyhedron $\mathcal{Q}'$ has 173 vertices. The distinct images of those vertices under the incidence matrix A are:

```
1 2 1 1 1 1 1 1 1 1 3  vertex of the shift polyhedron Q
2 1 1 1 1 1 1 1 1 1 3  vertex of Q
2 1 1 1 1 1 1 1 1 3 1  vertex of Q
1 1 1 1 1 1 1 1 1 1 2  vertex of Q
1 1 1 1 1 1 1 1 1 2 1  vertex of Q
1 1 1 1 1 1 1 1 2 1 1  vertex of Q
1 1 1 1 1 1 2 1 1 1 1  vertex of Q
2 1 1 1 1 1 1 1 1 2 2  not a vertex of Q but correspond to a
                       minimal generator of the canonical module
1 1 1 1 1 2 1 1 2 2 1  this and the following vectors correspond to
1 1 2 1 1 1 1 1 2 2 1  redundant monomials in the canonical module
.....................
```

By Theorem 11.3.7, $\omega_{K[G]}$ is minimally generated by the eight monomials corresponding to the first eight vectors above. Thus type$(K[G]) = 8$ and the rank of the last module in the graded free resolution of $K[G]$ is 8.

Remark 11.4.5 From Theorem 11.3.1 and Theorem 11.4.1 we obtain an effective method to compute the a-invariant of $K[G]$ that only requires a description of the shift polyhedron by linear inequalities and to solve a linear program. See Example 11.4.6.

Example 11.4.6 Consider the following bipartite graph G:

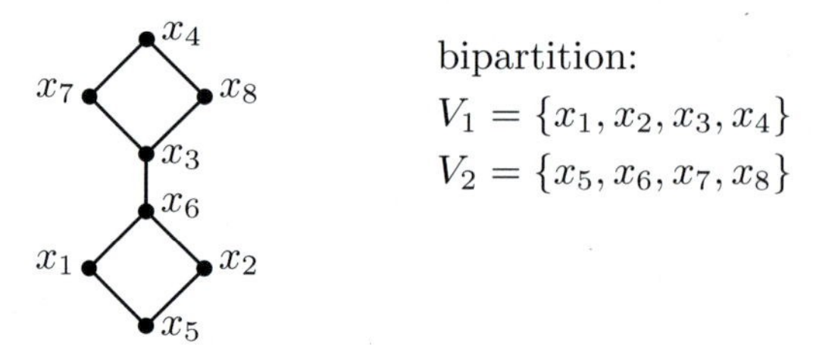

In order to estimate the a-invariant of G we set up the next linear program using the following input file for *Mathematica* [431]

```
vars:={x1,x2,x3,x4,x5,x6,x7,x8}
f:=(x1+x2+x3+x4+x5+x6+x7+x8)/2
ieq:={x1+x2+x3+x4-x5-x6-x7-x8 == 0, -x1<= -1, -x2<= -1,
     -x4<= -1, -x5<= -1, -x7<= -1,-x8 <= -1,x4-x7-x8 <= -1,
      -x3-x4+x7+x8 <= -1,x3+x4-x6-x7-x8 <= -1}
ConstrainedMin[f,ieq,vars]
```

where the set of inequalities was found using PORTA and it comes from an irreducible representation of the edge cone as in Eq. (11.5). The optimal value of this linear program is equal to 5 and is attained at the vertex $(1,1,2,1,1,2,1,1)$. Hence by Theorems 11.3.1 and 11.4.1 we get that the a-invariant of $K[G]$ is equal to -5.

Exercises

11.4.7 [189, Lemma 7.3.2] Let $G = G(m,n)$ be a connected bipartite graph. If $\mathcal{Q}$ is the shift polyhedron of G, with respect to C', and $x \in \mathcal{Q}$, then $x_i \geq 1$ for all $i = 1,\ldots,m+n$.

11.4.8 [189, Proposition 7.3.3] Let G be a connected bipartite graph. If $\mathcal{Q}$ is the shift polyhedron of G, with respect to C', and $x \in \mathcal{Q}$, then for $\mathfrak{A} \subsetneq V_1$ (resp. $\mathfrak{A} \subsetneq V_2$) such that $N(\mathfrak{A}) \subset \mathfrak{A}' \subset V_2$ (resp. $N(\mathfrak{A}) \subset \mathfrak{A}' \subset V_1$) one has

$$\sum_{x_i \in \mathfrak{A}} x_i - \sum_{x_i \in \mathfrak{A}'} x_i \leq -1.$$

11.5 Algebraic invariants of edge subrings

In this section we give upper bounds for the a-invariant of the edge subring $K[G]$ valid for any bipartite connected graph G. We examine the Gorenstein property and the type of $K[G]$. As a tool we use a technique, introduced in Section 10.1, based on decomposing a graph into blocks.

Proposition 11.5.1 *If G is a connected bipartite graph with a bipartition (V_1, V_2) and $|V_1| = m \leq |V_2| = n$, then $a(K[G]) \leq -n$.*

Proof. Let $V_1 = \{x_1,\ldots,x_m\}$. Take a monomial x^β in the canonical module of $K[G]$, that is, $\beta = (\beta_i)$ is an integral vector in $\mathrm{ri}(\mathbb{R}_+\mathcal{A})$. One may assume $m \geq 2$ because if $m = 1$ the graph G is a star and the result is clear. Thus $\beta_i \geq 1$ by Corollary 1.1.45, and $\sum_{i=1}^m \beta_i = \sum_{i=m+1}^{n+m} \beta_i$ because the edge cone lies in the hyperplane

$$\sum_{i=1}^{m} x_i = \sum_{i=m+1}^{m+n} x_i.$$

Hence $\deg(x^\beta)$, the normalized degree of x^β, is $\sum_{i=m+1}^{m+n}\beta_i$. Therefore $\deg(x^\beta)\geq n$, which proves the required inequality. □

Corollary 11.5.2 [63] *Let $\mathcal{K}_{m,n}$ be the complete bipartite graph. If $n\geq m$, then $a(K[\mathcal{K}_{m,n}])=-n$.*

Proof. By Exercise 10.7.26 and Proposition 11.5.1 it suffices to observe that the monomial x^β with

$$\beta=(\underbrace{n-m+1,1,\ldots,1}_{m\text{ entries}},\underbrace{1,\ldots,1}_{n\text{ entries}})$$

is in the canonical module of $K[G]$ and has degree equal to n. □

There are 2-connected graphs where the a-invariant does not reach the upper bound of Proposition 11.5.1 (see Exercise 11.5.14).

A formula for the type of a determinantal ring is due to Brennan; see [73, p. 115]. We show a special case of this formula.

Proposition 11.5.3 *Let $G=\mathcal{K}_{m,n}$ be the complete bipartite graph. If $n\geq m$, then $K[G]$ is a level algebra and its type is given by:*

$$\text{type}(K[G])=\binom{n-1}{m-1}.$$

Proof. From Exercise 10.7.26 it is seen that the canonical module of $K[G]$ is minimally generated by the set of monomials of the form $x_1^{a_1}\cdots x_m^{a_m}y_1\cdots y_n$, such that $a_1+\cdots+a_m=n$ and $a_i\geq 0$ for all i. Thus the number of such partitions is the type. As the canonical module is generated in a single degree the algebra $K[G]$ is level. □

Proposition 11.5.4 *If G is a bipartite connected graph with bipartition (V_1,V_2) and $|V_1|=m\leq|V_2|=n$, then $a(K[G])=-m$ if and only if $m=n$ and $\mathbf{1}=(1,\ldots,1)$ is in $\text{ri}(\mathbb{R}_+\mathcal{A})$.*

Proof. $\Rightarrow$) By Corollary 1.1.45, the canonical module $\omega_{K[G]}$ of the ring $K[G]$ is generated by monomials x^β such that all the entries of β are positive integers. As $n=m$ we derive that $x^{\mathbf{1}}=x_1x_2\cdots x_{m+n}$ is in the canonical module and it is the only monomial of normalized degree n in $\omega_{K[G]}$.

$\Leftarrow$) It follows from Proposition 11.5.1. □

Definition 11.5.5 Let G be a graph. A cycle containing all the vertices of G is said to be a *Hamilton cycle*, and a graph containing a Hamilton cycle is said to be *Hamiltonian*.

Corollary 11.5.6 *If G is a Hamiltonian bipartite graph with $2n$ vertices, then $a(K[G])=-n$.*

Proof. Let H be a Hamiltonian cycle of G and let $2n$ be the number of vertices of G. Note that $a(K[H]) = -n$ (because $K[H]$ is the ring of a hypersurface of degree n) and $\dim K[H] = \dim K[G]$. Hence one has $a(K[H]) \leq a(K[G])$ and consequently $a(K[G]) = -n$. $\Box$

Corollary 11.5.7 *If $G = \mathcal{K}_{n,n}$ is a complete bipartite graph, then the a-invariant of $K[G \setminus \{e\}]$ is equal to $-n$.*

Proof. Note that for $n \geq 3$ the graph $G \setminus \{e\}$ is Hamiltonian. $\Box$

Decomposing a bipartite graph into blocks Let G be a connected bipartite graph. Next we describe in a condensed form a technique that can be used to simplify the computation of the a-invariant and the type of $K[G]$. It is based in the graph theoretical notion of block (see Section 7.1) and the following result.

Proposition 11.5.8 *If B and C are two standard K-algebras, then*

$$a(B \otimes_K C) = a(B) + a(C).$$

If in addition B and C are Cohen–Macaulay, then

$$\text{type}(B \otimes_K C) = \text{type}(B) \cdot \text{type}(C).$$

Proof. The first equality follows from the proof of Proposition 5.1.11. For the second equality, see [171]. $\Box$

Proposition 11.5.9 *If $G_1, \ldots, G_r$ are the blocks of a connected bipartite graph, then*

$$a(K[G]) = \sum_{i=1}^{r} a(K[G_i]) \quad \textit{and} \quad \text{type}(K[G]) = \prod_{i=1}^{r} \text{type}(K[G_i]).$$

Proof. Since G is connected and bipartite one has

$$K[G] \simeq K[G_1] \otimes_K K[G_2] \otimes_K \cdots \otimes_K K[G_r].$$

Hence the result follows from Proposition 11.5.8. $\Box$

The next theorem is a combinatorial obstruction for an edge subring to be Gorenstein. Because of Proposition 11.5.9 it suffices to consider the class of Gorenstein subrings of 2-connected bipartite graphs.

Theorem 11.5.10 *Let G be a Gorenstein connected bipartite graph with bipartition (V_1, V_2), then G has a perfect matching and furthermore*

$$|\mathfrak{A}| < |N(\mathfrak{A})| \textit{ for all } \mathfrak{A} \subsetneq V_1.$$

Proof. Let x^β be the generator of $\omega_{K[G]}$. For each $1 \leq i \leq m+n$ choose a spanning tree of $G \setminus \{x_i\}$ and enlarge this to a spanning tree $\mathcal{T}_i$ of G. Note that the monomial x^{γ_i} with $\gamma_i = (\deg_{\mathcal{T}_i}(x_1), \ldots, \deg_{\mathcal{T}_i}(x_{n+m}))$ and $\deg_{\mathcal{T}_i}(x_i) = 1$ is the generator of the canonical module of $K[\mathcal{T}_i]$. Hence since $\mathrm{aff}(\mathbb{R}_+\mathcal{A}_i) = \mathrm{aff}(\mathbb{R}_+\mathcal{A})$, where $\mathbb{R}_+\mathcal{A}_i$ is the edge cone of $\mathcal{T}_i$, we get $x^{\gamma_i} \in \omega_{K[G]}$. Therefore $\beta_i = 1$, and consequently $\beta = \mathbf{1}$. Noting that β is in the affine space generated by the edge cone of G it follows $m = n$. Thus from Theorem 10.7.16 we obtain $|\mathfrak{A}| < |N(\mathfrak{A})|$ for all $\mathfrak{A} \subsetneq V_1$. $\square$

The converse of Theorem 11.5.10 does not hold.

Example 11.5.11 Consider the bipartite graph G:

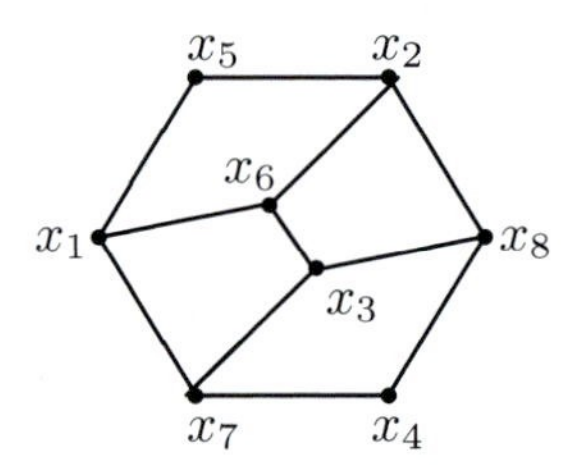

Using the following procedure in *Macaulay*2 [199], we rapidly obtain that $K[G]$ has type equal to three. On the other hand $\mathbf{1} = (1, \ldots, 1)$ is in the relative interior of the edge cone. In this example the shift polyhedron has vertices $\mathbf{1}$, $(2,1,1,1,1,1,1,2)$ and $(1,2,1,1,1,1,2,1)$.

```
KK=ZZ/31991
R=KK[x_1..x_8,t_1..t_11,MonomialOrder=>Eliminate 8]
I=ideal(t_1-x_1*x_5, t_2-x_1*x_6, t_3-x_1*x_7, t_4-x_2*x_5,
        t_5-x_2*x_6, t_6-x_2*x_8, t_7-x_3*x_6, t_8-x_3*x_7,
        t_9-x_3*x_8, t_10-x_4*x_7, t_11-x_4*x_8)
J= ideal selectInSubring(1,gens gb I)
M=coker gens J
res M
```

Corollary 11.5.12 *Let G be a bipartite connected graph with bipartition (V_1, V_2) and $|V_1| = m \leq |V_2| = n$. If G has no cut vertices and $K[G]$ is Gorenstein, then G has a perfect matching and $a(K[G]) = -n$.*

Proof. It follows from the proof of Theorem 11.5.10. $\square$

If one of the blocks of a graph G is a bridge e one can say a bit more about the relationship between G and $G \setminus \{e\}$.

Proposition 11.5.13 *Let G be a graph with a bridge e. If $G \setminus \{e\} = G_1 \cup G_2$ with G_1 and G_2 disjoint graphs such that $K[G_i]$ is Gorenstein for $i = 1, 2$, then $K[G]$ is Gorenstein.*

Proof. There is an isomorphism $K[G] \simeq (K[G_1] \otimes_K K[G_2])[t]$, where t is a variable. Since the tensor product of Gorenstein rings is Gorenstein and a polynomial ring over a Gorenstein ring is Gorenstein, the result follows. $\square$

Exercises

11.5.14 Consider the following bipartite graph G:

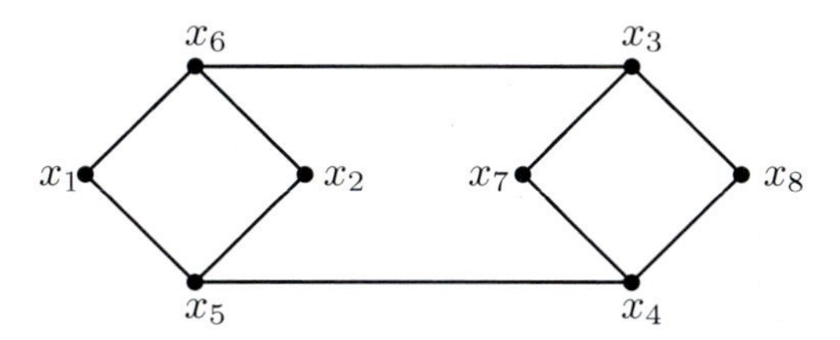

Using Theorems 11.3.1 and 11.4.1, show that $a(K[G]) = -5$ and that the shift polyhedron has four vertices.

11.5.15 Prove that the following are the equations of the edge cone of the graph G shown on the left.

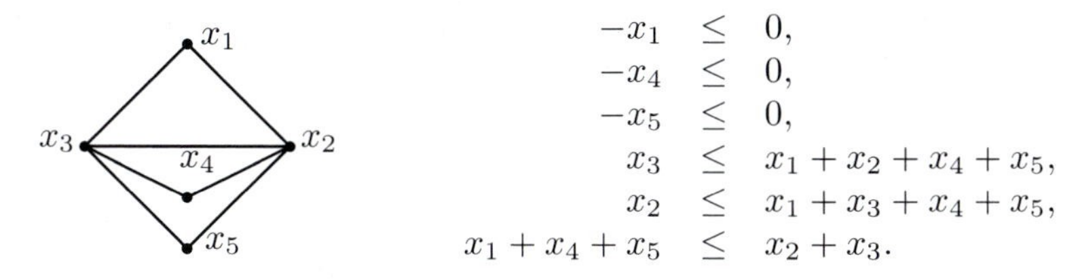

$$\begin{aligned} -x_1 &\leq 0, \\ -x_4 &\leq 0, \\ -x_5 &\leq 0, \\ x_3 &\leq x_1 + x_2 + x_4 + x_5, \\ x_2 &\leq x_1 + x_3 + x_4 + x_5, \\ x_1 + x_4 + x_5 &\leq x_2 + x_3. \end{aligned}$$

Show that the a-invariant of $K[G]$ is equal to -4.

11.5.16 Let G be the graph consisting of two squares joined by an edge. Prove that $K[G]$ is a Gorenstein ring, its a-invariant is -5, it has a perfect matching, and $\alpha = (1, \ldots, 1)$ is not in the relative interior of the edge cone of G. Prove that the monomial corresponding to $(1, 1, 2, 1, 1, 2, 1, 1)$ generates the canonical module of $K[G]$.

11.5.17 Let G be a graph with a bridge e. If $G \setminus \{e\} = G_1 \cup G_2$ with G_1 and G_2 disjoint graphs, then

$$a(G) = a(G_1) + a(G_2) - 1.$$

11.5.18 Let G be a graph and let $\mathcal{A}$ be the set of column vectors of its incidence matrix. If $K[G]$ is normal, then the canonical module of $K[G]$ is:

$$\omega_{K[G]} = (\{x^a \mid a \in \mathbb{N}\mathcal{A} \cap \mathrm{ri}(\mathbb{R}_+\mathcal{A})\}).$$

Use this formula to prove the following:

(a) The *join* $G_1 * G_2$ of two disjoint graphs G_1 and G_2 consists of $G_1 \cup G_2$ and all the lines joining G_1 with G_2. If G_1, G_2 are connected graphs on n vertices and $K[G_1]$, $K[G_2]$ are normal, then

$$a(K[G_1 * G_2]) = -n.$$

(b) Let G be a connected non-bipartite graph on n vertices. If $K[G]$ is normal, then

$$a(K[G]) - 1 \leq a(K[C(G)]) \leq -\left\lceil \frac{n+1}{2} \right\rceil.$$

(c) Let G_1, G_2 be two connected non-bipartite graphs. If $K[G_1]$ and $K[G_2]$ are normal, then

$$a(K[G_1]) + a(K[G_2]) \leq a(K[G_1 * G_2]).$$

Chapter 12

Normality of Rees Algebras of Monomial Ideals

Let $R = K[x_1, \ldots, x_n]$ be a polynomial ring over a field K and let I be an ideal of R generated by a finite set $F = \{x^{v_1}, \ldots, x^{v_q}\}$ of monomials. In this chapter we study the integral closure of I, and the normality and invariants of $R[It]$, the Rees algebra of I. The normalization of a Rees algebra is examined using the Danilov–Stanley formula, Carathéodory's theorem, and Hilbert bases of Rees cones.

Interesting classes of normal ideals, such as ideals of Veronese type and polymatroidal ideals, are introduced in the chapter. The divisor class group of a normal Rees algebra is computed using polyhedral geometry.

12.1 Integral closure of monomial ideals

Proposition 12.1.1 [275] *Let I be a monomial ideal of a polynomial ring R over a field K. Then $\overline{I}$, the integral closure of I, is a monomial ideal.*

Proposition 12.1.2 *Let $R = K[x_1, \ldots, x_n]$ be a polynomial ring over a field K and let $I \subset R$ be a monomial ideal. Then*

(a) $\overline{I} = (x^\alpha \mid x^{m\alpha} \in I^m$ *for some* $m \geq 1)$, *and*

(b) $\overline{I} = (x^\alpha \mid \alpha \in \operatorname{conv}(\log(I)) \cap \mathbb{Z}^n)$.

Proof. (a): If $x^{m\alpha} \in I^m$, then x^α satisfies a polynomial of the form $z^m + a_m$, with $a_m \in I^m$, hence x^α is in $\overline{I}$. On the other hand if $z = x^\alpha \in \overline{I}$, then there is an equation $z^r + a_1 z^{r-1} + \cdots + a_{r-1} z + a_r = 0$ with $a_i \in I^i$

for all i. Since I is a monomial ideal one obtains $z^m \in I^m$ for some $m \geq 1$. Observing that $\overline{I}$ is a monomial ideal the asserted equality follows.

(b): "$\supset$": Let $\alpha \in \operatorname{conv}(\log I) \cap \mathbb{Z}^n$. One can write $\alpha = \sum_{i=1}^q \mu_i \beta_i$, where $\mu_i \in \mathbb{Q}_+$, $\sum_{i=1}^q \mu_i = 1$ and $x^{\beta_i} \in I$. Pick $m \in \mathbb{N}_+$ such that $m\mu_i \in \mathbb{N}$ for all i. Then $x^{m\alpha} \in I^m$ and $x^\alpha \in \overline{I}$.

"$\subset$": Since I is generated by monomials for any $x^\alpha \in \overline{I}$ there is $m \in \mathbb{N}_+$ such that $x^{m\alpha} \in I^m$. It follows readily that $\alpha \in \operatorname{conv}(\log I)$. $\square$

Example 12.1.3 If $R = K[x_1, x_2, x_3]$ and $I = (x_1^3, x_2 x_3^2)$, then using the program *Normaliz* [68] we get $\overline{I} = (x_1^3, x_2 x_3^2, x_1^2 x_2 x_3)$.

A geometric description of the integral closure Let $\alpha \in \mathbb{Q}_+^n$, where $\mathbb{Q}_+$ is the set of nonnegative rational numbers. We define the *upper right corner* or *ceiling* of α as the vector $\lceil \alpha \rceil$ whose entries are given by

$$\lceil \alpha \rceil_i = \begin{cases} \alpha_i & \text{if } \alpha_i \in \mathbb{N}, \\ \lfloor \alpha_i \rfloor + 1 & \text{if } \alpha_i \notin \mathbb{N}, \end{cases}$$

where $\lfloor \alpha_i \rfloor$ stands for the integer part of α_i. Accordingly we can define the *ceiling* of any vector in $\mathbb{R}^n$ or the *ceiling* of any real number. Let $\operatorname{conv}(v_1, \ldots, v_q)$ be the convex hull (over the rationals), that is,

$$\operatorname{conv}(v_1, \ldots, v_q) = \left\{ \sum_{i=1}^q \lambda_i v_i \,\middle|\, \sum_{i=1}^q \lambda_i = 1,\ \lambda_i \in \mathbb{Q}_+ \right\}$$

is the set of all *convex combinations* of $v_1, \ldots, v_q$.

Proposition 12.1.4 *Let $R = K[x_1, \ldots, x_n]$ be a polynomial ring over a field K. If $I \subset R$ is an ideal generated by monomials $x^{v_1}, \ldots, x^{v_q}$, then*

$$\overline{I} = \left(\left\{ x^{\lceil \alpha \rceil} \,\middle|\, \alpha \in \operatorname{conv}(v_1, \ldots, v_q) \right\} \right)$$

and $\overline{I}$ is generated by all x^a with $a \in (\operatorname{conv}(v_1, \ldots, v_q) + [0,1)^n) \cap \mathbb{N}^n$.

Proof. "$\supset$": Let $\alpha = \sum_{i=i}^q \lambda_i v_i$ be a convex combination of $v_1, \ldots, v_q$ with $\lambda_i \in \mathbb{Q}_+$ for all i. Since $\lceil \alpha \rceil \geq \alpha$, there is $\beta \in \mathbb{Q}_+^n$ such that $\lceil \alpha \rceil = \beta + \alpha$. Hence there is $0 \neq p \in \mathbb{N}$ so that $p\beta \in \mathbb{N}^n$ and $p\lambda_i \in \mathbb{N}$ for all i. Therefore

$$x^{p\lceil \alpha \rceil} = x^{p\beta} x^{p\alpha} = x^{p\beta} (x^{v_1})^{p\lambda_1} \cdots (x^{v_q})^{p\lambda_q} \in I^p \ \Rightarrow\ x^{\lceil \alpha \rceil} \in \overline{I}.$$

"$\subset$": Let $x^\gamma \in \overline{I}$, that is, $x^{p\gamma} \in I^p$ for some $0 \neq p \in \mathbb{N}$. There are nonnegative integers $s_1, \ldots, s_q$ such that

$$x^{p\gamma} = x^\delta (x^{v_1})^{s_1} \cdots (x^{v_q})^{s_q} \text{ and } s_1 + \cdots + s_q = p.$$

Hence $\gamma = (\delta/p) + \sum_{i=1}^q (s_i/p) v_i$. We set $\alpha = \sum_{i=1}^q (s_i/p) v_i$. By dividing the entries of δ by p, we can write $\gamma = \theta + \beta + \alpha$, where $0 \leq \beta_i < 1$ for all i and $\theta \in \mathbb{N}^n$. Notice $\beta + \alpha \in \mathbb{N}^n$. It is seen that $\lceil \alpha \rceil = \beta + \alpha$. Thus $x^\gamma = x^\theta x^{\lceil \alpha \rceil}$ with $\alpha \in \operatorname{conv}(v_1, \ldots, v_q)$ as required. $\square$

Example 12.1.5 If $I = (x_1^3, x_2^4) \subset K[x_1, x_2]$, by Proposition 12.1.4, $\overline{I}$ is generated by the monomials of the points marked in the figure:

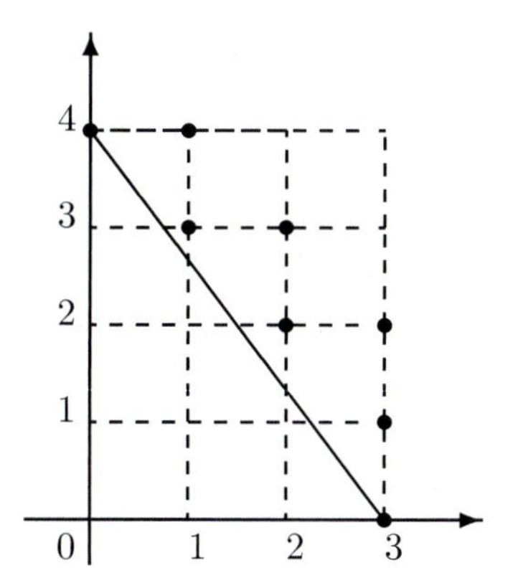

Thus $\overline{I} = (x_1^3, x_2^4, x_1x_2^3, x_1^2x_2^2)$.

Corollary 12.1.6 *Let $I \subset R = K[x_1, \ldots, x_n]$ be an ideal generated by monomials of degree at most k. The following hold.*

(a) *$\overline{I}$ is generated by monomials of degree at most $k + n - 1$.*

(b) *If $\operatorname{ht}(I) = n$, then $\overline{I}$ is generated by monomials of degree at most k.*

Proof. (a) Let $G(I) = \{x^{v_1}, \ldots, x^{v_q}\}$ be the minimal set of generators of I. Take $\alpha = (\alpha_i) \in \operatorname{conv}(v_1, \ldots, v_q)$ and set $\beta = (\beta_i) = \lceil \alpha \rceil$. As β_i is equal to $\alpha_i + \delta_i$, for some δ_i in $[0, 1)$, one has $\deg(x^\beta) < k + n$, as required.

(b) One has $\overline{(x_1^k, \ldots, x_n^k)} = (x_1, \ldots, x_n)^k \subset \overline{I}$ because $x_i^k \in I$ for all i. Therefore $\overline{I}$ does not require minimal generators of degree $k + 1$. $\square$

Complete and $\mathfrak{m}$-full ideals in dimension two Let $R = K[x_1, x_2]$ be a polynomial ring over an infinite field K, let $\mathfrak{m} = (x_1, x_2)$ and let I be an $\mathfrak{m}$-primary ideal of R minimally generated by q monomials, $q = \mu(I)$, that are listed lexicographically,

$$I = (x_1^{a_1}, x_1^{a_2}x_2^{b_{q-1}}, \ldots, x_1^{a_i}x_2^{b_{q-i+1}}, \ldots, x_1^{a_{q-1}}x_2^{b_2}, x_2^{b_1})$$

with $a_1 > a_2 > \cdots > a_{q-1} > a_q := 0$, $b_1 > b_2 > \cdots > b_{q-1} > b_q := 0$. We assume that $I \neq \mathfrak{m}^{q-1}$.

The ideal I is said to be $\mathfrak{m}$-*full* if $(\mathfrak{m}I\colon f) = I$ for some $f \in \mathfrak{m} \setminus \mathfrak{m}^2$ or equivalently I is $\mathfrak{m}$-full if $(I\colon \mathfrak{m}) = (I\colon f)$ [259, Proposition 14.1.6]. A useful characterization of $\mathfrak{m}$-full ideals is the following result.

Theorem 12.1.7 (Rees–Watanabe [425, Theorem 4]) *If I is an $\mathfrak{m}$-primary ideal of a regular local ring $(R, \mathfrak{m})$ of dimension two, then I is $\mathfrak{m}$-full if and only if for all ideals $I \subset J$, $\mu(J) \leq \mu(I)$.*

The *order* of I is $\mathrm{ord}(I) := \max\{k | I \subset \mathfrak{m}^k\}$. In regular local rings of dimension two one has $\mu(I) \leq \mathrm{ord}(I) + 1$ [259, Lemma 14.1.3].

Lemma 12.1.8 *If $\mu(I) = \mathrm{ord}(I) + 1$, then I is $\mathfrak{m}$-full.*

Proof. Let J be an ideal of R. Note that $I \subset J$ implies $\mathrm{ord}(J) \leq \mathrm{ord}(I)$. Since $\mu(J) \leq \mathrm{ord}(J) + 1 \leq \mu(I)$, I satisfies Theorem 12.1.7. □

The following result classifies all $\mathfrak{m}$-full monomial ideals of $K[x_1, x_2]$.

Theorem 12.1.9 [181] *I is $\mathfrak{m}$-full if and only if there is $1 \leq k \leq q$ such that the following conditions hold*

(i) $b_{q-i} - b_{q-i+1} = 1$ *for* $1 \leq i \leq k - 1$,

(ii) $k = q$ *or* $k < q$ *and* $b_{q-k} - b_{q-k+1} \geq 2$,

(iii) $a_i - a_{i+1} = 1$ *for* $k \leq i \leq q - 1$.

Proof. $\Rightarrow$) If I is $\mathfrak{m}$-full, $\mathrm{ord}(I) \leq q - 1$, as otherwise $I \subset (x_1, x_2)^q$, which has $q + 1$ minimal generators, which would violate Theorem 12.1.7. Thus $\mathrm{ord}(I) \leq q - 1$. Let $x_1^k x_2^{q-1-k}$ be a monomial of I of degree $q - 1$. This means that there are at most $q - 1 - k$ elements prior to $x_1^k x_2^{q-1-k}$ and at most k elements after. This gives

$$I = (x_1^{a_1}, x_1^{a_2}x_2, \ldots, x_1^{a_{k-1}}x_2^{q-2-k}, x_1^k x_2^{q-1-k}, x_1^{k-1}x_2^{b_{q-k}}, \ldots, x_1 x_2^{b_2}, x_2^{b_1}).$$

By choosing k as small as possible we achieve all three conditions.

$\Leftarrow$) By Lemma 12.1.8 we need only show $\mu(I) \geq \mathrm{ord}(I) + 1$. Since $q = \mu(I)$ its suffices to show that I has a monomial of degree $q - 1$. Notice that $b_{q-i} = i$ for $1 \leq i \leq k - 1$ and $a_i = q - i$ for $k \leq i \leq q - 1$. In particular $a_k = q - k$ and $b_{q-k+1} = k - 1$. Thus the monomial $x_1^{q-k}x_2^{k-1}$ belongs to I and has degree $q - 1$. □

If I is complete, then I is $\mathfrak{m}$-full [425, Theorem 5]. In polynomial rings in two variables any complete ideal is normal; more generally one has the following result of Zariski.

Theorem 12.1.10 [436, Appendix 5] *If $I_1, \ldots, I_r$ are complete ideals in a regular local ring R of dimension two, then $I_1 \cdots I_r$ is complete.*

In [101] Crispin Quiñonez studied the normality of $\mathfrak{m}$-primary monomial ideals in $K[x_1, x_2]$ and established a criterion in terms of certain partial blocks and associated sequences of rational numbers.

Example 12.1.11 The ideal $I = (x_1^3,\, x_1^2x_2^8,\, x_1x_2^{15},\, x_2^{21})$ is $\mathfrak{m}$-full and is not normal. The integral closure of I is $\overline{I} = (x_1^3,\, x_1^2x_2^7,\, x_1x_2^{14},\, x_2^{21})$.

Linear programming test We give a test to check whether or not a given monomial lies in the integral closure of a monomial ideal.

Proposition 12.1.12 (Membership test) *Let $v_1,\dots,v_q$ be a set of vectors in $\mathbb{N}^n$ and let A be the $n\times q$ matrix whose columns are the vectors $v_1,\dots,v_q$. If I is the ideal generated by $x^{v_1},\dots,x^{v_q}$, then a monomial x^b lies in the integral closure of I if and only if the linear program:*

Maximize $x_1+\cdots+x_q$

Subject to

$Ax\leq b$ and $x\geq 0$

has an optimal value greater than or equal to 1*, which is attained at a vertex of the rational polytope $\mathcal{P}=\{x\in\mathbb{R}^q\,|\,Ax\leq b \text{ and } x\geq 0\}$.*

Proof. $\Rightarrow$) Let $x^b\in\overline{I}$, that is, $x^{pb}\in I^p$ for some positive integer p. There are nonnegative integers $r_1,\dots,r_q$ such that $x^{pb}=x^{\delta}\,(x^{v_1})^{r_1}\cdots(x^{v_q})^{r_q}$ and $r_1+\cdots+r_q=p$. Hence the column vector c with entries $c_i=r_i/p$ satisfies $Ac\leq b$ and $c_1+\cdots+c_q=1$. Thus the linear program has an optimal value greater than or equal to 1.

$\Leftarrow$) Since the vertices of $\mathcal{P}$ have rational entries (see Proposition 1.1.46) and the maximum of $x_1+\cdots+x_q$ is attained at a vertex of the polytope $\mathcal{P}$ (see Proposition 1.1.41), there are $c_1,\dots,c_q$ in $\mathbb{Q}_+$ such that $c_1+\cdots+c_q\geq 1$ and $c_1v_1+\cdots+c_qv_q\leq b$. Hence there are $\epsilon_1,\dots,\epsilon_q$ in $\mathbb{Q}$ such that $0\leq\epsilon_i\leq c_i$ for all i and $\sum_{i=1}^q\epsilon_i=1$. Therefore there is a vector $\delta\in\mathbb{Q}_+^n$ such that $b=\delta+\sum_{i=1}^q\epsilon_iv_i$. Thus there is an integer $p>0$ such that

$$pb=\underbrace{p\delta}_{\in\mathbb{N}^n}+\underbrace{p\epsilon_1}_{\in\mathbb{N}}v_1+\cdots+\underbrace{p\epsilon_q}_{\in\mathbb{N}}v_q\ \Rightarrow\ x^b\in\overline{I}.\qquad\square$$

Remark 12.1.13 By Theorem 1.1.56 one can also use the dual problem

Minimize $b_1y_1+\cdots+b_ny_n$

Subject to

$yA\geq\mathbf{1}$ and $y\geq 0$

to check whether x^b is in $\overline{I}$. In this case one has a fixed polyhedron

$$\mathcal{Q}=\{y\in\mathbb{R}^n\,|\,yA\geq\mathbf{1}\text{ and }y\geq 0\}$$

that can be used to test membership of any monomial x^b, while in the primal problem the polytope $\mathcal{P}$ depends on b.

Example 12.1.14 Consider the ideal $I=(x_1^3,x_2^4,x_3^5,x_4^6,x_1x_2x_3x_4)$ and the exponent vector $b=(0,1,1,4)$. To check whether x^b is in $\overline{I}$ we use the following procedure in *Mathematica* [431]:

```
ieq:={3x1+x5<=0,4x2+x5<=1,5x3+x5<=1,6x4+x5<=4}
vars:={x1,x2,x3,x4,x5}
f:=x1+x2+x3+x4+x5
ConstrainedMax[f,ieq,vars]
```

The answer that *Mathematica* gives is:

```
{67/60, {x1 -> 0, x2 -> 1/4, x3 -> 1/5, x4 -> 2/3, x5 -> 0}}
```

where the first entry is the optimal value and the other entries correspond to a vertex of the polytope $\mathcal{P}$. By Proposition 12.1.12, we get $x^b \in \overline{I}$.

Using *PORTA* [84] one obtains that the vertices of the polyhedral set $\mathcal{Q}$ described in Remark 12.1.13 are the rows of the matrix:

$$M = \begin{bmatrix} 1/3 & 1/4 & 1/5 & 13/60 \\ 1/3 & 1/4 & 1/4 & 1/6 \\ 1/3 & 3/10 & 1/5 & 1/6 \\ 23/60 & 1/4 & 1/5 & 1/6 \end{bmatrix}.$$

This is a "membership test matrix" in the sense that a monomial x^b lies in $\overline{I}$ if and only if $Mb \geq \mathbf{1}$. If $x^b = x_2x_3x_4^4$, we get:

$$Mb = (79/60, 7/6, 7/6, 67/60)^\top \geq (1,1,1,1) \Rightarrow x^b \in \overline{I}.$$

Exercises

12.1.15 Let $R = K[x_1,\ldots,x_n]$ be a polynomial ring over a field K and F a finite set of monomials of the same degree. If $I = (F)$ is complete, then

$$\operatorname{conv}(\log(F)) \cap \mathbb{Z}^n = \log(F).$$

12.1.16 Let $R = K[x_1,\ldots,x_n]$ be a polynomial ring over a field K. If $\mathfrak{m} = (x_1,\ldots,x_n)$, then $\mathfrak{m}^d$ is the integral closure of $(x_1^d,\ldots,x_n^d)$.

12.1.17 Let $I = (x^{v_1},\ldots,x^{v_q})$ be a monomial ideal of a polynomial ring over a field K. Prove that $\overline{I^i} = \overline{(x^{iv_1},\ldots,x^{iv_q})}$ for $i \geq 1$.

12.1.18 If $I \subset K[x_1,x_2]$ is an ideal generated by monomials of degree k, then so is $\overline{I}$.

12.1.19 Prove that $\overline{I \cap J} \neq \overline{I} \cap \overline{J}$, where $I = (x_1^5, x_2^3)$ and $J = (x_2^5, x_3^2)$.

12.1.20 If $I = (x_1^2, x_2^3) \subset K[x_1,x_2]$ and $\mathcal{P} = \operatorname{conv}((2,0),(0,3))$, then $\mathcal{P} \cap \mathbb{Z}^2$ is equal to $\{(2,0),(0,3)\}$ and $\overline{I} = I + (x_1x_2^2)$.

12.1.21 If $I = (x_1^2, x_1x_2^2, x_2^3) \subset K[x_1,x_2]$, then I is normal.

12.2 Normality criteria

In this section we present a simple *normality criterion* which is suitable to study the normality of monomial ideals.

Proposition 12.2.1 *Let $(R,\mathfrak{m})$ be a Noetherian local ring and let I be a proper ideal of R. Then I is normal if and only if*

(a) *$IR_\mathfrak{p}$ is normal for any prime ideal $I\subset\mathfrak{p}\neq\mathfrak{m}$, and*

(b) *$\overline{I^r}\cap(I^r\colon\mathfrak{m})=I^r$ for all $r\geq 1$.*

Proof. $\Rightarrow$) That (a) is satisfied follows from the fact that integral closures and powers of ideals commute with localizations. Part (b) is clearly satisfied.

$\Leftarrow$) Assume $I^r\neq\overline{I^r}$ for some $r\geq 1$. Take $\mathfrak{p}\in\mathrm{Ass}_R(M)$, where $M=\overline{I^r}/I^r$. Since $\mathfrak{p}$ is in the support of M we get $M_\mathfrak{p}\neq(0)$ and consequently, by (a), we must have $\mathfrak{p}=\mathfrak{m}$. Hence there is an embedding

$$R/\mathfrak{m}\hookrightarrow M\qquad(\overline{1}\mapsto\overline{x_0}),$$

where $x_0\in\overline{I^r}\setminus I^r$ and $\mathfrak{m}=\mathrm{ann}(\overline{x_0})$. From (b) one concludes that $x_0\in I^r$, a contradiction. Thus $I^r=\overline{I^r}$ for all $r\geq 1$. $\Box$

Proposition 12.2.2 *Let $(R,\mathfrak{m})$ be a Noetherian local ring and let I be a proper ideal of R. Then I is integrally closed if and only if $IR_\mathfrak{p}$ is integrally closed for any prime ideal $I\subset\mathfrak{p}\neq\mathfrak{m}$ and $\overline{I}\cap(I\colon\mathfrak{m})=I$.*

Proof. It follows from the proof of Proposition 12.2.1. $\Box$

The next result proves that a normal monomial ideal of $K[x_1,\ldots,x_n]$ stays normal if we make any variable, say x_n, equal to 1 or 0.

Proposition 12.2.3 *Let $I=(x^{v_1},\ldots,x^{v_q})$ be a normal ideal. If each x^{v_i} is written as $x_n^{a_i}g_i$, where g_i is a monomial in $R'=K[x_1,\ldots,x_{n-1}]$, then*

(a) *$J=(g_1,\ldots,g_q)\subset R'$ is a normal ideal, and*

(b) *$L=(\{g_i|\,a_i=0\})\subset R'$ is a normal ideal.*

Proof. (a) We will show that $J^r=\overline{J^r}$ for all $r\geq 1$. Take a monomial x^α of R' that belong to $\overline{J^r}$. Then $(x^\alpha)^p\in J^{rp}$ for some $p\geq 1$ and we can write

$$(x^\alpha)^p=g_1^{b_1}\cdots g_q^{b_q}x^\beta,$$

where $\sum_{i=1}^q b_i=rp$ and $x^\beta\in R'$. Multiplying both sides of the equation above by x_n^{sp} with $s=\sum_{i=1}^q a_ib_i$, we get $(x_n^s x^\alpha)^p\in I^{rp}$. Hence $x_n^s x^\alpha\in I^r$ and we can write $x_n^s x^\alpha=(x_n^{a_1}g_1)^{c_1}\cdots(x_n^{a_q}g_q)^{c_q}x^\gamma$, where $\sum_{i=1}^q c_i=r$. Evaluating the last equality at $x_n=1$ gives that $x^\alpha\in J^r$, as required. Part (b) follows using similar arguments. $\Box$

Theorem 12.2.4 (Normality criterion) *Let $I = (x^{v_1}, \ldots, x^{v_q})$ be a monomial ideal of $R = K[x_1, \ldots, x_n]$ and let J_i be the ideal of R generated by all monomials obtained from $\{x^{v_1}, \ldots, x^{v_q}\}$ by making $x_i = 1$. Then I is normal if and only if*

(a) *J_i is normal for all i (resp. $I_\mathfrak{p}$ is normal for any prime ideal $\mathfrak{p} \neq \mathfrak{m}$),*

(b) *$\overline{I^r} \cap (I^r \colon \mathfrak{m}) = I^r$ for all $r \geq 1$, where $\mathfrak{m} = (x_1, \ldots, x_n)$.*

Proof. $\Rightarrow$) Condition (a) follows from Proposition 12.2.3 (resp. that $I_\mathfrak{p}$ is normal follows using that integral closures and powers of ideals commute with localizations; see Proposition 4.3.4). Part (b) is clearly satisfied.

$\Leftarrow$) To prove that $I^r = \overline{I^r}$ set $M = \overline{I^r}/I^r$. We proceed by contradiction assuming $I^r \subsetneq \overline{I^r}$. Take $\mathfrak{p} \in \mathrm{Ass}_R(M)$. There are embeddings

$$R/\mathfrak{p} \overset{\varphi}{\hookrightarrow} M \hookrightarrow R/I^r.$$

Thus $\mathfrak{p} \in \mathrm{Ass}_R(R/I^r)$. Consequently $\mathfrak{p}$ is generated by a subset of $\{x_1, \ldots, x_n\}$ and $\mathfrak{p} \subset \mathfrak{m}$. Note $\mathfrak{p} = \mathfrak{m}$, otherwise if $x_i \notin \mathfrak{p}$ for some i, then by (a) $IR_\mathfrak{p} = J_iR_\mathfrak{p}$ is normal and $M_\mathfrak{p} = (0)$, a contradiction because $\mathfrak{p}$ is in the support of M. Thus $\mathfrak{p} = \mathfrak{m}$. If $\overline{x_0} = \varphi(\overline{1})$, we get $x_0 \in \overline{I^r} \setminus I^r$ and $x_0 \in (I^r \colon \mathfrak{m})$, which contradicts (b). $\Box$

Proposition 12.2.5 *Let $R = \oplus_{i=0}^{\infty} R_i$ be an $\mathbb{N}$-graded ring and let $I = R_{\geq \alpha}$ be the ideal of R generated by the elements of degree at least α. If R is a domain, then I is integrally closed.*

Proof. For $f = f_s + f_{s+1} + \cdots + f_r \in R$, with $f_i \in R_i$ for all i and $f_s \neq 0$, we set $\deg(f) = s$. As R is a domain one has $\deg(gh) = \deg(g) + \deg(h)$ for $g, h \in R$ and $gh \neq 0$. Take $0 \neq f \in \overline{I}$ with $s = \deg(f)$. There is an equation

$$f^n + a_1 f^{n-1} + \cdots + a_{n-1} f + a_n = 0 \quad (a_i \in I^i).$$

Assume $s < \alpha$. If $a_i f^{n-i} \neq 0$, then $\deg(a_i f^{n-i}) \geq i\alpha + (n-i)s > ns$. Thus one can rewrite the equation above as

$$f_s^n + (\text{terms of degree greater than } ns) = 0,$$

a contradiction because $f_s \neq 0$ and R is graded. Hence $\deg(f) \geq \alpha$. $\Box$

Theorem 12.2.6 [153] *Let K be a field and let $R = K[x_1, \ldots, x_n]$ be a positively graded K-algebra. If R is a domain and c is the least common multiple of the weights of the x_i, then $I = R_{\geq nc}$ is a normal ideal. If R is normal, then the Rees ring $R[It]$ is normal.*

Proof. In [153] it is proved that I^p is equal to $R_{\geq pnc}$ for all $p \geq 1$, where the latter is integrally closed. $\Box$

Ideals of mixed products Let $R = K[x_1, \ldots, x_m, y_1, \ldots, y_n]$ be a ring of polynomials over a field K. Given k, r, s, t in $\mathbb{N}$ such that $k + r = s + t$, consider the square-free monomial ideal of *mixed products* given by

$$L = I_k J_r + I_s J_t,$$

where I_k (resp. J_r) is the ideal of R generated by the square-free monomials of degree k in the variables $x_1, \ldots, x_m$ (resp. $y_1, \ldots, y_n$). The Betti numbers and some of the invariants of L have been studied in [246, 260, 355]. The invariants of the symmetric algebra of L are studied in [288].

Next we present a complete classification of the normal ideals of mixed products that can be shown using the normality criterion of Theorem 12.2.4.

Theorem 12.2.7 [351] *If $L \neq R$, then L is normal if and only if it can be written (up to permutation of k, s and r, t) in one of the following forms:*

(a) $L = I_k J_r + I_{k+1} J_{r-1}$, $k \geq 0$ *and* $r \geq 1$.

(b) $L = I_k J_r$, $k \geq 1$ *or* $r \geq 1$.

(c) $L = I_k J_r + I_s J_t$, $0 = k < s = m$, *or* $0 = t < r = n$, *or* $k = t = 0$, $s = 1$.

Definition 12.2.8 Let G be a simple graph (resp. digraph) with vertices $x_1, \ldots, x_n$ and let $t \geq 2$ be an integer. The *path ideal* of G, denoted by $I_t(G)$, is the ideal of $K[x_1, \ldots x_n]$ generated by all square-free monomials $x_{i_1} \cdots x_{i_t}$ such that the x_{i_j} is adjacent to $x_{i_{j+1}}$ (resp. $e_j = (x_{i_j}, x_{i_{j+1}})$ is a directed edge from x_{i_j} to $x_{i_{j+1}}$) for all $1 \leq j \leq t - 1$.

Path ideals of digraphs were introduced by Conca and De Negri [89] and later studied in [50, 211]. Kubitzke and Olteanu [284] have studied the algebraic properties and invariants of path ideals of certain types of posets.

Corollary 12.2.9 *If G is a complete bipartite graph, then the path ideal $I_t(G)$ is normal for all $t \geq 2$.*

Proof. Let $x_1, \ldots, x_m, y_1, \ldots, y_n$ be the vertex set of G, one may assume that the edges of G are precisely the pairs of the form $\{x_i, y_j\}$. Therefore $I_t(G)$ is $I_d J_{d+1} + I_{d+1} J_d$ if $t = 2d + 1$ and $I_t(G)$ is $I_d J_d$ if $t = 2d$. Hence the result follows from Theorem 12.2.7. □

Exercises

12.2.10 Let $R = K[x_1, \ldots, x_m, y_1, \ldots, y_n]$ and $L = I_s J_t$, where I_s (resp. J_t) denote the ideal of R generated by the square-free monomials of degree s (resp. t) in the x_i (resp. y_i) variables. Prove that L is normal.

12.2.11 Let $R = K[x_1, \ldots x_n]$ and $R[x]$ be polynomial rings over a field K and let $L_t = I_t + xI_{t-2}$, where I_t is the ideal of R generated by the square-free monomials of degree $t \geq 2$. Then L_t is a normal ideal of $R[x]$.

12.2.12 Let I and J be two ideals of the polynomial rings $K[x_1, \ldots, x_m]$ and $K[y_1, \ldots, y_n]$, respectively. The *join* of I and J is:

$$I * J = I + J + (x_i y_j \,|\, 1 \leq i \leq m \text{ and } 1 \leq j \leq n).$$

If I and J are normal ideals generated by square-free monomials of the same degree $t \geq 2$, then their join $I * J$ is normal.

12.2.13 Let $X = \{x_1, \ldots, x_m\}$ and $\{y_1, \ldots, y_n\}$ be two sets of variables over a field K. Let I be a normal ideal of $K[X]$ generated by square-free monomials of degree t and let $L = I + (X)(Y)$. Then L is a normal ideal.

12.2.14 Let $R = K[x_1, \ldots, x_8]$ be a polynomial ring over a field K of characteristic zero and let I be the ideal of R generated by

$$x_1x_2x_3x_4,\ x_5x_6x_7x_8,\ x_1x_2x_4x_5,\ x_3x_4x_6x_8,\ x_1x_3x_7,\ x_2x_7x_8.$$

Prove that I is normal, whereas the ideal $J = I + (x_1y_1, \ldots, x_8y_1)$ is not.

12.2.15 Let $R = K[x_1, \ldots, x_6]$ be a polynomial ring over a field K. Prove that $I = (x_3x_5x_6, x_3x_4x_6, x_2x_5x_6, x_2x_4x_5, x_1x_3x_5, x_1x_2x_6)$ is not normal.

12.3 Rees cones and polymatroidal ideals

Let $R = K[x_1, \ldots, x_n]$ be a polynomial ring over a field K. In this part we prove that Rees algebras of polymatroidal ideals are normal domains. In particular ideals of Veronese type are normal.

Quasi-ideal Rees cones Let I be a monomial ideal of R minimally generated by the set $F = \{x^{v_1}, \ldots, x^{v_q}\}$. The *Rees cone* of I is the rational polyhedral cone in $\mathbb{R}^{n+1}$, denoted by $\mathbb{R}_+\mathcal{A}'$ or $\mathbb{R}_+(I)$, generated by

$$\mathcal{A}' := \{e_1, \ldots, e_n, (v_1, 1), \ldots, (v_q, 1)\} \subset \mathbb{R}^{n+1},$$

where e_i is the ith unit vector. Consider the index set

$$\mathcal{J} = \{1 \leq i \leq n |\ \langle e_i, v_j \rangle = 0 \text{ for some } j\} \cup \{n+1\}.$$

Notice that $\dim \mathbb{R}_+\mathcal{A}' = n + 1$. By Theorem 1.4.2, the Rees cone has a unique irreducible representation

$$\mathbb{R}_+\mathcal{A}' = \left(\bigcap_{i \in \mathcal{J}} H_{e_i}^+\right) \bigcap \left(\bigcap_{i=1}^{r} H_{\ell_i}^+\right) \tag{12.1}$$

such that, for each i, $0 \neq \ell_i \in \mathbb{Z}^{n+1}$, the non-zero entries of ℓ_i are relatively prime, the first n entries of ℓ_i are in $\mathbb{N}$ and the last entry of ℓ_i is negative.

Let $\mathcal{P}$ be the convex hull of $\mathcal{A} = \{v_1, \ldots, v_q\}$. Recall that the *Ehrhart ring* of $\mathcal{P}$ and the *Rees algebra* of I are the monomial subrings given by

$$A(\mathcal{P}) = K[x^a t^b |\, a \in b\mathcal{P} \cap \mathbb{Z}^n] \ \text{ and } \ R[It] = K[x_1, \ldots, x_n, x^{v_1}t, \ldots, x^{v_q}t]$$

respectively, where t is a new variable. Using Theorem 9.1.1, it follows readily that the *normalization* of $R[It]$ is

$$\overline{R[It]} = K[x^a t^b |\, (a,b) \in \mathbb{R}_+\mathcal{A}' \cap \mathbb{Z}^{n+1}].$$

Notation For use below we set $[n] = \{1, \ldots, n\}$.

Theorem 12.3.1 *If ℓ_k has the form $\ell_k = -d_k e_{n+1} + \sum_{i \in A_k} e_i$ for some $A_k \subset [n]$ for $k = 1, \ldots, r$, then $A(\mathcal{P})[x_1, \ldots, x_n] = \overline{R[It]}$.*

Proof. "$\subset$": This inclusion is clear because $A(\mathcal{P}) \subset \overline{R[It]}$.

"$\supset$": Let $x^a t^b = x_1^{a_1} \cdots x_n^{a_n} t^b \in \overline{R[It]}$ be a minimal generator, i.e., (a,b) cannot be written as a sum of two non-zero integral vectors in the Rees cone $\mathbb{R}_+\mathcal{A}'$. We may assume $a_i \geq 1$ for $1 \leq i \leq m$, $a_i = 0$ for $i > m$, and $b \geq 1$.

Case (I): $\langle (a,b), \ell_i \rangle > 0$ for all i. The vector $\gamma = (a,b) - e_1$ satisfies $\langle \gamma, \ell_i \rangle \geq 0$ for all i, that is $\gamma \in \mathbb{R}_+\mathcal{A}'$. Thus since $(a,b) = e_1 + \gamma$ we derive a contradiction.

Case (II): $\langle (a,b), \ell_i \rangle = 0$ for some i. We may assume

$$\{\ell_i |\, \langle (a,b), \ell_i \rangle = 0\} = \{\ell_1, \ldots, \ell_p\}.$$

Subcase (II.a): $e_i \in H_{\ell_1} \cap \cdots \cap H_{\ell_p}$ for some $1 \leq i \leq m$. It is not hard to verify that the vector $\gamma = (a,b) - e_i$ satisfies $\langle \gamma, \ell_k \rangle \geq 0$ for all k. Thus $\gamma \in \mathbb{R}_+\mathcal{A}'$, a contradiction because $(a,b) = e_i + \gamma$.

Subcase (II.b): $e_i \notin H_{\ell_1} \cap \cdots \cap H_{\ell_p}$ for all $1 \leq i \leq m$. Since the vector (a,b) belongs to the Rees cone it follows that we can write

$$(a,b) = \lambda_1(v_1, 1) + \cdots + \lambda_q(v_q, 1) \quad (\lambda_i \geq 0).$$

Therefore $a \in b\mathcal{P}$, i.e., $x^a t^b \in A(\mathcal{P})$. □

If $\ell_1, \ldots, \ell_r$ satisfy the condition of Theorem 12.3.1 (resp. $d_k = 1$ for all k), we say that the Rees cone of I is *quasi-ideal* (resp. *ideal*).

Corollary 12.3.2 *If the Rees cone of I is quasi-ideal and $K[Ft] = A(\mathcal{P})$, then the Rees algebra $R[It]$ is normal.*

Proof. By Theorem 12.3.1 we get

$$R[It] = K[Ft][x_1, \ldots, x_n] = A(\mathcal{P})[x_1, \ldots, x_n] = \overline{R[It]}. \qquad \square$$

Polymatroidal sets of monomials Let $\mathcal{A} = \{v_1, \ldots, v_q\} \subset \mathbb{N}^n$ be the set of bases of a discrete polymatroid of rank d, i.e., $|v_i| = d$ for all i and given any two $a = (a_i), c = (c_i)$ in $\mathcal{A}$, if $a_i > c_i$ for some index i, then there is an index j with $a_j < c_j$ such that $a - e_i + e_j$ is in $\mathcal{A}$. We refer to [221, 227] for the theory of discrete polymatroids.

The set $F = \{x^{v_1}, \ldots, x^{v_q}\}$ (resp. the ideal $I = (F) \subset R$) is called a *polymatroidal set of monomials* (resp. a *polymatroidal ideal*).

Lemma 12.3.3 *$F' = \{x^{v_i}/x_1 \colon x_1$ occurs in $x^{v_i}\}$ is also polymatroidal.*

Proof. It is left as an exercise. □

Lemma 12.3.4 *Let $\mathcal{A}' = \{e_1, \ldots, e_n, (v_1, 1), \ldots, (v_q, 1)\}$. If E is a facet of $\mathbb{R}_+\mathcal{A}'$, then $E = \mathbb{R}_+\mathcal{A}' \cap H_b$, where $b = e_i$ for some $1 \leq i \leq n+1$ or $b = (b_i) \in \mathbb{Z}^{n+1}$, $b_i \in \{0, 1\}$ for $i = 1, \ldots, n$ and $b_{n+1} \geq -d$.*

Proof. The proof is by induction on d. The case $d = 1$ is clear because the matrix whose columns are the vectors in $\mathcal{A}'$ is totally unimodular.

Assume that $d \geq 2$. For $1 \leq k \leq q$, we set $f_{v_i} = (v_i, 1)$, $B_i = \text{supp}(v_i)$ and $v_k = (v_{k1}, \ldots, v_{kn})$. Let E be a facet of $\mathbb{R}_+\mathcal{A}'$. There is a unique $0 \neq b = (b_i)$ in $\mathbb{Z}^{n+1}$ whose non-zero entries are relatively prime such that

(a) $E = \mathbb{R}_+\mathcal{A}' \cap H_b \neq \mathbb{R}_+\mathcal{A}'$, $\mathbb{R}_+\mathcal{A}' \subset H_b^+$, and

(b) there is a linearly independent set $A \subset H_b \cap \mathcal{A}'$ with $|A| = n$.

If $A = \{e_1, \ldots, e_n\}$ (resp. $A \subset \{f_{v_1}, \ldots, f_{v_q}\}$), then $b = e_{n+1}$ (resp. $b = (1, \ldots, 1, -d)$). Notice that $b_i \geq 0$ for $i = 1, \ldots, n$. Thus we may assume that A is the set $\{e_1, \ldots, e_s, f_{v_1}, \ldots, f_{v_t}\}$, where $1 \leq s \leq n-1$, $s + t = n$, and $\{1, \ldots, s\}$ is the set of all $i \in [n]$ such that $e_i \in H_b$. We may assume that $1 \in B_k$ for some $1 \leq k \leq q$. Indeed if 1 is not in $\cup_{i=1}^q B_i$, then the facets of $\mathbb{R}_+\mathcal{A}'$ different from $H_{e_1} \cap \mathbb{R}_+\mathcal{A}'$ are in one-to-one correspondence with the facets of

$$\mathbb{R}_+\{e_2, \ldots, e_n, f_{v_1}, \ldots, f_{v_q}\}.$$

Assume that $1 \notin B_1$. Since $v_{k1} > v_{11}$ for some k, by the symmetric exchange property of $\mathcal{A}$ [221, Theorem 4.1, p. 241] there is j with $v_{kj} < v_{1j}$ such that $v_1 + e_1 - e_j = v_i$ for some v_i in $\mathcal{A}$. Hence

$$f_{v_1} + e_1 = f_{v_i} + e_j \;\Rightarrow\; \langle f_{v_i}, b\rangle = -\langle e_j, b\rangle \;\Rightarrow\; \langle f_{v_i}, b\rangle = \langle e_j, b\rangle = 0,$$

i.e., f_{v_i} and e_j belong to H_b. Thus from the outset we may assume that $1 \in B_1$. Next assume that $t \geq 2$ and $1 \notin B_2$. Since $v_{11} > v_{21}$, by the symmetric exchange property of $\mathcal{A}$ there is j with $v_{1j} < v_{2j}$ such that $v_2 + e_1 - e_j = v_i$ for some v_i in $\mathcal{A}$. Hence

$$f_{v_2} + e_1 = f_{v_i} + e_j \;\Rightarrow\; \langle f_{v_i}, b\rangle = \langle e_j, b\rangle = 0,$$

i.e., f_{v_i} and e_j belong to H_b. Notice that $i \neq 1$. Indeed if $i = 1$, then from the equality above we get that $\{f_{v_2}, e_1, f_{v_1}, e_j\}$ is linearly dependent, a contradiction. Applying the arguments above repeatedly shows that we may assume that 1 belongs to B_i for $i = 1, \ldots, t$. We may also assume that $v_1, \ldots, v_r$ is the set of all vectors in $\mathcal{A}$ such that $1 \in B_i$, where $t \leq r$. For $1 \leq i \leq r$, we set $v_i' = v_i - e_1$. By Lemma 12.3.3 the set $\{v_1', \ldots, v_r'\}$ is again the set of basis of a discrete polymatroid of rank $d-1$. Consider the Rees cone $\mathbb{R}_+\mathcal{A}''$ generated by

$$\mathcal{A}'' = \{e_1, e_2, \ldots, e_n, f_{v_1'}, \ldots, f_{v_r'}\}.$$

Notice that $\mathbb{R}_+\mathcal{A}'' \subset H_b^+$ and that $\{e_1, e_2, \ldots, e_s, f_{v_1'}, \ldots, f_{v_t'}\}$ is a linearly independent set. Thus $\mathbb{R}_+\mathcal{A}'' \cap H_b$ is a facet of the cone $\mathbb{R}_+\mathcal{A}''$ and the result follows by induction. $\Box$

Theorem 12.3.5 [221] *If $\mathcal{P} = \mathrm{conv}(v_1, \ldots, v_q)$, then $K[Ft] = A(\mathcal{P})$.*

Theorem 12.3.6 [422] *If I is a polymatroidal ideal, then I is normal.*

Proof. By Lemma 12.3.4 and Theorem 12.3.5 $\mathbb{R}_+\mathcal{A}'$ is quasi-ideal and $K[Ft] = A(\mathcal{P})$. Thus, by Corollary 12.3.2, we get that $R[It]$ is normal. $\Box$

Corollary 12.3.7 *If I is a polymatroidal ideal, then I has the persistence property.*

Proof. By Theorems 12.3.6 and 7.7.3 I has the persistence property. $\Box$

Normality of an ideal of Veronese type Let $R = K[x_1, \ldots, x_n]$ be a ring of polynomials over an arbitrary field K. Given a sequence of integers $(s_1, s_2, \ldots, s_n; d)$ such that $1 \leq s_j \leq d \leq \sum_{i=1}^n s_i$ for all j, we define $\mathcal{A}$ as the set of partitions

$$\mathcal{A} = \{(a_1, \ldots, a_n) \in \mathbb{Z}^n |\, a_1 + \cdots + a_n = d;\ 0 \leq a_i \leq s_i \,\forall\, i\},$$

and F as the set of monomials $F = \{x^a |\, a \in \mathcal{A}\} = \{f_1, \ldots, f_q\}$.

Definition 12.3.8 The ideal $I = (F) \subset R$ is said to be of *Veronese type* of *degree* d with defining sequence $(s_1, \ldots, s_n; d)$. If $s_i = 1$ for all i we call I the dth *square-free Veronese ideal*, and if $s_i = d$ for all i we call I the dth *Veronese ideal*. Similar terminology may be applied to subrings.

The monomial subring $K[F] \subset R$ and its toric ideal have been studied by Sturmfels in [400]. In loc. cit. it is shown that the toric ideal of $K[F]$ has a quadratic Gröbner basis whose initial ideal is square-free.

Proposition 12.3.9 [147] *If $I \subset R$ is an ideal of Veronese type, then I is normal.*

Proof. The ideal I is a polymatroidal ideal. Then, the Rees algebra of I is normal by Theorem 12.3.6, i.e., I is a normal ideal. □

Proposition 12.3.10 [418] *If I is the* dth *square-free Veronese ideal of R, then I is normal.*

Proof. The result follows at once from Proposition 12.3.9. □

The basis monomial ring of a matroid Let M be a matroid on a finite set $X = \{x_1, \dots, x_n\}$ and let $\mathcal{B}$ be the collection of bases of M. The clutter with vertex set X and edge set $\mathcal{B}$ is called the *clutter of bases* of M.

Definition 12.3.11 The set of all $x_{i_1} \cdots x_{i_r} \in R$ such that $\{x_{i_1}, \dots, x_{i_r}\}$ is in $\mathcal{B}$ is denoted by F_M. The *basis monomial ring* of M is the ring $K[F_M]$. The square-free monomial ideal (F_M) is the *basis monomial ideal* of M.

It is well known [338] that all bases of a matroid M have the same number of elements; this common number is called the *rank* of the matroid. Thus $K[F_M]$ is a standard graded K-algebra and all monomials of F_M have the same degree; see Section 9.2.

Corollary 12.3.12 *If $I = (F_M)$ is the basis monomial ideal of a matroid, then $R[It]$ is normal.*

Proof. It follows from Theorem 12.3.6. □

Corollary 12.3.13 [429] *$K[F_M]$ is normal.*

Proof. As $K[F_M t] \simeq K[F_M]$, by Theorems 12.3.5 and 9.3.6, $K[F_M]$ is normal. □

Definition 12.3.14 The standard graded K-algebra $S = K[F_M]$ is said to be *Koszul* if the residue field K has a linear S-free resolution.

A sufficient condition for $K[F_M]$ to be Koszul is that its toric ideal has a quadratic Gröbner basis; see [46, 221]. In [430, Conjecture 12] White conjectured that the toric ideal of $K[F_M]$ is generated by quadrics. More recently Blum [46] asked whether $K[F_M]$ is Koszul. An open problem posed by Bøgvad is whether the toric ideal of a smooth projectively normal toric variety is generated by quadrics (see [61, 400]).

Exercises

12.3.15 Let $G = \{x^{u_1}, \dots, x^{u_s}\}$ and let $d = \max_i\{|u_i|\}$. Suppose that $F = \{x^{u_1}, \dots, x^{u_r}\}$ is the set of all x^{u_i} of degree d. If $\mathcal{Q} = \text{conv}(u_1, \dots, u_s)$ and $K[Gt] = A(\mathcal{Q})$, then $K[Ft] = A(\mathcal{P})$, where $\mathcal{P} = \text{conv}(u_1, \dots, u_r)$.

12.3.16 If $\mathcal{B} = \{\{x_1, x_2\}, \{x_2, x_3\}, \{x_3, x_4\}, \{x_1, x_4\}\}$, prove that $\mathcal{B}$ satisfies the bases exchange property of matroids.

12.4 Veronese subrings and the a-invariant

In this section we present an explicit generating set for the canonical module of a square-free Veronese subring and compute its a-invariant. A sharp upper bound for the a-invariant of the normalization of a subring generated by square-free monomials of the same degree will be presented, this bound is attained at a square-free Veronese subring.

Let $R = K[x_1,\ldots,x_n] = \oplus_{i=0}^{\infty} R_i$ be a polynomial ring over a field K with the standard grading and let $k \in \mathbb{N}_+$. An element in R_{ik} is said to have *normalized degree* i. The kth *Veronese subring* of R is given by:

$$R^{(k)} := \bigoplus_{i=0}^{\infty} R_{ik} \subset R,$$

and the kth *square-free Veronese subring* of R is given by:

$$K[V_k] := K[\{x_{i_1}\cdots x_{i_k} \,|\, 1 \leq i_1 < \cdots < i_k \leq n\}].$$

The kth *Veronese subring* of R is graded by $(R^{(k)})_i = R_{ki}$. Thus $R^{(k)}$ is a K-algebra generated by all the monomials of R of normalized degree 1 and $K[V_k]$ is a graded subring of $R^{(k)}$ with the *normalized grading*:

$$K[V_k] = \bigoplus_{i=0}^{\infty} (K[V_k])_i,$$

where $(K[V_k])_i = K[V_k] \cap (R^{(k)})_i$. In what follows we shall assume that $K[V_k]$ has the normalized grading.

Let F a finite set of square-free monomials of the same degree k. In this case there are embeddings

$$K[F] \subset K[V_k] \subset R^{(k)}.$$

Let us fix some of the notation that will be used throughout the rest of the section. Let $n \geq 2k \geq 4$ be two integers (this is not an essential restriction; see Remark 12.4.11). We set

$$\mathcal{V} = \{e_{i_1} + \cdots + e_{i_k} \,|\, 1 \leq i_1 < \cdots < i_k \leq n\},$$

where $e_1,\ldots,e_n$ are the canonical vectors in $\mathbb{R}^n$. The K-subring of R spanned by the set $\{x^a \,|\, a \in \mathcal{V}\}$ is equal to $K[V_k]$.

Remark 12.4.1 $\dim K[V_k] = n$. This follows from Corollary 8.2.21. The vector space generated by $\mathcal{V}$ is equal to $\mathbb{R}^n$ and $\dim \mathbb{R}_+\mathcal{V} = n$.

Lemma 12.4.2 [72] *Set* $N_1 = \{-e_1, \ldots, -e_n\}$, $N = N_1 \cup N_2$ *and*

$$N_2 = \{-e_1 - \cdots - e_{i-1} + (k-1)e_i - e_{i+1} - \cdots - e_n |\, 1 \leq i \leq n\}.$$

If H *is a supporting hyperplane of* $\mathbb{R}_+\mathcal{V}$ *such that* H *contains a set of* $n-1$ *linearly independent vectors in* $\mathcal{V}$, *then* $H = H_a$ *for some* $a \in N$.

Remark 12.4.3 The converse of Lemma 12.4.2 is also true because we are assuming $n \geq 2k \geq 4$, and this implies $n \geq k+2$. Note that if $n = 3$ and $k = 2$, then the cone $\mathbb{R}_+\mathcal{V}$ has only three facets.

Proposition 12.4.4 [72] *A point* $x = (x_1, \ldots, x_n) \in \mathbb{R}^n$ *is in* $\mathbb{R}_+\mathcal{V}$ *if and only if* x *is a feasible solution of the system of linear inequalities*

$$\begin{array}{rcll} -x_i & \leq & 0, & i = 1, \ldots, n \\ (k-1)x_i - \sum_{j \neq i} x_j & \leq & 0, & i = 1, \ldots, n. \end{array}$$

Proof. Let $\mathbb{R}_+\mathcal{V} = H_{b_1}^- \cap \cdots \cap H_{b_m}^-$ be the irreducible representation of $\mathbb{R}_+\mathcal{V}$. By Theorem 1.1.44 the set $H_{b_i} \cap \mathbb{R}_+\mathcal{V}$ is a facet of $\mathbb{R}_+\mathcal{V}$. Note that H_{b_i} is generated by a set of linearly independent vectors in $\mathcal{V}$; see Proposition 1.1.23. Hence, by Lemma 12.4.2, we get $H_{b_i} = H_a$ for some $a \in N$, where N is the set defined in Lemma 12.4.2. $\square$

Lemma 12.4.5 *Let* a *be a vector in* $C \cap \mathrm{ri}(\mathbb{R}_+\mathcal{V})$ *and set* $\mathcal{I} = \{i \,|\, a_i \geq 2\}$. *If* $|\mathcal{I}| \geq k$ *and* $i_1, \ldots, i_k$ *are distinct integers in* $\mathcal{I}$, *then* $a' = a - e_{i_1} - \cdots - e_{i_k}$ *also belongs to* $\mathbb{N}\mathcal{V} \cap \mathrm{ri}(\mathbb{R}_+\mathcal{V})$.

Proof. Without loss of generality one may assume $a_1 \geq a_2 \geq \cdots \geq a_n$, $a_k \geq 2$ and $a' = a - e_1 - \cdots - e_k$. As $\mathbb{R}_+\mathcal{V}$ has dimension n, it is seen that $a' \in \mathrm{ri}(\mathbb{R}_+\mathcal{V})$. By Proposition 12.3.10 the subring $K[V_k]$ is a normal domain, and therefore $\mathbb{N}\mathcal{V} = \mathbb{Z}\mathcal{V} \cap \mathbb{R}_+\mathcal{V}$. Since $a' \in \mathbb{Z}\mathcal{V}$, we conclude $a' \in \mathbb{N}\mathcal{V}$. $\square$

Theorem 12.4.6 [72] *Let* $\omega_{K[V_k]}$ *be the canonical module of* $K[V_k]$ *and let* $\mathfrak{B}$ *be the set of monomials* $x_1^{a_1} \cdots x_n^{a_n}$ *satisfying the following conditions:*

(a) $a_i \geq 1$ *and* $(k-1)a_i \leq -1 + \sum_{j \neq i} a_j$, *for all* i.

(b) $\sum_{i=1}^n a_i \equiv 0 \bmod (k)$.

(c) $|\{i \,|\, a_i \geq 2\}| \leq k-1$.

If $n \geq 2k \geq 4$, *then* $\mathfrak{B}$ *is a generating set for* $\omega_{K[V_k]}$.

Proof. By Theorem 9.1.5 we get $\omega_{K[V_k]} = (\{x^a |\, a \in \mathbb{N}\mathcal{V} \cap \mathrm{ri}(\mathbb{R}_+\mathcal{V})\})$. Using the arguments of the proof of Lemma 12.4.5 and by a repeated use of this lemma it is enough to prove that $\mathfrak{B} \subset \omega_{K[V_k]}$. Let $M \in \mathfrak{B}$; we may assume

$M = x_1^{a_1} \cdots x_{k-1}^{a_{k-1}} x_k \cdots x_n$, where $a_1 \geq \cdots \geq a_{k-1} \geq 1$. The monomial $N = x_1^{a_1} \cdots x_{k-1}^{a_{k-1}}$ can be factored as

$$N = \prod_{i=1}^{k-1} N_i, \text{ where } N_i = \begin{cases} (x_1 \cdots x_i)^{a_i - a_{i+1}}, & \text{if } 1 \leq i \leq k-2 \\ (x_1 \cdots x_{k-1})^{a_{k-1}}, & \text{if } i = k-1. \end{cases}$$

On the other hand, by (a) and (b), we can write $\prod_{i=k}^{n} x_i = N' \prod_{i=1}^{k-1} N_i'$, where $\deg(N_i') = (k-i)(a_i - a_{i+1})$ if $1 \leq i \leq k-2$, $\deg(N_i') = a_{k-1}$ if $i = k-1$, and $\deg(N') \equiv 0 \bmod (k)$. Hence $M = N' \prod_{i=1}^{k-1}(N_i N_i')$ is in $K[V_k]$, which readily implies $M \in \omega_{K[V_k]}$. $\Box$

Let $S = K[\{t_{i_1 \cdots i_k} \,|\, 1 \leq i_1 < \cdots < i_k \leq n\}]$ be a polynomial ring over a field K with one variable $t_{i_1 \cdots i_k}$ for each monomial $x_{i_1} \cdots x_{i_k}$. Here S has the standard grading. There is a graded homomorphism of K-algebras

$$\varphi \colon S \longrightarrow K[V_k], \text{ induced by } t_{i_1 \cdots i_k} \stackrel{\varphi}{\longmapsto} x_{i_1} \cdots x_{i_k},$$

the ideal $P = \ker(\varphi)$ is the *toric ideal* of $K[V_k]$. Thus the *Cohen–Macaulay type* of the ring $K[V_k]$, denoted by $\operatorname{type}(K[V_k])$, is the last Betti number in the minimal free resolution of S/P as an S-module; see Proposition 5.3.4.

Remark 12.4.7 To compute the type of $K[V_k]$ recall that this number is the minimal number of generators of the canonical module $\omega_{K[V_k]}$ of $K[V_k]$; see Corollary 5.3.6. Notice that a monomial $x_1^{a_1} \cdots x_{k-1}^{a_{k-1}} x_k \cdots x_n$ is in $\mathfrak{B}$ if and only if for all $1 \leq i \leq k-1$ one has $\sum_{j=1}^{k-1} a_j = mk - n + k - 1$ and $1 \leq a_i \leq m-1$, for some $m \geq 2$. Therefore $\frac{n}{k} \leq m \leq n - 2k + 2$. Hence, by Theorem 12.4.6, the computation of the type of $K[V_k]$ reduces to counting partitions of positive integers.

Corollary 12.4.8 *If $k = 2$, $n \geq 2k$ and n is odd, then*

$$\omega_{K[V_k]} = (\{x_1 \cdots x_{j-1} x_j^{2i} x_{j+1} \cdots x_n \,|\, 1 \leq j \leq n,\ 1 \leq i \leq (n-3)/2\}).$$

Corollary 12.4.9 *If $k = 2$ and $n \geq 2k$, then* $\operatorname{type}(K[V_k])$ *is $n(n-3)/2$ if n is odd, and* $\operatorname{type}(K[V_k])$ *is $(n^2 - 4n + 2)/2$ if n is even.*

Corollary 12.4.10 [72] *If $n \geq 2k \geq 4$, then the $a((K[V_k]) = -\lceil \frac{n}{k} \rceil$.*

Proof. Set $D = K[V_k]$ and $m = \lceil \frac{n}{k} \rceil$. To compute the a-invariant of D we use the formula of Lemma 9.1.7. It follows from Remark 12.4.7 that the degree of the generators in least degree of ω_D is at least m. To complete the proof we exhibit some generators of ω_D living in degree m. Write $n = qk + r$, $0 \leq r < k$; note $q \geq 2$. If $r \geq 1$, observe that the monomials

$$x_1^2 \cdots x_{k-r}^2 x_{k-r+1} \cdots x_{k-1} x_k \cdots x_n \text{ and } x_1 \cdots x_{n-k+r} x_{n-k+r+1}^2 \cdots x_{n-1}^2 x_n^2$$

belong to $(\omega_D)_m$. In particular, D cannot be a Gorenstein ring in this case. If $r=0$, then the monomial $M=x_1\cdots x_n$ satisfies $M\in(\omega_D)_m$. $\Box$

Remark 12.4.11 (Duality) Let $1\leq k\leq n-1$ be an integer and let

$$D_{n,k}=K[\{x_{i_1}\cdots x_{i_k}|\,1\leq i_1<\cdots<i_k\leq n\}],$$

be the K-subring of R spanned by the $x_{i_1}\cdots x_{i_k}$'s. Observe that there is a graded isomorphism of K-algebras of degree zero:

$$\rho\colon D_{n,k}\longrightarrow D_{n,n-k},\ \text{ induced by } \rho(x_{i_1}\cdots x_{i_k})=x_{j_1}\cdots x_{j_{n-k}},$$

where $\{j_1,\ldots,j_{n-k}\}=\{1,\ldots,n\}\setminus\{i_1,\ldots,i_k\}$. Thus if $n\leq 2k$, then

$$a(D_{n,k})=a(D_{n,n-k})=-\left\lceil\frac{n}{n-k}\right\rceil.$$

Because of this *duality* one may always assume that $n\geq 2k$.

Corollary 12.4.12 [108] *The subring $D_{n,k}$ is a Gorenstein ring if and only if $k\in\{1,n-1\}$ or $n=2k$.*

Proof. By duality one may assume $n\geq 2k\geq 4$. Set $D=D_{n,k}$. If D is Gorenstein, then by the proof of Corollary 12.4.10 we may assume $n=qk$. If $q\geq 3$, then $x_1\cdots x_n$ and $x_1^3x_2^2\cdots x_{k-1}^2x_k\cdots x_n$ belong to $(\omega_D)_q$ and $(\omega_D)_{q+1}$ respectively, which is impossible. Therefore $q=2$, as required.

Conversely assume $n=2k$. Let $\mathfrak{B}$ be as in Theorem 12.4.6. Take a monomial x^a in $\mathfrak{B}$; it suffices to verify that x^a is equal to $x_1\cdots x_n$. One may assume that $x^a=x_1^{a_1}\cdots x_{k-1}^{a_{k-1}}x_k\cdots x_n$, where $a_i\geq a_{i+1}\geq 1$. By hypothesis $\sum_{j=1}^{k-1}a_j=k(m-1)-1$ for some $m\geq 2$. On the other hand one has $ka_i\leq k+\sum_{j=1}^{k-1}a_j$ for all $1\leq i\leq k-1$. Hence $a_i\leq m-1$ for all i. Therefore using the previous equality again we rapidly derive $k(m-1)-1\leq(k-1)(m-1)$, which yields $m\leq 2$. As a consequence $a_i=1$ for all i, as required. $\Box$

Theorem 12.4.13 [72] *Let F be a finite set of square-free monomials of degree k in R such that* $\dim(K[F])=n$. *The following hold.*

(i) $a(\overline{K[F]})\leq a(K[V_k])$.

(ii) $a(\overline{K[F]})\leq-\lceil\frac{n}{k}\rceil$ *if $n\geq 2k$ and* $a(\overline{K[F]})\leq-\left\lceil\frac{n}{n-k}\right\rceil$ *if $n\leq 2k$, $n\neq k$.*

(iii) *If $n\geq 2k\geq 4$, then $\overline{K[F]}$ is generated as a K-algebra by elements of normalized degree less than or equal to $n-\lceil\frac{n}{k}\rceil$.*

Proof. (i): We set $\mathcal{V} = \{e_{i_1} + \cdots + e_{i_k} | 1 \leq i_1 < \cdots < i_k \leq n\}$ and $D = K[V_k]$. Let C, C_F be the semigroups of $\mathbb{N}^n$ generated by $\mathcal{V}$, $\log(F)$, respectively. By Proposition 12.3.10 D is normal. Hence, $\overline{K[F]} \subset D$. Let

$$M_1 = \{x^a | \, a \in \overline{C}_F \cap \mathrm{ri}(\mathbb{R}_+\overline{C}_F)\} \text{ and } M_2 = \{x^a | \, a \in C \cap \mathrm{ri}(\mathbb{R}_+C)\},$$

where $\overline{C}_F = \mathbb{Z}C_F \cap \mathbb{R}_+C_F$. Notice $\mathbb{R}_+\overline{C}_F = \mathbb{R}_+C_F$ and $\mathrm{aff}(\mathbb{R}_+C_F) = \mathbb{R}^n$. Therefore the relative interior of $\mathbb{R}_+\overline{C}_F$ equals its interior in $\mathbb{R}^n$. For similar reasons we have $\mathrm{ri}(\mathbb{R}_+C) = (\mathbb{R}_+C)^{\mathrm{o}}$. Hence $\mathrm{ri}(\mathbb{R}_+\overline{C}_F) \subset \mathrm{ri}(\mathbb{R}_+C)$. Altogether we obtain $M_1 \subset M_2$. Let x^b be an element of minimal degree in M_1 so that $\deg(x^b) = -a(\overline{K[F]})$. Set $r = -a(\overline{K[F]})$. Since x^b is in M_2 and $x^b \in \overline{K[F]}_r \subset D_r$, we conclude

$$-a(D) = \min\{\deg(x^a) | a \in M_2\} \leq r = -a(\overline{K[F]}).$$

(ii): It follows from Corollary 12.4.10, part (i), and the duality given in Remark 12.4.11.

(iii): It follows from the proof of part (i) and using part (ii). $\square$

Example 12.4.14 Let $K[F]$ be the subring of $R = K[x_1, \ldots, x_8]$ spanned by the monomials of R defining the edges of the graph shown below.

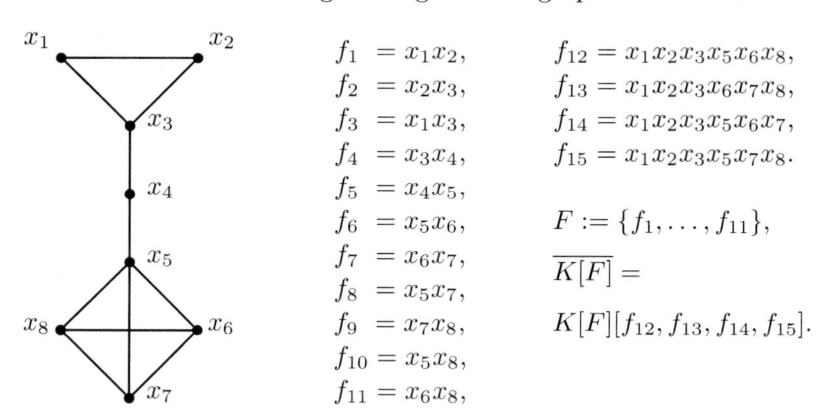

$$\begin{aligned} f_1 &= x_1x_2, & f_{12} &= x_1x_2x_3x_5x_6x_8,\\ f_2 &= x_2x_3, & f_{13} &= x_1x_2x_3x_6x_7x_8,\\ f_3 &= x_1x_3, & f_{14} &= x_1x_2x_3x_5x_6x_7,\\ f_4 &= x_3x_4, & f_{15} &= x_1x_2x_3x_5x_7x_8.\\ f_5 &= x_4x_5, & &\\ f_6 &= x_5x_6, & F &:= \{f_1, \ldots, f_{11}\},\\ f_7 &= x_6x_7, & \overline{K[F]} &=\\ f_8 &= x_5x_7, & &\\ f_9 &= x_7x_8, & & K[F][f_{12}, f_{13}, f_{14}, f_{15}].\\ f_{10} &= x_5x_8, & &\\ f_{11} &= x_6x_8, & & \end{aligned}$$

The generators of $\overline{K[F]}$ can be computed using the program *Normaliz* [68]. See also [413, Section 7.3]. A Noether normalization for $K[F]$ is given by

$$A_0 = K[h_1, \ldots, h_8] \hookrightarrow K[F] \hookrightarrow \overline{K[F]},$$

where $h_1 = f_1, h_3 = f_8 - f_{11}, h_5 = f_2 - f_3, h_7 = f_5 - f_7 - f_9 - f_{11}, h_2 = f_6$, $h_4 = f_9 - f_{10}, h_6 = f_3 - f_5, h_8 = f_1 - \sum_{i=2}^{11} f_i$.

As $\overline{K[F]}$ is Cohen–Macaulay, by Proposition 3.1.27, we get

$$\begin{aligned} \overline{K[F]} \;=\; & A_0 1 \oplus A_0 f_7 \oplus A_0 f_{10} \oplus a_0 f_{11} \oplus \\ & A_0 f_7^2 \oplus A_0 f_{10}^2 \oplus A_0 f_{10} f_{11} \oplus A_0 f_7 f_{11} \oplus \\ & A_0 f_7^3 \oplus A_0 f_7^2 f_{11} \oplus A_0 f_{10}^3 \oplus A_0 f_{10}^2 f_{11} \oplus \\ & A_0 f_{12} \oplus A_0 f_{13} \oplus A_0 f_{14} \oplus A_0 f_{15} \oplus A_0 f_{10}^3 f_{11}. \end{aligned}$$

Therefore the Hilbert series of $\overline{K[F]}$ is equal to

$$H(\overline{K[F]}, z) = \frac{1 + 3z + 4z^2 + 8z^3 + z^4}{(1 - z)^8} \quad \text{and} \quad a(\overline{K[F]}) = -4.$$

Exercises

12.4.15 Let P be the toric ideal of the Veronese subring $R^{(k)}$ or the toric ideal of the kth square-free Veronese subring $K[V_k]$. If $k \geq 2$, then P is generated by homogeneous binomials of degree two.

12.4.16 [197, 311] Prove the following: (a) $R^{(k)}$ is Gorenstein if and only if k divides n. (b) $a(K[V_k]) \leq a(R^{(k)})$. (c) If $n = 5$ and $k = 3$, then $a(K[V_k]) = -3$ and $a(R^{(k)}) = -2$. (d) $a(R^{(k)}) = -\left\lceil \frac{n}{k} \right\rceil$.

12.4.17 (a) If $n = 2k + 1 \geq 5$, then the canonical module of $K[V_k]$ is generated by the set of all x^a, $a = (a_1, \ldots, a_n)$, such that $|\{i|a_i = 2\}| = k-1$ and $|\{i|a_i = 1\}| = n - k + 1$.

(b) If $h_1, \ldots, h_n$ is a system of parameters for $K[V_k]$ with each h_i a form of degree 1, then $K[V_k]/(h_1, \ldots, h_n)$ is a *level algebra*, i.e., all the non-zero elements of its socle have the same degree.

12.4.18 If $n = 2k + 1 \geq 5$, then $\text{type}(K[V_k]) = \binom{n}{k-1}$.

12.4.19 Let F be the set of all monomials $x_i x_j$ such that x_i and x_j are connected by an edge of the following graph:

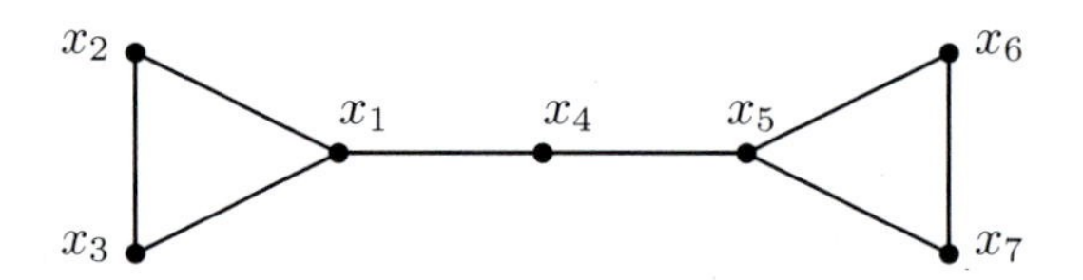

The monomial subring $A = K[F]$ is a graded subring of $\overline{A}$, where both rings are endowed with the normalized grading. Prove that the Hilbert series and Hilbert polynomial of A and $\overline{A}$ are given by:

$$F(A, t) = \frac{1 + t + t^2 + t^3 + t^4}{(1 - t)^7}, \quad F(\overline{A}, t) = \frac{1 + t + t^2 + 2t^3}{(1 - t)^7}$$

$$\varphi(A, t) = \frac{1}{144}t^6 + \frac{1}{16}t^5 + \frac{55}{144}t^4 + \frac{65}{48}t^3 + \frac{47}{18}t^2 + \frac{31}{12}t + 1,$$

$$\varphi(\overline{A}, t) = \frac{1}{144}t^6 + \frac{17}{240}t^5 + \frac{55}{144}t^4 + \frac{21}{16}t^3 + \frac{47}{18}t^2 + \frac{157}{60}t + 1.$$

In particular $e(A) = e(\overline{A}) = 5$, $a(A) = -3$ and $a(\overline{A}) = -4$.

12.5 Normalizations of Rees algebras

Let $R = K[x_1, \ldots, x_n]$ be a polynomial ring over a field K, and let I be a monomial ideal of R generated by $x^{v_1}, \ldots, x^{v_q}$. In this section we study the normality of $R[It]$ and the behavior of the filtration $\mathcal{F} = \{\overline{I^i}\}_{i \geq 0}$.

Proposition 12.5.1 *If $I = (x^{v_1}, \ldots, x^{v_q})$ is a monomial ideal of the ring of polynomials $R = K[x_1, \ldots, x_n]$, then $\overline{R[It]}$ is generated as a K-algebra by monomials of degree in t at most $n - 1$.*

Proof. If $\mathcal{H}$ is the minimal integral Hilbert basis of the Rees cone of I, then $\overline{R[It]} = K[\mathbb{N}\mathcal{H}]$ and the result follows from Proposition 1.5.4. □

Lemma 12.5.2 *Let I be a monomial ideal of $R = K[x_1, \ldots, x_n]$. If $\overline{R[It]}$ is generated as a K-algebra by monomials of degree in t at most s and I^i is integrally closed for $i \leq s$, then $R[It]$ is normal.*

Proof. Let $x^\alpha t^b$ be a generator of $\overline{R[It]}$ with $b \leq s$. Note that $x^\alpha \in \overline{I^b} = I^b$. Hence $x^\alpha t^b \in R[It]$, as required. □

Corollary 12.5.3 [345] *If I is a monomial ideal of a polynomial ring in n variables and I^i is integrally closed for $i \leq n - 1$, then I is normal.*

Proof. It follows from Proposition 12.5.1 and Lemma 12.5.2. □

Theorem 12.5.4 *Let $I = (x^{v_1}, \ldots, x^{v_q}) \subset K[x_1, \ldots, x_n]$ be a monomial ideal and let $r_0 = \operatorname{rank}(v_1, \ldots, v_q)$. The following hold.*

(i) $\overline{I^b} = I\overline{I^{b-1}}$ *for* $b \geq n$.

(ii) *If* $\mathcal{A} = \{v_1, \ldots, v_q\}$ *is homogeneous, then* $\overline{I^b} = I\overline{I^{b-1}}$ *for* $b \geq r_0$.

Proof. We set $\mathcal{A}' = \{(v_1, 1), \ldots, (v_q, 1), e_1, \ldots, e_n\}$. Assume $b \geq n$ (resp. $b \geq r_0$ and $\mathcal{A}$ homogeneous). Notice that we invariably have $I\overline{I^{b-1}} \subset \overline{I^b}$. To show the reverse inclusion take $x^\alpha \in \overline{I^b}$. Thus $x^\alpha t^b \in \overline{R[It]}$ and (α, b) is in the Rees cone $\mathbb{R}_+\mathcal{A}'$. By the proof of Proposition 1.5.4 and by Carathéodory's theorem (see Theorem 1.1.18), we can write

$$(\alpha, b) = \lambda_1(v_{i_1}, 1) + \cdots + \lambda_r(v_{i_r}, 1) + \mu_1 e_{j_1} + \cdots + \mu_s e_{j_s} \quad (\lambda_i, \mu_k \in \mathbb{Q}_+),$$

where $\{(v_{i_1}, 1), \ldots, (v_{i_r}, 1), e_{j_1}, \ldots, e_{j_s}\}$ is a set of vectors (resp. linearly independent set of vectors) contained in $\mathcal{A}'$ and $r \leq n$ (resp. $r \leq r_0$). Since $b = \lambda_1 + \cdots + \lambda_r$, we obtain that $\lambda_i \geq 1$ for some i. It follows readily that $x^\alpha \in I\overline{I^{b-1}}$, as required. □

Normalizations and Hilbert functions Let $R = K[x_1, \ldots, x_n]$ be a polynomial ring over a field K with $n \geq 2$ and let I be a zero dimensional monomial ideal of R.

The integral closure of I^i, denoted by $\overline{I^i}$, is again a monomial ideal for all $i \geq 1$ (see Proposition 12.1.2). The length of $R/\overline{I^i}$ will be denoted simply by $\ell(R/\overline{I^i})$. By Lemma 8.5.1 the length of $R/\overline{I^i}$ is $\dim_K(R/\overline{I^i})$. We are interested in computing the *Hilbert function*

$$H_{\mathcal{F}}(i) := \ell(R/\overline{I^i}) = \dim_K(R/\overline{I^i}); \quad i \in \mathbb{N} \setminus \{0\}; \;\; H_{\mathcal{F}}(0) = 0,$$

of the filtration $\mathcal{F} = \{\overline{I^i}\}_{i=0}^{\infty}$. The Hilbert function of $\mathcal{F}$ is a polynomial function of degree n:

$$H_{\mathcal{F}}(i) = c_n i^n + c_{n-1} i^{n-1} + \cdots + c_1 i + c_0 \quad (i \gg 0),$$

where $c_0, \ldots, c_n \in \mathbb{Q}$ and $c_n \neq 0$ (see the proof of Proposition 12.5.16). The polynomial $\varphi_{\mathcal{F}}(x) := c_n x^n + \cdots + c_0$ is the *Hilbert polynomial* of $\mathcal{F}$. By the results of [104], one has the equality $n!c_n = e(I)$, where $e(I)$ is the multiplicity of the ideal I in the sense of Section 4.4.

There is a unique set of monomials $\overline{F} = \{x^{u_1}, \ldots, x^{u_r}\}$ that minimally generate $\overline{I}$. For $1 \leq i \leq n$, we may assume that the ith entry of u_i is a_i and that all other entries of u_i are zero, where $a_1, \ldots, a_n$ are positive integers. We set $\alpha_0 = (1/a_1, \ldots, 1/a_n)$. Let $\{u_{n+1}, \ldots, u_s\}$ be the set of u_i such that $\langle u_i, \alpha_0 \rangle < 1$, and let $\{u_{s+1}, \ldots, u_r\}$ be the set of u_i such that $i > n$ and $\langle u_i, \alpha_0 \rangle \geq 1$. Consider the convex rational polyhedron

$$\begin{aligned} \mathcal{Q} &= \mathbb{Q}_+^n + \operatorname{conv}(u_1, \ldots, u_r) \\ &= \mathbb{Q}_+^n + \operatorname{conv}(u_1, \ldots, u_n, u_{n+1}, \ldots, u_s). \end{aligned}$$

The second equality follows from the finite basis theorem and the equality

$$\mathbb{Q}_+^n + \operatorname{conv}(u_1, \ldots, u_n) = \{x \,|\, x \geq 0; \, \langle x, \alpha_0 \rangle \geq 1\}.$$

Lemma 12.5.5 $\overline{I^i} = (\{x^a | \, a \in i\mathcal{Q} \cap \mathbb{Z}^n\})$ *for* $0 \neq i \in \mathbb{N}$.

Proof. Let $x^{v_1}, \ldots, x^{v_q}$ be a set of generators of I. Let $x^\alpha \in \overline{I^i}$, i.e., $x^{p\alpha} \in I^{ip}$ for some $0 \neq p \in \mathbb{N}$. Hence

$$\alpha/i \in \operatorname{conv}(v_1, \ldots, v_q) + \mathbb{Q}_+^n \subset \operatorname{conv}(u_1, \ldots, u_r) + \mathbb{Q}_+^n = \mathcal{Q}$$

and $\alpha \in i\mathcal{Q} \cap \mathbb{Z}^n$. Conversely let $\alpha \in i\mathcal{Q} \cap \mathbb{Z}^n$. Hence $x^{p\alpha} \in (\overline{I})^{ip}$ for some $0 \neq p \in \mathbb{N}$, this yields $x^\alpha \in \overline{(\overline{I})^i}$. Since $(\overline{I})^i \subset \overline{I^i}$ and the latter is complete we get $x^\alpha \in \overline{I^i}$. $\Box$

Corollary 12.5.6 $\ell(R/\overline{I^i}) = |\mathbb{N}^n \setminus i\mathcal{Q}|$ *for* $i \geq 1$.

Proof. It follows from Lemmas 8.5.1 and 12.5.5. $\Box$

In what follows we set $\mathcal{P} := \text{conv}(u_1, \ldots, u_s)$, $\mathcal{S} := \text{conv}(0, u_1, \ldots, u_n)$ and $d = \dim(\mathcal{P})$. The *Ehrhart function* of $\mathcal{P}$ is given by

$$\chi_{\mathcal{P}}(i) := |\mathbb{Z}^n \cap i\mathcal{P}|, \quad i \in \mathbb{N}.$$

This is a polynomial function of degree d. The *Ehrhart polynomial* of $\mathcal{P}$ is the unique polynomial $E_{\mathcal{P}}(x)$ of degree d such that $E_{\mathcal{P}}(i) = \chi_{\mathcal{P}}(i)$ for $i \gg 0$. Ehrhart functions are studied in Section 9.3.

Proposition 12.5.7 *$H_{\mathcal{F}}(i) = E_{\mathcal{S}}(i) - E_{\mathcal{P}}(i)$ for $i \in \mathbb{N}$.*

Proof. Since $E_{\mathcal{P}}(0) = E_{\mathcal{S}}(0) = 1$, we get $H_{\mathcal{F}}(0) = 0$. Assume $i \geq 1$. Notice that from the decomposition $\mathcal{Q} = (\mathbb{Q}_+^n \setminus \mathcal{S}) \cup \mathcal{P}$ we get

$$\begin{aligned} i\mathcal{Q} &= (\mathbb{Q}_+^n \setminus i\mathcal{S}) \cup i\mathcal{P}, \\ \mathbb{N}^n \setminus i\mathcal{Q} &= [\mathbb{N}^n \cap (i\mathcal{S})] \setminus [\mathbb{N}^n \cap (i\mathcal{P})]. \end{aligned}$$

Hence, by Corollary 12.5.6, $\ell(R/\overline{I^i}) = |\mathbb{N}^n \setminus i\mathcal{Q}| = E_{\mathcal{S}}(i) - E_{\mathcal{P}}(i)$. $\Box$

The Hilbert function of $\mathcal{F}$ and its Hilbert polynomial are equal for non-negative integer values:

Corollary 12.5.8 *$H_{\mathcal{F}}(i) = c_n i^n + \cdots + c_1 i + c_0$ for $i \in \mathbb{N}$ and $c_0 = 0$.*

Proof. By Lemma 9.3.8, $E_{\mathcal{P}}(i) = \chi_{\mathcal{P}}(i)$ and $E_{\mathcal{S}}(i) = \chi_{\mathcal{S}}(i)$ for $i \in \mathbb{N}$. Thus the result follows from Proposition 12.5.7. $\Box$

Example 12.5.9 If I is the monomial ideal (x_1^2, x_2^3), then $\overline{I} = I + (x_1 x_2^2)$. Notice that $\mathcal{P} = \text{conv}((2,0),(0,3))$ because $(1,2)$ lies above the hyperplane $x_1/2 + x_2/3 = 1$ and $\mathcal{S} = \text{conv}(0,(2,0),(0,3))$. Using *Normaliz* [68] we get $E_{\mathcal{S}}(i) = 3i^2 + 3i + 1$ and $E_{\mathcal{P}}(i) = i + 1$. Thus $H_{\mathcal{F}}(i) = 3i^2 + 2i$.

Example 12.5.10 If $I = (x_1^4, x_2^5, x_3^6, x_1 x_2 x_3^2)$, using *Normaliz* [68] we get:

$$\begin{aligned} \overline{I} = I + (&x_1 x_2^4, x_1^2 x_2^3, x_1^3 x_2^2, x_1^2 x_2^2 x_3, x_2^4 x_3^2, x_1 x_3^5, x_2^3 x_3^3, x_1^2 x_3^3, x_1^3 x_2 x_3, \\ &x_2^2 x_3^4, x_1 x_2^3 x_3, x_1^3 x_3^2, x_2 x_3^5). \end{aligned}$$

Notice that $\mathcal{P} = \text{conv}((4,0,0),(0,5,0),(0,0,6),(1,1,2))$ because $(1,1,2)$ is the only u_i lying strictly below the hyperplane $x_1/4 + x_2/5 + x_3/6 = 1$ and $\mathcal{S} = \text{conv}(0,(4,0,0),(0,5,0),(0,0,6))$. Using *Normaliz* [68] we get

$$\begin{aligned} H_{\mathcal{F}}(i) &= E_{\mathcal{S}}(i) - E_{\mathcal{P}}(i) \\ &= (1 + 6i + 19i^2 + 20i^3) - (1 + (1/6)i + (3/2)i^2 + (13/3)i^3) \\ &= (35/6)i + (35/2)i^2 + (47/3)i^3. \end{aligned}$$

Remark 12.5.11 For $i \geq n$, $\overline{I^i} = I\overline{I^{i-1}}$, see Theorem 12.5.4. Thus we can use polynomial interpolation and Corollary 12.5.8 to determine $c_1, \ldots, c_n$.

Example 12.5.12 Consider the ideal of Example 12.5.10. For $i = 0, 1, 2, 3$ the values of $H_{\mathcal{F}}(i) = \ell(R/\overline{I^i})$ are $0, 39, 207, 598$. To compute the Hilbert polynomial $\varphi_{\mathcal{F}}(x)$ of $\mathcal{F}$, using polynomial interpolation, one can use the following command in *Maple* [80]:

```
interp([0,1,2,3],[0,39,207,598],x);
```

to get that $\varphi_{\mathcal{F}}(x) = (47/3)x^3 + (35/2)x^2 + (35/6)x$.

Example 12.5.13 Consider the ideal $I = (x_1^{10}, x_2^8, x_3^5) \subset R$. The values of $H_{\mathcal{F}}(i)$ at $i = 0, 1, 2, 3$ are $0, 112, 704, 2176$. Using interpolation we obtain:

$$H_{\mathcal{F}}(i) = \ell(R/\overline{I^i}) = \frac{200}{3}i^3 + 40x^2 + \frac{16}{3}i.$$

Proposition 12.5.14 [424] *Let I_i be the ideal obtained from I by making $x_i = 0$ in the generators of I and let $e(I_i)$ be its multiplicity. Then*

$$2c_{n-1} \geq \sum_{i=1}^{n-1} \frac{e(I_i)}{(n-1)!}.$$

Let $e_0, e_1, \ldots, e_n$ be the Hilbert coefficients of $H_{\mathcal{F}}$. Recall that we have:

$$H_{\mathcal{F}}(i) = e_0\binom{i+n-1}{n} - e_1\binom{i+n-2}{n-1} + \cdots + (-1)^{n-1}e_{n-1}\binom{i}{1} + (-1)^n e_n,$$

where $e_0 = e(I)$ is the multiplicity of I and $c_n = e_0/n!$. Notice that $e_n = 0$ because $H_{\mathcal{F}}(0) = 0$.

Corollary 12.5.15 $e_0(n-1) - 2e_1 \geq e(I_1) + \cdots + e(I_{n-1}) \geq n-1$.

Proof. From the equality

$$c_{n-1} = \frac{1}{n!}\left[e_0\binom{n}{2} - ne_1\right] = \frac{1}{(n-1)!}\left[\frac{e_0(n-1)}{2} - e_1\right]$$

and using Proposition 12.5.14 we obtain the desired inequality. □

Proposition 12.5.16 $e_i \geq 0$ *for all* i.

Proof. Since the Rees algebra of $\mathcal{F}$ is $\overline{R[It]}$ and this algebra is normal and Cohen–Macaulay, we get that the ring $\mathrm{gr}_{\mathcal{F}}(R) = \oplus_{i\geq 0}\overline{I^i}/\overline{I^{i+1}}$ is Cohen–Macaulay (see [198, Corollary 2.1, p. 74]). Recall that $\mathrm{gr}_{\mathcal{F}}(R)$ is a finitely generated $(R/\overline{I})$-algebra, i.e., $\mathrm{gr}_{\mathcal{F}}(R) = (R/\overline{I})[\overline{z}_1, \ldots, \overline{z}_m]$, where $\overline{z}_1, \ldots, \overline{z}_m$

are homogeneous elements of positive degree (see [65, Proposition 4.5.4]). The canonical map $\mathrm{gr}_I(R) \to \mathrm{gr}_{\mathcal{F}}(R)$ is integral. Thus $\mathrm{gr}_{\mathcal{F}}(R)$ is a positively graded Noetherian module over the standard graded algebra $\mathrm{gr}_I(R)$. Hence, by the Hilbert–Serre theorem, we can write the Hilbert series of $\mathrm{gr}_{\mathcal{F}}(R)$ as $h(x)/(1-x)^n$, where $h(x)$ is a polynomial with nonnegative integer coefficients because $\mathrm{gr}_{\mathcal{F}}(R)$ is Cohen–Macaulay. To finish the proof notice that $e_i = h^{(i)}(1)/i!$ (see [65, Proposition 4.1.9]). □

Example 12.5.17 Let $\mathfrak{m} = (x_1, \ldots, x_n)$ be the irrelevant maximal ideal and let $I = \mathfrak{m}^k$ be the kth Veronese ideal of R. Then

$$H_{\mathcal{F}}(i) = \binom{ki+n-1}{n} = \frac{k^n}{n!}i^n + \frac{k^{n-1}}{(n-2)!2}i^{n-1} + \text{ terms of lower degree}$$

because I is normal. Here $e_0 = e(I) = k^n$, $e_1 = (n-1)(k^n - k^{n-1})/2$ and

$$c_{n-1} = \frac{1}{2}\sum_{i=1}^{n-1} \frac{e(I_i)}{(n-1)!}.$$

Exercises

12.5.18 Let $I = (x^{v_1}, \ldots, x^{v_q}) \subset R$ be a monomial ideal and let r_0 be the rank of $A = (v_1, \ldots, v_q)$. If $\mathcal{A} = \{v_1, \ldots, v_q\}$ is homogeneous and I^i is integrally closed for $i = 1, \ldots, r_0 - 1$, then $R[It]$ is a normal domain.

12.5.19 If $R = \mathbb{Q}[x_1, x_2]$ and $I = (x_1^2, x_2^3, x_1x_2)$, find a presentation of $\mathrm{gr}_I(R)$ and find the multiplicity of I using the CoCoA [88] procedure:

```
Use R::=Q[t[1..3],x[1..2]];
J:=Toric([[2,0,1,1,0],[0,3,1,0,1],[1,1,1,0,0]]);
I:=Ideal(x[1]^2,x[2]^3,x[1]*x[2]);
L:=I+J;
Multiplicity(R/LT(L));
```

12.5.20 If $I = (x_1^2, x_2^3, x_1x_2) \subset \mathbb{Q}[x_1, x_2]$, use Exercise 12.5.19 to prove:

$$\ell(I^i/I^{i+1}) = \begin{cases} 5i+4 & \text{if } \ 0 \le i \neq 2, \\ 27 & \text{if } \ i = 2. \end{cases}$$

Prove that the Samuel function of I is given by

$$\chi_R^I(i) := \ell(R/I^{i+1}) = \frac{5}{2}i^2 + \frac{13}{2}i + 4 \quad (i \ge 0).$$

12.5.21 If $R = \mathbb{Q}[x_1, x_2]$ and $I = (x_1x_2, x_1^2 + x_2^2)$, prove that $e(I) = 4$.

12.5.22 If $I = (x^2, y^3, z^5, w^6, xy, z^2w, xz^2, y^2z^2) \subset R = K[x, \ldots, w]$, use the methods of this section to prove:

$$\ell(R/\overline{I^i}) = (7/2)i^4 + (65/6)i^3 + (21/2)i^2 + (19/6)i; \quad \forall\, i \in \mathbb{N}.$$

12.6 Rees algebras of Veronese ideals

In this section we compute the a-invariant of the Rees algebra of a Veronese ideal. This will be used to show some degree bounds for the normalization of a uniform ideal; see Theorem 12.6.7.

The Rees algebra of a square-free Veronese ideal Let K be a field and let $R = K[x_1, \ldots, x_n]$ be a polynomial ring with coefficients in K. Given two positive integers k, n with $k \leq n$, we define

$$V_k = \{x_{i_1} \cdots x_{i_k} \,|\, 1 \leq i_1 < \cdots < i_k \leq n\}.$$

The ideal $L := (V_k)$ is called the kth *square-free Veronese ideal* of R. Recall that the *Rees algebra* of L is given by

$$R[Lt] = K[x_1, \ldots, x_n, f_1 t, \ldots, f_{\binom{n}{k}} t] \subset R[t],$$

where t is a new variable and $V_k = \{f_1, \ldots, f_{\binom{n}{k}}\}$. We can make $R[Lt]$ a standard K-algebra with the grading induced by setting $\delta(x_i) = 1$ and $\delta(t) = 1 - k$. In this grading $\delta(f_i t) = 1$ for all i.

The Rees algebra of L satisfies the hypothesis of Theorem 9.1.5 because: (i) $R[Lt]$ is a normal domain by Proposition 12.3.10, and (ii) $R[Lt]$ is a standard K-algebra.

The next expression for the a-invariant follows from Theorem 9.1.5.

Theorem 12.6.1 [3] *If* $1 \leq k < n$ *and* $n = pk + r$, *where* $0 \leq r < k$, *then*

$$a(R[Lt]) = \begin{cases} -(p+2) & \text{if } \; r \geq 2, \\ -(p+1) & \text{if } \; r = 1, \\ -(p+1) & \text{if } \; r = 0 \text{ and } k > 1, \\ -p & \text{if } \; r = 0 \text{ and } k = 1. \end{cases}$$

There is an alternative expression for a Rees algebra of a monomial ideal generated by monomials of the same degree k.

Proposition 12.6.2 *Let* $F = \{x^{v_1}, \ldots, x^{v_q}\}$ *be a set of monomials in* R *of fixed degree* $k \geq 1$. *If* $\mathbf{x} = \{x_1, \ldots, x_n\}$, *then* $R[Ft] \simeq K[F, \mathbf{x}t]$.

Proof. Consider the epimorphisms of K-algebras:

$$\phi\colon R[t_1, \ldots, t_q] \to R[tF], \text{ induced by } \phi(t_i) = x^{v_i} t \text{ and } \phi(x_i) = x_i,$$
$$\psi\colon R[t_1, \ldots, t_q] \to K[F, \mathbf{x}t], \text{ induced by } \psi(t_i) = x^{v_i} \text{ and } \psi(x_i) = x_i t.$$

It suffices to notice that $\ker(\phi) = \ker(\psi)$. This equality follows from the fact that $\ker(\phi)$ and $\ker(\psi)$ are binomial ideals (see Corollary 8.2.18). $\Box$

Corollary 12.6.3 [72] *The* a*-invariant of the edge subring of a complete graph* $\mathcal{K}_{n+1}$ *is given by* $a(K[\mathcal{K}_{n+1}]) = -\left\lceil \frac{n+1}{2} \right\rceil$.

Proof. It follows from Theorem 12.6.1 and Proposition 12.6.2. $\Box$

The Rees cone of a Veronese ideal In this part, we refer to Chapter 1 for standard terminology and notation on polyhedral cones. Setting

$$\begin{aligned}\mathcal{A} &= \{e_1,\ldots,e_n,ke_1+e_{n+1},\ldots,ke_n+e_{n+1}\},\\ \mathcal{A}' &= \{e_1,\ldots,e_n,a_1e_1+\cdots+a_ne_n+e_{n+1}|\, a_i\in\mathbb{N};\ |a|=k\},\end{aligned}$$

where e_i is the ith unit vector in $\mathbb{R}^{n+1}$ and $|a|:=a_1+\cdots+a_n$, one has the equality $\mathbb{R}_+\mathcal{A}=\mathbb{R}_+\mathcal{A}'$. This cone is the so-called *Rees cone*.

Lemma 12.6.4 *The irreducible representation of* $\mathbb{R}_+\mathcal{A}$ *is:*

$$\mathbb{R}_+\mathcal{A}=H_{e_1}^+\cap\cdots\cap H_{e_n}^+\cap H_{e_{n+1}}^+\cap H_a^+,\quad a=(1,\ldots,1,-k).$$

Proof. Set $N=\{e_1,\ldots,e_{n+1},a\}$. It suffices to prove that F is a facet of $\mathbb{R}_+\mathcal{A}$ if and only if $F=H_b\cap\mathbb{R}_+\mathcal{A}$ for some $b\in N$. Let $1\leq i\leq n+1$. Consider the following sets

$$\{e_1,e_2,\ldots,\widehat{e}_i,\ldots,e_n,e_{n+1}\}\ \text{ and }\ \{ke_1+e_{n+1},\ldots,ke_n+e_{n+1}\},$$

where $\widehat{e}_i$ means to omit e_i from the list. Since those sets are linearly independent it follows that $F=H_b\cap\mathbb{R}_+\mathcal{A}$ is a facet for $b\in N$. Conversely let F be a facet of $\mathbb{R}_+\mathcal{A}$. There are linearly independent vectors $\alpha_1,\ldots,\alpha_n\in\mathcal{A}$ and $0\neq b=(b_1,\ldots,b_{n+1})\in\mathbb{R}^{n+1}$ such that

(i) $F=\mathbb{R}_+\mathcal{A}\cap H_b$,

(ii) $\mathbb{R}\alpha_1+\cdots+\mathbb{R}\alpha_n=H_b$, and

(iii) $\mathbb{R}_+\mathcal{A}\subset H_b^+$.

Since $e_1,\ldots,e_n$ are in $\mathcal{A}$, by (iii), one has $\langle e_i,b\rangle=b_i\geq 0$ for $i=1,\ldots,n$. Set $B=\{\alpha_1,\ldots,\alpha_n\}$ and consider the matrix M_B whose rows are the vectors in B. Let us prove that there exists $c\in N$ such that $H_b=H_c$ and $\mathbb{R}_+\mathcal{A}\subset H_c^+$. Consider the following cases. Case (1): If the ith column of M_B is zero for some $1\leq i\leq n+1$, it suffices to take $c=e_i$. Case (2): If $B=\{ke_1+e_{n+1},\ldots,ke_n+e_{n+1}\}$, then take $c=a$. Case (3): If

$$B=\{e_{i_1},\ldots,e_{i_s},ke_{j_1}+e_{n+1},\ldots,ke_{j_t}+e_{n+1}\},$$

where $s,t>0$, $s+t=n$, $1\leq i_1<\cdots<i_s\leq n$, $1\leq j_1<\cdots<j_t\leq n$, and M_B has all its columns different from zero. Since $b_{i_1}=0$, using

$$\begin{aligned}\langle ke_{i_1}+e_{n+1},b\rangle &= b_{n+1}\geq 0,\\ \langle ke_{j_1}+e_{n+1},b\rangle &= kb_{j_1}+b_{n+1}=0,\end{aligned}$$

and $b_{j_1}\geq 0$, we obtain $b_{n+1}=0$. Then $e_{n+1}\in H_b$. It follows that H_b is generated by the set $\{e_1,e_2,\ldots,\widehat{e}_i,\ldots,e_n,e_{n+1}\}$ for some $1\leq i\leq n$ and we take $c=e_i$. $\square$

Theorem 12.6.5 *Let $R[Jt]$ be the Rees algebra of the kth Veronese ideal J.* (a) *If $2 \leq k < n$ and $n = pk + s$, where $0 \leq s < k$, then*

$$a(R[Jt]) = \begin{cases} -(p+2) & \text{if} \quad s \geq 2, \\ -(p+1) & \text{if} \quad s = 0 \text{ or } s = 1. \end{cases}$$

(b) *If $k \geq n$, then $a(R[Jt]) = -2$.*

Proof. Let $M = x^a x^b t^c$ be a monomial in the canonical module $\omega_{R[Jt]}$ of $R[Jt]$, where $x^b t^c = (f_1 t) \cdots (f_c t)$ and f_i is a monomial in R of degree k for all i. Note $\delta(M) = |a| + c$. Recall that $\log(M) = (a + b, c)$ is in the interior of $\mathbb{R}_+\mathcal{A}$ by Theorem 9.1.5. Therefore, using Lemma 12.6.4, one has $c \geq 1$, $a_i + b_i \geq 1$ for all i, and $|a| + |b| \geq kc + 1$. As $|b| = kc$, altogether we get:

$$|a| + |b| \geq n \text{ and } |a| \geq 1. \tag{12.2}$$

In particular $\delta(M) \geq 2$ and $a(R[Jt]) \leq -2$. To prove (b) note that by Lemma 12.6.4 the monomial $m = x_1^{k-n+2} x_2 \cdots x_n t$ is in $\omega_{R[Jt]}$ and $\delta(m)$ is equal to 2. Hence $a(R[Jt]) = -2$. To prove (a) there are the following three cases to consider.

Case $s \geq 2$: First we show $\delta(M) \geq p + 2$. If $c > p$, then $\delta(M) \geq p + 2$ follows from Eq.(12.2). On the other hand assume $c \leq p$. Observe:

$$k(p - c) + s \geq (p - c) + 2. \tag{$*$}$$

From Eq.(12.2) one has $|a| + |b| = |a| + kc \geq n = kp + s$. Consequently

$$\delta(M) = |a| + c \geq k(p - c) + s + c. \tag{$**$}$$

Hence from $(*)$ and $(**)$ we get $\delta(M) \geq p + 2$, as required. Therefore one has the inequality $a(R[Jt]) \leq -(p + 2)$, to show equality we claim that

$$m = x_1^2 x_2^2 \cdots x_{k-s+1}^2 x_{k-s+2} \cdots x_n t^{p+1}$$

is in $\omega_{R[Jt]}$ and $\delta(m) = p + 2$. An easy calculation shows that m is in $R[Jt]$ and $\delta(m) = p + 2$. Finally let us see that m is in $\omega_{R[Jt]}$ via Lemma 12.6.4. That the entries of $\log(m)$ satisfy $y_i > 0$ for all i is clear. The inequality

$$y_1 + y_2 + \cdots + y_n > k y_{n+1}, \tag{$***$}$$

after making y_i equal to the ith entry of $\log(m)$, transforms into

$$2(k - s + 1) + (n - (k - s + 1)) > k(p + 1)$$

but the left-hand side is $k(p + 1) + 1$, hence $\log(m)$ satisfies $(***)$.

Case $s = 1$: If $c \geq p$, then clearly $|a| + c \geq p + 1$, so let us suppose that $c < p$, then $n + c(1 - k) \geq n + p(1 - k)$ and using Eq. (12.2) we get

$$|a| + c = (|a| + |b| - n) + (n + c(1 - k)) \geq n + c(1 - k) \geq p + 1$$

so $a(R[Jt]) \leq -(p+1)$. Now observe that the monomial $m = x_1x_2\cdots x_nt^p$ has $\delta(m) = p+1$ and $m \in R[Jt]$. Finally, an argument similar to the previous case shows that m is in $\omega_{R[Jt]}$. Thus $a(R[Jt]) = -(p+1)$.

Case $s = 0$: First note that $p > 1$ because $n > k$. If $c \geq p$, then clearly $|a| + c \geq p+1$. So let us suppose $c < p$. Then, using Eq. (12.2), we get

$$|a| \geq k(p-c) > (p-c).$$

Then adding c we derive $|a| + c \geq p+1$. Hence $a(R[Jt]) \leq -(p+1)$. Note that the monomial $m = x_1^2x_2\cdots x_nt^p$ has $\delta(m) = p+1$ and belongs to $\omega_{R[Jt]}$. Thus, one has the required equality. □

Corollary 12.6.6 *Let I be an ideal of Veronese type of degree k. If $2 \leq k < n$ and $n = pk+s$, where $0 \leq s < k$, then*

$$a(R[It]) = \begin{cases} -(p+2) & \text{if} \quad s \geq 2, \\ -(p+1) & \text{if} \quad s = 0 \text{ or } s = 1. \end{cases}$$

Proof. Let L and J be the kth square-free Veronese ideal and the kth Veronese ideal, respectively. From Theorems 12.6.1 and 12.6.5 we obtain that $a(R[Lt])$ and $a(R[Jt])$ can be computed using the formula above. Since $R[It]$ is normal, using the proof of Theorem 12.4.13, we get

$$a(R[Lt]) \leq a(R[It]) \leq a(R[Jt]).$$

Therefore $a(R[It])$ is given by the formula above. □

Normalization of a uniform ideal Let $R = K[x_1,\ldots,x_n]$ be a ring of polynomials over a field K and let I be an ideal of R generated by monomials $x^{v_1},\ldots,x^{v_q}$ of degree $k \geq 2$. Ideals of this type are called *uniform*. In this case the Rees algebra of I is a standard graded K-algebra with the grading δ induced by $\delta(x_i) = 1$ and $\delta(t) = 1-k$. Notice that $\delta(x^{v_i}t) = 1$ for all i.

Notice that we can embed $R[It]$ in $R[Jt]$, where J is the kth Veronese ideal of R. The next result complements [72, Theorem 3.4] for the class of uniform ideals.

Theorem 12.6.7 *Let $\mathcal{R} = R[It]$ be the Rees algebra of I. If $2 \leq k < n$, then the normalization $\overline{\mathcal{R}}$ of I is generated as an $\mathcal{R}$-module by monomials $g \in R[t]$ of degree in t at most $n - \lfloor n/k \rfloor$.*

Proof. We set $\mathcal{A}' = \{e_1,\ldots,e_n,(v_1,1),\ldots,(v_q,1)\}$. As $\mathbb{Z}\mathcal{A}' = \mathbb{Z}^{n+1}$, according to Theorem 9.1.1, the normalization of I can be expressed as:

$$\overline{\mathcal{R}} = K[\{x^at^b \,|\, (a,b) \in \mathbb{Z}^{n+1} \cap \mathbb{R}_+\mathcal{A}'\}].$$

If $m = x^a t^b$ with $0 \neq (a, b) \in \mathbb{Z}^{n+1} \cap \mathbb{R}_+\mathcal{A}'$, then $\delta(m) \geq b$. To show this write $(a, b) = \sum_{i=1}^{n} \lambda_i e_i + \sum_{i=1}^{q} \mu_i(v_i, 1)$ for some $\lambda_i \geq 0, \mu_i \geq 0$. Hence

$$\begin{aligned} |a| &= \lambda_1 + \cdots + \lambda_n + (\mu_1 + \cdots + \mu_q)k \\ b &= \mu_1 + \cdots + \mu_q, \end{aligned}$$

and consequently $\delta(m) = |a| + (1-k)b = (\lambda_1 + \cdots + \lambda_n) + b \geq b$. Therefore $\overline{\mathcal{R}}$ is generated as a K-algebra by monomials of positive degree. There is a Noether normalization of $\mathcal{R}$

$$A = K[z_1, \ldots, z_{n+1}] \stackrel{\varphi}{\hookrightarrow} \mathcal{R} \stackrel{\psi}{\hookrightarrow} \overline{\mathcal{R}},$$

where $z_1, \ldots, z_{n+1} \in \mathcal{R}_1$. Notice that the composite $A \to \overline{\mathcal{R}}$ is a Noether normalization of $\overline{\mathcal{R}}$. By Theorem 9.1.6 $\overline{\mathcal{R}}$ is Cohen–Macaulay. Hence, by Proposition 3.1.27, $\overline{\mathcal{R}}$ is a free module over A and one can write

$$\overline{\mathcal{R}} = Am_1 \oplus \cdots \oplus Am_p, \tag{12.3}$$

where $m_i = x^{\beta_i} t^{b_i}$. Using that the length is additive one has the following expression for the Hilbert series

$$F(\overline{\mathcal{R}}, z) = \sum_{i=0}^{p} \frac{z^{\delta(m_i)}}{(1-z)^{n+1}} = \frac{h_0 + h_1 z + \cdots + h_s z^s}{(1-z)^{n+1}},$$

where $h_i = |\{j \mid \delta(m_j) = i\}|$. Note that $a(\overline{\mathcal{R}})$, the a-invariant of $\overline{\mathcal{R}}$, is equal to $s - (n+1)$. On the other hand one has

$$a(\overline{\mathcal{R}}) = -\min\{i \mid (\omega_{\overline{\mathcal{R}}})_i \neq 0\},$$

where $\omega_{\overline{\mathcal{R}}}$ is the canonical module of the normalization $\overline{\mathcal{R}}$. Using the proof of Theorem 12.4.13 together with Theorem 12.6.5 yields

$$a(\overline{\mathcal{R}}) = s - (n+1) \leq a(R[Jt]) \leq -\left\lfloor \frac{n}{k} \right\rfloor - 1$$

and $s \leq n - \lfloor n/k \rfloor$. Altogether if $m_i = x^{\beta_i} t^{b_i}$ one has $b_i \leq \delta(m_i) \leq n - \lfloor n/k \rfloor$, that is, the degree in t of m_i is at most $n - \lfloor n/k \rfloor$, as required. □

Proposition 12.6.8 $\overline{I^i} = I\overline{I^{i-1}}$ *for* $i \geq n + 2 + a(R[Jt])$. *Furthermore equality holds for* $i \geq n - \lfloor n/k \rfloor + 1$.

Proof. It follows for the proof of Theorem 12.6.7 and using Eq.(12.3). □

Definition 12.6.9 The *index of normalization* of I, denoted by $s(I)$, is the smallest integer s such that $\overline{I^i} = I\overline{I^{i-1}}$ for $i > s$.

Corollary 12.6.10 $s(I) \leq n - \lfloor n/k \rfloor$.

Exercises

12.6.11 Let $B = K[x_1, \ldots, x_n, t_1, \ldots, t_n]$ be a polynomial ring, let A be the following matrix of indeterminates

$$A = \begin{pmatrix} x_1 & x_2 & \cdots & x_n \\ t_1 & t_2 & \cdots & t_n \end{pmatrix},$$

and let $I_2(A)$ be the ideal generated by the 2-minors of A. If R is the polynomial ring $K[x_1, \ldots, x_n]$ and $\mathfrak{m} = (x_1, \ldots, x_n)$, then the toric ideal of $R[\mathfrak{m}t]$ is $I_2(A)$ and $a(R[\mathfrak{m}t]) = a(B/I_2(A)) = -n$.

12.6.12 [72] Let $K[F] \subset R$ be a subring generated by a set F of monomials of degree k. Prove that $\overline{K[F]}$ is generated as a $K[F]$-module by elements $g \in R$ of normalized degree at most $\dim(K[F]) - 1$.

12.7 Divisor class group of a Rees algebra

In this section we determine the divisor class group of the Rees algebra of a normal monomial ideal.

Let $R = K[x_1, \ldots, x_n]$ be a polynomial ring over a field K and let I be an ideal of R of height $g \geq 2$ minimally generated by monomials $x^{v_1}, \ldots, x^{v_q}$ of degree at least two. For technical reasons we shall assume that each variable x_i occurs in at least one monomial x^{v_j}.

The *Rees cone* of the point configuration $\mathcal{A} = \{v_1, \ldots, v_q\}$ is the rational polyhedral cone in $\mathbb{R}^{n+1}$, denoted by $\mathbb{R}_+\mathcal{A}'$, consisting of the nonnegative linear combinations of the set

$$\mathcal{A}' := \{e_1, \ldots, e_n, (v_1, 1), \ldots, (v_q, 1)\} \subset \mathbb{R}^{n+1},$$

where e_i is the ith unit vector. Notice that $\dim(\mathbb{R}_+\mathcal{A}') = n + 1$. Thus according to Corollary 1.1.50 there is a unique irreducible representation

$$\mathbb{R}_+\mathcal{A}' = H^+_{e_1} \cap \cdots \cap H^+_{e_{n+1}} \cap H^+_{\ell_1} \cap \cdots \cap H^+_{\ell_r} \tag{12.4}$$

such that $0 \neq \ell_i \in \mathbb{Z}^{n+1}$ and the non-zero entries of ℓ_i are relatively prime for all i.

Lemma 12.7.1 *If $\ell_j = (a_1, \ldots, a_n, -a_{n+1})$, then $a_i \geq 0$ for $i = 1, \ldots, n$ and $a_{n+1} > 0$.*

Proof. Clearly $a_i \geq 0$ because $e_i \in \mathbb{R}_+\mathcal{A}'$ for $i = 1, \ldots, n$. Notice that $a_i > 0$ for some $1 \leq i \leq n$. Assume $a_{n+1} \leq 0$. Then

$$H^+_{e_1} \cap \cdots \cap H^+_{e_{n+1}} \subset H^+_{e_1} \cap \cdots \cap H^+_{e_{n+1}} \cap H^+_{\ell_j}.$$

To show the inclusion take $x = (x_i)$ is in the left-hand side, then $x_i \geq 0$ for all i and $a_1x_1 + \cdots + a_nx_n \geq 0 \geq a_{n+1}x_{n+1}$. Thus $x \in H_{\ell_j}^+$. Hence $H_{\ell_j}^+$ can be deleted from Eq. (12.4), a contradiction. Thus $a_{n+1} > 0$. □

Lemma 12.7.2 *If* $1 \leq i \leq r$, *then the following equality holds*

$$\mathbb{Z} = \langle e_1, \ell_i\rangle\mathbb{Z} + \cdots + \langle e_n, \ell_i\rangle\mathbb{Z} + \langle (v_1, 1), \ell_i\rangle\mathbb{Z} + \cdots + \langle (v_q, 1), \ell_i\rangle\mathbb{Z}. \quad (12.5)$$

Proof. Fix an integer $1 \leq i \leq r$. Since $H_{\ell_i} \cap \mathbb{R}_+\mathcal{A}'$ is a facet of the Rees cone there is $(v_j, 1)$ such that $\langle (v_j, 1), \ell_i\rangle = 0$. Setting $v_j = (c_1, \ldots, c_n)$ and $\ell_i = (a_1, \ldots, a_n, -a_{n+1})$, we get $c_1a_1 + \cdots + c_na_n - a_{n+1} = 0$. Notice that the right-hand side of Eq. (12.5) is equal to

$$a_1\mathbb{Z} + \cdots + a_n\mathbb{Z} + \langle (v_1, 1), \ell_i\rangle\mathbb{Z} + \cdots + \langle (v_q, 1), \ell_i\rangle\mathbb{Z}.$$

Thus we need only show the equality $\mathbb{Z} = a_1\mathbb{Z} + \cdots + a_n\mathbb{Z}$. By Lemma 12.7.1 $a_{n+1} > 0$. Thus, from the equality $\sum_{i=1}^{n} c_ia_i = a_{n+1}$, and using that the non-zero entries of ℓ_j are relatively prime, we get that the non-zero elements in $a_1, \ldots, a_n$ are relatively prime, as desired. □

Next we prove a result about the torsion freeness of the divisor class group of $R[It]$. It shows that the rank of this group can be read off from the irreducible representation of the Rees cone given in Eq (12.4).

Proposition 12.7.3 [380, Theorem 1.1] *If* $R[It]$ *is normal, then its divisor class group* $\mathrm{Cl}(R[It])$ *is a free abelian group of rank* r.

Proof. Set $p = n + 1 + r$. Consider the map $\varphi\colon \mathbb{Z}^{n+1} \to \mathbb{Z}^p$ given by

$$\varphi(v) = \langle v, e_1\rangle e_1 + \cdots + \langle v, e_{n+1}\rangle e_{n+1} + \langle v, \ell_1\rangle e_{n+2} + \cdots + \langle v, \ell_r\rangle e_p$$

and consider the matrix B whose rows are the vectors in the set

$$\mathcal{B} = \varphi(\mathcal{A}') = \{\varphi(e_1), \ldots, \varphi(e_n), \varphi(v_1, 1), \ldots, \varphi(v_q, 1)\}.$$

The matrix B is given by:

$$B = \begin{bmatrix} e_1 & 0 & \ell_{11} & \cdots & \ell_{r1} \\ e_2 & 0 & \ell_{12} & \cdots & \ell_{r2} \\ \cdots & \cdots & \cdots & \cdots & \cdots \\ e_n & 0 & \ell_{1n} & \cdots & \ell_{rn} \\ v_1 & 1 & \langle (v_1, 1), \ell_1\rangle & \cdots & \langle (v_1, 1), \ell_r\rangle \\ \cdots & \cdots & \cdots & \cdots & \cdots \\ v_q & 1 & \langle (v_q, 1), \ell_1\rangle & \cdots & \langle (v_q, 1), \ell_r\rangle \end{bmatrix},$$

where $\ell_i = (\ell_{i1}, \ldots, \ell_{i(n+1)})$. It is not hard to see that the Smith normal form of B is equal to $\mathrm{diag}(1, \ldots, 1, 0, \ldots, 0)$, where the number of 1's is $n + 1$. Hence, using Theorem 9.8.19, we get $\mathrm{Cl}(R[It]) \simeq \mathbb{Z}^p/\mathbb{Z}\mathcal{B} \simeq \mathbb{Z}^r$. □

Exercise

12.7.4 Let $I = (x_1x_2, x_2x_3, x_3x_4, x_1x_4)$ be the edge ideal of a cycle of length 4. Prove that $\mathrm{Cl}(R[It])$ is a free abelian group of rank 2.

12.8 Stochastic matrices and Cremona maps

In this section we discuss a conjecture on the normality of ideals arising from doubly stochastic matrices and study Cremona monomial maps.

Let $A = (a_{ij})$ be a non-singular matrix of order n with entries in $\mathbb{N}$. The matrix A is called *doubly stochastic* of degree d if

$$\sum_{k=1}^{n} a_{ki} = \sum_{k=1}^{n} a_{jk} = d \geq 2 \text{ for all } 1 \leq i, j \leq n.$$

If $\sum_{k=1}^{n} a_{ki} = d \geq 2$ for all i, we say that A is a *d-stochastic matrix* by columns. The set

$$\mathcal{A} = \{v_1, \ldots, v_n\}$$

will denote the set of columns of A and I will denote the monomial ideal of the polynomial ring $R = K[x_1, \ldots, x_n]$ generated by $x^{v_1}, \ldots, x^{v_n}$.

Proposition 12.8.1 *If $A = (a_{ij})$ is a d-stochastic matrix by columns and $R[It]$ is normal, then* $\det(A) = \pm d$.

Proof. By Proposition 9.3.33 we have $H = \mathbb{Z}^n/(v_1, \ldots, v_n) \simeq \mathbb{Z}_d$. By Lemma 1.3.17 the order of H is $|\det(A)|$. Thus $d = |\det(A)|$. $\square$

Conjecture 12.8.2 The Rees algebra $R[It]$ is normal for every doubly stochastic matrix A of degree d with entries in $\{0, 1\}$ and $\det(A) = \pm d$.

To prove this conjecture one could try to use the Birkhoff von Neumann theorem: "A matrix is doubly stochastic if and only if it is a convex combination of permutation matrices multiplied by d" [372, Theorem 8.6].

Proposition 12.8.3 *If $A = (a_{ij})$ is a d-stochastic matrix by columns, then* $\det(A)$ *is a multiple of* d.

Proof. Let $F_1, \ldots, F_n$ be the rows of A and let B be the matrix with rows $F_1, \ldots, F_{n-1}, \sum_{i=1}^{n} F_i$. Since $\det(B) = \det(A)$ and $\sum_{i=1}^{n} F_i = (d, \ldots, d)$, the assertion follows. $\square$

Proposition 12.8.4 *Let $A = (a_{ij})$ be a 2-stochastic matrix by columns with entries in $\{0, 1\}$. If $\det(A) = \pm 2$, then $R[It]$ is normal.*

Proof. Since $d = 2$, A is the incidence matrix of a simple graph $G = (V, E)$ with vertex set $V = \{x_1, \ldots, x_n\}$ and edge set E, where $\{x_i, x_j\} \in E$ if and only if $e_i + e_j \in \mathcal{A}$. Notice that $|V| = |E|$ because A is a square matrix.

We claim that G is a connected. Let $G_1, \ldots, G_k$ be the connected components of G. By re-ordering the vertices of G we can write

$$A = \text{diag}(M_1, \ldots, M_k),$$

where M_i is the incidence matrix of G_i. Let n_i be the number of vertices of G_i. By Lemma 10.2.6, $\text{rank}(M_i) = n_i$ if G_i is non-bipartite and $\text{rank}(M_i) = n_i - 1$ if G_i is bipartite. As G is a disjoint union of $G_1, \ldots, G_k$, we get $n = n_1 + \cdots + n_k$. Thus, using that $\text{rank}(A)$ is $\sum_{i=1}^{k} \text{rank}(M_i)$, it follows that G_i is a connected non-bipartite graph with at least n_i edges for any i. Since the number of edges of G is n we obtain that M_i is a square matrix of order n_i for all i. Then $\det(A) = \det(M_1) \cdots \det(M_k)$. By Proposition 12.8.3 we get $\det(M_i) = 2m_i$ for $1 \leq i \leq k$, where $m_i \in \mathbb{Z}$ for all i. Hence $k = 1$.

Let $T = (V_T, E_T)$ be a spanning tree of G (see Exercise 10.1.65). Then $|E_T| + 1 = |V_T| = |V| = |E|$ and G is obtained from T by adding one edge. This means that G has a unique odd cycle and some branches. Using Corollary 10.3.12 we get $K[G] = K[\{x^a | \, a \in \mathcal{A}\}]$ is normal and according to Corollary 10.5.6 we have that the Rees algebra $R[It]$ is normal. □

Example 12.8.5 Consider the following doubly stochastic matrix

$$A^t = \begin{bmatrix} 0 & 0 & 1 & 1 & 1 & 1 \\ 0 & 1 & 0 & 1 & 1 & 1 \\ 1 & 1 & 1 & 0 & 1 & 0 \\ 1 & 1 & 1 & 1 & 0 & 0 \\ 1 & 0 & 0 & 1 & 1 & 1 \\ 1 & 1 & 1 & 0 & 0 & 1 \end{bmatrix}$$

whose rows are $v_1, \ldots, v_6$. Notice that $\det(A) = \pm 8 \neq \pm 4$. The ideal I is *minimally non-normal*, that is I is not normal and I' is normal for every proper minor I' of I (see Definition 14.2.23 for the notion of minor).

Lemma 12.8.6 *If A is a d-stochastic matrix by columns and* $\det(A) = \pm d$, *then $A^{-1}(e_i - e_j) \in \mathbb{Z}^n$ for all i, j.*

Proof. Fixing indices i, j, we can write $A^{-1}(e_i - e_j) = \sum_{k=1}^{n} \lambda_k e_k$ for some $\lambda_1, \ldots, \lambda_n$ in $\mathbb{Q}$. Notice that $A^{-1}(e_i)$ is the ith column of A^{-1}. Since $\mathbf{1}A = d\mathbf{1}$, we get $\mathbf{1}/d = \mathbf{1}A^{-1}$. Therefore $|A^{-1}(e_i)| = |A^{-1}(e_j)| = 1/d$ and $\sum_k \lambda_k = 0$. Then we can write

$$A^{-1}(e_i - e_j) = \sum_{k=2}^{n} \lambda_k (e_i - e_1) \implies e_i - e_j = \sum_{k=2}^{n} \lambda_k (v_k - v_1).$$

Thus there is $0 \neq s \in \mathbb{N}$ such that $s(e_i - e_j)$ belong to $\mathbb{Z}\{v_1 - v_k\}_{k=2}^n$. By Proposition 10.8.7, the group $\mathbb{Z}^n/\mathbb{Z}\{v_1 - v_k\}_{k=2}^n$ is free. Then we can write $e_i - e_j = \sum_{k=2}^n \eta_k(v_k - v_1)$, for some η_i's in $\mathbb{Z}$. Since $\mathbb{Z}\{v_1 - v_k\}_{k=2}^n$ is also free (of rank $n-1$), the vectors $v_2 - v_1, \ldots, v_n - v_1$ are linearly independent. Thus $\lambda_k = \eta_k \in \mathbb{Z}$ for all $k \geq 2$, hence ultimately $A^{-1}(e_i - e_j) \in \mathbb{Z}^n$. $\square$

Theorem 12.8.7 [386] *If A is a d-stochastic matrix by columns such that* $\det(A) = \pm d$, *then there are unique vectors $\beta_1, \ldots, \beta_n, \gamma \in \mathbb{N}^n$ such that the following two conditions hold:*

(a) *$A\beta_i = \gamma + e_i$ for all i, where β_i, γ and e_i are column vectors;*

(b) *The matrix B with columns $\beta_1, \ldots, \beta_n$ has at least one zero entry in every row.*

Moreover, $\det(B) = \pm(|\gamma| + 1)/d = \pm|\beta_i|$ *for all i.*

Proof. First we show the uniqueness. Assume that $\beta_1', \ldots, \beta_n', \gamma'$ is a set of vectors in $\mathbb{N}^n$ such that: (a$'$) $A\beta_i' = \gamma' + e_i$ for all i, and (b$'$) The matrix B' whose column vectors are $\beta_1', \ldots, \beta_n'$ has at least one zero entry in every row. Let $\Delta = (\Delta_i)$ and $\Delta' = (\Delta_i')$ be nonnegative vectors such that $A^{-1}(\gamma - \gamma') = \Delta' - \Delta$. Then from (a) and (a$'$) we get

$$\beta_i - \beta_i' = A^{-1}(\gamma - \gamma') = \Delta' - \Delta, \ \forall i \implies \beta_{ik} - \beta_{ik}' = \Delta_k' - \Delta_k, \ \forall i, k, \quad (12.6)$$

where $\beta_i = (\beta_{i1}, \ldots, \beta_{in})$ and $\beta_i' = (\beta_{i1}', \ldots, \beta_{in}')$. It suffices to show that $\Delta = \Delta'$. If $\Delta_k' > \Delta_k$ for some k; then, by Eq. (12.6), we obtain $\beta_{ik} > 0$ for $i = 1, \ldots, n$, which contradicts (b). Similarly if $\Delta_k' < \Delta_k$ for some k; then, by Eq. (12.6), we obtain $\beta_{ik}' > 0$ for $i = 1, \ldots, n$, which contradicts (b$'$). Thus $\Delta_k = \Delta_k'$ for all k, i.e., $\Delta = \Delta'$.

Next we prove the existence of $\beta_1, \ldots, \beta_n$ and γ. By Lemma 12.8.6, for $i \geq 2$ we can write

$$0 \neq \alpha_i = A^{-1}(e_1 - e_i) = \alpha_i^+ - \alpha_i^-$$

where α_i^+ and α_i^- are in $\mathbb{N}^n$. Notice that $\alpha_i^+ \neq 0$ and $\alpha_i^- \neq 0$. Indeed the sum of the entries of $A^{-1}(e_i)$ is equal to $1/d$. Thus $|\alpha_i| = |\alpha_i^+| - |\alpha_i^-| = 0$, and consequently the positive and negative part of α_i are both non-zero for $i \geq 2$. The vector α_i^+ can be written as $\alpha_i^+ = (\alpha_{i1}^+, \ldots, \alpha_{in}^+)$ for $i \geq 2$. For $1 \leq k \leq n$ consider the integers given by $m_k = \max_{2 \leq i \leq n}\{\alpha_{ik}^+\}$ and set $\beta_1 = (m_1, \ldots, m_n)$. Since $\beta_1 \geq \alpha_i^+$, for each $i \geq 2$ there is $\theta_i \in \mathbb{N}^n$ such that $\beta_1 = \theta_i + \alpha_i^+$. Therefore

$$\alpha_i = A^{-1}(e_1 - e_i) = \alpha_i^+ - \alpha_i^- = \beta_1 - (\theta_i + \alpha_i^-).$$

We set $\beta_i = \theta_i + \alpha_i^-$ for $i \geq 2$. Since we have $A\beta_1 - e_1 = A\beta_i - e_i$ for $i \geq 2$, it follows readily that $A\beta_1 - e_1 \geq 0$ (make $i = 2$ in the last equality and compare entries). Thus, setting $\gamma := A\beta_1 - e_1$, it follows that $\beta_1, \ldots, \beta_n$ and γ satisfy (a). If each row of B has some zero entry the proof of the existence is complete. If every entry of a row of B is positive we subtract the vector $\mathbf{1} = (1, \ldots, 1)$ from that row and change γ accordingly so that (a) is still satisfied. Applying this argument repeatedly we get a sequence $\beta_1, \ldots, \beta_n, \gamma$ satisfying (a) and (b).

We now prove the last part of the assertion. If β_{ij} denotes the j-entry of β_i, then $A\beta_i = \gamma + e_i$ is equivalent to $\beta_{i1}v_1 + \cdots + \beta_{in}v_n = \gamma + e_i$. Thus $|\beta_i|d = |\gamma| + 1$. Condition (a) is equivalent to $AB = \Gamma + I$, where Γ is the matrix all of whose columns are equal to γ. Since $\det(B) = \pm\det(\Gamma + I)/d$ it suffices to show that $\det(\Gamma + I) = |\gamma| + 1$. By Exercise 12.8.16, if Γ has rank at most one and D is the identity, we get $\det(\Gamma + I) = \text{trace}(\Gamma) + 1$. $\square$

Example 12.8.8 Consider the following matrix A and its inverse:

$$A = \begin{pmatrix} d & d-1 & 0 \\ 0 & 1 & d-1 \\ 0 & 0 & 1 \end{pmatrix}; \qquad A^{-1} = \frac{1}{d}\begin{pmatrix} 1 & 1-d & (d-1)^2 \\ 0 & d & d(1-d) \\ 0 & 0 & d \end{pmatrix}.$$

To compute the β_i's and γ we follow the proof of Theorem 12.8.7. Then $\beta_1 = (2, d, 0)$, $\beta_2 = (1, d+1, 0)$, $\beta_3 = (d, 1, 1)$, $\gamma = (d^2 + d - 1, d, 0)$, and

$$B = \begin{pmatrix} 2 & 1 & d \\ d & d+1 & 1 \\ 0 & 0 & 1 \end{pmatrix}.$$

By subtracting the vector $(1, 1, 1)$ from rows 1 and 2, we get

$$B' = \begin{pmatrix} 1 & 0 & d-1 \\ d-1 & d & 0 \\ 0 & 0 & 1 \end{pmatrix}.$$

The column vectors $\beta_1' = (1, d-1, 0)$, $\beta_2' = (0, d, 0)$, $\beta_3' = (d-1, 0, 1)$, $\gamma' = (d^2 - d, d-1, 0)$ satisfy (a) and (b).

Cremona maps Let $R = K[x_1, \ldots, x_n]$ be a polynomial ring over a field K. In what follows we assume that A is a d-stochastic matrix by columns and that the corresponding set of monomials $F = \{x^{v_1}, \ldots, x^{v_n}\} \subset R$ have no non-trivial common factor. We also assume throughout that every x_i divides at least one member of F, a harmless condition.

Definition 12.8.9 F defines a rational (monomial) map $\mathbb{P}^{n-1} \dashrightarrow \mathbb{P}^{n-1}$ denoted again by F and written as a tuple $F = (x^{v_1}, \ldots, x^{v_n})$. F is called a *Cremona map* if it admits an inverse rational map with source $\mathbb{P}^{n-1}$.

A rational monomial map F is defined everywhere if and only if the defining monomials are pure powers of the variables, in which case it is a Cremona map if and only if $F = (x_{\sigma(1)}, \ldots, x_{\sigma(n)})$ for some permutation σ.

Proposition 12.8.10 [385, Proposition 2.1] *F is a Cremona map if and only if* $\det(A) = \pm d$.

Thus Conjecture 12.8.2 has the following reformulation: "the rational map $F\colon \mathbb{P}^{n-1} \dashrightarrow \mathbb{P}^{n-1}$ is a Cremona map if and only if $R[It]$ is normal." For $d = 2$ this conjecture was proved in [385]; see Proposition 12.8.4. Cremona maps defined by monomials of degree $d = 2$ are thoroughly analyzed and classified via integer arithmetic and graph combinatorics in [98].

Theorem 12.8.11 *If $F\colon \mathbb{P}^{n-1} \dashrightarrow \mathbb{P}^{n-1}$ is a Cremona map then its inverse is also defined by monomials of fixed degree.*

Proof. We set $f_i = x^{v_i}$ for $i = 1, \ldots, n$. By Proposition 12.8.10, A has determinant $\pm d$. Therefore Theorem 12.8.7 implies the existence of an $n \times n$ matrix B such that $AB = \Gamma + I$, where Γ is a matrix with repeated column γ throughout. Let $g_1, \ldots, g_n$ denote the monomials defined by the columns of B and call G the corresponding rational monomial map. The above matrix equality translates into the equality

$$(f_1(g_1, \ldots, g_n), \ldots, f_n(g_1, \ldots, g_n)) = (x^\gamma \cdot x_1, \ldots, x^\gamma \cdot x_n).$$

Thus the left-hand side is proportional to the vector $(x_1, \ldots, x_n)$ which means that the composite map $F \circ G$ is the identity map wherever the two are defined. On the other hand Theorem 12.8.7 also says that G is also a Cremona map. Therefore G has to be the inverse of F, as required. Notice that the proof of Theorem 12.8.7 provides an algorithm to compute B and γ. The input for this algorithm is the matrix A. $\square$

Exercises

12.8.12 If B is a real non-singular matrix such that the sum of the elements in each row is $d \neq 0$, then the sum of the elements in each row of B^{-1} is equal to $1/d$

12.8.13 Let $A = (a_{ij})$ be a d-stochastic matrix by columns of order n. If B is the matrix $(b_{ij}) = (1 - a_{ij})$, then

$$(n - d)\det(A) = (-1)^{n-1} d \det(B).$$

12.8.14 Let $A = (a_{ij})$ be a doubly stochastic matrix of order n with

$$\sum_{i=1}^{n} a_{ik} = \sum_{j=1}^{n} a_{\ell j} = d \quad (\forall\, k, \ell).$$

If $|\det(A)| = d \geq 1$, then $\gcd\{n, d\} = 1$.

12.8.15 Let F be a Cremona map of the form $F = (x_1^{a_1}x_2^{a_2}, x_2^{b_2}x_3^{b_3}, x_1^{c_1}x_3^{c_3})$. Prove that, up to permutation of the variables and the monomials, F is one of the following two kinds:

$$F = (x_1x_2, x_2x_3, x_1x_3) \quad \text{or} \quad F = (x_1^d, x_2x_3^{d-1}, x_1^{d-1}x_3).$$

12.8.16 Let $\Gamma = (\gamma_{i,j})$ be an $n \times n$ matrix over a commutative ring, and let $D = \operatorname{diag}(d_1, \ldots, d_n)$ be a diagonal matrix. Then

$$\begin{aligned} \det(\Gamma + D) &= \det(\Gamma) \\ &+ \sum_i d_i \Delta_{[n]\setminus\{i\}} + \sum_{1 \le i_1 < i_2 \le n} d_{i_1} d_{i_2} \Delta_{[n]\setminus\{i_1,i_2\}} \\ &+ \cdots + \sum_{1 \le i_1 < \cdots < i_{n-1} \le n} d_{i_1} \cdots d_{i_{n-1}} \Delta_{[n]\setminus\{i_1,\ldots,i_{n-1}\}} \\ &+ \det D, \end{aligned}$$

where $\Delta_{[n]\setminus\{i_1,\ldots,i_k\}}$ denotes the principal $(n-k) \times (n-k)$-minor of Γ with rows and columns $[n] \setminus \{i_1, \ldots, i_k\}$. Here $[n] = \{1, \ldots, n\}$.

Chapter 13

Combinatorics of Symbolic Rees Algebras of Edge Ideals of Clutters

In this chapter we give a description—using notions from combinatorial optimization and polyhedral geometry— of the minimal generators of the symbolic Rees algebra of the edge ideal of a clutter and show a complete graph theoretical description of the minimal generators of the symbolic Rees algebra of the ideal of covers of a graph. For a connected non-bipartite graph G whose edge subring is normal, we give conditions for G to have a perfect matching.

The notion of a indecomposable graph is related to the strong perfect graph theorem. We give a description—in terms of cliques—of the symbolic Rees algebra and the Simis cone of the edge ideal of a perfect graph.

13.1 Vertex covers of clutters

In this section we characterize the notion of a minimal vertex cover of a clutter in algebraic and combinatorial terms and relate the matching number and the vertex covering number of a clutter to the algebraic invariants of the corresponding edge ideal.

Let $\mathcal{C}$ be a clutter with vertex set $X = \{x_1, \ldots, x_n\}$ and let $R = K[X]$ be a polynomial ring over a field K. We denote the incidence matrix of $\mathcal{C}$ by A and denote the column vectors of A by $v_1, \ldots, v_q$. The *edge ideal* of $\mathcal{C}$, denoted by $I(\mathcal{C})$, is the ideal of R generated by $x^{v_1}, \ldots, x^{v_q}$.

Definition 13.1.1 A subset $C \subset X$ is a *minimal vertex cover* of a clutter $\mathcal{C}$ if: (c_1) every edge of $\mathcal{C}$ contains at least one vertex of C, and (c_2) there is no proper subset of C with the first property. If C only satisfies condition (c_1), then C is called a *vertex cover* of $\mathcal{C}$.

Proposition 13.1.2 *Let $\mathcal{Q}(A) = \{x|\, x \geq 0;\, xA \geq \mathbf{1}\}$ be the set covering polyhedron of $\mathcal{C}$. The following are equivalent:*

(a) $\mathfrak{p} = (x_1, \ldots, x_r)$ *is a minimal prime of* $I = I(\mathcal{C})$.

(b) $C = \{x_1, \ldots, x_r\}$ *is a minimal vertex cover of* $\mathcal{C}$.

(c) $\alpha = e_1 + \cdots + e_r$ *is a vertex of* $\mathcal{Q}(A)$.

Proof. (a) $\Leftrightarrow$ (b): This follows at once from Lemma 6.3.37.

(b) $\Rightarrow$ (c): Fix $1 \leq i \leq r$. To make notation simpler fix $i = 1$. We may assume that there is s_1 such that $x^{v_j} = x_1 m_j$ for $j = 1, \ldots, s_1$ and $x_1 \notin \operatorname{supp}(x^{v_j})$ for $j > s_1$. Notice that $\operatorname{supp}(m_{k_1}) \cap (C \setminus \{x_1\}) = \emptyset$ for some $1 \leq k_1 \leq s_1$, otherwise $C \setminus \{x_1\}$ is a vertex cover of $\mathcal{C}$ strictly contained in C, a contradiction. Hence for each $1 \leq i \leq r$ there is v_{k_i} in $\{v_1, \ldots, v_q\}$ such that $x^{v_{k_i}} = x_i m_{k_i}$ and $\operatorname{supp}(m_{k_i}) \subset \{x_{r+1}, \ldots, x_n\}$. The vector α is clearly in $\mathcal{Q}(A)$, and since $\{e_i\}_{i=r+1}^{n} \cup \{v_{k_1}, \ldots, v_{k_r}\}$ is linearly independent, and

$$\langle \alpha, e_i \rangle = 0 \quad (i = r+1, \ldots, n), \qquad \langle \alpha, v_{k_i} \rangle = 1 \quad (i = 1, \ldots, r),$$

we get that the vector α is a basic feasible solution of the linear system $x \geq 0; xA \geq \mathbf{1}$. Therefore α is a vertex of $\mathcal{Q}(A)$ by Corollary 1.1.49.

(c) $\Rightarrow$ (b): If $C' \subsetneq C$ is a vertex cover of $\mathcal{C}$, then $\alpha' = \sum_{x_i \in C'} e_i$ satisfies $\alpha' A \geq \mathbf{1}$ and $\alpha' \geq 0$. Using that α is a basic feasible solution of the linear system $x \geq 0; xA \geq \mathbf{1}$, it is not hard to verify that α' is also a vertex of $\mathcal{Q}(A)$. If V is the vertex set of $\mathcal{Q}(A)$, by Theorem 1.1.42, we can write

$$\mathcal{Q}(A) = \mathbb{R}_+^n + \operatorname{conv}(V).$$

Since $\alpha = \beta + \alpha'$, for some $\beta \in \mathbb{R}_+^n$, we obtain $\mathcal{Q}(A) = \mathbb{R}_+^n + \operatorname{conv}(V \setminus \{\alpha\})$, a contradiction (see Propositions 1.1.36 and 1.1.39). Thus C is a minimal vertex cover, as required. □

Corollary 13.1.3 *A vector $\alpha \in \mathbb{R}^n$ is an integral vertex of $\mathcal{Q}(A)$ if and only if $\alpha = e_{i_1} + \cdots + e_{i_s}$ for some minimal vertex cover $\{x_{i_1}, \ldots, x_{i_s}\}$ of $\mathcal{C}$.*

Proof. By Proposition 13.1.2 it suffices to observe (see Exercise 11.3.8) that any integral vertex of $\mathcal{Q}(A)$ has entries in $\{0, 1\}$. □

Let $\mathcal{C}$ be a clutter. A set of edges of $\mathcal{C}$ is called *independent* or a *matching* if no two of them have a common vertex. We denote the smallest number of vertices in any minimal vertex cover of $\mathcal{C}$ by $\alpha_0(\mathcal{C})$ and the maximum number of independent edges of $\mathcal{C}$ by $\beta_1(\mathcal{C})$. We call $\alpha_0(\mathcal{C})$ the *vertex covering number* and $\beta_1(\mathcal{C})$ the *matching number* of $\mathcal{C}$. These numbers are related to min-max problems (see Exercise 13.1.7).

Definition 13.1.4 If $\alpha_0(\mathcal{C}) = \beta_1(\mathcal{C})$ we say that the clutter $\mathcal{C}$ (resp. the edge ideal $I(\mathcal{C})$) has the *König property*.

Definition 13.1.5 The *monomial grade* of $I(\mathcal{C})$, denoted by $\text{mgrade}(I(\mathcal{C}))$, is the maximum integer r such that there exists a regular sequence of monomials $x^{\alpha_1}, \ldots, x^{\alpha_r}$ in $I(\mathcal{C})$.

The combinatorial invariants $\alpha_0(\mathcal{C})$ and $\beta_1(\mathcal{C})$ can be interpreted in terms of algebraic invariants of $I(\mathcal{C})$.

Proposition 13.1.6 $\text{ht}\, I(\mathcal{C}) = \alpha_0(\mathcal{C})$ *and* $\beta_1(\mathcal{C}) = \text{mgrade}(I(\mathcal{C}))$.

Proof. The first equality follows at once from Proposition 6.5.4. The second equality follows readily from Exercise 13.1.10. $\square$

Exercises

13.1.7 Let A be the incidence matrix of a clutter $\mathcal{C}$. Prove that the vertex covering number and the matching number of $\mathcal{C}$ satisfy:

$$\begin{aligned}\alpha_0(\mathcal{C}) &\geq \min\{\langle \mathbf{1}, x\rangle \,|\, x \geq 0; xA \geq \mathbf{1}\} \\ &= \max\{\langle y, \mathbf{1}\rangle \,|\, y \geq 0; Ay \leq \mathbf{1}\} \geq \beta_1(\mathcal{C}).\end{aligned}$$

Prove that $\alpha_0(\mathcal{C}) = \beta_1(\mathcal{C})$ if and only if both sides of the equality have integral optimum solutions.

13.1.8 If $\mathcal{C}$ is a clutter, prove that the following conditions are equivalent:

(a) $\alpha_0(\mathcal{C}) = \beta_1(\mathcal{C})$.

(b) $x_1 \cdots x_n t^g$ belongs to $R[It]$, where $g = \text{ht}(I)$ and $I = I(\mathcal{C})$.

13.1.9 Let $\underline{f} = \{f_1, \ldots, f_r\}$ be a sequence of elements of a ring R and let $f_0 = 0$. Prove that $\underline{f}$ is a regular sequence if and only if

$$((f_1, \ldots, f_{i-1})\colon f_i) = (f_1, \ldots, f_{i-1}) \quad \text{for all} \quad i \geq 1.$$

13.1.10 A sequence $x^{\alpha_1}, \ldots, x^{\alpha_r}$ of monomials of R is a regular sequence if and only if x^{α_i} and x^{α_j} have no common variables for all $i \neq j$.

13.1.11 Let $\mathcal{C}$ be a clutter and let C be a minimal vertex cover of $\mathcal{C}$. Prove that $|C \cap e| = 1$ for some edge e of $\mathcal{C}$.

13.2 Symbolic Rees algebras of edge ideals

Let $\mathcal{C}$ be a clutter with vertex set $X = \{x_1, \ldots, x_n\}$ and edge set $E(\mathcal{C})$, let $R = K[X]$ be a polynomial ring over a field K, and let $I = I(\mathcal{C})$ be the edge ideal of $\mathcal{C}$. The *blowup algebra* studied here is the *symbolic Rees algebra*

$$R_s(I) = R \oplus I^{(1)}t \oplus \cdots \oplus I^{(i)}t^i \oplus \cdots \subset R[t]$$

of I, where t is a new variable and $I^{(i)}$ is the ith symbolic power of I. The main theorem of this section is a description—in combinatorial optimization terms—of the minimal set of generators of $R_s(I)$ as a K-algebra.

As usual, we denote by $\mathcal{C}^\vee$ the clutter whose edges are the minimal vertex covers of $\mathcal{C}$. If C is a subset of X, its *characteristic vector* is the vector $v = \sum_{x_i \in C} e_i$, where e_i is the ith unit vector in $\mathbb{R}^n$. Let $C_1, \ldots, C_s$ be the minimal vertex covers of $\mathcal{C}$ and let u_k be the characteristic vector of C_k for $1 \leq k \leq s$. In our situation, according to Propositions 4.3.24 and 6.5.4, the bth symbolic power of I has a simple expression:

$$I^{(b)} = \mathfrak{p}_1^b \cap \cdots \cap \mathfrak{p}_s^b = (\{x^a \mid \langle a, u_k\rangle \geq b \text{ for } k = 1, \ldots, s\}), \tag{13.1}$$

where $\mathfrak{p}_k$ is the prime ideal of R generated by C_k. In particular, if $b = 1$, we obtain the primary decomposition of I because $I^{(1)} = I$.

Definition 13.2.1 The *Simis cone* of $I = I(\mathcal{C})$ is the polyhedral cone:

$$\mathrm{Cn}(I) = H_{e_1}^+ \cap \cdots \cap H_{e_{n+1}}^+ \cap H_{(u_1,-1)}^+ \cap \cdots \cap H_{(u_s,-1)}^+.$$

The term Simis cone was coined in [147] to recognize the pioneering work of Aron Simis on symbolic powers of monomial ideals [377]. The Simis cone is a finitely generated rational cone (Theorem 1.1.29). Hence, by Theorem 1.3.9, there is a unique minimal finite set of integral vectors $\mathcal{H} \subset \mathbb{Z}^{n+1}$ such that $\mathbb{Z}^{n+1} \cap \mathbb{R}_+\mathcal{H} = \mathbb{N}\mathcal{H}$ and $\mathrm{Cn}(I) = \mathbb{R}_+\mathcal{H}$ (minimal relative to taking subsets). The set $\mathcal{H}$ is called the *minimal Hilbert basis* of $\mathrm{Cn}(I)$. See Section 1.3 for a detailed study of Hilbert bases.

Theorem 13.2.2 *$\mathcal{H}$ is the set of all integral vectors $0 \neq \alpha \in \mathrm{Cn}(I)$ such that α is not the sum of two other non-zero integral vectors in $\mathrm{Cn}(I)$.*

Proof. It follows at once from Theorem 1.3.9. □

Theorem 13.2.3 [147] *Let $\mathcal{H} \subset \mathbb{N}^{n+1}$ be a Hilbert basis of $\mathrm{Cn}(I)$. If $K[\mathbb{N}\mathcal{H}]$ is the semigroup ring of $\mathbb{N}\mathcal{H}$, then $R_s(I) = K[\mathbb{N}\mathcal{H}]$.*

Proof. Recall that $K[\mathbb{N}\mathcal{H}] = K[\{x^a t^b \mid (a, b) \in \mathbb{N}\mathcal{H}\}]$. Take $x^a t^b \in R_s(I)$, that is, $x^a \in \mathfrak{p}_i^b$ for all i. Hence $\langle (a, b), (u_i, -1)\rangle \geq 0$ for all i or equivalently $(a, b) \in \mathrm{Cn}(I)$. Thus $(a, b) \in \mathbb{N}\mathcal{H}$ and $x^a t^b \in K[\mathbb{N}\mathcal{H}]$. Conversely take $x^a t^b \in K[\mathbb{N}\mathcal{H}]$, then (a, b) is in $\mathrm{Cn}(I)$ and $\langle (a, b), (u_i, -1)\rangle \geq 0$ for all i. Hence $x^a \in \mathfrak{p}_i^b$ for all i and $x^a \in I^{(b)}$, as required. □

Corollary 13.2.4 [301] *$R_s(I)$ is a finitely generated K-algebra.*

Proof. Let $\mathrm{Cn}(I)$ be the Simis cone of I and let $\mathcal{H} \subset \mathbb{N}^{n+1}$ be a Hilbert basis of $\mathrm{Cn}(I)$. Applying Theorem 13.2.3, we get $R_s(I) = K[\mathbb{N}\mathcal{H}]$. Thus, $R_s(I)$ is a finitely generated K-algebra. □

Definition 13.2.5 Let $a = (a_i) \neq 0$ be a vector in $\mathbb{N}^n$ and let $b \in \mathbb{N}$. If a, b satisfy $\langle a, u_k\rangle \geq b$ for $k = 1, \ldots, s$, we say that a is a *b-cover* of $\mathcal{C}^\vee$.

Lemma 13.2.6 *$x^a t^b$ is in $R_s(I)$ if and only if a is a b-cover of $\mathcal{C}^\vee$.*

Proof. It follows from the description of $I^{(b)}$ given in Eq. (13.1). □

The notion of a b-cover occurs in combinatorial optimization (see for instance [373, Chapter 77, p. 1378] and the references therein) and algebraic combinatorics [147, 225].

Definition 13.2.7 A b-cover a of $\mathcal{C}^\vee$ is called *reducible* if there exists an i-cover c and a j-cover d of $\mathcal{C}^\vee$ such that $a = c + d$ and $b = i + j$. If a is not reducible, we call a *irreducible*.

The irreducible 0 and 1 covers of $\mathcal{C}^\vee$ are the unit vectors $e_1, \ldots, e_n$ and the characteristic vectors $v_1, \ldots, v_q$ of the edges of $\mathcal{C}$, respectively.

Lemma 13.2.8 *A monomial $x^a t^b$ is a minimal generator of $R_s(I)$, as a K-algebra, if and only if a is an irreducible b-cover of $\mathcal{C}^\vee$.*

Proof. It follows from the discussion above, by decomposing any b-cover into irreducible ones. □

Let S be a set of vertices of a clutter $\mathcal{C}$. The *induced subclutter* on S, denoted by $\mathcal{C}[S]$, is the maximal subclutter of $\mathcal{C}$ with vertex set S. Thus the vertex set of $\mathcal{C}[S]$ is S and the edges of $\mathcal{C}[S]$ are exactly the edges of $\mathcal{C}$ contained in S. Notice that $\mathcal{C}[S]$ may have isolated vertices, i.e., vertices that do not belong to any edge of $\mathcal{C}[S]$. If $\mathcal{C}$ is a discrete clutter, i.e., all the vertices of $\mathcal{C}$ are isolated, we set $I(\mathcal{C}) = 0$ and $\alpha_0(\mathcal{C}) = 0$.

Let $\mathcal{C}$ be a clutter and let X_1, X_2 be a partition of its vertex set $V(\mathcal{C})$ into non-empty sets. Clearly, one has the inequality

$$\alpha_0(\mathcal{C}) \geq \alpha_0(\mathcal{C}[X_1]) + \alpha_0(\mathcal{C}[X_2]). \tag{13.2}$$

If $\mathcal{C}$ is a graph and equality occurs, Erdös and Gallai [144] call $\mathcal{C}$ a *decomposable graph*. This motivates the following similar notion for clutters.

Definition 13.2.9 A clutter $\mathcal{C}$ is called *decomposable* if there are non-empty vertex sets X_1, X_2 such that $X = V(\mathcal{C})$ is the disjoint union of X_1 and X_2, and $\alpha_0(\mathcal{C}) = \alpha_0(\mathcal{C}[X_1]) + \alpha_0(\mathcal{C}[X_2])$. If $\mathcal{C}$ is not decomposable, it is called *indecomposable*.

Examples of indecomposable graphs include complete graphs, odd cycles and complements of odd cycles of length at least five (see Lemma 13.3.1).

Definition 13.2.10 (Schrijver [373]) The *duplication* of a vertex x_i of a clutter $\mathcal{C}$ means extending its vertex set X by a new vertex x_i' and replacing the edge set $E(\mathcal{C})$ by

$$E(\mathcal{C}) \cup \{(e \setminus \{x_i\}) \cup \{x_i'\} | \, x_i \in e \in E(\mathcal{C})\}.$$

The *deletion* of x_i, denoted by $\mathcal{C} \setminus \{x_i\}$, is the clutter formed from $\mathcal{C}$ by deleting the vertex x_i and all edges containing x_i. A clutter obtained from $\mathcal{C}$ by a sequence of deletions and duplications of vertices is called a *parallelization*.

It is not difficult to verify that these two operations commute. If $a = (a_i)$ is a vector in $\mathbb{N}^n$, we denote by $\mathcal{C}^a$ the clutter obtained from $\mathcal{C}$ by successively deleting any vertex x_i with $a_i = 0$ and duplicating $a_i - 1$ times any vertex x_i if $a_i \geq 1$ (for graphs cf. [191, p. 53]).

Example 13.2.11 Let G be the graph whose only edge is $\{x_1, x_2\}$ and let $a = (3, 3)$. We set $x_i^1 = x_i$ for $i = 1, 2$. The parallelization G^a is a complete bipartite graph with bipartition $V_1 = \{x_1^1, x_1^2, x_1^3\}$ and $V_2 = \{x_2^1, x_2^2, x_2^3\}$. Note that x_i^k is a vertex, i.e., k is an index not an exponent.

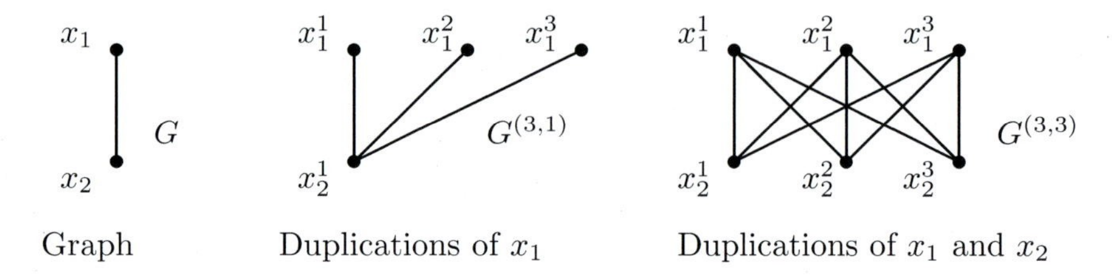

Lemma 13.2.12 *Let $\mathcal{C}$ be a clutter and let A be its incidence matrix. If $a = (a_i)$ is a vector in $\mathbb{N}^n$, then*

$$\beta_1(\mathcal{C}^a) \leq \max\{\langle y, \mathbf{1}\rangle | \, y \in \mathbb{N}^q; \, Ay \leq a\}.$$

Proof. We may assume that $a = (a_1, \ldots, a_m, 0, \ldots, 0)$, where $a_i \geq 1$ for $i = 1, \ldots, m$. Recall that for each $1 \leq i \leq m$ the vertex x_i is duplicated $a_i - 1$ times and for each $i > m$ the vertex x_i is deleted. We denote the duplications of x_i by $x_i^2, \ldots, x_i^{a_i}$ and set $x_i^1 = x_i$. Thus, we can write

$$V(\mathcal{C}^a) = \{x_1^1, \ldots, x_1^{a_1}, \ldots, x_i^1, \ldots, x_i^{a_i}, \ldots, x_m^1, \ldots, x_m^{a_m}\}.$$

There are $f_1, \ldots, f_{\beta_1}$ independent edges of $\mathcal{C}^a$, where $\beta_1 = \beta_1(\mathcal{C}^a)$. Each f_i has the form

$$f_k = \{x_{k_1}^{j_{k_1}}, x_{k_2}^{j_{k_2}}, \ldots, x_{k_r}^{j_{k_r}}\} \qquad (1 \leq k_1 < \cdots < k_r \leq m; \; 1 \leq j_{k_i} \leq a_{k_i}).$$

We set $g_k = \{x_{k_1}^1, x_{k_2}^1, \ldots, x_{k_r}^1\} = \{x_{k_1}, x_{k_2}, \ldots, x_{k_r}\}$. By definition of $\mathcal{C}^a$ we get that $g_k \in E(\mathcal{C})$ for all k. We may re-order the f_i's so that

$$\underbrace{g_1 = g_2 = \cdots = g_{s_1}}_{s_1}, \underbrace{g_{s_1+1} = \cdots = g_{s_2}}_{s_2 - s_1}, \ldots, \underbrace{g_{s_{r-1}+1} = \cdots = g_{s_r}}_{s_r - s_{r-1}}$$

and $g_{s_1}, \ldots, g_{s_r}$ distinct, where $s_r = \beta_1$. Let v_i be the characteristic vector of g_{s_i}. Set $y = s_1 e_1 + (s_2 - s_1)e_2 + \cdots + (s_r - s_{r-1})e_r$. We may assume that the incidence matrix A of $\mathcal{C}$ has column vector $v_1, \ldots, v_q$. Then y satisfies $\langle y, \mathbf{1} \rangle = \beta_1$. For each k_i the number of variables of the form $x_{k_i}^\ell$ that occur in $f_1, \ldots, f_{\beta_1}$ is at most a_{k_i} because the f_i are pairwise disjoint. Hence for each k_i the number of times that the variable $x_{k_i}^1$ occurs in $g_1, \ldots, g_{\beta_1}$ is at most a_{k_i}. Then

$$Ay = s_1 v_1 + (s_2 - s_1)v_2 + \cdots + (s_r - s_{r-1})v_r \leq a. \qquad \Box$$

Definition 13.2.13 The clutter of minimal vertex covers of $\mathcal{C}$, denoted by $b(\mathcal{C})$ or $\mathcal{C}^\vee$, is called the *blocker* of $\mathcal{C}$ or the *Alexander dual* of $\mathcal{C}$.

Lemma 13.2.14 ([116, Lemma 2.15], [373, p. 1385, Eq. (78.6)]) *Let $\mathcal{C}$ be a clutter and let $\mathcal{C}^\vee$ be the blocker of $\mathcal{C}$. If $a = (a_i) \in \mathbb{N}^n$, then*

$$\min\left\{ \sum_{x_i \in C} a_i \,\middle|\, C \in \mathcal{C}^\vee \right\} = \alpha_0(\mathcal{C}^a).$$

Proof. We may assume that $a = (a_1, \ldots, a_m, a_{m+1}, \ldots, a_{m_1}, 0, \ldots, 0)$, where $a_i \geq 2$ for $i = 1, \ldots, m$, $a_i = 1$ for $i = m+1, \ldots, m_1$, and $a_i = 0$ for $i > m_1$. Thus for $i = 1, \ldots, m$ the vertex x_i is duplicated $a_i - 1$ times. We denote the duplications of x_i by $x_i^2, \ldots, x_i^{a_i}$ and set $x_i^1 = x_i$.

We prove first the inequality "$\leq$". Let C^a be a minimal vertex cover of $\mathcal{C}^a$ with α_0 elements, where α_0 is equal to $\alpha_0(\mathcal{C}^a)$. We may assume that $C^a \cap \{x_1, \ldots, x_{m_1}\} = \{x_1, \ldots, x_s\}$. Note that $x_i^1, \ldots, x_i^{a_i}$ are in C^a for $i = 1, \ldots, s$. Indeed since C^a is a minimal vertex cover of $\mathcal{C}^a$, there exists an edge e of $\mathcal{C}^a$ such that $e \cap C^a = \{x_i^1\}$. Then $(e \setminus \{x_i^1\}) \cup \{x_i^j\}$ is an edge of C^a for $j = 1, \ldots, a_i$. Consequently $x_i^j \in C^a$ for $j = 1, \ldots, a_i$. Hence

$$a_1 + \cdots + a_s \leq |C^a| = \alpha_0. \tag{13.3}$$

On the other hand the set $C' = \{x_1, \ldots, x_s\} \cup \{x_{m_1+1}, \ldots, x_n\}$ is a vertex cover of $\mathcal{C}$. Let D be a minimal vertex cover of $\mathcal{C}$ contained in C'. Let e_D denote the characteristic vector of D. Then, since $a_i = 0$ for $i > m_1$, using Eq. (13.3) we get

$$\langle a, e_D \rangle = \sum_{x_i \in D} a_i = \sum_{x_i \in D \cap \{x_1, \ldots, x_s\}} a_i \leq \sum_{x_i \in \{x_1, \ldots, x_s\}} a_i \leq \alpha_0.$$

This completes the proof of the inequality "$\leq$". Next we show the inequality "$\geq$". Let C be a minimal vertex cover of $\mathcal{C}$. Note that the set

$$C' = \cup_{x_i \in C}\{x_i^1, \ldots, x_i^{a_i}\}$$

is a vertex cover of $\mathcal{C}^a$. Indeed any edge e^a of the clutter $\mathcal{C}^a$ is of the form $e^a = \{x_{i_1}^{j_1}, \ldots, x_{i_r}^{j_r}\}$ for some edge $e = \{x_{i_1}, \ldots, x_{i_r}\}$ of $\mathcal{C}$ and since e is covered by C, we have that e^a is covered by C'. Therefore one has that $\alpha_0(\mathcal{C}^a) \leq |C'| = \sum_{x_i \in C} a_i$. As C was an arbitrary vertex cover of $\mathcal{C}$ we get the asserted inequality. $\Box$

Theorem 13.2.15 [307] *Let $\mathcal{C}$ be a clutter with vertex set $X = \{x_1, \ldots, x_n\}$ and let $0 \neq a \in \mathbb{N}^n$, $b \in \mathbb{N}$. Then $x^a t^b$ is a minimal generator of $R_s(I(\mathcal{C}))$, as a K-algebra, if and only if $\mathcal{C}^a$ is indecomposable and $b = \alpha_0(\mathcal{C}^a)$.*

Proof. We may assume that $a = (a_1, \ldots, a_m, 0, \ldots, 0)$, where $a_i \geq 1$ for $i = 1, \ldots, m$. For each $1 \leq i \leq m$ the vertex x_i is duplicated $a_i - 1$ times, and the vertex x_i is deleted for each $i > m$. We denote the duplications of x_i by $x_i^2, \ldots, x_i^{a_i}$ and set $x_i^1 = x_i$ for $1 \leq i \leq m$. The vertex set of $\mathcal{C}^a$ is

$$X^a = \{x_1^1, \ldots, x_1^{a_1}, \ldots, x_i^1, \ldots, x_i^{a_i}, \ldots, x_m^1, \ldots, x_m^{a_m}\} = X^{a_1} \cup \cdots \cup X^{a_m},$$

where $X^{a_i} = \{x_i^1, \ldots, x_i^{a_i}\}$ for $1 \leq i \leq m$ and $X^{a_i} \cap X^{a_j} = \emptyset$ for $i \neq j$.

$\Rightarrow$) Assume that $x^a t^b$ is a minimal generator of $R_s(I(\mathcal{C}))$. Then, by Lemma 13.2.8, a is an irreducible b-cover of $\mathcal{C}^\vee$. First we prove that b is equal to $\alpha_0(\mathcal{C}^a)$. There is k such that $a_k \neq 0$. We may assume that $a - e_k \neq 0$. By Lemma 13.2.14 we need only show the equality

$$b = \min\left\{ \sum_{x_i \in C} a_i \,\middle|\, C \in \mathcal{C}^\vee \right\}.$$

As a is a b-cover of $\mathcal{C}^\vee$, the minimum is greater than or equal to b. If the minimum is greater than b, then we can write $a = (a - e_k) + e_k$, where $a - e_k$ is a b-cover and e_k is a 0-cover of $\mathcal{C}^\vee$, a contradiction.

Next we show that $\mathcal{C}^a$ is indecomposable. We proceed by contradiction. Assume that $\mathcal{C}^a$ is decomposable. Then there is a partition X_1, X_2 of X^a such that $\alpha_0(\mathcal{C}^a) = \alpha_0(\mathcal{C}^a[X_1]) + \alpha_0(\mathcal{C}^a[X_2])$. For $1 \leq i \leq n$, we set

$$\ell_i = |X^{a_i} \cap X_1| \quad \text{and} \quad p_i = |X^{a_i} \cap X_2|$$

if $1 \leq i \leq m$, and $\ell_i = p_i = 0$ if $i > m$. Consider the vectors $\ell = (\ell_i)$ and $p = (p_i)$. Notice that a has a decomposition $a = \ell + p$ because one has a partition $X^{a_i} = (X^{a_i} \cap X_1) \cup (X^{a_i} \cap X_2)$ for $1 \leq i \leq m$. To derive a contradiction we now claim that ℓ (resp. p) is an $\alpha_0(\mathcal{C}^a[X_1])$-cover (resp. $\alpha_0(\mathcal{C}^a[X_2])$-cover) of $\mathcal{C}^\vee$. Take an arbitrary C in $\mathcal{C}^\vee$. The set

$$C_a = \bigcup_{x_i \in C} \{x_i^1, \ldots, x_i^{a_i}\} = \bigcup_{x_i \in C} X^{a_i}$$

is a vertex cover of $\mathcal{C}^a$. Indeed, if f_k is any edge of $\mathcal{C}^a$, then f_k has the form

$$f_k = \{x_{k_1}^{j_{k_1}}, \ldots, x_{k_r}^{j_{k_r}}\} \quad (1 \leq k_1 < \cdots < k_r \leq m;\ 1 \leq j_{k_i} \leq a_{k_i}) \tag{13.4}$$

for some edge $\{x_{k_1}, \ldots, x_{k_r}\}$ of $\mathcal{C}$. Since $\{x_{k_1}, \ldots, x_{k_r}\} \cap C \neq \emptyset$, we get $f_k \cap C_a \neq \emptyset$. Thus C_a is a vertex cover of $\mathcal{C}^a$. Therefore $C_a \cap X_1$ and $C_a \cap X_2$ are vertex covers of $\mathcal{C}^a[X_1]$ and $\mathcal{C}^a[X_2]$, respectively, because $E(\mathcal{C}^a[X_i])$ is contained in $E(\mathcal{C}^a)$ for $i = 1, 2$. Hence using the partitions

$$C_a \cap X_1 = \bigcup_{x_i \in C} (X^{a_i} \cap X_1) \quad \text{and} \quad C_a \cap X_2 = \bigcup_{x_i \in C} (X^{a_i} \cap X_2)$$

we obtain

$$\alpha_0(\mathcal{C}^a[X_1]) \leq |C_a \cap X_1| = \sum_{x_i \in C} \ell_i \quad \text{and} \quad \alpha_0(\mathcal{C}^a[X_2]) \leq |C_a \cap X_2| = \sum_{x_i \in C} p_i.$$

This completes the proof of the claim. Consequently a is a reducible b-cover of $\mathcal{C}^\vee$, where $b = \alpha_0(\mathcal{C}^a)$, a contradiction to the irreducibility of a.

$\Leftarrow$) Assume that $\mathcal{C}^a$ is an indecomposable clutter and $b = \alpha_0(\mathcal{C}^a)$. We set $a = (a_1, \ldots, a_n)$. To show that $x^a t^b$ is a minimal generator of $R_s(I(\mathcal{C}))$ we need only show that a is an irreducible b-cover of $\mathcal{C}^\vee$. To begin with, notice that a is a b-cover of $\mathcal{C}^\vee$ by Lemma 13.2.14. We proceed by contradiction assuming that there is a decomposition $a = \ell + p$, where $\ell = (\ell_i)$ is a c-cover of $\mathcal{C}^\vee$, $p = (p_i)$ is a d-cover of $\mathcal{C}^\vee$, and $b = c + d$. Each X^{a_i} can be decomposed as $X^{a_i} = X^{\ell_i} \cup X^{p_i}$, where $X^{\ell_i} \cap X^{p_i} = \emptyset$, $\ell_i = |X^{\ell_i}|$, and $p_i = |X^{p_i}|$. We set

$$X^\ell = X^{\ell_1} \cup \cdots \cup X^{\ell_m} \quad \text{and} \quad X^p = X^{p_1} \cup \cdots \cup X^{p_m}.$$

Then one has a decomposition $X^a = X^\ell \cup X^p$ of the vertex set of $\mathcal{C}^a$. We now show that $\alpha_0(\mathcal{C}^a[X^\ell]) \geq c$ and $\alpha_0(\mathcal{C}^a[X^p]) \geq d$. By symmetry, it suffices to prove the first inequality. Take an arbitrary minimal vertex cover C_ℓ of $\mathcal{C}^a[X^\ell]$. Then $C_\ell \cup X^p$ is a vertex cover of $\mathcal{C}^a$ because if f is an edge of $\mathcal{C}^a$ contained in X^ℓ, then f is covered by C_ℓ, otherwise f is covered by X^p. Hence there is a minimal vertex cover C_a of $\mathcal{C}^a$ such that $C_a \subset C_\ell \cup X^p$. We set $V_m = \{x_1, \ldots, x_m\}$. Since $\mathcal{C}[V_m]$ is a subclutter of $\mathcal{C}^a$, there is a minimal vertex cover C_1 of $\mathcal{C}[V_m]$ contained in C_a. Then the set $C_1 \cup \{x_i | i > m\}$ is a vertex cover of $\mathcal{C}$. Therefore there is a minimal vertex cover C of $\mathcal{C}$ such that $C \cap V_m \subset C_a$. Altogether one has:

$$C \cap V_m \subset C_a \subset C_\ell \cup X^p \Rightarrow C \cap V_m \subset C_a \cap V_m \subset (C_\ell \cup X^p) \cap V_m. \tag{13.5}$$

We may assume that $C_a \cap V_m = \{x_1, \ldots, x_s\}$. Next we claim that $X^{a_i} \subset C_a$ for $1 \leq i \leq s$. Take an integer i between 1 and s. Since C_a is a minimal vertex cover of $\mathcal{C}^a$, there exists an edge e of $\mathcal{C}^a$ such that $e \cap C_a = \{x_i^1\}$.

Then $(e \setminus \{x_i^1\}) \cup \{x_i^j\}$ is an edge of $\mathcal{C}^a$ for $j = 1, \ldots, a_i$, this follows using that the edges of $\mathcal{C}^a$ are of the form described in Eq. (13.4). Consequently $x_i^j \in C_a$ for $j = 1, \ldots, a_i$. This completes the proof of the claim. Thus one has $X^{\ell_i} \subset X^{a_i} \subset C_a$ for $1 \leq i \leq s$. Hence, by Eq. (13.5), and noticing that $X^{\ell_i} \cap X^p = \emptyset$, we get $X^{\ell_i} \subset C_\ell$ for $1 \leq i \leq s$. So, using that $\ell_i = 0$ for $i > m$, we get

$$\alpha_0(\mathcal{C}^a[X^\ell]) \geq |C_\ell| \geq \sum_{i=1}^{s} \ell_i \geq \sum_{x_i \in C \cap V_m} \ell_i = \sum_{x_i \in C} \ell_i \geq c.$$

Therefore $\alpha_0(\mathcal{C}^a[X^\ell]) \geq c$. Similarly $\alpha_0(\mathcal{C}^a[X^p]) \geq d$. Thus

$$\alpha_0(\mathcal{C}^a[X^\ell]) + \alpha_0(\mathcal{C}^a[X^p]) \geq c + d = b = \alpha_0(\mathcal{C}^a),$$

and consequently, by Eq. (13.2), we have equality. Thus we have shown that $\mathcal{C}^a$ is a decomposable clutter, a contradiction. □

Lemma 13.2.16 *If $\mathcal{C}$ is an indecomposable clutter with the König property, then either $\mathcal{C}$ has no edges and has exactly one isolated vertex or $\mathcal{C}$ has only one edge and no isolated vertices.*

Proof. Let $f_1, \ldots, f_g$ be a set of independent edges and let $X' = \cup_{i=1}^g f_i$, where $g = \alpha_0(\mathcal{C})$. Note that $g = 0$ if $\mathcal{C}$ has no edges. Then $V(\mathcal{C})$ has a partition

$$V(\mathcal{C}) = (\cup_{i=1}^g f_i) \cup \left(\cup_{x_i \in V(\mathcal{C}) \setminus X'} \{x_i\}\right).$$

As $\mathcal{C}$ is indecomposable, we get that either $g = 0$ and $V(\mathcal{C}) = \{x_i\}$ for some vertex x_i or $g = 1$ and $V(\mathcal{C}) = f_i$ for some i. Thus in the second case, as $\mathcal{C}$ is a clutter, we get that $\mathcal{C}$ has exactly one edge and no isolated vertices. □

Corollary 13.2.17 *Let $\mathcal{C}$ be a clutter and let $I = I(\mathcal{C})$ be its edge ideal. Then all indecomposable parallelizations of $\mathcal{C}$ satisfy the König property if and only if $I^i = I^{(i)}$ for $i \geq 1$.*

Proof. $\Rightarrow$) It suffices to prove that $R[It] = R_s(I)$. Clearly $R[It] \subset R_s(I)$. To prove the reverse inclusion take a minimal generator $x^a t^b$ of $R_s(I)$. If $b = 0$, then $a = e_i$ for some i and $x^a t^b = x_i$ is in $R[It]$. Assume $b \geq 1$. By Theorem 13.2.15 $\mathcal{C}^a$ is an indecomposable clutter such that $b = \alpha_0(\mathcal{C}^a)$. As $\mathcal{C}^a$ satisfies the König property, using Lemma 13.2.16, it is not hard to see that $b = 1$ and that $E(\mathcal{C}^a) = \{e\}$ consists of a single edge e of $\mathcal{C}$, i.e., $x^a t^b = x_e t$, where $x_e = \prod_{x_i \in e} x_i$. Thus $x^a t^b \in R[It]$.

$\Leftarrow$) Since $R[It] = R_s(I)$, by Theorem 13.2.15, we obtain that the only indecomposable parallelizations are either induced subclutters of $\mathcal{C}$ with exactly one edge and no isolated vertices or subclutters consisting of exactly one isolated vertex. Thus in both cases they satisfy the König property. □

Corollary 13.2.18 *Let $\mathcal{C}$ be a clutter with vertex set $X = \{x_1, \ldots, x_n\}$ and let $S \subset X$. Then the induced clutter $H = \mathcal{C}[S]$ is indecomposable if and only if the monomial $(\prod_{x_i \in S} x_i)t^{\alpha_0(H)}$ is a minimal generator of $R_s(I(\mathcal{C}))$.*

Proof. We set $a = \sum_{x_i \in S} e_i$. Since $\mathcal{C}^a = \mathcal{C}[S]$, the result follows from Theorem 13.2.15. $\Box$

Corollary 13.2.19 *Let $\mathcal{H}$ be the Hilbert basis of* $\mathrm{Cn}(I(\mathcal{C}))$. *Then*

$$\begin{aligned} \mathcal{H} &= \{(a,b) \mid x^a t^b \text{ is a minimal generator of } R_s(I(\mathcal{C}))\} \\ &= \{(a, \alpha_0(\mathcal{C})) \mid \mathcal{C}^a \text{ is an indecomposable parallelization of } \mathcal{C}\} \\ &= \{(a,b) \mid a \text{ is an irreducible } b\text{-cover of } \mathcal{C}^\vee\}. \end{aligned}$$

Proof. The first and third equalities follow readily from Lemma 13.2.8 and Theorem 13.2.2. The second equality follows from Theorem 13.2.15. $\Box$

This result allows us to compute all indecomposable parallelizations of $\mathcal{C}$ and all indecomposable induced subclutters of $\mathcal{C}$ using Hilbert bases (see also Exercise 13.2.31).

Computing the Hilbert basis of a Simis cone Let $x^{v_1}, \ldots, x^{v_q}$ be the minimal set of generators of the edge ideal $I(\mathcal{C})$ of a clutter $\mathcal{C}$ and let $\mathbb{R}_+\mathcal{A}'$ be the Rees cone of $\mathcal{A} = \{v_1, \ldots, v_q\}$.

Lemma 13.2.20 *If $\ell_i = (u_i, -1)$ for $i = 1, \ldots, s$, then the halfspace $H_{\ell_i}^+$ occurs in the irreducible representation of $\mathbb{R}_+\mathcal{A}'$.*

Proof. By Corollary 13.1.3 the integral vertices of $\mathcal{Q}(A)$ are $u_1, \ldots, u_s$, where A is the incidence matrix of the clutter $\mathcal{C}$. Hence the result follows from Theorem 1.4.2. $\Box$

Procedure 13.2.21 To determine the Hilbert basis of $\mathrm{Cn}(I)$ and the generators of $R_s(I(\mathcal{C}))$ we proceed as follows. Using *Normaliz*, one computes the irreducible representation of $\mathbb{R}_+\mathcal{A}'$ and apply Lemma 13.2.20 to obtain the set of "inequalities" $\Gamma = \{e_1, \ldots, e_{n+1}, \ell_1, \ldots, \ell_s\}$ that define $\mathrm{Cn}(I)$. Then one can use Γ as input for *Normaliz* [68] to obtain the Hilbert basis. Then one uses Corollary 13.2.19 to obtain the generators of $R_s(I(\mathcal{C}))$.

This procedure is based on the fact that the program *Normaliz* [68] is able to compute the minimal Hilbert basis of a pointed cone given by a linear system of inequalities (cf. [189, Procedure 4.6.10]).

Example 13.2.22 Using Procedure 13.2.21 and *Normaliz* [68], we compute the minimal generators of the symbolic Rees algebra of the edge ideal

$$I = I(\mathcal{C}) = (x_1x_2, x_2x_3, x_1x_3).$$

The input file for *Normaliz* is:

```
3
3
1 1 0
0 1 1
1 0 1
3
```

Using *Normaliz*, with *mode* 3, and Lemma 13.2.20, we obtain that $\mathrm{Cn}(I)$ is defined by the "inequalities" given by the rows of the matrix:

```
7
4
0 0 1  0
0 0 0  1
1 0 1 -1
1 0 0  0
0 1 1 -1
0 1 0  0
1 1 0 -1
4
```

Using this input file for *Normaliz*, with *mode* 4, to compute the Hilbert basis of $\mathrm{Cn}(I)$, we get:

```
7 Hilbert basis elements:
 0 1 1 1
 1 0 1 1
 1 1 1 2
 0 0 1 0
 0 1 0 0
 1 0 0 0
 1 1 0 1
```

Thus, by Corollary 13.2.19, the symbolic Rees algebra of the edge ideal I is minimally generated as a K-algebra by x_1, x_2, x_3, It, $x_1x_2x_3t^2$.

Algebras of vertex covers In this part we will further examine minimal sets of generators of symbolic Rees algebras of edge ideals using polyhedral geometry and vertex covers.

Let $I = I(\mathcal{C})$ be the edge ideal of a clutter $\mathcal{C}$ and let $v_1, \ldots, v_q$ be the characteristic vectors of the edges of $\mathcal{C}$. Recall that *Rees cone* of I, denoted by $\mathbb{R}_+(I)$ or $\mathbb{R}_+\mathcal{A}'$, is the rational cone generated by the set

$$\mathcal{A}' = \{e_1, \ldots, e_n, (v_1, 1), \ldots, (v_q, 1)\} \subset \mathbb{R}^{n+1},$$

where e_i is the ith unit vector. By Proposition 1.1.51 and Theorem 1.4.2, the Rees cone has a unique irreducible representation

$$\mathbb{R}_+(I) = H^+_{e_1} \cap H^+_{e_2} \cap \cdots \cap H^+_{e_{n+1}} \cap H^+_{\ell_1} \cap H^+_{\ell_2} \cap \cdots \cap H^+_{\ell_r} \tag{13.6}$$

such that, for each k, $\ell_k \in \mathbb{Z}^{n+1}$, the non-zero entries of ℓ_k are relatively prime, the first n entries of ℓ_k are in $\mathbb{N}$, the last entry of ℓ_k is negative, and none of the closed halfspaces $H_{e_1}^+, \ldots, H_{e_{n+1}}^+, H_{\ell_1}^+, \ldots, H_{\ell_r}^+$ can be omitted from the intersection.

The *facets* (i.e., the proper faces of maximum dimension or equivalently the faces of dimension n) of the Rees cone are exactly:

$$H_{e_1} \cap \mathbb{R}_+(I), \ldots, H_{e_{n+1}} \cap \mathbb{R}_+(I), H_{\ell_1} \cap \mathbb{R}_+(I), \ldots, H_{\ell_r} \cap \mathbb{R}_+(I).$$

Let $C_1, \ldots, C_s$ be the minimal vertex covers of $\mathcal{C}$ and let $u_1, \ldots, u_s$ be their characteristic vectors, i.e., $u_k = \sum_{x_i \in C_k} e_i$ for $k = 1, \ldots, s$. By Lemma 13.2.20 we may always assume that

$$\ell_k = (u_k, -1) \text{ for } 1 \leq k \leq s,$$

and that every facet $H_{\ell_k} \cap \mathbb{R}_+(I)$, with $k > s$, satisfies $\langle \ell_k, e_{n+1} \rangle < -1$. Thus one has:

Remark 13.2.23 The primary decomposition of $I(\mathcal{C})$ as well as all the minimal vertex covers of $\mathcal{C}$ can be read off from the irreducible representation of the Rees cone $\mathbb{R}_+(I(\mathcal{C}))$.

The *ideal of covers* of $\mathcal{C}$, denoted by $I_c(\mathcal{C})$, is the ideal of R generated by $x^{u_1}, \ldots, x^{u_s}$. We also denote $I_c(\mathcal{C})$ by $I(\mathcal{C})^\vee$ and call $I(\mathcal{C})^\vee$ the Alexander dual of $I(\mathcal{C})$. Notice that $I_c(\mathcal{C}) = I(\mathcal{C}^\vee) = I(\mathcal{C})^\vee$.

The symbolic Rees algebra of $I_c(\mathcal{C})$ can also be interpreted in terms of "b-vertex covers" because $R_s(I_c(\mathcal{C}))$ is the symbolic Rees algebra of the edge ideal of $\mathcal{C}^\vee$ and $(\mathcal{C}^\vee)^\vee = \mathcal{C}$. Let $a = (a_1, \ldots, a_n) \neq 0$ be a vector in $\mathbb{N}^n$ and let $b \in \mathbb{N}$. Recall that a is a *b-vertex cover* of $\mathcal{C}$ if $\langle v_i, a \rangle \geq b$ for $i = 1, \ldots, q$.

Definition 13.2.24 The *algebra of vertex covers* of I, denoted by $R_c(I)$, is the K-subalgebra of $R[t]$ generated by all monomials $x^a t^b$ such that a is a b-cover of $\mathcal{C}$. This algebra turns out to be equal to $R_s(I_c(\mathcal{C}))$.

The irreducible 0 and 1 covers of $\mathcal{C}$ are the unit vector $e_1, \ldots, e_n$ and the vectors $u_1, \ldots, u_s$. The minimal generators of $R_c(I)$, as a K-algebra, correspond to the irreducible covers of $\mathcal{C}$ (see Lemma 13.2.8). Notice the following dual descriptions:

$$\begin{aligned} I^{(b)} &= (\{x^a \mid \langle a, u_i \rangle \geq b \text{ for } i = 1, \ldots, s\}), \\ I_c(\mathcal{C})^{(b)} &= (\{x^a \mid \langle a, v_i \rangle \geq b \text{ for } i = 1, \ldots, q\}). \end{aligned}$$

Thus, we have the following duality between symbolic Rees algebras.

Theorem 13.2.25 $R_c(I) = R_s(I_c(\mathcal{C}))$ *and* $R_c(I_c(\mathcal{C})) = R_s(I)$.

Theorem 13.2.26 *$R_c(I)$ is a finitely generated K-algebra.*

Proof. It follows from Corollary 13.2.4 because $R_c(I)$ is the symbolic Rees algebra of the edge ideal $I_c(\mathcal{C})$. $\square$

Lemma 13.2.27 *If $\ell_k = (a_k, -d_k)$ is any of the vectors of Eq. (13.6), where $a_k \in \mathbb{N}^n$, $d_k \in \mathbb{N}_+$, then a_k is an irreducible d_k-cover of $\mathcal{C}$ and $x^{a_k}t^{d_k}$ is a minimal generator of $R_s(I_c(\mathcal{C}))$.*

Proof. We proceed by contradiction and assume that there is a d_k'-cover a_k' and a d_k''-cover a_k'' such that $a_k = a_k' + a_k''$ and $d_k = d_k' + d_k''$. Set $F' = H_{(a_k', -d_k')} \cap \mathbb{R}_+(I)$ and $F'' = H_{(a_k'', -d_k'')} \cap \mathbb{R}_+(I)$. Clearly F', F'' are proper faces of $\mathbb{R}_+(I)$ and $F = \mathbb{R}_+(I) \cap H_{\ell_k} = F' \cap F''$. Applying Theorem 1.1.44(d) to F' and F'' it is seen that $F' \subset F$ or $F'' \subset F$, i.e., $F = F'$ or $F = F''$. We may assume $F = F'$. Hence $H_{(a_k', -d_k')} = H_{\ell_k}$. Taking orthogonal complements we get that $(a_k', -d_k') = \lambda(a_k, -d_k)$ for some $\lambda \in \mathbb{Q}_+$, because the orthogonal complement of H_{ℓ_k} is generated by ℓ_k. Since the non-zero entries of ℓ_k are relatively prime, we may assume that $\lambda \in \mathbb{N}$. Thus $d_k' = \lambda d_k \geq d_k \geq d_k'$ and λ must be 1. Hence $a_k = a_k'$ and a_k'' must be zero, a contradiction. $\square$

Remark 13.2.28 Let F_{n+1} be the facet of the Rees cone $\mathbb{R}_+(I)$ determined by the hyperplane $H_{e_{n+1}}$. Thus, by Lemma 13.2.27, we have a map ψ:

$$\begin{array}{rcl} \{\text{Facets of } \mathbb{R}_+(I(\mathcal{C}))\} \setminus \{F_{n+1}\} & \stackrel{\psi}{\longrightarrow} & R_s(I_c(\mathcal{C})) \\ H_{\ell_k} \cap \mathbb{R}_+(I) & \stackrel{\psi}{\longrightarrow} & x^{a_k}t^{d_k}, \text{ where } \ell_k = (a_k, -d_k) \\ H_{e_i} \cap \mathbb{R}_+(I) & \stackrel{\psi}{\longrightarrow} & x_i \end{array}$$

whose image is a good approximation for the minimal set of generators of $R_s(I_c(\mathcal{C}))$ as a K-algebra. Likewise, from the facets of $\mathbb{R}_+(I_c(\mathcal{C}))$, we obtain an approximation for the minimal set of generators of $R_s(I(\mathcal{C}))$.

The next example shows a connected graph G for which the image of the map ψ does not generate $R_s(I_c(G))$.

Example 13.2.29 Consider the following graph G:

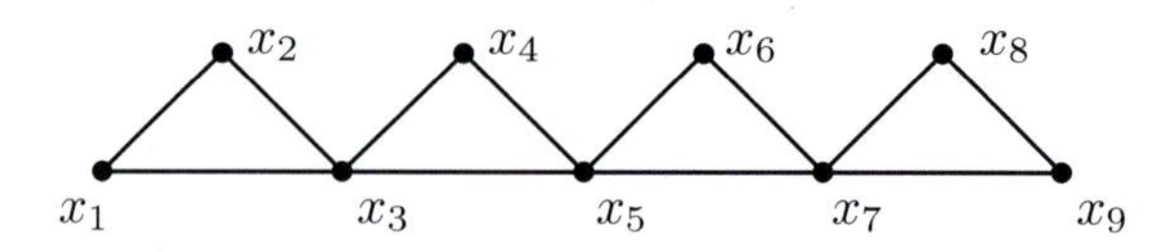

Using *Normaliz* [68] it is seen that the vector $a = (1, 1, 2, 0, 2, 1, 1, 1, 1)$ is an irreducible 2-cover of G such that the supporting hyperplane $H_{(a,-2)}$ does not define a facet of the Rees cone of $I(G)$.

For balanced clutters, the image of the map ψ generates $R_s(I_c(\mathcal{C}))$. This follows from the next result (see Corollary 14.3.11). In particular the image of the map ψ generates $R_s(I_c(\mathcal{C}))$ when $\mathcal{C}$ is a bipartite graph.

Proposition 13.2.30 [183, 185] *If $\mathcal{C}$ is a balanced clutter, then*

$$R_s(I_c(\mathcal{C})) = R[I_c(\mathcal{C})t].$$

This result was first shown for bipartite graphs in [183, Corollary 2.6] and later generalized to balanced clutters in [185].

Exercises

13.2.31 Let $\mathcal{C}$ be a clutter with vertices $x_1, \ldots, x_n$ and $\alpha = (a_1, \ldots, a_n, b)$ a vector in $\{0, 1\}^n \times \mathbb{N}$. Then α is in the minimal Hilbert basis of $\mathrm{Cn}(I(\mathcal{C}))$ if and only if the induced subclutter $H = \mathcal{C}[\{x_i | \, a_i = 1\}]$ is indecomposable with $b = \alpha_0(H)$.

13.2.32 Let $\mathcal{C}$ be a clutter. Prove that $(1, \ldots, 1, \alpha_0(\mathcal{C}))$ is in the minimal Hilbert basis of $\mathrm{Cn}(I(\mathcal{C}))$ if and only if $\mathcal{C}$ is indecomposable.

13.3 Blowup algebras in perfect graphs

Harary and Plummer [209] studied indecomposable graphs. To the best of our knowledge there is no structure theorem for indecomposable graphs. In this section we relate these graphs to the theory of perfect graphs and to symbolic Rees algebras. Then we present some general properties of indecomposable clutters.

Lemma 13.3.1 *Let $C_n = \{x_1, \ldots, x_n\}$ be a cycle of length n.* (a) *If n is odd and $n \geq 5$, then the complement $\overline{C}_n$ of C_n is indecomposable,* (b) *if n is odd, then C_n is indecomposable, and* (c) *any complete graph is indecomposable.*

Proof. (a) Assume that $G = \overline{C}_n$ is decomposable. Then there are disjoint sets X_1, X_2 such that $V(G) = X_1 \cup X_2$ and $\alpha_0(G) = \alpha_0(G[X_1]) + \alpha_0(G[X_2])$. Since $\beta_0(G) = 2$, it is seen that $G[X_i]$ is a complete graph for $i = 1, 2$. We may assume that $x_1 \in X_1$. Then x_2 must be in X_2, otherwise $\{x_1, x_2\}$ is an edge of $G[X_1]$, a contradiction. By induction it follows that $x_1, x_3, x_5, \ldots, x_n$ are in X_1. Hence, $\{x_1, x_n\}$ is an edge of $G[X_1]$, a contradiction. (b) This is left as an exercise. (c) It follows readily from the fact that the covering number of a complete graph in r vertices is $r - 1$. $\square$

For graphs, we can use this lemma together with Corollary 13.2.19 and Exercise 13.2.31 to locate all *induced odd cycles* (*odd holes*) and all *induced complements of odd cycles* (*odd antiholes*) of length at least five. Notice that, by Theorem 13.2.15, odd holes and odd antiholes correspond to minimal generators of the symbolic Rees algebra of the edge ideal of the graph.

A graph G is called a *Berge graph* if and only if G has no odd holes or odd antiholes of length at least five.

Theorem 13.3.2 (Strong perfect graph theorem [85]) *A graph G is perfect if and only if G is a Berge graph.*

In commutative algebra odd holes occurred for the first time in [377], and later in the description of $I(G)^{\{2\}}$, the join of an edge ideal of a graph G with itself [381], and in the description of the associated primes of powers of ideals of vertex covers of graphs [164]. The expository paper [163] surveys algebraic techniques for detecting odd cycles and odd holes in a graph and for computing the chromatic number of a hypergraph (see Theorem 7.7.19).

Theorem 13.3.3 [164] *If G is a graph and $\mathfrak{p}$ is an associated prime of $I_c(G)^2$, then $\mathfrak{p} = (x_i, x_j)$ for some edge $\{x_i, x_j\}$ of G or $\mathfrak{p} = (x_i | \in S)$, where S is a set of vertices of G that induces an odd hole.*

Example 13.3.4 Let G be the graph below. Computing the generators of $R_s(I(G))$ via Procedure 13.2.21 and Corollary 13.2.19, gives that the indecomposable parallelizations of G are: seven vertices, nine edges, one induced triangle, three induced pentagons, and the duplication shown below.

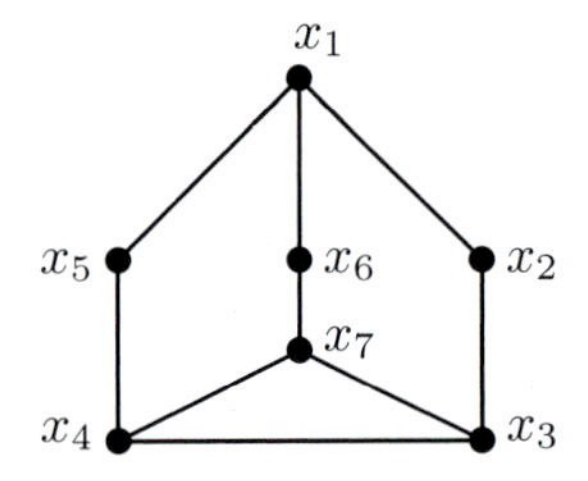

Decomposable graph G

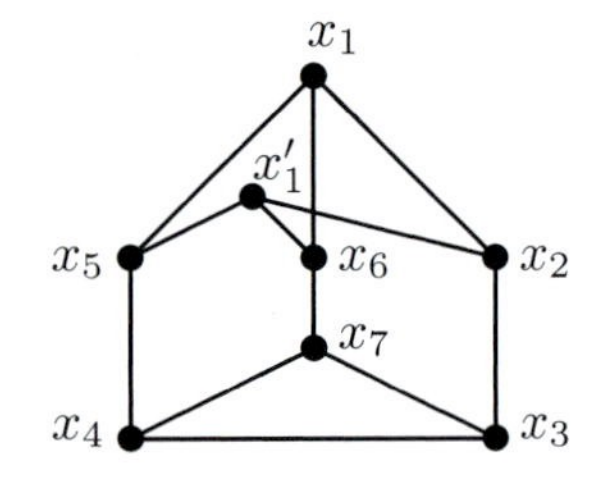

Indecomposable graph $G^{(2,1,1,1,1,1,1)}$

The next result shows that indecomposable graphs occur naturally in the theory of perfect graphs.

Proposition 13.3.5 [117, Proposition 2.13] *A graph G is perfect if and only if the indecomposable parallelizations of G are exactly the complete subgraphs of G*

The following was one of the first deep results in the study of symbolic powers of edge ideals from the viewpoint of graph theory.

Corollary 13.3.6 [383, Theorem 5.9] *Let G be a graph and let I be its edge ideal. Then G is bipartite if and only if $I^i = I^{(i)}$ for $i \geq 1$.*

Proof. $\Rightarrow$) If G is a bipartite graph, then any parallelization of G is again a bipartite graph. This means that any parallelization of G satisfies the König property because bipartite graphs satisfy this property; see Theorem 7.1.8. Thus $I^i = I^{(i)}$ for all i by Corollary 13.2.17.

$\Leftarrow$) Assume that $I^i = I^{(i)}$ for $i \geq 1$. Thanks to Corollary 13.2.17 all indecomposable induced subgraphs of G have the König property. If G is not bipartite, then G has an induced odd cycle, a contradiction because induced odd cycles are indecomposable by Lemma 13.3.1 and do not satisfy the König property. $\Box$

Basic properties of indecomposable clutters If e is a edge of a clutter $\mathcal{C}$, we denote by $\mathcal{C} \setminus \{e\}$ the spanning subclutter of $\mathcal{C}$ obtained by deleting e and keeping all the vertices of $\mathcal{C}$.

Definition 13.3.7 A clutter $\mathcal{C}$ is called *vertex critical* (resp. *edge critical*) if $\alpha_0(\mathcal{C} \setminus \{x_i\}) < \alpha_0(\mathcal{C})$ (resp. $\alpha_0(\mathcal{C} \setminus \{e\}) < \alpha_0(\mathcal{C})$) for all $x_i \in V(\mathcal{C})$ (resp. for all $e \in E(\mathcal{C})$).

Lemma 13.3.8 Let x_i be a vertex and let e be an edge of a clutter $\mathcal{C}$.

(a) If $\alpha_0(\mathcal{C} \setminus \{x_i\}) < \alpha_0(\mathcal{C})$, then $\alpha_0(\mathcal{C} \setminus \{x_i\}) = \alpha_0(\mathcal{C}) - 1$.

(b) If $\alpha_0(\mathcal{C} \setminus \{e\}) < \alpha_0(\mathcal{C})$, then $\alpha_0(\mathcal{C} \setminus \{e\}) = \alpha_0(\mathcal{C}) - 1$.

Proof. Part (a) is left as an exercise (cf. Proposition 7.2.11). Part (b) follows from Proposition 7.2.13. $\Box$

Definition 13.3.9 A clutter $\mathcal{C}$ is called *connected* if there is no $U \subset V(\mathcal{C})$ such that $\emptyset \subsetneq U \subsetneq V(\mathcal{C})$ and such that $e \subset U$ or $e \subset V(\mathcal{C}) \setminus U$ for each edge e of $\mathcal{C}$.

Proposition 13.3.10 *If a clutter $\mathcal{C}$ is indecomposable, then it is connected and vertex critical.*

Proof. Assume that $\mathcal{C}$ is disconnected. Then there is a partition X_1, X_2 of $V(\mathcal{C})$ such that

$$E(\mathcal{C}) \subset E(\mathcal{C}[X_1]) \cup E(\mathcal{C}[X_2]). \tag{13.7}$$

For $i = 1, 2$, let C_i be a minimal vertex cover of $\mathcal{C}[X_i]$ with $\alpha_0(\mathcal{C}[X_i])$ vertices. Then, by Eq. (13.7), $C_1 \cup C_2$ is a minimal vertex cover of $\mathcal{C}$. Hence $\alpha_0(\mathcal{C}[X_1]) + \alpha_0(\mathcal{C}[X_2])$ is greater than or equal to $\alpha_0(\mathcal{C})$. So $\alpha_0(\mathcal{C})$ is equal to $\alpha_0(\mathcal{C}[X_1]) + \alpha_0(\mathcal{C}[X_2])$, a contradiction to the indecomposability of $\mathcal{C}$. Thus $\mathcal{C}$ is connected.

We now show that $\alpha_0(\mathcal{C} \setminus \{x_i\}) < \alpha_0(\mathcal{C})$ for all i. If $\alpha_0(\mathcal{C} \setminus \{x_i\}) = \alpha_0(\mathcal{C})$, then $V(\mathcal{C}) = X_1 \cup X_2$, where $X_1 = V(\mathcal{C}) \setminus \{x_i\}$ and $X_2 = \{x_i\}$. Note that $\mathcal{C}[X_1] = \mathcal{C} \setminus \{x_i\}$. As $\alpha_0(\mathcal{C}[X_1]) = \alpha_0(\mathcal{C})$ and $\alpha_0(\mathcal{C}[X_2]) = 0$, we contradict the indecomposability of $\mathcal{C}$. Thus $\alpha_0(\mathcal{C} \setminus \{x_i\}) < \alpha_0(\mathcal{C})$. $\Box$

Proposition 13.3.11 *If $\mathcal{C}$ is a connected edge critical clutter, then $\mathcal{C}$ is indecomposable.*

Proof. Assume that $\mathcal{C}$ is decomposable. Then there is a partition X_1, X_2 of $V(\mathcal{C})$ into non-empty vertex sets such that $\alpha_0(\mathcal{C}) = \alpha_0(\mathcal{C}[X_1]) + \alpha_0(\mathcal{C}[X_2])$. Since $\mathcal{C}$ is connected, there is an edge $e \in E(\mathcal{C})$ intersecting both X_1 and X_2. Pick a minimal vertex cover C of $\mathcal{C} \setminus \{e\}$ with less than $\alpha_0(\mathcal{C})$ vertices. As $E(\mathcal{C}[X_i])$ is a subset of $E(\mathcal{C} \setminus \{e\}) = E(\mathcal{C}) \setminus \{e\}$ for $i = 1, 2$, we get that C covers all edges of $\mathcal{C}[X_i]$ for $i = 1, 2$. Hence C must have at least $\alpha_0(\mathcal{C})$ vertices, a contradiction. □

Exercises

13.3.12 Let $\mathcal{D}$ be a clutter obtained from $\mathcal{C}$ by adding a new vertex v and some new edges containing v and some vertices of $V(\mathcal{C})$. If $a = \mathbf{1} \in \mathbb{N}^n$ is an irreducible $\alpha_0(\mathcal{C})$-cover of $\mathcal{C}^\vee$ such that $\alpha_0(\mathcal{D}) = \alpha_0(\mathcal{C}) + 1$, then $a' = (a, 1)$ is an irreducible $\alpha_0(\mathcal{D})$-cover of $\mathcal{D}^\vee$.

13.3.13 [12] If G is a complete graph and H is the graph obtained by taking a cone over a pentagon, then

$$\begin{aligned} R_s(I(G)) &= K[\{x^a t^b \mid x^a \text{ is square-free ; } \deg(x^a) = b + 1\}], \\ R_s(I(H)) &= R[I(H)t][x_1 \cdots x_5 t^3, x_1 \cdots x_6 t^4, x_1 \cdots x_5 x_6^2 t^5]. \end{aligned}$$

13.3.14 [209] Prove that any odd cycle is indecomposable.

13.3.15 Using Procedure 13.2.21 and Corollary 13.2.19, show that the graph G below has exactly 103 indecomposable parallelizations, 92 of which correspond to subgraphs. The only indecomposable parallelization G^a, with $a_i > 0$ for all i, is that obtained by duplication of the five outer vertices, i.e., $a = (2, 2, 2, 2, 2, 1, 1, 1, 1, 1)$ and $\alpha_0(G^a) = 11$.

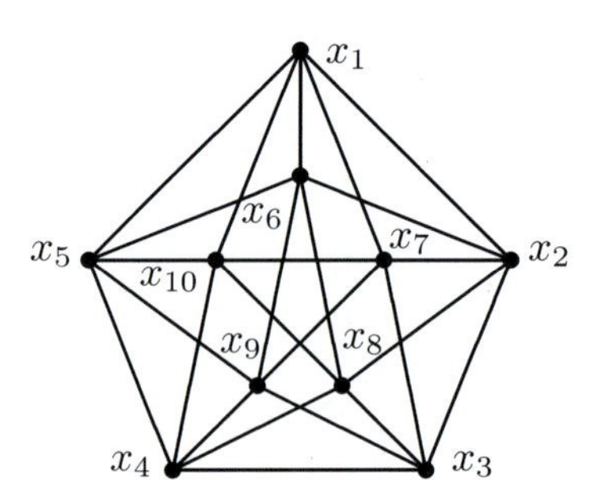

Decomposable graph G

13.4 Algebras of vertex covers of graphs

Let G be a graph and let $I_c(G)$ be its ideal of covers. In this section we give a graph theoretical description of the irreducible b-covers of G, i.e., we describe the minimal generators of the symbolic Rees algebra of $I_c(G)$.

Lemma 13.4.1 *If $a = (a_1, \ldots, a_n) \in \mathbb{N}^n$ is an irreducible b-cover of G, then $0 \leq b \leq 2$ and $0 \leq a_i \leq 2$ for $i = 1, \ldots, n$.*

Proof. Recall that a is a b-cover of G if and only if $a_i + a_j \geq b$ for any edge $\{x_i, x_j\}$ of G. If $b = 0$ or $b = 1$, then by the irreducibility of a it is seen that either $a = e_i$ for some i or $a = e_{i_1} + \cdots + e_{i_r}$ for some minimal vertex cover $\{x_{i_1}, \ldots, x_{i_r}\}$ of G. Thus we may assume that $b \geq 2$.

Case (I): $a_i \geq 1$ for all i. Clearly $\mathbf{1} = (1, \ldots, 1)$ is a 2-cover. If $a - \mathbf{1} \neq 0$, then $a - \mathbf{1}$ is a $b-2$ cover and $a = \mathbf{1} + (a - \mathbf{1})$, a contradiction. Hence $a = \mathbf{1}$. Pick any edge $\{x_i, x_j\}$ of G. Since a is a b-cover, we get $2 = a_i + a_j \geq b$ and b must be equal to 2.

Case (II): $a_i = 0$ for some i. We may assume $a_i = 0$ for $1 \leq i \leq r$ and $a_i \geq 1$ for $i > r$. Notice that the set $S = \{x_1, \ldots, x_r\}$ is independent because if $\{x_i, x_j\}$ is an edge and $1 \leq i < j \leq r$, then $0 = a_i + a_j \geq b$, a contradiction. Consider the neighbor set $N_G(S)$ of S. We may assume that $N_G(S) = \{x_{r+1}, \ldots, x_s\}$. Observe that $a_i \geq b \geq 2$ for $i = r+1, \ldots, s$, because a is a b-cover. Write

$$a = (0, \ldots, 0, a_{r+1} - 2, \ldots, a_s - 2, a_{s+1} - 1, \ldots, a_n - 1) + \\ (\underbrace{0, \ldots, 0}_{r}, \underbrace{2, \ldots, 2}_{s-r}, \underbrace{1, \ldots, 1}_{n-s}) = c + d.$$

Clearly d is a 2-cover. If $c \neq 0$, using that $a_i \geq b \geq 2$ for $r + 1 \leq i \leq s$ and $a_i \geq 1$ for $i > s$ it is not hard to see that c is a $(b-2)$-cover. This gives a contradiction, because $a = c + d$. Hence $c = 0$. Therefore $a_i = 2$ for $r < i \leq s$, $a_i = 1$ for $i > s$, and $b = 2$. $\square$

Corollary 13.4.2 *$R_s(I_c(G))$ is generated, as a K-algebra, by monomials of degree in t at most two and total degree at most $2n$.*

Proof. Let $x^a t^b$ be a minimal generator of $R_s(I_c(G))$. Then $a = (a_i)$ is an irreducible b-cover of G. By Lemma 13.4.1, we get $0 \leq b \leq 2$ and $0 \leq a_i \leq 2$ for all i. If $b = 0$ or $b = 1$, we get that the degree of $x^a t^b$ is at most n because when $b = 0$ or 1 one has $a = e_i$ or $a = \sum_{x_i \in C_k} e_i$ for some minimal vertex cover C_k of G, respectively. If $b = 2$, by the proof of Lemma 13.4.1, either $a = \mathbf{1}$ or $a_i = 0$ for some i. Thus $\deg(x^a) \leq 2(n-1)$. $\square$

For use below consider the vectors $\ell_1, \ldots, \ell_r$ that occur in the irreducible representation of $\mathbb{R}_+(I(G))$ given in Eq. (13.6).

Corollary 13.4.3 *If* $\ell_i = (\ell_{i1}, \ldots, \ell_{in}, -\ell_{i(n+1)})$, *then* $0 \leq \ell_{ij} \leq 2$ *for* $j = 1, \ldots, n$ *and* $1 \leq \ell_{i(n+1)} \leq 2$.

Proof. It suffices to observe that $(\ell_{i1}, \ldots, \ell_{in})$ is an irreducible $\ell_{i(n+1)}$-cover of G and to apply Lemma 13.4.1. $\square$

Lemma 13.4.4 $a = (1, \ldots, 1)$ *is an irreducible* 2*-cover of* G *if and only if* G *is non-bipartite.*

Proof. $\Rightarrow$) We proceed by contradiction. Assume that G is bipartite. Then G has a bipartition (V_1, V_2). Set $a' = \sum_{x_i \in V_1} e_i$ and $a'' = \sum_{x_i \in V_2} e_i$. Since V_1 and V_2 are minimal vertex covers of G, we can decompose a as $a = a' + a''$, where a' and a'' are 1-covers, which is impossible.

$\Leftarrow$) Notice that a cannot be the sum of a 0-cover and a 2-cover. Indeed if $a = a' + a''$, where a' is a 0-cover and a'' is a 1-cover, then a'' has an entry a_i equal to zero. Pick an edge $\{x_i, x_j\}$ incident with x_i, then $\langle a'', e_i + e_j \rangle \leq 1$, a contradiction. Thus we may assume that $a = c + d$, where c, d are 1-covers. Let C_r be an odd cycle of G of length r. Notice that any vertex cover of C_r must contain a pair of adjacent vertices. Hence the vertex covers of G corresponding to c and d must contain a pair of adjacent vertices, a contradiction because c and d are complementary vectors and the complement of a vertex cover is an independent set. $\square$

Theorem 13.4.5 *Let* $0 \neq a = (a_i) \in \mathbb{N}^n$ *and let* $G^\vee$ *be the clutter of minimal vertex covers of* G. *Then the following hold.*

(i) *If* G *is bipartite, then* a *is an irreducible* b*-cover of* G *if and only if* $b = 0$ *and* $a = e_i$ *for some* $1 \leq i \leq n$ *or* $b = 1$ *and* $a = \sum_{x_i \in C} e_i$ *for some* $C \in E(G^\vee)$.

(ii) *If* G *is non-bipartite, then* a *is an irreducible* b*-cover if and only if* a *has one of the following forms*:

(a) (0*-covers*) $b = 0$ *and* $a = e_i$ *for some* $1 \leq i \leq n$,

(b) (1*-covers*) $b = 1$ *and* $a = \sum_{x_i \in C} e_i$ *for some* $C \in E(G^\vee)$,

(c) (2*-covers*) $b = 2$ *and* $a = (1, \ldots, 1)$,

(d) (2*-covers*) $b = 2$ *and up to permutation of vertices*

$$a = (\underbrace{0, \ldots, 0}_{|\mathfrak{A}|}, \underbrace{2, \ldots, 2}_{|N_G(\mathfrak{A})|}, 1, \ldots, 1)$$

for some independent set of vertices $\mathfrak{A} \neq \emptyset$ *of* G *such that*

(d_1) $N_G(\mathfrak{A})$ *is not a vertex cover of* G *and* $V \neq \mathfrak{A} \cup N_G(\mathfrak{A})$,

(d_2) *the induced subgraph* $G[V \setminus (\mathfrak{A} \cup N_G(\mathfrak{A}))]$ *has no isolated vertices and is not bipartite.*

Proof. (i) $\Rightarrow$) Since G is bipartite, by Proposition 13.2.30, we have the equality $R_s(I_c(G)) = R[I_c(G)t]$. Thus the minimal set of generator of $R_s(I_c(G))$ as a K-algebra is $\{x_1, \ldots, x_n, x^{u_1}t, \ldots, x^{u_r}t\}$, where $u_1, \ldots, u_r$ are the characteristic vectors of the minimal vertex covers of G. As a is an irreducible b-cover of G, $x^a t^b$ is a minimal generator of $R_s(I_c(\mathcal{C}))$. Therefore either $a = e_i$ for some i and $b = 0$ or $a = u_i$ for some i and $b = 1$. The converse follows readily and is valid for any graph or clutter.

(ii) $\Rightarrow$) By Lemma 13.4.1, $0 \leq b \leq 2$ and $0 \leq a_i \leq 2$ for all i. If $b = 0$ or $b = 1$, then clearly a has the form indicated in (a) or (b), respectively.

Assume $b = 2$. If $a_i \geq 1$ for all i, then $a_i = 1$ for all i, otherwise if $a_i = 2$ for some i, then $a - e_i$ is a 2-cover and $a = e_i + (a - e_i)$, a contradiction. Hence $a = \mathbf{1}$. Thus we may assume that a has the form

$$a = (0, \ldots, 0, 2, \ldots, 2, 1, \ldots, 1).$$

We set $\mathfrak{A} = \{x_i | \, a_i = 0\} \neq \emptyset$, $B = \{x_i | \, a_i = 2\}$, and $C = V \setminus (\mathfrak{A} \cup B)$. Observe that $\mathfrak{A}$ is an independent set because a is a 2-cover and $B = N_G(\mathfrak{A})$ because a is irreducible. Hence it is seen that conditions (d_1) and (d_2) are satisfied. By Lemma 13.4.4, the proof of the converse is straightforward. $\square$

If $I \subset R$ is an ideal, the ring $\mathcal{F}_s(I) = R_s(I)/\mathfrak{m}R_s(I)$ is called the *symbolic special fiber* of I (cf. Definition 14.2.10). If G is a graph, the algebraic properties of the ring $\mathcal{F}_s(I)$ are studied in [91], when G is bipartite the symbolic special fiber of $I_c(G)$ turns out to be Koszul [91, 356] and in this case, by Proposition 13.2.30, one has $\mathcal{F}_s(I_c(G)) = R[I_c(G)t]/\mathfrak{m}R[I_c(G)t]$.

Exercises

13.4.6 Let G be a graph with vertex set $V(G) = \{x_1, \ldots, x_n\}$ and let H_α^+ be any of the closed halfspaces that occur in the irreducible representation of $\mathbb{R}_+(I_c(G))$ with $\alpha = (a_1, \ldots, a_n, -b)$, $a_i \in \mathbb{N}$ for all i, $0 \neq b \in \mathbb{N}$, and the non-zero entries of α are relatively prime.

(a) Let H be the cone over G. If $a_i \geq 1$ for all i and

$$\beta = (a_1, \ldots, a_n, (\textstyle\sum_{i=1}^n a_i) - b, -\sum_{i=1}^n a_i) = (\beta_1, \ldots, \beta_{n+1}, -\beta_{n+2}),$$

then $H_\beta \cap \mathbb{R}_+(I_c(H))$ is a facet of $\mathbb{R}_+(I_c(H))$ and $x_1^{\beta_1} \cdots x_{n+1}^{\beta_{n+1}} t^{\beta_{n+2}}$ is a minimal generator of $R_s(I(H))$.

(b) Let $G_0 = G$ and let G_r be the cone over G_{r-1} for $r \geq 1$. If α is equal to $(1, \ldots, 1, -g)$ and $b = n + (r-1)(n-g)$, then

$$(\underbrace{1, \ldots, 1}_{n}, \underbrace{n-g, \ldots, n-g}_{r})$$

is an irreducible b-cover of G_r and $R_s(I(G_r))$ has a generator of degree in t equal to b.

(c) Let $G = C_s$ be an odd cycle of length $s = 2k+1$. Then, $\alpha_0(C_s)$ is equal to $(s+1)/2 = k+1$ and

$$x_1 \cdots x_s x_{s+1}^k \cdots x_{s+r}^k t^{rk+k+1}$$

is a minimal generator of $R_s(I(G_r))$. This proves that the degree in t of the minimal generators of $R_s(I(G_r))$ is much larger than the number of vertices of the graph G_r [225].

13.5 Edge subrings in perfect matchings

In this section we present membership criteria and a generalized version of the marriage theorem. For a connected non-bipartite graph whose edge subring is normal, we give conditions for the existence of a perfect matching.

Let G be a connected graph and let $\mathcal{A} = \{v_1, \ldots, v_q\}$ be the set of all vectors $e_i + e_j$ such that $\{x_i, x_j\}$ is an edge of G. The *edge cone* of G, denoted by $\mathbb{R}_+\mathcal{A}$, is defined as the cone generated by $\mathcal{A}$. Below we recover an explicit combinatorial description of the edge cone (see Section 10.7).

Let $\mathfrak{A}$ be an *independent set* of vertices of G. The supporting hyperplane of the edge cone of G defined by

$$\sum_{x_i \in \mathfrak{A}} x_i - \sum_{x_i \in N_G(\mathfrak{A})} x_i = 0$$

is denoted by $H_{\mathfrak{A}}$. If $a_{\mathfrak{A}} = \sum_{x_i \in \mathfrak{A}} e_i - \sum_{x_i \in N_G(\mathfrak{A})} e_i$, then $H_{\mathfrak{A}} = H_{a_{\mathfrak{A}}}$.

Edge cones and their representations by closed halfspaces are a useful tool to study a-invariants of edge subrings [419, 405]; see Chapter 11. The following result is a prototype of these representations (cf. Theorem 10.7.8 and Exercise 10.7.24). As an application we give a direct proof of the next result using Rees cones and Theorem 13.4.5.

Corollary 13.5.1 [406, Corollary 2.8] *A vector $a = (a_1, \ldots, a_n) \in \mathbb{R}^n$ is in $\mathbb{R}_+\mathcal{A}$ if and only if a satisfies the following system of linear inequalities*

$$\begin{aligned} a_i &\geq 0, \quad i = 1, \ldots, n; \\ \textstyle\sum_{x_i \in N_G(\mathfrak{A})} a_i - \sum_{x_i \in \mathfrak{A}} a_i &\geq 0, \quad \textit{for all independent sets } \mathfrak{A} \subset V(G). \end{aligned}$$

Proof. We set $\mathcal{B} = \{(v_1, 1), \ldots, (v_q, 1)\}$ and $I = I(G)$. Notice the equality

$$\mathbb{R}_+(I) \cap \mathbb{R}\mathcal{B} = \mathbb{R}_+\mathcal{B}, \tag{13.8}$$

where $\mathbb{R}\mathcal{B}$ is the $\mathbb{R}$-vector space spanned by $\mathcal{B}$. Consider the irreducible representation of $\mathbb{R}_+(I)$ given in Eq. (13.6) and write $\ell_i = (a_i, -d_i)$, where $0 \neq a_i \in \mathbb{N}^n$, $0 \neq d_i \in \mathbb{N}$. Next we show the equality:

$$\mathbb{R}_+\mathcal{A} = \mathbb{R}\mathcal{A} \cap \mathbb{R}_+^n \cap H^+_{(2a_1/d_1 - \mathbf{1})} \cap \cdots \cap H^+_{(2a_r/d_r - \mathbf{1})}, \tag{13.9}$$

where $\mathbf{1} = (1, \ldots, 1)$. Take $\alpha \in \mathbb{R}_+\mathcal{A}$. Clearly $\alpha \in \mathbb{R}\mathcal{A} \cap \mathbb{R}_+^n$. We can write

$$\alpha = \lambda_1 v_1 + \cdots + \lambda_q v_q \ \Rightarrow\ |\alpha| = 2(\lambda_1 + \cdots + \lambda_q) = 2b.$$

Thus $(\alpha, b) = \lambda_1(v_1, 1) + \cdots + \lambda_q(v_q, 1)$, i.e., $(\alpha, b) \in \mathbb{R}_+\mathcal{B}$. Hence, from Eq. (13.8), we get $(\alpha, b) \in \mathbb{R}_+(I)$ and

$$\langle(\alpha, b), (a_i, -d_i)\rangle \geq 0 \ \Rightarrow\ \langle\alpha, a_i\rangle \geq bd_i = (|\alpha|/2)d_i = |\alpha|(d_i/2).$$

Writing $\alpha = (\alpha_1, \ldots, \alpha_n)$ and $a_i = (a_{i1}, \ldots, a_{in})$, the last inequality gives:

$$\alpha_1 a_{i_1} + \cdots + \alpha_n a_{in} \geq (\alpha_1 + \cdots + \alpha_n)(d_i/2) \ \Rightarrow\ \langle\alpha, a_i - (d_i/2)\mathbf{1}\rangle \geq 0.$$

Then $\langle\alpha, 2a_i/d_i - \mathbf{1}\rangle \geq 0$ and $\alpha \in H^+_{(2a_i/d_i - \mathbf{1})}$ for all i. This proves the inclusion "$\subset$" of Eq. (13.9). The other inclusion follows similarly. Now, by Lemma 13.2.27, a_i is an irreducible d_i-cover of G. Therefore, using Theorem 13.4.5, we readily get the equality

$$\mathbb{R}_+\mathcal{A} = \left(\bigcap_{\mathfrak{A}\in\mathcal{F}} H^-_{\mathfrak{A}}\right) \bigcap \left(\bigcap_{i=1}^{n} H^+_{e_i}\right),$$

where $\mathcal{F}$ is the collection of all the independent sets of vertices of G. From this equality the assertion follows at once. □

Proposition 13.5.2 [419] *Let G be a connected non-bipartite graph with n vertices. If $K[G]$ is a normal domain, then a monomial $x_1^{\beta_1} \cdots x_n^{\beta_n}$ belongs to $K[G]$ if and only if the following two conditions hold*

(i) *$\beta = (\beta_1, \ldots, \beta_n)$ is in the edge cone of G, and*

(ii) *$\sum_{i=1}^n \beta_i$ is an even integer.*

Proof. $\Leftarrow$) Let $\mathcal{A} = \{v_1, \ldots, v_q\}$ be the set of column vector of the incidence matrix of G. Assume that $\beta \in \mathbb{R}_+\mathcal{A}$ and $\deg(x^\beta)$ even. We proceed by induction on $\deg(x^\beta)$. Using Corollary 10.2.11 one has the isomorphism

$$\mathbb{Z}^n/(v_1, \ldots, v_q) \simeq \mathbb{Z}_2.$$

Hence $2\beta \in \mathbb{R}_+\mathcal{A} \cap \mathbb{Z}\mathcal{A} = \mathbb{N}\mathcal{A}$ and one can write

$$2\beta = 2\sum_{i=1}^{q} s_i v_i + \sum_{i=1}^{q} \epsilon_i v_i, \quad s_i \in \mathbb{N} \text{ and } \epsilon_i \in \{0, 1\},$$

by induction one may assume $\sum_{i=1}^q s_i v_i = 0$. Therefore, from the equality above, one concludes that the subgraph whose edges are defined by the set $\{v_i |\, \epsilon_i = 1\}$ is an edge disjoint union of cycles $C_1, \ldots, C_r$. By induction one may further assume that all the C_i's are odd cycles. Note $r \geq 2$, because $\deg(x^b)$ is even. As G is connected, using Corollary 10.3.12, it follows that x^β is in $K[G]$.

$\Rightarrow$) It follows readily by Corollary 9.1.3 because $K[G]$ is normal. □

Proposition 13.5.3 *Let G be a connected graph with n vertices. If n is even and $K[G]$ is normal of dimension n, then $x_1 \cdots x_n$ is in $K[G]$ if and only if $|\mathfrak{A}| \leq |N(\mathfrak{A})|$ for every independent set of vertices $\mathfrak{A}$ of G.*

Proof. $\Rightarrow$) Let $\mathcal{A} = \{v_1, \ldots, v_q\}$ be the set of column vector of the incidence matrix of G. Since $K[G]$ is normal: $\mathbb{R}_+\mathcal{A} \cap \mathbb{Z}\mathcal{A} = \mathbb{N}\mathcal{A}$. Hence the vector $a = (a_1, \ldots, a_n) = \mathbf{1}$ is in $\mathbb{R}_+\mathcal{A}$. Using Corollary 13.5.1 we get that a satisfies the inequalities:

$$|\mathfrak{A}| = \sum_{x_i \in \mathfrak{A}} a_i \leq \sum_{x_i \in N(\mathfrak{A})} a_i = |N(\mathfrak{A})|$$

for every independent set of vertices $\mathfrak{A}$ of G, as required.

$\Leftarrow$) First we use Corollary 13.5.1 to conclude that a is in the edge cone, then apply Proposition 13.5.2 to get $x^a \in K[G]$. $\Box$

Corollary 13.5.4 (Marriage theorem) *Let G be a graph with an even number of vertices. If G is connected and satisfies the odd cycle condition, then the following are equivalent*

(a) *G has a perfect matching.*

(b) $|\mathfrak{A}| \leq |N(\mathfrak{A})|$ *for all $\mathfrak{A}$ independent set of vertices of G.*

Proof. The proof follows using Theorem 7.1.9 and Proposition 13.5.3. $\Box$

Exercises

13.5.5 Let G be a connected non-bipartite graph with n vertices and let $v_1, \ldots, v_q$ be the columns of its incidence matrix. If n is even, prove that the vector $(1, \ldots, 1)$ is in $\mathcal{L} = \mathbb{Z}v_1 + \cdots + \mathbb{Z}v_q$.

13.5.6 Let G be a connected non-bipartite graph with n vertices. If n is even, prove that the monomial $x_1 \cdots x_n$ is in the field of fractions of $K[G]$.

13.5.7 Let G be a connected non-bipartite graph with n vertices. If $K[G]$ is a normal domain, then $x_1 \cdots x_n$ is in $K[G]$ if and only if $(1, \ldots, 1)$ is in the edge cone of G and n is even.

13.5.8 Prove that the following graph does not have a perfect matching by exhibiting an independent set $\mathfrak{A}$ such that $|\mathfrak{A}| \not\leq |N(\mathfrak{A})|$.

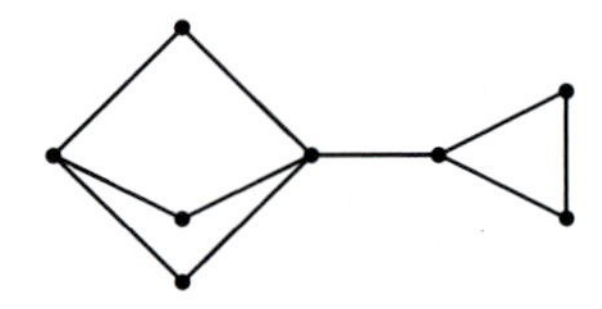

13.5.9 Consider the graph $G = G_1 \cup G_2$ with six vertices $x_1 \cdots x_6$ consisting of two disjoint triangles:

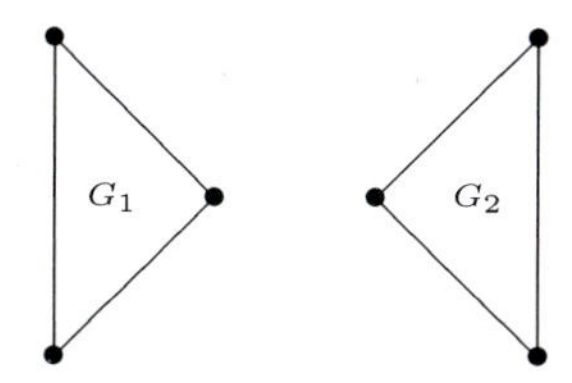

Prove that the vector $\mathbf{1} = (1, \ldots, 1)$ belongs to the edge cone of G and that $x^{\mathbf{1}} = x_1 \cdots x_6$ is not in $K[G]$.

13.6 Rees cones and perfect graphs

Let G be a graph and let $I_c(G)$ be its ideal of covers. In this section we characterize when G is perfect in terms of the irreducible representation of the Rees cone of $I_c(G)$ and show an application to Ehrhart rings.

Let G be a graph with vertex set $X = \{x_1, \ldots, x_n\}$. In what follows we shall always assume that G has no isolated vertices. We denote a complete subgraph of G with r vertices by $\mathcal{K}_r$. The empty set is regarded as an independent set whose characteristic vector is the zero vector.

Theorem 13.6.1 [86, 296] *The following statements are equivalent:*

(a) *G is a perfect graph.*

(b) *The complement of G is perfect.*

(c) *The independence polytope of G, i.e., the convex hull of the incidence vectors of the independent sets of G, is given by:*

$$\left\{(a_i) \in \mathbb{R}^n_+ \mid \textstyle\sum_{x_i \in \mathcal{K}_r} a_i \leq 1;\ \forall \mathcal{K}_r \subset G\right\}.$$

The equivalence between (a) and (b) is due to Lovász [296] and this is called the *weak perfect graph theorem*. That (b) and (c) are equivalent is due to Fulkerson and Chvátal [86].

Lemma 13.6.2 *Let G be a graph. Then the set*

$$F = \{(a_i) \in \mathbb{R}^{n+1} \mid \textstyle\sum_{x_i \in V(\mathcal{K}_r)} a_i = (r-1)a_{n+1}\} \cap \mathbb{R}_+(I_c(G))$$

is a facet of $\mathbb{R}_+(I_c(G))$, where $\mathcal{K}_r$ is a complete subgraph of G.

Proof. If $\mathcal{K}_r = \emptyset$, then $r = 0$ and $F = H_{e_{n+1}} \cap \mathbb{R}_+(I_c(G))$, which is a facet because $e_1, \ldots, e_n \in F$. If $r = 1$, then $F = H_{e_i} \cap \mathbb{R}_+(I_c(G))$ for some $1 \leq i \leq n$, which is a facet because $e_j \in F$ for $j \notin \{i, n+1\}$ and there is at least one minimal vertex cover of G not containing x_i. We may assume that $V(\mathcal{K}_r)=\{x_1,\ldots,x_r\}$ and $r\geq 2$. For each $1\leq i\leq r$ there is a minimal vertex cover C_i of G not containing x_i. Notice that C_i contains $V(\mathcal{K}_r)\setminus\{x_i\}$. Let u_i be the characteristic vector of C_i. Since $\operatorname{rank}(u_1,\ldots,u_r)$ is r, the set $\{(u_1,1),\ldots,(u_r,1),e_{r+1},\ldots,e_n\}$ is linearly independent and contained in F, i.e., $\dim(F)=n$. Hence F is a facet of $\mathbb{R}_+(I_c(G))$ because the hyperplane that defines F is a supporting hyperplane. $\Box$

We regard $\mathcal{K}_0$ as the empty set with zero elements. A sum over an empty set is defined to be 0.

Proposition 13.6.3 *Let $J=I_c(G)$ be the ideal of covers of G. Then G is perfect if and only if the following equality holds*

$$\mathbb{R}_+(J)=\left\{(a_i)\in\mathbb{R}^{n+1}\middle|\ \textstyle\sum_{x_i\in\mathcal{K}_r}a_i\geq(r-1)a_{n+1};\ \forall\,\mathcal{K}_r\subset G\right\}. \quad (13.10)$$

Moreover this is the irreducible representation of $\mathbb{R}_+(J)$ if G is perfect.

Proof. $\Rightarrow$) The inclusion "$\subset$" is clear because any minimal vertex cover of G contains at least $r-1$ vertices of any $\mathcal{K}_r$. To show the reverse inclusion take a vector $a=(a_i)$ satisfying $b=a_{n+1}\neq 0$ and

$$\textstyle\sum_{x_i\in\mathcal{K}_r}a_i\geq(r-1)b;\ \forall\,\mathcal{K}_r\subset G\ \Rightarrow\ \sum_{x_i\in\mathcal{K}_r}(a_i/b)\geq r-1;\ \forall\,\mathcal{K}_r\subset G.$$

This implication follows because by making $r=0$ we get $b>0$. We may assume that $a_i\leq b$ for all i. Indeed if $a_i>b$ for some i, say $i=1$, then we can write $a=e_1+(a-e_1)$. From the inequality

$$\sum_{\substack{x_i\in\mathcal{K}_r\\ x_1\in\mathcal{K}_r}}a_i=a_1+\sum_{x_i\in\mathcal{K}_{r-1}}a_i\geq a_1+(r-2)b\geq 1+(r-1)b$$

it is seen that $a-e_1$ belongs to the right-hand side of Eq. (13.10). Thus, if necessary, we may apply this observation again to $a-e_1$ and so on till we get that $a_i\leq b$ for all i. Hence, by Theorem 13.6.1(c), the vector $\gamma=\mathbf{1}-(a_1/b,\ldots,a_n/b)$ belongs to the independence polytope of G. Thus we can write

$$\gamma=\lambda_1w_1+\cdots+\lambda_sw_s;\quad(\lambda_i\geq 0;\ \textstyle\sum_i\lambda_i=1),$$

where $w_1,\ldots,w_s$ are characteristic vectors of independent sets of G. Hence

$$\gamma=\lambda_1(\mathbf{1}-u_1')+\cdots+\lambda_s(\mathbf{1}-u_s'),$$

where $u_1',\ldots,u_s'$ are characteristic vectors of vertex covers of G. For each i we can write $u_i'=u_i+\epsilon_i$, where u_i is the characteristic vector of a minimal vertex cover of G and $\epsilon_i\in\{0,1\}^n$. Therefore

$$\begin{aligned}&\mathbf{1}-\gamma=\lambda_1u_1'+\cdots+\lambda_su_s'\Longrightarrow\\ &\qquad a=b\lambda_1(u_1,1)+\cdots+b\lambda_s(u_s,1)+b\lambda_1\epsilon_1+\cdots+b\lambda_s\epsilon_s.\end{aligned}$$

Thus $a \in \mathbb{R}_+(J)$. If $b = 0$, clearly $a \in \mathbb{R}_+(J)$. Hence we get equality in Eq. (13.10), as required. The converse follows using similar arguments. To finish the proof notice that, by Lemma 13.6.2, the decomposition of Eq. (13.10) is irreducible. $\Box$

Remark 13.6.4 *Normaliz* [68] determines the irreducible representation of a Rees cone. Thus Proposition 13.6.3 can be used to check whether a given graph is perfect.

Definition 13.6.5 The *clique clutter* of a graph G, denoted by $\mathrm{cl}(G)$, is the clutter on $V(G)$ whose edges are the maximal cliques of G (maximal with respect to inclusion).

Definition 13.6.6 The incidence matrix A of a clutter is called *perfect* if the polytope defined by the system $x \geq 0$; $xA \leq \mathbf{1}$ is integral.

Theorem 13.6.7 ([24], [373, Corollary 83.1a(vii)]) *Let $\mathcal{C}$ be a clutter and let A be its incidence matrix. Then $\mathcal{C}$ is balanced if and only if every submatrix of A is perfect.*

The *vertex-clique matrix* of a graph G is the incidence matrix of $\mathrm{cl}(G)$, the clique clutter of G.

Theorem 13.6.8 ([296], [86]) *Let A be the incidence matrix of a clutter. Then the following are equivalent:*

(a) *The system $x \geq 0$; $xA \leq \mathbf{1}$ is* TDI.

(b) *A is perfect.*

(c) *A is the vertex-clique matrix of a perfect graph.*

Let $v_1, \ldots, v_q$ be a set of points in $\mathbb{N}^n$ and let $\mathcal{P} = \mathrm{conv}(v_1, \ldots, v_q)$. The *Ehrhart ring* of the lattice polytope $\mathcal{P}$ is the K-subring of $R[t]$ given by

$$A(\mathcal{P}) = K[\{x^a t^b \,|\, a \in b\mathcal{P} \cap \mathbb{Z}^n\}].$$

Corollary 13.6.9 *Let A be a perfect matrix with column vectors $v_1, \ldots, v_q$. If there is $x_0 \in \mathbb{R}^n$ such that all the entries of x_0 are positive and $\langle v_i, x_0 \rangle = 1$ for all i, then $A(\mathcal{P}) = K[x^{v_1}t, \ldots, x^{v_q}t]$.*

Proof. The inclusion "$\supset$" is clear. To show the other inclusion take $x^a t^b$ in $A(\mathcal{P})$. Then we can write $(a, b) = \sum_{i=1}^{q} \lambda_i (v_i, 1)$, where $\lambda_i \geq 0$ for all i. Hence $\langle a, x_0 \rangle = b$. By Theorem 13.6.8 the system $x \geq 0$; $xA \leq \mathbf{1}$ is TDI. Therefore, applying Proposition 1.3.28, we have:

$$(a, b) = \eta_1(v_1, 1) + \cdots + \eta_q(v_q, 1) - \delta_1 e_1 - \cdots - \delta_n e_n \quad (\eta_i \in \mathbb{N};\ \delta_i \in \mathbb{N}).$$

Then $b = \langle a, x_0 \rangle = b - \delta_1 \langle x_0, e_1 \rangle - \cdots - \delta_n \langle x_0, e_n \rangle$. Using that $\langle x_0, e_i \rangle > 0$ for all i, we conclude that $\delta_i = 0$ for all i, i.e., $x^a t^b \in K[x^{v_1}t, \ldots, x^{v_q}t]$. $\Box$

Exercises

13.6.10 Let A be an $n \times q$ integer matrix with column vectors $v_1, \ldots, v_q$. If the polyhedron $\{x | x \geq 0; xA \leq \mathbf{1}\}$ is integral and $\mathbb{R}_+\mathcal{B} \cap \mathbb{Z}^{n+1} = \mathbb{N}\mathcal{B}$, where $\mathcal{B} = \{(v_i, 1)\}_{i=1}^q \cup \{-e_i\}_{i=1}^n$, then the system $x \geq 0;\ xA \leq \mathbf{1}$ is TDI.

13.6.11 Let G be a graph and let $\overline{G}$ be its complement. Then:

(a) $I_c(\overline{G}) = (\{x^a | X \setminus \mathrm{supp}(x^a)$ is a maximal clique of $G\})$.

(b) If G is perfect, then $\mathbb{R}_+(I_c(\overline{G}))$ is equal to

$$\left\{(a_i) \in \mathbb{R}^{n+1} |\ \textstyle\sum_{x_i \in S} a_i \geq (|S| - 1)a_{n+1};\ \forall\, S \text{ independent set of } G\right\}.$$

13.6.12 Let G be a perfect graph with vertex set X and let β_0 be its vertex independence number. Then there is a partition $X_1, \ldots, X_{\beta_0}$ of X such that X_i is a clique of G for all i. If G is unmixed, then $\mathrm{cl}(\overline{G})^\vee$ has a perfect matching, where $\overline{G}$ is the complement of G.

13.7 Perfect graphs and algebras of covers

Let G be a graph with vertex set $X = \{x_1, \ldots, x_n\}$ and let $I = I(G)$ be its edge ideal. The main purpose of this section is to study the symbolic Rees algebra of I and the Simis cone of I when G is a perfect graph, i.e., we study the algebra of vertex covers of $G^\vee$. We show that the cliques of a perfect graph G completely determine both the Hilbert basis of the Simis cone and the symbolic Rees algebra of $I(G)$.

The Simis cone of I is denoted by $\mathrm{Cn}(I)$ (see Definition 13.2.1). Let $C_1, \ldots, C_s$ be the minimal vertex covers of G. For $1 \leq k \leq s$ let u_k be the characteristic vector of C_k, i.e., $u_k = \sum_{x_i \in C_k} e_i$.

If $\mathcal{H}$ is the minimal integral Hilbert basis of $\mathrm{Cn}(I)$, then $R_s(I(G))$ is equal to $K[\mathbb{N}\mathcal{H}]$, the semigroup ring of $\mathbb{N}\mathcal{H}$ (see Corollary 13.2.19). Next we describe $\mathcal{H}$ when G is perfect.

Theorem 13.7.1 *Let $\omega_1, \ldots, \omega_p$ be the characteristic vectors of the non-empty cliques of a perfect graph G. If*

$$\mathcal{H} = \{(\omega_1, |\omega_1| - 1), \ldots, (\omega_p, |\omega_p| - 1)\}.$$

Then $\mathbb{N}\mathcal{H} = \mathrm{Cn}(I) \cap \mathbb{Z}^{n+1}$, that is, $\mathcal{H}$ is the integral Hilbert basis of $\mathrm{Cn}(I)$.

Proof. The inclusion $\mathbb{N}\mathcal{H} \subset \mathrm{Cn}(I) \cap \mathbb{Z}^{n+1}$ is clear because any clique of size r intersects any minimal vertex cover in at least $r - 1$ vertices. Let us show the reverse inclusion. Let (a, b) be a minimal generator of $\mathrm{Cn}(I) \cap \mathbb{Z}^{n+1}$, where $0 \neq a = (a_i) \in \mathbb{N}^n$ and $b \in \mathbb{N}$. Then

$$\textstyle\sum_{x_i \in C_k} a_i = \langle a, u_k \rangle \geq b, \tag{13.11}$$

for all k. If $b = 0$ or $b = 1$, then $(a, b) = e_i$ for some $i \leq n$ or $(a, b) = (e_i + e_j, 1)$ for some edge $\{x_i, x_j\}$, respectively. In both cases $(a, b) \in \mathcal{H}$. Thus we may assume that $b \geq 2$ and $a_j \geq 1$ for some j. Using Eq. (13.11) we obtain

$$\sum_{x_i \in C_k} a_i + \sum_{x_i \in X \setminus C_k} a_i = |a| \geq b + \sum_{x_i \in X \setminus C_k} a_i = b + \langle \mathbf{1} - u_k, a \rangle, \tag{13.12}$$

for all k. Set $c = |a| - b$. Notice that $c \geq 1$ because $a \neq 0$. Indeed if $c = 0$, from Eq. (13.12) we get $\sum_{x_i \in X \setminus C_k} a_i = 0$ for all k, i.e., $a = 0$, a contradiction. Consider the vertex-clique matrix of G':

$$A' = (\mathbf{1} - u_1 \cdots \mathbf{1} - u_s),$$

where $\mathbf{1} - u_1, \ldots, \mathbf{1} - u_s$ are regarded as column vectors. From Eq. (13.12) we get $(a/c)A' \leq \mathbf{1}$. Hence by Theorem 13.6.1(c) we obtain that a/c belongs to $\mathrm{conv}(\omega_0, \omega_1, \ldots, \omega_p)$, where $\omega_0 = 0$, i.e., we can write $a/c = \sum_{i=0}^{p} \lambda_i \omega_i$, where $\lambda_i \geq 0$ for all i and $\sum_i \lambda_i = 1$. Thus we can write

$$(a, c) = c\lambda_0(\omega_0, 1) + \cdots + c\lambda_p(\omega_p, 1).$$

Using Theorem 13.6.8(a) it follows that the subring $K[\{x^{\omega_i} t | 0 \leq i \leq p\}]$ is normal. Hence there are $\eta_0, \ldots, \eta_p$ in $\mathbb{N}$ such that

$$(a, c) = \eta_0(\omega_0, 1) + \cdots + \eta_p(\omega_p, 1).$$

Thus $|a| = \eta_0|\omega_0| + \cdots + \eta_p|\omega_p|$ and $c = \eta_0 + \cdots + \eta_p = |a| - b$, consequently:

$$(a, b) = \eta_0(\omega_0, |\omega_0| - 1) + \eta_1(\omega_1, |\omega_1| - 1) + \cdots + \eta_p(\omega_p, |\omega_p| - 1).$$

Notice that there is u_ℓ such that $\langle a, u_\ell \rangle = b$; otherwise since $a_j \geq 1$, by Eq. (13.11) the vector $(a, b) - e_j$ would be in $\mathrm{Cn}(I) \cap \mathbb{Z}^{n+1}$, contradicting the minimality of (a, b). Therefore from the equality

$$0 = \langle (a, b), (u_\ell, -1) \rangle = \eta_0 + \textstyle\sum_{i=1}^{p} \eta_i \langle (\omega_i, |\omega_i| - 1), (u_\ell, -1) \rangle$$

we conclude that $\eta_0 = 0$, i.e., $(a, b) \in \mathbb{N}\mathcal{H}$, as required. $\square$

Corollary 13.7.2 *If G is a perfect graph, then*

$$R_s(I(G)) = K[x^a t^r \,|\, x^a \text{ is square-free } ; \langle \mathrm{supp}(x^a) \rangle = \mathcal{K}_{r+1};\ 0 \leq r < n].$$

Proof. By Theorem 13.2.3, we have the equality $R_s(I(G)) = K[\mathbb{N}\mathcal{H}]$, thus the formula follows from Theorem 13.7.1. $\square$

Lemma 13.7.3 *Let G be a graph and let $0 \neq a = (a_i) \in \mathbb{N}^n$. If $a_i \in \{0, 1\}$ for all i and $G[\{x_i | a_i > 0\}] = \mathcal{K}_{b+1}$, then a is an irreducible b-cover of $G^\vee$.*

Proof. By Lemma 13.6.2, the closed halfspace $H^+_{(a,-b)}$ must occur in the irreducible representation of $\mathbb{R}_+(I_c(G))$. Hence a is an irreducible b-cover of $G^\vee$ by Lemma 13.2.27. □

Corollary 13.7.4 [423] *If G is a graph, then*

$$K[x^a t^r \,|\, x^a \textit{ square-free };\ \langle \mathrm{supp}(x^a)\rangle = \mathcal{K}_{r+1};\ 0 \le r < n] \subset R_s(I(G))$$

with equality if and only if G is a perfect graph.

Proof. The inclusion follows from Lemma 13.7.3. If G is a perfect graph, then by Corollary 13.7.2 the equality holds. Conversely if the equality holds, then by Lemmas 13.2.27 and 13.6.2 we have

$$\mathbb{R}_+(I_c(G)) = \left\{(a_i) \in \mathbb{R}^{n+1} \middle|\ \textstyle\sum_{x_i\in\mathcal{K}_r} a_i \ge (r-1)a_{n+1};\ \forall\,\mathcal{K}_r \subset G\right\}.$$

Hence an application of Proposition 13.6.3 gives that G is perfect. □

Exercises

13.7.5 Let G be a graph with clique number $\omega(G)$ and chromatic number $\chi(G)$. Notice that the chromatic number is given by

$$\chi(G) = \min\{k \,|\, \exists\, X_1,\ldots,X_k \text{ independent sets of } G \text{ whose union is } X\}.$$

Prove that $\omega(G) \le \chi(G)$.

13.7.6 If G is a perfect graph and $\mathcal{C} = G^\vee$, show that the image of ψ given in Remark 13.2.28 generates $R_s(I_c(G^\vee))$.

Hint Use that the irreducible b-covers of $G^\vee$ correspond to cliques of the graph G (see Corollary 13.7.4). Notice that $I_c(G^\vee)$ is equal to $I(G)$.

Chapter 14

Combinatorial Optimization and Blowup Algebras

In this chapter we relate commutative algebra and blowup algebras with combinatorial optimization to gain insight on these research areas. A main goal here is to connect algebraic properties of blowup algebras associated to edge ideals with combinatorial and optimization properties of clutters and polyhedra. A conjecture of Conforti and Cornuéjols about packing problems is examined from an algebraic point of view. We study max-flow min-cut problems of clutters, packing problems, and integer rounding properties of systems of linear inequalities—and their underlying polyhedra—to analyze algebraic properties of blowup algebras and edge ideals (e.g., normality, normally torsion freeness, Cohen–Macaulayness, unmixedness). Systems with integer rounding properties and clutters with the max-flow min-cut property come from linear optimization problems [372, 373].

The study of algebraic and combinatorial properties of edge ideal of clutters and hypergraphs is of current interest; see [119, 155, 206, 285, 326] and the references therein. A comprehensive reference for combinatorial optimization and hypergraph theory is the 3-volume book of Schrijver [373]. For a thorough study of clutters—that includes 18 conjectures in the area—from the point of view of combinatorial optimization, see [93].

In this chapter we make use of polyhedral geometry and combinatorial optimization to study blowup algebras and vice versa. We refer the reader to Chapter 1 for the undefined terminology and notation regarding these areas. As a handy reference in Section 14.1 we introduce and fix some of the notation and definitions that will be used throughout this chapter.

14.1 Blowup algebras of edge ideals

In this section we introduce some of the notation and definitions that will be used throughout this chapter. In particular the irreducible representation of a Rees cone is introduced here.

Let $\mathcal{C}$ be a *clutter* with vertex set $X = \{x_1, \ldots, x_n\}$ and let $R = K[X]$ be a polynomial ring over a field K. The set of vertices and edges of $\mathcal{C}$ are denoted by $V(\mathcal{C})$ and $E(\mathcal{C})$, respectively. Let $f_1, \ldots, f_q$ be the edges of $\mathcal{C}$ and let $v_k = \sum_{x_i \in f_k} e_i$ be the *characteristic vector* of f_k for $1 \leq k \leq q$. Recall that the *edge ideal* of $\mathcal{C}$, denoted by $I = I(\mathcal{C})$, is the ideal of R generated by all monomials $x_e = \prod_{x_i \in e} x_i$ such that e is an edge of $\mathcal{C}$. Thus, I is minimally generated by $F = \{x^{v_1}, \ldots, x^{v_q}\}$. In what follows we shall always assume that the height of I is at least 2.

The *blowup algebras* and *monomial algebras* studied in this chapter are the following: (a) *Rees algebra*

$$R[It] := R \oplus It \oplus \cdots \oplus I^i t^i \oplus \cdots \subset R[t],$$

where t is a new variable, (b) *extended Rees algebra*

$$R[It, t^{-1}] := R[It][t^{-1}] \subset R[t, t^{-1}],$$

(c) *symbolic Rees algebra*

$$R_s(I) := R + I^{(1)}t + I^{(2)}t^2 + \cdots + I^{(i)}t^i + \cdots \subset R[t],$$

where $I^{(i)}$ is the ith symbolic power of I, (d) *associated graded ring*

$$\mathrm{gr}_I(R) := R/I \oplus I/I^2 \oplus \cdots \oplus I^i/I^{i+1} \oplus \cdots \simeq R[It] \otimes_R (R/I),$$

with multiplication

$$(a + I^{i+1})(b + I^{j+1}) = ab + I^{i+j+1} \qquad (a \in I^i,\ b \in I^j),$$

(e) *monomial subring*

$$K[F] = K[x^{v_1}, \ldots, x^{v_q}] \subset R$$

spanned by $F = \{x^{v_1}, \ldots, x^{v_q}\}$, (f) *homogeneous monomial subring*

$$K[Ft] = K[x^{v_1}t, \ldots, x^{v_q}t] \subset R[t]$$

spanned by Ft, (g) *homogeneous monomial subring*

$$K[Ft \cup \{t\}] = K[x^{v_1}t, \ldots, x^{v_q}t, t] \subset R[t]$$

spanned by $Ft \cup \{t\}$, (h) *homogeneous monomial subring*

$$S = K[x^{w_1}t, \ldots, x^{w_r}t] \subset R[t],$$

where $\{w_1,\ldots,w_r\}$ is the set of all $\alpha \in \mathbb{N}^n$ such that $0 \leq \alpha \leq v_i$ and we allow x^{v_i} to be an arbitrary monomial, (i) *Ehrhart ring*

$$A(\mathcal{P}) = K[\{x^a t^i \,|\, a \in \mathbb{Z}^n \cap i\mathcal{P}; i \in \mathbb{N}\}] \subset R[t]$$

of the lattice polytope $\mathcal{P} = \mathrm{conv}(v_1,\ldots,v_q)$, and (j) *Stanley–Reisner ring*

$$K[\Delta_{\mathcal{C}}] = R/I(\mathcal{C}).$$

of the independence complex $\Delta_{\mathcal{C}}$. This ring is also called the *edge ring* of $\mathcal{C}$.

The *incidence matrix* of $\mathcal{C}$ will be denoted by A; it is the $n \times q$ matrix with column vectors $v_1,\ldots,v_q$. We shall always assume that the rows and columns of A are different from zero. Consider the set

$$\mathcal{A}' = \{e_1,\ldots,e_n,(v_1,1),\ldots,(v_q,1)\} \subset \mathbb{N}^{n+1},$$

where e_i is the ith unit vector. The *Rees cone* of I, will be denoted by $\mathbb{R}_+(I)$, it is the rational polyhedral cone generated by $\mathcal{A}'$. The closed halfspace $H_{e_i}^+$ occurs in the irreducible representation of $\mathbb{R}_+(I)$ for $i = 1,\ldots,n+1$ (see Exercise 14.2.34). Thus, by Theorem 1.4.2, there is a unique irreducible representation

$$\mathbb{R}_+(I) = H_{e_1}^+ \cap \cdots \cap H_{e_{n+1}}^+ \cap H_{\alpha_1}^+ \cap \cdots \cap H_{\alpha_r}^+$$

such that $0 \neq \alpha_i \in \mathbb{Q}^{n+1}$, $\langle \alpha_i, e_{n+1}\rangle = -1$ for all i, and none of the closed halfspaces can be omitted from the intersection.

Let $\mathfrak{p}_1,\ldots,\mathfrak{p}_s$ be the minimal primes of the edge ideal $I = I(\mathcal{C})$ and let

$$C_k = \{x_i \,|\, x_i \in \mathfrak{p}_k\} \quad (k = 1,\ldots,s)$$

be the corresponding minimal vertex covers of $\mathcal{C}$ (see Proposition 13.1.2). The set of column vectors of A will be denoted by $\mathcal{A} = \{v_1,\ldots,v_q\}$ and $\mathcal{Q}(A)$ will denote the *set covering polyhedron*

$$\mathcal{Q}(A) := \{x \,|\, x \geq 0; xA \geq \mathbf{1}\}.$$

Notation Let d_k be the unique positive integer such that $d_k\alpha_k$ has relatively prime integral entries. We set $\ell_k = d_k\alpha_k$ for $k = 1,\ldots,r$.

Theorem 14.1.1 *The irreducible representation of the Rees cone is:*

$$\mathbb{R}_+(I) = H_{e_1}^+ \cap \cdots \cap H_{e_{n+1}}^+ \cap H_{\ell_1}^+ \cap \cdots \cap H_{\ell_r}^+ \tag{14.1}$$

where $\ell_k = -e_{n+1} + \sum_{x_i \in C_k} e_i$ for $1 \leq k \leq s$. Moreover all the vertices of $\mathcal{Q}(A)$ are integral if and only if $r = s$.

Proof. By Proposition 13.1.2 $u_k = \sum_{x_i \in C_k} e_i$ is a vertex of $\mathcal{Q}(A)$. Since $\ell_k + e_{n+1} = (u_k, 0)$ for $k = 1, \ldots, s$, by Theorem 1.4.3, the result follows. □

Notation In the sequel we shall always assume that $u_1, \ldots, u_s$ are the vectors given by $u_k = \sum_{x_i \in C_k} e_i$, $\ell_1, \ldots, \ell_s$ are the vectors defined as

$$\ell_k = -e_{n+1} + \sum_{x_i \in C_k} e_i,$$

and that $\ell_1, \ldots, \ell_r$ are the vectors of Eq. (14.1). Notice that $d_k = 1$ for $k = 1, \ldots, s$ and $d_k = -\langle \ell_k, e_{n+1} \rangle$ for $1 \leq k \leq r$.

The *Simis cone* of I is the polyhedral cone:

$$\mathrm{Cn}(I) = H_{e_1}^+ \cap \cdots \cap H_{e_{n+1}}^+ \cap H_{\ell_1}^+ \cap \cdots \cap H_{\ell_s}^+,$$

see Definition 13.2.1. The Simis and Rees cones are related to the normality and torsion freeness of I (Theorem 14.2.2 and Proposition 14.2.3).

14.2 Rees algebras and polyhedral geometry

In this section we describe when the integral closure of a Rees algebra of an edge ideal is equal to the symbolic Rees algebra in combinatorial and algebraic terms. If the Rees algebra is normal, we describe the primary decomposition of the zero ideal of the associated graded ring. We show that certain properties of clutters and edge ideals are closed under taking minors and Alexander duals.

Lemma 14.2.1 *If $1 \leq j \leq r$ and $\langle \ell_j, e_{n+1} \rangle = -1$, then $1 \leq j \leq s$. In particular $r = s$ if and only if $\langle \ell_j, e_{n+1} \rangle = -1$ for $i = 1, \ldots, r$*

Proof. Let $\ell_j = (a_1, \ldots, a_n, -1)$. Since $\mathbb{R}_+(I) \subset H_{\ell_j}^+$ we get $a_i \geq 0$ for $i = 1, \ldots, n$. Consider the ideal $\mathfrak{p}$ of R generated by the set of x_i such that $a_i > 0$. Note that $I \subset \mathfrak{p}$ because $\langle \ell_j, (v_i, 1) \rangle \geq 0$ for all $1 \leq i \leq n$. For simplicity of notation assume that $\ell_j = (a_1, \ldots, a_m, 0, \ldots, 0, -1)$, where $a_i > 0$ for all i. Take an arbitrary vector $v \in \mathcal{A}'$ such that $v \in H_{\ell_j}$, then v satisfies the equation

$$a_1 y_1 + \cdots + a_m y_m = y_{n+1}.$$

Observe that v also satisfies the equation $y_1 + \cdots + y_m = y_{n+1}$. Using that H_{ℓ_j} contains n linearly independent vectors from $\mathcal{A}'$ we conclude that $H_{\ell_j} = H_b$, where $b = e_1 + \cdots + e_m - e_{n+1}$. Hence $\ell_j = b$ and consequently $a_i = 1$ for all i. It remains to show that $\mathfrak{p}$ is a minimal prime of I, and this follows from the irreducibility of Eq. (14.1). □

Theorem 14.2.2 $\overline{R[It]} = R_s(I) \iff \langle \ell_i, e_{n+1}\rangle = -1$ *for* $i = 1, \ldots, r$.

Proof. The equality $\overline{R[It]} = R_s(I)$ holds if and only if $\mathbb{Z}^{n+1} \cap \mathbb{R}_+(I)$ is equal to $\mathbb{Z}^{n+1} \cap \mathrm{Cn}(I)$ if and only if the Rees and Simis cones are equal. By the irreducibility of the representations of both cones it follows that $\overline{R[It]} = R_s(I)$ if and only if $r = s$. Now, to finish the proof recall that by Lemma 14.2.1, $r = s$ if and only if $\langle \ell_i, e_{n+1}\rangle = -1$ for $i = 1, \ldots, r$. $\square$

Proposition 14.2.3 *If I is a monomial ideal, then $R[It]$ is normal if and only if any of the following two equivalent conditions hold:*

(a) *$\mathcal{A}'$ is a Hilbert basis, i.e., $\mathbb{N}\mathcal{A}' = \mathbb{Z}^{n+1} \cap \mathbb{R}_+(I)$.*

(b) *I is normal, i.e., $I^i = \overline{I^i}$ for all $i \geq 1$.*

Proof. The Rees algebra of I can be written as

$$\begin{aligned} R[It] &= K[\{x_1, \ldots, x_n, x^{v_1}t, \ldots, x^{v_q}t\}] \\ &= K[\{x^a t^b \mid (a, b) \in \mathbb{N}\mathcal{A}'\}] \\ &= R \oplus It \oplus I^2t^2 \oplus \cdots \oplus I^i t^i \oplus \cdots . \end{aligned}$$

By Theorem 9.1.1 and Exercise 4.3.49, the integral closure of $R[It]$ in its field of fractions can be expressed as

$$\begin{aligned} \overline{R[It]} &= K[\{x^a t^b \mid (a, b) \in \mathbb{Z}^{n+1} \cap \mathbb{R}_+(I)\}] \\ &= R \oplus \overline{I}t \oplus \overline{I^2}t^2 \oplus \cdots \oplus \overline{I^i}t^i \oplus \cdots , \end{aligned}$$

where $\overline{I^i}$ is the integral closure of I^i. Thus, $R[It]$ is normal if and only if $\mathcal{A}'$ is a Hilbert basis if and only if $I^i = \overline{I^i}$ for all $i \geq 1$. $\square$

Let $\mathfrak{p}$ be a prime ideal of R. In what follows we denote the localization of $R[It]$ at the multiplicative set $R \setminus \mathfrak{p}$ by $R[It]_{\mathfrak{p}}$

Lemma 14.2.4 *If $\mathfrak{p} \in \mathrm{Spec}(R)$, then $\mathfrak{p}R[It]_{\mathfrak{p}} \cap R = \mathfrak{p}$.*

Proof. It is left as an exercise. $\square$

Lemma 14.2.5 *If $B = R[It]$ and $\mathfrak{p}$ is a minimal prime of I, then $\mathfrak{p}B_{\mathfrak{p}} \cap B$ is a minimal prime of IB.*

Proof. From the equality $\mathfrak{p}B_{\mathfrak{p}} = \mathfrak{p}(R_{\mathfrak{p}}[\mathfrak{p}R_{\mathfrak{p}}t]) = \mathfrak{p}R_{\mathfrak{p}} + \mathfrak{p}^2R_{\mathfrak{p}}t + \cdots$ we get that $\mathfrak{p}B_{\mathfrak{p}}$ is a prime ideal. Hence its contraction $\mathfrak{p}B_{\mathfrak{p}} \cap B$ is also a prime ideal. To prove the minimality take $P \in \mathrm{Spec}(B)$ such that $IB \subset P \subset \mathfrak{p}B_{\mathfrak{p}} \cap B$. Contracting to R and using Lemma 14.2.4 we get $P \cap R = \mathfrak{p}$. From this equality it follows that $PB_{\mathfrak{p}} \cap B = P$. Therefore

$$\mathfrak{p}B_{\mathfrak{p}} \cap B \subset PB_{\mathfrak{p}} \cap B = P \implies P = \mathfrak{p}B_{\mathfrak{p}} \cap B. \qquad \square$$

Notation In what follows $J_k^{(d_k)}$ will denote the ideal of $R[It]$ given by

$$J_k^{(d_k)} = (\{x^a t^b \mid \langle (a,b), \ell_k\rangle \geq d_k\}) \cap R[It] \quad (k = 1, \ldots, r)$$

and J_k will denote the ideal of $R[It]$ given by

$$J_k = (\{x^a t^b \mid \langle (a,b), \ell_k\rangle > 0\}) \cap R[It] \quad (k = 1, \ldots, r),$$

where $d_k = -\langle \ell_k, e_{n+1}\rangle$. In general the ideal $J_k^{(d_k)}$ might not be equal to the d_k-th symbolic power of J_k; see [415]. If $d_k = 1$ we have $J_k^{(1)} = J_k$. By Lemma 14.2.1 we have that $d_k = 1$ if and only if $1 \leq k \leq s$.

Proposition 14.2.6 *$J_1, \ldots, J_r$ are distinct height one prime ideals of $R[It]$ that contain $IR[It]$ and J_k is equal to $\mathfrak{p}_k R[It]_{\mathfrak{p}_k} \cap R[It]$ for $k = 1, \ldots, s$. If $\mathcal{Q}(A)$ is integral, then*

$$\mathrm{rad}(IR[It]) = J_1 \cap \cdots \cap J_s.$$

Proof. $IR[It]$ is clearly contained in J_k for all k by definition of J_k. To show that J_k is a prime ideal of height one it suffices to notice that the right-hand side of the isomorphism:

$$R[It]/J_k \simeq K[\{x^a t^b \in R[It] \mid \langle (a,b), \ell_k\rangle = 0\}]$$

is an n-dimensional integral domain, because $F_k = \mathbb{R}_+(I) \cap H_{\ell_k}$ is a facet of the Rees cone for all k. The prime ideals $J_1, \ldots, J_r$ are distinct because the dimension of $\mathbb{R}_+(I)$ is $n+1$ and $F_1, \ldots, F_r$ are distinct facets of the Rees cone. Set $P_k = \mathfrak{p}_k R[It]_{\mathfrak{p}_k} \cap R[It]$ for $1 \leq k \leq s$. By Lemma 14.2.5 this ideal is a minimal prime of $IR[It]$ and admits the following description

$$\begin{aligned} P_k &= \mathfrak{p}_k R_{\mathfrak{p}_k}[\mathfrak{p}_k R_{\mathfrak{p}_k} t] \cap R[It] \\ &= \mathfrak{p}_k + (\mathfrak{p}_k^2 \cap I)t + (\mathfrak{p}_k^3 \cap I^2)t^2 + \cdots + (\mathfrak{p}_k^{i+1} \cap I^i)t^i + \cdots \end{aligned}$$

Notice that $x^a \in \mathfrak{p}_k^{b+1}$ if and only if $\langle a, \sum_{x_i \in C_k} e_i\rangle \geq b+1$. Hence $J_k = P_k$.

Assume that $\mathcal{Q}(A)$ is integral, i.e., $r = s$. Take $x^\alpha t^b \in J_k$ for all k. Using Eq. (14.1) it is not hard to see that $(\alpha, b+1) \in \mathbb{R}_+(I)$; that is, $x^\alpha t^{b+1}$ is in $\overline{R[It]}$ and $x^\alpha t^{b+1} \in \overline{I^{b+1}} t^{b+1}$. It follows that $x^\alpha t^b$ is a monomial in the radical of $IR[It]$. This proves the asserted equality. $\square$

The following formulas, pointed out to us by Vasconcelos, show the difference between the symbolic Rees algebra and the normalization of I.

$$R_s(I) = \bigcap_{k=1}^{s} R[It]_{\mathfrak{p}_k} \cap R[t]; \qquad \overline{R[It]} = \bigcap_{k=1}^{r} \overline{R[It]}_{\mathfrak{p}_k} \cap R[t],$$

where $\mathfrak{p}_k = J_k \cap R$ for $k = 1, \ldots, r$. These representations are related to the so-called Rees valuations of I; see [414, Chapter 8].

Theorem 14.2.7 [185, 188] *The following are equivalent:*

(a) $\mathcal{Q}(A)$ *is integral.*

(b) $\mathbb{R}_+(I) = H^+_{e_1} \cap \cdots \cap H^+_{e_{n+1}} \cap H^+_{\ell_1} \cap \cdots \cap H^+_{\ell_s}$, *i.e.,* $r = s$.

(c) $R_s(I) = \overline{R[It]}$.

(d) *The minimal primes of* $IR[It]$ *are of the form* $\mathfrak{p}R[It]_{\mathfrak{p}} \cap R[It]$, *where* $\mathfrak{p}$ *is a minimal prime of* I.

Proof. (a) $\Leftrightarrow$ (b): It follows at once from Theorem 14.1.1. (b) $\Leftrightarrow$ (c): It follows from Theorem 14.2.2 and Lemma 14.2.1. (a) $\Rightarrow$ (d): It follows from Proposition 14.2.6. (d) $\Rightarrow$ (b): From Proposition 14.2.6, $J_1, \ldots, J_r$ are minimal primes of $IR[It]$ and J_k is equal to $\mathfrak{p}_k R[It]_{\mathfrak{p}_k} \cap R[It]$ for $k = 1, \ldots, s$. Thus $r = s$. $\Box$

Corollary 14.2.8 *If* $\mathcal{Q}(A)$ *has only integral vertices and* $\mathcal{C}$ *has* n *vertices, then* $\alpha_0(\mathcal{C}^a) \leq n - 1$ *for all indecomposable parallelizations* $\mathcal{C}^a$ *of* $\mathcal{C}$.

Proof. By Theorem 14.2.7 we have the equality $\overline{R[It]} = R_s(I)$. Take any indecomposable parallelization $\mathcal{C}^a$ of $\mathcal{C}$ and consider the monomial $m = x^a t^b$, where $b = \alpha_0(\mathcal{C}^a)$. By Theorem 13.2.15 m is a minimal generator of $R_s(I)$. Now, according to Proposition 12.5.1, a minimal generator of $\overline{R[It]}$ has degree in t at most $n - 1$, i.e., $b \leq n - 1$. $\Box$

Example 14.2.9 Let $I = I(\mathcal{C}) = (x_1x_2x_5, x_1x_3x_4, x_2x_3x_6, x_4x_5x_6)$ be the edge ideal associated to the clutter:

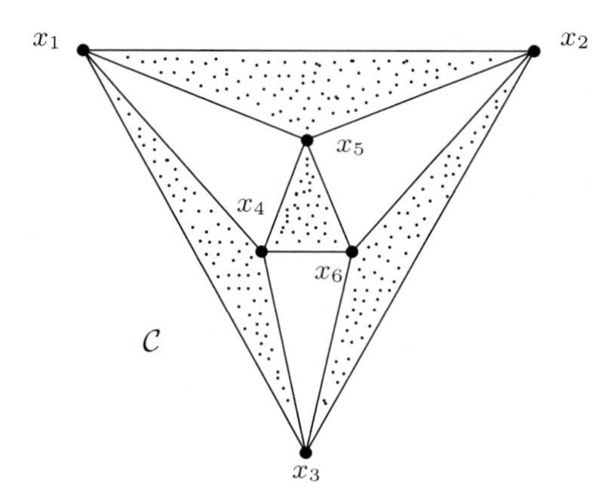

This clutter, usually denoted by $\mathcal{Q}_6$ in the literature, plays an important role in combinatorial optimization [93]. Using *Normaliz* [68], Theorems 1.4.9, 14.2.7, and Remark 13.2.23, we obtain that $\mathcal{Q}(A)$ is integral,

$$R[It] \subsetneq R_s(I) = \overline{R[It]} = R[It][x_1 \cdots x_6 t^2],$$

$R[It]$ is not normal, the minimal primes of I are

$$\begin{array}{llll} \mathfrak{p}_1 = (x_1, x_6), & \mathfrak{p}_2 = (x_2, x_4), & \mathfrak{p}_3 = (x_3, x_5), & \\ \mathfrak{p}_4 = (x_1, x_2, x_5), & \mathfrak{p}_5 = (x_1, x_3, x_4), & \mathfrak{p}_6 = (x_2, x_3, x_6), & \mathfrak{p}_7 = (x_4, x_5, x_6), \end{array}$$

and $\mathcal{C}^\vee$, the blocker of $\mathcal{C}$, has the max-flow min-cut property.

The *dual hypergraph* $\mathcal{H}^\top$ of a hypergraph $\mathcal{H} = (V, E)$ is the hypergraph with vertex set E and edges all sets $\{e \in E | v \in e\}$ for $v \in V$. So the incidence matrix of $\mathcal{H}^\top$ is the transpose of the incidence matrix of $\mathcal{H}$. The clutter $\mathcal{Q}_6$ is the dual hypergraph of $\mathcal{K}_6$, the complete graph on 6 vertices.

Recall that the *analytic spread* of I, denoted by $\ell(I)$, is given by

$$\ell(I) = \dim R[It]/\mathfrak{m}R[It].$$

This number satisfies $\mathrm{ht}(I) \leq \ell(I) \leq \dim R$ [412, Corollary 5.1.4]. The analytic spread of a monomial ideal I can be computed in terms of the Newton polyhedron of I [42].

Definition 14.2.10 The ring $\mathcal{F}(I) = R[It]/\mathfrak{m}R[It]$ is called the *special fiber* of I.

Corollary 14.2.11 *If $\mathcal{Q}(A)$ is integral, then $\ell(I) < n$.*

Proof. By Theorem 14.2.7 we have $r = s$. If $\ell(I) = n$, then the height of $\mathfrak{m}R[It]$ is equal to 1. Hence there is a height one prime ideal P of $R[It]$ such that $IR[It] \subset \mathfrak{m}R[It] \subset P$. By Proposition 14.2.6 the ideal P has the form $\mathfrak{p}_k R[It]_{\mathfrak{p}_k} \cap R[It]$, this readily yields a contradiction. $\square$

This corollary also follows directly from [312, Theorem 3].

Proposition 14.2.12 *If $\mathcal{C}$ is a uniform clutter, then $\ell(I) = \mathrm{rank}(A)$.*

Proof. Since $\deg(x^{v_i}) = d$ for all i. There are isomorphisms

$$R[It]/\mathfrak{m}R[It] \simeq K[x^{v_1}t, \ldots, x^{v_q}t] \simeq K[x^{v_1}, \ldots, x^{v_q}].$$

Thus, by Corollary 8.2.21, we get that $\ell(I)$ is the rank of A. $\square$

Theorem 14.2.13 $\inf_i\{\mathrm{depth}(R/I^i)\} \leq \dim(R) - \ell(I)$ *with equality if the associated graded ring* $\mathrm{gr}_I(R)$ *is Cohen–Macaulay.*

The inequality is due to Burch [76]. If $\mathrm{gr}_I(R)$ is C–M, the equality comes from [131]. By a result of Brodmann [55], $\mathrm{depth}\, R/I^k$ is constant for $k \gg 0$. Brodmann improved the Burch inequality by showing that the constant value is bounded by $\dim(R) - \ell(I)$. For a study of the initial and limit behavior of the numerical function $f(k) = \mathrm{depth}\, R/I^k$, see [223].

Lemma 14.2.14 *If* $x^{v_1} = x_1x^{v'_1}, \ldots, x^{v_p} = x_1x^{v'_p}$ *and* $x_1 \notin \operatorname{supp}(x^{v_i})$ *for* $i > p$*, where* x_1 *is a variable in* C_k *for some* $1 \leq k \leq s$*, then there is* $x^{v'_j}$ *such that* $\operatorname{supp}(x^{v'_j}) \cap C_k = \emptyset$.

Proof. If $\operatorname{supp}(x^{v'_j}) \cap C_k \neq \emptyset$ for all j, then $C_k \setminus \{x_1\}$ is a vertex cover of $\mathcal{C}$, a contradiction because C_k is a minimal vertex cover. □

Proposition 14.2.15 *If* $1 \leq i \leq s$*, then* J_i *is the* J_i*-primary component of* $IR[It]$.

Proof. We set $P = J_i$. It suffices to show the equality $R[It] \cap IR[It]_P = P$. In general the left-hand side is contained in the right-hand side. To show the reverse inclusion we may assume that P can be written as

$$P = (x_1, \ldots, x_m, x^{v_1}t, \ldots, x^{v_p}t)R[It],$$

where $\mathfrak{p} = (x_1, \ldots, x_m)$ is a minimal prime of I. Set $C = \{x_1, \ldots, x_m\}$.

Case (I): Consider x_k with $1 \leq k \leq m$. By Lemma 14.2.14 there is j such that $x^{v_j} = x_kx^{\alpha}$ and $\operatorname{supp}(x^{\alpha}) \cap C = \emptyset$. Thus since x^{α} is not in P (because of the second condition) we obtain $x_k \in R[It] \cap IR[It]_P$.

Case (II): Now, consider $x^{v_k}t$ with $1 \leq k \leq p$. Since

$$\langle (v_k, 1), e_1 + \cdots + e_m - e_{n+1} \rangle \geq 1,$$

the monomial x^{v_k} contains at least two variables in C. Thus we may assume that x_1, x_2 are in the support of x^{v_k}. Again by Lemma 14.2.14 there are j, ℓ such that $x^{v_j} = x_1x^{\alpha}$, $x^{v_\ell} = x_2x^{\gamma}$, and the support of x^{α} and x^{γ} disjoint from C. Hence the monomial $x^{v_k}x^{\alpha+\gamma}t$ belongs to I^2t and $x^{\alpha+\gamma}$ is not in P. Writing $x^{v_k}t = (x^{v_k}x^{\alpha+\gamma}t)/x^{\alpha+\gamma}$, we get $x^{v_k}t \in R[It] \cap IR[It]_P$. □

Lemma 14.2.16 $\operatorname{rad}(J_k^{(d_k)}) = J_k$ *for* $1 \leq k \leq r$.

Proof. By construction one has $\operatorname{rad}(J_k^{(d_k)}) \subset J_k$. The reverse inclusion follows by noticing that if $x^at^b \in J_k$, then $(x^at^b)^{d_k} \in J_k^{(d_k)}$. □

Theorem 14.2.17 [258, Theorem 1.11] *If the height of* $I \geq 2$*, then the following are equivalent:*

(i) $\operatorname{gr}_I(R)$ *is torsion-free over* R/I.

(ii) $\operatorname{gr}_I(R)$ *is reduced.*

(iii) $R[It]$ *is normal and* $\operatorname{Cl}(R[It])$*, the divisor class group of* $R[It]$*, is a free abelian group whose rank is the number of minimal primes of* I.

Proposition 14.2.18 ([70], [188]) *If* $R[It]$ *is normal, then*

$$IR[It] = J_1^{(d_1)} \cap J_2^{(d_2)} \cap \cdots \cap J_r^{(d_r)}.$$

Proof. "$\subset$": Assume $x^\alpha t^b \in IR[It]$. Since $x^\alpha \in I^{b+1}$ we readily obtain $(\alpha, b+1) \in \mathbb{N}\mathcal{A}'$. In particular we get $(\alpha, b+1) \in \mathbb{R}_+(I)$. Therefore

$$0 \leq \langle(\alpha, b+1), \ell_k\rangle = \langle(\alpha, b), \ell_k\rangle - d_k$$

and consequently $x^\alpha t^b \in J^{(d_k)}$ for $1 \leq k \leq r$.

"$\supset$": Assume $x^\alpha t^b \in J^{(d_k)}$ for all k. Since the element $(\alpha, b+1)$ is in $\mathbb{R}_+(I) \cap \mathbb{Z}^{n+1}$ and using that $R[It]$ is normal yields $(\alpha, b+1) \in \mathbb{N}\mathcal{A}'$. It follows that $x^\alpha t^b \in I^{b+1}t^b \subset IR[It]$. □

Since the program *Normaliz* [68] computes the irreducible representation of the Rees cone and the integral closure of $R[It]$, the following result is an effective criterion for the reducedness of the associated graded ring.

Theorem 14.2.19 ([258], [147]) *The following are equivalent:*

(a) $R[It] = R_s(I)$.

(b) $\mathrm{gr}_I(R)$ *is reduced.*

(c) $R[It]$ *is normal and* $\langle \ell_i, e_{n+1}\rangle = -1$ *for* $i = 1, \ldots, r$.

Proof. (a) ⇒ (b) By Corollary 4.3.26 the Rees algebra $R[It]$ is normal. Thus using Theorem 14.2.7 we obtain that $r = s$ and that the minimal primes of $IR[It]$ are $J_1, \ldots, J_s$. Hence from Proposition 14.2.18 we have

$$IR[It] = J_1^{(d_1)} \cap J_2^{(d_2)} \cap \cdots \cap J_s^{(d_s)}.$$

Therefore $IR[It]$ is a radical ideal because $d_i = 1$ for $i = 1, \ldots, s$. Since $\mathrm{gr}_I(R) \simeq R[It]/IR[It]$ we get that $\mathrm{gr}_I(R)$ is reduced.

(b) ⇒ (c) By Theorem 14.2.17, the ring $R[It]$ is normal and $\mathrm{Cl}(R[It])$, the divisor class group of $R[It]$, is a free abelian group whose rank is the number of minimal primes of I, i.e., the rank of $\mathrm{Cl}(R[It])$ is equal to s. On the other hand by Proposition 12.7.3, the rank of $\mathrm{Cl}(R[It])$ is equal to r. Thus $s = r$ and $\langle \ell_i, e_{n+1}\rangle = -1$ for $i = 1, \ldots, r$.

(c) ⇒ (a) By Lemma 14.2.1, $r = s$. Hence, by Theorem 14.2.7, we get $\overline{R[It]} = R_s(I)$. Thus $R[It] = R_s(I)$. □

Example 14.2.20 Let $I = (x_1x_5, x_2x_4, x_3x_4x_5, x_1x_2x_3)$. Using *Normaliz* [68] with the input file:

```
4
5
1 0 0 0 1
0 1 0 1 0
0 0 1 1 1
1 1 1 0 0
3
```

we get the output file:

```
9 generators of integral closure of Rees algebra:
  1  0  0  0  0  0
  0  1  0  0  0  0
  0  0  1  0  0  0
  0  0  0  1  0  0
  0  0  0  0  1  0
  1  0  0  0  1  1
  0  1  0  1  0  1
  0  0  1  1  1  1
  1  1  1  0  0  1

10 support hyperplanes:
   0   0   1   1   1  -1
   1   0   0   0   0   0
   0   1   0   0   0   0
   0   0   0   0   0   1
   0   0   1   0   0   0
   1   0   0   1   0  -1
   0   0   0   1   0   0
   0   0   0   0   1   0
   0   1   0   0   1  -1
   1   1   1   0   0  -1
```

The first block of the output file shows the list of generators of $\overline{R[It]}$. Thus $R[It]$ is normal. The second block gives the irreducible representation of the Rees cone of I. This means that all the ℓ_i's that occur in the irreducible representation of the Rees cone $\mathbb{R}_+(I)$ (see Eq. (14.1)) have its last entry equal to -1. Thus, by Theorem 14.2.19, the ring $\mathrm{gr}_I(R)$ is reduced.

Proposition 14.2.21 *If* $\overline{R[It]} = R_s(I)$, *then* $\overline{R[I_c(\mathcal{C})t]} = R_s(I_c(\mathcal{C}))$.

Proof. We set $J = I_c(\mathcal{C})$. Using Theorem 14.2.7 one has

$$\mathbb{R}_+(I) = H_{e_1}^+ \cap \cdots \cap H_{e_{n+1}}^+ \cap H_{\ell_1}^+ \cap \cdots \cap H_{\ell_s}^+,$$

where $\ell_i = (u_i, -1)$ for all i. Then, by Corollary 1.1.30, we get:

$$\left(\bigcap_{i=1}^{n} H_{e_i}^+\right) \cap \left(\bigcap_{i=1}^{q} H_{(v_i,1)}^+\right) = \left(\sum_{i=1}^{n+1} \mathbb{R}_+ e_i\right) + \left(\sum_{i=1}^{s} \mathbb{R}_+(u_i, -1)\right). \quad (14.2)$$

By Theorem 14.2.7 we need only show the equality

$$\mathbb{R}_+(J) = \left(\bigcap_{i=1}^{n+1} H_{e_i}^+\right) \cap \left(\bigcap_{i=1}^{q} H_{(v_i,-1)}^+\right).$$

Clearly the left-hand side is contained in the right-hand side. To show the reverse inclusion take (α, b) in the right-hand side. Then

$$0 \leq \langle(\alpha, b), (v_i, -1)\rangle = \langle(\alpha, -b), (v_i, 1)\rangle$$

for all i. Hence using Eq. (14.2) we can write

$$(\alpha, -b) = \sum_{i=1}^{s} \lambda_i(u_i, -1) + \sum_{i=1}^{n+1} \mu_i e_i \quad (\lambda_i; \mu_j \geq 0). \tag{14.3}$$

Since $\lambda_1 + \cdots + \lambda_s - b = \mu_{n+1} \geq 0$, there are $\lambda_1', \ldots, \lambda_s'$ such that

$$0 \leq \lambda_i' \leq \lambda_i;\ b = \lambda_1' + \cdots + \lambda_s';\ \mu_{n+1} = (\lambda_1 - \lambda_1') + \cdots + (\lambda_s - \lambda_s').$$

Therefore using Eq. (14.3) we conclude

$$(\alpha, b) = \sum_{i=1}^{s} \lambda_i'(u_i, 1) + \sum_{i=1}^{n} \mu_i e_i + \sum_{i=1}^{s}(\lambda_i - \lambda_i')u_i \in \mathbb{R}_+(J). \qquad \square$$

Corollary 14.2.22 [93, Theorem 1.17] *If $\mathcal{Q}(A)$ is integral and $A^\vee$ is the incidence matrix of $\mathcal{C}^\vee$, then $\mathcal{Q}(A^\vee)$ is integral.*

Proof. It follows at once from Theorem 14.2.7 and Proposition 14.2.21. $\square$

The following notion of minor of an edge ideal was introduced in [188]. It is inspired by the notion of a minor in combinatorial optimization [93] and is consistent with the terminology of Section 6.5 (Definition 6.5.3).

Definition 14.2.23 A *minor* of I is an ideal $(0) \subsetneq I' \subsetneq R$ obtained from I by making any sequence of variables equal to 1 or 0 in $F = \{x^{v_1}, \ldots, x^{v_q}\}$. The ideal I is considered itself a minor. A *minor* of $\mathcal{C}$ is a clutter $\mathcal{C}'$ that corresponds to a minor $(0) \subsetneq I' \subsetneq R$ of I. The clutter $\mathcal{C}$ is itself a minor. If $\mathcal{C}'$ is a minor of $\mathcal{C}$ and $E(\mathcal{C}') \neq E(\mathcal{C})$, we call $\mathcal{C}'$ a *proper minor.*

If $I' \subset R$ is a minor of I, notice that $\mathcal{C}'$ is obtained from I' by considering $G(I')$, the unique set of square-free monomials that minimally generate I'. Indeed if $G(I') = \{x^{\alpha_1}, \ldots, x^{\alpha_m}\}$, then I' defines a clutter $\mathcal{C}'$ (resp. matrix A') whose vertices are the variables of R and whose edges (resp. columns) are the supports of the monomials x^{α_i} (resp. the column vectors α_i). We can also take as vertex set of $\mathcal{C}'$, the set of variables that occur in $x^{\alpha_1}, \ldots, x^{\alpha_m}$. This choice of vertex set is useful in induction arguments, where one needs to reduce the number of variables.

Proposition 14.2.24 *If $\overline{I^i} = I^{(i)}$ for some $i \geq 2$ and $J = I'$ is a minor of I, then $\overline{J^i} = J^{(i)}$.*

Proof. Assume that J is the minor obtained from I by making $x_1 = 0$. Take $x^\alpha \in J^{(i)}$. Then $x^\alpha \in I^{(i)} = \overline{I^i}$ because $J \subset I$. Thus $x^\alpha \in \overline{I^i}$. Since $x_1 \notin \text{supp}(x^\alpha)$ it follows that $x^\alpha \in \overline{J^i}$. This proves $J^{(i)} \subset \overline{J^i}$. The other inclusion is clear because $J^{(i)}$ is integrally closed and $J^i \subset J^{(i)}$.

Assume that J is the minor obtained from I by making $x_1 = 1$. Take $x^\alpha \in J^{(i)}$. Notice that $x_1^i x^\alpha \in I^{(i)} = \overline{I^i}$. Indeed if $x_1 \in \mathfrak{p}_k$, then $x_1^i \in \mathfrak{p}_k^i$, and if $x_1 \notin \mathfrak{p}_k$, then $J \subset \mathfrak{p}_k$ and $x^\alpha \in \mathfrak{p}_k^i$. Thus $x_1^i x^\alpha \in I^{(i)}$. Hence $(x_1^i x^\alpha)^p \in I^{ip}$ for some $p \in \mathbb{N}_+$. Since $x_1 \notin \text{supp}(x^\alpha)$ it follows that $x^\alpha \in \overline{J^i}$. $\square$

Corollary 14.2.25 (a) *If $R_s(I) = \overline{R[It]}$, then $R_s(I') = \overline{R[I't]}$ for any minor I' of I.* (b) *The integrality of $\mathcal{Q}(A)$ is preserved under taking minors and parallelizations.*

Proof. (a): It follows from Proposition 14.2.24. (b): This part follows from part (a), Theorem 14.2.7, and [373, p. 1390]. $\square$

Proposition 14.2.26 *Let $\mathcal{C}^\vee$ be the blocker of $\mathcal{C}$. If $\overline{R[It]}$ is equal to $R_s(I)$ and $|C \cap B| \leq 2$ for $C \in \mathcal{C}$ and $B \in \mathcal{C}^\vee$, then $R[It]$ is normal.*

Proof. Let $x^a t^b = x_1^{a_1} \cdots x_n^{a_n} t^b \in \overline{R[It]}$ be a minimal generator, that is (a, b) cannot be written as a sum of two non-zero integral vectors in the Rees cone $\mathbb{R}_+(I)$. We may assume $a_i \geq 1$ for $1 \leq i \leq m$, $a_i = 0$ for $i > m$, and $b \geq 1$.

Case (I): $\langle (a, b), \ell_i \rangle > 0$ for all i. The vector $\gamma = (a, b) - e_1$ satisfies $\langle \gamma, \ell_i \rangle \geq 0$ for all i, that is $\gamma \in \mathbb{R}_+(I)$. Thus since $(a, b) = e_1 + \gamma$ we derive a contradiction.

Case (II): $\langle (a, b), \ell_i \rangle = 0$ for some i. We may assume

$$\{\ell_i | \langle (a, b), \ell_i \rangle = 0\} = \{\ell_1, \dots, \ell_p\}.$$

Subcase (II.a): $e_i \in H_{\ell_1} \cap \cdots \cap H_{\ell_p}$ for some $1 \leq i \leq m$. It is not hard to verify that the vector $\gamma = (a, b) - e_i$ satisfies $\langle \gamma, \ell_k \rangle \geq 0$ for all $1 \leq k \leq s$. Thus $\gamma \in \mathbb{R}_+(I)$, a contradiction because $(a, b) = e_i + \gamma$.

Subcase (II.b): $e_i \notin H_{\ell_1} \cap \cdots \cap H_{\ell_p}$ for all $1 \leq i \leq m$. Since the vector (a, b) belongs to the Rees cone it follows that we can write

$$(a, b) = \lambda_1 (v_1, 1) + \cdots + \lambda_q (v_q, 1) \quad (\lambda_i \geq 0). \tag{$*$}$$

By the choice of $x^a t^b$ we may assume $0 < \lambda_1 < 1$. Set $\gamma = (a, b) - (v_1, 1)$ and notice that by Eq. $(*)$ this vector has nonnegative entries. We claim that γ is in the Rees cone. Since by hypothesis one has that $\ell_j = u_j$ (see Theorem 14.2.7) and $0 \leq \langle (v_1, 1), \ell_j \rangle \leq 1$ for all j we readily obtain

$$\langle \gamma, \ell_k \rangle = \begin{cases} \langle (a, b), \ell_k \rangle - \langle (v_1, 1), \ell_k \rangle & = & 0 & \text{if } 1 \leq k \leq p, \\ \langle (a, b), \ell_k \rangle - \langle (v_1, 1), \ell_k \rangle & \geq & 0 & \text{otherwise.} \end{cases}$$

Thus $\gamma \in \mathbb{R}_+(I)$ and $(a, b) = (v_1, 1) + \gamma$. As a result $\gamma = 0$ and $(a, b) \in R[It]$, as desired. $\square$

Normality criteria and Rees cones For each proper face $\mathcal{G}$ of the rational polyhedral cone $\mathbb{R}_+\mathcal{A}$ we set

$$\mathcal{A}_\mathcal{G} = \{v_i \,|\, v_i \in \mathcal{G}\} \ \text{ and } \ F_\mathcal{G} = \{x^{v_i} \,|\, v_i \in \mathcal{G}\}.$$

If $\mathcal{G}$ is a proper face of $\mathbb{R}_+\mathcal{A}$, there is a supporting hyperplane H_a in $\mathbb{R}^n$ such that $\mathcal{G} = \mathbb{R}_+\mathcal{A} \cap H_a \neq \emptyset$, $\mathbb{R}_+\mathcal{A} \not\subset H_a$, and $\mathbb{R}_+\mathcal{A} \subset H_a^+$. Notice that

$$\mathcal{A}_\mathcal{G} = \{v_i \in \mathcal{A} \,|\, \langle v_i, a\rangle = 0\} \neq \emptyset \ \text{ and } \ F_\mathcal{G} = \{x^{v_i} \,|\, \langle v_i, a\rangle = 0\} \neq \emptyset.$$

Lemma 14.2.27 *If $\mathbb{N}\mathcal{A}$ is a normal semigroup (resp. $\mathcal{A}$ is a Hilbert basis), then $\mathbb{N}\mathcal{A}_\mathcal{G}$ is a normal semigroup (resp. $\mathcal{A}_\mathcal{G}$ is a Hilbert basis) for each proper face $\mathcal{G}$ of $\mathbb{R}_+\mathcal{A}$.*

Proof. Assume that $\mathbb{N}\mathcal{A}$ is normal. Let $\alpha \in \mathbb{R}_+\mathcal{A}_\mathcal{G} \cap \mathbb{Z}\mathcal{A}_\mathcal{G}$. We may assume without loss of generality that $\langle v_i, a\rangle = 0$ for $1 \leq i \leq p$ and $\langle v_i, a\rangle > 0$ for $p+1 \leq i \leq q$. Using that $\alpha \in \mathbb{R}_+\mathcal{A} \cap \mathbb{Z}\mathcal{A} = \mathbb{N}\mathcal{A}$ we can write:

$$\alpha = \lambda_1 v_1 + \cdots + \lambda_q v_q = a_1 v_1 + \cdots + a_p v_p,$$

where $\lambda_i \in \mathbb{N}$ for $1 \leq i \leq q$ and $a_i \in \mathbb{R}_+$ for $1 \leq i \leq p$. Taking inner products we get the equality:

$$0 = a_1\langle v_1, a\rangle + \cdots + a_p\langle v_p, a\rangle = \lambda_{p+1}\langle v_{p+1}, a\rangle + \cdots + \lambda_q\langle v_q, a\rangle.$$

Since the λ_i's are nonnegative and $\langle v_i, a\rangle > 0$ for $p+1 \leq i \leq q$, one has $\lambda_{p+1} = \cdots = \lambda_q = 0$. Therefore $\alpha \in \mathbb{N}\mathcal{A}_\mathcal{G}$. This shows that $\mathbb{N}\mathcal{A}_\mathcal{G}$ is normal. The other assertion is shown similarly. □

Lemma 14.2.28 *If $K[F]$ is normal, then $K[F_\mathcal{G}]$ is normal for any proper face $\mathcal{G}$ of $\mathbb{R}_+\mathcal{A}$.*

Proof. Notice $K[F] = K[\mathbb{N}\mathcal{A}]$ and $K[F_\mathcal{G}] = K[\mathbb{N}\mathcal{A}_\mathcal{G}]$. By Corollary 9.1.3 $K[F]$ is normal if and only if $\mathbb{N}\mathcal{A}$ is normal. Thanks to Lemma 14.2.27 the semigroup is $\mathbb{N}\mathcal{A}_\mathcal{G}$ is normal. Therefore $K[F_\mathcal{G}]$ is normal. □

The converse of Lemma 14.2.27 fails as the next example shows.

Example 14.2.29 If $\mathcal{A} = \{(1,1,0),(1,0,1),(0,1,1)\}$ and $\mathbf{1} = (1,1,1)$, then $\mathbf{1}$ is in $\mathbb{R}_+\mathcal{A} \cap \mathbb{Z}^3$ and $\mathbf{1} \notin \mathbb{N}\mathcal{A}$. Thus $\mathcal{A}$ is not a Hilbert basis. By Proposition 1.1.23 any proper face $\mathcal{G}$ of $\mathbb{R}_+\mathcal{A}$ is a cone generated by a proper subset of $\mathcal{A}$. It follows that $\mathcal{A}_\mathcal{G}$ is a Hilbert basis.

Proposition 14.2.30 *If $K[\mathbb{N}\mathcal{A}'_\mathcal{G}]$ is normal for every proper face $\mathcal{G}$ of the Rees cone $\mathbb{R}_+(I)$ and there is a fixed integer $1 \leq i \leq n$ such that the ith entry of ℓ_j is either 0 or 1 for all $1 \leq j \leq r$, then $R[It]$ is normal.*

Proof. Recall that $R[It] = K[\mathbb{N}\mathcal{A}']$. Let (α, b) be an element in the minimal integral Hilbert basis of the affine semigroup $\overline{\mathbb{N}\mathcal{A}'}$, where $\alpha \in \mathbb{N}^n$ and $b \in \mathbb{N}$. It suffices to prove that (α, b) belongs to $\mathbb{N}\mathcal{A}'$.

Case (I): Assume $(\alpha, b) \in \mathrm{rb}(\mathbb{R}_+(I))$. By Theorem 1.1.44(b) the relative boundary of $\mathbb{R}_+(I)$ is given by:

$$\mathrm{rb}(\mathbb{R}_+(I)) = \left(\bigcup_{i=1}^{n+1} \mathbb{R}_+(I) \cap H_{e_i}\right) \cup \left(\bigcup_{i=1}^{r} \mathbb{R}_+(I) \cap H_{\ell_i}\right).$$

Hence (α, b) is in some facet of the Rees cone and by hypothesis it follows readily that (α, b) is in $\mathbb{N}\mathcal{A}'$.

Case (II): If $(\alpha, b) \in \mathrm{ri}(\mathbb{R}_+(I))$, then $\langle(\alpha, b), \ell_j\rangle > 0$ for $1 \leq j \leq r$ and $\langle(\alpha, b), e_k\rangle > 0$ for $1 \leq k \leq n+1$. For simplicity of notation we may assume $i = 1$. Write $(\alpha, b) = (\alpha_1, \ldots, \alpha_n, b)$. Notice $(\alpha, b) = \beta + e_1$, where $\beta = (\alpha_1 - 1, \alpha_2, \ldots, \alpha_n, b)$. Therefore

$$\langle \beta, \ell_j\rangle = \langle(\alpha, b), \ell_j\rangle - \langle e_1, \ell_j\rangle \geq 0$$

for all j, that is, $\beta \in \overline{\mathbb{N}\mathcal{A}'}$. Since $e_1 \in \overline{\mathbb{N}\mathcal{A}'}$, we obtain a contradiction. □

Theorem 14.2.31 *If $\mathcal{Q}(A)$ is integral, then $R[It]$ is normal if and only if $K[\mathbb{N}\mathcal{A}'_{\mathcal{G}}]$ is normal for any facet $\mathcal{G}$ of $\mathbb{R}_+(I)$.*

Proof. By Theorem 14.1.1 $r = s$ and the vectors $\ell_1, \ldots, \ell_r$ have their first n entries in $\{0, 1\}$ because $\ell_i = u_i$ for $i = 1, \ldots, s$. Hence the result follows from Lemma 14.2.28 and Proposition 14.2.30. □

Exercises

14.2.32 Prove that $\mathrm{gr}_I(R)$ is reduced if and only if $\mathrm{Cn}(I) \cap \mathbb{Z}^{n+1} = \mathbb{N}\mathcal{A}'$.

14.2.33 Consider a sequence $1 \leq i_1 < \cdots < i_s \leq n$ of integers. Prove that the following two conditions are equivalent:

(a) $e_{i_1} + \cdots + e_{i_s}$ is a vertex of $\mathcal{Q}(A)$.

(b) $F = \mathbb{R}_+(I) \cap H_{(a,-1)}$ is a facet of the Rees cone for some vector $a = (a_i)$ in $\mathbb{Z}^n$ such that $\{i \mid a_i \neq 0\} = \{i_1, \ldots, i_s\}$.

14.2.34 If height of $I \geq 2$, prove the following assertions:

(a) $\dim(\mathbb{R}_+(I)) = n + 1$.

(b) $H_{e_i} \cap \mathbb{R}_+(I)$ is a facet of $\mathbb{R}_+(I)$ for $i = 1, \ldots, n+1$.

(c) $H^+_{e_1}, \ldots, H^+_{e_{n+1}}$ occur in the irreducible representation of $\mathbb{R}_+(I)$.

14.2.35 Let A be a matrix with entries in $\mathbb{N}^n$ and let A' be the matrix obtained from A by adjoining a row of 1's. Prove that the vertices of $\mathcal{Q}(A)$ and $\mathcal{Q}(A')$ are related by

$$\text{vert}(\mathcal{Q}(A')) = \{e_{n+1}\} \cup \{(\beta, 0) | \, \beta \in \text{vert}(\mathcal{Q}(A))\}.$$

In particular $\mathcal{Q}(A)$ is integral if and only if $\mathcal{Q}(A')$ is integral.

14.2.36 Let $\mathcal{C}$ be a uniform clutter with n vertices and let A be its incidence matrix. If A has rank n and $I = I(\mathcal{C})$, then $\text{gr}_I(R)$ is not reduced.

14.2.37 Let $I = (x^{v_1}, \ldots, x^{v_4})$, where $v_1, \ldots, v_4$ denote the column vectors of the matrix:

$$A = \begin{pmatrix} 2 & 1 & 2 & 1 \\ 9 & 6 & 8 & 4 \\ 0 & 0 & 1 & 1 \\ 1 & 6 & 1 & 2 \end{pmatrix}.$$

Prove that $R[It]$ is not normal and $\det(A) = \pm 1$.

14.2.38 Let L_1, L_2 be monomial ideals with disjoint sets of variables. If L_1, L_2 are generated by monomials of degrees d_1 and d_2, respectively, then $\ell(L_1 + L_2) = \ell(L_1) + \ell(L_2)$, where $\ell(L_i)$ is the analytic spread of L_i.

14.2.39 Let $I = I(\mathcal{Q}_6) = (x_1x_2x_5, x_1x_3x_4, x_2x_3x_6, x_4x_5x_6)$ be the edge ideal of the clutter $\mathcal{Q}_6$. Use *Macaulay*2 to show the equality

$$\overline{I^2} = (I^2, x_1x_2 \cdots x_6).$$

Prove that $\overline{I^3}/I\overline{I^2}$ has four generators and prove that $I^{(i)} = II^{(i-1)}$ for $i \geq \ell(I) = 4$, where $\ell(I)$ is the analytic spread of I.

14.2.40 Let $I = (x_6x_7x_8, x_2x_4x_7, x_1x_2x_6, x_1x_3x_8, x_1x_3x_5)$ be the edge ideal of a clutter $\mathcal{C}$. Prove that $R[It] \subsetneq \overline{R[It]} = R_s(I)$, $\text{ht}(I) = 3$ and $\mathcal{C}$ satisfies the König property.

14.2.41 Let $I = (x^{v_1}, x^{v_2}, x^{v_3}, x^{v_4}, x^{v_5})$, where the v_i's are:

$$\begin{aligned} v_1 &= (1,1,1,1,0,0,0,0,0,0), \quad v_2 = (1,0,0,0,1,1,1,0,0,0), \\ v_3 &= (0,1,0,0,1,0,0,1,1,0), \quad v_4 = (0,0,1,0,0,1,0,1,0,1), \\ v_5 &= (0,0,0,1,0,0,1,0,1,1). \end{aligned}$$

Prove that the Rees algebra $R[It]$ is not normal, whereas for every proper face $\mathcal{G}$ of the Rees cone $K[\mathbb{N}\mathcal{A}'_{\mathcal{G}}]$ is normal. See Proposition 14.2.30.

14.2.42 Consider the matrix A whose transpose is:

$$A^\top = \begin{bmatrix} 1&0&0&1&0&0&1&0&0\\ 1&0&0&0&1&0&0&1&0\\ 1&0&0&0&0&1&0&0&1\\ 0&1&0&1&0&0&0&0&1\\ 0&1&0&0&1&0&0&1&0\\ 0&1&0&0&0&1&1&0&0\\ 0&0&1&0&0&1&1&0&0\\ 0&0&1&0&1&0&0&1&0\\ 0&0&1&1&0&0&0&0&1 \end{bmatrix}$$

This is a doubly stochastic singular matrix of rank 6. Prove that $\mathcal{Q}(A)$ is integral, $R[It]$ is not normal, $A(P) = \overline{K[Ft]}$ and $\overline{K[Ft]} \neq K[Ft]$.

14.3 Packing problems and blowup algebras

In this section, we present several characterizations of the max-flow min-cut property of a clutter in algebraic and combinatorial terms and prove that certain properties are closed under taking minors and parallelizations. We study packing problems and analyze the Conforti–Cornuéjols conjecture about the packing property [90].

The following basic result about the nilpotent elements of the associated graded ring holds for arbitrary monomial ideals.

Lemma 14.3.1 *If I is a monomial ideal of R, then the nilradical of the associated graded ring of I is given by*

$$\begin{aligned} \mathrm{nil}(\mathrm{gr}_I(R)) &= (\{\overline{g} \in I^i/I^{i+1} \,|\, g^s \in I^{si+1}; i \geq 0; s \geq 1\}) \\ &= (\{\overline{x^\alpha} \in I^i/I^{i+1} \,|\, x^{s\alpha} \in I^{si+1}; i \geq 0; s \geq 1\}). \end{aligned}$$

Proof. The first equality holds because the radical of any graded algebra is generated by homogeneous elements.

To prove the second equality take an equivalence class $\overline{0} \neq \overline{g} \in I^i/I^{i+1}$ such that $g^s \in I^{is+1}$. There are non-zero scalars $\lambda_1, \ldots, \lambda_r$ in the field K and monomials $m_1, \ldots, m_r$ in I^i such that

$$g = \lambda_1 m_1 + \cdots + \lambda_r m_r$$

and $m_1 \prec \cdots \prec m_r$, where $\prec$ is the lexicographical ordering. We may assume that $m_j \notin I^{i+1}$ for all j. Since m_r^s is the leading term of g^s and since I^{is+1} is a monomial ideal, we get $m_r^s \in I^{is+1}$. Thus $\overline{m_r}$ is in the nilradical of $\mathrm{gr}_I(R)$. By induction it follows that $\overline{m_j}$ is in $\mathrm{nil}(\mathrm{gr}_I(R))$ for all j, as required. $\Box$

Proposition 14.3.2 *If* $\mathrm{gr}_I(R)$ *is reduced (resp.* $R[It]$ *is normal) and* I' *is a minor of* I*, then* $\mathrm{gr}_{I'}(R)$ *is reduced (resp.* $R[I't]$ *is normal).*

Proof. Assume that $\mathrm{gr}_I(R)$ is reduced. Notice that we need only show that $\mathrm{gr}_{I'}(R)$ is reduced when I' is a minor obtained from I by making $x_1 = 0$ or $x_1 = 1$. Using Lemma 14.3.1 both cases are quite easy to prove. We leave the details as an exercise. If $R[It]$ is normal, then $R[I't]$ is normal by Proposition 12.2.3. $\Box$

Theorem 14.3.3 *Let* $\mathcal{C}$ *be a clutter and let* $\mathcal{C}'$ *be a parallelization of* $\mathcal{C}$*. If* $I(\mathcal{C})$ *is normal, then* $I(\mathcal{C}')$ *is normal.*

Proof. By Proposition 14.3.2 normality is closed under taking minors. Thus we need only show that the normality of $I = I(\mathcal{C})$ is preserved when we duplicate a vertex of $\mathcal{C}$. Let $V(\mathcal{C}) = \{x_2, \ldots, x_n\}$ be the vertex set of $\mathcal{C}$ and let $\mathcal{C}'$ be the duplication of the vertex x_2. We denote the duplication of x_2 by x_1. We may assume that

$$I = I(\mathcal{C}) = (x_2x^{w_1}, \ldots, x_2x^{w_r}, x^{w_{r+1}}, \ldots, x^{w_q}),$$

where $x^{w_i} \in K[x_3, \ldots, x_n]$ for all i. We must show that the ideal

$$I(\mathcal{C}') = I + (x_1x^{w_1}, \ldots, x_1x^{w_r})$$

is normal. Consider the sets $\mathcal{B}_0 = \{(0, 1, w_1, 1), \ldots, (0, 1, w_r, 1)\}$,

$$\begin{aligned} \mathcal{B} &= \{e_2, \ldots, e_n\} \cup \mathcal{B}_0 \cup \{(0, 0, w_{r+1}, 1), \ldots, (0, 0, w_q, 1)\}, \\ \mathcal{B}' &= \mathcal{B} \cup \{e_1, (1, 0, w_1, 1), \ldots, (1, 0, w_r, 1)\}. \end{aligned}$$

As I is normal, by Proposition 14.2.3, we have $\mathbb{Z}^{n+1} \cap \mathbb{R}_+\mathcal{B} = \mathbb{N}\mathcal{B}$ and we need only show the equality $\mathbb{Z}^{n+1} \cap \mathbb{R}_+(I) = \mathbb{N}\mathcal{B}'$. It suffices to show the inclusion "$\subset$". Take an integral vector $w = (a, b, c, d)$ in $\mathbb{R}_+(I)$, where $a, b, d \in \mathbb{Z}$ and $c \in \mathbb{Z}^{n-2}$. Then

$$\begin{aligned} w = (a, b, c, d) &= \sum_{i=1}^{r} \alpha_i(0, 1, w_i, 1) + \sum_{i=r+1}^{q} \alpha_i(0, 0, w_i, 1) \\ &\quad + \sum_{i=1}^{r} \beta_i(1, 0, w_i, 1) + \sum_{i=1}^{n} \gamma_i e_i \end{aligned}$$

for some α_i, β_i, γ_i in $\mathbb{R}_+$. Comparing entries it follows that

$$\begin{aligned} (0, a + b, c, d) &= \sum_{i=1}^{r} (\alpha_i + \beta_i)(0, 1, w_i, 1) + \sum_{i=r+1}^{q} \alpha_i(0, 0, w_i, 1) \\ &\quad + (\gamma_1 + \gamma_2)e_2 + \sum_{i=3}^{n} \gamma_i e_i, \end{aligned}$$

that is, the vector $(0, a+b, c, d)$ is in $\mathbb{Z}^{n+1} \cap \mathbb{R}_+\mathcal{B} = \mathbb{N}\mathcal{B}$. Thus there are λ_i, μ_i in $\mathbb{N}$ such that.

$$(0, a+b, c, d) = \sum_{i=1}^{r} \mu_i(0, 1, w_i, 1) + \sum_{i=r+1}^{q} \mu_i(0, 0, w_i, 1) + \sum_{i=2}^{n} \lambda_i e_i.$$

Comparing entries we obtain the equalities $a + b = \mu_1 + \cdots + \mu_r + \lambda_2$,

$$\begin{aligned} c &= \mu_1 w_1 + \cdots + \mu_q w_q + \lambda_3 e_3 + \cdots + \lambda_n e_n, \\ d &= \mu_1 + \cdots + \mu_q. \end{aligned}$$

Case (I): $b \leq \sum_{i=1}^{r} \mu_i$. If $b < \mu_1$, we set $b = \mu_1'$, $\mu_1' < \mu_1$, and define $\mu_1'' = \mu_1 - \mu_1'$. Otherwise pick $s \geq 2$ such that

$$\mu_1 + \cdots + \mu_{s-1} \leq b \leq \mu_1 + \cdots + \mu_s.$$

Then $b = \mu_1 + \cdots + \mu_{s-1} + \mu_s'$, where $\mu_s' \leq \mu_s$. Set $\mu_s'' = \mu_s - \mu_s'$. Hence

$$\begin{aligned} a + b &= \mu_1 + \cdots + \mu_r + \lambda_2 = a + \mu_1 + \cdots + \mu_{s-1} + \mu_s', \\ a &= \mu_s + \cdots + \mu_r + \lambda_2 - \mu_s' = \mu_{s+1} + \cdots + \mu_r + \mu_s'' + \lambda_2, \\ w &= \sum_{i=1}^{s-1} \mu_i(0, 1, w_i, 1) + \mu_s'(0, 1, w_s, 1) + \sum_{i=r+1}^{q} \mu_i(0, 0, w_i, 1) \\ &\quad + \mu_s''(1, 0, w_s, 1) + \sum_{i=s+1}^{r} \mu_i(1, 0, w_i, 1) + \lambda_2 e_1 + \sum_{i=3}^{n} \lambda_i e_i, \end{aligned}$$

that is, $w = (a, b, c, d) \in \mathbb{N}\mathcal{B}'$.

Case (II): $b > \sum_{i=1}^{r} \mu_i$. Then $b = \sum_{i=1}^{r} \mu_i + \lambda_2'$. Since

$$a + b = \mu_1 + \cdots + \mu_r + \lambda_2 = a + \mu_1 + \cdots + \mu_r + \lambda_2'$$

we get $a = \lambda_2 - \lambda_2'$. In particular $\lambda_2 \geq \lambda_2'$. Then

$$w = \sum_{i=1}^{r} \mu_i(0, 1, w_i, 1) + \sum_{i=r+1}^{q} \mu_i(0, 0, w_i, 1) + a e_1 + \lambda_2' e_2 + \sum_{i=3}^{n} \lambda_i e_i,$$

that is, $w = (a, b, c, d) \in \mathbb{N}\mathcal{B}'$. □

Definition 14.3.4 The clutter $\mathcal{C}$ satisfies the *max-flow min-cut* (MFMC) property if both sides of the LP-duality equation

$$\min\{\langle \alpha, x\rangle \,|\, x \geq 0; xA \geq \mathbf{1}\} = \max\{\langle y, \mathbf{1}\rangle \,|\, y \geq 0; Ay \leq \alpha\} \qquad (14.4)$$

have integral optimum solutions x and y for each nonnegative integral vector α. The system $x \geq 0; xA \geq \mathbf{1}$ is called *totally dual integral* (TDI) if the maximum in Eq. (14.4) has an integral optimum solution y for each integral vector α with finite maximum.

Definition 14.3.5 A clutter $\mathcal{C}$ is called *Mengerian* if $\beta_1(\mathcal{C}^a) = \alpha_0(\mathcal{C}^a)$ for all $a \in \mathbb{N}^n$, i.e., $\mathcal{C}$ is *Mengerian* if all its parallelizations have the König property.

Theorem 14.3.6 [147, 188, 258, 373] *Let $\mathcal{C}$ be a clutter and let A be its incidence matrix. The following are equivalent:*

(i) $\mathrm{gr}_I(R)$ *is reduced, where $I = I(\mathcal{C})$ is the edge ideal of $\mathcal{C}$.*

(ii) $R[It]$ *is normal and $\mathcal{Q}(A)$ is an integral polyhedron.*

(iii) $x \geq 0$; $xA \geq \mathbf{1}$ *is a* TDI system.

(iv) $\mathcal{C}$ *has the max-flow min-cut property.*

(v) $I^i = I^{(i)}$ *for $i \geq 1$.*

(vi) I *is normally torsion-free, i.e.,* $\mathrm{Ass}(R/I^i) \subset \mathrm{Ass}(R/I)$ *for $i \geq 1$.*

(vii) $\mathcal{C}$ *is Mengerian, i.e., $\beta_1(\mathcal{C}^a) = \alpha_0(\mathcal{C}^a)$ for all $a \in \mathbb{N}^n$.*

Proof. By Theorem 14.2.19, $\mathrm{gr}_I(R)$ is reduced if and only if $R[It]$ is normal and the vectors $\ell_1, \ldots, \ell_r$ occurring in Eq. (14.1) are integral. Thus (i) and (ii) are equivalent by Theorem 1.4.3.

(ii)$\Rightarrow$(iii): Let α be a vector such that Eq. (14.4) has a finite maximum equal to b. We may assume $b > 0$, otherwise the maximum in Eq. (14.4) is attained at $y = 0$. Note $\alpha \geq 0$. As $\mathcal{Q}(A)$ is integral, $b \in \mathbb{N}$ and there is $0 \neq s \in \mathbb{N}$ such that $x^{s\alpha} \in I^{sb}$. Hence $x^\alpha \in \overline{I^b} = I^b$ and this implies that the maximum in Eq. (14.4) has an integral optimum solution y.

(iii)$\Rightarrow$ (i): Let $\overline{x^\alpha} \in I^i/I^{i+1}$ be a nilpotent element and let $m = x^\alpha$; that is, m is a monomial such that: (i) $\overline{m} \in I^i/I^{i+1}$ and (ii) $m^s \in I^{is+1}$, for some $0 \neq s \in \mathbb{N}$. By Lemma 14.3.1 it suffices to prove that $m \in I^{i+1}$. From (ii) there are $a_1, \ldots, a_q$ in $\mathbb{N}$ and $\delta \in \mathbb{N}^n$ such that

$$x^{s\alpha} = (x^{v_1})^{a_1} \cdots (x^{v_q})^{a_q} x^\delta \ \text{ and } \ \textstyle\sum_{j=1}^q a_j = is + 1.$$

The rational vector $y_0 = (a_1/s, \ldots, a_q/s)$ satisfies $Ay_0 \leq \alpha$, $y_0 \geq 0$ and the sum of its entries is equal to z_0. Hence the linear program

Maximize $f(y) = \sum_{i=1}^q y_i$

Subject to $y \geq 0; Ay \leq \alpha$

has an optimal value greater than or equal to $z_0 = i + 1/s$. Note that the polyhedron $\mathcal{Q} = \{y |\, Ay \leq \alpha; y \geq 0\}$ is bounded. Indeed any $y = (y_1, \ldots, y_q)$ in $\mathcal{Q}$ must satisfy $y_j \leq \max\{\alpha_1, \ldots, \alpha_q\}$ for all j, where α_j is the jth entry of α. Since the system $x \geq 0; xA \leq \mathbf{1}$ is TDI, the optimum value of the linear program above is attained by an integral vector $b = (b_1, \ldots, b_q)$. Thus $b_1 + \cdots + b_q \geq i + 1$. As $b \in \mathcal{Q}$, we can write

$$x^\alpha = (x^{v_1})^{b_1} \cdots (x^{v_q})^{b_q} x^\gamma,$$

for some $\gamma \in \mathbb{N}^n$. This proves $m \in I^{i+1}$, as required.

(iii)$\Leftrightarrow$ (iv): This is Proposition 1.4.8. (i)$\Leftrightarrow$ (v): This was shown in Theorem 14.2.19. (v) $\Leftrightarrow$ (vi): This was shown in Proposition 4.3.29.

(iv) $\Rightarrow$ (vii): Let A' be the incidence matrix of $\mathcal{C}' = \mathcal{C}^a$. We have shown that (ii) is equivalent to (iv). Thus $I(\mathcal{C})$ is normal and $\mathcal{Q}(A)$ is integral. By Theorem 14.3.3 the ideal $I(\mathcal{C}')$ is normal, and since the integrality of $\mathcal{Q}(A)$ is closed under minors and duplications (see Corollary 14.2.25), we get that $\mathcal{Q}(A')$ is also integral. Thus, using once again that (ii) is equivalent to (iv), we get that $\mathcal{C}'$ satisfies the max-flow min-cut property. Therefore the LP-duality equation

$$\min\{\langle \mathbf{1}, x\rangle \,|\, x \geq 0; xA' \geq \mathbf{1}\} = \max\{\langle y, \mathbf{1}\rangle \,|\, y \geq 0; A'y \leq \mathbf{1}\}$$

has optimum integral solutions x, y. Observe that the left-hand side of this equality is $\alpha_0(\mathcal{C}')$ and the right-hand side is $\beta_1(\mathcal{C}')$.

(vii) $\Rightarrow$ (iv): Conversely if $\mathcal{C}^a$ has the König property for all $a \in \mathbb{N}^n$, then by Lemmas 13.2.12 and 13.2.14 both sides of the LP-duality equation

$$\min\{\langle a, x\rangle \,|\, x \geq 0; xA \geq \mathbf{1}\} = \max\{\langle y, \mathbf{1}\rangle \,|\, y \geq 0; Ay \leq a\}$$

have integral optimum solutions x and y for each nonnegative integral vector a, i.e., $\mathcal{C}$ has the max-flow min-cut property. □

Corollary 14.3.7 *Let $\mathcal{C}'$ be a parallelization or a minor of a clutter $\mathcal{C}$. If $\mathcal{C}$ has the max-flow min-cut property, then so does $\mathcal{C}'$.*

Proof. The max-flow min-cut property of a clutter is closed under taking minors; this follows from Proposition 14.3.2 and Theorem 14.3.6. That the max-flow min-cut property is closed under parallelizations follows from Theorem 14.3.6. □

Corollary 14.3.8 *If a clutter $\mathcal{C}$ has the max-flow min-cut property, then all parallelizations and minors of $\mathcal{C}$ have the König property.*

Proof. Assume that the clutter $\mathcal{C}$ has the max-flow min-cut property. By Corollary 14.3.7 it suffices to prove that $\mathcal{C}$ has the König property. Let A be the incidence matrix of $\mathcal{C}$. By hypothesis the LP-duality equation

$$\min\{\langle \mathbf{1}, x\rangle \,|\, x \geq 0; xA \geq \mathbf{1}\} = \max\{\langle y, \mathbf{1}\rangle \,|\, y \geq 0; Ay \leq \mathbf{1}\}$$

has optimum integral solutions x, y. To complete the proof notice that the left-hand side of this equality is $\alpha_0(\mathcal{C})$ and the right-hand side is $\beta_1(\mathcal{C})$. □

Definition 14.3.9 A clutter $\mathcal{C}$ is called *dyadic* if $|C \cap B| \leq 2$ for $C \in \mathcal{C}$ and $B \in \mathcal{C}^\vee$.

Corollary 14.3.10 [94, Theorem 1.3] *If $\mathcal{Q}(A)$ is integral and $\mathcal{C}$ is dyadic, then $x \geq 0$; $xA \geq \mathbf{1}$ is a* TDI *system.*

Proof. By Proposition 14.2.26 the Rees algebra $R[It]$ is normal. Thus the proof follows from Theorem 14.3.6. $\Box$

Corollary 14.3.11 [185] *If $\mathcal{C}$ is a balanced clutter, then*

(i) $R[I(\mathcal{C})t] = R_s(I(\mathcal{C}))$ *and* $R[I_c(\mathcal{C})t] = R_s(I_c(\mathcal{C}))$.

(ii) $\mathrm{gr}_{I(\mathcal{C})}(R)$ *is reduced and all minors of $\mathcal{C}$ satisfy the König property.*

Proof. By [373, Corollary 83.1a(iv), p. 1441], we get that both $\mathcal{C}$ and $\mathcal{C}^\vee$ are Mengerian. Thus the result follows at once from Theorem 14.3.6. $\Box$

Corollary 14.3.12 [156, Theorem 5.3] *If $\mathcal{C}$ is the clutter of facets of a simplicial forest, then $\mathcal{C}$ has the König property.*

Proof. We set $I = I(\mathcal{C})$. By Theorem 6.5.17, $\mathcal{C}$ is totally balanced. Hence, by Corollary 14.3.11, the associated graded ring $\mathrm{gr}_I(R)$ is reduced and all minors of $\mathcal{C}$ satisfy the König property. $\Box$

Remark 14.3.13 Let $\mathcal{C}$ be the clutter of facets of an unmixed simplicial forest. By Theorem 6.5.17, $\mathcal{C}$ has no special cycles. As $\mathcal{C}$ is balanced, it has the König property (Corollary 14.3.11), and by Lemma 6.5.7 it has a perfect matching of König type. Thus, Theorem 6.5.18 holds for $\mathcal{C}$ [156].

Corollary 14.3.14 *If A is totally unimodular, then $\mathrm{gr}_I(R)$ is reduced.*

Proof. If A is totally unimodular, then A is balanced. Thus we can apply Corollary 14.3.11. $\Box$

Corollary 14.3.15 [383, Theorem 5.9] *If $\mathcal{C}$ is a bipartite graph and I is its edge-ideal, then I is normally torsion-free and $R[It]$ is normal.*

Proof. By Proposition 10.2.2, the incidence matrix of a bipartite graph is totally unimodular. Thus the result follows from Corollary 14.3.14 and Theorem 14.3.6. $\Box$

Definition 14.3.16 A clutter $\mathcal{C}$ is said to satisfy the *packing property* (PP for short) if $\alpha_0(\mathcal{C}') = \beta_1(\mathcal{C}')$ for any minor $\mathcal{C}'$ of $\mathcal{C}$; that is, all minors of $\mathcal{C}$ satisfy the König property.

Theorem 14.3.17 (A. Lehman [290], [93, Theorem 1.8]) *If a clutter $\mathcal{C}$ has the packing property, then $\mathcal{Q}(A)$ is integral.*

The converse of this result is not true. A famous example is the clutter $\mathcal{Q}_6$ of Example 14.2.9. It does not pack and $\mathcal{Q}(A)$ is integral.

Corollary 14.3.18 [93] *If a clutter $\mathcal{C}$ has the max-flow min-cut property, then $\mathcal{C}$ has the packing property.*

Proof. It follows at once from Corollary 14.3.8. □

Conforti and Cornuéjols [90] conjectured that the converse is also true; see also [93, Conjecture 1.6].

Conjecture 14.3.19 (Packing problem, Conforti–Cornuéjols) If a clutter $\mathcal{C}$ has the packing property, then $\mathcal{C}$ has the max-flow min-cut property.

The following is an equivalent version of this conjecture [93].

Conjecture 14.3.20 (Duplication Conjecture, [93, Conjecture 4.24]) If $\mathcal{C}$ has the packing property and $\mathcal{C}'$ is the clutter obtained from $\mathcal{C}$ by duplicating a vertex x_i, then $\mathcal{C}'$ has the König property.

The Conforti–Cornuéjols conjecture can be studied via blowup algebras. By Theorem 14.3.6, the following is an algebraic version of this conjecture.

Conjecture 14.3.21 If $\alpha_0(\mathcal{C}') = \beta_1(\mathcal{C}')$ for all minors $\mathcal{C}'$ of $\mathcal{C}$, then the ring $\mathrm{gr}_I(R)$ is reduced.

Using Theorems 14.3.17 and 14.3.6 this conjecture reduces to:

Conjecture 14.3.22 If $\alpha_0(\mathcal{C}') = \beta_1(\mathcal{C}')$ for all minors $\mathcal{C}'$ of $\mathcal{C}$, then $R[It]$ is normal.

Proposition 14.3.23 *Let J_i be the ideal obtained from I by making $x_i = 1$ in F. If $\mathcal{Q}(A)$ is integral, then I is normal if and only if J_i is normal for $i = 1, \ldots, n$ and* $\mathrm{depth}(R/I^k) \geq 1$ *for all $k \geq 1$.*

Proof. $\Rightarrow$) The normality of an edge ideal is closed under taking minors (Proposition 12.2.3), hence J_i is normal for all i. Since $R[It]$ is normal, we get that $R[It]$ is C–M thanks to Theorem 9.1.6. Then by Theorem 4.3.19 the ring $\mathrm{gr}_I(R)$ is C–M. Hence using Theorem 14.2.13 and Corollary 14.2.11 we get that $\mathrm{depth}(R/I^i) \geq 1$ for all i.

$\Leftarrow$) It follows readily adapting the arguments given in the proof of the normality criterion presented in Theorem 12.2.4. □

By Proposition 14.3.23, we get that Conjecture 14.3.21 also reduces to:

Conjecture 14.3.24 If $\alpha_0(\mathcal{C}') = \beta_1(\mathcal{C}')$ for any minor $\mathcal{C}'$ of $\mathcal{C}$, then

$$\mathrm{depth}(R/I^i) \geq 1 \quad \text{for all } i \geq 1.$$

Definition 14.3.25 A clutter $\mathcal{C}$ is called *minimally non-packing* if it does not pack but all its proper minors do.

The next conjecture implies Conjecture 14.3.19 [94].

Conjecture 14.3.26 [94] *If $\mathcal{C}$ is a minimally non-packing clutter and the polyhedron $\mathcal{Q}(A)$ is integral, then $\alpha_0(\mathcal{C}) = 2$.*

The next result essentially says that, for uniform clutters, it suffices to prove Conjecture 14.3.19 for Cohen–Macaulay clutters.

Theorem 14.3.27 [116] *Let $\mathcal{C}$ be a d-uniform clutter. If $\mathcal{C}$ satisfies* PP *(resp. max-flow min-cut), then there is a d-uniform Cohen–Macaulay clutter $\mathcal{C}'$ satisfying* PP *(resp. max-flow min-cut) such that $\mathcal{C}$ is a minor of $\mathcal{C}'$.*

Theorem 14.3.28 *If $\overline{R[It]} = R_s(I)$ and $K[Ft] = A(\mathcal{P})$, then $R[It]$ is normal.*

Proof. By Theorem 14.2.7 the Rees cone of I is quasi-ideal. Thus by Corollary 12.3.2 the Rees algebra of I is normal. □ □

Proposition 14.3.29 *If $\mathcal{C}$ is a d-uniform clutter such that $\overline{I^b} = I^{(b)}$ for all b and $d \geq 2$, then $\overline{I^b}$ is generated by monomials of degree bd for $b \geq 1$.*

Proof. The monomial ideal $\overline{I^b}$ has a unique minimal set of generators consisting of monomials. Take x^a in this minimal set. Notice that (a, b) is in $\mathbb{R}_+(I)$. Thus we may proceed as in the proof of Theorem 12.3.1 to obtain that (a, b) is in the cone generated by $\{(v_1, 1), \ldots, (v_q, 1)\}$. This yields $\deg(x^a) = bd$. □

Proposition 14.3.30 [185] *If $\mathcal{C}$ is a d-uniform clutter and $d \geq 2$, then $I^i = I^{(i)}$ for all $i \geq 1$ if and only if $\mathcal{Q}(A)$ is integral and $K[Ft] = A(\mathcal{P})$.*

Proof. $\Rightarrow$) By Theorem 14.2.7 the polyhedron $\mathcal{Q}(A)$ is integral. Since $I^{(i)}$ is integrally closed (see Corollary 4.3.26), we get that $R[It]$ is normal. Therefore applying Theorem 9.3.31, we obtain $K[Ft] = A(\mathcal{P})$. Here the hypothesis on the degrees of x^{v_i} is essential.

$\Leftarrow$) By Theorem 14.2.7 $\overline{I^i} = I^{(i)}$ for all i, thus applying Theorem 14.3.28 gives that $R[It]$ is normal and we get the required equality. In this part the hypothesis on the degrees of x^{v_i} is not needed. □

For uniform clutters, using Proposition 14.3.30 and Theorem 14.3.17, we obtain another algebraic version of the Conforti–Cornuéjols conjecture:

Conjecture 14.3.31 If $\mathcal{C}$ is a uniform clutter with the packing property, then one has the equality $K[Ft] = A(\mathcal{P})$.

Corollary 14.3.32 *Let $\mathcal{C}$ be a uniform clutter and let A be its incidence matrix. If the polyhedra*

$$\{x|\, x \geq 0;\, xA \leq \mathbf{1}\} \quad \text{and} \quad \{x|\, x \geq 0;\, xA \geq \mathbf{1}\}$$

are integral, then $\mathcal{C}$ has the max-flow min-cut property.

Proof. By Proposition 14.3.30, the clutter $\mathcal{C}$ has the max-flow min-cut property if and only if $\mathcal{Q}(A)$ is integral and $K[Ft] = A(\mathcal{P})$. Thus, the result follows from Corollary 13.6.9. $\square$

Notation For an integral matrix $B \neq (0)$, the greatest common divisor of all the non-zero $r \times r$ subdeterminants of B will be denoted by $\Delta_r(B)$.

Corollary 14.3.33 *Let B be the matrix obtained from A by adjoining a row of 1's. If $x^{v_1}, \ldots, x^{v_q}$ have degree $d \geq 2$ and $\mathcal{C}$ has the max-flow min-cut property, then $\Delta_r(B) = 1$ with $r = \mathrm{rank}(B)$.*

Proof. By Theorem 14.3.6 and Proposition 14.3.30, we obtain the equality $A(\mathcal{P}) = K[Ft]$. Hence a direct application of Theorem 9.3.25 gives that $\Delta_r(B)$ is equal to 1. $\square$

Definition 14.3.34 Let M be a matroid on X and let $\mathcal{C}$ be the clutter of bases of M. The edge ideal $I(\mathcal{C})$ is called the *basis monomial ideal* of M.

Proposition 14.3.35 *Let $\mathcal{C}$ be the clutter of bases of a matroid M on X. If $\mathcal{C}$ satisfies the packing property, then $\mathrm{gr}_{I(\mathcal{C})}(R)$ is reduced.*

Proof. Since $\mathcal{Q}(A)$ is integral (Theorem 14.3.17) and $R[I(\mathcal{C})t]$ is normal (Corollary 12.3.12), we get that $\mathrm{gr}_{I(\mathcal{C})}(R)$ is reduced by Theorem 14.3.6. $\square$

Definition 14.3.36 Let $X_1, \ldots, X_d$ be a partition of X. The matroid $\mathcal{M}$ whose collection of bases is

$$\{\{y_1, \ldots, y_d\} | \, y_i \in X_i \text{ for } i = 1, \ldots, d\}$$

is called the *transversal matroid* defined by $X_1, \ldots, X_d$.

Proposition 14.3.37 *Let $X_1, \ldots, X_d$ be a partition of X and let $\mathcal{M}$ be the transversal matroid on X defined by $X_1, \ldots, X_d$. If I is the basis monomial ideal of $\mathcal{M}$, then $\mathrm{gr}_I(R)$ is reduced.*

Proof. Let $\mathcal{C}$ be the clutter of bases of $\mathcal{M}$ and let A be the incidence matrix of $\mathcal{C}$. Then $I = I(\mathcal{C})$. Notice that the edges of $\mathcal{C}^\vee$ are $X_1, \ldots, X_d$, therefore the incidence matrix of $\mathcal{C}^\vee$ is totally unimodular. Hence $\mathcal{C}^\vee$ is a balanced clutter and $(\mathcal{C}^\vee)^\vee = \mathcal{C}$. Thus, by Corollary 14.3.11, $\mathrm{gr}_I(R)$ is reduced. $\square$

Another property which is closed under taking minors is being the basis monomial ideal of a matroid.

Proposition 14.3.38 *Let I be the basis monomial ideal of a matroid M on X of rank d. If I' is a minor of I, then I' is again the basis monomial ideal of a matroid M'.*

Proof. Let I' be the minor obtained from I by making $x_n = 1$ and let $G(I')$ be the unique minimal generating set of I' consisting of monomials.

First we prove that I' is generated by monomials of degree $d-1$. Take $f \in G(I')$. We proceed by contradiction. Assume that $f = x_1x_2\cdots x_d$ with $x_n \notin \text{supp}(f)$. Pick a monomial in I of the form $g = x_{i_1}x_{i_2}\cdots x_{i_{d-1}}x_n$. If $\text{supp}(g) \setminus \{x_n\} \subset \{x_1,\ldots,x_d\}$ we obtain a contradiction because of the minimality of $G(I')$. Thus we may assume $x_{i_1} \notin \text{supp}(f)$. By the exchange property (Theorem 1.9.14) we may assume that the monomial

$$g_1 = x_1x_{i_2}x_{i_3}\cdots x_{i_{d-1}}x_n$$

belongs to I. If $\text{supp}(g_1) \setminus \{x_n\} \subset \{x_1,\ldots,x_d\}$ we obtain a contradiction. Thus we may assume $x_{i_2} \notin \text{supp}(f)$. Again by the exchange property we may assume that $g_2 = x_1x_2x_{i_3}x_{i_4}\cdots x_{i_{d-1}}x_n$ is in I. Repeating this procedure we get $x_1x_2\cdots x_{d-1}x_n \in I$, a contradiction. Thus I' is generated by monomials of degree $d-1$, as claimed.

We set $G(I') = \{x^{\alpha_1},\ldots,x^{\alpha_r}\}$, $B_i' = \text{supp}(x^{\alpha_i})$, and $B_i = B_i' \cup \{x_n\}$. The collection of bases $\{B_1,\ldots,B_r\}$ satisfies the exchange property, hence so does the collection $\mathcal{B}' = \{B_1',\ldots,B_r'\}$. Thus, by Theorem 1.9.14, there is a matroid M' whose collection of bases is $\mathcal{B}'$. Hence I' is the basis monomial ideal of M', as required. The case $x_n = 0$ is also not hard to show. □

Proposition 14.3.39 *If $\mathcal{C}$ is a graph, then the following are equivalent:*

(a) $\text{gr}_I(R)$ *is reduced.*

(b) $\mathcal{C}$ *is bipartite.*

(c) $\mathcal{Q}(A)$ *is integral.*

(d) $\mathcal{C}$ *has the packing property.*

Proof. (a) $\Rightarrow$ (d): By Theorem 14.3.6, $\mathcal{C}$ has the max-flow min-cut property. Then, by Corollary 14.3.18, $\mathcal{C}$ has the packing property. (d) $\Rightarrow$ (b): It suffices to show that $\mathcal{C}$ has no induced odd cycles (Proposition 7.1.2). If $C_r = \{x_1,\ldots,x_r\}$ is an induced cycle, then deleting all vertices outside C_r we obtain that C_r is a minor of $\mathcal{C}$. Thus C_r has the König property. It follows that r must be even. (b) $\Rightarrow$ (c): By Proposition 10.2.2, the matrix A is totally unimodular. Then by a result of Hoffman and Kruskal (see Theorem 1.5.6), $\mathcal{Q}(A)$ is integral. (c) $\Rightarrow$ (a): Since any graph is a dyadic clutter, using Corollary 14.3.10, we get that $x \geq 0$; $xA \geq \mathbf{1}$ is a TDI system. Thus by Theorem 14.3.6, the ring $\text{gr}_I(R)$ is reduced. □

Definition 14.3.40 A clutter is *binary* if its edges and its minimal vertex covers intersect in an odd number of vertices.

Theorem 14.3.41 [375] *A binary clutter $\mathcal{C}$ has the max-flow min-cut property if and only if $\mathcal{Q}_6$ is not a minor of $\mathcal{C}$.*

Corollary 14.3.42 *If $\mathcal{C}$ is a binary clutter with the packing property, then $\mathcal{C}$ has the max-flow min-cut property.*

Proof. Any clutter with the packing property cannot have $\mathcal{Q}_6$ as a minor because $\mathcal{Q}_6$ does not satisfy the König property. Thus $\mathcal{C}$ has the max-flow min-cut property by Theorem 14.3.41. $\Box$

Lemma 14.3.43 *If I is an unmixed ideal and $\mathcal{C}$ satisfies the König property, then $x^{\mathbf{1}} = x_1x_2\cdots x_n$ belongs to the subring $K[x^{v_1}, \ldots, x^{v_q}]$.*

Proof. We may assume $x^{\mathbf{1}} = x^{v_1}\cdots x^{v_g}x^{\delta}$ for some x^{δ}, where g is the height of I. If $\delta \neq 0$ pick $x_n \in \mathrm{supp}(x^{\delta})$. Since the variable x_n occurs in some monomial minimal generator of I, there is a minimal prime $\mathfrak{p}$ containing x_n. Thus using that $x^{v_1}, \ldots, x^{v_g}$ have disjoint support we conclude that $\mathfrak{p}$ contains at least $g+1$ variables, a contradiction. $\Box$

Proposition 14.3.44 *Let $I_i = I \cap K[X \setminus \{x_i\}]$. If I is an unmixed ideal such that the following conditions hold*

(a$_1$) *$\mathcal{Q}(A)$ is integral,*

(a$_2$) *I_i is normal for $i = 1, \ldots, n$, and*

(a$_3$) *$\mathcal{C}$ has the König property,*

then $R[It]$ is normal.

Proof. Take $x^at^b = x_1^{a_1}\cdots x_n^{a_n}t^b \in \overline{R[It]}$ a minimal generator; that is, x^at^b cannot be written as a product of two non-constant monomials of $\overline{R[It]}$. By condition (a$_2$) we may assume $a_i \geq 1$ for all i. Notice that $x_1\cdots x_nt^g$ is always in $\overline{R[It]}$ if $\mathcal{Q}(A)$ is integral, where g is the height of I. We claim that $b \leq g$. If $b > g$, consider the decomposition

$$x^at^b = (x_1\cdots x_nt^g)(x_1^{a_1-1}\cdots x_n^{a_n-1}t^{b-g}).$$

To derive a contradiction consider the irreducible representation of the Rees cone $\mathbb{R}_+(I)$ (see Eq. (14.1)). Observe that using condition (a$_1$) gives

$$\sum_{i\in C_k} a_i \geq b \qquad (k = 1, \ldots, s)$$

because $(a, b) \in \mathbb{R}_+(I)$. Now since I is unmixed we get

$$\sum_{i\in C_k} (a_i - 1) \geq b - g \qquad (k = 1, \ldots, s),$$

and consequently $x_1^{a_1-1}\cdots x_n^{a_n-1}t^{b-g} \in \overline{R[It]}$, a contradiction to the choice of x^at^b. Thus $b \leq g$. Using condition (a$_3$) we get $x_1\cdots x_nt^g \in I^gt^g \subset R[It]$, which readily implies $x^at^b \in R[It]$. $\Box$

Corollary 14.3.45 *Let $\mathcal{C}$ be an unmixed clutter with the packing property and let $I = I(\mathcal{C})$ be its edge ideal. If the ideal $I_i = I \cap K[X \setminus \{x_i\}]$ is normal for all i, then $\mathcal{C}$ satisfies the max-flow min-cut property.*

Proof. By Theorem 14.3.17, the set covering polyhedron $\mathcal{Q}(A)$ is integral. Then $R[It]$ is normal by Proposition 14.3.44. Using Theorem 14.3.6, we get that $\mathcal{C}$ satisfies the max-flow min-cut property. □

Corollary 14.3.46 [204] *If a minimal counterexample $\mathcal{C}$ to the Conforti–Cornuéjols conjecture exists, then $\mathcal{C}$ cannot be unmixed.*

Proof. Assume that $\mathcal{C}$ is unmixed. Let $I = I(\mathcal{C})$ be the edge ideal of $\mathcal{C}$, let $I_i = I \cap K[X \setminus \{x_i\}]$, and let $\mathcal{C}_i$ be the clutter that corresponds to I_i. Then $\mathcal{C}_i$ satisfies the max-flow min-cut property, and by Theorem 14.3.6 we get that I_i is normal. Hence by Corollary 14.3.45, the clutter $\mathcal{C}$ has the max-flow min-cut property, a contradiction. □

Proposition 14.3.47 *Let $Y \subset X$ and let $I_Y = I \cap K[Y]$. If I_Y has the König property for all Y and $\overline{R[It]}$ is generated as a K-algebra by monomials of the form $x^a t^b$, with x^a square-free, then $R[It]$ is normal.*

Proof. Take $x^a t^b$ a generator of $\overline{R[It]}$, with x^a square-free. By induction we may assume $x^a t^b = x_1 \cdots x_n t^b$. Hence, since $(1, \ldots, 1, b)$ is in the Rees cone, we get $|C_k| \geq b$ for $k = 1, \ldots, s$. In particular $g = \mathrm{ht}(I) \geq b$. As I has the König property, we get that the monomial $x_1 \cdots x_n$ is in I^g and consequently $x^a t^b \in R[It]$. □

Proposition 14.3.48 *Let $I_i = I \cap K[X \setminus \{x_i\}]$. If I_i is a normal ideal for $i = 1, \ldots, n$ and*

$$H_{\ell_1} \cap H_{\ell_2} \cap \cdots \cap H_{\ell_r} \cap \mathbb{R}_+^{n+1} \neq (0), \tag{14.5}$$

then $R[It]$ is normal.

Proof. Let $x^a t^b = x_1^{a_1} \cdots x_n^{a_n} t^b \in \overline{R[It]}$ be a minimal generator. It suffices to prove that $0 \leq b \leq 1$ because this readily implies that x^a is either a variable or a monomial in F. Assume $b \geq 2$. Since I_i is normal we may assume that $a_i \geq 1$ for all i. As each variable occurs in at least one monomial of F, using Eq. (14.5) it follows that there is $(v_k, 1)$ such that $\langle (v_k, 1), \ell_i \rangle = 0$ for $i = 1, \ldots, r$. Therefore

$$\langle (a - v_k, b - 1), \ell_i \rangle \geq 0 \qquad (i = 1, \ldots, r).$$

Thus $(a, b) - (v_k, 1) \in \mathbb{R}_+(I)$, a contradiction to the choice of $x^a t^b$. □

Rees algebras and their a-invariants In this part we compute some a-invariants in terms of combinatorial invariants of graphs and clutters, and exhibit some families of Gorenstein Rees algebras.

Theorem 14.3.49 *Let $\mathcal{C}$ be a d-uniform clutter and let $a(R[It])$ be the a-invariant of $R[It]$ with respect to the grading of $R[It]$ induced by* $\deg(x_i) = 1$ *and* $\deg(t) = -(d-1)$. *If* $\mathrm{gr}_I(R)$ *is reduced, then*

$$a(R[It]) \geq -\left[n - (d-1)(\alpha_0(\mathcal{C}) - 1)\right],$$

with equality if I is unmixed.

Proof. By assigning $\deg(x_i) = 1$ and $\deg(t) = -(d-1)$, the Rees ring $R[It]$ becomes a standard graded K-algebra. By Theorem 14.3.6 the Rees algebra $R[It]$ is a normal domain. Then according to a formula of Danilov–Stanley (see Theorem 9.1.5) its *canonical module* is the ideal of $R[It]$ given by

$$\omega_{R[It]} = (\{x_1^{a_1} \cdots x_n^{a_n} t^{a_{n+1}} \,|\, a = (a_i) \in (\mathbb{R}_+(I))^{\rm o} \cap \mathbb{Z}^{n+1}\}).$$

Set $\alpha_0 = \alpha_0(\mathcal{C})$. By Theorem 14.3.6, the Rees algebra of I is normal and $\mathcal{Q}(A)$ is integral. The ring $R[It]$ is Cohen–Macaulay by Theorem 9.1.6. Then by Proposition 5.2.3, the a-invariant can be expressed as

$$a(R[It]) = -\min\{\, i \,|\, (\omega_{R[It]})_i \neq 0\}.$$

Using Eq. (14.1) it is seen that $(1, \ldots, 1, \alpha_0 - 1) \in (\mathbb{R}_+(I))^{\rm o}$. Thus the inequality follows by computing the degree of $x_1 \cdots x_n t^{\alpha_0 - 1}$.

Assume that I is unmixed. Take any monomial $x^a t^b = x_1^{a_1} \cdots x_n^{a_n} t^b$ in the ideal $\omega_{R[It]}$, that is, $(a, b) \in (\mathbb{R}_+(I))^{\rm o}$. By Theorem 14.2.7 the vector (a, b) has positive entries and satisfies

$$-b + \textstyle\sum_{x_i \in C_k} a_i \geq 1 \quad (k = 1, \ldots, s). \tag{14.6}$$

If $\alpha_0 \geq b + 1$ we readily obtain the inequality

$$\deg(x^a t^b) = a_1 + \cdots + a_n - b(d-1) \geq n - (d-1)(\alpha_0 - 1). \tag{14.7}$$

Now assume $\alpha_0 \leq b$. Using the normality of $R[It]$ and Eqs. (14.1) and (14.6) it follows that the monomial

$$m = x_1^{a_1 - 1} \cdots x_n^{a_n - 1} t^{b - \alpha_0 + 1}$$

belongs to $R[It]$. Since $x^a t^b = m x_1 \cdots x_n t^{\alpha_0 - 1}$, the inequality (14.7) also holds in this case. Altogether we conclude the desired equality. $\Box$

Corollary 14.3.50 [183] *Let $\mathcal{C}$ be a d-uniform clutter. If I is unmixed with* $\alpha_0(\mathcal{C}) = 2$ *and* $\mathrm{gr}_I(R)$ *is reduced, then $R[It]$ is a Gorenstein ring and*

$$a(R[It]) = -(n - d + 1).$$

Proof. From the proof of Theorem 14.3.49 it follows that the monomial $x_1 \cdots x_n t$ generates the canonical module. $\square$

Remark 14.3.51 If $\alpha_0(\mathcal{C}) \geq 3$, then $R[It]$ is not a Gorenstein ring because $x_1 \cdots x_n t^{\alpha_0 - 1}$ and $x_1 \cdots x_n t$ are distinct minimal generators of $\omega_{R[It]}$. This holds in a more general setting (see Proposition 4.3.39).

Corollary 14.3.52 [183] *Let $J = I_c(\mathcal{C})$ be the ideal of covers of $\mathcal{C}$. If $\mathcal{C}$ is a bipartite graph and $I = I(\mathcal{C})$ is unmixed, then $R[Jt]$ is Gorenstein and*

$$a(R[Jt]) = -(n - \alpha_0(\mathcal{C}) + 1).$$

Proof. Notice that $R[Jt]$ has the grading induced by assigning $\deg(x_i) = 1$ and $\deg(t) = 1 - \alpha_0(\mathcal{C})$. Thus the formula follows from Corollary 14.3.50 once we recall that $\mathrm{gr}_J(R)$ is reduced according to Corollary 14.3.11. $\square$

Corollary 14.3.53 *Let $\mathcal{C}$ be a bipartite graph. Then the Rees cone $\mathbb{R}_+(I)$ is the intersection of the closed halfspaces given by the linear inequalities*

$$\begin{array}{rl} x_i \geq 0 & i = 1, \ldots, n+1, \\ -x_{n+1} + \sum_{v_i \in C} x_i \geq 0 & C \text{ is a minimal vertex cover of } \mathcal{C}, \end{array}$$

and none of those halfspaces can be omitted from the intersection.

Proof. Let A be the incidence matrix of $\mathcal{C}$. By Proposition 10.2.2, the matrix A is totally unimodular. Then the vertices of $\mathcal{Q}(A)$ are integral by Theorem 1.5.6. Hence the result follows from Theorem 14.1.1. $\square$

Corollary 14.3.54 *Let $\mathcal{C}$ be a bipartite graph. Then F is a facet of the Rees cone $\mathbb{R}_+(I)$ if and only if*

(a) $F = \mathbb{R}_+(I) \cap H_{e_i}$ *for some* $1 \leq i \leq n+1$, *or*

(b) $F = \mathbb{R}_+(I) \cap \{x \in \mathbb{R}^{n+1} | -x_{n+1} + \sum_{v_i \in C} x_i = 0\}$ *for some minimal vertex cover C of $\mathcal{C}$.*

Proof. The result follows from Theorem 1.1.44 and Corollary 14.3.53. $\square$

Theorem 14.3.55 [187] *Let $\mathcal{C}$ be a bipartite graph. If $a(R[It])$ is the a-invariant of $R[It]$ with respect to the grading of $R[It]$ induced by* $\deg(x_i) = 1$ *and* $\deg(t) = -1$, *then*

$$a(R[It]) = -(\beta_0(\mathcal{C}) + 1).$$

Proof. By Corollary 14.3.15, the Rees algebra $R[It]$ is a normal domain. Clearly $R[It]$ is a standard graded K-algebra with the grading induced by

$\deg(x_i) = 1$ and $\deg(t) = -1$. Then according to a formula of Danilov–Stanley (see Theorem 9.1.5) the canonical module of $R[It]$ is the ideal of $R[It]$ given by

$$\omega_{R[It]} = (\{x_1^{a_1} \cdots x_n^{a_n} t^{a_{n+1}} \,|\, a = (a_i) \in (\mathbb{R}_+(I))^{\text{o}} \cap \mathbb{Z}^{n+1}\}),$$

where $(\mathbb{R}_+(I))^{\text{o}}$ is the topological interior of the Rees cone. The ring $R[It]$ is Cohen–Macaulay by Theorem 9.1.6. Therefore the a-invariant can be expressed as

$$a(R[It]) = -\min\{\, i \,|\, (\omega_{R[It]})_i \neq 0\},$$

see Theorem 5.2.3. Let $a = (a_i)$ be an arbitrary vector in $(\mathbb{R}_+(I))^{\text{o}} \cap \mathbb{Z}^{n+1}$. By Corollary 14.3.53 a satisfies $a_i \geq 1$ for $1 \leq i \leq n+1$ and

$$-a_{n+1} + \textstyle\sum_{v_i \in C} a_i \geq 1$$

for any minimal vertex cover C of $\mathcal{C}$. Let C be a vertex cover of $\mathcal{C}$ with $\alpha_0(\mathcal{C})$ elements and let $A = V \setminus C$. Note $\beta_0(\mathcal{C}) = |A|$. Hence if m denotes the monomial $x_1^{a_1} \cdots x_n^{a_n} t^{a_{n+1}}$, then

$$\begin{aligned} \deg(m) &= a_1 + \cdots + a_n - a_{n+1} \\ &= \sum_{v_i \in A} a_i + \sum_{v_i \in C} a_i - a_{n+1} \geq \beta_0(\mathcal{C}) + 1. \end{aligned}$$

This proves the inequality $a(R[It]) \leq -(\beta_0(\mathcal{C}) + 1)$. On the other hand using Corollary 14.3.53 and the assumption $\alpha_0(\mathcal{C}) \geq 2$ we get that the monomial $m_1 = x_1 \cdots x_n t^{\alpha_0 - 1}$ is in $\omega_{R[It]}$ and has degree $\beta_0(\mathcal{C}) + 1$. Thus $a(R[It]) \geq -(\beta_0(\mathcal{C}) + 1)$. □

Corollary 14.3.56 *If $\mathcal{C}$ is a bipartite graph, then* $\text{type}(R[It]) \geq \alpha_0(\mathcal{C}) - 1$.

Proof. By Corollary 14.3.53 it is seen that the monomials $x_1 \cdots x_n t^i$, with $i = 1, \ldots, \alpha_0(\mathcal{C}) - 1$, are distinct minimal generators of $\omega_{R[It]}$. □

Exercises

14.3.57 Prove that $\mathcal{Q}_6$ is a binary non-dyadic clutter.

14.3.58 If a clutter $\mathcal{C}$ has the packing property and $I = I(\mathcal{C})$, then $I^2 = \overline{I^2}$.

14.3.59 Let v_1 and v_2 be two vectors in $\mathbb{R}^n$ with entries in $\{0, 1\}$ and let $I = (x^{v_1}, x^{v_2}) \subset R$. Prove that the Rees algebra $R[It]$ is normal.

14.3.60 Prove that the ideal $I = (x_1^2, x_2^2) \subset K[x_1, x_2]$ is not normal.

14.3.61 Let $I = (x_1x_4x_7, x_3x_5x_8, x_2x_6x_9, x_1x_4x_8, x_3x_4x_7, x_1x_5x_7)$ be the edge ideal of a clutter $\mathcal{C}$ and let A be its incidence matrix. Prove that (a) $\text{gr}_I(R)$ is reduced, and (b) A is not totally unimodular.

14.3.62 Let I and I' be monomial ideals of R and $R' = K[x_2, \ldots, x_n]$, respectively. If I is square-free and $I = x_1 I'$, prove that $\mathrm{gr}_I(R)$ is reduced if and only if $\mathrm{gr}_{I'}(R')$ is reduced.

14.3.63 If $\mathcal{C}$ is a minimally non-packing clutter, then $\mathcal{C}$ is either ideal or minimally non-ideal. Recall that $\mathcal{C}$ is called *ideal* if $\mathcal{Q}(A)$ is integral, where A is the incidence matrix of $\mathcal{C}$, and that $\mathcal{C}$ is called *minimally non-ideal* if $\mathcal{C}$ is not ideal but all its proper minors are ideal [94].

14.3.64 If G is an odd cycle, then G does not satisfy the König property.

14.3.65 Let $\mathcal{C}$ be a clutter with vertex set $X = \{x_1, \ldots, x_n\}$. Consider the clutter $\mathcal{C}'$ with vertex set $X \cup \{y_1, \ldots, y_n\}$ and whose edges are the edges of $\mathcal{C}$ together with $\{x_i, y_i\}$ for $i = 1, \ldots, n$. Prove that $\mathcal{C}$ satisfies the max-flow min-cut property if and only if $\mathcal{C}'$ does.

14.3.66 Let $\mathcal{C}$ be a clutter with vertex set $X = \{x_1, \ldots, x_n\}$ such that all its edges have $d \geq 2$ vertices. Construct a clutter $\mathcal{C}'$ satisfying the following properties: (a) all the edges of $\mathcal{C}'$ have d vertices. (b) $\mathcal{C}'$ is unmixed. (c) $\mathcal{C}$ satisfies the max-flow min-cut property if and only if $\mathcal{C}'$ does.

14.3.67 Prove that condition (a_3) of Proposition 14.3.44 is not needed when I is generated by monomials of the same degree (cf. Corollary 14.4.7).

14.3.68 Let $R = K[x_1, \ldots, x_n]$ and $R[z_1, \ldots, z_\ell]$ be polynomial rings over a field K. If I is an ideal of R generated by square-free monomials and I is normal, then $J = (I, x_1 z_1 \cdots z_\ell)$ is a normal ideal of $R[z_1, \ldots, z_\ell]$.

14.3.69 If $\mathrm{gr}_I(R)$ is reduced and $\mathcal{C}$ is d-uniform, then any square submatrix B of A of order m such that the sum of the entries in any row or column of B is equal to d satisfies $m \equiv 0 \bmod (d)$ and $\det(B) = 0$.

14.3.70 Let $x^{a_1}, \ldots, x^{a_r}$ be a set of generators of I and let B be the matrix whose columns are $a_1, \ldots, a_r$. Then both sides of the equation

$$\min\{\langle \mathbf{1}, x\rangle \,|\, x \geq 0; xA \geq \mathbf{1}\} = \max\{\langle y, \mathbf{1}\rangle \,|\, y \geq 0; Ay \leq \mathbf{1}\}$$

have integral optimum solutions if and only if both sides of the equation

$$\min\{\langle \mathbf{1}, x\rangle \,|\, x \geq 0; xB \geq \mathbf{1}\} = \max\{\langle y, \mathbf{1}\rangle \,|\, y \geq 0; By \leq \mathbf{1}\}$$

have integral optimum solutions. If any of the two systems have integral optimum solutions x, y their optimal values are equal.

14.4 Uniform ideal clutters

A clutter is called *ideal* if its set covering polyhedron is integral. In this section we study the combinatorial structure of *uniform ideal clutters*. The max-flow min-cut property of clutters is studied here from a linear algebra perspective. If a uniform clutter has the max-flow min-cut property, we prove that its incidence matrix is equivalent over $\mathbb{Z}$ to an "identity matrix" and that its column vectors form a Hilbert basis. The structure of normally torsion-free Cohen–Macaulay edge ideals of height two is also studied here.

Lemma 14.4.1 [185] *If $\mathcal{C}$ is a d-uniform clutter and $\mathcal{Q}(A)$ is integral, then there exists a minimal vertex cover of $\mathcal{C}$ intersecting every edge of $\mathcal{C}$ in exactly one vertex.*

Proof. Let B be the incidence matrix of $\mathcal{C}^\vee$. Using Corollary 14.2.22, we get that $\mathcal{Q}(B)$ is an integral polyhedron. We proceed by contradiction. Assume that for each column α_k of B there exists v_{i_k} in $\{v_1,\ldots,v_q\}$ such that $v_{i_k}B \geq \mathbf{1}+e_k$. Consider the vector $\alpha = v_{i_1}+\cdots+v_{i_s}$. From the inequality

$$\alpha B \geq (s+1,\ldots,s+1)$$

we obtain that $\alpha/(s+1) \in \mathcal{Q}(B)$. Notice that $\mathcal{Q}(B) = \mathbb{R}_+^n + \mathrm{conv}(v_1,\ldots,v_q)$; see Theorem 1.1.42. Thus, we can write

$$\alpha/(s+1) = \lambda_1 v_1+\cdots+\lambda_q v_q+\mu_1 e_1+\cdots+\mu_n e_n \quad (\lambda_i,\mu_j \geq 0;\ \textstyle\sum \lambda_i = 1).$$

Therefore, taking inner product with $\mathbf{1}$, we get $|\alpha| = sd \geq (s+1)d$, a contradiction. $\square$

Definition 14.4.2 A graph G is called *strongly perfect* if every induced subgraph H of G has a maximal independent set of vertices $\mathfrak{A}$ such that $|\mathfrak{A}\cap K| = 1$ for any maximal clique K of H.

Bipartite and chordal graphs are strongly perfect. If A is the vertex-clique matrix of G, then G being strongly perfect implies that the clique polytope of G, $\{x|\, x \geq 0;\, xA \leq \mathbf{1}\}$, has a vertex that intersects every maximal clique. In this sense, *uniform ideal clutters* can be thought of as being analogous to strongly perfect graphs.

Proposition 14.4.3 *Let $\mathcal{C}$ be a d-uniform clutter. If $\overline{R[It]} = R_s(I)$ and I is unmixed, then $H_{\ell_1}\cap\cdots\cap H_{\ell_r}\cap\mathbb{R}_+^{n+1} \neq (0)$.*

Proof. By Proposition 14.2.21 one has $\overline{R[I_c(\mathcal{C})t]} = R_s(I_c(\mathcal{C}))$. Thus, by Lemma 14.4.1, there is v_k such that $|\mathrm{supp}(v_k)\cap C_i| = 1$ for $i=1,\ldots,r$. This means that $(v_k,1)$ is in the intersection of $H_{\ell_1},\ldots,H_{\ell_r}$. $\square$

Proposition 14.4.4 *If $\mathcal{C}$ is a d-uniform clutter whose covering polyhedron $\mathcal{Q}(A)$ is integral, then there are $X_1,\ldots,X_d$ mutually disjoint minimal vertex covers of $\mathcal{C}$ such that $X=\cup_{i=1}^d X_i$.*

Proof. By induction on d. By Lemma 14.4.1 there is a minimal vertex cover X_1 of $\mathcal{C}$ such that $|\text{supp}(x^{v_i})\cap X_1|=1$ for all i. Consider the ideal I' obtained from I by making $x_i=1$ for $x_i\in X_1$. Let $\mathcal{C}'$ be the clutter corresponding to I' and let A' be the incidence matrix of $\mathcal{C}'$. The integrality of $\mathcal{Q}(A)$ is preserved under taking minors (Corollary 14.2.25), so $\mathcal{Q}(A')$ is integral. Then $\mathcal{C}'$ is a $(d-1)$-uniform clutter with $\mathcal{Q}(A')$ integral and $V(\mathcal{C}')=X\setminus X_1$. Therefore by induction hypothesis there are $X_2,\ldots,X_d$ pairwise disjoint minimal vertex covers of $\mathcal{C}'$ such that $X\setminus X_1=X_2\cup\cdots\cup X_d$.

To complete the proof we now show that $X_2,\ldots,X_d$ are minimal vertex covers of $\mathcal{C}$. If e is an edge of $\mathcal{C}$ and $2\leq k\leq d$, then $e\cap X_1=\{x_i\}$ for some i. Since $e\setminus\{x_i\}$ is an edge of $\mathcal{C}'$, we get $(e\setminus\{x_i\})\cap X_k\neq\emptyset$. Hence X_k is a vertex cover of $\mathcal{C}$. Furthermore if $x\in X_k$, then by the minimality of X_k there is an edge e' of $\mathcal{C}'$ disjoint from $X_k\setminus\{x\}$. Since $e=e'\cup\{y\}$ is an edge of $\mathcal{C}$ for some $y\in X_1$, we obtain that e is an edge of $\mathcal{C}$ disjoint from $X_k\setminus\{x\}$. Therefore X_k is a minimal vertex cover of $\mathcal{C}$, as required. $\square$

Corollary 14.4.5 *If $\mathcal{C}$ is a d-uniform clutter and $\mathcal{Q}(A)$ is integral, then there is a partition $X_1,\ldots,X_d$ of X such that $\mathcal{C}$ is a subclutter of the clutter of bases of the transversal matroid $\mathcal{M}$ defined by $X_1,\ldots,X_d$.*

Proof. It suffices to notice that, by Proposition 14.4.4, any edge of $\mathcal{C}$ intersects X_i in exactly one vertex for any i. $\square$

Theorem 14.4.6 [205] *Let $\mathcal{C}$ be a d-uniform clutter. Assume that $I(\mathcal{C})$ is normally torsion-free of height two. Then $\mathcal{C}$ is Cohen–Macaulay if and only if* (i) *$\mathcal{C}$ is unmixed, and* (ii) *there is a partition*

$$X^1=\{x_1^1,x_2^1\},\ldots,X^d=\{x_1^d,x_2^d\}$$

of X and a perfect matching $e_1=\{x_1^1,\ldots,x_1^d\}$, $e_2=\{x_2^1,\ldots,x_2^d\}$ of $\mathcal{C}$ so that all edges of $\mathcal{C}$ have the form $\{x_{i_1}^1,\ldots,x_{i_d}^d\}$ for some non-decreasing sequence $1\leq i_1\leq\cdots\leq i_d\leq 2$.

Proof. $\Rightarrow$) By Corollary 3.1.17 Cohen–Macaulay rings are unmixed, so (i) holds true. We shall prove (ii) by induction on d.

We claim that $\mathcal{C}$ has a free vertex. By Theorem 14.3.6 $\mathcal{Q}(A)$ is integral. Hence, by Proposition 14.4.4, there are minimal vertex covers $Z_1,\ldots,Z_d$ of $\mathcal{C}$ such that $Z_1,\ldots,Z_d$ partition X and $|Z_i\cap e|=1$ for all $e\in E(\mathcal{C})$ and $i=1,\ldots,d$. Since $\mathcal{C}$ is unmixed, one has $|Z_i|=2$ for all i. By Theorem 14.3.6, $\mathcal{C}$ has the König property. Then, by Lemma 6.5.7, $\mathcal{C}$ has a perfect matching of König type, i.e., there are edges e_1,e_2 of $\mathcal{C}$ such that

$e_1 \cap e_2 = \emptyset$ and $e_1 \cup e_2 = X$. We may assume that $e_1 = \{x_1, \ldots, x_d\}$, $e_2 = \{y_1, \ldots, y_d\}$, and $Z_i = \{x_i, y_i\}$ for all i. Any minimal vertex cover of $\mathcal{C}$ has the form $C = \{x, y\}$ for some $x \in e_1$ and $y \in e_2$. Let $G = \mathcal{C}^\vee$ be the blocker of $\mathcal{C}$, which, in this case, is a graph. The graph G is bipartite with bipartition e_1, e_2. As $R/I(\mathcal{C})$ is Cohen–Macaulay, $I(G) = I(\mathcal{C}^\vee)$ has a linear resolution (Theorem 6.3.41). Then, by Theorem 7.6.7, the complement graph G' of G is chordal. By Lemma 7.5.4, G' has a simplicial vertex z. We may assume that $z = x_k$ for some k. The induced subgraphs $G'[e_1]$ and $G'[e_2]$ of G' are complete graphs on d vertices. Next we prove that $x_k \notin N_{G'}(e_2)$ for any k. If $x_k \in N_{G'}(e_2)$ for some k, then $\{x_k, y_\ell\}$ is an edge of G' for some ℓ. Then y_ℓ would have to be adjacent in G' to any x_i in e_1, in particular $\{x_\ell, y_\ell\} \in E(G')$, a contradiction. Thus $\{x_k, y_i\} \in E(G)$ for all i. Note that y_k is a free vertex of $\mathcal{C}$. Indeed let e be any edge of $\mathcal{C}$ containing y_k, then x_k is not in e because $|e \cap Z_k| = 1$. Hence since $\{x_k, y_i\}$ is a vertex cover of $\mathcal{C}$ for any i we get that $y_i \in e$ for any i, i.e., $e = e_2$.

Let I' be the edge ideal obtained from $I(\mathcal{C})$ by making $x_k = 1$ and $y_k = 1$ and let $\mathcal{C}'$ be the clutter on $V' = X \setminus \{x_k, y_k\}$ such that $I' = I(\mathcal{C}')$. The ideal I' is Cohen–Macaulay of height two, normally torsion-free, and is generated by monomials of degree $d-1$. Therefore, by the induction hypothesis, there is a partition $X^2 = \{x_1^2, x_2^2\}, \ldots, X^d = \{x_1^d, x_2^d\}$ of V' such that all edges of $\mathcal{C}'$ have the form $\{x_{i_2}^2, \ldots, x_{i_d}^d\}$ for some $1 \leq i_1 \leq \cdots \leq i_d \leq 2$. To complete the proof we set $x_1^1 = x_k$, $x_2^1 = y_k$, and $X^1 = \{x_1^1, x_2^1\}$.

$\Leftarrow$) It is left as an exercise. Here, the assumption that $I(\mathcal{C})$ is normally torsion-free is not needed. $\square$

Corollary 14.4.7 *Let $\mathcal{C}$ be a d-uniform clutter. If $\overline{R[It]} = R_s(I)$ and I is unmixed, then $\mathcal{C}$ and $\mathcal{C}^\vee$ have the König property.*

Proof. According to Proposition 14.2.21 one has $\overline{R[I_c(\mathcal{C})]} = R_s(I_c(\mathcal{C}))$. Thus, by duality (Theorem 6.3.39), we need only show that $\mathcal{C}$ has the König property, and this follows readily from Proposition 14.4.4. $\square$

Theorem 14.4.8 *Let $\mathcal{C}$ be a d-uniform clutter and let $I_i = I \cap K[X \setminus \{x_i\}]$. If I is unmixed and $\mathcal{Q}(A)$ is integral, then $\mathrm{gr}_I(R)$ is reduced if and only if I_i is normal for $i = 1, \ldots, n$.*

Proof. It follows from Corollary 14.4.7 and Proposition 14.3.44. $\square$

Proposition 14.4.9 *If $\mathcal{C}$ is a d-uniform clutter with a perfect matching and $\mathcal{Q}(A)$ is integral, then $\mathcal{C}$ has the König property and is vertex critical.*

Proof. By Theorem 1.1.42, we can write $\mathcal{Q}(A) = \mathbb{R}_+^n + \mathrm{conv}(u_1, \ldots, u_s)$, where the u_i's are the characteristic vectors of the minimal vertex covers of $\mathcal{C}$. As $\mathbf{1}/d \in \mathcal{Q}(A)$, using this equality, we get

$$\mathbf{1}/d = \delta + \lambda_1 u_1 + \cdots + \lambda_s u_s; \quad (\delta \in \mathbb{R}_+^n;\ \lambda_i \geq 0;\ \textstyle\sum_i \lambda_i = 1).$$

Therefore $n \geq gd$, where $g = \alpha_0(\mathcal{C})$. By hypothesis there are mutually disjoint edges $f_1, \ldots, f_r$ such that $X = f_1 \cup \cdots \cup f_r$ and $n = rd$. Thus $n = rd \geq gd$ and $r \geq g$. On the other hand $g = \alpha_0(\mathcal{C}) \geq \beta_1(\mathcal{C}) \geq r$. Thus $r = g$ and $n = gd$. Hence $\mathcal{C}$ has the König property. We now prove that $\mathcal{C}$ is vertex critical. By Proposition 14.4.4 there are $X_1, \ldots, X_d$ mutually disjoint minimal vertex covers of $\mathcal{C}$ such that $X = \cup_{i=1}^{d} X_i$. Hence

$$n = gd = |X_1| + \cdots + |X_d|.$$

As $|X_i| \geq g$ for all i, we get $|X_i| = g$ for all i. It follows easily that $\mathcal{C}$ is vertex critical. Indeed notice that each vertex x_i belongs to a minimal vertex cover of $\mathcal{C}$ with g vertices. Hence $\alpha_0(G \setminus x_i) < \alpha_0(G)$. □

Proposition 14.4.10 *Let $\mathcal{C}$ be a d-uniform clutter with a perfect matching $f_1, \ldots, f_r$. If $\mathcal{Q}(A)$ is integral, then $r = \alpha_0(\mathcal{C})$ and there are $X_1, \ldots, X_d$ disjoint minimal vertex covers of $\mathcal{C}$ of size $\alpha_0(\mathcal{C})$ such that $X = \cup_{i=1}^{d} X_i$.*

Proof. It follows from the proof of Proposition 14.4.9. □

Theorem 14.4.11 [119] *If $\mathcal{C}$ is a d-uniform clutter with the max-flow min-cut property, then*

(a) $\Delta_r(A) = 1$, *where* $r = \operatorname{rank}(A)$.

(b) $\mathbb{N}\mathcal{A} = \mathbb{R}_+\mathcal{A} \cap \mathbb{Z}^n$, *where* $\mathcal{A} = \{v_1, \ldots, v_q\}$.

(c) *A diagonalizes over $\mathbb{Z}$ to an identity matrix.*

Proof. (a): Let B be the matrix with column vectors $(v_1, 1), \ldots, (v_q, 1)$. Since the clutter is uniform, the last row vector of B, i.e., the vector $\mathbf{1}$, is a $\mathbb{Q}$-linear combination of the first n rows of B. Thus A and B have the same rank. By Proposition 14.3.30 we obtain $K[Ft] = A(\mathcal{P})$. In particular $\overline{K[Ft]} = A(\mathcal{P})$ because $A(\mathcal{P})$ is normal. Hence, by Theorem 9.3.25, we have $\Delta_r(B) = 1$. Recall that $\Delta_r(A) = 1$ if and only if A is equivalent over $\mathbb{Z}$ to an "identity matrix." We can regard A as a matrix of size $(n+1) \times q$ by adding a row of zeros. Thus it suffices to prove that B is equivalent to A over $\mathbb{Z}$. Thanks to Proposition 14.4.4, there are $X_1, \ldots, X_d$ mutually disjoint minimal vertex covers of $\mathcal{C}$ such that $X = \cup_{i=1}^{d} X_i$ and

$$|\operatorname{supp}(x^{v_i}) \cap X_k| = 1 \quad \forall\, i, k.$$

By permuting the variables we may assume that X_1 is equal to $\{x_1, \ldots, x_r\}$. Hence the last row of B, which is the vector $\mathbf{1}$, is the sum of the first $|X_1|$ rows of B, i.e., B is equivalent to A over $\mathbb{Z}$.

(b): It suffices to prove the inclusion "$\supset$". Let a be an integral vector in $\mathbb{R}_+\mathcal{A}$. Then $a = \lambda_1 v_1 + \cdots + \lambda_q v_q$, $\lambda_i \geq 0$ for all i. Thus, setting $b = \sum_i \lambda_i$, one has $|a| = bd$. We claim that $(a, \lceil b \rceil)$ belongs to $\mathbb{R}_+(I)$. Let $u_1, \ldots, u_s$

be the characteristic vectors of the minimal vertex covers of $\mathcal{C}$. Since $\mathcal{Q}(A)$ is integral, by Theorem 14.1.1, we can write

$$\mathbb{R}_+(I) = H_{e_1}^+ \cap H_{e_2}^+ \cap \cdots \cap H_{e_{n+1}}^+ \cap H_{\ell_1}^+ \cap H_{\ell_2}^+ \cap \cdots \cap H_{\ell_s}^+, \tag{14.8}$$

where $\ell_i = (u_i, -1)$ for $1 \leq i \leq s$. Notice that $(a, b) \in \mathbb{R}_+(I)$, thus using Eq. (14.8) we get that $\langle a, u_i\rangle \geq b$ for all i. Hence $\langle a, u_i\rangle \geq \lceil b\rceil$ for all i because $\langle a, u_i\rangle$ is an integer for all i. Using Eq. (14.8) again we get that $(a, \lceil b\rceil) \in \mathbb{R}_+(I)$, as claimed. By Theorem 14.3.6 the Rees algebra $R[It]$ is normal. Then, applying Proposition 14.2.3, we obtain that $(a, \lceil b\rceil)$ is in $\mathbb{N}\mathcal{A}'$. There are nonnegative integers $\eta_1, \ldots, \eta_q$ and $\rho_1, \ldots, \rho_n$ such that

$$(a, \lceil b\rceil) = \eta_1(v_1, 1) + \cdots + \eta_q(v_q, 1) + \rho_1 e_1 + \cdots + \rho_n e_n.$$

Hence it is seen that $|a| = \lceil b\rceil d + \sum_i \rho_i = bd$ and consequently $\rho_i = 0$ for all i and $b = \lceil b\rceil$. It follows at once that $a \in \mathbb{N}\mathcal{A}$ as required.

(c): By part (a) one has $\Delta_r(A) = 1$. Thus the invariant factors of A are all equal to 1 (see Theorem 1.3.15), i.e., the Smith normal canonical form of A is an identity matrix. □

This result and the Conforti–Cornuéjols conjecture suggest the following weaker conjecture.

Conjecture 14.4.12 [185] If $\mathcal{C}$ is a uniform clutter with the packing property, then either of the following equivalent conditions hold:

(a) $\mathbb{Z}^n/\mathbb{Z}\mathcal{A}$ is a free group.

(b) A diagonalizes over $\mathbb{Z}$ to an identity matrix.

This conjecture will be proved later for d-uniform clutters with a perfect matching and $\alpha_0(\mathcal{C}) = 2$ (see Theorem 14.4.15).

Corollary 14.4.13 *Let $\mathcal{C}$ be a uniform clutter. Then $\mathcal{C}$ satisfies the max-flow min-cut property if and only if $\mathcal{Q}(A)$ is integral and $\mathbb{N}\mathcal{A} = \mathbb{R}_+\mathcal{A} \cap \mathbb{Z}^n$.*

Proof. $\Rightarrow$) It follows from Theorems 14.4.11 and 14.3.6.

$\Leftarrow$) By Proposition 14.3.30 we need only show $K[Ft] = A(\mathcal{P})$. The inclusion "$\subset$" is clear. To show the inclusion "$\supset$" take $x^a t^b \in A(\mathcal{P})$. Then $a \in b\mathcal{P} \cap \mathbb{Z}^n$. From $\mathbb{N}\mathcal{A} = \mathbb{R}_+\mathcal{A} \cap \mathbb{Z}^n$ it is seen that $a = \sum_{i=1}^q \eta_i v_i$ for some $\eta_i \in \mathbb{N}$ such that $\sum_i \eta_i = b$. Thus $x^a t^b \in K[Ft]$, as required. □

Corollary 14.4.14 *Let $\mathcal{C}$ be a d-uniform clutter. If $\mathcal{C}$ has the max-flow min-cut property, then $\mathcal{C}$ has a perfect matching if and only if $n = d\alpha_0(\mathcal{C})$.*

Proof. $\Rightarrow$) As $\mathcal{Q}(A)$ is integral, from the proof of Proposition 14.4.9 we obtain the equality $n = d\alpha_0(\mathcal{C})$.

$\Leftarrow$) We set $g = \alpha_0(\mathcal{C})$. Let B be the incidence matrix of $\mathcal{C}^\vee$. As $\mathcal{Q}(A)$ is integral, by Corollary 14.2.22, we get that $\mathcal{Q}(B)$ is an integral polyhedron. Consequently, by Theorem 1.1.42, we can write

$$\mathcal{Q}(B) = \mathbb{R}_+^n + \mathrm{conv}(v_1, \ldots, v_q),$$

Therefore, since the rational vector $\mathbf{1}/g$ is in $\mathcal{Q}(B)$, we can write

$$\mathbf{1}/g = \delta + \mu_1 v_1 + \cdots + \mu_q v_q \quad (\delta \in \mathbb{R}_+^n;\ \mu_i \geq 0;\ \mu_1 + \cdots + \mu_q = 1).$$

Hence $n/g = |\delta| + (\sum_{i=1}^q \mu_i)d = |\delta| + d$. Since $n = dg$, we obtain that $\delta = 0$. Thus the vector $\mathbf{1}$ is in $\mathbb{R}_+\mathcal{A} \cap \mathbb{Z}^n$. By Theorem 14.4.11(b) this intersection is equal to $\mathbb{N}\mathcal{A}$. Then we can write $\mathbf{1} = \eta_1 v_1 + \cdots + \eta_q v_q$, for some $\eta_1, \ldots, \eta_q$ in $\mathbb{N}$. Hence it is readily seen that $\mathcal{C}$ has a perfect matching. $\square$

The next result gives some support to Conjecture 14.3.19.

Theorem 14.4.15 [119] *Let $\mathcal{C}$ be a d-uniform clutter with a perfect matching such that $\mathcal{C}$ has the packing property and $\alpha_0(\mathcal{C}) = 2$. Then:*

(a) *A diagonalizes over $\mathbb{Z}$ to an "identity matrix."*

(b) *If $K[Ft]$ is normal, then $\mathcal{C}$ has the max-flow min-cut property.*

(c) *If $\mathcal{A}$ is linearly independent, then $\mathcal{C}$ has the max-flow min-cut property.*

Proof. (a): By Lehman theorem $\mathcal{Q}(A)$ is integral; see Theorem 14.3.17. Thus by Proposition 14.4.10 there is a perfect matching f_1, f_2 of X with $X = f_1 \cup f_2$ and there is a partition $X_1, \ldots, X_d$ of X such that X_i is a minimal vertex cover of $\mathcal{C}$ for all i, $|X_i| = 2$ for all i, and

$$|\mathrm{supp}(x^{v_i}) \cap X_k| = 1 \quad \forall\, i, k. \tag{14.9}$$

Thus we may assume that $X_i = \{x_{2i-1}, x_{2i}\}$ for $i = 1, \ldots, d$. Notice that $n = 2d$ because $X = f_1 \cup f_2$.

By induction on $r = \mathrm{rank}(A)$. Since $\mathbf{1}$ is the sum of the first two rows of A it suffices to prove that 1 is the only invariant factor of $A^\top$ or equivalently that the Smith normal form of $A^\top$ is the "identity." Let $w_1, \ldots, w_{2d}$ be the columns of $A^\top$ and let V_i be the linear space generated by $w_1, \ldots, w_{2i}$. For k odd, one has $w_k + w_{k+1} = \mathbf{1}$. Hence if k is odd and we remove columns w_k and w_{k+1} from $A^\top$ we obtain a submatrix whose rank is greater than or equal to $r - 1$. Thus after permuting columns we may assume

$$\dim(V_i) = \begin{cases} i+1 & \text{if} \quad 1 \leq i \leq r-1, \\ r & \text{if} \quad r \leq i. \end{cases} \tag{14.10}$$

Let J be the monomial ideal defined by the rows of $[w_1, \ldots, w_{2(r-1)}]$. Then a minimal set of generators of J consists of monomials of degree $r-1$ and will still satisfy condition in Eq. (14.9) with d replaced by $r-1$. Furthermore J satisfies the packing property because it is a minor of I. If $[w_1, \ldots, w_{2(r-1)}]$ diagonalizes (over the integers) to the identity matrix so does $A^\top$. Therefore we may assume $d = r-1$ and $I = J$.

Let B be the matrix $[w_1, \ldots, w_{2(d-1)}]$ and let I' be the monomial ideal defined by the rows of B; that is, I' is obtained from I making $x_{2d-1} = 1$ and $x_{2d} = 1$. Notice that by induction B diagonalizes to an identity matrix $[I_{r-1}, \mathbf{0}]$ of order $r-1$. Since I has the König property we may permute rows and columns and assume that the matrix $A^\top$ is written as:

$$
\underbrace{\begin{array}{ccccc}
10 & 10 & 10 & \cdots & 10 \\
01 & 01 & 01 & \cdots & 01 \\
\bigcirc & \bigcirc & \bigcirc & \cdots & \bigcirc \\
\vdots & \vdots & & & \vdots \\
\bigcirc & \bigcirc & \bigcirc & \cdots & \bigcirc \\
\bigcirc & \bigcirc & \bigcirc & \cdots & \bigcirc \\
\vdots & \vdots & & & \vdots \\
\bigcirc & \bigcirc & \bigcirc & \cdots & \bigcirc
\end{array}}_{B}
\begin{array}{cc}
10 & \leftarrow \\
01 & \\
10 & \leftarrow \\
\vdots & \vdots \\
10 & \leftarrow \\
01 & \\
\vdots & \\
01 &
\end{array}
$$

where either a pair $1\,0$ or $0\,1$ must occur in the places marked with a circle and such that the number of $1's$ in the last column is greater than or equal to the number of $1's$ in any other column. Consider the matrix C obtained from A by removing the rows whose penultimate entry is equal to 1 (these are marked above with an arrow) and removing the last column. Let K be the monomial ideal defined by the rows of C; that is, K is obtained from I by making $x_{2d-1} = 0$ and $x_{2d} = 1$. By the choice of the last column and because of Eq. (14.10) it is seen that K has height two. Since K is a minor of I it has the König property. As a consequence using row operations $A^\top$ can be brought to the form:

$$
\underbrace{\begin{array}{ccccc}
10 & 10 & 10 & \cdots & 10 \\
10 & 10 & 10 & \cdots & 10 \\
\bigcirc & \bigcirc & \bigcirc & \cdots & \bigcirc \\
\vdots & \vdots & & & \vdots \\
\bigcirc & \bigcirc & \bigcirc & \cdots & \bigcirc \\
\bigcirc & \bigcirc & \bigcirc & \cdots & \bigcirc \\
\vdots & \vdots & & & \vdots \\
\bigcirc & \bigcirc & \bigcirc & \cdots & \bigcirc
\end{array}}_{B_1}
\begin{array}{c}
10 \\
01 \\
10 \\
\vdots \\
10 \\
01 \\
\vdots \\
01
\end{array}
$$

where B_1 has rank $r-1$ and diagonalizes to an identity. Therefore it is readily seen that this matrix reduces to

$$\begin{matrix} I_{r-1} & 0 & 0 & \cdots & 0 & 0 & 0 \\ \mathbf{0} & 0 & 0 & \cdots & 0 & 1 & -1 \\ \mathbf{0} & 0 & 0 & \cdots & 0 & a_1 & b_1 \\ \vdots & \vdots & & & \vdots & \vdots & \vdots \\ \mathbf{0} & 0 & 0 & \cdots & 0 & a_r & b_r \end{matrix}$$

To finish the proof observe that by rank considerations (see Eq. (14.10)) this matrix reduces to $[I_r, \mathbf{0}]$, as required.

(b): By part (a) we have $\Delta_r(A) = 1$. Consider the matrix B with column vectors $(v_1, 1), \ldots, (v_q, 1)$. Since $|v_i| = d$ for all i, one has $\Delta_r(B) = 1$ (see Exercise 14.4.23). Then by Theorem 9.3.25 this condition is equivalent to the equality $\overline{K[Ft]} = A(P)$. Hence $K[Ft] = A(\mathcal{P})$ because $K[Ft]$ is a normal domain. By Lehman theorem (Theorem 14.3.17), $\mathcal{Q}(A)$ is integral. Thus $\mathcal{C}$ satisfies max-flow min-cut by Proposition 14.3.30 and Theorem 14.3.6.

(c): $x^{v_1}t, \ldots, x^{v_q}t$ are algebraically independent and $K[Ft]$ is normal because it is a polynomial ring. Hence the result follows from (b). $\Box$

Recall that I is called *minimally non-normal* if I is not normal and all its proper minors are normal.

Theorem 14.4.16 *Let $\mathcal{C}$ be a d-uniform clutter and $\{X_i\}_{i=1}^d$ a partition of X such that $X_i = \{x_{2i-1}, x_{2i}\}$ is a minimal vertex cover of $\mathcal{C}$ for all i. Then*

(a) $\text{rank}(A) \leq d + 1$.

(b) *If C is a minimal vertex cover of $\mathcal{C}$, then* $2 \leq |C| \leq d$.

(c) *If $\mathcal{C}$ satisfies the König property and there is a minimal vertex cover C of $\mathcal{C}$ with $|C| = d \geq 3$, then* $\text{rank}(A) = d + 1$.

(d) *If $I = I(\mathcal{C})$ is minimally non-normal and $\mathcal{C}$ has the packing property, then* $\text{rank}(A) = d + 1$.

Proof. (a): For each odd integer k the sum of rows k and $k+1$ of the matrix A is equal to $\mathbf{1}$. Thus the rank of A is bounded by $d+1$.

(b): By the pigeon hole principle, any $C \in \mathcal{C}^\vee$ satisfies $2 \leq |C| \leq d$.

(c): Notice that C contains exactly one vertex of each X_j because $X_j \not\subset C$. Thus we may assume $C = \{x_1, \ldots, x_{2d-1}\}$. Consider $x^\alpha := x_2 \cdots x_{2d}$ and notice that $x_k x^\alpha \in I$ for each $x_k \in C$ because $x_k x^\alpha$ is in every minimal prime of I. Writing $x_k = x_{2i-1}$ with $1 \leq i \leq d$ we conclude that the monomial $x^{\alpha_i} = x_2 \cdots x_{2(i-1)} x_{2i-1} x_{2(i+1)} \cdots x_{2d}$ is a minimal generator of I. Thus we may assume $x^{\alpha_i} = x^{v_i}$ for $i = 1, \ldots, d$. The vector $\mathbf{1}$ belongs to $\mathbb{Q}\mathcal{A}$ because $\mathcal{C}$ has the König property. It follows that the matrix with rows $v_1, \ldots, v_d, \mathbf{1}$ has rank $d+1$, as required.

(d): Let $x^\alpha t^b$ be a minimal generator of $\overline{R[It]}$ not in $R[It]$ and let $m = x^\alpha$. Using Proposition 14.3.29, one has $\deg(m) = bd$. By hypothesis $\mathfrak{m}$ is the only associated prime of $N = \overline{I^b}/I^b$. Hence, by Corollary 2.1.29, we have

$$\mathrm{rad}(\mathrm{ann}(N)) = \bigcap_{\mathfrak{p}\in\mathrm{Ass}(N)} \mathfrak{p} = \mathfrak{m} = (x_1, \ldots, x_n)$$

and $\mathfrak{m}^r \subset \mathrm{ann}(N)$ for some $r > 0$. Thus for i odd we can write

$$x_i^r x^\alpha = (x^{v_1})^{a_1} \cdots (x^{v_q})^{a_q} x^\delta,$$

where $a_1 + \cdots + a_q = b$ and $\deg(x^\delta) = r$. If we write $x^\delta = x_i^{s_1} x_{i+1}^{s_2} x^\gamma$ with x_i, x_{i+1} not in the support of x^γ, making $x_j = 1$ for $j \notin \{i, i+1\}$, it is not hard to see that $r = s_1 + s_2$ and $\gamma = 0$. Thus we get an equation:

$$x_i^{s_2} x^\alpha = (x^{v_1})^{a_1} \cdots (x^{v_q})^{a_q} x_{i+1}^{s_2}$$

with $s_2 > 0$. Using a similar argument we obtain an equation:

$$x_{i+1}^{w_1} x^\alpha = (x^{v_1})^{b_1} \cdots (x^{v_q})^{b_q} x_i^{w_1} \quad \text{with} \quad w_1 > 0.$$

Hence, $x_{i+1}^{s_2+w_1}(x^{v_1})^{a_1} \cdots (x^{v_q})^{a_q} = x_i^{s_2+w_1}(x^{v_1})^{b_1} \cdots (x^{v_q})^{b_q}$. As $\mathbb{Z}^n/\mathbb{Z}\mathcal{A}$ is torsion-free (Theorem 14.4.15), we get $e_i - e_{i+1} \in \mathbb{Z}\mathcal{A}$ for i odd. Finally to conclude that $\mathrm{rank}(A) = d + 1$ notice that $\mathbf{1} \in \mathbb{Z}\mathcal{A}$. □

The next theorem gives an interesting class of uniform clutters, coming from combinatorial optimization, that satisfy Conjecture 9.6.22.

Theorem 14.4.17 *If A is a balanced matrix and $\mathcal{C}$ is a d-uniform clutter, then any regular triangulation of the cone $\mathbb{R}_+\mathcal{A}$ is weakly unimodular.*

Proof. Let $\mathcal{A}_1, \ldots, \mathcal{A}_m$ be the elements of a regular triangulation of $\mathbb{R}_+\mathcal{A}$. Then $\dim \mathbb{R}_+\mathcal{A}_i = \dim \mathbb{R}_+\mathcal{A}$ and $\mathcal{A}_i$ is linearly independent for all i. Let $\mathcal{C}_i$ be the subclutter of $\mathcal{C}$ whose edges correspond to the vectors in $\mathcal{A}_i$, and let A_i be the incidence matrix of $\mathcal{C}_i$. As A_i is a balanced matrix, using Corollary 14.3.11 and Theorem 14.3.6, we get that the clutter $\mathcal{C}_i$ has the max-flow min-cut property. Hence by Theorem 14.4.11 one has $\Delta_r(A_i) = 1$, where r is the rank of A. Thus the invariant factors of A_i are all equal to 1 (see Theorem 1.3.15). Therefore by the fundamental structure theorem for finitely generated abelian groups (see Theorem 1.3.16) the quotient group $\mathbb{Z}^n/\mathbb{Z}\mathcal{A}_i$ is torsion-free for all i. Notice that

$$\dim \mathbb{R}_+\mathcal{A} = \mathrm{rank}\, \mathbb{Z}\mathcal{A} \quad \text{and} \quad \dim \mathbb{R}_+\mathcal{A}_i = \mathrm{rank}\, \mathbb{Z}\mathcal{A}_i$$

for all i. Since r is equal to $\dim \mathbb{R}_+\mathcal{A}$. It follows that the quotient group $\mathbb{Z}\mathcal{A}/\mathbb{Z}\mathcal{A}_i$ is torsion-free and has rank 0 for all i, consequently $\mathbb{Z}\mathcal{A} = \mathbb{Z}\mathcal{A}_i$ for all i, i.e., the triangulation is weakly unimodular. □

Definition 14.4.18 Let $K[Ft]=K[x^{v_1}t,\ldots,x^{v_q}t]$ be a homogeneous subring with the standard grading induced by $\deg(x^at^b)=b$. Let

$$K[t_1,\ldots,t_q]/I_{\mathcal{A}}\simeq K[Ft],\qquad \overline{t}_i\mapsto x^{v_i}t,$$

be a presentation of $K[Ft]$. The *regularity* of $K[Ft]$, denoted $\text{reg}(K[Ft])$, is defined as the regularity of $K[t_1,\ldots,t_q]/I_{\mathcal{A}}$ as an $K[t_1,\ldots,t_q]$-module.

Theorem 14.4.19 *Let $\mathcal{C}$ be a d-uniform unmixed clutter with $g=\alpha_0(\mathcal{C})$. If $\mathcal{C}$ has the max-flow min-cut property, then $K[Ft]=A(\mathcal{P})$, the a-invariant of $A(\mathcal{P})$ is bounded from above by $-g$ and* $\text{reg}(A(\mathcal{P}))\leq(d-1)(g-1)$.

Proof. Setting $\mathcal{B}=\{(v_i,1)\}_{i=1}^q$ and $\mathcal{A}'=\mathcal{B}\cup\{e_i\}_{i=1}^n$, one has the equality

$$\mathbb{R}_+\mathcal{B}=\mathbb{R}\,\mathcal{B}\cap\mathbb{R}_+\mathcal{A}'. \tag{14.11}$$

By Theorem 14.3.6, $R[I(\mathcal{C})t]$ is normal and $\mathcal{Q}(A)$ is integral. Hence, using Theorem 9.3.31, we obtain $K[Ft]=A(\mathcal{P})$ and $A(\mathcal{P})$ becomes a standard graded K-algebra. By Theorems 14.1.1 and 14.2.7, we obtain the equality

$$\mathbb{R}_+(I)=H_{e_1}^+\cap\cdots\cap H_{e_{n+1}}^+\cap H_{(u_1,-1)}^+\cap\cdots\cap H_{(u_s,-1)}^+, \tag{14.12}$$

where $u_1,\ldots,u_s$ are the characteristic vectors of the minimal vertex covers of $\mathcal{C}$. Hence, by Proposition 14.4.4, there are $X_1,\ldots,X_d$ mutually disjoint minimal vertex covers of $\mathcal{C}$ of size g such that $X=\cup_{i=1}^d X_i$. Notice that $|X_i\cap f|=1$ for $1\leq i\leq d$ and $f\in E(\mathcal{C})$. We may assume that $X_i=C_i$ for $1\leq i\leq d$. Therefore, using Eqs. (14.11) and (14.12), we get

$$\mathbb{R}_+\mathcal{B}=\mathbb{R}\,\mathcal{B}\cap H_{e_1}^+\cap\cdots\cap H_{e_n}^+\cap H_{e_{n+1}}^+\cap\left(\bigcap_{i\in\mathcal{I}}H_{(u_i,-1)}^+\right), \tag{14.13}$$

where $i\in\mathcal{I}$ if and only if $H_{(u_i,-1)}^+$ defines a proper face of $\mathbb{R}_+\mathcal{B}$. The Ehrhart ring $A(\mathcal{P})$ is Cohen–Macaulay (Theorem 9.1.6). Then, by Theorem 9.1.5 and Eq. (14.13), the canonical module of $A(\mathcal{P})$ is the ideal given by

$$\omega_{A(\mathcal{P})}=((xt)^a\,|\,a\in\mathbb{R}\,\mathcal{B};\,a_i\geq 1\,\forall\,i;\,\langle(u_k,-1),a\rangle\geq 1\ \text{ for }\ k\in\mathcal{I}), \tag{14.14}$$

where $(xt)^a=x_1^{a_1}\cdots x_n^{a_n}t^{a_{n+1}}$. The a-invariant of $A(\mathcal{P})$ is given by

$$a(A(\mathcal{P}))=-\min\{\,i\,|\,(\omega_{A(\mathcal{P})})_i\neq 0\}, \tag{14.15}$$

see Proposition 5.2.3. Take an arbitrary monomial $x^at^b=x_1^{a_1}\cdots x_n^{a_n}t^b$ in the ideal $\omega_{A(\mathcal{P})}$. By Eqs. (14.13) and (14.14), the vector (a,b) is in $\mathbb{R}_+\mathcal{B}$ and $a_i\geq 1$ for all i. Thus we can write $(a,b)=\sum_{i=1}^q\lambda_i(v_i,1)$ with $\lambda_i\geq 0$ for all i. Since $\langle v_i,u_k\rangle=1$ for all i,k, we obtain

$$g=|u_k|\leq\sum_{x_i\in C_k}a_i=\langle a,u_k\rangle=\sum_{i=1}^q\lambda_i\langle v_i,u_k\rangle=\sum_{i=1}^q\lambda_i=b$$

for $1 \leq k \leq d$. This means that $\deg(x^a t^b) \geq g$. Thus $-a(A(\mathcal{P})) \geq g$, as required. Next we show that $\operatorname{reg}(A(\mathcal{P})) \leq (d-1)(g-1)$. Since $A(\mathcal{P})$ is Cohen–Macaulay, by Theorem 6.4.1, we have

$$\operatorname{reg}(A(\mathcal{P})) = \dim(A(\mathcal{P})) + a(A(\mathcal{P})) \leq \dim(A(\mathcal{P})) - g. \tag{14.16}$$

Using that $\langle v_i, u_k \rangle = 1$ for all i, k, by induction on d, it is seen that $\operatorname{rank}(A) \leq g+(d-1)(g-1)$. Thus, using the fact that $\dim(A(\mathcal{P})) = \operatorname{rank}(A)$ and Eq. (14.16), we get $\operatorname{reg}(A(\mathcal{P})) \leq (d-1)(g-1)$. $\square$

Exercises

14.4.20 Let $\mathcal{C}$ be the clutter with vertex set $X = \{x_1, \ldots, x_9\}$ and edges:

$$\begin{array}{llll} f_1 = \{x_1, x_2\}, & f_2 = \{x_3, x_4, x_5, x_6\}, & f_3 = \{x_7, x_8, x_9\}, & \\ f_4 = \{x_1, x_3\}, & f_5 = \{x_2, x_4\}, & f_6 = \{x_5, x_7\}, & f_7 = \{x_6, x_8\}. \end{array}$$

Prove the following: the incidence matrix of $\mathcal{C}$ is a balanced matrix, $\mathcal{Q}(A)$ is integral, $|C \cap f_i| \geq 2$ for any $C \in \mathcal{C}^\vee$ and for any i, $\mathcal{C}$ has a perfect matching, $\alpha_0(\mathcal{C} \setminus \{x_9\}) = \alpha_0(\mathcal{C}) = 4$, $\mathcal{C}$ is not vertex critical, and the uniformity hypothesis is essential in Propositions 14.4.4 and 14.4.9.

14.4.21 Let A be the incidence matrix of a cycle of length 3 and let B be the matrix obtained from A by adjoining the row $\mathbf{1}$. Prove that $\det(A) = 2$ and $\Delta_3(B) = 1$. In particular A and B are not equivalent over $\mathbb{Z}$.

14.4.22 Consider the clutter $\mathcal{C}$ whose incidence matrix is

$$\begin{bmatrix} 1 & 0 & 0 & 1 \\ 0 & 1 & 0 & 1 \\ 0 & 0 & 1 & 1 \\ 0 & 1 & 1 & 0 \\ 1 & 0 & 1 & 0 \end{bmatrix}.$$

Prove that $\mathcal{C}$ satisfies max-flow min-cut, A is not equivalent over $\mathbb{Z}$ to an identity matrix, and the columns of A do not form a Hilbert basis for the cone they generate. Thus the uniformity hypothesis is essential in the two statements of Theorem 14.4.11.

14.4.23 Let $\mathcal{A} = \{v_1, \ldots, v_q\} \subset \mathbb{Z}^n$. If $|v_i| = d \neq 0$ for all i and $\mathbb{Z}^n/\mathbb{Z}\mathcal{A}$ is a torsion-free $\mathbb{Z}$-module, then $\mathbb{Z}^{n+1}/\mathbb{Z}\{(v_1, 1)), \ldots, (v_q, 1)\}$ is torsion-free.

14.4.24 Let G be an unmixed bipartite graph, let $\mathcal{A}^\vee = \{u_1, \ldots, u_s\}$ be the set of column vectors of the incidence matrix of $G^\vee$, and let $\mathcal{P}^\vee = \operatorname{conv}(\mathcal{A}^\vee)$. Then $K[x^{u_1}t, \ldots, x^{u_s}t] = A(\mathcal{P}^\vee)$ and $\operatorname{reg}(A(\mathcal{P}^\vee)) \leq (|V(G)|/2) - 1$.

14.4.25 If we do not require that $|v_i| = d$ for all i in Theorem 14.4.17, give an example to show that this theorem is false even if $K[F]$ is homogeneous.

14.4.26 [119] Let $\{u_1, \ldots, u_r\}$ be the set of all characteristic vectors of the collection of bases $\mathcal{B}$ of a transversal matroid $\mathcal{M}$. Then the polyhedral cone $\mathbb{R}_+\{u_1, \ldots, u_r\}$ has a weakly unimodular regular triangulation.

14.5 Clique clutters of comparability graphs

In this section we prove that the clique clutter of a comparability graph satisfies the max-flow min-cut property. We also present some classical combinatorial results on posets, e.g., Dilworth decomposition theorem and a min-max theorem of Menger.

Let $P = (X, \prec)$ be a *partially ordered set* (*poset* for short) on the finite vertex set $X = \{x_1, \ldots, x_n\}$ and let G be its *comparability graph*. Recall that the vertex set of G is X and the edge set of G is the set of all unordered pairs $\{x_i, x_j\}$ such that x_i and x_j are comparable. A *clique* of G is a subset of the set of vertices that induces a complete subgraph.

Lemma 14.5.1 *If G^1 (resp. $\mathrm{cl}(G)^1$) is the graph (resp. clutter) obtained from G (resp. $\mathrm{cl}(G)$) by duplicating the vertex x_1, then $\mathrm{cl}(G)^1 = \mathrm{cl}(G^1)$.*

Proof. Let y_1 be the duplication of x_1. We now prove the inclusion $E(\mathrm{cl}(G)^1) \subset E(\mathrm{cl}(G^1))$. The other inclusion follows readily using similar arguments. Take $e \in E(\mathrm{cl}(G)^1)$.

Case (i): Assume $y_1 \notin e$. Then $e \in E(\mathrm{cl}(G))$. Clearly e is a clique of G^1. If $e \notin E(\mathrm{cl}(G^1))$, then e can be extended to a maximal clique of G^1. Hence $e \cup \{y_1\}$ must be a clique of G^1. Note that $x_1 \notin e$ because $\{x_1, y_1\}$ is not an edge of G^1. Then $e \cup \{x_1\}$ is a clique of G, a contradiction.

Case (ii): Assume $y_1 \in e$. Then there is $f \in E(\mathrm{cl}(G))$, with $x_1 \in f$, such that $e = (f \setminus \{x_1\}) \cup \{y_1\}$. Since $\{x, x_1\} \in E(G)$ for any x in $f \setminus \{x_1\}$, one has that $\{x, y_1\} \in E(G^1)$ for any x in $f \setminus \{x_1\}$. Then e is a clique of G^1. If e is not a maximal clique of G^1, there is $x \notin e$ which is adjacent in G to any vertex of $f \setminus \{x_1\}$ and x is adjacent to y_1 in G^1. In particular $x \neq x_1$. Then x is adjacent in G to x_1 and consequently x is adjacent in G to any vertex of f, a contradiction because f is a maximal clique of G. $\square$

Let $\mathcal{D}$ be a *digraph*; that is, $\mathcal{D}$ consists of a finite set $V(\mathcal{D})$ of vertices and a set $E(\mathcal{D})$ of ordered pairs of distinct vertices called edges. Let A, B be two sets of vertices of $\mathcal{D}$. For use below recall that a (directed) path of $\mathcal{D}$ is called an *A–B path* if it runs from a vertex in A to a vertex in B. A common vertex of A and B is also an A–B path. A set C of vertices is called an *A–B disconnecting* set if C intersects each A–B path. Menger [316] gave a min-max theorem for the maximum number of disjoint A–B paths in an undirected graph. This theorem also holds for digraphs.

Theorem 14.5.2 (Menger's theorem, see [373, Theorem 9.1]) *Let $\mathcal{D}$ be a digraph and let A, B be two subsets of $V(\mathcal{D})$. Then the maximum number of vertex-disjoint A–B paths is equal to the minimum size of an A–B disconnecting vertex set.*

Proof. Let k be the minimum size of an A–B disconnecting vertex set and let ℓ be the maximum number of vertex-disjoint A–B paths. Clearly $\ell \leq k$. Equality will be shown by induction on $E(\mathcal{D})$, the number of edges of $\mathcal{D}$. The case $E(\mathcal{D}) = \emptyset$ is trivial. Pick an edge $a = (u, v)$ of $\mathcal{D}$. If each A–B disconnecting vertex set of $\mathcal{D} \setminus a$ has size at least k, then inductively there are k vertex-disjoint A–B paths in $\mathcal{D} \setminus a$, hence in $\mathcal{D}$. Thus we may assume that there exists an A–B disconnecting vertex set C of $\mathcal{D} \setminus a$ of size less than or equal to $k - 1$. Then $C \cup \{u\}$ and $C \cup \{v\}$ are A–B disconnecting vertex sets of $\mathcal{D}$ of size k.

Now, each A–$(C \cup \{u\})$ disconnecting vertex set $\mathfrak{A}$ of $\mathcal{D} \setminus a$ has size at least k, as $\mathfrak{A}$ is an A–B disconnecting vertex set of $\mathcal{D}$. Indeed, if $\mathcal{P}$ is an A–B path of $\mathcal{D}$, then $\mathcal{P}$ intersects $C \cup \{u\}$, and hence $\mathcal{P}$ contains an A–$(C \cup \{u\})$ path. So $\mathcal{P}$ intersects $\mathfrak{A}$. Therefore by induction $\mathcal{D} \setminus a$ contains k vertex-disjoint A–$(C \cup \{u\})$ paths $\mathcal{P}_0, \ldots, \mathcal{P}_{k-1}$. Similarly $\mathcal{D} \setminus a$ contains k vertex-disjoint $(C \cup \{v\})$–B paths $\mathcal{Q}_0, \ldots, \mathcal{Q}_{k-1}$. If p_i is the last vertex of $\mathcal{P}_i$, we may assume—by reducing the size of $\mathcal{P}_i$ if necessary—that any other vertex of $\mathcal{P}_i$ is not in $C \cup \{u\}$. Similarly if q_i is the first vertex of $\mathcal{Q}_i$, we may assume that all other vertices of $\mathcal{Q}_i$ are not in $C \cup \{v\}$. Since $|C| = k - 1$, we have

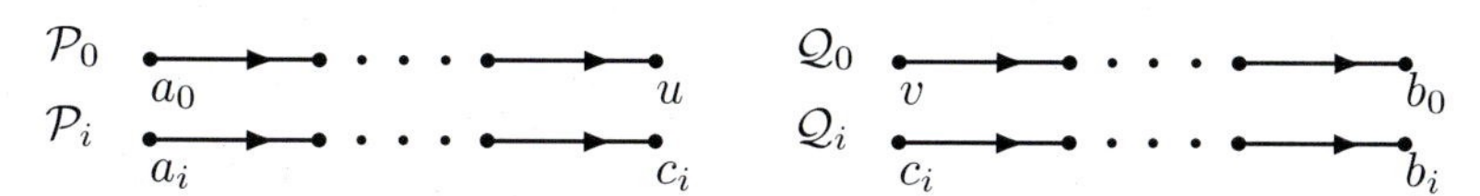

where $C = \{c_1, \ldots, c_{k-1}\}$, $a_i \in A$, $b_i \in B$ for $i = 0, \ldots, k - 1$. Any path in the first collection intersects any path in the second collection only in C, since otherwise $\mathcal{D} \setminus a$ contains an A–B path avoiding C. Therefore we can pairwise concatenate the paths to obtain k vertex-disjoint A–B paths. This proof was adapted from [195]. $\square$

Theorem 14.5.3 [118] *Let $P = (X, \prec)$ be a poset on the vertex set X and let G be its comparability graph. If $\mathcal{C} = \mathrm{cl}(G)$ is the clutter of maximal cliques of G, then $\mathcal{C}$ satisfies the max-flow min-cut property.*

Proof. We can regard P as a transitive digraph without cycles of length two with vertex set X and edge set $E(P)$, i.e., the edges of P are ordered pairs (a, b) of distinct vertices with $a \prec b$ such that:

(i) $(a, b) \in E(P)$ and $(b, c) \in E(P) \Rightarrow (a, c) \in E(P)$, and
(ii) $(a, b) \in E(P) \Rightarrow (b, a) \notin E(P)$.

Note that by (i) P is acyclic; that is, it has no directed cycles. We set $X = \{x_1, \ldots, x_n\}$. Let x_1 be a vertex of P and let y_1 be a new vertex. Consider the digraph P^1 with vertex set $X^1 = X \cup \{y_1\}$ and edge set

$$E(P^1) = E(P) \cup \{(y_1, x) |\, (x_1, x) \in E(P)\} \cup \{(x, y_1) |\, (x, x_1) \in E(P)\}.$$

The digraph P^1 is transitive. Indeed let (a, b) and (b, c) be two edges of P^1. If $y_1 \notin \{a, b, c\}$, then $(a, c) \in E(P) \subset E(P^1)$ because P is transitive. If $y_1 = a$, then (x_1, b) and (b, c) are in $E(P)$. Hence $(x_1, c) \in E(P)$ and $(y_1, c) \in E(P^1)$. The cases $y_1 = b$ and $y_1 = c$ are treated similarly. Thus P^1 defines a poset $P^1 = (V(P^1), \prec^1)$. The comparability graph H of P^1 is precisely the graph G^1 obtained from G by duplicating the vertex x_1 by the vertex y_1. From Lemma 14.5.1 we get that $\mathrm{cl}(G)^1 = \mathrm{cl}(G^1)$, where $\mathrm{cl}(G)^1$ is the clutter obtained from $\mathrm{cl}(G)$ by duplicating the vertex x_1 by the vertex y_1. Altogether we obtain that the clutter $\mathrm{cl}(G)^1$ is the clique clutter of the comparability graph G^1 of the poset P^1.

By Theorem 14.3.6(vii) it suffices to prove that $\mathrm{cl}(G)^w$ has the König property for all $w \in \mathbb{N}^n$. Since duplications commute with deletions we may assume that $w = (w_1, \ldots, w_r, 0, \ldots, 0)$, where $w_i \geq 1$ for $i = 1, \ldots, r$. Consider the clutter $\mathcal{C}_1$ obtained from $\mathrm{cl}(G)$ by duplicating $w_i - 1$ times the vertex x_i for $i = 1, \ldots, r$. We denote the vertex set of $\mathcal{C}_1$ by X_1. By successively applying the fact that $\mathrm{cl}(G)^1 = \mathrm{cl}(G^1)$, we conclude that there is a poset P_1 with comparability graph G_1 and vertex set X_1 such that $\mathcal{C}_1 = \mathrm{cl}(G_1)$. As before we regard P_1 as a transitive acyclic digraph.

Let A and B be the set of minimal and maximal elements of the poset P_1, i.e., the elements of A and B are the sources and sinks of P_1, respectively. We set $S = \{x_{r+1}, \ldots, x_n\}$. Consider the digraph $\mathcal{D}$ whose vertex set is $V(\mathcal{D}) = X_1 \setminus S$ and whose edge set is defined as follows. A pair (x, y) in $V(\mathcal{D}) \times V(\mathcal{D})$ is in $E(\mathcal{D})$ if and only if $(x, y) \in E(P_1)$ and there is no vertex z in X_1 with $x \prec z \prec y$. Notice that $\mathcal{D}$ is a sub-digraph of P_1 which is not necessarily the digraph of a poset. We set $A_1 = A \setminus S$ and $B_1 = B \setminus S$. Note that $\mathcal{C}^w = \mathcal{C}_1 \setminus S$, the clutter obtained from $\mathcal{C}_1$ by removing all vertices of S and all edges sharing a vertex with S. If every edge of $\mathcal{C}_1$ intersects S, then $E(\mathcal{C}^w) = \emptyset$ and there is nothing to prove. Thus we may assume that there is a maximal clique K of G_1 disjoint form S. Note that by the maximality of K and by the transitivity of P_1 we get that K contains at least one source and one sink of P_1, i.e., $A_1 \neq \emptyset$ and $B_1 \neq \emptyset$ (see argument below).

The maximal cliques of G_1 not containing any vertex of S correspond exactly to the A_1–B_1 paths of $\mathcal{D}$. Indeed let $c = \{v_1, \ldots, v_s\}$ be a maximal clique of G_1 disjoint from S. Consider the sub-poset P_c of P_1 induced by c. Note that P_c is a tournament, i.e., P_c is an oriented graph (no-cycles of length two) such that any two vertices of P_c are comparable. By Exercise 7.1.21 any tournament has a Hamiltonian path, i.e., a spanning oriented path. Therefore we may assume that

$$v_1 \prec v_2 \prec \cdots \prec v_{s-1} \prec v_s$$

By the maximality of c we get that v_1 is a source of P_1, v_s is a sink of P_1, and (v_i, v_{i+1}) is an edge of $\mathcal{D}$ for $i = 1, \ldots, s-1$. Thus c is an A_1–B_1 path of $\mathcal{D}$, as required. Conversely let $c = \{v_1, \ldots, v_s\}$ be an A_1–B_1 path of $\mathcal{D}$. Clearly c is a clique of P_1 because P_1 is a poset. Assume that c is not a maximal clique of G_1. Then there is a vertex $v \in X_1 \setminus c$ such that v is related to every vertex of c. Since v_1, v_s are a source and a sink of P_1, respectively, we get $v_1 \prec v \prec v_s$. We claim that $v_i \prec v$ for $i = 1, \ldots, s$. By induction assume that $v_i \prec v$ for some $1 \leq i < s$. If $v \prec v_{i+1}$, then $v_i \prec v \prec v_{i+1}$, a contradiction to the fact that (v_i, v_{i+1}) is an edge of $\mathcal{D}$. Thus $v_{i+1} \prec v$. Making $i = s$ we get that $v_s \prec v$, a contradiction. This proves that c is a maximal clique of G_1. Therefore, since the maximal cliques of G_1 not containing any vertex in S are exactly the edges of $\mathcal{C}^w = \mathcal{C}_1 \setminus S$, by Menger's theorem (see Theorem 14.5.2) we obtain that $\mathcal{C}^w$ satisfies the König property. □

Corollary 14.5.4 *Let $P = (X, \prec)$ be a poset on the vertex X and let G be its comparability graph. Then the maximum number of disjoint maximal cliques of G is equal to the minimum size of a set intersecting all maximal cliques of G.*

Proof. By Theorem 14.5.3, the clique clutter of G satisfies the max-flow min-cut property. Then by Theorem 14.3.6, the clique clutter of G satisfies the König property. □

Let $P = (X, \prec)$ be a poset with vertex set X. A subset L of X is called a *chain* of P if $x \prec y$ or $y \prec x$ for all $x \neq y$ in L. A set A of X is called an *anti-chain* of P if x and y are not related for all $x \neq y$ in A.

Proposition 14.5.5 *Let $P = (X, \prec)$ be a poset with vertex set X. Then the maximum size of a chain of P equals the minimum number of disjoint anti-chains into which X can be decomposed (covered).*

Proof. Let L be a chain of maximum size and let $m = |L|$. Clearly max $\leq$ min, because if $X_1, \ldots, X_k$ is a decomposition of X into disjoint anti-chains, then $|L \cap X_i| \leq i$ for all i, so $|L| \leq k$. Let $x \in X$ and let $h(x)$ be the size of the longest chain of P with end vertex x. Since any two vertices x, y with $h(x) = h(y)$ are incomparable, we get a decomposition

$$X = \{x|\, h(x) = 1\} \cup \{x|\, h(x) = 2\} \cup \cdots \cup \{x|\, h(x) = m\}$$

into at most m disjoint anti-chains, so min $\leq$ max. □

Corollary 14.5.6 *If G is the comparability graph of a poset $P = (X, \prec)$, then G is perfect.*

Proof. Let $\omega(G)$ and $\chi(G)$ be the clique number and the chromatic number of G, respectively. By Proposition 14.5.5, we have $\omega(G) = \chi(G)$. As the class of comparability graphs is closed under taking induced subgraphs, we get that G is a perfect graph. $\Box$

Theorem 14.5.7 (Dilworth decomposition theorem) *Let $P = (X, \prec)$ be a poset on X. Then the maximum size of an anti-chain is equal to the minimum number of disjoint chains into which X can be decomposed.*

Proof. Let G be the comparability graph of P. By Corollary 14.5.6, G is perfect. Hence by the weak perfect graph theorem (see Theorem 13.6.1), the complement $\overline{G}$ of G is also perfect. In particular $\omega(\overline{G}) = \chi(\overline{G})$ and the result follows readily. $\Box$

Recall that the *vertex-clique matrix* of a graph G is the incidence matrix of the clique clutter of G.

Corollary 14.5.8 *Let G be a comparability graph and let A be the vertex-clique matrix of G. Then the following polytopes are integral:*

$$\mathcal{P}(A) = \{x | \, x \geq 0; \, xA \leq \mathbf{1}\} \quad \textit{and} \quad \mathcal{Q}(A) = \{x | \, x \geq 0; \, xA \geq \mathbf{1}\}$$

Proof. G is a perfect graph by Corollary 14.5.6. Hence the polytope $\mathcal{P}(A)$ is equal to the independence polytope of G (Theorem 13.6.1). In particular $\mathcal{P}(A)$ is integral. By Theorem 14.5.3 the clique clutter $\mathrm{cl}(G)$ has the max-flow min-cut property. Thus applying Theorem 14.3.6, we get that $\mathcal{Q}(A)$ has only integral vertices. $\Box$

Conjecture 14.5.9 *Let G be a perfect graph and let A be the vertex-clique matrix of G. If the polyhedron $\mathcal{Q}(A) = \{x | \, x \geq 0; xA \geq \mathbf{1}\}$ is integral, then the system $x \geq 0; xA \geq \mathbf{1}$ is TDI.*

The conjecture holds when the clique clutter of a perfect graph G is uniform (see Corollary 14.3.32).

Corollary 14.5.10 *If G is a comparability graph and $\mathrm{cl}(G)$ is its clique clutter, then the edge ideal of $\mathrm{cl}(G)$ is normally torsion-free and normal.*

Proof. It follows from Theorems 14.5.3 and 14.3.6. $\Box$

Definition 14.5.11 Let $d \geq 2$, $g \geq 2$ be two integers and let

$$X^1 = \{x_1^1, \ldots, x_g^1\}, \, X^2 = \{x_1^2, \ldots, x_g^2\}, \, \ldots, X^d = \{x_1^d, \ldots, x_g^d\}$$

be disjoint sets of variables. A clutter $\mathcal{C}$ with vertex set $X = X^1 \cup \cdots \cup X^d$ and edge set $E(\mathcal{C}) = \{\{x_{i_1}^1, x_{i_2}^2, \ldots, x_{i_d}^d\} | \, 1 \leq i_1 \leq i_2 \leq \cdots \leq i_d \leq g\}$ is called a *complete admissible d-uniform clutter* (cf. Definition 6.6.4).

Complete admissible uniform clutters were first studied in [160] and more recently in [205, 325] (see Section 6.6).

Theorem 14.5.12 *If $\mathcal{C}$ is a complete admissible d-uniform clutter, then its edge ideal $I(\mathcal{C})$ is normally torsion-free and normal.*

Proof. Let $P = (X, \prec)$ be the poset with vertex set X and partial order given by $x_k^\ell \prec x_p^m$ if and only if $1 \leq \ell < m \leq d$ and $1 \leq k \leq p \leq g$. We denote the comparability graph of P by G. We claim that $E(\mathcal{C}) = E(\text{cl}(G))$, where $\text{cl}(G)$ is the clique clutter of G. Let $f = \{x_{i_1}^1, x_{i_2}^2, \ldots, x_{i_d}^d\}$ be an edge of $\mathcal{C}$, i.e., we have $1 \leq i_1 \leq i_2 \leq \cdots \leq i_d \leq g$. Clearly f is a clique of G. If f is not maximal, then there is a vertex x_k^ℓ not in f which is adjacent in G to every vertex of f. In particular x_k^ℓ must be comparable to $x_{i_\ell}^\ell$, which is impossible. Thus f is an edge of $\text{cl}(G)$. Conversely let f be an edge of $\text{cl}(G)$. We can write $f = \{x_{i_1}^{k_1}, x_{i_2}^{k_2}, \ldots, x_{i_s}^{k_s}\}$, where $k_1 < \cdots < k_s$ and $i_1 \leq \cdots \leq i_s$. By the maximality of f we get that $s = d$ and $k_i = i$ for $i = 1, \ldots, d$. Thus f is an edge of $\mathcal{C}$. Hence by Corollary 14.5.10 we obtain that $I(\mathcal{C})$ is normally torsion-free and normal. □

Exercises

14.5.13 Let G be a graph. Let $G^1 = G \setminus x_1$ (resp. $\text{cl}(G)^1 = \text{cl}(G) \setminus x_1$) be the graph (resp. clutter) obtained from G (resp. $\text{cl}(G)$) by deleting the vertex x_1. If G is a cycle of length three, prove that $\text{cl}(G)^1 \neq \text{cl}(G^1)$.

14.5.14 Let $\text{cl}(G)$ be the clique clutter of a perfect graph G. If $\text{cl}(G)$ is d-uniform and satisfies the packing property, prove that $\text{cl}(G)$ satisfies the max-flow min-cut property.

14.6 Duality and integer rounding problems

Throughout this section we keep the notation of Section 14.1 but allow I to be a monomial ideal and A to be a matrix with entries in $\mathbb{N}$.

In this section we characterize the normality of a monomial ideal in terms of integer rounding properties of linear systems and show a duality theorem for monomial subrings. We show that the Rees algebra of the ideal of covers of a perfect graph is normal.

Let I be a monomial ideal of R generated by $x^{v_1}, \ldots, x^{v_q}$, and let A be the $n \times q$ matrix with column vectors $v_1, \ldots, v_q$. In what follows consider the homogeneous monomial subring

$$S = K[x^{w_1}t, \ldots, x^{w_r}t] \subset R[t],$$

where $\{w_1, \ldots, w_r\}$ is the set of all $\alpha \in \mathbb{N}^n$ such that $0 \leq \alpha \leq v_i$ for some i.

Definition 14.6.1 Given a polyhedron $\mathcal{Q}$ in $\mathbb{R}^n$, its *blocking polyhedron*, denoted by $B(\mathcal{Q})$, is defined as:

$$B(\mathcal{Q}) := \{z \in \mathbb{R}^n |\, z \geq 0;\ \langle z, x\rangle \geq 1 \text{ for all } x \text{ in } \mathcal{Q}\}.$$

Lemma 14.6.2 *If $\mathcal{Q} = \mathcal{Q}(A)$, then $B(\mathcal{Q}) = \mathbb{R}^n_+ + \text{conv}(v_1, \ldots, v_q)$.*

Proof. "$\subset$": Take z in $B(\mathcal{Q})$, then $\langle z, x\rangle \geq 1$ for all $x \in \mathcal{Q}$ and $z \geq 0$. Let $u_1, \ldots, u_r$ be the vertex set of $\mathcal{Q}$. In particular $\langle z, u_i\rangle \geq 1$ for all i. Then $\langle (z, 1), (u_i, -1)\rangle \geq 0$ for all i. From Theorem 1.4.2, we get that $(z, 1)$ is in the cone generated by $\mathcal{A}'$. Thus z is in $\mathbb{R}^n_+ + \text{conv}(v_1, \ldots, v_q)$.

"$\supset$": This inclusion is clear. □

Definition 14.6.3 A rational polyhedron $\mathcal{Q}$ has the *integer decomposition property* if for each natural number k and for each integer vector a in $k\mathcal{Q}$, a is the sum of k integer vectors in $\mathcal{Q}$, where $k\mathcal{Q}$ is equal to $\{ka |\, a \in \mathcal{Q}\}$.

Theorem 14.6.4 *I is normal if and only if the blocking polyhedron $B(\mathcal{Q})$ of $\mathcal{Q} = \mathcal{Q}(A)$ has the integer decomposition property and all minimal integer vectors of $B(\mathcal{Q})$ are columns of A (minimal with respect to $\leq$).*

Proof. By Lemma 14.6.2 we get $B(\mathcal{Q}) \cap \mathbb{Q}^n = \mathbb{Q}^n_+ + \text{conv}_{\mathbb{Q}}(v_1, \ldots, v_q)$, because $B(\mathcal{Q})$ is a rational polyhedron. From this equality we readily obtain

$$\overline{I^k} = (\{x^a |\, a \in kB(\mathcal{Q}) \cap \mathbb{Z}^n\}) \ \text{ for } \ 0 \neq k \in \mathbb{N}. \tag{14.17}$$

$\Rightarrow$) Assume that I is normal, i.e., $\overline{I^k} = I^k$ for $k \geq 1$. Let a be an integral vector in $kB(\mathcal{Q})$. Then, by Eq. (14.17), $x^a \in I^k$ and a is the sum of k integral vectors in $B(\mathcal{Q})$; that is, $B(\mathcal{Q})$ has the integer decomposition property. Take a minimal integer vector a in $B(\mathcal{Q})$. Then $x^a \in \overline{I} = I$ and we can write $a = \delta + v_i$ for some v_i and for some $\delta \in \mathbb{N}^n$. Thus $a = v_i$ by the minimality of a.

$\Leftarrow$) Assume that $B(\mathcal{Q})$ has the integer decomposition property and all minimal integer vectors of $B(\mathcal{Q})$ are columns of A. Take $x^a \in \overline{I^k}$, i.e., a is an integral vector of $kB(\mathcal{Q})$. Hence a is the sum of k integral vectors $\alpha_1, \ldots, \alpha_k$ in $B(\mathcal{Q})$. Since any minimal vector of $B(\mathcal{Q})$ is a column of A we may assume that $\alpha_i = c_i + v_i$ for $i = 1, \ldots, k$. Hence $x^a \in I^k$. □

Theorem 14.6.5 *If $I = \overline{I}$ and $\mathcal{Q} = \mathcal{Q}(A)$, then I is normal if and only if the blocking polyhedron $B(\mathcal{Q})$ has the integer decomposition property.*

Proof. $\Rightarrow$) If I is normal, by Theorem 14.6.4, the blocking polyhedron $B(\mathcal{Q})$ has the integer decomposition property.

$\Leftarrow$) Take $x^a \in \overline{I^k}$. By Eq. (14.17), a is an integral vector of $kB(\mathcal{Q})$. Hence a is the sum of k integral vectors $\alpha_1, \ldots, \alpha_k$ in $B(\mathcal{Q})$. By Eq. (14.17), with $k = 1$, we get that $\alpha_1, \ldots, \alpha_k$ are in $\overline{I} = I$. Hence $x^a \in I^k$. □

Corollary 14.6.6 *If I is the edge ideal of a clutter, then I is normal if and only if the blocking polyhedron $B(\mathcal{Q})$ has the integer decomposition property.*

Proof. Recall that I is an intersection of prime ideals. Thus it is seen that $\overline{I} = I$ and the result follows from Theorem 14.6.5. □

Definition 14.6.7 Let A be a matrix with entries in $\mathbb{N}$. The linear system $x \geq 0; xA \geq \mathbf{1}$ has the *integer rounding property* if

$$\max\{\langle y, \mathbf{1}\rangle \,|\, Ay \leq w; y \in \mathbb{N}^q\} = \lfloor \max\{\langle y, \mathbf{1}\rangle \,|\, y \geq 0; Ay \leq w\} \rfloor \tag{14.18}$$

for each integral vector w for which the right-hand side is finite.

Systems with the integer rounding property have been widely studied; see [372, Chapter 22], [373, Chapter 5], and the references therein.

Theorem 14.6.8 ([19], [373, p. 82]) *The system $x \geq 0; xA \geq \mathbf{1}$ has the integer rounding property if and only if the blocking polyhedron $B(\mathcal{Q})$ of $\mathcal{Q} = \mathcal{Q}(A)$ has the integer decomposition property and all minimal integer vectors of $B(\mathcal{Q})$ are columns of A (minimal with respect to $\leq$).*

Corollary 14.6.9 *Let $I = (x^{v_1}, \ldots, x^{v_q})$ be a monomial ideal and let A be the matrix with column vectors $v_1, \ldots, v_q$. Then I is a normal ideal if and only if the system $xA \geq \mathbf{1}; x \geq 0$ has the integer rounding property.*

Proof. According to Theorem 14.6.8, the system $xA \geq \mathbf{1}; x \geq 0$ has the integer rounding property if and only if the blocking polyhedron $B(\mathcal{Q})$ of $\mathcal{Q} = \mathcal{Q}(A)$ has the integer decomposition property and all minimal integer vectors of $B(\mathcal{Q})$ are columns of A (minimal with respect to $\leq$). Thus the result follows at once from Theorem 14.6.4. □

This corollary was first observed by N. V. Trung when I is the edge ideal of a hypergraph.

Lemma 14.6.10 *The following equation holds*:

$$\mathrm{conv}(w_1, \ldots, w_r) = \mathbb{R}^n_+ \cap (\mathrm{conv}(w_1, \ldots, w_r) + \mathbb{R}_+\{-e_1, \ldots, -e_n\}).$$

Proof. The inclusion "$\subset$" is clear. To show the inclusion "$\supset$" take a vector z such that $z \geq 0$ and

$$z = \lambda_1 w_1 + \cdots + \lambda_r w_r - \delta_1 e_1 - \cdots - \delta_n e_n, \tag{14.19}$$

where $\lambda_i \geq 0$, $\lambda_1 + \cdots + \lambda_r = 1$, and $\delta_j \geq 0$ for all i, j. Consider the vector $z' = \lambda_1 w_1 + \cdots + \lambda_r w_r - \delta_1 e_1$. We set $T = \mathrm{conv}(w_1, \ldots, w_r)$ and $w_i = (w_{i1}, \ldots, w_{in})$. We claim that z' is in T. We may assume that $\delta_1 > 0$,

$\lambda_i > 0$ for all i, and that the first entry w_{i1} of w_i is positive for $1 \leq i \leq s$ and is equal to zero for $i > s$. From Eq. (14.19) we get $\lambda_1 w_{11} + \cdots + \lambda_s w_{s1} \geq \delta_1$.

Case (I): $\lambda_1 w_{11} \geq \delta_1$. Then we can write

$$z' = \frac{\delta_1}{w_{11}}(w_1 - w_{11}e_1) + \left(\lambda_1 - \frac{\delta_1}{w_{11}}\right) w_1 + \lambda_2 w_2 + \cdots + \lambda_r w_r.$$

Notice that $w_1 - w_{11}e_1$ is again in $\{w_1, \ldots, w_r\}$. Thus z' is a convex combination of $w_1, \ldots, w_r$, i.e., $z' \in T$.

Case (II): $\lambda_1 w_{11} < \delta_1$. Let m be the largest integer less than or equal to s such that $\lambda_1 w_{11} + \cdots + \lambda_{m-1} w_{(m-1)1} < \delta_1 \leq \lambda_1 w_{11} + \cdots + \lambda_m w_{m1}$. Then

$$\begin{aligned} z' \;=\; & \sum_{i=1}^{m-1} \lambda_i (w_i - w_{i1}e_1) + \left[\frac{\delta_1}{w_{m1}} - \left(\sum_{i=1}^{m-1} \frac{\lambda_i w_{i1}}{w_{m1}}\right)\right](w_m - w_{m1}e_1) + \\ & \left[\lambda_m - \frac{\delta_1}{w_{m1}} + \left(\sum_{i=1}^{m-1} \frac{\lambda_i w_{i1}}{w_{m1}}\right)\right] w_m + \sum_{i=m+1}^{r} \lambda_i w_i. \end{aligned}$$

Notice that $w_i - w_{i1}e_1$ is again in $\{w_1, \ldots, w_r\}$ for $i = 1, \ldots, m$. Thus z' is a convex combination of $w_1, \ldots, w_r$, i.e., $z' \in T$. This proves the claim. We can apply the argument above to any entry of z or z', thus we obtain that $z' - \delta_2 e_2 \in T$. Thus by induction we obtain that $z \in T$, as required. $\square$

Definition 14.6.11 [372, p. 117] Let $\mathcal{P}$ be a rational polyhedron in $\mathbb{R}^n$. The *antiblocking polyhedron* of $\mathcal{P}$ is defined as:

$$T(\mathcal{P}) := \{z \mid z \geq 0; \langle z, x\rangle \leq 1 \text{ for all } x \in \mathcal{P}\}.$$

Lemma 14.6.12 *If* $\mathcal{P} = \{x \mid x \geq 0; xA \leq \mathbf{1}\}$, *then*

$$T(\mathcal{P}) = \text{conv}(w_1, \ldots, w_r).$$

Proof. One has the equality $\mathcal{P} = \{z \mid z \geq 0; \langle z, w_i\rangle \leq 1\, \forall i\}$ because for each w_i there is v_j such that $w_i \leq v_j$. Hence, by Corollary 1.1.34, we can write

$$\mathcal{P} = \{z \mid z \geq 0; \langle z, w_i\rangle \leq 1\, \forall i\} = \text{conv}(\beta_0, \beta_1, \ldots, \beta_m) \tag{14.20}$$

for some $\beta_1, \ldots, \beta_m$ in $\mathbb{Q}_+^n$ and $\beta_0 = 0$. From Eq. (14.20) we get

$$\{z \mid z \geq 0; \langle z, \beta_i\rangle \leq 1\, \forall i\} = T(\mathcal{P}). \tag{14.21}$$

Using Eq. (14.20) and noticing that $\langle \beta_i, w_j\rangle \leq 1$ for all i, j, we get

$$\mathbb{R}_+^n \cap (\text{conv}(\beta_0, \ldots, \beta_m) + \mathbb{R}_+\{-e_1, \ldots, -e_n\}) = \{z \mid z \geq 0; \langle z, w_i\rangle \leq 1\, \forall i\}.$$

Hence using this equality and [372, Theorem 9.4] we obtain

$$\mathbb{R}_+^n \cap (\text{conv}(w_1, \ldots, w_r) + \mathbb{R}_+\{-e_1, \ldots, -e_n\}) = \\ \{z \mid z \geq 0; \langle z, \beta_i\rangle \leq 1\, \forall i\}. \tag{14.22}$$

Therefore, by Lemma 14.6.10 together with Eqs. (14.21) and (14.22), we conclude that $T(\mathcal{P})$ is equal to $\text{conv}(w_1,\ldots,w_r)$, as required. $\Box$

Definition 14.6.13 Let A be a matrix with entries in $\mathbb{N}$. The linear system $x \geq 0; xA \leq \mathbf{1}$ has the *integer rounding property* if

$$\lceil \min\{\langle y, \mathbf{1}\rangle \,|\, y \geq 0;\ Ay \geq a\}\rceil = \min\{\langle y, \mathbf{1}\rangle \,|\, Ay \geq a;\ y \in \mathbb{N}^q\}$$

for each integral vector a for which $\min\{\langle y, \mathbf{1}\rangle \,|\, y \geq 0;\ Ay \geq a\}$ is finite.

Remark 14.6.14 Let A be a matrix with entries in $\mathbb{N}$. Then the linear system $x \geq 0; xA \leq \mathbf{1}$ has the *integer rounding property* if and only if

$$\lceil \min\{\langle y, \mathbf{1}\rangle \,|\, y \geq 0;\ Ay \geq a\}\rceil = \min\{\langle y, \mathbf{1}\rangle \,|\, Ay \geq a;\ y \in \mathbb{N}^q\}$$

for each vector $a \in \mathbb{N}^n$ for which $\min\{\langle y, \mathbf{1}\rangle \,|\, y \geq 0;\ Ay \geq a\}$ is finite. This follows decomposing an integral vector a as $a = a_+ - a_-$ and noticing that for $y \geq 0$ we have that $Ay \geq a$ if and only if $Ay \geq a_+$

Theorem 14.6.15 ([19], [373, p. 82]) *If $\mathcal{P} = \{x \,|\, x \geq 0;\ xA \leq \mathbf{1}\}$, then the system $xA \leq \mathbf{1}; x \geq 0$ has the integer rounding property if and only if $T(\mathcal{P})$ has the integer decomposition property and all maximal integer vectors of $T(\mathcal{P})$ are columns of A (maximal with respect to $\leq$).*

Theorem 14.6.16 *The system $x \geq 0;\ xA \leq \mathbf{1}$ has the integer rounding property if and only if the subring $K[x^{w_1}t,\ldots,x^{w_r}t]$ is normal.*

Proof. Let $\mathcal{P} = \{x \,|\, x \geq 0;\ xA \leq \mathbf{1}\}$ and let $T(\mathcal{P})$ be its antiblocking polyhedron. According to Lemma 14.6.12 one has

$$T(\mathcal{P}) = \text{conv}(w_1,\ldots,w_r). \tag{14.23}$$

Setting $\mathcal{B} = \{(w_i,1)\}_{i=1}^r$, by Theorem 14.6.15 and Corollary 9.1.3, it suffices to prove that the equality $\mathbb{R}_+\mathcal{B} \cap \mathbb{Z}\mathcal{B} = \mathbb{N}\mathcal{B}$ holds true if and only if $T(\mathcal{P})$ has the integer decomposition property and all maximal integer vectors of $T(\mathcal{P})$ are columns of A.

Assume that $\mathbb{R}_+\mathcal{B} \cap \mathbb{Z}\mathcal{B} = \mathbb{N}\mathcal{B}$. Let b be a natural number and let a be an integer vector in $bT(\mathcal{P})$. Then using Eq. (14.23) it is seen that (a,b) is in $\mathbb{R}_+\mathcal{B}$. In our situation one has $\mathbb{Z}\mathcal{B} = \mathbb{Z}^{n+1}$. Hence $(a,b) \in \mathbb{N}\mathcal{B}$ and a is the sum of b integer vectors in $T(\mathcal{P})$. Thus $T(\mathcal{P})$ has the integer decomposition property. Assume that a is a maximal integer vector of $T(\mathcal{P})$. It is not hard to see that $(a,1)$ is in $\mathbb{R}_+\mathcal{B}$, i.e., $(a,1) \in \mathbb{N}\mathcal{B}$. Thus $(a,1)$ is a linear combination of vectors in $\mathcal{B}$ with coefficients in $\mathbb{N}$. Hence $(a,1)$ is equal to $(w_j,1)$ for some j. There exists v_i such that $a = w_j \leq v_i$. Therefore by the maximality of a, we get $a = v_i$ for some i. Thus a is a column of A.

Conversely assume that $T(\mathcal{P})$ has the integer decomposition property and that all maximal integer vectors of $T(\mathcal{P})$ are columns of A. Let (a,b) be an integral vector in $\mathbb{R}_+\mathcal{B}$ with $a\in\mathbb{N}^n$ and $b\in\mathbb{N}$. Hence, using Eq. (14.23), we get $a\in bT(\mathcal{P})$. Thus $a=\alpha_1+\cdots+\alpha_b$, where α_i is an integral vector of $T(\mathcal{P})$ for all i. Since each α_i is less than or equal to a maximal integer vector of $T(\mathcal{P})$, we get that $\alpha_i\in\{w_1,\ldots,w_r\}$. Then $(a,b)\in\mathbb{N}\mathcal{B}$. $\Box$

Corollary 14.6.17 *Let A be the incidence matrix of a uniform clutter. If either linear system $x\geq 0; xA\leq \mathbf{1}$ or $x\geq 0; xA\geq\mathbf{1}$ has the integer rounding property and $\mathcal{P}=\mathrm{conv}(v_1,\ldots,v_q)$, then*

$$K[x^{v_1}t,\ldots,x^{v_q}t]=A(\mathcal{P}).$$

Proof. In general the inclusion "$\subset$" holds. Assume that $x\geq 0;xA\leq\mathbf{1}$ has the integer rounding property and that every edge of $\mathcal{C}$ has d elements. Let $w_1,\ldots,w_r$ be the set of all $\alpha\in\mathbb{N}^n$ such that $\alpha\leq v_i$ for some i. Then by Theorem 14.6.16 the subring $K[x^{w_1}t,\ldots,x^{w_r}t]$ is normal. Using that $v_1,\ldots,v_q$ is the set of w_i with $|w_i|=d$, it is not hard to see that $A(\mathcal{P})$ is contained in $K[x^{v_1}t,\ldots,x^{v_q}t]$.

Assume that $x\geq 0;xA\geq\mathbf{1}$ has the integer rounding property. Let $I=I(\mathcal{C})$ be the edge ideal of $\mathcal{C}$ and let $R[It]$ be its Rees algebra. By Corollary 14.6.9, $R[It]$ is a normal domain. Since the clutter $\mathcal{C}$ is uniform the required equality follows at once from Theorem 9.3.31. $\Box$

Corollary 14.6.18 *If G is a perfect unmixed graph and $v_1,\ldots,v_q$ are the characteristic vectors of the maximal stable sets of G, then the subring $K[x^{v_1}t,\ldots,x^{v_q}t]$ is normal.*

Proof. The minimal vertex covers of G are exactly the complements of the maximal stable sets of G. Thus $|v_i|=d$ for all i, where $d=\dim(R/I(G))$. On the other hand the maximal stable sets of G are exactly the maximal cliques of $\overline{G}$. Thus, by Theorem 13.6.8 and Corollary 14.6.17, the subring $K[x^{v_1}t,\ldots,x^{v_q}t]$ is an Ehrhart ring, and consequently it is normal. $\Box$

Proposition 14.6.19 *Let $I=(x^{v_1},\ldots,x^{v_q})$ be a monomial ideal and let $v_i^*=\mathbf{1}-v_i$. Then $R[It]$ is normal if and only if the set*

$$\Gamma=\{-e_1,\ldots,-e_n,(v_1^*,1),\ldots,(v_q^*,1)\}$$

is a Hilbert basis.

Proof. Let $\mathcal{A}'=\{e_1,\ldots,e_n,(v_1,1),\ldots,(v_q,1)\}$. Assume that $R[It]$ is normal. Then $\mathcal{A}'$ is a Hilbert basis. Let (a,b) be an integral vector in $\mathbb{R}_+\Gamma$, with $a\in\mathbb{Z}^n$ and $b\in\mathbb{Z}$. Then we can write

$$(a,b)=\mu_1(-e_1)+\cdots+\mu_n(-e_n)+\lambda_1(v_1^*,1)+\cdots+\lambda_q(v_q^*,1),$$

where $\mu_i \geq 0$ and $\lambda_j \geq 0$ for all i, j. Therefore

$$-(a, -b) + b\mathbf{1} = \mu_1 e_1 + \cdots + \mu_n e_n + \lambda_1(v_1, 1) + \cdots + \lambda_q(v_q, 1),$$

where $\mathbf{1} = e_1 + \cdots + e_n$. As $\mathcal{A}'$ is a Hilbert basis we can write

$$-(a, -b) + b\mathbf{1} = \mu_1' e_1 + \cdots + \mu_n' e_n + \lambda_1'(v_1, 1) + \cdots + \lambda_q'(v_q, 1),$$

where $\mu_i' \in \mathbb{N}$ and $\lambda_j' \in \mathbb{N}$ for all i, j. Thus $(a, b) \in \mathbb{N}\Gamma$. This proves that Γ is a Hilbert basis. The converse can be shown using similar arguments. □

Definition 14.6.20 *Let $A = (a_{ij})$ be a matrix with entries in $\{0, 1\}$. Its dual is the matrix $A^* = (a_{ij}^*)$, where $a_{ij}^* = 1 - a_{ij}$.*

The following duality is valid for incidence matrices of clutters. It will be used later to establish a duality theorem for monomial subrings.

Theorem 14.6.21 *If A is the incidence matrix of a clutter and A^* is its dual matrix, then $x \geq 0; xA \geq \mathbf{1}$ has the integer rounding property if and only if $x \geq 0; xA^* \leq \mathbf{1}$ has the integer rounding property.*

Proof. We set $\mathcal{Q} = \{x | x \geq 0; xA \geq \mathbf{1}\}$ and $\mathcal{P}^* = \{x | x \geq 0; xA^* \leq \mathbf{1}\}$. If $v_i^* = \mathbf{1} - v_i$ for all i, then A^* is the matrix with column vectors $v_1^*, \ldots, v_q^*$. Let $w_1^*, \ldots, w_s^*$ be the set of all $\alpha \in \mathbb{N}^n$ such that $\alpha \leq v_i^*$ for some i. Then, using Lemmas 14.6.2 and 14.6.12, we obtain

$$B(\mathcal{Q}) = \mathbb{R}_+^n + \operatorname{conv}(v_1, \ldots, v_q) \ \text{ and } \ T(\mathcal{P}^*) = \operatorname{conv}(w_1^*, \ldots, w_s^*).$$

$\Rightarrow$) Thanks to Theorem 14.6.15 we need only show that $T(\mathcal{P}^*)$ has the integer decomposition property and all maximal integer vectors of $T(\mathcal{P}^*)$ are columns of A^*. Let $b \in \mathbb{N}_+$ and let a be an integer vector in $bT(\mathcal{P}^*)$. Then we can write $a = b(\lambda_1 w_1^* + \cdots + \lambda_s w_s^*)$ with $\sum_i \lambda_i = 1$ and $\lambda_i \geq 0$. For each $1 \leq i \leq s$ there is $v_{j_i}^*$ in $\{v_1^*, \ldots, v_q^*\}$ such that $w_i^* \leq v_{j_i}^*$. Thus for each i we can write $\mathbf{1} - w_i^* = v_{j_i} + \delta_i$, where $\delta_i \in \mathbb{N}^n$. Therefore

$$\mathbf{1} - a/b = \lambda_1(v_{j_1} + \delta_1) + \cdots + \lambda_s(v_{j_s} + \delta_s).$$

This means that $\mathbf{1} - a/b \in B(\mathcal{Q})$, i.e., $b\mathbf{1} - a$ is an integer vector in $bB(\mathcal{Q})$. Hence by Theorem 14.6.8 we can write $b\mathbf{1} - a = \alpha_1 + \cdots + \alpha_b$ for some $\alpha_1, \ldots, \alpha_b$ integer vectors in $B(\mathcal{Q})$, and for each α_i there is v_{k_i} in $\{v_1, \ldots, v_q\}$ such that $v_{k_i} \leq \alpha_i$. Thus $\alpha_i = v_{k_i} + \epsilon_i$ for some $\epsilon_i \in \mathbb{N}^n$ and consequently:

$$a = (\mathbf{1} - v_{k_1}) + \cdots + (\mathbf{1} - v_{k_b}) - c = v_{k_1}^* + \cdots + v_{k_b}^* - c,$$

where $c = (c_1, \ldots, c_n) \in \mathbb{N}^n$. Notice that $v_{k_1}^* + \cdots + v_{k_b}^* \geq c$ because $a \geq 0$. If $c_1 \geq 1$, then the first entry of $v_{k_i}^*$ is non-zero for some i and we can write

$$a = v_{k_1}^* + \cdots + v_{k_{i-1}}^* + (v_{k_i}^* - e_1) + v_{k_{i+1}}^* + \cdots + v_{k_b}^* - (c - e_1).$$

Since $v_{k_i}^* - e_1$ is again in $\{w_1^*, \ldots, w_s^*\}$, we can apply this argument recursively to obtain that a is the sum of b integer vectors in $\{w_1^*, \ldots, w_s^*\}$. This proves that $T(\mathcal{P}^*)$ has the integer decomposition property. Let a be a maximal integer vector of $T(\mathcal{P}^*)$. As the vectors $w_1^*, \ldots, w_s^*$ have entries in $\{0, 1\}$, we get $T(\mathcal{P}^*) \cap \mathbb{Z}^n = \{w_1^*, \ldots, w_s^*\}$. Then $a = w_i^*$ for some i. As $w_i^* \leq v_j^*$ for some j, we get that $a = v_j^*$, i.e., a is a column of A^*.

$\Leftarrow$) Thanks to Corollary 14.6.9, the system $x \geq 0; xA \geq \mathbf{1}$ has the integer rounding property if and only if $R[It]$ is normal. Thus by Proposition 14.6.19 we need only show that the set $\Gamma = \{-e_1, \ldots, -e_n, (v_1^*, 1), \ldots, (v_q^*, 1)\}$ is a Hilbert basis. Let (a, b) be an integral vector in $\mathbb{R}_+\Gamma$, with $a \in \mathbb{Z}^n$ and $b \in \mathbb{Z}$. Then we can write

$$(a, b) = \mu_1(-e_1) + \cdots + \mu_n(-e_n) + \lambda_1(v_1^*, 1) + \cdots + \lambda_q(v_q^*, 1),$$

where $\mu_i \geq 0$, $\lambda_j \geq 0$ for all i, j. Hence $A^*\lambda \geq a$, where $\lambda = (\lambda_i)$. By hypothesis the system $x \geq 0; xA^* \leq \mathbf{1}$ has the integer rounding property. Then one has

$$b \geq \lceil \min\{\langle y, \mathbf{1}\rangle \,|\, y \geq 0;\ A^*y \geq a\}\rceil = \min\{\langle y, \mathbf{1}\rangle \,|\, A^*y \geq a;\ y \in \mathbb{N}^q\} = \langle y_0, \mathbf{1}\rangle$$

for some $y_0 = (y_i) \in \mathbb{N}^q$ such that $|y_0| = \langle y_0, \mathbf{1}\rangle \leq b$ and $a \leq A^*y_0$. Then

$$a = y_1v_1^* + \cdots + y_qv_q^* - \delta_1e_1 - \cdots - \delta_ne_n,$$

where $\delta_1, \ldots, \delta_n$ are in $\mathbb{N}$. Hence we can write

$$(a, b) = y_1(v_1^*, 1) + \cdots + y_{q-1}(v_{q-1}^*, 1) + (y_q + b - |y_0|)(v_q^*, 1) - (b - |y_0|)v_q^* - \delta,$$

where $\delta = (\delta_i)$. As the entries of A^* are in $\mathbb{N}$, the vector $-v_q^*$ can be written as a nonnegative integer combination of $-e_1, \ldots, -e_n$. Thus $(a, b) \in \mathbb{N}\Gamma$. This proves that Γ is a Hilbert basis. □

The following result is a duality theorem for monomial subrings.

Theorem 14.6.22 [52] *Let A be the incidence matrix of a clutter and let v_i^* be the vector $\mathbf{1} - v_i$. If $w_1^*, \ldots, w_s^*$ is the set of all $\alpha \in \mathbb{N}^n$ such that $\alpha \leq v_i^*$ for some i, then the following conditions are equivalent:*

(a) *$R[It]$ is normal, where $I = (x^{v_1}, \ldots, x^{v_q})$.*

(b) *$S^* = K[x^{w_1^*}t, \ldots, x^{w_s^*}t]$ is normal.*

(c) *$\{-e_1, \ldots, -e_n, (v_1^*, 1), \ldots, (v_q^*, 1)\}$ is a Hilbert basis.*

(d) *$x \geq 0; xA \geq \mathbf{1}$ has the integer rounding property.*

(e) *$x \geq 0; xA^* \leq \mathbf{1}$ has the integer rounding property.*

Proof. (a) $\Leftrightarrow$ (c): This was shown in Proposition 14.6.19. (a) $\Leftrightarrow$ (d): This was shown in Corollary 14.6.9. (b) $\Leftrightarrow$ (e): This part was shown in Theorem 14.6.16. (d) $\Leftrightarrow$ (e): This follows from Theorem 14.6.21. □

Definition 14.6.23 Let I be the edge ideal of a clutter $\mathcal{C}$. The *dual* of I, denoted by I^*, is the ideal of R generated by all monomials $x_1 \cdots x_n / x_e$ such that e is an edge of $\mathcal{C}$, where $x_e = \prod_{x_i \in e} x_i$.

Corollary 14.6.24 *Let $\mathcal{C}$ be a clutter and let A be its incidence matrix. If $\mathcal{P} = \{x | x \geq 0; xA \leq \mathbf{1}\}$ is an integral polytope and $I = I(\mathcal{C})$, then $R[I^*t]$ and $S = K[x^{w_1}t, \ldots, x^{w_r}t]$ are normal.*

Proof. Since $\mathcal{P}$ has only integral vertices, by a theorem of Lovász (see Theorem 13.6.8) the system $x \geq 0; xA \leq \mathbf{1}$ is totally dual integral, i.e., the minimum in the LP-duality equation

$$\max\{\langle a, x\rangle \,|\, x \geq 0; xA \leq \mathbf{1}\} = \min\{\langle y, \mathbf{1}\rangle \,|\, y \geq 0; Ay \geq a\} \qquad (14.24)$$

has an integral optimum solution y for each integral vector a with finite minimum. In particular the system $x \geq 0; xA \leq \mathbf{1}$ satisfies the integer rounding property. Therefore $R[I^*t]$ and $K[x^{w_1}t, \ldots, x^{w_r}t]$ are normal by Theorem 14.6.22. □

Corollary 14.6.25 [423] *If G is a perfect graph, then $R[I_c(G)t]$ is normal.*

Proof. Let $\mathcal{C}$ be the clique clutter of $\overline{G}$, the complement of G, and let A be its incidence matrix. The graph $\overline{G}$ is perfect by the weak perfect graph theorem (see Theorem 13.6.1). Hence, by Theorem 13.6.8, the polytope $\mathcal{P}(A) = \{x | \, x \geq 0; \, xA \leq \mathbf{1}\}$ is integral. If I is the edge ideal of $\mathcal{C}$, by Corollary 14.6.24, the ideal I^* is normal. To complete the proof notice that I^* is equal to $I_c(G)$. □

Corollary 14.6.26 [165, Corollary 5.11] *If $I_c(G)$ is the ideal of covers of a perfect graph G, then $\operatorname{Ass}(R/I_c(G)^k)$ form an ascending chain.*

Proof. We set $J = I_c(G)$. By Corollary 14.6.25, $R[Jt]$ is normal. Thus $J^k = \overline{J^k}$ for all k, so by Theorem 7.7.3, J has the persistence property. □

Proposition 14.6.27 *Let $\mathcal{C}$ be a balanced clutter and let I be its edge ideal. Then the Rees algebra $R[I^*t]$ of I^* is normal.*

Proof. Let A be the incidence matrix of $\mathcal{C}$. This matrix is balanced by hypothesis. By Theorem 13.6.7 A is balanced if and only if every submatrix of A is perfect. Thus, the result follows from Corollary 14.6.24. □

Corollary 14.6.28 *If G is a perfect and unmixed graph, then $R[I_c(G)t]$ is a Gorenstein standard graded K-algebra.*

Proof. We set $g = \operatorname{ht}(I(G))$ and $J = I_c(G)$. By assigning $\deg(x_i) = 1$ and $\deg(t) = -(g-1)$, the Rees algebra $R[Jt]$ becomes a standard graded

K-algebra. The ring $R[Jt]$ is normal by Theorem 14.6.25. Then, according to Theorem 9.1.5, its canonical module is the ideal of $R[Jt]$ given by

$$\omega_{R[Jt]} = (\{x_1^{a_1} \cdots x_n^{a_n} t^{a_{n+1}} \,|\, a = (a_i) \in \mathbb{R}_+(J)^{\rm o} \cap \mathbb{Z}^{n+1}\}).$$

By Theorem 9.1.6 the ring $R[Jt]$ is Cohen–Macaulay. Using Eq. (13.10) of Proposition 13.6.3 it is seen that $x_1 \cdots x_n t$ belongs to $\omega_{R[Jt]}$. Take any monomial $x^a t^b = x_1^{a_1} \cdots x_n^{a_n} t^b$ in the ideal $\omega_{R[Jt]}$, that is $(a, b) \in \mathbb{R}_+(J)^{\rm o}$. Hence the vector (a, b) has positive integer entries and satisfies

$$\textstyle\sum_{x_i \in \mathcal{K}_r} a_i \geq (r-1)b + 1 \qquad (14.25)$$

for every complete subgraph $\mathcal{K}_r$ of G. If $b = 1$, clearly $x^a t^b$ is a multiple of $x_1 \cdots x_n t$. Now assume $b \geq 2$. Using the normality of $R[Jt]$ and Eqs. (13.10) and (14.25) it follows that the monomial $m = x_1^{a_1 - 1} \cdots x_n^{a_n - 1} t^{b-1}$ belongs to $R[Jt]$. Since $x^a t^b = m x_1 \cdots x_n t$, we obtain that $\omega_{R[Jt]}$ is generated by $x_1 \cdots x_n t$ and thus $R[Jt]$ is a Gorenstein ring. □

Corollary 14.6.29 *Let $B_1, \ldots, B_q$ be the collection of basis of a matroid M with vertex set X and let $v_1, \ldots, v_q$ be their characteristic vectors. If A is the matrix with column vectors $v_1, \ldots, v_q$, then all systems*

$$x \geq 0; xA \geq \mathbf{1}, \quad x \geq 0; xA^* \geq \mathbf{1}, \quad x \geq 0; xA \leq \mathbf{1}, \quad x \geq 0; xA^* \leq \mathbf{1}$$

have the integer rounding property.

Proof. Consider the basis monomial ideal $I = (x^{v_1}, \ldots, x^{v_q})$ of the matroid M. By [338, Theorem 2.1.1], the collection of basis of the dual matroid M^* of M is given by $X \setminus B_1, \ldots, X \setminus B_q$. Now, the basis monomial ideal of a matroid is normal (Corollary 12.3.12). Thus the result follows at once from the duality of Theorem 14.6.22. □

Linear systems of the form xA $\leq$ 1 We now turn our attention to study the integer rounding property of linear systems of inequalities of the form $xA \leq \mathbf{1}$ (see Definition 1.3.23).

Proposition 14.6.30 *Let $v_1, \ldots, v_q$ be the column vectors of a nonnegative integer matrix A and let $A(\mathcal{P})$ be the Ehrhart ring of $\mathcal{P} = \mathrm{conv}(0, v_1, \ldots, v_q)$. Then the system $xA \leq \mathbf{1}$ has the integer rounding property if and only if*

$$K[x^{v_1}t, \ldots, x^{v_q}t, t] = A(\mathcal{P}).$$

Proof. By Theorem 1.3.24, we have that the system $xA \leq \mathbf{1}$ has the integer rounding property if and only if the set $\mathcal{B} = \{(v_1, 1), \ldots, (v_q, 1), (0, 1)\}$ is an integral Hilbert basis. Thus the proposition follows by noticing the equality

$$A(\mathcal{P}) = K[\{x^a t^b | (a, b) \in \mathbb{R}_+\mathcal{B} \cap \mathbb{Z}^{n+1}\}]$$

and the inclusion $K[x^{v_1}t, \ldots, x^{v_q}t, t] \subset A(\mathcal{P})$. □

Theorem 14.6.31 *Let $\mathcal{A} = \{v_1, \ldots, v_q\}$ be the set of column vectors of a matrix A with entries in $\mathbb{N}$. If the system $xA \leq \mathbf{1}$ has the integer rounding property, then*

(a) *$K[F]$ is normal, where $F = \{x^{v_1}, \ldots, x^{v_q}\}$, and*

(b) *$\mathbb{Z}^n/\mathbb{Z}\mathcal{A}$ is a torsion-free group.*

The converse holds if $|v_i| = d$ for all i.

Proof. If $xA \leq \mathbf{1}$ has the integer rounding property, then (a) and (b) follow from Corollary 1.3.32. Conversely assume that $|v_i| = d$ for all i and that (a) and (b) hold. We need only show that $\mathcal{B} = \{(v_i, 1)\}_{i=1}^q \cup \{e_{n+1}\}$ is a Hilbert basis. Let (a, b) be an integral vector in $\mathbb{R}_+\mathcal{B}$, where $a \in \mathbb{N}^n$ and $b \in \mathbb{N}$. Then we can write

$$(a, b) = \lambda_1(v_1, 1) + \cdots + \lambda_q(v_q, 1) + \mu(0, 1), \tag{14.26}$$

for some $\lambda_1, \ldots, \lambda_q, \mu$ in $\mathbb{Q}_+$. Hence using (b) gives that a is in $\mathbb{R}_+\mathcal{A} \cap \mathbb{Z}\mathcal{A}$. Hence $x^a \in \overline{K[F]} = K[F]$, i.e., $a \in \mathbb{N}\mathcal{A}$. Then we can write $a = \sum_{i=1}^q \eta_i v_i$ for some $\eta_1, \ldots, \eta_q$ in $\mathbb{N}^n$. Since $|v_i| = d$ for all i, one has $\sum_i \lambda_i = \sum_i \eta_i$. Therefore using Eq. (14.26), we get $\mu \in \mathbb{N}$. Therefore we get

$$(a, b) = \eta_1(v_1, 1) + \cdots + \eta_q(v_q, 1) + \mu(0, 1) \ \Rightarrow \ (a, b) \in \mathbb{N}\mathcal{B}. \qquad \square$$

Corollary 14.6.32 *Let A be the incidence matrix of a connected graph G. Then the system $xA \leq \mathbf{1}$ has the integer rounding property if and only if G is a bipartite graph.*

Proof. $\Rightarrow$) Let $\mathcal{A} = \{v_1, \ldots, v_q\}$ be the set of columns of A. If G is not bipartite, then according to Corollary 10.2.11 one has $\mathbb{Z}^n/\mathbb{Z}\mathcal{A} \simeq \mathbb{Z}_2$, a contradiction to Theorem 14.6.31(b).

$\Leftarrow$) By Corollary 10.2.11 and Proposition 10.3.1 the ring $K[x^{v_1}, \ldots, x^{v_q}]$ is normal and $\mathbb{Z}^n/\mathbb{Z}\mathcal{A} \simeq \mathbb{Z}$. Thus, by Theorem 14.6.31, the system $xA \leq \mathbf{1}$ has the integer rounding property, as required. $\square$

Integer rounding properties in simple graphs Let G be a simple graph with vertex set $X = \{x_1, \ldots, x_n\}$, let $v_1, \ldots, v_q$ be the column vectors of the incidence matrix A of G, let $R = K[x_1, \ldots, x_n]$ be a polynomial ring over a field K and let $I = I(G)$ be the edge ideal of G. Recall that the *extended Rees algebra* of I is the subring

$$R[It, t^{-1}] := R[It][t^{-1}] \subset R[t, t^{-1}],$$

where $R[It]$ is the Rees algebra of I.

Lemma 14.6.33 $R[It, t^{-1}] \simeq K[t, x_1t, \ldots, x_nt, x^{v_1}t, \ldots, x^{v_q}t]$.

Proof. We set $S = K[t, x_1t, \ldots, x_nt, x^{v_1}t, \ldots, x^{v_q}t]$. Note that S and $R[It, t^{-1}]$ are integral domains of the same dimension. This follows from the dimension formula given in Corollary 8.2.21. Thus it suffices to prove that there is an epimorphism $\overline{\psi}\colon S \to R[It, t^{-1}]$ of K-algebras.

Let $u_0, u_1, \ldots, u_n, t_1, \ldots, t_q$ be a new set of variables and let φ, ψ be the maps of K-algebras defined by the diagram

$$\begin{array}{ccc} K[u_0, u_1, \ldots, u_n, t_1, \ldots, t_q] & \xrightarrow{\ \psi\ } & R[It, t^{-1}] \\ \downarrow{\scriptstyle\varphi} & \overset{\overline{\psi}}{\nearrow} & \\ S & & \end{array} \qquad \begin{array}{l} u_0 \overset{\varphi}{\longmapsto} t, \\ u_i \overset{\varphi}{\longmapsto} x_it, \\ t_i \overset{\varphi}{\longmapsto} x^{v_i}t, \end{array} \qquad \begin{array}{l} u_0 \overset{\psi}{\longmapsto} t^{-1}, \\ u_i \overset{\psi}{\longmapsto} x_i, \\ t_i \overset{\psi}{\longmapsto} x^{v_i}t. \end{array}$$

As $\ker(\varphi)$ is a binomial ideal (Corollary 8.2.18) we get $\ker(\varphi) \subset \ker(\psi)$. Hence there is an epimorphism $\overline{\psi}$ of K-algebras that makes the diagram commutative, i.e., $\psi = \overline{\psi}\varphi$. $\Box$

Theorem 14.6.34 *Let G be a connected graph and let A be its incidence matrix. Then $K[G] := K[x^{v_1}, \ldots, x^{v_q}]$ is normal if and only if the system $x \geq 0; xA \leq \mathbf{1}$ has the integer rounding property.*

Proof. Let $I = I(G)$ be the edge ideal of G. By Corollary 10.5.6 the subring $K[G]$ is normal if and only if $R[It]$ is normal. Using Theorem 4.3.17, we get that $R[It]$ is normal if and only if $R[It, t^{-1}]$ is normal. By Lemma 14.6.33, $R[It, t^{-1}]$ is normal if and only if the subring

$$S = K[t, x_1t, \ldots, x_nt, x^{v_1}t, \ldots, x^{v_q}t]$$

is normal. Finally applying Theorem 14.6.16 we get that S is normal if and only if the system $x \geq 0; xA \leq \mathbf{1}$ has the integer rounding property. $\Box$

Theorem 14.6.35 *Let G be a connected graph and let A be its incidence matrix. Then the system $x \geq 0; xA \leq \mathbf{1}$ has the integer rounding property if and only if any of the following equivalent conditions hold.*

(a) *$x \geq 0; xA \geq \mathbf{1}$ is a system with the integer rounding property.*

(b) *$R[It]$ is a normal domain, where $I = I(G)$ is the edge ideal of G.*

(c) *$K[x^{v_1}t, \ldots, x^{v_q}t]$ is normal, where $v_1, \ldots, v_q$ are the columns of A.*

(d) *$K[t, x_1t, \ldots, x_nt, x^{v_1}t, \ldots, x^{v_q}t]$ is normal.*

Proof. According to Corollary 14.6.9, the system $x \geq 0; xA \geq \mathbf{1}$ has the integer rounding property if and only if the Rees algebra $R[It]$ is normal. Thus the result follows from the proof of Theorem 14.6.34. $\Box$

Corollary 14.6.36 *Let G be a graph and let $I = I(G)$ be its edge ideal. Then $R[It]$ is normal if and only if $R[I^*t]$ is normal.*

Proof. Recall that: (i) the Rees algebra $R[It]$ is normal if and only if the extended Rees algebra $R[It, t^{-1}]$ is normal (Theorem 4.3.17), and (ii) $R[It, t^{-1}]$ is isomorphic to

$$K[t, x_1t, \ldots, x_nt, x^{v_1}t, \ldots, x^{v_q}t],$$

where I is the edge ideal of G (Lemma 14.6.33). Then the result follows applying Theorem 14.6.22. $\Box$

The next example shows that Corollary 14.6.36 does not extend to arbitrary uniform clutters.

Example 14.6.37 Consider the clutter $\mathcal{C}$ whose incidence matrix A is the transpose of the matrix:

$$\begin{bmatrix}
0 & 0 & 1 & 1 & 0 & 1 & 1 & 1 & 1 & 1 \\
0 & 0 & 1 & 0 & 1 & 1 & 1 & 1 & 1 & 1 \\
0 & 1 & 1 & 0 & 0 & 1 & 1 & 1 & 1 & 1 \\
1 & 1 & 0 & 0 & 0 & 1 & 1 & 1 & 1 & 1 \\
0 & 1 & 1 & 0 & 1 & 0 & 1 & 1 & 1 & 1 \\
1 & 1 & 1 & 1 & 1 & 0 & 0 & 1 & 1 & 0 \\
1 & 1 & 1 & 1 & 1 & 0 & 0 & 1 & 0 & 1 \\
1 & 1 & 1 & 1 & 1 & 0 & 1 & 1 & 0 & 0 \\
1 & 1 & 1 & 1 & 1 & 1 & 1 & 0 & 0 & 0 \\
1 & 1 & 1 & 1 & 0 & 0 & 1 & 1 & 0 & 1
\end{bmatrix}.$$

Let $I = I(\mathcal{C})$ be the edge ideal of $\mathcal{C}$. Note that $\mathcal{C}$ is uniform. Using *Normaliz* [68] it is seen that $R[It]$ is normal and that $R[I^*t]$ is not normal.

Proposition 14.6.38 *Let G be a graph without isolated vertices and let $\overline{G}$ be its complement. Then $I(\overline{G})^\vee = I(G)^*$ if and only if G is triangle free.*

Proof. $\Rightarrow$) Assume that G has a triangle $\mathcal{C}_3 = \{x_1, x_2, x_3\}$. We may assume $n \geq 4$. Notice that $\overline{C} = \{x_4, \ldots, x_n\}$ is a vertex cover of $\overline{G}$, i.e., $x_4 \cdots x_n$ belongs to $I(\overline{G})^\vee$ and consequently it belongs to $I(G)^*$, a contradiction because $I(G)^*$ is generated by monomials of degree $n-2$.

$\Leftarrow$) Let $x^a = x_1 \cdots x_r$ be a minimal generator of $I(\overline{G})^\vee$. Then the set $C = \{x_1, \ldots, x_r\}$ is a minimal vertex cover of $\overline{G}$. Hence $X \setminus C$ is a maximal complete subgraph of G. Thus by hypothesis $X \setminus C$ is an edge of G, i.e., $x^a \in I(G)^*$. This proves the inclusion "$\subset$". Conversely, let x^a be a minimal generator of $I(G)^*$. There is an edge $\{x_1, x_2\}$ of G such that $x^a = x_3 \cdots x_n$. Every edge of $\overline{G}$ must intersect $C = \{x_3, \ldots, x_n\}$, i.e., $x^a \in I(\overline{G})^\vee$. $\Box$

Corollary 14.6.39 *Let G be a triangle free graph without isolated vertices. Then $R[I(G)t]$ is normal if and only if $R[I(\overline{G})^\vee t]$ is normal.*

Proof. It follows from Corollary 14.6.36 and Proposition 14.6.38. $\Box$

Example 14.6.40 Let G be the graph consisting of two vertex disjoint odd cycles of length 5 and let $\overline{G}$ be its complement. By Proposition 10.5.8 and Corollary 14.6.39 the Rees algebras $R[I(G)t]$ and $R[I(\overline{G})^{\vee}t]$ are not normal.

Exercises

14.6.41 Let G be a perfect graph with vertex set $X = \{x_1, \ldots, x_n\}$ and let S be the subring generated by all $x^a t$ such that $\text{supp}(x^a)$ is a clique of G. Prove that S is normal.

14.6.42 Let G be a bipartite graph without isolated vertices and let $\overline{G}$ be its complement. Then $I(\overline{G})^{\vee} = I(G)^*$.

14.6.43 If $I \subset R$ is a Cohen–Macaulay square-free monomial ideal of height two, then $R[It]$ is normal.

14.6.44 Let A be the vertex-clique matrix of a graph G and let $v_1, \ldots, v_q$ be the column vectors of A. Consider the ideal $I^* = (x^{v_1^*}, \ldots, x^{v_q^*})$, where $v_i^* = \mathbf{1} - v_i$ for all i. Prove the duality: I^* is the ideal of covers of $\overline{G}$.

14.6.45 If G is a pentagon, then the Rees algebra of $I_c(G)$ is normal and G is not a perfect graph.

14.6.46 Let A be the incidence matrix of a clutter $\mathcal{C}$. If $\mathcal{C}$ is uniform and has the max-flow min-cut property, then the system $xA \leq \mathbf{1}$ has the integer rounding property.

14.7 Canonical modules and integer rounding

Here we give a description of the canonical module and the a-invariant for subrings arising from systems with the integer rounding property.

Let A be a matrix of size $n \times q$ with entries in $\mathbb{N}$ such that A has non-zero rows and non-zero columns. Let $v_1, \ldots, v_q$ be the columns of A. For use below consider the set $w_1, \ldots, w_r$ of all $\alpha \in \mathbb{N}^n$ such that $\alpha \leq v_i$ for some i. Let $R = K[x_1, \ldots, x_n]$ be a polynomial ring over a field K and let

$$S = K[x^{w_1}t, \ldots, x^{w_r}t] \subset R[t]$$

be the subring of $R[t]$ generated by $x^{w_1}t, \ldots, x^{w_r}t$, where t is a new variable.

The ring S is a standard graded K-algebra such that $\deg(x^a t^b) = b$ for any monomial $x^a t^b$ of S. If S is normal, then according to Theorem 9.1.5 the canonical module of S is the ideal given by

$$\omega_S = (\{x^a t^b \mid (a, b) \in \mathbb{N}\mathcal{B} \cap (\mathbb{R}_+\mathcal{B})^{\circ}\}), \tag{14.27}$$

where $\mathcal{B} = \{(w_1, 1), \ldots, (w_r, 1)\}$ and $(\mathbb{R}_+\mathcal{B})^{\mathrm{o}}$ is the interior of $\mathbb{R}_+\mathcal{B}$ relative to $\mathrm{aff}(\mathbb{R}_+\mathcal{B})$, the affine hull of $\mathbb{R}_+\mathcal{B}$. In our case $\mathrm{aff}(\mathbb{R}_+\mathcal{B}) = \mathbb{R}^{n+1}$.

Let $\{\beta_0, \beta_1, \ldots, \beta_m\} \subset \mathbb{Q}_+^n$ be the set of vertices of the polytope

$$\mathcal{P} = \{x \mid x \geq 0; xA \leq \mathbf{1}\},$$

where $\beta_0 = 0$, and let $\beta_1, \ldots, \beta_p$ be the set of all maximal elements of $\beta_0, \beta_1, \ldots, \beta_m$ (maximal with respect to $\leq$).

The following lemma is not hard to prove.

Lemma 14.7.1 *For each $1 \leq i \leq p$ there is a unique positive integer δ_i such that the non-zero entries of $(-\delta_i\beta_i, \delta_i)$ are relatively prime.*

Notation In what follows $\{\beta_1, \ldots, \beta_p\}$ is the set of maximal elements of $\{\beta_0, \ldots, \beta_m\}$ and $\delta_1, \ldots, \delta_p$ are the unique positive integers in Lemma 14.7.1.

Theorem 14.7.2 *If the system $x \geq 0; xA \leq \mathbf{1}$ has the integer rounding property, then S is normal, the canonical module of S is given by*

$$\omega_S = \left(x^a t^b \,\middle|\, (a, b) \begin{pmatrix} -\delta_1\beta_1 & \cdots & -\delta_p\beta_p & e_1 & \cdots & e_n \\ \delta_1 & \cdots & \delta_p & 0 & \cdots & 0 \end{pmatrix} \geq \mathbf{1} \right), \tag{14.28}$$

and the a-invariant of S is equal to $-\max_i\{\lceil 1/\delta_i + |\beta_i| \rceil\}$.

Proof. Note that in Eq. (14.28) we regard $(-\delta_i\beta_i, \delta_i)$ and e_j as column vectors for all i, j. The normality of S follows from Theorem 14.6.16. Recall that we have the following duality:

$$\begin{aligned} \mathcal{P} = \{x \mid x \geq 0; \langle x, w_i \rangle \leq 1 \,\forall i\} &= \mathrm{conv}(\beta_0, \beta_1, \ldots, \beta_m), \\ T(\mathcal{P}) = \{x \mid x \geq 0; \langle x, \beta_i \rangle \leq 1 \forall i\} &= \mathrm{conv}(w_1, \ldots, w_r), \end{aligned}$$

see Lemma 14.6.12. Therefore, by the maximality of $\beta_1, \ldots, \beta_p$, we obtain

$$\mathrm{conv}(w_1, \ldots, w_r) = \{x \mid x \geq 0; \langle x, \beta_i \rangle \leq 1, \,\forall\, i = 1, \ldots, p\}.$$

Hence, setting $\mathcal{B} = \{(w_i, 1)\}_{i=1}^r$ and noticing $\mathbb{Z}\mathcal{B} = \mathbb{Z}^{n+1}$, it is seen that

$$\mathbb{R}_+\mathcal{B} = H_{e_1}^+ \cap \cdots \cap H_{e_n}^+ \cap H_{(-\delta_1\beta_1, \delta_1)}^+ \cap \cdots \cap H_{(-\delta_p\beta_p, \delta_p)}^+. \tag{14.29}$$

Notice that $H_{e_i} \cap \mathbb{R}_+\mathcal{B}$ and $H_{(-\delta_j\beta_j, \delta_j)} \cap \mathbb{R}_+\mathcal{B}$ are proper faces of $\mathbb{R}_+\mathcal{B}$ for any i, j. Hence from Eq. (14.29) we get that a vector (a, b), with $a \in \mathbb{Z}^n$, $b \in \mathbb{Z}$, is in the relative interior of $\mathbb{R}_+\mathcal{B}$ if and only if the entries of a are positive and $\langle (a, b), (-\delta_i\beta_i, \delta_i) \rangle \geq 1$ for all i. Thus the required expression for ω_S follows using the normality of S and Eq. (14.27).

It remains to prove the formula for $a(S)$, the a-invariant of S. Consider the vector $(\mathbf{1}, b_0)$, where $b_0 = \max_i\{\lceil 1/\delta_i + |\beta_i| \rceil\}$. Using Eq. (14.28), it is not hard to see (by direct substitution of $(\mathbf{1}, b_0)$), that the monomial $x^{\mathbf{1}}t^{b_0}$ is in ω_S. Thus, from Eq. (9.2) of Lemma 9.1.7, we get $a(S) \geq -b_0$. Conversely if $x^a t^b$ is in ω_S, then again from Eq. (14.28) we get $\langle(-\delta_i\beta_i, \delta_i), (a, b)\rangle \geq 1$ for all i and $a_i \geq 1$ for all i, where $a = (a_i)$. Hence

$$b\delta_i \geq 1 + \delta_i\langle a, \beta_i\rangle \geq 1 + \delta_i\langle \mathbf{1}, \beta_i\rangle = 1 + \delta_i|\beta_i|.$$

Since b is an integer we obtain $b \geq \lceil 1/\delta_i + |\beta_i|\rceil$ for all i. Therefore $b \geq b_0$, i.e., $\deg(x^a t^b) = b \geq b_0$. As $x^a t^b$ was an arbitrary monomial in ω_S, by the formula for the a-invariant of S given in Eq. (9.2) of Lemma 9.1.7, we obtain that $a(S) \leq -b_0$. Altogether one has $a(S) = -b_0$, as required. $\Box$

Theorem 14.7.3 *Assume that the system $x \geq 0$; $xA \leq \mathbf{1}$ has the integer rounding property. If S is Gorenstein and $c_0 = \max\{|\beta_i|\colon 1 \leq i \leq p\}$ is an integer, then $|\beta_k| = c_0$ for each $1 \leq k \leq p$ such that β_k has integer entries.*

Proof. We proceed by contradiction. Assume that $|\beta_k| < c_0$ for some integer $1 \leq k \leq p$ such that β_k is integral. We may assume that β_k is $(1, \ldots, 1, 0, \ldots, 0)$ and $|\beta_k| = s$. From Eq. (14.29) $x^{\beta_k}t^{s-1}$ cannot be in S because $(\beta_k, s-1)$ does not belong to $H^+_{(-\delta_k\beta_k, \delta_k)}$. Consider $x^a t^b$, where $a = \beta_k + \mathbf{1}$, $b = b_0 + s - 1$ and $b_0 = -a(S)$. We claim that $x^a t^b$ is in ω_S. By Theorem 14.7.2 it suffices to show that $\langle(a, b), (-\delta_j\beta_j, \delta_j)\rangle \geq 1$ for $1 \leq j \leq p$. Thus we need only show that $\langle(a, b), (-\beta_j, 1)\rangle > 0$ for $1 \leq j \leq p$. From the proof of Theorem 14.7.2, it is seen that $-a(S) = \max_i\{\lfloor|\beta_i|\rfloor\} + 1$. Hence we get $b_0 = c_0 + 1$. One has the following equality

$$\langle(a, b), (-\beta_j, 1)\rangle = -|\beta_j| - \langle\beta_k, \beta_j\rangle + c_0 + s.$$

Set $\beta_j = (\beta_{j1}, \ldots, \beta_{jn})$. From Eq. (14.29) we get that the entries of each β_j are less than or equal to 1. Case (I): If $\beta_{ji} < 1$ for some $1 \leq i \leq s$, then $s - \langle\beta_k, \beta_j\rangle > 0$ and $c_0 \geq |\beta_j|$. Case (II): $\beta_{ji} = 1$ for $1 \leq i \leq s$. Then $\beta_j \geq \beta_k$. Thus by the maximality of β_k we obtain $\beta_j = \beta_k$. In both cases we obtain $\langle(a, b), (-\beta_j, 1)\rangle > 0$, as required. Hence the monomial $x^a t^b$ is in ω_S. Since S is Gorenstein and ω_S is generated by $x^{\mathbf{1}}t^{b_0}$, we obtain that $x^a t^b$ is a multiple of $x^{\mathbf{1}}t^{b_0}$, i.e., $x^{\beta_k}t^{s-1}$ must be in S, a contradiction. $\Box$

Theorem 14.7.4 *Assume that $x \geq 0; xA \leq \mathbf{1}$ has the integer rounding property. If $-a(S) = 1/\delta_i + |\beta_i|$ for $i = 1, \ldots, p$, then S is Gorenstein.*

Proof. We set $b_0 = -a(S)$ and $\mathcal{B} = \{(w_1, 1), \ldots, (w_r, 1)\}$. The ring S is normal by Theorem 14.6.16. Since the monomial $x^{\mathbf{1}}t^{b_0} = x_1 \cdots x_n t^{b_0}$ is in ω_S, we need only show that $\omega_S = (x^{\mathbf{1}}t^{b_0})$. Take $x^a t^b \in \omega_S$. It suffices to prove that $x^{a-\mathbf{1}}t^{b-b_0}$ is in S. Using Theorem 14.6.16, one has the equality $\mathbb{R}_+\mathcal{B} \cap \mathbb{Z}^{n+1} = \mathbb{N}\mathcal{B}$. Thus we need only show that $(a - \mathbf{1}, b - b_0)$ is in $\mathbb{R}_+\mathcal{B}$.

From Eq. (14.29), the proof reduces to showing that $(a-\mathbf{1}, b-b_0)$ is in $H^+_{(-\beta_i,1)}$ for $i=1,\ldots,p$. As $(a,b)\in\omega_S$, from Theorem 14.7.2, we get

$$\langle (a,b),(-\delta_i\beta_i,\delta_i)\rangle = -\langle a,\delta_i\beta_i\rangle + b\delta_i \geq 1 \implies -\langle a,\beta_i\rangle \geq -b+1/\delta_i$$

for $i=1,\ldots,p$. Therefore

$$\langle (a-\mathbf{1},b-b_0),(-\beta_i,1)\rangle = -\langle a,\beta_i\rangle + |\beta_i| + b - b_0 \geq -b+1/\delta_i+|\beta_i|+b-b_0 = 0$$

for all i, as required. □

Corollary 14.7.5 *If $\mathcal{P}=\{x|\, x\geq 0; xA\leq \mathbf{1}\}$ is an integral polytope, then S is Gorenstein if and only if $a(S)=-(|\beta_i|+1)$ for $i=1,\ldots,p$.*

Proof. Notice that if $\mathcal{P}$ is integral, then β_i has entries in $\{0,1\}$ for $1\leq i\leq p$ and consequently $\delta_i = 1$ for $1\leq i\leq p$. Thus the result follows from Theorems 14.7.3 and 14.7.4. □

Conjecture 14.7.6 If A is the incidence matrix of a connected graph and the system $x\geq 0$; $xA\leq \mathbf{1}$ has the integer rounding property, then S is Gorenstein if and only if $-a(S)=1/\delta_i+|\beta_i|$ for $i=1,\ldots,p$.

The answer to this conjecture is positive if A is the incidence matrix of a bipartite graph because in this case $\mathcal{P}$ is an integral polytope and we may apply Corollary 14.7.5.

For use below we consider the empty set as a clique whose vertex set is empty. Note that $\text{supp}(x^a)=\emptyset$ if and only if $a=0$.

Theorem 14.7.7 *Let G be a perfect graph and let $S=K[x^{\omega_1}t,\ldots,x^{\omega_r}t]$ be the subring generated by all square-free monomials x^at such that $\text{supp}(x^a)$ is a clique of G. Then the canonical module of S is given by*

$$\omega_S = \left(\left\{x^at^b \,\middle|\, (a,b)\begin{pmatrix} -\beta_1 & \cdots & -\beta_s & e_1 & \cdots & e_n \\ 1 & \cdots & 1 & 0 & \cdots & 0\end{pmatrix} \geq \mathbf{1}\right\}\right),$$

where $\beta_1,\ldots,\beta_s$ are the characteristic vectors of the maximal independent sets of G, and the a-invariant of S is equal to $-(\max_i\{|\beta_i|\}+1)$.

Proof. Let $v_1,\ldots,v_q$ be the set of characteristic vectors of the maximal cliques of G. Note that $w_1,\ldots,w_r$ is the set of all $\alpha\in\mathbb{N}^n$ such that $\alpha\leq v_i$ for some i. Since G is a perfect graph, by Theorem 13.6.1, we have

$$\mathcal{P}=\{x|x\geq 0; xA\leq \mathbf{1}\} = \text{conv}(\beta_0,\beta_1,\ldots,\beta_p),$$

where $\beta_0=0$ and $\beta_1,\ldots,\beta_p$ are the characteristic vectors of the independent sets of G. We may assume that $\beta_1,\ldots,\beta_s$ correspond to the maximal independent sets of G. Furthermore, since $\mathcal{P}$ has only integral vertices, by a result of Lovász (see Theorem 13.6.8) the system $x \geq 0; xA \leq \mathbf{1}$ is totally dual integral, i.e., the minimum in the LP-duality equation

$$\max\{\langle \alpha, x\rangle |\, x \geq 0; xA \leq \mathbf{1}\} = \min\{\langle y, \mathbf{1}\rangle |\, y \geq 0; Ay \geq \alpha\} \qquad (14.30)$$

has an integral optimum solution y for each integral vector α with finite minimum. In particular the system $x \geq 0; xA \leq \mathbf{1}$ satisfies the integer rounding property. Therefore the result follows readily from Theorem 14.7.2. $\Box$

Proposition 14.7.8 *Let G be a perfect graph, let $w_1, \ldots, w_r$ be the characteristic vectors of the cliques of G, and let $\beta_1, \ldots, \beta_s$ be the characteristic vectors of the maximal independent sets of G. Then the set*

$$\Gamma = \{(-\beta_1, 1), \ldots, (-\beta_s, 1), e_1, \ldots, e_n\}$$

is a Hilbert basis of $(\mathbb{R}_+\mathcal{B})^*$, where $\mathcal{B} = \{(w_1, 1), \ldots, (w_r, 1)\}$.

Proof. The characteristic vector of the empty set is set to be equal to zero. As G is perfect we have

$$\text{conv}(w_1, \ldots, w_r) = \{x |\, x \geq 0; x(\beta_1 \cdots \beta_s) \leq \mathbf{1}\}.$$

Therefore it is seen that

$$\mathbb{R}_+(w_1, 1) + \cdots + \mathbb{R}_+(w_r, 1) = H_{e_1}^+ \cap \cdots \cap H_{e_n}^+ \cap \mathbb{R}_+(-\beta_1, 1) \cap \cdots \mathbb{R}_+(-\beta_s, 1).$$

Thus, by duality (see Corollary 1.1.30), we obtain that $(\mathbb{R}_+\mathcal{B})^* = \mathbb{R}_+\Gamma$. Using that the system $x \geq 0$; $x(\beta_1, \ldots, \beta_s) \leq \mathbf{1}$ is TDI it follows that $\mathbb{Z}^n \cap \mathbb{R}_+\Gamma = \mathbb{N}\Gamma$, i.e., Γ is a Hilbert basis, as required. $\Box$

Corollary 14.7.9 *Let G be a connected bipartite graph and let $I = I(G)$ be its edge ideal. Then the extended Rees algebra $R[It, t^{-1}]$ is a Gorenstein standard K-algebra if and only if G is unmixed.*

Proof. Let ω_S be the canonical module of $S = R[It, t^{-1}]$. As S is Cohen–Macaulay, recall that S is Gorenstein if and only if ω_S is a principal ideal. Since G is a perfect graph. The result follows using Lemma 14.6.33 together with the description of the canonical module given in Theorem 14.7.7. $\Box$

Exercise

14.7.10 If A is the incidence matrix of a graph, then $\delta_i = 1$ or $\delta_i = 1/2$ for each i (see Lemma 14.7.1).

14.8 Clique clutters of Meyniel graphs

In this section it is shown that clique clutters of Meynel graphs are Ehrhart in the sense of [306] (see Definition 9.3.27).

Recall that a *Meyniel graph* is a simple graph in which every odd cycle of length at least five has at least two chords.

Theorem 14.8.1 [247] *A graph G is Meyniel if and only if for each induced subgraph H and for each vertex v of H, there exists an independent set in H containing v and intersecting all maximal cliques of H (maximal with respect to inclusion).*

Theorem 14.8.2 *Let G be a Meyniel graph and let $\mathcal{A}=\{v_1,\ldots,v_q\}$ be the set of columns of the vertex-clique matrix of G. If $A(\mathcal{P})$ is the Ehrhart ring of $\mathcal{P}=\text{conv}(v_1,\ldots,v_q)$, then $K[x^{v_1}t,\ldots,x^{v_q}t]=A(\mathcal{P})$.*

Proof. The inclusion "$\subset$" is clear. To prove the other inclusion take $x^at^b\in A(\mathcal{P})$, i.e., $a\in b\mathcal{P}\cap\mathbb{Z}^n$ and $b\in\mathbb{N}$. Using that $\mathcal{P}$ is the convex hull of the v_i's, it is not hard to see that we can write

$$(a,b)=\lambda_1(v_1,1)+\cdots+\lambda_q(v_q,1),\quad \lambda_i\geq 0\,\forall\, i. \tag{14.31}$$

Let A be the vertex-clique matrix of G. Any Meyniel graph is perfect [373, Theorem 66.6]. Thus, by Theorem 13.6.8, the system $x\geq 0; xA\leq \mathbf{1}$ is TDI. Therefore, by Proposition 1.3.28, we get that the set

$$\mathcal{H}=\{(v_1,1),\ldots,(v_q,1),-e_1,\ldots,-e_n\}$$

is a Hilbert basis, i.e., $\mathbb{R}_+\mathcal{H}\cap\mathbb{Z}^{n+1}=\mathbb{N}\mathcal{H}$. As (a,b) is an integral vector in $\mathbb{R}_+\mathcal{H}$, we can write

$$(a,b)=\eta_1(v_1,1)+\cdots+\eta_q(v_q,1)-\mu_1e_1-\cdots-\mu_ne_n, \tag{14.32}$$

$\eta_i\in\mathbb{N},\mu_j\in\mathbb{N}$ for all i,j. By Theorem 14.8.1, for each x_k in $V(G)=\{x_1,\ldots,x_n\}$ there exists an independent set B_k of G containing x_k and intersecting all maximal cliques of G. We set $\beta_k=\sum_{x_i\in B_k}e_i$ for $1\leq k\leq n$. Notice that a clique of G and an independent set of G can meet in at most one vertex. Then $\beta_1,\ldots,\beta_n$ are vectors in $\{0,1\}^n$ such that $\langle v_j,\beta_i\rangle=1$ and $\langle e_i,\beta_i\rangle=1$ for all i,j. Hence, using Eqs. (14.31) and (14.32), we obtain

$$\begin{aligned}
\langle a,\beta_i\rangle &= \lambda_1\langle v_1,\beta_i\rangle+\cdots+\lambda_q\langle v_q,\beta_i\rangle=b,\quad \forall\, i.\\
\langle a,\beta_i\rangle &= \eta_1\langle v_1,\beta_i\rangle+\cdots+\eta_q\langle v_q,\beta_i\rangle-\mu_1\langle e_1,\beta_i\rangle-\cdots-\mu_n\langle e_n,\beta_i\rangle\\
&= b-\mu_1\langle e_1,\beta_i\rangle-\cdots-\mu_n\langle e_n,\beta_i\rangle,\quad \forall\, i.
\end{aligned}$$

Therefore for $i=1,\ldots,n$, we get $\sum_{j=1}^n\mu_j\langle e_j,\beta_i\rangle=0$. Since $\langle e_i,\beta_i\rangle=1$ for all i, we get $\mu_i=0$ for all i. Hence the vector (a,b) belongs to the semigroup $\mathbb{N}\mathcal{A}$. Thus $x^at^b\in K[x^{v_1}t,\ldots,x^{v_q}t]$. $\square$

Corollary 14.8.3 *Let G be a Meyniel graph, let A be the vertex-clique matrix of G, and let $v_1, \ldots, v_q$ be the column vectors of A. The following conditions are equivalent:*

(i) $x \geq 0; xA \geq \mathbf{1}$ *is a TDI system.*

(ii) $I^i = I^{(i)}$ *for* $i \geq 1$, *where* $I = (x^{v_1}, \ldots, x^{v_q})$.

(iii) $\mathcal{Q}(A) = \{x | x \geq 0; xA \geq \mathbf{1}\}$ *is an integral polyhedron.*

Proof. (i) $\Rightarrow$ (ii) and (ii) $\Rightarrow$ (iii): Follow from Theorem 14.3.6. (iii) $\Rightarrow$ (i): By Theorem 14.8.2 we get $K[x^{v_1}t, \ldots, x^{v_q}t] = A(\mathcal{P})$. As $\mathcal{Q}(A)$ is integral, a direct application of Proposition 14.3.30 and Theorem 14.3.6 gives that the system $x \geq 0; xA \geq \mathbf{1}$ is TDI. □

Exercises

14.8.4 Let $v_1, \ldots, v_6$ be the characteristic vectors of the maximal cliques of the graph below and let $\mathcal{P}$ be its convex hull.

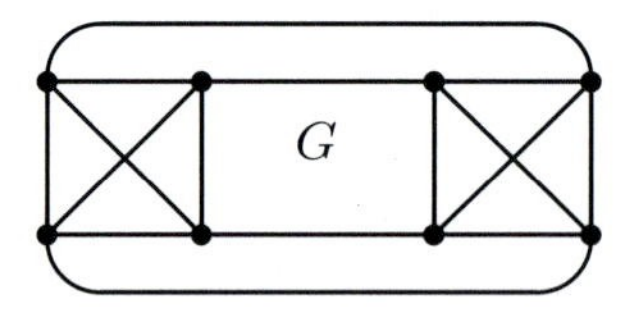

Use *Normaliz* [68] to show that $K[x^{v_1}t, \ldots, x^{v_6}t] \subsetneq A(\mathcal{P})$. Show that the incidence matrix of cl(G), the clique clutter of G, is totally unimodular by showing that the incidence matrix of cl(G) is the transpose of the incidence matrix of the complete bipartite graph $\mathcal{K}_{2,4}$.

14.8.5 Consider the following graph G:

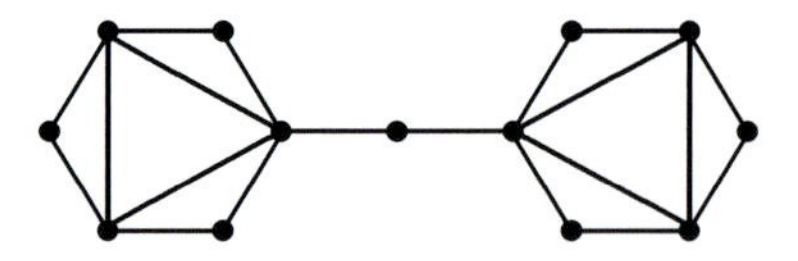

Prove that the edge ideal $I = I(\text{cl}(G))$ of the clique clutter of G is not normal. This graph is chordal, hence perfect. Thus edge ideals of clique clutters of perfect graphs are in general not normal.

Appendix

Graph Diagrams

For convenience we will display some Cohen–Macaulay graphs and unmixed graphs with small number of vertices.

A.1 Cohen–Macaulay graphs

The complete list of Cohen–Macaulay connected graphs with at most six vertices is:

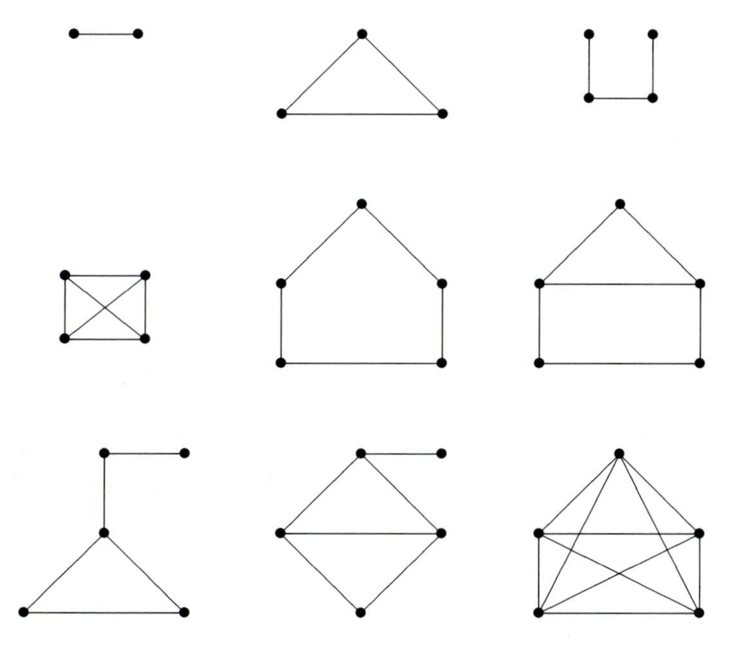

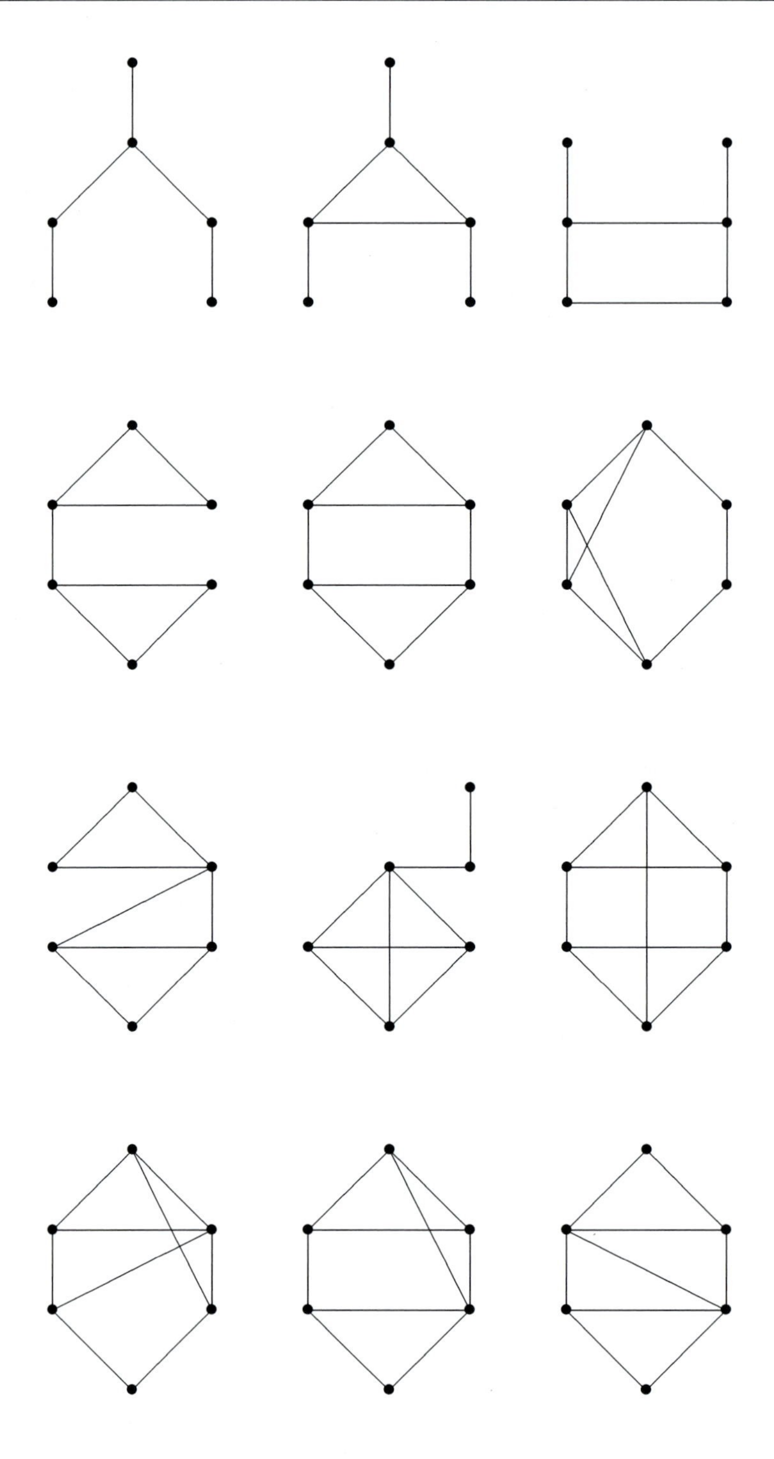

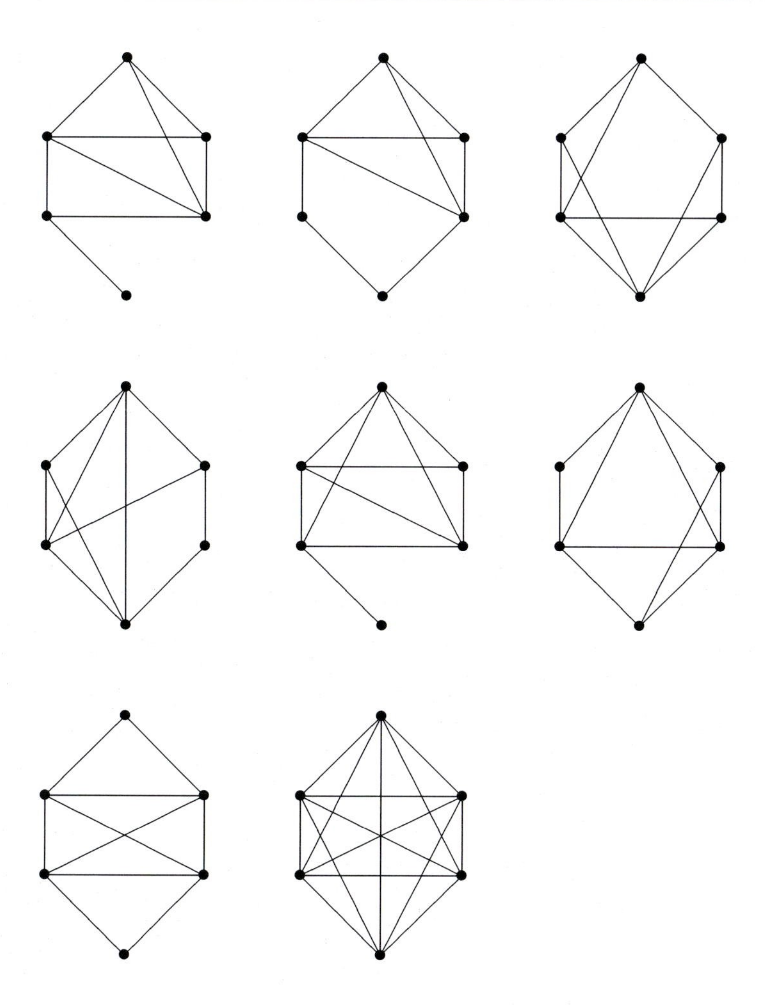

A.2 Unmixed graphs

Next we display the set of all unmixed non-Cohen–Macaulay connected graphs with at most six vertices.

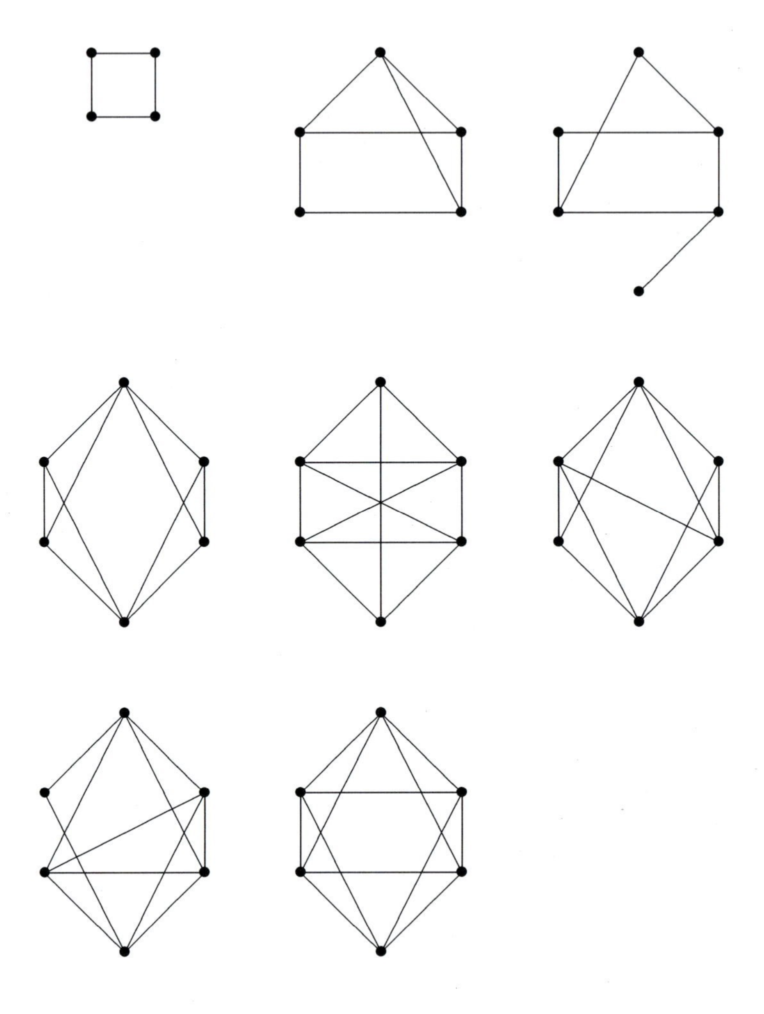

Bibliography

[1] W. W. Adams and P. Loustaunau, *An Introduction to Gröbner Bases*, Graduate Studies in Mathematics **3**, American Mathematical Society, Providence, RI, 1994.

[2] M. Aigner, *Combinatorial Theory*, Springer, 1997.

[3] A. Alcántar, Rees algebras of square-free Veronese ideals and their a-invariants, *Discrete Math.* **302** (2005), 7–21.

[4] A. Alcántar and R. H. Villarreal, Critical binomials of monomial curves, *Comm. Algebra* **22** (1994), 3037–3052.

[5] C. Alfaro and C. E. Valencia, On the sandpile group of the cone of a graph, *Linear Algebra Appl.* **436** (2012), no. 5, 1154–1176.

[6] A. Alilooee and S. Faridi, When is a squarefree monomial ideal of linear type? Preprint, 2013, `arXiv:1309.1771`.

[7] N. Alon, Combinatorial Nullstellensatz, *Combin. Probab. Comput.* **8** (1999), no. 1-2, 7–29.

[8] R. P. Anstee, Hypergraphs with no special cycles, *Combinatorica* **3** (1983), no. 2, 141–146.

[9] M. F. Atiyah and I. G. Macdonald, *Introduction to Commutative Algebra*, Addison-Wesley, Reading, MA, 1969.

[10] D. Avis and K. Fukuda, A pivoting algorithm for convex hulls and vertex enumeration of arrangements and polyhedra, *Discrete Comput. Geom.* **8** (1992), no. 3, 295–313.

[11] D. Avis and K. Fukuda, Reverse search for enumeration, *Discrete Appl. Math.* **65** (1996), no. 1-3, 21–46.

[12] C. Bahiano, Symbolic powers of edge ideals, *J. Algebra* **273** (2004), no. 2, 517–537.

[13] A. Banerjee, The regularity of powers of edge ideals, *J. Algebraic Combin.*, to appear. DOI: 10.1007/s10801-014-0537-2.

[14] J. Bang-Jensen and G. Gutin, *Digraphs. Theory, Algorithms and Applications*, Springer Monographs in Mathematics, Springer, 2006.

[15] M. Barile, On the arithmetical rank of the edge ideals of forests, *Comm. Algebra* **36** (2008), no. 12, 4678–4703.

[16] M. Barile, D. Kiani, F. Mohammadi, and S. Yassemi, Arithmetical rank of the cyclic and bicyclic graphs, *J. Algebra Appl.* **11** (2012), no. 2, Article Number: 1250039.

[17] M. Barile, M. Morales, and A. Thoma, Set-theoretic complete intersections on binomials, *Proc. Amer. Math. Soc.* **130** (2002), 1893–1903.

[18] H. Bass, On the ubiquity of Gorenstein rings, *Math. Z.* **82** (1963), 8–28.

[19] S. Baum and L. E. Trotter, Integer rounding for polymatroid and branching optimization problems, *SIAM J. Algebraic Discrete Methods* **2** (1981), no. 4, 416–425.

[20] D. Bayer and M. Stillman, Computation of Hilbert functions, *J. Symbolic Comput.* **14** (1992), 31–50.

[21] M. Beck and S. Robins, *Computing the Continuous Discretely*, Springer, New York, 2007.

[22] E. Becker, R. Grobe, and M. Niermann, Radicals of binomial ideals, *J. Pure Appl. Algebra* **117** (1997), 41–79.

[23] M. Beintema, A note on Artinian Gorenstein algebras defined by monomials, *Rocky Mountain J. Math.* **23** (1993), 1–3.

[24] C. Berge, Balanced matrices, *Math. Programming* **2** (1972), 19–31.

[25] C. Berge, *Graphs and Hypergraphs*, North-Holland Mathematical Library, Vol. 6, North-Holland Publishing Co., Amsterdam-London; American Elsevier Publishing Co., Inc., New York, 1976.

[26] C. Berge, *Some Common Properties for Regularizable Graphs, Edge-critical Graphs and B-graphs*, North-Holland Math. Stud. **60**, North-Holland, Amsterdam, 1982, pp. 31–44.

[27] C. Berge, *Hypergraphs Combinatorics of Finite Sets*, Mathematical Library **45**, North-Holland, 1989.

[28] A. Berman and R. J. Plemmons, *Nonnegative Matrices in the Mathematical Sciences*, Classics in Applied Mathematics **9**, Society for Industrial and Applied Mathematics (SIAM), Philadelphia, PA, 1994.

[29] I. Bermejo, I. García-Marco, and E. Reyes, Graphs and complete intersection toric ideals, *J. Algebra Appl.*, to appear.

[30] I. Bermejo, I. García-Marco, and J. J. Salazar-González, An algorithm for checking whether the toric ideal of an affine monomial curve is a complete intersection, *J. Symbolic Comput.* **(42)** (2007), 971–991.

[31] I. Bermejo and P. Gimenez, On Castelnuovo–Mumford regularity of projective curves, *Proc. Amer. Math. Soc.* **128** (2000), 1293–1299.

[32] I. Bermejo and P. Gimenez, Saturation and Castelnuovo–Mumford regularity, *J. Algebra* **303** (2006), no. 2, 592–617.

[33] I. Bermejo, P. Gimenez, E. Reyes, and R. H. Villarreal, Complete intersections in affine monomial curves, *Bol. Soc. Mat. Mexicana* (3) **11** (2005), 191–203.

[34] I. Bermejo, P. Gimenez, and A. Simis, Polar syzygies in characteristic zero: the monomial case, *J. Pure Appl. Algebra* **213** (2009), 1–21.

[35] D. Bertsimas and J. N. Tsitsiklis, *Introduction to Linear Optimization*, Athena Scientific, Massachusetts, 1997.

[36] A. Bhat, J. Biermann, and A. Van Tuyl, Generalized cover ideals and the persistence property, *J. Pure Appl. Algebra* **218** (2014), no. 9, 1683–1695.

[37] A. Bigatti, Computation of Hilbert–Poincaré series, *J. Pure Appl. Algebra* **119** (1997), 237–253.

[38] A. Bigatti, R. La Scala, and L. Robbiano, Computing toric ideals, *J. Symbolic Comput.* **27** (1999), 351–365.

[39] A. Bigatti and L. Robbiano, Borel sets and sectional matrices, *Annals of Combinatorics* **1** (1997), 197–213.

[40] N. Biggs, *Algebraic Graph Theory*, Cambridge University Press, Cambridge, 1993.

[41] L. J. Billera, Polyhedral theory and commutative algebra, in *Mathematical Programming: The State of the Art* (A. Bachem, M. Grötschel, and B. Korte, Eds.), Springer-Verlag, 1983, pp. 57–77.

[42] C. Bivià-Ausina, The analytic spread of monomial ideals, *Comm. Algebra* **31** (2003), no. 7, 3487–3496.

[43] A. Björner, *Topological Methods, Handbook of Combinatorics*, Vol. 1, 2, 1819–1872, Elsevier, Amsterdam, 1995.

[44] A. Björner and M. Wachs, Shellable nonpure complexes and posets I, *Trans. Amer. Math. Soc.* **348** (1996), no. 4, 1299–1327.

[45] A. Björner and M. Wachs, Shellable nonpure complexes and posets II, *Trans. Amer. Math. Soc.* **349** (1997), no. 10, 3945–3975.

[46] S. Blum, Base-sortable matroids and Koszulness of semigroup rings, *European J. Combin.* **22** (2001), no. 7, 937–951.

[47] M. Boij and J. Söderberg, Betti numbers of graded modules and the multiplicity conjecture in the non-Cohen–Macaulay case, *Algebra Number Theory* **6** (2012), no. 3, 437–454.

[48] B. Bollobás, *Modern Graph Theory*, Graduate Texts in Mathematics **184**, Springer-Verlag, New York, 1998.

[49] V. Bonanzinga, C. Escobar, and R. H. Villarreal, On the normality of Rees algebras associated to totally unimodular matrices, *Results Math.* **41**, 3/4, (2002), 258–264.

[50] R. R. Bouchat, H. T. Hà, and A. O'Keefe, Path ideals of rooted trees and their graded Betti numbers, *J. Combin. Theory Ser. A* **118** (2011), no. 8, 2411–2425.

[51] P. Bouwknegt, q-identities and affinized projective varieties I. Quadratic monomial ideals, *Commun. Math. Phys.* **210** (2000), no. 3, 641–661.

[52] J. P. Brennan, L. A. Dupont, and R. H. Villarreal, Duality, a-invariants and canonical modules of rings arising from linear optimization problems, *Bull. Math. Soc. Sci. Math. Roumanie (N.S.)* **51** (2008), no. 4, 279–305.

[53] H. Bresinsky, Symmetric semigroups of integers generated by 4 elements, *Manuscripta Math.* **17** (1975), 205–219.

[54] H. Bresinsky, Monomial space curves in $\mathbb{A}^3$ as set-theoretic complete intersections, *Proc. Amer. Math. Soc.* **75** (1979), 23–24.

[55] M. Brodmann, The asymptotic nature of the analytic spread, *Math. Proc. Cambridge Philos. Soc.* **86** (1979), 35–39.

[56] M. Brodmann, Asymptotic stability of $\text{Ass}(M/I^nM)$, *Proc. Amer. Math. Soc.* **74** (1979), 16–18.

[57] A. Brøndsted, *Introduction to Convex Polytopes*, Graduate Texts in Mathematics **90**, Springer-Verlag, 1983.

[58] H. Bruggesser and P. Mani, Shellable decompositions of cells and spheres, *Math. Scand.* **29** (1971), 197–205.

[59] P. Brumatti, A. Simis, and W. V. Vasconcelos, Normal Rees algebras, *J. Algebra* **112** (1988), 26–48.

[60] W. Bruns and J. Gubeladze, Normality and covering properties of affine semigroups, *J. reine angew. Math.* **510** (1999), 161–178.

[61] W. Bruns and J. Gubeladze, *Polytopes, Rings, and K-Theory*, Springer Monographs in Mathematics, Springer, Dordrecht, 2009.

[62] W. Bruns, J. Gubeladze, and N. V. Trung, Normal polytopes, triangulations, and Koszul algebras, *J. reine angew. Math.* **481** (1997), 123–160.

[63] W. Bruns and J. Herzog, On the computation of a-invariants, *Manuscripta Math.* **77** (1992), 201–213.

[64] W. Bruns and J. Herzog, On multigraded resolutions, *Math. Proc. Cambridge Philos. Soc.* **118** (1995), 245–257.

[65] W. Bruns and J. Herzog, *Cohen–Macaulay Rings*, Cambridge University Press, Cambridge, Revised Edition, 1998.

[66] W. Bruns and T. Hibi, Stanley–Reisner rings with pure resolutions. *Comm. Algebra* **21** (1995), 1201–1217.

[67] W. Bruns and T. Hibi, Cohen–Macaulay partially ordered sets with pure resolutions, *European J. Combin.* **19** (1998), 779–785.

[68] W. Bruns, B. Ichim, T. Römer, and C. Söger: *Normaliz*. Algorithms for rational cones and affine monoids.
Available from `http://www.math.uos.de/normaliz`.

[69] W. Bruns and R. Koch, Computing the integral closure of an affine semigroup, *Univ. Iagel. Acta Math.* **39** (2001), 59–70.

[70] W. Bruns and G. Restuccia, Canonical modules of Rees algebras, *J. Pure Appl. Algebra* **201** (2005), 189–203.

[71] W. Bruns, A. Simis, and N. V. Trung, Blow-up of straightening-closed ideals in ordinal Hodge algebras, *Trans. Amer. Math. Soc.* **326** (1991), 507–528.

[72] W. Bruns, W. V. Vasconcelos, and R. H. Villarreal, Degree bounds in monomial subrings, *Illinois J. Math.* **41** (1997), 341–353.

[73] W. Bruns and U. Vetter, *Determinantal Rings*, Lecture Notes in Mathematics **1327**, Springer-Verlag, 1988.

[74] B. Buchberger, An algorithmic method in polynomial ideal theory, in *Recent Trends in Mathematical Systems Theory* (N.K. Bose, Ed.), Reidel, Dordrecht, 1985, 184–232.

[75] D. Buchsbaum and D. Eisenbud, Algebra structures for finite free resolutions, and some structure theorems for ideals of codimension 3, *Amer. J. Math.* **99** (1977), 447–485.

[76] L. Burch, Codimension and analytic spread, *Proc. Camb. Phil. Soc.* **72** (1972), 369–373.

[77] A. H. Busch, F. F. Dragan, and R. Sritharan, *New Min-Max Theorems for Weakly Chordal and Dually Chordal Graphs*, Lecture Notes in Computer Science **6509**, Springer-Verlag, 207–218, 2010.

[78] G. Carrá Ferro and D. Ferrarello, Cohen–Macaulay graphs arising from digraphs. Preprint, 2007, `arXiv:0703417`.

[79] M. Cavaliere, M. Rossi, and G. Valla, *On Short Graded Algebras* (Proc. Workshop in Commutative Algebra, Salvador, Brazil), Lecture Notes in Mathematics **1430**, Springer-Verlag, Berlin, 1990, pp. 21–31.

[80] B.W. Char, K.G. Geddes, G.H. Gonnet, and S.M. Watt, *Maple V Language Reference Manual*, Springer-Verlag, Berlin, 1991.

[81] J. Chen, S. Morey, and A. Sung, The stable set of associated primes of the ideal of a graph, *Rocky Mountain J. Math.* **32** (2002), 71–89.

[82] J. Chifman and S. Petrović, Toric ideals of phylogenetic invariants for the general group-based model on claw trees, *Lecture Notes in Computer Science* **4545**, Springer-Verlag, 307–321, 2007.

[83] L. Chouinard, Krull semigroups and divisor class groups, *Canad. J. Math.* **33** (1981), 1459–1468.

[84] T. Christof, revised by A. Löbel and M. Stoer, *PORTA: A Polyhedron Representation Transformation Algorithm*, 1997. `http://www.zib.de/Optimization/Software/Porta`.

[85] M. Chudnovsky, N. Robertson, P. Seymour, and R. Thomas, The strong perfect graph theorem, *Ann. of Math.* **(2) 164** (2006), no. 1, 51–229.

[86] V. Chvátal, On certain polytopes associated with graphs, *J. Combinatorial Theory Ser. B* **18** (1975), 138–154.

[87] V. Chvátal, *Linear Programming*, W.H. Freeman and Company, New York, 1983.

[88] CoCoATeam, CoCoA: a system for doing Computations in Commutative Algebra. Available at `http://cocoa.dima.unige.it`.

[89] A. Conca and E. De Negri, M-sequences, graph ideals, and ladder ideals of linear type, *J. Algebra* **211** (1999), 599–624.

[90] M. Conforti and G. Cornuéjols, Clutters that pack and the max-flow min-cut property, *The Fourth Bellairs Workshop on Combinatorial Optimization* (W. R. Pulleyblank, F. B. Shepherd, Eds.), 1993.

[91] A. Constantinescu and M. Varbaro, Koszulness, Krull dimension, and other properties of graph-related algebras, *J. Algebraic Combin.* **34** (2011), no. 3, 375–400.

[92] S. M. Cooper, R. J. D. Embree, H. T. Hà, and A. H. Hoefel, Symbolic powers of monomial ideals. Preprint, 2014, `arXiv:1309.5082v3`.

[93] G. Cornuéjols, *Combinatorial Optimization: Packing and Covering*, CBMS-NSF Regional Conference Series in Applied Mathematics **74**, SIAM (2001).

[94] G. Cornuéjols, B. Guenin, and F. Margot, The packing property, *Math. Programming* **89** (2000), no. 1, Ser. A, 113–126.

[95] H. Corrales and C. E. Valencia, On the critical ideals of graphs, *Linear Algebra Appl.* **439** (2013), no. 12, 3870–3892.

[96] A. Corso and U. Nagel, Monomial and toric ideals associated to ferrers graphs, *Trans. Amer. Math. Soc.* **361** (2009), no. 3, 1371–1395.

[97] A. Corso, W. V. Vasconcelos, and R. H. Villarreal, Generic Gaussian ideals, *J. Pure Appl. Algebra* **125** (1998), 117–127.

[98] B. Costa and A. Simis, Cremona maps defined by monomials, *J. Pure Appl. Algebra* **216** (2012), no. 1, 202–215.

[99] D. Cox, J. Little, and D. O'Shea, *Ideals, Varieties, and Algorithms*, Springer-Verlag, 1992.

[100] D. Cox, J. Little, and H. Schenck, *Toric Varieties*, Graduate Studies in Mathematics **124**, American Mathematical Society, Providence, RI, 2011.

[101] V. Crispin Quiñonez, Integral closure and other operations on monomial ideals, *J. Commut. Algebra* **2** (2010), no. 3, 359–386.

[102] V. I. Danilov, The geometry of toric varieties, *Russian Math. Surveys* **33** (1978), 97–154.

[103] H. Dao, C. Huneke, and J. Schweig, Bounds on the regularity and projective dimension of ideals associated to graphs, *J. Algebraic Combin.* **38** (2013), no. 1, 37–55.

[104] D. Delfino, A. Taylor, W. V. Vasconcelos, R. H. Villarreal, and N. Weininger, Monomial ideals and the computation of multiplicities, *Commutative Ring Theory and Applications* (Fez, 2001), pp. 87–106, Lecture Notes in Pure and Appl. Math. **231**, Dekker, New York, 2003.

[105] J. A. De Loera, R. Hemmecke and M. Köppe, *Algebraic and Geometric Ideas in the Theory of Discrete Optimization*, MOS-SIAM Series on Optimization **14**, Society for Industrial and Applied Mathematics (SIAM), Philadelphia, PA, 2013.

[106] J. A. De Loera, J. Rambau, and F. Santos, *Triangulations*, Algorithms and Computation in Mathematics **25**, Springer, 2010.

[107] C. Delorme, Sous-monoides d'intersection complète de ℕ, *Ann. Sci. École Norm. Sup.* **9** (1976), 145–154.

[108] E. De Negri and T. Hibi, Gorenstein algebras of Veronese type, *J. Algebra* **193** (1997), 629–639.

[109] F. Di Biase and R. Urbanke, An algorithm to calculate the kernel of certain polynomial ring homomorphism, *Experiment. Math.* **4** (1995), 227–234.

[110] A. Dickenstein and E. A. Tobis, Independent sets from an algebraic perspective, *Internat. J. Algebra Comput.* **22** (2012), no. 2, 83–98.

[111] R. Diestel, *Graph Theory*, Graduate Texts in Mathematics **173**, Springer-Verlag, New York, 1997.

[112] G. A. Dirac, On rigid circuit graphs, *Abh. Math. Sem. Univ. Hamburg* **38** (1961), 71–76.

[113] A. Dochtermann and A. Engstrom, Algebraic properties of edge ideals via combinatorial topology, *Electron. J. Combin.* **16** (2009), no. 2, R2.

[114] L. Doering and T. Gunston, Algebras arising from planar bipartite graphs, *Comm. Algebra* **24** (1996), 3589–3598.

[115] A. Dress, A new algebraic criterion for shellability, *Beiträge Algebra Geom.* **34** (1993), 45–55.

[116] L. A. Dupont, E. Reyes, and R. H. Villarreal, Cohen–Macaulay clutters with combinatorial optimization properties and parallelizations of normal edge ideals, *São Paulo J. Math. Sci.* **3** (2009), no. 1, 61–75.

[117] L. A. Dupont and R. H. Villarreal, Symbolic Rees algebras, vertex covers and irreducible representations of Rees cones, *Algebra Discrete Math.* **10** (2010), no. 2, 64–86.

[118] L. A. Dupont and R. H. Villarreal, Edge ideals of clique clutters of comparability graphs and the normality of monomial ideals, *Math. Scand.* **106** (2010), no. 1, 88–98.

[119] L. A. Dupont and R. H. Villarreal, Algebraic and combinatorial properties of ideals and algebras of uniform clutters of TDI systems, *J. Comb. Optim.* **21** (2011), no. 3, 269–292.

[120] I. M. Duursma, C. Rentería, and H. Tapia-Recillas, Reed–Muller codes on complete intersections, *Appl. Algebra Engrg. Comm. Comput.* **11** (2001), no. 6, 455–462.

[121] A. M. Duval, Algebraic shifting and sequentially Cohen–Macaulay simplicial complexes, *Electron. J. Combin.* **3** (1996), no. 1, R21.

[122] J. A. Eagon and M. Hochster, R-sequences and indeterminates, *Quart. J. Math. Oxford* **25** (1974), 61–71.

[123] J. A. Eagon and D.G. Northcott, Ideals defined by matrices and a certain complex associated with them, *Proc. Royal Soc.* **269** (1962), 188–204.

[124] J. A. Eagon and V. Reiner, Resolutions of Stanley–Reisner rings and Alexander duality, *J. Pure Appl. Algebra* **130** (1998), 265–275.

[125] J. Edmonds and R. Giles, A min-max relation for submodular functions on graphs, *Ann. of Discrete Math.* **1**, North-Holland, Amsterdam, 1977, pp. 185–204.

[126] H. Edwards, *Divisor Theory*, Birkhäuser, Boston, 1990.

[127] E. Ehrhart, Démonstration de la loi de réciprocité du polyedre rationnel, *C. R. Acad. Sci. Paris Ser. A* **265** (1967), 91–94.

[128] D. Eisenbud, *Commutative Algebra with a View toward Algebraic Geometry*, Graduate Texts in Mathematics **150**, Springer-Verlag, 1995.

[129] D. Eisenbud, *The Geometry of Syzygies: A Second Course in Commutative Algebra and Algebraic Geometry*, Graduate Texts in Mathematics **229**, Springer, New York, 2005.

[130] D. Eisenbud and S. Goto, Linear free resolutions and minimal multiplicity, *J. Algebra* **88** (1984), 89–133.

[131] D. Eisenbud and C. Huneke, Cohen–Macaulay Rees algebras and their specialization, *J. Algebra* **81** (1983), 202–224.

[132] D. Eisenbud and B. Mazur, Evolutions, symbolic squares, and Fitting ideals, *J. reine angew. Math.* **488** (1997), 189–201.

[133] D. Eisenbud and F.-O. Schreyer, Betti numbers of graded modules and cohomology of vector bundles, *J. Amer. Math. Soc.* **22** (2009), no. 3, 859–888.

[134] D. Eisenbud and B. Sturmfels, Binomial ideals, *Duke Math. J.* **84** (1996), 1–45.

[135] S. Eliahou, *Courbes monomiales et algèbre de Rees symbolique*, PhD thesis, Université de Genève, 1983.

[136] S. Eliahou, Idéaux de définition des courbes monomiales, *Lecture Notes in Mathematics* **1092**, Springer-Verlag, 1984, pp. 229–240.

[137] S. Eliahou, Minimal syzygies of monomial ideals and Gröbner bases, Reporte Técnico No. 6, CINVESTAV–IPN, 1989.

[138] S. Eliahou and R. H. Villarreal, The second Betti number of an edge ideal, Aportaciones Matemáticas, Serie Comunicaciones **25** (1999), Soc. Mat. Mex., pp. 115–119.

[139] S. Eliahou and R. H. Villarreal, On systems of binomials in the ideal of a toric variety, *Proc. Amer. Math. Soc.* **130** (2002), 345–351.

[140] J. Elias, L. Robbiano, and G. Valla, Numbers of generators of perfect ideals, *Nagoya Math. J.* **123** (1991), 39–76.

[141] E. Emtander, Betti numbers of hypergraphs, *Comm. Algebra* **37** (2009), no. 5, 1545–1571.

[142] V. Ene and J. Herzog, *Gröbner Bases in Commutative Algebra*, Graduate Studies in Mathematics **130**, American Mathematical Society, Providence, RI, 2012.

[143] V. Ene, O. Olteanu, and N. Terai, Arithmetical rank of lexsegment edge ideals, *Bull. Math. Soc. Sci. Math. Roumanie (N.S.)* **53** (2010), no. 4, 315–327.

[144] P. Erdös and T. Gallai, On the minimal number of vertices representing the edges of a graph, *Magyar Tud. Akad. Mat. Kutató Int. Közl.* **6** (1961), 181–203.

[145] P. Erdös, A. Hajnal, and J.W. Moon, A problem in graph theory, *Amer. Math. Monthly* **71** (1964), 1107–1110.

[146] C. Escobar, J. Martínez-Bernal, and R. H. Villarreal, Relative volumes and minors in monomial subrings, *Linear Algebra Appl.* **374** (2003), 275–290.

[147] C. Escobar, R. H. Villarreal, and Y. Yoshino, Torsion freeness and normality of blowup rings of monomial ideals, *Commutative Algebra, Lect. Notes Pure Appl. Math.* **244**, Chapman & Hall/CRC, Boca Raton, FL, 2006, pp. 69–84.

[148] M. Estrada and A. López, A note on symmetric semigroups and almost arithmetic sequences, *Comm. Algebra* **22** (1994), 3903–3905.

[149] M. Estrada and R. H. Villarreal, Cohen–Macaulay bipartite graphs, *Arch. Math.* **68** (1997), 124–128.

[150] E. G. Evans and P. Griffith, *Syzygies*, London Math. Soc., Lecture Note Series **106**, Cambridge University Press, Cambridge, 1985.

[151] G. Ewald, *Combinatorial Convexity and Algebraic Geometry*, Graduate Texts in Mathematics **168**, Springer-Verlag, New York, 1996.

[152] M. Farber, Characterizations of strongly chordal graphs, *Discrete Math.* **43** (1983), no. 2-3, 173–189.

[153] S. Faridi, Normal ideals of graded rings, *Comm. Algebra* **28** (2000), 1971–1977.

[154] S. Faridi, The facet ideal of a simplicial complex, *Manuscripta Math.* **109** (2002), 159–174.

[155] S. Faridi, Simplicial trees are sequentially Cohen–Macaulay, *J. Pure Appl. Algebra* **190** (2004), 121–136.

[156] S. Faridi, Cohen–Macaulay properties of square-free monomial ideals, *J. Combin. Theory Ser. A* **109** (2005), no. 2, 299–329.

[157] S. Faridi, *Monomial Ideals via Square-free Monomial Ideals*, Lecture Notes in Pure and Applied Math. **244**, Taylor & Francis, Philadelphia, 2005, pp. 85–114.

[158] K. Fischer, W. Morris, and J. Shapiro, Affine semigroup rings that are complete intersections, *Proc. Amer. Math. Soc.* **125** (1997), 3137–3145.

[159] H. Flaschka and L. Haine, Torus orbits on G/P, *Pacific J. Math.* **149** (1991), 251–292.

[160] G. Fløystad and J. E. Vatne, (Bi-)Cohen–Macaulay simplicial complexes and their associated coherent sheaves, *Comm. Algebra* **33** (2005), no. 9, 3121–3136.

[161] R. M. Fossum, *The Divisor Class Group of a Krull Domain*, Springer-Verlag, 1973.

[162] L. Fouli and K.-N. Lin, Rees algebras of square-free monomial ideals, *J. Commut. Algebra*, to appear.

[163] C. Francisco, H. T. Hà, and J. Mermin, Powers of square-free monomial ideals and combinatorics, in *Commutative Algebra* (I. Peeva Ed.), 373–392, Springer, New York, 2013.

[164] C. A. Francisco, H. T. Hà, and A. Van Tuyl, Associated primes of monomial ideals and odd holes in graphs, *J. Algebraic Combin.* **32** (2010), no. 2, 287–301.

[165] C. Francisco, H.T. Hà, and A. Van Tuyl, Colorings of hypergraphs, perfect graphs, and associated primes of powers of monomial ideals, *J. Algebra* **331** (2011), no. 1, 224–242.

[166] C. Francisco, A. Hoefel, and A. Van Tuyl, EdgeIdeals: a package for (hyper)graphs, *J. Softw. Algebra Geom.* **1** (2009), 1–4.

[167] C. A. Francisco and A. Van Tuyl, Sequentially Cohen–Macaulay edge ideals, *Proc. Amer. Math. Soc.* **135** (2007), 2327–2337.

[168] P. Frankl, Extremal set systems, in *Handbook of Combinatorics II*, Elsevier, 1995, pp. 1293–1329.

[169] R. Fröberg, A study of graded extremal rings and of monomial rings, *Math. Scand.* **51** (1982), 22–34.

[170] R. Fröberg, Rings with monomial relations having linear resolutions, *J. Pure Appl. Algebra* **38** (1985), 235–241.

[171] R. Fröberg, A note on the Stanley–Reisner ring of a join and of a suspension, *Manuscripta Math.* **60** (1988), 89–91.

[172] R. Fröberg, On Stanley–Reisner rings, in *Topics in Algebra*, Part 2. Polish Scientific Publishers, 1990, pp. 57–70.

[173] R. Fröberg, The Frobenius number of some semigroups, *Comm. Algebra* **22** (1994), 6021–6024.

[174] R. Fröberg and D. Laksov, *Compressed Algebras*, Lecture Notes in Mathematics **1092** (1984), Springer-Verlag, pp. 121–151.

[175] D. R. Fulkerson, A. J. Hoffman, and M.H. McAndrew, Some properties of graphs with multiple edges, *Canad. J. Math.* **17** (1965), 166–177.

[176] W. Fulton, *Introduction to Toric Varieties*, Princeton University Press, 1993.

[177] E. Gawrilow, and M. Joswig, polymake: a framework for analyzing convex polytopes. Polytopes—combinatorics and computation (Oberwolfach, 1997), 43–73, DMV Sem., 29, Birkhäuser, Basel, 2000.

[178] A. V. Geramita, M. Kreuzer, and L. Robbiano, Cayley–Bacharach schemes and their canonical modules, *Trans. Amer. Math. Soc.* **339** (1993), no. 1, 163–189.

[179] A. M. H. Gerards and A. Sebö, Total dual integrality implies local strong unimodularity, *Math. Programming* **38** (1987), no. 1, 69–73.

[180] R. Gilmer, *Commutative Semigroup Rings*, Chicago Lectures in Math., Univ. of Chicago Press, Chicago, 1984.

[181] P. Gimenez, A. Simis, W. V. Vasconcelos, and R. H. Villarreal, On complete monomial ideals, *J. Commut. Algebra*, to appear.

[182] I. Gitler, E. Reyes, and J. A. Vega, Complete intersection toric ideals of oriented graphs and chorded-theta subgraphs, *J. Algebraic Combin.* **38** (2013), no. 3, 721–744.

[183] I. Gitler, E. Reyes, and R. H. Villarreal, Blowup algebras of ideals of vertex covers of bipartite graphs, *Contemp. Math.* **376** (2005), 273–279.

[184] I. Gitler, E. Reyes, and R. H. Villarreal, Ring graphs and complete intersection toric ideals, *Discrete Math.* **310** (2010), no. 3, 430–441.

[185] I. Gitler, E. Reyes, and R. H. Villarreal, Blowup algebras of square–free monomial ideals and some links to combinatorial optimization problems, *Rocky Mountain J. Math.* **39** (2009), no. 1, 71–102.

[186] I. Gitler and C. Valencia, Bounds for invariants of edge-rings, *Comm. Algebra* **33** (2005), 1603–1616.

[187] I. Gitler, C. Valencia, and R. H. Villarreal, A note on the Rees algebra of a bipartite graph, *J. Pure Appl. Algebra* **201** (2005), 17–24.

[188] I. Gitler, C. Valencia, and R. H. Villarreal, A note on Rees algebras and the MFMC property, *Beiträge Algebra Geom.* **48** (2007), no. 1, 141–150.

[189] I. Gitler and R. H. Villarreal, *Graphs, Rings and Polyhedra*, Aportaciones Mat. Textos, **35**, Soc. Mat. Mexicana, México, 2011.

[190] C. Godsil and G. Royle, *Algebraic Graph Theory*, Graduate Texts in Mathematics **207**, Springer, 2001.

[191] M. C. Golumbic, *Algorithmic Graph Theory and Perfect Graphs*, Second Edition, Annals of Discrete Mathematics **57**, Elsevier Science B.V., Amsterdam, 2004.

[192] M. González-Sarabia, C. Rentería, and A. J. Sánchez, Minimum distance of some evaluation codes, *Appl. Algebra Engrg. Comm. Comput.* **24** (2013), no. 2, 95–106.

[193] A. Goodarzi, On the Hilbert series of monomial ideals, *J. Combin. Theory Ser. A* **120** (2013), no. 2, 315–317.

[194] P. Gordan, Über die Auflösung linearer Gleichungen mit reellen Coefficienten, *Math. Ann.* **6** (1873), no. 1, 23–28.

[195] F. Göring, Short proof of Menger's theorem, *Discrete Math.* **219** (2000), no. 1-3, 295–296.

[196] E. Gorla, J. C. Migliore, and U. Nagel, Gröbner bases via linkage, *J. Algebra* **384** (2013), 110–134.

[197] S. Goto, The Veronesean subrings of Gorenstein rings, *J. Math. Kyoto Univ.* **16** (1976), 51–55.

[198] S. Goto and K. Nishida, The Cohen–Macaulay and Gorenstein Rees algebras associated to filtrations, *Mem. Amer. Math. Soc.* **110** (1994), no. 526, 1–134.

[199] D.R. Grayson and M.E. Stillman, *Macaulay*2, a software system for research in algebraic geometry.
Available at `http://www.math.uiuc.edu/Macaulay2/`.

[200] G. M. Greuel and G. Pfister, *A Singular Introduction to Commutative Algebra*, 2nd extended edition, Springer, Berlin, 2008.

[201] J. Grossman, D. M. Kulkarni, and I. Schochetman, On the minors of an incidence matrix and its Smith normal form, *Linear Algebra Appl.* **218** (1995), 213–224.

[202] A. Grothendieck, *Local Cohomology*, Lecture Notes in Mathematics **41**, Springer-Verlag, Berlin, 1967.

[203] H. T. Hà and K.-N. Lin, Normal 0-1 polytopes. Preprint, 2013, `arXiv:1309.4807`.

[204] H. T. Hà and S. Morey, Embedded associated primes of powers of square-free monomial ideals, *J. Pure Appl. Algebra* **214** (2010), no. 4, 301–308.

[205] H. T. Hà, S. Morey, and R. H. Villarreal, Cohen–Macaulay admissible clutters, *J. Commut. Algebra* **1** (2009), no. 3, 463–480.

[206] H. T. Hà and A. Van Tuyl, Monomial ideals, edge ideals of hypergraphs, and their graded Betti numbers, *J. Algebraic Combin.* **27** (2008), 215–245.

[207] H. Haghighi, N. Terai, S. Yassemi, and R. Zaare-Nahandi, Sequentially S_r simplicial complexes and sequentially S_2 graphs, *Proc. Amer. Math. Soc.* **139** (2011), no. 6, 1993–2005.

[208] F. Harary, *Graph Theory*, Addison-Wesley, Reading, MA, 1972.

[209] F. Harary and M. D. Plummer, On indecomposable graphs, *Canad. J. Math.* **19** (1967), 800–809.

[210] R. Hartshorne, *Algebraic Geometry*, Springer-Verlag, New York, 1977.

[211] J. He and A. Van Tuyl, Algebraic properties of the path ideal of a tree, *Comm. Algebra* **38** (2010), no. 5, 1725–1742.

[212] B. Hedman, The maximum number of cliques in dense graphs, *Discrete Math.* **54** (1985), 161–166.

[213] W. Heinzer, A. Mirbagheri, and L.J. Ratliff, Jr., Parametric decomposition of monomial ideals II, *J. Algebra* **187** (1997), 120–149.

[214] W. Heinzer, L.J. Ratliff, Jr., and K. Shah, Parametric decomposition of monomial ideals I, *Houston J. Math.* **21** (1995), 29–52.

[215] R. Hemmecke, On the computation of Hilbert bases of cones, Mathematical software (Beijing, 2002), 307–317, World Sci. Publ., River Edge, NJ, 2002.

[216] J. Herzog, Generators and relations of abelian semigroups and semigroup rings, *Manuscripta Math.* **3** (1970), 175–193.

[217] J. Herzog, Ein Cohen–Macaulay–Kriterium mit Anwendungen auf den Konormalenmodul und den Differentialmodul, *Math. Z.* **163** (1978), 149–162.

[218] J. Herzog, Alexander duality in commutative algebra and combinatorics, *Algebra Colloq.* **11** (2004), no. 1, 21–30.

[219] J. Herzog, A generalization of the Taylor complex construction, *Comm. Algebra* **35** (2007), no. 5, 1747–1756.

[220] J. Herzog and T. Hibi, Componentwise linear ideals, *Nagoya Math. J.* **153** (1999), 141–153.

[221] J. Herzog and T. Hibi, Discrete polymatroids, *J. Algebraic Combin.* **16** (2002), no. 3, 239–268.

[222] J. Herzog and T. Hibi, Distributive lattices, bipartite graphs and Alexander duality, *J. Algebraic Combin.* **22** (2005), no. 3, 289–302.

[223] J. Herzog and T. Hibi, The depth of powers of an ideal, *J. Algebra* **291** (2005), 534–550.

[224] J. Herzog and T. Hibi, *Monomial Ideals*, Graduate Texts in Mathematics **260**, Springer-Verlag, 2011.

[225] J. Herzog, T. Hibi, and N. V. Trung, Symbolic powers of monomial ideals and vertex cover algebras, *Adv. Math.* **210** (2007), 304–322.

[226] J. Herzog, T. Hibi, N. V. Trung, and X. Zheng, Standard graded vertex cover algebras, cycles and leaves, *Trans. Amer. Math. Soc.* **360** (2008), 6231–6249.

[227] J. Herzog, T. Hibi, and M. Vladoiu, Ideals of fiber type and polymatroids, *Osaka J. Math.* **42** (2005), 1–23.

[228] J. Herzog, T. Hibi, and X. Zheng, Dirac's theorem on chordal graphs and Alexander duality. *European J. Combin.* **25** (2004), no. 7, 949–960.

[229] J. Herzog, T. Hibi, and X. Zheng, Monomial ideals whose powers have a linear resolution, *Math. Scand.* **95** (2004), 23–32.

[230] J. Herzog and M. Kühl, On the Betti numbers of finite pure and linear resolutions, *Comm. Algebra* **12** (1984), 1627–1646.

[231] J. Herzog and D. Popescu, Finite filtrations of modules and shellable multicomplexes, *Manuscripta Math.* **121** (2006), no. 3, 385–410.

[232] J. Herzog, A. Rauf, and M. Vladoiu, The stable set of associated prime ideals of a polymatroidal ideal, *J. Algebraic Combin.* **37** (2013), no. 2, 289–312.

[233] J. Herzog, A. Simis, and W. V. Vasconcelos, *Koszul Homology and Blowing-Up Rings*, Lecture Notes in Pure and Applied Math. **84**, Marcel Dekker, New York, 1983, pp. 79–169.

[234] J. Herzog, A. Simis, and W. V. Vasconcelos, On the arithmetic and homology of algebras of linear type, *Trans. Amer. Math. Soc.* **283** (1984), 661–683.

[235] J. Herzog, A. Simis, and W. V. Vasconcelos, On the canonical module of the Rees algebra and the associated graded ring of an ideal, *J. Algebra* **105** (1987), 285–302.

[236] J. Herzog, A. Simis, and W. V. Vasconcelos, Arithmetic of normal Rees algebras, *J. Algebra* **143** (1991), 269–294.

[237] J. Herzog and H. Srinivasan, Bounds for multiplicities, *Trans. Amer. Math. Soc.* **350** (1998), 2879–2902.

[238] J. Herzog, W. V. Vasconcelos, and R. H. Villarreal, Ideals with sliding depth, *Nagoya Math. J.* **99** (1985), 159–172.

[239] T. Hibi, Union and glueing of a family of Cohen–Macaulay partially ordered sets, *Nagoya Math. J.* **107** (1987), 91–119.

[240] T. Hibi, *Algebraic Combinatorics on Convex Polytopes*, Carslaw Publications, Australia, 1992.

[241] T. Hibi, A. Higashitani, A. Tsuchiya, and K. Yoshida Ehrhart polynomials with negative coefficients. Preprint, 2013, `arXiv:1312.7049`.

[242] T. Hibi and H. Ohsugi, Normal polytopes arising from finite graphs, *J. Algebra* **207** (1998), 409–426.

[243] T. Hibi and N. Terai, Betti numbers of minimal free resolutions of Stanley–Reisner rings, Semigroups, formal languages and combinatorial on words (Japanese) (Kyoto 1994), Sūrikaisekikenkyūsho Kōkyūroku **910** (1995), 98–107.

[244] T. Hibi and N. Terai, Alexander duality theorem and second Betti numbers of Stanley–Reisner rings, *Adv. Math.* **124** (1996), 332–333.

[245] P.J. Hilton and U. Stammbach, *A Course in Homological Algebra*, Graduate Texts in Mathematics **4**, Springer-Verlag, 1971.

[246] L. T. Hoa and N. D. Tam, On some invariants of a mixed product of ideals, *Arch. Math.* **94** (2010), no. 4, 327–337.

[247] C. T. Hoáng, On a conjecture of Meyniel, *J. Combin. Theory Ser. B* **42** (1987), no. 3, 302–312.

[248] M. Hochster, Rings of invariants of tori, Cohen–Macaulay rings generated by monomials, and polytopes, *Ann. of Math.* **96** (1972), 318–337.

[249] M. Hochster, Criteria for equality of ordinary and symbolic powers of prime ideals, *Math. Z.* **133** (1973), 53–65.

[250] M. Hochster, Cohen–Macaulay rings, combinatorics and simplicial complexes, in *Ring Theory II*, Lecture Notes in Pure and Applied Math. **26**, Marcel Dekker, New York, 1977, pp. 171–223.

[251] S. Hoşten and J. Shapiro, Primary decomposition of lattice basis ideals, *J. Symbolic Comput.* **29** (2000), no. 4-5, 625–639.

[252] S. Hoşten and R. R. Thomas, Gomory integer programs, *Math. Programming* **96** (2003), no. 2, Ser. B, 271–292.

[253] C. Huneke, Linkage and the Koszul homology of ideals, *Amer. J. Math.* **104** (1982), 1043–1062.

[254] C. Huneke, On the associated graded ring of an ideal, *Illinois J. Math.* **26** (1982), 121–137.

[255] C. Huneke, Hyman Bass and ubiquity: Gorenstein rings, *Contemp. Math.* **243** (1999), 55–78.

[256] C. Huneke and M. Miller, A note on the multiplicity of Cohen–Macaulay algebras with pure resolutions, *Canad. J. Math.* **37** (1985), 1149–1162.

[257] C. Huneke and J. Ribbe, Symbolic squares in regular local rings, *Math. Z.* **229** (1998), 31–44.

[258] C. Huneke, A. Simis, and W. V. Vasconcelos, Reduced normal cones are domains, *Contemp. Math.* **88** (1989), 95–101.

[259] C. Huneke and I. Swanson, *Integral Closure of Ideals Rings, and Modules*, London Math. Soc., Lecture Note Series **336**, Cambridge University Press, Cambridge, 2006.

[260] C. Ionescu and G. Rinaldo, Some algebraic invariants related to mixed product ideals, *Arch. Math.* **91** (2008), 20–30.

[261] N. Jacobson, *Basic Algebra I*, Second Edition, W. H. Freeman and Company, New York, 1996.

[262] N. Jacobson, *Basic Algebra II*, W. H. Freeman and Company, New York, 1989.

[263] P. M. Johnson, A weighted graph problem from commutative algebra. Preprint, 2007, `arXiv:0704.3696`.

[264] T. Jòzefiak, Ideals generated by minors of a symmetric matrix, *Comment. Math. Helvetici* **53** (1978), 595–607.

[265] L. Juan, Some results about symmetric semigroups, *Comm. Algebra* **21** (1993), 3637–3645.

[266] T. Kahle, Decompositions of binomial ideals, *J. Softw. Algebra Geom.* **4** (2012), 1–5.

[267] T. Kahle and E. Miller, Decompositions of commutative monoid congruences and binomial ideals, *Algebra Number Theory*, to appear.

[268] T. Kaiser, M. Stehlík, and R. Škrekovski, Replication in critical graphs and the persistence of monomial ideals, *J. Combin. Theory Ser. A* **123** (2014), no. 1, 239–251.

[269] G. Kalai and R. Meshulam, Unions and intersections of Leray complexes, *J. Combin. Theory Ser. A* **113** (2006), 1586–1592.

[270] A. Katsabekis, M. Morales, and A. Thoma, Binomial generation of the radical of a lattice ideal, *J. Algebra* **324** (2010), no. 6, 1334–1346.

[271] A. Katsabekis and A. Thoma, Toric sets and orbits on toric varieties, *J. Pure Appl. Algebra* **181** (2003), 75–83.

[272] A. Katsabekis and A. Thoma, Parametrizations of toric varieties over any field, *J. Algebra* **308** (2007), no. 2, 751–763.

[273] M. Katzman, Bipartite graphs whose edge algebras are complete intersections, *J. Algebra* **220** (1999), 519–530.

[274] M. Katzman, Characteristic-independence of Betti numbers of graph ideals, *J. Combin. Theory Ser. A* **113** (2006), no. 3, 435–454.

[275] G. Kempf, F. Knudsen, D. Mumford, and B. Saint-Donat, *Toroidal Embeddings* I, Lecture Notes in Mathematics **339**, Springer, 1973.

[276] K. Kimura, Non-vanishingness of Betti numbers of edge ideals, in: *Harmony of Gröbner Bases and the Modern Industrial Society*, World Scientific, 2012, pp. 153–168.

[277] K. Kimura, Non-vanishingness of Betti numbers of edge ideals and complete bipartite subgraphs. Preprint, 2013, `arXiv:1306.1333`.

[278] K. Kimura and N. Terai, Binomial arithmetical rank of edge ideals of forests, *Proc. Amer. Math. Soc.* **141** (2013), no. 6, 1925–1932.

[279] K. Kimura, N. Terai, and K. Yoshida, Arithmetical rank of squarefree monomial ideals of small arithmetic degree, *J. Algebraic Combin.* **29** (2009), 389–404.

[280] S. Kleiman and B. Ulrich, Gorenstein algebras, symmetric matrices, self-linked ideals, and symbolic powers, *Trans. Amer. Math. Soc.* **349** (1997), 4973–5000.

[281] B. Korte and J. Vygen, *Combinatorial Optimization Theory and Algorithms*, *Algorithms and Combinatorics* **21**, Springer, 2000.

[282] M. Kreuzer and L. Robbiano, *Computational Commutative Algebra* 2, Springer-Verlag, Berlin, 2005.

[283] J. Kruskal, The number of simplices in a complex, in *Mathematical Optimization Techniques* (R. Bellman, Ed.), University of California Press, 1963, pp. 251–278.

[284] M. Kubitzke and A. Olteanu, Algebraic properties of classes of path ideals of posets, *J. Pure Appl. Algebra* **218** (2014), no. 6, 1012–1033.

[285] M. Kummini, Regularity, depth and arithmetic rank of bipartite edge ideals, *J. Algebraic Combin.* **30** (2009), no. 4, 429–445.

[286] J. P. S. Kung, Basis-exchange properties, *Theory of matroids*, 62–75, *Encyclopedia Math. Appl.* **26**, Cambridge Univ. Press, 1986.

[287] E. Kunz, The value-semigroup of a one dimensional Gorenstein ring, *Proc. Amer. Math. Soc.* **25** (1970), 748–751.

[288] M. La Barbiera and G. Restuccia, Mixed product ideals generated by s-sequences, *Algebra Colloq.* **18** (2011), no. 4, 553–570.

[289] M. D. Larsen, *Multiplicative Theory of Ideals*, Pure and Applied Mathematics **43**, Academic Press, 1971.

[290] A. Lehman, On the width-length inequality and degenerate projective planes, in *Polyhedral Combinatorics* (W. Cook and P. Seymour Eds.) DIMACS Series in Discrete Mathematics and Theoretical Computer Science **1**, Amer. Math. Soc., 1990, pp. 101–105.

[291] M. Llabres and F. Rossello, A new family of metrics for biopolymer contact structures, *Comput. Biol. Chem.* **28** (2004), no. 1, 21–37.

[292] H. H. López, C. Rentería, and R. H. Villarreal, Affine cartesian codes, *Des. Codes Cryptogr.* **71** (2014), no. 1, 5–19.

[293] H. H. López and R. H. Villarreal, Complete intersections in binomial and lattice ideals, *Internat. J. Algebra Comput.* **23** (2013), no. 6, 1419–1429.

[294] H. H. López and R. H. Villarreal, Computing the degree of a lattice ideal of dimension one, *J. Symbolic Comput.* **65** (2014), 15–28.

[295] D. Lorenzini, Smith normal form and Laplacians, *J. Combin. Theory Ser. B* **98** (2008), 1271–1300.

[296] L. Lovász, Normal hypergraphs and the perfect graph conjecture, *Discrete Math.* **2** (1972), no. 3, 253–267.

[297] L. Lovász and M. D. Plummer, *Matching Theory*, Annals of Discrete Mathematics **29**, Elsevier Science B.V., Amsterdam, 1986.

[298] C. L. Lucchesi and D. H. Younger, A minimax theorem for directed graphs, *J. London Math. Soc.* (2) **17** (1978), no. 3, 369–374.

[299] G. Lyubeznik, *Set Theoretic Intersections and Monomial Ideals*, PhD thesis, Columbia University, 1984.

[300] G. Lyubeznik, The minimal non-Cohen–Macaulay monomial ideals, *J. Pure Appl. Algebra* **51** (1988), 261–266.

[301] G. Lyubeznik, On the arithmetical rank of monomial ideals, *J. Algebra* **112** (1988), 86–89.

[302] G. Lyubeznik, A survey of problems and results on the number of defining equations, in *Commutative Algebra* (M. Hochster et. al., Eds.), MSRI Publications **15**, Springer-Verlag, 1989, 375–390.

[303] M. Mahmoudi, A. Mousivand, M. Crupi, G. Rinaldo, N. Terai, and S. Yassemi, Vertex decomposability and regularity of very well-covered graphs, *J. Pure Appl. Algebra* **215** (2011), no. 10, 2473–2480.

[304] W. H. Marlow, *Mathematics for Operation Research*, New York, Wiley, 1978.

[305] J. Martínez-Bernal, S. Morey, and R. H. Villarreal, Associated primes of powers of edge ideals, *Collect. Math.* **63** (2012), no. 3, 361–374.

[306] J. Martínez-Bernal, E. O'Shea, and R. H. Villarreal, Ehrhart clutters: regularity and max-flow min-cut, *Electron. J. Combin.* **17** (2010), no. 1, R52.

[307] J. Martínez-Bernal, C. Rentería, and R. H. Villarreal, Combinatorics of symbolic Rees algebras of edge ideals of clutters, *Contemp. Math.* **555** (2011), 151–164.

[308] J. Martínez-Bernal and R. H. Villarreal, Toric ideals generated by circuits, *Algebra Colloq.* **19** (2012), no. 4, 665–672.

[309] H. Matsumura, *Commutative Algebra*, Benjamin-Cummings, Reading, MA, 1980.

[310] H. Matsumura, *Commutative Ring Theory*, Cambridge Studies in Advanced Mathematics **8**, Cambridge University Press, 1986.

[311] T. Matsuoka, On an invariant of Veronesean rings, *Proc. Japan Acad.* **50** (1974), 287–291.

[312] S. McAdam, Asymptotic prime divisors and analytic spreads, *Proc. Amer. Math. Soc.* **80** (1980), 555–559.

[313] S. McAdam, *Asymptotic Prime Divisors*, Lecture Notes in Mathematics **103**, Springer–Verlag, New York, 1983.

[314] P. McMullen, The maximum numbers of faces of a convex polytope, *Mathematika* **17** (1970), 179–184.

[315] P. McMullen and G. Shephard, *Convex Polytopes and the Upper Bound Conjecture*, London Mathematical Society, Lect. Note Ser. **3**, Cambridge University Press, 1971.

[316] K. Menger, Zur allgemeinen Kurventheorie, *Fundamenta Mathematicae* **10** (1927), 96–115.

[317] E. Miller and B. Sturmfels, *Combinatorial Commutative Algebra*, Graduate Texts in Mathematics **227**, Springer, 2004.

[318] M. Miller and R. H. Villarreal, A note on generators of least degree in Gorenstein ideals, *Proc. Amer. Math. Soc.* **124** (1996), 377–382.

[319] T. T. Moh, Set-theoretic complete intersections, *Proc. Amer. Math. Soc.* **94** (1985), 217–220.

[320] B. Mohar and C. Thomassen, *Graphs on Surfaces*, Johns Hopkins University Press, Baltimore, Maryland, 1997.

[321] S. Moradi and D. Kiani, Bounds for the regularity of edge ideal of vertex decomposable and shellable graphs, *Bull. Iranian Math. Soc.* **36** (2010), 267–277.

[322] M. Morales and A. Thoma, Complete intersection lattice ideals, *J. Algebra* **284** (2005), 755–770.

[323] S. Morey, Stability of associated primes and equality of ordinary and symbolic powers of ideals, *Comm. Algebra* **27** (1999), 3221–3231.

[324] S. Morey, Depths of powers of the edge ideal of a tree, *Comm. Algebra* **38** (2010), no. 11, 4042–4055.

[325] S. Morey, E. Reyes, and R. H. Villarreal, Cohen–Macaulay, shellable and unmixed clutters with a perfect matching of König type, *J. Pure Appl. Algebra* **212** (2008), no. 7, 1770–1786.

[326] S. Morey and R. H. Villarreal, Edge ideals: algebraic and combinatorial properties, in *Progress in Commutative Algebra, Combinatorics and Homology, Vol. 1* (C. Francisco, L. C. Klingler, S. Sather-Wagstaff, and J. C. Vassilev, Eds.), De Gruyter, Berlin, 2012, pp. 85–126.

[327] T. S. Motzkin, Comonotone curves and polyhedra, Abstract, *Bull. Amer. Math. Soc.* **63** (1957), 35.

[328] U. Nagel and S. Petrović, Properties of cut ideals associated to ring graphs, *J. Commut. Algebra* **1** (2009), no. 3, 547–565.

[329] W. Narkiewicz, *Elementary and Analytic Theory of Algebraic Numbers*, Second Edition, Springer-Verlag, Berlin; PWN—Polish Scientific Publishers, Warsaw, 1990.

[330] J. Neves, M. Vaz Pinto, and R. H. Villarreal, Regularity and algebraic properties of certain lattice ideals, *Bull. Braz. Math. Soc. (N.S.)* **45** (2014), no. 4, 777–806.

[331] E. Nevo, Regularity of edge ideals of C_4-free graphs via the topology of the lcm-lattice, *J. Combin. Theory Ser. A.* **118** (2011), no. 2, 491–501.

[332] M. Newman, *Integral Matrices*, Pure and Applied Mathematics **45**, Academic Press, New York, 1972.

[333] K. Nishida, Powers of ideals in Cohen–Macaulay rings, *J. Math. Soc. Japan* **48** (1996), 409–426.

[334] D. G. Northcott, A generalization of a theorem on the contents of polynomials, *Proc. Camb. Phil. Soc.* **55** (1959), 282–288.

[335] L. O'Carroll, F. Planas-Vilanova, and R. H. Villarreal, Degree and algebraic properties of lattice and matrix ideals, *SIAM J. Discrete Math.* **28** (2014), no. 1, 394–427.

[336] M. Ohtani, Graphs and ideals generated by some 2-minors, *Comm. Algebra* **39** (2011), no. 3, 905–917.

[337] I. Ojeda and R. Peidra, Cellular binomial ideals. Primary decomposition of binomial ideals, *J. Symbolic Comput.* **30** (2000), 383–400.

[338] J. Oxley, *Matroid Theory*, Oxford University Press, Oxford, 1992.

[339] C. Peskine and L. Szpiro, Liaison des variétés algébriques, *Invent. Math.* **26** (1974), 271–302.

[340] M. D. Plummer, Some covering concepts in graphs, *J. Combinatorial Theory* **8** (1970), 91–98.

[341] J. L. Ramírez Alfonsín, *The Diophantine Frobenius Problem*, Oxford Lecture Series in Mathematics and its Applications, **30**, Oxford University Press, Oxford, 2005.

[342] L. J. Ratliff, Jr., On prime divisors of I^n, n large, *Michigan Math. J.* **23** (1976), no. 4, 337–352.

[343] L. J. Ratliff, Jr., On asymptotic prime divisors, *Pacific J. Math.* **111** (1984), no. 2, 395–413.

[344] G. Ravindra, Well-covered graphs, *J. Combinatorics Information Syst. Sci.* **2** (1977), no. 1, 20–21.

[345] L. Reid, L. G. Roberts, and M. A. Vitulli, Some results on normal homogeneous ideals, *Comm. Algebra* **31** (2003), no. 9, 4485–4506.

[346] G. Reisner, Cohen–Macaulay quotients of polynomial rings, *Adv. Math.* **21** (1976), 31–49.

[347] P. Renteln, The Hilbert series of the face ring of a flag complex, *Graphs Combin.* **18** (2002), no. 3, 605–619.

[348] C. Rentería, A. Simis, and R. H. Villarreal, Algebraic methods for parameterized codes and invariants of vanishing ideals over finite fields, *Finite Fields Appl.* **17** (2011), no. 1, 81–104.

[349] C. Rentería and H. Tapia, Reed–Muller codes: an ideal theory approach, *Comm. Algebra* **25** (1997), 401–413.

[350] C. Rentería and R. H. Villarreal, Koszul homology of Cohen–Macaulay rings with linear resolutions, *Proc. Amer. Math. Soc.* **115** (1992), 51–58.

[351] G. Restuccia and R. H. Villarreal, On the normality of monomial ideals of mixed products, *Comm. Algebra* **29** (2001), 3571–3580.

[352] E. Reyes, C. Tatakis, and A. Thoma, Minimal generators of toric ideals of graphs, *Adv. in Appl. Math.* **48** (2012), no. 1, 64–78.

[353] E. Reyes and J. Toledo, On the strong persistence property for monomial ideals, preprint, 2013.

[354] E. Reyes, R. H. Villarreal, and L. Zárate, A note on affine toric varieties, *Linear Algebra Appl.* **318** (2000), 173–179.

[355] G. Rinaldo, Betti numbers of mixed product ideals, *Arch. Math.* **91** (2008), no. 5, 416–426.

[356] G. Rinaldo, Koszulness of vertex cover algebras of bipartite graphs, *Comm. Algebra* **39** (2011), no. 7, 2249–2259.

[357] R. T. Rockafellar, The elementary vectors of a subspace of R^N, in *Combinatorial Mathematics and its Applications*, Proc. Chapel Hill Conf., Univ. North Carolina Press, 1969, pp. 104–127.

[358] R. T. Rockafellar, *Convex Analysis*, Princeton University Press, 1970.

[359] Ö. J. Rödseth, On a linear Diophantine problem of Frobenius, *J. reine angew. Math.* **301** (1978), 171–178.

[360] J. C. Rosales and P. A. García-Sánchez, *Numerical Semigroups*, Developments in Mathematics **20**, Springer, New York, 2009.

[361] M. E. Rossi and G. Valla, Multiplicity and t-isomultiple ideals, *Nagoya Math. J.* **110** (1988), 81–111.

[362] M. Roth and A. Van Tuyl, On the linear strand of an edge ideal, *Comm. Algebra* **35** (2007), no. 3, 821–832.

[363] J. J. Rotman, *An Introduction to Homological Algebra*, Academic Press, 1979.

[364] H. Sabzrou and F. Rahmati, Matrices defining Gorenstein lattice ideals, *Rocky Mountain J. Math.* **35** (2005), no. 3, 1029–1042.

[365] H. Sabzrou and F. Rahmati, The Frobenius number and a-invariant, *Rocky Mountain J. Math.* **36** (2006), no. 6, 2021–2026.

[366] J. Sally, Cohen–Macaulay rings of maximal embedding dimension, *J. Algebra* **56** (1979), 168–183.

[367] E. Sarmiento, M. Vaz Pinto, and R. H. Villarreal, The minimum distance of parameterized codes on projective tori, *Appl. Algebra Engrg. Comm. Comput.* **22** (2011), no. 4, 249–264.

[368] H. Scarf and D. Shallcross, The Frobenius problem and maximal lattice free bodies, *Math. Oper. Res.* **18** (1993), 511–515.

[369] H. Schenck, *Computational Algebraic Geometry*, London Math. Soc., Student Texts **58**, Cambridge University Press, Cambridge, 2003.

[370] P. Schenzel, Uber die freien auflösungen extremaler Cohen–Macaulay-ringe, *J. Algebra* **64** (1980), 93–101.

[371] A. Schrijver, On total dual integrality, *Linear Algebra Appl.* **38** (1981), 27–32.

[372] A. Schrijver, *Theory of Linear and Integer Programming*, John Wiley & Sons, New York, 1986.

[373] A. Schrijver,*Combinatorial Optimization*, Algorithms and Combinatorics **24**, Springer-Verlag, Berlin, 2003.

[374] E. Selmer, On the linear diophantine problem of Frobenius, *Journal für Mathematik*, band 293/294 (1977), 1–17.

[375] P. D. Seymour, The matroids with the max-flow min-cut property, *J. Combinatorial Theory Ser. B* **23** (1977), 189–222.

[376] Y. H. Shen, On a class of squarefree monomial ideals of linear type, *J. Commut. Algebra*, to appear.

[377] A. Simis, *Combinatoria Algebrica*, XVIII Coloquio Brasileiro de Matematica, IMPA, 1991 (Apendice. Palimpsesto 2: Potencias simbolicas, 2.1).

[378] A. Simis, Effective computation of symbolic powers by Jacobian matrices, *Comm. Algebra* **24** (1996), 3561–3565.

[379] A. Simis, On the Jacobian module associated to a graph, *Proc. Amer. Math. Soc.* **126** (1998), 989–997.

[380] A. Simis and N.V. Trung, The divisor class group of ordinary and symbolic blow-ups, *Math. Z.* **198** (1988), 479–491.

[381] A. Simis and B. Ulrich, On the ideal of an embedded join, *J. Algebra* **226** (2000), no. 1, 1–14.

[382] A. Simis and W. V. Vasconcelos, The syzygies of the conormal module, *Amer. J. Math.* **103** (1981), 203–224.

[383] A. Simis, W. V. Vasconcelos, and R. H. Villarreal, On the ideal theory of graphs, *J. Algebra* **167** (1994), 389–416.

[384] A. Simis, W. V. Vasconcelos, and R. H. Villarreal, The integral closure of subrings associated to graphs, *J. Algebra* **199** (1998), 281–289.

[385] A. Simis and R. H. Villarreal, Constraints for the normality of monomial subrings and birationality, *Proc. Amer. Math. Soc.* **131** (2003), 2043–2048.

[386] A. Simis and R. H. Villarreal, Combinatorics of Cremona monomial maps, *Math. Comp.* **81** (2012), no. 279, 1857–1867.

[387] J. M. S. Simões-Pereira, On matroids on edge sets of graphs with connected subgraphs as circuits II, *Discrete Math.* **12** (1975), 55–78.

[388] D. E. Smith, On the Cohen–Macaulay property in commutative algebra and simplicial topology, *Pacific J. Math.* **141** (1990), 165–196.

[389] A. Soleyman Jahan, and X. Zheng, Pretty clean monomial ideals and linear quotients. Preprint, 2007, `arXiv:0707.2914`.

[390] E. Sperner, Ein satz über untermengen einer endlichen menge, *Math. Z.* **27** (1928), 544–548.

[391] R. Stanley, The upper bound conjecture and Cohen–Macaulay rings, *Stud. Appl. Math.* **54** (1975), 135–142.

[392] R. Stanley, Hilbert functions of graded algebras, *Adv. Math.* **28** (1978), 57–83.

[393] R. Stanley, The number of faces of simplicial polytopes and spheres, *Ann. New York Acad. Sci.* **440** (1985), pp. 212–223.

[394] R. Stanley, *Enumerative Combinatorics I*, Wadsworth-Brooks/Cole, Monterey, California, 1986.

[395] R. Stanley, *Combinatorics and Commutative Algebra*, Birkhäuser Boston, 2nd ed., 1996.

[396] R. Stanley, *Algebraic Combinatorics*, Springer, New York, 2013.

[397] I. Stewart and D. Tall, *Algebraic Number Theory*, Chapman and Hall Mathematics Series, 1979.

[398] B. Sturmfels, Gröbner bases and Stanley decompositions of determinantal rings, *Math. Z.* **205** (1990), 137–144.

[399] B. Sturmfels, Gröbner bases of toric varieties, *Tôhoku Math. J.* **43** (1991), 249–261.

[400] B. Sturmfels, *Gröbner Bases and Convex Polytopes*, University Lecture Series **8**, American Mathematical Society, Rhode Island, 1996.

[401] N. Terai, Alexander duality theorem and Stanley–Reisner rings, *Sūrikaisekikenkyūsho Kōkyūroku* **1078** (1999), 174–184.

[402] A. Thoma, On the set-theoretic complete intersection problem for monomial curves in $\mathbb{A}^n$ and $\mathbb{P}^n$, *J. Pure Appl. Algebra* **104** (1995), 333–344.

[403] B. Toft, Colouring, stable sets and perfect graphs, in *Handbook of Combinatorics I*, Elsevier, 1995, pp. 233–288.

[404] B. Ulrich, Gorenstein rings and modules with high numbers of generators, *Math. Z.* **188** (1984), 23–32.

[405] C. Valencia and R. H. Villarreal, Canonical modules of certain edge subrings, *European J. Combin.* **24** (2003), no. 5, 471–487.

[406] C. Valencia and R. H. Villarreal, Explicit representations of the edge cone of a graph, *Int. Journal Contemp. Math. Sciences* **1** (2006), 53-66.

[407] A. Van Tuyl, Sequentially Cohen–Macaulay bipartite graphs: vertex decomposability and regularity, *Arch. Math.* **93** (2009), 451–459.

[408] A. Van Tuyl, A beginner's guide to edge and cover ideals, in *Monomial Ideals, Computations and Applications*, Lecture Notes in Mathematics **2083**, Springer, 2013, pp. 63–94.

[409] A. Van Tuyl and R. H. Villarreal, Shellable graphs and sequentially Cohen–Macaulay bipartite graphs, *J. Combin. Theory Ser. A* **115** (2008), no.5, 799-814.

[410] W. V. Vasconcelos, *Divisor Theory in Module Categories*, Mathematics Studies **14**, North Holland, 1974.

[411] W. V. Vasconcelos, On the equations of Rees algebras, *J. reine angew. Math.* **418** (1991), 189–218.

[412] W. V. Vasconcelos, *Arithmetic of Blowup Algebras*, London Math. Soc., Lecture Note Series **195**, Cambridge University Press, Cambridge, 1994.

[413] W. V. Vasconcelos, *Computational Methods in Commutative Algebra and Algebraic Geometry*, Springer-Verlag, 1998.

[414] W. V. Vasconcelos, *Integral Closure*, Springer Monographs in Mathematics, Springer-Verlag, New York, 2005.

[415] R. H. Villarreal, Rees algebras and Koszul homology, *J. Algebra* **119** (1988), 83–104.

[416] R. H. Villarreal, Cohen–Macaulay graphs, *Manuscripta Math.* **66** (1990), 277–293.

[417] R. H. Villarreal, Rees algebras of edge ideals, *Comm. Algebra* **23** (1995), 3513–3524.

[418] R. H. Villarreal, Normality of subrings generated by square free monomials, *J. Pure Appl. Algebra* **113** (1996), 91–106.

[419] R. H. Villarreal, On the equations of the edge cone of a graph and some applications, *Manuscripta Math.* **97** (1998), 309–317.

[420] R. H. Villarreal, Monomial algebras and polyhedral geometry, in *Handbook of Algebra, Vol. 3* (M. Hazewinkel Ed.), Elsevier, Amsterdam, 2003, pp. 257–314.

[421] R. H. Villarreal, Unmixed bipartite graphs, *Rev. Colombiana Mat.* **41** (2007), no. 2, 393–395.

[422] R. H. Villarreal, Rees cones and monomial rings of matroids, *Linear Algebra Appl.* **428** (2008), 2933–2940.

[423] R. H. Villarreal, Rees algebras and polyhedral cones of ideals of vertex covers of perfect graphs, *J. Algebraic Combin.* **27** (2008), 293-305.

[424] R. H. Villarreal, Normalization of monomial ideals and Hilbert functions, *Proc. Amer. Math. Soc.* **136** (2008), 1933–1943.

[425] J. Watanabe, $\mathfrak{m}$-full ideals, *Nagoya Math. J.* **106** (1987), 101–111.

[426] A. Watkins, *Hilbert Functions of Face Rings Arising from Graphs*, M.S. Thesis, Rutgers University, 1992.

[427] R. Webster, *Convexity*, Oxford University Press, Oxford, 1994.

[428] C. Weibel, *An Introduction to Homological Algebra*, Cambridge University Press, Cambridge, 1994.

[429] N. L. White, The basis monomial ring of a matroid, *Adv. Math.* **24** (1977), 292–297.

[430] N. L. White, A unique exchange property for bases, *Linear Algebra Appl.* **31** (1980), 81–91.

[431] S. Wolfram, *Mathematica. A System for Doing Mathematics by Computer*, Addison Wesley, 1989.

[432] R. Woodroofe, Vertex decomposable graphs and obstructions to shellability, *Proc. Amer. Math. Soc.* **137** (2009), 3235–3246.

[433] R. Woodroofe, Chordal and sequentially Cohen–Macaulay clutters, *Electron. J. Combin.* **18** (2011), no. 1, R208.

[434] R. Woodroofe, Matchings, coverings, and Castelnuovo–Mumford regularity, *J. Commut. Algebra* **6** (2014), no. 2, 287–304.

[435] R. Zaare-Nahandi, Cohen–Macaulayness of bipartite graphs, revisited, *Bull. Malays. Math. Sci. Soc.* (2), to appear.

[436] O. Zariski and P. Samuel, *Commutative Algebra*, Vol. II, Van Nostrand, Princeton, 1960.

[437] X. Zheng, Resolutions of facet ideals, *Comm. Algebra* **32** (2004), no. 6, 2301–2324.

[438] G. M. Ziegler, *Lectures on Polytopes*, Graduate Texts in Mathematics **152**, Springer-Verlag, 1994.

Notation Index

$\alpha_0(G)$, vertex covering number, 266
$\alpha_1(G)$, edge covering number, 267
$(I\colon J^\infty)$, saturation, 130
$(I\colon f^\infty)$, saturation, 130
$(N_1\colon {}_RN_2)$, ideal quotient, 64
$(a_i) \leq (b_i)$, $a_i \leq b_i$ for all i, 3
$A(\mathcal{P})$, Ehrhart ring, 381
$A[F]$, subring generated by F, 86
A^t or $A^\top$, transpose of A, 3
A^{o}, interior of A, 2
$A_{\mathcal{P}}$, normalized Ehrhart ring, 375
$C(G)$, cone of a graph G, 458
C^*, dual cone, 9
$F(M,t)$, Hilbert series, 171
$F_{\mathcal{G}} = \{x^{v_i} \mid v_i \in \mathcal{G}\}$, 580
G' or $\overline{G}$, complement of G, 280
$G(I)$, monomial gens of I, 202
$G[V(H)]$, induced subgraph, 262
$G \setminus \{v\}$, deletion of a vertex, 264
G^a, parallelization of G, 303
$G_1 * G_2$, join of two graphs, 498
$H(M,i)$, Hilbert function, 77
$H(y,c)$, hyperplane, 2
$H^+(y,c)$, halfspace, 2
H_I^a, affine Hilbert function, 342
$H_{\mathfrak{A}}$, supporting hyperplane, 468
$I(X)$, vanishing ideal, 123
I^*, dual of I, 623
$I^\vee$, Alexander dual of I, 221
$I^{(n)}$, symbolic power, 152
I_Δ, Stanley–Reisner ideal, 212
$I_{\mathcal{A}}$, toric ideal, 326
$I_\rho(\mathcal{L})$, lattice ideal, 316
$I_c(\mathcal{C})$, ideal of covers, 221
K, a field, 62
$K[F]$, monomial subring, 321
$K[G]$, edge subring, 423
$K[\Delta]$, Stanley–Reisner ring, 212
$K[\mathbf{x}^{\pm 1}]$, Laurent polynomials, 320
$K[x_1^{\pm 1}, \ldots, x_n^{\pm 1}]$, Laurent ring, 320
$K^* = K \setminus \{0\}$, 127
$K_\star(\underline{x})$, Koszul complex, 105
M^*, dual of M, 99
$N_G(A)$, neighbor set, 264
$N_G(v)$, neighbor set of v, 264
$R(-a)$, shift, 135
$R/I(G)$, edge-ring, 268
$R[It]$ or $\mathcal{R}(I)$, Rees algebra, 142
$R^{(k)}$, kth Veronese subring, 513
$R_+ = \oplus_{i=1}^\infty R_i$, $R = \oplus_{i=0}^\infty R_i$, 78
$R_s(I)$, symbolic Rees algebra, 568
$R_{\mathfrak{p}}$, localization at a prime, 64
$S^{-1}(M)$, localization, 63
$T(M)$, torsion subgroup, 22
$V(G)$, V_G, vertex set, 261
$V(I)$, affine variety of I, 123
$V(I)$, prime ideals over I, 62
$V(I)$, projective variety of I, 133
$[[r+1,q]] := \{r+1,\ldots,q\}$, 356
$[q] := \{1,\ldots,q\}$, 355
$\mathrm{Ass}(I)$, associated primes, 65
$\mathrm{Ass}_R(M)$, associated primes, 65
Δ^q, q-skeleton of Δ, 215
$\Delta_i(B)$, g.c.d. of ith minors, 388
$\Delta_r(A)$, 32
Δ_I, Stanley–Reisner complex, 212

Spec(R), spectrum of a ring, 62
Supp(M), support of M, 66
$\|x\|$, Euclidean norm, 2
α_+, positive part of α, 55
α_-, negative part of α, 55
$\alpha_0(\mathcal{C})$, covering number, 221
ht (I), height of an ideal, 64
ann $_R(M)$, annihilator of M, 64
$\beta_0'(\mathcal{C})$, 230
$\beta_0(G)$, vertex indep. number, 266
$\beta_0(\mathcal{C})$, 221
$\beta_1(\mathcal{C})$, matching number, 538
$\beta_1(G)$, matching number, 266
$\chi(G)$, chromatic number, 263
$\ell_R(M)$, length of a module, 71
gcd, greatest common divisor, 367
$\langle\, ,\, \rangle$, usual inner product, 2
$\lceil\alpha\rceil$, ceiling of α, 500
lcm, least common multiple, 128
lk(F), link of a face, 213
log(F), log set of F, 370
log(x^a), log of a monomial, 370
$\mathbb{A}^q_K$, affine space, 123
$\mathbb{N}=\{0,1,2,\dots\}$, 5
$\mathbb{N}\Gamma$, semigroup generated by Γ, 29
$\mathbb{N}_+=\{1,2,\dots\}$, 38
$\mathbb{N}\mathcal{A}'$, Rees semigroup of $\mathcal{A}$, 46
$\mathbb{N}\mathcal{H}$, semigroup spanned by $\mathcal{H}$, 28
$\mathbb{P}^q_K$, projective space, 133
$\mathbb{Q}_+=\{x\in\mathbb{Q}\,|\,x\geq 0\}$, 7
$\mathbb{R}\mathcal{A}$, vector space gen. by $\mathcal{A}$, 5
$\mathbb{R}_+(I)$, Rees cone, 508
$\mathbb{R}_+=\{x\in\mathbb{R}\,|\,x\geq 0\}$, 5
$\mathbb{R}_+v$, 5
$\mathbb{R}_+\mathcal{A}'$, Rees cone, 39
$\mathbb{Z}\mathcal{A}$, subgroup spanned by $\mathcal{A}$, 7
$\mathbf{1}=(1,\dots,1)$, 40
$\mathbf{x}=\{x_1,\dots,x_n\}$, variables, 326
$\mathcal{C}/x_i$, contraction, 234
$\mathcal{C}[S]$, induced subclutter, 228
$\mathcal{C}\setminus x_i$, deletion, 234
$\mathcal{C}^\vee$, Alexander dual, 221, 543
$\mathcal{C}^\vee$, blocker, 221
$\mathcal{C}^a$, parallelization of $\mathcal{C}$, 542
$\mathcal{P}$, polytope, 2
$\mathcal{Q}(A)$, covering polyhedron, 39
$\mathfrak{F}_+$, positive part, 417
PP, packing property, 588
S(f,g), the S-polynomial, 128
w.r.t, with respect to, 77
C–M, Cohen–Macaulay, 269
$\mu(M)$, number of generators, 71
$\omega(G)$, clique number, 263
ω_S, canonical module, 178
$\overline{I}$ or I_a, integral closure, 145
$\overline{\mathbb{N}\mathcal{A}}$, normalization, 46
$\sqrt{I}$ or rad (I), radical of I, 65
$\varphi_M(t)$, Hilbert polynomial, 77
$|x|$, value of x, 29
$\widetilde{A}$, almost integral closure, 101
$\widetilde{\chi}(\Delta)$, reduced Euler charac., 211
$b(\mathcal{C})$, blocker, 221, 543
c_0, bipartite components, 467
$e(M)$, multiplicity of M, 77
$e(\mathfrak{q})$, multiplicity of an ideal, 161
$e(\mathfrak{q},M)$, multiplicity of $\mathfrak{q}$, 161
e_i, ith unit vector, 21
$f(\Delta)$, the f-vector of Δ, 251
$f\rightarrow_F g$, reduction w.r.t F, 127
$g(\mathcal{S})$, Frobenius number, 348
$h(\Delta)$, the h-vector of Δ, 252
$h^{\langle j\rangle}$, Macaulay symbol, 191
k, a field, 62
$k[\mathbf{x}]$, polynomial ring, 62
$r(I)$, reduction number, 463
$x\cdot y$, inner product of x and y, 2
x^a, monomial, 326
C_n, cycle, 262
$\chi(\Delta)$, Euler characteristic, 211
$\mathbb{R}_+\mathcal{A}$, cone generated by $\mathcal{A}$, 5
$\mathcal{A}'=\{(v_i,1)\}_{i=1}^q\cup\{e_i\}_{i=1}^n$, 46
$\mathcal{A}=\{v_1,\dots,v_q\}$, 46
$\mathcal{A}_\mathcal{G}=\{v_i\,|\,v_i\in\mathcal{G}\}$, 580
$\mathcal{G}_P$, Graver basis, 326
$\mathcal{K}_n$, complete graph, 262
$\mathcal{K}_{m,n}$, complete bipartite, 263

$\mathcal{Q}_6$, clutter, 573
$\mathcal{Z}(M)$, zero divisors, 67
$\mathfrak{N}_R$ or nil(R), nilradical of R, 65
Cn(I), Simis cone of I, 540
Div(A), divisorial ideals, 101
Pic(A), Picard group, 98
Prin(A), principal ideals, 101
Soc(M), socle of M, 85
Sym$_R(M)$, symmetric algebra, 141
adj(H), adjoint of a matrix, 70
aff(A), affine space, 1
arith-deg$(I(\mathcal{C}))$, 232
bight$(I(\mathcal{C}))$, big height, 230
char(R), characteristic of R, 63
conv(A), convex hull of A, 1
deg(M), degree of M, 77
deg(v), degree of a vertex, 261
del$_\Delta(v)$, deletion of a vertex, 219
depth(M), depth of a module, 80
dim(M), Krull dimension, 64
frank(G), primitive cycles, 430
gr$_I(R)$, associated graded ring, 149
iff, if and only if, 62
in$_\prec(I)$, initial ideal, 127
mgrade(I), monomial grade, 539
rank(G), cycle rank, 429
rb(A), relative boundary, 2
ri(A) relative interior, 2
s.o.p, system of parameters, 79
star(σ), star of a face, 224
supp(α), support of a vector, 55
supp(x^a), support of x^a, 203
trdeg, transcendence degree, 114
type(M), type of M, 85, 184, 495
vol$(\mathcal{P})$, relative volume, 21
$\widetilde{H}_q(\Delta; A)$, reduced homology, 210

Index

a-invariant, 173
active column, 34
acyclic
  digraph, 291
  graph, 262
adjacency matrix, 301
adjacent, 261
admissible grading, 368
affine
  k-algebra, 111
  combination, 1
  Hilbert function, 342
  hyperplane, 2
  map, 1
  space, 1, 123
    generated by a set, 1
affinely independent, 5
Alexander dual
  of a clutter, 221, 543
  of a complex, 222
  of an edge ideal, 549
  of an ideal, 221
algebra, 86
  finite, 87
  finitely generated, 87
  of finite type, 87
  of vertex covers, 549
almost integral, 101
  closure, 101
analytic spread, 574
annihilator, 64
anti-chain, 613
antiblocking polyhedron, 618
arithmetic
  degree, 232
  rank, 233
arrows, 52
Artin–Rees lemma, 162
Artinian
  module, 71
  reduction, 195
ascending chain condition, 61
associated
  graded ring, 149, 568
  matrix
    of a monomial subring, 322
  prime
    of a module, 65
    of an ideal, 65
  valuation of $\mathfrak{p}$, 95

b-cover, 541
B-graph, 270
balanced
  binomial, 411
  clutter, 239
    König property, 588
  matrix, 239
basic feasible solution, 14
basis
  of a matroid, 59
  of a polymatroid, 510
basis monomial
  ideal of a matroid, 512
  ideal of a polymatroid, 510
  ring of a matroid, 512
Berge graph, 552

Betti numbers, 137
  initial, 298
  initial virtual, 298
big height, 230
  of a monomial ideal, 231
binary clutter, 592
binomial, 311
  expansion, 191
  ideal, 311, 312
  non-pure, 311
  of a closed walk, 424
  set theoretic
    complete intersection, 324
  with a square-free term, 377
  with nonsquare-free terms, 411
binomial ideal
  non-pure, 311
bipartite graph, 263
bipartition, 263
birational extension, 477
Björner and Wachs, 216
block, 264
blocker
  of a clutter, 221, 543
blocking polyhedron, 488
blowup algebras, 568
boundary complex, 256
  of a polytope, 4
bow-tie, 449
bridge, 264
Bruggesser–Mani, 5
Buchberger
  algorithm, 128
  criterion, 129
Burch inequality, 574

C–M, 81, 269
C–M filtration, 219
c-minor
  of a clutter, 235
  of an ideal, 235
canonical module, 178
Carathéodory's theorem, 5
Castelnuovo–Mumford, 226
catenary ring, 84
Cayley–Hamilton, 86
ceiling of a vector, 500
ceiling or a real number, 500
chain
  in a poset, 613
chain in a poset, 268
change of variable
  formula, 21
characteristic
  cone, 18
  vector, 568
characteristic of a ring, 63
Chinese remainder theorem, 74
chord, 240, 280, 427
chordal graph, 240, 280
  is shellable, 285
chromatic number, 263
  computation, 308
circuit, 57
  ideal, 332
  of a graph, 446
  of a matroid, 59
  of a multigraph, 452
  of a toric ideal, 327
clique, 240, 263, 610
  number, 263
clique clutter, 240, 563
closed halfspace, 3
clutter, 43, 568
  of bases of matroid, 512
co-chordal graph, 297
co-CM clutter, 230
CoCoA
  see computer algebra, 169
codimension
  of a module, 64
  of an ideal, 64
Cohen–Macaulay
  clutter, 234
  direct product, 84, 86
  graph, 269
  ideal, 82

module, 81
localization, 81
ring, 82
simplicial complex, 213
colon ideal, 64
coloration, 267
comparability graph, 610
complement of a graph, 280
complementary slackness, 16, 18
complete
bipartite graph, 263
graph, 262
ideal, 145
complete admissible
clutter, 247
is unmixed, 247
uniform clutter, 614
complete intersection, 83
generically, 143
graph, 437
set theoretic, 83
completely normal, 101
semigroup, 417
componentwise linear, 222
composition series, 71
computer algebra systems
CoCoA, 169
Macaulay2, 130
Maple, 522
Mathematica, 493
Normaliz, 500
PORTA, 10
cone
finitely generated, 7
generated by a set, 467
over a graph, 458
pointed, 30
polyhedral, 6
Conforti–Cornuéjols conjecture, xiii, 567, 589, 590, 604
connected
clutter, 553
component, 262
graph, 262
connector of a binomial, 411
conormal module, 143
constructible complex, 218
content of a polynomial, 462
contraction
of a clutter, 234
of and edge ideal, 234
convex
combination, 1
cone, 6
hull, 1
polyhedron, 3
polytope, 2
set, 1
covering number, 221, 538
edge, 267
vertex, 266
Cremona map, 534
critical binomials, 352
full set, 352
critical group, 401
cutpoint, 264
cutvertex, 264
cycle, 245, 262
basis, 429
rank, 429
space, 429
vector, 429

d-tree, 301
Danilov–Stanley formula, 368
decomposition theorem
for polyhedra, 9
Dedekind ring, 103
Dedekind–Mertens formula, 462
degree
is additive, 344
of a module, 77
of a vertex, 261
of an affine algebra, 343
Dehn–Sommerville equations, 257
deletion
of a clutter, 234

of a face, 219
of a vertex, 219, 303
of an edge ideal, 234
dependent set of matroid, 59
depth
lemma, 81
of a module, 80
descending chain condition, 71
Dickson's lemma, 127
digraph, 52, 267, 610
strongly connected, 397
Dilworth decomposition, 614
dimension
of a monomial subring, 322
of a set, 1
of a simplicial complex, 209
of an ideal, 64
theorem, 79
directed
cut, 52
graph, 267
discrete
graph, 262
valuation, 417
valuation ring, 93
distance, 263
division algorithm, 128
divisor class group, 96, 103
divisorial fractional ideal, 98
dual
cone, 9, 369
hypergraph, 574
of an edge ideal, 623
dual problem
linear programming, 15
duality of Veronese subrings, 516
duality theorem
for cones, 8
for edge ideals, 222
for linear programming, 15
for normal Rees algebras, 622
for PB matrices, 397
for symbolic Rees algebras, 549
duplication
of vertices, 542
dyadic clutter, 587

Eagon–Reiner criterion, 222
edge, 261
cone, 467, 484
cover, 267
covering number, 267
critical clutter, 553
critical graph, 273
generator, 424
polytope, 454
ring, 268, 569
edge ideal
of a clutter, 220, 568
of a graph, 268
of a hypergraph, 220
edge subring
of a graph, 423, 454
of a multigraph, 448
Ehrhart
clutter, 389
configuration, 389
function, 383
polynomial, 383
normalized, 376
reciprocity law, 384
ring, 381, 569
normalized, 375
series, 383
elementary
integral vector, 57
vector, 55
zonotope, 447
elimination order, 130
embedded prime, 68
Euler
characteristic, 211
reduced, 211
formula, 256, 436
identity, 267
even cycle, 262
exact sequence, 66, 296

short, 66, 296
exchange property
bases, 60
exponent vector, 370
extended Rees algebra, 148, 568
extremal
Cohen–Macaulay ring, 188
Gorenstein ring, 187
extremal rays, 15
of a cone, 15

f-vector
of a complex, 251
of a polytope, 256
face, 3, 209
ideal, 202
ring, 212
face-lattice, 4
facet
of a polyhedral set, 3
of a simplicial complex, 213
facet ideal, 245
faithfully flat, 90
Farkas's lemma, 6
feasible solutions, 16, 18
field of fractions, 70
filtration of a ring, 157
fine grading, 250
finite
algebra, 87
basis theorem, 9
homomorphism, 87
length, 71
type algebra, 87
finitely generated
semigroup, 29
flag complexes, 280
flat homomorphism, 90
forest, 262
forms, 76
four colour theorem, 310
fractional ideal, 98
free
variable, 237
vertex, 237
vertex property, 240
Frobenius number, 348
full semigroup, 422
fundamental theorem
for f.g. abelian groups, 32
of linear inequalities, 6

Gauss lemma, 462
geometrically linked, 106
Gershgorin circle theorem, 399
going
down, 89
up, 89
Gordan's lemma, 38
version 1, 29
version 2, 29
Gorenstein
graded ideal, 184
ideal, 85
Rees algebras, 156
ring, 85
Gröbner basis, 128
of toric ideals, 322
reduced, 128
graded
algebra, 78
ideal, 77
map, 76
module, 76
ring, 76
submodule, 76
Gram–Schmidt process, 19
graph, 261
2-connected, 264
connected, 262
connected components, 262
invariant, 262
planar graph, 435
regular, 302
well-covered, 269
Graver basis, 326
GRevLex order, 130, 343
Grothendieck theorem, 109

group of differences, 416

H-path, 431
h-vector
  of a C–M complex, 254
  of a complex, 252
  of a polytope, 256
  of a standard algebra, 179
Hamilton cycle, 494
Heger's theorem, 49
height of an ideal, 64
Herzog–Kühl formulas, 139
Hilbert
  basis, 28
    existence, 30
    minimal, 28
    of a cone, 28
    theorem, 62
  coefficients, 384, 522
  function, 77, 171
    affine, 342
  polynomial
    of a face ring, 252
    over an Artinian local ring, 77
  series, 171
    of a face ring, 252
Hilbert–Burch theorem, 138
Hilbert–Serre theorem, 172
Hochster
  configuration, 457
  local cohomology, 214
  theorem on normality, 368
Hoffman and Kruskal, 47
homeomorphic graphs, 436
homogeneous
  element, 76
  ideal, 77
  monomial subring, 372
  point configuration, 372, 380
  resolution, 137
  ring, 77
homogenization of an ideal, 132
homomorphism
  integral, 87
  of algebras, 87
  of graphs, 262
  of rings, 63
Hull of an ideal, 401
hypergraph, 220

ideal
  of covers, 221
  of covers of a clutter, 549
  of linear type, 143
  of mixed products, 507
  of relations, 114
  of Veronese type, 511
  quotient, 64
ideal clutter, 598, 599
implicit representation, 8
improper face
  of a polyhedral set, 3
incidence matrix
  of a clutter, 43
  of a digraph, 52
    is totally unimodular, 55
  of a graph, 440
  Smith form, 444
inclusion diagram, 4
independence
  complex
    of a clutter, 221
  number
    of a graph, 266
    small, 230
  polynomial of a graph, 272
  polytope, 561
independent edges, 538
independent set
  of edges, 235, 264
  of vertices, 221, 265
index
  of normalization, 528
  of regularity, 174
  of stability, 303
induced matching
  number, 229

of a clutter, 229
induced subclutter, 228
induced subgraph, 262
initial
degree, 183
form, 403
ideal, 127
integer
decomposition, 391, 616
rounding property
of $x \geq 0; xA \geq \mathbf{1}$, 617
of $x \geq 0; xA \leq \mathbf{1}$, 619
of $xA \leq w$, 35
integral
basis, 369
element, 87, 145
extension, 87
hull, 17
polyhedron, 16
integral closure, 87
commute with localizations, 146, 147
of a domain, 89
of a semigroup, 46
of an edge subring, 450
of ideals, 145
of monomial ideals, 499
of monomial subrings, 366, 367
integrally closed
ideal, 145
ring, 89
semigroup, 46
invariant factors, 20, 444
formula, 32
invariant ring, 90
invertible fractional ideal, 98
irreducible
components
of a variety, 125
ideal, 206
representation, 9, 12
of a cone, 14
submodule, 68
vertex cover, 541
irredundant
representation, 9
irredundant decomposition
of modules, 69
irrelevant maximal ideal, 78, 114
isolated
prime, 68
subgroup, 93
vertex, 261
of a clutter, 234, 235
isomorphic graphs, 262

Jacobian
criterion, 167
ideal, 167
matrix, 167
Jacobson radical, 65
join
of graphs, 498
of ideals, 508
of simplicial complexes, 216

König
property, 235, 539
theorem, 266
Koszul algebra, 512
Koszul complex, 105
Krull
dimension, 64
intersection theorem, 159
principal ideal theorem, 82
ring, 94
semigroup, 417
Krull–Akizuki, 95
Kruskal–Katona criterion, 252

Laplacian ideal, 401
Laplacian matrix, 400
lattice, 4, 207
complete, 4
distributive, 207
ideal, 316
isomorphism, 4

of monomial ideals, 207
point, 20
polytope, 20, 365, 380
simplex, 25
subgroup, 316
Laurent
monomial, 320
subring, 321
toric ideal, 321
polynomial, 148
leading
coefficient, 127
term, 127
leaf, 240
length of a module, 71
level algebra, 518
lex order, 111, 127
lexicographical order, 92, 127, 327
graded, 343
line, 6
line graph, 299
of an ideal, 428
lineality space, 18
linear
quotients, 223
resolution, 139
variety, 1
link of a face, 213
linkage
class, 106
even, 106
of ideals, 106
local ring, 63
localization, 63
at a prime, 64
preserves associated primes, 75, 307
locally a complete intersection, 154
log set, 370
log-matrix, 477
loop, 261
lower facet, 408
LP, 15
LP duality theorem, 15
Lucchesi–Younger theorem, 52
Lying over condition, 88

m-full ideal, 501
Macaulay
symbol, 191
theorem, 191
Macaulay2
see computer algebra systems, 130
Maple, 22, 522
marriage problem
a generalization, 560
marriage theorem, 266, 475
matching, 264
number, 266, 538
matrix
totally unimodular, 47
matrix ideal, 396
Matrix-Tree Theorem, 401
matroid, 59
bases of, 60
basis, 59
basis monomial ideal, 591
basis monomial ring, 512
rank of, 60, 512
representable, 59
max-flow min-cut, 43, 585
characterization, 43, 586
comparability graphs, 611
example, 45
if and only if Mengerian, 586
integer rounding property, 628
of graphs, 592
of uniform clutter, 603
sufficient condition, 590
maximal
clique, 240, 263, 563
independent cover, 270
independent set, 265
Menger theorem, 611
Mengerian
clutter, 586

Meyniel graph, 633
MFMC, 43, 585
min-max problem, 538
minimal
  face, 16
  generating set, 29
  prime
    of a ring, 62
    of an ideal, 62
    of a module, 68
    of an edge ideal, 221
  resolution, 137
  vertex cover
    of a clutter, 221, 537
    of a graph, 265
minimally
  non-ideal, 598
  non-normal, 532, 606
  non-packing, 589
minimum number of gens, 134
Minkowski representation, 8
minor
  of a clutter, 235, 578
  of a matrix, 48
  of an ideal, 578
mixed products, 507
mode, 548
module
  of finite length, 71
  of fractions, 63
monomial, 126
  algebra, 568
  grade, 539
  ideal, 202
  order, 127
  ring, 202
  space curve, 352
  subring, 321
  walk, 424
Mori–Nagata, 94
multi Rees Algebra, 157
multigraph, 262, 448
multiplicity
  additivity, 162
  of a face ring, 255
  of a graded module, 77
  of a graph, 298
  of an edge-ring, 233
  of an ideal, 161

Nakayama's lemma
  general version, 70
  graded version, 78
neighbor set
  of a subset, 264
  of a vertex, 264
nilpotent element, 65
nilradical, 65
Noether normalization
  homogeneous, 118
  lemma, 113
    homogeneous, 117
  of a monomial ring, 118
  of an edge subring, 463
Noetherian
  module, 61
  ring, 62
non-pure binomial ideal, 311
non-pure shellable, 216
normal
  ideal, 145, 571
  point configuration, 380
  polytope, 380
  ring, 89
  semigroup, 46, 417
normality
  closed under
    minors, 584
    parallelizations, 584
  criterion, 506
  descent, 157
Normaliz, 22, 44, 48, 422, 500, 576
  an overview, 50
normalization
  of a Rees algebra, 509
  of a ring, 89
  of a semigroup, 46

of an edge subring, 450
of monomial subrings, 366
normalized
degree, 513
grading, 513
valuation, 417
volume, 25, 383
normally torsion-free, 153, 586
versus normal, 154, 155
Nullstellensatz, 125
numerical semigroup, 347

objective function, 15
odd antiholes, 551
odd component, 262, 266
odd cycle condition, 451
odd holes, 551
optimal
solution, 15
value, 15
order complex, 268
order of a graph, 261
order of an ideal, 502
ordered group, 92
oriented
cycle, 289
graph, 267
outerplanar graph, 436

packing problem, 589
packing property, 588
parallelization
of a clutter, 542
of a graph, 303
parallelotope, 31
volume, 32
partial character, 316
partially ordered set, 4
partition, 255
Pascal's triangle, 160
path, 262
ideal, 507
PB matrix, 397
perfect
graph, 263
matching, 475
of a clutter, 235
of a graph, 266
of König type, 235
matrix, 563
Perron–Frobenius, 398
persistence property, 303
Picard group, 98
Pick's formula, 385
point configuration, 380
pointed
cone, 30
characterization, 38
polyhedron, 15, 17, 18
polarization
of a monomial ideal, 203
polyhedral
cone, 6
set, 3
face, 3
facet, 3
polyhedron, 3
integral, 16
polymatroidal ideal, 510
polynomial
function, 165
quasi-homogeneous, 77
Polyprob, 22
polytopal
complex, 4
subring, 364, 380
polytope, 2
lattice, 380
PORTA, 10
poset, 4
positive part, 55
positively graded algebra, 114
PP, 588
presentation
ideal, 114, 142, 319
of a graded algebra, 114
primal problem, 15

primary
ideal, 68
extension of, 91
submodule, 68
primary decomposition
irredundant, 69
of a graded module, 78
of a module, 69
of an ideal, 69
of monomial ideals, 204
prime avoidance
general version, 74
graded version, 120
prime spectrum, 62
primitive
binomial, 326
cycle, 427
cycle property, 432
principal fractional ideal, 98
projective
closure, 133
of a monomial curve, 134
dimension, 137
space, 133
proper face
of a polyhedral set, 3
proper minor, 578
pure resolution, 139
pure shellable complex, 216
pure simplicial complex, 213

quasi-forest, 297
quasi-ideal Rees cone, 509
quasi-regular sequence, 159
quotient ideal, 64

radical ideal, 151
radical of an ideal, 65
rank
of a matroid, 60
of a module, 74
of a valuation ring, 93
of an ordered group, 93
rational hyperplane, 2
rational polyhedron, 3
reciprocity law, 384
reduced ring, 65
reduced simplicial homology, 210
reduction number
of an algebra, 466
of an ideal, 463
reduction of a polynomial, 127
Rees
algebra, 142
of a filtration, 158
cone, 39, 508, 529, 548, 569
configuration, 43
semigroup, 46
reflexive module, 99
regular
element, 67
graph, 267
ring, 82
is Cohen–Macaulay, 165
sequence, 79
triangulation, 408
regularity, 226
regularity of an edge subring, 608
Reisner theorem, 214
relative
boundary, 2
point, 2
interior, 2
point, 2
volume, 21, 383
example, 22
formula, 22
of a simplex, 25
remainder, 128
residue field, 63
residue group, 417
revlex order, 127
ring extension, 86
ring graph, 431

S-polynomial, 128
Samuel function, 161
sandpile group, 401

saturation
of a lattice, 317
of an ideal, 130
Segre
product, 177, 466
variety, 341
semigroup, 347
affine, 46
finitely generated, 29
ring, 321
semigroup ring, 417, 419
semilocal ring, 74
sequentially Cohen–Macaulay
clutter, 232, 285
complex, 220
graph, 232, 285
module, 219
Serre's normality criterion, 90
set covering polyhedron, 39, 538
set theoretic
complete intersection, 83, 233
binomial, 324
variety, 341
shedding vertex, 219, 226
shellable
clutter, 232, 234
complex
non-pure, 217
graph, 282
polytope, 5
shelling, 216, 217
shift in the graduation, 135
shift polyhedron, 370, 484
with respect to C, 490
Simis cone, 540
simple
components, 313
polynomial, 313
simple module, 71
simplex
lattice, 25
of a simplicial complex, 209
oriented, 209
unimodular, 25
simplicial complex, 209
connected in codim 1, 213
Alexander dual, 222
Cohen–Macaulay, 213
constructible, 218
geometric realization, 255
strongly connected, 213
simplicial forest, 240
simplicial sphere, 256
Gorenstein property, 257
simplicial vertex, 293
sink, 290
skeleton, 230
skeleton of a complex, 215
sliding depth, 105
Smith normal form, 20
example, 22
socle
of a graded algebra, 183
of a module, 85
source, 290
spanning
subgraph, 262
tree
existence, 439
special cycle, 239, 245
special fiber, 574
symbolic, 557
spectrum of a ring, 62
square, 262
square-free
binomial, 333
monomial, 202
monomial ideal, 202
term, 377
of a binomial, 411
Veronese ideal, 524
Veronese subring, 513
stability number, 266
stable set, 265
stable vertex set, 221
standard

algebra, 114
grading, 77
monomial, 129
Stanley–Reisner
complex, 212
ideal, 212
ring, 212, 569
star, 263
star of a face, 224
stochastic matrix, 531
strong perfect graph theorem, 552
strongly
chordal graph, 240
Cohen–Macaulay ideal, 105
connected complex, 213
perfect graph, 599
subdivision of a graph, 431
subgraph, 262
support
of a binomial, 354
of a module, 66
of a monomial, 203, 445
of a vector, 55, 468
supporting hyperplane, 3
suspension of a graph, 277
symbolic
power, 152
of a prime ideal, 95
Rees algebra, 568
symmetric
algebra, 141, 142
exchange property, 60
semigroup, 348
system of parameters, 81
for modules, 79
homogeneous, 115
syzygetic ideal, 143
syzygy
module, 129, 137
computation of, 130
theorem, 136

t-unimodular, 48
TDI system
$xA \leq b$, 31
$x \geq 0;\ Bx \geq b$, 52
$x \geq 0; xA \geq 1$, 585
$x \geq 0;\ xA \geq 1$, 42
$x \geq 0; xA \leq w$, 36
and Hilbert bases, 34, 45
tensor algebra, 141
term order, 127
terms, 126
toric
ideal, 321, 326, 403
of a monomial curve, 352
of a point configuration, 326
of an edge subring, 424
set, 335
parameterized by x^{v_i}, 335
variety, 336
torsion
free, 22
subgroup, 22
total quotient group, 416
total ring of fractions, 70
totally balanced clutter, 239
totally dual integral, 42, 585
system $x \geq 0;\ Bx \geq b$, 52
system $x \geq 0;\ xA \geq 1$, 42
system $x \geq 0; xA \leq w$, 36
system $xA \leq b$, 31
totally unimodular, 47
tournament, 267
transcendence
basis, 114
degree, 114
transitive
digraph, 292
directed graph, 289
transpose of a matrix, 3
transversal
matroid, 591
number, 266
tree, 262, 496
spanning, 496
triangle, 262

trivial character, 316
twists of a graded module, 137
type, 184, 495, 515
    Cohen–Macaulay, 184
    of a module, 85

underlying digraph, 397
unicyclic graph, 450
uniform
    clutter, 223
    ideal, 527
unimodular
    cone, 364
    covering, 381
        weakly, 26
        with support in $\mathcal{A}$, 26
    matrix, 48, 444
        cf. t-unimodular, 48
    regular triangulations, 408
    simplex, 25
universal Gröbner basis, 327
unmixed
    clutter, 234
    graph, 269
    ideal, 83
Unmixedness theorem, 84
Upper bound conjecture, 258
usual grading, 77

valuation, 92
valuation ring, 92
    of a valuation, 93
valuation semigroup, 417
value group, 93
value of a vector, 29
vanishing ideal, 123, 345
variety
    affine, 123
    coordinate ring, 123
    dimension of, 123
    irreducible, 124
    irreducible components, 125
        over $\mathcal{C}$, 126
    projective, 133
    toric, 336
vector configuration, 380
vector matroid, 59
Veronese subring, 513
    square-free, 513
Veronese variety, 341
vertex, 10
    cover
        minimal, 265
    cover of a clutter, 538
    cover of a graph, 265
    covering number, 221, 266, 538
    critical clutter, 553
    critical graph, 269
    decomposable clutter, 285
    decomposable complex, 219
    independence number, 221
vertex-clique matrix, 563, 614
very well-covered graph, 295
volume
    of a parallelotope, 32
    of a polytope, 21
    of a simplex, 24

walk, 262
    closed, 262
weak perfect graph theorem, 561
weakly chordal graph, 295
weakly unimodular
    covering, 26
    regular triangulation, 408
    triangulation
        example, 409
well-covered graph, 269, 288
whisker, 209, 277, 284
whisker graph, 277, 284

Zariski
    closure, 123
    topology, 123
        is Artinian, 126
        of the prime spectrum, 62
zero divisor, 67
zero set of an ideal, 123